D1591594

Squire's Fundamentals of Radiology

Squire's Fundamentals of Radiology

FIFTH EDITION

Robert A. Novelline, M.D.

Professor of Radiology, Harvard Medical School

Director, Emergency Radiology, and Director, Undergraduate Radiology Education, Massachusetts General Hospital, Boston, Massachusetts

Harvard University Press

Cambridge, Massachusetts, and London, England · 1997

Printed in the United States of America

Library of Congress Cataloging-in-Publication Data

Novelline, Robert A.
Squire's fundamentals of radiology / Robert A. Novelline.
—5th ed.
p. cm.
Rev. ed. of Fundamentals of radiology / Lucy Frank Squire,
Robert A. Novelline. 4th ed. 1988.
Includes index.
ISBN 0-674-83339-2 (hardcover : alk. paper)
1. Diagnosis, Radioscopic. 2. Diagnostic imaging. I. Squire,
Lucy Frank, 1915– Fundamentals of radiology. II. Title.
[DNLM: 1. Diagnostic Imaging. WN 180 N938s 1997]
RC78.S69 1997
616.07′572—dc20
DNLM/DLC
for Library of Congress 96-36574

To the memory of
Dr. Richard Schatzki,
with grateful appreciation
for his inspiration
and example

Preface

In the preparation of this textbook, my first goal has been to maintain the format and spirit of the previous editions, conceived by my colleague, mentor, and friend Lucy Frank Squire, M.D. For over a quarter of a century her *Fundamentals of Radiology* has introduced medical students to the basic principles of diagnostic radiology in a logical and entertaining fashion. In response to the significant advances in imaging technology of the past decade, and the remarkable changes in health care delivery during that period, I have added to this edition hundreds of state-of-the-art diagnostic images and included discussions of many new topics. Students preparing for a career in clinical medicine need to know about the wide range of diagnostic imaging examinations available today, the indications for these examinations, and the role of these examinations in the orderly imaging workup of their patients. Nearly all the new illustrative cases in this edition have been student-tested in my own radiology courses at Harvard Medical School.

Like its predecessors, this edition is designed to be the medical student's introduction to radiology. It can also serve as a radiology instruction manual for those in the allied health sciences. The increase in size of the fifth edition is primarily due to the inclusion of many new teaching cases illustrated with computed tomography, ultrasound, magnetic-resonance, and computer-reformatted images. During the past decade, many traditional plain film x-ray, tomographic, fluoroscopic, and angiographic procedures have been replaced by faster and more accurate cross-sectional imaging procedures, which, in many cases, cost less than the examinations they replaced and are easier and more comfortable for patients.

Readers of previous editions had to refer to other textbooks to gain the foundation in plain film anatomy and cross-sectional imaging anatomy necessary for radiology instruction. In this edition I have provided a new chapter on plain film x-ray and cross-sectional anatomy. Additional instruction in vascular anatomy and neuroanatomy is included in new chapters on vascular radiology and neuroradiology. Today medical educators are emphasizing the importance of primary care instruction for future physicians, and to meet that need the range of subjects in this book has been expanded to include women's imaging, men's imaging, pediatric radiology, and other primary care topics. And a chapter on interventional radiology has been added to introduce readers to the variety of biopsy procedures and therapeutic interventional procedures performed by radiologists today. Recognizing the importance of illustrating the various radiological procedure rooms, I have included in this edition a number of photographs of patients undergoing imaging examinations. Also included are discussions of the patients' experiences during the different procedures, to help students better prepare their patients. To test students' advancing knowledge, unknowns are provided in most of the chapters, with the answers given at the back of the book. It is best to read this book from cover to cover, since the concepts, principles, and methods are treated cumulatively.

My greatest hope is that readers will learn the fundamentals of diagnostic radiology from this textbook, and that what they learn will help them in the care of their patients. As with previous editions, I give my address below, in the hope that readers will communicate their critical responses to this book and tell me how their needs may be better served.

Robert A. Novelline, M.D.
Department of Radiology
Massachusetts General Hospital
Boston, Massachusetts 02114-2698

Contents

Squire's Fundamentals of Radiology

What is it?

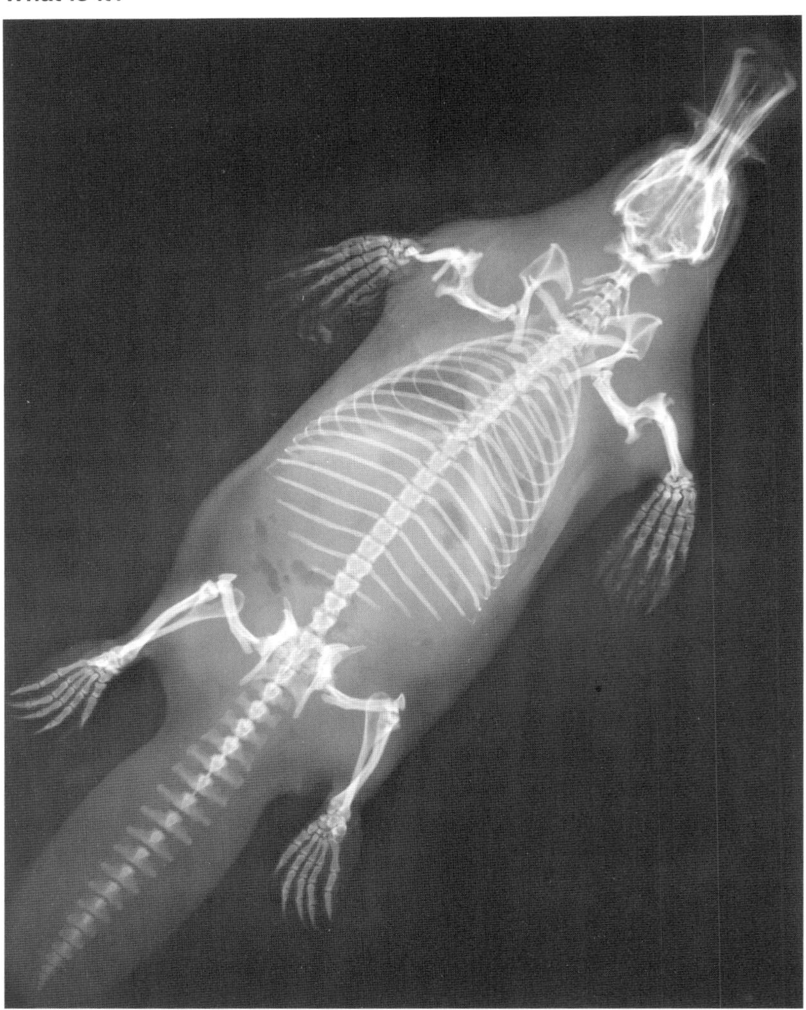

Figure 1.1. Its name is *Ornithorhynchus anatinus,* and you have never seen an x-ray portrait of it before. Nevertheless, there is one creature and one only in the animal kingdom which could give this x-ray appearance, and you can reason out its identity.

1 Basic Concepts

As you probably already know, x-rays are produced by bombarding a tungsten target with an electron beam. They are a form of radiant energy similar in several respects to visible light. For example, they radiate from the source in all directions unless stopped by an absorber. As with light rays, a very small part of the beam of x-rays will be absorbed by air, whereas all of the beam will be absorbed by a sheet of thick metal. The fundamental difference between x-rays and light rays is their range of wavelengths, the wavelengths of all x-rays being shorter than those of ultraviolet light. The useful science of diagnostic radiology is based on this difference, since many substances that are opaque to light are penetrated by x-rays. It was this attractive property that caught the attention of Professor Wilhelm Conrad Roentgen of the University of Würzburg on a cold November night in 1895, when he first observed certain physical phenomena he could not explain.

Figure 1.2. Staged version of the discovery of the roentgen ray.

Roentgen had been experimenting with an apparatus that, unknown to him, caused the emission of x-rays as a by-product. Accustomed to the darkened laboratory, he observed that whenever the apparatus was working, a chemical-coated piece of cardboard lying on the table glowed with a pale green light. We know now that fluorescence, or the emission of visible light, can be produced in a variety of ways by complex nuclear energy exchanges. But in 1895 Roentgen recognized at first only the fact that he had unintentionally produced *a hitherto unknown form of radiant energy that was invisible, could cause fluorescence, and passed through objects opaque to light.* When he placed his hand between the source of the beam and the glowing cardboard, he could see the bones inside his fingers within the shadow of his hand. He found that the new rays, which he named x-rays, penetrated wood. Using photographic paper instead of a fluorescing material, he made an "x-ray picture" of a hand through the door of his laboratory.

Six years later the first Nobel Prize in physics was awarded to Roentgen for his discovery, and by then this remarkably systematic investigator had explored most of the basic physical and medical applications of the new ray.

The idea of being able to see through opaque objects caught the public fancy all over the world, and a great deal of nonsense was written on the subject in many languages within the first decade after its discovery. There is a fascinating file of cartoons and articles in the Library of Congress documenting this fever.

Are you quite sure you can imagine exactly what Roentgen saw when he first observed that he could "see through his hand" with the help of the new ray? In order to grasp this clearly, you must first understand the important difference between what Roentgen saw

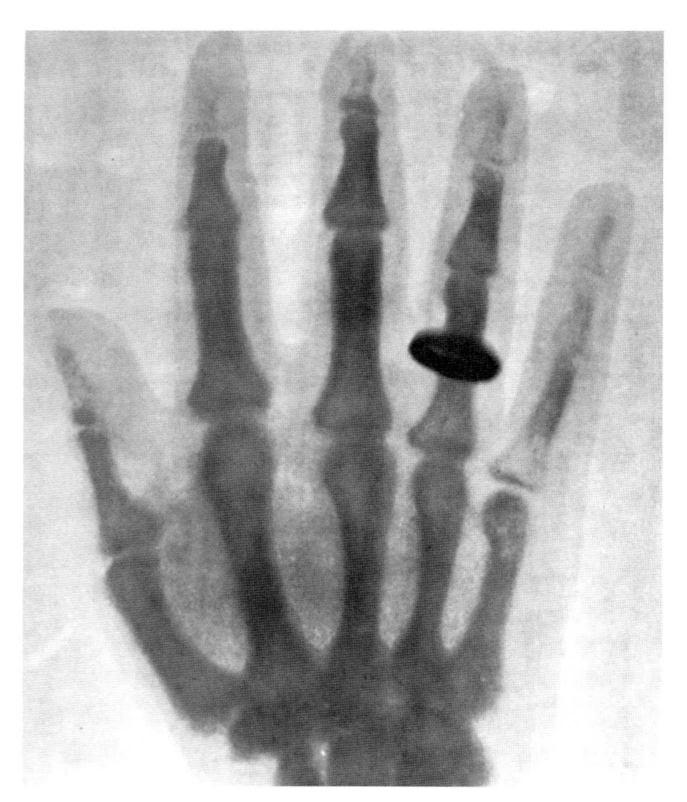

Figure 1.3. What Roentgen saw when he placed his hand between the tube and the fluorescing cardboard on the laboratory table. Areas not in the shadow of his hand fluoresced vigorously and appear white. Fewer x-rays reached the areas under the bones, so they were not so luminous and appear dark.

fluoroscopically (Figure 1.3) and what you see today in an x-ray film of the hand, such as the one in Figure 1.4. You should call this a *radiograph,* not an x-ray.

When light hits photographic film, a photochemical process takes place in which metallic silver is precipitated in fine particles within the gelatin emulsion, rendering the film black when it is developed chemically. Places on the film that are not exposed to light remain clear. When a "positive" paper print is made of this "negative" film, the values are reversed: the black, silver-bearing areas prevent light from reaching the photosensitive paper, while clear areas in the film permit the paper to be blackened (Figure 1.4).

The x-ray film you will see in medical school is equivalent to the negative film you may have worked with in your own photographic darkroom. X-rays, like light rays, precipitate silver in a photographic film, but they do so much less rapidly than light rays. A patient cannot be expected to hold still long enough for films to be made by using x-rays alone, and too much exposure to radiation is both dangerous and technically undesirable. Therefore, an ingenious reinforcing technique has been worked out using a special film container, or *cassette.*

The cassette (Figure 1.5) contains two fluorescent intensifying screens, which are activated by the x-rays and in turn emit light rays that reinforce the photochemical effects of the x-rays themselves on the film. In this way the silver-precipitating effect of the x-rays combined with that of the light rays they generate work together to blacken the film. The use of x-ray cassettes with specialized intensifying screens and films permits diagnostic x-ray imaging with less radiation of the patient and faster exposure times. When an object interposed between the x-ray beam source and the cassette has absorbed the rays, no light activation of the fluorescent screen will take place; neither x-rays nor light rays will reach the film, and no silver will be precipitated.

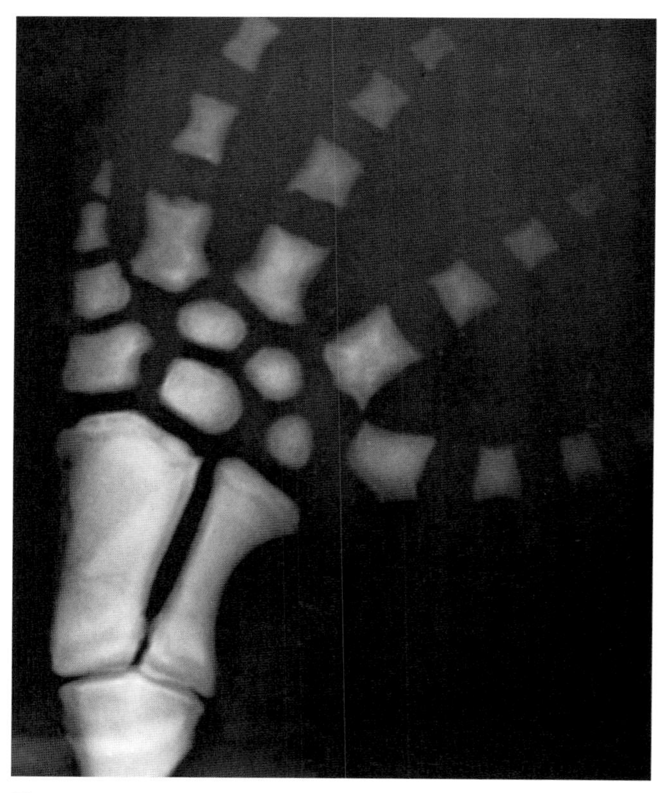

Figure 1.4. Radiograph of a hand . . . well, actually not a hand. Guess what it might be.

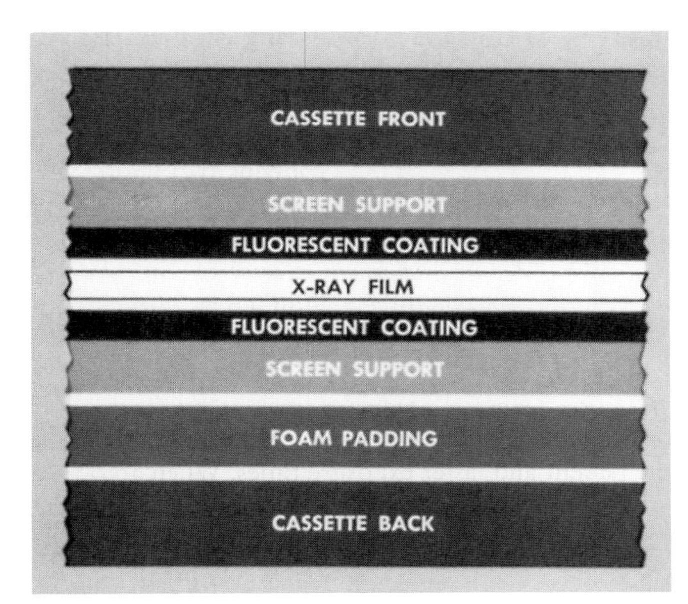

Figure 1.5A. Cross section of a cassette, or modern film holder. The x-ray film in use today consists of an acetate sheet coated on both sides with photographic emulsion, and the cassette is constructed so that plastic fluorescent screens are applied in contact with each side of the film. In this way light rays reinforce the photochemical effect on the film of the x-rays themselves. This effectively speeds up the exposure, shortening the time, and reduces the blurring effect of the patient's motion.

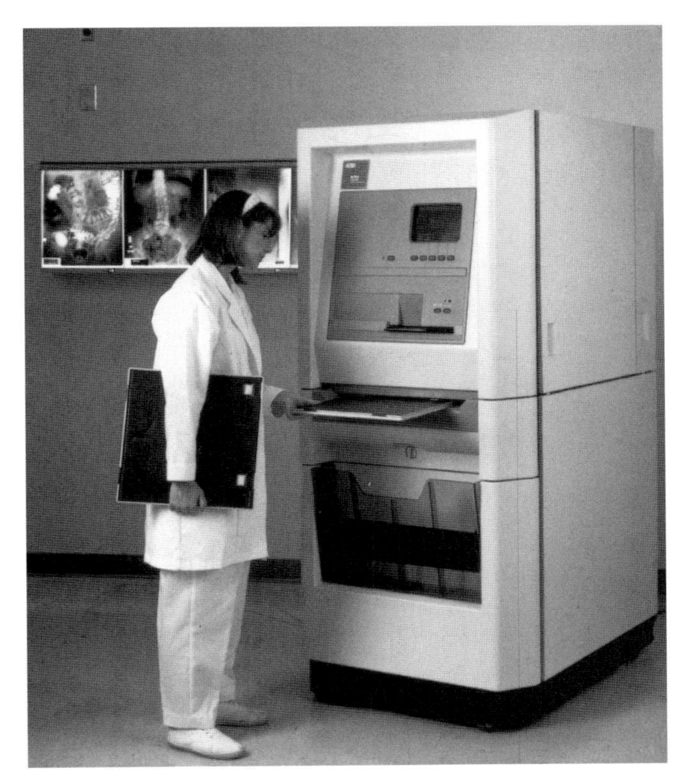

Figure 1.5B. Technologist holding a film cassette in her right hand while inserting another cassette into a film processor with her left hand.

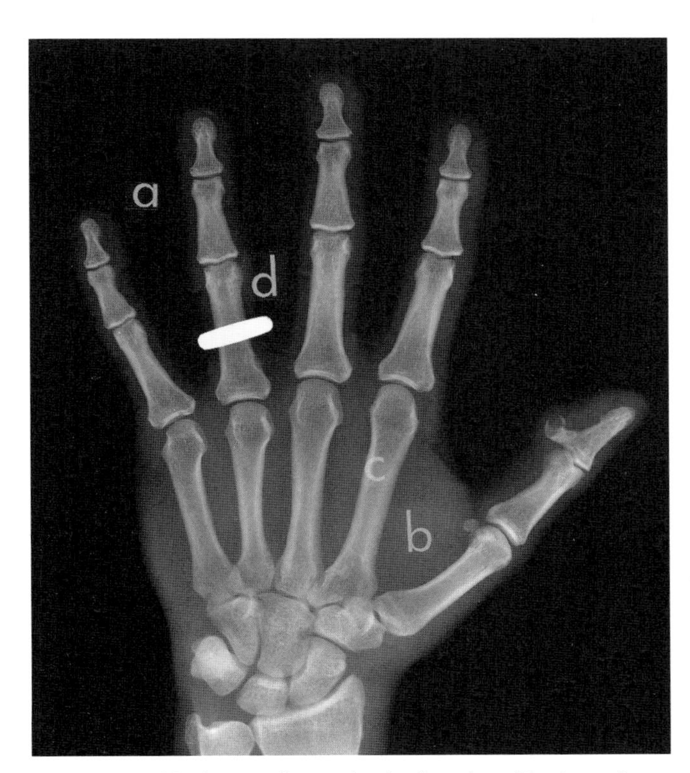

Figure 1.6. Modern radiograph of a hand. *a:* Blackened area where only air is interposed between beam and film. *b:* Soft tissues absorb part of the beam before it reaches the film. *c:* Calcium salts in bone absorb even more x-rays, leaving the film only lightly exposed and relatively little silver precipitated in the emulsion. *d:* The dense metal of the ring absorbs all rays; no silver is precipitated. (Note: This, like most x-ray illustrations in textbooks and periodicals published in this country, is a doubly reversed print, so that what you see here is what you will see whenever you hold a film of the hand against the light.)

In Figure 1.6 a man's left hand has been placed over the cassette and exposed to a beam of x-rays. Notice that the film not covered by any part of the hand has been intensely blackened because very little of the beam was absorbed by the air, which was the only absorber interposed between x-ray tube and film. The fleshy parts of the hand (or *soft tissues,* as they are called by the radiologist) absorbed a good deal of the beam, so that the film appears gray. Very few x-rays reached that part of the film directly under the *bones,* because bones contain large amounts of calcium, which absorb more x-rays than the soft tissues. Every *metal* absorbs x-rays to an extent depending on its atomic number and thickness. No x-rays at all were able to pass through the

Figure 1.4 is not a human hand but a whale's flipper reduced photographically.

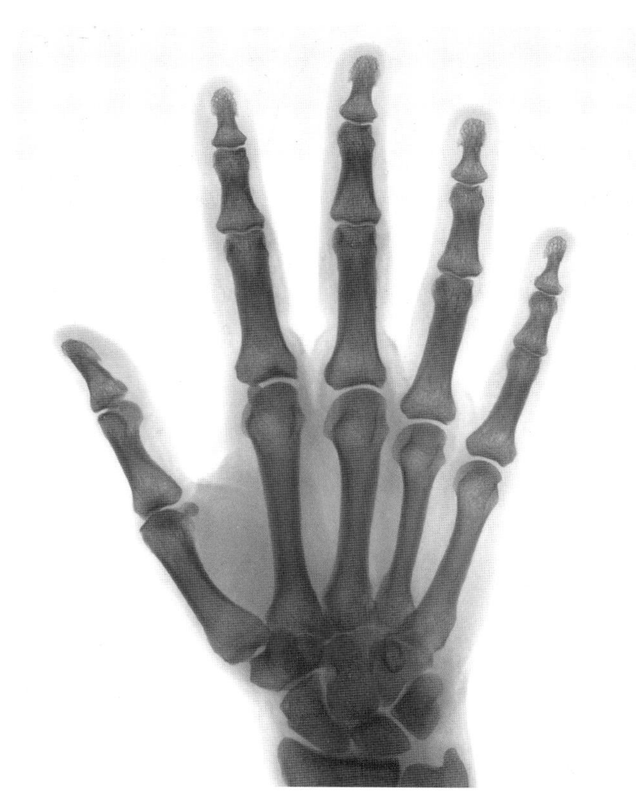

Figure 1.7. Positive print of a radiograph of the hand, made by singly reversing the values. This is also approximately as the hand would look on the fluoroscopic screen.

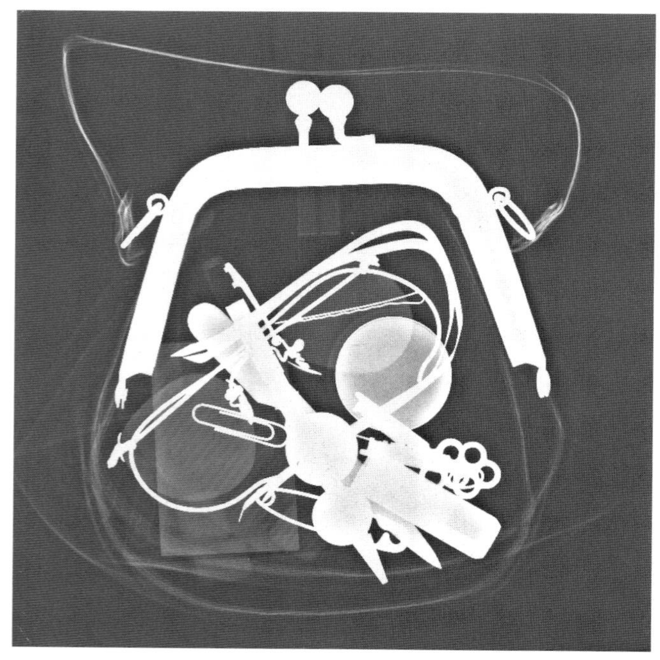

Figure 1.8

gold ring, and the film underneath it was not altered photographically and appears completely white.

What Roentgen saw, on the contrary, was the reverse of all the light-dark values you have been looking at in the film of the hand. X-rays reached the coated cardboard in abundance all around his hand so that the *background* fluoresced vigorously, while the shadow of his hand emitted less light and appeared gray-green. The cardboard underneath the bones of his fingers appeared darkest of all, since it received almost no activating rays. (Compare Figure 1.3.)

Fluoroscopic light is very faint unless it is amplified electronically. Today all fluoroscopic rooms are equipped with such "image-intensification" machines, functioning in a lighted room. You may not see much fluoroscopy in your lifetime, but you will see many thousands of x-ray films. For this reason we suggest that you make a practice of thinking in terms of the white and black values that relate to the usual x-ray film as you saw them in the hand in Figure 1.6. Think of dense objects as white and of those more easily penetrated as gray or black. All the illustrations in this book are printed like Figure 1.6, and you will find that most journals and books reproduce x-ray illustrations in this way.

While it is essential to understand which are the more dense (or *radiopaque*) substances and which the more transparent (or *radiolucent*) ones, your concern, even as you first begin looking at radiographs, should not be only with density. You will often make quite reasonable and useful deductions from the *form and shape* of radiographic shadows. If you figured out that Figure 1.1 was, and could only be, a radiograph of a duck-billed platypus, you have experienced the sort of educated guessing used all the time in radiology. You guess imaginatively and then subject your guess to a rigorous logical analysis based on radiological and medical data. By putting together expected density and expected form, you will soon find that you can predict the appearance of the radiograph of an object or structure.

Begin, then, by applying imagination and judgment to a variety of nonmedical objects. Try to predict the type of shadow that would appear on the film if you x-rayed an egg.

Figure 1.8 is a radiograph of a woman's purse. Although the cloth from which the purse was made offered almost no obstruction to the x-rays, anything made of metal inside it, including the frame of the purse, absorbed the rays and left a white profile on the film. You will be able to identify from their outlines

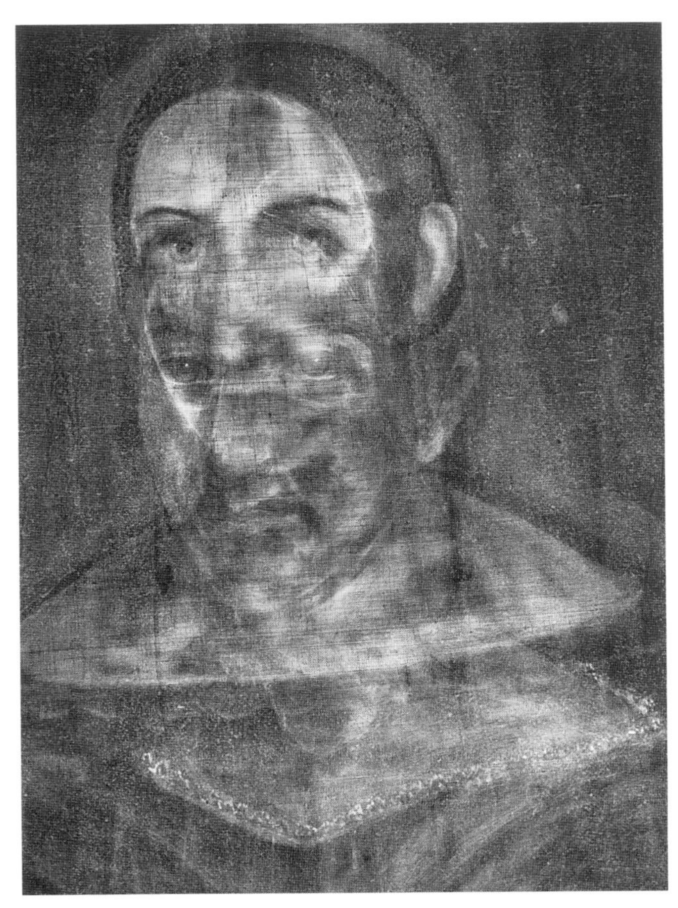

Figure 1.9. Radiograph of a portrait in oils painted over an earlier portrait. Only the woman with the pale eyes is visible as one looks at the painting.

alone a paper clip, a bobby pin, a safety pin, a pair of rimless glasses, coins, a lipstick case, two locker keys (overlapped), a nail file, and a metal pencil. You can almost construe the woman: a poverty-stricken, myopic individual who is taking two lab courses and wears makeup. You might, of course, be mistaken about the state of her finances; folding money would be quite radiolucent.

Pure metals are relatively radiopaque and so are their salts. Consequently, so also are the mixtures of oil and brilliantly colored metallic salts responsible for the whole field of oil painting. The radiography of paintings and other works of art is a fascinating and technically useful branch of the science. Frauds, inept reconstructions, and masterpieces painted over by amateurs may sometimes be detected by x-ray studies.

In Figure 1.9 two painters have used the same canvas—or, dissatisfied with the portrait of the man whose eyes appear as the lower pair, the same painter may have done the portrait of the woman with the light eyes and severely dressed hair, covering the earlier portrait. Only the woman is visible as one looks at the painting.

Variations in the precise metallic composition of artists' colors used at different times in history may help in the identification and dating of such works of art. The pigments in use since about 1800 have been made of the salts of metals with much lower atomic numbers than those of the older pigments, and for that reason they x-ray quite differently.

Thus a modern forgery of an old master, no matter how adroit a copy, will yield a radiograph entirely different from a radiograph of the original. But a copy made by a pupil of the master or another artist of the same school, painting at about the same time in history with the same hand-ground, earth-mineral colors, can be expected to x-ray in about the same way.

The characteristic use of brush strokes, which, even better than the signature, often stamps the work of a great artist, may also help to identify a concealed

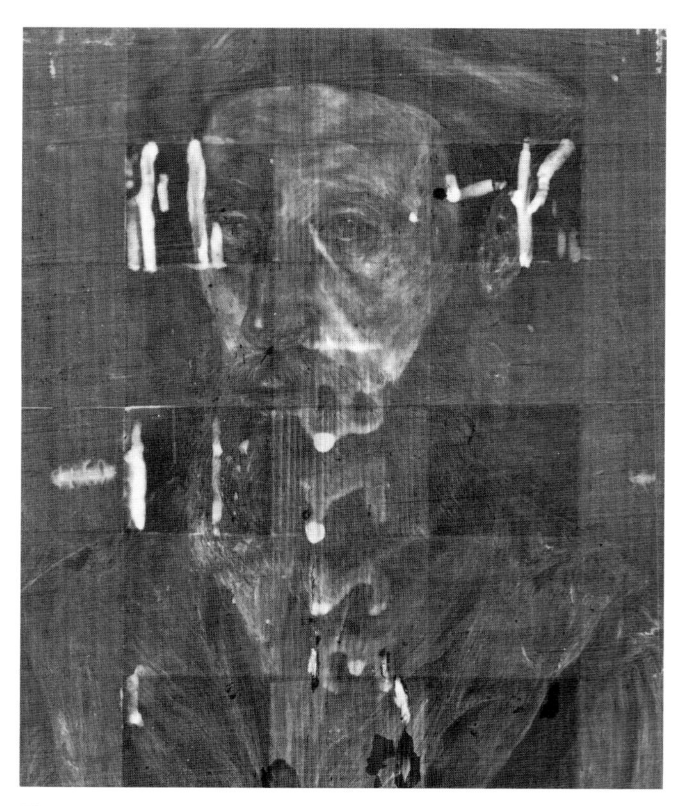

Figure 1.10

painting covered by a lesser artist. You would be able to imagine the radiograph of a contemporary canvas with the vigorous, heavy brush strokes of Van Gogh showing through, for example.

Remember too that any radiograph of a painting represents the summation not only of the various paint densities but also of the x-ray shadows of the canvas itself and the supporting structures. The wooden frame on which the canvas is stretched will cast some shadow, and if there are any nails in the wood they will appear in the radiograph also. Figure 1.10 shows a radiograph of a painting supported on wooden strips. The curious white areas are wormholes that have been filled with white lead. The x-rays have been completely absorbed, you notice, by the white-lead *casts* of the wormholes, and under them no x-rays have reached the film to blacken it. The white areas on the film are actually, therefore, *shadow-profiles* of these white-lead casts. Remember this! It has an important parallel in barium work in medical x-ray studies of the gastrointestinal tract.

The industrial uses of x-rays are many and important. Flaws, cracks, and fissures in heavy steel can be shown by x-raying big equipment or building materials. Especially powerful machines are needed for this sort of work, ones that will produce a more penetrating beam of x-rays of very short wavelength, often called hard x-rays. X-rays of long wavelength, or soft x-rays, are used to study thin or delicate objects. Very soft x-rays are used to study tissue sections of bone 1 or 2 microns in thickness (microradiography), while very hard x-rays are used to penetrate deep into the body and destroy malignant tumor cells (radiation therapy). Between these two extremes fall the wavelengths that are used in medical x-ray diagnosis.

The *electromagnetic spectrum* is a scaled arrangement of all types of radiant energy according to wavelength. Within the range used in diagnostic radiology, x-ray technologists are trained to select and use the particular wavelength suited to the density and thickness of the part they are filming. They do this by varying the kilovoltage of the machine: the higher the kilovoltage, the harder or more penetrating the beam of rays produced. They can also vary the amount of radiation in the beam by altering the milliamperage used, and, finally, they can control the time of exposure. Thus, for instance, for a thin object like the hand they use a soft beam for a short time, and for a dense object like the pelvis, a hard beam and a long exposure.

Radiodensity as a Function of Thickness

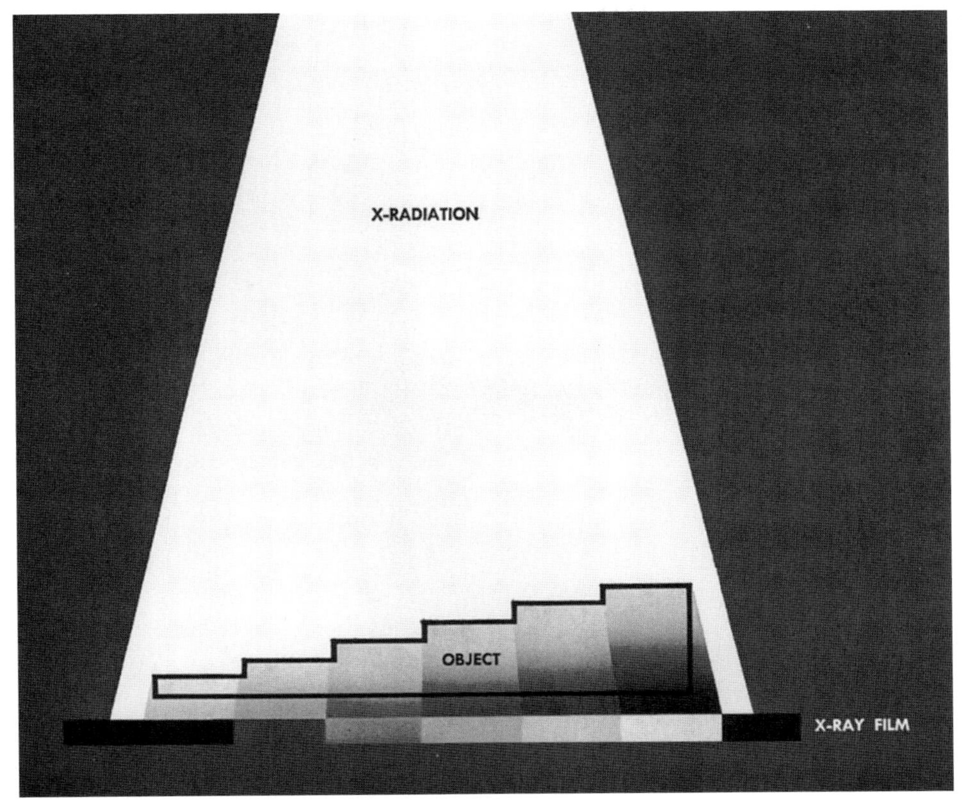

Figure 1.11. Radiodensity as a function of the thickness of the object. Here the object to be filmed is of homogeneous composition and has a stepwise range of thickness. Gray shading indicates the degree of absorption of the x-rays.

Radiodensity as a Function of Composition, with Thickness Kept Constant

Having reasoned through all this, you must consider in greater detail the *relative radiodensities* of various substances and tissues. To do this most easily, let us eliminate thickness completely for the moment. Consider an imaginary row of 1-centimeter cubes of lead, air, butter, bone, liver, blood, muscle, subcutaneous fat, and barium sulfate. Can you arrange them in order of their radiodensity, decreasing from left to right?

If they were all pure elemental chemicals, you certainly could arrange them in order by looking up their atomic numbers. Only one of them is quite as simple as that, and a judicious guess will surely place first to your left as most dense the cube of lead, with an atomic number of 82. Are you hesitating between bone and barium sulfate? Barium has an atomic number of 56, and calcium in the bone cube has an atomic number of 20. Bone, however, is not even pure calcium salt. It has a functioning physiological structure with holes and spaces to accommodate body fluids and marrow. It is composed of an organic matrix into which the complex bone mineral is precipitated. All such organic substances reduce the radiodensity of the cube of bone, and it will consequently have even less radiodensity than a similar cube of packed bone dust. The cube of barium sulfate must be placed next to the lead cube, therefore, and after it, the cube of bone.

As to the most radiolucent of all, you can have no trouble with that: surely you will have put the cube of air far to the right, at the opposite end of the scale from the lead. The film under the air cube will be black, since the sparse scattering of air molecules offers

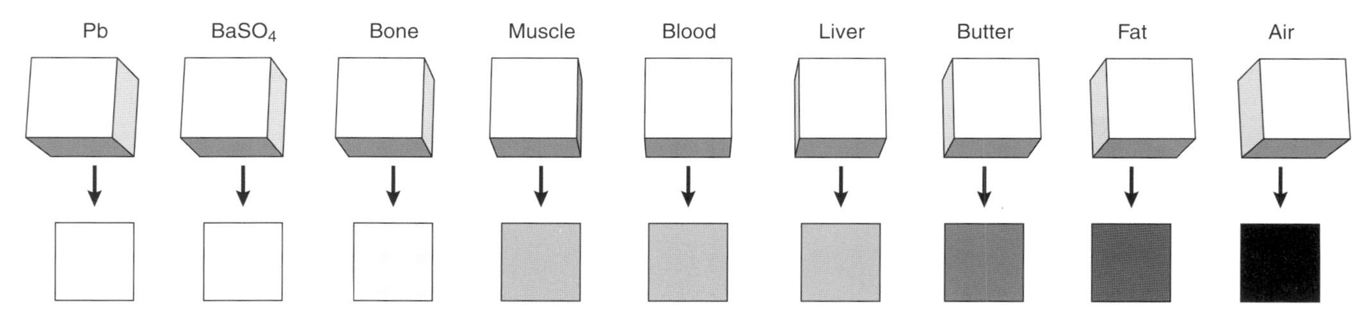

Figure 1.12. Radiodensity as a function of composition, with thickness kept constant.

almost no obstacle to the rays. The square of film under the lead, unaltered because no rays penetrated the cube to reach it, will be clear white, whereas that under the bone will show a tinge of gray.

Butter and subcutaneous fat have very similar x-ray densities. They are extremely radiolucent and must be placed next to air in the scale we are considering. Neither butter nor fatty tissue is homogeneous, since the first is never quite free of water and the second contains both circulating fluids and a supporting network of fibrous connective tissue. Their squares on the radiograph would be almost the same very dark gray.

Between the three very dense cubes and the three very lucent ones there remain to be arranged the three cubes of blood, muscle, and liver. These will all x-ray an almost identical medium gray, and you should remember that all moist solid or fluid-filled organs and tissue masses will have about the same radiodensity, greater than fat or air but considerably less than bone or metal. Thus the muscular heart with its blood-filled chambers could be expected to x-ray as you see it on the chest film, a homogeneous mass much denser than the air-containing lung on both sides of it, but showing no differentiation between muscular ventricle wall and blood within the ventricle.

Remember that in this discussion of relative radiodensities, we have kept thickness and form constant, as well as such technical factors as kilovoltage and time of exposure. We have planned this deliberately so that you can more easily construct a working concept of the relative densities of different tissues. In practice, the radiologist adjusts the technical factors to accentuate these differences. Upon this useful spectrum of differing radiodensities of human tissues is based the whole field of medical radiography.

Once these basic facts are learned, radiology becomes an exercise in logical deduction and an absorbing habit of mind. More important for you, it is also a delightful extra dimension in learning, a sort of custom-tailored illustrative tool related to nearly everything you will study in medical school. If you wish, you can use it to help you learn from the first day you begin to study anatomy, through your courses in physical diagnosis, pathology, medicine, and surgery, as a means of comprehending and remembering medical facts.

How Roentgen Shadows Instruct You about Form

Figure 1.13

Consider now the contribution of *form*. Figure 1.13 is a radiograph of three roses, which we can use as an example of the basic logic of the roentgen shadows of complex objects. Flowers require only a very soft beam, of course, because they are both thin and delicate. A glance will tell you that one rose is full-blown and the other two more recently opened. You can deduce a great deal of information from the form, outline, shape, and structure of roentgen shadows. This is so true that in time you will learn to recognize with confidence the *identity* of certain shadows in medical radiographs because of their shape or form.

Now study the density of various parts of a single petal and compare the radiodensity, or whiteness, of the petals with that of the leaves. The leaves look less dense than the flowers and stems. Notice too that the veins within each leaf are denser than the rest of it. Veins of leaves have, of course, a structure independent of the cells composing the flatter part of the leaf. The stems are thicker and they also convey fluid. In both medical and nonmedical radiographs you can, in general, anticipate added density wherever there is fluid.

Radiographs as Summation Shadowgrams

Another reason for the denser appearance of the petals compared with the leaves in Figure 1.13 is that they do not lie flat against the film but are curved and folded and overlap one another. This gives you a clue to a very important facet of radiological interpretation. A sheet of any uniform composition, if it lies flat and parallel to the film, will have a uniform x-ray density and cast a homogeneous shadow. If it is curved, however, those parts which lie perpendicular to the plane of the film will radiograph as though they were much more dense.

This is perfectly simple. X-rays pass through a complex object and render upon the film not a picture at all but a "composite shadowgram," representing the sum of the densities interposed between beam source and film. Thus a sheet of rose petal that lies perpendicular to the film, or in the plane of the ray, is equivalent to many thicknesses of petal laid one upon another and, quite logically, is much more dense than a single sheet lying flat. Find the leaf that is turned on edge.

Curved sheets, considered geometrically, arrange themselves into groups of planes, and should be so considered in imagination when you are interpreting an x-ray film. Of course, in nature, and consequently in medicine, the curved plane is common and the symmetrical plane rare. In the radiograph of any curved plane structure, therefore, learn to think in terms of those parts of it that are *relatively parallel to the film* and those that are *roughly perpendicular to it.*

Observe, finally, that the shadow of the stem of the rose in Figure 1.13 has a form you will find characteristic of any *tubular structure* of uniform composition. The margins are relatively dense because they represent long, curved planes radiographed tangentially, and the center area between them appears as a darker, more radiolucent streak. Rose stems are not truly hollow as one looks at them with the naked eye, but the central core, like that of tubular bones, is filled with a structure having less radiodensity than the margins. Hence the rose stem looks hollow and tubular on the film, just as a hollow tube containing air would look.

By this time you have several important principles clearly in mind, although you have learned them largely from examples. *First,* you know that x-rays are radiant energy of very short wavelength, beyond light in the electromagnetic spectrum, and that they penetrate, differently according to their wavelengths, substances opaque to light.

Second, you know that a beam of x-rays penetrates a complex object like the hand in accordance with the relative radiodensities of the materials which compose the object. You know that the beam produces on the film a composite shadowgram representing the sum of those radiodensities, layer for layer and part for part. You know that radiodensity is a function of atomic number and of thickness.

Third, you have realized that the parts of an object may become recognizable as to form, and their structure may be deduced, according to whether they are constructed most like solid or hollow spheres, cubes, or cylinders, or like plane sheets lying flat or curved upward away from the film.

Because we believe that the working of problems and puzzles will greatly increase your enjoyment of this book, we have included some in almost every chapter. They are geared to the chapter in question both in subject matter and in difficulty. In general, they are presented with a few details about the patient, and you should imagine yourself the intern or practicing physician in charge of that patient. Often, especially in the early chapters, you are asked not for a diagnosis but rather for an impression of variation from the normal of a particular structure. You will see that solving these puzzles will help you gauge as you go along just how roentgen shadows can be reasoned out and used as a mnemonic device in learning medicine. We think it will also persuade you that you know more and can reason better than you had realized.

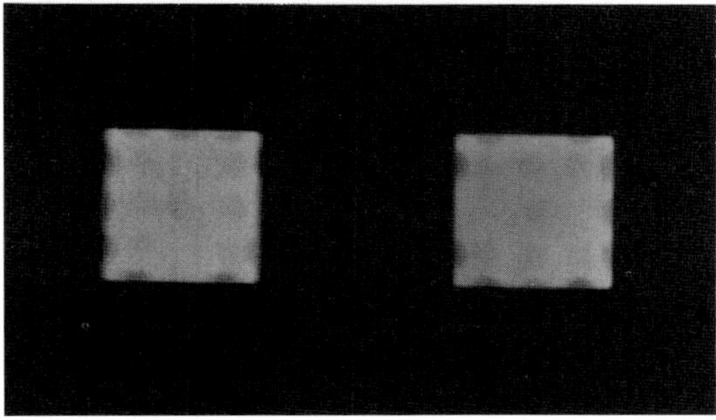

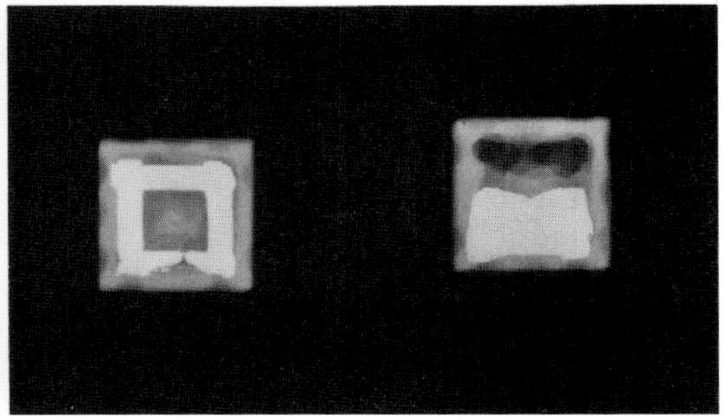

Figure 1.14 ***(Unknown 1.1).*** Sometimes the radiologist figures in criminology as an adjunctive source of information. The lucky throw you see in the innocent-looking pair of dice in the photograph was actually not luck at all but planned economy. Below are two radiographs, one of a pair of loaded dice and one of a pair of unloaded dice for which they could be switched. It is simple enough to decide which are the loaded dice, but can you figure out precisely what has been done to them?

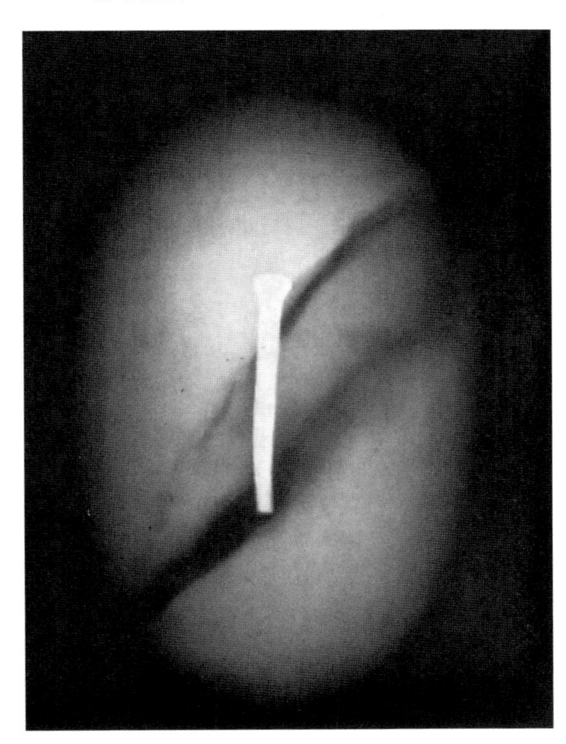

Figure 1.15 ***(Unknown 1.2).*** This is not a familiar object, and though you can figure out what its structure is from this, its radiograph, you will be very gifted indeed if you can say where it was when found.

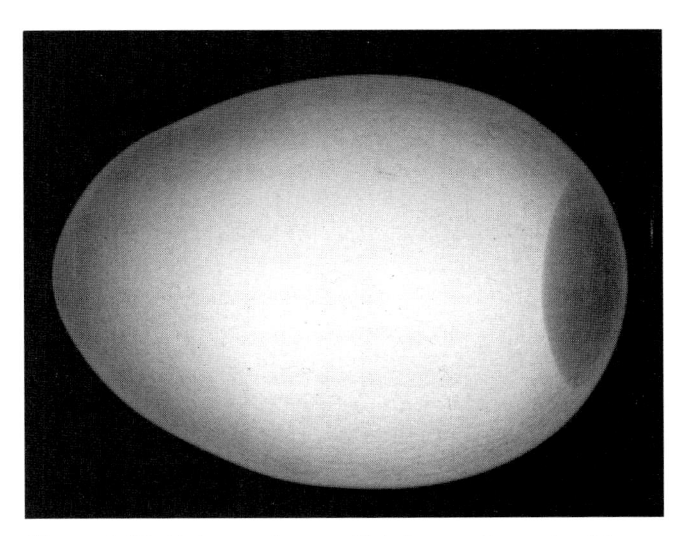

Figure 1.16. Not an unknown. This is a radiograph of the egg referred to earlier. You no doubt anticipated its shape, but did you predict the air pocket at the blunted end? Note that the density of the shell increases peripherally, as with any hollow sphere.

2 The Imaging Techniques

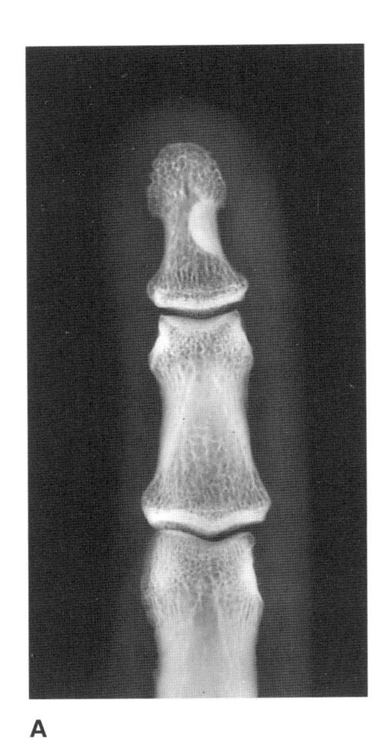

A

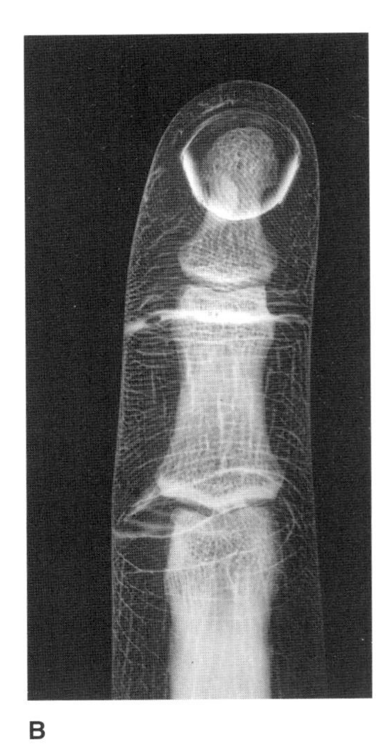

B

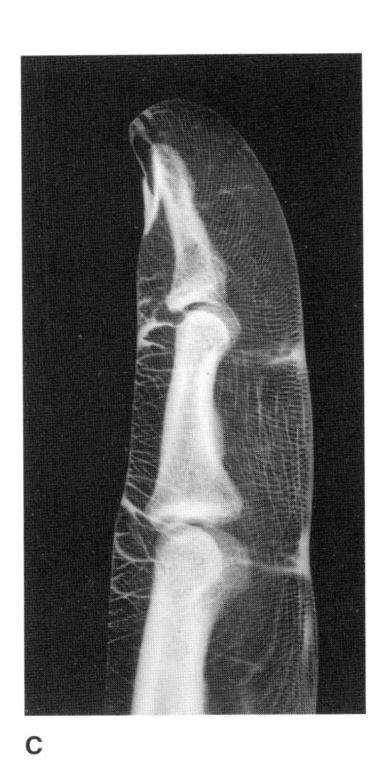

C

Figure 2.1

In this chapter we will teach you several important facts about plain films and introduce you to the other diagnostic imaging techniques in clinical use today. You will learn more about these procedures in later chapters. Remember to think three-dimensionally when looking at x-rays and other diagnostic images.

Thinking Three-Dimensionally about Plain Films

The three radiographs of a finger in Figure 2.1 illustrate at once how important it is for you to learn to think in three dimensions about x-ray shadows, reconstructing form from two views at right angles. Note that the soft tissues in A are seen as a faint uniform gray outline encompassing the bones. In B and C, however, the skin with its wrinkles and folds and the crevice between the cuticle and nail all seem to become visible. They appear so because they have been coated with a creamy substance containing a metallic salt.

Actually, the skin itself is no more visible than it was before, but the radiopaque cream collecting on its patterned, irregular surface forms a visible coating that marks the position of the skin. A and B were made in the *frontal projection;* C is made from the side and is called a *lateral view.*

Although A probably looks very flat to you and B and C give an illusion of depth, you will have realized that you can look at a medical x-ray film and *think about it three-dimensionally* even though you do not see it that way. The radiograph is a composite shadowgram and represents the added densities of many layers of tissue. You must think in layers when looking at any radiograph.

The most striking contrasts in radiodensity exist in the region of the chest, where air-filled lungs (radiolucent) on both sides of the muscular fluid-filled heart (relatively opaque) occupy the inside of a bony cage (a fretwork of crossed radiopaque strips).

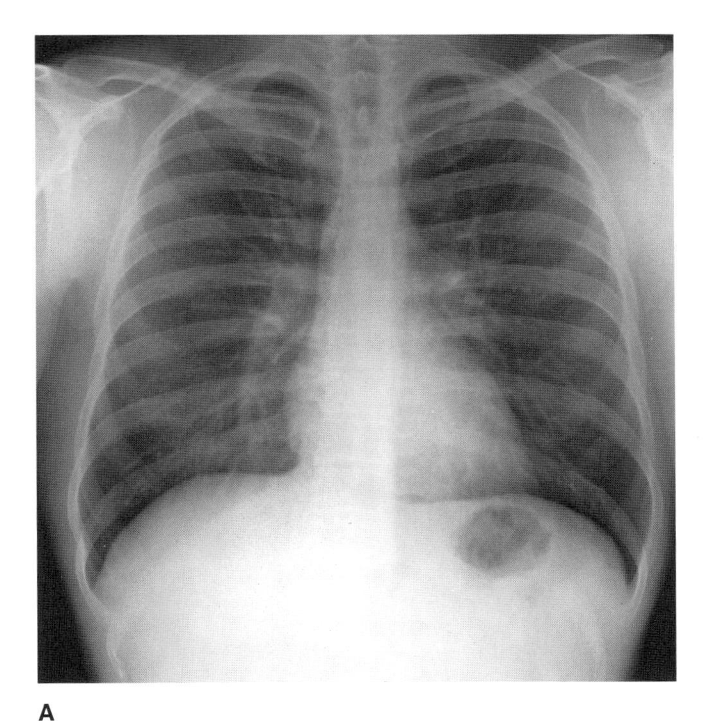

A

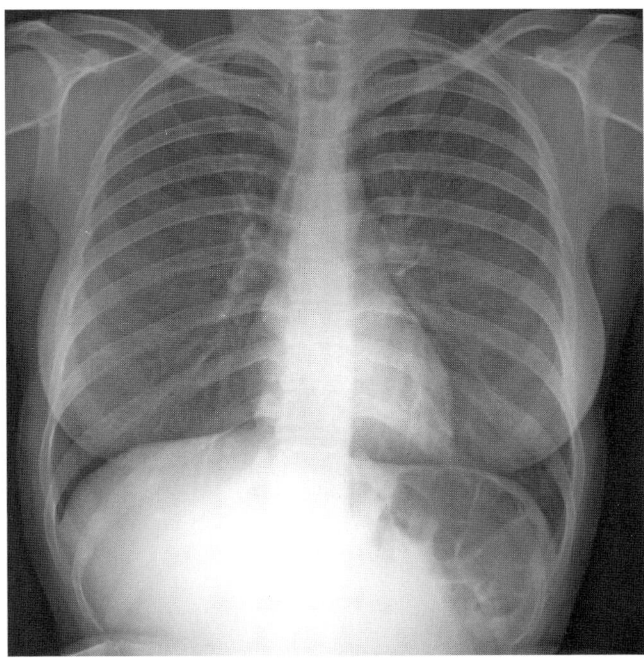

B

Figure 2.2

The Routine Posteroanterior (PA) Film

In Figure 2.2A imagine the structures through which the x-ray beam has passed from back to front: the skin of the back; subcutaneous fat; lots of muscle encasing the flat blades of the scapulae, the vertebral column, and the posterior shell of the rib cage; then the lungs with the heart and other mediastinal structures between them; the sternum and the anterior shell of the ribs; the pectoral muscles and subcutaneous fat; breast tissue and, finally, skin.

Note the crescents of density that are added in Figure 2.2B, where the x-rays have had to traverse the female breasts in addition to all the other tissue layers. Below the shadow of the breasts and above that of the diaphragm the film is blacker where more rays have reached it.

One of the problems that will worry you as you begin looking at chest films is how to put them up on the light boxes against which they are viewed: since they are transparent, you can look through them from either side. *Always place them so that you seem to be facing the patient.* Naturally this is only possible with PA and AP (anteroposterior) views.

X-ray films are usually marked by the technologist to indicate which was the patient's right side—or, in the case of films of the extremities, whether it was the right or left leg, for example. In chest films you can usually be somewhat independent of the marker because the left ventricle and the arch of the aorta cast more prominent shadows on the left side of the patient's spine. Always view a chest film, then, so that the patient is facing you with the patient's left on your right, and remember that "left" in regard to a finding on the film invariably means the *patient's* left. When you read "the right breast is missing" you are going to check, automatically, the breast shadow to your left.

Most of the chest films you see will have been made with the beam passing in a sagittal direction "posteroanteriorly," the x-ray tube behind and the film in front of the patient. This is the standard PA chest film, and films of all kinds are called *PA views if the beam passes through the patient from back to front.* It is customary to make a PA chest film of any patient who is able to stand and be positioned.

PA and AP Chest Films Compared

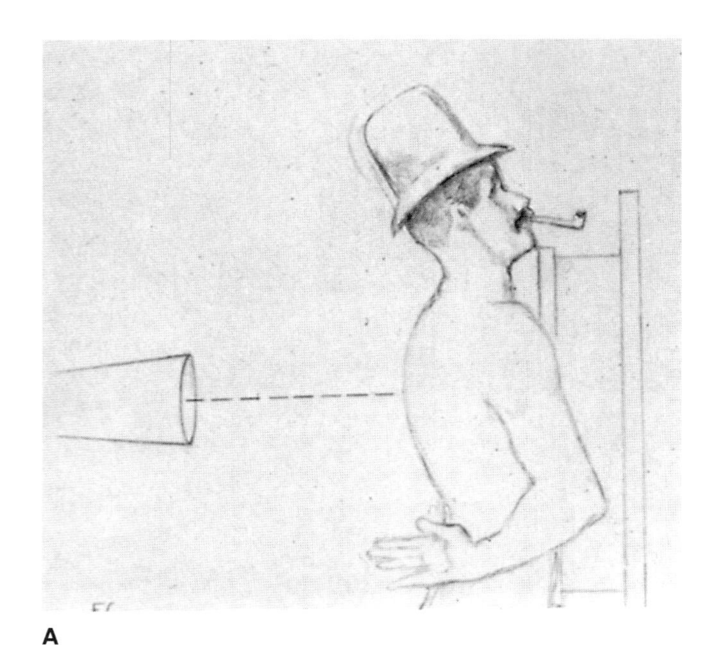

A

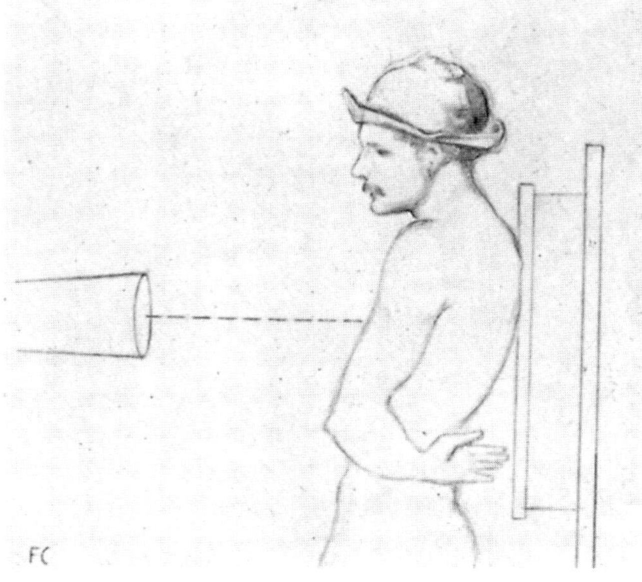

B

C

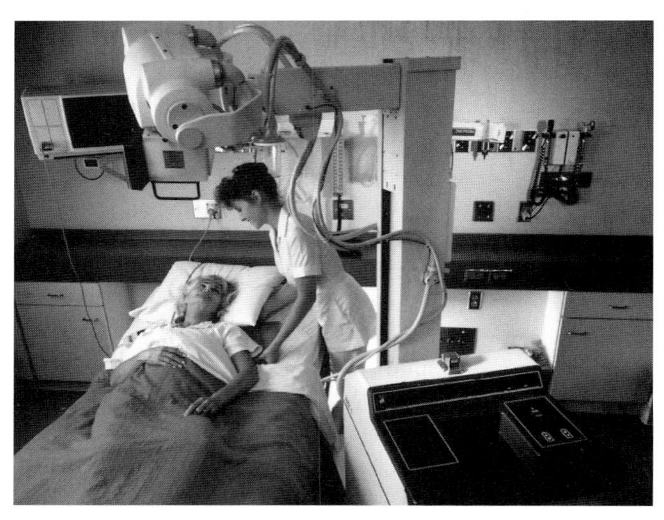

D

Figure 2.3. A: Posteroanterior beam produces a PA chest film, the conventional view you see most often. (Drawing after Cézanne.) B: Anteroposterior beam produces an AP film. Note that the film is named for the direction the beam takes through the patient. (Drawing after Cézanne.) C: Patient standing for a PA chest film; the x-ray tube is behind the technologist and the x-ray cassette in front of the patient's chest. D: Patient positioned for an AP chest film with portable equipment in the patient's hospital room. The technologist slides an x-ray cassette under the patient's chest; the x-ray tube is suspended from above.

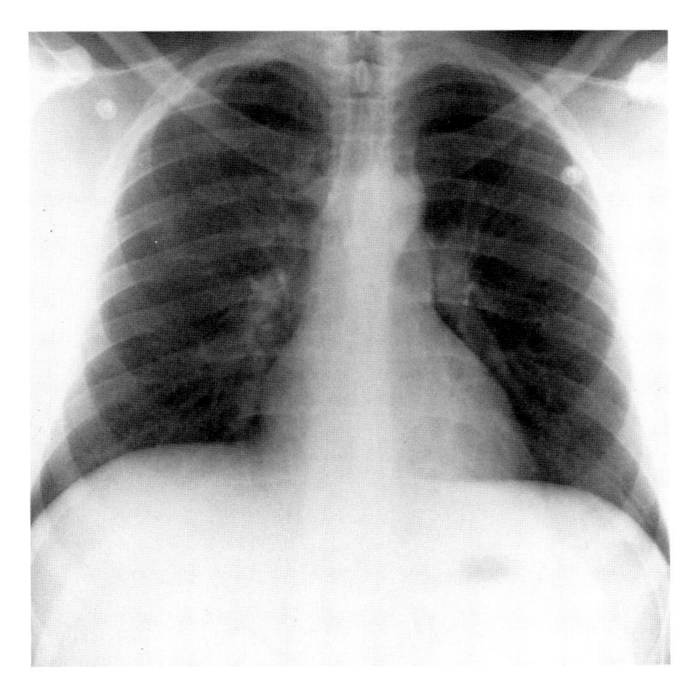

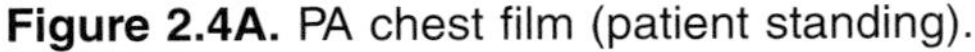

Figure 2.4A. PA chest film (patient standing).

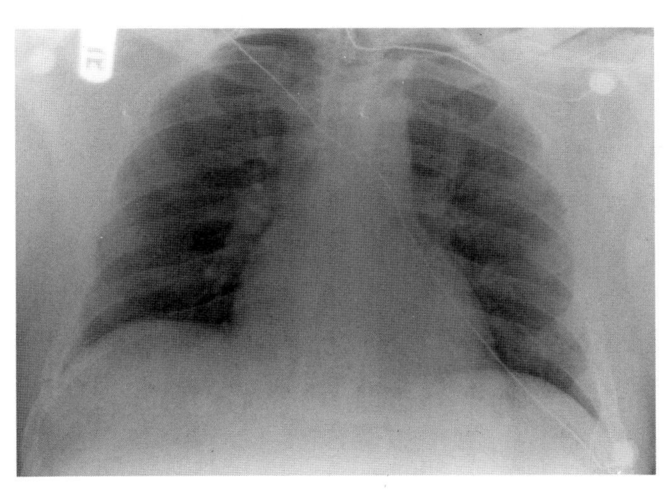

Figure 2.4B. AP chest film (patient is supine; same patient as in Figure 2.4A).

Less satisfactory but often valuable AP films of the chest are made with a portable x-ray machine when the patient is too sick to leave bed. The patient is propped up against pillows and the film is placed behind the patient, the exposure being made with the x-ray tube over the bed. Thus the ray passes through the patient "anteroposteriorly." You will be seeing such portable AP films of your very sick patients, especially those in intensive care units. Although they do not compare in quality with the PA films made with better technical facilities in the x-ray department, they do offer important information about the progress of the patient's disease. Sometimes patients who cannot stand are not too sick to be taken in their beds to the x-ray department and filmed AP with the equipment available there, a better film being obtained in this way than is possible with the portable machine.

After looking at many normal PA chest films, you will have formed a mental image of what a normal chest film looks like. A normal AP chest film, however, looks different for the following reasons. The divergence of the rays enlarges the shadow of the heart, which is far anterior in the chest, and the position of the patient, leaning back, makes the posterior ribs look more horizontal. These differences are even more marked at the shorter tube-film distances used in portable radiography at the bedside. Remember that, in addition, the diaphragm will be higher and the lung volumes less than in a standing patient.

The Lateral Chest Film

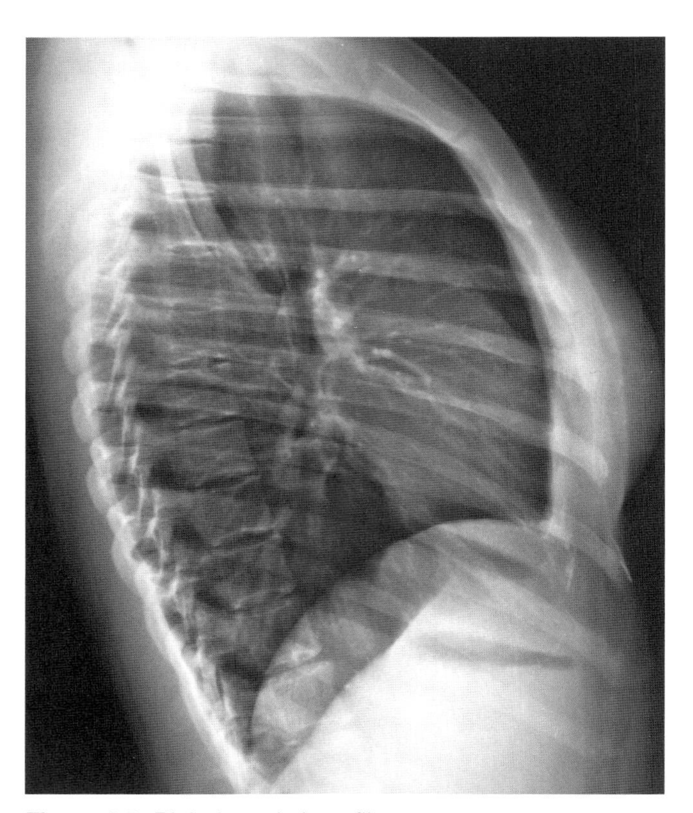

Figure 2.5. Right lateral chest film.

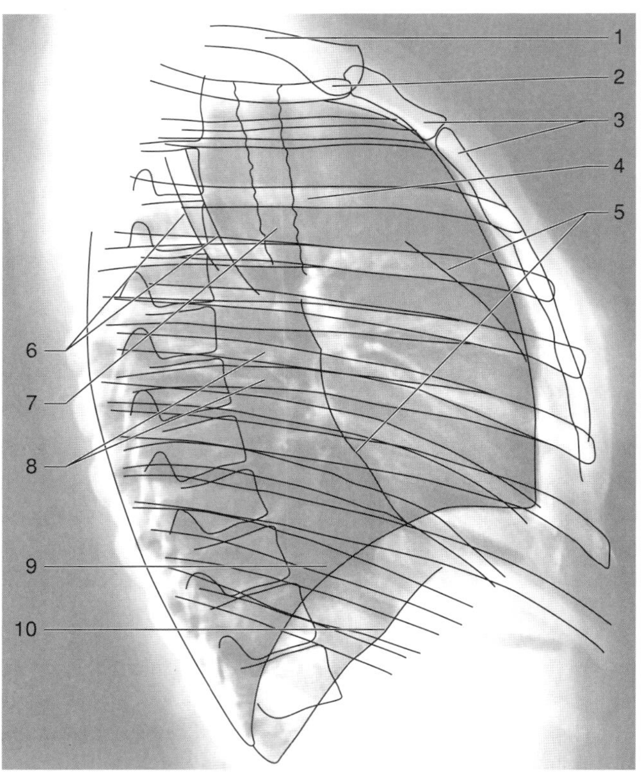

Figure 2.6. Labeled drawing of Figure 2.5: *1,* clavicle; *2,* medial end of the first rib; *3,* junction of the manubrium with the body of the sternum; *4,* pair of third ribs superimposed; *5,* anterior and posterior surfaces of the heart; *6,* scapulae; *7,* air in the trachea; *8,* pair of sixth ribs not superimposed; *9* and *10,* right and left hemidiaphragms.

After the standard PA film, the next most common view of the chest is the "lateral." It is marked with an R or an L *according to whether the right or the left side of the patient was against the film.* Most often a *left* lateral is made, because the heart is closer to the film and less magnified. Note in Figure 2.5 that the ribs all seem roughly parallel, some pairs superimposed by the beam, forming a single denser white shadow. Mark how far the vertebral column projects into the chest. Large segments of lung extending farther back on either side of the spine are superimposed on it in the lateral view. You may not be able to tell whether you have a right or left lateral in your hand if the technician has forgotten the marker.

The Lordotic View

Consider a patient who presented to the emergency ward with a persistent cough, one episode of blood-streaked sputum, weight loss, and a daily fever. The routine PA chest film in Figure 2.7 is not strikingly abnormal at first glance, but there was a strong clinical suspicion of pulmonary tuberculosis, so a special projection called a *lordotic view* was made. Because the patient stands leaning backward in exaggerated lordosis, the horizontal beam of AP x-rays foreshortens the chest by penetrating it at such an oblique angle that the anterior and posterior segments of the same ribs are superimposed. The result of this maneuver is, of course, to project the clavicles upward so that by looking between the ribs you can much more effectively visualize the lung tissue of the apex. Note that this case is a good exercise in the use of *bilateral symmetry* in examining films made with a sagittal beam. Now you are able to see that there *is* a fluffy white shadow in the upper part of the left lung, best seen in the second interspace. Note that there is nothing like it in the same interspace on the other side. Analysis of the patient's sputum confirmed the diagnosis of active tuberculosis.

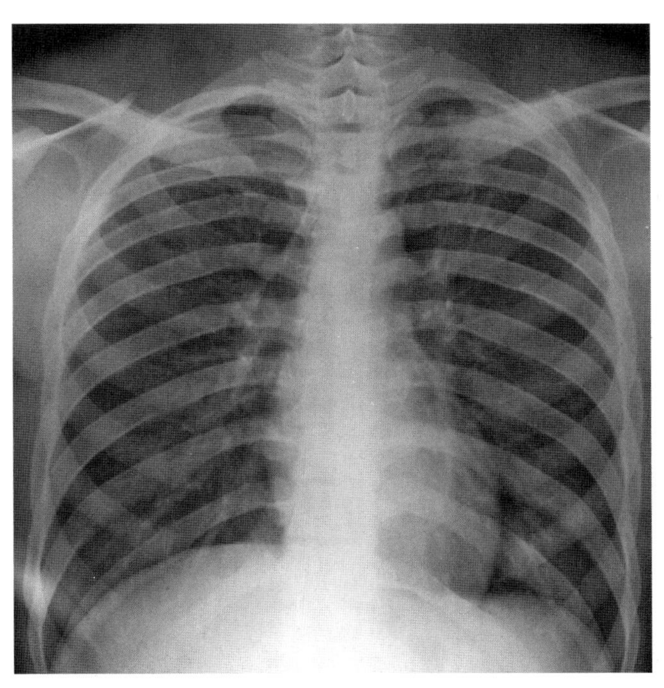

Figure 2.7. Standard PA view of the chest of a patient with cough, fever, weight loss, and hemoptysis.

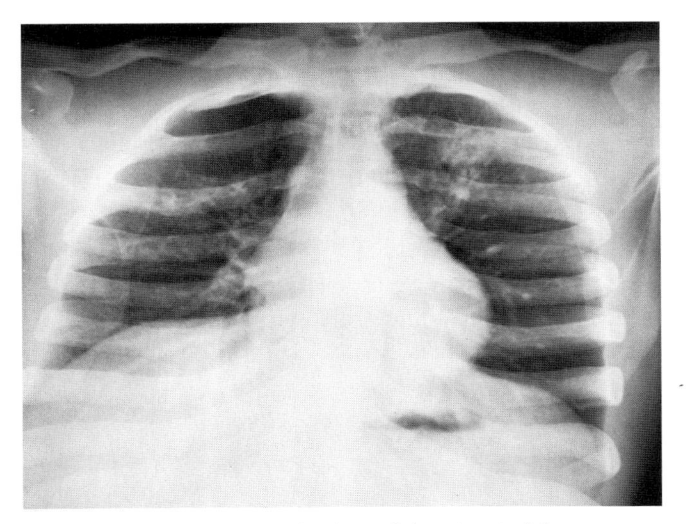

Figure 2.8. Special lordotic view of the chest of the same patient.

Figure 2.9. Position in which Figure 2.8 was made. (After Seurat.)

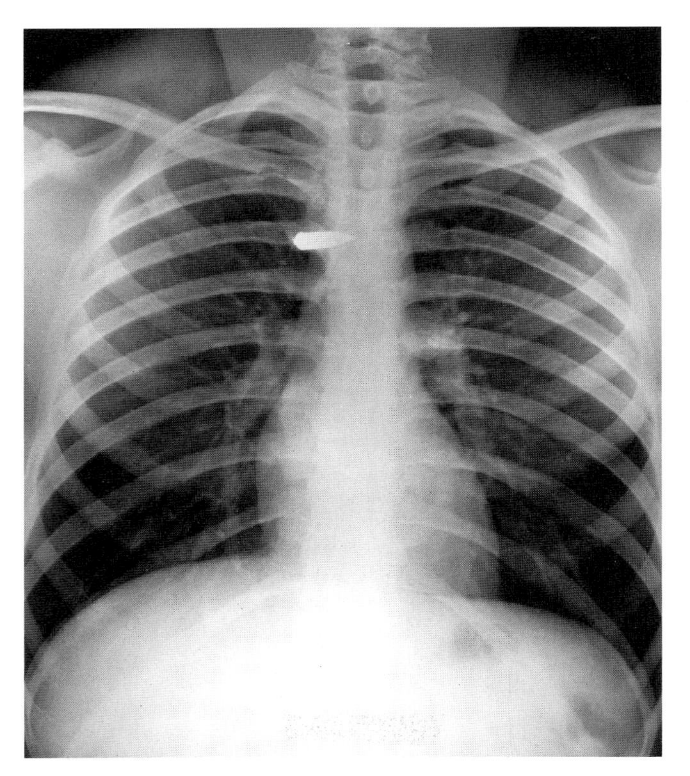

Figure 2.10

The chest film in Figure 2.10 offers you an opportunity to test your progress in three-dimensional thinking. There is an obvious metallic radiodensity. The shape of the metal object suggests it might be a bullet. This, in fact, is a film made of a soldier wounded in the Sicilian campaign during World War II. He was transferred to a hospital, where the surgeons observed what you observe. They requested, as you are about to do, a lateral view to determine the location of the bullet. It might, of course, be in any of the structures whose roentgen shadows superimpose in this view on the origin of the fifth rib.

The importance of localizing a bullet is illustrated by the cross-sectional drawing shown in Figure 2.11. If the bullet is lodged in the spinal cord or the trachea, or in one of the major vascular structures at this level, there may be less hope of saving the patient. In fact, the bullet was located harmlessly in the anterior mediastinum, had not injured any vital structure, and was removed without incident. (For the lateral view see Figure 2.12.)

You can never know precisely where a foreign body is located from a single radiograph. A film made at right angles to the first is essential, and minute metallic foreign bodies in the soft tissues of the extremities are localized very accurately by a refinement of this procedure. Fractured bones can appear to be in good position, end to end, in one film, while a second film made at right angles shows that the fragments are separated and do not align. Often the making of such supplementary lateral films is a routine matter. At other times you will have to ask that they be made on your patients. Always ask for a right lateral chest film if you think the lesion is on the right, so that the structure to be studied is as close as possible to the film. In most medical centers in this country a chest film series includes a PA chest film and a *left* lateral. Can you decide why?

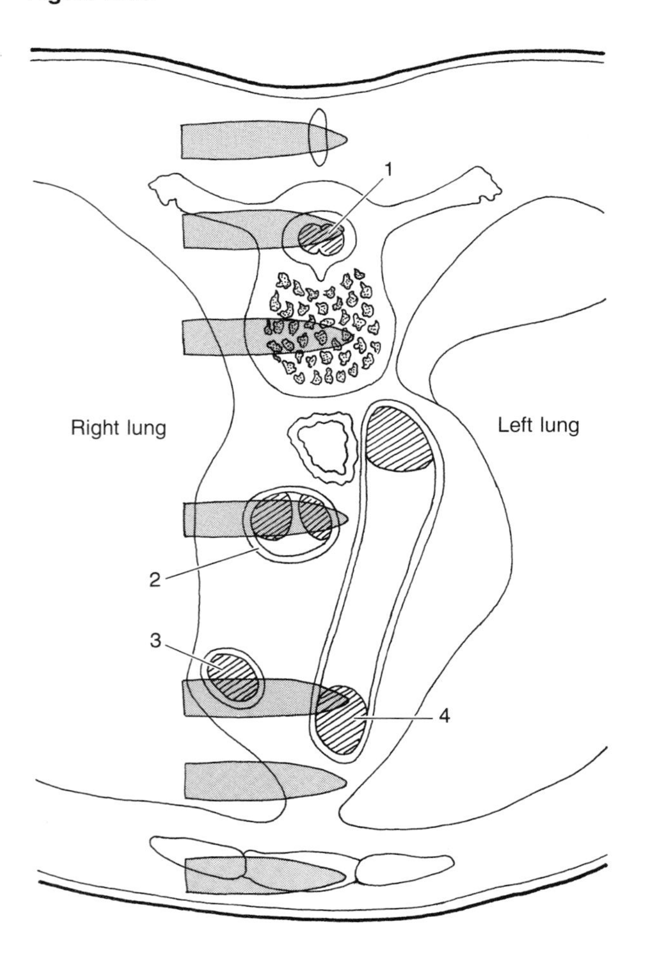

Figure 2.11. All the places the bullet could reside. *1,* spinal cord; *2,* trachea at bifurcation; *3,* superior vena cava; *4,* ascending aorta.

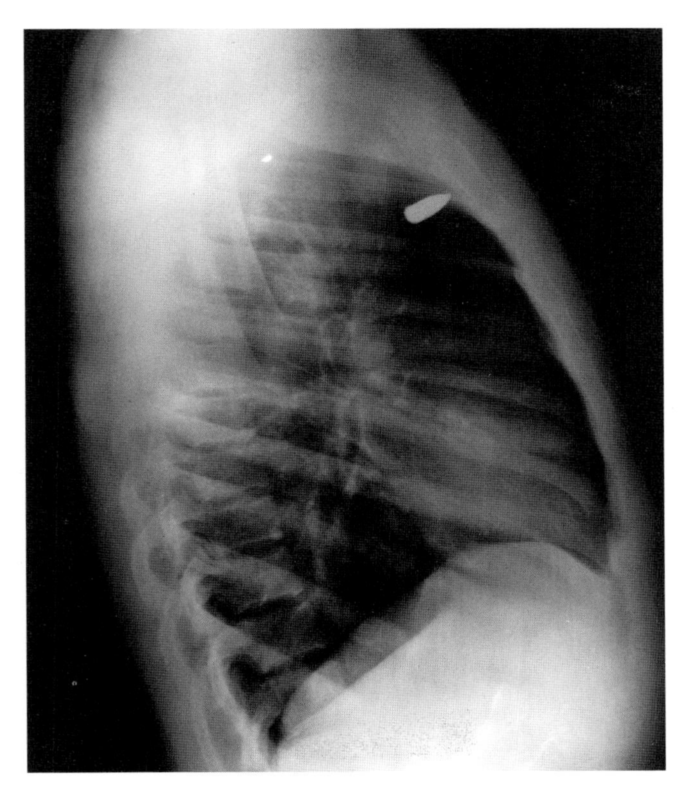

Figure 2.12

At about this point you will begin to say to yourself, "How am I going to know which views are important for me to understand and learn to use?" If, in the course of your training in medicine, you can familiarize yourself with the chest structures and their shadows *as seen in the standard PA and lateral views,* you will have built yourself a very useful and satisfying tool, and you should have no trouble in doing so. Do not feel confused or defeated if occasionally you see chest films that look like nothing you have ever seen before. Some of these will in fact be films of grossly abnormal chests. Others, however, will turn out to be films made by special or rarely used x-ray projections with which you are not yet familiar. You should rely comfortably on your acquaintance with the standard views, but not be incurious about or resistant to the possibilities of other modes of examination.

There are all sorts of ingenious obliquities of projection and many fascinating special procedures in the armamentarium of the radiologist that you will want to know about. Two of them, the posteroanterior obliques of the chest, are sometimes used in studying the heart or hila of the lung. Detailed study of the ribs is obtained by obliques made anteroposteriorly. Other procedures, designed for visualizing a particular structure in a particular way, also offer anatomic information not otherwise available. Sometimes the radiologist decides which views or procedures to obtain. At other times you will ask specifically for certain projections, or, better yet, discuss with the radiologist the advantages of their being used in the study of your patient's particular problem.

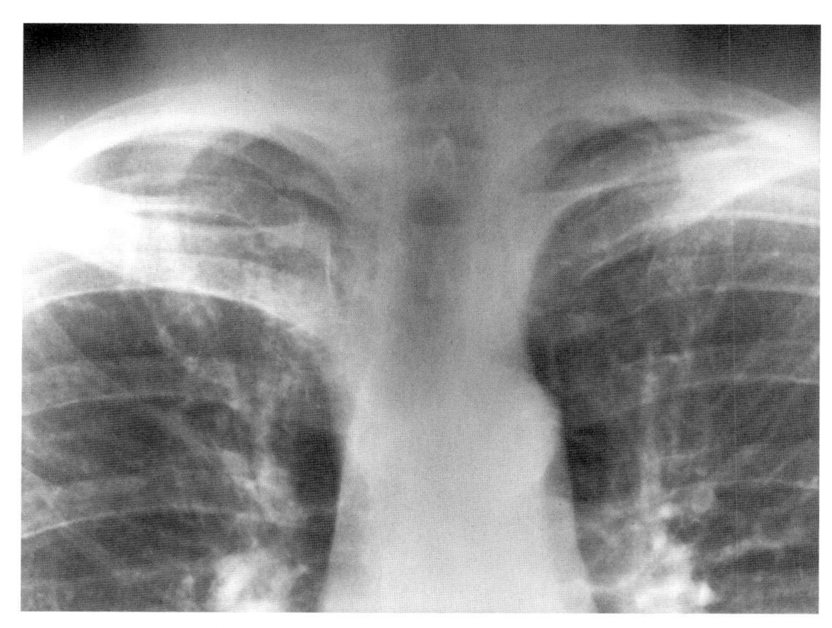

Figure 2.13A. PA film of upper bony thorax.

Conventional Tomography

The two films on this page spread were made of the same patient. Figure 2.13A is an ordinary PA radiograph; Figure 2.13B is a special-procedure film called a tomogram (or a conventional tomogram). Tomography will help you to visualize better the shadows that must be added together to make up the usual x-ray film. It is important, therefore, to understand how such radiographs are made.

Imagine that a frozen cadaver is sawed into *coronal* slices about 1 inch thick and that you then make a radiograph of each slice. Each film will have on it only the shadows cast by the densities of the structures in that slice. There will be no confusing superimposition of the shadows of structures from other slices to trouble you. How much simpler it would be, for example, to be able to study the manubrium and medial halves of the clavicles if they were not superimposed upon the shadow of the thoracic spine as they are in the standard PA chest film. On the next few pages you will find some radiographed cadaver slices to study. They are arranged in order from front to back, the first slice having been omitted. (It included the anterior chest wall, rib cartilages, and sternum.) You will find it helpful to refer back to these slices as you learn the x-ray appearance of various organs and structures. Now notice how well you can see in Figure 2.13B the shadows cast by the clavicles where they join the manubrium.

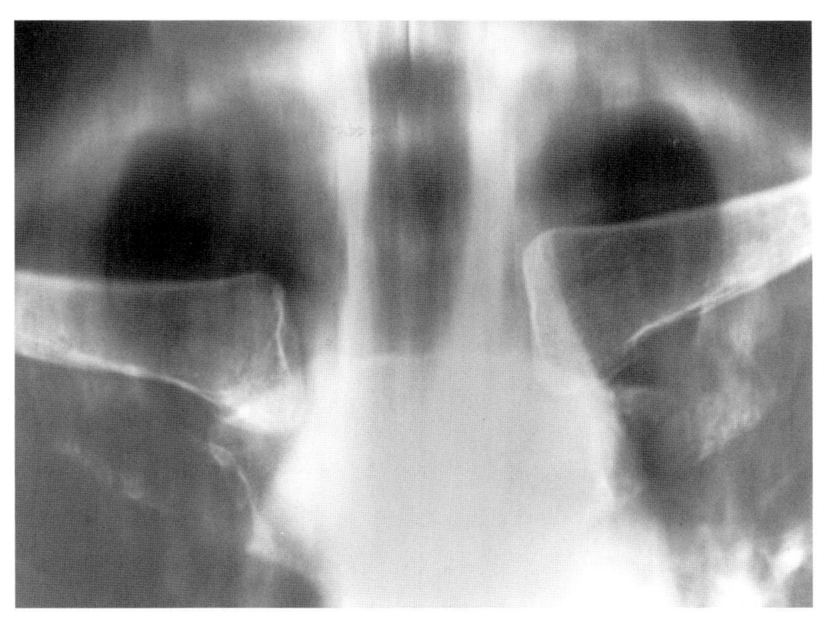

Figure 2.13B. Coronal conventional tomogram of the anterior part of the bony thorax showing the junction of the clavicles with the manubrium.

Conventional tomographic studies effectively slice the living patient so that you can study the shadows cast by certain structures free of superimposed shadows. The term "tomogram" is a general one and there are different types of tomographic studies, the techniques of which depend upon the result desired, that is, the shadows intended for study and those you wish to distort.

Conventional tomograms are less likely to be requested today than in the past. For many procedures conventional tomography has been replaced by computed tomography, which you will learn about later.

On first acquaintance tomograms may look blurred and confusing to you. Whenever you are puzzled by one you see in this book, try coming back to the cadaver slices in Figures 2.14–2.17 to get your bearings, remembering that *only the structures in one plane will be in focus in the tomogram.* Remember too that the thickness of these particular cadaver slices may not match perfectly the chosen plane of the study you happen to be looking at, since the pivot point determining the plane of a tomographic study is calculated arbitrarily for a certain distance in centimeters from the surface of the body.

Radiographs of Coronal Slices of a Frozen Cadaver

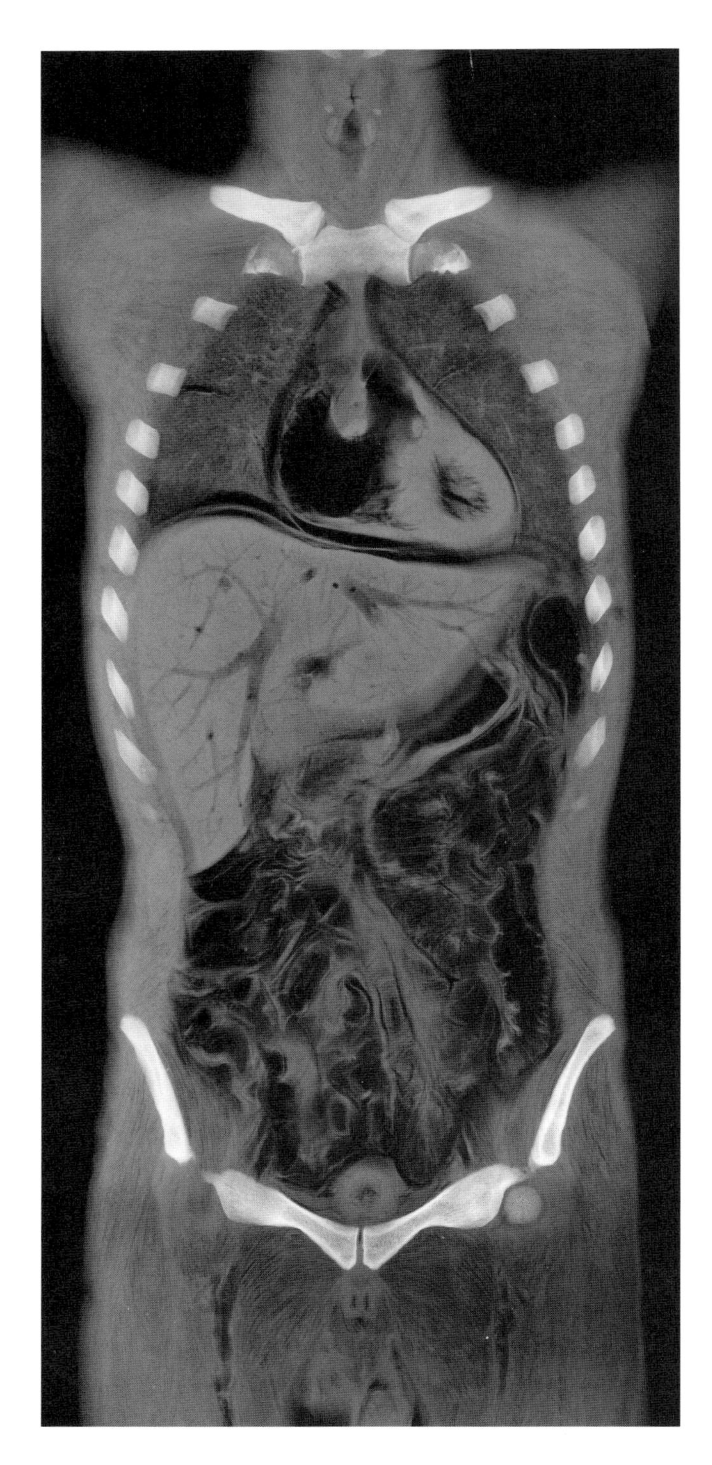

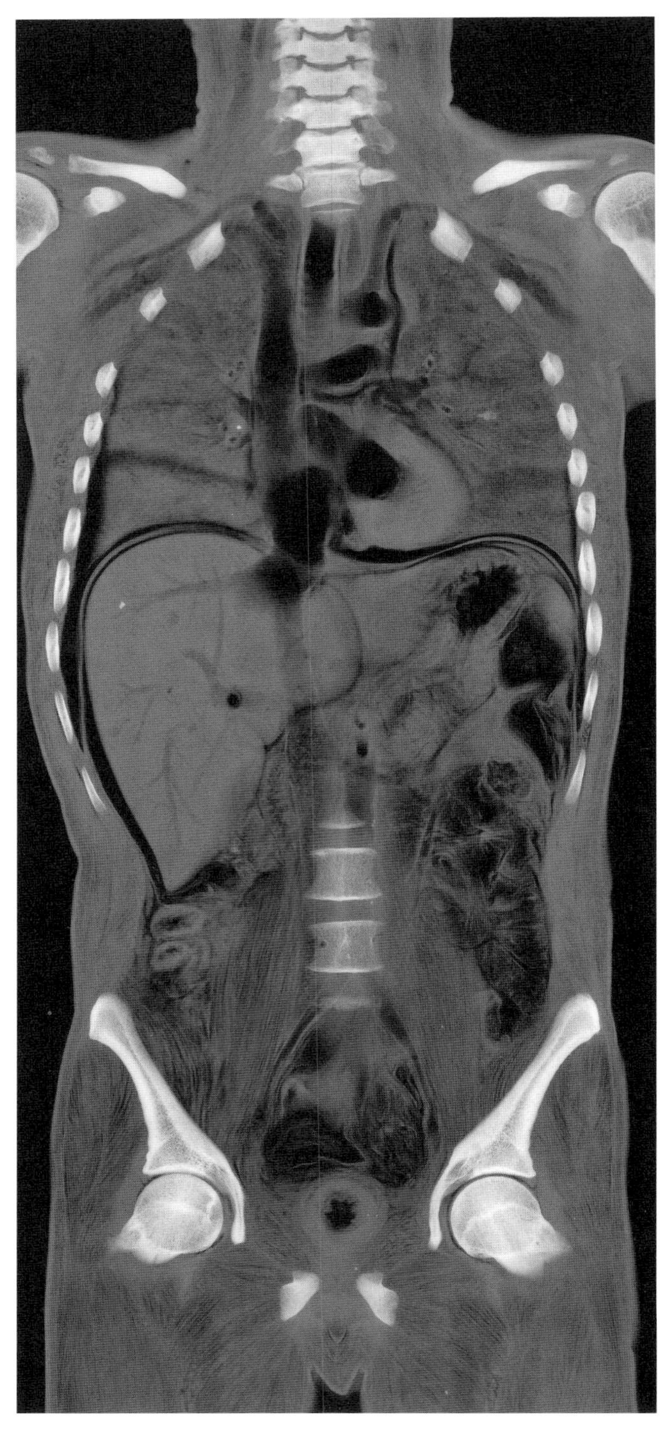

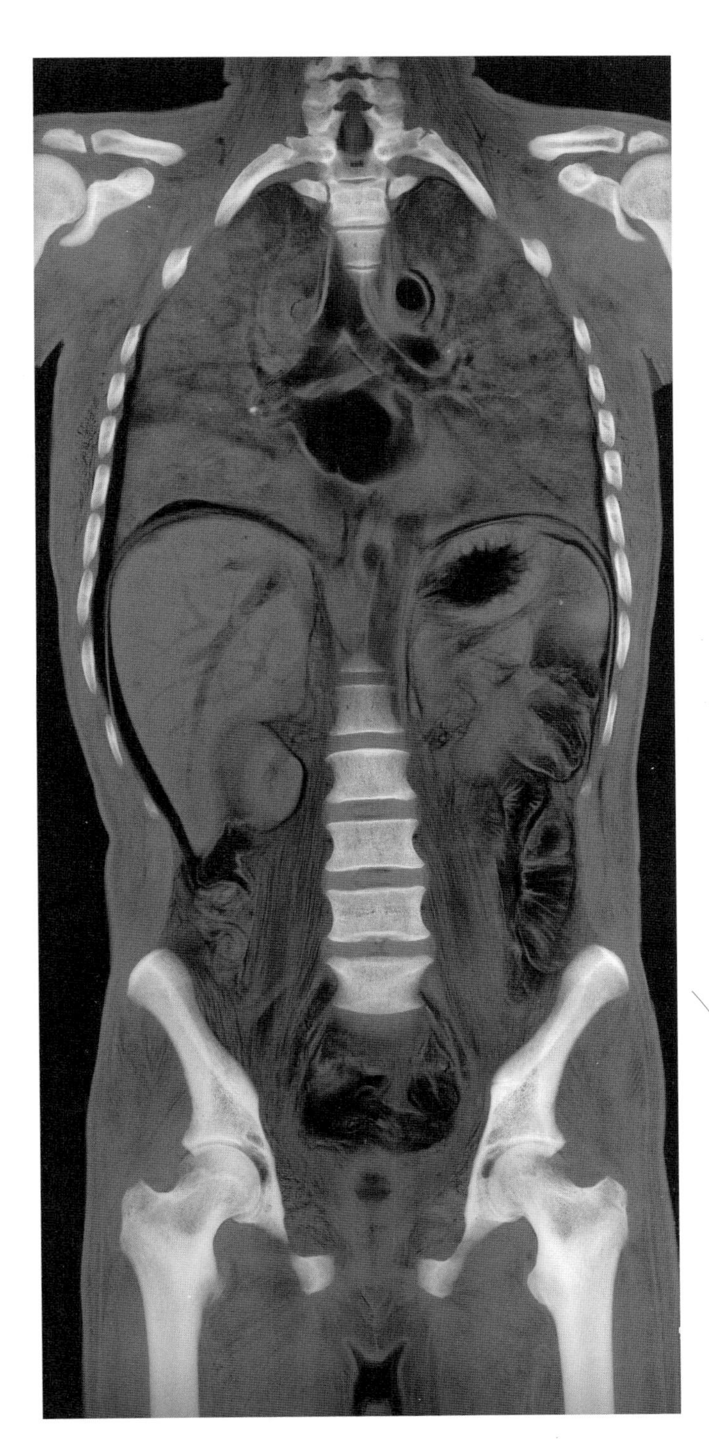

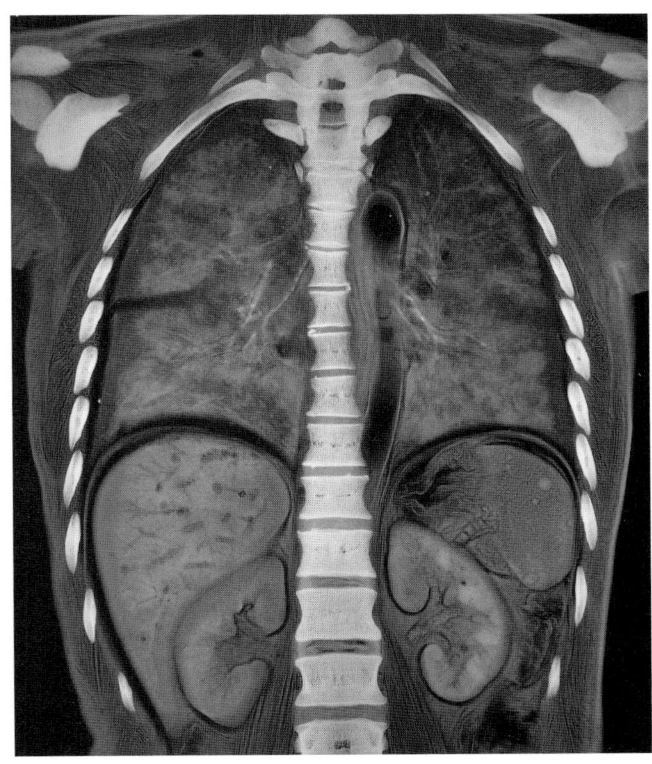

Figures 2.14 to 2.17. Radiographs of a series of coronal slices of a cadaver, arranged from front to back. Identify the following:

Junction of manubrium and clavicles
Superior vena cava (empty and filled with air)
Fundus of stomach
(Each locates the level of the slice just as a body-section tomographic study would identify the level of the slice by including certain structures and excluding others)
Symphysis pubis
Empty cavity of left ventricle
Trachea, carina, and major bronchi with air-filled left atrium immediately below

Note the change in shape of the liver from section to section. Note too that these radiographs are *unlike* tomograms because there are no blurred images of structures in other slices.

Problems

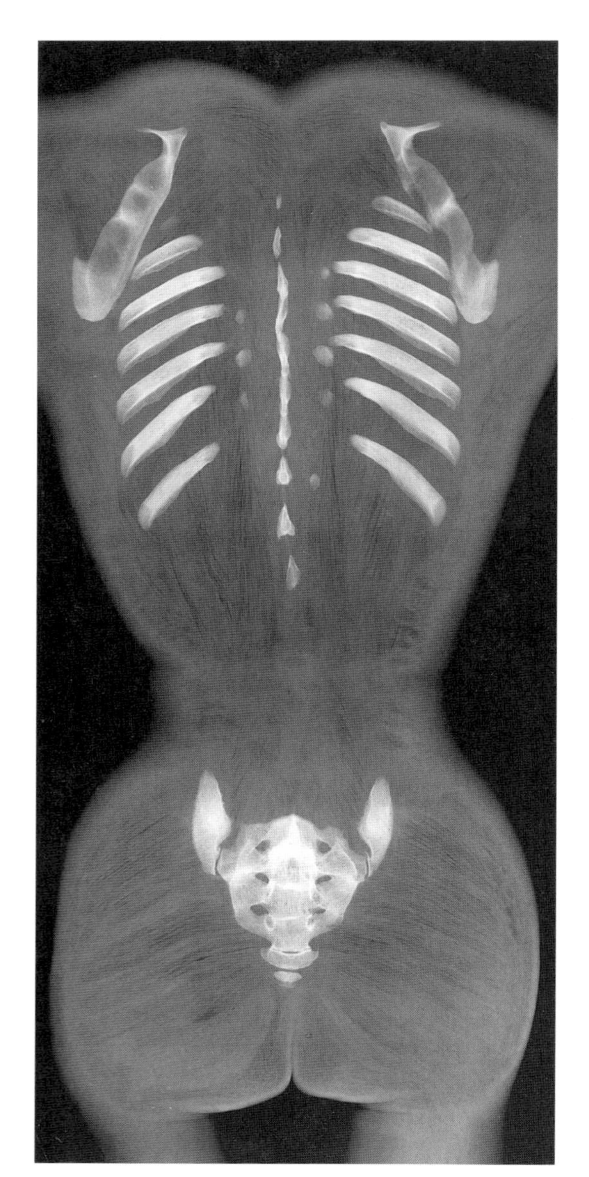

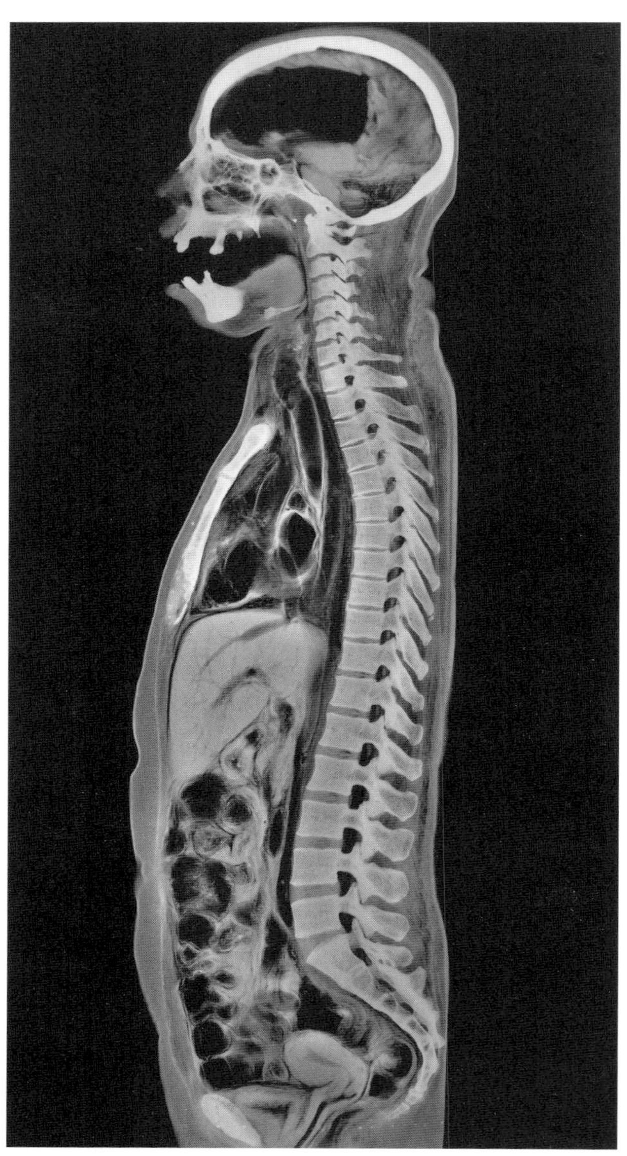

Figure 2.18 ***(Unknown 2.1) (left)*** **and Figure 2.19** ***(Unknown 2.2) (right).*** Figure out precisely what has been radiographed.

Conventional Tomograms of the Living Patient in the Coronal Plane

Conventional tomograms are made by *moving both the x-ray tube and the film around the patient during the exposure,* as in Figure 2.20. They are moved about a pivot point calculated to fall in the plane of the object to be studied. In this way the shadows of all the structures *not* in the plane selected for study are *intentionally blurred* because they move relative to the film. Thus in the diagram (Figure 2.21) the object to be studied, *b,* will be "in focus" in the film, while the shadow of an object at *a* will be magnified, blurred, and distorted to lie between *a′* and *a″* on the film. Only the structures in the plane of the pivot point will be recognizable (as in Figure 2.22C); the shadows representing organs in front of or behind it are distorted in such a way that shape and form are no longer recognizable and the blurred images are easy for your eye to ignore. Conventional tomography is used as an adjunctive study whenever detail is needed of a structure superimposed on and obscured by other structures in the line of the x-ray beam.

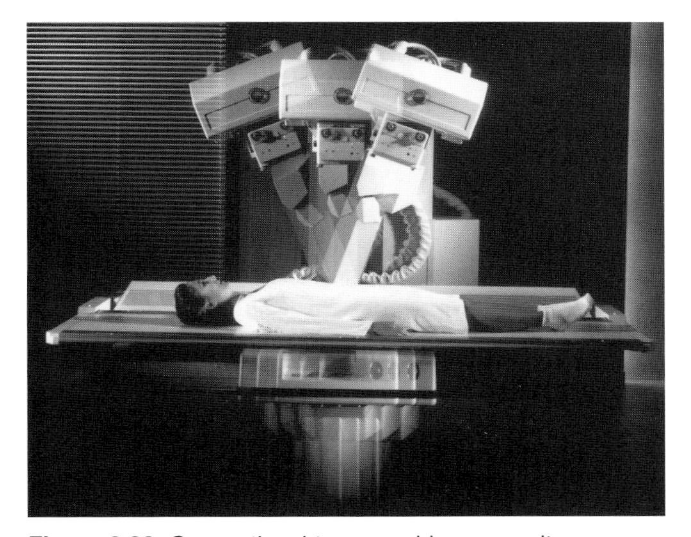

Figure 2.20. Conventional tomographic x-ray unit showing range of tube motion.

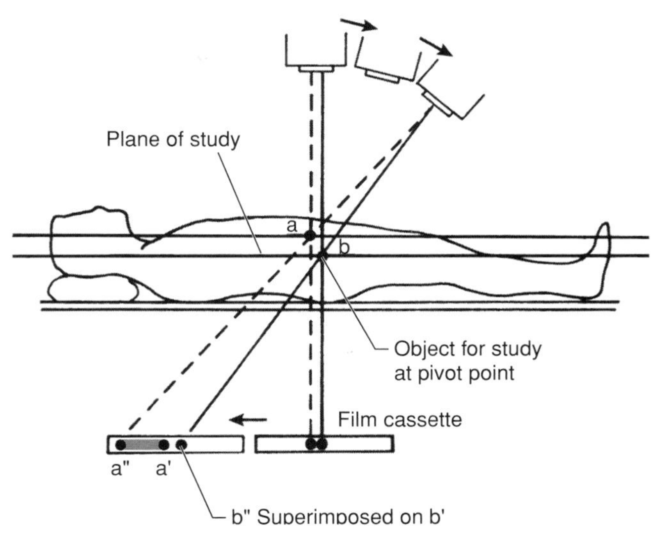

Figure 2.21. How conventional tomography obtains a coronal slice.

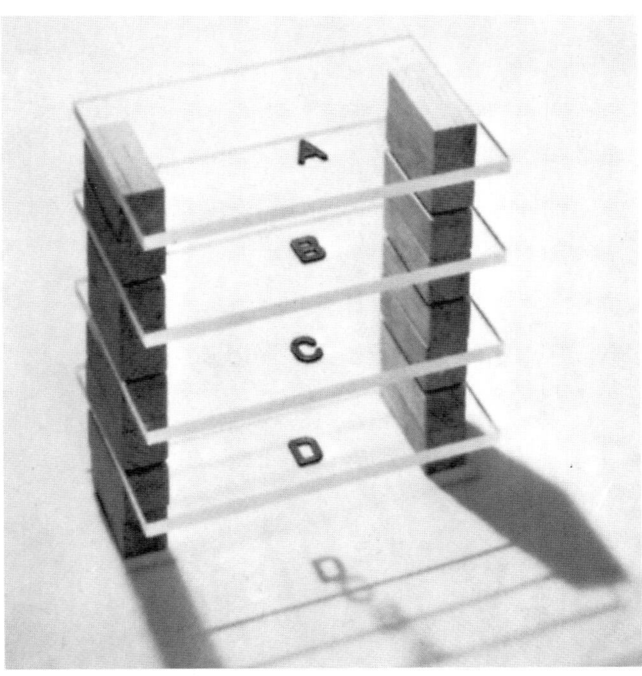

A

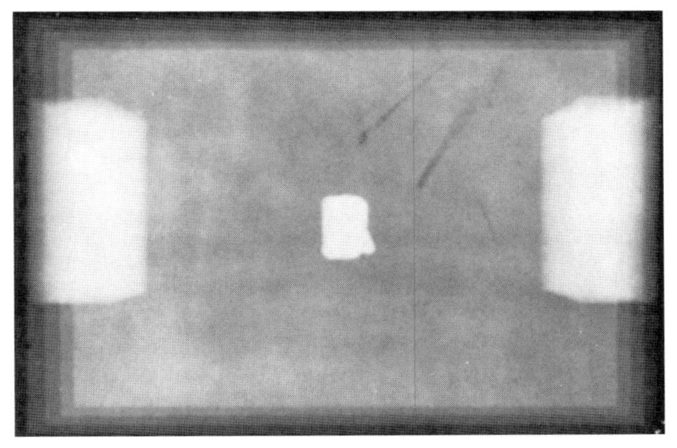

B

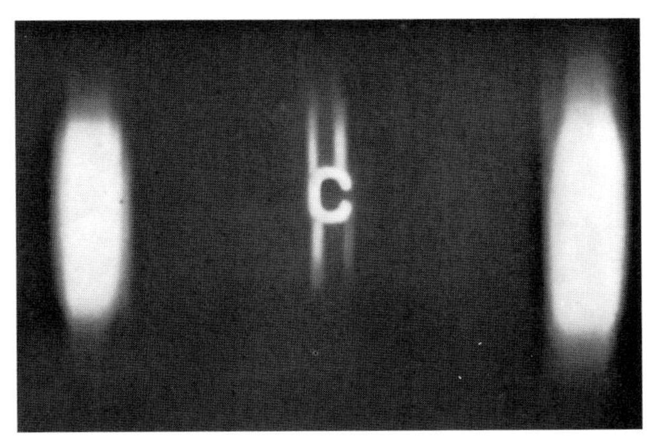

C

Figure 2.22. Demonstration of the effect obtained with conventional tomography. A: Series of plastic shelves, each holding a lead letter superimposed vertically. B: Conventional radiograph superimposes the shadows of the letters. C: Conventional tomogram at the level of *C* shows that letter clearly but distorts and blurs the others.

Fluoroscopy

Fluoroscopy is a common radiological technique that allows real-time visualization of the patient. You may already be familiar with the use of fluoroscopy during contrast examinations of the gastrointestinal tract to follow the course of barium through the esophagus, stomach, and bowel. Fluoroscopy is also used to guide the radiologist performing selective arterial and venous catheter placement for angiographic procedures. In addition, most interventional radiological procedures require fluoroscopic guidance.

During fluoroscopy a continuous beam of x-rays passes through the patient to cast an image on a fluorescing screen, which is amplified by an electronic image intensifier and viewed on a high-resolution television screen. Figure 2.23 shows an angiographic suite in which a coronary arteriogram is being performed. The x-ray tube is located underneath the patient and the large cylindrical image intensifier above; the beam of x-rays is passing through the patient from below. The angiographer observes a coronary artery contrast injection on the television monitors. Note that on the fluoroscopic image, black and white are reversed so that bone and contrast agents appear dark and radiolucent structures such as the lungs appear light.

Figure 2.24A shows a simpler fluoroscopic room, such as might be used for gastrointestinal examinations. Again the image intensifier is located above the patient; the x-ray tube is concealed below within the x-ray table. Films taken by the radiologist during the fluoroscopic portion of the procedure are called "spot films." They detail small areas of special interest observed during the fluoroscopic segment (Figure 2.24B). On completion of fluoroscopy the x-ray technologist obtains a set of larger conventional x-rays (Figure 2.24C), which are called "overhead" films because they are taken with a second x-ray tube that is rolled along a track suspended above the patient.

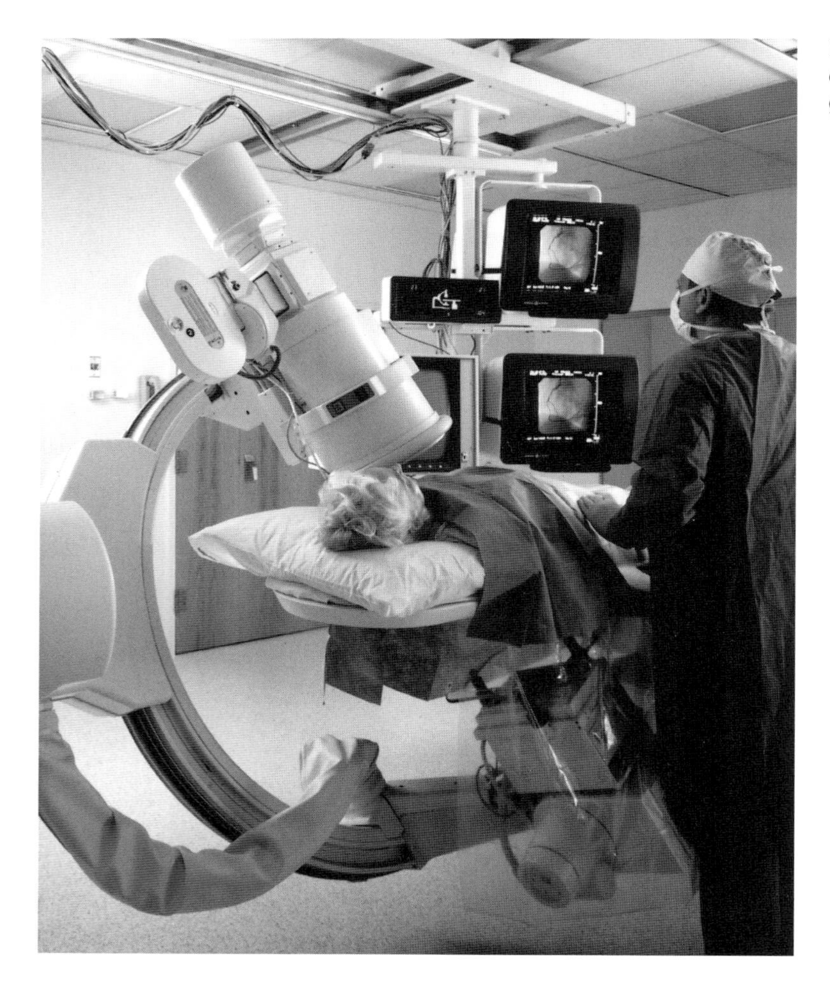

Figure 2.23. Fluoroscopy equipment in an angiographic procedure room.

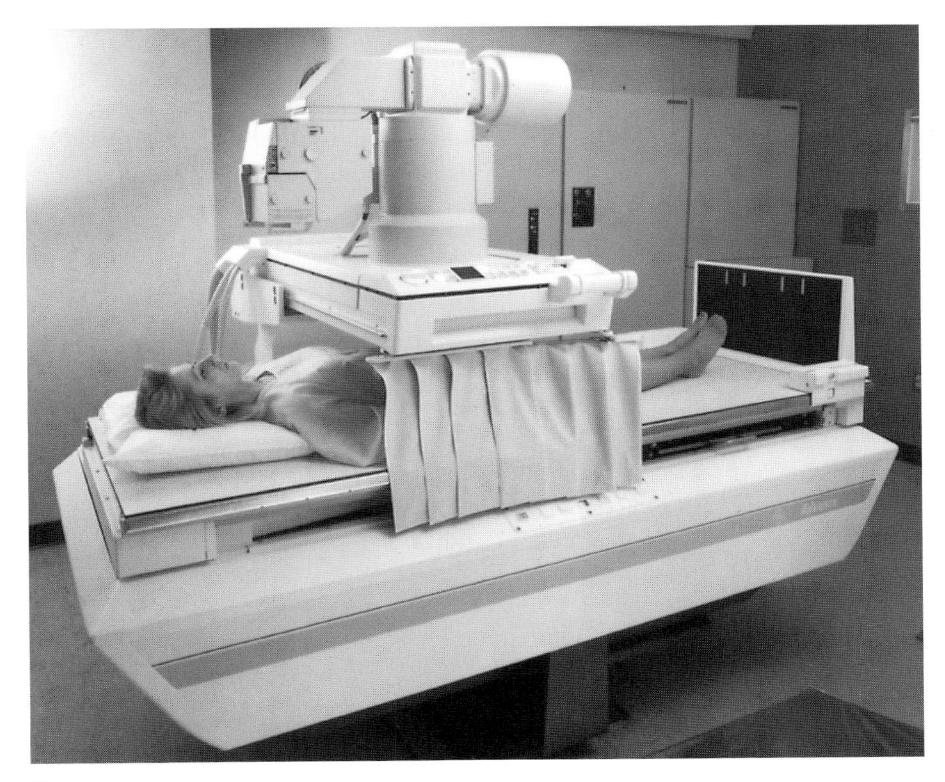

A

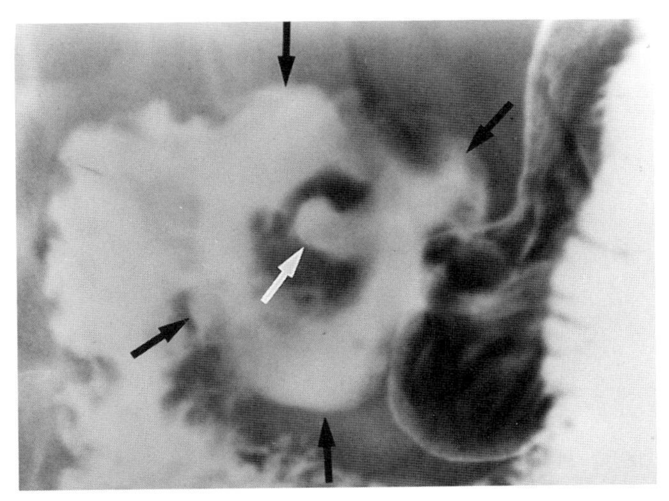

B

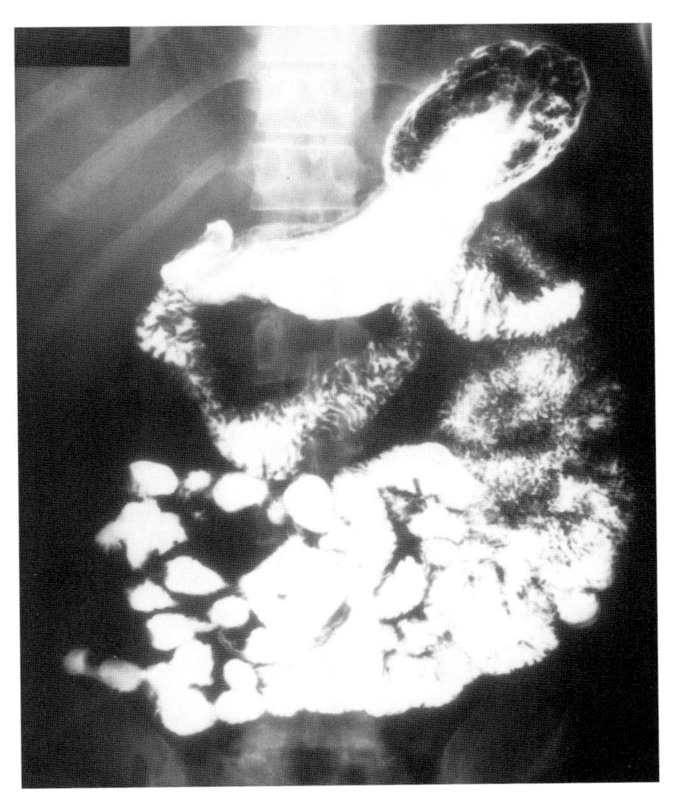

C

Figure 2.24. A: Fluoroscopy equipment for gastrointestinal examinations. B: Spot film of the duodenum taken during fluoroscopy. The *white arrow* indicates an ulcer crater within the duodenal bulb *(black arrows)*. C: Overhead film of the contrast-filled upper gastrointestinal tract taken after fluoroscopy. Contrast material opacifies the stomach, duodenum, and most of the small bowel; none has yet passed into the colon.

Angiography

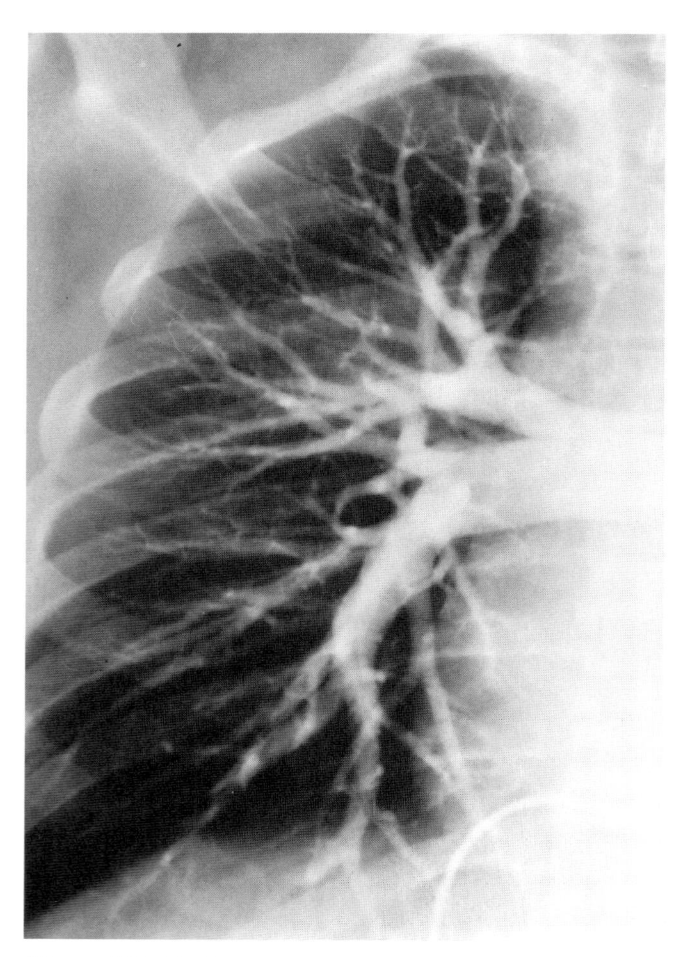

Figure 2.25. Right pulmonary arteriogram.

Angiography includes a variety of procedures in which the vascular system is imaged by x-ray during intravascular injection of iodinated contrast agents. Images of arterial structures are called *arteriograms,* and images of venous structures are called *venograms.* The arterial system is usually opacified by contrast injection into a percutaneously placed small-caliber, flexible, arterial catheter, usually introduced through a femoral artery. Under fluoroscopic guidance the catheter is manipulated through the arterial system until its tip is in position within the artery under examination. A wide variety of catheter shapes and styles are available, as well as sophisticated directional equipment that permits selective catheterization of virtually every major artery in the body. Once the catheter tip is positioned, contrast material is infused by a power injector at a controlled rate and volume, while x-ray filming is obtained with a rapid film changer, digital radiography system, or movie camera (cineradiography).

Venography of major veins such as the superior vena cava, inferior vena cava, and renal veins is performed by similar techniques, using a femoral vein approach. Some venous structures, such as the pulmonary veins and portal vein, do not anatomically lend themselves to direct catheterization techniques and are usually imaged by contrast injection into the supplying arteries (pulmonary artery, superior mesenteric artery) and filming through the venous phases of the arterial contrast injection. Venous studies of the extremities (leg venogram, arm venogram) do not require actual catheter techniques. These studies require only a simple injection of contrast material into the peripheral veins of the foot or hand.

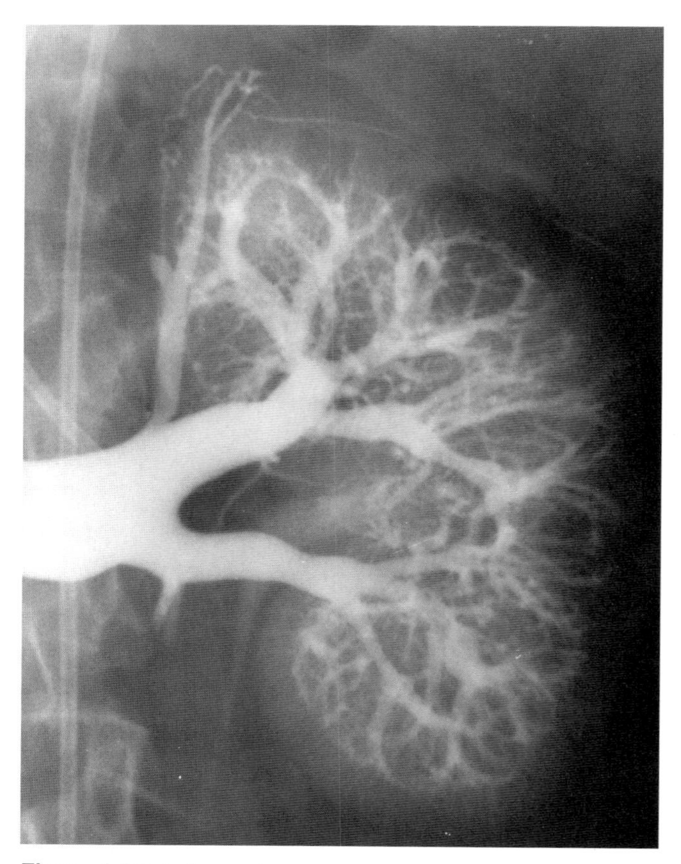

Figure 2.26. Left renal venogram.

Computed Tomography

Computed tomography (CT) gives you a whole new way of looking at the body because it provides the equivalent of cross-sectional slice radiographs of the living body. These are what the lay public calls CAT (computerized axial tomography) scans, and they are a vital source of radiological information in medicine.

You can begin by understanding the difference between plain radiographs, conventional (or plain) tomography, and CT. Remember that ordinary plain x-ray films are superimposition shadowgrams: the images of all superimposed structures appear on the film. *Conventional* tomography gives you sharply focused radiographic images of one plane of the patient, upon which are superimposed (unfortunately) the *blurred* images of structures in slices on both sides of the plane chosen for study.

A CT scan, on the other hand, gives you focused radiographic information about one cross-sectional slice of the patient only, without any confusing superimposed images. Thus a CT scan gives you a range of density values for a particular chosen slice of the patient, which should be studied with regional cross-sectional anatomy in mind. You will be able to learn the relationships between structures in the body much more accurately with the help of the added dimension CT provides.

An awareness of the relative x-ray densities of different tissues and organs and their interfaces with fat planes in the body helps you as you look at CT scans. In computed tomography a pencil-thin collimated beam of x-rays passes through the body in the axial plane chosen for study as the x-ray tube moves in a continuous arc around the patient. Carefully aligned and placed directly opposite the x-ray tube are special electronic detectors, a hundred times more sensitive than ordinary x-ray film. These detectors convert the exiting beam on the other edge of the body slice into amplified electrical pulses, the intensity of which depends upon the amount of the remaining beam of x-rays that has not been absorbed by the intervening tissues. Thus if the beam has passed mainly through dense areas of the body (such as bone), fewer x-rays will emerge than when the beam traverses mainly low-density tissue (such as lung). The x-ray tube and detectors are housed in the gantry, the doughnut-shaped structure through which the patient passes during scanning. The gantry can actually be tilted to take slices at an angle to the long axis of the patient.

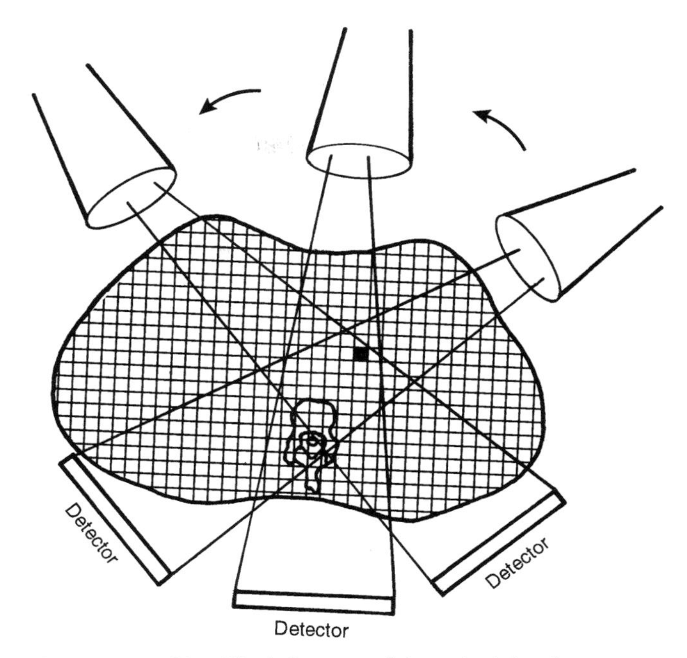

Figure 2.27. Simplified diagram of the principle of computed tomography.

If you conceive of a body slice as a mosaic of unit volumes, or voxels (see Figure 2.27), forming a geometric grid, you can see that a single denser unit volume (perhaps calcium containing, like the small black square in Figure 2.27) will absorb more of the beam than other, less dense neighboring voxels.

As fast as it is received by the detectors, this information is conveyed to a computer, which then calculates the x-ray absorption for each voxel in the mosaic. The pictorial arrangement of absorption values makes up the final CT image. The absorption value is expressed in Hounsfield units (after one of the inventors of CT). Water was arbitrarily assigned the value of zero, while denser values range upward to bone (which can be +500 or more). Less dense structures range downward through fat to air (which can be −500 or less).

The attenuation number so obtained for each voxel in the mosaic matrix slice is converted into a dot on a television monitor screen, the brightness of which depends on the density of that unit volume and thus reflects its anatomic structure. Denser tissues (such as bone) appear white; less dense tissues appear darker; and air appears black. The "picture" produced is equivalent to a radiograph of that cross-sectional slice of the living patient.

It is conventional to view the CT scan so produced as though you were looking up at it from the patient's

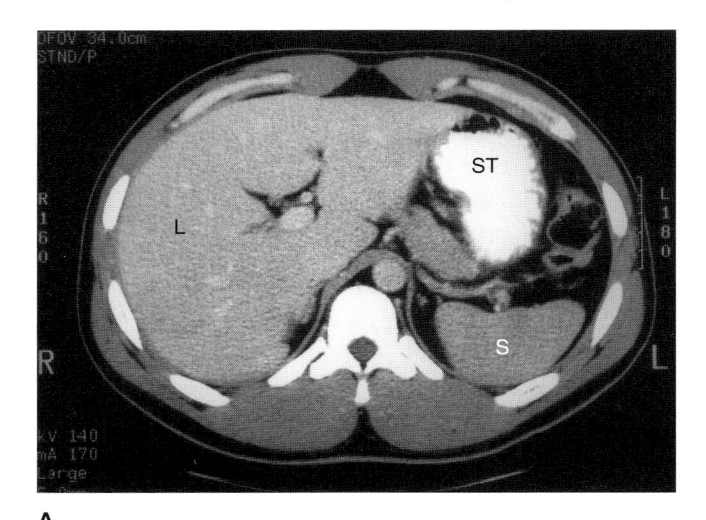

A

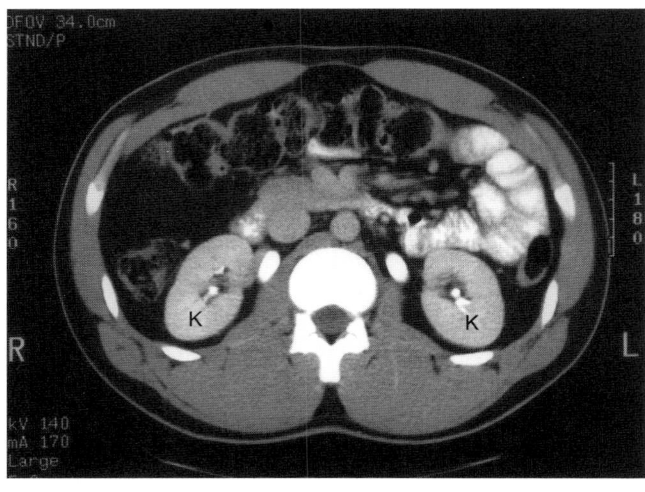

B

Figure 2.28. A: CT scan of the upper abdomen showing the liver *(L)* filling the right upper quadrant and the smaller spleen *(S)* posteriorly on the left. Note the oral contrast–filled stomach *(ST),* anteriorly on the left. B: CT scan at a lower level showing both kidneys *(K's)* opacified with intravenous contrast media. Loops of small bowel opacified with oral contrast appear to the left. Across the anterior abdomen at this level is a portion of the transverse colon containing air and fecal material.

feet (Figures 2.28A and 2.28B), and it is important to remember that therefore the structures seen on your right are those on the left side of the patient's body, just as they are when you view an ordinary chest film. Permanent images are produced by transferring the images onto x-ray film with a laser camera. For each CT slice obtained, the imaging settings (window and level) can be altered on the CT scanner controls to better show individual tissues (bone versus lung versus heart and great vessels). You will find that most CT scans are filmed at more than one setting. For example, a chest CT (Figure 2.29) is generally filmed with "lung windows" to optimally show the lung parenchyma and "soft-tissue windows" to best show the heart, blood vessels, and other structures in the mediastinum and chest wall. A head CT in a trauma patient is filmed with soft-tissue windows to show any brain injuries and also bone windows to show any fractures.

The CT scans in your patient's film envelope are documented on 14 × 17 inch x-ray films (called "hard copies"), each such film having from 6 to 12 to 20 scan slices in sequence, so that you can look from one slice to the next, above or below, for additional information about the form of a structure or organ. The actual CT images are usually also stored in computer form (magnetic tape, laser disk), so that additional hard copies can be obtained, if necessary, at a later date.

The usual CT series of scans for examining the chest and abdomen consists of contiguous 5- or 10-millimeter-thick slices, but slices as thin as 3.0 millimeters or even 1.0 millimeter can be obtained when finer detail is needed for diagnosis. In most radiology departments CT protocols are written and followed that detail the most optimal CT technique for examining various body regions or for evaluating various clinical conditions. The protocols describe not only the slice thickness and extent of study (location of first and last slices), but gantry angle (0 degrees for a true cross section versus tilted to better show structures in other angles of section), whether any oral, intravenous, or other contrast material is required, and whether any computer rearrangements (special reformations) of the axial slices are required. The protocols also describe what kind of "windows" (bone, soft-tissue, lung, liver, brain, etc.) should be hard copied for each slice. The x-ray dose per slice of a CT scan varies from 1 to 4 rads (but only to the slice being imaged) and is comparable to the exposure for conventional x-ray studies of the area.

High-density materials such as barium or metal (a hip prosthesis or metal surgical clips) may produce artifacts like bright stars with sharply geometric radiating white lines that may degrade the image obtained and interfere with the information available from it. Motion also degrades the image, but this effect is minimized with newer, high-speed scanners.

Conventional CT scanners require only 1 to 2 seconds to complete a slice, but patients who cannot hold their breath may have motion artifacts on their scans. This may be a problem with unconscious, very ill, or dyspneic patients and small children requiring CT studies. But the newer, high-speed scanners can virtually eliminate respiratory motion. A conventional CT scan may take 10 to 20 minutes for completion of slices; with a high-speed scanner (helical or spiral scanner) an entire chest or abdomen can be scanned in 90 seconds, or the time equivalent of one breath hold.

Mild sedation and reassurance by you as well as by the radiologist may help calm an anxious patient. The gantry is huge and may be frightening (Figure 2.30). It behooves you to inspect the CT rooms in your x-ray department so that you can explain to your patient beforehand that, unless intravenous (IV) contrast material is required, the procedure is as painless as having a photograph taken, in spite of the look of the machine. As compared to magnetic-resonance scanning (discussed later in this chapter), in which the patient's entire body is placed within the bore of a superconducting magnet, only a portion of the body is surrounded by the more doughnut-shaped gantry of a CT scanner.

Body CT scans in the axial plane can be produced with the patients supine or prone or lying on their side. Other planes of imaging, especially of the head and extremities, are possible, but you will need to learn to think of most CT body scans as transaxial and supine, since patients are most comfortable and relaxed lying on their back.

It is important for you to realize that CT should be considered most of the time as a sophisticated study for special problems, usually arranged following consultation with the radiologist. Other, less expensive, procedures like plain films and ultrasound should be used when the information obtained is comparable.

The important exceptions to this principle are in traumatized patients and central nervous system emergencies. In head trauma the superior capacity of CT to recognize intracranial hemorrhage often makes ordinary skull films a dangerous waste of time. Patients with abdominal trauma too should be taken straight to the CT suite for serial scans from the diaphragm through the pelvis, supplying a wealth of emergency information about hemorrhage and organ injury that can save lives.

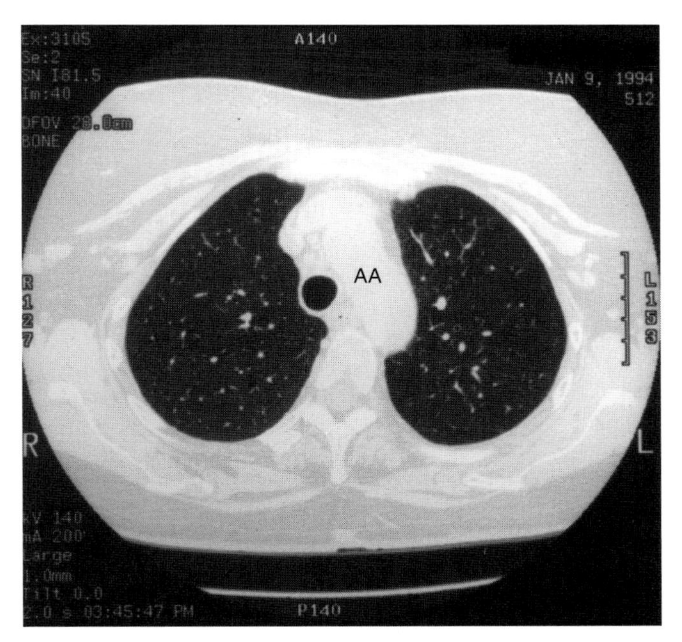

A

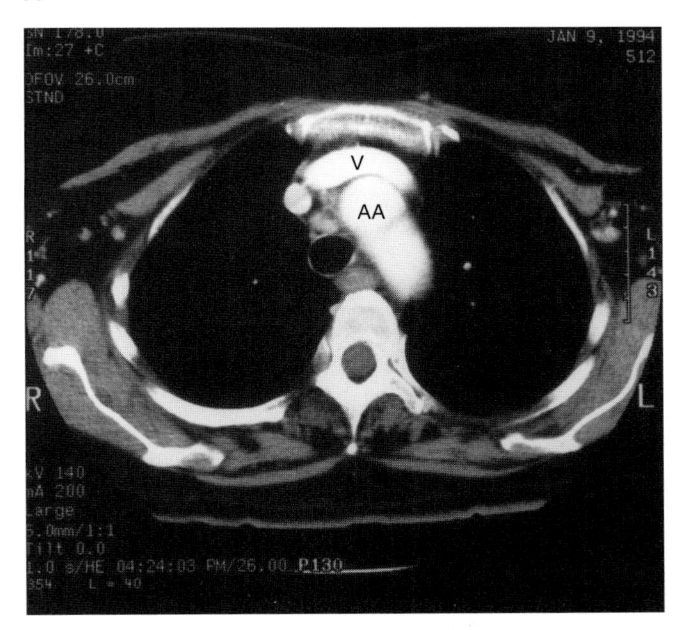

B

Figure 2.29. A: Chest CT scan at the level of the aortic arch *(AA)* filmed with "lung window" settings; note the pulmonary vessels and bronchi shown within the lung parenchyma. B: Chest CT scan at the same level filmed with "soft-tissue window settings," which show the structures of the mediastinum and chest wall better. The aortic arch *(AA)* is opacified with intravenous contrast media, as is the left subclavian vein *(V)* shown coursing anterior to the arch.

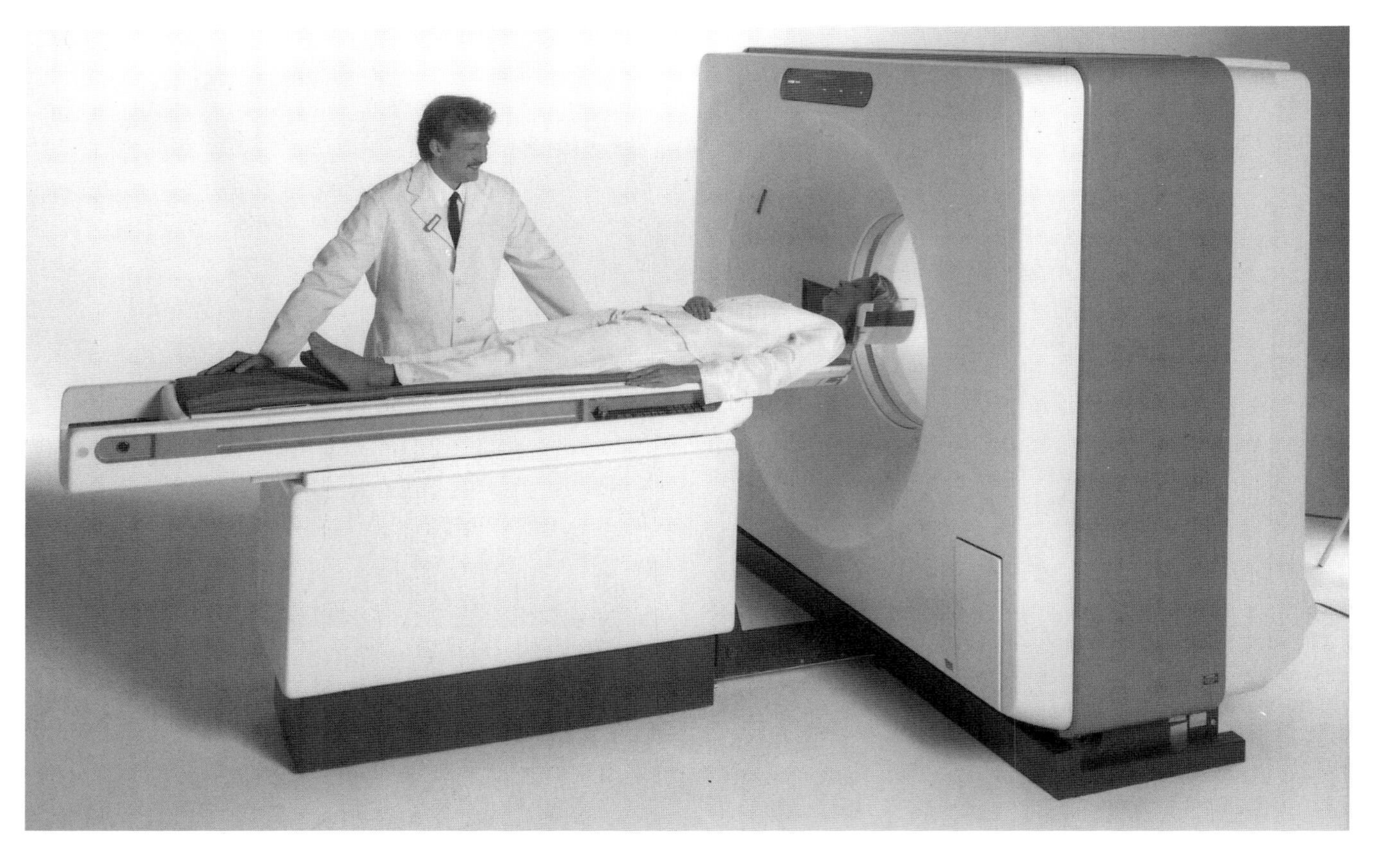

A

Figure 2.30. A: Patient on a CT table about to be moved within the gantry of the scanner. B: The CT scanner console. The radiologist is adjusting window and level settings to optimally visualize various types of tissues.

B

Depending on the clinical condition under investigation, contrast media may be used during CT scanning to enhance the difference in density of various structures. The gastrointestinal (GI) tract can be illuminated by giving the patient diluted water-soluble oral contrast material, which will help to distinguish the stomach and bowel from other soft-tissue structures and masses. Intravenous administration of water-soluble contrast material will produce a temporary increase in the density of vascular structures and highly vascularized organs. This effect is referred to as enhancement and is extremely useful. For example, a great vessel and the tumor mass encasing and constricting it will appear as one homogeneously dense mass unless the vessel is enhanced with contrast material, when its narrowing will be apparent.

Three-Dimensional CT

Three-dimensional CT images may be produced by computer stacking of a series of contiguous CT slices. Figure 2.31 shows the 3DCT of a patient with a facial fracture. A 3DCT image can provide the surgeon with an image that most realistically displays the position and orientation of displaced fracture fragments. Although the fracture line and fracture fragments were clearly apparent on the individual axial slices, it is much easier to perceive the "big picture" by looking at the 3D image than by mentally stacking all the individual axial CT slices. No additional scanning is required to produce a 3DCT image. The CT scanner computer or a free-standing computer is directed to make a 3D model from the series of axial CT scans. After the model is generated, it can be rotated in real time to be viewed from every side and even sliced to reveal the interior three-dimensional anatomy.

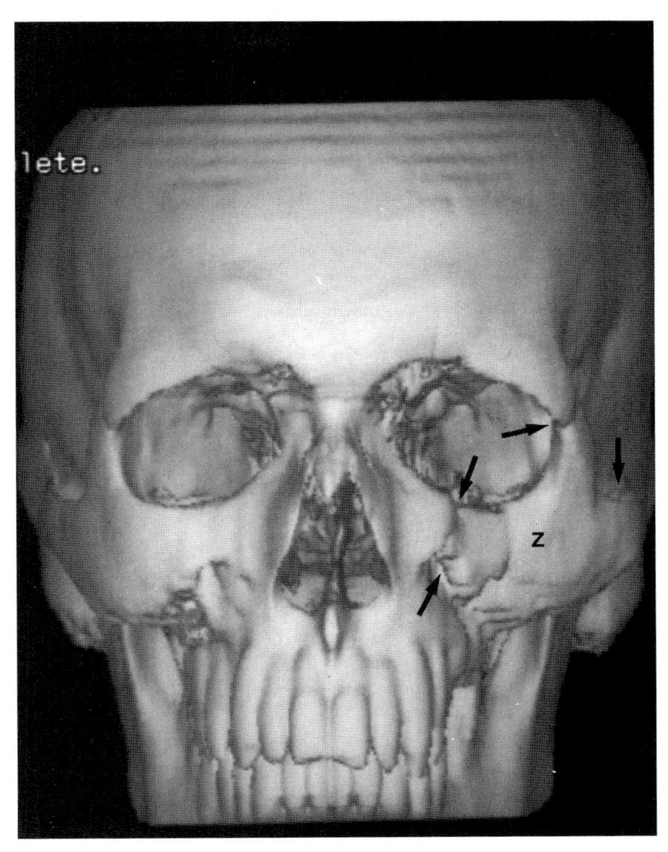

Figure 2.31. 3DCT of a patient with a left zygoma (cheekbone) fracture. The left zygoma *(Z)* is fractured along its articulations with the frontal bone, maxillary bone, and temporal bone *(arrows),* and it is displaced downward and posteriorly. Note that the left orbit appears ovoid and increased in size, compared with the right.

High-Speed (Helical or Spiral) CT and CT Angiography

Conventional CT scanning is performed by obtaining a series of individual axial scans during suspended respiration. The x-ray tube and detector assembly rotate about the patient while scanning each slice. Between scans, motion of the tube and detector ceases, and the patient is allowed to breathe during a 5–10-second delay, during which the scanner table moves the patient to the next scanning position. The recent introduction of improved mechanics in helical or spiral high-speed CT gantries allows for uninterrupted scanning in such a way that the patient moves through the scanner at a constant rate during continuous rotation of the x-ray tube–detector assembly. Scanning is so fast that an entire CT examination of the head, chest, or abdomen may be performed within 90 seconds, often with only a single breath hold required by the patient. Consequently, motion artifact is virtually eliminated with high-speed scanning. While scanning, the x-ray tube traverses a helical or spiral path around the patient.

As you no doubt have already realized, high-speed scanning is helpful not only when motion is a problem, for example during imaging of pediatric and sick adult patients, but also when scanning is done to evaluate structures that move within the patient, such as blood vessels (arterial pulsation) and lungs (respiratory motion). High-speed CT is capable of producing detailed three-dimensional displays of blood vessels (Figure 2.32). Referred to as *CT angiography,* this technique may be used to evaluate aortic aneurysms and aortic dissections, renal artery stenoses, and a variety of other vascular conditions.

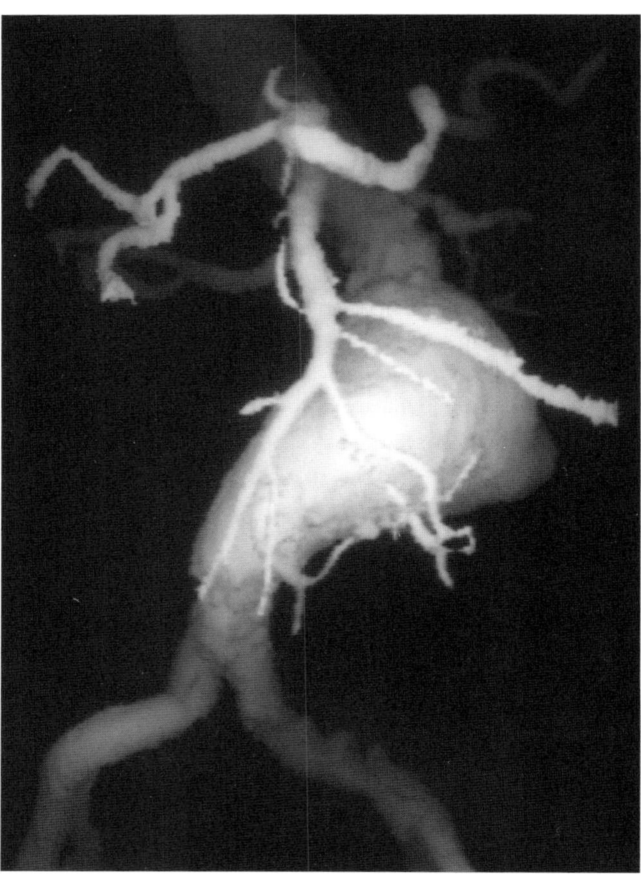

Figure 2.32. 3DCT angiography of a patient with an abdominal aortic aneurysm. The structures closer to the viewer (celiac artery, superior mesenteric artery, anterior wall of the aneurysm) are more brightly illuminated. The renal arteries and iliac arteries, which course posteriorly, are less well illuminated.

Ultrasound

Ultrasound, or *ultrasonography,* also gives you an image of a slice of the body, by directing a narrow beam of high-frequency sound waves into the body and recording the manner in which sound is reflected from organs and structures. The ultrasonographer uses a hand-held transducer (Figure 2.33) containing piezoelectric crystals, which change electrical energy into high-frequency sound waves. The sound beam is directed into the region of interest and then reflected back toward the transducer at interfaces between tissues of different *acoustic impedance* (which is determined by the physical density of the tissue and the velocity of the sound). As the acoustic impedance mismatch between two tissues increases at any given interface, the reflected sound, or *echo,* becomes stronger. When the reflected sounds return to the transducer, they are converted to electrical signals, which are then computer analyzed to produce the ultrasound images.

These images are viewed in "real time" and can be used to display motion of the heart and blood vessels.

At ultrasound *solid organs* (Figure 2.34) appear as *echogenic* structures because they consist of tissues with multiple acoustic interfaces, whereas *cysts* and *fluid collections* (Figure 2.35) appear echo-free (echolucent or *anechoic*) because they lack internal acoustic reflectors. Air and bone cannot be adequately visualized with ultrasound because the acoustic impedance mismatch between these structures and the adjacent soft tissues is very great; most of the sound energy is reflected, so that little is left to visualize the structures beyond the interface.

Ultrasound does not produce an image that is as sharp and clear as CT, but it has five singular advantages. First, ultrasound is a safe procedure that does not employ ionizing radiation and that produces no biological injury. Consequently, it has found wide applications in the imaging of obstetrical, gynecological, pediatric, and testicular conditions. Second, ultrasound can be employed in the transaxial plane or sagittally or at any obliquity required to show the anatomic region being investigated. Third, it is far less expensive than either CT or magnetic-resonance imaging. Fourth, ultrasound can be performed portably at the bedside of very sick patients. Fifth, real-time ultrasound can provide moving images of the heart, fetus, and other structures.

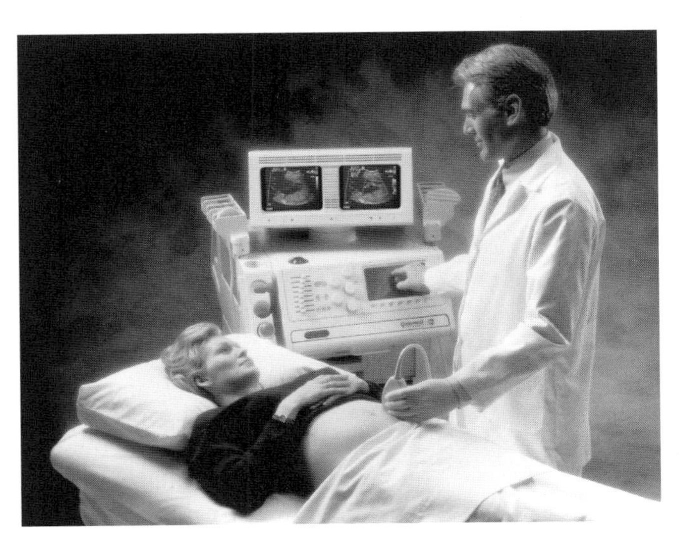

Figure 2.33. Patient undergoing an abdominal ultrasound examination.

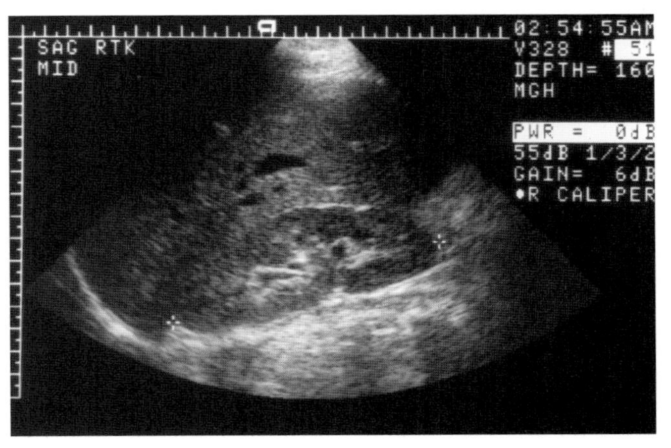

Figure 2.34. Sagittal ultrasound of the right abdomen. The right kidney has been marked with "cursors" by the ultrasonographer. The liver can be seen anterior to the kidney. Note the bright white echogenic diaphragm margining the upper liver. The sinus fat within the kidney is also echogenic.

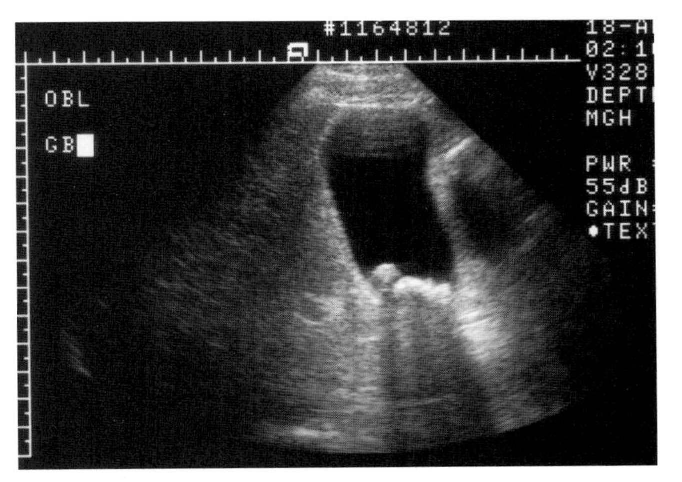

Figure 2.35. Ultrasound of a gallbladder containing multiple stones. The fluid within the gallbladder is anechoic (no echoes): no reflected sound waves are returned to the transducer by the fluid filling the gallbladder. The stones, however, are very echogenic. Note the dense white echoes reflected from the stones, which also block the transmission of ultrasound waves from above, producing dark acoustic shadows behind the stones.

Magnetic-Resonance Imaging

Like ultrasound, *magnetic-resonance imaging* (MRI) or magnetic-resonance scanning (MR scanning) does not use ionizing radiation as ordinary x-rays and CT do. This technique for imaging places the patient within the bore of a powerful magnet and passes radio waves through the body in a particular sequence of very short pulses (Figure 2.36). Each pulse causes a responding pulse of radio waves to be emitted from the patient's tissues. The location from which the signals have originated is recorded by a detector and sent to a computer, which then produces a two-dimensional picture representing a predetermined section or slice of the patient.

The specific principles involved in magnetic-resonance imaging are quite complex and beyond the scope of this book. But a basic understanding of this imaging technique will give you a better understanding of its clinical applications and the patient's experience. MRI uses very powerful magnets, ranging in field strength from 0.3 to 1.5 Tesla. By comparison, 1 Tesla is equivalent to 10,000 gauss, and the earth's magnetic field is only 0.5 gauss. Consequently, patients with cardiac pacemakers and certain metallic implants cannot be examined with MR scanning.

Current diagnostic magnetic-resonance scanning is based on imaging hydrogen atoms in fat and water molecules. In a magnetic field, the hydrogen atoms, which are small magnets themselves, align themselves with the magnetic field, in much the same way that a compass aligns itself with the earth's magnetic field. During scanning, pulsed radiowaves of a particular radiofrequency (RF) are directed at the patient, causing these small atomic magnets to be knocked out of alignment. The hydrogen atoms will eventually reestablish the previous equilibrium with the surrounding magnet, and when they do so, they will emit the absorbed radiofrequency waves. The distribution of the emitted radiofrequency waves is analyzed by computer to produce the image. The time required by the hydrogen atoms to regain the equilibrium state is referred to as the relaxation time. Two relaxation times are recognized with MR scanning: the T1, or longitudinal relaxation time; and the T2, or transverse relaxation time.

A wide variety of MR techniques are available to optimally visualize different tissues and disease processes. The most commonly used are spin-echo sequences. Two other MR terms you will hear about are

Figure 2.36. Magnetic-resonance imaging suite. Looking from the control room where the technologists and radiologists direct the scan, you can see a patient being moved into the bore of the magnet. The scanning room is shielded from external radiofrequency (RF) waves. Ferromagnetic materials cannot be brought into the scanning room because of the powerful magnetic field.

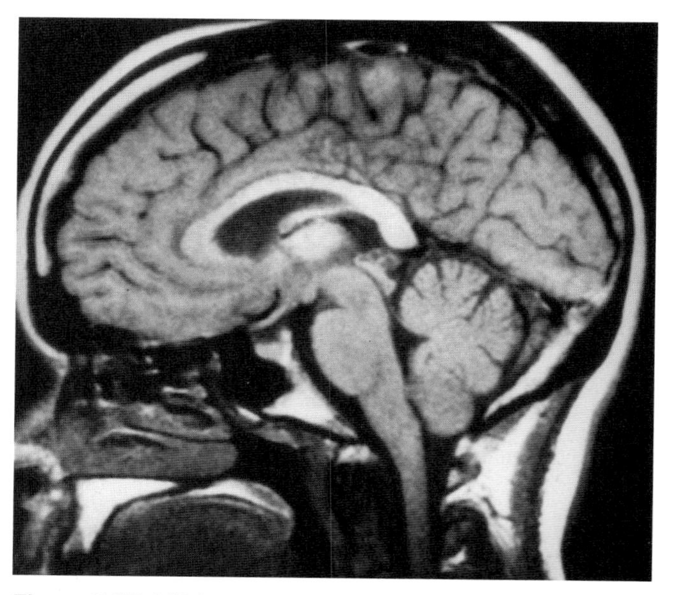

Figure 2.37. MR image made midsagittally through the brain. Well shown are the medial surface of a cerebral hemisphere, the corpus callosum, cerebellum, midbrain, and upper spinal cord.

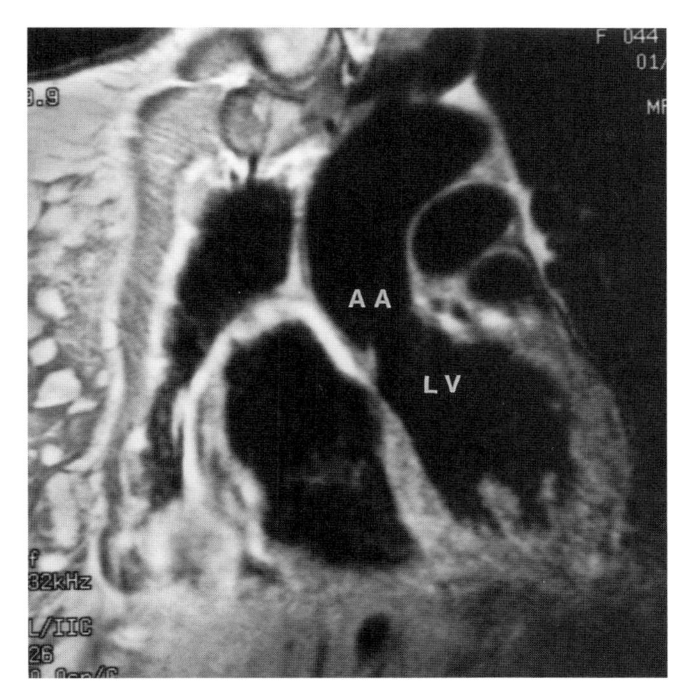

A

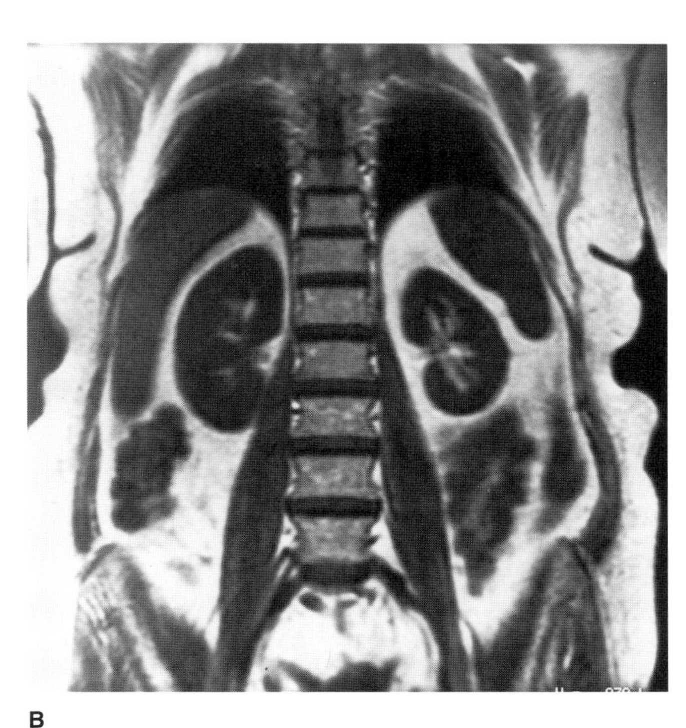

B

Figure 2.38. A: Coronal MR image of the heart. Blood within the cardiac chambers and blood vessels has almost no MR signal and appears black at the MR settings used to obtain this scan. Cardiac muscle appears gray, and fat (which has a strong MR signal) within the mediastinum appears white. This scan was obtained at the level of the aortic valve shown open between the left ventricular cavity *(LV)* and the ascending aorta *(AA)*. B: Coronal MR image of the posterior abdomen in another patient. This obese individual has a generous amount of high MR signal fat (appearing white) in the peritoneal cavity, in the retroperitoneum (shown around the kidneys), and in the subcutaneous fascia (shown between the skin and the abdominal wall musculature). Find the spine, psoas muscles, kidneys, liver, and spleen. Note the high position of the diaphragm and the low lung volumes.

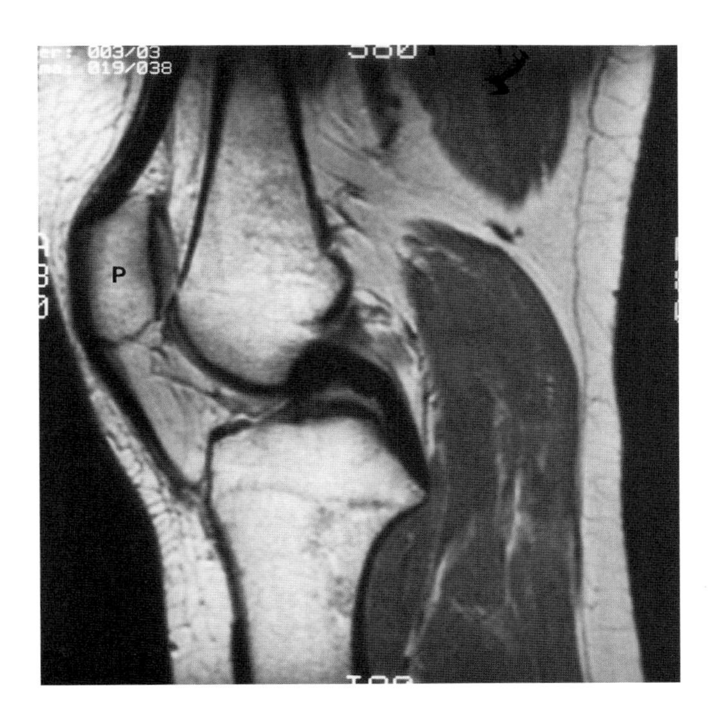

Figure 2.39. Sagittal MR image of the knee. Fat within the bone marrow and soft tissues has a strong signal at the MR settings used, and consequently structures composed of fatty tissue appear white. Compact bone and the tendons have little to no MR signal and appear black. Muscle tissue with a weak signal appears dark gray. Note that the patella *(P)* is suspended in front of the distal femur by the quadriceps tendon superiorly and the patella tendon inferiorly.

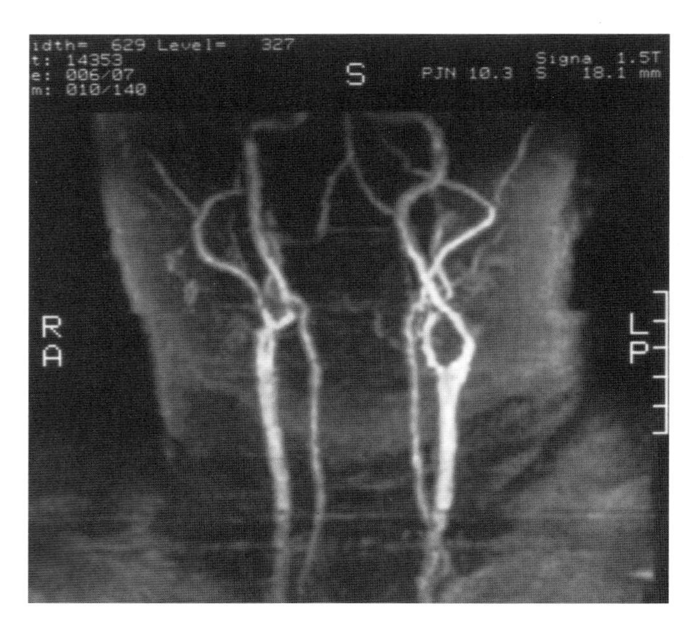

Figure 2.40. 3DMR angiography of the neck arteries. No intravenous contrast material was needed to produce these images of the carotid and vertebral arteries.

the repetition time (TR) and the echo time (TE). The repetition time is the time between successive RF pulses; the echo time is the time between the RF pulse that excites the hydrogen atoms and the arrival of the return signal at the detector. Longer TR and TE values will produce images that are more dependent upon the T2 values of the tissues; shorter TR and TE values will produce images that are more T1 dependent. By changing the TR and TE values one can alter the relative *signal intensities* of different tissues to better visualize the organ or clinical condition under investigation.

Various body tissues emit characteristic MR signals, which determine whether they will appear white, gray, or black on the final scans. Tissues that emit strong MR signals appear white in MR scans, whereas those emitting little or no signal appear black. Note that the x-ray terms radiolucent and radiodense do not apply to MRI; instead, structures that appear white on MR scans are said to have *high signal strength,* whereas dark gray or black objects are said to have a *low signal,* or no signal at all. Compact bone will generally appear black. Fat will appear bright on a T1-weighted image, but decrease in intensity slightly on a T2-weighted image. Most tumors and inflammatory masses appear bright on T2-weighted images. With most MR techniques, rapidly moving blood appears black because the blood moves out of the anatomic section being imaged before the emission of the RF signal from the excited protons.

A great advantage of MR over CT scanning is that direct multiplanar scanning is possible. MRI can produce primary images in almost any imaging plane, including the axial, coronal, sagittal, or any specially chosen oblique plane. In addition, greater differentiation of soft-tissue structures is possible with MRI than with CT. A disadvantage of MRI is the longer *acquisition time* (time to collect data for imaging) of several minutes, which results in greater motion artifact. This is a problem with MR scans of the thorax and abdomen because of respiratory motion, but not with MR scans of the head and extremities. As compared with a 10- to 20-minute CT scan, an MR examination may take 30 to 45 minutes to acquire the data to produce the scan images. Three-dimensional reformations can be produced of both CT and MR images; but since MR shows blood vessels without contrast media, it permits contrast-free 3D vascular imaging (Figure 2.40).

You should be aware that many patients feel claustrophobic within the bore of an MR scanner, some so severely that the examination may have to be discontinued before completion. These symptoms are often alleviated by a reassuring discussion before the examination, as well as by sedation to relax anxious patients. We strongly recommend that in addition to observing MR scanning, you observe as many radiological procedures as possible so that you are in a position to describe them accurately and to prepare your patients for what to expect.

Radioisotope Scanning

Finally, *nuclear imaging*, another branch of radiology, offers important physiological information that you must be familiar with. This branch of radiology is based on the visualization of particular living organs and tissues through the injection of a radioactive isotope (radionuclide) that takes up residence there briefly. It does so because the selected chemical substance to which the isotope has been attached (radionuclide-labeled substance) is normally involved in the physiologic metabolism of that organ or will remain there long enough to be imaged. An image is obtained because the radioactive isotope emits gamma rays for a brief period of time. The emitted rays are recorded by a *gamma camera* (Figure 2.41) or, less commonly, by a *rectilinear scanner* during the period of gamma emission. A few hours or days later, the isotope will stop emitting detectable rays as it returns to a stable state. Its return to stability is measured in terms of its *half-life:* the period until it is seen to be emitting half as much radiation as it did initially. Isotopes chosen for tagging are those that will remain in the organ to be studied long enough to produce a usable image but with relatively short half-lives, so as to minimize radiation to the patient's tissues.

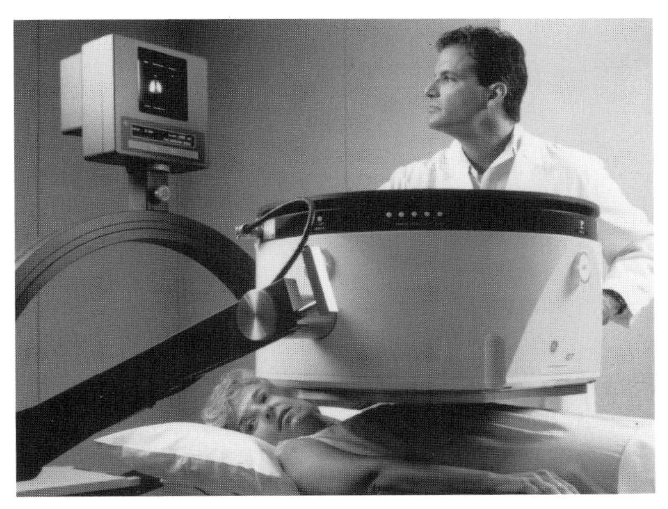

Figure 2.41. Gamma camera positioned over a patient for an anterior view perfusion lung scan. Images will also be taken with the camera adjacent to the patient's back (posterior view) and against both sides of the chest (lateral views). Oblique and other views are also possible.

Technetium-99m (Tc99m) has proved to be the most useful radioactive tracer; it is relatively inexpensive, has a short but useful half-life, and is readily available from portable generators. Tc99m is linked to various physiological substances that will seek different organs. Technetium-99m-pertechnetate is trapped by the thyroid gland and can be used for thyroid imaging. Two other examples of other useful Tc99m compounds are Tc99m-macroaggregated albumin (which is trapped in pulmonary capillaries) for lung scanning (Figure 2.42) and Tc-99m-methylene diphosphonate for bone scanning (Figure 2.43). Other radionuclides are also used for diagnostic imaging. You may be already familiar, for example, with the use of thallium-201 scanning in the evaluation of myocardial blood flow.

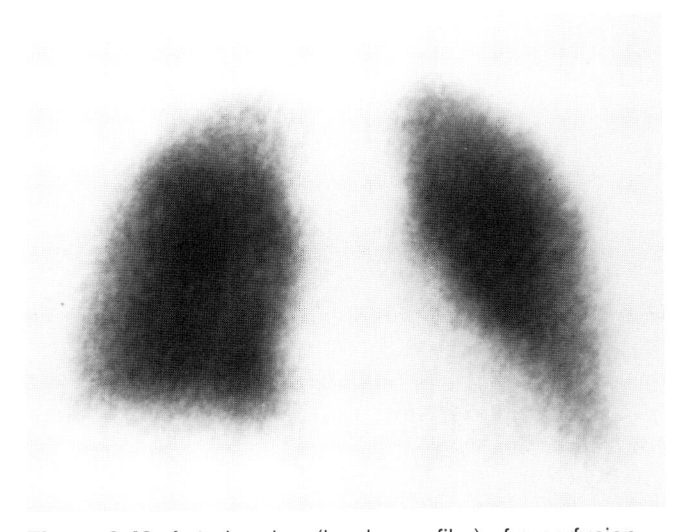

Figure 2.42. Anterior view (hard copy film) of a perfusion lung scan. The detected radioactivity was emitted by macroaggregates of intravenously injected, radioisotope-labeled albumin, which had become trapped in pulmonary capillaries. Between the lungs there is no activity overlying the silhouette of the heart and mediastinum.

A commonly requested radioisotope examination is the bone scan. The image obtained shows areas of more or less intensity of radiation related to portions of the bone having increased turnover. Thus "hot spots" showing markedly increased activity of bone will be seen as dense black areas on a gamma camera or rectilinear scan of the whole skeleton (see bone scan, Figure 2.44). Unfortunately, these are very nonspecific and do not tell us the cause of the increased

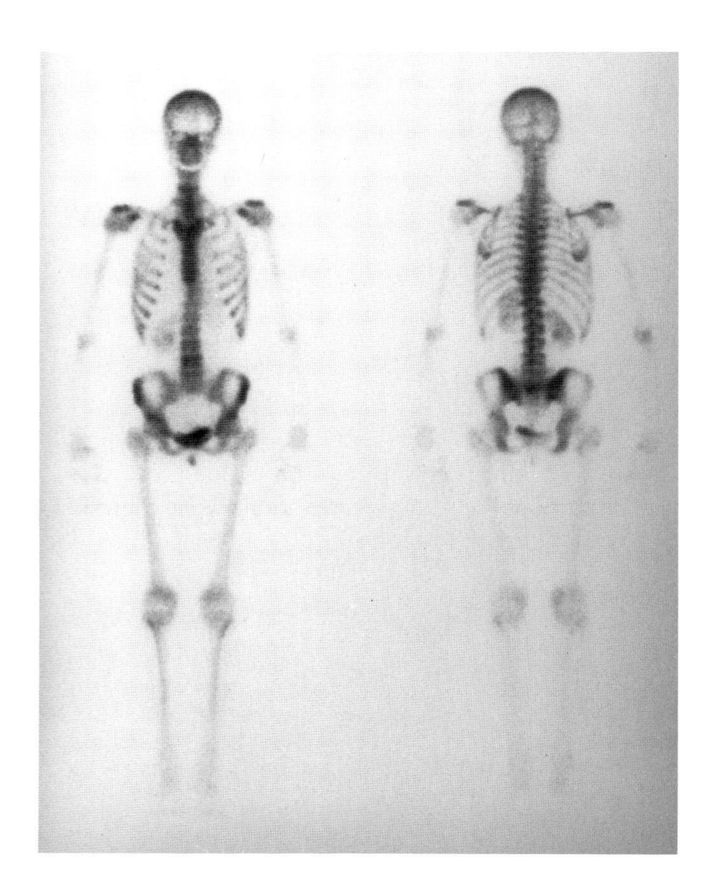

Figure 2.43. Normal technetium bone scan obtained with a rectilinear scan, which can cover the entire body in one scanning sweep. Both anterior and posterior views were obtained. You no doubt correctly guessed that the anterior scan is to the left (the anteriorly located "y-shaped" sternum and facial bones are better seen) and the posterior scan to the right (the posteriorly located back of the skull and spine are better seen).

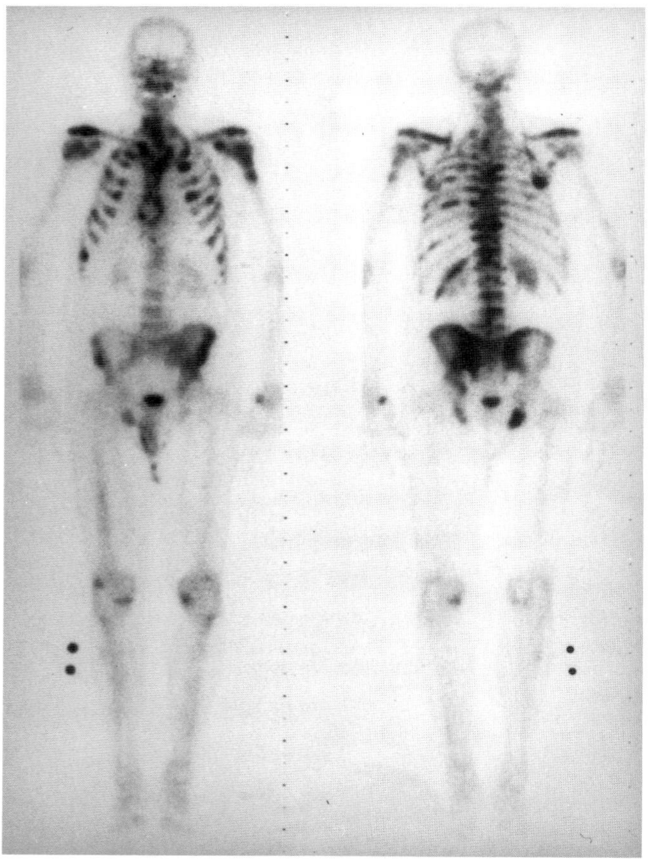

Figure 2.44. Technetium bone scan of a middle-aged woman with metastatic breast cancer. Multiple bone metastases are shown as areas of increased radioisotope uptake *(blacker areas)* in the spine, ribs, shoulders, and pelvis.

bone turnover. If they are located in symmetrical joint areas, for example, they may be being caused by acute arthritis, and if they are located eccentrically like those in Figure 2.44, they may be assumed to indicate the location of bone metastases from the patient's known or suspected cancer. A new technique you may hear about is *SPECT* imaging, which uses a gamma camera that rotates around the patient to produce tomographic-like nuclear images.

As we proceed through this book, other important procedures using radioactive isotopes will be described. Remember for now that nuclear medicine gives you less precise anatomic information but much more important physiological information, which will help you to understand and remember metabolic processes, both normal and abnormal.

Realize that in the usual isotope scan, the image obtained is produced by gamma radiation from the *entire thickness* of the organ, not from a single slice of it as in CT, MR, and ultrasonography. Realize also that just as *fluoroscopy* in plain radiography consists of continuous or intermittent observation of tissues penetrated by x-rays and produces *dynamic* radiographic information, so too any of the other imaging methods we have been discussing can be used dynamically. The motion of the fetal heart is routinely monitored by real-time ultrasound as evidence that a quiet fetus is, in fact, alive. Dynamic studies using rapidly sequenced CT scans during the intravenous injection of contrast material produce time-lapse information about the vascularity of a liver mass. Similarly, sequential isotope scans are in use to document flow patterns such as blood flow through the heart chambers in a patient suspected of having a congenital heart anomaly.

As a student, you certainly can learn to recognize some of the basic changes imaged on plain films of the chest, abdomen, and bones. You cannot expect to learn to recognize all of the innumerable more subtle plain radiographic changes the radiologist identifies. Neither will you be able to interpret the findings the radiologist recognizes in the many supplementary techniques such as CT, MR, and ultrasound. A four-year residency in radiology is scarcely enough time in which to learn to do that.

It is important, though, for you to learn while you are in medical school how to use the help the radiologist can give you in planning which procedures ought to be included in your diagnostic workup plan, and the order in which they should be undertaken.

3 Normal Radiological Anatomy

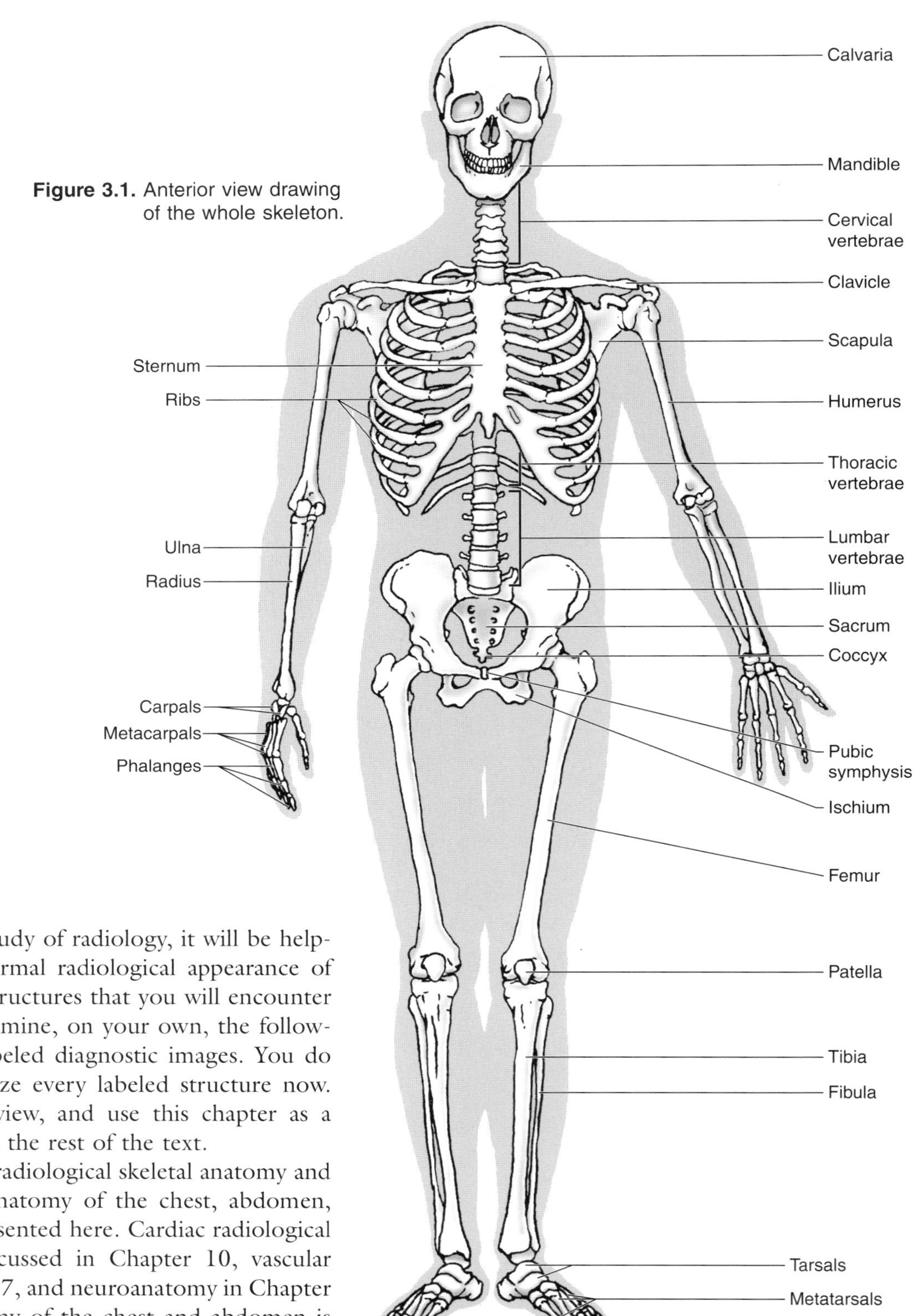

Figure 3.1. Anterior view drawing of the whole skeleton.

As you begin your study of radiology, it will be helpful to review the normal radiological appearance of common anatomic structures that you will encounter in later chapters. Examine, on your own, the following drawings and labeled diagnostic images. You do not need to memorize every labeled structure now. Rather, get an overview, and use this chapter as a reference as you read the rest of the text.

Normal plain film radiological skeletal anatomy and cross-sectional CT anatomy of the chest, abdomen, and pelvis will be presented here. Cardiac radiological anatomy will be discussed in Chapter 10, vascular anatomy in Chapter 17, and neuroanatomy in Chapter 18. Plain film anatomy of the chest and abdomen is detailed in the chest and abdomen chapters.

Although CT may be thought of as a higher-tech and more sophisticated imaging technique than plain films, it is often easier for the beginner to master the plain film anatomy of the chest and abdomen after becoming familiar with cross-sectional anatomy as displayed by CT.

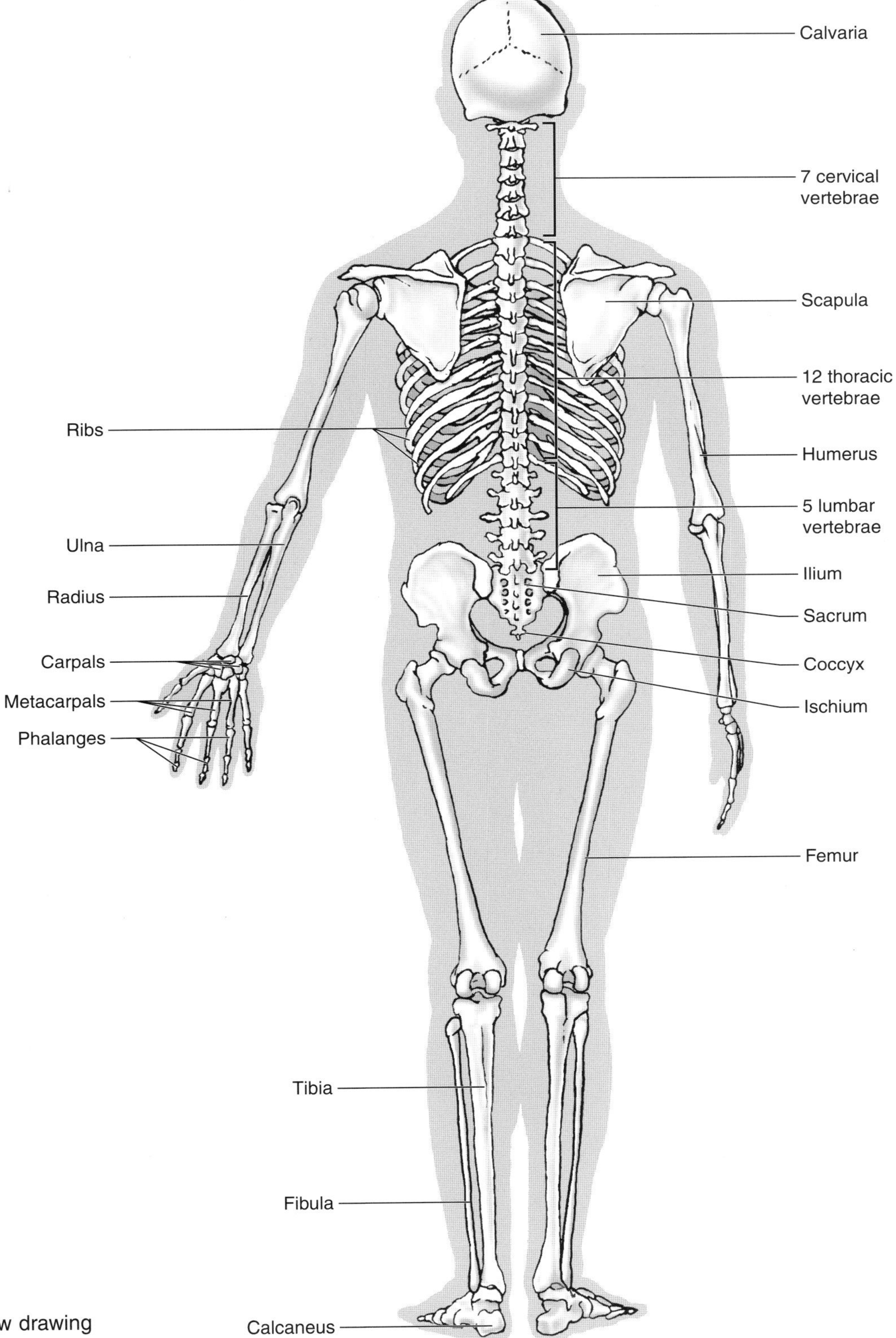

Figure 3.2. Posterior view drawing of the whole skeleton.

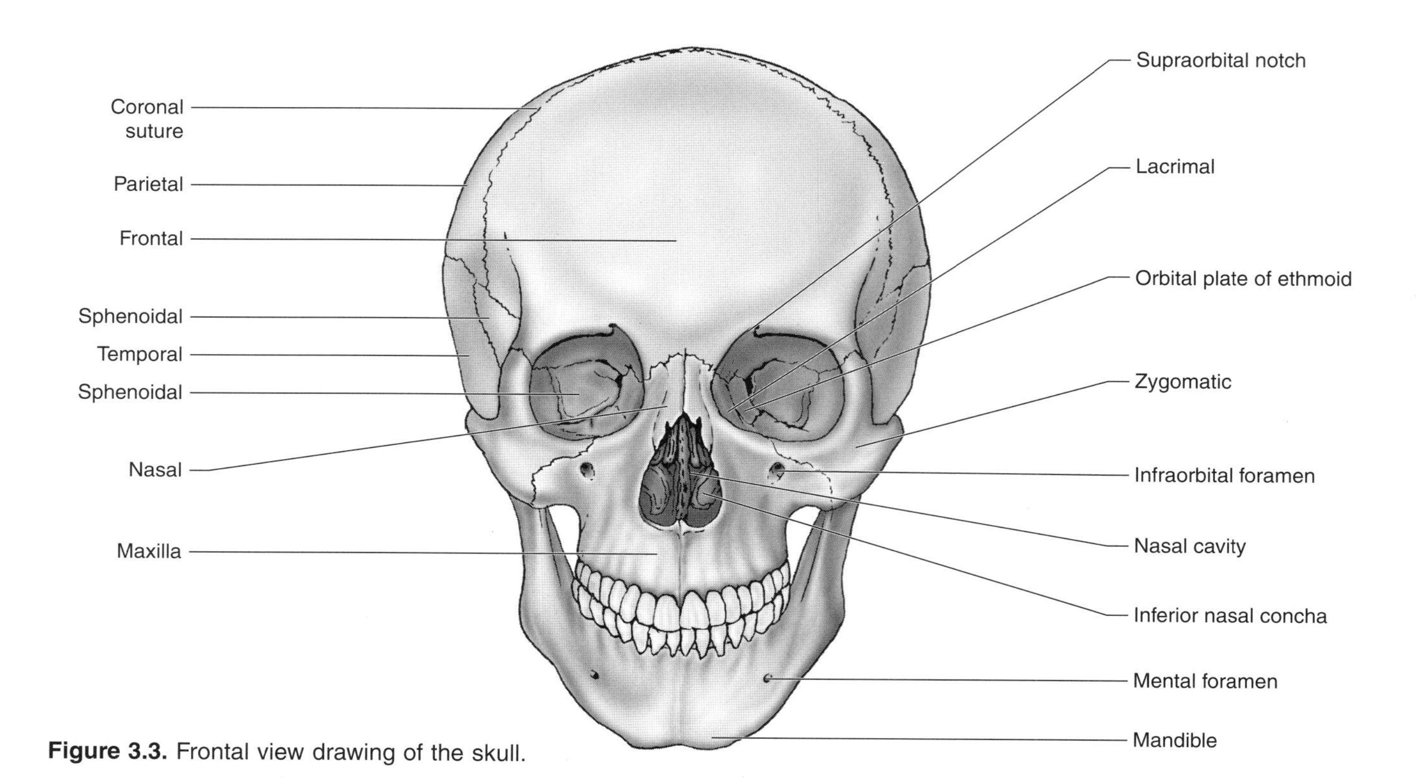

Figure 3.3. Frontal view drawing of the skull.

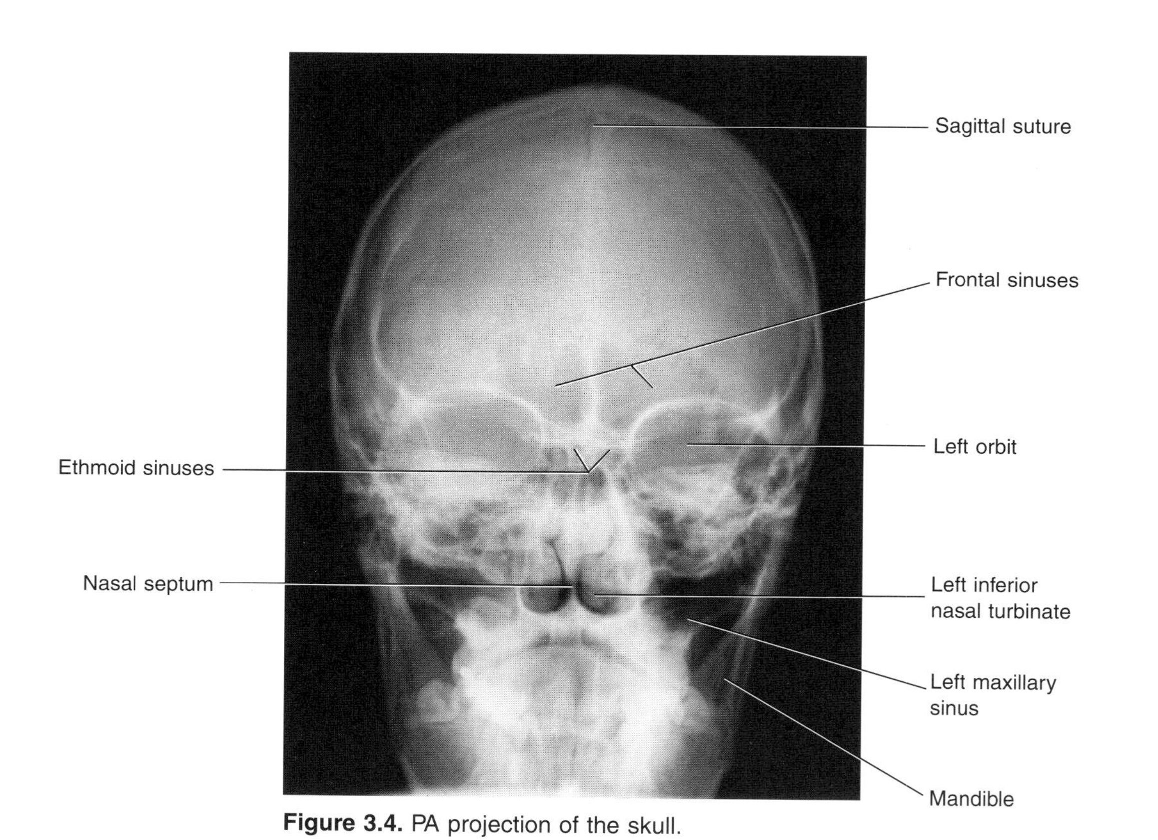

Figure 3.4. PA projection of the skull.

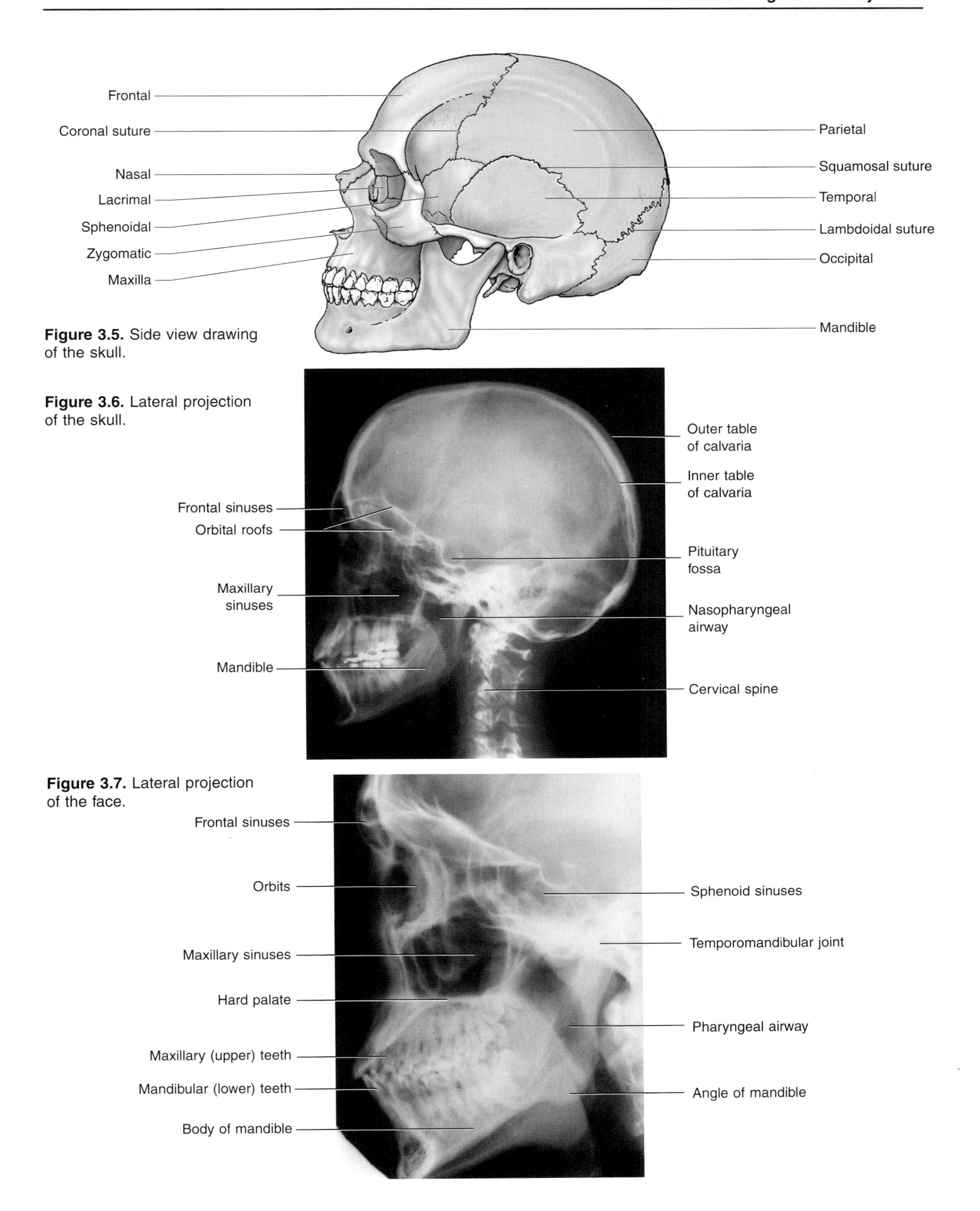

Figure 3.5. Side view drawing of the skull.

Figure 3.6. Lateral projection of the skull.

Figure 3.7. Lateral projection of the face.

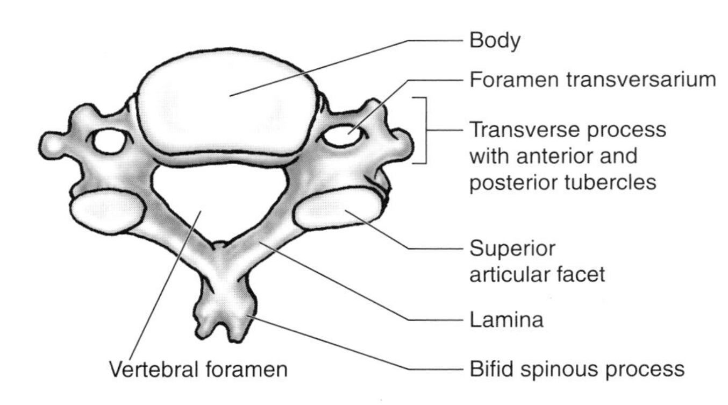

Figure 3.8. Superior view drawing of a typical cervical vertebra (C3–C7 vertebra).

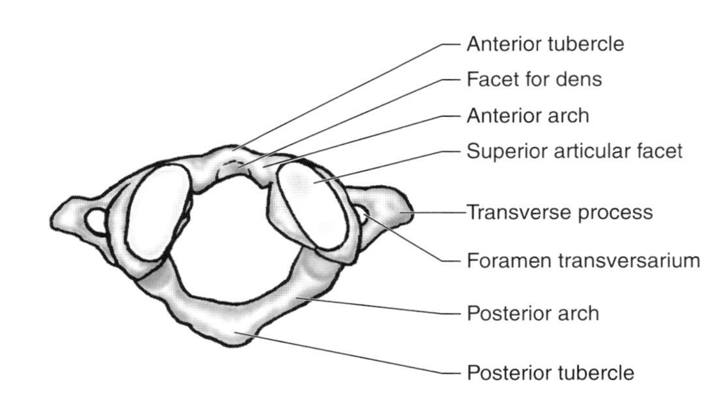

Figure 3.9. Superior view drawing of the atlas (C1 vertebra).

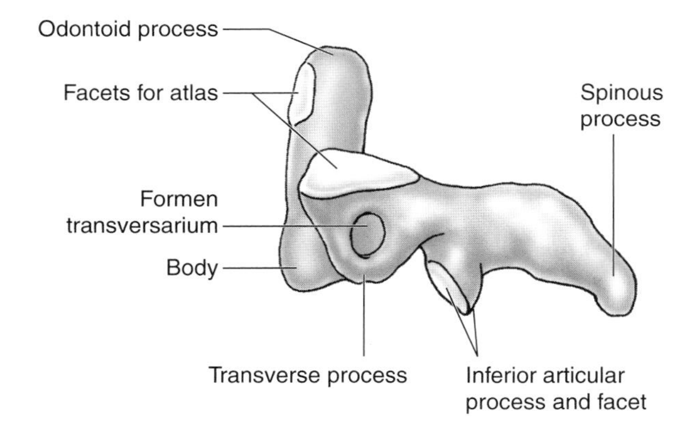

Figure 3.10. Lateral view drawing of the axis (C2 vertebra).

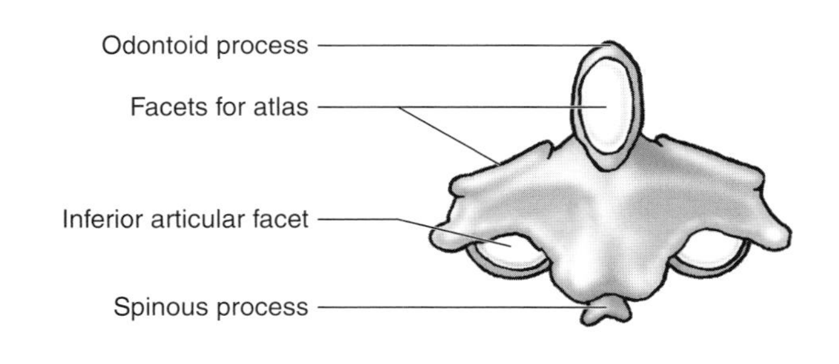

Figure 3.11. Frontal view drawing of the axis (C2 vertebra).

Figure 3.12. Lateral projection of the cervical spine.

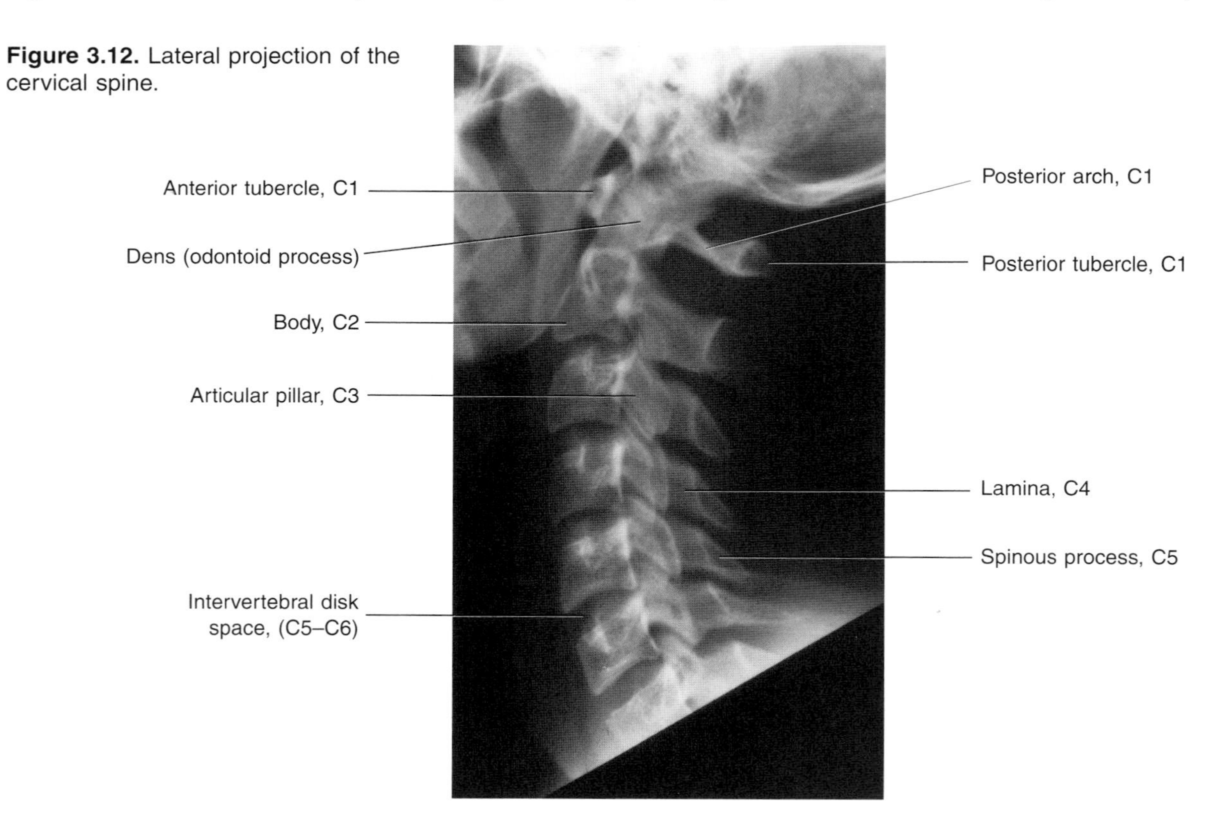

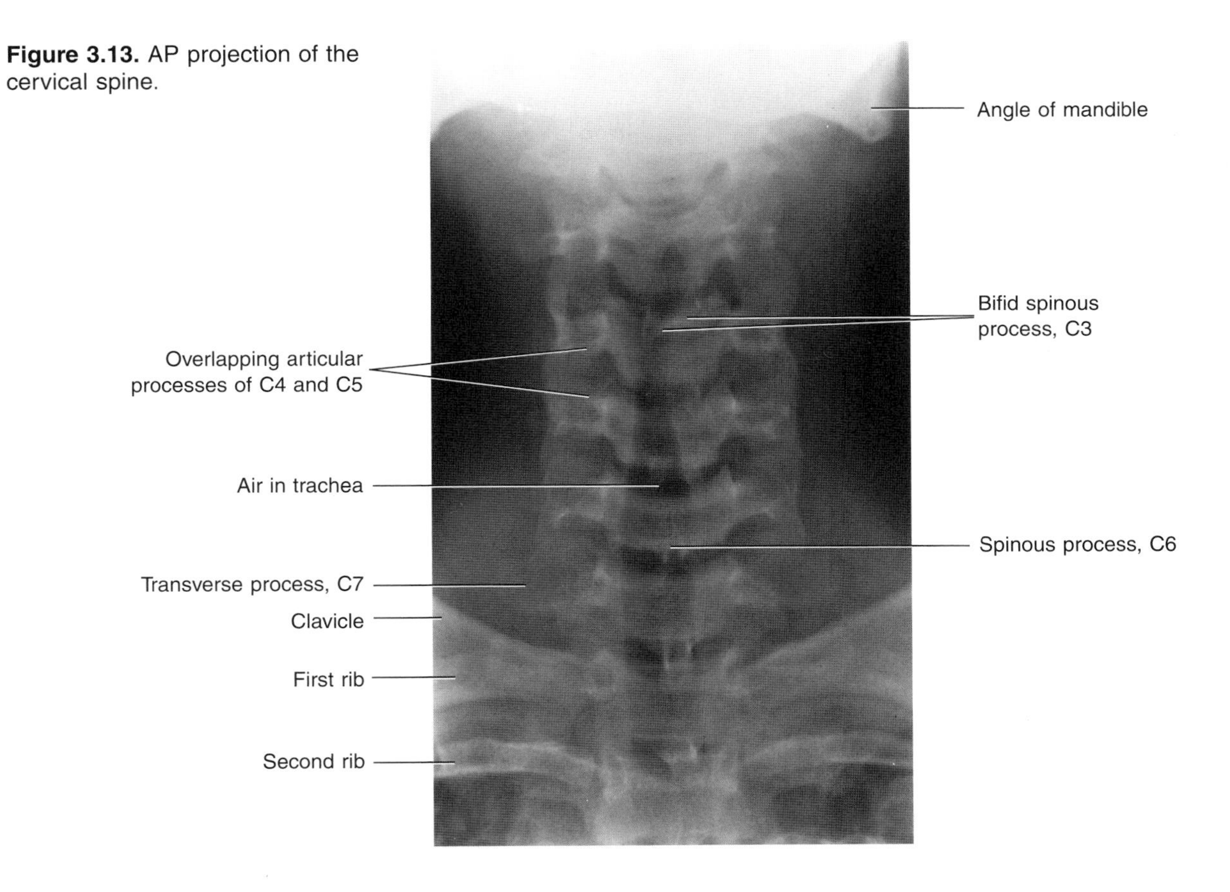

Figure 3.13. AP projection of the cervical spine.

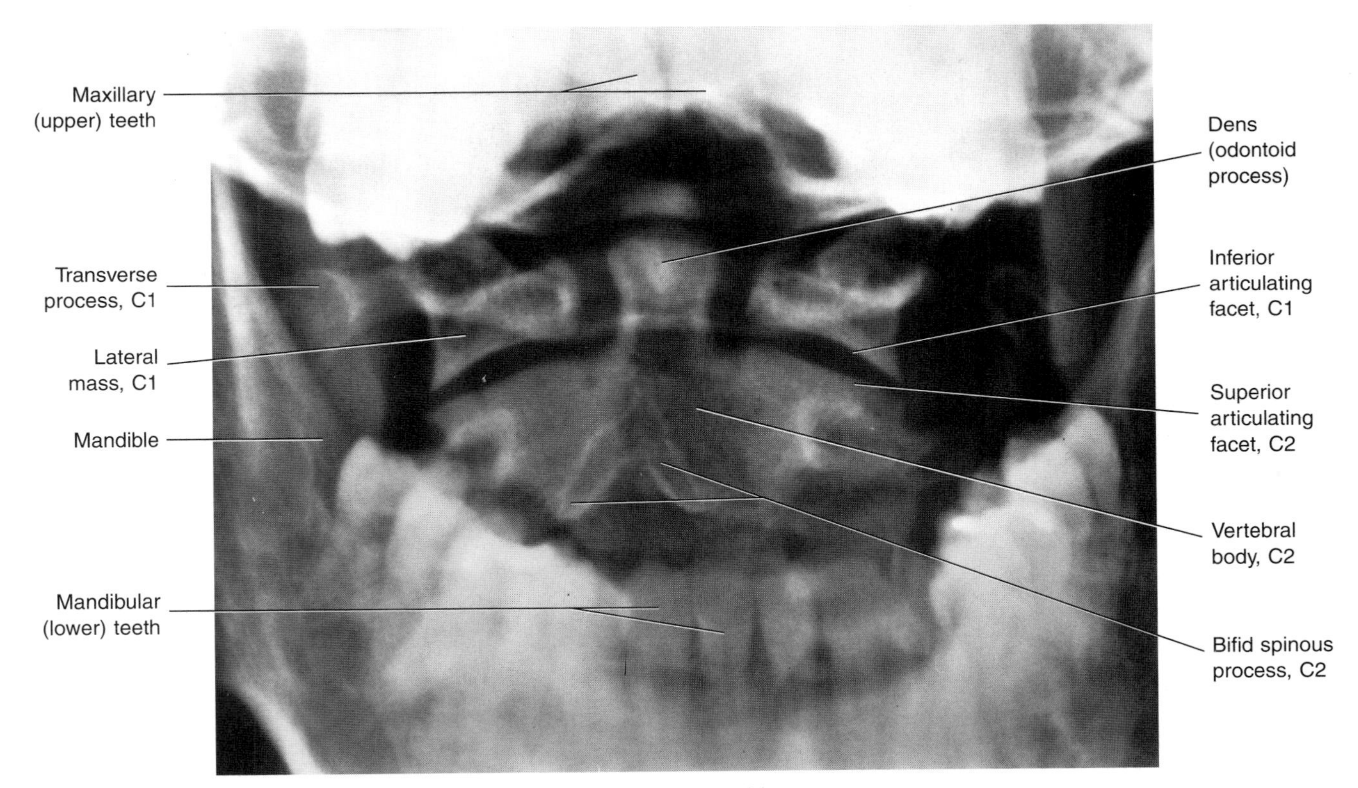

Figure 3.14. Detail of AP projection of the odontoid process.

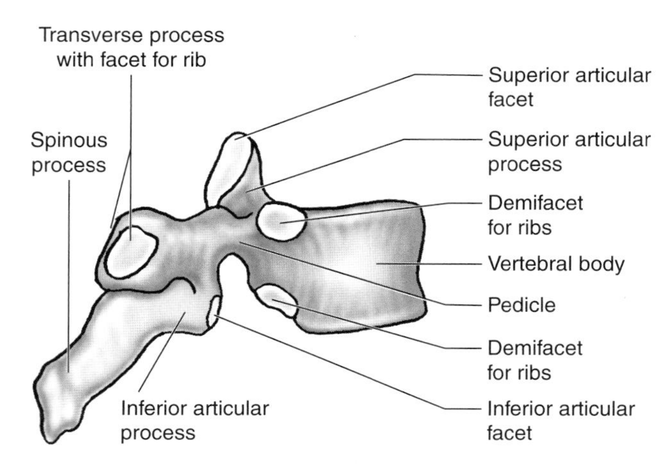

Figure 3.15. Lateral view drawing of a typical thoracic vertebra.

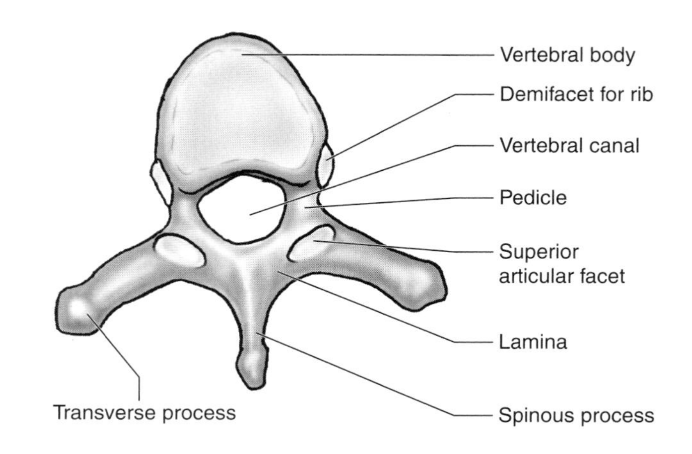

Figure 3.16. Superior view drawing of a typical thoracic vertebra.

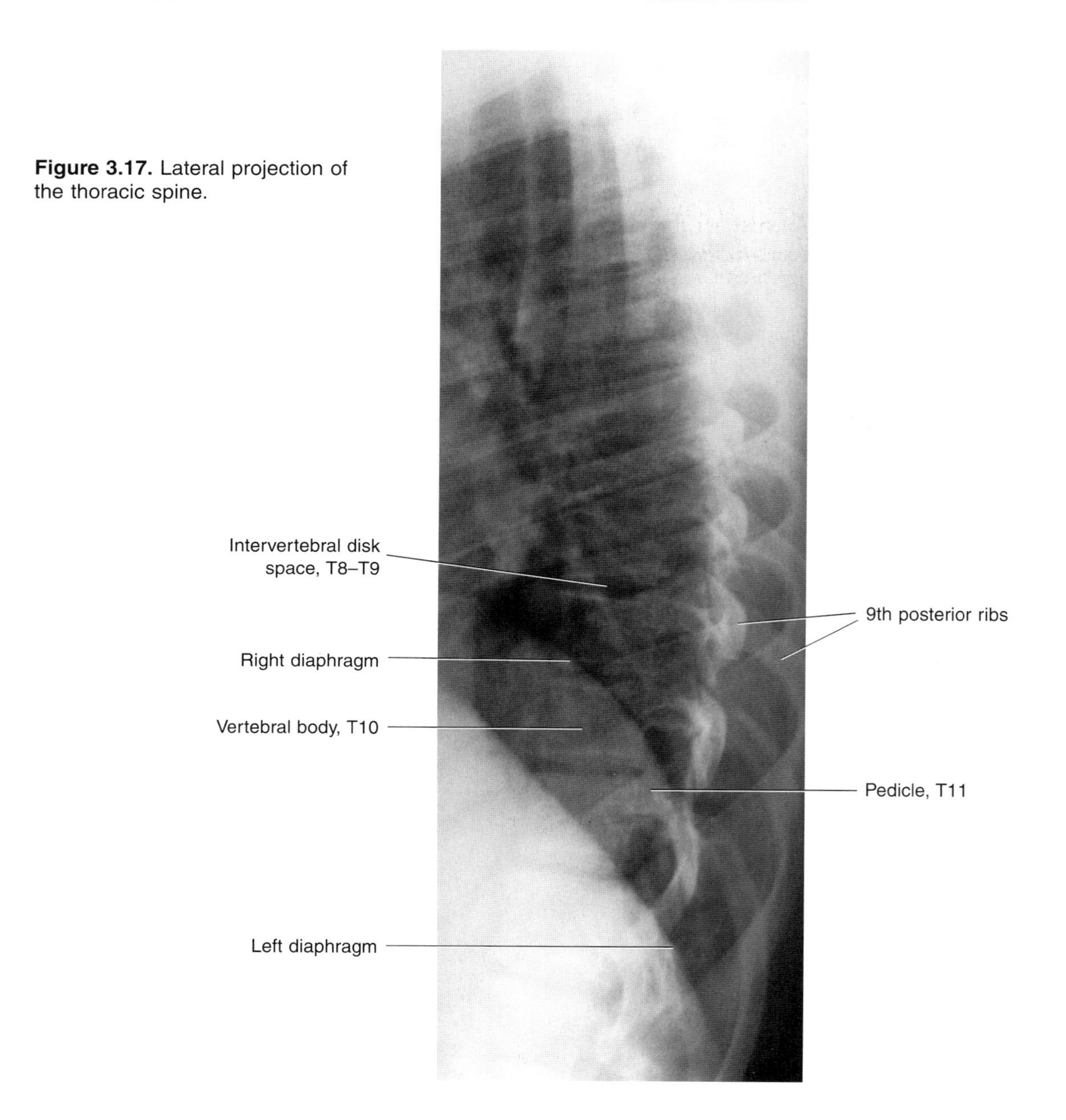

Figure 3.17. Lateral projection of the thoracic spine.

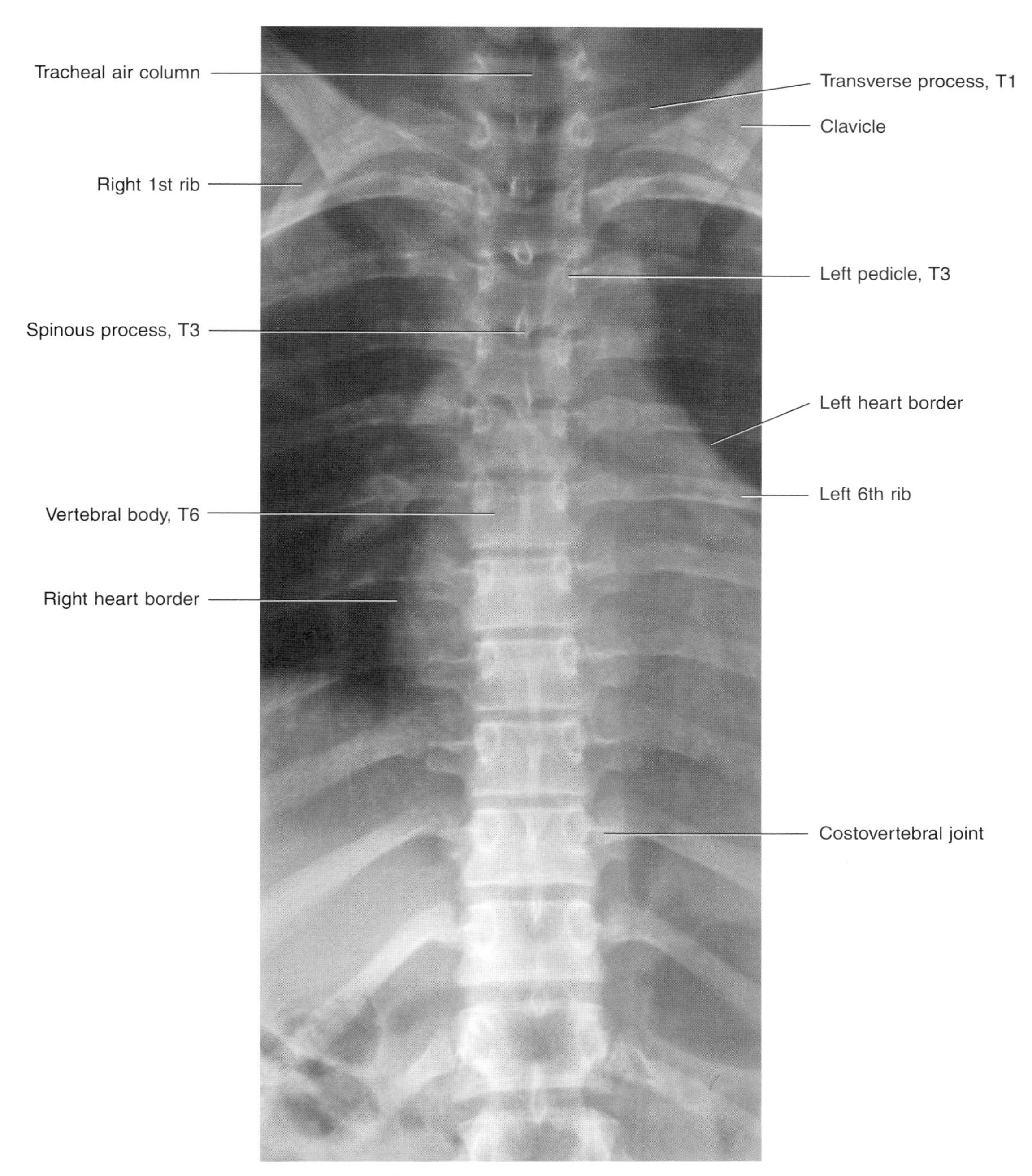

Figure 3.18. AP projection of the thoracic spine.

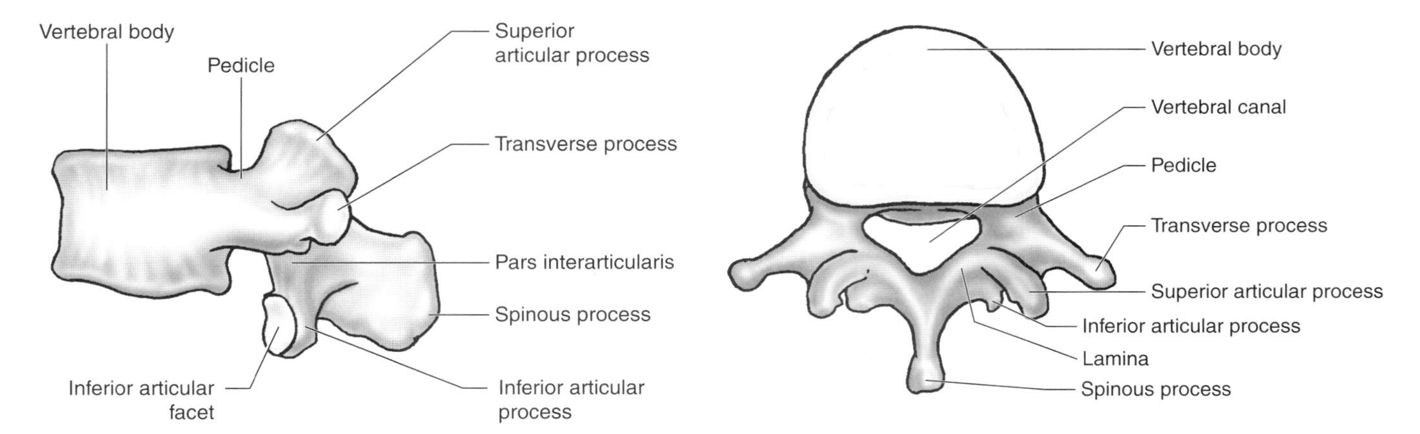

Figure 3.19. Lateral view drawing of a typical lumbar vertebra.

Figure 3.20. Superior view drawing of a typical lumbar vertebra.

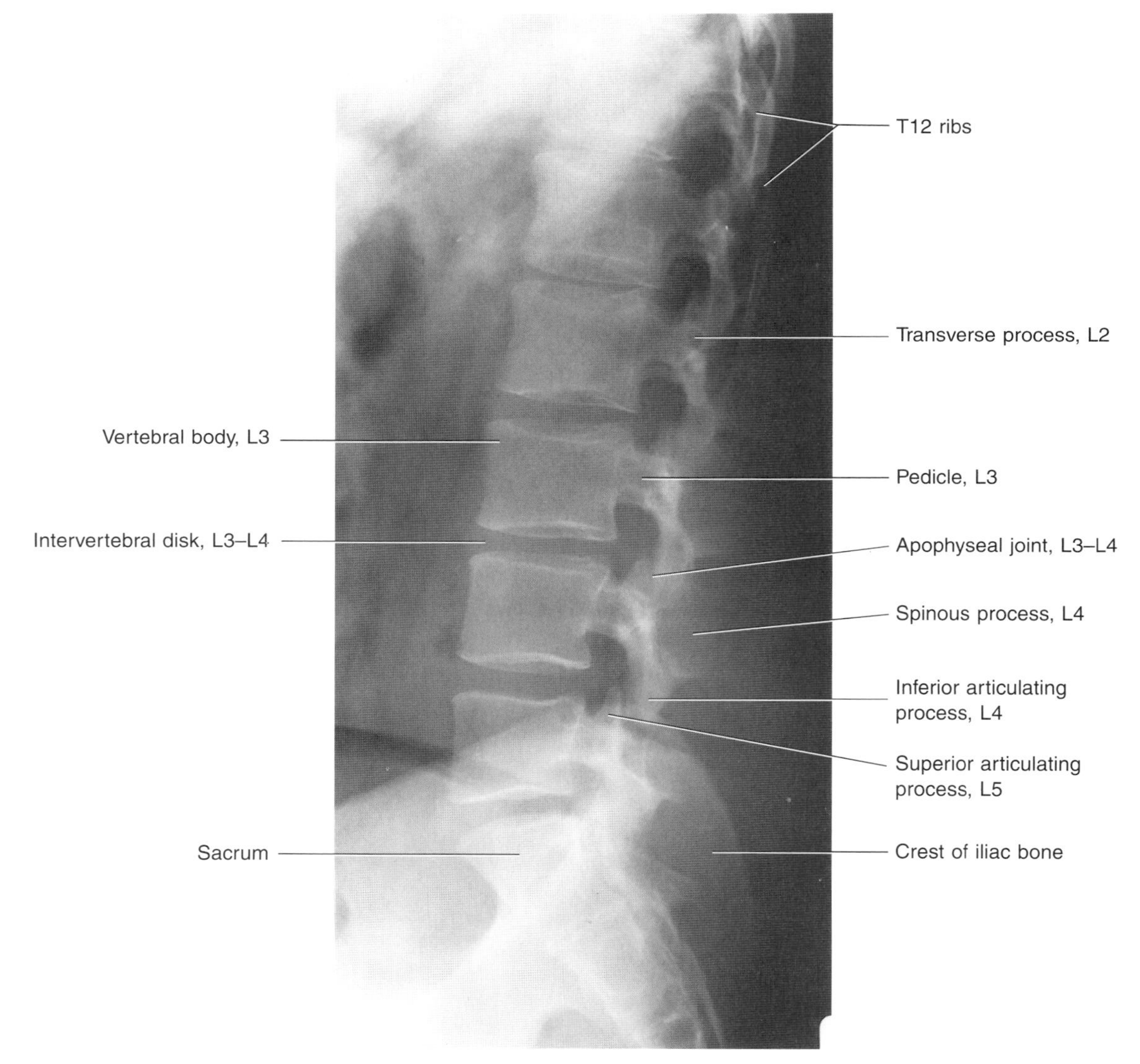

Figure 3.21. Lateral projection of the lumbar spine.

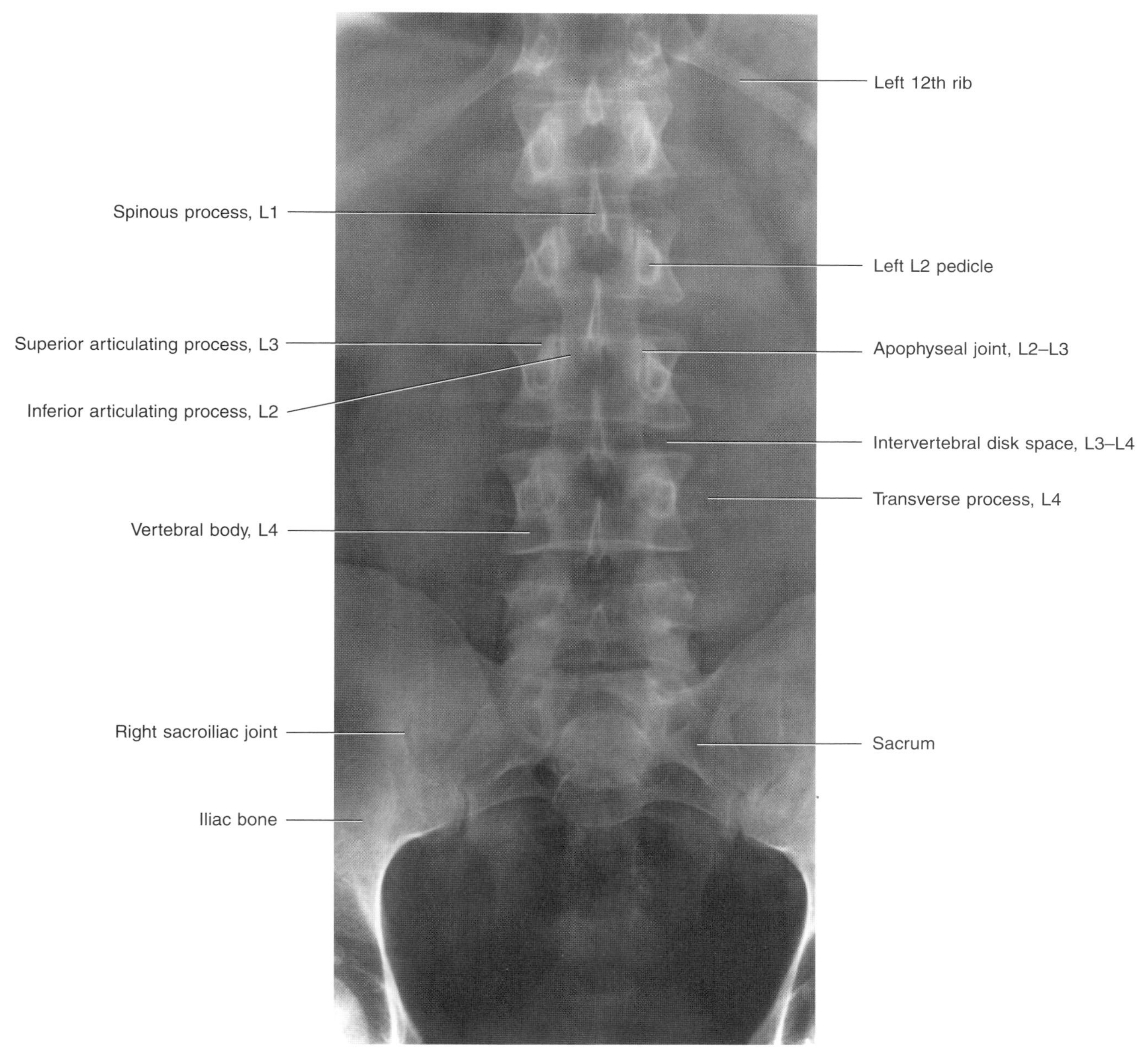

Figure 3.22. AP projection of the lumbar spine.

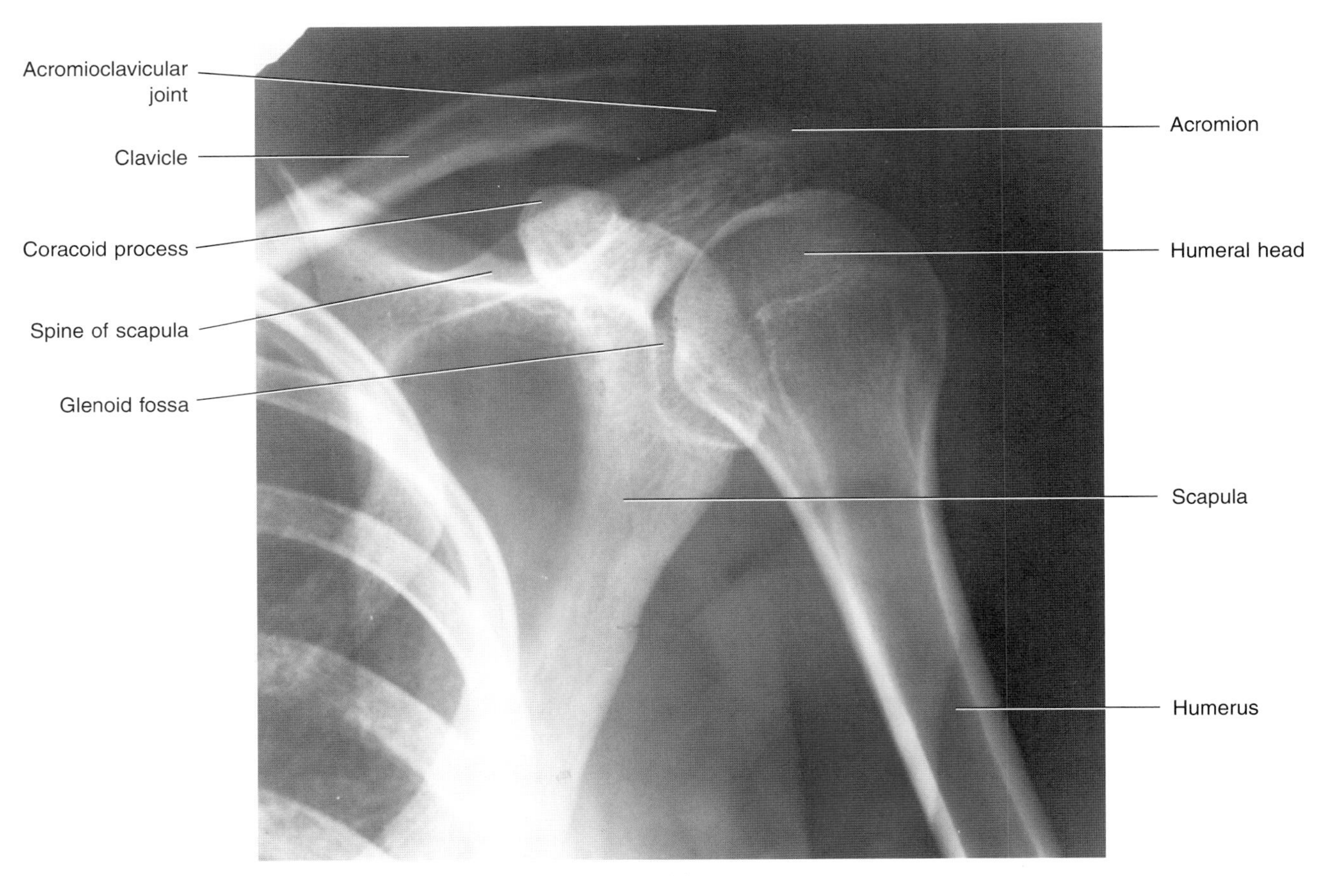

Figure 3.23. AP projection of the shoulder.

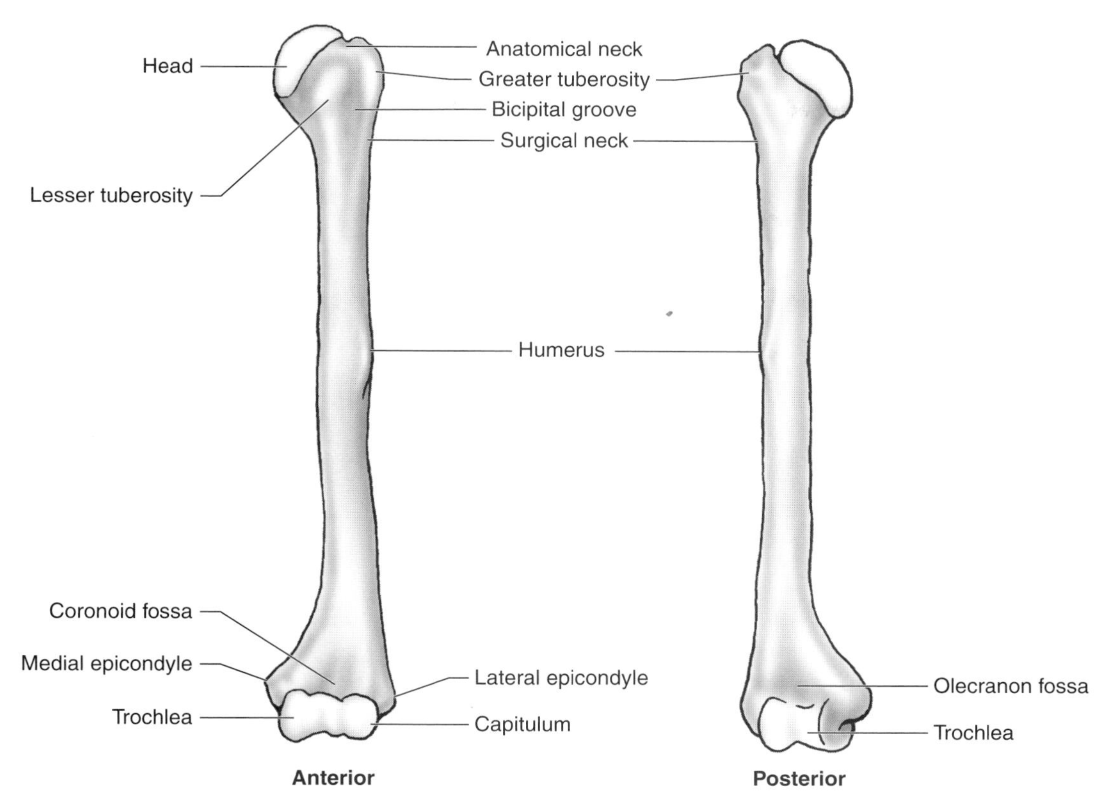

Figure 3.24. Anterior and posterior view drawings of the humerus.

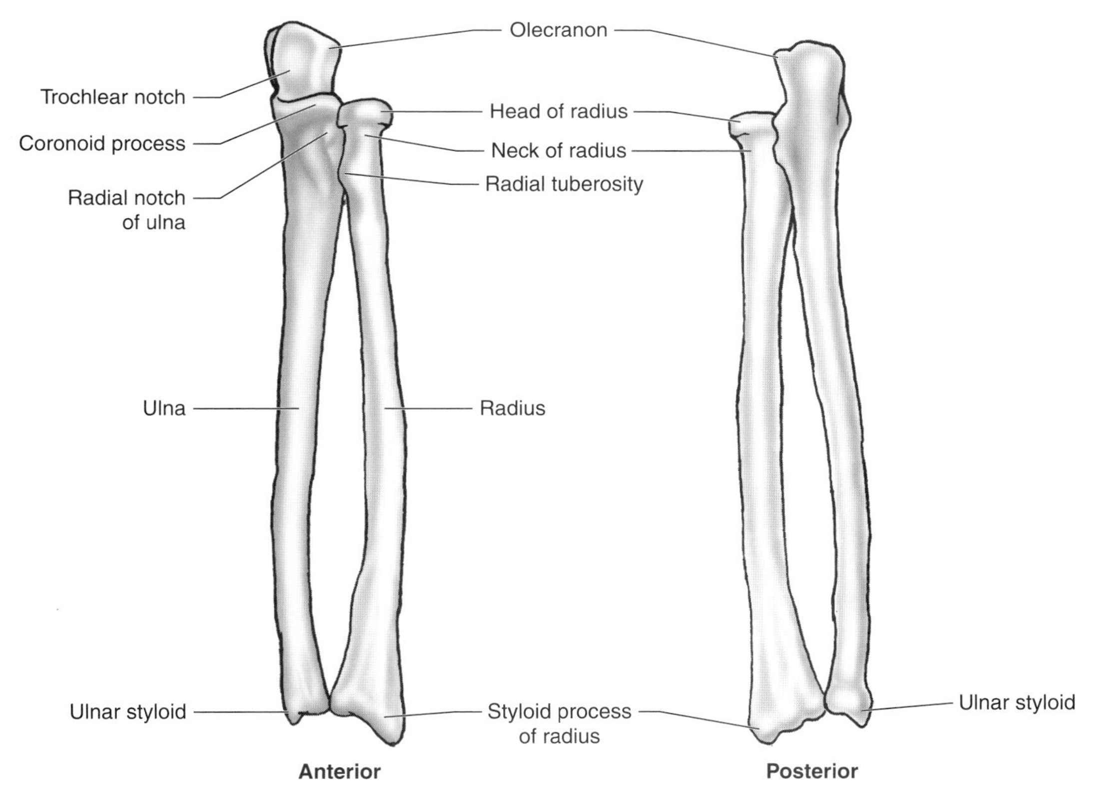

Figure 3.25. Anterior and posterior view drawings of the radius and ulna.

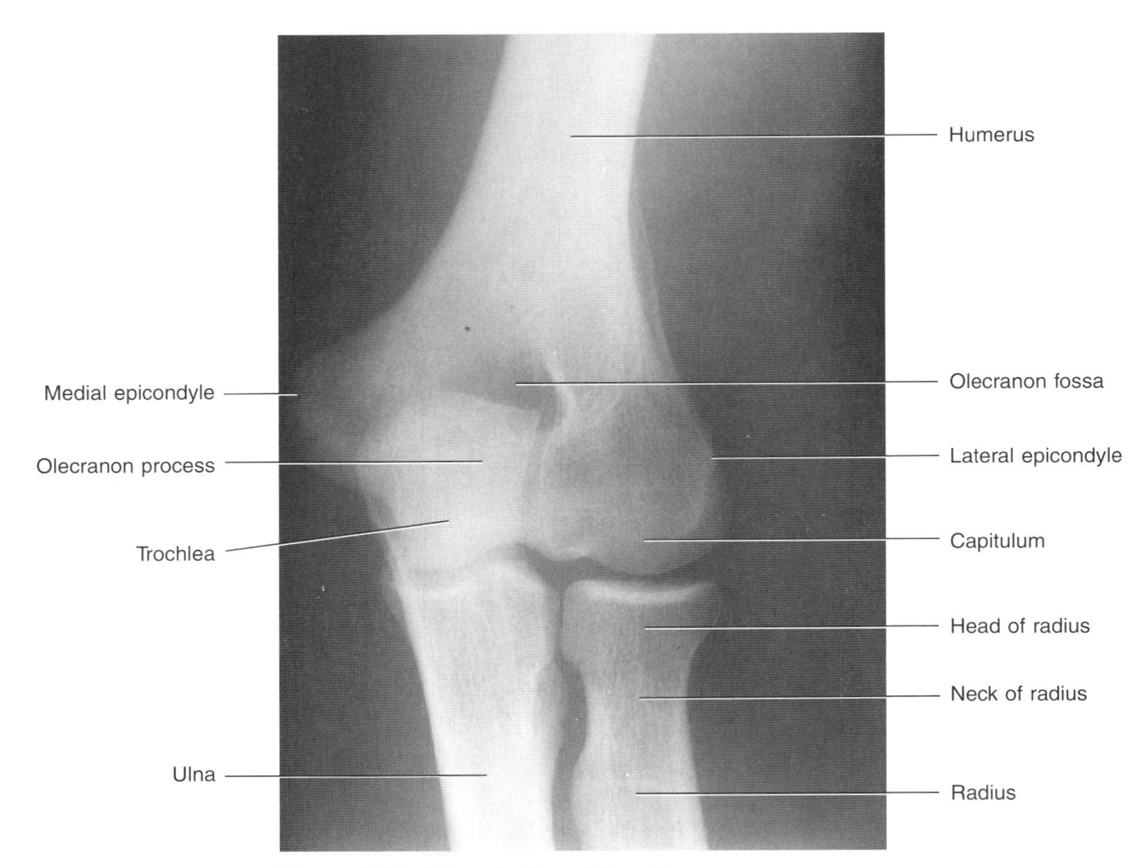

Figure 3.26. AP projection of the elbow.

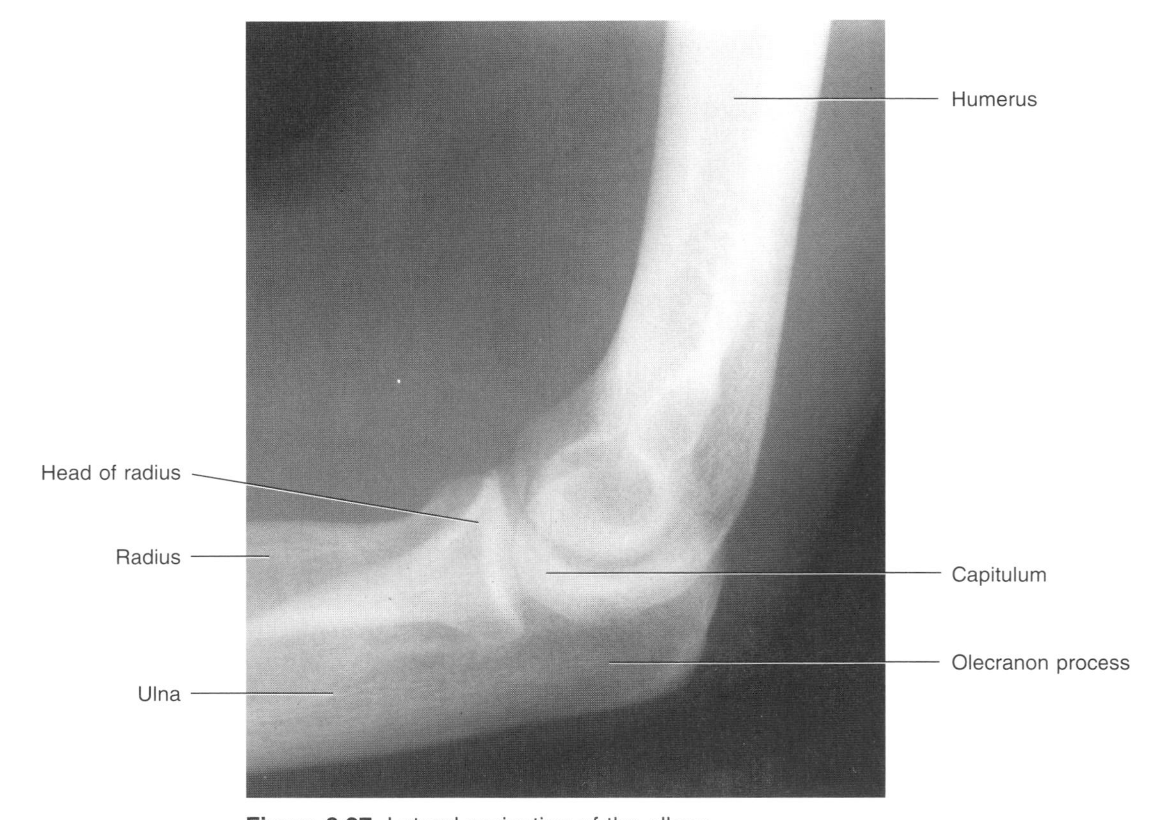

Figure 3.27. Lateral projection of the elbow.

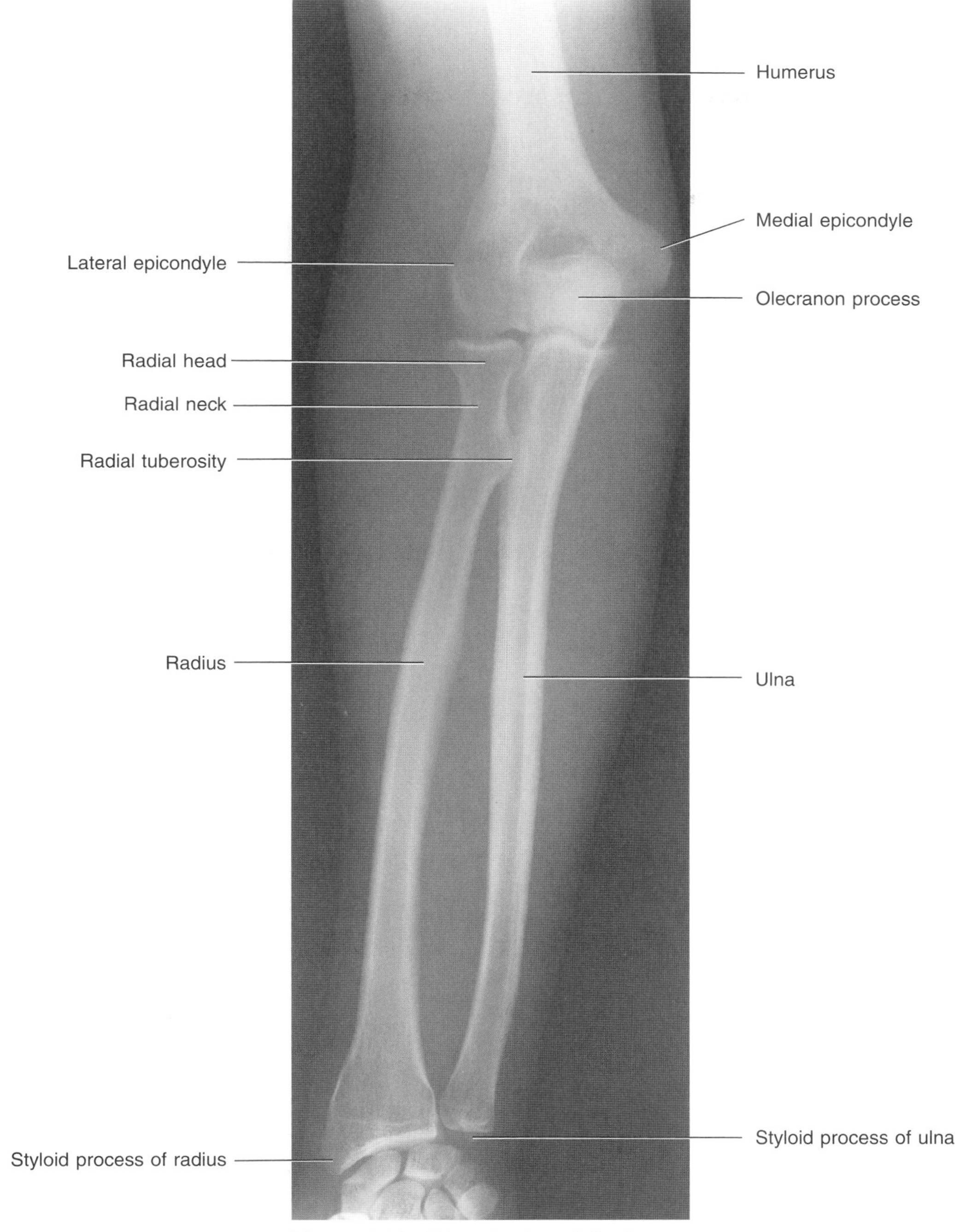

Figure 3.28. AP projection of the forearm.

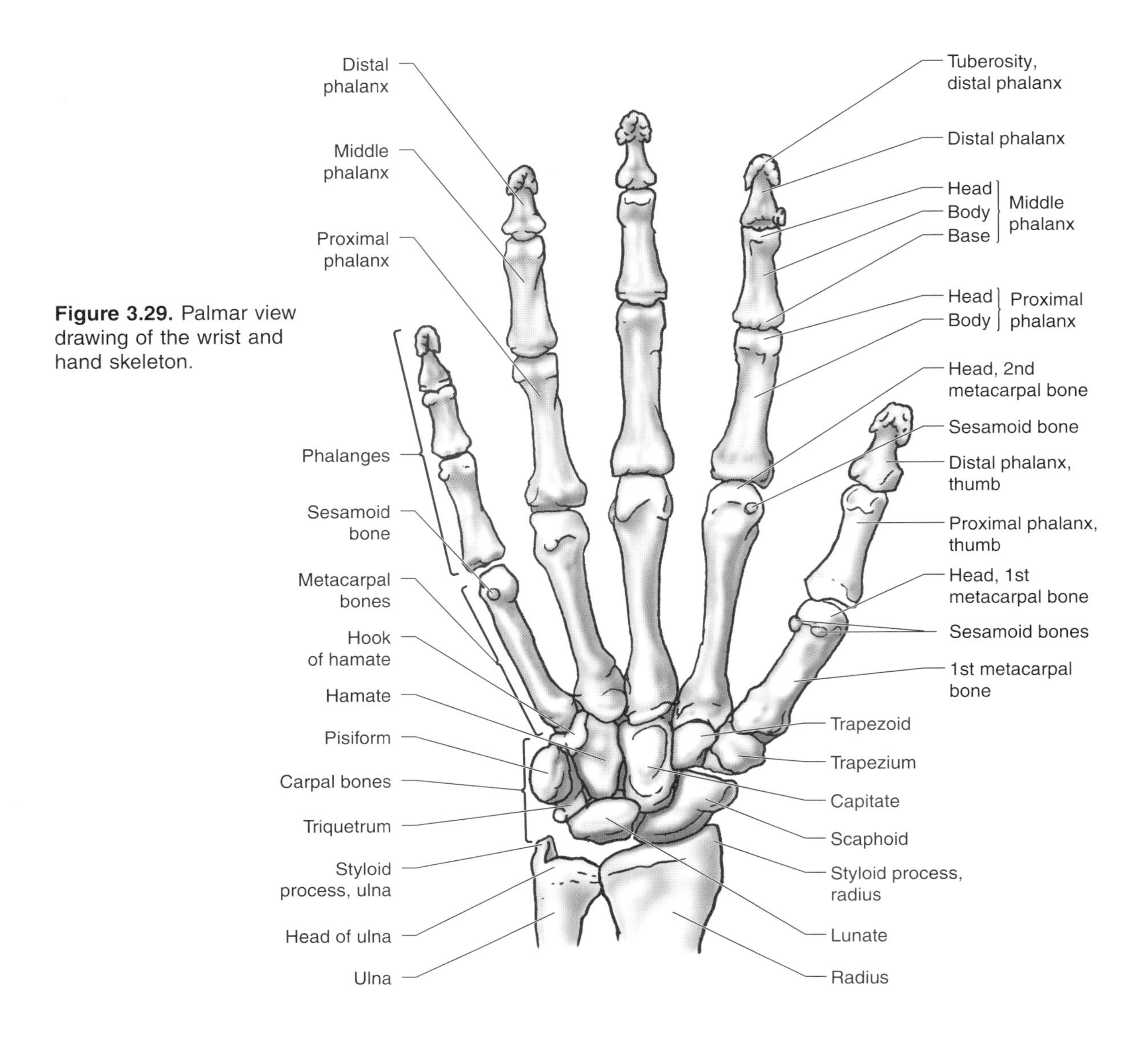

Figure 3.29. Palmar view drawing of the wrist and hand skeleton.

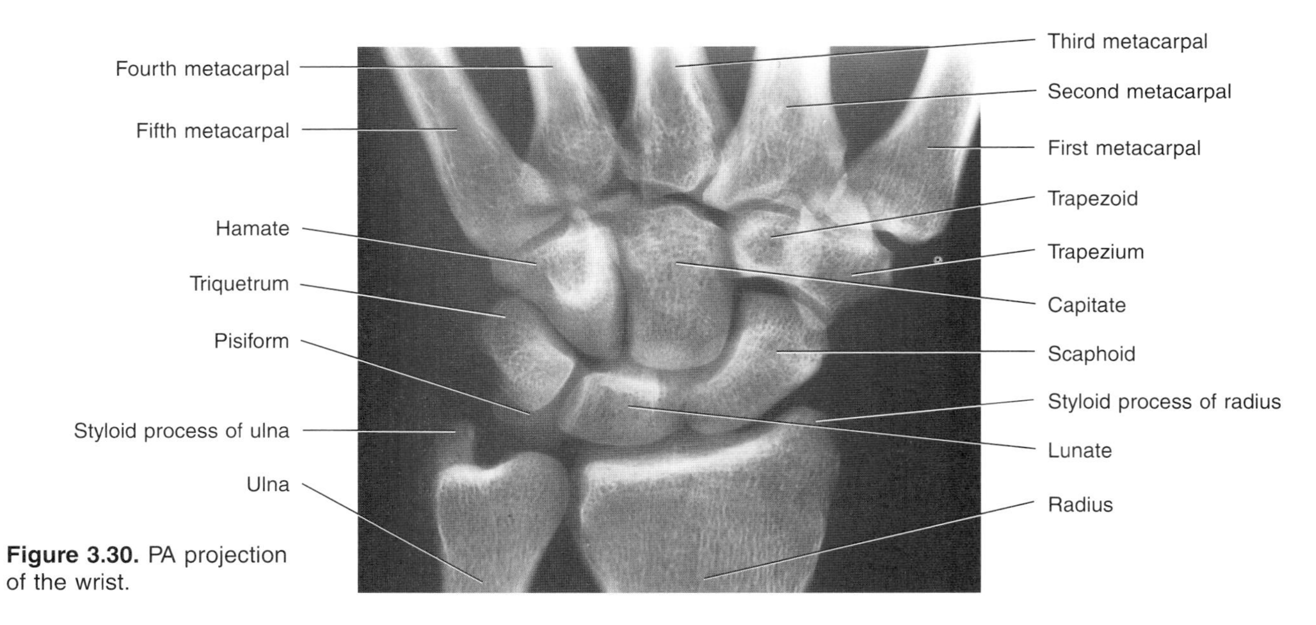

Figure 3.30. PA projection of the wrist.

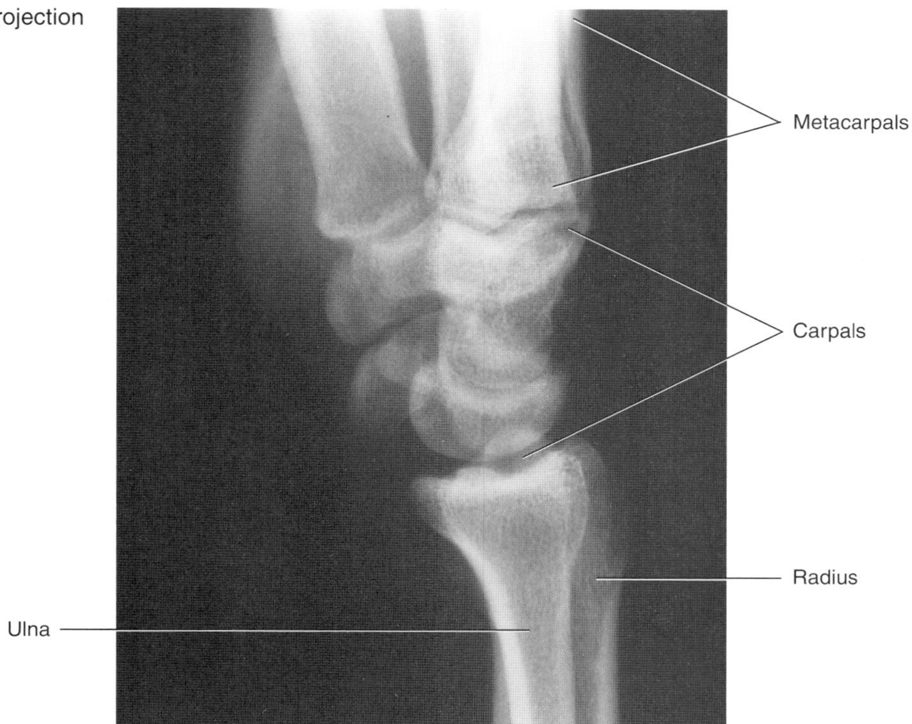

Figure 3.31. Lateral projection of the wrist.

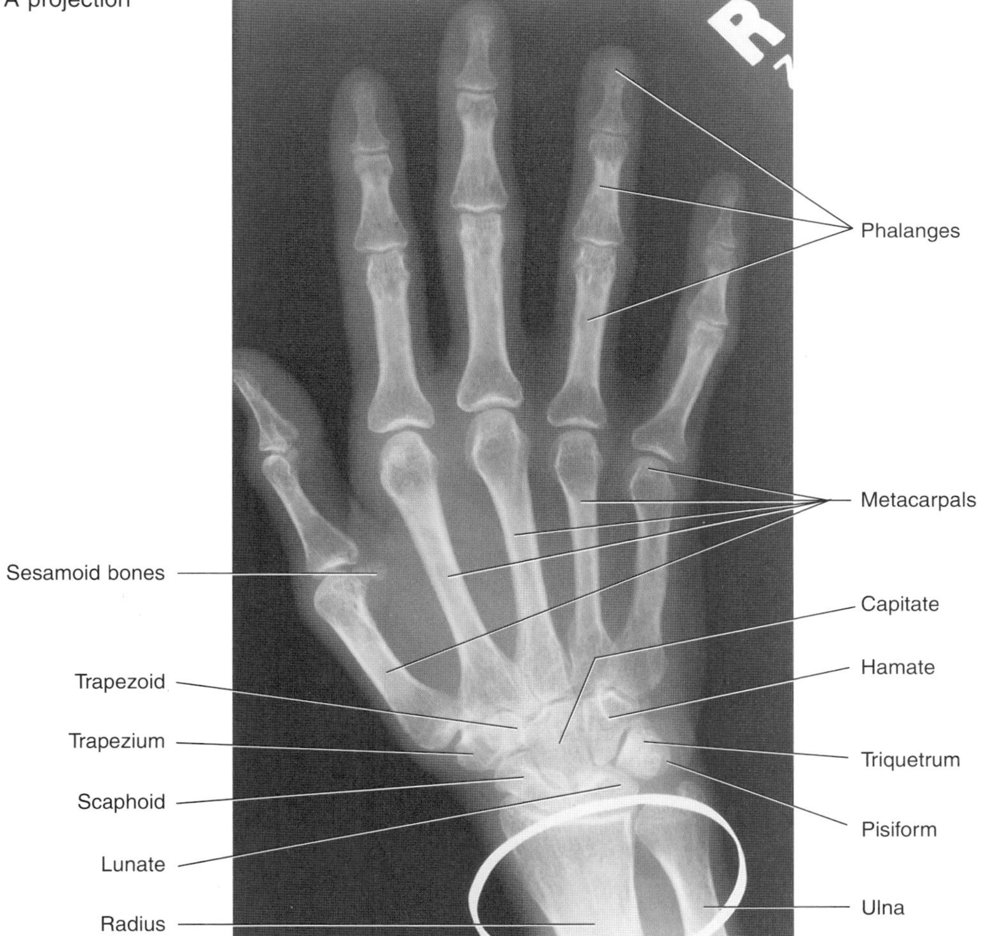

Figure 3.32. PA projection of the hand.

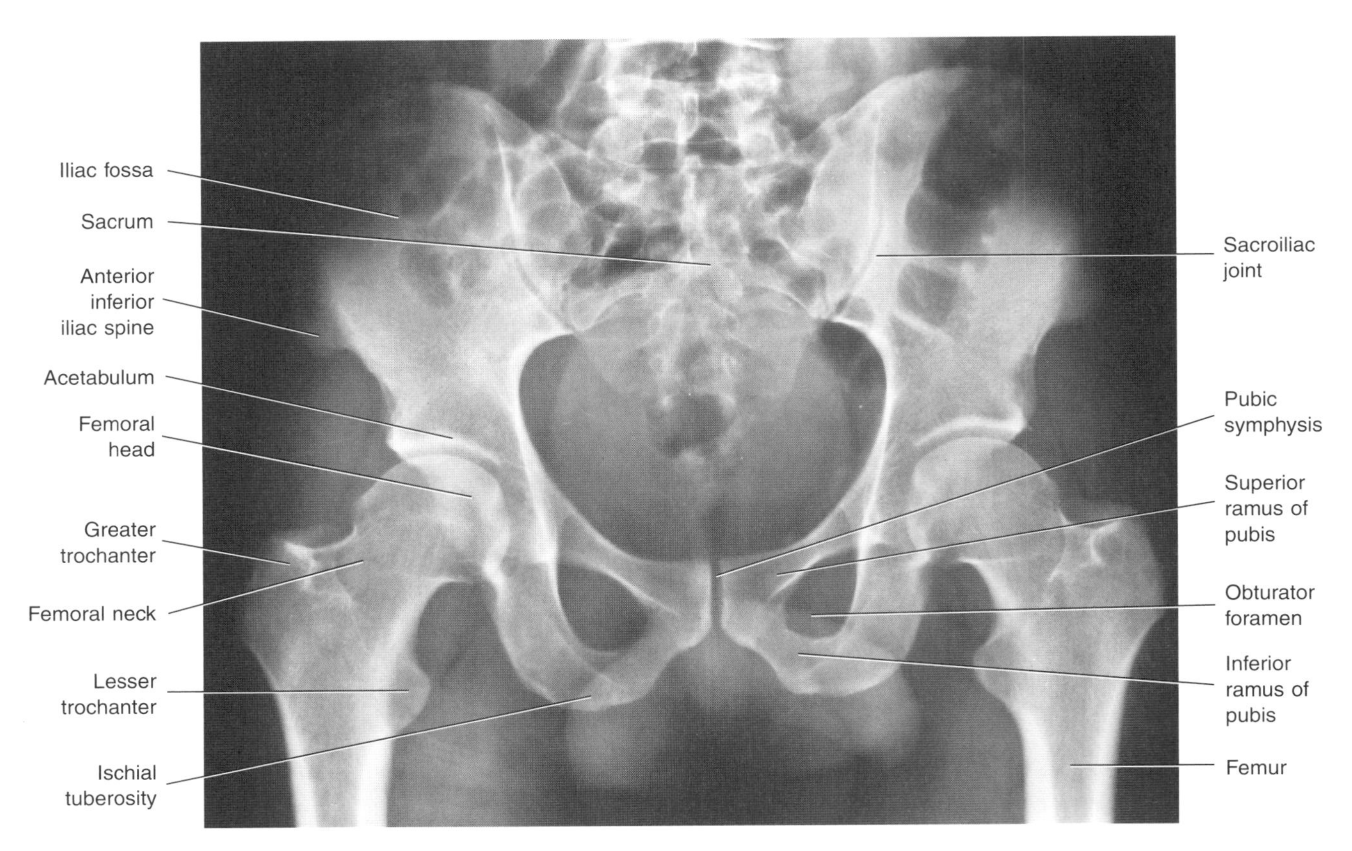

Figure 3.33. AP projection of the pelvis.

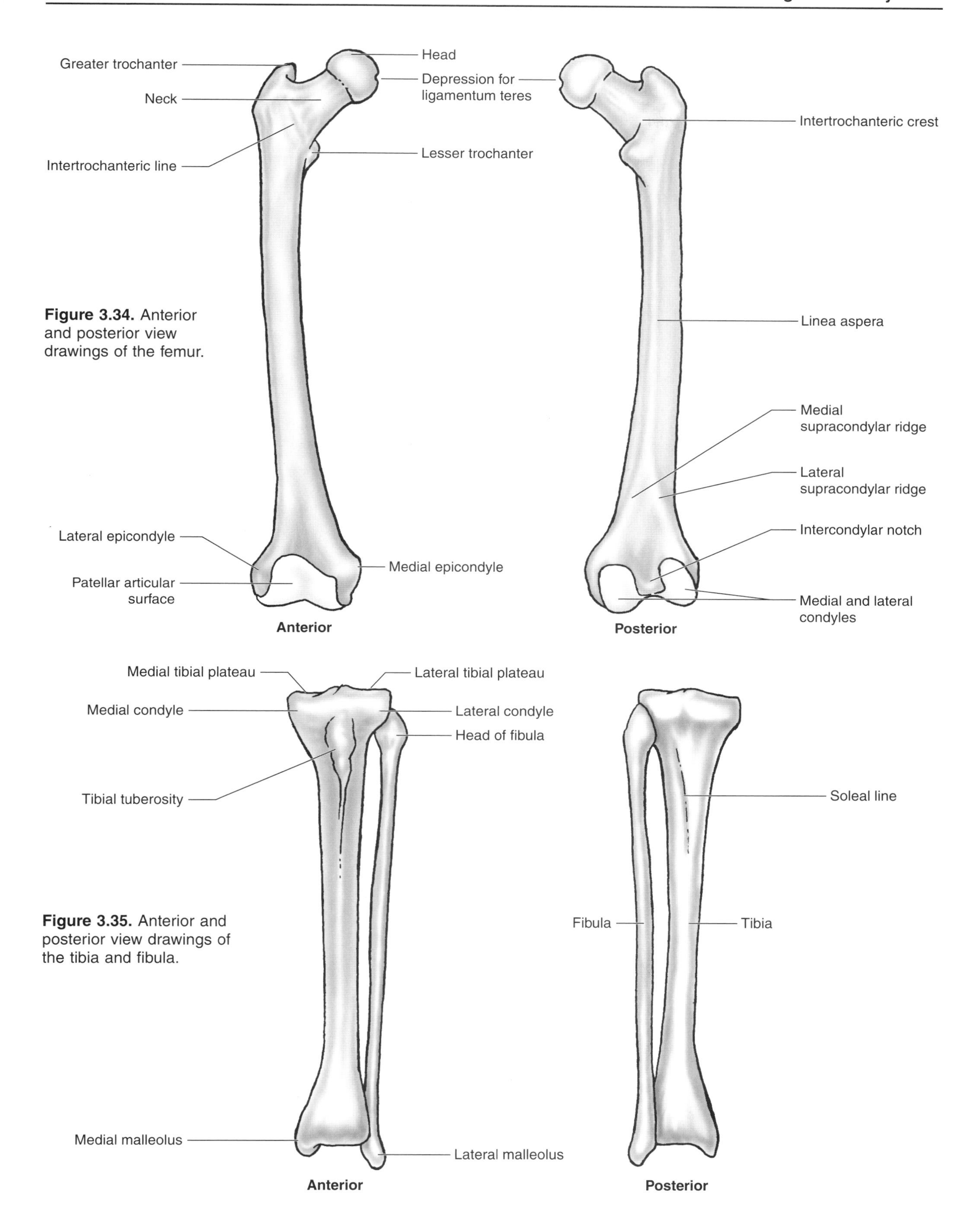

Figure 3.34. Anterior and posterior view drawings of the femur.

Figure 3.35. Anterior and posterior view drawings of the tibia and fibula.

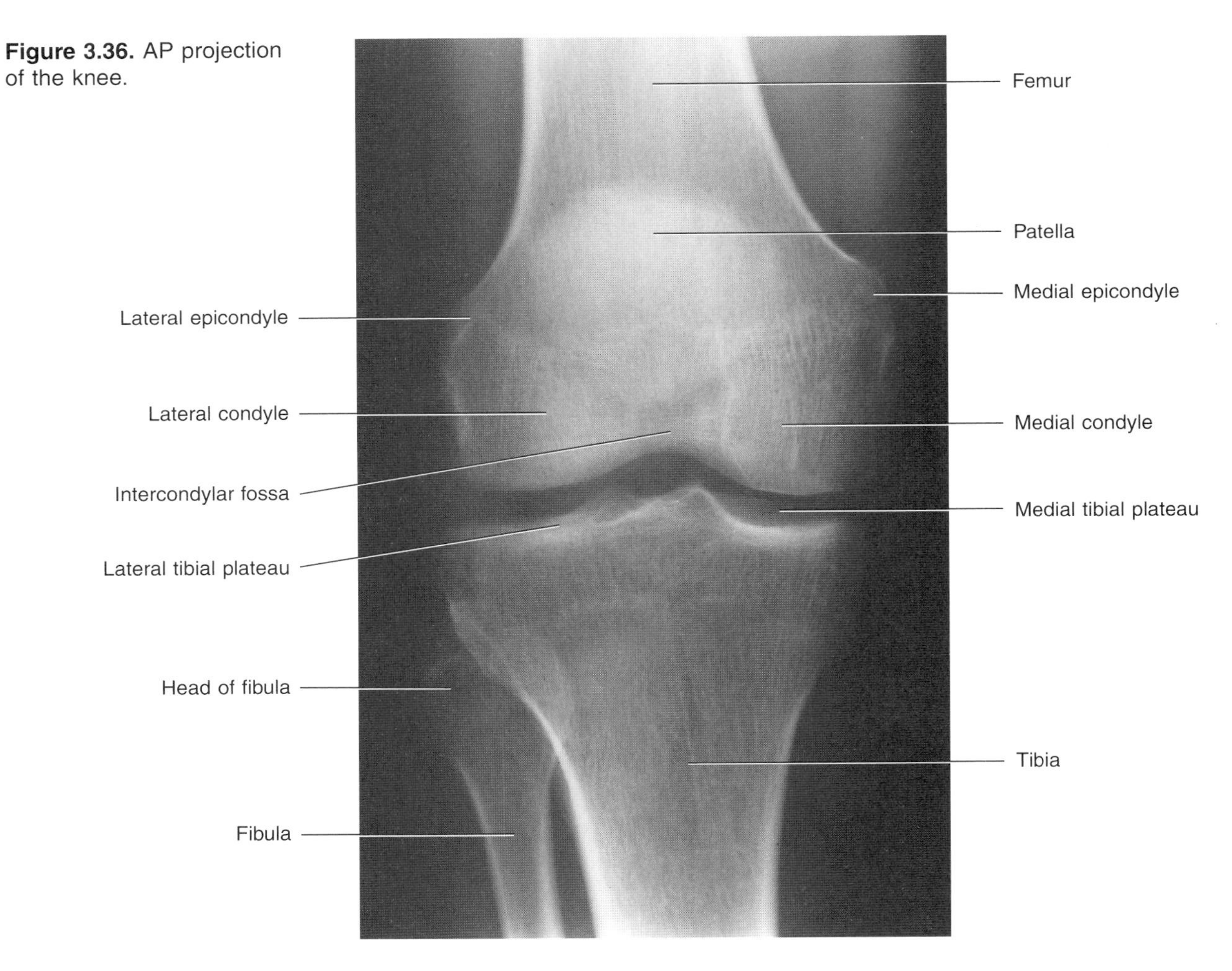

Figure 3.36. AP projection of the knee.

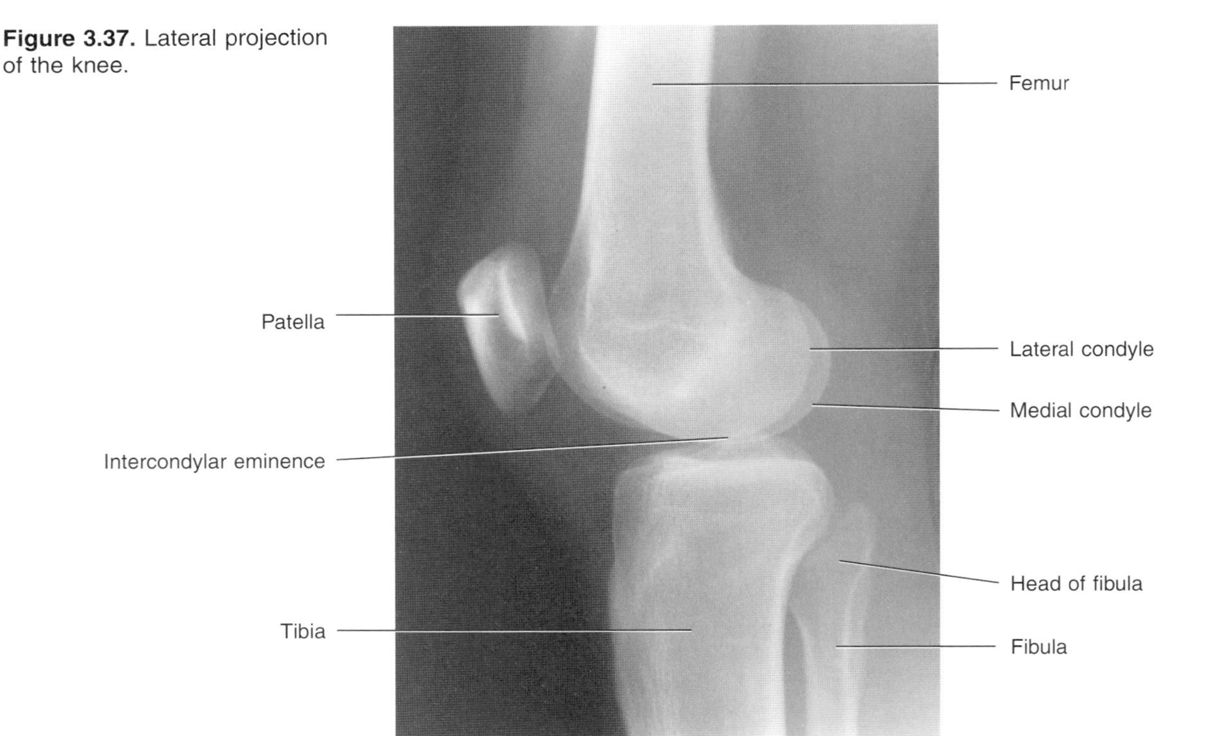

Figure 3.37. Lateral projection of the knee.

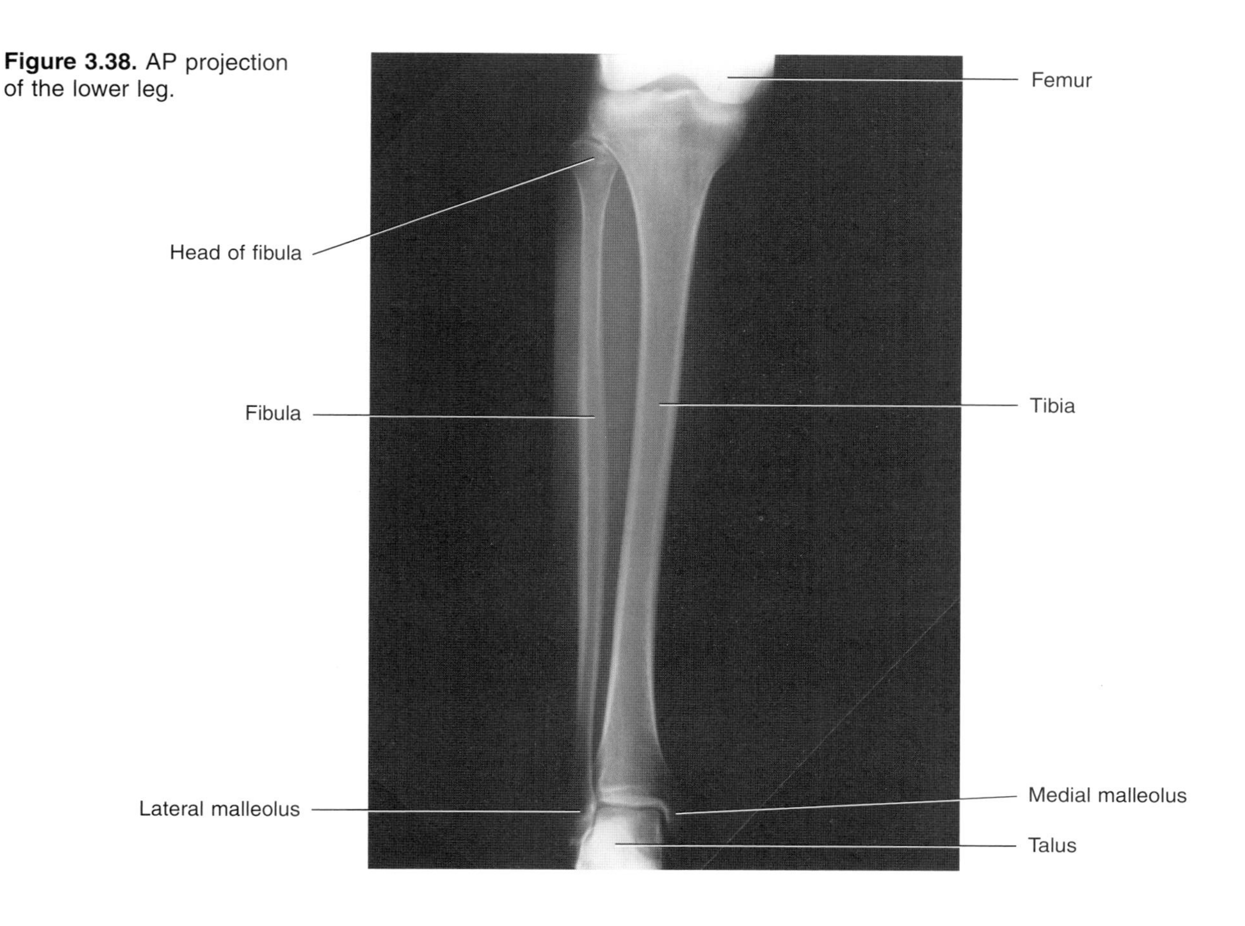

Figure 3.38. AP projection of the lower leg.

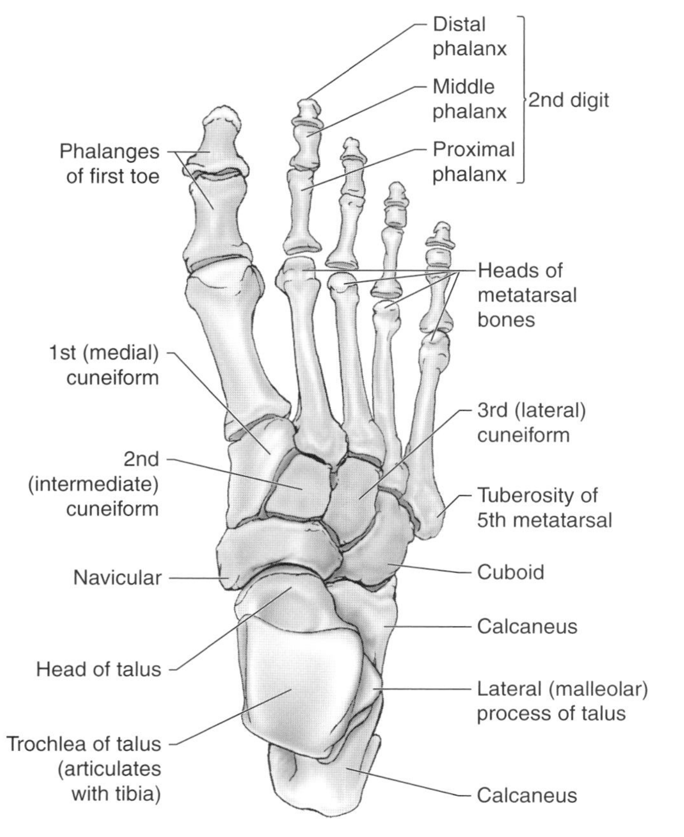

Figure 3.39. Dorsal view drawing of the foot skeleton.

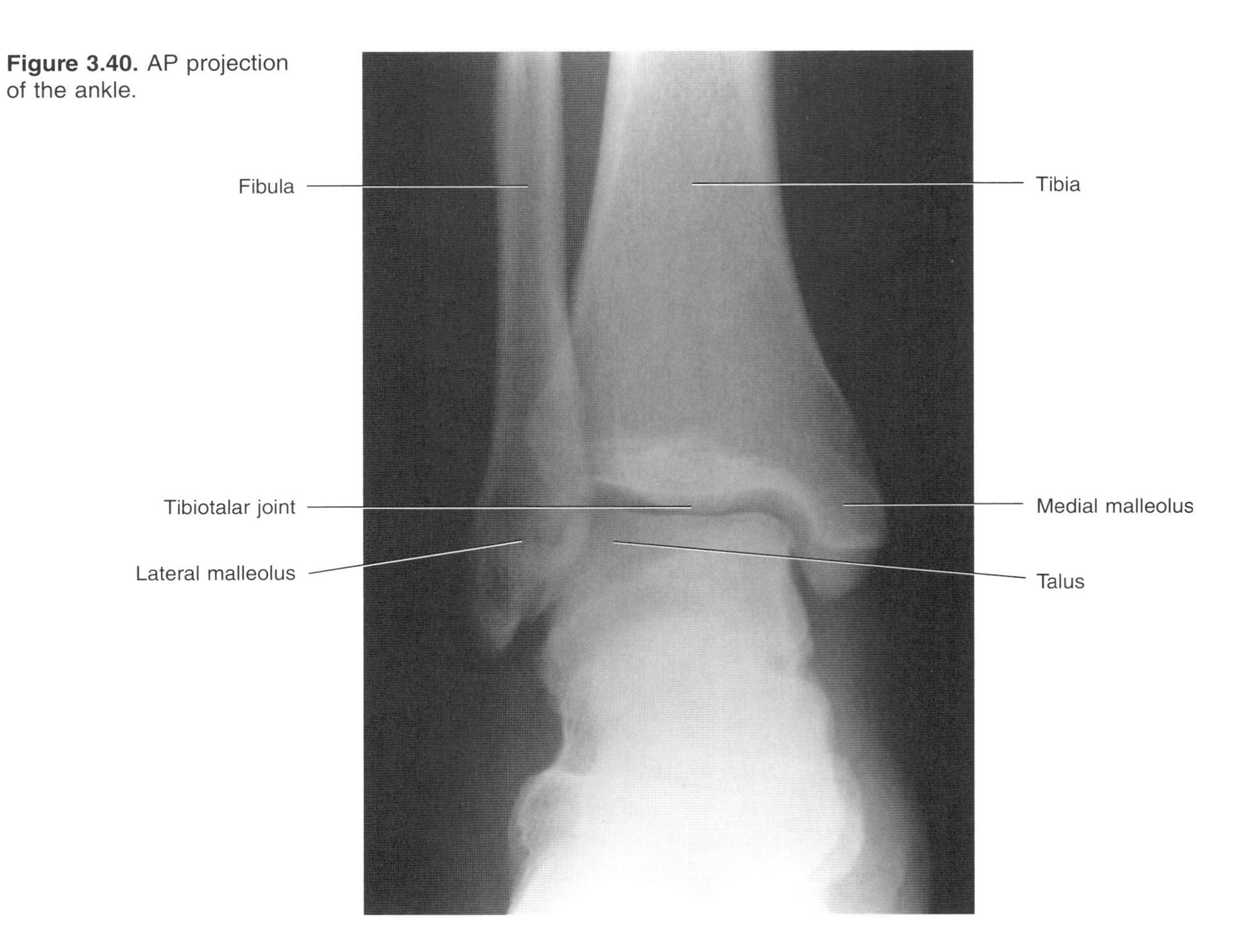

Figure 3.40. AP projection of the ankle.

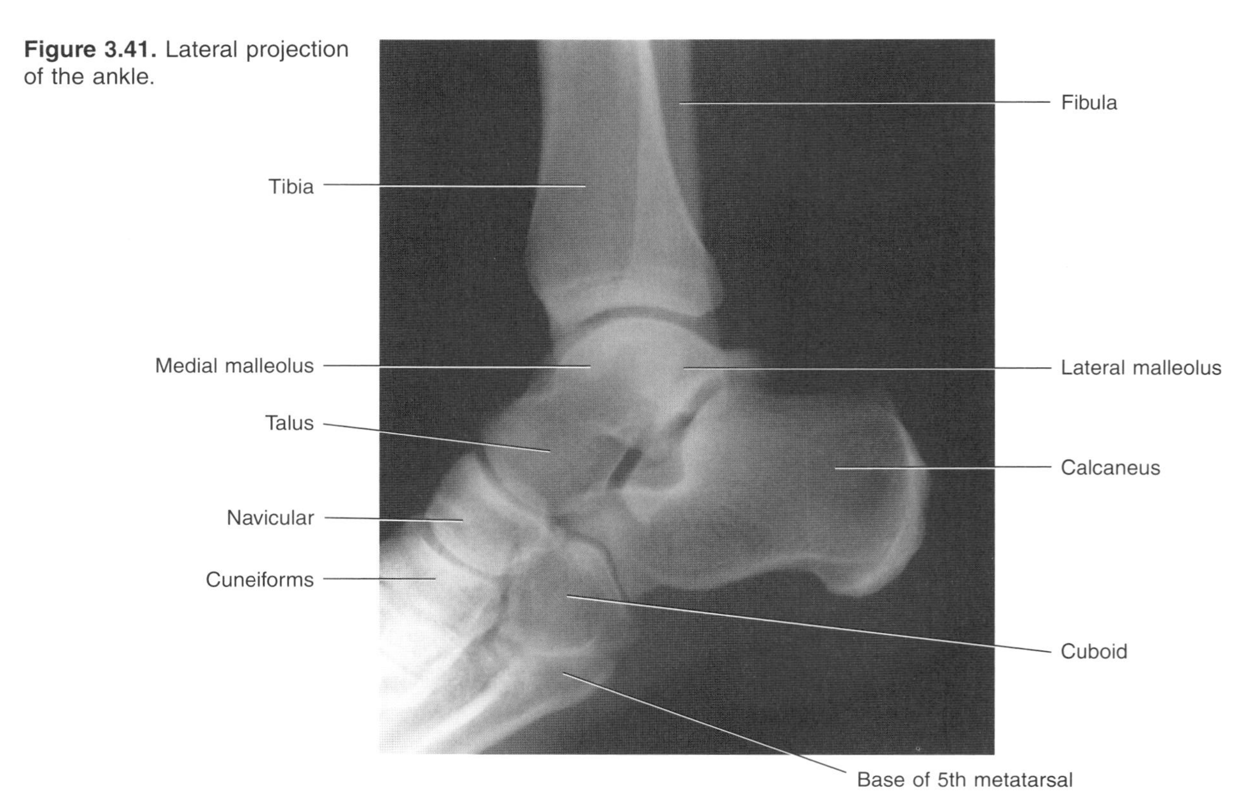

Figure 3.41. Lateral projection of the ankle.

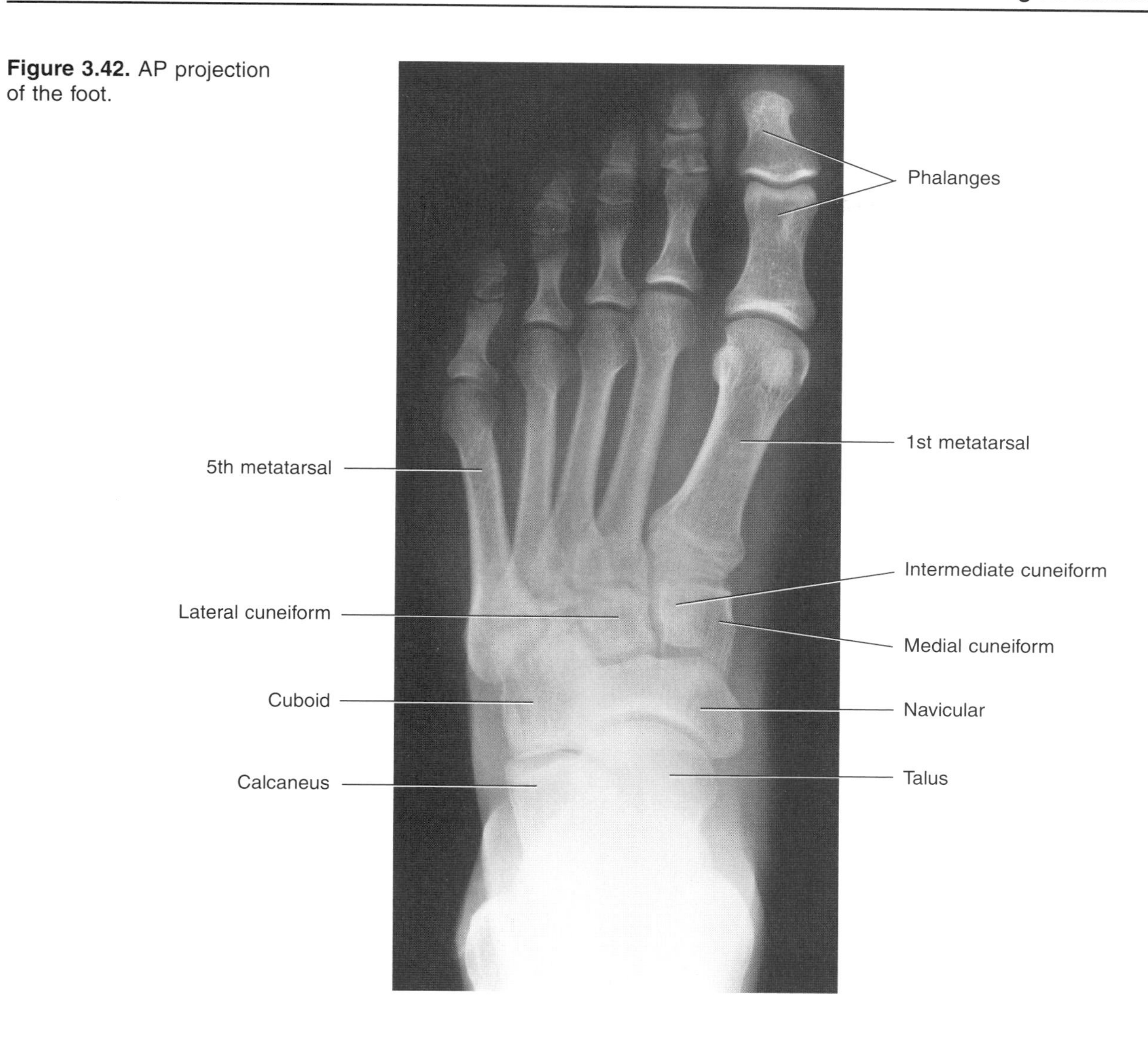

Figure 3.42. AP projection of the foot.

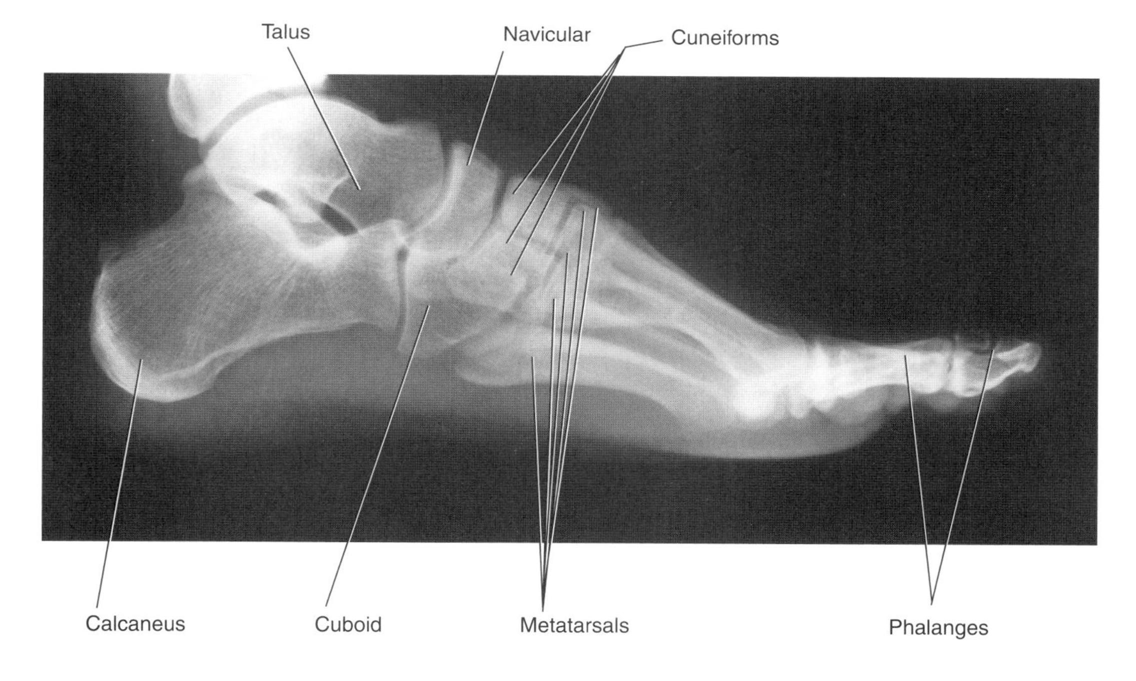

Figure 3.43. Lateral projection of the foot.

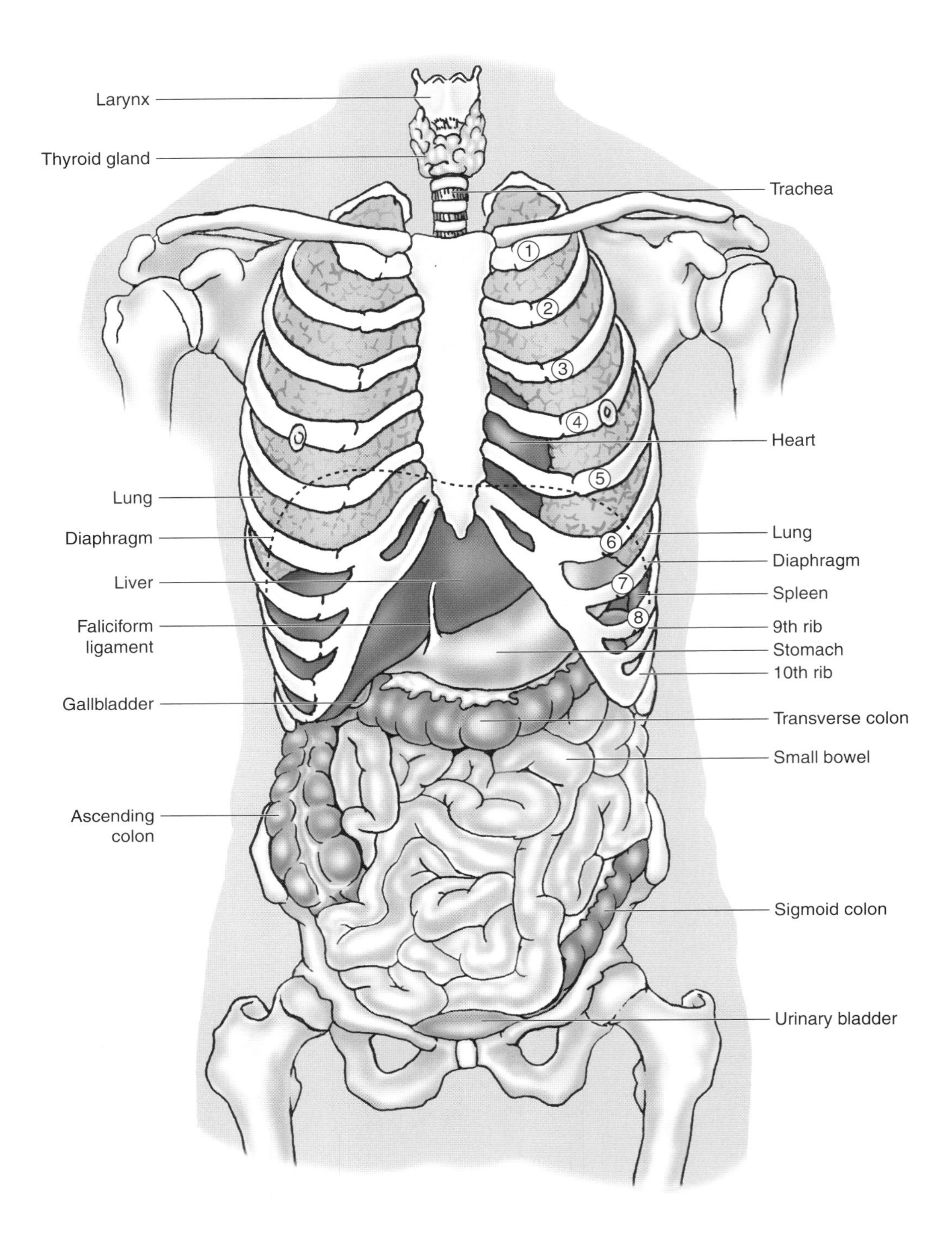

Figure 3.44. Anterior view drawing of the thoracic and abdominal viscera.

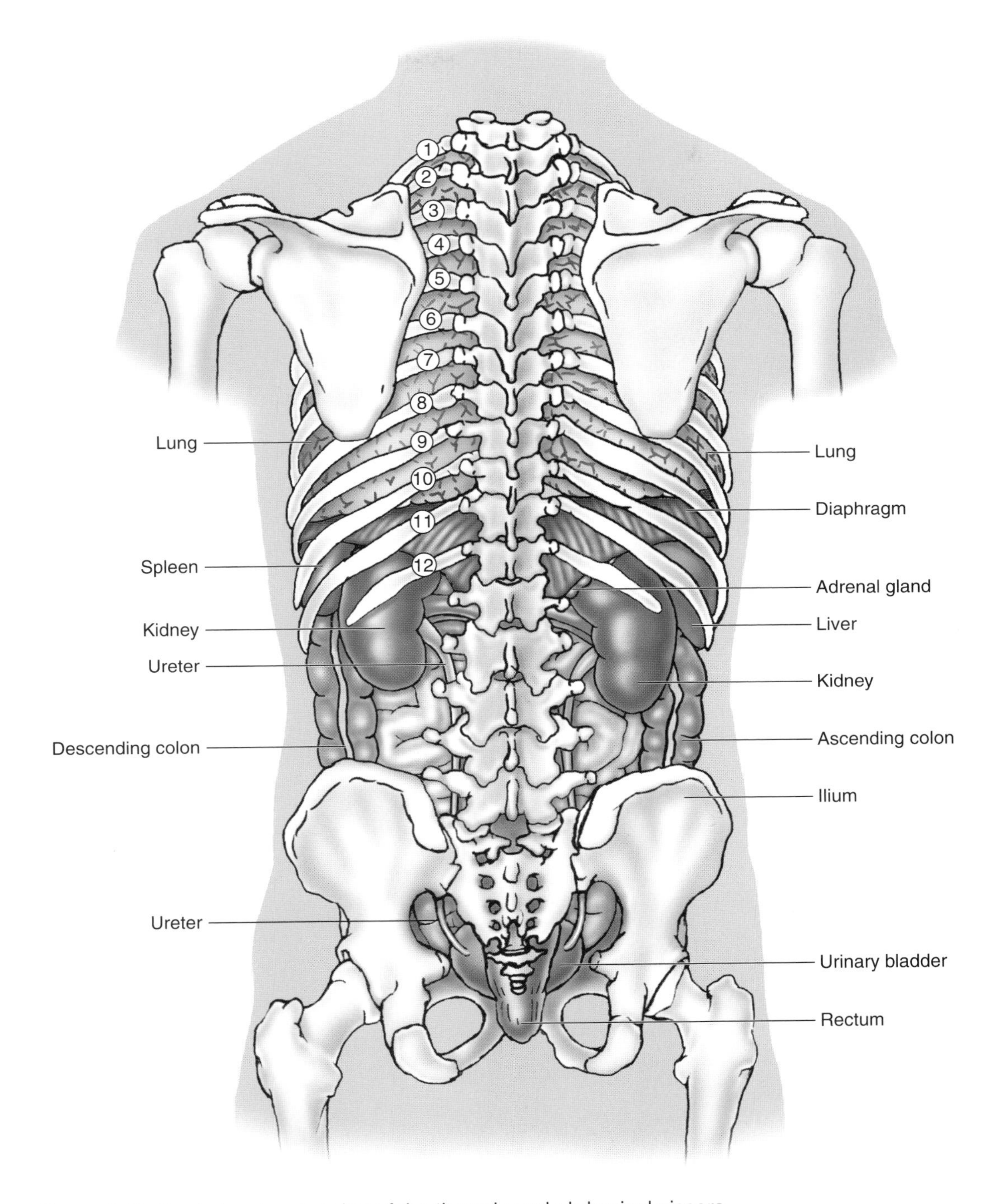

Figure 3.45. Posterior view drawing of the thoracic and abdominal viscera.

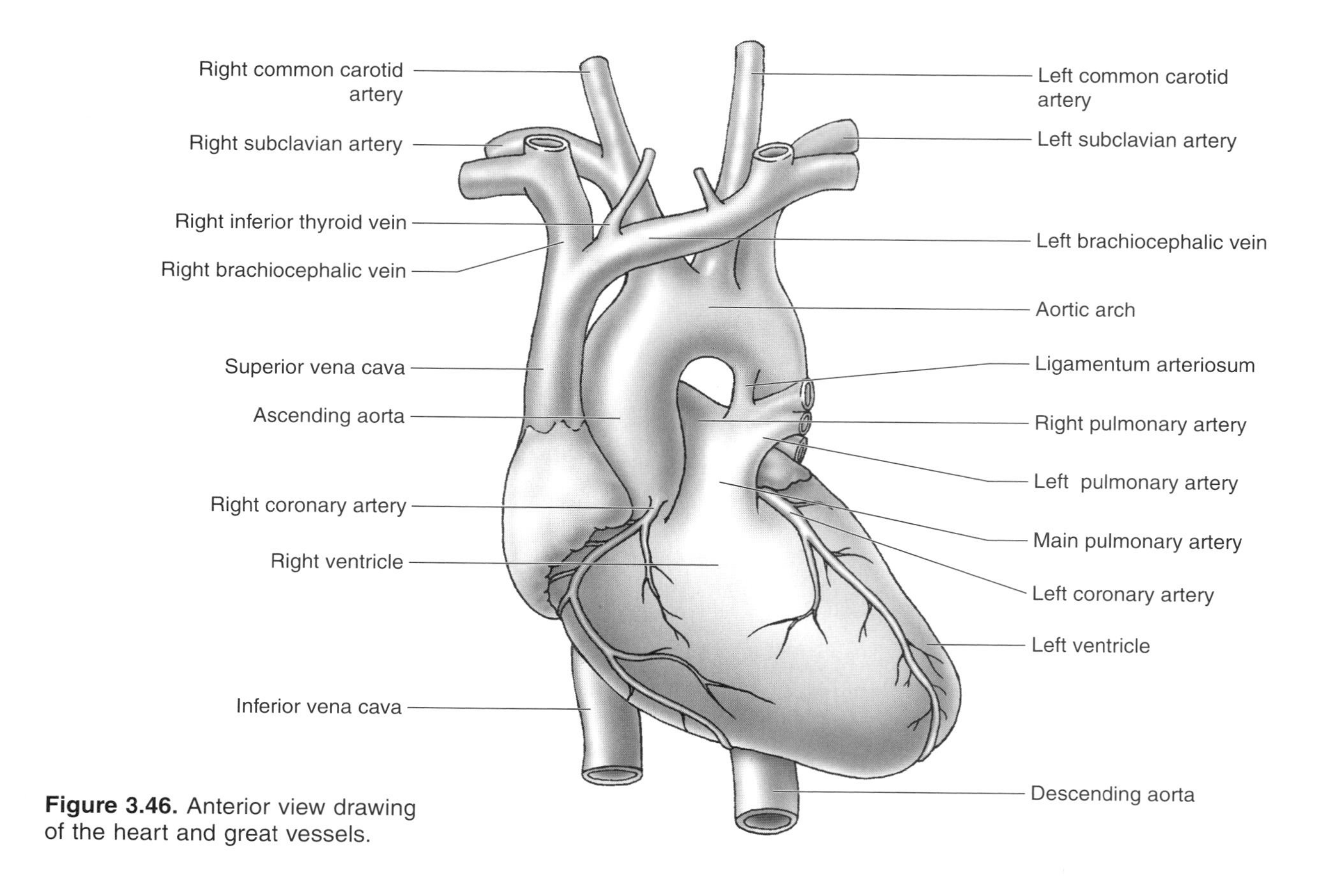

Figure 3.46. Anterior view drawing of the heart and great vessels.

Figures 3.47 to 3.57. Series of chest CT scans with intravenous contrast material, filmed with soft-tissue windows.

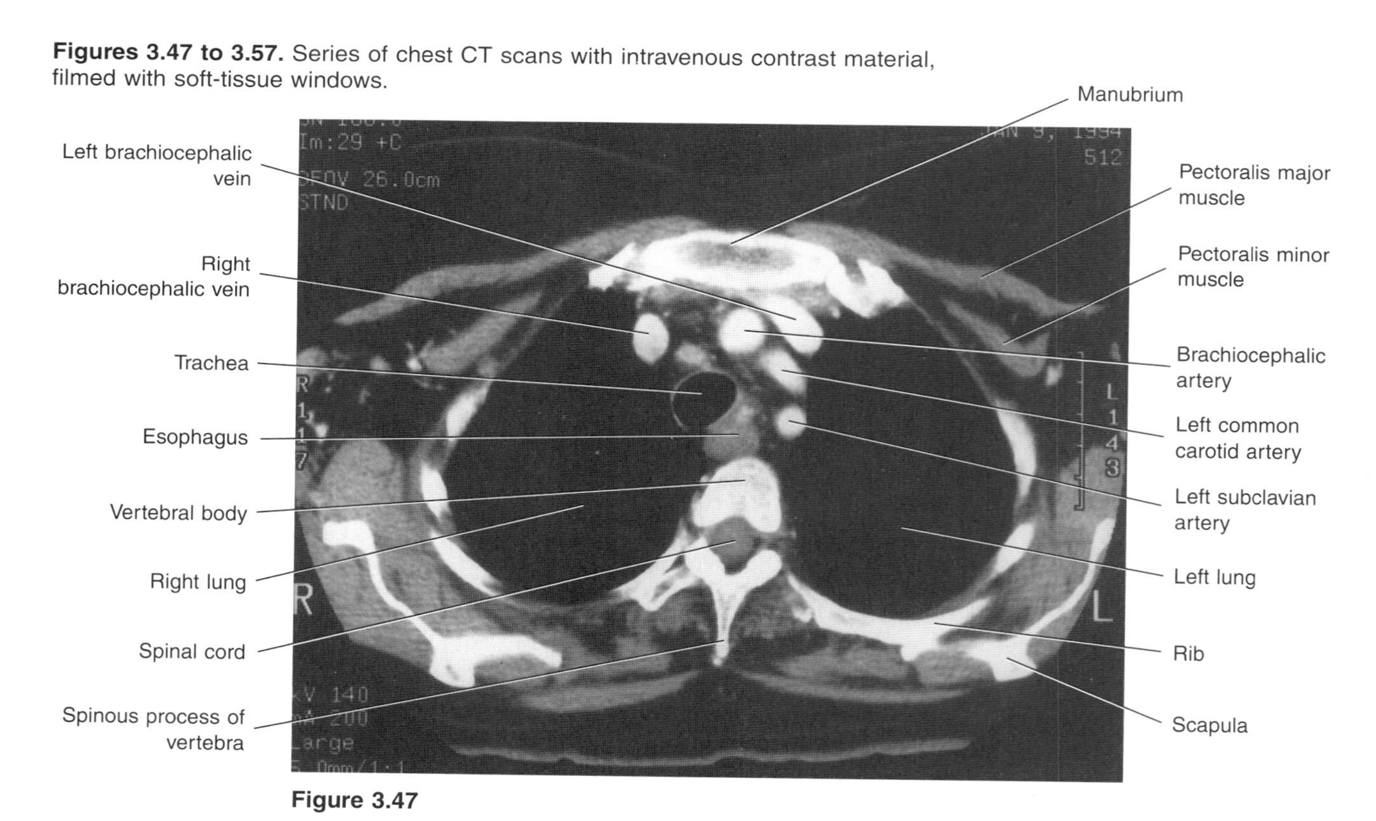

Figure 3.47

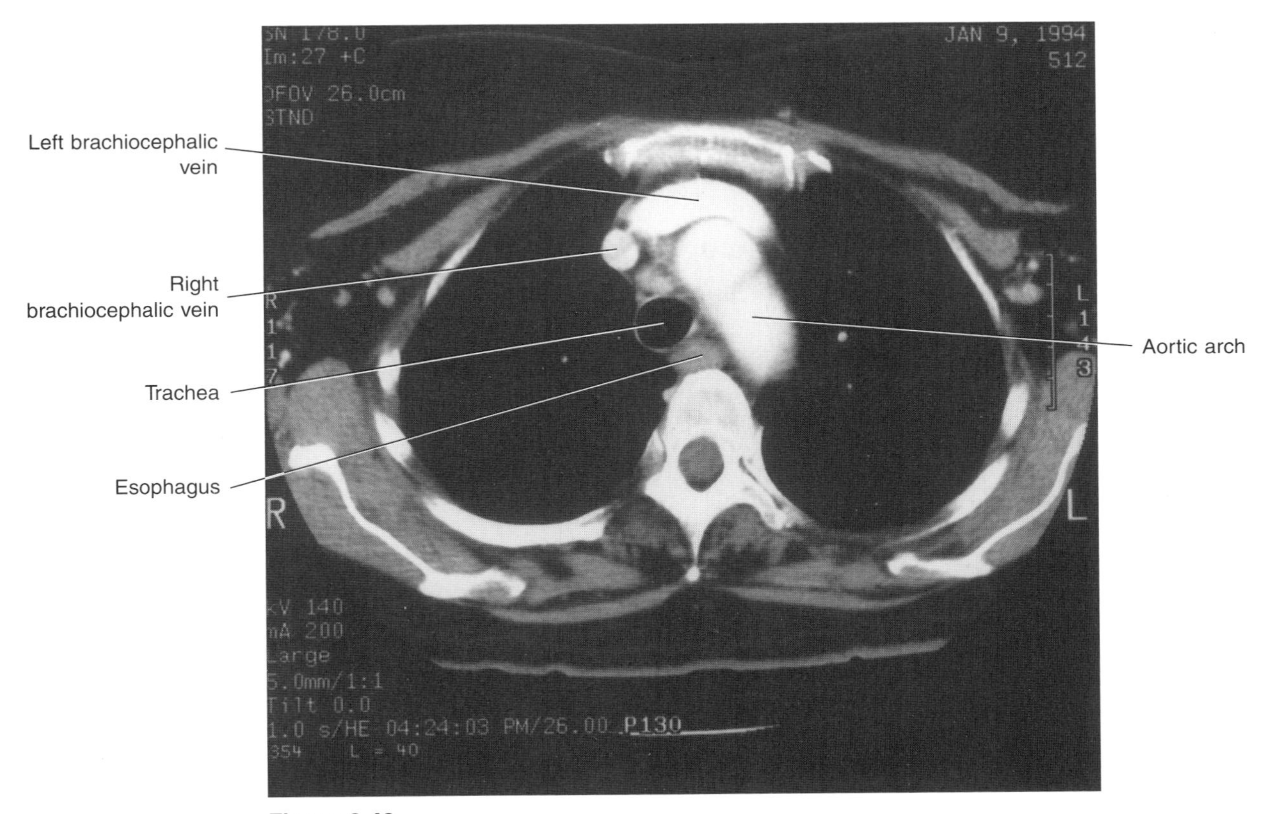

Figure 3.48

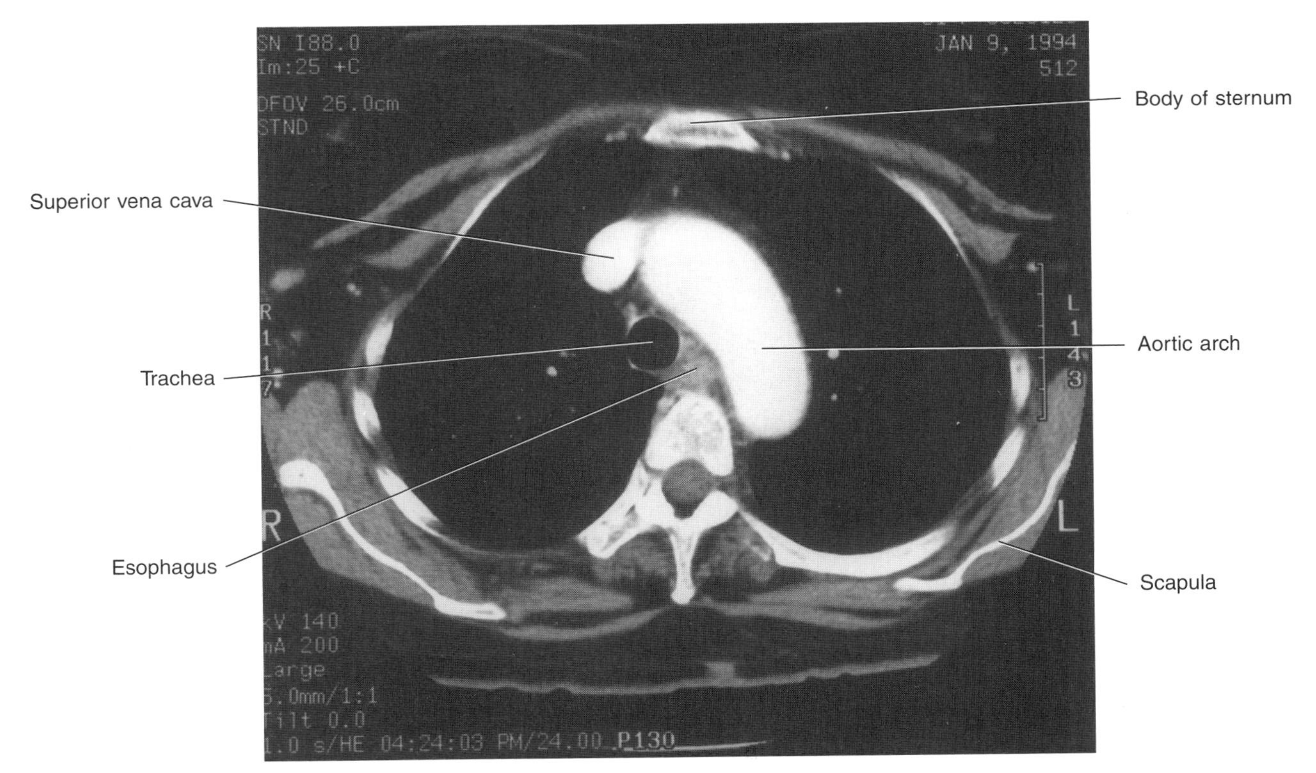

Figure 3.49

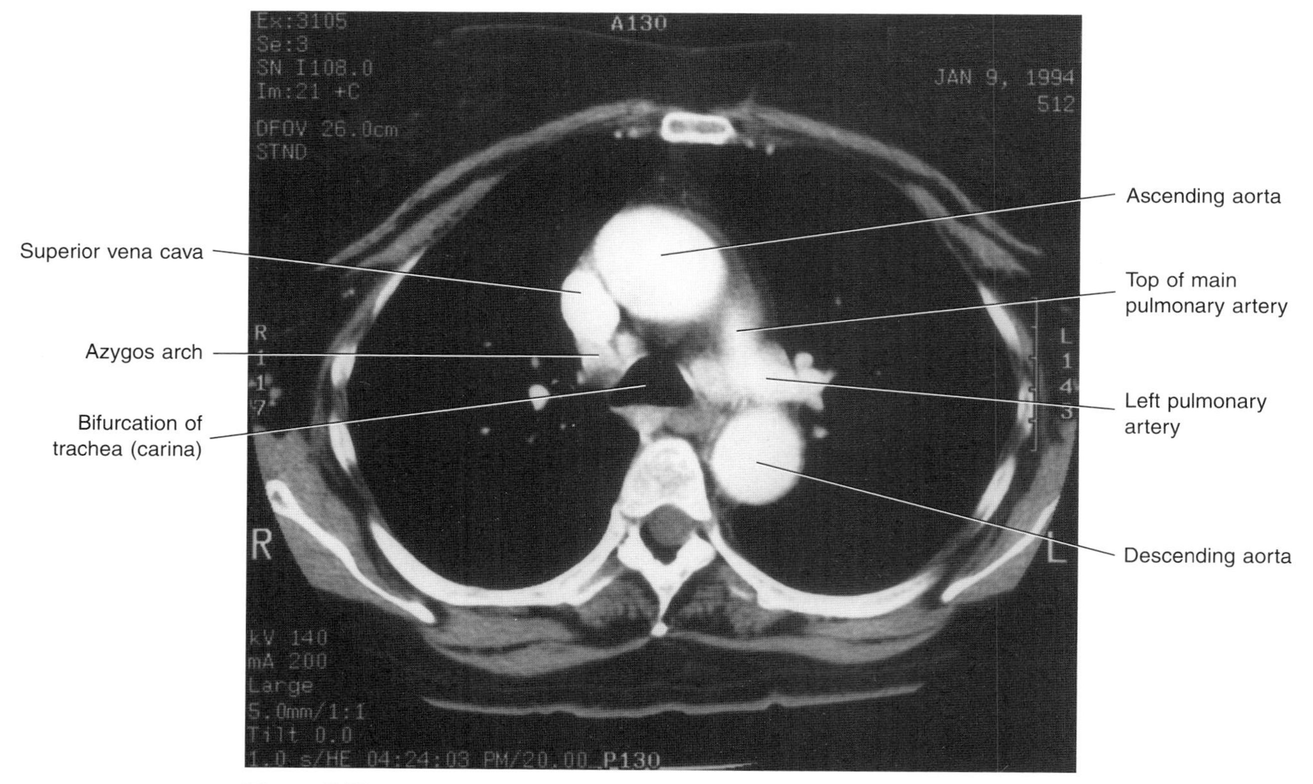

Figure 3.50

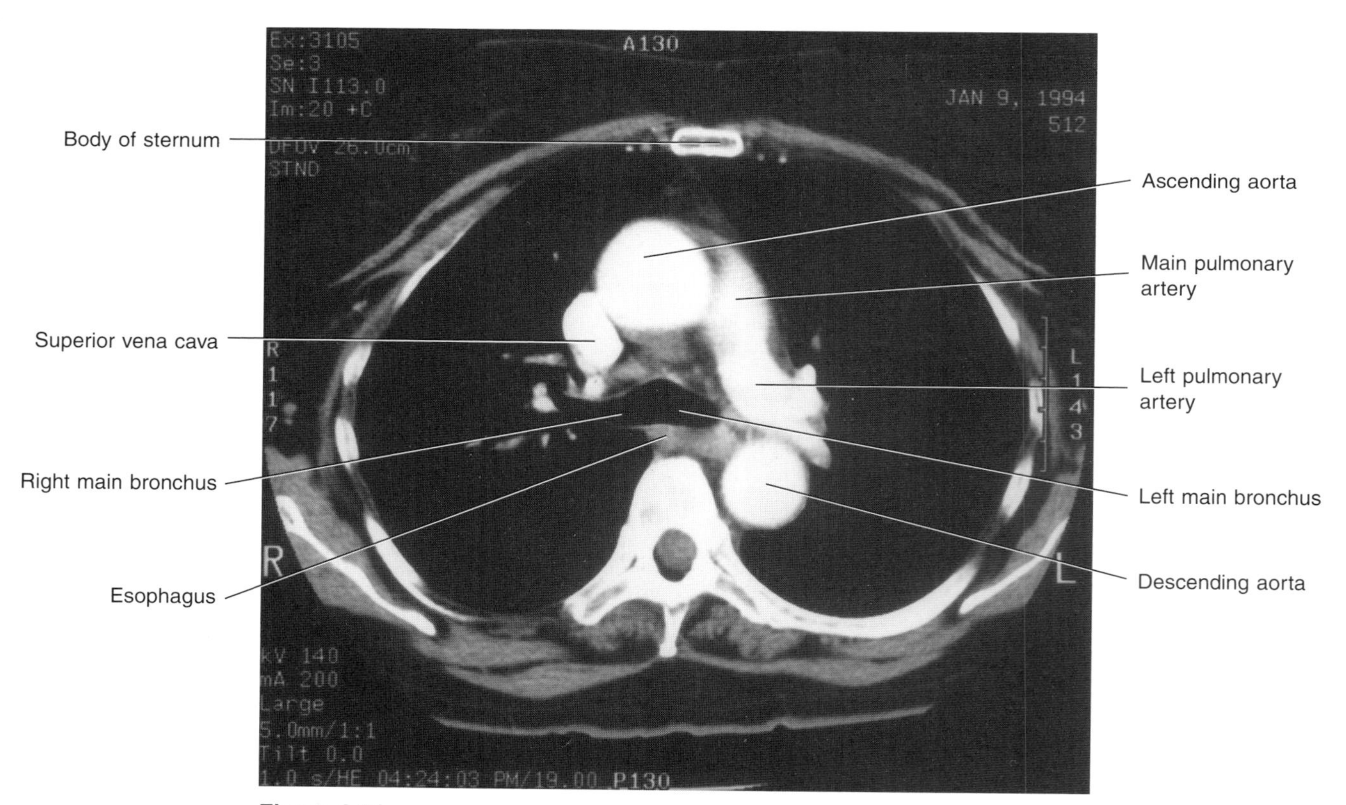

Figure 3.51

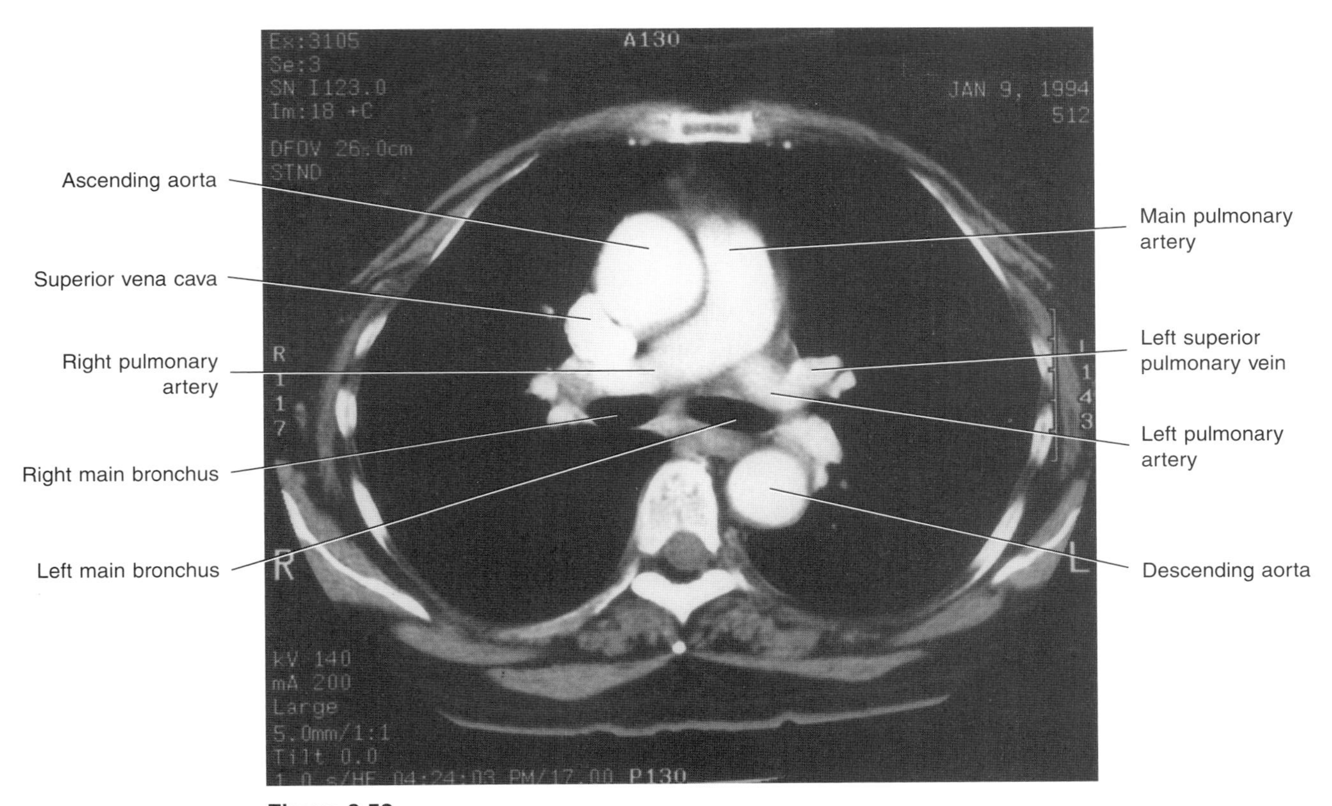

Figure 3.52

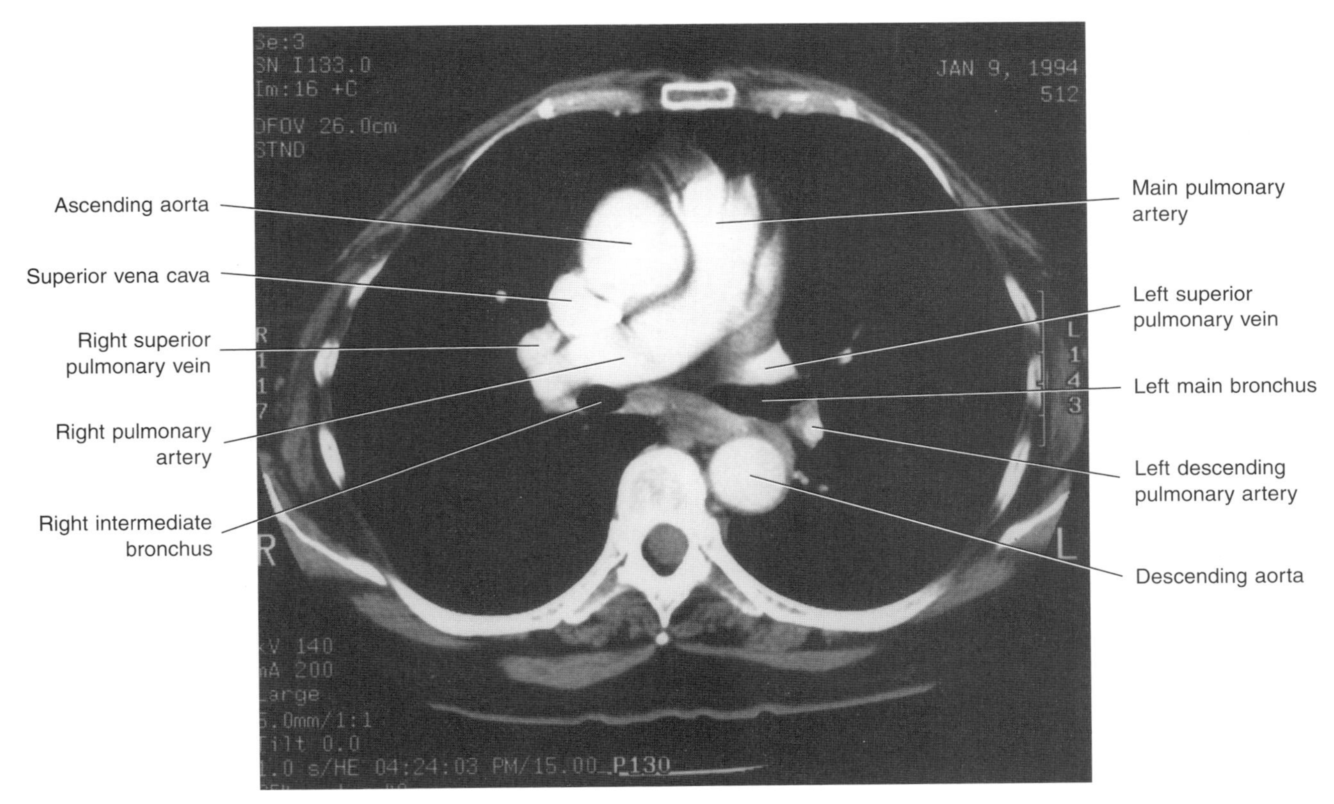

Figure 3.53

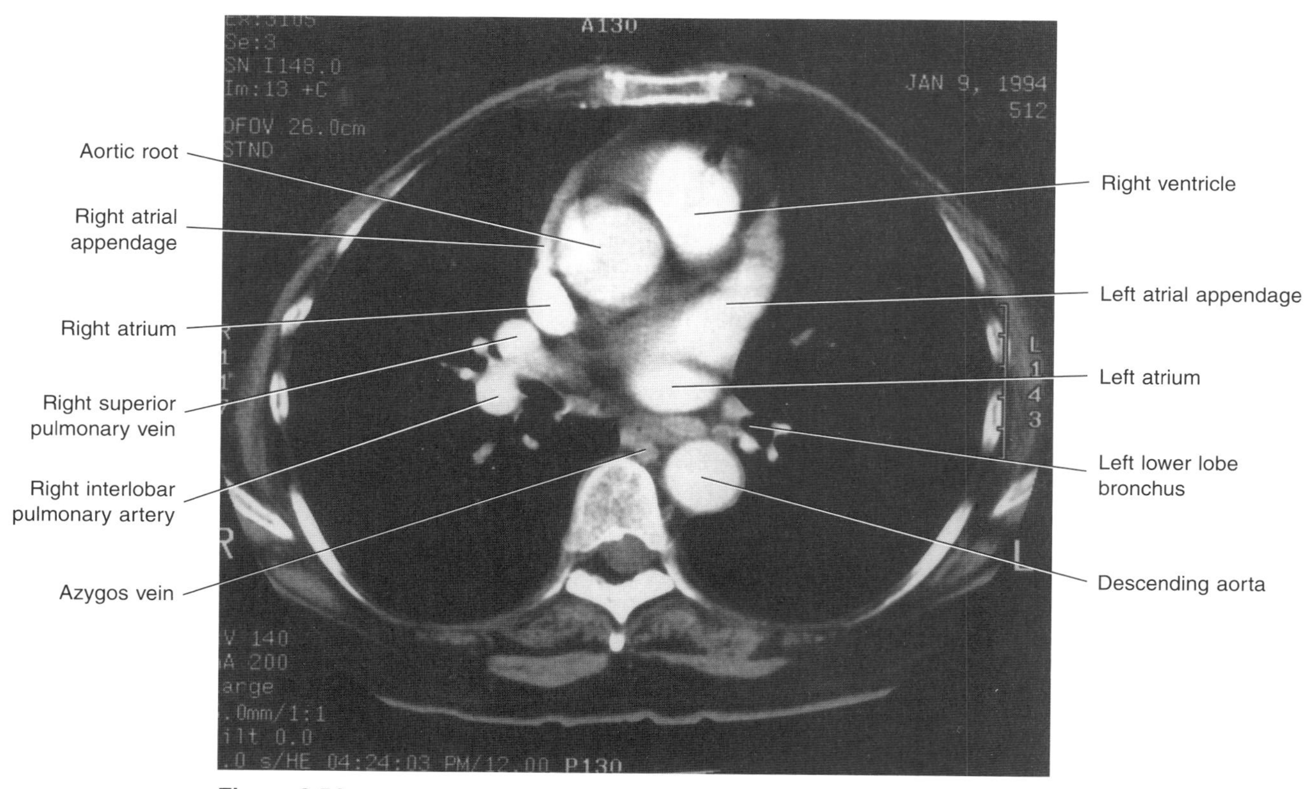

Figure 3.54

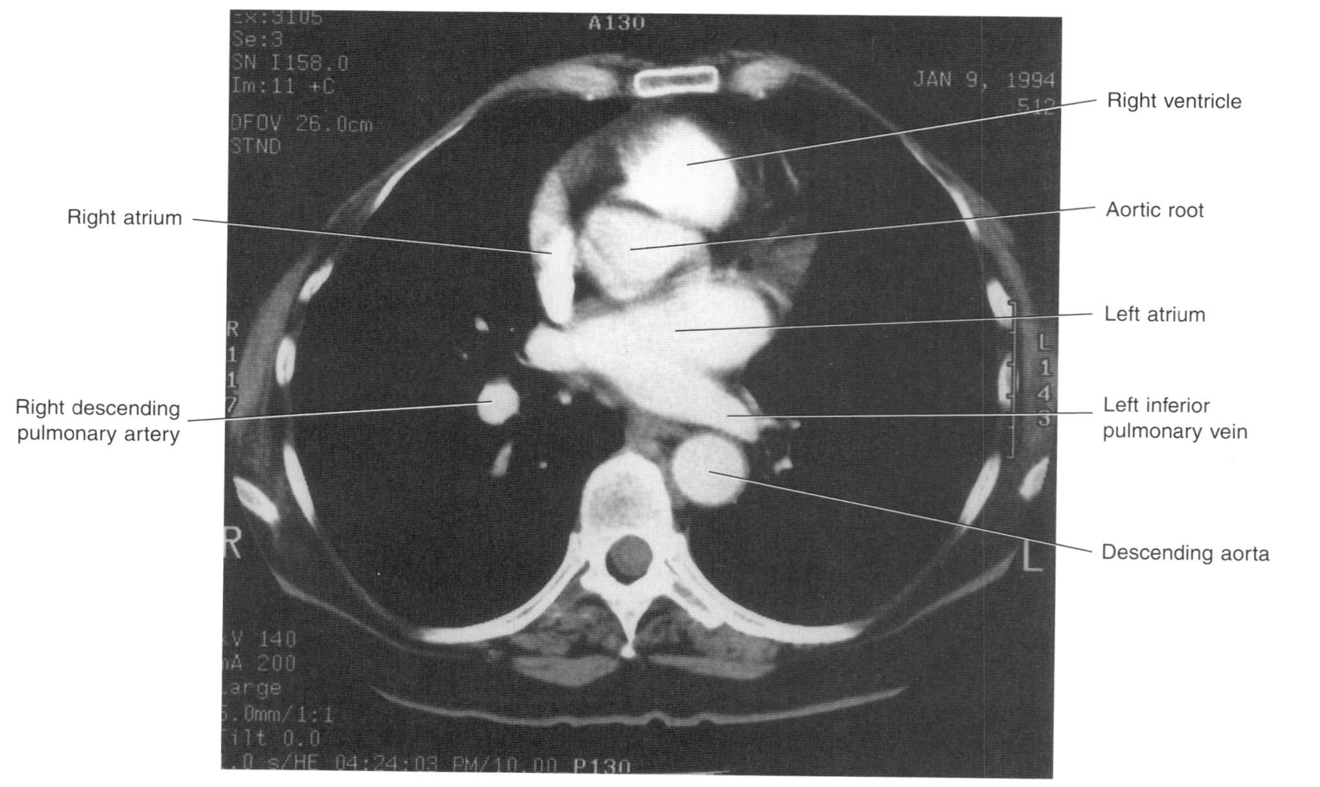

Figure 3.55

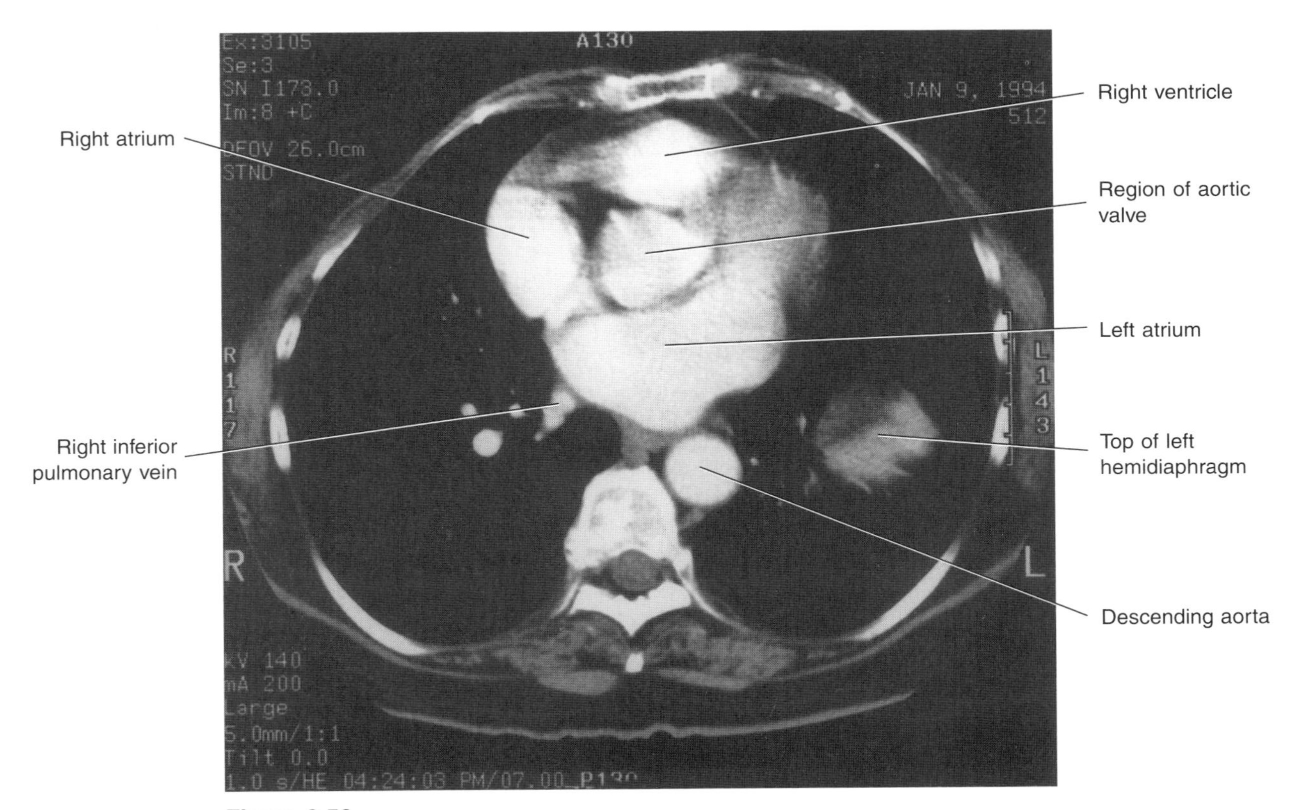

Figure 3.56

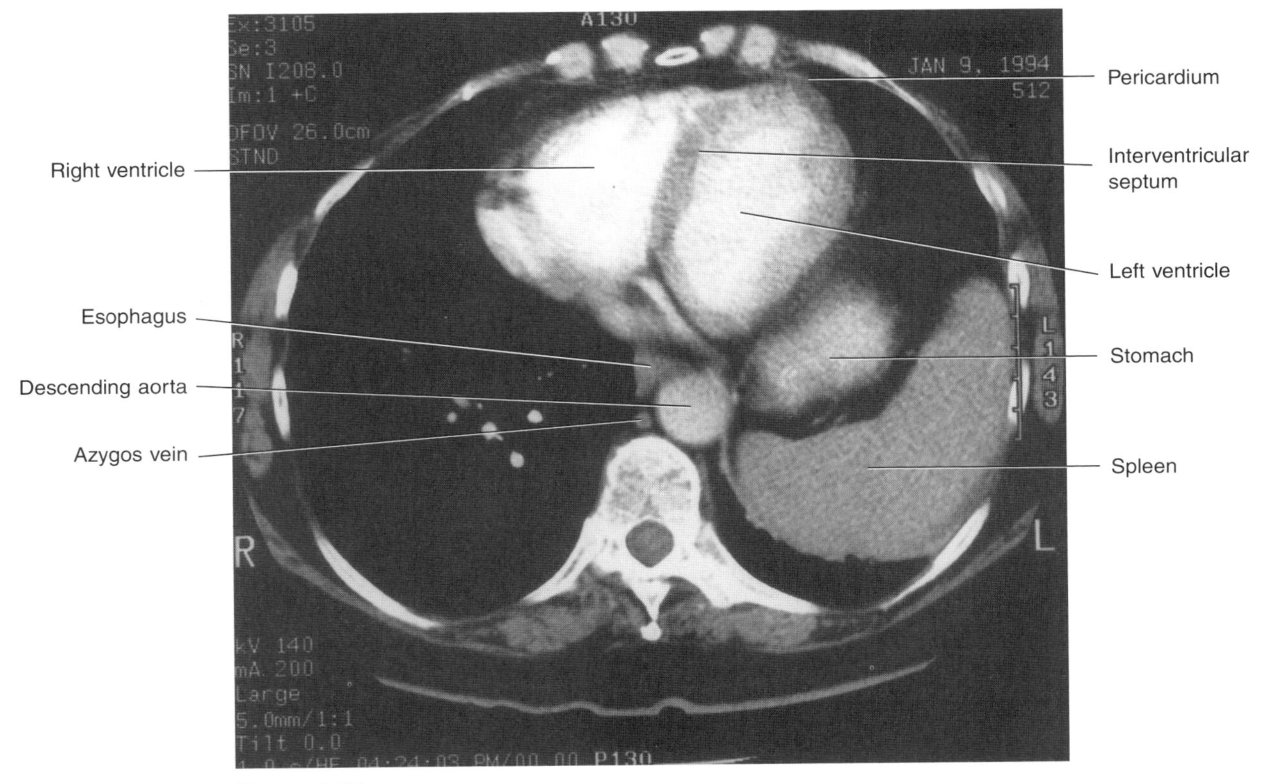

Figure 3.57

Figures 3.58 to 3.61. Series of chest CT scans filmed with lung windows.

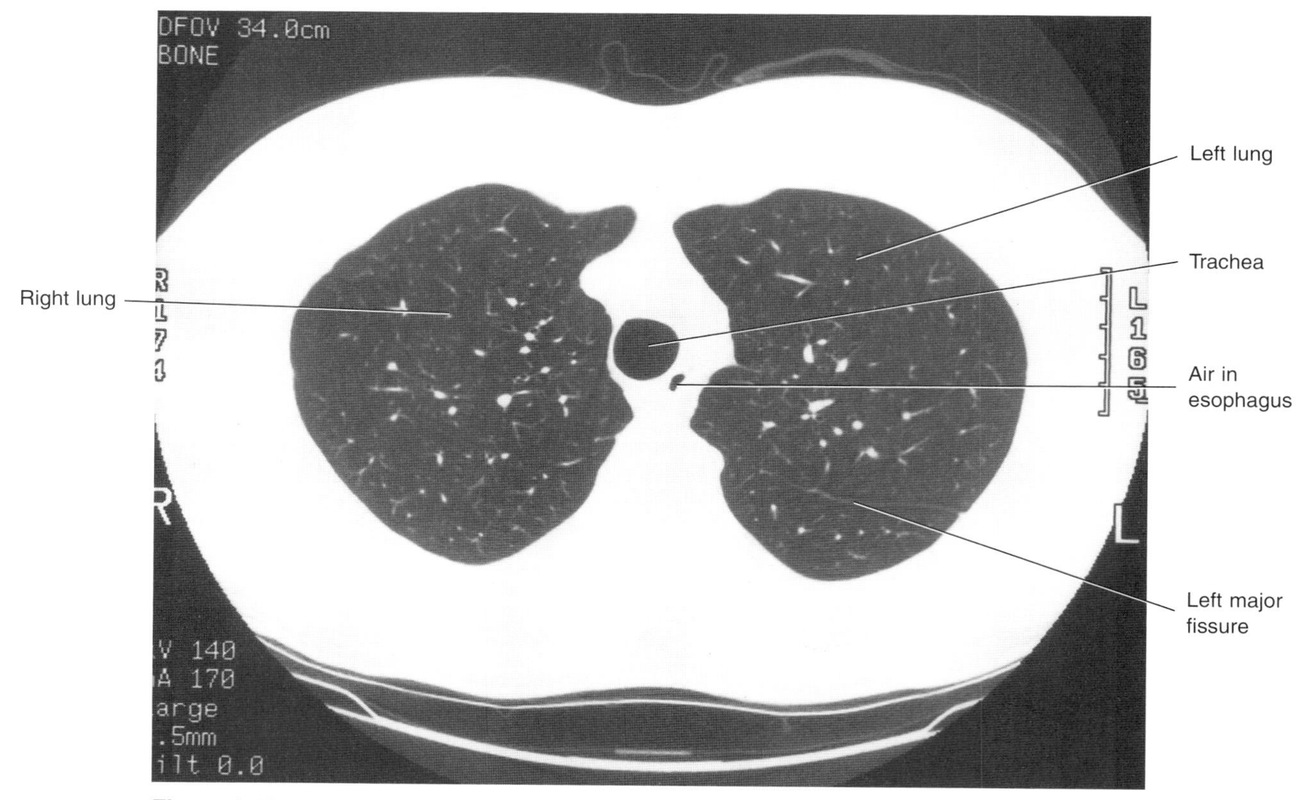

Figure 3.58

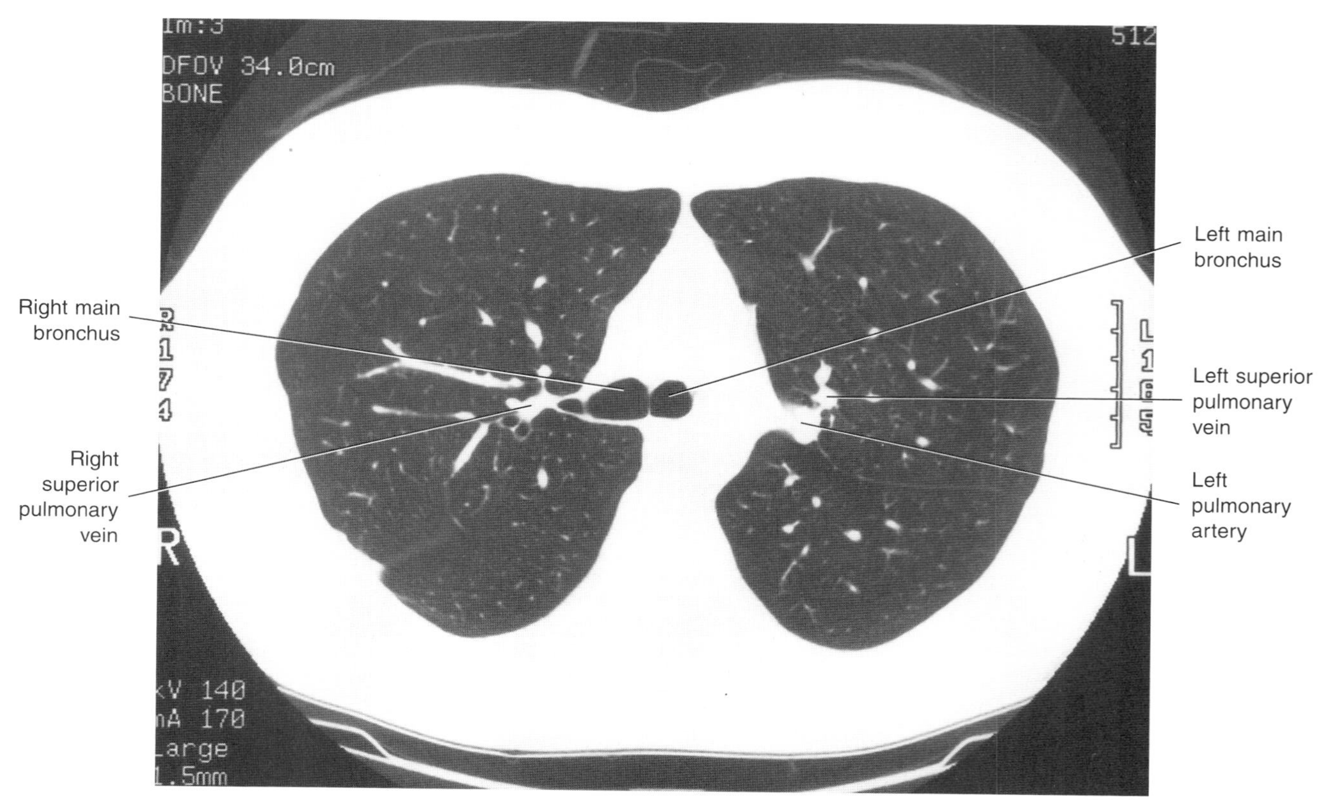

Figure 3.59

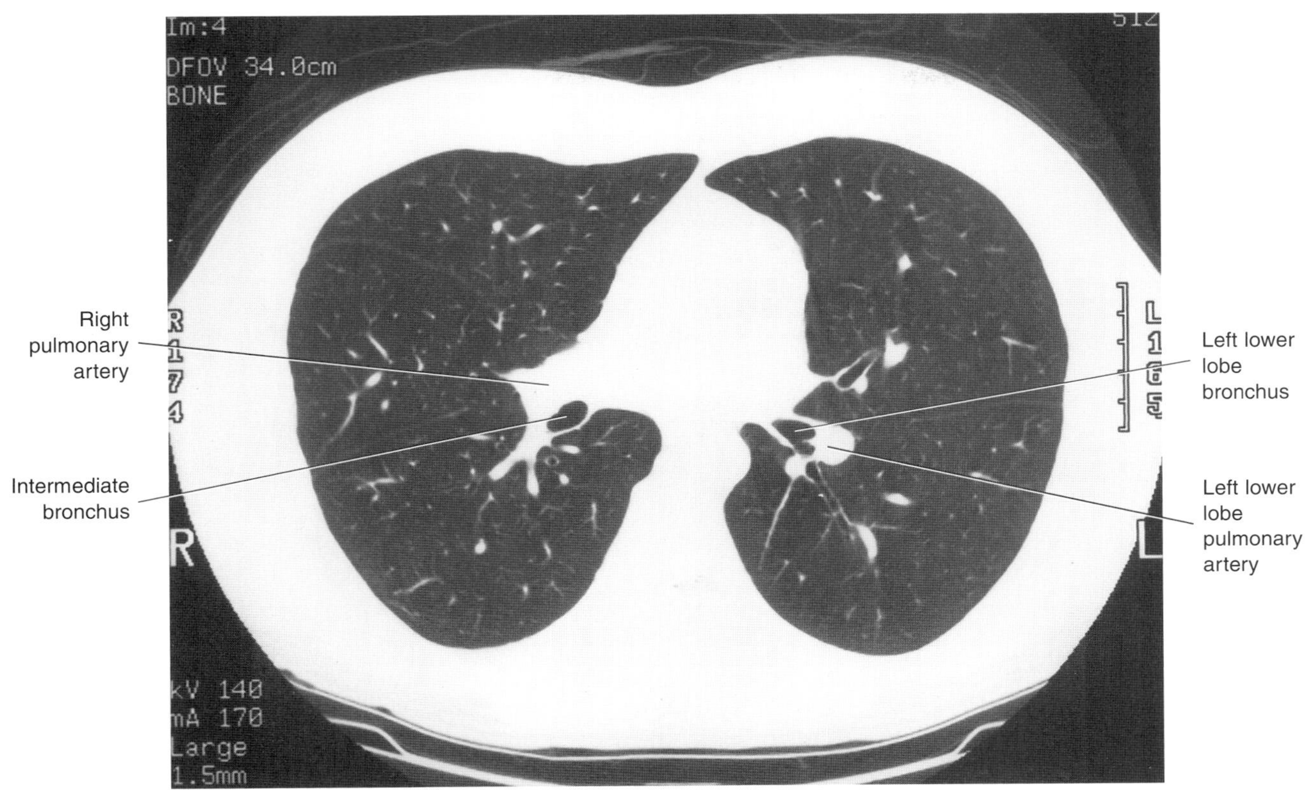

Figure 3.60

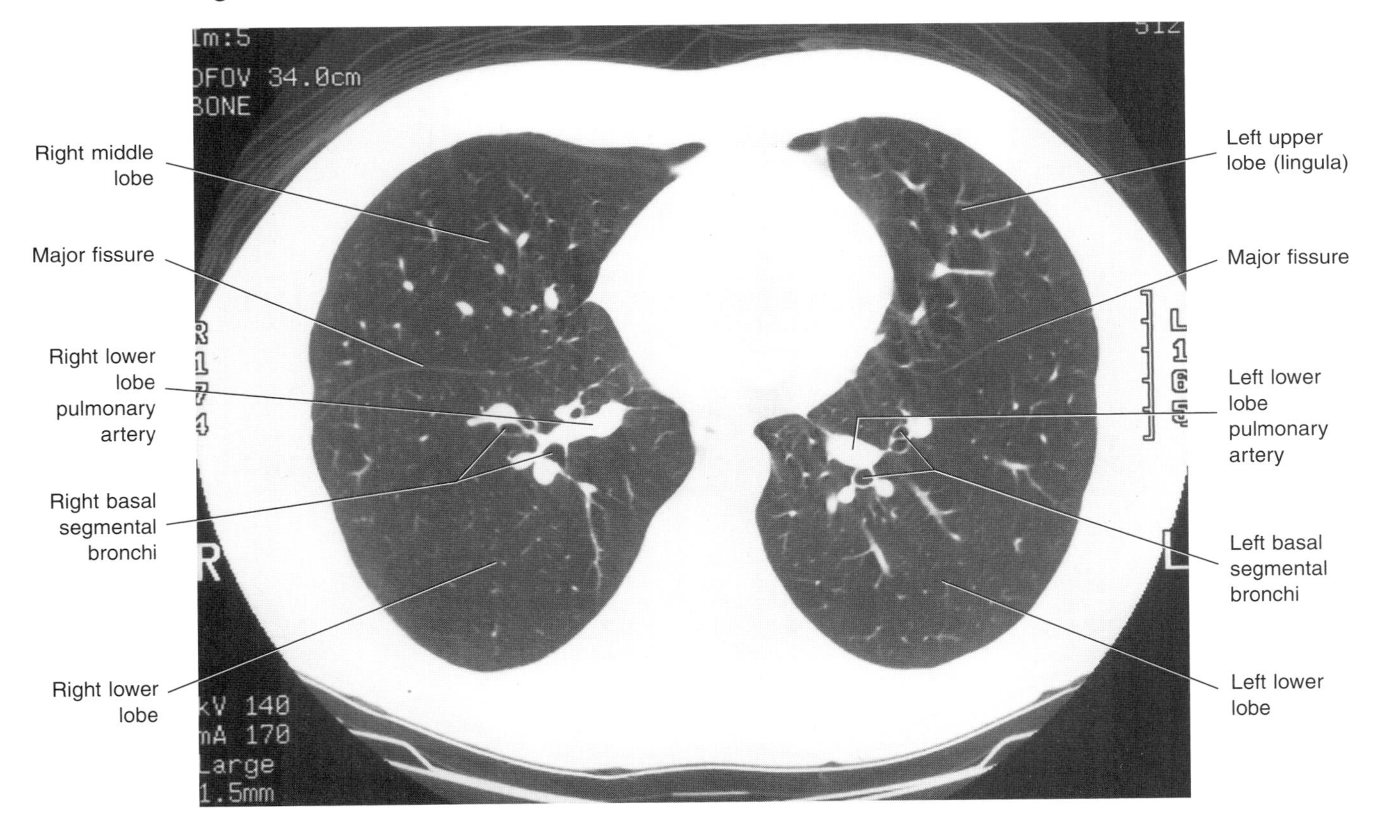

Figure 3.61

Figures 3.62 to 3.73. Series of abdominal CT scans with oral and intravenous contrast material, filmed with soft-tissue windows.

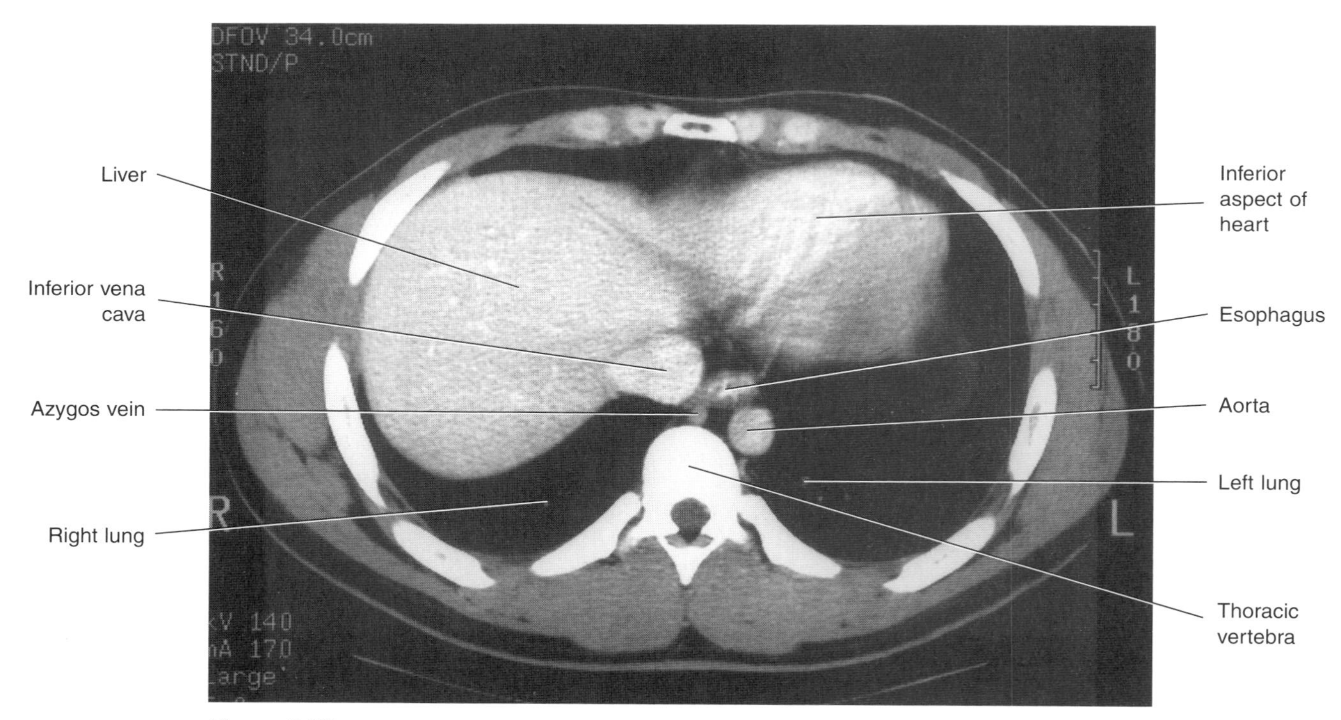

Figure 3.62

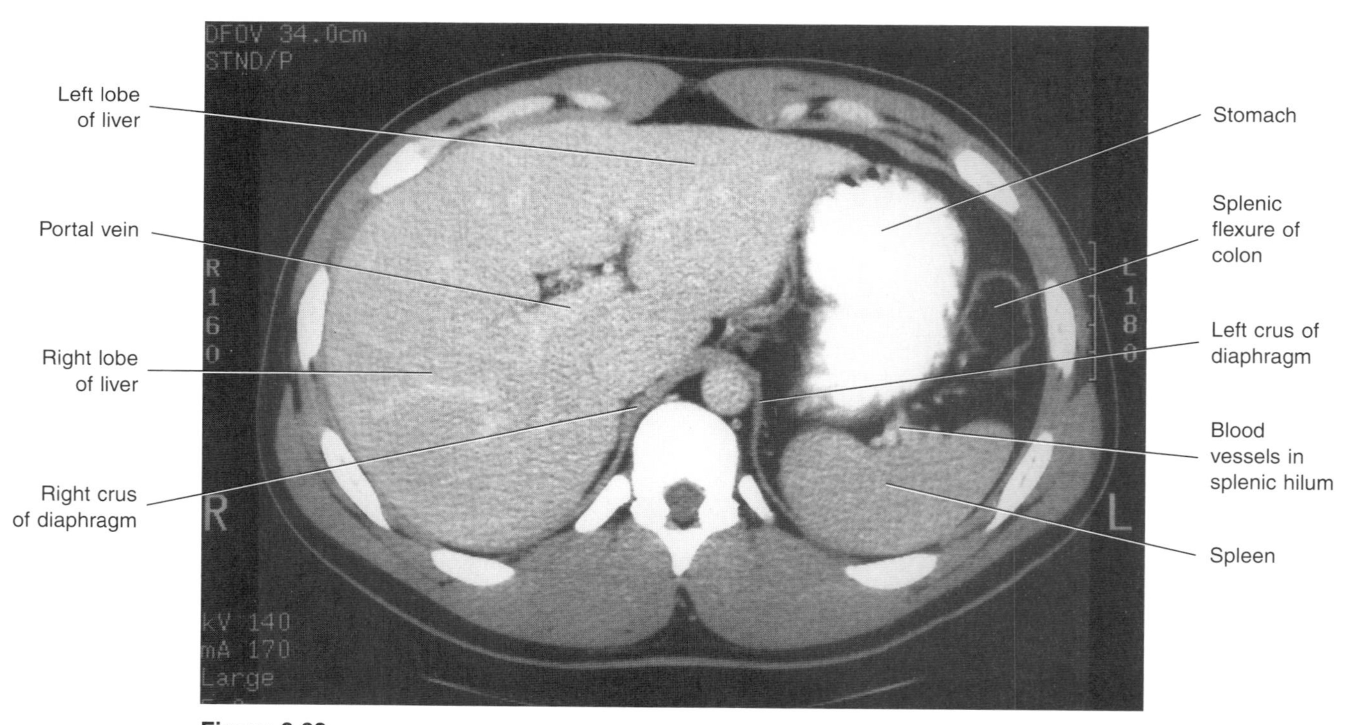

Figure 3.63

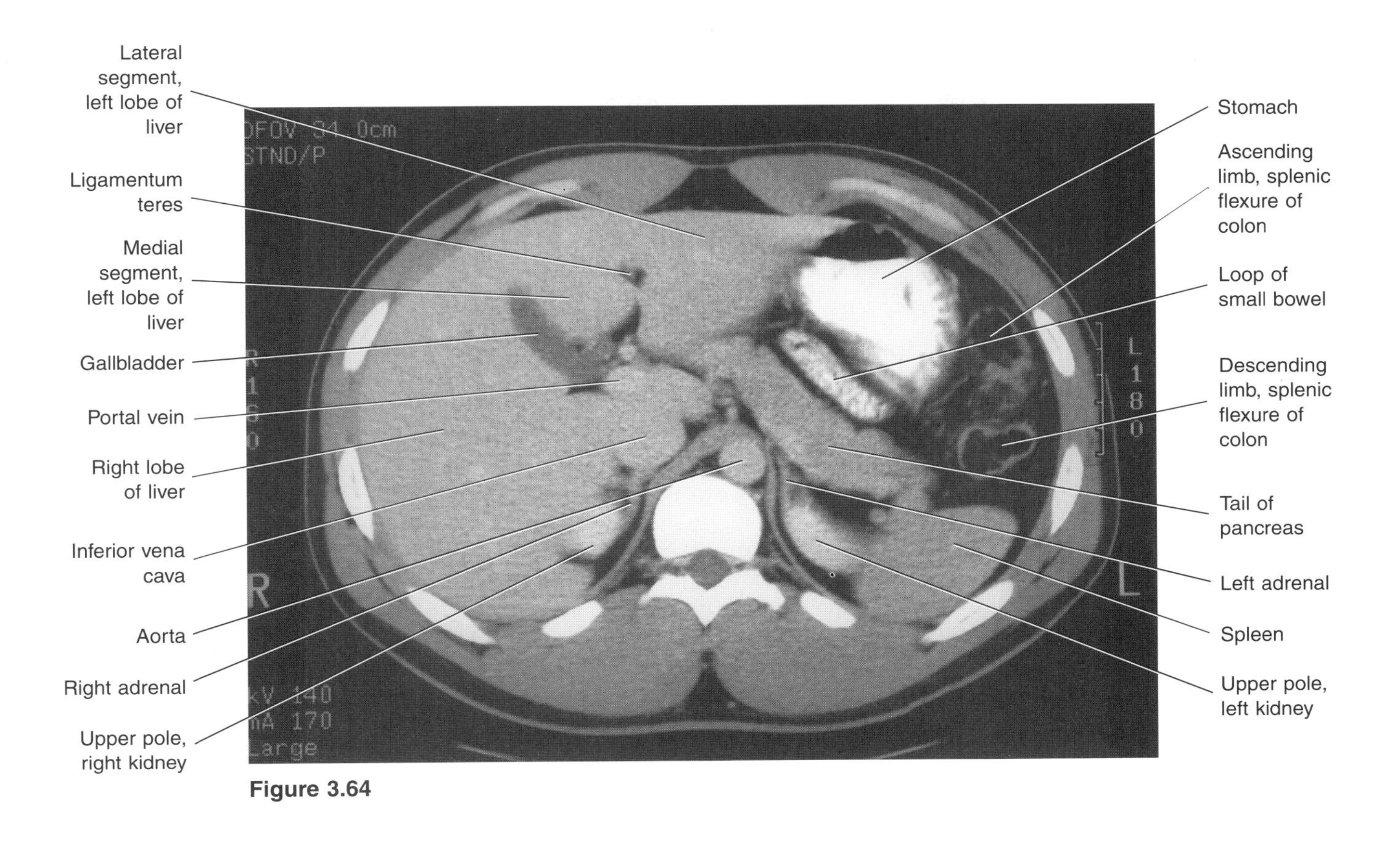

Figure 3.64

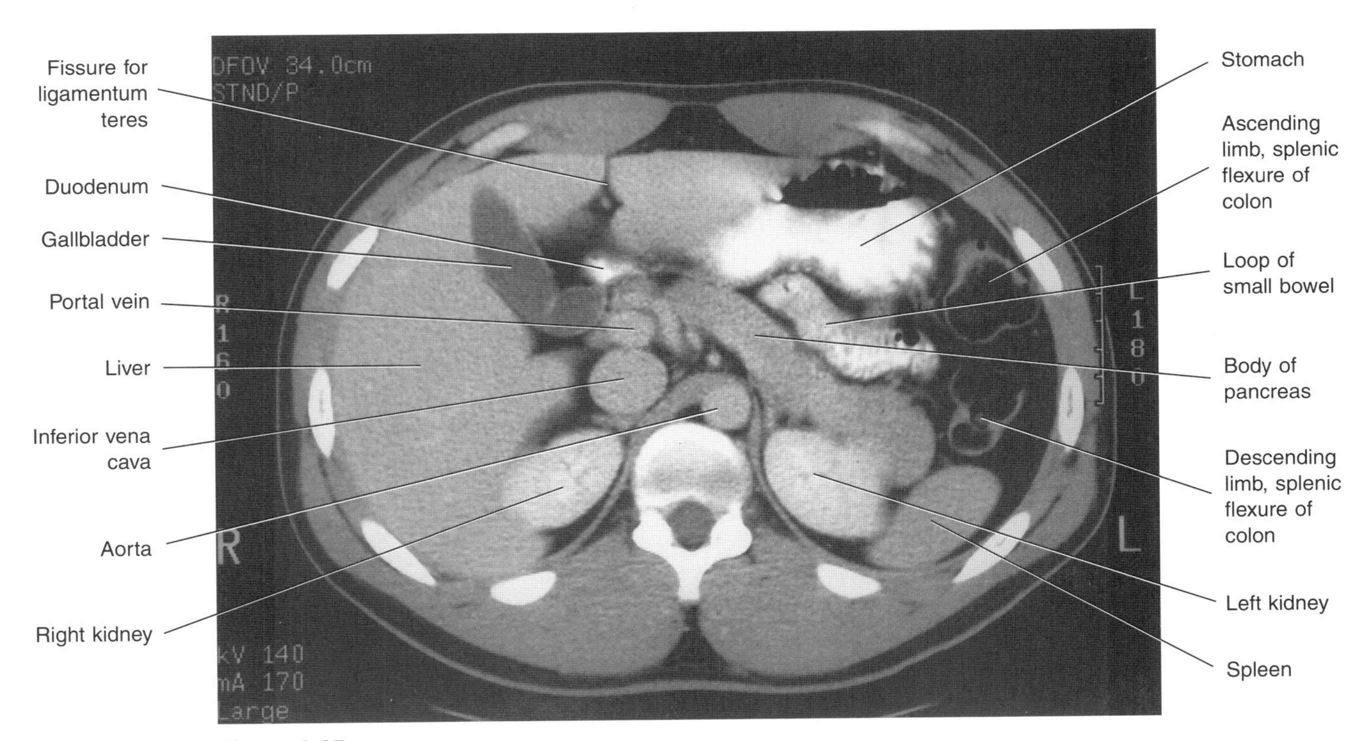

Figure 3.65

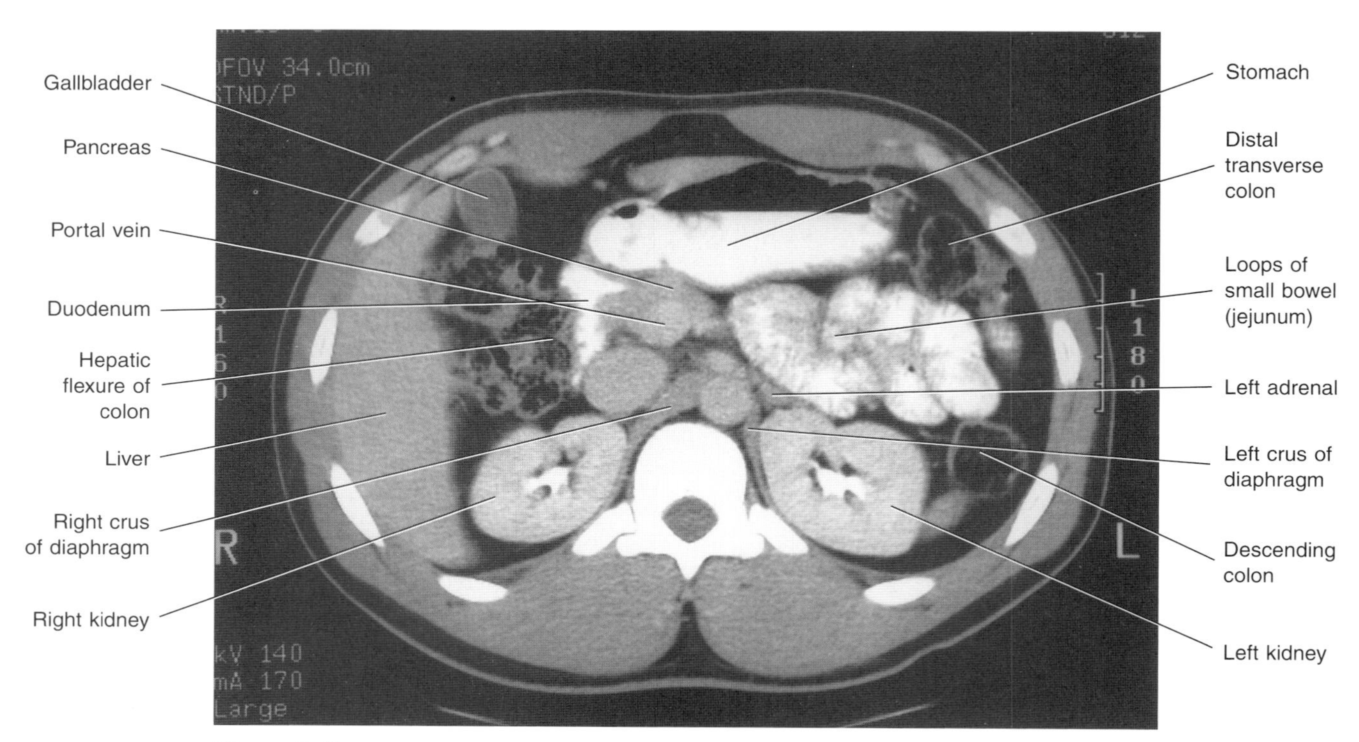

Figure 3.66

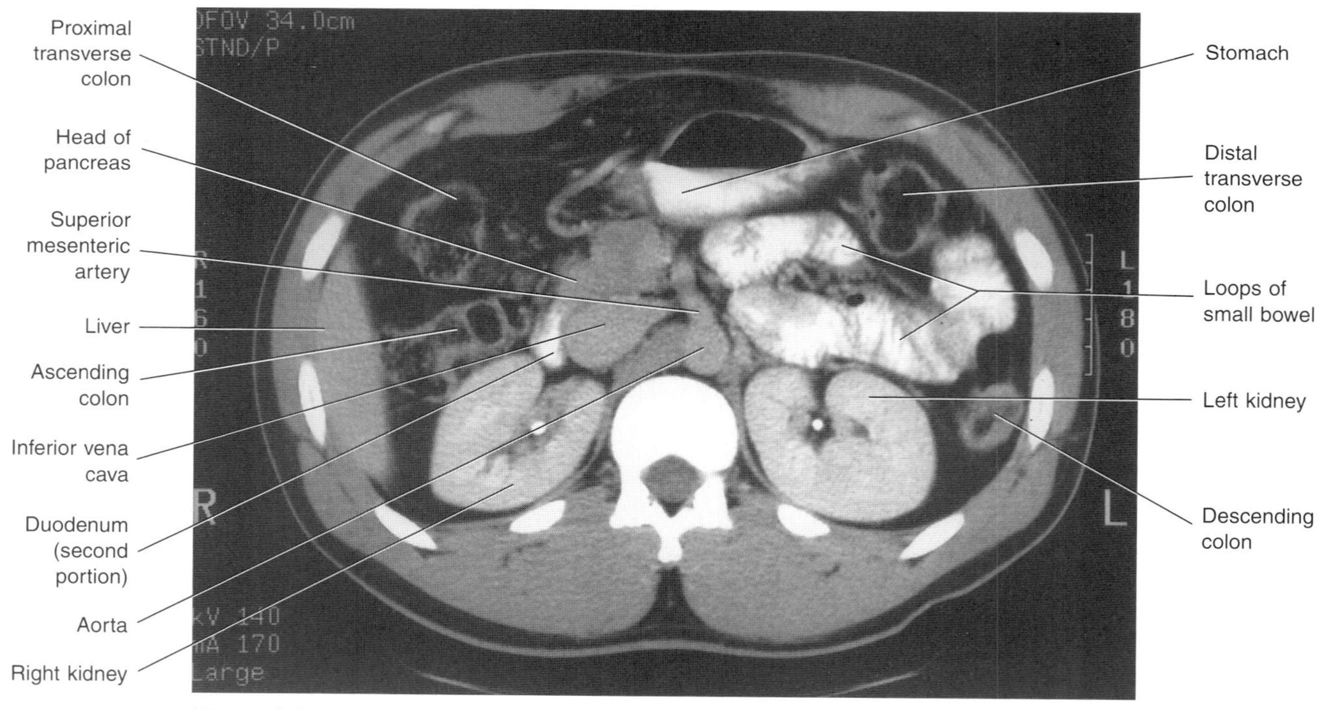

Figure 3.67

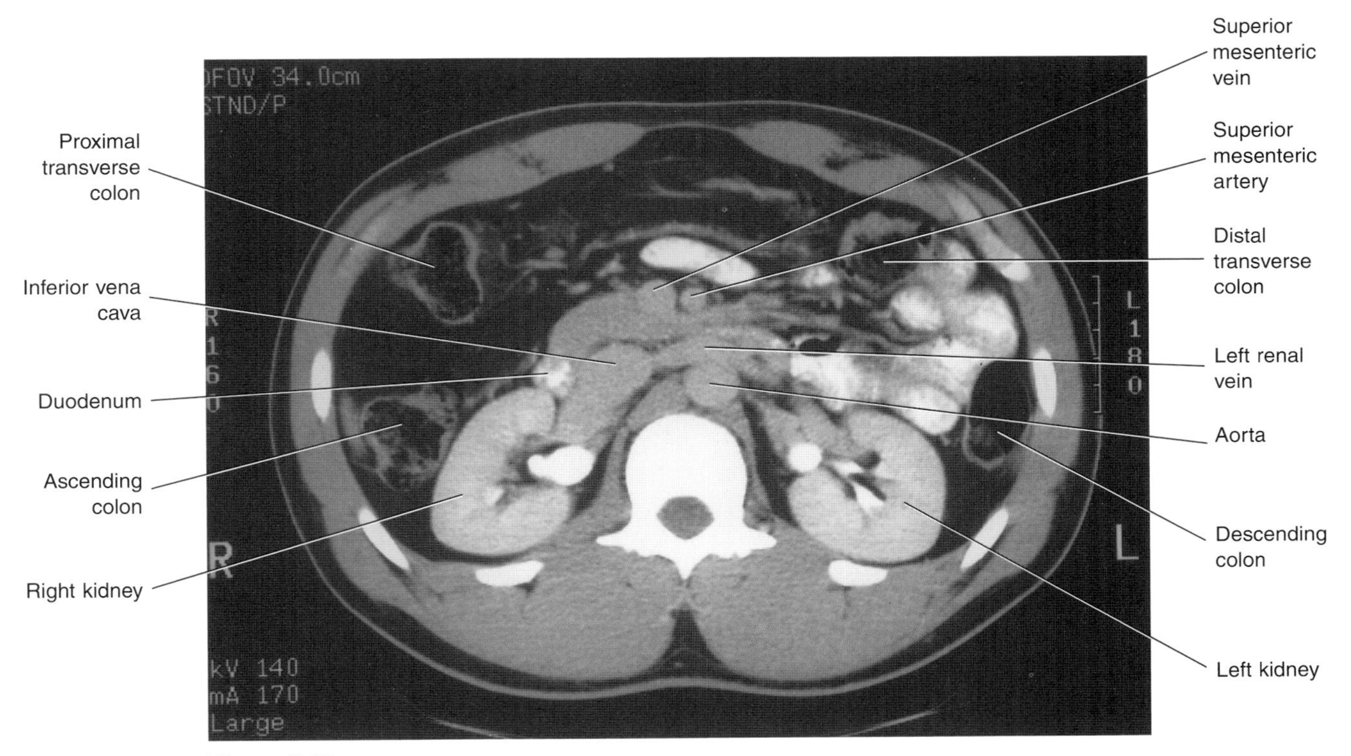

Figure 3.68

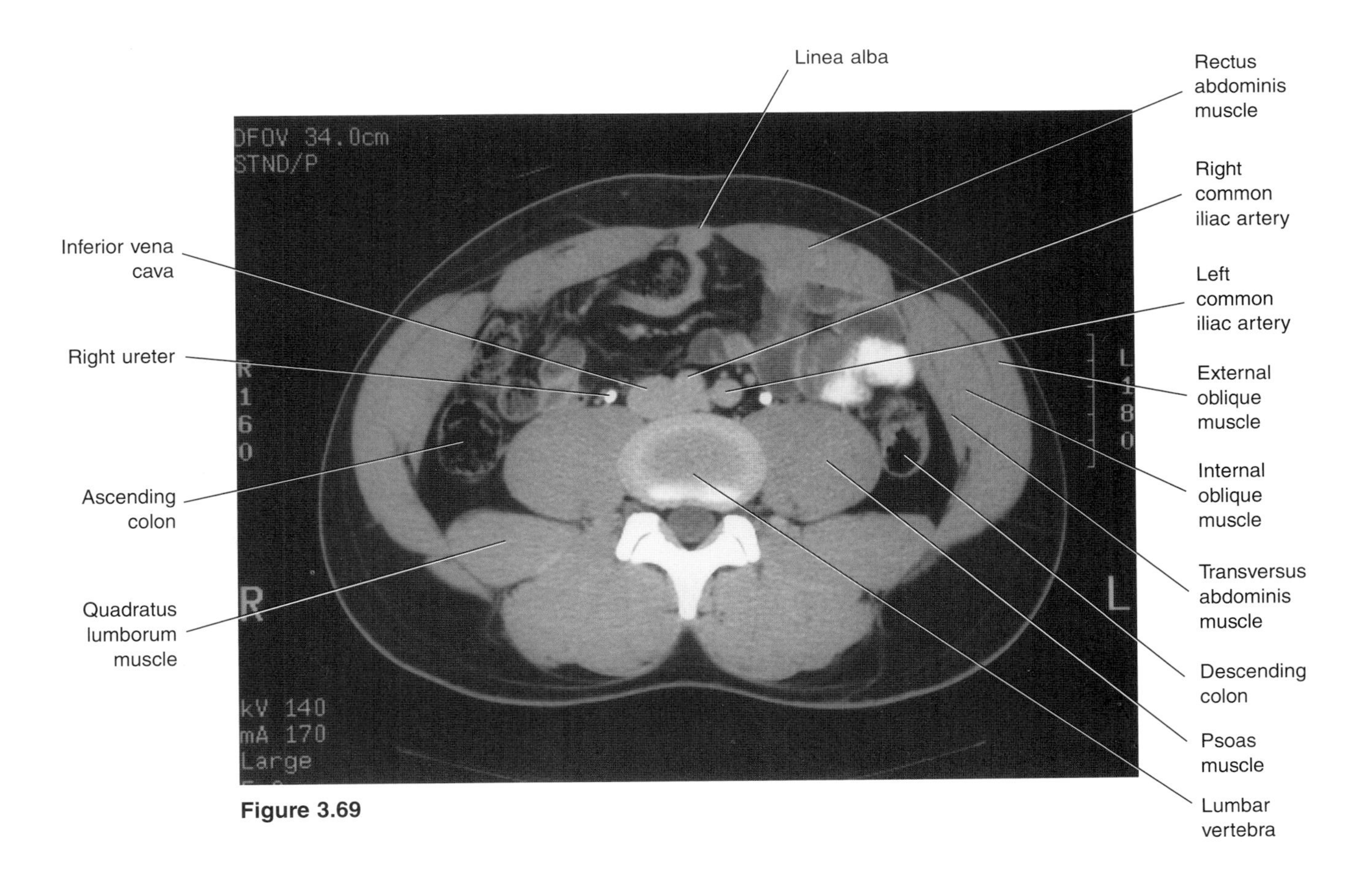

Figure 3.69

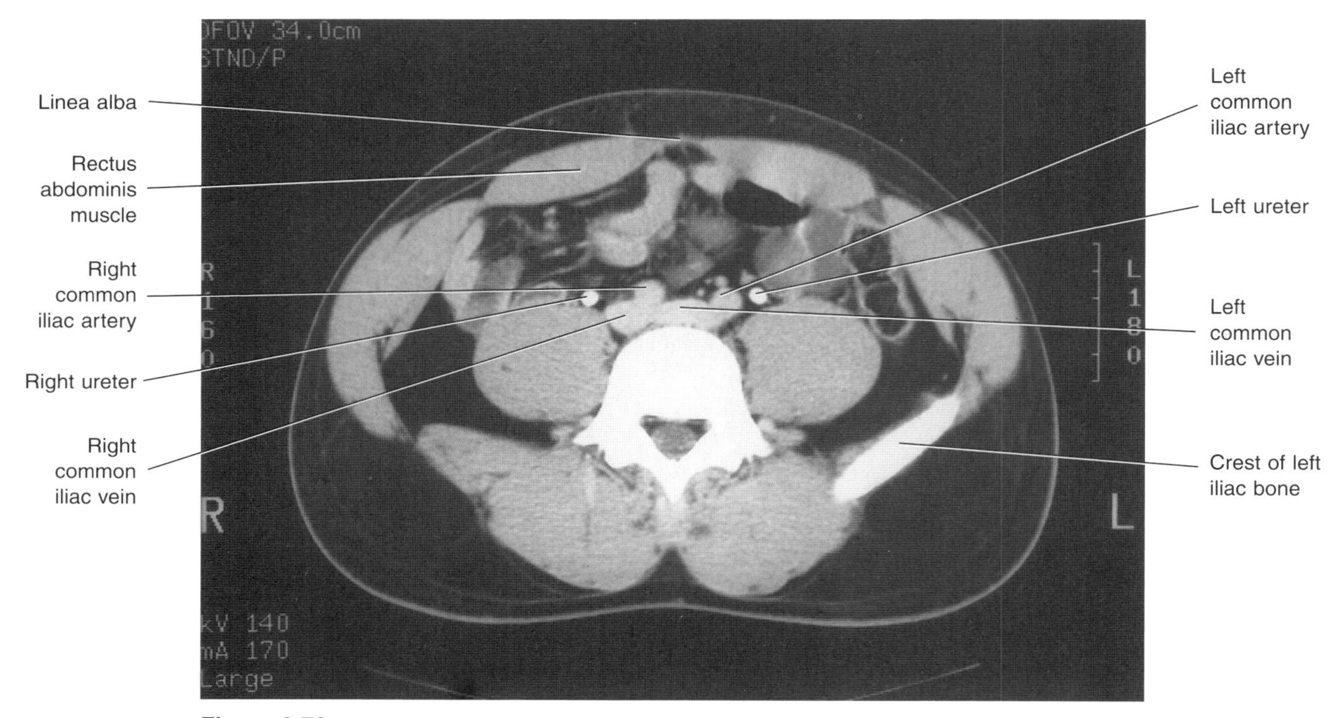

Figure 3.70

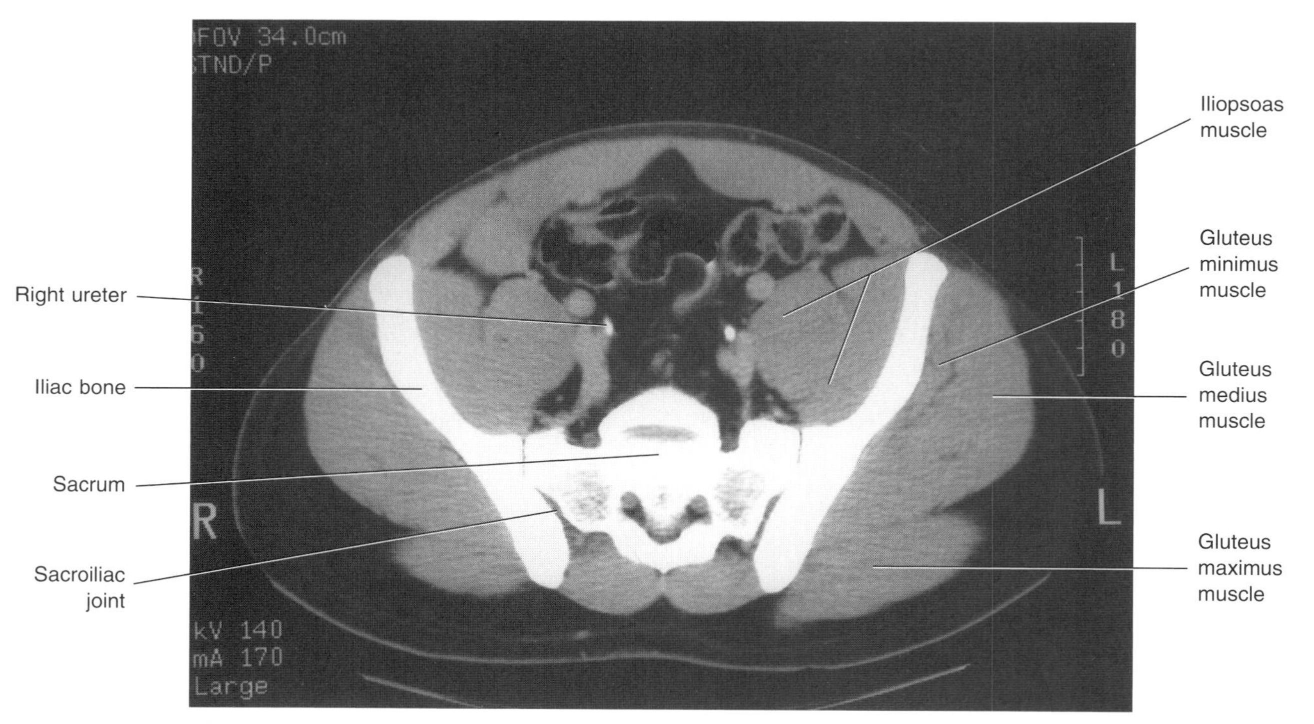

Figure 3.71

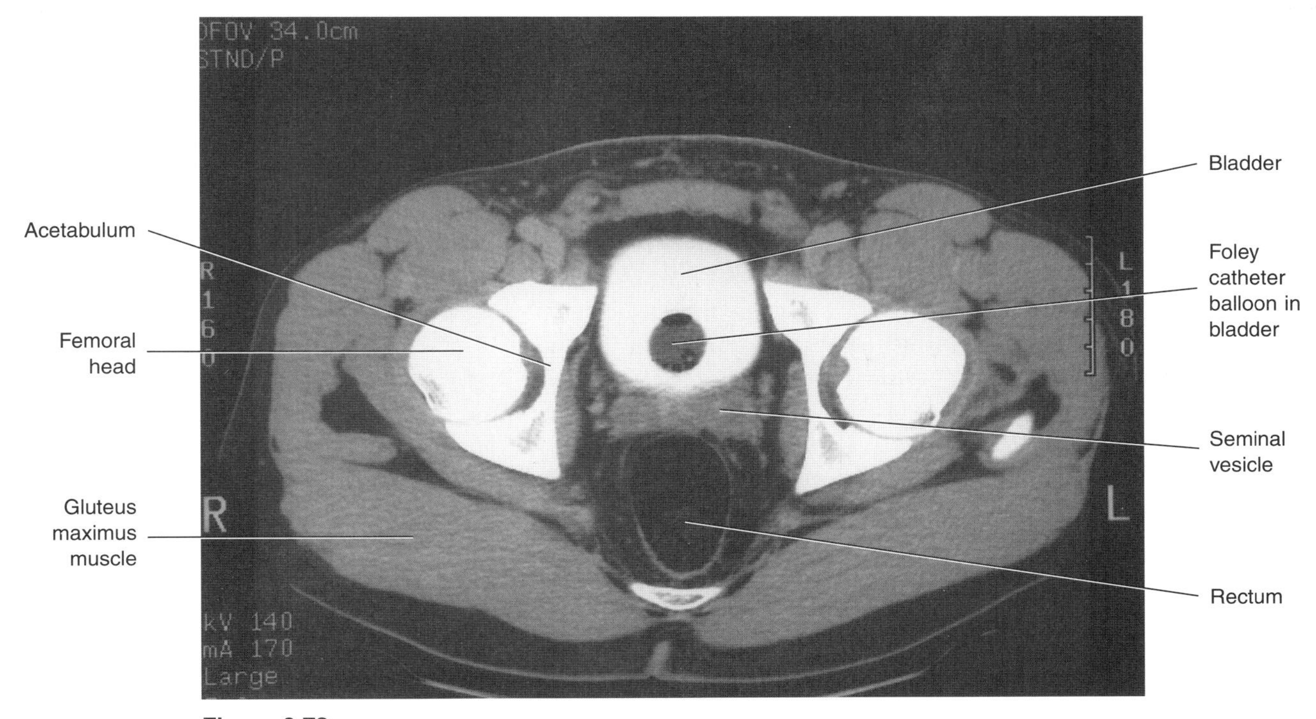

Figure 3.72

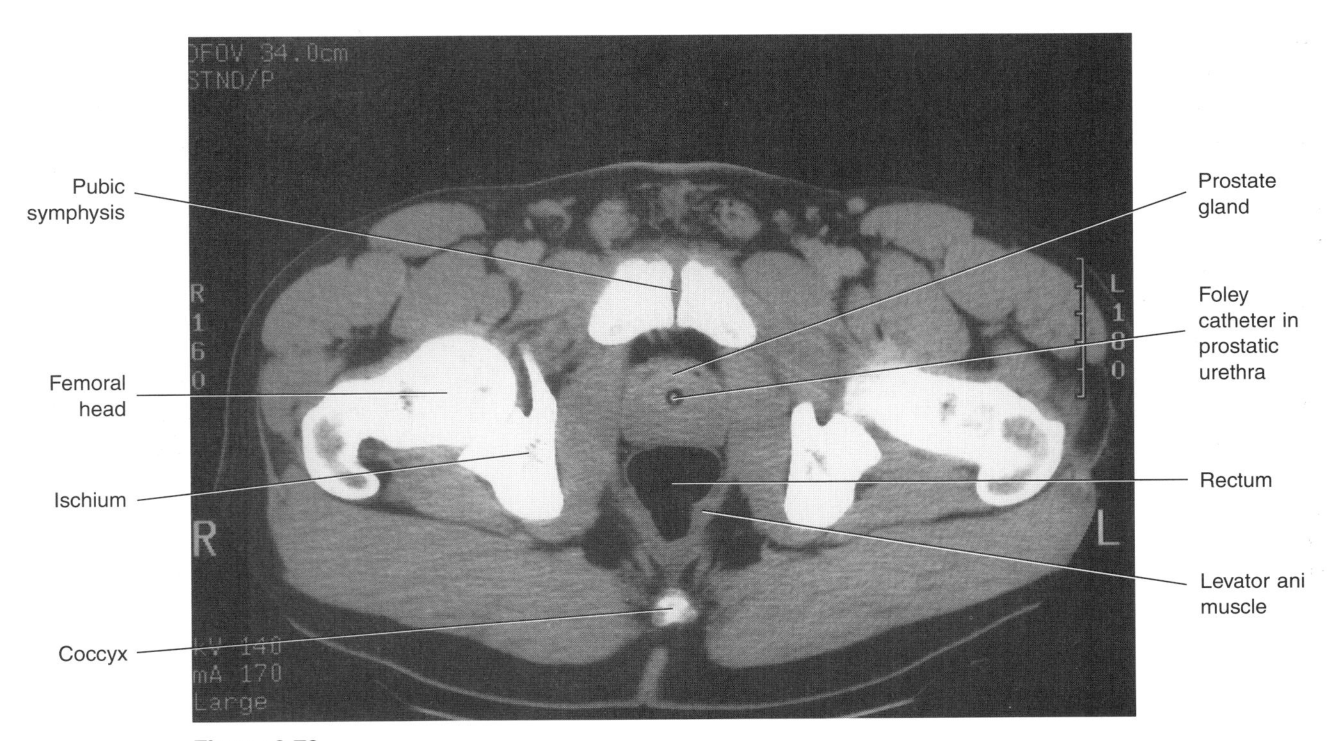

Figure 3.73

4 How to Study the Chest

Now that you have gained an overview of normal radiological anatomy, you are ready to begin your study of imaging fundamentals. In the process you will also learn how to use the radiologist as a valued consultant early in the patient's workup. The chest film, important in the diagnosis of a variety of diseases, offers you the opportunity to develop your skills in reading x-rays and, eventually, other diagnostic images. As you proceed through this chapter and the rest of the book, concentrate on looking at every shadow on an x-ray or other image and asking yourself, "*Why* does it look like that?"

One glance at a chest film often is enough to "see" a very striking abnormality. Having seen it, you must reason out its structural identity, seldom quite so obvious, and attempt to deduce its nature in accordance with a knowledge of the patient's illness. While the single-glance approach has its value, it may be dangerous for the patient, because the presence of a very obvious abnormality tends to psychologically suppress your search for more subtle changes. And the subtler changes are quite often more important to the patient than the obvious ones.

Let us say that you correctly interpret the shadow of a large mass in the lung on Mr. B's chest film (Figure 4.2A) as being consistent with the cancer you thought he might have when you questioned and examined him. You will have failed him if you neglect a deliberate quest for any possible secondary involvement of his bones, since quite a different program of treatment may then become appropriate. Figure 4.2B illustrates the point by showing in more detail, with a more penetrating x-ray beam, the extent of his bone destruction.

The system generally employed by the radiologist is to *look at* various structures in a deliberate order, concentrating on the anatomy of each while excluding the superimposed shadows of other structures. Even as an exercise in intellectual discipline, this is not as difficult as it sounds. Prove it to your own satisfaction by trying to *look at* one clavicle or one rib on any of the chest films in this chapter, thinking of its normal anatomic proportions and excluding other shadows overlying it that you know are not part of the bone you are studying.

The best way to be systematic about studying any film is to adopt a definite order in which you look at the structures whose shadows appear there. For a chest film you will *look at* the bony framework and then, just as deliberately, *look through it* at lung tissue and the heart.

Begin with the scapulae. Then look at the portions of humerus and shoulder joint often visible on the chest film. Inspect the clavicles, and then finally study the ribs, quickly but in pairs from top to bottom. When you can, always compare the two sides for symmetry. The spine and sternum are, of course, superimposed upon each other and upon the dense shadows of the mediastinal structures in the PA view, so that, at the kilovoltages used for lung study, little of the beam penetrates and the film remains less well exposed down the midline.

Remember that the technique used for chest films has been designed for study of the lung; what you see of the bones is incidental. Ideal techniques for studying these same bones will be quite different. In a PA chest film, for example, the scapulae and posterior ribs are as far as possible from the film. Therefore, they are enlarged and distorted to some extent. In addition, on the chest film the scapulae have been intentionally rotated to the sides as much as they can be by placing the hands on the hips, palms out, with the elbows forward. Try it. In the PA view of the chest this maneuver prevents the superimposition of the scapular shadows upon the upper lung fields, and only the medial margin of the scapula will be seen overlapping the axillary portions of the upper ribs. Decide whether the scapulae were properly rotated in Figure 4.1.

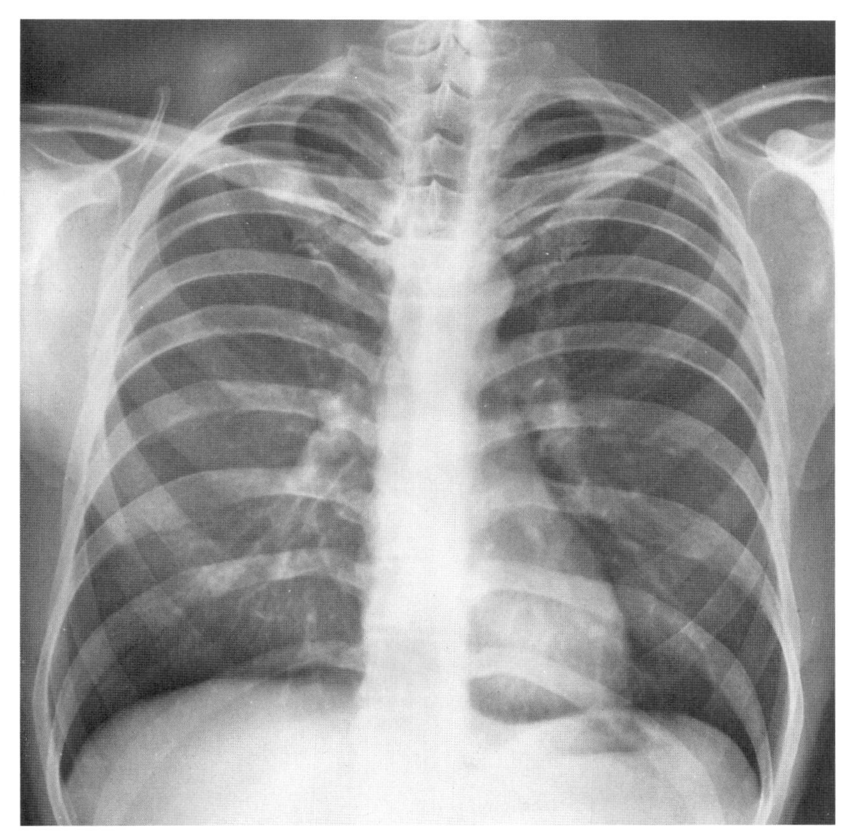

Figure 4.1. Normal PA chest film.

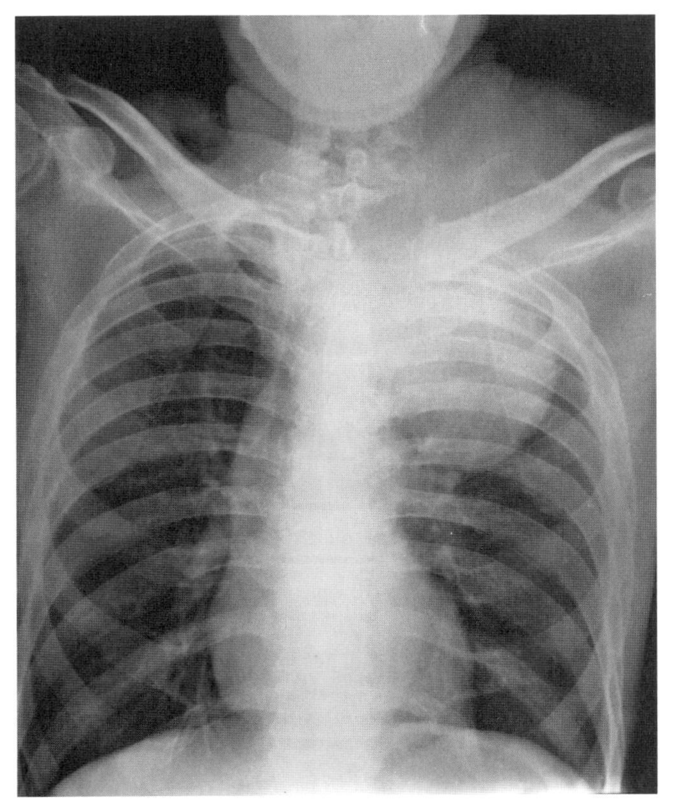

A

Figure 4.2. A *(left):* PA chest film of Mr. B, admitted with a cough, chest pain, hoarseness, and a fist-sized mass in the left supraclavicular region. B *(below):* Detail study of the thoracic inlet made AP with a more penetrating beam. The left posterior first rib and parts of the first two thoracic vertebrae have been destroyed by the tumor.

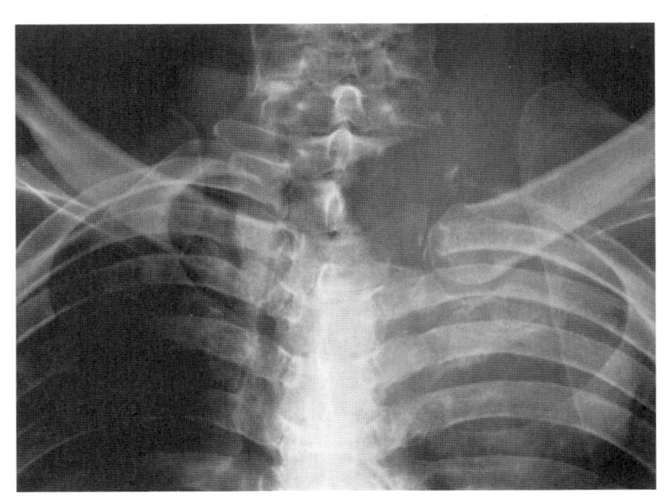

B

Projection

Do not discount the factor of *projection* in altering the appearance of structures far away from the film. You need not be confused by such changes, however, once you are familiar with them. In Figure 4.3 you have a diagram illustrating the effect of projection, in which you can equate the "object" with the scapula in any AP and PA chest film. You can equate it also with the anteriorly placed heart in the AP film in Figure 4.4. Note that the heart in this film appears to be larger, with less sharp margins, than the hearts in the PA chest films you have seen up to now. Note also the slight difference in the width and shape of the posterior rib interspaces compared with those on the usual PA film.

The routine chest film measures 14 × 17 inches, and its cassette film holder is placed with the long dimension vertical. In broad-chested people little of the *shoulder girdle* and *humerus* will be seen, but in slender, smaller individuals you may actually have all of the shoulder and most of the upper arm to study. Figure 4.5 is a radiograph of the shoulder made AP. Figure 4.6, next to it so that you can look back and forth, is a photograph of the bones of the shoulder. Notice that you seem to see the coracoid *through* the spine of the scapula because they superimpose, just as you see the head of the humerus and the acromion additively.

The man in Figure 4.7 had fallen from a horse. Since there are several fracture lines and fracture fragments, this is a *comminuted* fracture.

The woman in Figure 4.8 could not comb her hair

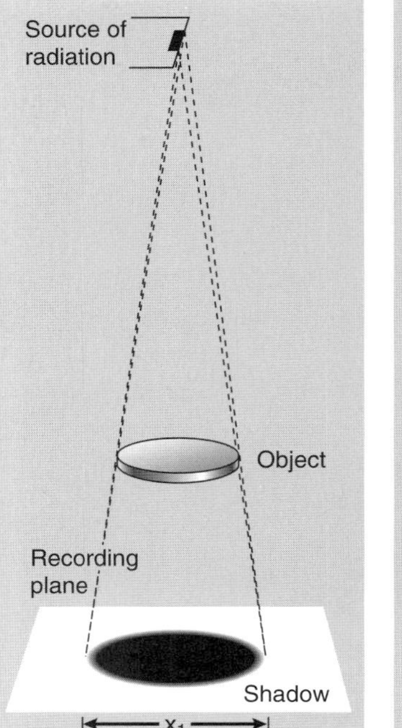

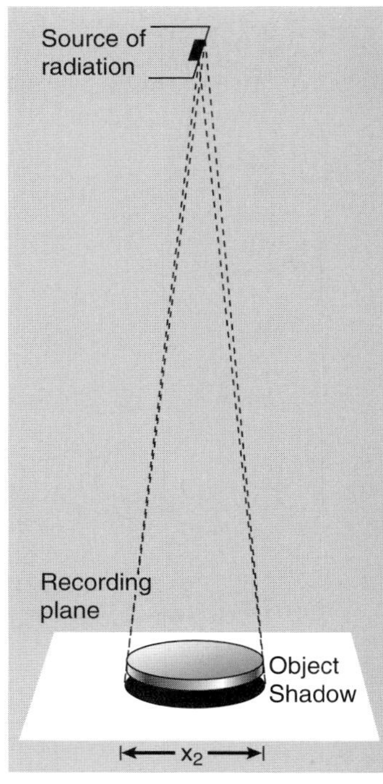

Figure 4.3. The effect of projection in enlarging the roentgen shadows of objects far away from the film.

without intense pain in her shoulder. She had tenderness over the insertion of the supraspinatus tendon, and, as you see, she has *calcification* in that area (*b*) and around the shoulder joint—dense white shadows not present on any of the x-rays of normal shoulders you have seen so far. These findings (*b*) are typical of "bursitis," or calcific tendinitis of the shoulder.

Note the dense white triangular shadow medial to the midhumerus (*a*) in Figure 4.8. It is common in

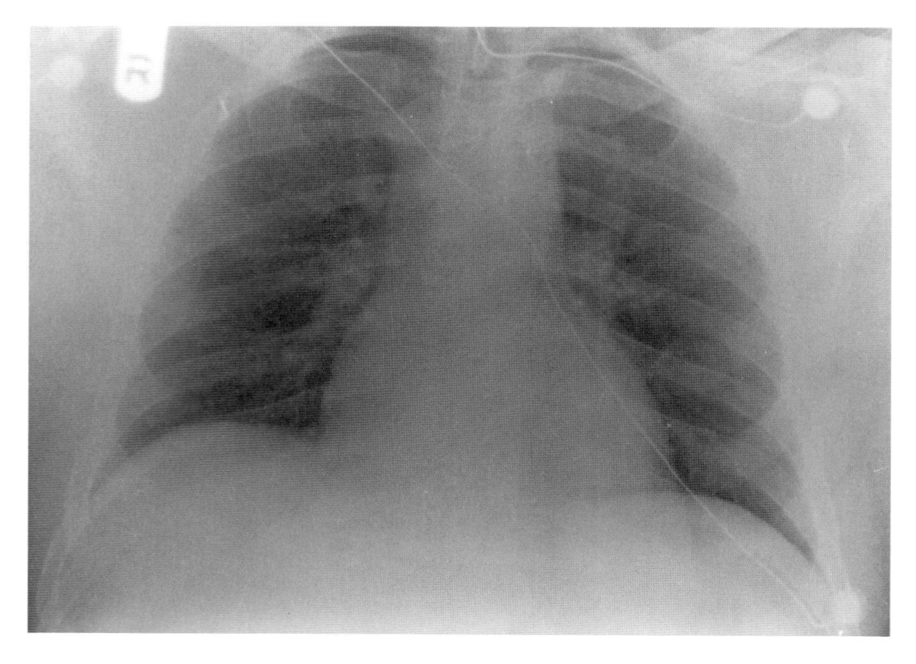

Figure 4.4. AP view of the chest with the posterior ribs close to the film. The heart, far from the film, is projected and looks larger than normal. Compare this view with Figures 2.4A and 2.4B. The metallic wires and round devices shown overlying the chest represent cardiac monitoring electrodes. This patient had suffered a myocardial infarction and was being monitored in a cardiac intensive care unit. Consequently, his chest film had to be made with a portable machine using an AP technique.

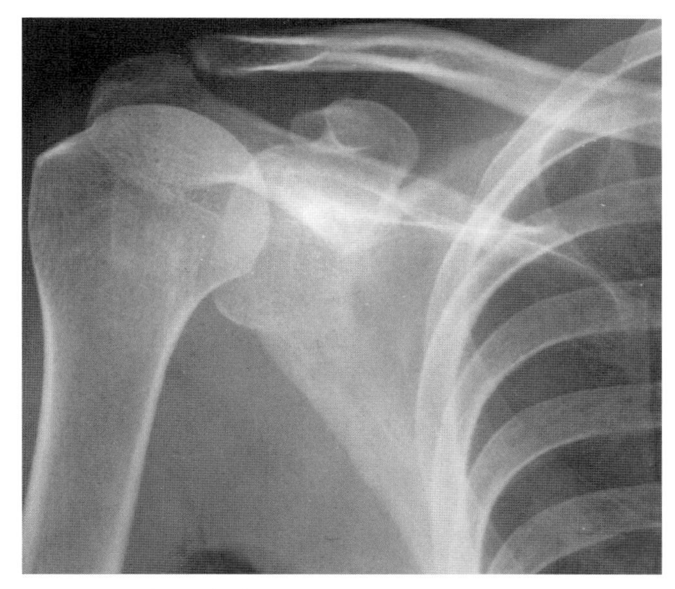

Figure 4.5. Identify the clavicle, acromioclavicular joint, head and greater tuberosity of the humerus, glenoid fossa, acromion, and coracoid process.

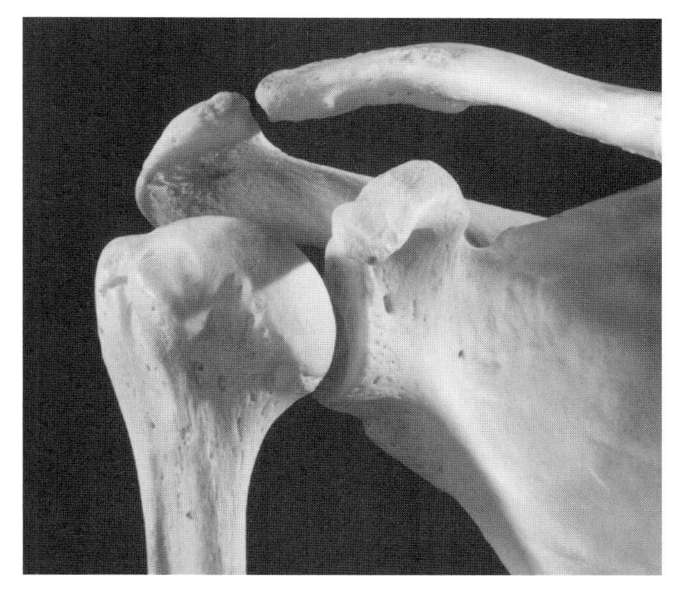

Figure 4.6. Photograph of the bones of the shoulder.

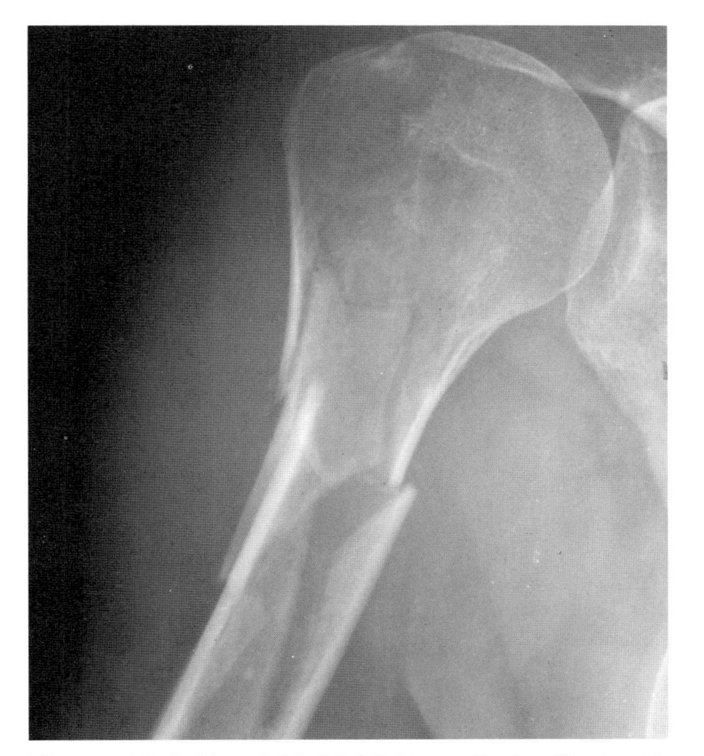

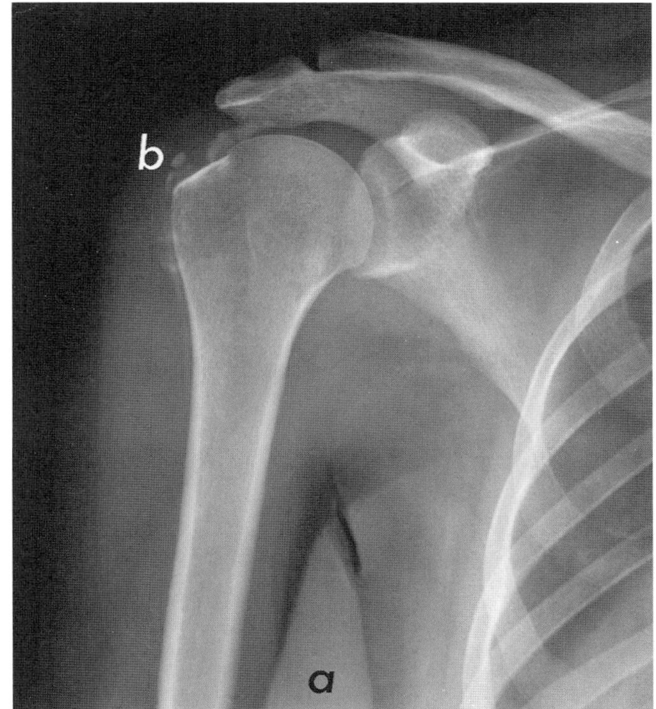

Figures 4.7 *(left)* and 4.8 *(right)*. Two patients with shoulder pain.

radiology and is called an *overlap shadow.* In this instance it is created by the added densities of heavy breast and soft tissues of the upper arm. The confusion arising from the unexpected density of the shadow at *a* will be easy for you to resolve if you remember that *thickness* as well as *composition* determines radiodensity. Although fat, skin, and muscle ought to be less radiodense than bone, the shadow cast by a thick mass of these tissues will approach that of bone, as you see in this figure. Note, on the other hand, that a small amount of air imprisoned in the axilla is black on the film, probably because it was a long pocket of air x-rayed end-on.

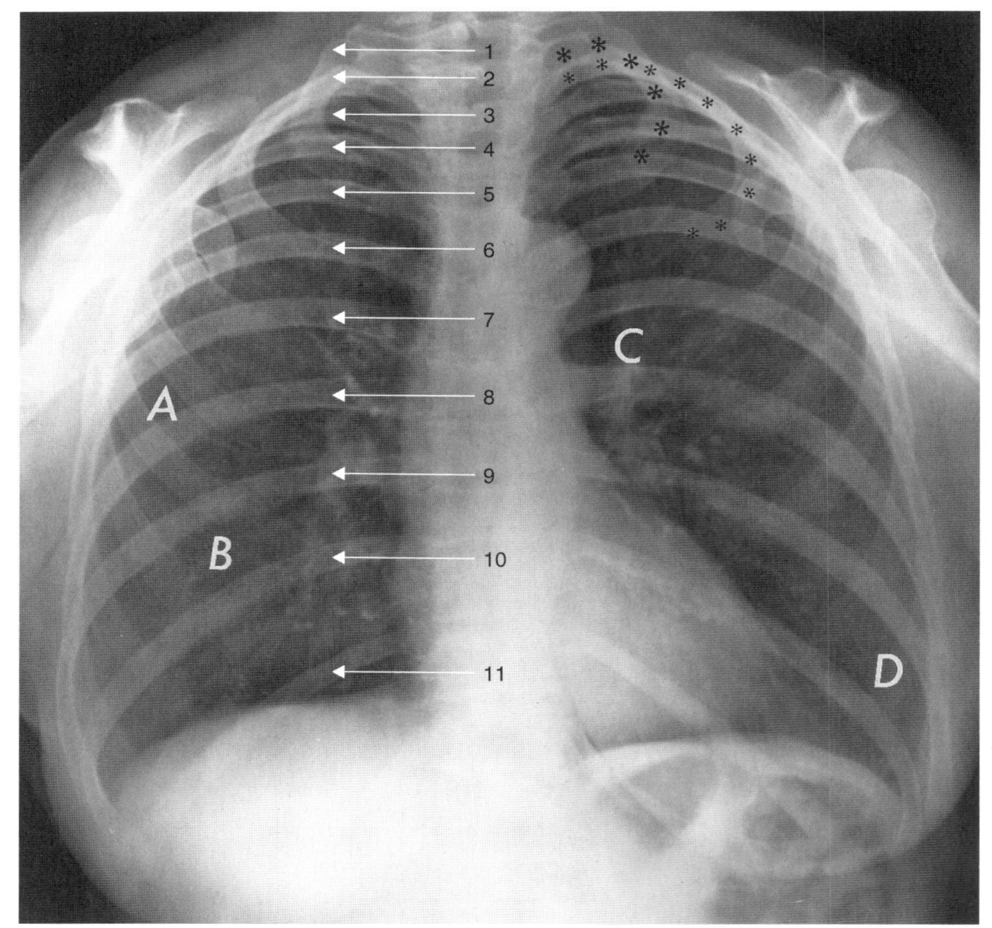

Figure 4.9. Counting and identifying ribs and rib interspaces is an important part of the systematic chest film survey. (See text for instructions.) Note, incidentally, that the breast shadows in this patient come well below the level of the diaphragm and do not obscure the lower lung field; in many female patients they do.

The Rib Cage

Using the bilateral symmetry of each pair of ribs in Figure 4.9, and beginning at the origin of the first rib at its junction with the first thoracic vertebra, trace each rib as far as you can anteriorly to the beginning of the radiolucent (and hence invisible) costal cartilage. The ribs are useful to radiologists because they locate an abnormal shadow by its proximity to a particular rib or interspace on a film they are describing. Anyone reading the radiologist's written report can identify in this way the precise shadow being discussed. Thus *A* in Figure 4.9 could be described as lying in the seventh interspace on the right, close to the axilla (that is, the outer third of the space between the posterior halves of the right seventh and eighth ribs). If you do not locate it there, count again: you are probably getting lost in the overlap tangle of ribs 1, 2, and 3. To avoid this, identify the first rib carefully by finding its anterior junction with the manubrium and following this rib *backward* to the spine. Then count down the posterior ribs. *B* would be said to be located in the ninth interspace on the right. Note that the word "interspace" always implies the space between *posterior* segments of adjoining ribs, unless the anterior is specified. Try your hand at designating the location of *C* and *D*, covering the left half of this figure and the spine with all its numbers. (Have you noticed anything peculiar about this film? Is there anything missing? Compare it with Figure 4.1.)

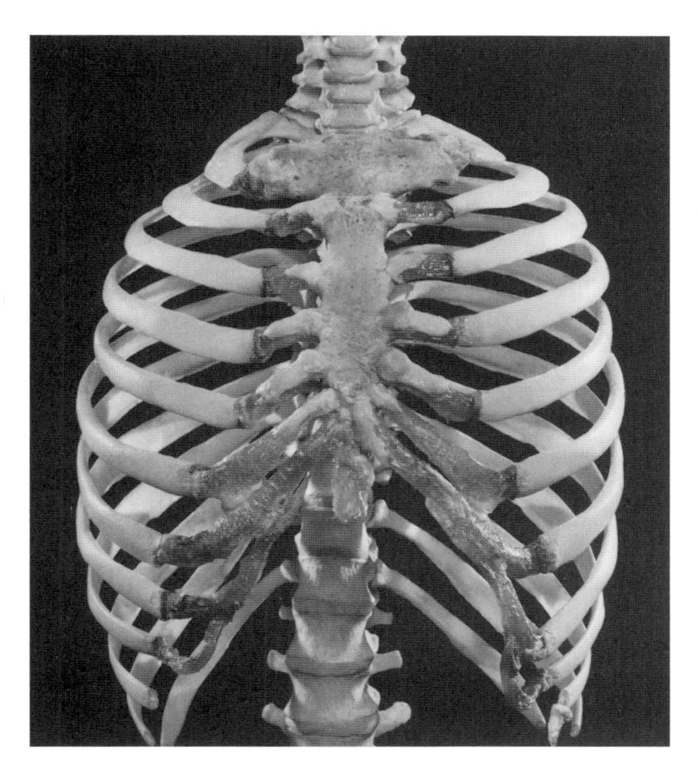

A. Anterior and posterior ribs.

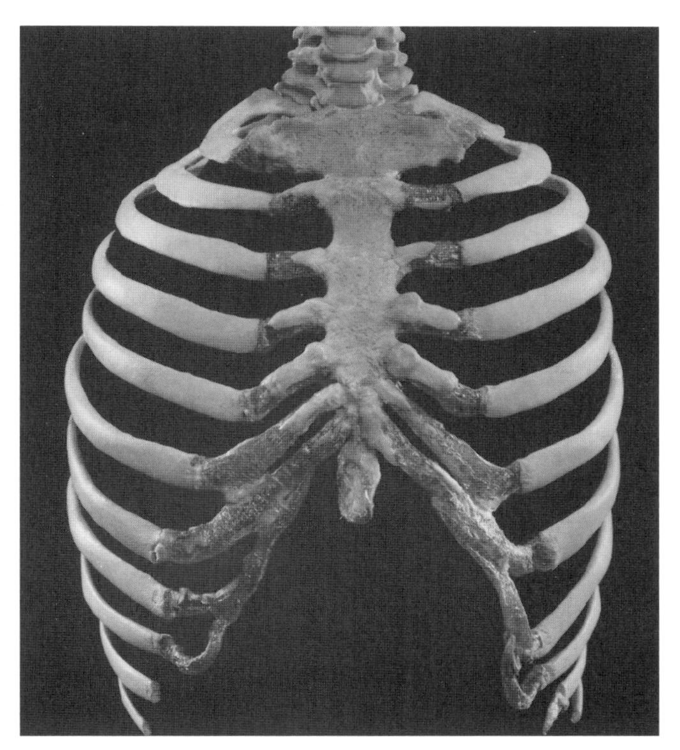

B. Anterior ribs only.

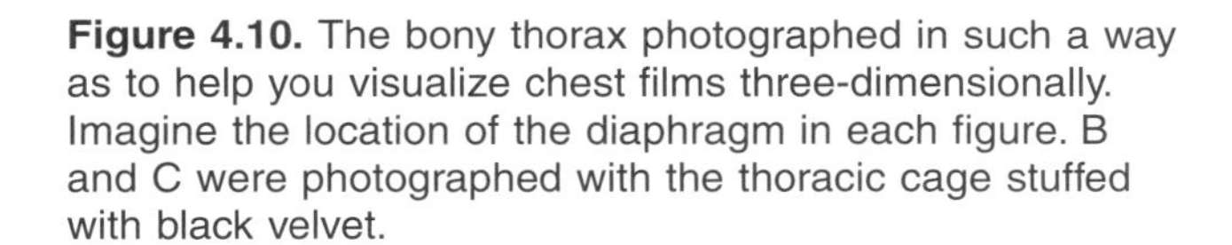

Figure 4.10. The bony thorax photographed in such a way as to help you visualize chest films three-dimensionally. Imagine the location of the diaphragm in each figure. B and C were photographed with the thoracic cage stuffed with black velvet.

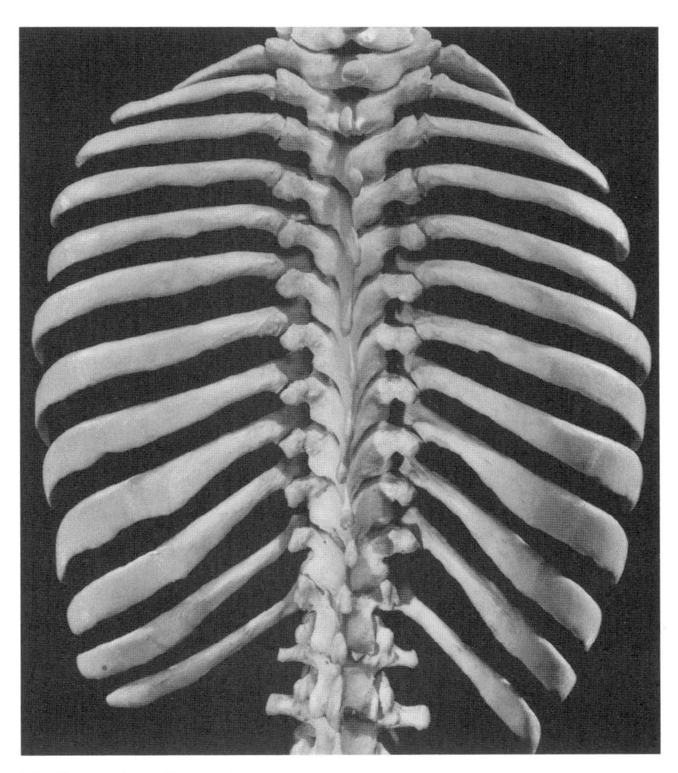

C. Posterior ribs only.

The *ribs* confuse everyone beginning to look at chest films. The miracle is that one can discern anything useful about the heart and lungs through such a cross-hatched pattern of shadows. Three-dimensional thinking will be easier if you try to concentrate first on the posterior halves of the ribs and then on the anterior. In Figure 4.10B and C the same thorax was photographed from the front and from the back after the cavity had been stuffed with black velvet to give you the illusion you seek in trying to study the posterior ribs while excluding from your mind the anterior ones.

Remember to think in terms of *coronal* slices and the summation shadows they produce, as in the cadaver in Chapter 2. The transaxial images used in CT are useful too in three-dimensional thinking, and we will return to them presently.

Confusing Shadows Produced by Rotation

Because of their curiously curved shape, the shadows of the two *clavicles* will appear symmetrical on the chest film only if there is no rotation of the chest. In a perfectly true PA film the beam passes straight through the midsagittal plane. The arms and shoulders of the patient are arranged symmetrically, either hands on hips or arms overhead, and the technologist checks for rotation and corrects it before making the exposure. Turned even a few degrees, the clavicles will exhibit a remarkable degree of asymmetry. This fact will prove very useful to you, because a glance at the clavicles will tell you whether or not the beam has passed through the sagittal plane and whether you are therefore looking at a true PA or AP film without rotation.

Even slight rotation is undesirable in a chest film, because the heart and mediastinum are then radiographed obliquely and their shadows appear enlarged and distorted. If you think of the mediastinum as a disc of denser structures flattened between the two inflated lungs and normally x-rayed end-on in a PA chest film, it is easy to see how rotation of this disc will produce a wider shadow. If it were a valid finding, enlargement of the heart or widening of the mediastinal shadow would be an important piece of roentgen evidence for disease. You have to be able to disregard apparent enlargement due to rotation, therefore, and the best clue to rotation is asymmetry of the shadows of the two clavicles. Learn to watch them, mentally noting their symmetry or lack of it, in your systematic survey of the chest film.

Now look back at Figure 4.9. Did you notice that there were no clavicles? The patient was born without them and is an ideal subject for you to use in learning to count ribs. Compare this film with any normal chest film and observe that you can mentally subtract the shadow of the clavicle when you want to in order to study or count the first three ribs.

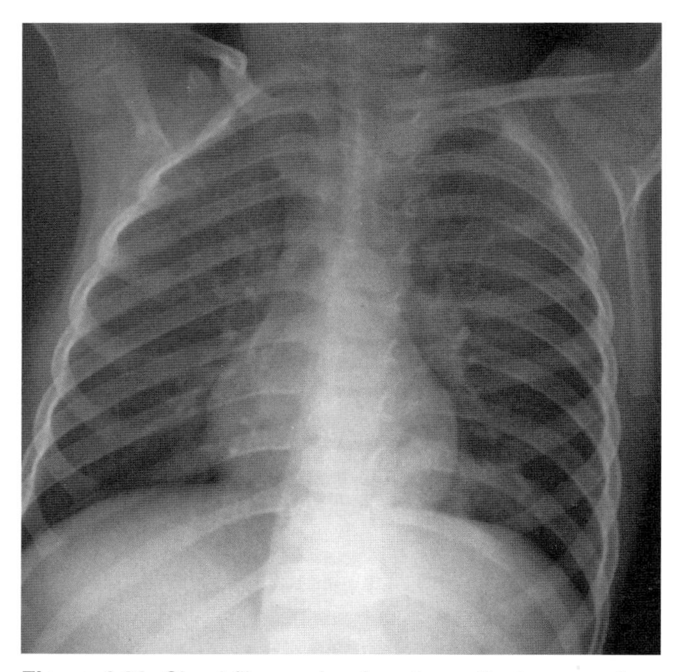

Figure 4.11. Chest film made when the patient was accidentally somewhat rotated. Note the marked asymmetry of the clavicles.

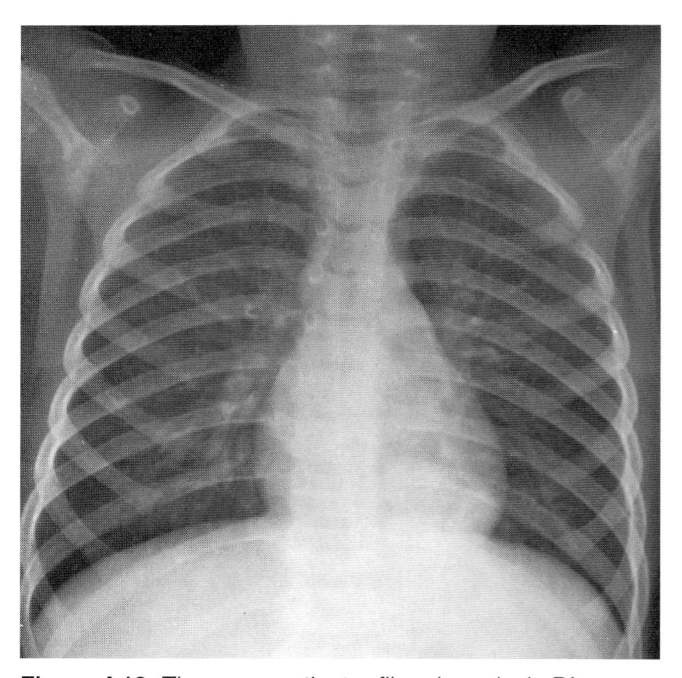

Figure 4.12. The same patient refilmed precisely PA.

Problems in Studying Ribs and Clavicles

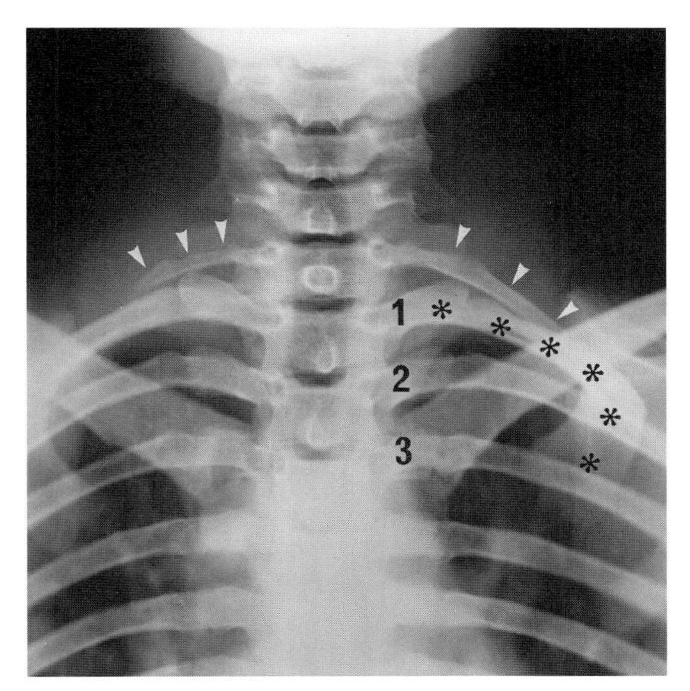

Figure 4.13 ***(Unknown 4.1).*** If you think the ribs are correctly labeled here, how do you account for the structures indicated by the white arrows?

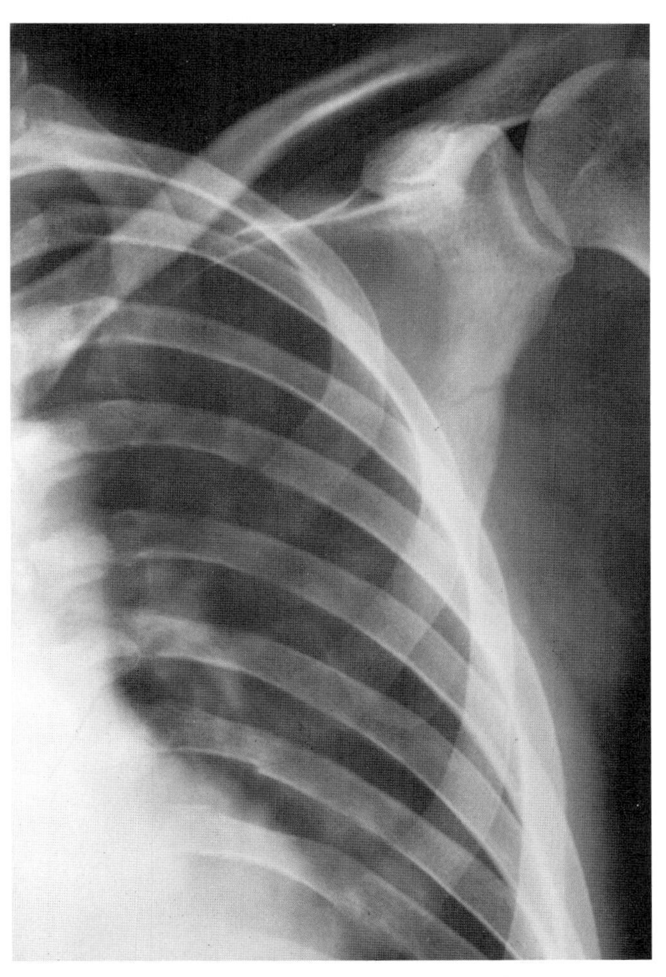

Figure 4.14 ***(Unknown 4.2) (right).*** This patient has been filmed after an automobile accident. Which rib is fractured?

Figure 4.15 ***(Unknown 4.3) (below).*** This is not a PA film, for the very good reason that the patient was in a great deal of pain and was x-rayed on a stretcher. Study the bones using any normal shoulder girdle for comparison. Then study the soft tissues outside the chest cage around the shoulder girdle. How can you account for the dark streaks?

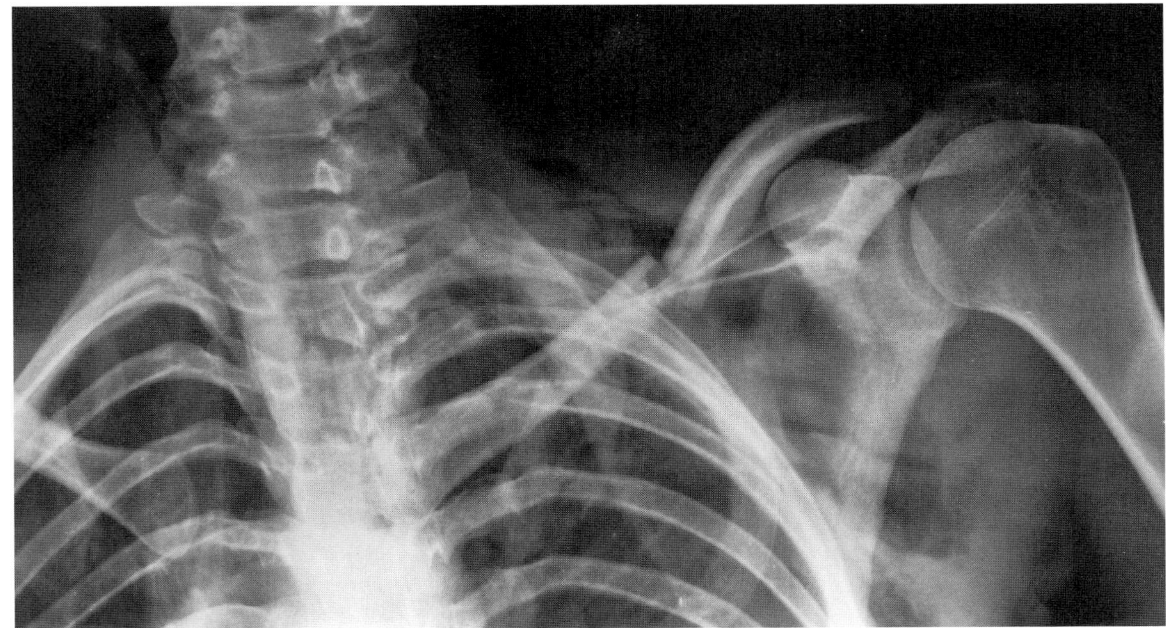

The Importance of Exposure

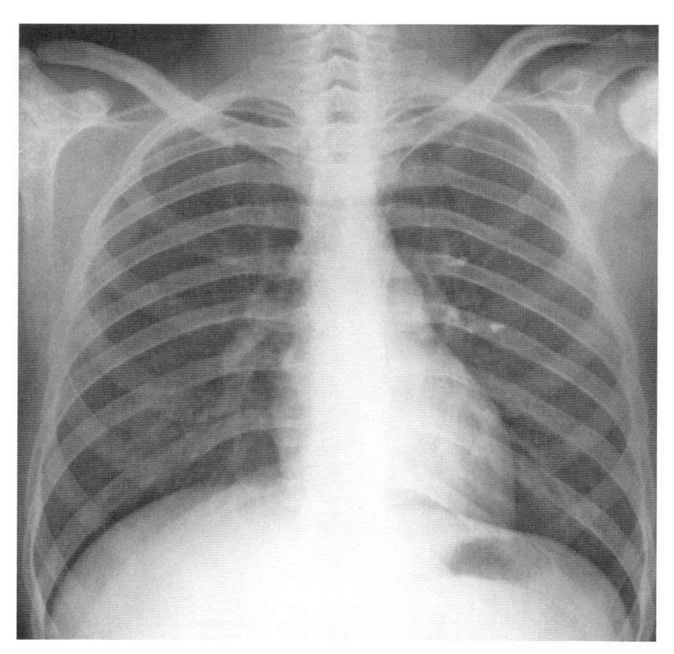

Figure 4.16. Regular chest film made PA at 6 feet with the patient standing.

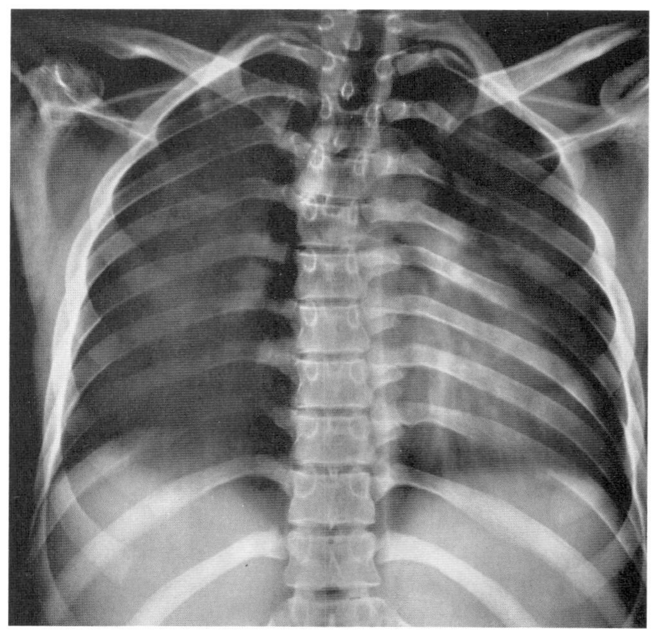

Figure 4.17. AP film of the chest exposed for study of the spine. The patient is lying down; a Bucky grid is used. The ideal exposure for a chest film today is somewhere between these two and allows faint visualization of the intervertebral spaces through the heart.

The thoracic spine is not well seen in the chest films you have been looking at because its density added to those of the mediastinal structures and sternum together absorb almost all the rays and few reach the film to blacken it. This is true of the techniques commonly used for studying the lung. Seeing detail through very dense parts of the body requires a different technique. More penetration can be achieved in several ways. One is by increasing the exposure factors (kilovoltage, milliamperage, and time) to produce a beam of x-rays of shorter wavelength, so-called harder rays. A film made in this way is often called an overexposed film, intentionally overexposed in order to increase the penetration of dense structures.

Unfortunately, when x-rays impinge on matter of any kind, secondary x-rays are generated that radiate in all directions as if from many point sources of light. These rays are called scattered radiation and they are, of course, added photographically to the primary beam, additionally blackening the film. Since there are thus multiple sources of radiation, a blurred and distorted image is produced. Moreover, increasing the exposure factors to produce a harder and more penetrating beam also increases the amount of scattered radiation. Thus a simple overexposed film will usually lack contrast and sharpness.

Scattered radiation can be eliminated by an ingenious device called a *Bucky grid.* Interposed between the patient and the film, it is a flat grid composed of alternating very thin strips of radiolucent and radiopaque material (plastic and lead, for example). Only the most perpendicular rays pass through the lucent plastic strips. The oblique rays, representing most of the scattered radiation, strike the sides of the lead strips and are absorbed.

If the interposed grid is motionless, of course, the lead strips will appear on the film as fine white lines. To prevent this, it is only necessary to move the grid across the film throughout the exposure; no lines will appear.

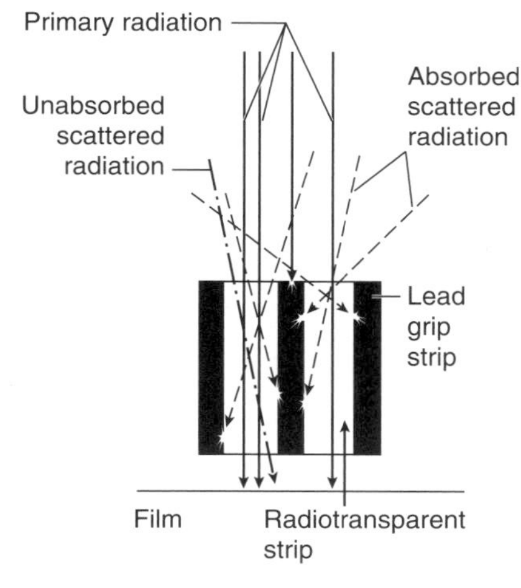

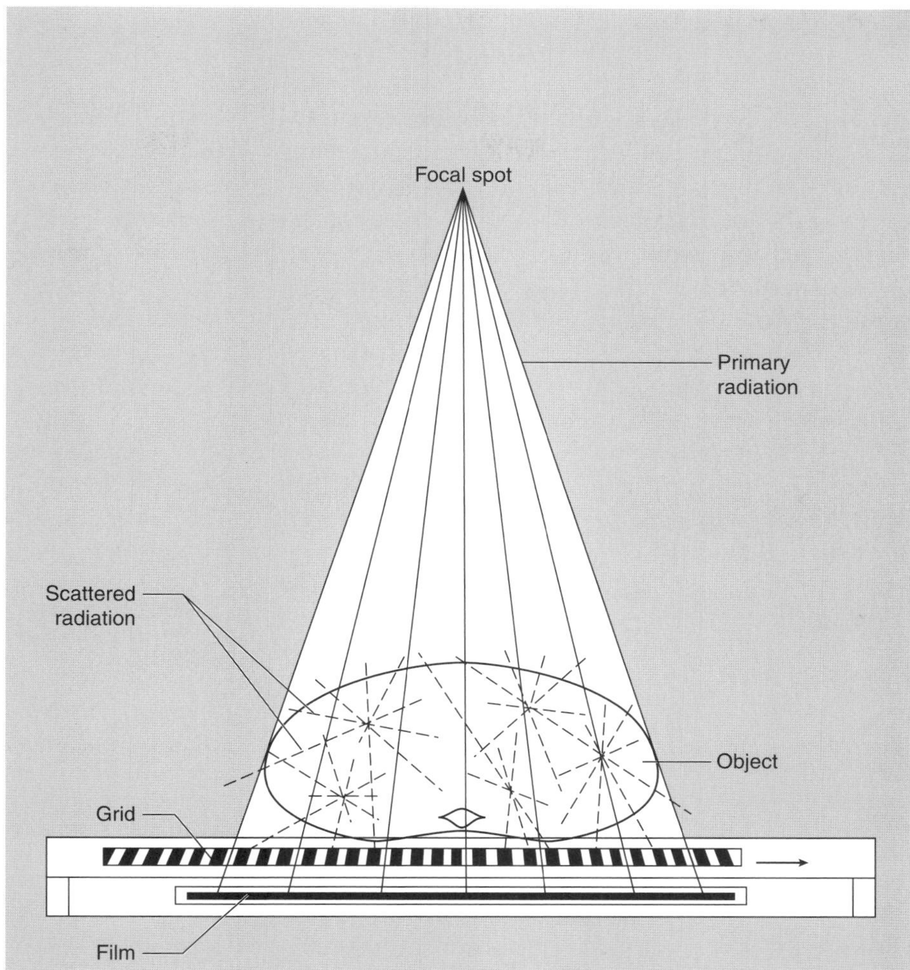

Figure 4.18. The Bucky grid, plus increased kilovoltage and time, gives the desired increased penetration and clearer detail to radiographs of thick parts of the body (see text).

You will find that in an obese patient, or in any patient whose spine, mediastinum, pelvis, or heavy long bones must be studied by x-ray, Bucky-technique films will have been made automatically by the technologist. Every film you see of the abdomen will have been made in this way also.

The PA chest film in Figure 4.16 was made expressly for the purpose of studying the lung. Figure 4.17 was made AP (so that the spine toward which the study was directed would be close to the film), and a Bucky grid and the appropriate bone exposure techniques were used to produce it. Notice how well you can see the structure of the vertebrae with their interposed disc spaces, which should be just visible in a routine chest film. Note also that here you can see the ribs below the diaphragm, scarcely visible in most regular chest films. This film would be useless for studying the lung, all the delicate detail being lost.

Therefore, as you look at any chest film you should try to estimate whether it is exposed correctly or overexposed or underexposed. Correct PA exposure for routine chest films allows you to see the more radiolucent intervertebral spaces but not the detailed anatomy of the vertebrae. This is important for the following reasons: if a film is underexposed you will be tempted to overinterpret vascular shadows in the lungs, whereas in an overexposed film you may miss minute densities of importance because they are "burned out."

Soft Tissues

Just as the lung detail is burned out with these techniques, so also are the soft tissues of the chest lying outside the thoracic cage. Having completed your survey of the bones, you should now *look at* these soft tissues, studying breast tissues, supraclavicular areas, axillae, and the subcutaneous tissue and muscles along the sides of the chest where it is seen in tangent. You will be able to study the soft tissues in any film exposed for study of the lungs, and they often give you vital information about the patient. Are the soft tissues scanty, indicating perhaps that the patient has lost weight? Are the normally symmetrical triangles of dark fat in the supraclavicular region disturbed in any way? Look back at Figure 4.2A and at Figure 4.15. Always be sure to check whether there are two breasts: a chest film showing one missing breast often means that the patient is being studied for recurrence of cancer, and you should scrutinize the bones and lung field for evidence of metastases. The lung field under a missing breast appears a little darker than the other lung field because of the missing tissue of the breast and sometimes missing pectoral muscles that were removed at the time of the mastectomy.

So much, then, for the first step in studying a chest film: a systematic survey of the bones and soft tissues. You are ready now to look past the bones and soft tissues at the shadow of the lung itself.

Problems

Unknown 4.4 (Figure 4.19)

Analyze this film of an asymptomatic man and decide whether the film is normal.

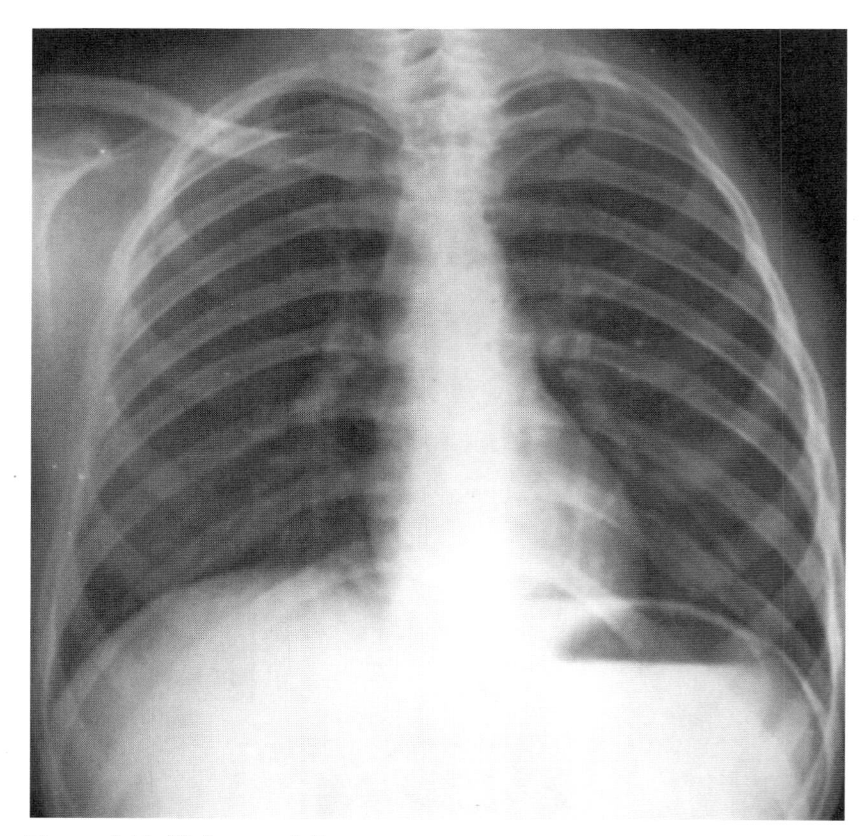

Figure 4.19 ***(Unknown 4.4)***

Unknowns 4.5–4.7 (Figures 4.20–4.22A)

Compare the bones and soft tissues in the films of these three asymptomatic women. How do you account for the differences?

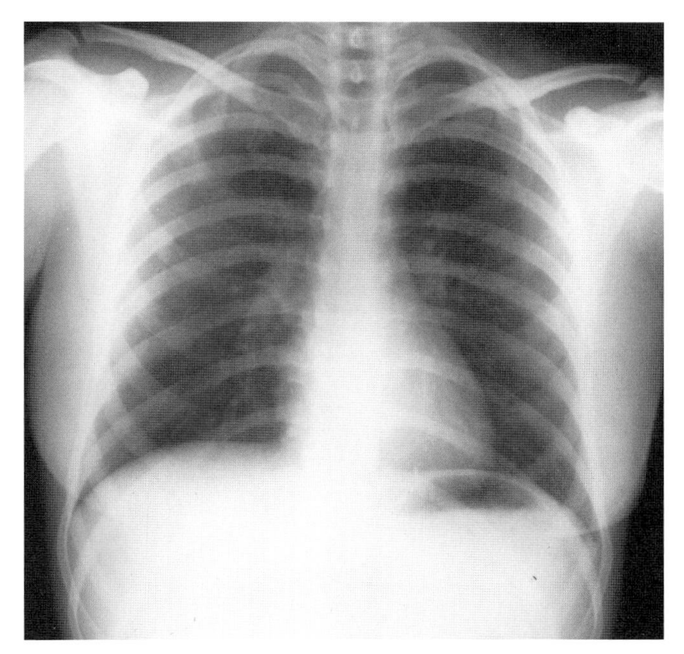

Figure 4.20 ***(Unknown 4.5)***

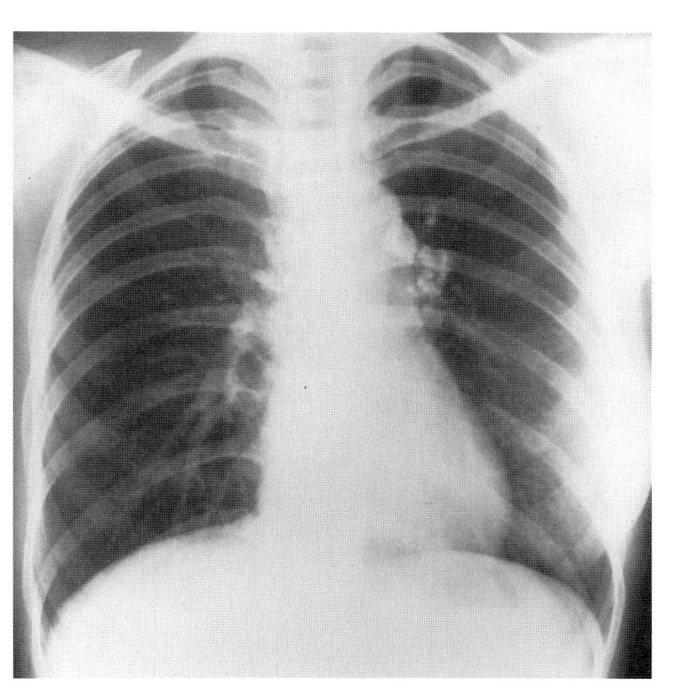

Figure 4.21 ***(Unknown 4.6)***

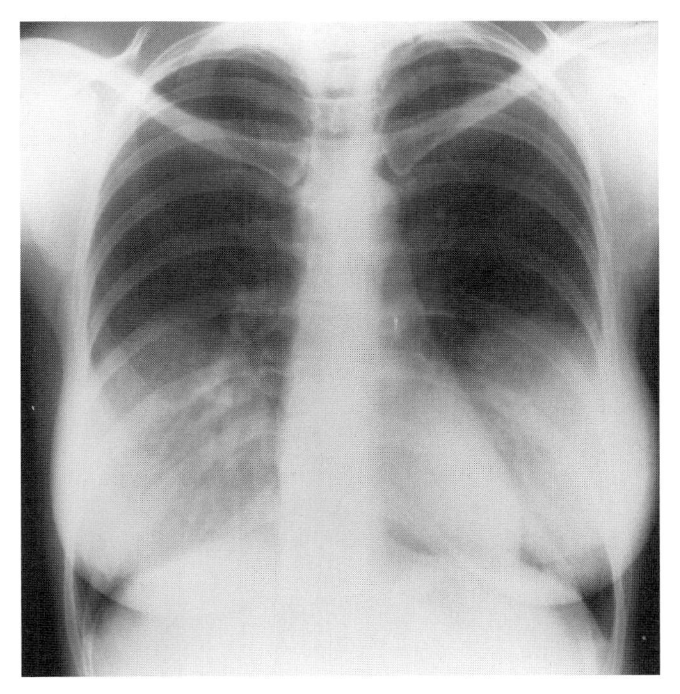

Figure 4.22A ***(Unknown 4.7)***

5 The Lung

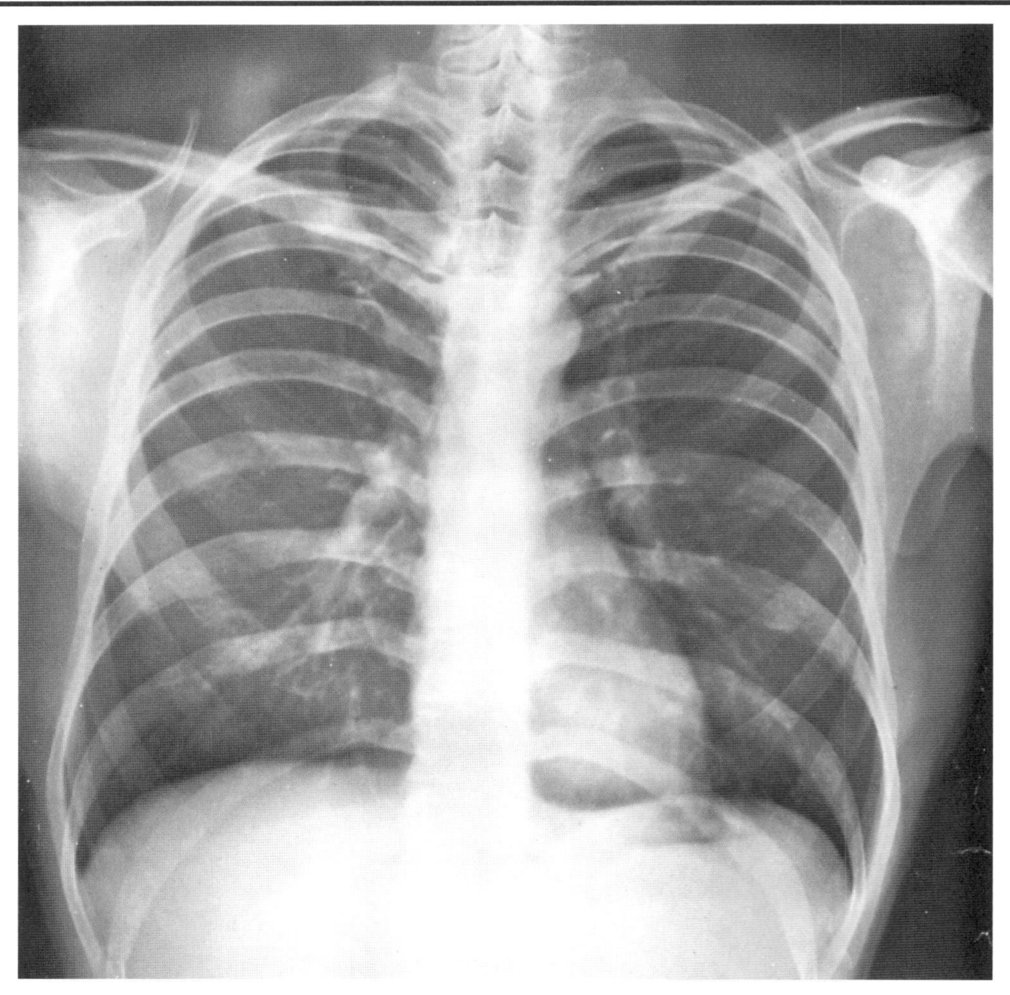

Figure 5.1. A normal chest film.

The Normal Lung

Look now past all those distracting shadows of the ribs and soft tissues at the roentgen images of the lung itself. Because of its contained air the normally expanded lung is largely radiolucent, as we have said, but nevertheless you do see in Figure 5.1 traceries of branching gray linear shadows. What precisely are they? You can reason it out.

Reflect first that logically any structure of greater radiodensity suspended in the middle of a radiolucent structure like the lung will absorb some of the x-rays and cast a gray shadow on the film, a patch of film where less silver has been precipitated.

If this structure is spherical and of uniform composition, it will cast a round shadow. If the surface of the mass is knobby and irregular, knobs will be present in the outline, as in Figure 5.2. If the suspended dense structure is cylindrical, like a blood-filled vessel traversing lung substance, a tapering linear gray shadow results; and if the vessel branches, the shadow will branch.

When a vessel passes through the lung in a direction roughly parallel to the film (and perpendicular to the ray), its tapering and its branching will be accurately rendered on the PA film. But if it passes through the lung in a more nearly sagittal direction, it will then line up with the beam, absorbing more x-rays so that its shadow will appear as a dense round spot. The situation is analogous to the rose leaf on edge in the first chapter. You can find such end-on vessels in Figure 5.1, or in any chest film.

It is quite logical, then, that the normal "lung markings" are indeed blood vessels and not bronchi and bronchioles. The bronchial tree, because it is filled with air and has thin walls, casts little or no shadow when it is normal. It is therefore practical to think of the normal "lung markings" as wholly vascular.

The tracheobronchial tree may be *rendered visible*, of course, with relatively harmless radiopaque fluids instilled via a tracheal catheter into the lung of a living patient, who later coughs up or absorbs and excretes the opaque substance. This procedure, called *bronchography*, is rarely used clinically today.

Much less dramatic but nonetheless important information is available from any simple PA chest film as you study the vascular shadows in the lung, marked only by the blood they contain. Notice first that the largest vessels at the hilum of the lung cast the heaviest and widest shadows, just as you would expect. This Medusa-like tangle of arteries and veins on either side of the heart shadow is referred to by the radiologist in reports as the "hilum" or "lung root." The right hilar vessels seem to extend out farther than those on the left, but that is only because a part of the left hilum is obscured by the shadow of the more prominent left side of the heart. Measured from the center of the vertebral column, the vessels will be found to be symmetrical except for the slightly higher takeoff of the left pulmonary artery, which curves over the left main bronchus rather abruptly (Figure 5.7). For this reason the left hilum on any normal chest film is a little higher than the right one.

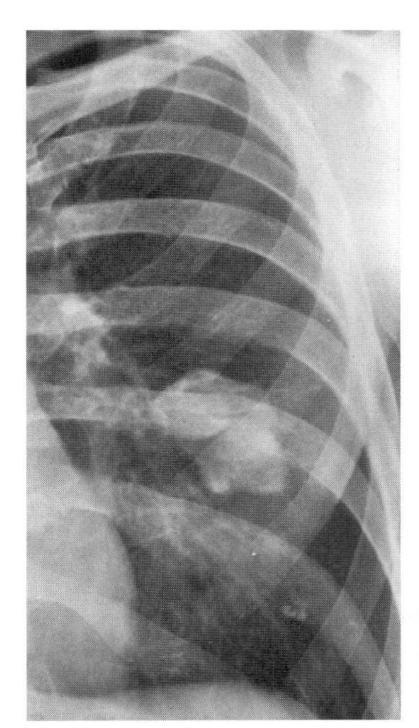

Figure 5.2. The mass in this air-filled lung casts a shadow recording its knobby outline.

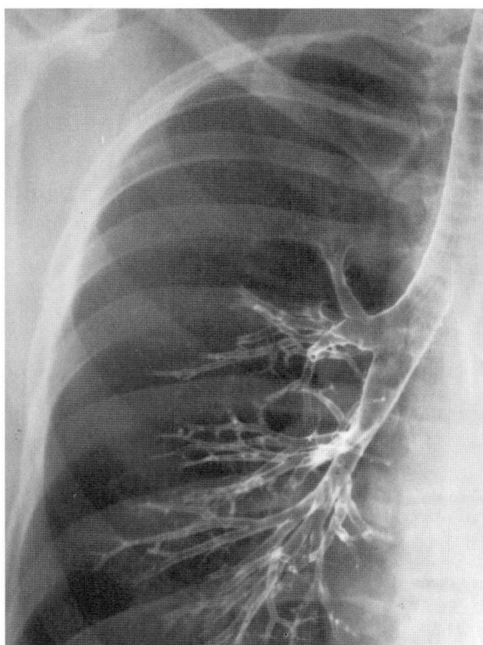

Figure 5.3A. Bronchogram showing the tracheobronchial tree coated with opaque material. Vessels filled with blood are only faintly visible.

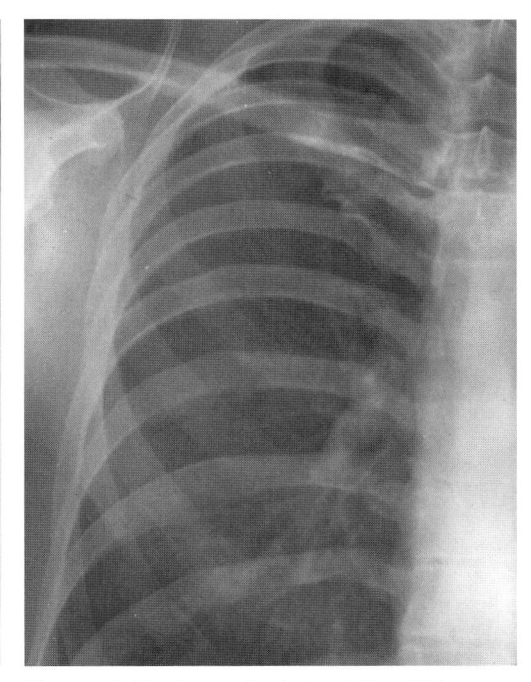

Figure 5.3B. A standard chest film. Note that the air column is seen in the trachea, but the rest of the tracheobronchial tree is not visible at all. Note the faint vascular shadows.

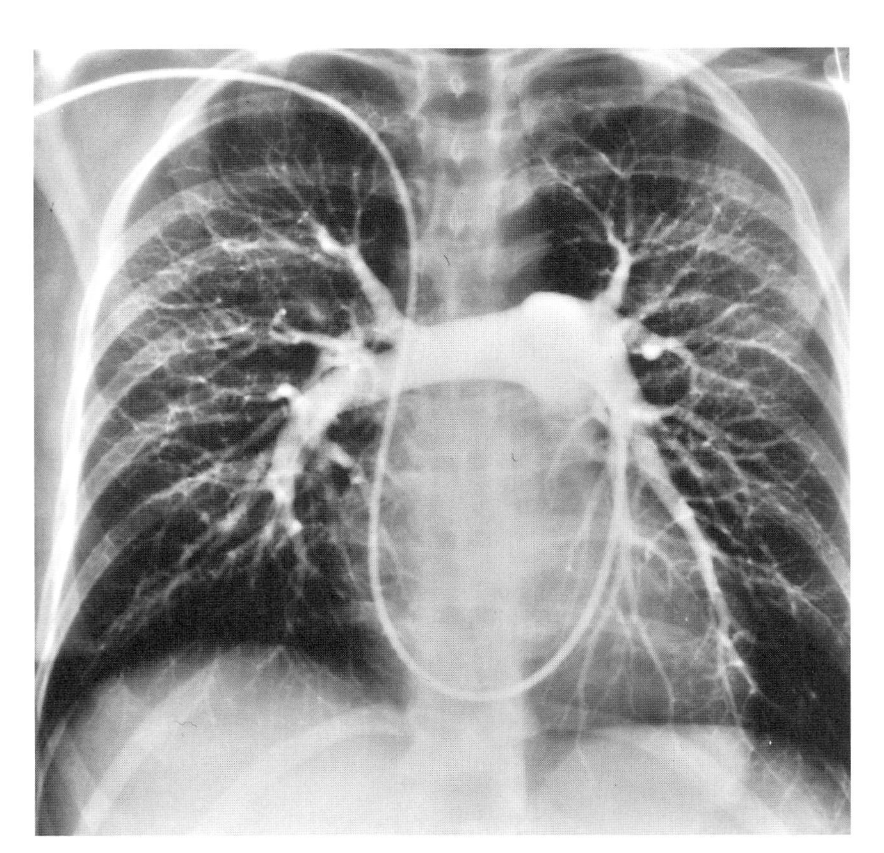

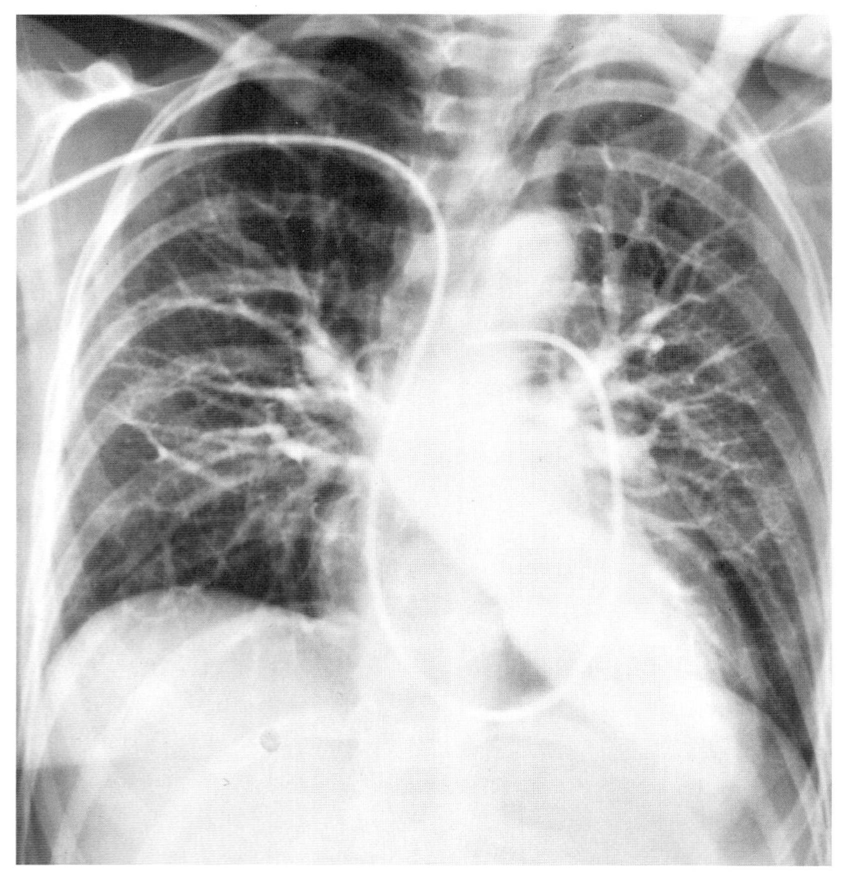

Figures 5.4 and 5.5. The vascular tree within the lung may be opacified in the living patient, so that the blood vessels are more clearly seen than they are on the plain chest film. Radiopaque contrast media may be injected through a vascular catheter positioned in the pulmonary artery tree to opacify the pulmonary arteries and veins while imaging is performed with a film changer taking a series of x-ray exposures. Figure 5.4 *(top)* was made during the early phase of filming; it shows opacification of the pulmonary arteries. Figure 5.5 *(bottom)*, a film made a few seconds later, shows opacification of the pulmonary veins as well as opacification of the left atrium, left ventricle, and aorta. Note that the pulmonary veins drain into the left atrium at a level below where the pulmonary arteries leave the mediastinum.

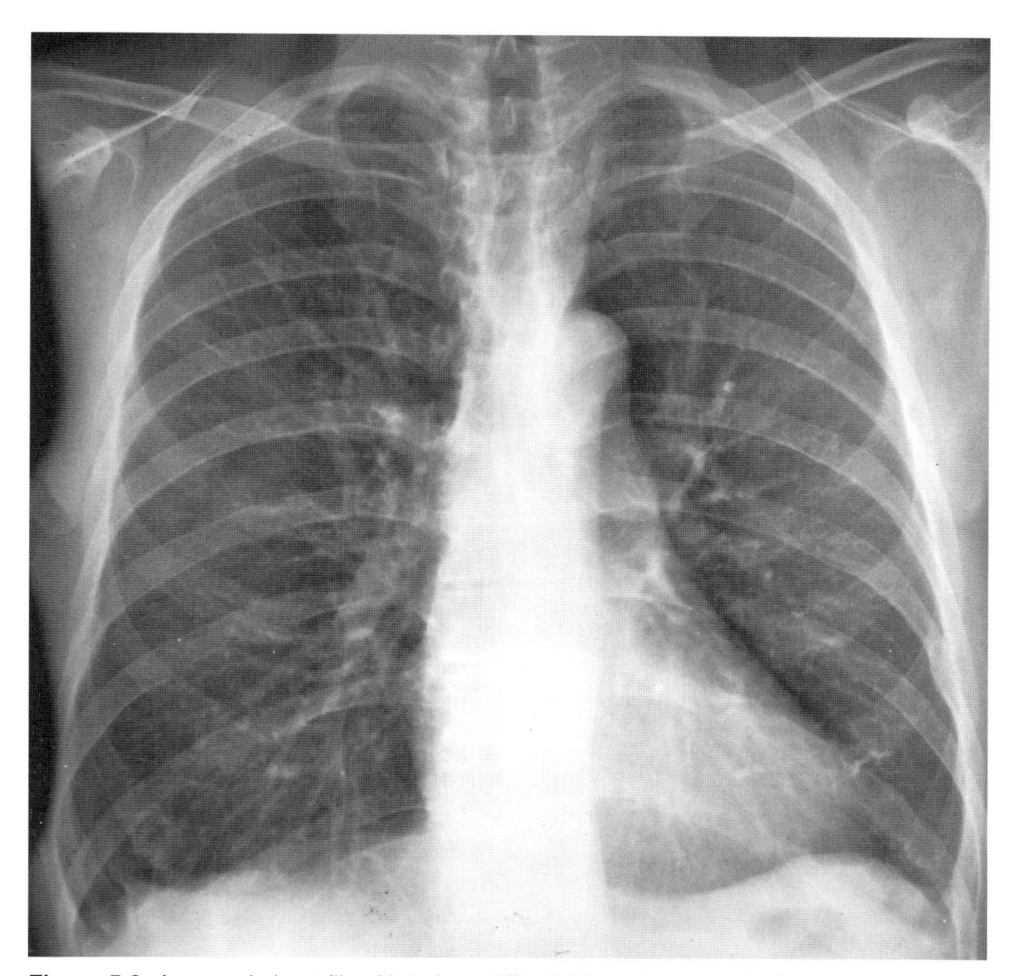

Figure 5.6. A normal chest film. Note how difficult it is to be sure which vessels are arteries and which are veins on this "plain film" without added contrast material.

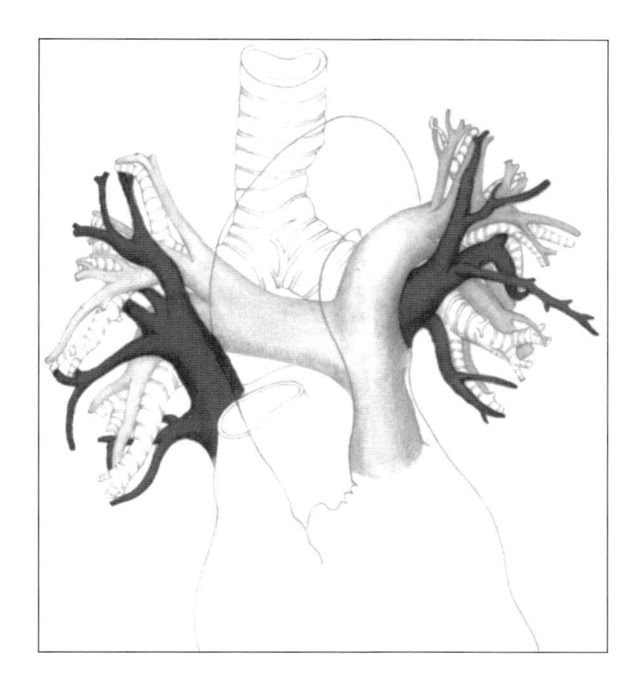

Figure 5.7. The anatomical composition of the hila. The aorta has been rendered as though transparent (find it). The tracheobronchial tree is indicated with cartilage rings; the pulmonary arteries are light and the pulmonary veins dark.

Variations in Pulmonary Vascularity

Compare the normal hila in Figure 5.1 with a few abnormal hilar shadows. The lung root may be enlarged because of engorgement of its *veins,* for example, in any condition in which there is obstruction to the return of oxygenated blood from the lung to the left side of the heart. Such a condition exists in acute left heart failure after a myocardial infarction. More chronically, the same condition exists with mitral stenosis, where the gradual narrowing of the mitral valve results in back pressure in the pulmonary veins. Figure 5.8 shows the appearance of the hila in moderately advanced mitral stenosis. Note the enlargement of the lung root and its obviously fat and tortuous branches, compared with the slim, straight vessels in Figure 5.1.

Dilatation of the *arteries* in the hila will also become familiar to you in types of congenital heart disease in which an abnormal opening in the septum reroutes blood from the left chambers back into the right chambers and to the pulmonary circulation, thus overloading the right heart and pulmonary arteries. A patent ductus arteriosus with a shunt of blood from the aorta to the pulmonary artery, and septal defects between the atria, commonly give this picture. Figure 5.9 shows an example of the marked hilar arterial engorgement seen in congenital heart disease of this type.

There is often actually some enlargement of both veins and arteries, and it is not usually possible for you to say from the plain radiograph which vessels predominate. In judging the appearance of the hila in the patient whose film you see for the first time, you will decide simply that you are looking at vascular trunks of normal caliber or that they are enlarged.

The vascular trunks of the hila normally branch and taper out into the lungs in all directions. They are so fine in the far peripheral lung close to the chest wall that you can sometimes no longer see them, depending on the technique used. If you mask off between two pieces of paper first the hila and medial half of the lung and then the lateral half, you will be struck by the decreased number of trunks laterally. But this will

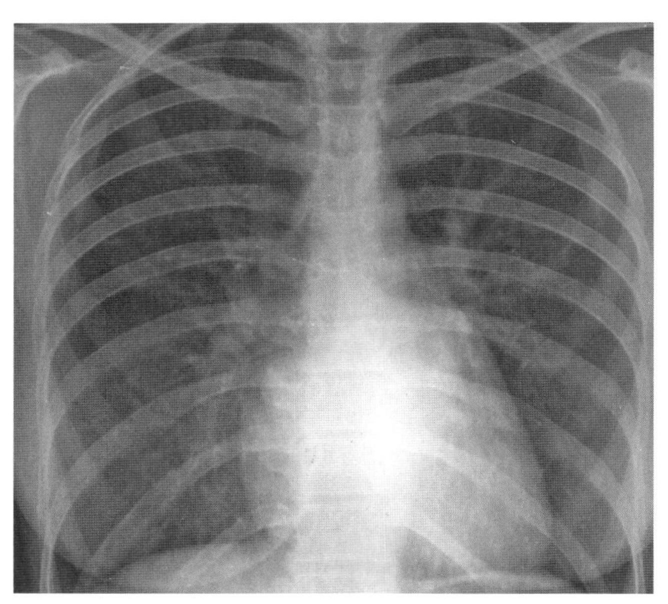

Figure 5.8. Engorged hilar shadows in mitral stenosis. Note the upper lobe vessels, which are enlarged in mitral disease. Compare the upper lobe vessels in Figure 5.1.

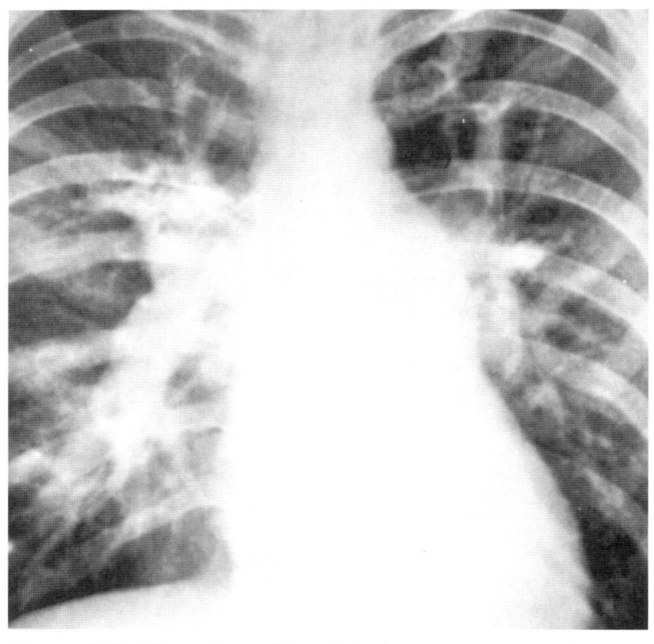

Figure 5.9. Hila enlarged by dilated pulmonary arteries and veins in a patient with interatrial septal defect. The normal quantity of blood returning to the right atrium from the vena cavae is augmented by blood shunted through the defect from the left atrium. The result is recirculation of blood through the lungs and overload of the pulmonary circulation.

not surprise you when you recall that the lung is much thicker medially where it borders the mediastinum than at its lateral extremity, and that there are many more vessels superimposed on each other in the medial half of the lung on the radiograph. If you similarly divide the lung on the x-ray film into upper and lower halves, you can see at once that there are many more branching vascular trunks in the lower half of the lung than there are in the upper half. This too is a function of thickness, and to think three-dimensionally about the vascular tree within the lung at this point is to recall the pyramidal shape of the lung with its broad base against the diaphragm and its apex coming to a point under the arch of the first rib. In the upright patient there is also increased flow in the lower lobe vessels, which increases their caliber.

You will be disturbed from time to time by the juxtacardiac portion of the lower right lung (the right cardiophrenic angle). Many vascular trunks overlap there in the PA view, because those for the anteriorly placed middle lobe are superimposed on those for the posteriorly placed lower lobe. You can easily be misled into supposing that there is some increased opacity in this area, when in fact none exists. You can prove this to your own satisfaction by reviewing in films on this page and preceding ones the appearance of the portion of the right lung lying just above the diaphragm and to the right of the heart. Observe that even in normal films (see Figure 5.1) the area looks more heavily traversed by vessel trunks than you expect it to be. Part of the difficulty is the visual trick your eyes play; you are probably comparing the lung on the two sides of the heart, but because of the shape of the heart, the lung tissue just beyond the left border of the heart is not actually comparable to the problem area on the right, which we have been discussing. The point is easily proved by measuring from the midline: the truly comparable part of the left lung lies closer to the midline, obscured by the shadow of the heart itself. In Figures 5.10 and 5.11 you have abnormal and normal cardiophrenic angles to compare.

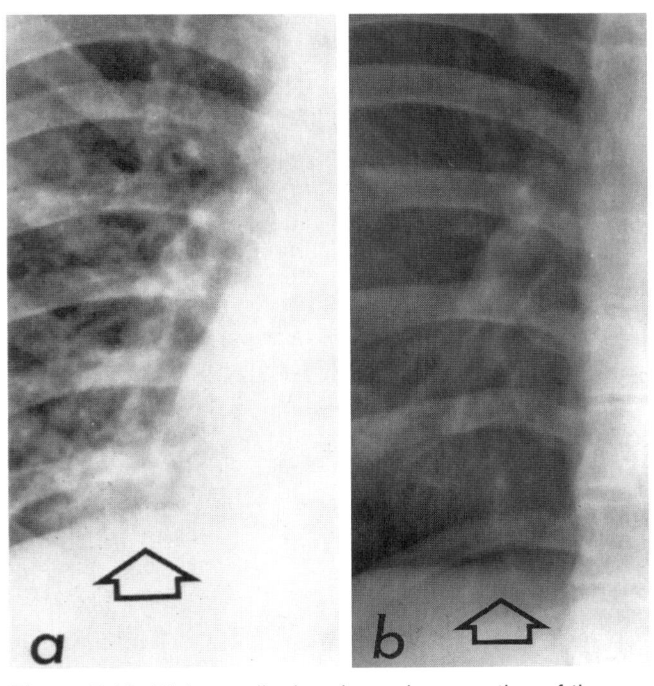

Figure 5.10. Right cardiophrenic angle, a section of the lung often difficult to assess because of the large numbers of vessels superimposed. In *a,* a fluffy opacity is filling in the area that is clear in the film of the normal lung, *b.* The man in *a* had pneumonia.

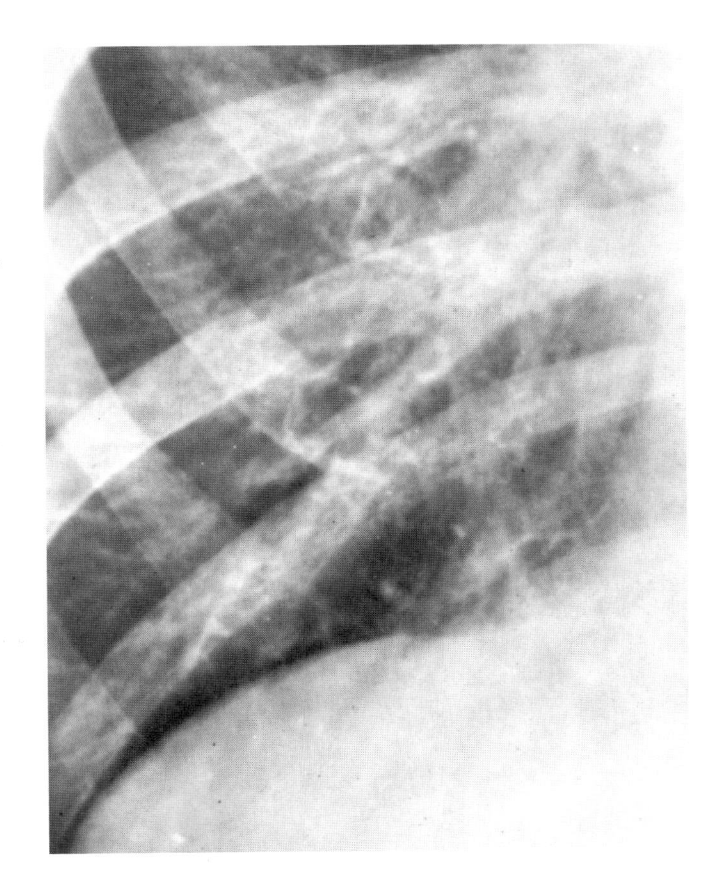

Figure 5.11. Cardiophrenic angle and lower right lung in mild cardiac failure. The shadows of the engorged veins in the hilum superimposed on those of the arteries give a matted, thickened look to the hilum and lung. Note Kerley's B-lines: horizontal, laterally placed linear opacities that represent edema in the interlobular septa.

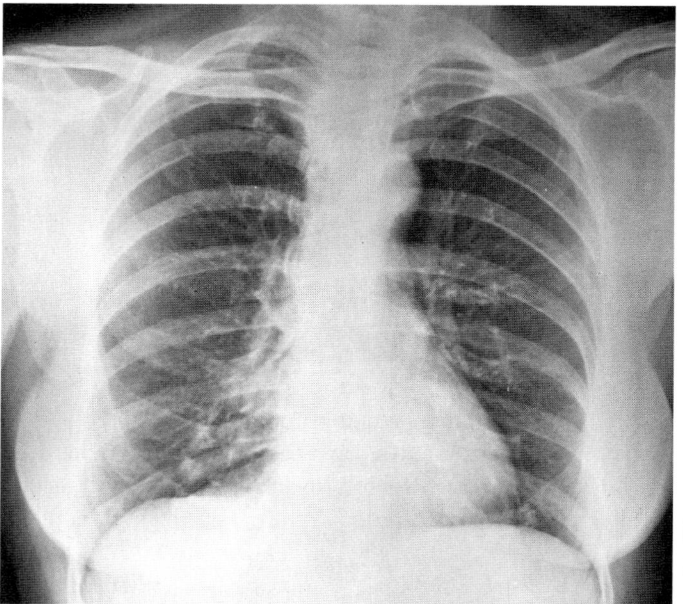

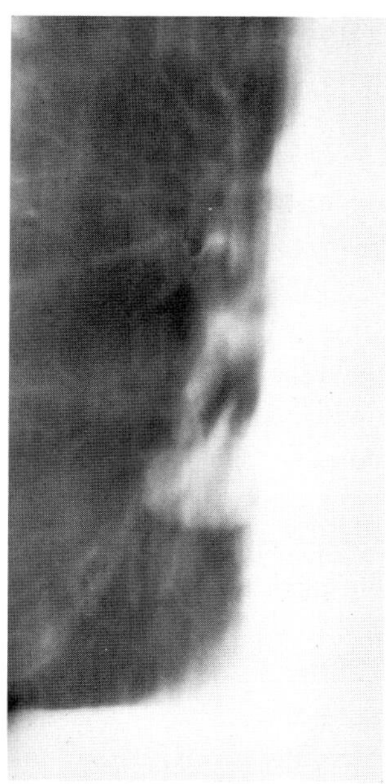

Figures 5.12 and 5.13. Ill-defined thickness in the lower right hilum in a patient with a cough and bloody sputum (Figure 5.12) proved on conventional tomography, Figure 5.13 *(right),* to be a smooth round mass below a clearly normal right pulmonary artery. At surgery a benign tumor was found.

Figure 5.14. Patient with a cough and hemoptysis. A: The PA chest film shows an upper right hilar mass that could be located in front of or behind the hilum. B: Lung window CT scan at the level of the carina shows a posterior mass that abuts the hilum. C *(facing page):* Soft-tissue window CT scan at the same level again shows the mass as well as pre-carinal adenopathy in the mediastinum, just anterior to the carina *(arrows).* This mass proved to be a bronchogenic carcinoma with mediastinal metastases.

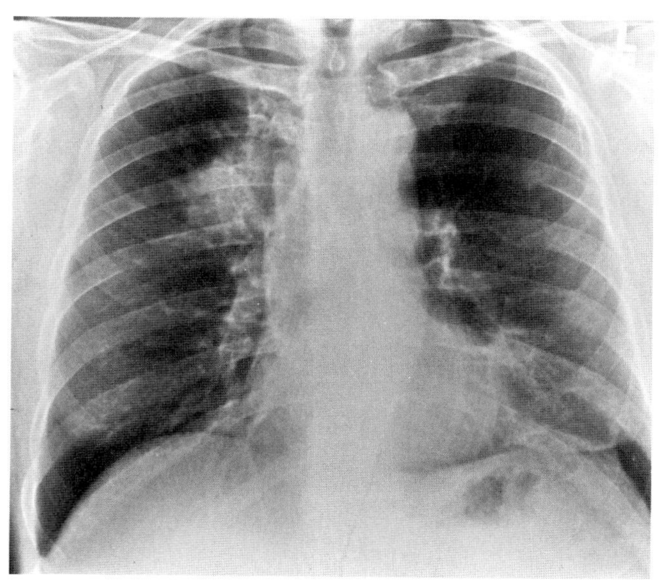

A

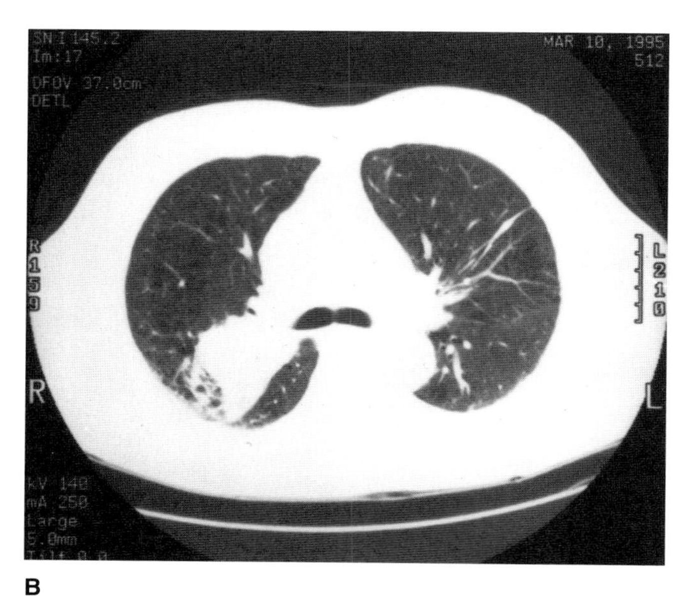

B

Often enlargement of the hilum is not vascular in nature. There are many *lymph nodes* in the hilum and mediastinum, lost among the heavier shadows of vessels and normally too small to be seen. They may enlarge, however, and become visible, either singly or in groups (when they respond to an inflammatory process in the lung, for example, or are secondarily invaded by tumor). They may be seen as overlapping round shadows or, when they are matted together, they may cast a confluent shadow.

Primary tumor masses occurring near the hilum are common. If you are thinking three-dimensionally about the lung root on the radiograph, you will also realize that tumor masses in the peripheral lung tissue in front of or behind the hilum may cast shadows which superimpose on that of the hilum in the PA chest film. Figures 5.12 and 5.14 are examples of this sort of problem. In Figure 5.12 the mass is just below the right hilum, and in Figure 5.14 it is just above the right hilum. A variety of special procedures can help to distinguish the nature of such masses. Conventional tomograms were immensely useful in the past, and you should think of those in the illustrations used here as radiographs of a slice of the patient made through the level of the hilum in the coronal plane. Computed tomography is the preferred method of evaluating the hilar structures today. CT, offering you a cross-sectional slice, will show you whether such a mass lies anterior to or posterior to the hilum.

Hilar enlargements due to tumor tend to be round and smooth in outline and are frequently unilateral. Masses that prove to be clusters of enlarged nodes, you will find, look like a radiographed bunch of grapes, with many overlapping round shadows. Vascular hilar enlargements, on the other hand, taper into the lung and are almost invariably bilateral. You are going to see exceptions, of course, but these very rough generalizations will provide you with a temporary working rule. With contrast CT, hilar enlargement due to tumors or nodes can be routinely distinguished from vascular hilar enlargement.

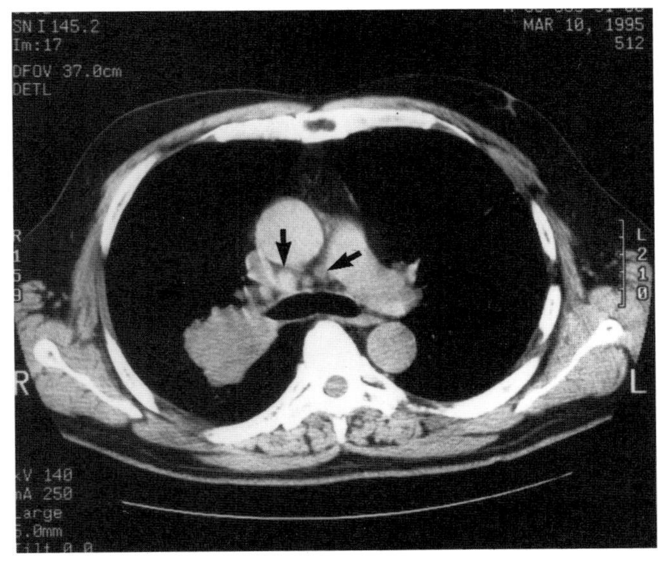

C

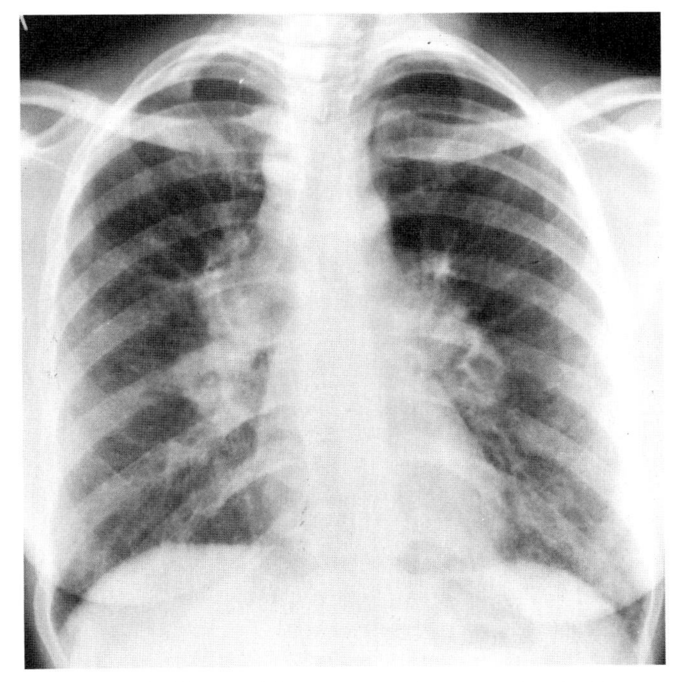

Figure 5.15. Bilateral hilar adenopathy in a patient with sarcoidosis.

Limitations and Fallibility

Finally, remember that sometimes a very innocent-looking hilum is not actually normal and may conceal among its vascular shadows tumor-involved nodes not yet large enough to be seen on the films. Retrospective studies of chest films in large groups of patients who were found to have tumor-involved hilar nodes at CT have shown a number of perfectly normal-appearing hilar shadows. The radiologist can do nothing about this except to report that your patient *seems* to have normal hilar shadows on the chest film, and both you and the radiologist have to consider that the presence of normal-sized but tumor-involved nodes would make a difference in the treatment of the patient.

You learn the limitations of other modes of inquiry constantly in medical school, and you must be aware that despite its great usefulness the radiographic inquiry also has some limitations, even in the hands of the most expert interpreters. When radiologists can help you solve a specific problem, they will do so. When they know they cannot help you, *or that a simple negative report is likely to be misinterpreted as a clean bill of health for the structure in question,* it is their obligation to warn you of that fact and to suggest other studies when they may produce additional information. Today, for example, radiologists will suggest CT for the patient with lung cancer to determine whether there are enlarged mediastinal nodes present. It is this type of problem which, more than anything else, makes it imperative that you not rely entirely on the written report but supplement it with a personal consultation with the radiologist while viewing the films yourself. Without sufficient clinical information the radiologist cannot truly serve the best interests of the patient, cannot offer you a report framed around the difficulties of a particular human being with a particular set of symptoms.

The Pulmonary Microcirculation

In Figures 5.4 and 5.5 you have seen the entire pulmonary arterial and venous systems visualized at arteriography. By advancing the arterial catheter farther toward the lung periphery, we can visualize the microcirculation. The display on the next page shows normal circulation with injected contrast material, as well as three abnormal arterial patterns. The abnormal patterns show dramatic changes in the terminal arterial branches. In Figure 5.17 you see pulmonary arterial hypertension with occlusion of multiple small terminal arterial branches, producing a pruned-tree appearance. By contrast, in Figure 5.18 you see hypervascularity in a patient with a left-to-right cardiac shunt and marked enlargement of all the terminal branches. This is a result of the increased volume of blood passing through the lung. You could have predicted these changes from your knowledge of pathology and physiology, just as you can predict that the chest film of the patient in Figure 5.17 will show fewer vessels than normal, and that of the patient in Figure 5.18 an engorged pulmonary bed with many more vessels. Figure 5.19 shows multiple small pulmonary emboli blocking the terminal arteries.

Variations in the Pulmonary Microcirculation

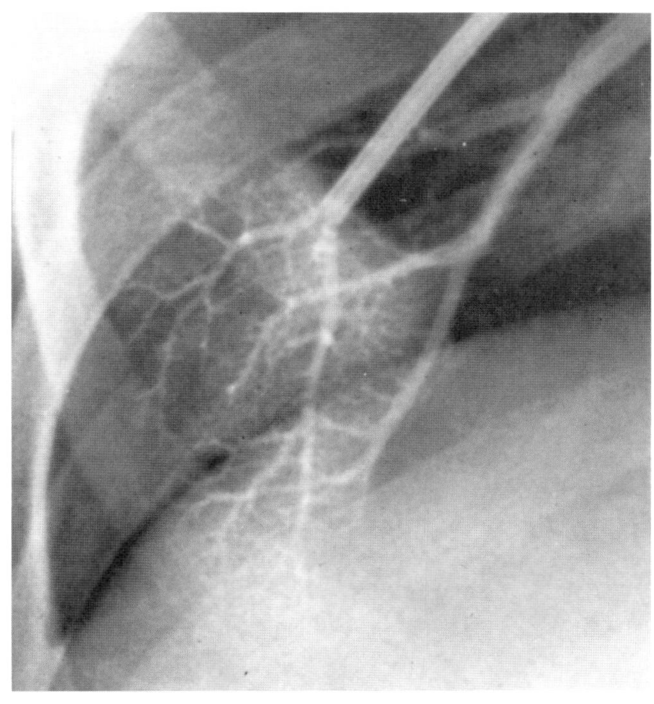

Figure 5.16. Normal wedge arteriogram showing the capillary bed in the lung.

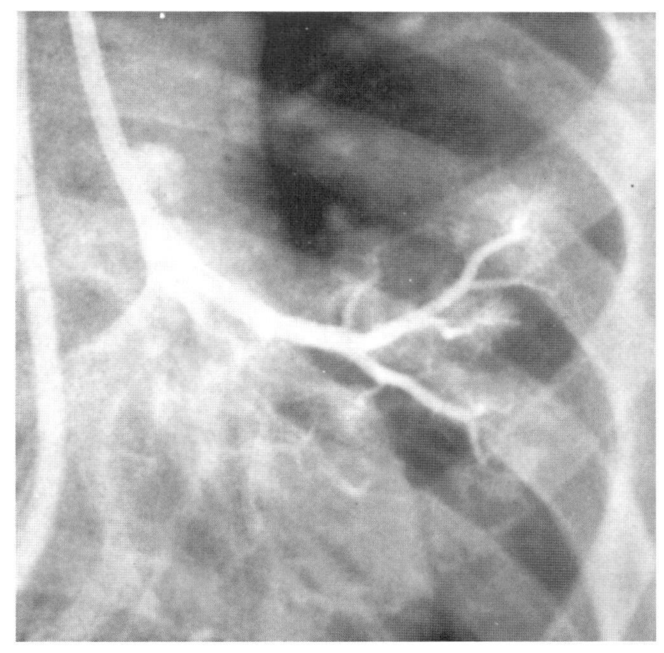

Figure 5.17. Patient with pulmonary arterial hypertension. From the narrowing of small end-arteries and sparse branching, this has been called the pruned-tree arteriogram. Pulmonary flow is reduced 50 percent.

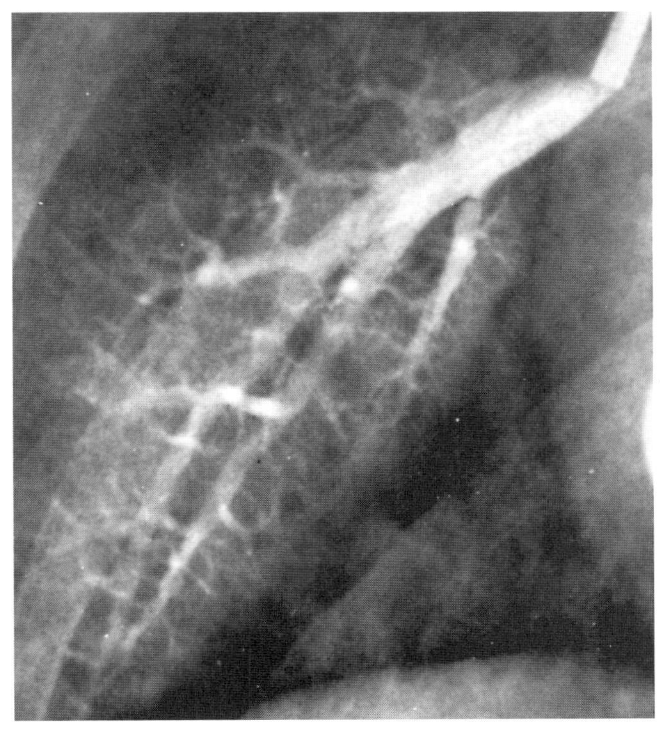

Figure 5.18. Wedge arteriogram of a patient with interatrial septal defect and a left-to-right shunt so extensive that the pulmonary flow was increased to 470 percent of the systemic flow.

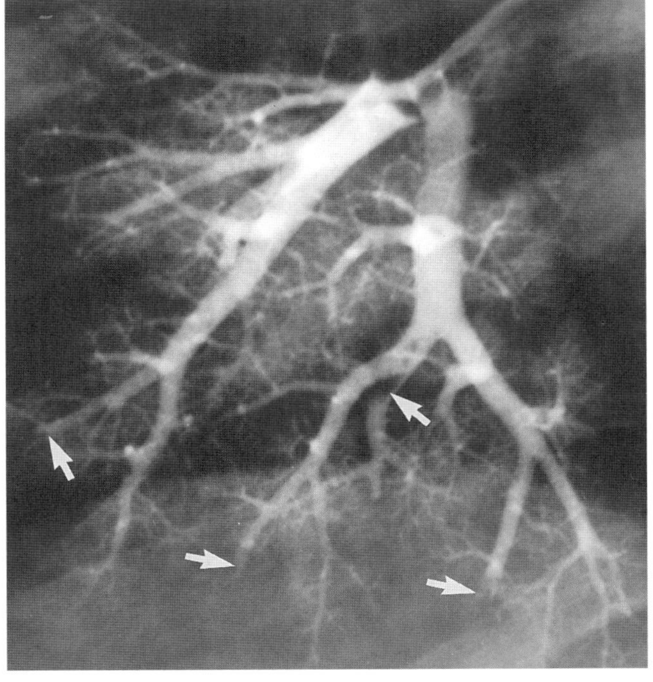

Figure 5.19. Balloon occlusion pulmonary arteriogram (BOPA) at the base shows multiple small emboli as intraluminal filling defects or sharp cutoffs *(arrows).*

Solitary and Disseminated Lesions in the Lung

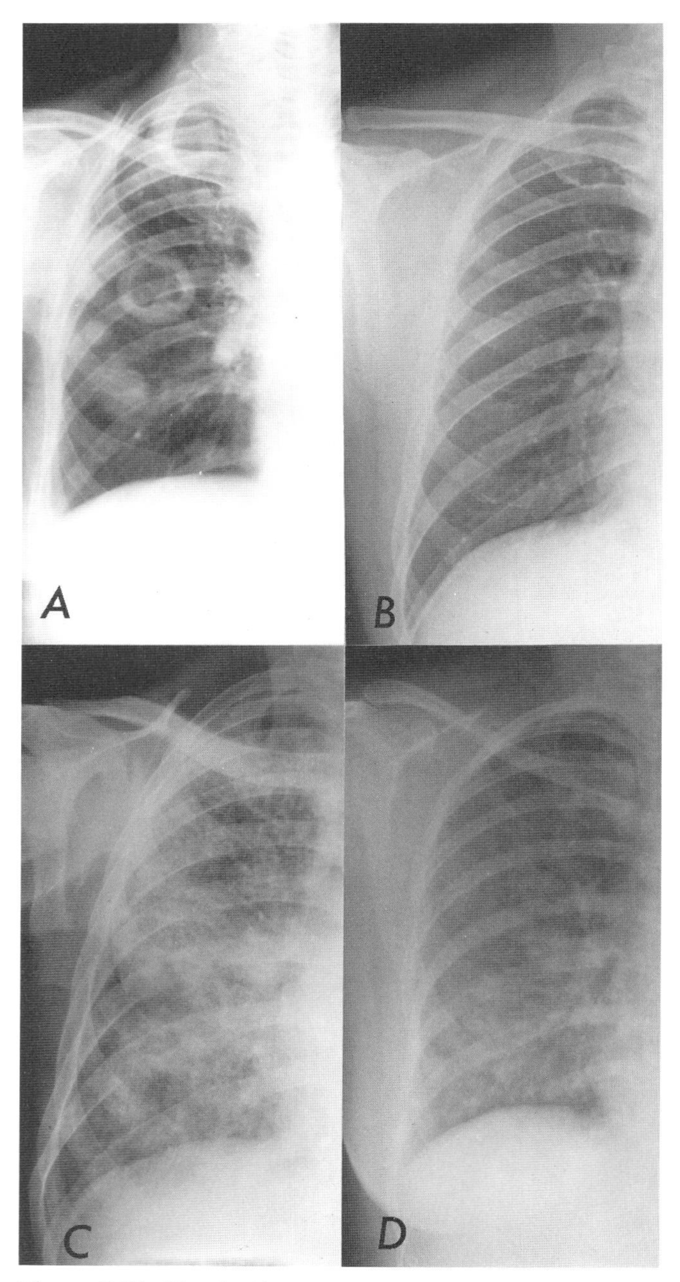

Figure 5.20. (See text.)

Imagine now the *alveolar portion of the lung* folded like a conical cuff around three sides of the hilum. You have looked at the peripheral lung when it was normal and seemed completely radiolucent in the lateral third of the PA film, where it is not superimposed on hilar trunks. What shadows will be added, then, if the vascular tree is normal but the alveolar lung is sprinkled with minute tumor nodules, patches of pneumonia, or small areas of atalectasis or collapse, or where it is threaded and reefed in by scar tissue from old infections, or flooded with interstitial fluid?

Look briefly over the eight lungs on these two pages and then come back to the text. Are any of them normal? Which ones seem to have changes in the lung so widespread that you think at once of some generalized process involving *all* lung tissue? Which show fewer than six isolated areas of abnormality?

Now consider them one by one. In A, several round shadows appear in an otherwise normal lung, their margins smooth and sharp, since they are surrounded by well-aerated lung tissue on all sides. One of them appears circular with a darker central area because it has a hollow, air-filled cavity inside it. (More x-rays pass through this part than through the shell tangentially.) The radiologist may not be able to tell you whether these are tumor masses growing in the lung with central cavitation or multiple abscesses with breakdown. But the radiologist *can* help you assess the probabilities for one or the other diagnosis on the basis of their appearance, their growth rate from film to film over a period of time, and the clinical history.

B, the right lung of the patient seen in E, is normal and can be used as a norm for studying the others.

Both C and D show innumerable patches of increased opacity, which on the original film involved both lungs. (G is the left half of D.) Unfortunately, many different conditions produce a picture similar to that shown on these two films. Some are common, others rare. From the film alone, without any knowledge of the acuteness of the patient's illness or of occupational background, or even of the tentative clinical diagnosis, the radiologist cannot arrive at the most probable diagnosis.

When you know, however, that the man in C had inhaled beryllium salts in a fluorescent-lamp factory over a period of time, you *can* say that his chest film shows shadows just like those seen in autopsy-proven cases of berylliosis, where myriad small granulomas and a lacework of scar tissue produce such roentgen shadows in the lung. And if you know that the woman in D and G was pulled out of the water several hours before being x-rayed, half drowned, her film becomes intelligible because this is a picture often seen after such mishaps. Many small areas of collapse from inhaled water and bronchial secretions produce this sort of patchy opacity. In addition, from the violent struggle in the water there is usually some pulmonary edema, with extra fluid in the interstitium of the lung about the vessels, and extensive hemorrhage.

Any chest film is only a point on a curve in the course of the patient's disease. *Change* from film to film in a day or a week or a year often alters the whole spectrum of diagnostic possibilities considered on viewing the original film. You may still not be able to be sure what the patient has, but you can then be sure of a good many things that he or she has not. To know that the man in C had shown the changes you see there for several months before this film was made, and that his lung picture did not change appreciably before he died, would strongly affirm your conviction that he had a chronic lung injury, probably related to his known industrial exposure. The half-drowned woman got well in a few days, and you could predict she would. In fact, H shows her left lung two days after G. Slightly enlarged vessel shadows seem the only remaining abnormality.

In case you have not been able to find anything wrong with E, look again at the ninth interspace. This solitary nodule had not changed since a chest film one year before, and at surgery it proved to be a benign tumor. In F the entire lung is sprinkled with minute areas of an opacity rightly suggesting calcium and representing the healed scars of an old infection with varicella pneumonia unchanged for many years. It has been estimated that such lesions must be at least 2 millimeters in size to be visible by x-ray.

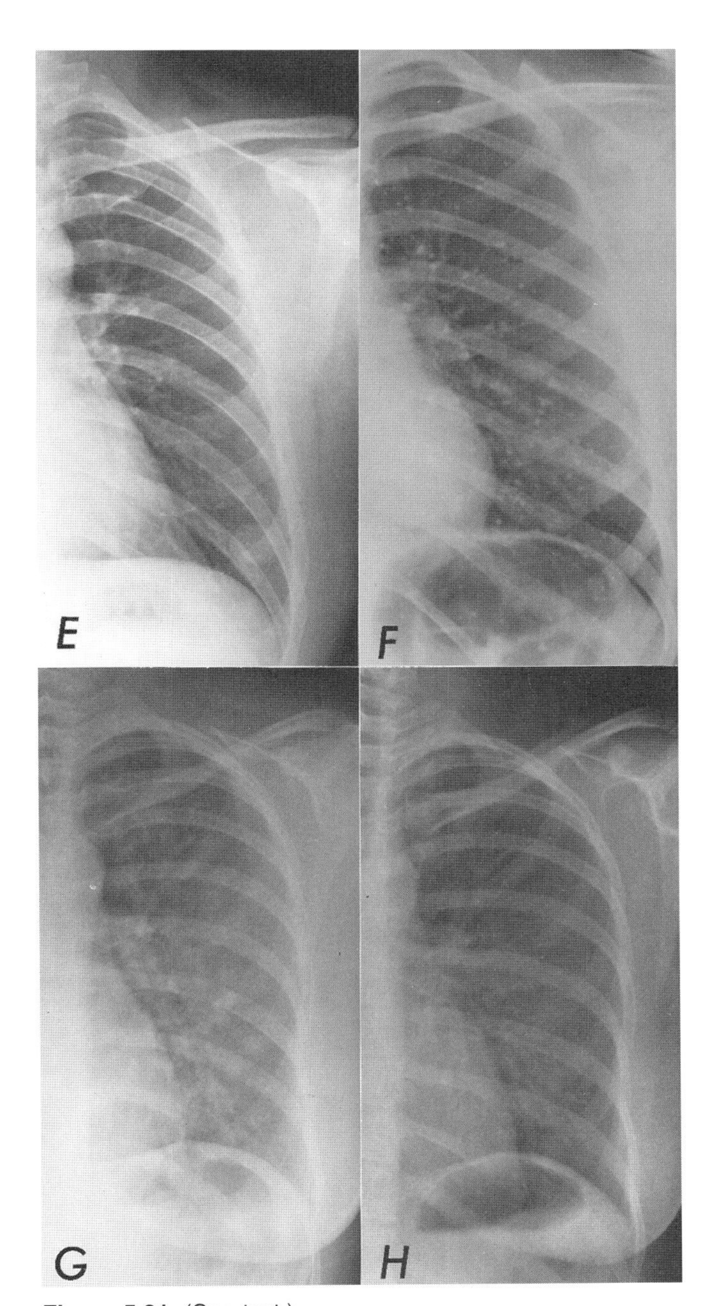

Figure 5.21. (See text.)

Air-Space and Interstitial Disease

Two basic patterns of disease in the lung do differ significantly in their roentgen appearance. *Air-space disease* involves the alveoli, which fill with fluid or exudate that displaces the air in them. These areas of alveolar filling (whether large and single, multiple or coalescing as the disease progresses) appear white and radiopaque on the chest film. Various kinds of pneumonia do show air-space disease, but it is by no means pathognomonic as a picture of pneumonia, and many other disease processes can produce this picture.

Interstitial disease is distributed through lung tissue that is otherwise well aerated. Any interstitial process will ultimately produce either linear strands of density or spherical densities, small or large, seen to be superimposed upon the normal radiating pattern of vessel trunks. You have already seen some examples of interstitial disease (Figures 5.20 and 5.21) in this chapter, and in the next chapter you will be studying air-space disease as an illustration of lung anatomy.

Unfortunately, neither process can be interpreted as a pathognomonic finding independent of clinical information about the patient, and positive diagnosis usually depends on other modes of investigation. Taken together with the patient's history, physical findings, and laboratory data, confirmatory radiological findings will help you arrive at a presumptive or working diagnosis.

Beware of leaning too heavily on your decision that the chest film shows either air-space or interstitial roentgen findings, because in a number of pathological states they may coexist. For example, in cardiac failure with or without pulmonary edema, the abnormal opacities on the chest film are initially produced by the presence of interstitial fluid surrounding the vascular trunks and advancing along interstitial planes into the lung. In time this fluid spills over into the alveoli and produces clusters of opacities that gradually coalesce. Thus in cardiac failure interstitial and air-space disease are often seen together.

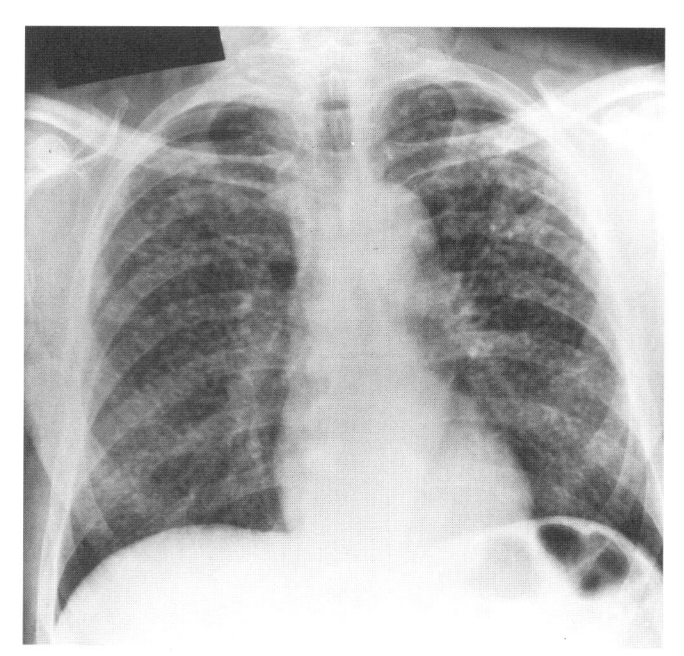

Figure 5.22

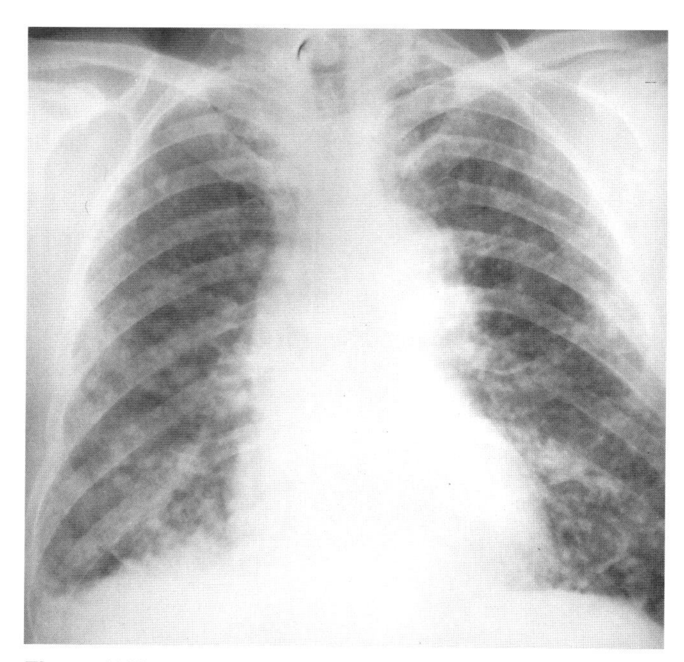

Figure 5.23

Figure 5.24

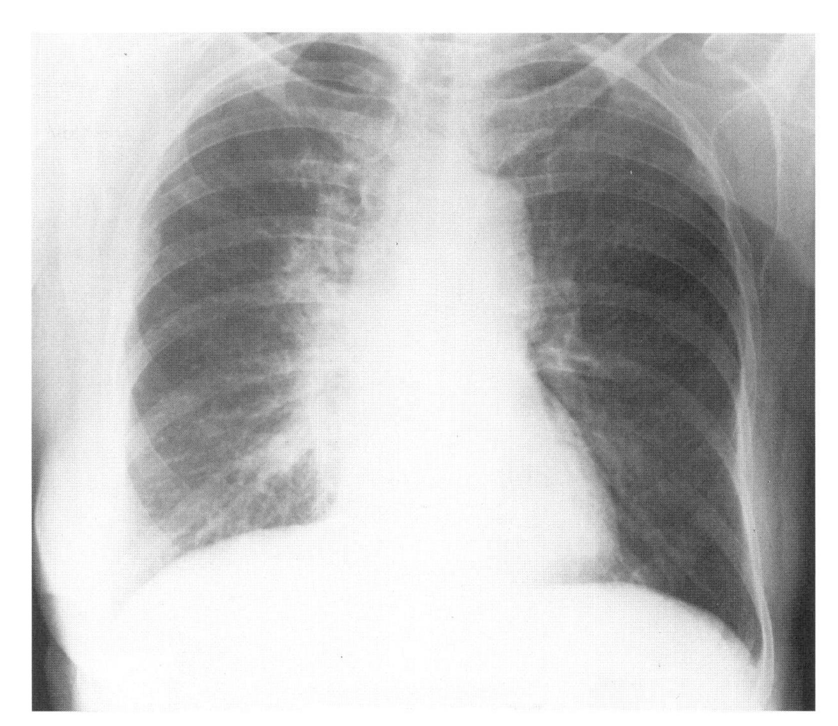

Figure 5.25

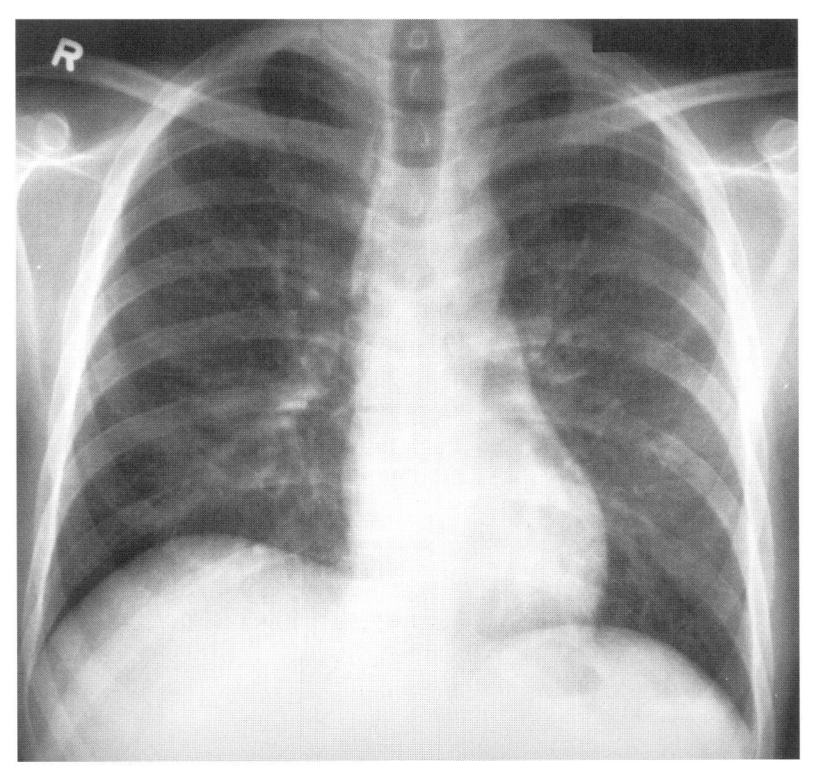

Remember that *bilateral disseminated interstitial disease* is a very nonspecific pattern on the chest film. In fact, there are over a hundred disease conditions that may produce it. Often the picture and the clinical data together add up to a high probability that one specific disease process is involved. The radiologist needs as much relevant clinical information as you can provide on the requisition for a chest film.

Now try to decide on a most probable diagnosis for the figures on this page spread *before you go on to the next page*.

The Importance of Clinical Findings

The clinical history in patients with *bilateral disseminated interstitial disease* helps to narrow the differential diagnosis to a few probable entities. All four of the patients on the preceding page spread have bilateral disseminated interstitial disease in their lungs. While they do show differences, none has a pathognomonic pattern. Figures 5.22 and 5.25 show similar streaky linear lesions; Figure 5.24 also has linear opacities, mainly extending outward from the hila, and Figure 5.23 shows more nodular interstitial lesions. None of this helps you much toward a diagnosis, and the radiologist interpreting the films can only describe the findings.

When you inform the radiologist, however, that the man in Figure 5.22 had worked in a silicon-laden atmosphere without respiratory protection for years, silicosis becomes by far the most likely explanation.

You ought to have informed the radiologist on your requisition for the patient in Figure 5.23 that he has been treated for several years for carcinoma of the prostate. When you know that (and that earlier chest films were normal), metastases to the lung become the almost certain explanation.

There *is* a clue on the film of the patient in Figure 5.24: the left breast is absent. This radiating perihilar pattern is sometimes seen in lymphangitic spread of breast cancer from the mediastinum outward into the lung, choking the lymphatics with tumor. It is *not* a pathognomonic pattern, since carcinoma of either the stomach or the pancreas can produce the same findings by extension to the mediastinum first and then outward along the lymphatics into the lung. These patients often present with sudden extreme dyspnea. This was the presenting complaint of the patient in Figure 5.24, and although she had her mastectomy five years ago, she has no symptoms suggesting either stomach or pancreatic cancer, so that lymphangitic spread from breast cancer becomes the most probable diagnosis.

Finally, the young man in Figure 5.25 also has bilateral disseminated interstitial disease. He has no industrial history of importance, since he is a salesman in a brokerage office. But he is acutely ill with cough and fever. When we learn, however, that he is in a high-risk group for exposure to the human immunodeficiency virus (HIV), the diagnosis of *Pneumocystis carinii* pneumonia occurring with acquired immune deficiency syndrome (AIDS) then becomes very probable. That diagnosis was made presumptively and confirmed clinically.

Remember to include *all* the clinical information you have when you write out the requisition, so that the radiologist can best advise you.

High-Resolution CT of the Lung

Computed tomography of the lung itself (lung parenchyma) has proven to be a most useful addition to the plain film diagnosis of diffuse lung disease. Because of CT's resolution for fine detail, and its ability to sort out otherwise superimposed parenchymal structures, CT may show evidence of definite lung abnormality when the plain films of the chest are normal or equivocal; when the plain films are abnormal, CT may assist the radiologist in making a more specific diagnosis of the patient's exact lung condition.

The CT technique used to examine the lung is called *high-resolution CT,* or *HRCT.* Very thin slices (from 1.0 millimeters to 1.5 millimeters in thickness) are obtained, after the patient has taken a deep breath. Figure 5.26 shows a normal example. Note that HRCT yields images that are vastly superior in visualization to the images provided by plain films of the bronchi and blood vessels, the interstitial connective tissue, and the air spaces. Figure 5.27 reproduces the chest film and HRCT of a woman with ideopathic interstitial fibrosis. Note the thickening of the bronchial walls and interstitial opacities on the CT scans, which produce the interstitial patterns on her chest film.

Four different lung patterns are better seen with HRCT than with plain films. *Reticular opacities* are produced by conditions that thicken the interstitial fiber network with fluid, fibrous tissue, inflammatory cells, or tumor cells. A common appearance at CT is thickening of the tissues around airways and blood vessels, called bronchovascular thickening. *Nodular opacities,* especially small ones, are also better shown by HRCT. These may represent inflammatory diseases, such as sarcoidosis and TB, or metastatic disease. *Pulmonary consolidations* that opacify air spaces can also be detected earlier with HRCT than with plain films. A common CT appearance is that of

Figure 5.26. Normal high-resolution lung CT scans.

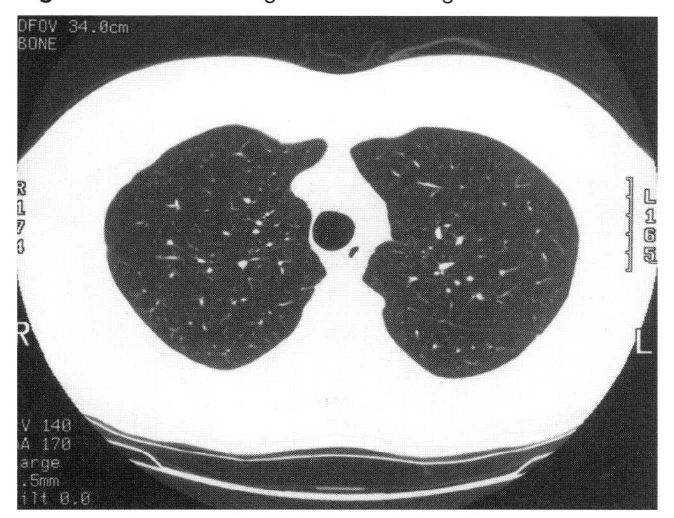

A. Upper lung.

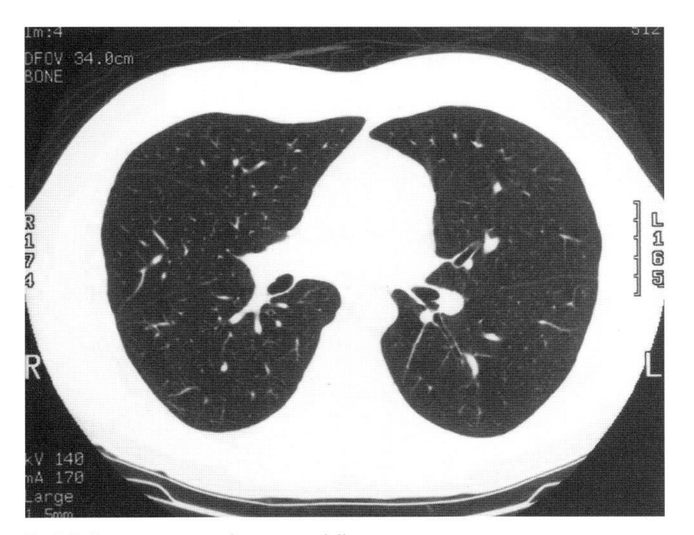

B. Midlung near pulmonary hila.

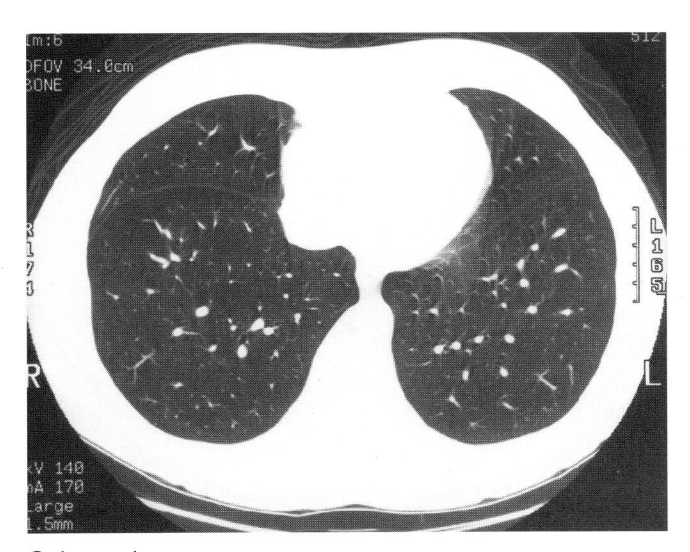

C. Lower lung.

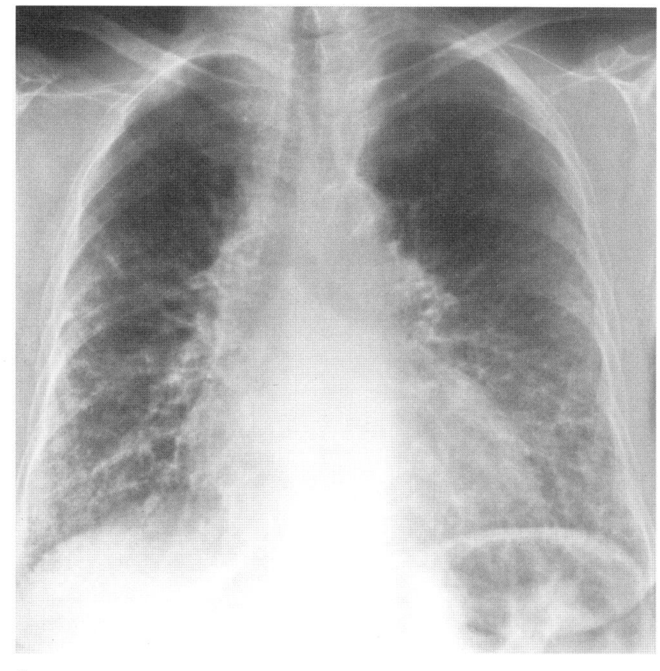

A

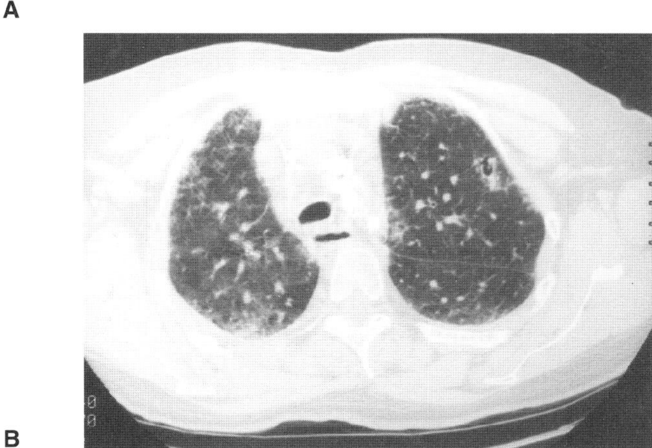

B

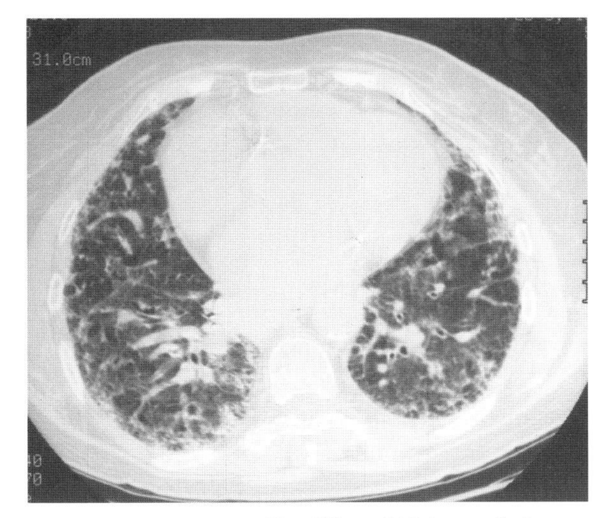

C

Figure 5.27. Abnormal chest film (A) and high-resolution CT scans (B and C) of a woman with idiopathic pulmonary fibrosis.

"ground-glass" opacities that produce a hazy increase in lung density without obscuring the blood vessels. Ground-glass opacities are very important to recognize because they may represent ongoing, acute conditions such as pneumonias and pulmonary edema. *Focal areas of decreased lung opacity* may be seen in patients with air trapping, or in patients who have suffered lung destruction as the result of lung disease. Emphysema is characterized by abnormal enlargement of the distal air spaces, often associated with destruction of the air-space walls. Lung cysts may be seen that measure 1.0 centimeter or more in diameter, with walls less than 3.0 millimeters thick.

Figures 5.28 to 5.31. High-resolution CT examinations of four patients with diffuse lung disease.

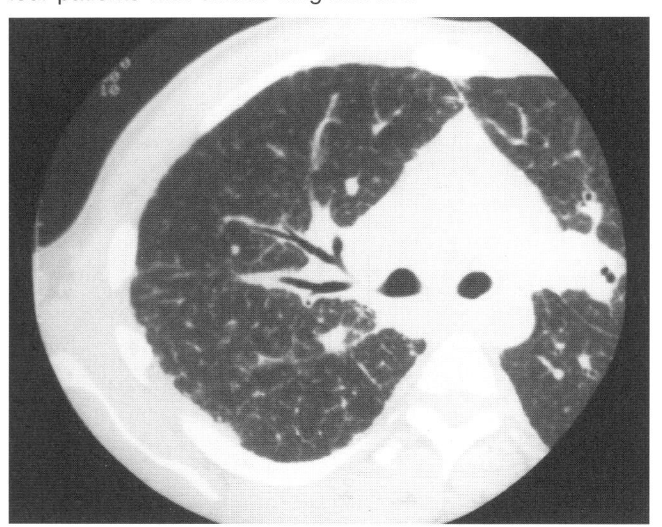

A

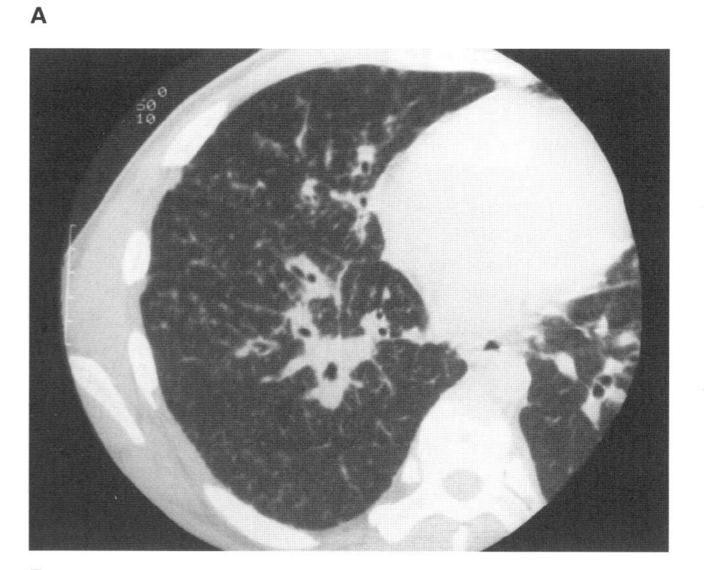

B

Figure 5.28. Exam 1.

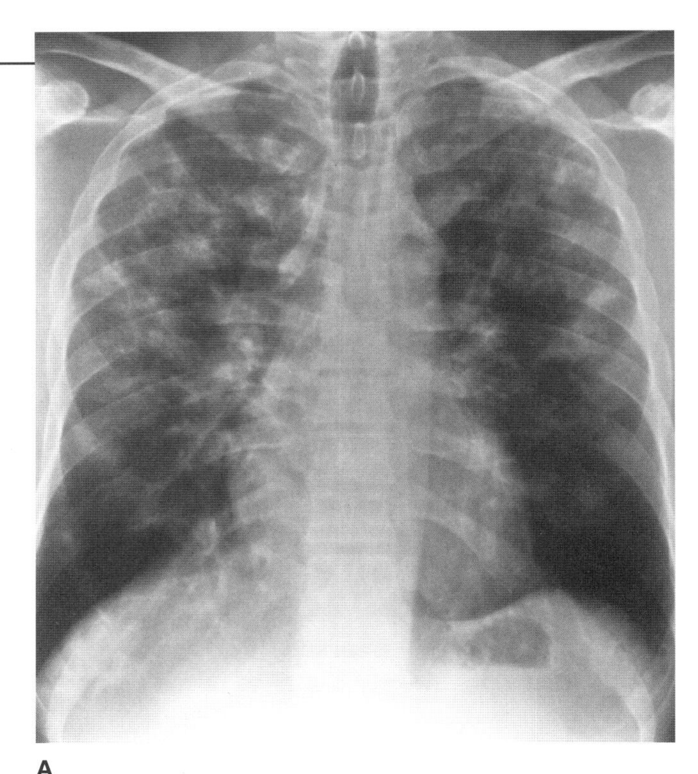

A

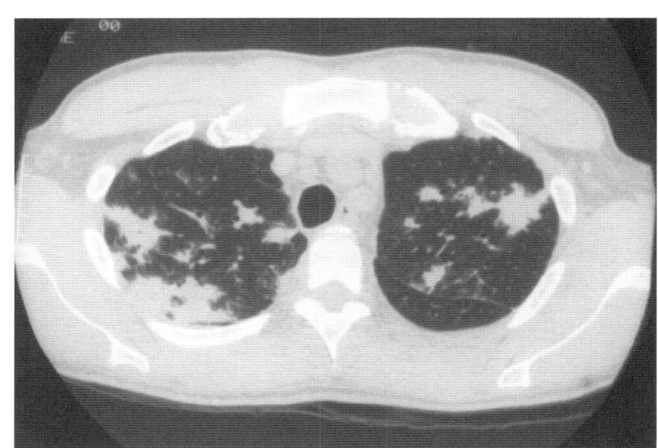

B

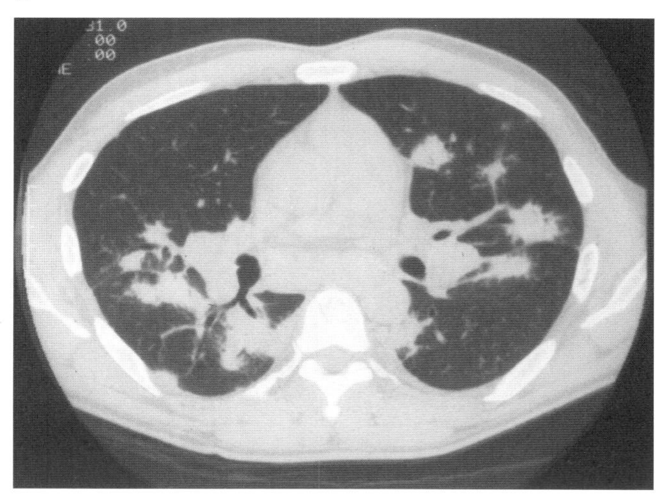

C

Figure 5.29. Exam 2.

Review the HRCT scans of four patients with diffuse lung disease shown in Figures 5.28–5.31 and compare them with the normal HRCT. Scans at two different levels are reproduced for each patient, and a plain film is also shown for the patients in Figures 5.29 and 5.31. One patient has an acute pneumonia with ground-glass opacities; one has reticular opacities characterized by marked peribronchial thickening; one has nodular opacities with hilar adenopathy; and one has end-stage lung disease with fibrosis and focal areas of decreased lung opacity (areas of lung destruction and lung cysts). Decide which CT examination matches each of these four patients *before* turning the page.

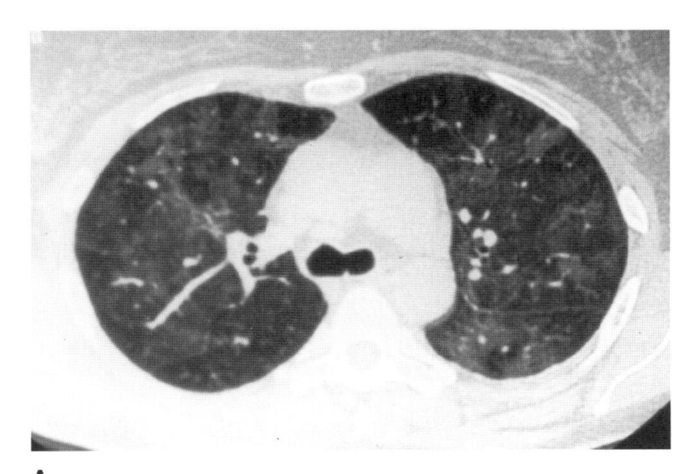

A

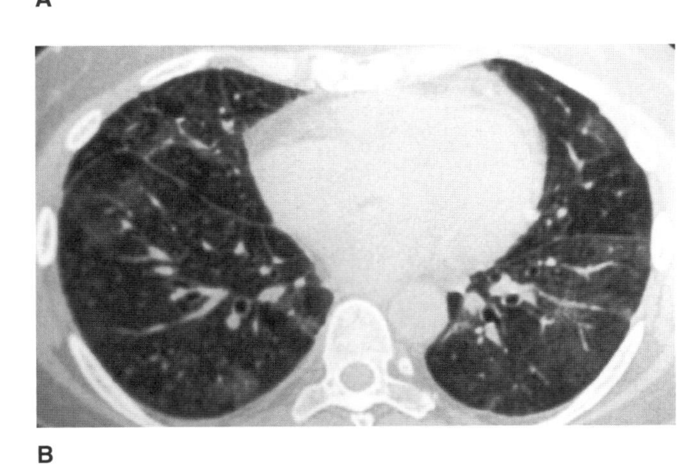

B

Figure 5.30. Exam 3.

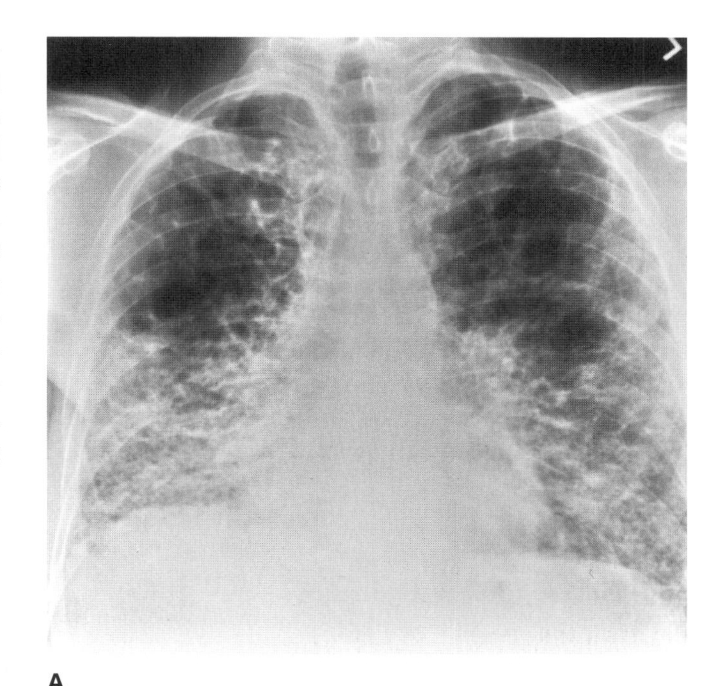

A

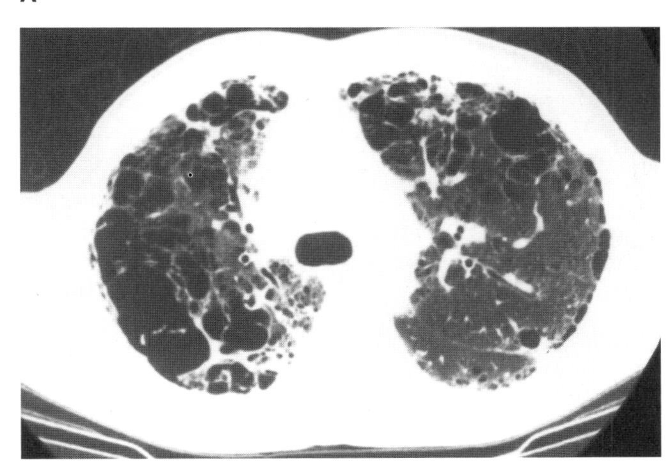

B

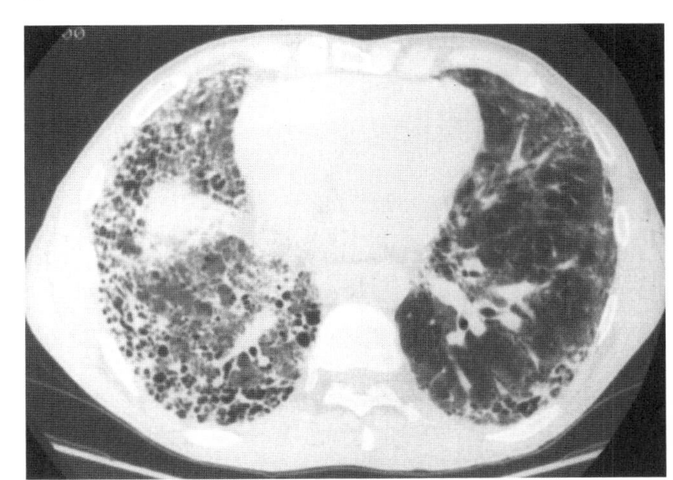

C

Figure 5.31. Exam 4.

CT exam 1 (Figure 5.28) belongs to the patient with reticular opacities characterized by marked peribronchial thickening. Compare the increased width of the bronchial walls with the normal HRCT, shown in Figure 5.26. This patient proved to have sarcoid, characterized by an interstitial pattern on plain chest films.

CT exam 2 belongs to the patient with nodular opacities and hilar adenopathy. Figure 5.29A is the PA chest film of the patient. The nodules are much better seen on the CT exam (Figures 5.29B and C) because they are not superimposed on one another as on the plain chest film. This patient had metastatic disease.

CT exam 3 belongs to the patient with an acute pneumonia characterized by ground-glass pulmonary opacities. Note the ill-defined gray opacities scattered throughout both lungs (Figures 5.30A and B). These areas represent inflammatory fluid and cells within air spaces; they appear gray instead of black because the density of the affected air spaces is increased by the presence of inflammatory fluid. This patient proved to have pneumocystis carinii pneumonia associated with AIDS. Your clinical suspicion for pneumocystis pneumonia should be high in patients who are at high risk for HIV exposure, because their chest films may be normal or almost normal when this pneumonia is present. A much more dramatic picture can be obtained with an HRCT.

And finally, CT exam 4 (Figure 5.31) belongs to the patient with end-stage lung disease characterized by extensive fibrosis and lung destruction. As also shown on the patient's PA chest film (Figure 5.31A), areas of interstitial abnormality (reticular opacities) are interspersed with areas of decreased lung opacification (areas of lung destruction and lung cysts). This patient has been chronically ill with idiopathic pulmonary fibrosis for many years.

Now that you have seen examples of CT's superior ability to image pulmonary parenchymal abnormalities, you may wonder how often you should be requesting this examination. Obviously HRCT should be requested in cases of patients with normal or nearly normal chest films whose clinical findings are compatible with pulmonary disease. In addition, HRCT is helpful when the chest film is abnormal but further characterization of the exact pulmonary pattern is needed for diagnosis.

Problems

Unknown 5.1 (Figure 5.32)

A young man known to be an intravenous drug abuser (heroin) is admitted with high fever, sweats, and coughing. What are your conclusions?

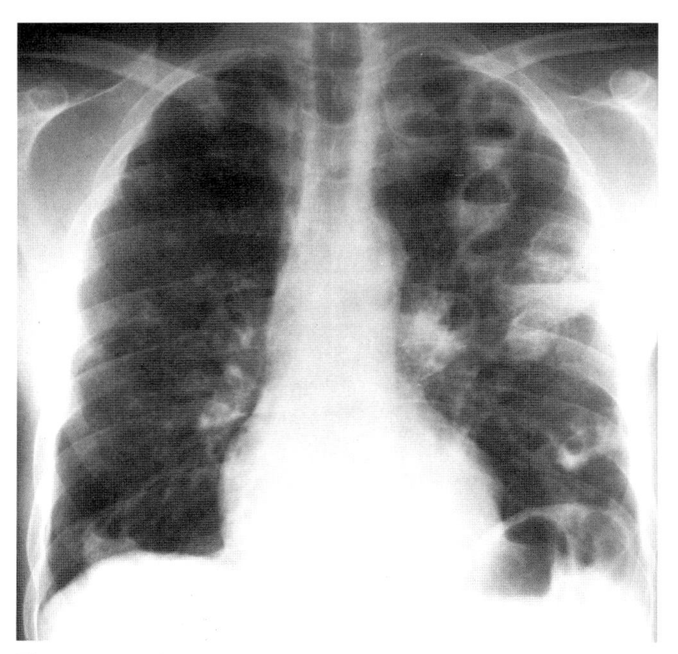

Figure 5.32 ***(Unknown 5.1)***

Unknown 5.2 (Figure 5.33)

A 65-year-old cigarette-smoking retired naval officer seeks medical help for exertional dyspnea. What conditions are you considering?

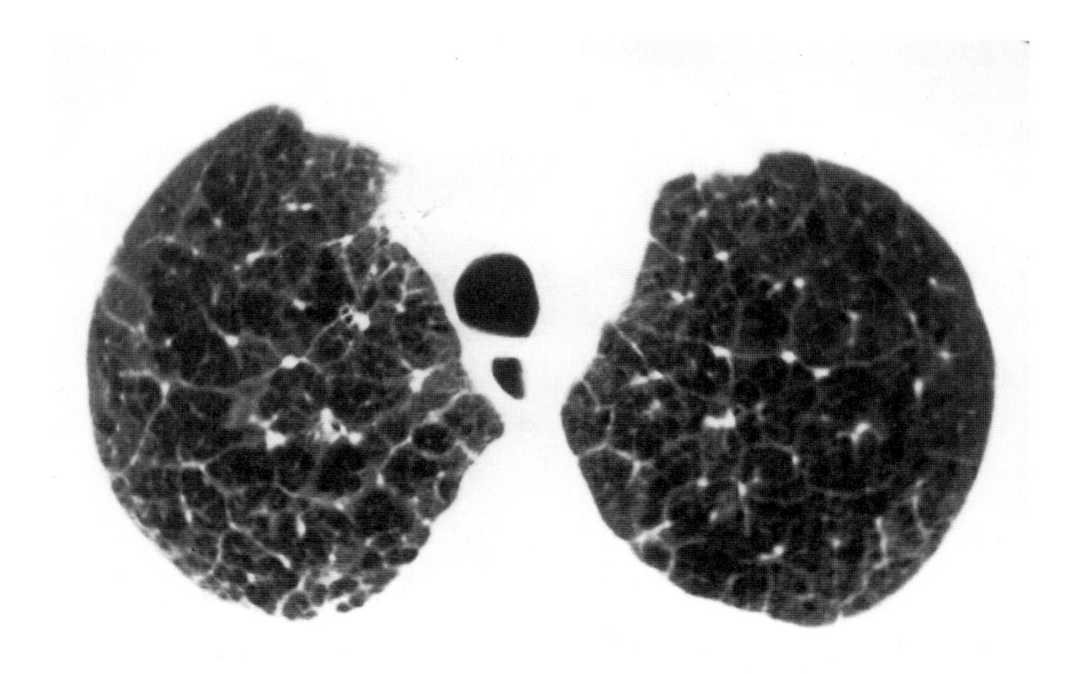

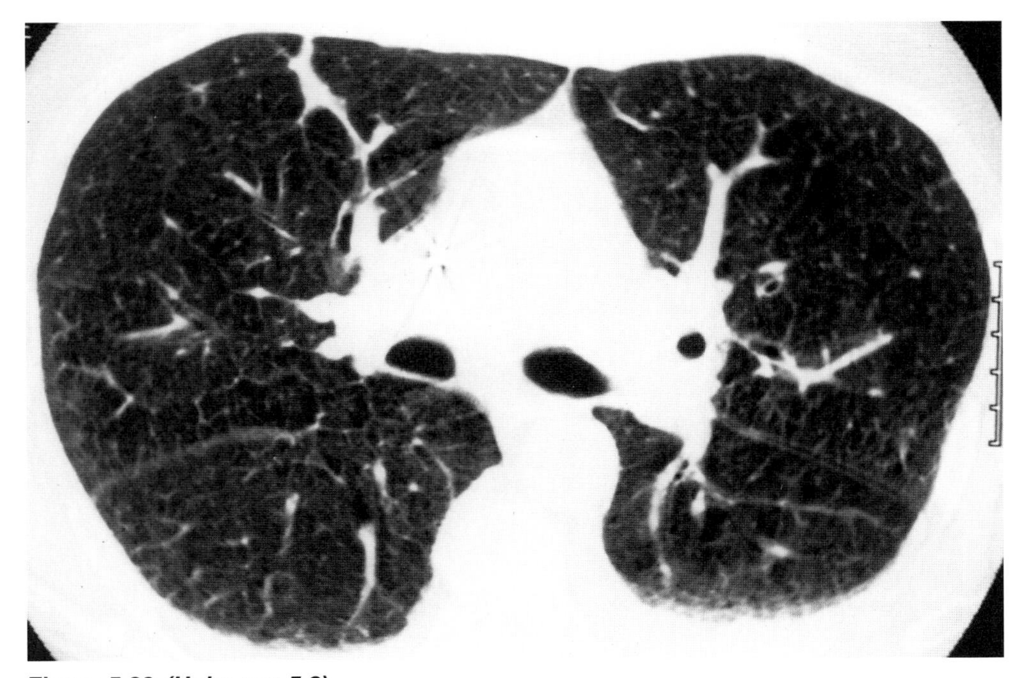

Figure 5.33 ***(Unknown 5.2)***

6 Lung Consolidations and Pulmonary Nodules

Consolidation of a Whole Lung

By massive lung consolidation we mean, for practical purposes, that a whole lung, a whole lobe, or at least one entire bronchopulmonary segment is solid in that it is almost entirely airless. The solid part will cast a uniform shadow of approximately the same density as the heart shadow, and its projection shadow will relate to the shape of the part involved. Although this sounds like a theoretical situation, the fact is that in everyday radiological practice shadows of this kind are common. Hardly a day goes by in a big general hospital without there turning up, for example, a radiograph in which, through a logical analysis of the abnormal shadows on the chest film, one can recognize a consolidated lobe in a patient with clinical lobar pneumonia. Likewise a shadow that can only represent a solid right upper lobe may be recognized in a patient already suspected of having lung cancer. To teach you to understand first the roentgen appearance of whole-lung consolidation and then that of consolidation of only one lobe is the main purpose of this chapter.

In Figure 6.1 we have diagrammed for you the roentgen findings you must anticipate when one or the other *whole lung* becomes solid but does not change in size or shape. Begin by noticing that in A the normal heart shadow *(white)* is thrown into relief by the normally aerated lung *(black)* on either side of it. So also the two domed diaphragmatic shadows covering the liver and spleen are seen in relief because there is air in the lung above them. The stomach bubble under the medial half of the left hemidiaphragm may be seen in the upright patient as the shadow of radiolucent air imprisoned in the fundus of the stomach above a horizontal fluid level.

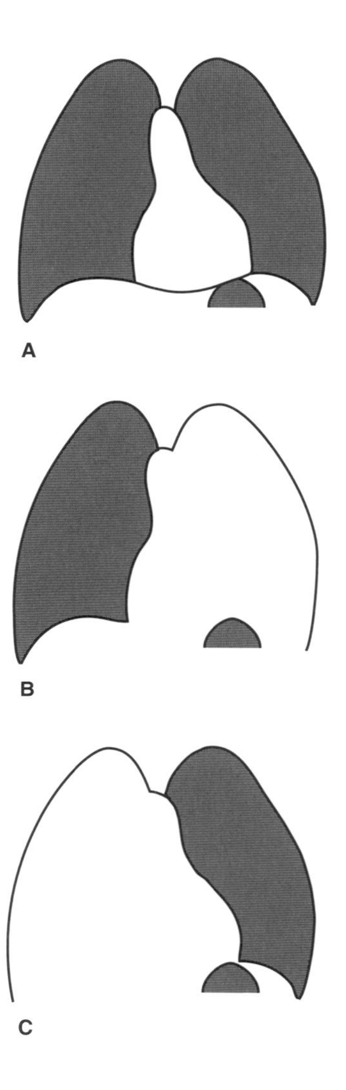

Figure 6.1

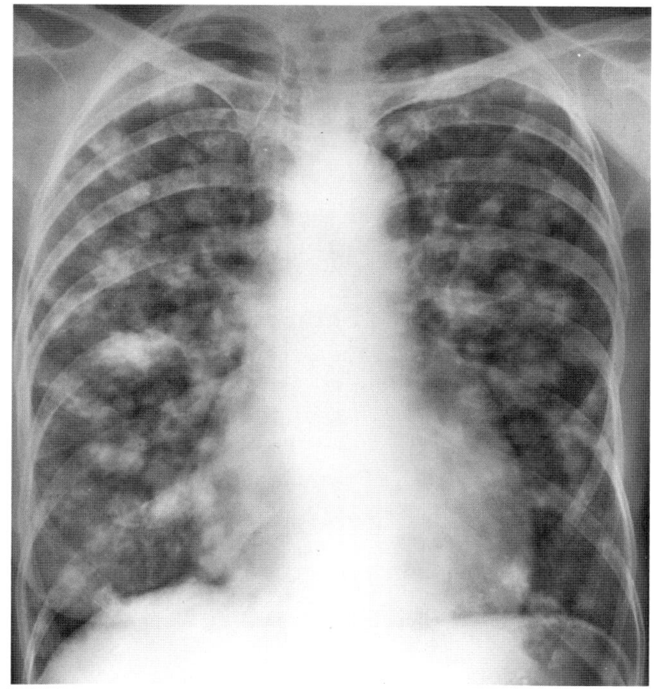

Figure 6.2

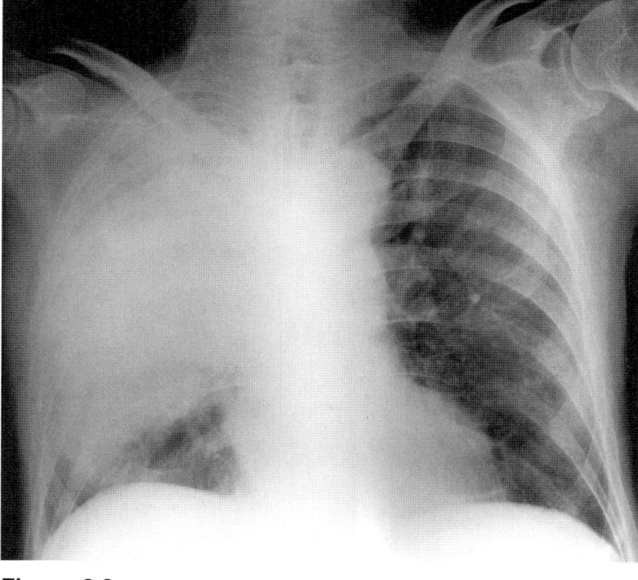

Figure 6.3

Now suppose that the entire left lung becomes consolidated, as in B. The heart, mediastinal structures, and dense lung are all of the same density *(white)* and their shadows merge into one, so that the left side of the heart profile disappears. They also merge with the shadows of the spleen and left lobe of the liver, and the outline of the left diaphragm is therefore lost, its location indicated only by the rays that reach the film through the air in the stomach. Look back at the cadaver sections in Chapter 2 to check these relations of stomach, spleen, and diaphragm.

If the left lung remains normal and the right solidifies, the chest film will look like C. The liver, right lung, and heart being nearly identical in density, their shadows now merge.

Disappearance of profiles or interfaces normally seen on a chest film of this sort implies solid change in the lung next to them because the usual air-solid roentgen interface no longer exists. It is interesting to realize that even a lung that is riddled with small disseminated nodules of tumor will still contain enough air to behave like a well-aerated lung with respect to profiles. The patient in Figure 6.2 proved at autopsy to have both lungs generously sprinkled with metastatic interstitial tumor nodules; yet you do see the heart shadow and both hemidiaphragms because of the air in the alveoli around the disseminated lesions.

Contrast with that film Figure 6.3, where pneumonia consolidating the entire upper part of the right lung results in loss of the upper part of the mediastinal and heart shadows. But the shadow of the right hemidiaphragm and the lower right heart is preserved.

Any consolidation against the mediastinum will result in loss of a part of the mediastinal border, therefore, and any consolidation of the base of the lung will erase the shadow of the diaphragm or a segment of it. Because the heart is in the anterior half of the chest, consolidation that erases the border of the heart must, of course, be located in the anterior part of the lung. Thus you will not be surprised the first time you observe for yourself that although the diaphragmatic shadow on one side is absent and the lower part of the lung on that side appears dense, the border of the heart is seen clearly through it, thrown into relief by the juxtaposed air-filled *anterior* lung. When you see this, you will reason accurately that the *lower* lobe, in contact with the diaphragm, is solid, whereas the rest of the lung is normal.

Consolidation of One Lobe

In the drawings of seemingly transparent lung on this page you can see exactly why a solid lower lobe erases the shadow of the diaphragm and how the upper lobes, which do not touch the diaphragm, apply themselves, full of air, against the heart in the anterior part of the chest and preserve its profile on the PA film. You can also see at once that, with consolidation of the upper and middle lobes on the right, the right heart border would disappear but the profile of the diaphragm would be preserved by the well-aerated lower lobe.

The oblique planes of the two major fissures are important to remember, for their location will often be visible to you in the lateral view and bear a significant relationship to an area of abnormality. The fissures normally contain two layers of visceral pleura in contact and are seen in the normal chest film only when the pleura, x-rayed tangentially, appears as a thin line of density outlined on both sides by lung. The *minor fissure* on the right should be thought of as roughly horizontal, extending forward and laterally from the middle of the major fissure to form the floor of the right upper lobe and the roof of the right middle lobe. On the PA view it is frequently seen as a thin line extending straight laterally from the hilum.

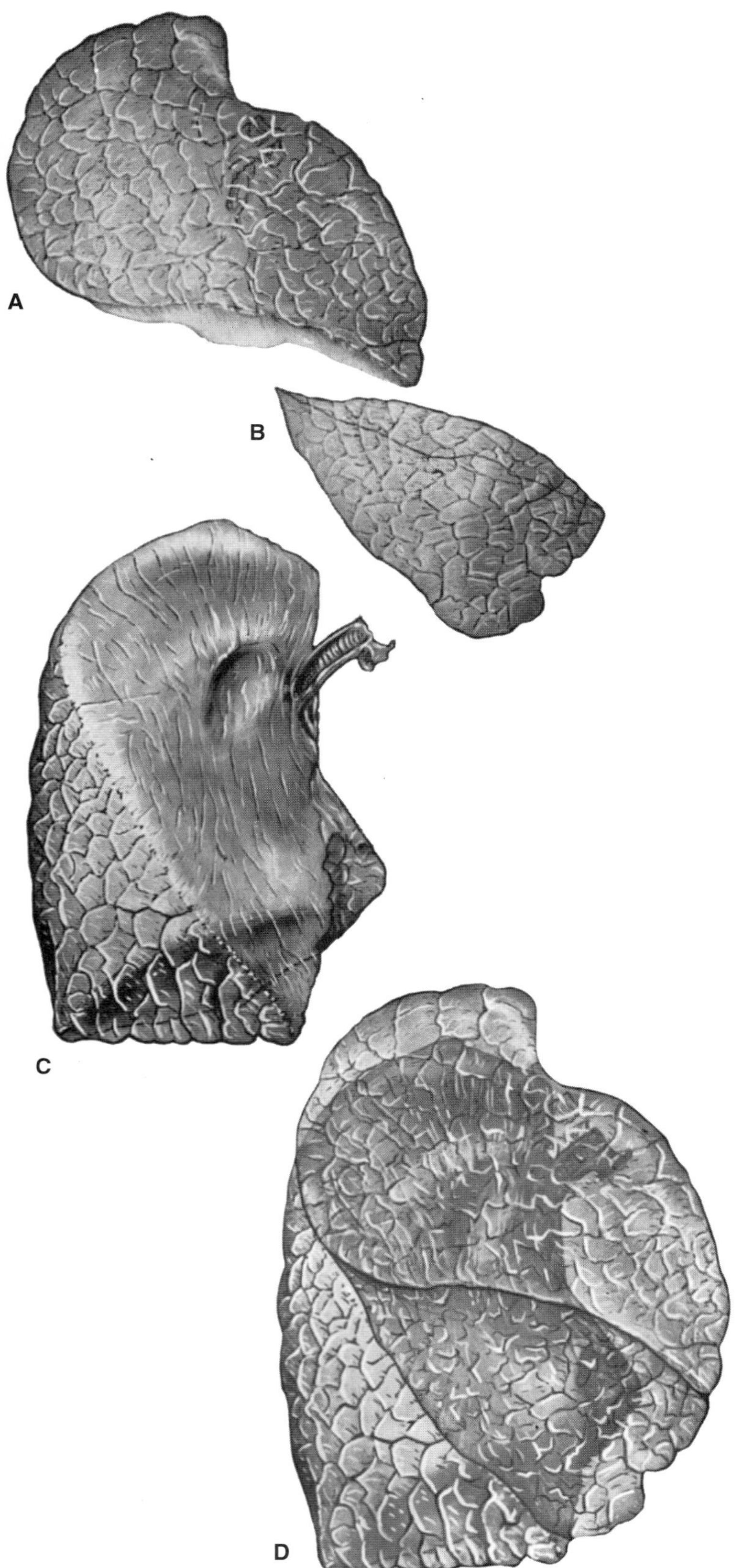

Figure 6.4. Transparent drawings of the right lung, seen from the lateral surface, separated into upper, middle, and lower lobes (A, B, C) and reassembled (D). The patient faces to your right.

Facing page: Transparent drawings of individual segments of the right lung.

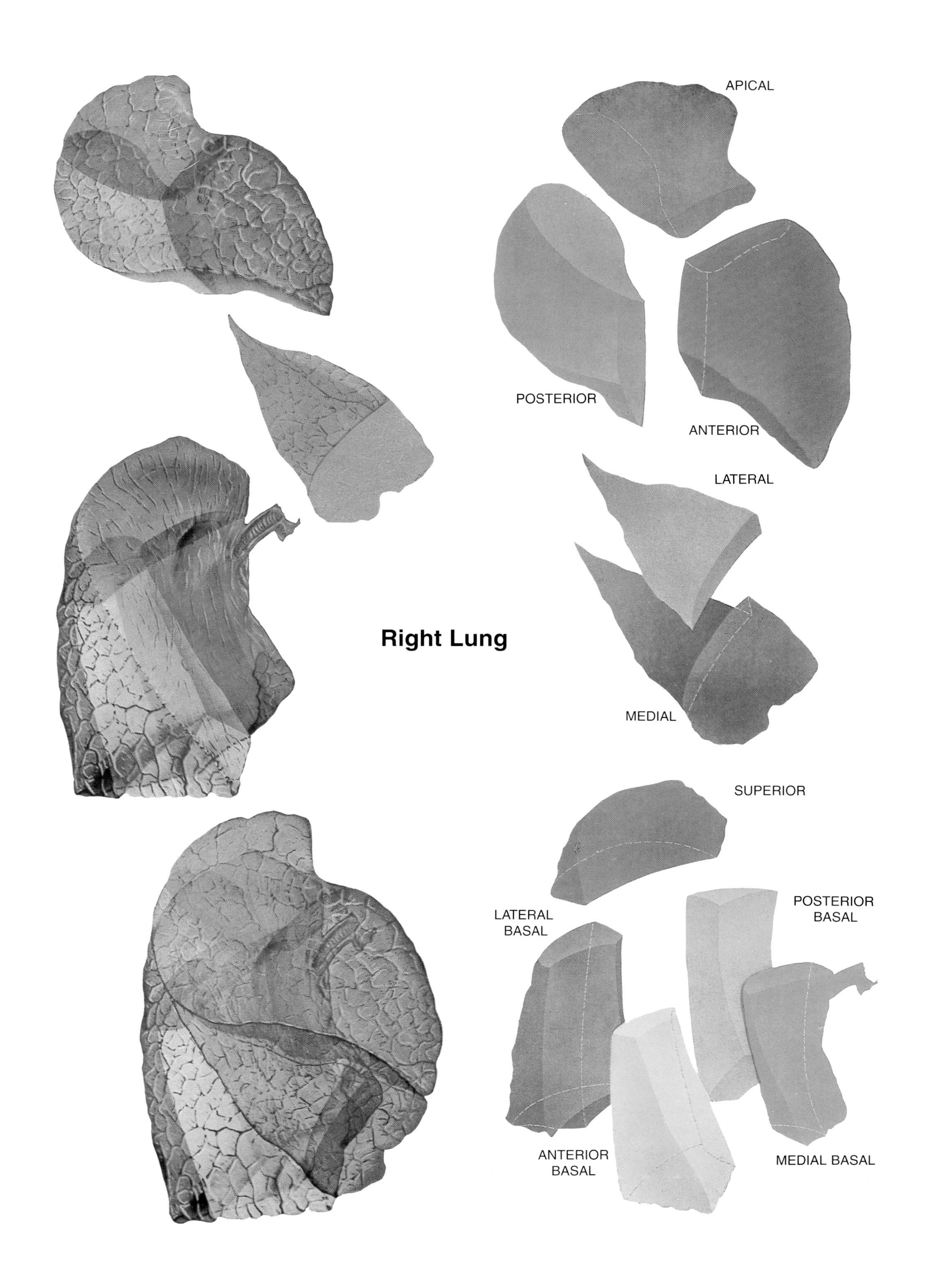
APICAL
POSTERIOR
ANTERIOR
LATERAL
Right Lung
MEDIAL
SUPERIOR
LATERAL
BASAL
POSTERIOR
BASAL
ANTERIOR
BASAL
MEDIAL BASAL

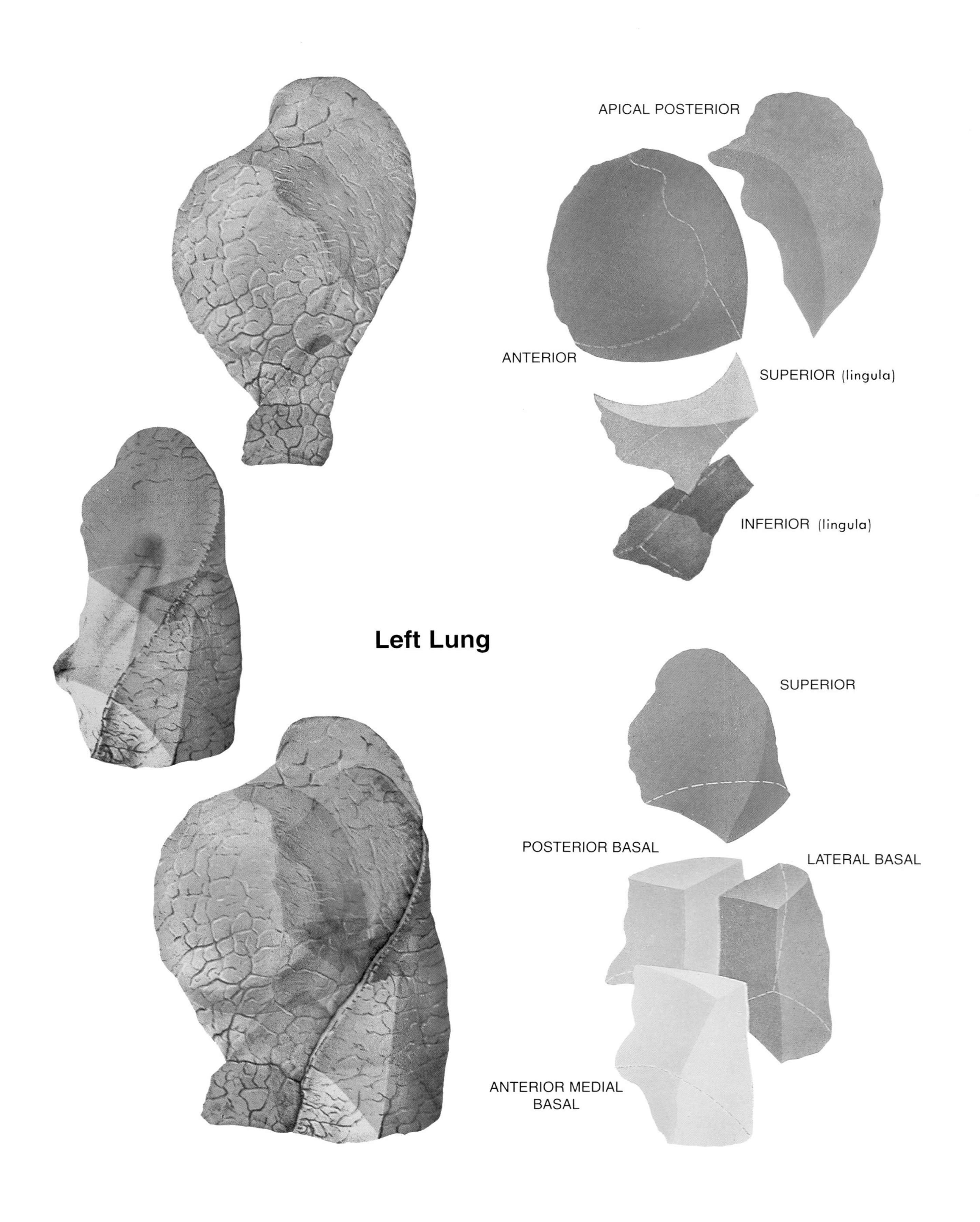
APICAL POSTERIOR
ANTERIOR
SUPERIOR (lingula)
INFERIOR (lingula)
Left Lung
SUPERIOR
POSTERIOR BASAL
LATERAL BASAL
ANTERIOR MEDIAL BASAL

Exercise

Take pencil and paper and reproduce the outline of Figure 6.1A several times. Make diagrams predicting the block of density you would expect to see on the PA film if each of the five lobes became consolidated. Then try to predict the appearance of the block of density you would see on the *lateral view* to go with each PA. This will be easier if you begin with the right upper lobe, producing a density extending from the horizontal plane of the minor fissure upward to the apex. Work out for yourself these predictable shadow profiles, and check them, so that you will never forget them. As you reason out each one, note which borders of the heart, diaphragm, and mediastinum can be expected to disappear with each block of density.

Remember that a dense sphere within the lung will project as a circular shadow on either the PA or the lateral view, but that *asymmetrical wedges* of different sorts will project quite differently according to the direction of the ray passing through them. The middle lobe best illustrates this point, since it is a long wedge x-rayed end-on in the PA view and appears as a much smaller shadow than it does when its full length is seen in the lateral view of the chest. The shadow profile in each instance is to be learned as an exercise in reasoning, independent of anatomic *surface* markings. Decide whether the consolidated middle lobe is going to be more dense-appearing on the PA or the lateral chest film. Think in terms of summation shadowgrams and of the shape and location of the lobe.

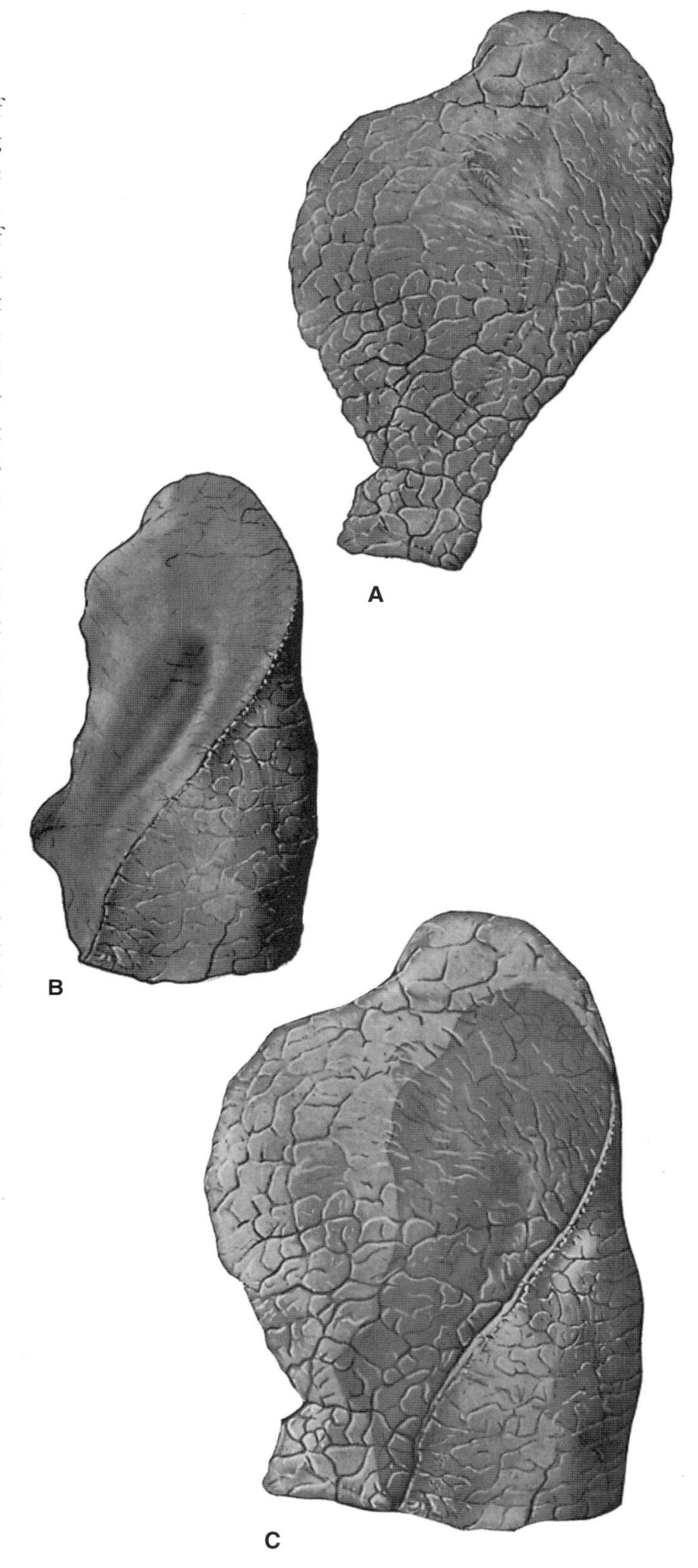

Figure 6.5. Transparent drawings of the left lung, seen from the lateral surface, separated into upper and lower lobes (A, B) and reassembled (C). The patient faces to your left. Note the similarities and differences between the middle lobe on the right and its analogue, the lingular segment of the upper lobe on the left. Opacity in either will obscure the lower part of the heart profile in the PA view.

Facing page: Transparent drawings of individual segments of the left lung.

The Diagrams You Should Have Drawn for the Lobes of the Right Lung

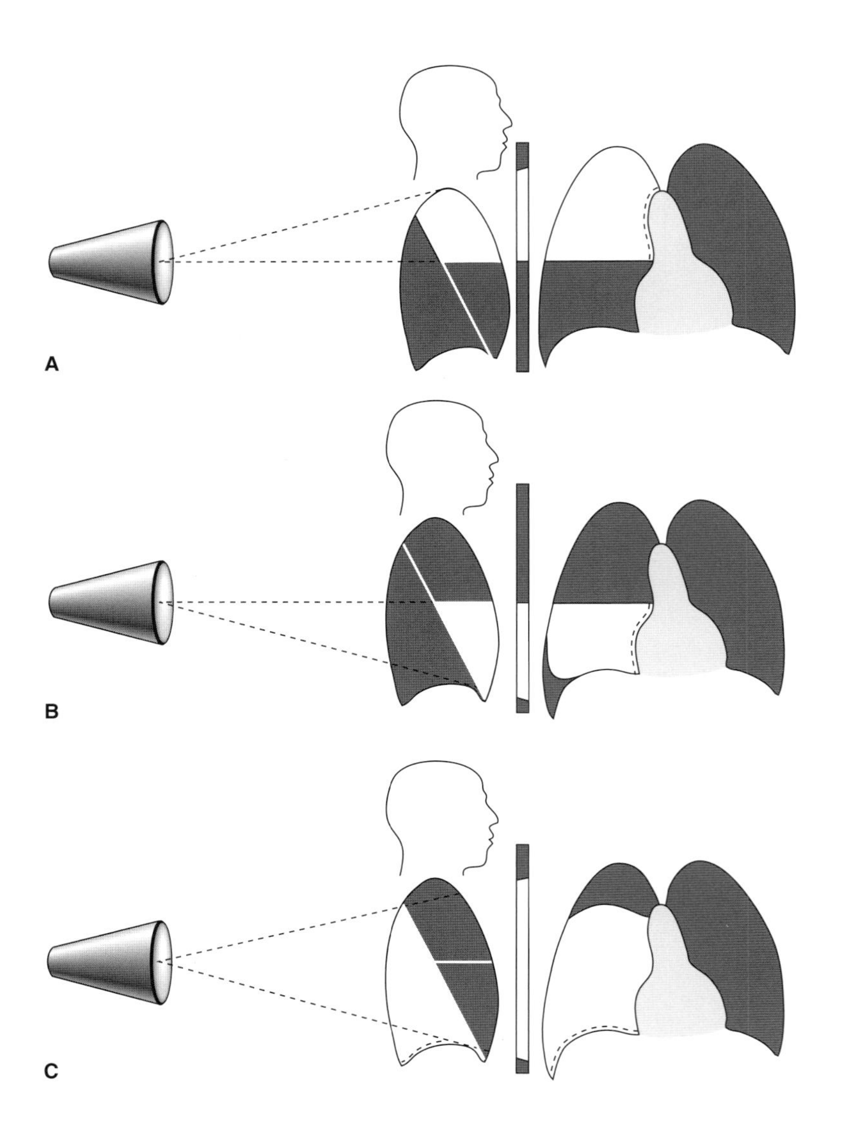

Figure 6.6. Projection shadows of each of the three lobes on the right. In A, the right upper lobe is consolidated, its inferior margin outlined by the air-filled middle lobe lying beneath the minor fissure. In B, the middle lobe alone is opaque; note that in the PA view it does not extend into the costophrenic sinus against the lateral insertion of the diaphragm. In C, on the contrary, the lung tissue filling the right costophrenic sinus is seen to be opaque because the right lower lobe is opaque. The heart shadow has been rendered in gray in these diagrams in order to clarify the shape of the lung mass shadows, but in an actual radiograph the heart shadow's opacity would merge with that of the middle lobe in B. Disappearing borders are outlined by dotted lines.

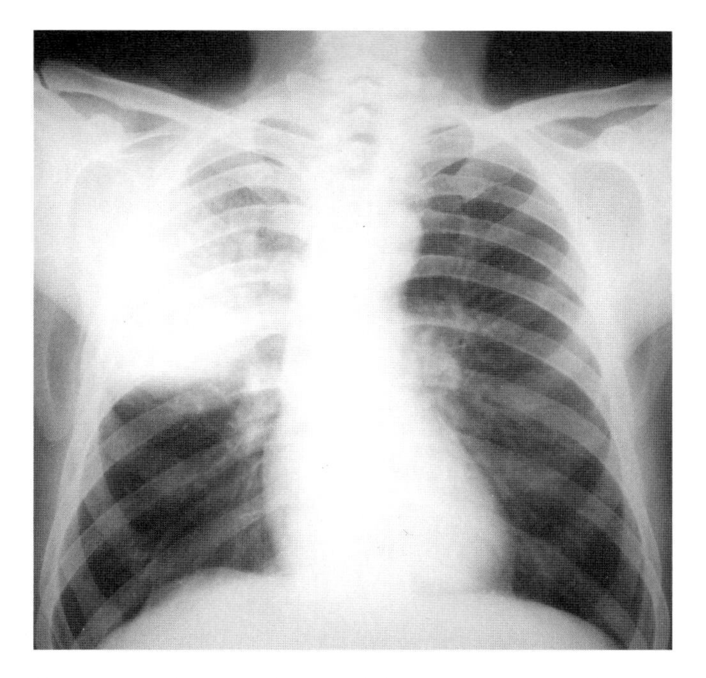

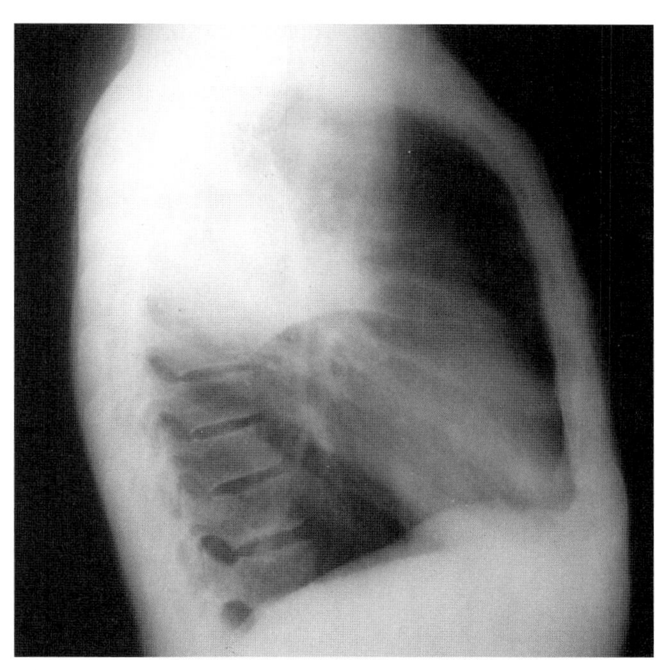

Figure 6.7. PA and right lateral views of a patient with right upper lobe pneumonia. The anterior segment is incompletely consolidated.

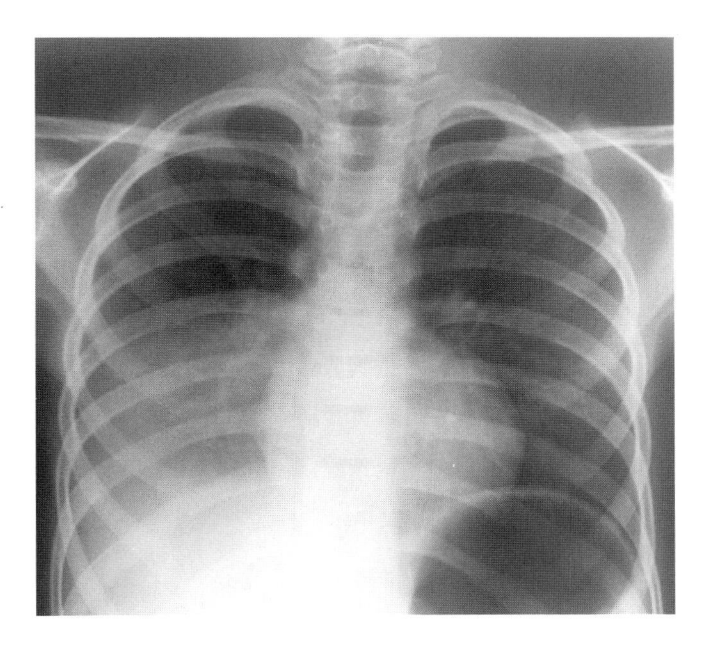

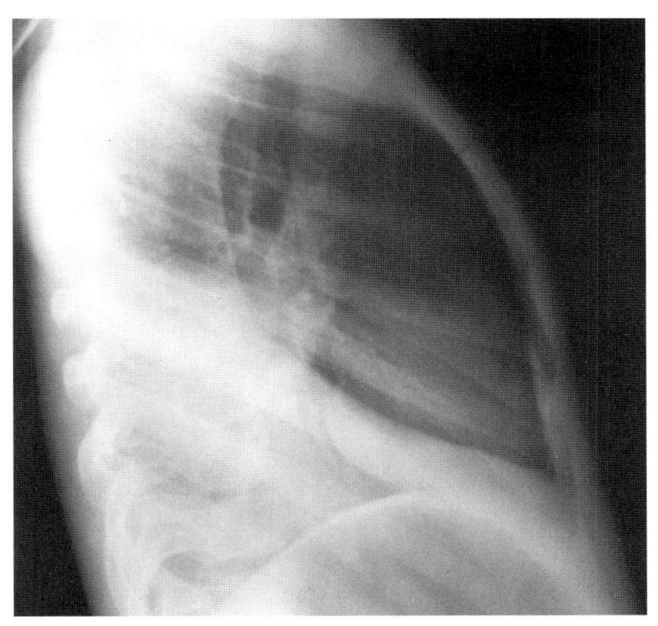

Figure 6.8. PA and right lateral views of a patient with right lower lobe consolidation. Note the preservation of the heart profile in the PA and the absence of the right diaphragmatic profile in the lateral. The wedge of opacity in the lateral view represents overlap of the consolidated lower lobe on the heart.

The Diagrams You Should Have Drawn for the Lobes of the Left Lung

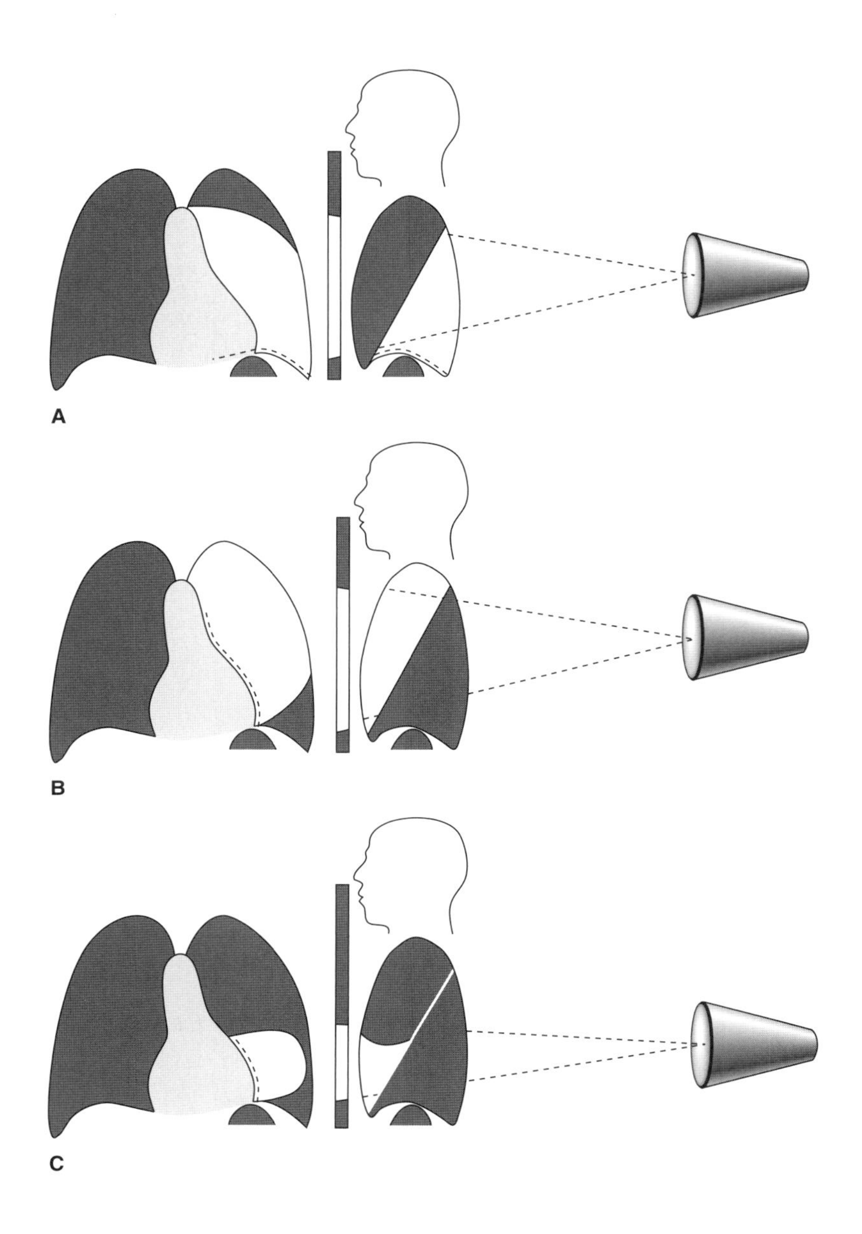

Figure 6.9. Projection shadows of the two lobes on the left. In A, the left lower lobe is consolidated; the diaphragmatic shadow disappears *(dotted lines),* since the roentgen shadows of the lower lobe and the spleen are merged; the approximate location of the left hemidiaphragm may be indicated by the stomach bubble if one is present; the left heart border does not disappear but is seen through the opaque lower lobe because air in the lingula of the left upper lobe anteriorly still throws it into relief. In B, the entire left upper lobe is consolidated and the left heart border is lost, but the diaphragm is seen because of lower lobe air above it. In C, only the lingular portion of the left upper lobe is solid, erasing the left heart border. Predict the radiographic appearance of the PA and lateral views in a patient with left upper consolidation that spared the lingula.

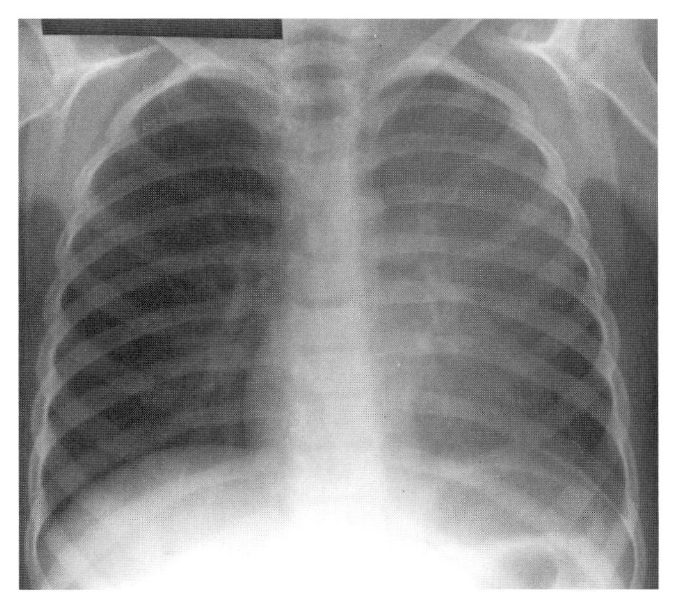

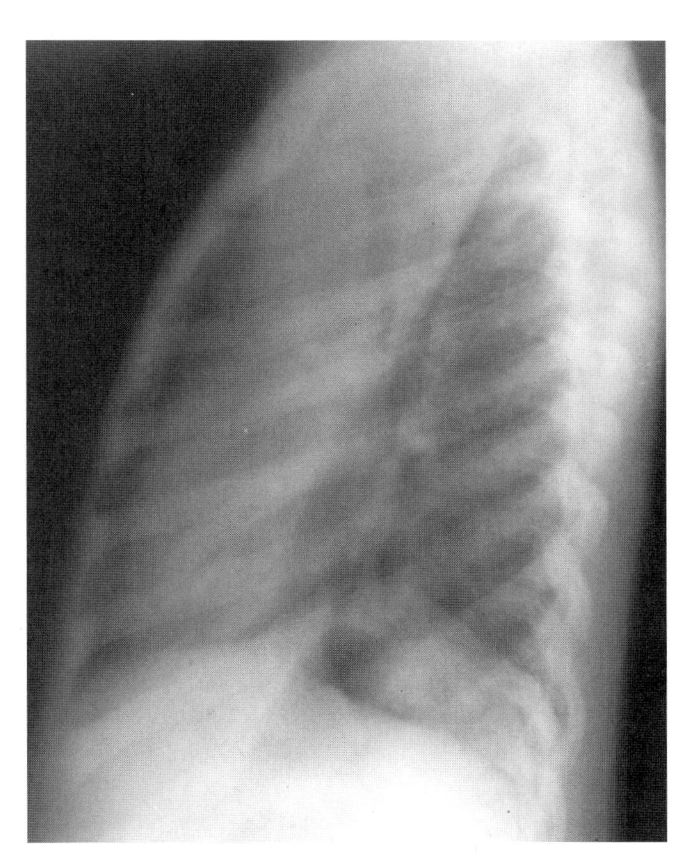

Figure 6.10. PA and lateral views of a patient with left upper lobe consolidation in pneumonia.

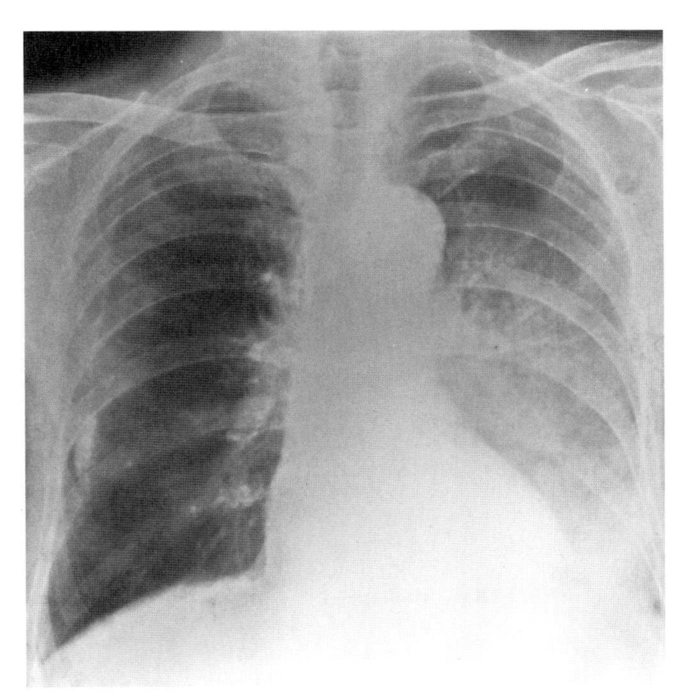

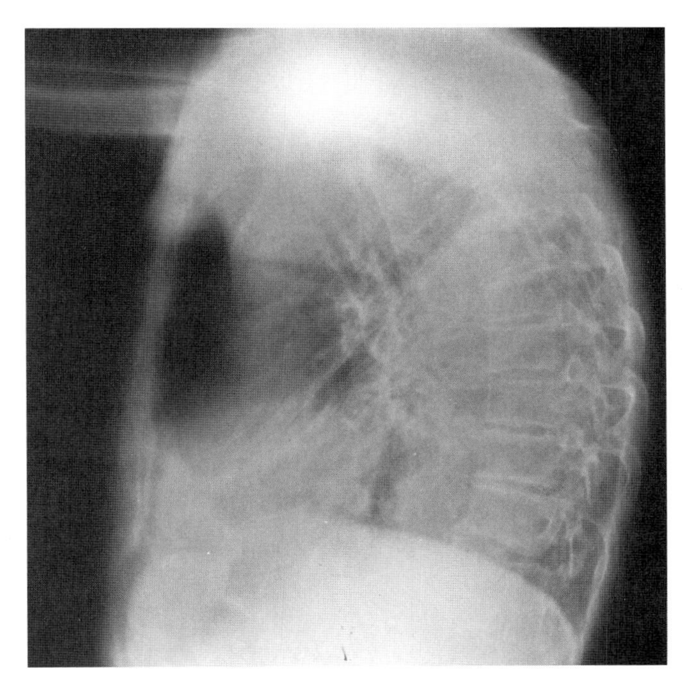

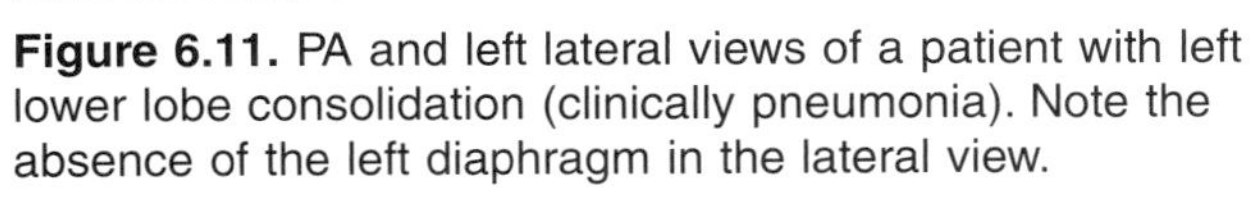

Figure 6.11. PA and left lateral views of a patient with left lower lobe consolidation (clinically pneumonia). Note the absence of the left diaphragm in the lateral view.

Consolidation of Only a Part of One Lobe

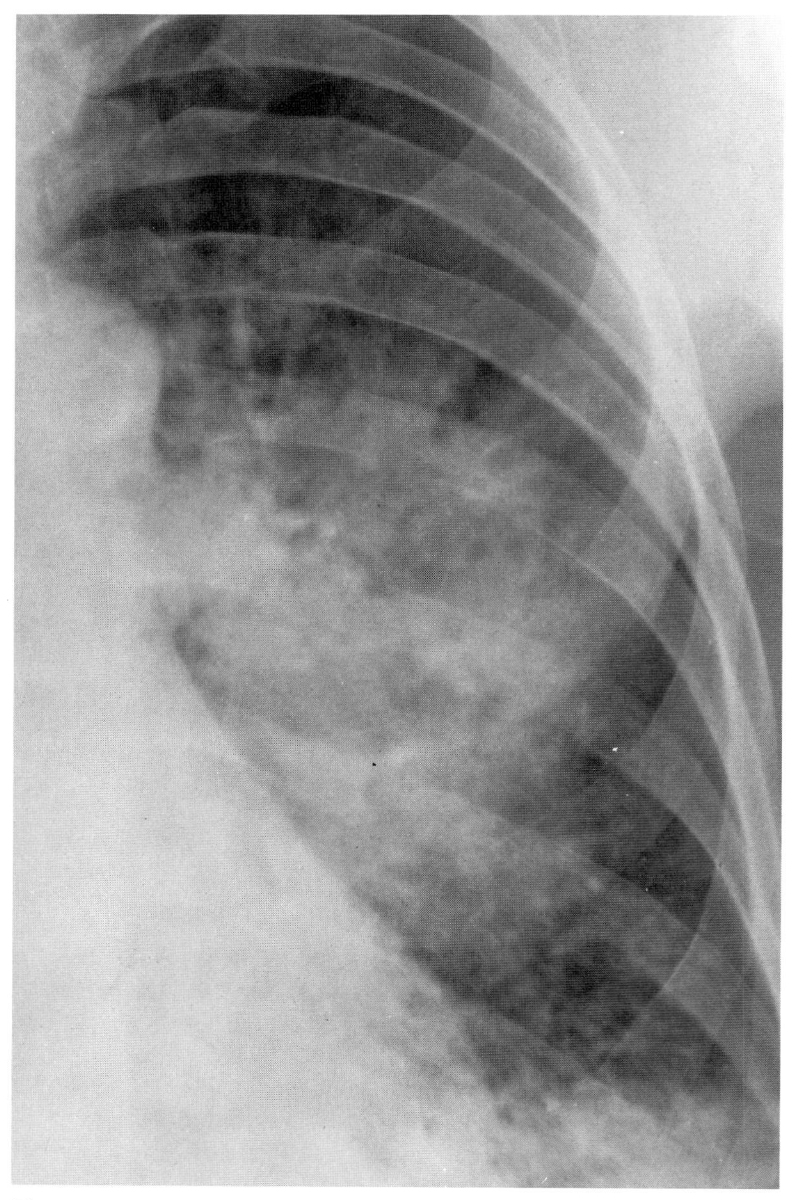

Figure 6.12. Patient with only a part of the left lower lobe involved. There is a patch of pneumonia not in juxtaposition with the diaphragm. You know that this patch of consolidation is not anterior because you see the heart profile so well.

When massive opacities involve an entire lung or an entire lobe, you have certain clues which tell you how much of the lung is consolidated, and sometimes these same clues help you determine the location of a patch of consolidation in the lung that does not occupy an entire lobe or bronchopulmonary segment, but occupies only a part of one. Consolidation of all the lung tissue against the diaphragm will cause the outline of the diaphragm to disappear entirely, but a patch of dense lung against the lateral half of the diaphragm will cause the disappearance of only the lateral half of its outline, leaving the medial half visible.

One day you may see a chest film in which the lower left hemithorax appears opaque, but if you can nevertheless see the entire diaphragmatic profile, you will know that there must be air filling the lower lobe. If, however, you cannot see any part of the diaphragm, but can still see the left heart border through dense lung, you will know that there must be air in the upper lobe. You will not call either picture consolidation of the entire lung.

Neither will you be prompted to call a patch of partial consolidation like that in Figure 6.12 involvement of a whole lobe; instead you will think of it as involving most of a bronchopulmonary segment in the posterior part of the chest, since the heart border is so well preserved.

Now that you can recognize and locate anatomically areas of massive consolidation in the lung, as contrasted with the scattered small areas of opacity you saw in the last chapter, you are probably somewhat impatient to know how to label them. As we have said, pneumonia and tumor can both produce solid areas in the lung giving the findings outlined above. It is current parlance to speak of air-space disease as opposed to interstitial disease, and to attempt to differentiate them from each other on the PA chest film. It is true that the consolidation of lobar pneumonia *should* be thought of as pure air-space disease. It is also true that other abnormalities of the lung in which the entire pathological change is interstitial *do* produce linear strands of density on the radiograph. Yet anyone who rigidly attempts to classify all disease as either air-space or interstitial from the roentgen appearance is in for a very disappointing and frustrating experience, because only some diseases pathologically show pure air-space or pure interstitial change. Other pulmonary processes (tuberculosis is one) may involve both the interstitium and the alveoli concurrently.

Lung that is airless because it has collapsed can produce much the same appearance as consolidation except that there will be evidence for change in the size and shape of the lung part involved. Collections of fluid in the pleural space can also produce opaque areas in the thoracic cavity, of course, obscuring the otherwise healthy lung it envelops and causing the disappearance of the diaphragmatic outline.

In both pneumonia and tumor, moreover, some atelectasis and pleural fluid are common in addition to the primary consolidation in the lung itself. You have to remember that these processes go together pathologically and, since they may cast very similar shadows, are often impossible to differentiate from each other by studying only the initial films. In the next chapters you will learn how to determine the presence of pleural effusion and how to analyze the particular signals indicating that a lobe has collapsed. Then you will add them to the signs of consolidation we have covered above and interpret chest films systematically on each level.

In Chapter 4 you set up a system for beginning to study a chest film by surveying the bony structures and the soft tissues. In Chapter 5 you added the systematic survey of the hila and their tapering vessels and the parenchyma of the lung itself. In this chapter you have added a survey to make sure that no large patches of lung appear opaque and that the heart borders and both hemidiaphragmatic outlines are present, checking for *disappearance of profiles normally seen*. You are building gradually the sort of careful analysis of a chest film that will help you now in using roentgen data and serve you all your life in understanding the films on your own patients. Do not be impatient for diagnostic labels.

Problems

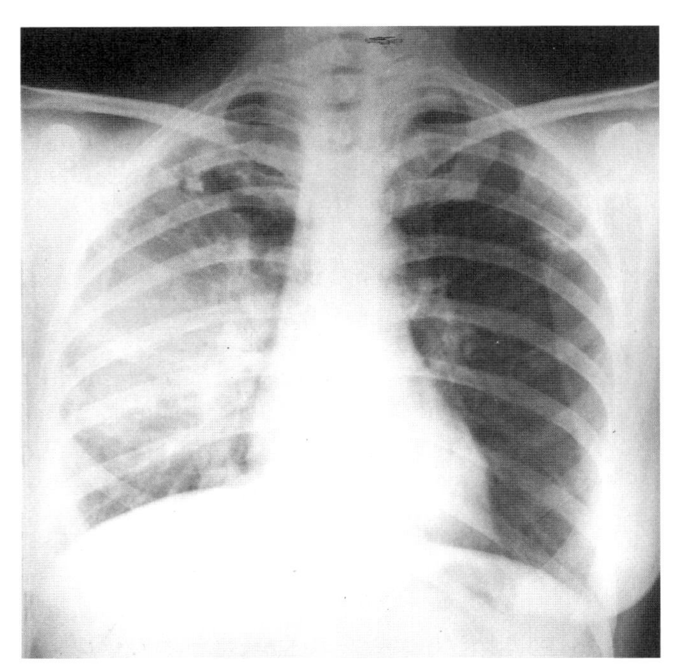

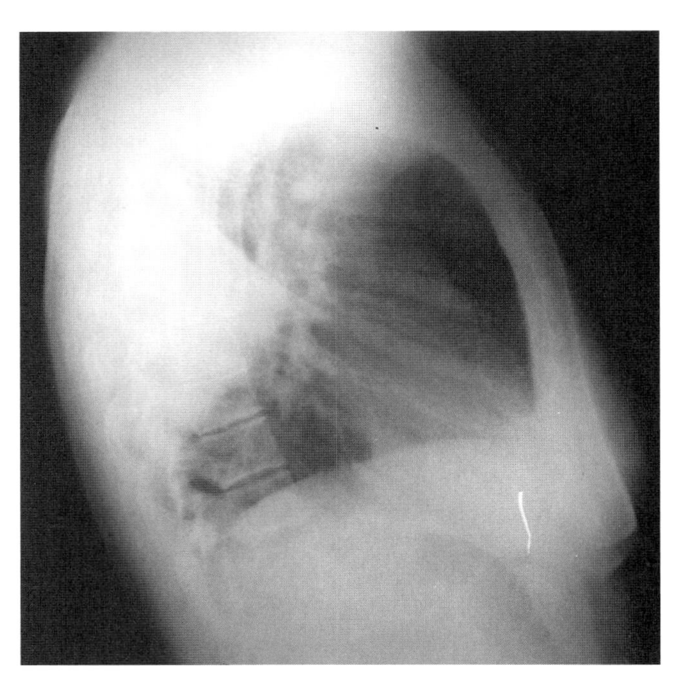

Figure 6.13 ***(Unknown 6.1).*** Clinically this young woman had pneumonia. Precisely what part of the right lung is consolidated? Is this air-space or interstitial disease?

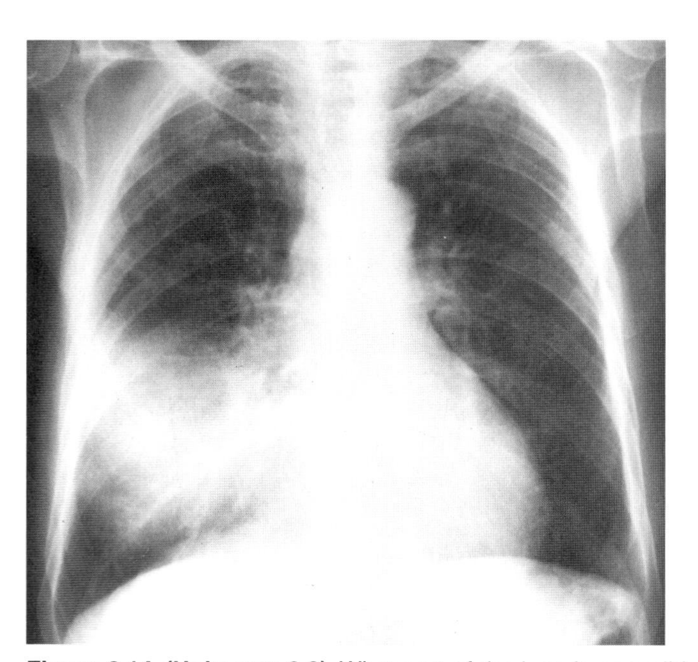

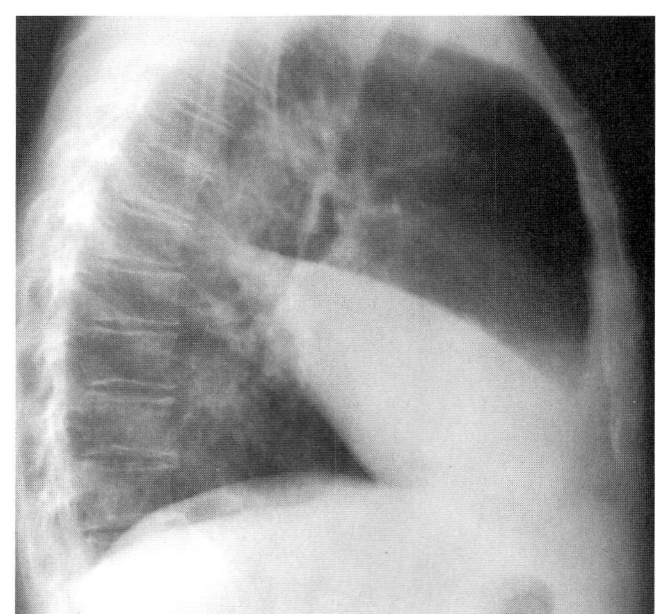

Figure 6.14 ***(Unknown 6.2).*** What part of the lung is consolidated here?

Solitary and Multiple Pulmonary Nodules

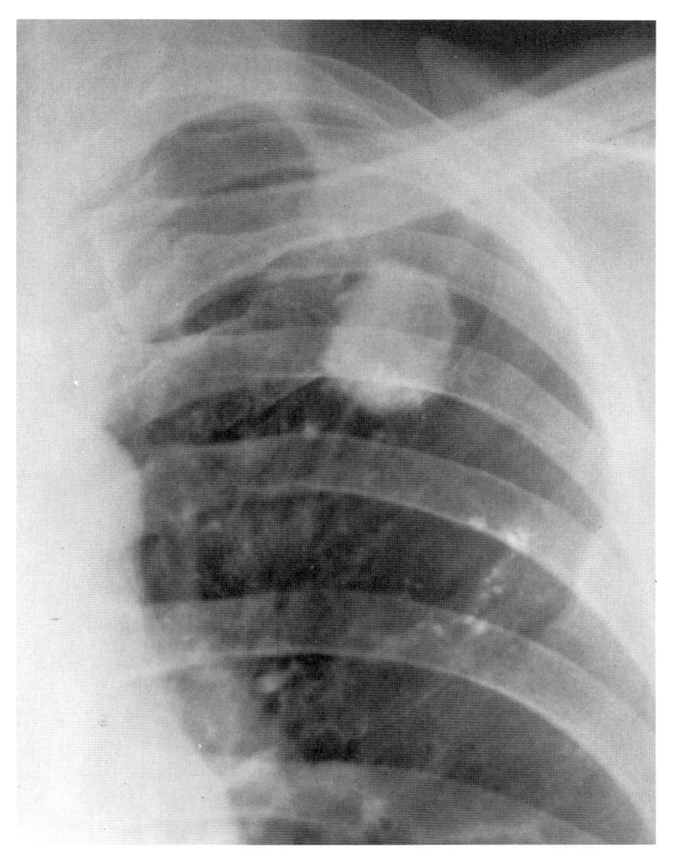

A special management problem arises when a solitary mass density (pulmonary nodule) is incidentally identified on a chest film examination. You have seen one such mass in Figure 5.21E, which, when resected, proved to be a benign tumor. Any such finding on a patient's chest x-rays must be subjected to serious consideration and diagnostic workup. A solitary pulmonary nodule could be something as unimportant as an old granuloma from prior histoplasmosis or TB, or something as serious as a malignant neoplasm. And if a nodule is neoplastic, it could be either a benign or malignant tumor; and if malignant it could be either a primary lung cancer or a single metastasis from any of a number of distant primaries.

The most cost-effective and efficient way to proceed when you see a single pulmonary nodule on a chest film is to retrieve the patient's prior chest films for comparison. If the pulmonary nodule was present on prior films dating back 2 years or more, and the nodule

Figure 6.15 ***(left).*** A densely calcified solitary pulmonary nodule that is most probably a granuloma and might safely be watched. Contrast its density with that in the patient in Figure 6.16.

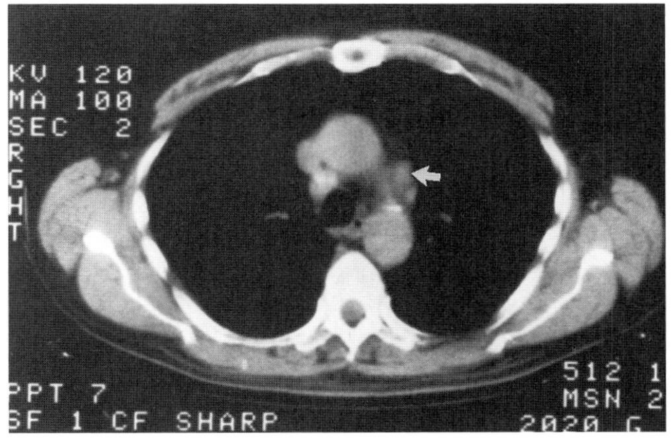

Figure 6.16B. CT scan of the patient in Figure 6.16A made at a slightly different level does not show the pulmonary nodule but does show enlarged lymph nodes *(arrow)* in the space between the ascending and descending aorta on the left, not present on normal CT scans at this level. This patient proved to have a lung cancer with mediastinal metastases.

Figure 6.16A. Patient with a solitary pulmonary nodule *(arrow).* A film made one year before was normal. The nodule does not appear to be calcified.

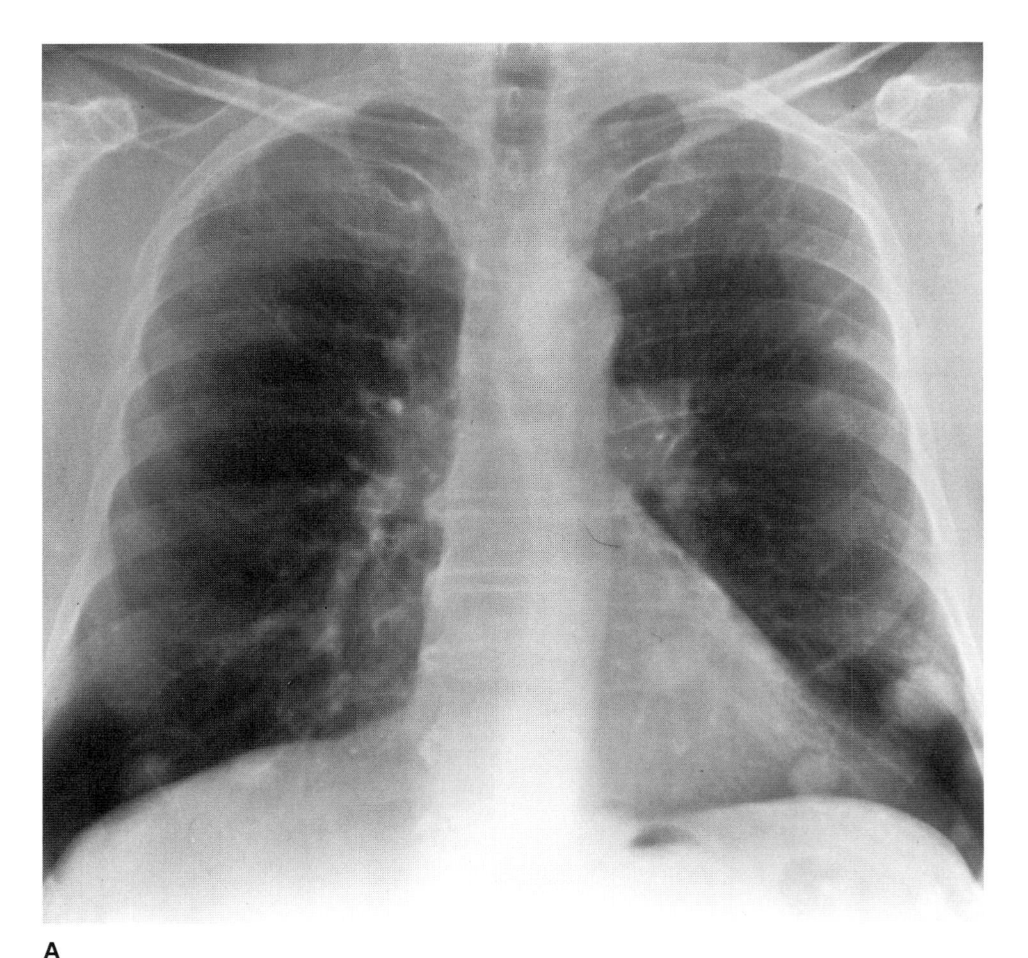

A

Figure 6.17. Multiple pulmonary nodules in an elderly man with metastases from a lower extremity soft-tissue sarcoma. A and B: PA and lateral chest films showing numerous pulmonary nodules. C, D, and E: CT scans at three different levels showing the nodules more clearly than the chest films. The nodules in the left costophrenic sulci are especially well shown by CT.

is unchanged, you can assume that the nodule represents a benign, old granuloma, which can be serially observed at intervals of three to six months. And if the mass contains calcium that is central and densely particulate, it is also likely to be a benign granuloma. If there are no old films, or if the nodule did not appear on the earlier films, then a diagnostic evaluation is mandatory.

The first order of business is to ascertain whether the mass is indeed a *single* pulmonary nodule or not, because if *multiple* nodules are present, then the clinical workup must focus on a search for metastatic disease or inflammatory conditions associated with multiple nodules, such as sarcoidosis or histoplasmosis. The chest films themselves should be scrutinized for other nodules, and if none are found then a chest CT should be requested. CT can show nodules too small to be shown by plain films, or nodules in locations difficult or impossible to identify on plain films. In addition, when the nodule is a primary lung cancer, CT may demonstrate metastases to the mediastinum that are not seen on chest films (Figures 6.16A and B). If no additional nodules are seen at CT and no mediastinal nodes are identified, a tissue diagnosis is indicated. If the nodule is located in the lung periphery, a tissue sample may be obtained by percutaneous needle aspiration biopsy under CT guidance. Material can be aspirated for both cytological and bacteriologic examination. This procedure will be discussed in greater detail in Chapter 19.

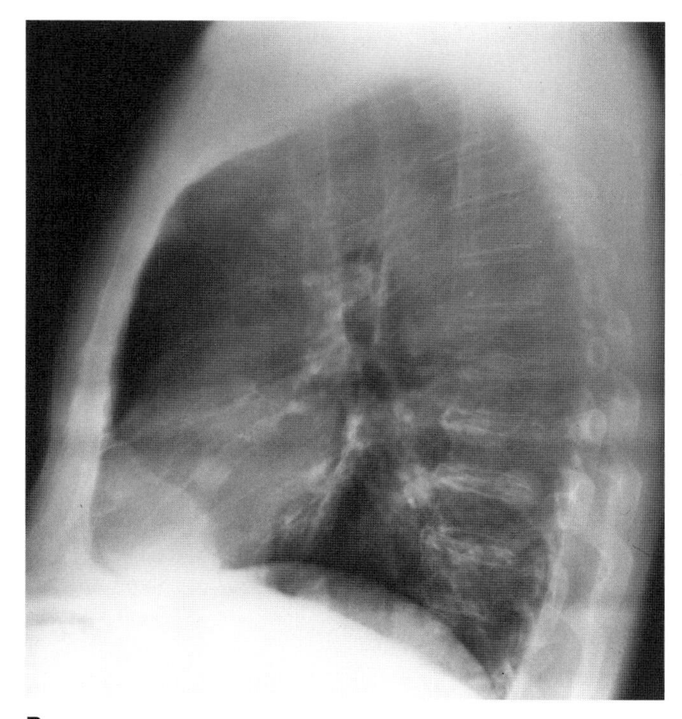

B

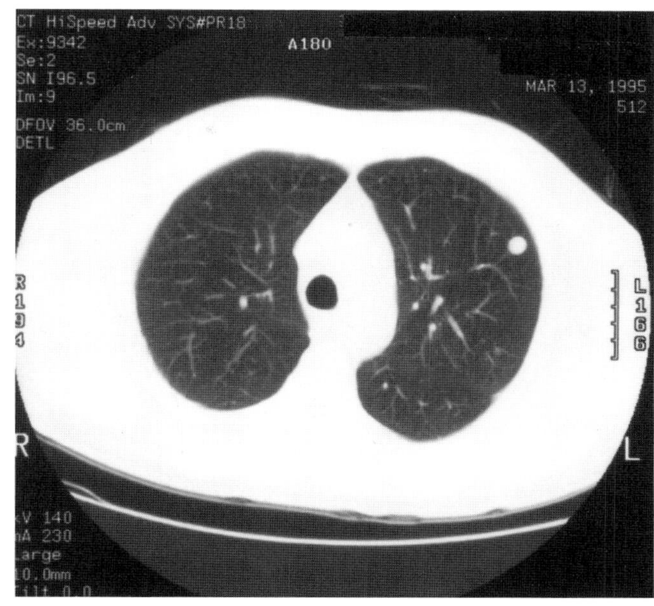
CT HiSpeed Adv SYS#PR18
Ex:9342
A180
Se:2
SN I96.5
Im:9
MAR 13, 1995
512
DFOV 36.0cm
DETL
R
1
9
4
L
1
6
6
R
L
kV 140
mA 230
Large
10.0mm

C

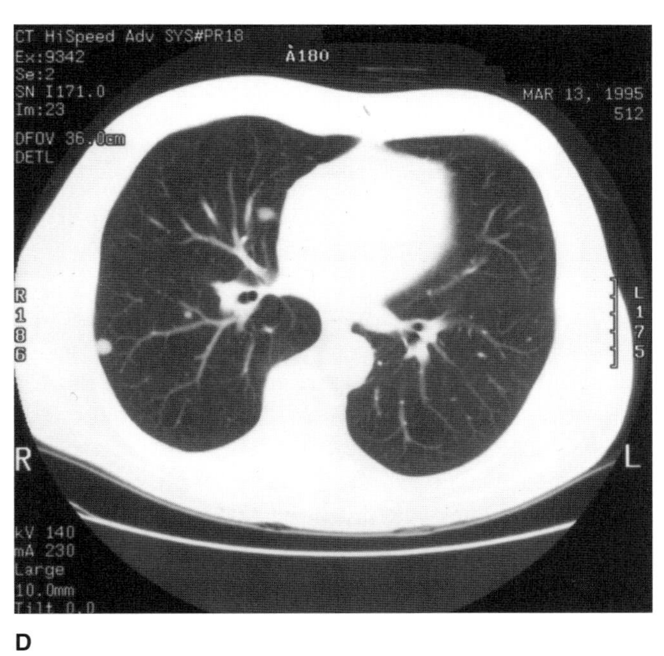
CT HiSpeed Adv SYS#PR18
Ex:9342
A180
Se:2
SN I171.0
Im:23
MAR 13, 1995
512
DFOV 36.0cm
DETL
R
1
8
6
L
1
7
5
R
L
kV 140
mA 230
Large
10.0mm

D

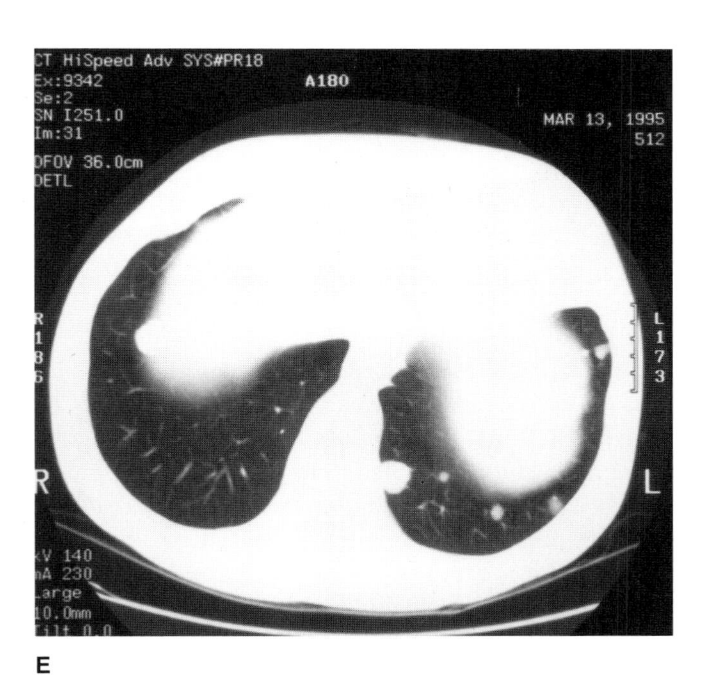
CT HiSpeed Adv SYS#PR18
Ex:9342
A180
Se:2
SN I251.0
Im:31
MAR 13, 1995
512
DFOV 36.0cm
DETL
R
1
8
6
L
1
7
3
R
L
kV 140
mA 230
Large
10.0mm

E

7 The Diaphragm, the Pleural Space, and Pulmonary Embolism

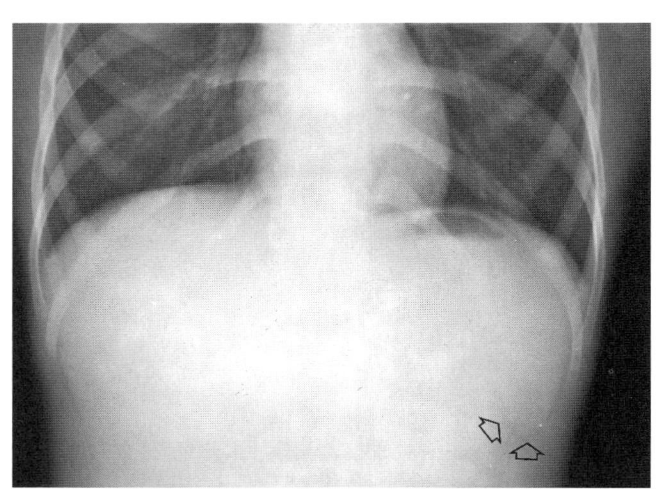

Figure 7.1. Normal hemidiaphragms. Note the stomach bubble with fluid level under the left hemidiaphragm. The *arrows* mark the tip of the spleen. The spleen and fluid-filled stomach form a continuous shadow. Only the crown of the left hemidiaphragm, tangential to the x-ray beam, is seen. It has the wall of the stomach plastered against its inferior surface. The sloped curving anterior and posterior parts of the diaphragm are oblique to the beam and hence are not seen at all. The right hemidiaphragm is closely applied against the liver. Only the interface for its superior surface is seen.

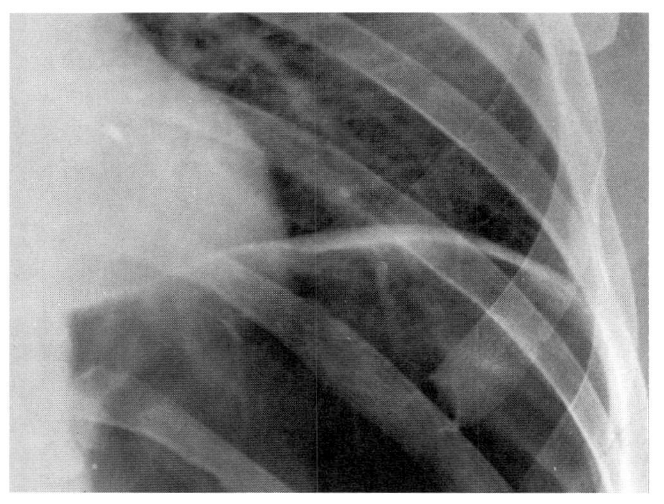

Figure 7.2. Left diaphragm filmed with the patient sitting upright after blunt trauma and bowel rupture caused a large amount of air to be admitted to the peritoneal space. The spleen and stomach are displaced downward. Note the vascular trunks in the lower lobe posteriorly extending well below the level of the crown of the diaphragm and superimposed on the subdiaphragmatic air. These lower lobe vessels are not seen in Figure 7.1 because they are behind the spleen.

The region of the diaphragm on the chest film affords some fine exercises in the logic of roentgen shadows. Just as you see the profile of the heart, dense between two lucent lungs, so you see the dome of the diaphragm because of a change in the sum of all superimposed densities. The sum of the shadows just below the *level* of the dome of the diaphragm on the radiograph includes a part of the lung posteriorly and the dense liver or spleen, solidly applied against its inferior concave surface. Above the level of the dome of the diaphragm the sum of all the densities is dominated by that of the lung, which offers little obstruction to the beam. Hence on the chest film the diaphragm and its subtended organs are silhouetted, white against the lucency of the lung field above, *even though their shadows are added to that piece of lung which dips into the posterior sulcus.*

Anatomically composed of a thin sheet of muscle attached to the xiphoid, lower six costal cartilages, ribs, and upper lumbar vertebrae, the diaphragm itself contributes little to the white shadow on the chest film that we mean when we refer to the "diaphragm." If free air in the peritoneal space interposes between the spleen and the diaphragm, as it did in the patient in Figure 7.2, the thin sheet of muscle alone is seen with air both above and below it. As usual when a curved, shell-like structure is x-rayed, what you see is that part of the diaphragm which is traversed in tangent by the beam. Although it appears to be linear, you will think in terms of roentgen densities and know it is a domed shell dividing the chest from the abdomen. Under the fluoroscope you could see it contract downward and flatten with inspiration and relax upward as the patient breathes out.

On most chest films made with the patient standing, the fundus of the stomach will be seen high against the diaphragm, usually containing swallowed air and fluid gastric juice (or lunch). A typical stomach bubble (Figure 7.1) shows a straight line marking the fluid level, above which air provides a radiolucent pocket

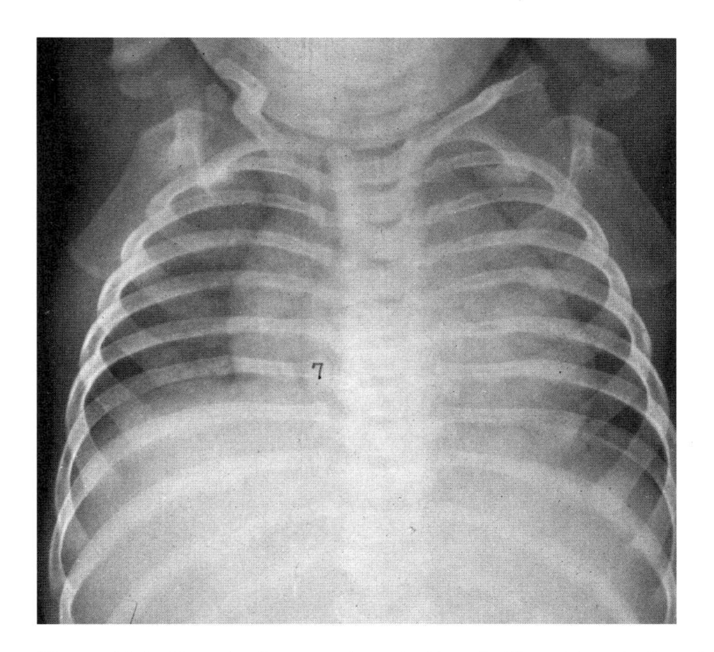

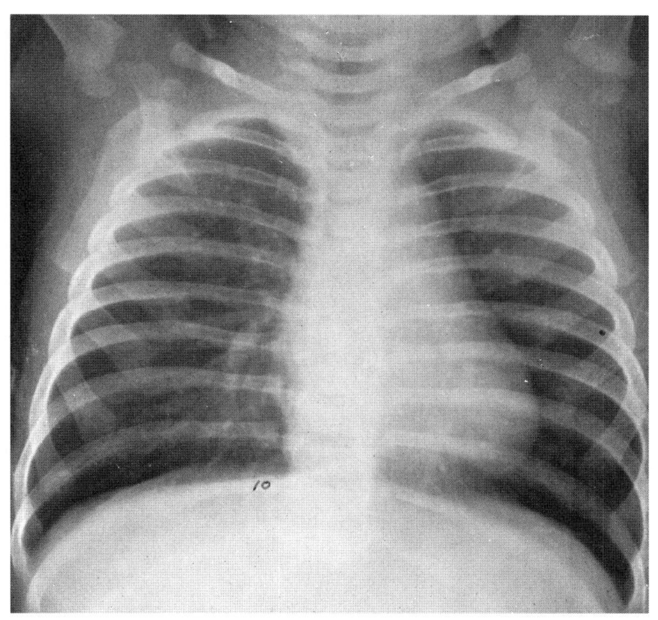

Figure 7.3. The diaphragm at expiration *(left)* and inspiration *(right)* in a child.

through which more rays may pass. The same film made with the patient lying down and the beam still directed sagittally will show no fluid level because the beam will strike the fluid level perpendicularly. Notice that in Figure 7.1 you seem to see the thickness of the diaphragm itself because there is air above and below it, but what you are actually looking at is the diaphragm plus the wall of the stomach.

A potentially hollow viscus that is completely filled with fluid and surrounded by dense viscera will not be distinguishable radiologically, but any hollow viscus that contains air will appear on the film as a dark shadow. While you are about it, take time to consider what you can do with air and a fluid level in radiography. Any hollow structure, normal or abnormal, that contains or can safely be made to contain both a gas and a fluid will show a fluid level provided the beam crosses the plane of that level. Thus by tilting the patient in several directions and always projecting the beam horizontally across the surface of the air-fluid interface, the radiologist can visualize the entire inside of a cavity piece by piece. This technique can be applied to the inside of the stomach, the inside of an abscess cavity, or the inside of the pleural space when it contains both fluid and air. Air thus becomes a useful *contrast substance,* forming a radiolucent cast of the hollow structure containing it, just as barium sulfate and other safely inert substances form radiopaque fluid casts of the hollow structures into which they are introduced.

Compare the chest films made at expiration with those made when the same patients have taken a deep breath (Figures 7.3 and 7.4). Poorly aerated alveoli and crowded-together vessels naturally decrease the radiolucency of the lung to some extent. Note too that the flexible mediastinum and fluid-filled heart have been compressed upward by the high diaphragms in expiration, so that they cast an appreciably wider shadow and appear to be enlarged. This will be true of any film made at expiration and is an additional reason why it is important for you to determine the level of the diaphragm in the course of your survey of any chest film. The patient must be cajoled into taking a deep breath if he is at all capable of it, and before you attempt to draw any conclusions from his chest film, you must check the position of his diaphragm and decide whether he has done so. If a chest film made at expiration in not recognized *as an expiration film,* an erroneous diagnosis of cardiomegaly could be made in a patient with a normal-sized heart!

Because the stomach bubble (present only if there is air in the stomach) normally lies close against the undersurface of the left diaphragm, it should be included in your systematic survey of a chest film. The interposition of anything between the diaphragm and the fundus of the stomach will displace the bubble downward. It may be deformed by the presence of tumor in the stomach. Both its appearance and its location are significant.

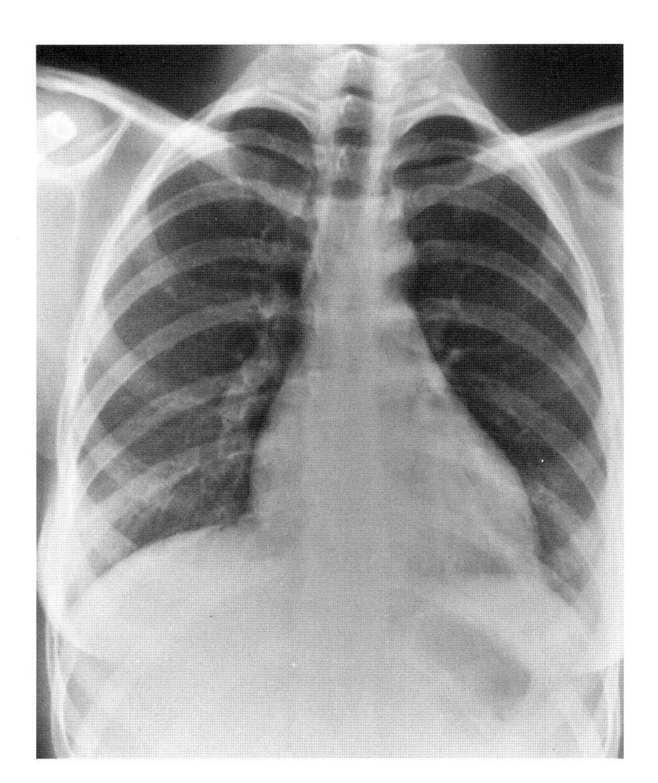

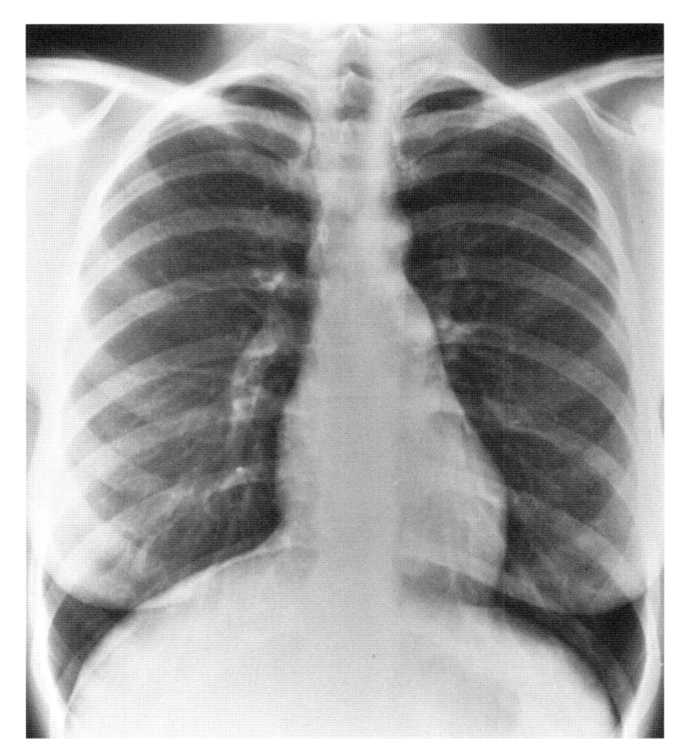

Figure 7.4. The diaphragm at expiration *(left)* and inspiration *(above)* in an adult.

For example, when any massive density in the chest just above the left diaphragm causes the disappearance of the normal diaphragmatic outline, as you have seen in the last chapter, the location of the stomach bubble may tell you where the diaphragm is. In the lateral chest film the presence of the stomach bubble close under one diaphragmatic shadow determines which is the left diaphragm (Figure 7.6).

Although anatomists think of and see the diaphragm as a single sheet of muscle and tendon, dividing chest from abdomen, radiologists see it on the PA chest film and at fluoroscopy as two curved shadows on either side of the heart. Radiologists speak of the "left and right hemidiaphragms," in spite of the fact that they know that usage to be anatomically and semantically imprecise. It is convenient to refer to the two halves of the diaphragm in this way because they often respond independently to unilateral disease in the chest above or in the abdomen below.

The two hemidiaphragms, then, as seen on PA adult chest films, normally are smooth curves taking off *from the midline at the origin of the tenth or eleventh rib.* You should make a practice of counting down the posterior ribs close to the spine to determine the level of the diaphragm. Try it on a few of the chest films

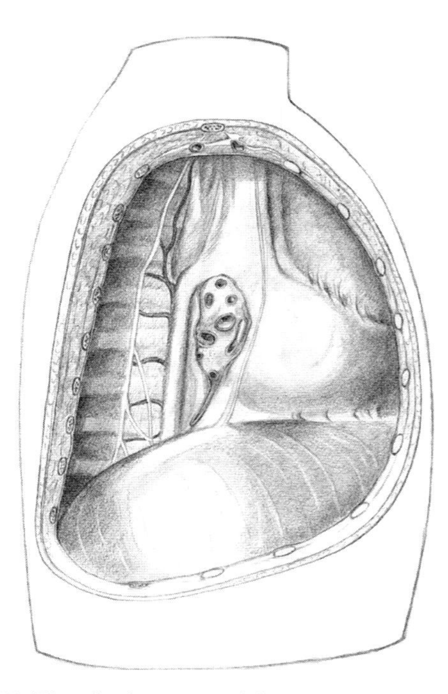

Figure 7.5. The diaphragm and the costophrenic sulcus with the lateral chest wall cut away.

you have seen (being sure to identify the first rib by tracing it backward from the sternoclavicular junction). If you determine the level of the diaphragm on a few actual patients' films seen in the course of your clinical day, you will find that hospitalized patients tend to show considerable variation in that level. The well person efficiently obeys the request of the technologist to "take a deep breath"; but the anxious, tired, pain-beset patient may fail to do so, even though he has a fractured ankle requiring open surgical reduction and nothing at all wrong with his chest or abdomen. In addition, the well patient usually stands upright for x-rays, whereas the sick patient may not be able to do so; it is much easier to take a deep breath when you are upright than when you are lying supine on an x-ray table. The result of not taking a deep breath is of course that the lower part of the lung close to the diaphragm is poorly inflated with air and consequently more dense on the chest film, giving an appearance of abnormality where, in fact, none exists.

The diaphragm may be elevated by large collections of fluid in the peritoneal space, as in a patient with ascites. With distention of many loops of the large or small bowel in intestinal obstruction, the diaphragm is usually high and may also be limited in its downward motion, responding reflexly to abdominal pain. For the same reason it is normally high and "splinted" in its motion for a few days after abdominal surgery. You would expect it to be high in the third trimester of pregnancy and it is.

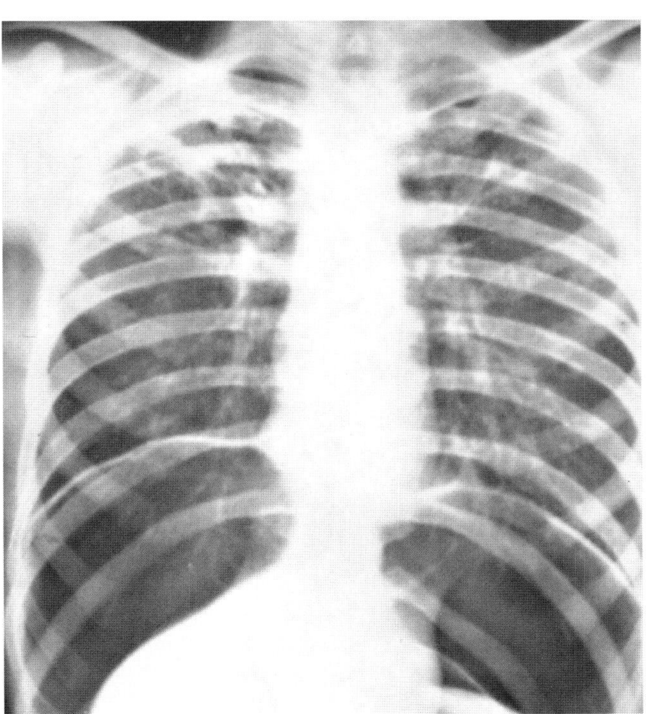

Figure 7.7 *(Unknown 7.1).* Determine the level of the diaphragm in this patient with bilateral upper lobe tuberculosis. Did the patient take a deep breath?

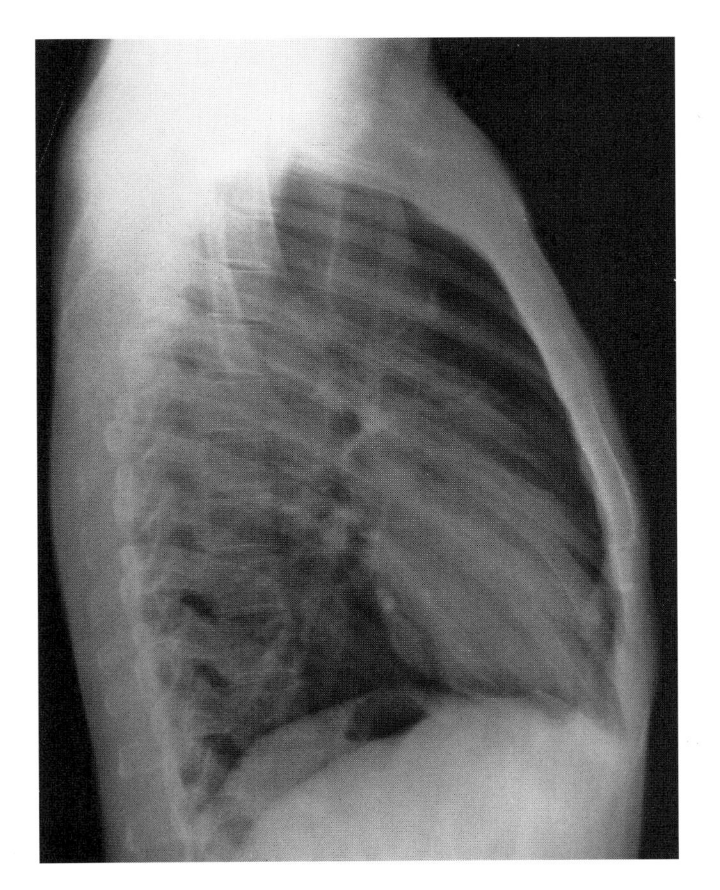

Figure 7.6 *(left).* A normal right lateral chest film. Note the two diaphragmatic profile shadows curving down posteriorly. Which is the left diaphragm? Note the vascular trunks curving down posteriorly below the level of the crown of the diaphragm. Here it is easy to see that the left is the higher diaphragm since the stomach bubble is close underneath it. You do not see its anterior segment: the heart, in contact with it, ablates its interface. What appears to be the continuing left diaphragm is actually the anterior segment of the right diaphragm, always seen continuously from the posterior sulcus to the anterior chest wall.

On the other hand, the diaphragm may be depressed and flattened in any condition that greatly increases the volume of the structures within the thoracic cage. Thus in emphysema, with irreversible trapping of air in the lung and gradually increasing overexpansion, the diaphragm is low and flat. It may show serrated margins because then the insertions into the lower ribs become visible. The diaphragm may also be depressed by the added volume of large collections of pleural fluid or of tumor masses in the lung.

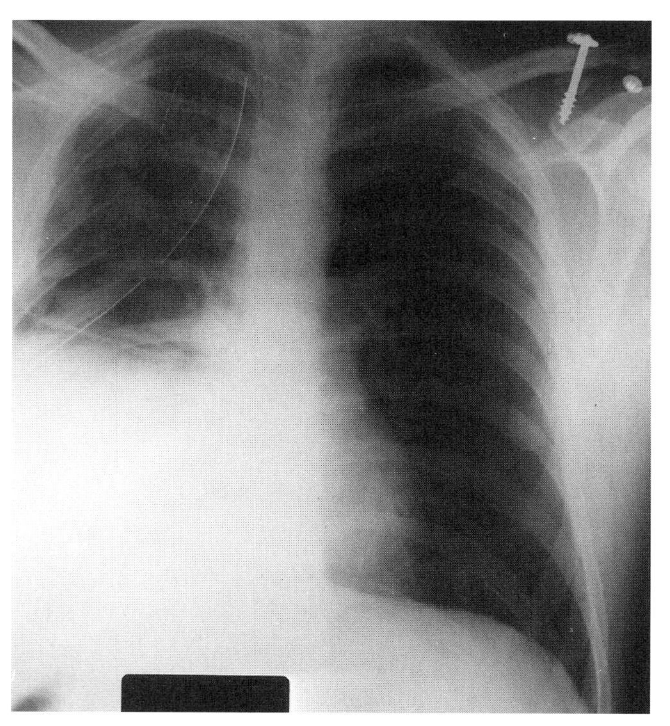

A

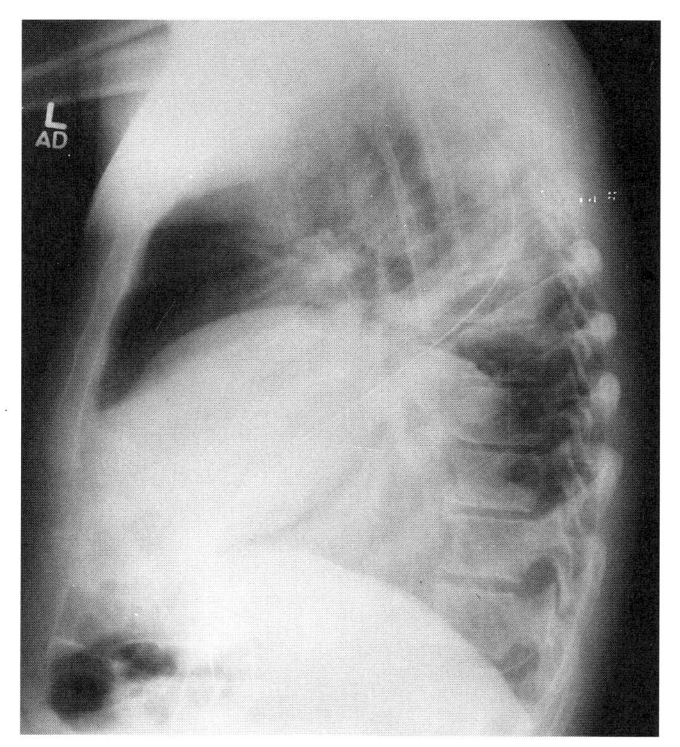

B

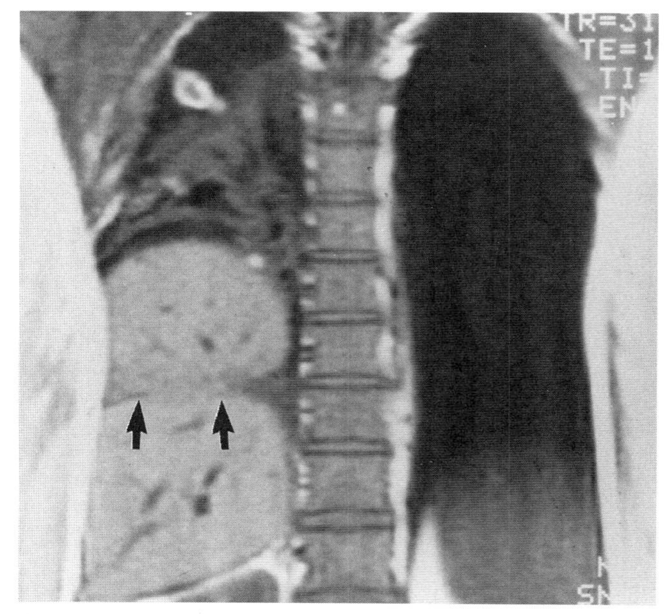

C

Figure 7.8. "Apparent" diaphragm elevation resulting from hemidiaphragm rupture and herniation of the liver into the lower right chest. This 36-year-old construction worker was struck in the right chest with an I-beam, which caused multiple right rib fractures, a right pneumothorax, and a right lung contusion. After a portable film (not shown) identified the pneumothorax, he was taken to radiology for PA (A) and lateral (B) chest films. They showed what appeared to be marked elevation of the right hemidiaphragm. Did you note the many right rib fractures? Because of the possibility of diaphragm rupture, coronal MR scanning (C) was done; it showed a break *(arrows)* in the right hemidiaphragm (appears black at MR) with herniation of a large portion of the right lobe of the liver. Note the lower position of the left hemidiaphragm and the increased signal in the right lung due to contusion and atelectasis. The bright nodular signal at the right apex represents a pulmonary hematoma.

Pleural Effusion

The pleura is a closed empty envelope, one side of which (visceral pleura) invests the surface of the lung, dipping into its fissures. The other side (parietal pleura) is applied against the inner surface of the thoracic cage. Too delicate to be seen radiographically under normal circumstances, the pleura may become visible when it has been thickened by inflammation and catches the beam of x-rays tangentially against the chest wall. The two thicknesses of pleura in the *minor fissure* may frequently be seen as a thin white line extending straight laterally from the right hilum, because the minor fissure is normally horizontal. Both the left and right *major fissures* and the minor fissure on the right may be seen on the lateral chest film whenever they happen to line up with the beam. The major fissures are too oblique to be seen on the PA chest film.

The pleural space, although normally empty and collapsed, *may come to contain either fluid or air or both,* any of which will alter the appearance of the chest film. A massive collection of fluid on one side can displace the mediastinum toward the opposite side, depress the diaphragm, partially collapse the lung, and render the entire hemithorax dense and white. Air in large or small amounts may gain access to the pleural space by rupture through the pleural surface of the lung, or after trauma when the lung is punctured by the ends of fractured ribs. If the amount of pleural air is large, the lung will be seen partially collapsed against the mediastinum. Any amount of air in the pleural space allows you to see some part of the surface of the lung that you do not see in the normal chest film because the lung lies closely in contact with the chest wall. Detection of a small pneumothorax depends on seeing the veil-like pleural margin of the lung, beyond which no lung markings extend.

Large amounts of pleural air or fluid are easy to visualize; small amounts are much more difficult. If you look again at any normal diaphragm shadow in the PA projection, you will see that it dips laterally to form a sharp angle with the chest wall. The base of the lower lobe, cupped convexly over the diaphragm, dips into this recess at the sides and dips deep posteriorly. The *costophrenic sulcus* (or *sinus*), of which only the lateral part appears in the PA chest film, is a continuous ditch formed between the chest wall and the diaphragm. The lowest part of this ditch, when

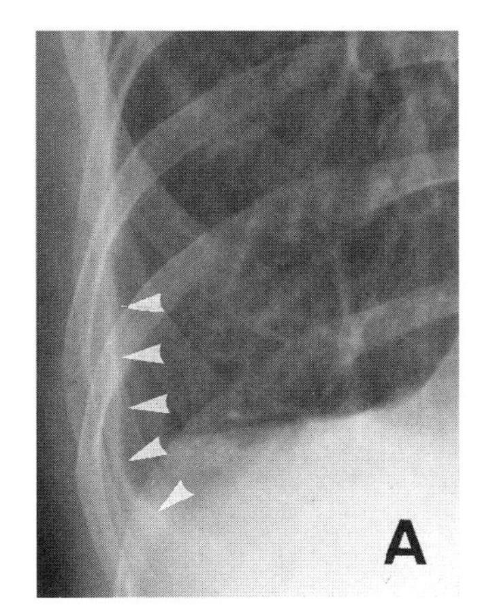

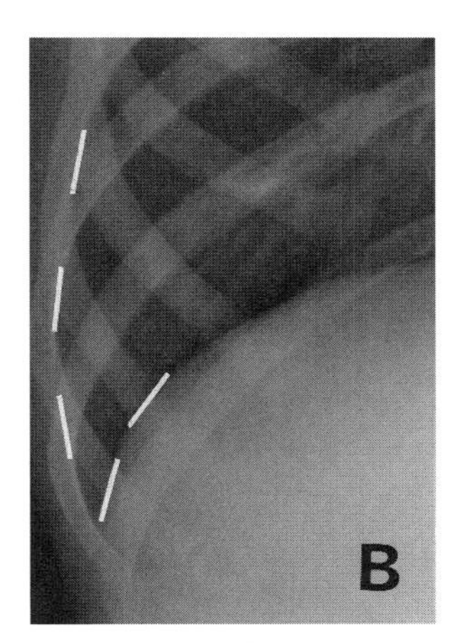

Figure 7.9. A: The old thickened pleura, caught tangentially by the beam of x-rays here, is actually a cuff of scarred tissue curving away from you and toward you over the surface of the lung, and represents parietal and visceral pleura densely adherent to each other. The pleural space in this area is obliterated, and with it the costophrenic sulcus. This is not to be confused with a small pleural effusion. B: A normal costophrenic sulcus, for comparison.

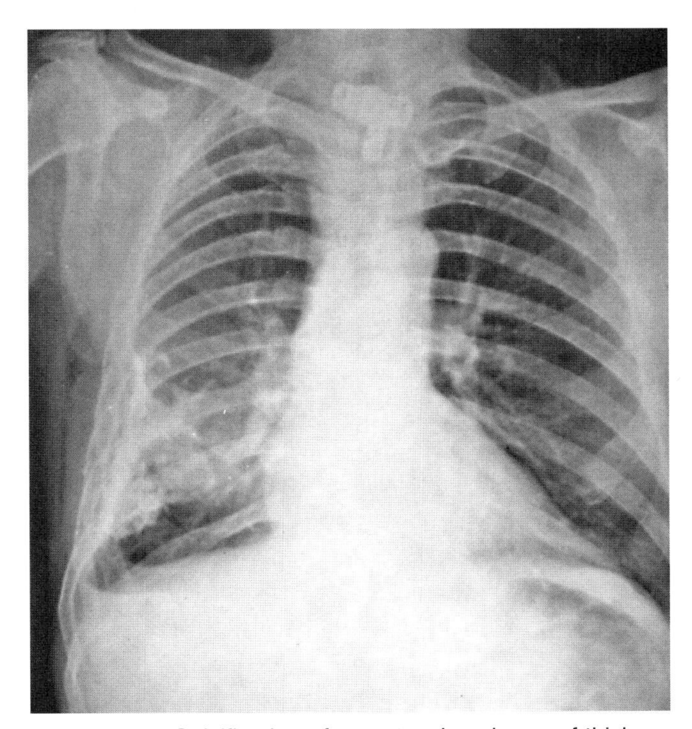

Figure 7.10. Calcification of an extensive plaque of thickened pleura. Note the air in the soft tissues in the lateral chest wall. This was a draining sinus in this patient with chronic empyema, which explains the obliterated costophrenic sinus.

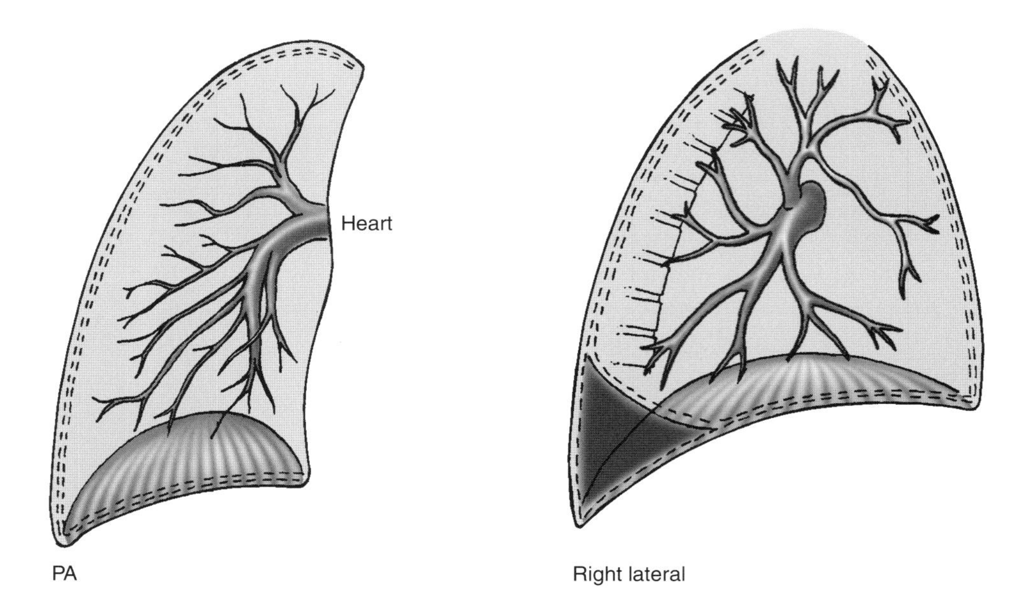

Figure 7.11. The small hydrothorax not seen in the PA view is seen in the lateral view. The *broken lines* indicate layers of parietal and visceral pleura, between which the fluid lies. What will happen to the fluid if, in preparing for a diagnostic thoracentesis posteriorly, the intern asks the patient to sit on a chair and lean forward? The fluid will flow forward into the lateral sulcus when the patient leans forward, making it difficult or impossible to withdraw fluid.

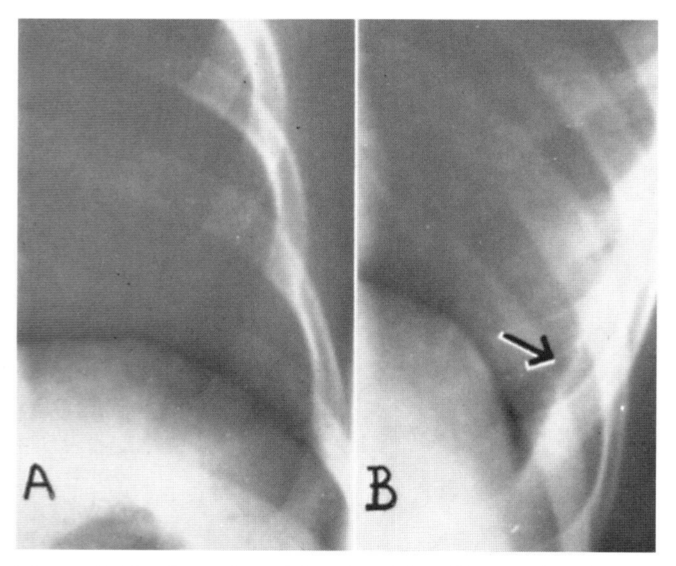

Figure 7.12. The value of the decubitus film for showing small pleural effusions. A: Patient erect. Pleural fluid is not seen. B: Patient lying tilted on his left side. Fluid *(arrow)* that is free in the pleural space flows into the lateral costophrenic sinus, where it is easy to see.

the patient sits or stands, is located far posteriorly on either side of the spine, as you have already appreciated from the lateral chest film. Into this ditch extends the base of each lower lobe against the posterior segment of the diaphragm, and pleural fluid gravitates into it. Thus the first hundred milliliters of pleural fluid that accumulate may not be visible in the lateral costophrenic sulcus on the adult PA chest film, but *would* be seen in a lateral chest film obscuring the posterior portion of the diaphragm. Confirmation of a pleural effusion this small is usually made today with the help of ultrasound, which can also direct placement of a thoracentesis needle.

When enough fluid is present to fill the posterior sulcus, the lateral part of the sulcus begins to fill, and this will be noted on the PA chest film as a blunting or obliteration of the costophrenic sinus on that side.

As a greater amount of fluid collects, the density of it obscures the rounded shadow of the diaphragm entirely and it will be seen as an upward-curving

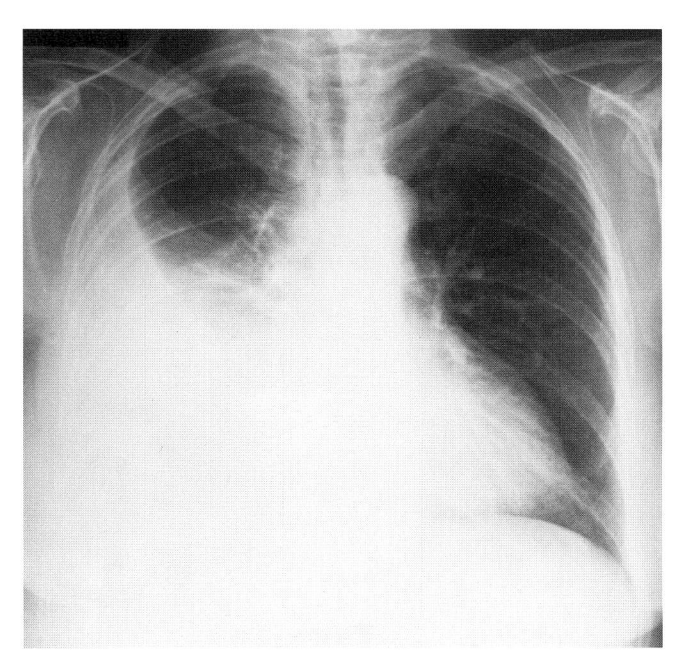

Figure 7.13. Pleural fluid engulfing the right diaphragm, surrounding the lung, and compressing the middle and lower lobes toward the hilum.

shadow against the chest wall (Figure 7.13). It never forms a horizontal fluid level unless there is also air present opening up the pleural space.

Whenever you see an obscured diaphragm, you must wonder whether there is fluid above it and look closely for an upward curve of density against the lateral chest wall or for a fluid level. Do not call a curved fluid line in simple effusion a "fluid level." Whenever you do see a true horizontal fluid level, you must check carefully for the margin of the lung, which is certainly present with pneumothorax and always visible somewhere. You must not forget to wonder what may be going on in that part of the lung which is concealed by the fluid shadow and to plan its better visualization. Massive effusions are more likely to be malignant in origin than small effusions (Figure 7.14).

Figure 7.14. Large accumulations of fluid show a crescentic line ascending along the lateral chest wall in the PA view *(black arrows).* Remember that the same amount is sloping up the anterior and posterior chest wall. Note the displacement of the trachea *(open arrow)* to the opposite side (mediastinal shift).

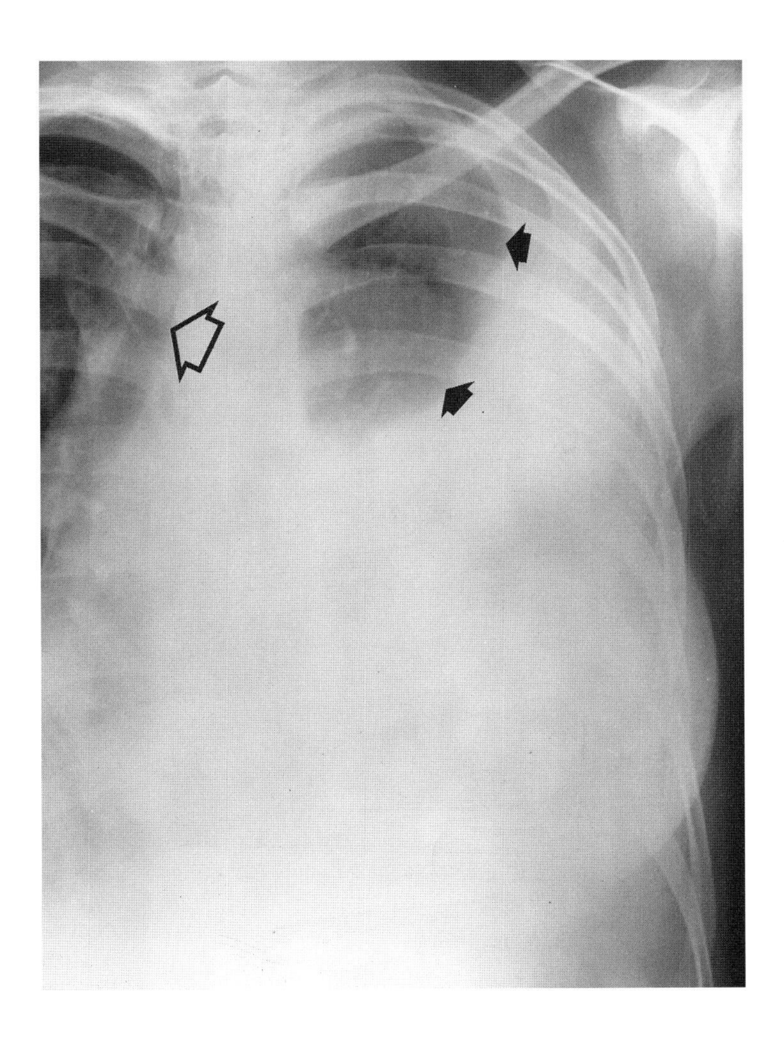

Pneumothorax

Although small amounts of pleural fluid collect far posteriorly against the diaphragm, small amounts of pleural air collect high over the cupola of the lung apex and against the upper lateral chest wall. Air may be very difficult to see there because of the overlapping tangle of bones, and you can quite easily miss minimal pneumothorax unless you are looking carefully for it. It is more obvious when the lung is less well aerated; thus a film made *at full expiration* may show clearly the margin of slightly denser lung outlined by darker pleural air against the chest wall, or a film may be made PA with the patient lying on his good side. The air in the pleural space will then collect between the lung and the chest wall, where it is easy to see. This is called a *lateral decubitus film* (Figure 7.16).

The lateral decubitus chest film is also useful in determining the presence of subpulmonic fluid. When the patient is placed on the affected side, fluid trapped beneath the base of the lung against the diaphragm can be dumped out into the pleural space against the lateral, now dependent, chest wall, as in Figure 7.17C. Without this maneuver the PA film might be interpreted as a "high right diaphragm." The lateral decubitus chest film is also useful for diagnosing small effusions in the patient who cannot be filmed upright.

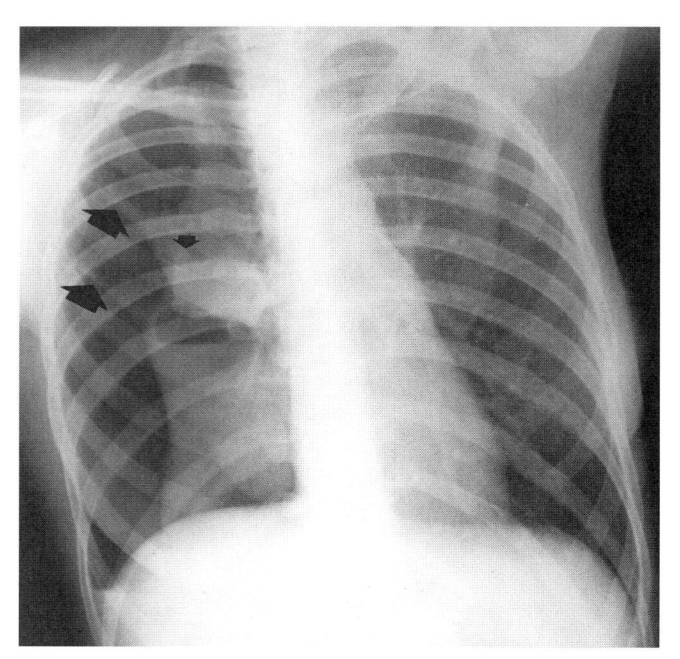

Figure 7.15. Large pneumothorax, small hydrothorax. The right lung is only partly collapsed because of the dense pleural adhesions retaining it *(large arrows)*. The *small arrow* points to the smaller of two cavities in the diseased lung. The larger cavity has a fluid level in it. You can see the three lobes collapsing separately in the pleural air. Note the small horizontal fluid-air interface in the right costophrenic angle, sure evidence that this is hydropneumothorax.

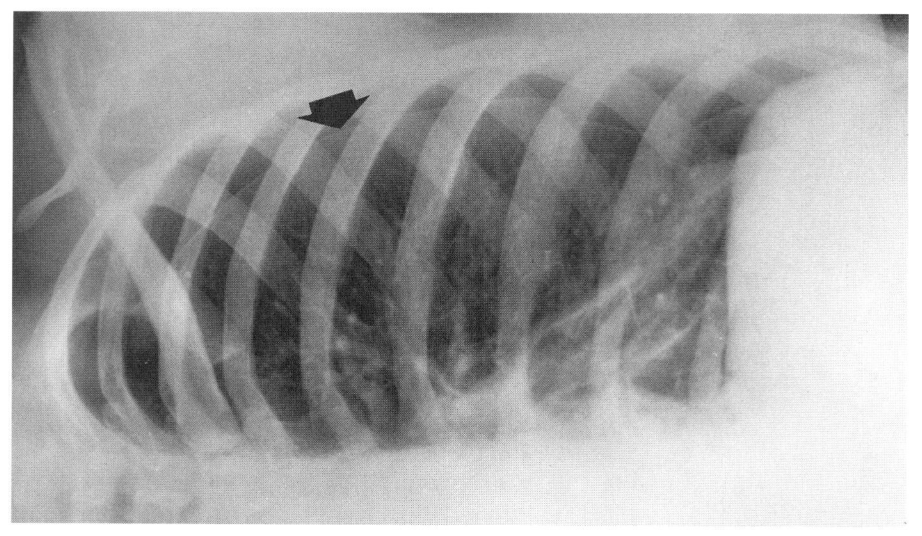

Figure 7.16. Pneumothorax seen PA in the right lateral decubitus position (patient lying on his right side, horizontal x-ray beam). Air rises to the highest point in the chest. Due to elastic recoil of the lung, there is passive atelectasis of the lung, which falls away from the chest wall in the presence of pneumothorax. The *arrow* indicates the lung margin.

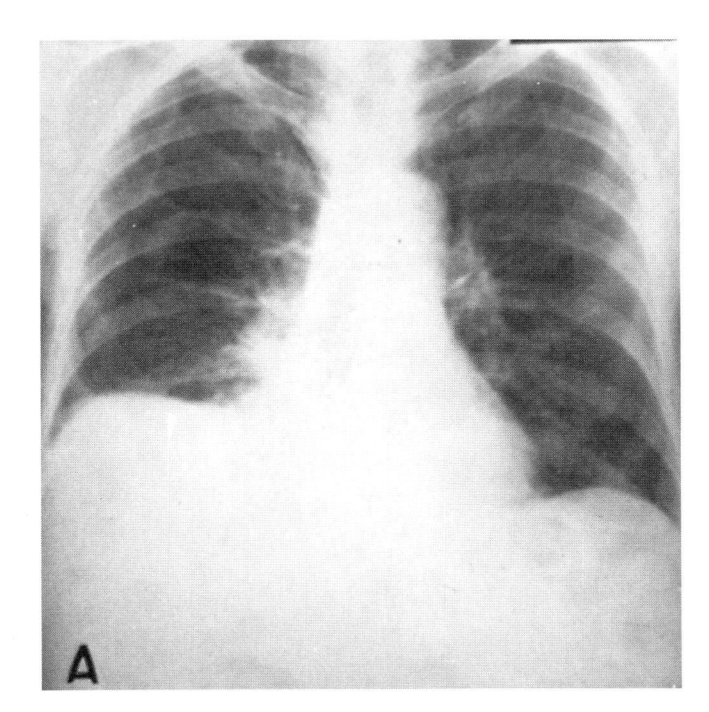

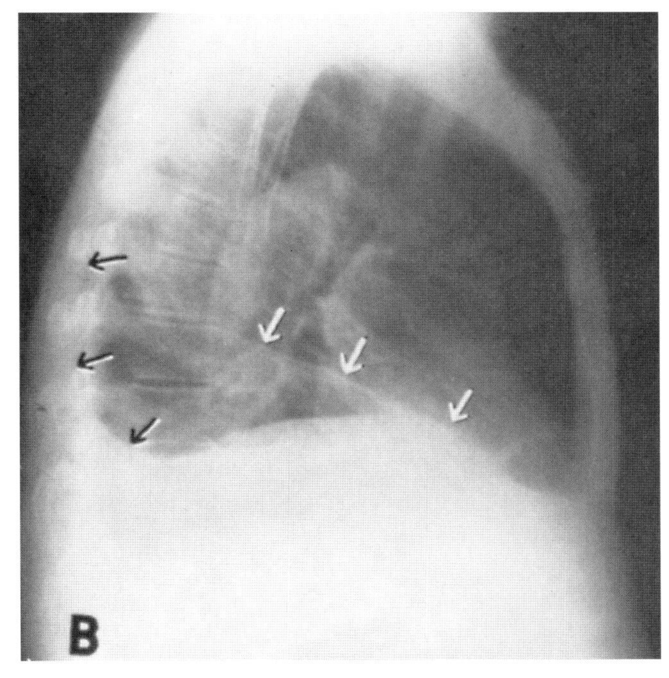

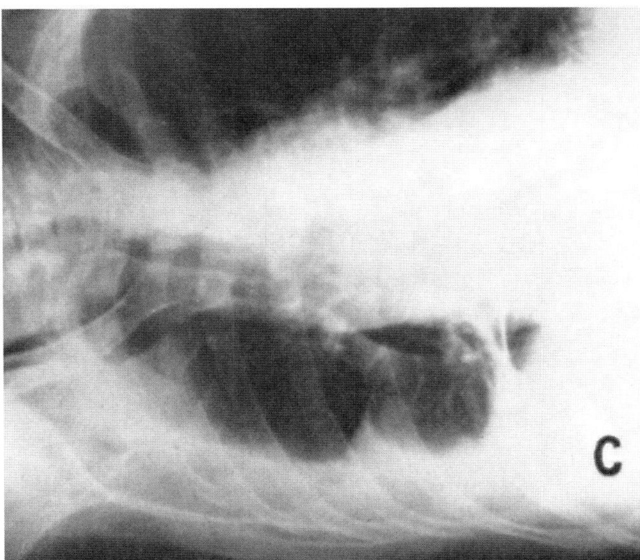

Figure 7.17. A: Subpulmonic collection of pleural fluid imitates a high right diaphragm in the PA view, although the dome of what appears to be the diaphragm is more lateral than usual. B: Lateral film shows fluid curving up along the posterior chest wall *(black arrows)* and extending into the major fissure *(white arrows).* C: Decubitus film shifts the fluid against the lateral chest wall, where it is seen to flow into the minor fissure.

Fluid may also be seen in the fissures. Remember that the planes of the two major fissures descend obliquely from high against the posterior chest wall to a point low against the anterior chest wall. Similarly, the plane of the minor fissure on the right is normally horizontal, extending forward and laterally from the middle of the major fissure at a level opposite the right hilum. Collections of fluid within these fissures will lie in the same planes, and you can look for them there. Frequently, when enough free fluid has accumulated, you can see it dipping into both major and minor fissures, as you do in Figures 7.17B and C.

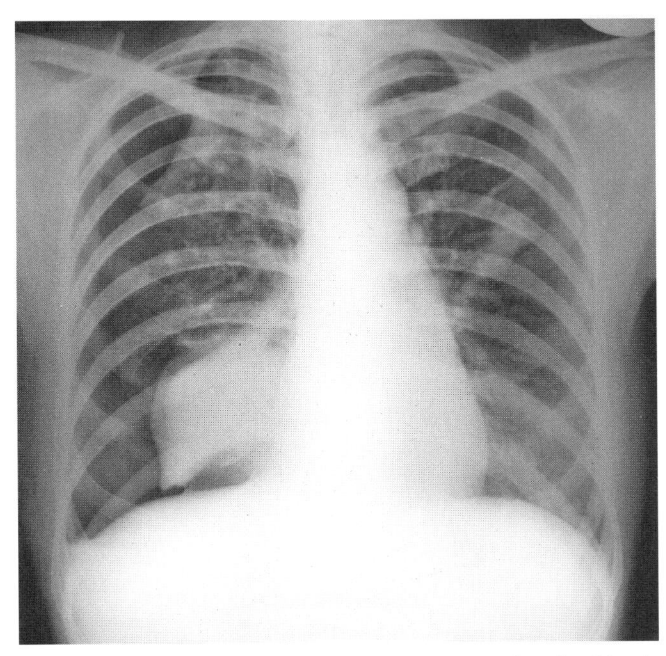

Figure 7.18 ***(Unknown 7.2).*** Clinically this patient had had fever and a cough with hemoptysis for six months but had refused medical attention. Tuberculosis was the working diagnosis. Analyze the film.

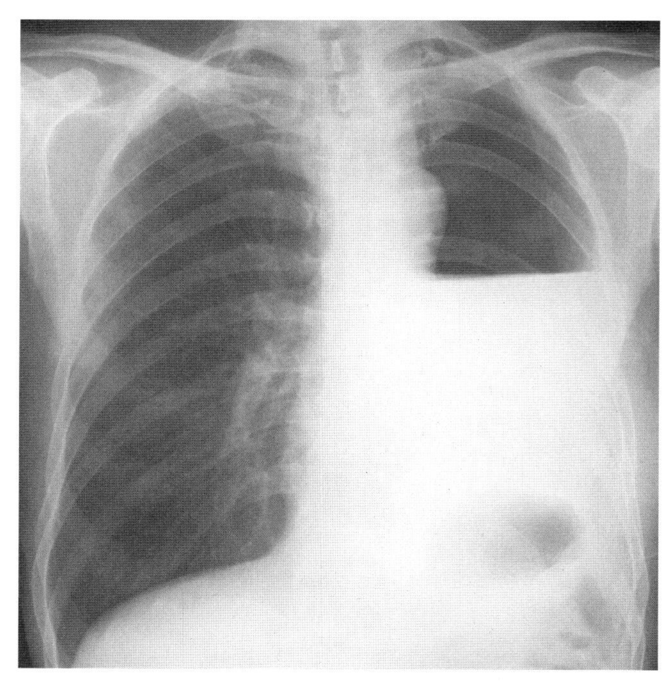

Figure 7.19 ***(Unknown 7.3).*** Locate the fluid level.

Pulmonary Embolism

The mortality rate of patients suffering acute pulmonary embolism can be significantly diminished with early diagnosis and optimum therapy. Radiological examinations such as ventilation/perfusion radioisotope lung scanning, pulmonary arteriography, and venous ultrasound play a central role in the diagnostic workup of this condition. But chest films are usually the first imaging procedure performed, and you should be familiar with the appearance of pulmonary embolism on those films.

The majority of patients with this condition have abnormal chest films, although a perfectly normal film may be seen with acute pulmonary embolism. The most common finding, when findings are present, is signs of diminished lung volume (Figure 7.20), such as linear or patchy segments of atelectasis or simply an elevated hemidiaphragm on the affected side. Localized areas of peripheral oligemia with or without distended proximal pulmonary arteries may also be seen (Westermark's sign). When the chest film *is normal* in a patient with new chest pain and dyspnea suggesting pulmonary embolism, the major contribution of the film is ruling out other causes of acute chest pain and dyspnea, such as pneumothorax, pneumonia, and rib fracture.

In approximately 10 percent of cases, pulmonary embolism results in pulmonary infarction, which may manifest radiologically as an air-space opacity or pleural effusion. Since infarction is always ischemia of pleura-based lung, you can anticipate that the opacity will appear on the chest film, at the lateral periphery of the lung, when intraalveolar hemorrhage and, later, organization occur. Remember, though, that there is a good deal of pleura-based lung which abuts the fissures or even the mediastinum, so that infarcts do not always appear laterally on the chest film. They *are* more common there, and often present as rounded opacities ("Hampton's hump") near the costophrenic sulcus above an elevated diaphragm. This infarcted piece of lung, when it is close to the diaphragm, may have its shape obscured by the presence of some pleural affiliation too, of course.

The patient with clinically suspected pulmonary embolism *usually* requires a radioisotope ventilation/perfusion lung scan and sometimes pulmonary arteriography for a definitive diagnosis.

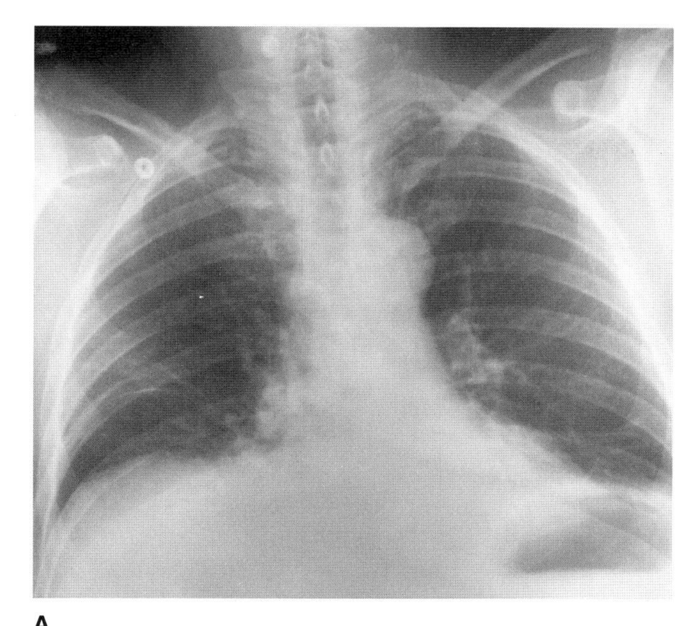

A

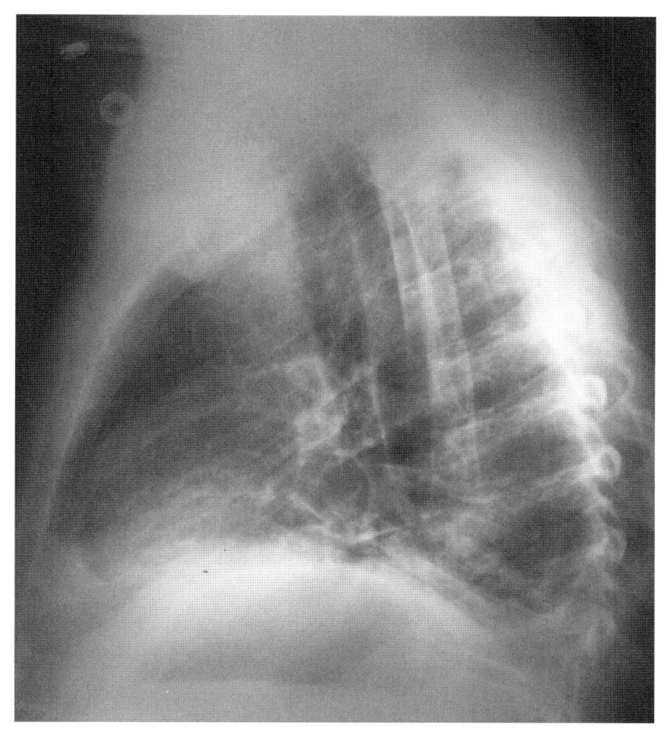

B

Figure 7.20. Plain films showing the most common chest abnormality in patients with acute pulmonary embolism: evidence of decreased lung volume, usually in the form of atelectasis. Note the low lung volumes and linear-like densities at the lung bases, apparent on both the PA film (A) and the lateral film (B).

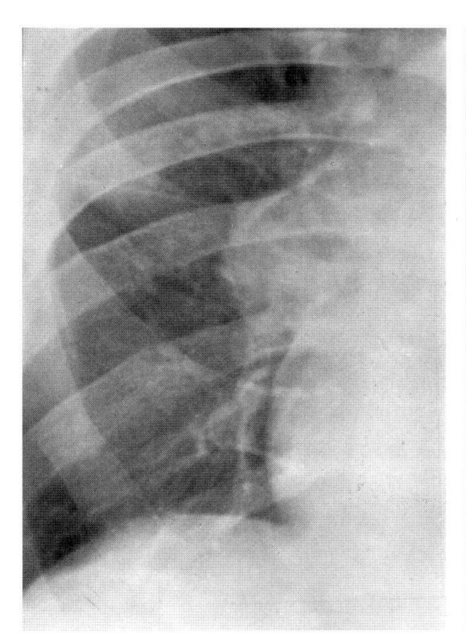

A

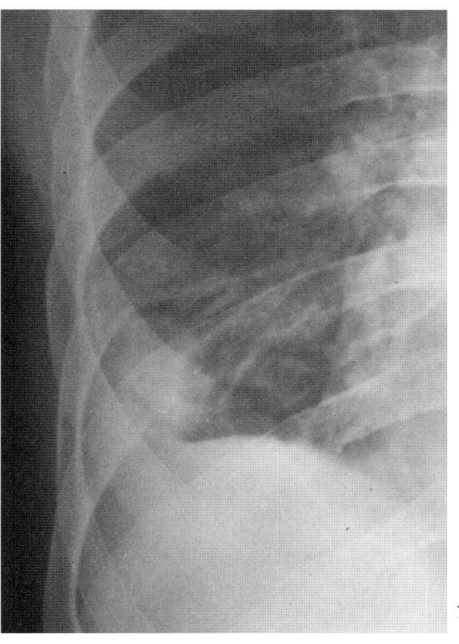

B

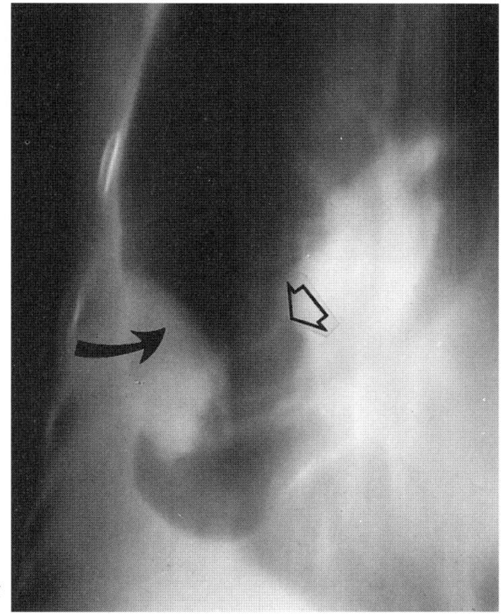

C

Figure 7.21. A and B show the early and late development of characteristic roentgen findings in pulmonary infarction of the right base against the lateral chest wall in the right lower lobe. Remember that the middle lobe does not occupy the lateral sulcus. C is a tomogram obtained at the same time as the film in B. *Curved arrow:* mound of infarcted lung (Hampton's hump). *Open arrow:* pruned hilar pulmonary artery. The small vessels seen below it are doubtless in the right middle lobe, which is still normally perfused.

Radioisotope Perfusion and Ventilation Lung Scans

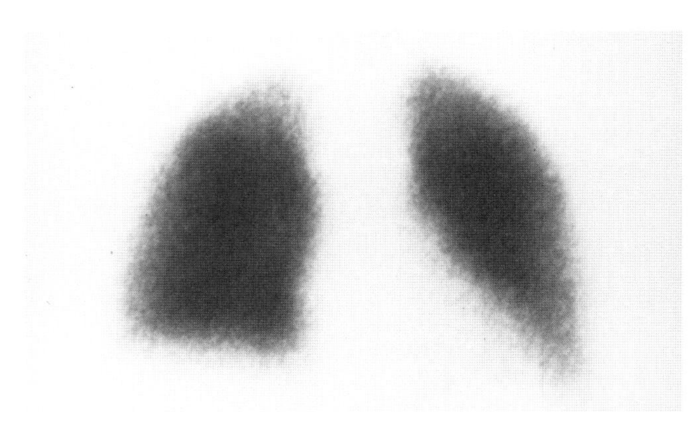

Figure 7.22. Normal perfusion scan, frontal projection.

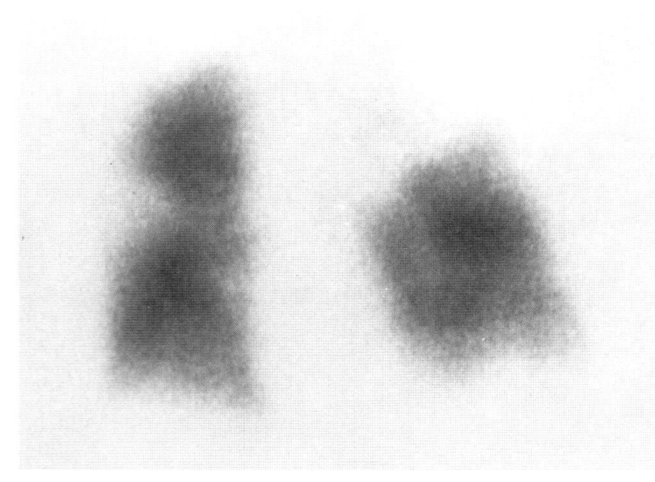

Figure 7.23. Perfusion scan in pulmonary embolism showing multiple perfusion defects.

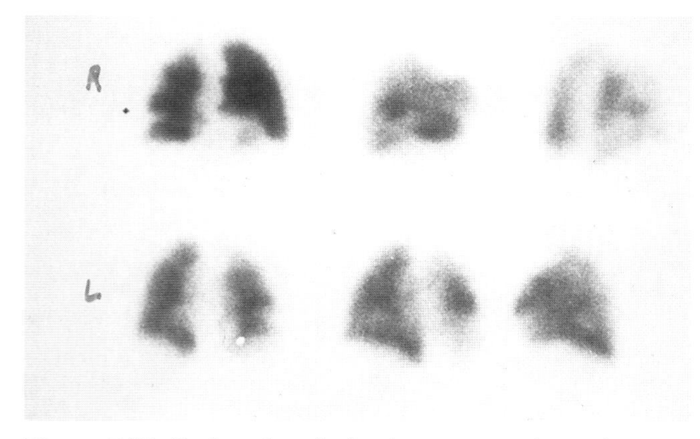

Figure 7.24. Series of perfusion lung scans of a patient clinically suspected of pulmonary embolism. *Top row:* anterior, right lateral, right posterior oblique. *Bottom row:* posterior, left posterior oblique, left lateral. Note the numerous defects in the perfusion of the peripheral lung.

Radioisotope perfusion lung scanning is performed with an intravenous injection of macroaggregated particles of human serum albumin that have been "tagged" with technetium-99m. The particles are slightly larger than erythrocytes and have the same lumen diameter as pulmonary capillaries (greater than 8 microns). Consequently, as they perfuse the lung, they are trapped in capillary branches throughout the pulmonary arterial tree, where the emission of gamma rays continues until disintegration is complete. The particles become inactive in a matter of hours. Scanning techniques carried out during this interval produce an image of the lungs (in terms of their arterial beds) that is recorded by a gamma camera from the anterior, posterior, lateral, and oblique aspects of the patient. The distribution of the trapped emitting particles is normally uniform throughout the lungs; hence the image produced should be that of two blackened lung-shaped shadows with the slightly asymmetrical heart shadow between them—white because it has no trapped particles emitting rays (Figure 7.22).

This procedure is useful in the diagnosis of pulmonary embolism because emboli block pulmonary artery branches. The lung tissue peripheral to that block is therefore *not* perfused with the isotope, and a "defect" (nonblackened area) is produced on the scan (Figure 7.23).

If the scan of a patient suspected of pulmonary embolism shows no perfusion defect, the scan is said to be normal and the patient can be presumed not to have an embolism. When the perfusion scan is abnormal, a ventilation scan is performed. This procedure is carried out by the inhalation of a radioactive gas (usually xenon-133). In this way the degree of ventilation of all parts of the lung can be imaged. A number of disease conditions of the lung do cause alterations in ventilation (pneumonia, emphysema, tumors), but uncomplicated pulmonary embolism does not. Thus a patient clinically suspected of having a pulmonary embolus who has a *perfusion scan (Q) defect* (as in Figure 7.24) *but a normal ventilation scan (V)* (Figure 7.25) very probably *has* an embolus *(V/Q mismatch).*

In older patients the presence of emphysema, often not apparent on the chest film, creates special problems in the diagnosis of pulmonary embolism. In these patients, *matching scans (V/Q match)* generally indicate a segment of abnormal lung causing defects on both scans and indicating both underperfusion and underventilation of that segment, probably not an

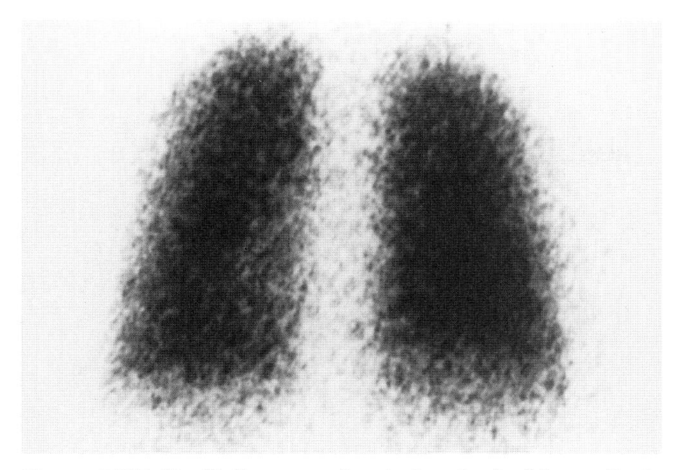

Figure 7.25. Ventilation scan (posterior view) of the patient in Figure 7.24 is normal.

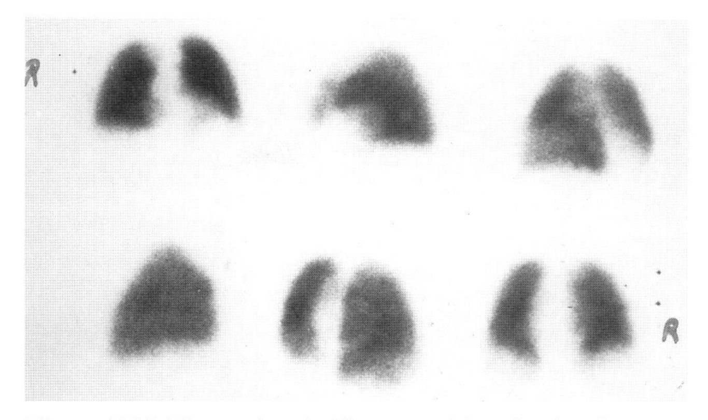

Figure 7.26. The patient in Figures 7.24 and 7.25 after 16 days of anticoagulation. The perfusion scans (obtained in a sequence different from that in Figure 7.24) are normal. The emboli have resolved.

embolus. But even patients with emphysema can have pulmonary emboli. In such patients, when the clinical suspicion of embolus is high and life-threatening, a diagnosis of embolism must be established with the help of angiography. The embolus itself can then be visualized radiographically as a lucent filling defect in a blocked artery (Figure 7.27), and the patient can be appropriately treated in order to prevent recurrent embolism, which may be fatal.

When you refer a patient for a lung scan, four different scan results are possible to assist you in planning the next step in your patient's workup. A completely *normal* perfusion scan rules out pulmonary embolism and makes a ventilation scan unnecessary. A *negative* or *low-probability* scan shows one or more minor perfusion abnormalities or defects with ventilation/perfusion matches (V/Q matches), which are thought to represent not emboli, but abnormalities related to other pulmonary conditions. A *positive* or *high-probability* scan shows two or more large or moderate-sized segmental perfusion/ventilation mismatches (V/Q mismatches). An *intermediate* (or *indeterminate*) scan shows features of both low- and high-probability scans.

Pulmonary arteriography is a more accurate method than radioisotope lung scanning for diagnosing pulmonary embolism, but it is a more invasive and more expensive procedure. It is usually indicated today when (1) the results of the lung scan are uncertain (indeterminate scan), (2) when the lung scan is interpreted as positive (high-probability scan) in a patient at high risk for anticoagulation, or (3) when the lung scan is interpreted as negative (low-probability scan) in a patient with overwhelming clinical evidence of pulmonary embolism. If the lung scan is interpreted as "normal," arteriography is not needed. Remember, a *normal* lung scan rules out pulmonary embolism and should be differentiated from a *negative* lung scan, which is not completely normal; the minor abnormalities on a negative lung scan could be caused by small emboli.

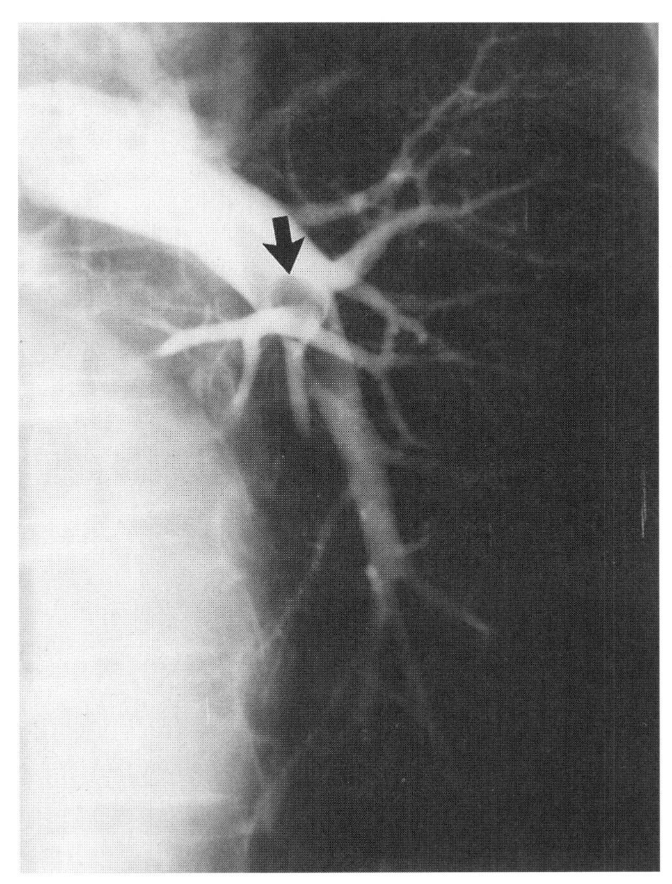

Figure 7.27. Pulmonary arteriogram of a patient with a large embolus *(arrow)* in the left lower lobe artery.

8 Lung Overexpansion, Lung Collapse, and Mediastinal Shift

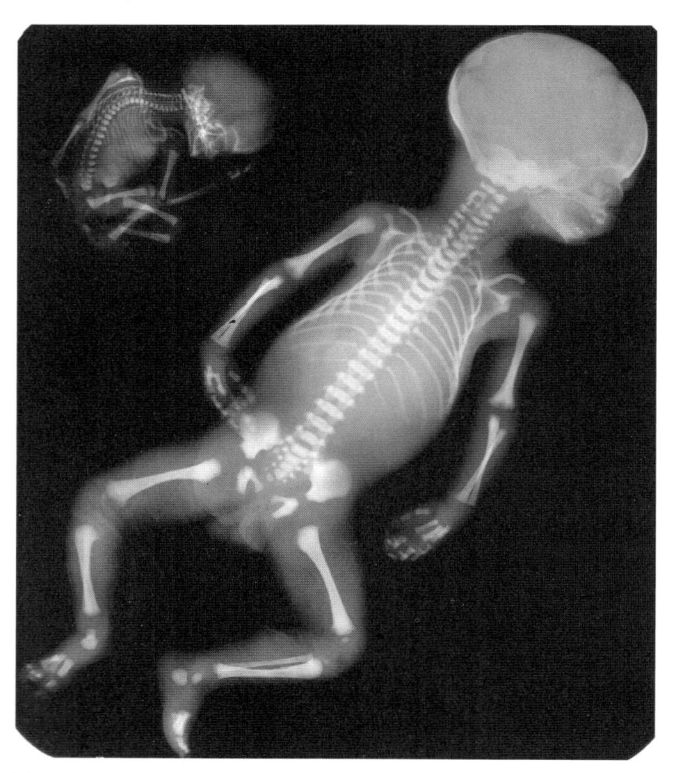

Figure 8.1. Stillborn twins, one dead at three months, one born dead at term. Note that the lungs, heart, and abdominal structures blend into one uniform shadow.

When lung tissue is inflated with more than its normal content of air, it becomes more radiolucent than usual. No matter how clearly radiolucent the normal tissue on a chest film now looks to you, remember that it is quite logical that a cube of air should be more radiolucent than the same cube of air traversed by blood-filled capillaries. As you would expect, then, an x-ray of an overexpanded lung will show *increased radiolucency,* the affected lung appearing too dark with the exposures used for chest work. In addition, the blood vessels will be spread apart as they are separated farther and farther by the ballooned alveoli. In obstructive emphysema localized to one segment of lung, this appearance may be so exaggerated as to be confused with pneumothorax.

Atelectasis, on the other hand, causes the lung to appear less radiolucent than usual, and atelectasis of one lobe will be seen first as a difference in density between the two sides of the chest film. Thus you will be looking for an unaccountably dense area in the lung. You have already seen diffusely increased density due to the high position of the diaphragm in a film made at peak expiration, and you realize that that amount of decreased radiolucency at some phase of the respiratory cycle goes with every breath your patient takes.

You see in Figure 8.1 the film of a stillborn infant who has never breathed at all. The lungs and bony thorax are collapsed about the heart and mediastinal structures as one uniformly dense shadow within the rib cage, and they blend continuously with the shadows of the dense abdominal structures. The tracheobronchial tree is filled with amniotic fluid.

In Figure 8.2 the arterial tree of a segment of lung has been injected with opaque fluid and sealed off, inflation and deflation of the lung being carried out through a tube tied into the bronchus. In Figure 8.2A you see the lung collapsed around its arterial tree in just about the same degree of hypoaeration that would exist near the diaphragm at full expiration. In Figure 8.2B the specimen has been inflated to approximate the lung near the diaphragm at deep inspiration.

In Figure 8.3 you see the lungs of a woman at expiration and inspiration. At expiration the right hemidiaphragm is at the level of the ninth rib, and at inspiration at the level of the tenth. Note the improved aeration of the lungs on the inspiration film, especially apparent at the lung bases. In this chapter you will learn the changes in the appearance of the chest film when both lungs are underexpanded and when both are overexpanded. Then we will consider the changes seen when the volume of one hemithorax is altered enough to produce a shift of the mediastinum from its midline position.

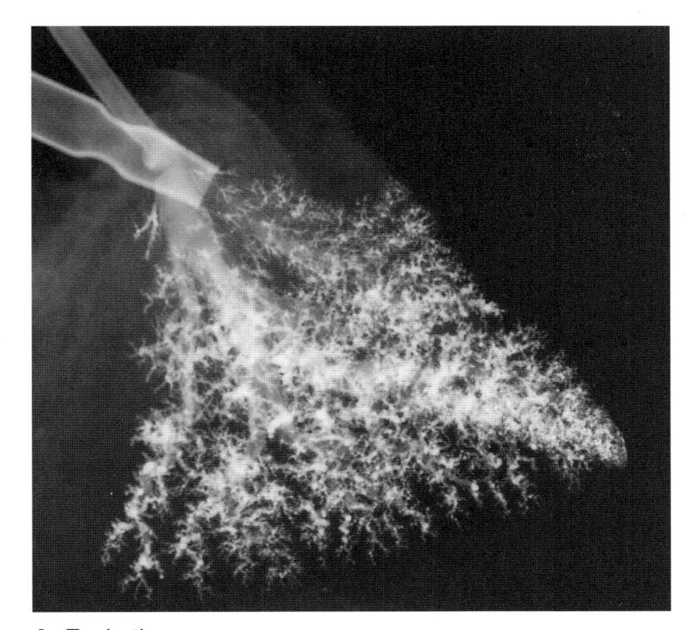

A. Expiration.

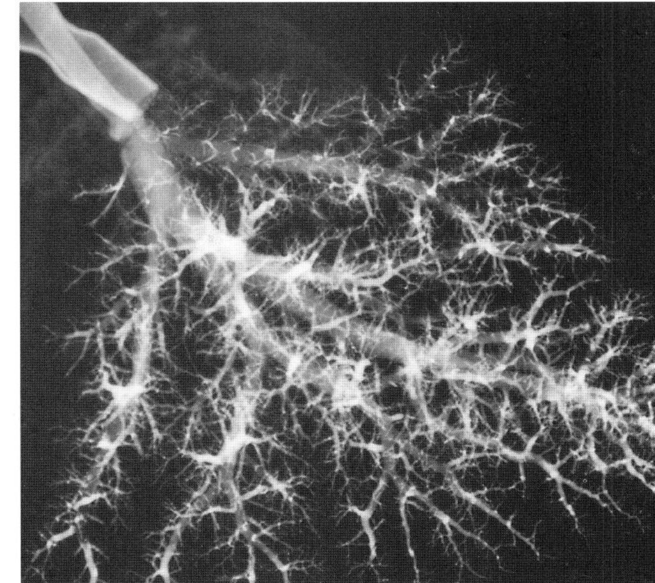

B. Inspiration.

Figure 8.2

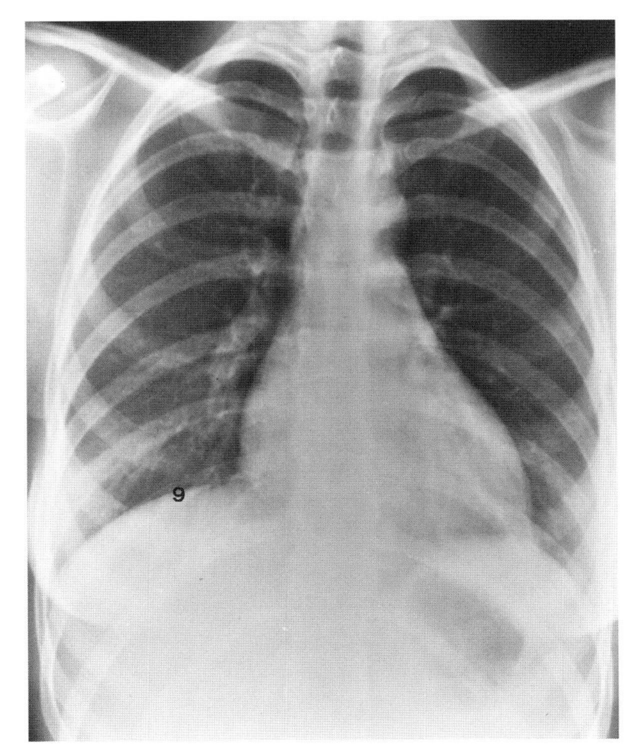

A. Expiration.

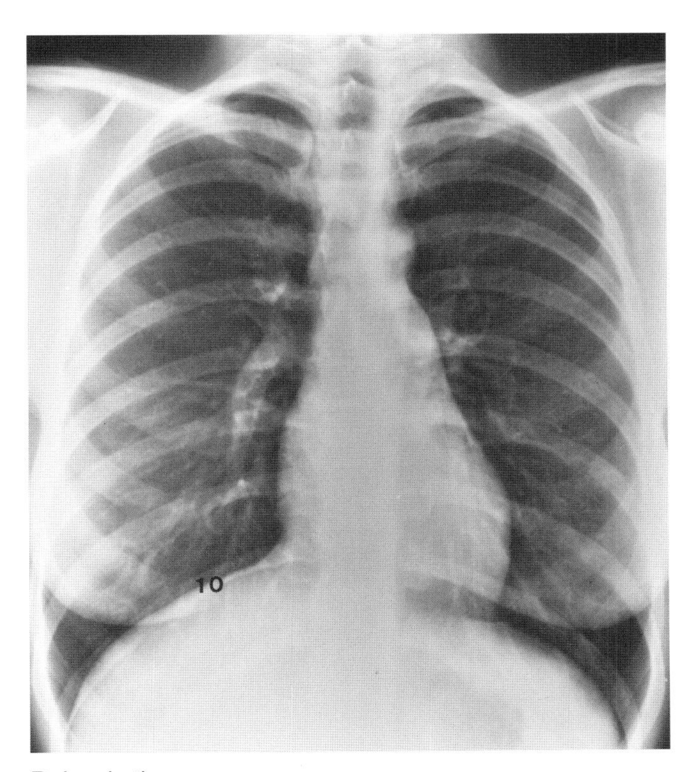

B. Inspiration.

Figure 8.3

Emphysema

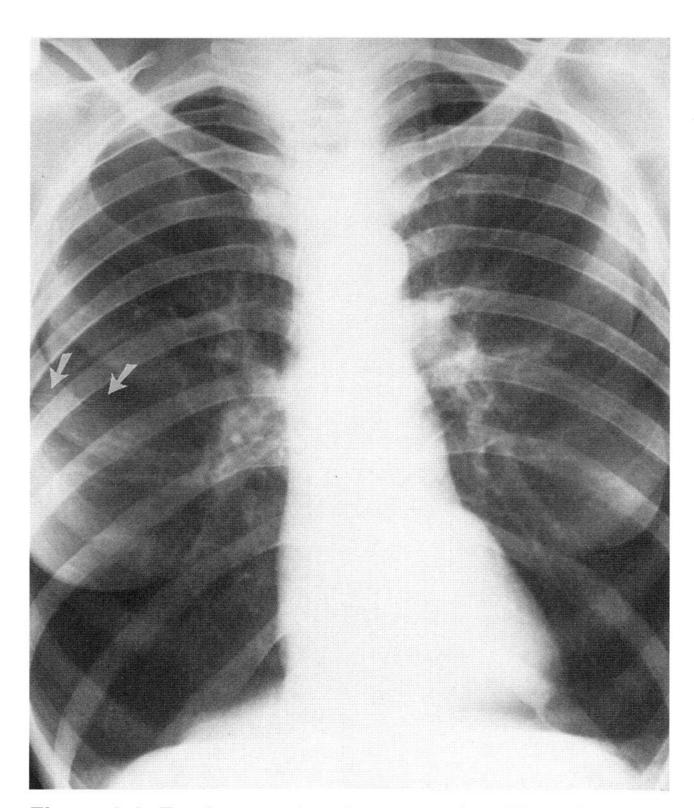

Figure 8.4. Emphysema in a heavy smoker. There is marked overaeration and hyperexpansion of both lungs. Note the very low position of the diaphragm; the right hemidiaphragm is below the level of the eleventh rib (last rib seen in the right lower chest). The heart appears normal in size, but remember that right ventricular enlargement (cor pulmonale in this case) is difficult (or may be impossible) to recognize in the PA view alone. In the lateral view the lower half of the anterior clear space may be obliterated by forward extension of the massive right ventricle. The minor fissure *(arrows)* is slightly elevated and should be distinguished from the walls of the blebs seen in Figure 8.5.

With chronic emphysema the lungs usually are both overexpanded, the diaphragm low, flattened, and often serrated. Many cases will be obvious to you from the increased radiolucency you will see at the usual roentgen exposures for chest work. Lesser degrees of generalized emphysema may be much less obvious. If such patients underwent fluoroscopy, the radiologist would see a diaphragm that moves down only slightly on inspiration and returns only very slowly on forced expiration. In many patients with emphysema the concomitant development of pulmonary fibrosis adds the shadow of a web of filamentous strands of increased

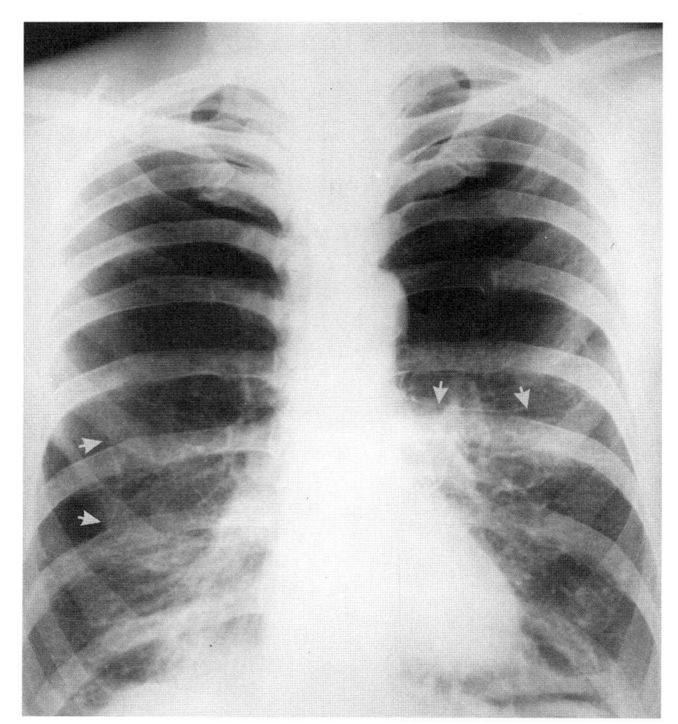

Figure 8.5. Widespread emphysema in a patient with chronic bronchitis. Multiple bullae in the apices increase the lucency of the upper lungs. The *arrows* mark walls of bullae.

opacity. These strands radiate outward from the hilum through the lung. Localized emphysematous bullae may be seen anywhere in the lung, like huge air cysts bordered by dense thin walls that enclose them. Rupture of such bullae, producing spontaneous pneumothorax, is not unusual.

Sometimes it is difficult to be certain from the PA chest film that the lungs are overinflated. When you look at the lateral, however, the situation becomes much more convincing and shows hyperinflation with flattening of the diaphragm and increase in the AP diameter of the chest. Thin-section (high-resolution) CT, which has become available in recent years, can beautifully delineate lung parenchyma and may be used to confirm the chest films (as in Figure 8.6B), to identify bullae, and to classify the type of emphysema.

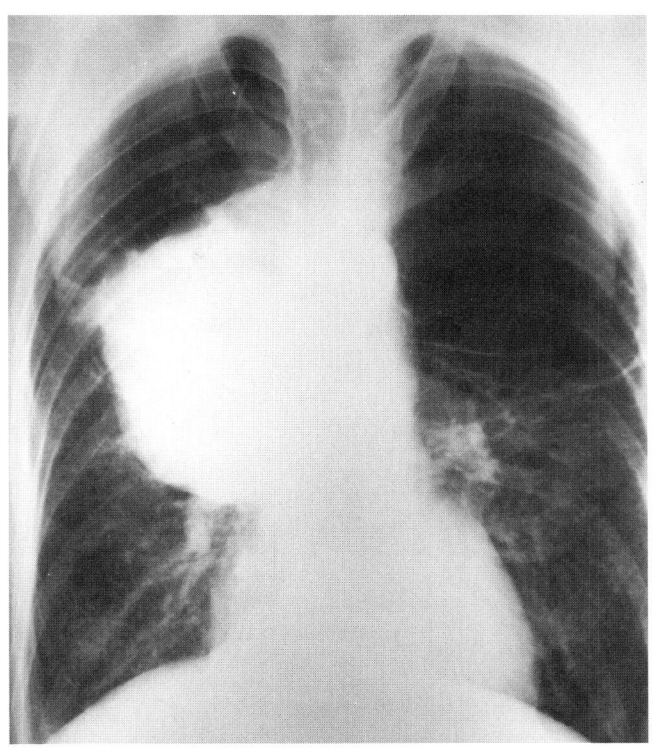

A. Bullous emphysema in a patient who smoked for years and now has also developed a lung cancer (the mass projecting to the right).

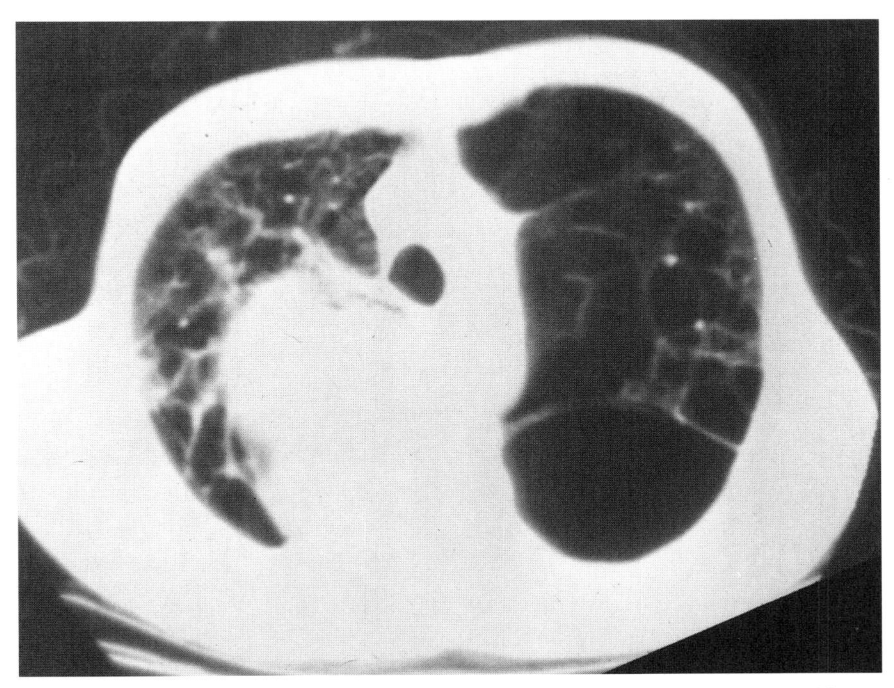

B. CT scan, same patient, with lung window settings. Note the bullous destruction of the left upper lung, which is useless. Imagine what perfusion and ventilation scans might show on this patient.

Figure 8.6

Normal Mediastinal Position

Some describe the mediastinum as a region. We prefer to think of it as a bundle of structures sandwiched between the two inflated lungs. With the exception of the air-filled trachea and main bronchi, all these structures have the same radiodensity and merge into a homogeneous shadow superimposed upon that of the spine in the PA projection. Thus the shadows of the mediastinal structures cannot be separated from one another except by a variety of special procedures employing contrast substances, computed tomography, and magnetic-resonance imaging. On the routine chest film, only the lateral margins of the mediastinum outlined by air in the lungs on either side can be identified.

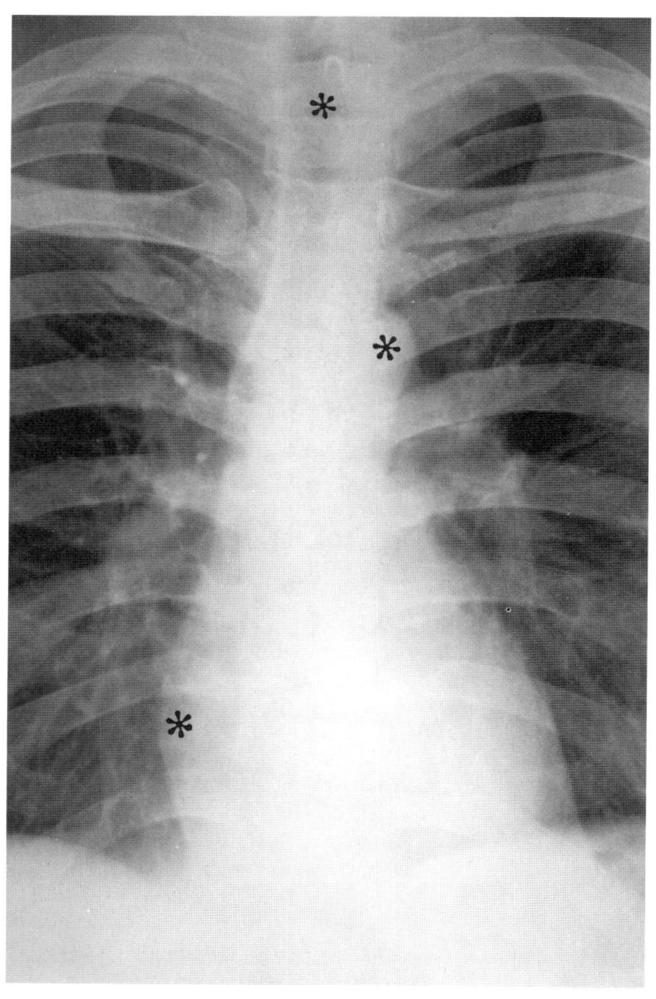

Figure 8.7. Well-exposed film showing the tracheobronchial tree. Markers indicate the three tag points by which you can determine the position of the mediastinum: the tracheal air column, the aortic arch, and the right heart border.

With changes in the air content of either lung, or with unilateral large accumulations of pleural air or fluid, the mediastinum will bow to one side like an elastic diaphragm. You will need to identify a few anatomic points along the margins of the mediastinal shadow and know their normal locations if you are to be able to recognize mediastinal displacement on a routine PA chest film. There are three of these signal points that ought to be included in your systematic chest survey.

The first is the column of air in the trachea, visible as a dark vertical shadow on the PA chest film, normally a little to the right of the midline as it approaches the carina. The second is the white knob you see to the left of the spine at about the fifth rib posteriorly. This knob is the shadow margin created by the arch of the aorta as it swings posteriorly and turns downward to become the descending aorta. You see it, of course, only because there is radiolucent lung tissue abutted against it, and it will disappear if that lung tissue becomes airless or if a dense mass lies against the aortic arch.

Finding the trachea and the aortic arch in their usual locations, then, will tell you that the upper mediastinum is where it ought to be. When there is a marked decrease in the amount of air in the right upper lobe, for example, the trachea will be found shifted toward that side. The arch of the aorta will be pulled with it toward the midline, its shadow disappearing as it becomes superimposed upon that of the spine. In just the same way, the trachea and the aortic arch may be displaced to the left when there is a decrease in the air content of the left upper lobe. If you review the anatomic relations of the aortic arch, you will find that normally both upper and lower lobes adjoin it. For this reason, if there is no air at all in the left upper lobe it will lie, dense and much decreased in size, against the anterior mediastinum, and you may actually still see the aorta through it, illumined by the overexpanded superior segment of the lower lobe. Check the position of the trachea and the aortic arch on some of the chest films you have seen so far.

The third tag point in determining the position of the mediastinum is the shadow of the right heart border. Major changes in the size of either *lower* lobe will swing the heart to one side, and it will look displaced. You may think that since we have not yet discussed the heart and cardiac enlargement, you will not be able to use the right heart border with much

Figures 8.8 and 8.9. CT sections illustrating shift of the mediastinum.

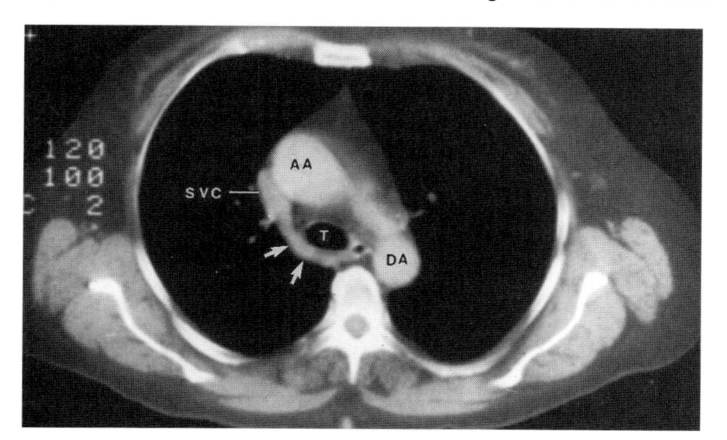

Figure 8.8. A normal scan. Both this scan and the one in Figure 8.9 were made just below the level of the aortic arch. You can identify the ascending *(AA)* and descending *(DA)* aorta, which have been opacified with intravenous contrast material. The superior vena cava *(SVC)* lies against the ascending aorta on the right and is receiving blood from the azygous vein, via the forward-swinging azygous arch *(arrows)* encircling the air-containing radiolucent trachea *(T)*. The trachea and lungs are much more radiolucent than the fat-containing triangle of anterior mediastinum extending forward to the sternum in the midline. The vascular structures have all been enhanced by intravenous contrast medium.

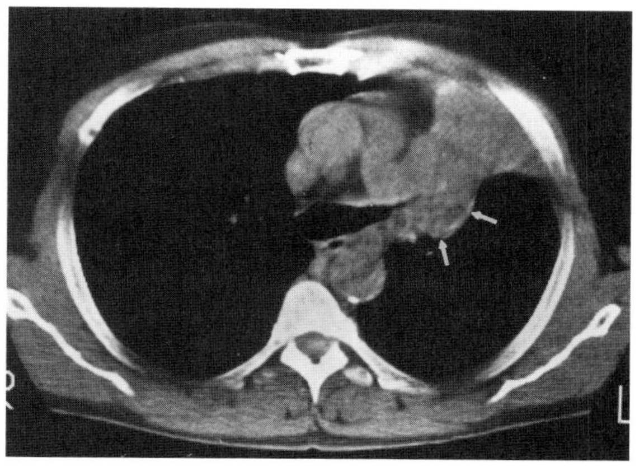

Figure 8.9. Scan of another patient, whose ascending aorta and superior vena cava (unopacified in this scan) have swung to the left. This patient has a bronchogenic carcinoma that has obstructed the left upper lobe bronchus, causing atelectasis of the left upper lobe. The collapse of the upper lobe has greatly diminished the volume of the left lung, so that the mediastinal structures are pulled to the left, the right lung expanding to fill the space. Note that the ascending aorta has rotated to the left from its normal position in the right chest. You can see the rounded mass of the tumor *(arrows)* and the concave margin of the major fissure beyond it.

sense of security, but be reassured. Up to now we have shown you very few abnormal hearts and a good many normal chest films. Look back over some of them at the right border of the heart as it curves down toward the diaphragm. By the time you have looked at a dozen or so, you should be convinced that the border of the normal right heart shadow is generally about a fingerbreadth beyond the right border of the spine on a full chest film of 14 × 17 inches (and proportionately less on these reductions). Of course, this is a very rough working rule; you will learn how to modify it as you look at more and more films and appraise more and more enlarged hearts.

Obviously, elevation of the diaphragm compressing the liquid-filled heart from below will exaggerate the lateral projection of both heart borders. Accuracy about mediastinal position, therefore, will depend on your having counted the ribs so that you are sure the diaphragm is drawn down well. Obviously too, if the right middle lobe lying against the right heart is consolidated, that border will disappear and cannot be used in tagging the position of the lower mediastinum. Also, the tag points will appear displaced in a patient with even minor degrees of scoliosis.

If, however, you are satisfied that the diaphragms are well down, that the clavicles and ribs are symmetrical and show no rotation, and that the right heart border appears to be in about its usual position, you can then say that the lower mediastinum is not appreciably displaced.

If the *whole lung* on one side collapses, all three tag points will show a shift in position, since the whole mediastinum swings to that side. If *only an upper lobe* is involved, you may find that the trachea and the aortic arch are shifted while the right heart border is not.

Mediastinal Shift

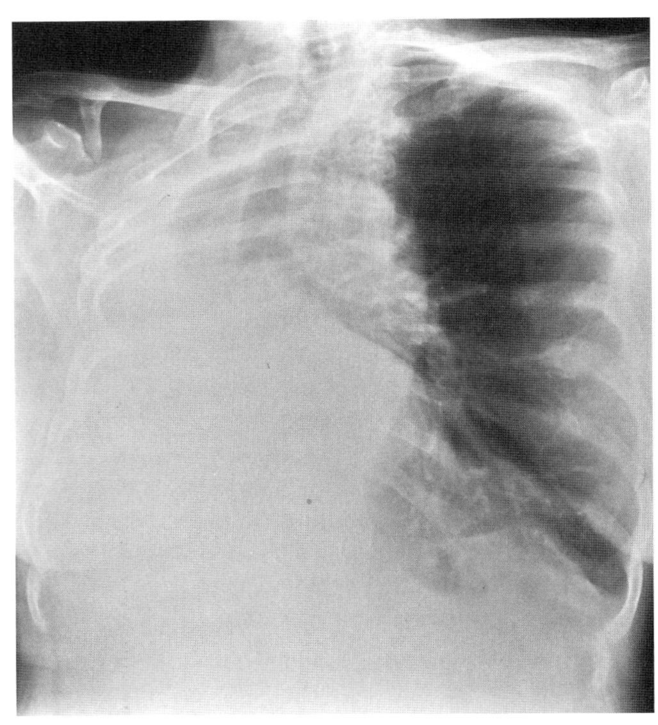

A. PA chest film shows marked shift of the mediastinum and heart into the right chest. All three tag points are deviated to the right. The right heart margin (not seen) is against the right chest wall; the trachea and aortic arch have swung to the right. The upper portion of the descending aorta still remains on the left.

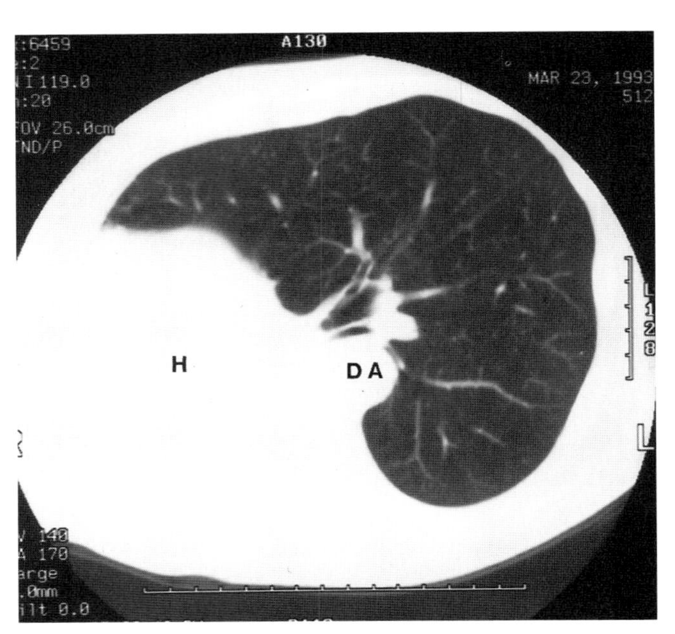

B. Lung window CT scan at the level of the heart shows the heart *(H)* entirely within the right chest. Note the hyperexpansion of the left lung to compensate for the increased volume of the left hemithorax. The anterior portions of the left lung swing across the midline, "herniating" into the right chest. The descending aorta *(DA)* is in its normal position.

Figure 8.10. Mediastinal shift after total right pneumonectomy.

The mediastinum may be displaced *permanently* (as by surgical removal of the whole lung, as shown in Figure 8.10); *temporarily* (as when a large pleural effusion develops); or *transiently* (as when a foreign body in one major bronchus interferes with the inflation or deflation of that part of the lung during each inspiration). Permanent and temporary displacements of the mediastinum will usually be appreciable from the PA chest film, and you will be looking for them every time you check the three tag points in your systematic chest survey.

Because any single chest film represents only the state of affairs in your patient's chest at one particular fraction of a second in time, it is quite possible to expose the film when the mediastinum is in midposition, even though there is a definite mediastinal shift during some other phase of respiration. The single PA film made routinely at inspiration, which you hold in your hand later that day, may give no indication at all that there was a *transient shift of the mediastinum* at expiration. If you suspect there may be one, you ought to ask for *films made at inspiration and expiration,* a commonly used device for documenting transient mediastinal shift.

The mediastinum may be *pushed* to one side by pressure from an overexpanded lung, as in obstructive emphysema, when air is drawn into that part of the lung with every breath but incompletely expelled. A check-valve foreign body may do this (see Figures 8.11A and B).

Obstructive emphysema probably also exists at some point in the natural history of any endobronchial tumor, although the obstruction is soon completed as the tumor grows, and the lung beyond it

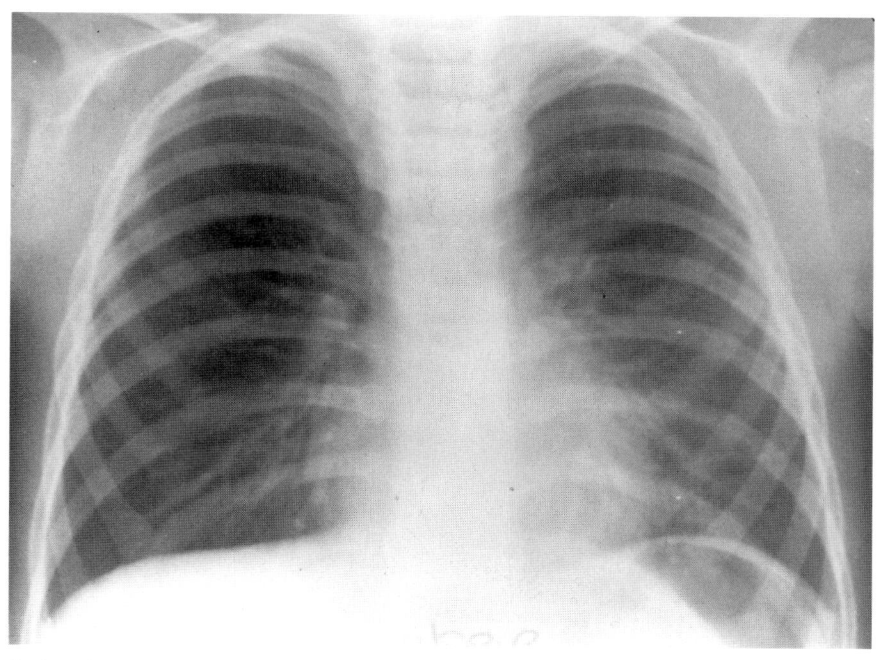

A. Inspiration.

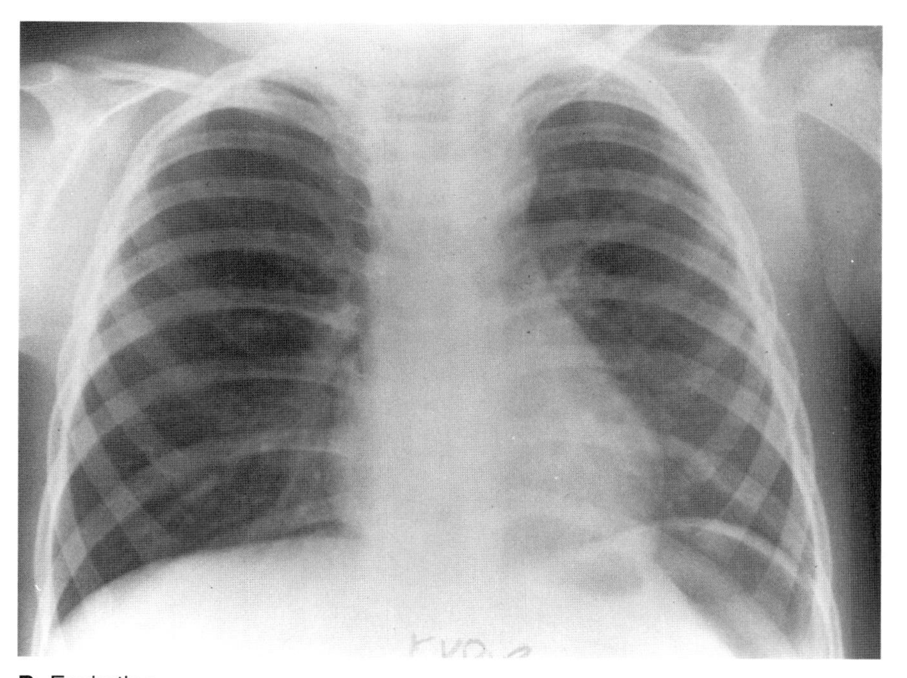

B. Expiration.

Figure 8.11. Transient mediastinal shift with a radiolucent endobronchial foreign body in the right main bronchus. At inspiration (A) the tracheal air shadow lies normally over the spine to the right of midline. The aorta and right heart border are in normal position. At expiration (B) the right lung is unable to deflate; it is the left lung which *can* deflate, so that the trachea, aortic shadow, and right heart border all shift to the left as the mediastinum moves away from the side where there is air trapping (greater relative radiolucency of the right lung). The right lung does not change size between the two films, but the right heart border moves closer to the spine at expiration. A piece of chewing gum was recovered from the right main bronchus.

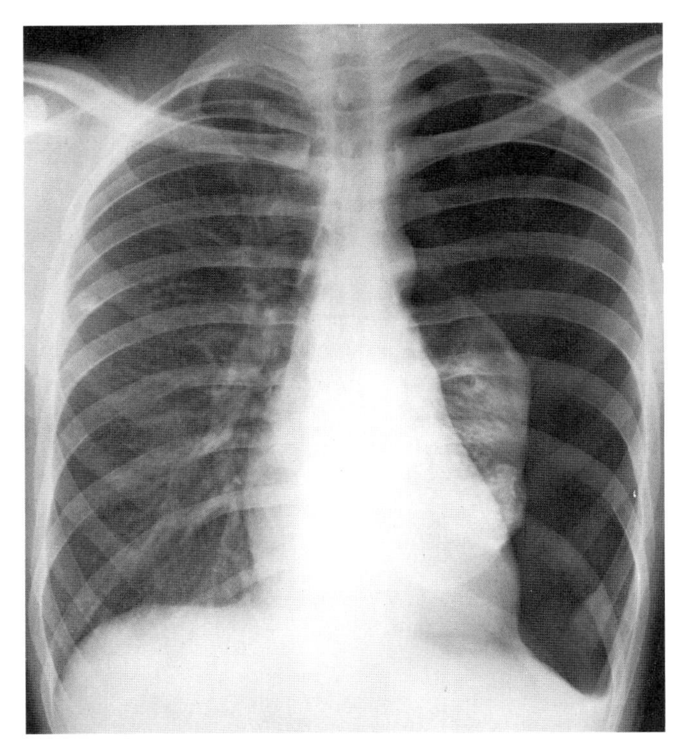

Figure 8.12. Pneumothorax with the mediastinum in the midline. The volume of pleural air is compensated for here by the degree of collapse of the left lung.

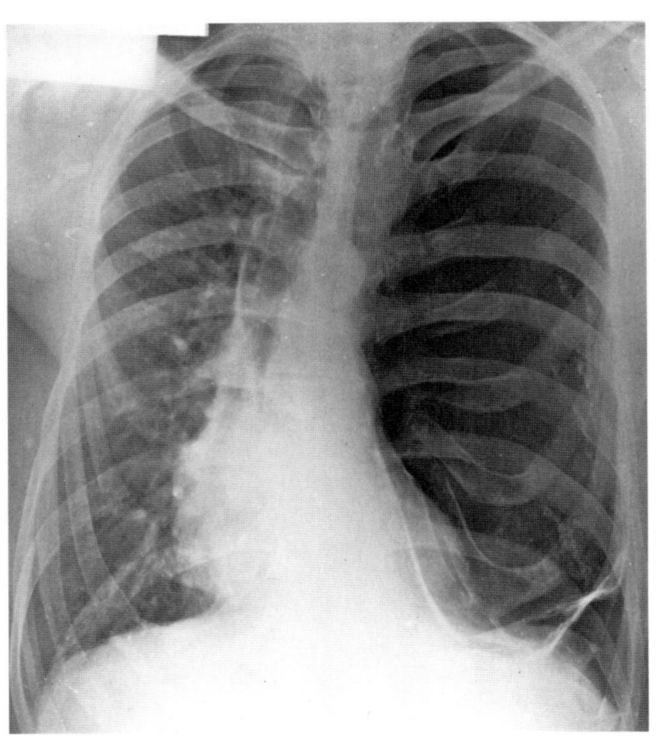

Figure 8.13. Obstructive bullous emphysema of part of the left lung, causing shift of the mediastinum to the right. No pneumothorax was present. Note that you do not see the outer surface of the lung, as you did in Figure 8.12.

collapses as its air is absorbed or escapes to other segments. On serial films you would be able to observe the mediastinum at first displaced *away* from the side of the lesion by the overexpanded lung and, a week or so later, displaced *toward* the side of the lesion as the affected lobe collapses—another illustration of the value of serial films, of the proper evaluation of changing roentgen signs, and of constant discussion with the radiologist.

If you make a practice of thinking of the mediastinum as a flexible disc held in the midline *whenever the volumes of the two hemithoraces are equal,* you will not find it difficult to understand and remember how mediastinal shift occurs. The mediastinum *must* shift whenever there is a significant change in the volume on one side. Massive pleural effusion shifts the mediastinum to the opposite side. After pneumonectomy, or in massive collapse of one whole lung, pronounced mediastinal shift also occurs (Figure 8.10). (Unilateral bullous emphysema may shift the mediastinum, compressing the good lung, as in Figure 8.13.)

The mediastinum may *not* shift if the various additions and subtractions in volume on one side cancel each other out, so that the volume of the abnormal hemithorax remains equal to that of the normal side. For example, in Figure 8.12 you see a patient with a large pneumothorax. Air trapped in the pleural space has added to the volume of the left hemithorax, but at the same time the left lung has collapsed to one-third its normal volume. Note that the mediastinum remains in the midline. Of course, with tension pneumothorax the mediastinum shifts away from that side.

In addition, the mediastinum may not shift if it has become fixed as a result of adhesions subsequent to inflammation, or because of tumor invasion. Furthermore, it may be checked in its displacement by pleural adhesions that prevent the full collapse of one lung.

Mediastinal Shift due to Abnormality of One Whole Lung

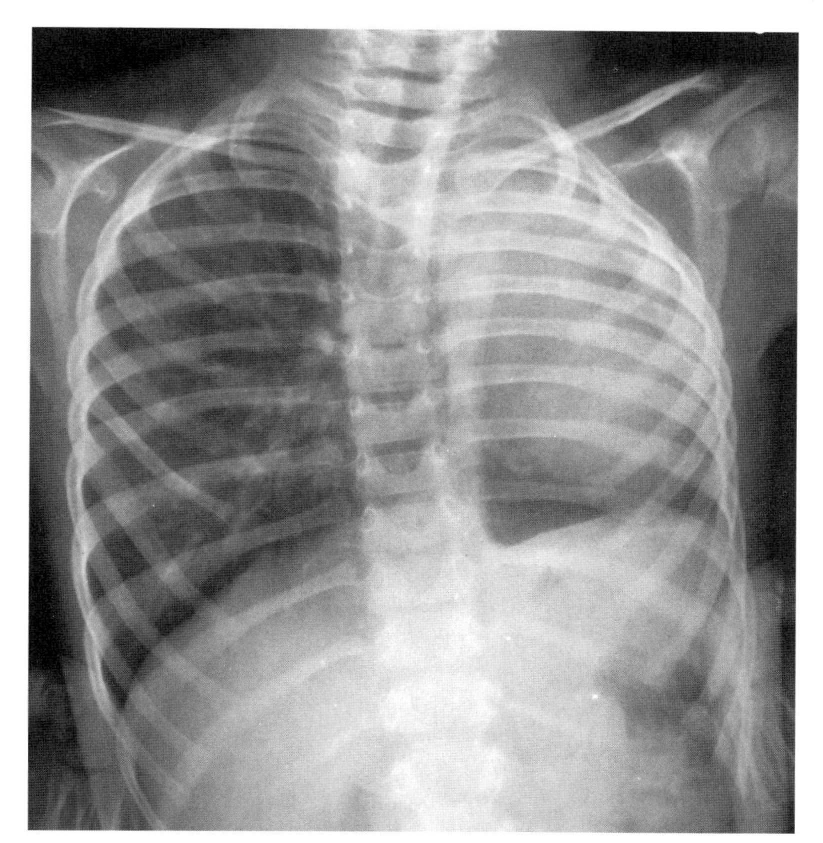

A

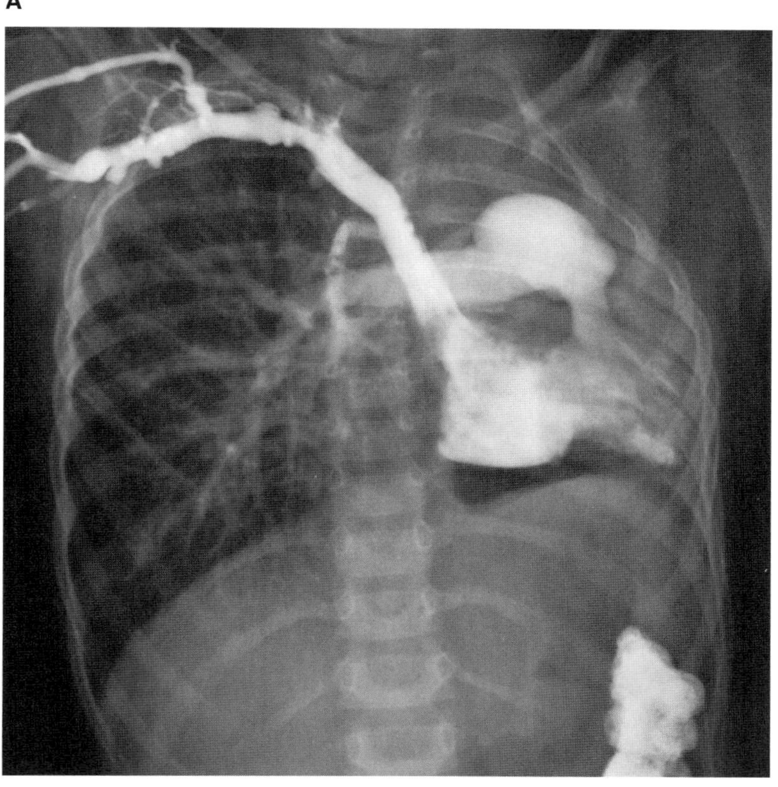

B

Figure 8.14. Permanently displaced mediastinum in a 4-year-old who had agenesis of the left lung and compensatory overexpansion of the right lung coming across the midline anteriorly. A: The dense mass in the left hemithorax is the heart. B: Angiographic contrast material injected into a right arm vein opacifies the arm veins, the right subclavian vein, superior vena cava, right atrium, right ventricle, and pulmonary artery circulation. Note the absence of the left pulmonary artery.

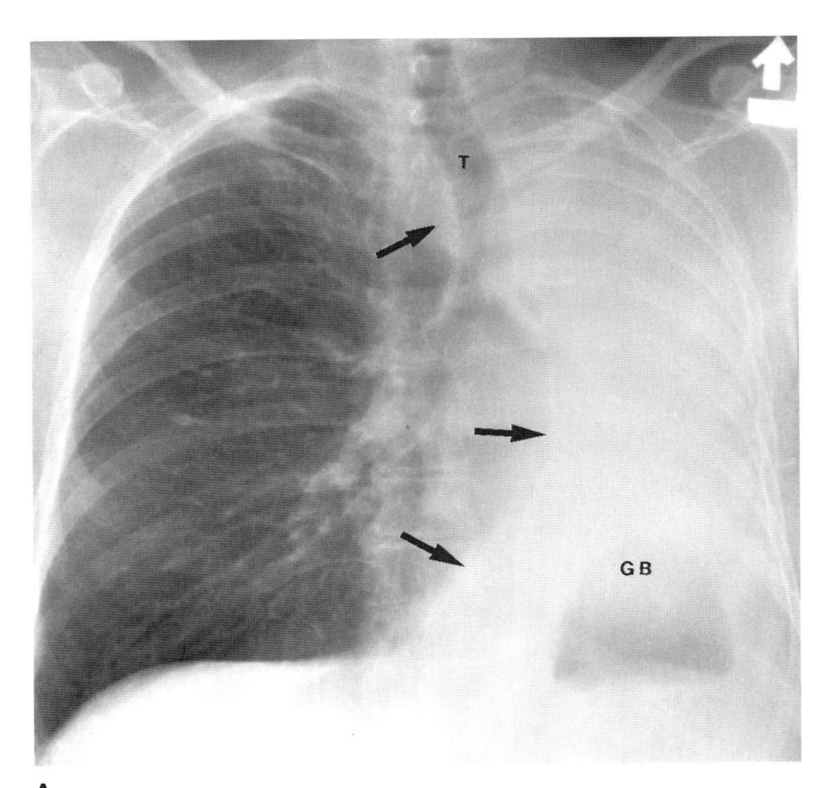

A

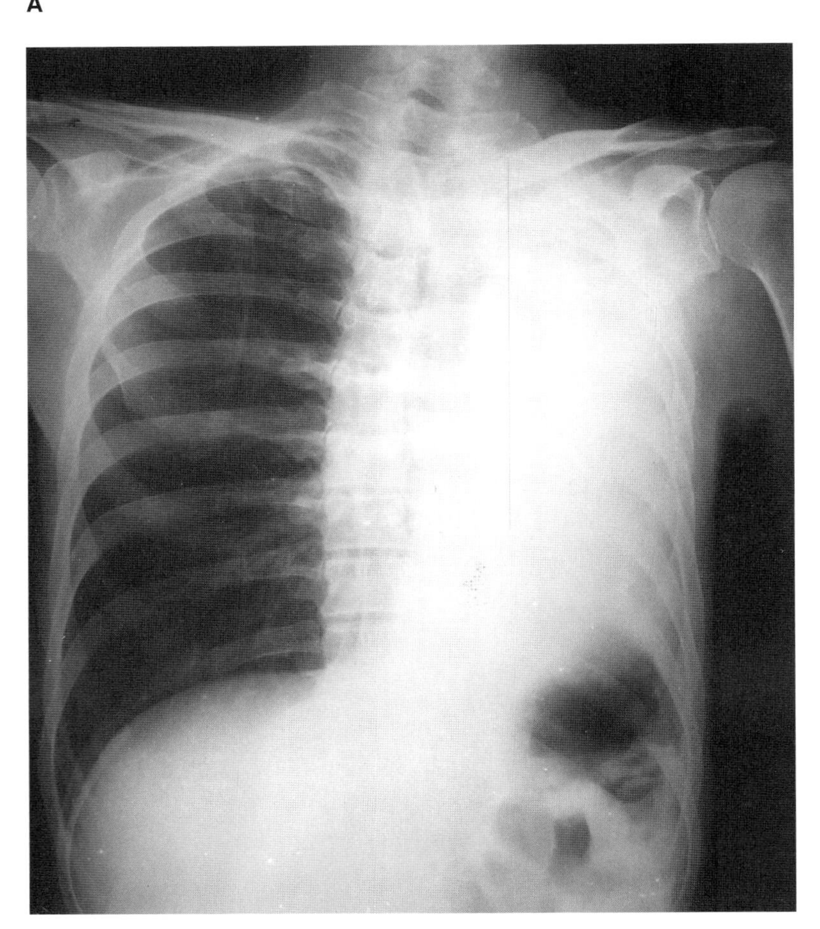

B

Figures 8.15A and B. Two patients with collapse of the entire left lung with compensatory overexpansion of the right lung. Note the elevation of the gastric bubble *(GB)*, indicating a high left hemidiaphragm. There is pronounced displacement of all mediastinal structures to the left with herniation to the left of the right lung *(arrows)*. *T* is the trachea. Bronchoscopy showed a carcinoma obstructing the left main bronchus in both patients.

Mediastinal Shift due to Collapse of One Lobe on the Right

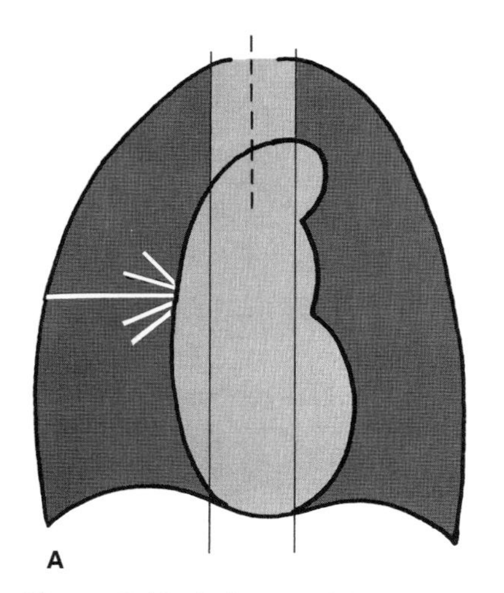

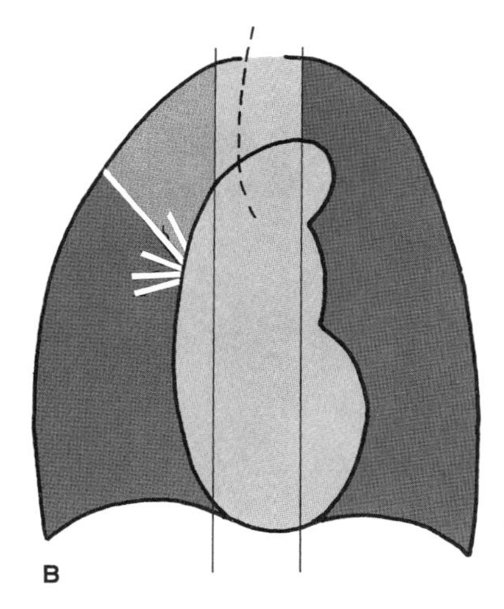

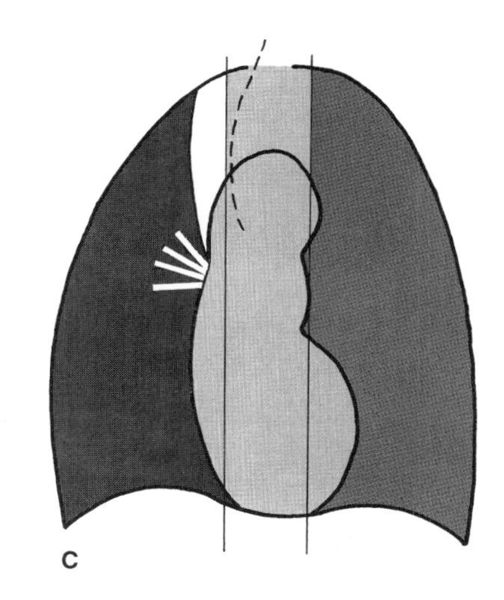

Figure 8.16. Collapse of the right upper lobe. A: Normal chest with normal position of the minor (horizontal) fissure, right hilum, trachea, aortic arch, and right heart border with equal aeration of all lobes. B: Right upper lobe collapsed 50 percent. Minor fissure deflected upward, trachea pulled slightly to the right. No change in the right heart border. C: Major collapse of the right upper lobe, which is now a flat wedge of opacity against the superior mediastinum. The trachea and the aortic arch are tilted to the right. The right hilum is drawn upward. Overaeration in the lower and middle lobes is compensatory. Note the upward displacement of the hilum.

A collapsing lobe tends to fold up fanwise against the mediastinum in a characteristic manner, and a dynamic concept of these collapse patterns and the roentgen signs by which they are to be recognized is, again, nothing more than an exercise in the logic of radiodensities applied anatomically. You would anticipate, for example, that with atelectasis of the *right upper lobe* you could see the location of the minor fissure separating it from the middle lobe better and better as the contrast increased between the poorly aerated lung tissue above it and the well-aerated lung tissue below it. Moreover, since the fissure is fixed at the hilum, it is natural that it would be seen to tilt upward from that fixed point as the upper lobe collapsed. When completely collapsed, the pancake-flat upper lobe would apply itself against the upper mediastinum and merge its shadow with that of other mediastinal structures. The shadow of the trachea would be drawn to the right, and the aortic arch would be drawn with it. The right hilum would be drawn slightly upward.

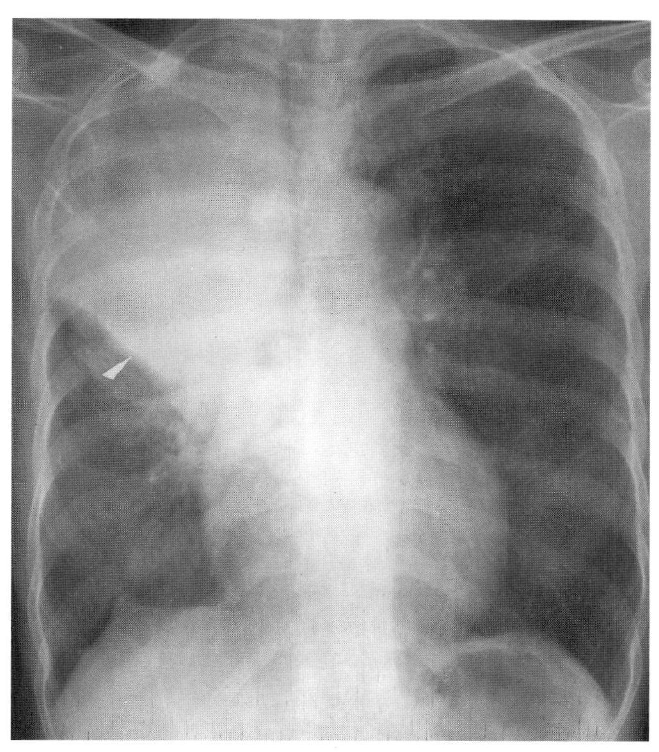

Figure 8.17. Patient with right upper lobe atelectasis distal to a bronchogenic carcinoma. The *arrow* indicates the elevated minor fissure.

Collapsed Lobes on the Right

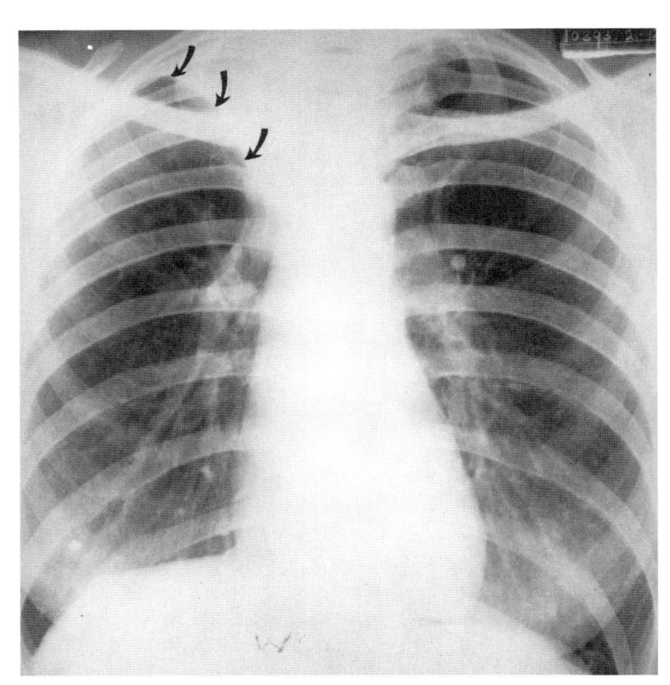

Figure 8.18. Nearly complete collapse of the right upper lobe in a patient with a 10-year history of symptoms of a cough and occasional hemoptysis. The *arrows* indicate the curving margin of an elevated minor fissure. At surgery obstruction of the right upper lobe bronchus by a bronchial adenoma was discovered and a lobectomy was performed. Note the high right takeoff of vessels to the lower lobe.

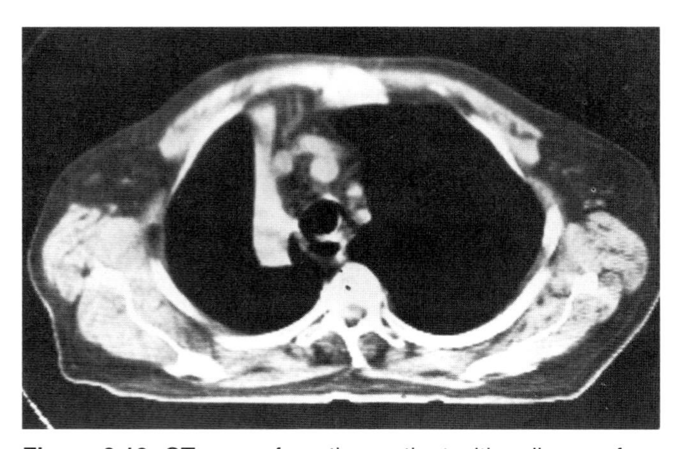

Figure 8.19. CT scan of another patient with collapse of the right upper lobe. The section is just above the aortic arch. You can identify the small flat upper lobe lying against the mediastinal structures, all of which are displaced to the right. Identify the left and right brachiocephalic veins (farthest anterior), then the trachea, and, between it and the brachiocephalic veins, the brachiocephalic artery. To the left of the trachea lie the left common carotid and left subclavian arteries.

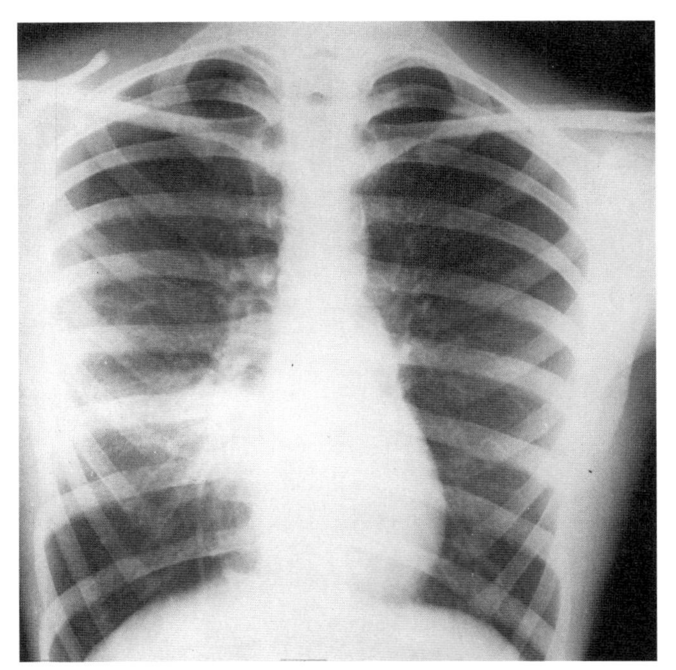

A

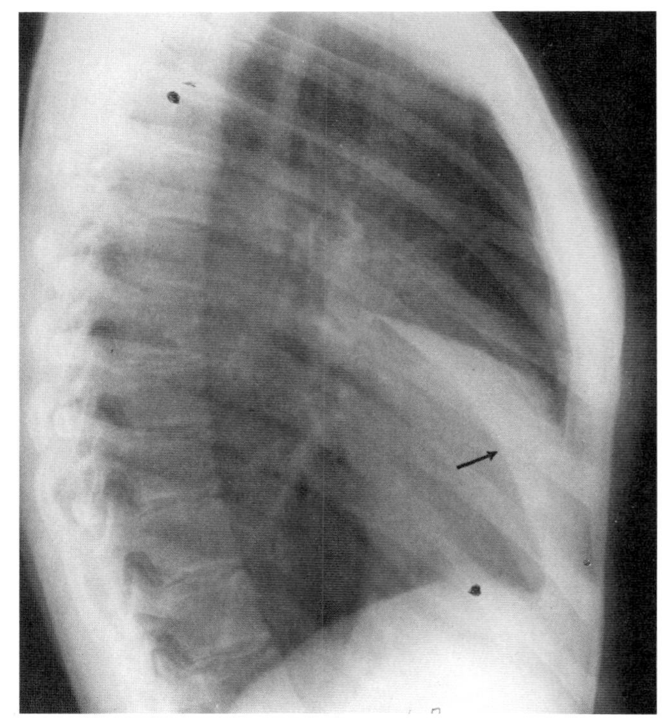

B

Figure 8.20. Collapse of the right middle lobe. A line drawn between the two black dots in the lateral view (B) would indicate the normal location of the major (oblique) fissure. Hence the lower part must be bowed forward *(arrow)*. The minor fissure is depressed. The mediastinum is not displaced, because the volume loss in the right middle lobe is too small. Even if the right middle lobe totally collapses, there is not sufficient change in right lung volume to alter the position of the mediastinum.

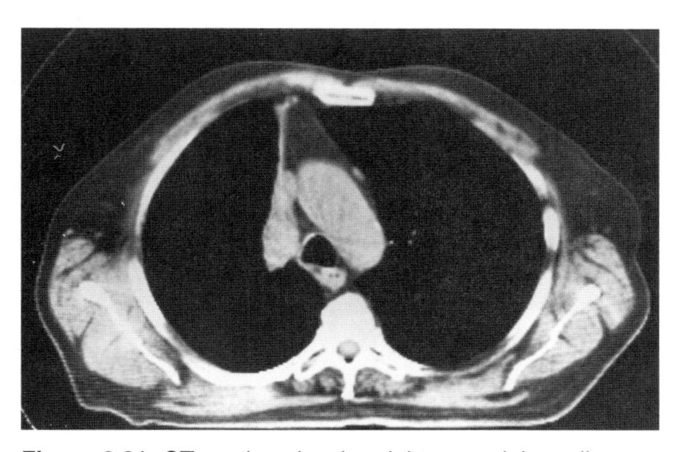

Figure 8.21. CT section showing right upper lobe collapse. This scan is made at the level of the aortic arch, lower than that in Figure 8.19. The aorta and the fatty anterior mediastinal triangle are displaced to the right.

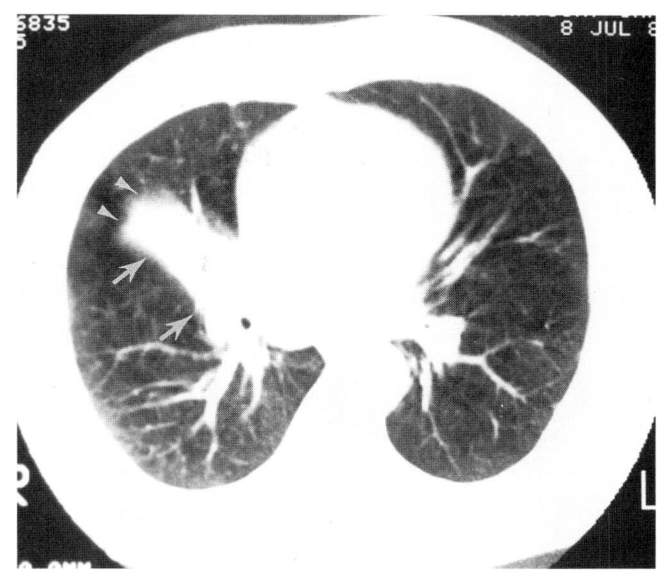

Figure 8.22. CT section of a collapsing right middle lobe filmed with lung window settings. This section is much lower than that in Figure 8.21 and through the heart. Compare Figures 8.19 and 8.21. The right heart border would be lost on a PA film of the patient in Figure 8.22.

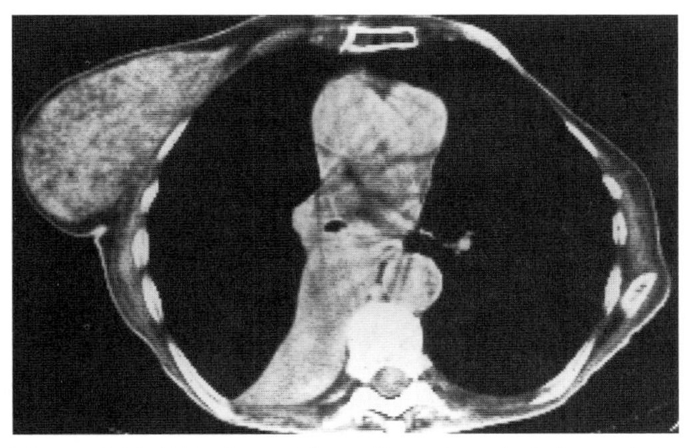

Figure 8.23. Right lower lobe collapsed posteriorly against the vertebral column and posterior ribs. This patient had had a left mastectomy. There is a dense mass of metastatic nodes between the vertebral body and the flattened, anteriorly displaced bronchus intermedius *(small black oval).*

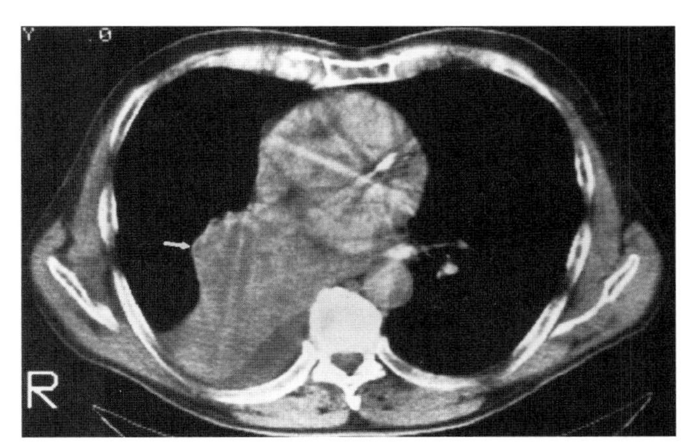

Figure 8.24. Right lower lobe collapse in obstructing bronchogenic carcinoma. Note the rounded tumor mass *(arrow)* bulging to the right, anterior to the collapsed lower lobe. Note too the pleural fluid collection (less dense) posterior to the collapsed lobe.

Mediastinal Shift due to Collapse of One Lobe on the Left

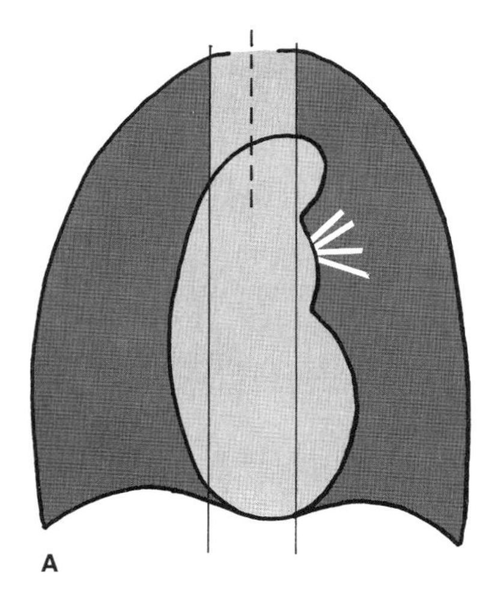

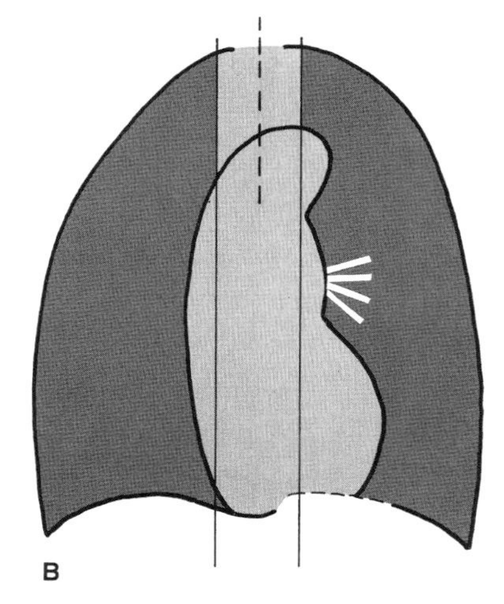

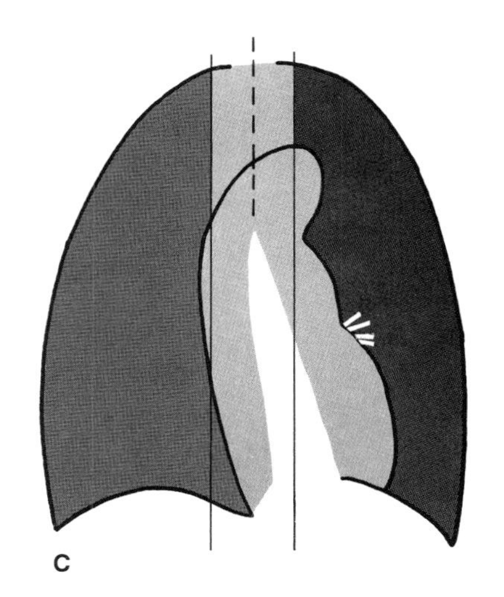

Figure 8.25. Collapse of the left lower lobe. A: Normal profiles and mediastinal tag points. B: Early signs of left lower lobe collapse: less heart shadow to the right of the spine, vague decrease in the lucency of the lower left lung with preservation of the left hemidiaphragm, which, however, becomes slightly elevated medially. C: Massive collapse of the left lower lobe. Little or no heart shadow is seen to the right of the spine. The medial half of the profile of the left diaphragm is missing. The left lower lobe is now a wedge of opacity seen through the heart and against the spine. The left hilum is depressed.

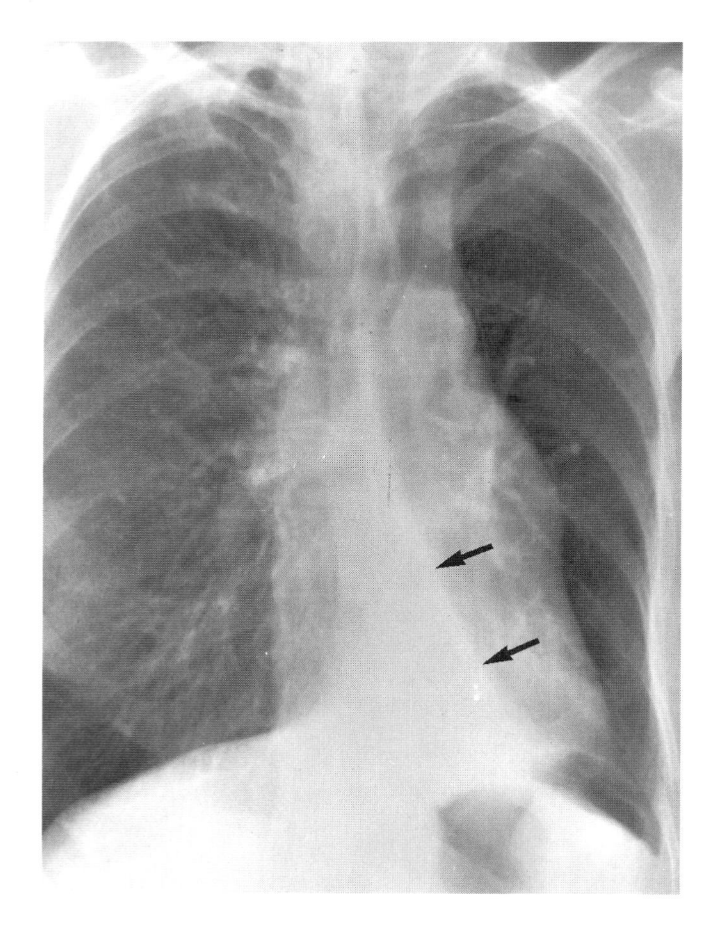

Now put together the signs of *left lower lobe* collapse in just the same deliberately logical way. Since the major fissure lines up with the beam of x-rays only in the lateral view and is always quite oblique to the ray in the PA view, no clear-cut margin between normal and atelectatic lung tissue is to be seen on the PA film as the left lower lobe begins to collapse. The heart gradually shifts to the left, however, so that you see less and less of the heart border to the right of the spine. You watch the left diaphragm become slightly more elevated and less and less clearly seen medially as the left lower lobe collapses against it, although the *lateral* half of the shadow profile of the diaphragm remains clear because of the compensatory expansion of the lingula of the left upper lobe, now touching it. The left hilum is depressed, gradually disappearing behind the left border of the heart, an important and often missed roentgen sign of left lower lobe collapse.

Figure 8.26. Left lower lobe collapse, seen as a wedge of density through the heart. The *arrows* indicate the margin of the collapsed lobe. The left upper lobe is overinflated, and the medial part of the left hemidiaphragm is lost. The spine is revealed and the low left hilum is obscured by displacement of the heart to the left.

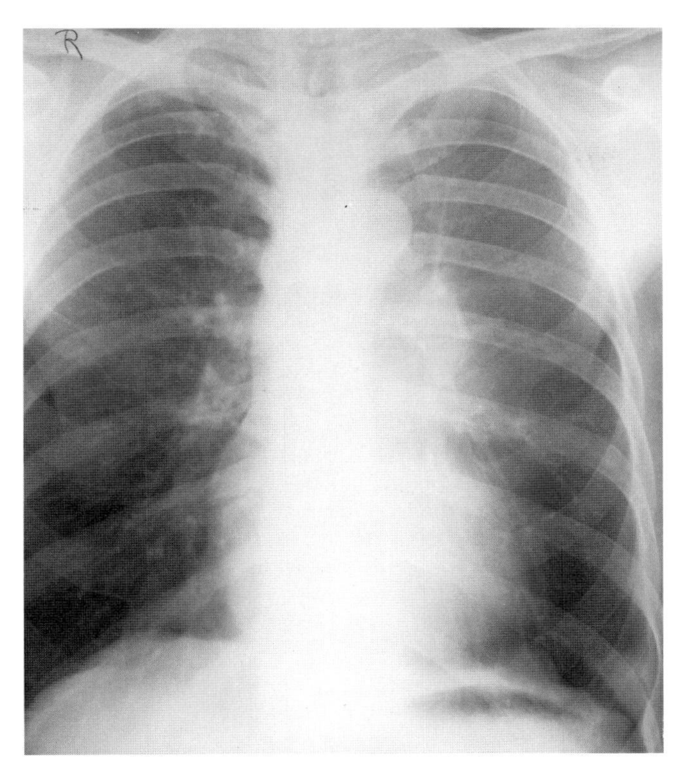

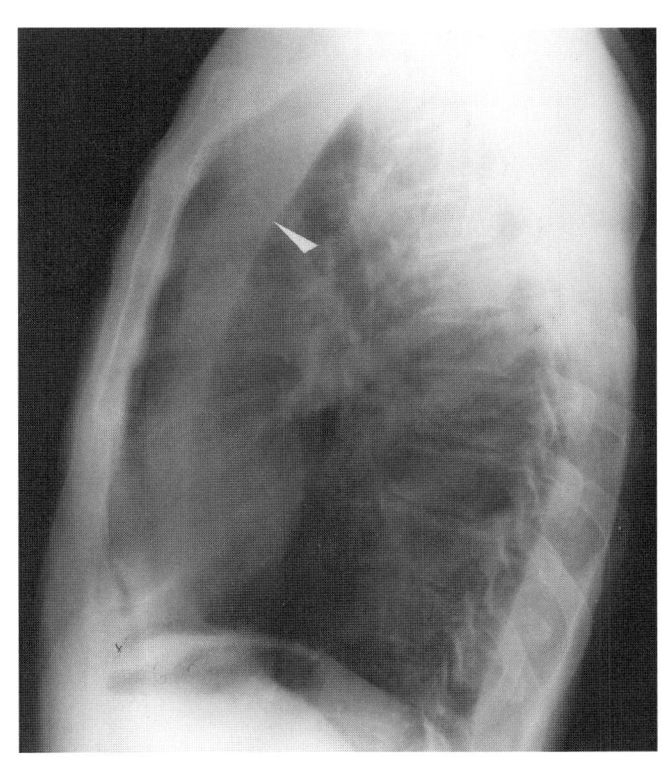

Figure 8.27. Left upper lobe collapse due to obstructing carcinoma. Note the veil-like density of the left upper lung and the obscured left heart border. The trachea is pulled to the midline and the aorta is too prominent, implying upper mediastinal shift. In the lateral film (B) the *arrow* indicates the major fissure bowed forward as the left upper lobe collapses. It is seen here as a slender wedge of opacity along the anterior part of the chest.

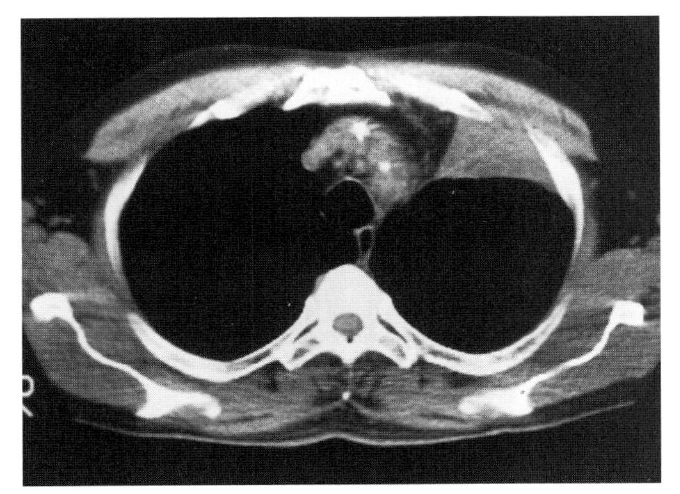

Figure 8.28. CT scan showing left upper lobe collapse in another patient, who had chest films similar to those in Figure 8.27. Note that the aortic arch is drawn forward and to the left, shortening the fatty wedge of the anterior mediastinum. The left upper lobe is seen as a wedge of opacity against the left anterior chest wall, with the hyperexpanded left lower lobe bulging into its posteroinferior surface.

The vessels of the left upper lobe appear spread apart and the lung tissue more lucent than that in comparable interspaces on the right. The totally collapsed lower lobe appears, finally, as a wedge-shaped shadow against the mediastinum posteriorly. Its outer margin is visible through the heart shadow, thrown into contrast by air in the normal lung tissue against it laterally, that is, in the overexpanded upper lobe (Figure 8.26). On the next page you will find CT scans to help you remember how the left upper and lower lobes collapse.

CT Scans of Three Patients with Lobar Collapse on the Left

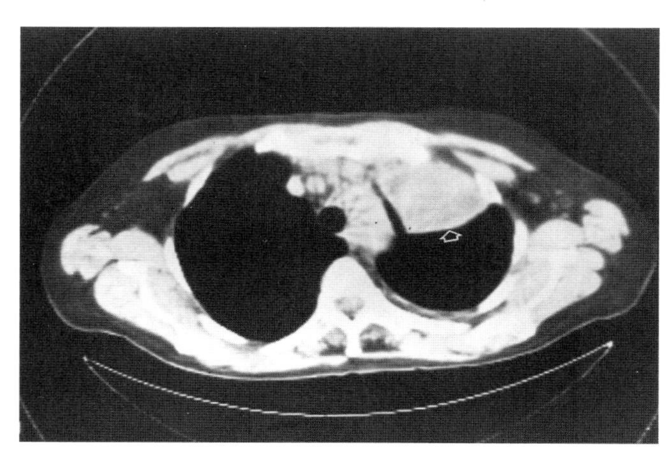

Figure 8.29. CT scan showing left upper lobe collapse. The section is at the level of the top of the aortic arch. The overexpanded superior segment of the lower lobe is pinched as a sliver of black between the atelectatic left upper lobe and the mediastinum. Here the posterior surface of the left upper lobe bulges backward because of a tumor mass within it. The *open arrow* indicates the major fissure interface between the left upper and left lower lobes.

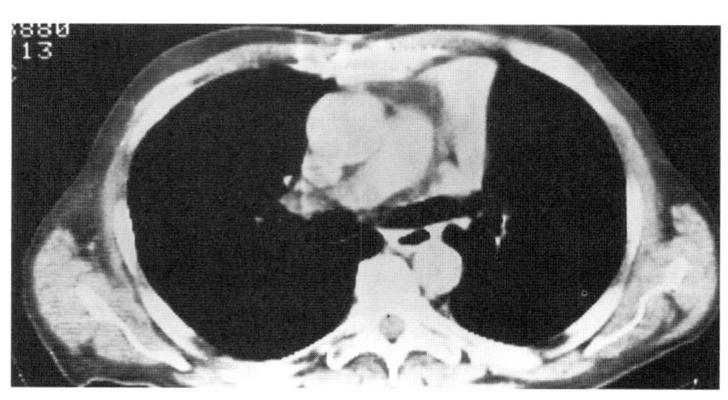

Figure 8.30. Left upper lobe collapse in the patient shown in Figure 8.29, but scanned lower down, at the level of the tracheal bifurcation. Here the lingula segment of the left upper lobe is seen flattened against the main pulmonary artery. The upper left cardiac profile would be lost in a PA film.

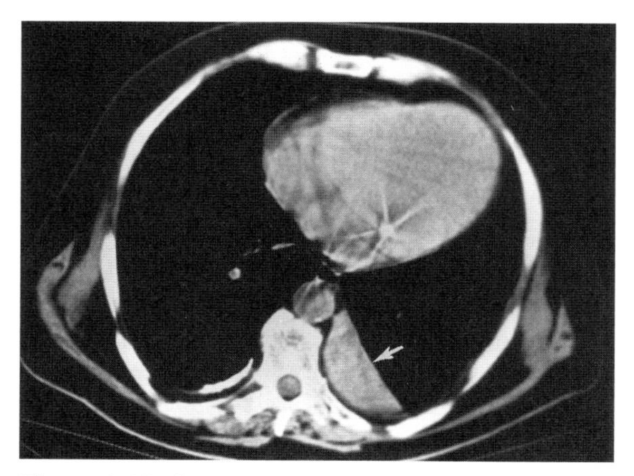

Figure 8.31. Collapse of the left lower lobe. Compare the plain film example in Figure 8.26. Note the mediastinal shift to the left (the section level is through the heart). The left lower lobe always collapses posteromedially against the spine and posterior ribs. The *arrow* shows a displaced major fissure between the upper and lower lobes.

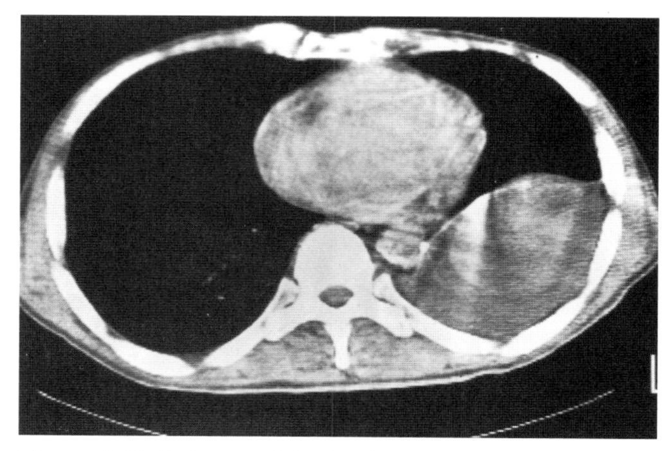

Figure 8.32. Collapse of the left lower lobe in another patient. Here the mediastinal shift is much less pronounced because part of the lost volume is compensated for by a collection of pleural fluid engulfing the lobe. If you discount the streak artifacts, you will see the crescent of less dense pleural fluid between the collapsing denser left lower lobe and the posterior chest wall. Figure out the CT scan for a patient with collapse of the whole left lung.

Tallying Roentgen Findings with Clinical Data

Roentgen observations that imply the presence of extensive emphysema or massive atelectasis will serve to remind you of distortions in architecture and aberrations of function that you might otherwise forget to consider in a particular patient. An elderly man with chronic emphysema is most concerned with his respiratory difficulties, but you will not be able to look at those overexpanded lungs and fibrous traceries of shadow on his radiograph without thinking of the increased work being done by the right side of his heart.

You must not forget, in looking at chest films, that changes in volume within the thorax may compensate for other changes and obscure them. If the mediastinum seems in its normal midline position despite the fact that one entire lung is opaque, you know only that the volumes of the two hemithoraces *are* equal. Underneath that white opacity there may be just enough collapse to compensate for the added volume of tumor or pleural effusion, and the radiologist will remind you that CT can differentiate fluid in the pleural space from tumor or atelectatic lung.

Remember that the radiograph is a shadowgram. Although you know that inflammation and tumor both render the lung opaque, you must anticipate, for example, that in either condition some atelectasis is likely to be present as well, adding to the opacity of the already involved lung. Either tumor or inflammation may cause collapse of a lobe even though the entire lobe is not actually involved in the primary process.

The consolidated lobe in pneumonia often remains normal in size, as you saw it in Chapter 6; but equally often such a lobe is distinctly decreased in size. In a good many patients with the clinical signs of pneumonia you may see some additional roentgen indications of atelectasis, resulting from poor aeration. As you increase your knowledge of medicine, you will learn to evaluate the roentgen signs of lobar collapse according to whether, for example, your patient was admitted with clear-cut pneumonia or has only a minor degree of fever and cough the day following surgery. In the former instance you must think in terms of pneumonia-plus-atelectasis and treat accordingly. In the latter you must think in terms of atelectasis primarily, perhaps due to a post-anesthesia mucous plug, and consider the possibility of infection developing in the collapsed lobe. That they may look the same on the films should not disturb you, since you are using the radiographic findings as part of the entire clinical analysis rather than as oracular information. A respectable percentage of lung cancers are at first thought to be pneumonia with atelectasis, and only the unaccountable failure of the lungs to reexpand fully with proper treatment eventually raises the question of an airway-obstructing tumor.

It is immensely important for you to realize that the type of analysis of roentgen findings you have been learning offers you an improved understanding of the dynamic pathological changes within the thoracic cage of your patient. It is much more useful to you than any collection of diagnostic tags and labels. When you see mediastinal shift on a chest film or appreciate exaggerated radiolucency or the disappearance of normal profiles, you are recognizing roentgen findings rather than diagnoses. Such findings are heavy with implication about what is going on inside your patient. Their presence will often go a long way toward confirming, expanding, or exploding an original working diagnosis based on the patient's history and physical examination.

9 The Mediastinum

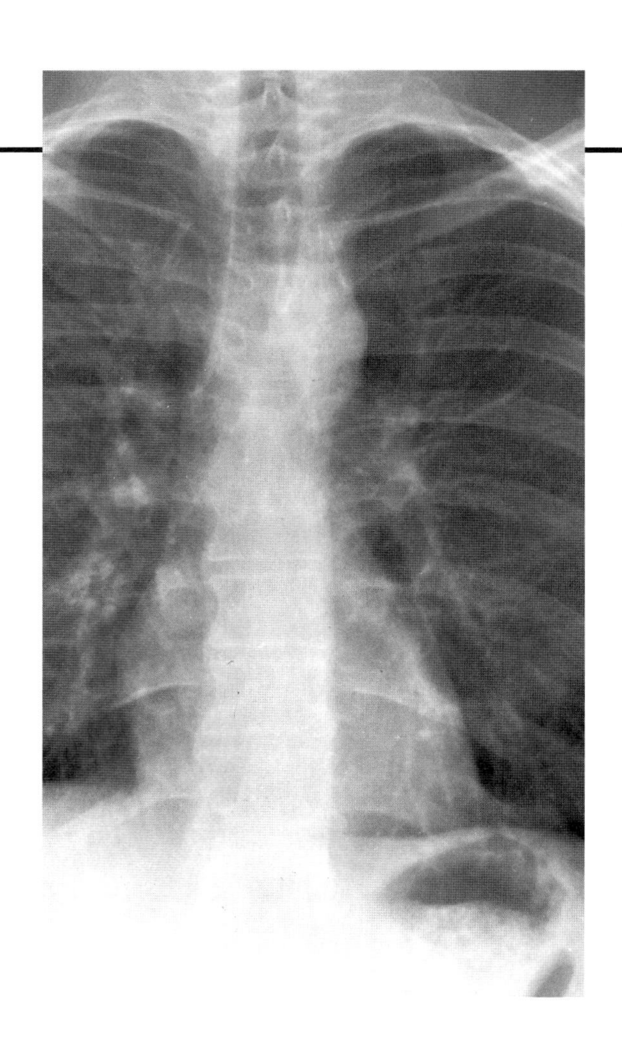

The heart is the largest of the mediastinal structures, and all of the profiles that bulge beyond the shadow of the spine on both sides in the PA chest film represent parts of the heart or of its great vessels. You can think of these profiles as nine intersecting arcs (Figure 9.2). Justify their identity on the basis of the angiocardiograms on the opposite page. Remember that some of the structures producing these shadows are more posterior in the chest *(6)* and others far anterior *(2, 7)*. Note also that the margins of the heart and great vessels appear to extend more laterally on the angiocardiograms than on the plain film in Figure 9.1. They appear that way because the angiocardiograms were filmed AP, with the patient lying on an angiographic table, and consequently the more anteriorly positioned heart and vessels are magnified, in comparison to the mediastinal outlines of the PA chest film.

Remember too that when contrast medium mixed with blood fills a particular chamber of the heart, its shadow may seem to you quite different in shape from what you have learned about that chamber based on a gross examination of the heart and its surface markings. Where a chamber is thickest its shadow will be most dense in the angiogram, and where it tapers off and becomes very thin a much less dense shadow is produced. Look at the shadow of the right ventricle, for example, in Figure 9.3, the dextrocardiogram. The slender, flattened part of the chamber, which extends far to the left against the interventricular septum in the PA view, hardly seems to belong to the dense massive shadow of the rest of the ventricle. Note also that you appreciate only vaguely the location of the tricuspid valve in this view, because the right atrium and right ventricle are partly superimposed. In the levocardiogram notice that you see the dense upper margin of the crab-shaped left atrium through the shadow of the ascending aorta, in spite of the fact that you know the left atrium is on the posterior surface of the heart and that the ascending aorta arises anteriorly. Their opaque-filled cavities have cast separate shadows, outlining them for you, and the two shadows overlap in this view.

The plain film of the chest made PA, then, shows you a number of mediastinal bulges seen in profile

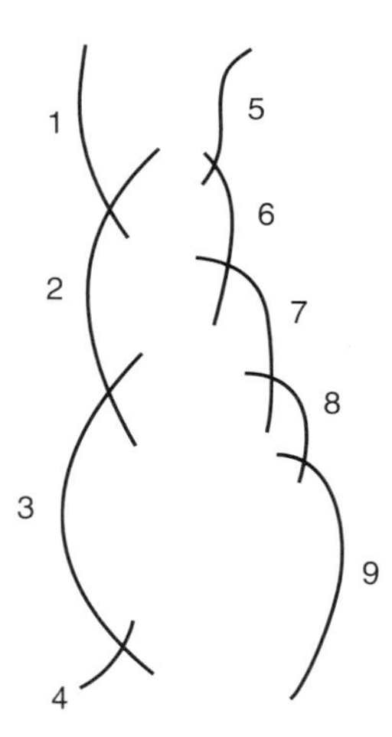

Figures 9.1 and 9.2. The normal mediastinal profiles (Figure 9.1, *top*) are all vascular and resolve into a series of nine intersecting arcs, as shown in Figure 9.2 *(bottom): 1,* right brachiocephalic vessels; *2,* ascending aorta and superimposed superior vena cava; *3,* right atrium; *4,* inferior vena cava; *5,* left brachiocephalic vessels; *6,* aortic arch; *7,* pulmonary trunk; *8,* left atrial appendage; *9,* left ventricle.

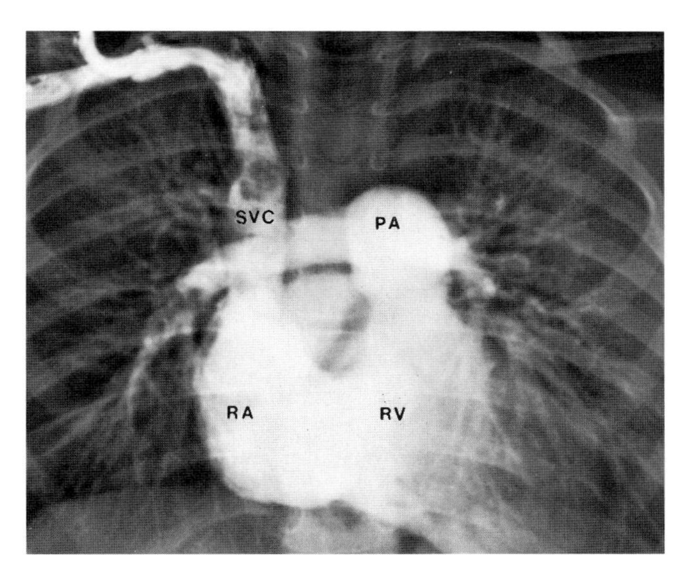

Dextrocardiogram.

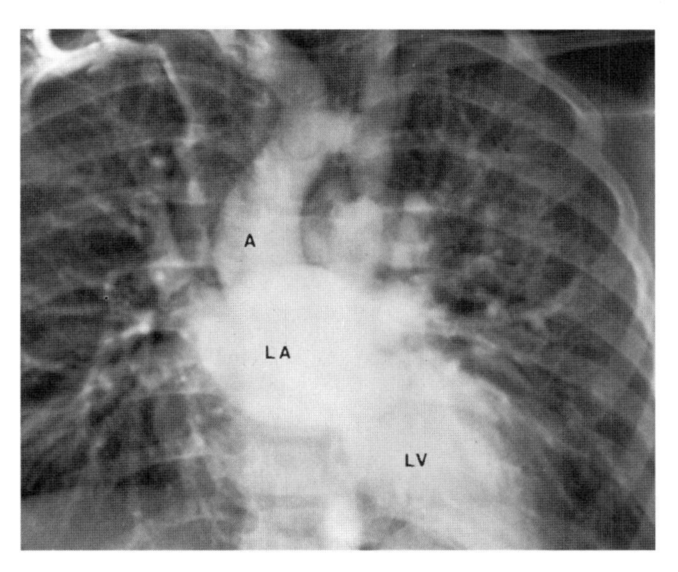

Levocardiogram.

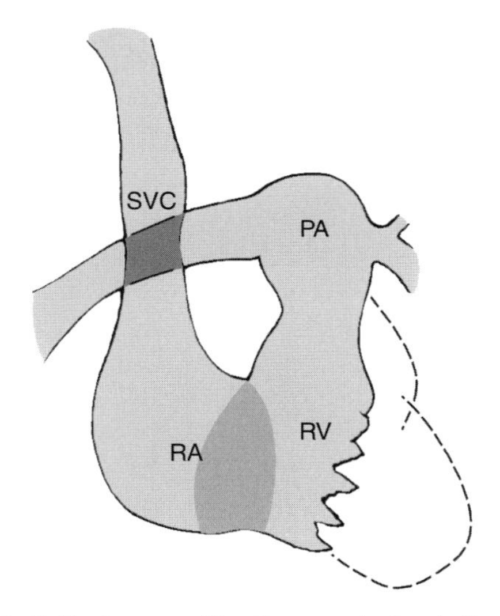

Figure 9.3 ***(left above, with diagram below).*** Dextrocardiogram, taken from a series of films made after a bolus of contrast material that was injected through a vein in the right arm passed through the heart. The time during which the right heart is opacified is called the *dextrophase.* You see the right atrium *(RA)* and right ventricle *(RV)*, as well as the pulmonary artery *(PA). SVC* indicates the superior vena cava.

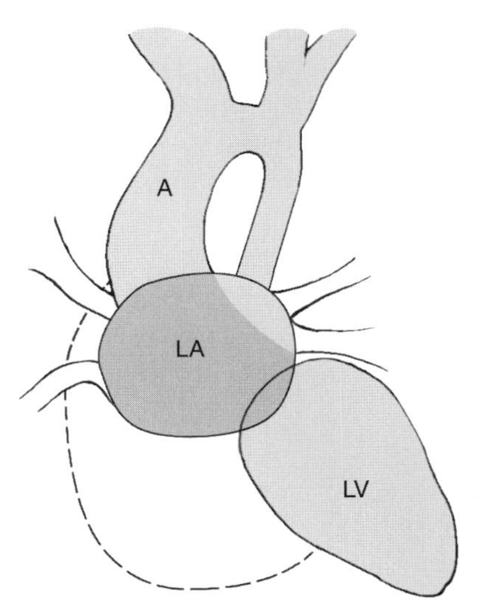

Figure 9.4 ***(right above, with diagram below).*** Levocardiogram of the same patient, taken 3 seconds later, when the right side of the heart had been cleared of contrast material. You now see opacification of the pulmonary veins entering the left atrium *(LA),* the left ventricle *(LV),* and the aorta *(A).* It is easy to see that the *venous* hila are lower than the arterial hila. This child had coarctation of the aorta (shown on the levocardiogram but not on the diagram); note the narrowing (stenosis) of the proximal descending aorta, just below the aortic arch.

against the radiolucent lung on either side of the spine, all of them vascular shadows. In addition, you can usually see air in the trachea, but all other mediastinal structures merge with one another and their shadows are superimposed upon those of the spine, the heart, and the sternum. You cannot account for the shadow of the esophagus or distinguish lymph nodes, thymus, or nerves; the thoracic duct merges with the shadows of other soft tissues and fluid-carrying vessels. Except for their marginal profiles and their branches entering the lucent lungs, even the great vessels are merged with other shadows.

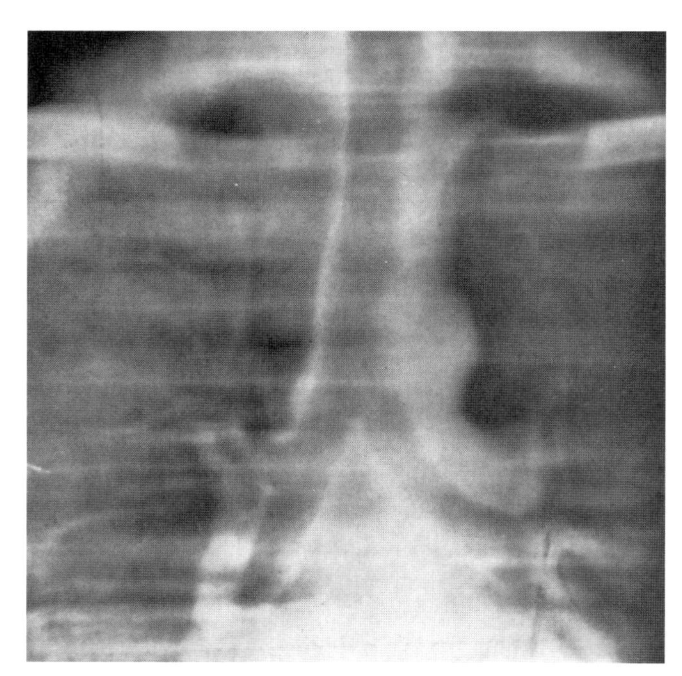

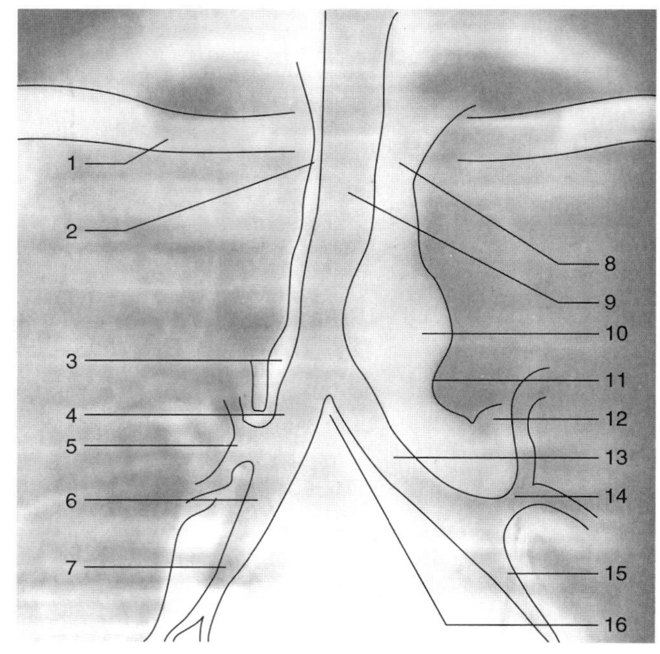

Figure 9.5 ***(above, with diagram at right).*** Coronal conventional tomogram of a normal tracheobronchial tree.
(1) Clavicle
(2) Normal opacity of great vessels and tracheal wall
(3) Azygos vein
(4) Right main bronchus
(5) Right upper lobe bronchus
(6) Site of origin of right middle lobe bronchus
(7) Right lower lobe bronchus
(8) Normal opacity of blood vessels and tracheal wall
(9) Trachea
(10) Aortic arch
(11) Concave profile between aortic arch and left pulmonary artery
(12) Left pulmonary artery
(13) Left main bronchus
(14) Left upper lobe bronchus
(15) Left lower lobe bronchus
(16) Carina

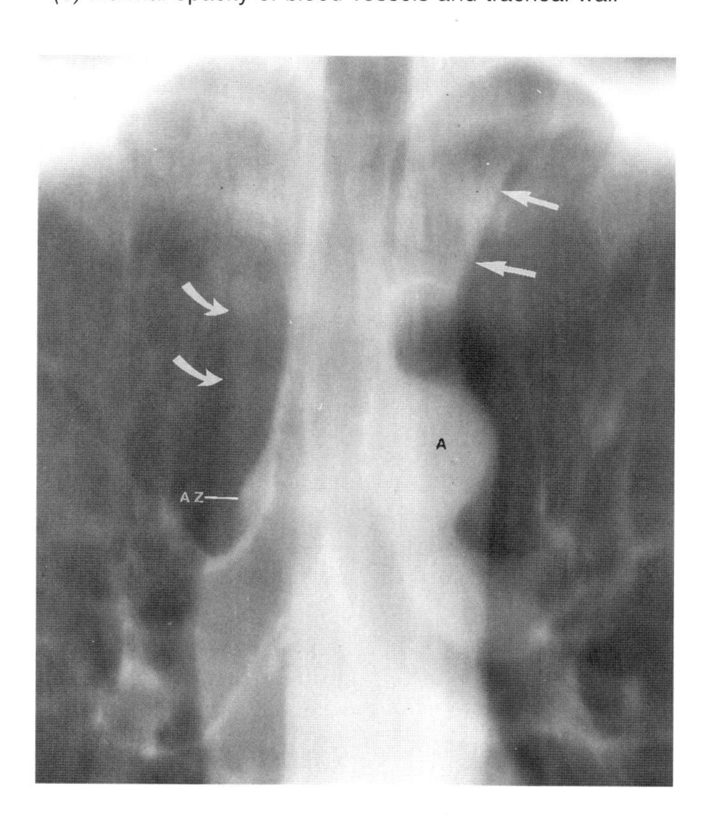

On this page spread you are using the normal dark air column in the trachea and bronchi as a contrast medium—which you can do, of course, up to a point in any well-exposed chest film. In Figures 9.5 and 9.6 you have conventional tomograms to study. These eliminate the superimposed images of other structures in front of, or in back of, the coronal plane of study, and afford you a more precise view of the tracheobronchial anatomy. Identify all the parts of the upper mediastinal structures visible in these conventional tomograms and apply what you have now learned to recognize to the well-exposed chest films you will be seeing in your own patients. Upper mediastinal shift will be appreciable, as you learned in the last chapter,

Figure 9.6. This coronal conventional tomogram visualizes many of the structures you have just identified in Figure 9.5. Note the bulge in the right paratracheal stripe, the azygos arch *(AZ)* about to empty into the superior vena cava, whose shadow is seen lateral and posterior to it *(curved arrows).* The left subclavian artery *(straight arrows)* is seen arising from the aortic arch *(A).*

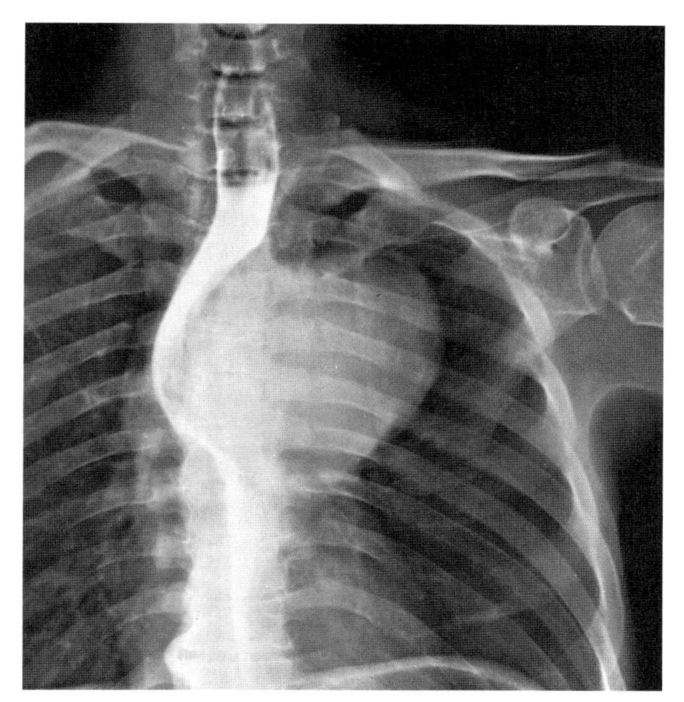

Figure 9.7. A patient with an aneurysm of the aortic arch causing displacement of the trachea and the barium-filled esophagus.

Figure 9.8 ***(below, with diagram at right).*** Coronal midthoracic conventional tomographic study of a man aged 65 with dyspnea and weight loss. Note the irregular narrowing of the trachea *(1)* and right main bronchus *(2)*. The mass *(X)* would appear on the regular chest film as an abnormal bulge on the right opposite the aortic arch. Note the downward displacement of a branch of the right upper lobe bronchus *(3)*. The pathological diagnosis was bronchogenic carcinoma with metastases to mediastinal lymph nodes.

from displacement of the trachea in the PA film. The trachea is normally located slightly to the right of the midline because it is closely applied against the mass of the arch of the aorta. In older patients in whom the mass of the aortic arch becomes larger (ectatic) as a result of aging, the trachea may often be seen a little farther to the right without indicating mediastinal shift.

The esophagus, which lies just behind the trachea, is often deflected with the trachea by masses like the aortic aneurysm in Figure 9.7, in which the esophagus is visualized because the patient has swallowed barium. Note that only certain mediastinal structures (the esophagus and trachea, not the entire upper mediastinum) have shifted, whereas the mass of the aneurysm extends to the left. Aeration of the two upper lobes of the right and left lungs is equal.

In questions of tracheal compression or invasion, conventional tomography is useful, as you see in the man in Figure 9.8, who was admitted with dyspnea and weight loss and soon developed stridor and great difficulty in breathing. Tumor is seen invading and narrowing the right main bronchus in this unfortunate patient, for whom radiotherapy would be only palliative. Both the mass of metastatic nodes and the narrowing of the trachea would also be well visualized at CT.

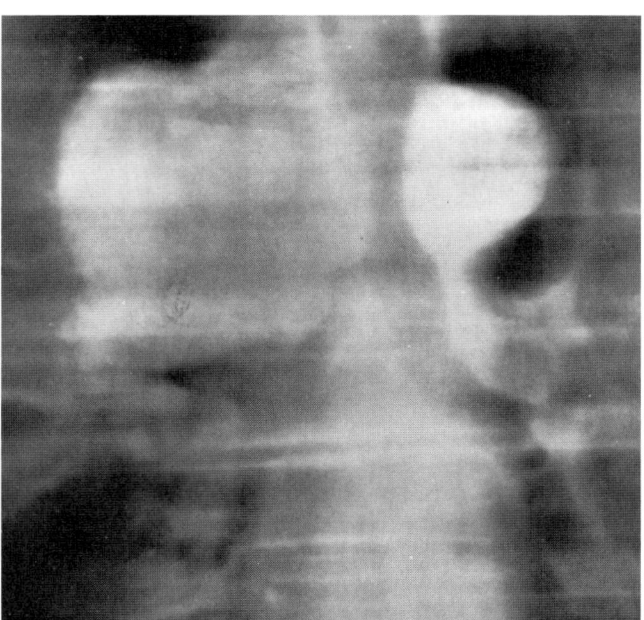

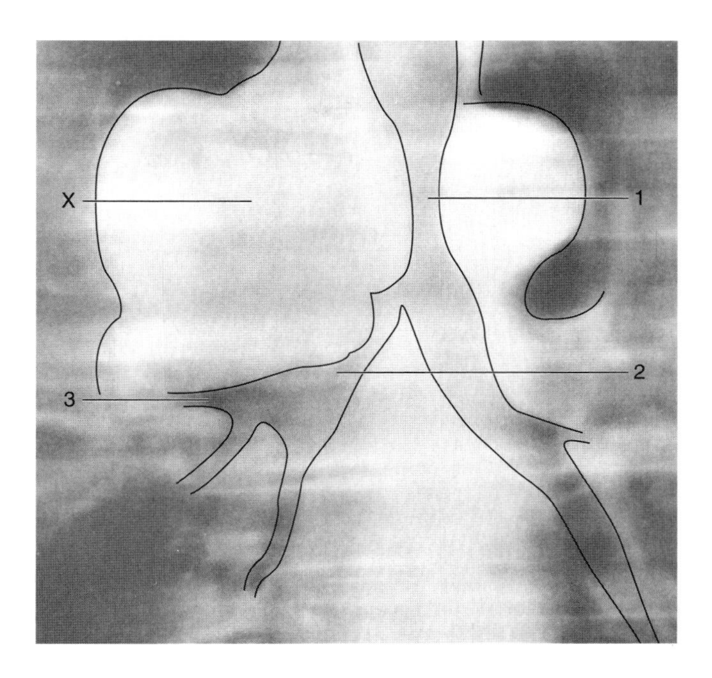

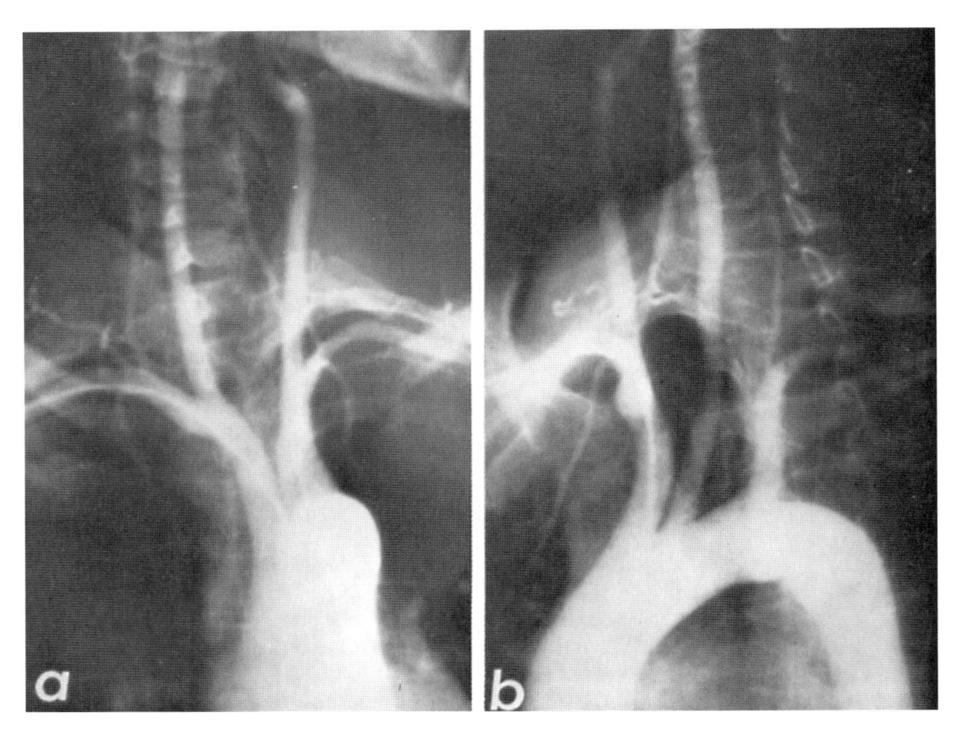

Figure 9.9. The normal aortic arch and brachiocephalic arteries. Contrast material was injected into the ascending aorta through a catheter that had been inserted via a right brachial artery route. In *a* the patient is almost AP (very slightly rotated to the left). In *b* he has been sharply rotated to his right, positioning the arch in profile, so that its branches no longer overlap. Remember that both arterial and venous structures account for arcs 1 and 5 in Figure 9.2, but that here only the arteries are contrast opacified.

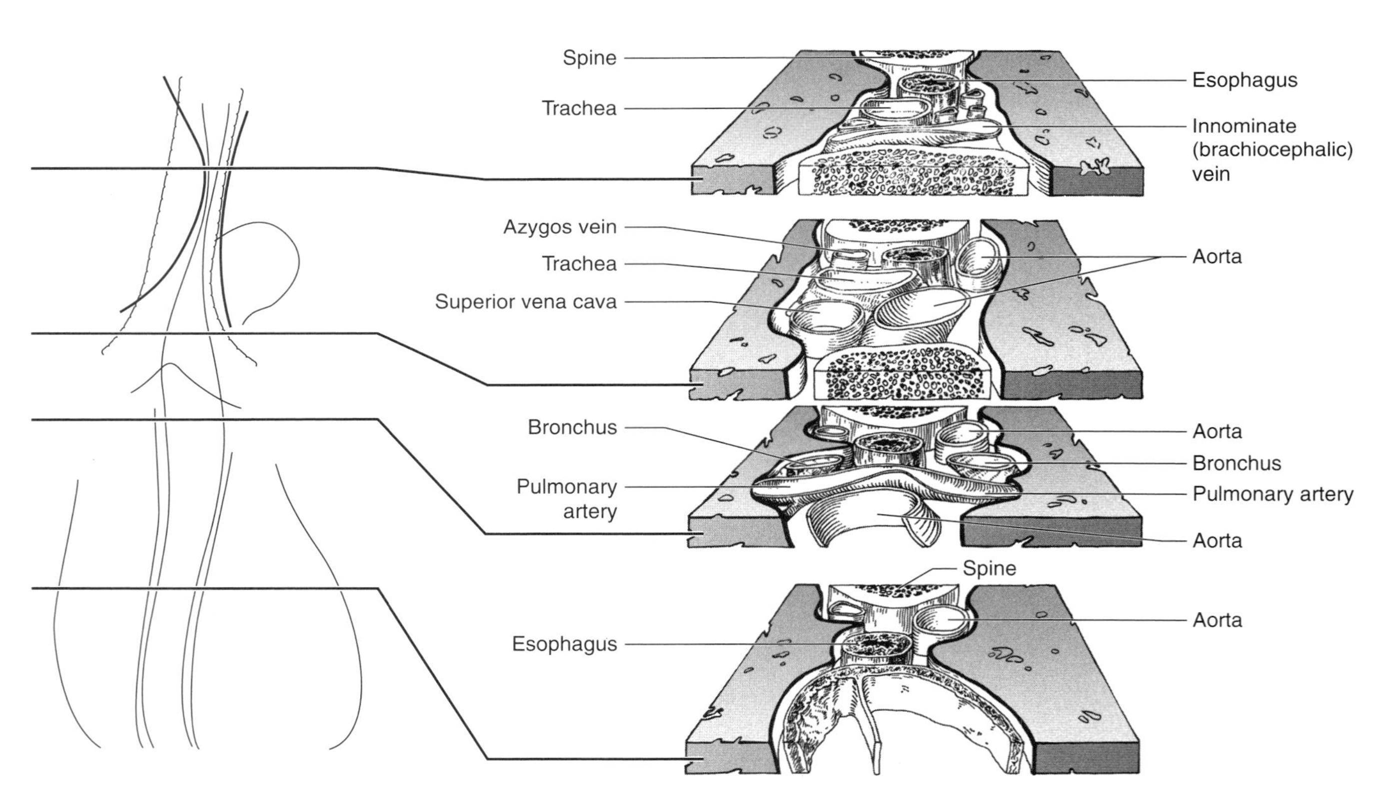

Figure 9.10. Mediastinal relationships. The heavy lines on the sectional drawings are pleural reflections. Remember that here you are looking at the *top* of the sections from front to back, whereas on all CT scans you are looking up at the scan from the patient's feet, with the spine toward you.

CT Sections: Four Levels, Four Patients

For decades angiography has been able to document alterations in the heart and brachiocephalic vessels. Figure 9.9 is a thoracic aortogram. The bolus of contrast material has been injected into the ascending aorta via a catheter inserted into the brachial artery. In *a* the patient is lying flat and in *b* he has been turned to his right. Imagine the superior vena cava lying against the arch on the right. Note that in *b* you can identify the brachiocephalic artery arising first and then dividing into the right subclavian and right common carotid arteries. Next, the left common carotid artery and the left subclavian artery arise in turn from the arch of the aorta.

Review the serial thoracic CT scans (in different patients) reproduced to your right; they show four different levels of the mediastinum opacified with intravenous contrast. In Figure 9.11, made just above the arch of the aorta through the brachiocephalic vessels, you can easily identify the anteriorly placed right *(RBV)* and left *(LBV)* brachiocephalic veins, which in the scan just below have joined to form the superior vena cava *(SVC)*. Identify the arterial branches of the aortic arch *(A)*.

Figures 9.9a and 9.10 afford you a *frontal* view with the vertebrae away from you, whereas the CT scans are conventionally displayed so that you are looking *up from the patient's feet* with the vertebrae down.

Now look back from Figure 9.12, in which you see the sectioned arch of the aorta *(A)*, to Figure 9.11, in which you can identify the branches it gives off: the brachiocephalic (or innominate) artery and then in turn the left common carotid and left subclavian arteries. Note the trachea to the right of the aortic arch in Figure 9.12 and its bifurcation *(B)* just posterior to the arch in Figure 9.13, as well as the branching right and left main bronchi in Figure 9.14. The ascending *(AA)* and descending *(DA)* limbs of the aorta and the main pulmonary artery *(MPA)* branching into the left *(LPA)* and right *(RPA)* pulmonary arteries are all clearly seen in Figures 9.13 and 9.14. The density *(arrow)* in the anterior mediastinum in Figure 9.12 is normal thymus in an adolescent.

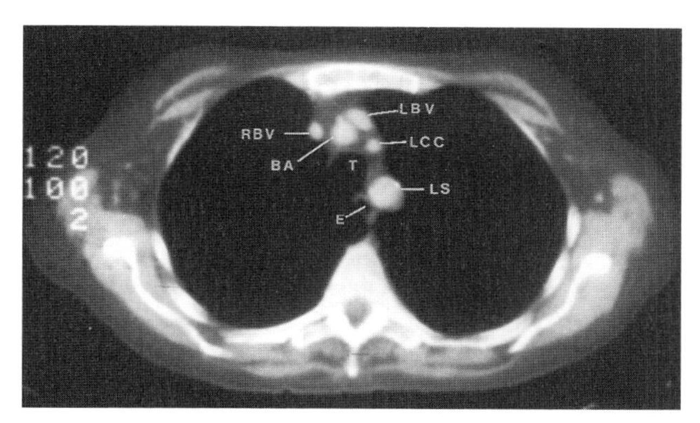

Figure 9.11

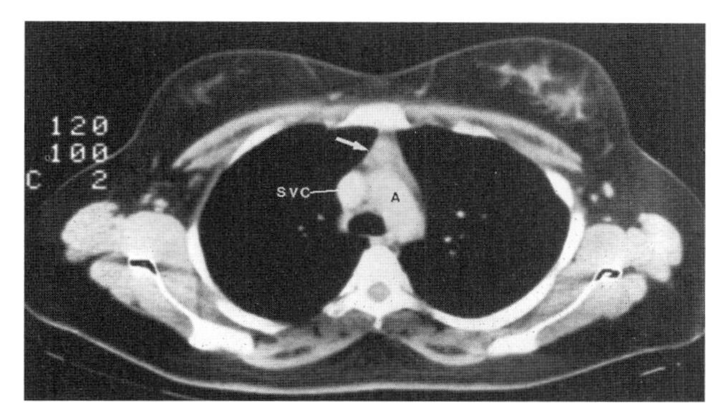

Figure 9.12

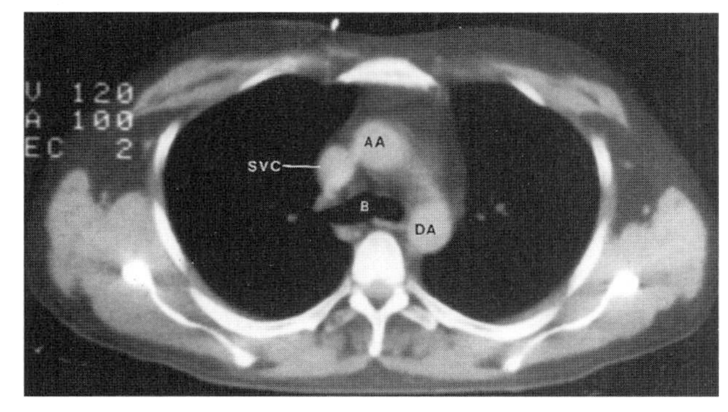

Figure 9.13

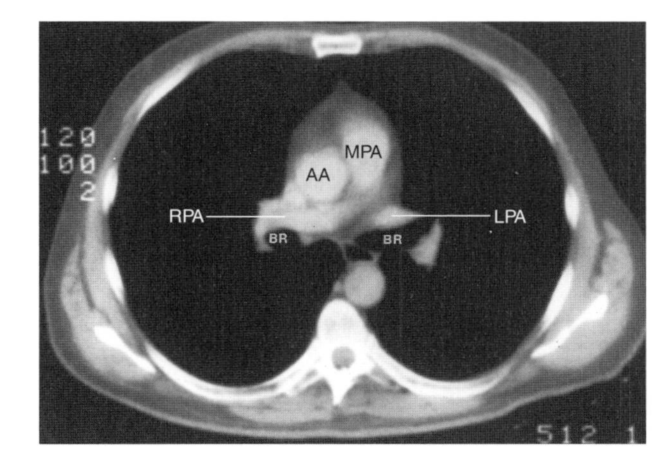

Figure 9.14

Mediastinal Compartments and Masses Arising within Them

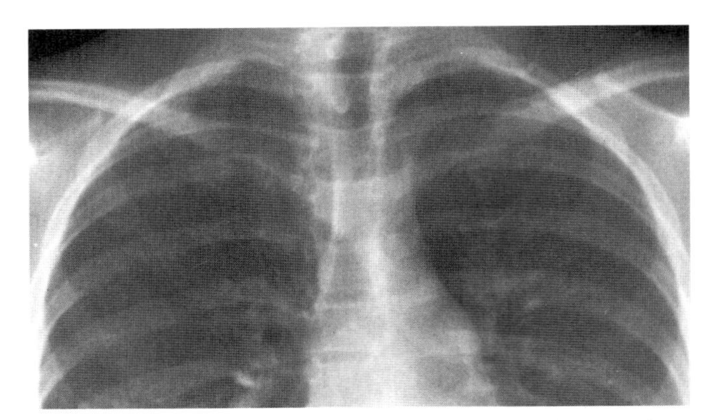

Figure 9.15. The normal superior mediastinum flattened between the two inflated upper lobes.

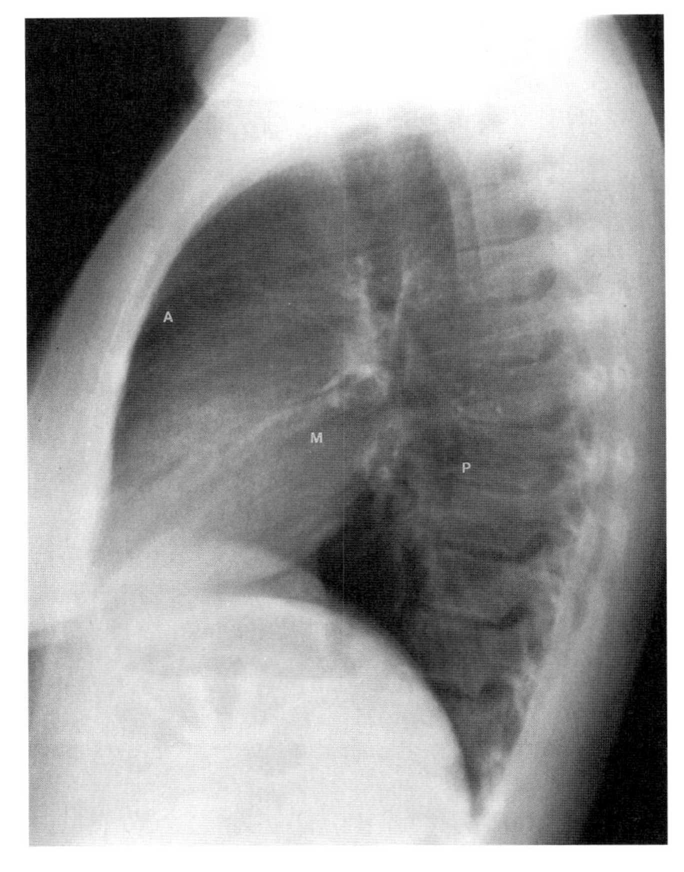

Figure 9.16. The mediastinal zones: *A,* anterior mediastinum; *M,* middle mediastinum; *P,* posterior mediastinum and paraspinal area.

The mediastinum, considered as a disc of structures compressed between the two inflated lungs in the PA chest film, is seen en face in the lateral chest film instead of in tangent. One can divide it into anterior, middle, and posterior sections and discuss the structures and masses in each. Unfortunately, there is some difference of opinion in labeling the various compartments of the mediastinum. Some writers place the heart in the anterior mediastinum; some place it in the middle mediastinum. For our purposes one need not put too fine a point on it, since *anterior mediastinal masses* generally arise from the region anterior to the heart and are seen, in the lateral chest film, to fill in the radiolucent anterior clear space in front of the heart. Such masses include goiter extending down from the thoracic inlet, thymoma, teratoma, and lymphoma. The last often extends back to occupy also the middle compartment and of course may involve any part of the mediastinum, because lymph nodes are located in all compartments.

Middle mediastinal masses generally arise from structures posterior to the heart: the esophagus (carcinoma as well as dilatation of the esophagus itself in achalasia and scleroderma), the tracheobronchial tree (bronchogenic carcinoma and cysts), and lymph nodes located there. *Posterior mediastinal masses* are often neural in origin (ganglioneuromas, neurofibromas), and of course aneurysms of the posterior part of the aortic arch and the descending aorta are posterior mediastinal masses.

Near the diaphragm is where you will also see masses related to the herniation of abdominal structures through the diaphragm (hernias of Morgagni, anteriorly, and Bochdalek, posteriorly, and those paraesophageal hernias we call hiatus hernias, which are often symptomatic). Pericardial cysts occur most often in the right paracardiac angle. They are seen on the frontal chest film to ablate part of the right heart shadow, and on the lateral film to superimpose on the heart shadow.

Masses in the superior part of the mediastinum or thoracic inlet are often (but not always) *goiters,* in which the thyroid mass extends down into the mediastinum. Compare Figure 9.17 with the appearance of the normal thoracic inlet in the frontal view in Figure 9.15. Note that the trachea *(arrows)* in Figure 9.17 is displaced.

Now analyze Figure 9.18. Does this CT section belong to the patient in Figure 9.17? Note that the trachea is displaced by a large mass in the anterior mediastinum. Compare with Figure 9.11 on the last page spread by tipping the page so that you can look at Figures 9.18 and 9.11 together. Yes, the mass in Figure 9.18 proved to be a goiter, but the CT section is not of the same patient as in Figure 9.17: the trachea is displaced to the left, not to the right. Note also that it is compressed, with an oval-shaped air column.

Analyze Figure 9.19. There is a mass in the superior mediastinum, extending to both sides, which proved to be a mass of tumor-invaded lymph nodes metastatic from a distant primary site.

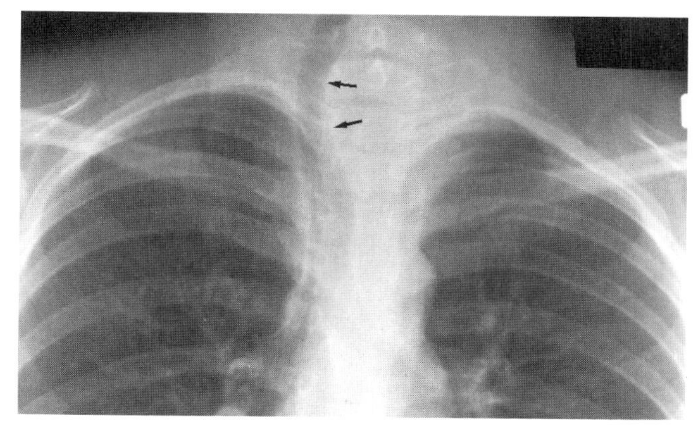

Figure 9.17

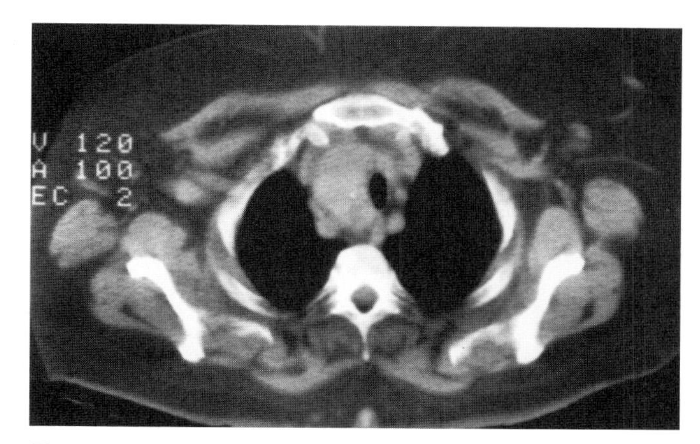

Figure 9.18

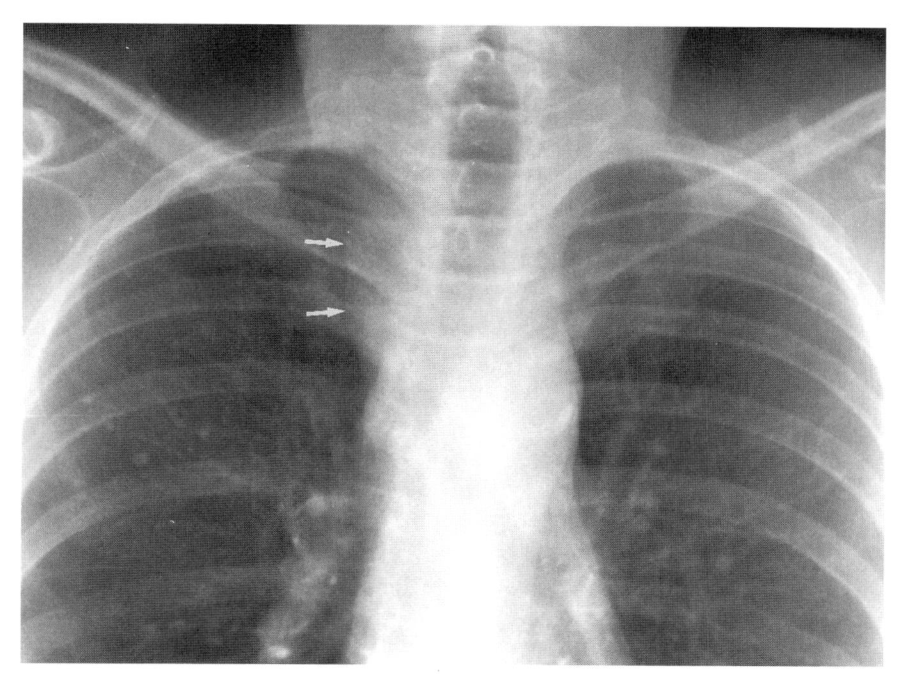

Figure 9.19

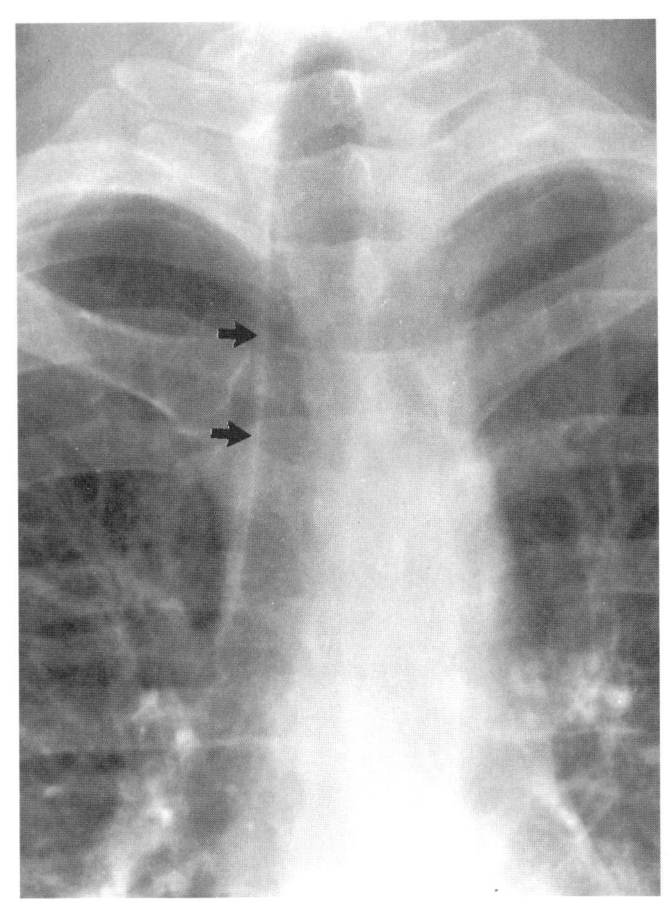

Figure 9.20. Normal mediastinum for comparison with Figures 9.21 and 9.22. The *arrows* indicate the right paratracheal space/stripe.

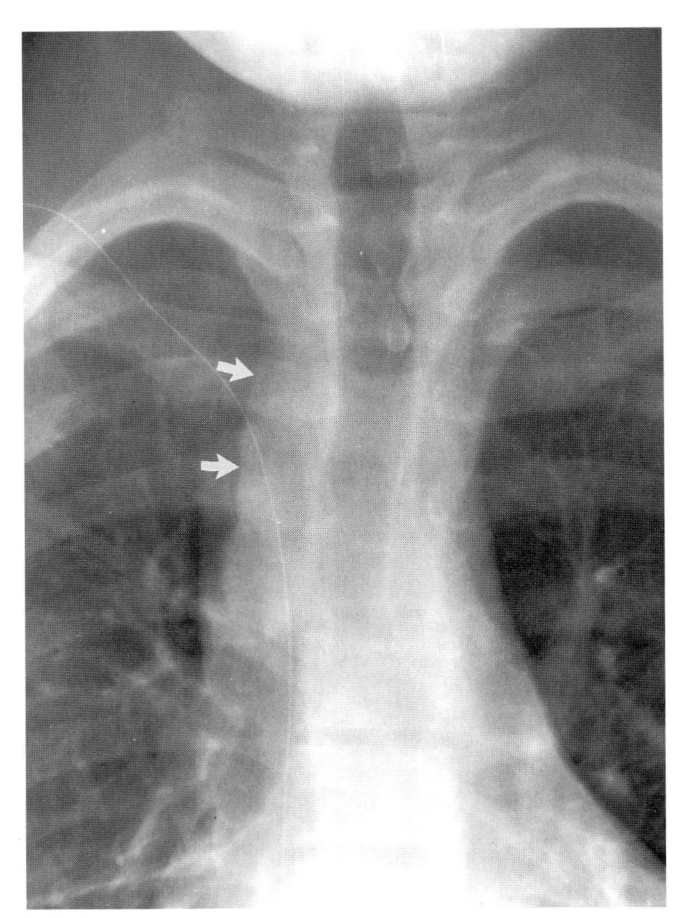

Figure 9.21. Widened right paratracheal space due to hemorrhage following trauma (see text).

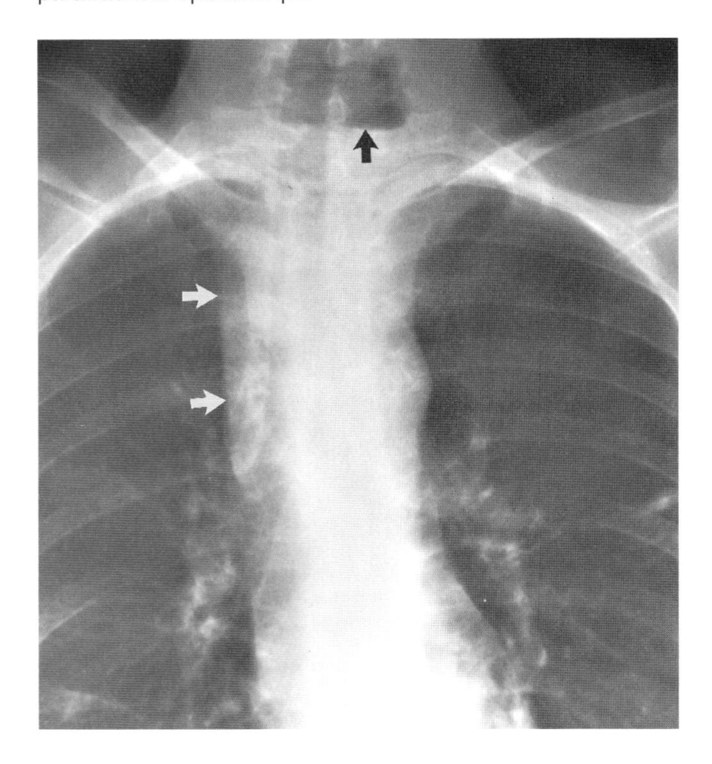

Figure 9.22. Mediastinitis widening the shadow of the superior mediastinum (see text).

By no means all obvious thickening or widening of the upper mediastinum is caused by tumor mass of one kind or another. Compare the three illustrations on the page to your left. Figure 9.20 is normal and shows normal thickness of the right paratracheal space (or paratracheal stripe) and the slightly-to-the-right-of-midline location of the trachea that you have learned to expect.

Figures 9.21 and 9.22, however, show definite widening of this region. The patient in Figure 9.21 had a mediastinal hematoma *(arrows)* following traumatic insertion of a subclavian venous catheter. The catheter was removed and replaced with a right arm central venous pressure line, which you can see in the illustration. The patient in Figure 9.22 had difficulty in swallowing, chest pain, and high fever before admission; she has a retropharyngeal abscess (air-fluid level indicated by the *black arrow*) and extension downward into the mediastinum with development of mediastinitis *(white arrows)*. Note the mottled appearance and indenting of the upper mediastinum. This case serves as a reminder that the mediastinum is a continuation of the soft tissues of the neck. Infection and other processes may extend from one region to the other in either direction.

The *thymus* is normally large in infancy and gradually regresses as it is replaced with fat in the adult. You have in Figure 9.23 the appearance on the chest film of the normal infant thymus, a sail-shaped shadow extending from the mediastinal border and seen somewhat better at expiration than at inspiration. In Figure 9.12 you have already seen the normal, rather dense thymus extending from the arch of the aorta anteriorly to the sternum as a triangular wedge at CT. Look back at that figure and compare it with the less dense anterior mediastinal region of the somewhat older patient in Figure 9.13, in whom the usual fatty decreased density of the anterior mediastinum is well seen.

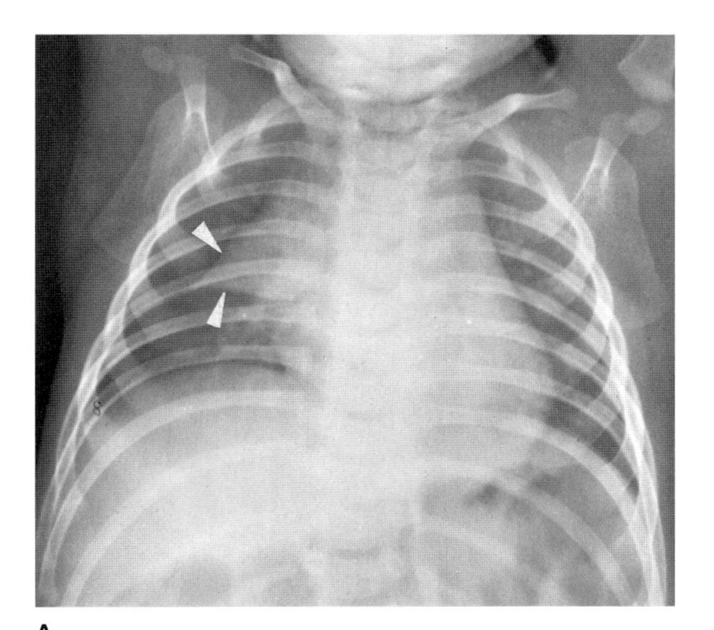
A

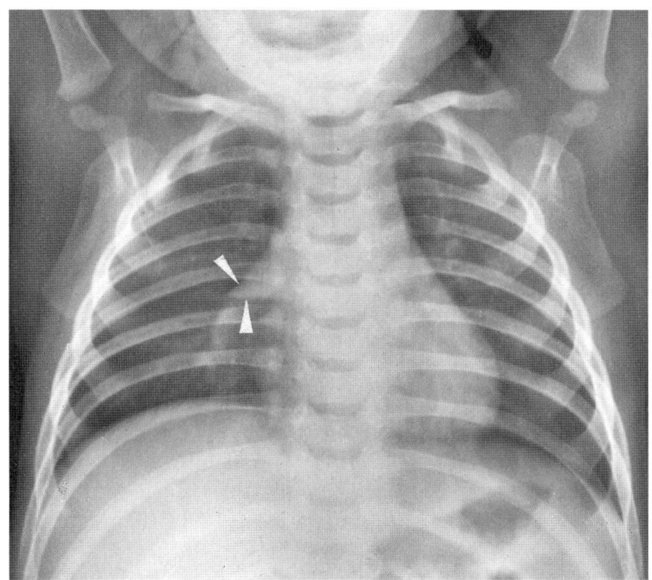
B

Figure 9.23. The normally enlarged thymus of an infant seen as a triangular, sail-shaped shadow overlapping the hilum as it projects laterally from the anterior mediastinum. It is better visualized at expiration (A) than at inspiration (B) a moment or so later.

Anterior Mediastinal Masses

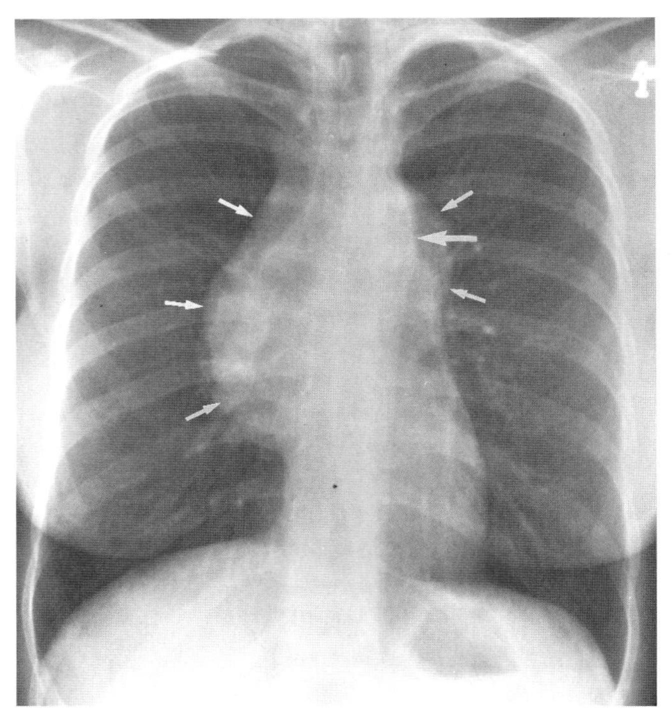

A. Plain film.

Figure 9.24. Teratoma in the anterior mediastinum (see text).

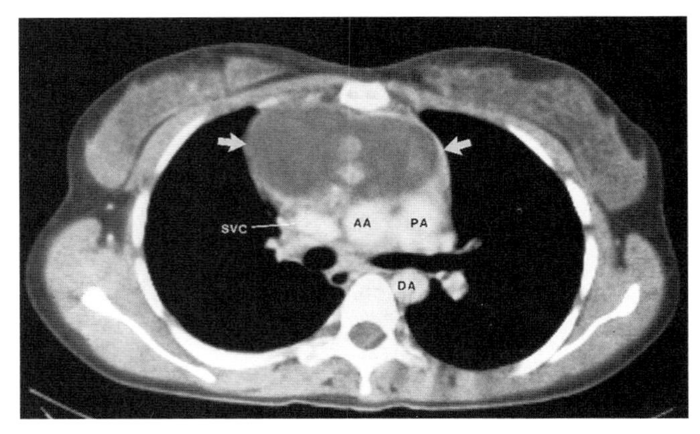

B. CT scan.

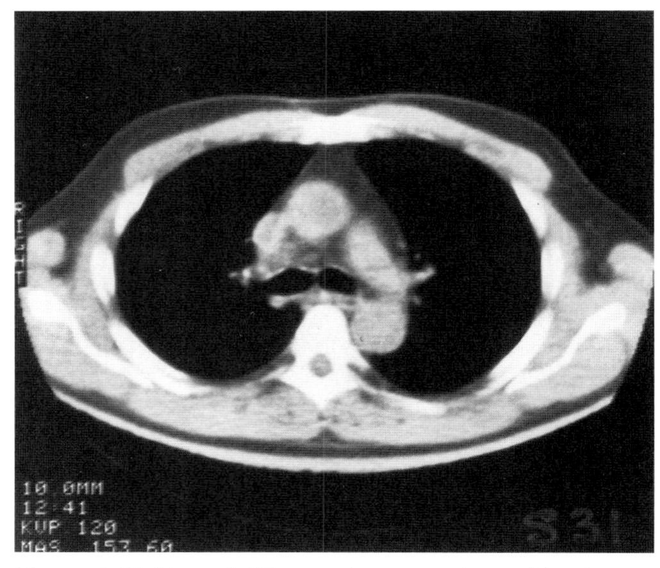

Figure 9.25. Normal CT scan, for comparison with others on this page spread.

Masses in the mediastinum anterior to the heart are generally one of four kinds: ectopic thyroids (goiters extending substernally), teratomas (benign or malignant), thymomas, or lymphomas. The mass you see in the chest film of the patient in Figure 9.24A *(small arrows)* was asymptomatic. Note that it is plastered against the right side of the heart and ablates the right heart shadow, just as pneumonia or large segments of collapsed lung do. This mass, however, lies across the location of the horizontal fissure and bulges outward so that it does not conform anatomically to either of those diagnoses. You know even from the PA view alone that the mass must be located anteriorly against the heart, since the upper part of the heart border is lost. You can also see its left margin as distinct from the far posterior arch and descending aorta *(large arrow)*. It is, therefore, most likely to be one of the four masses listed above.

Now look at the CT scan in Figure 9.24B and decide what the tissue composition of the mass is likely to be. The density of the vascular structures tells you that intravenous enhancement has been used, and this throws into contrast the lucency of the well-encapsulated mass, which must be largely composed of fat. It has within it, however, several round calcific densities, so it is not entirely fatty. It must logically be a teratoma, therefore, as indeed it proved to be on resection. The dense areas were cartilage partly calcified.

Notice the posterior displacement of the superior vena cava, the ascending aorta, and the pulmonary artery (compare with Figure 9.25, a normal scan made

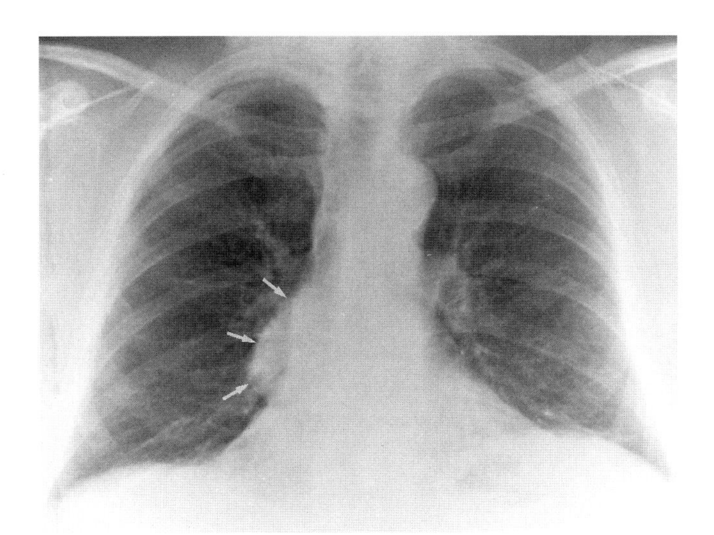

A

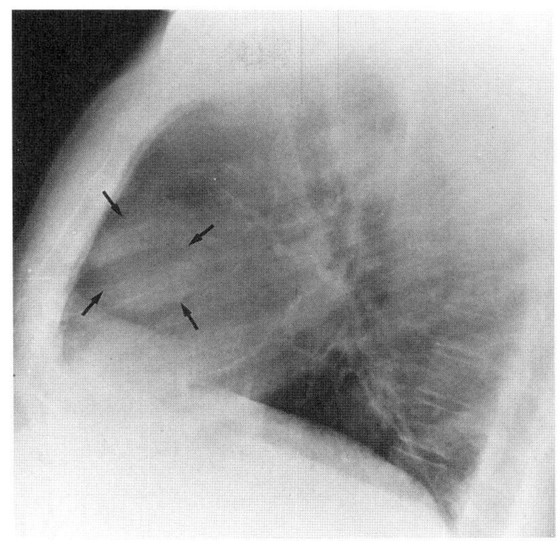

B

Figure 9.26. Anterior mediastinal thymoma in PA and lateral chest films (A and B) and in a CT scan (C).

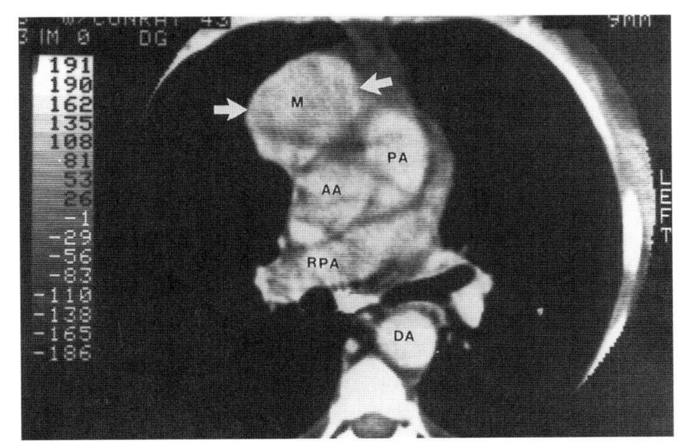

C

at a very slightly higher level, as you can tell from the tracheal bifurcation). The descending aorta is normally located against the vertebra.

Thymomas also appear as anterior mediastinal masses, often asymptomatic, sometimes occurring in conjunction with the symptoms and signs of myasthenia gravis. The mass *(arrows)* in Figure 9.26A was asymptomatic and might therefore have been any of the four entities, although it is low in the chest for thyroid.

Note that with the two views the plain films tell you that the mass is smaller than the one you have just been studying in Figure 9.24. For this reason it does not completely ablate the heart margin; small masses seldom do. Compare the matching CT scan, Figure 9.26C, with the other two on this page spread. Again the anteriorly placed mass deflects the aorta and superior vena cava posteriorly. At surgery it proved to be a thymoma.

Note that while Figure 9.25 was made at the level of the tracheal bifurcation, both Figure 9.24 and Figure 9.26 must have been made a bit lower, because the air-filled structures representing the major bronchi are now widely separated.

Anterior and Middle Mediastinal Masses

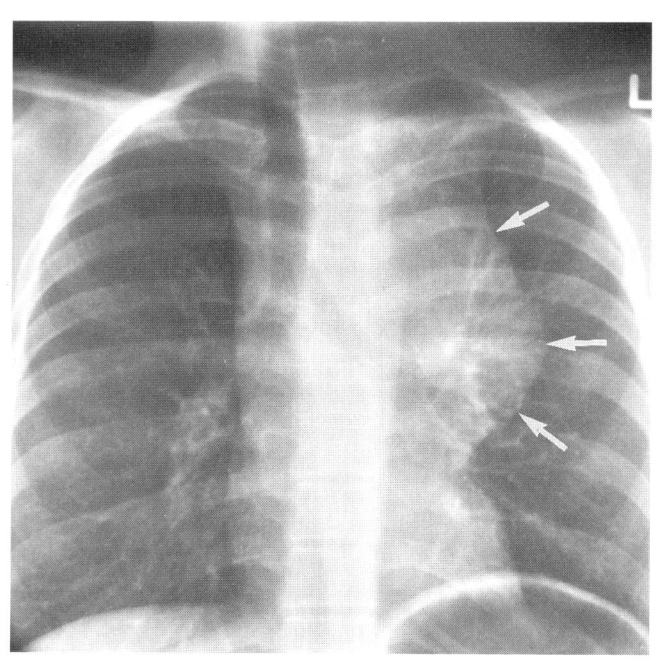

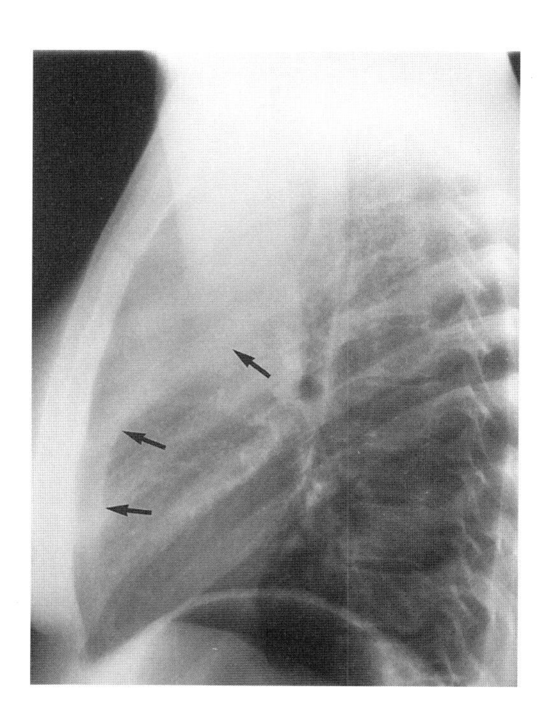

Figure 9.27. PA *(left)* and lateral *(right)* views of an anterior and superior mediastinal mass, which proved to be a lymphoma. Note the density filling in the anterior clear space in the lateral view.

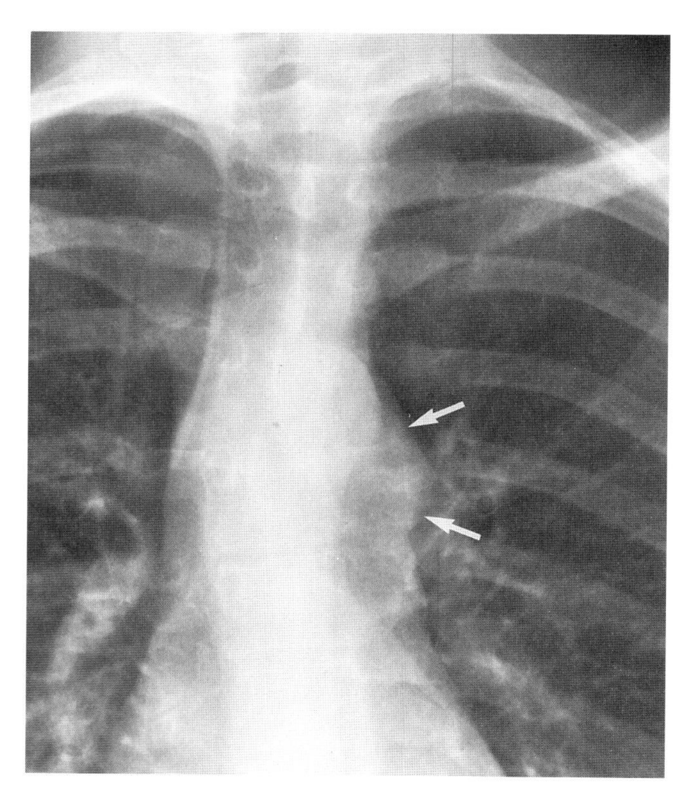

A

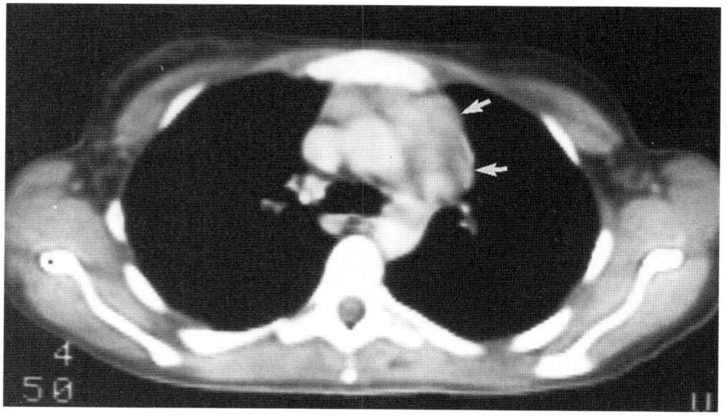

B

Figure 9.28. Hodgkin's lymphoma. Posteroanterior radiograph (A) and CT scan (B) of the chest. The findings are in effect those of anterior mediastinal lymphadenopathy.

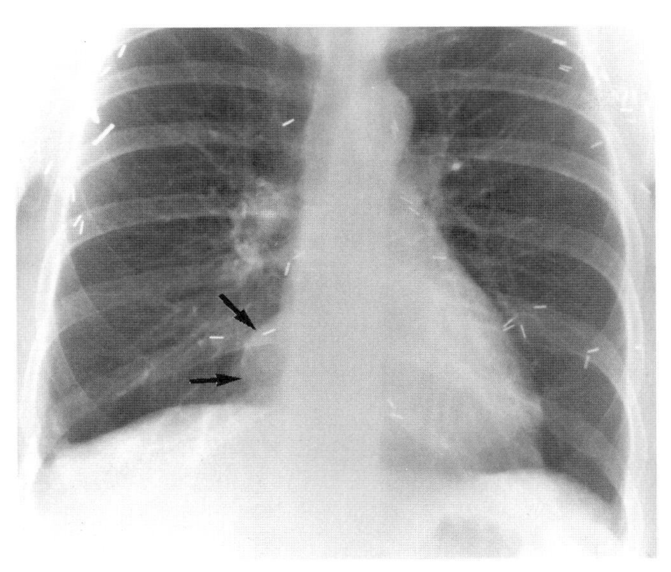

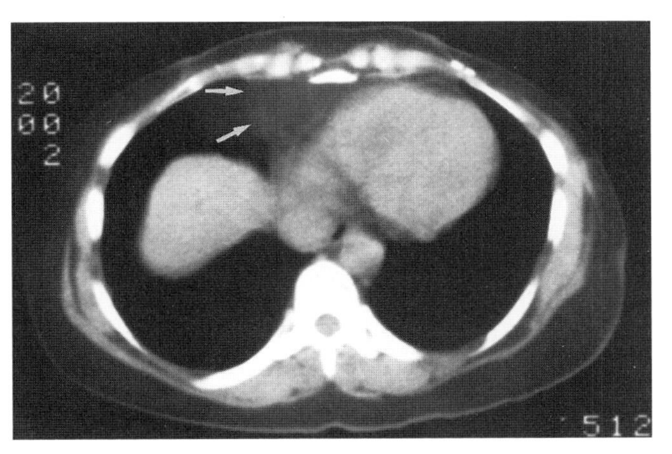

B

A

Figure 9.29. This patient had a history of bilateral mastectomies for carcinoma; note the multiple metallic surgical clips. At her annual checkup, although she was asymptomatic, her physician was concerned about the right paracardiac mass seen in A. However, on a CT scan (B) made low in the chest, the anteriorly placed mass is seen to be low in density—in fact, fat density of −136 Hounsfield units—and it is therefore typical of a paracardiac fat pad or lipoma rather than a recurrent tumor.

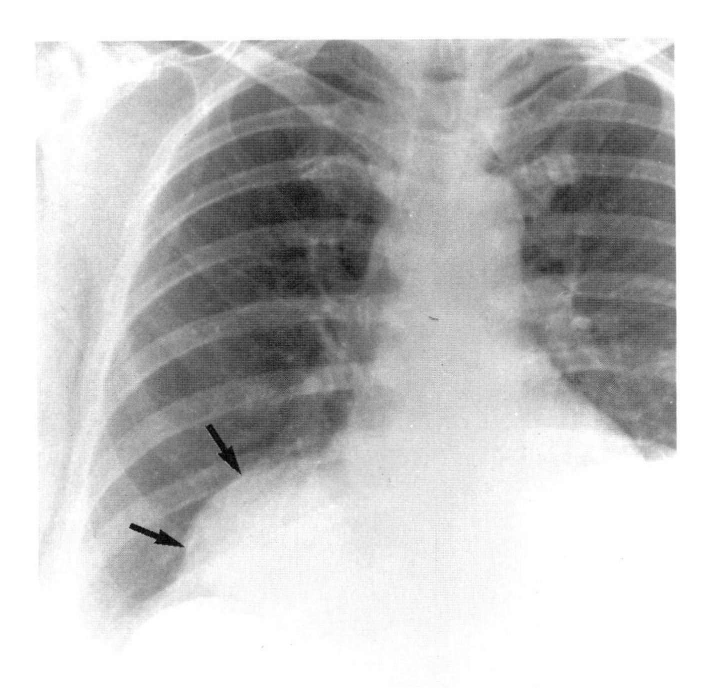

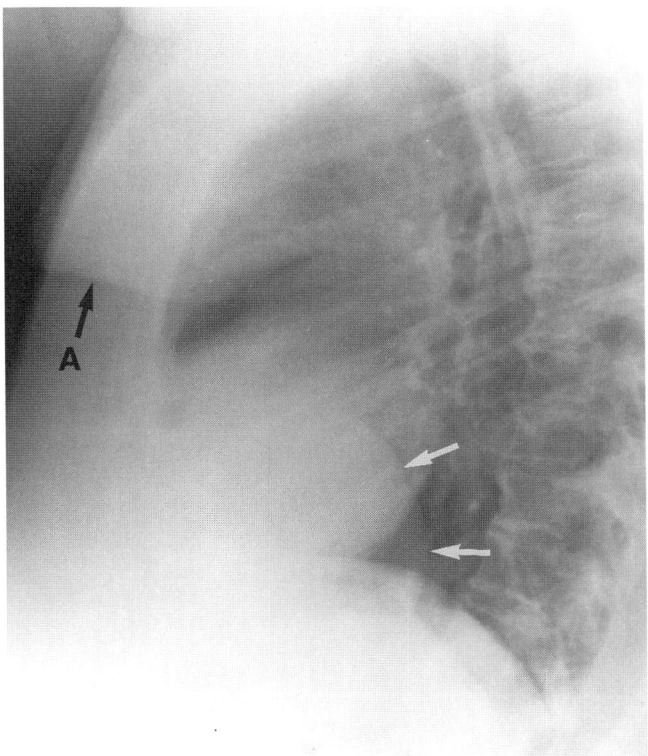

Figure 9.30. Pericardial cyst. The PA and lateral films demonstrate a round mass in the right cardiophrenic angle *(arrows)*. CT would indicate its cystic character. The patient was asymptomatic. *A* indicates air against the undersurface of the patient's arm.

Posterior Mediastinal Masses

Masses occurring in the paraspinal or posterior mediastinal areas are usually neural masses arising in nerves as they issue from the spinal cord. Because they apply themselves snugly against the ribs and spinal column, they often cause erosion of bone and back pain. In Figure 9.31 an oval mass is seen extending to the right of the spine across four rib spaces. Note that there is erosion of the fourth and fifth posterior ribs inferiorly *(white arrows)* with dense reactive bone along the inferior medial margin of each. The spaces between the fourth and fifth and the fifth and sixth ribs are wider than the contralateral spaces on the other side of the spine. The margin of the mass can also be seen extending down behind the heart and displacing the barium-filled esophagus to the left *(black arrows)*. You know that this is not an anterior mediastinal mass because it does not obliterate the heart border. In addition, it is eroding the *posterior* ribs. It proved to be a ganglioneuroma.

The CT scan on another patient, seen in Figure 9.32, shows a partly calcified mass lying against the posterior ribs on the left. This too proved to be a ganglioneuroma. Benign posterior mediastinal masses are more common than malignant ones.

Other posterior mediastinal masses include aneurysms of the descending aorta. They can generally be seen through the heart on the PA chest film.

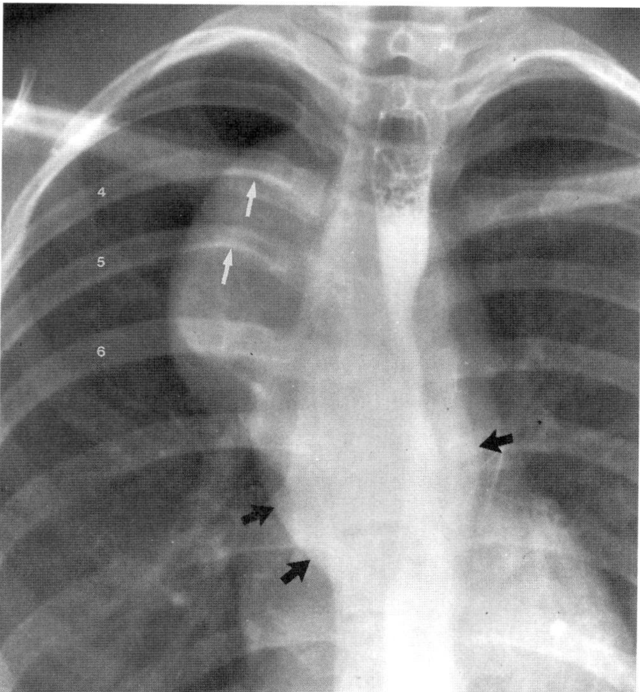

Figure 9.31. Posterior mediastinal ganglioneuroma. How would the CT scan look?

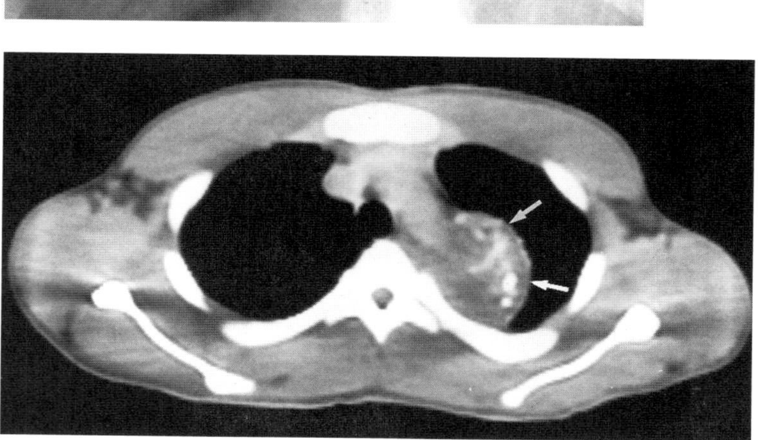

Figure 9.32. Posterior mediastinal mass in another patient. How would the PA chest film look?

Problems

Unknown 9.1 (Figure 9.33A)

Analyze the film and decide whether the lesion in the right upper lobe *(single arrow)* could be related to the mediastinal mass *(two arrows)*. The patient was admitted with superior vena cava syndrome.

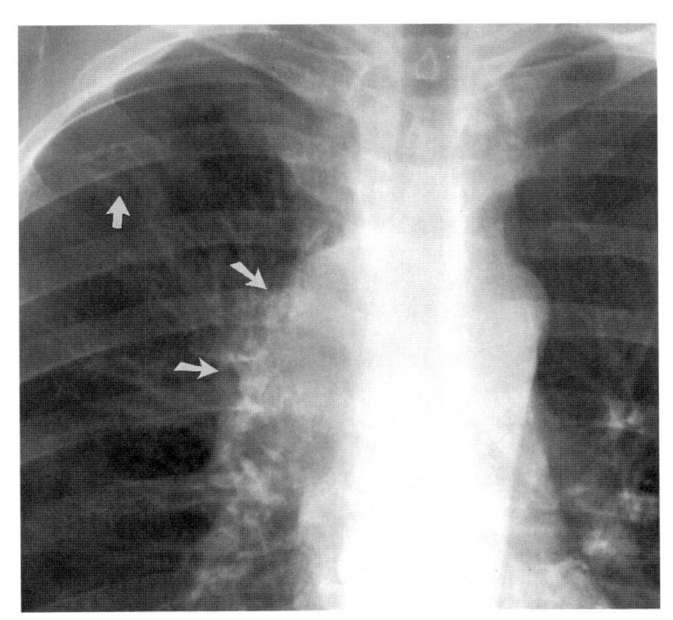

Figure 9.33A ***(Unknown 9.1)***

10 The Heart

Today a variety of simple and sophisticated imaging procedures are available to provide detailed information about the structure and function of the heart. These include chest films, fluoroscopy, echocardiography, computed tomography, radioisotope examinations, angiography, and magnetic-resonance imaging.

Generally the *plain film* diagnosis of heart disease is limited to the determination of cardiac enlargement (overall size and specific chamber enlargement), pulmonary vascular abnormalities, cardiac calcifications, and congestive failure. The other imaging techniques can give detailed information about myocardial thickness and motion, precise chamber size, and the presence of valvular disease, coronary artery disease, or pericardial disease. Several of these techniques also provide information about cardiac function.

In this chapter we propose to discuss first the plain film findings in heart disease, and then to give you examples of help provided by more complex studies.

Measurement of Heart Size

We are a race of measurers, partly because it is easier to measure than to think. Accurate measurement and an analysis of the significance of the findings form the basis of good science. Before angiocardiography and cross-sectional imaging techniques were developed, making possible the study of the individual cardiac chambers, the overall size of the heart was measured in its every dimension and some of these measurements proved useful. Since the development of specialized cardiac imaging, however, much less reliance is placed on plain film measurements in a cardiac patient. Nevertheless, in day-to-day routine patient problems, evaluation of the shadow of the heart on the plain film of the chest will prove useful to you. You can develop for your own daily use a ***rough*** estimate of the size of the heart, using only the measure-

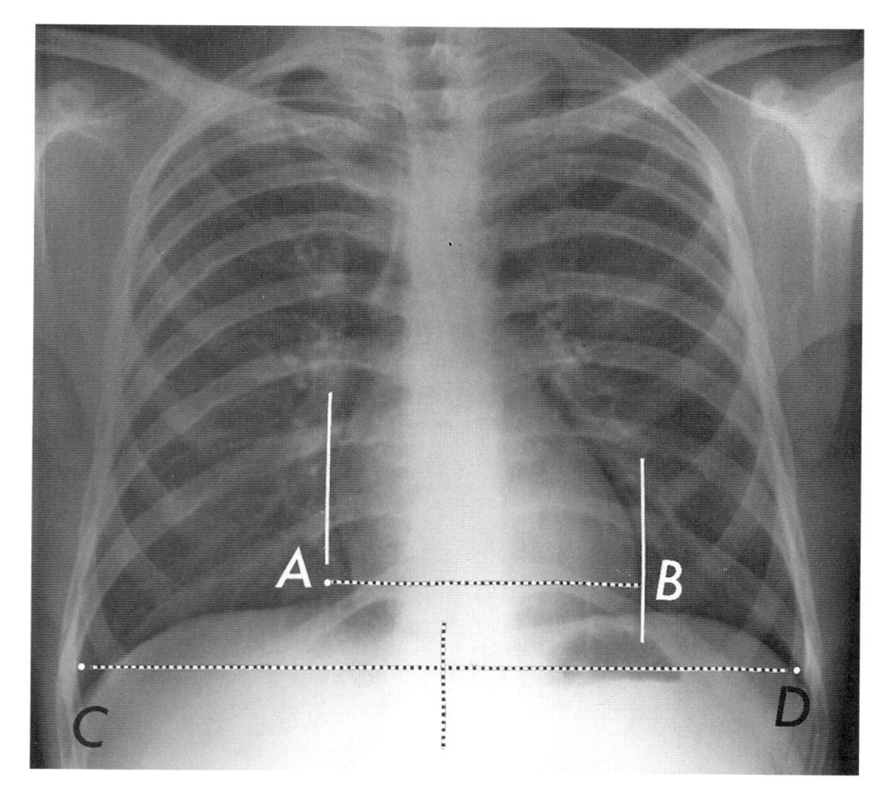

Figure 10.1. This young man was filmed because a heart murmur was detected on routine physical examination. Is his heart enlarged?

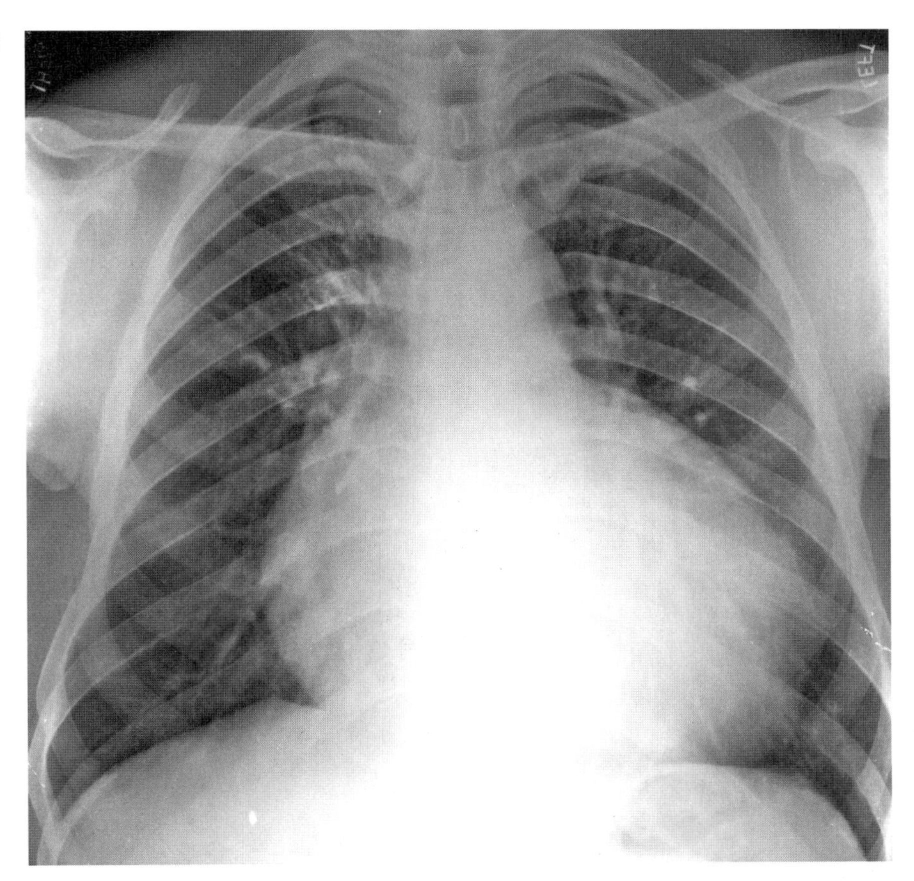

Figure 10.2. Measure this patient's heart for enlargement.

ment you make on a PA film (which enlarges the heart by projection less than 5 percent). If you know one easy-to-carry-out measuring system and employ it on every PA chest film you study, you will soon develop an ability to *estimate* heart size. A left lateral film, now a routine part of the chest film study, will enhance the accuracy of plain film assay of heart size.

You must add to this mode of assaying the status of the heart an awareness of *the ways in which cardiac enlargement may be (1) simulated (as on a poor inspiration film) or (2) masked (as in a large left pleural effusion)*. Even more important, you must have some familiarity with changes in the shape of the heart due to specific chamber enlargements, since a change in shape either with or without enlargement may sometimes indicate the type of heart disease that is present.

In sum, then, you must be able to estimate overall cardiac size while accepting the important limitations of that estimate, to discount conditions which may simulate enlargement, and to be aware of the changes produced in the shape of the cardiac profile by various disease processes. These things every physician ought to know.

The simplest method of measuring the heart is to determine its relation to the width of the chest at its widest part near the level of the diaphragm. This is called the *cardiothoracic ratio*, and it is calculated from the PA chest film only. Measure between two vertical lines drawn tangential to the most prominent point on the right and the left cardiac profiles. The prominence of the bulge on the right is usually a little higher than the apex of the left profile. In adults the width of the heart should be less than half the greatest thoracic diameter, measured from inside the rib cage at its widest point.

No ruler is necessary for this measurement, nor do you need to remember anything more than the 50 percent figure. Using any handy piece of paper with a straight edge (the handiest will often be the margin of the patient's own film envelope), determine the width of the heart. Then decide whether this width exceeds the distance from the midpoint (spine) to the inside of the rib cage (half the transthoracic diameter). Still more simply, you can measure from the midline to the *right* heart border and see whether that distance will fit into the piece of lung field to the *left* of the heart, something you can do from the back row at clinical rounds! For example, in Figure 10.1 is the distance from *A* to the midline less than or greater than that from *B* to *D*?

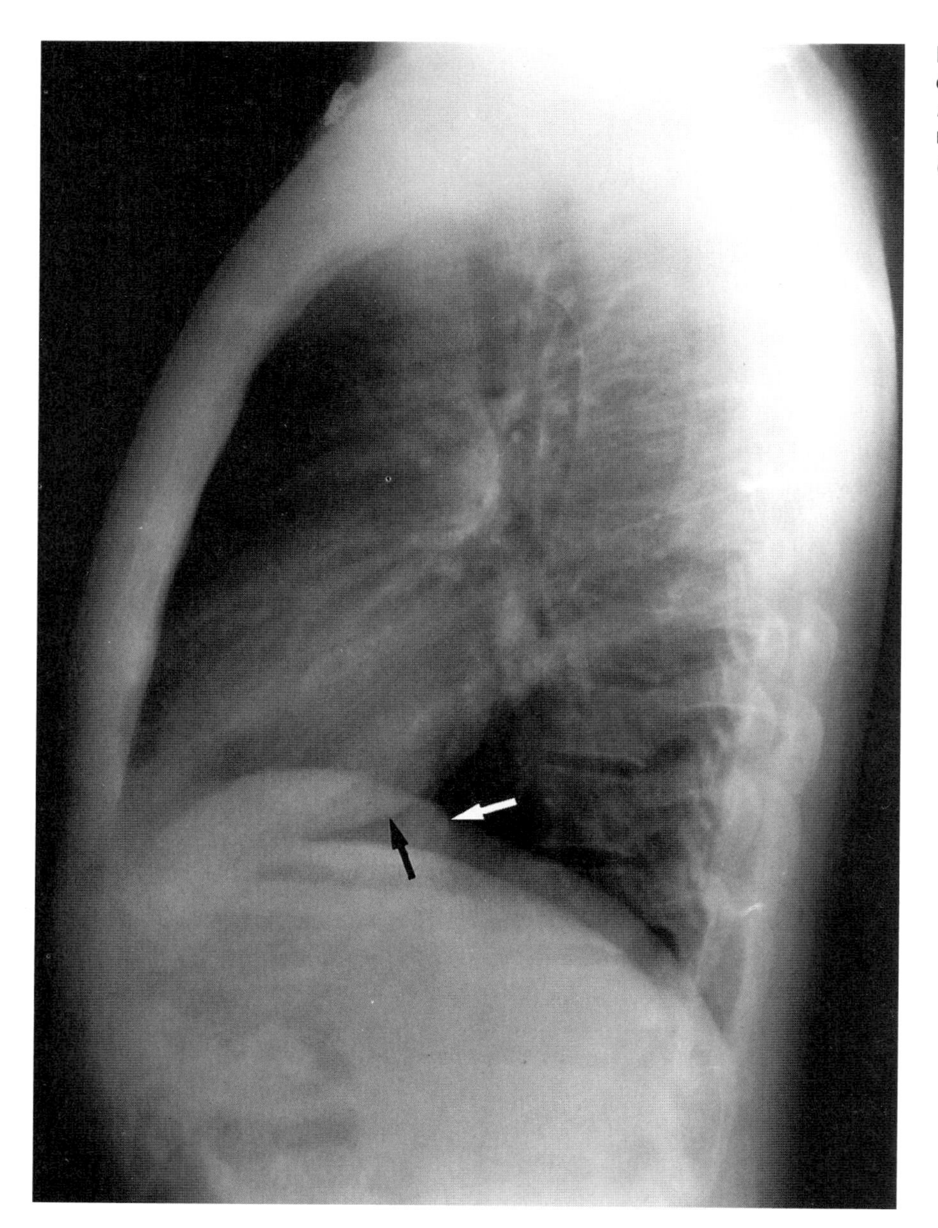

Figure 10.3. Normal left lateral film; the convex posterior heart border *(black arrow)* does not extend beyond the posterior margin of the inferior vena cava *(white arrow).*

The left lateral film (Figure 10.3) is an excellent check on the PA appearance of the heart. When apparent enlargement to the left in the PA view is checked on the left lateral chest film, any increase in the *mass* of the left ventricle will extend the border of the heart posteriorly and low against the diaphragm. Conversely, increase in the mass of the right ventricle will be seen in the lateral film to fill in the lower part of the anterior clear space behind the sternum but will not extend the heart posteriorly.

No doubt you measured the heart in Figure 10.1 and found it normal in size; but compare it now with almost any of the chest films in the preceding chapters and you will certainly be struck by the flat, almost absent aortic arch. This young patient had been discovered to have hypertension. The possibility of coarctation of the aorta was suggested in view of unobtainable pressure in his legs. Reinspection of his chest film revealed the saucered erosions of the undersurface of his ribs, where dilated intercostal arteries had developed as collateral pathways. His coarctation was successfully revised surgically, and he was restored to normal health. Figure 10.5 is a detail of the chest film in Figure 10.1. The radiological manifestations of coarctation are seldom present in children younger than 10. You must remember too that a number of other conditions can cause rib notching. Neurofibromatosis is one.

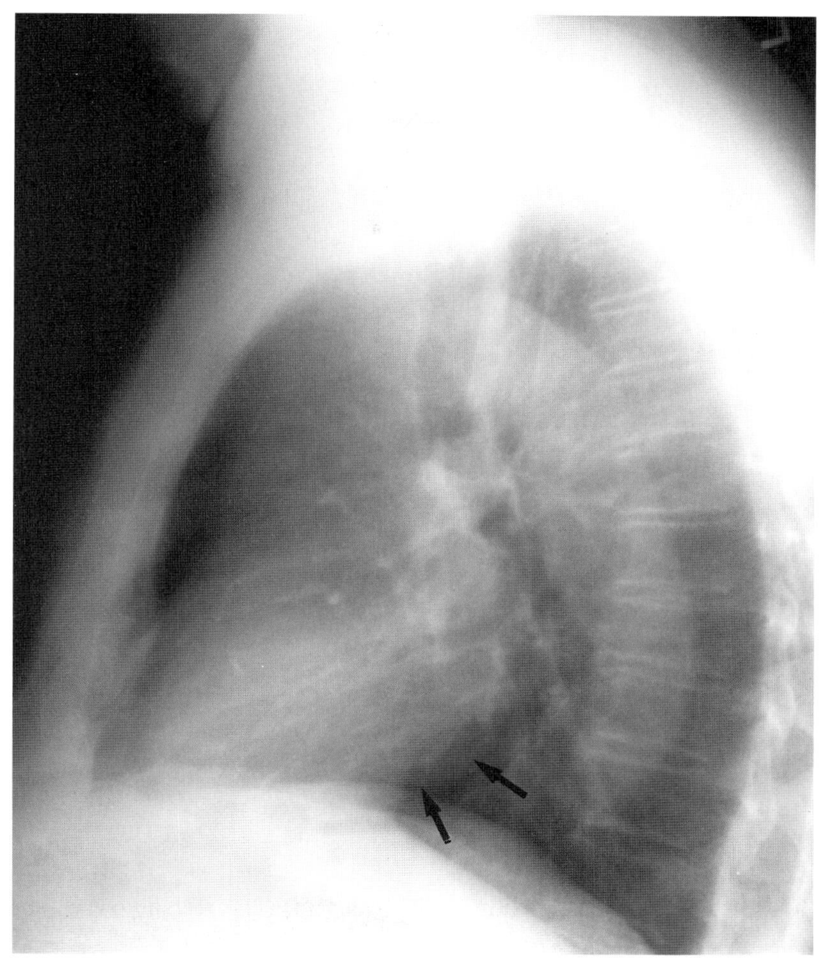

Figure 10.4. Left lateral film of a patient with cardiomegaly. The *black arrows* indicate the posterior heart border.

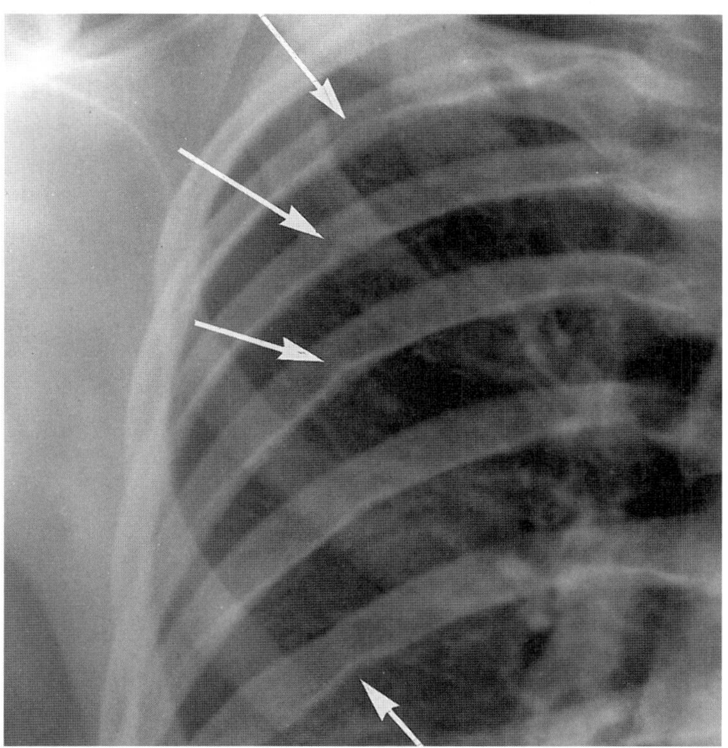

Figure 10.5. Detail from Figure 10.1. The *arrows* indicate rib notching.

Factors Limiting Information Obtained by Measurement

It is valuable to realize that heart shadows may be abnormal in shape even though normal in size. They may also be enlarged with or without a distinctive change in shape; hearts that decompensate may be enlarged and shapeless.

Hearts may also be only *apparently enlarged* for a variety of reasons that you must be able to discount. You already know some of the ways in which cardiac enlargement may be simulated. You have seen it in chest films made at *expiration* (Figure 10.6A), and it is logical to expect that a high diaphragm will tilt the heart upward, bringing its apex closer to the lateral chest wall. In addition, the flare of the ribs is greater at inspiration and decreases at expiration, further altering the apparent cardiothoracic ratio. Any time that you have reason to expect a patient's diaphragm is high, you should anticipate an *apparently* enlarged heart shadow. Whenever *any kind of abdominal distention* (late pregnancy, ascites, intestinal obstruction) is present, you may not be able to estimate heart size.

Remember too that portable chest films are usually made AP and result in an appreciable enlargement of the heart shadow by projection, since the heart is farther away from the film. If the patient is filmed supine, the diaphragm is likely to be higher. Valuable data may be obtained from bedside chest films on the very sick patient, but an estimate of heart size is not among them.

The next point to be checked is that there is *no rotation* of the patient. You have already seen in an earlier chapter the degree to which rotation may produce an appearance of widening of the heart and mediastinal shadows, and you know that symmetry of the clavicles and ribs gives you assurance that no rotation is present. We will be discussing intentionally rotated oblique films in more detail later, at which time you will be able to study more closely the precise effect of rotation on the cardiac shadow.

Deformity of the thoracic cage will, of course, often render impossible any attempt to measure the size of the heart, and you would not expect to be able to do so in a patient with severe scoliosis, for example. The solitary (but symmetrical) deformity of a depressed sternum (pectus excavatum) usually displaces the heart to the left, and your suspicions will be aroused when you find no right heart border from which to measure. A lateral film will settle the matter.

You might wonder whether the size of the heart shadow would be increased if the film happened to be taken at full diastole, and decreased if it were made at

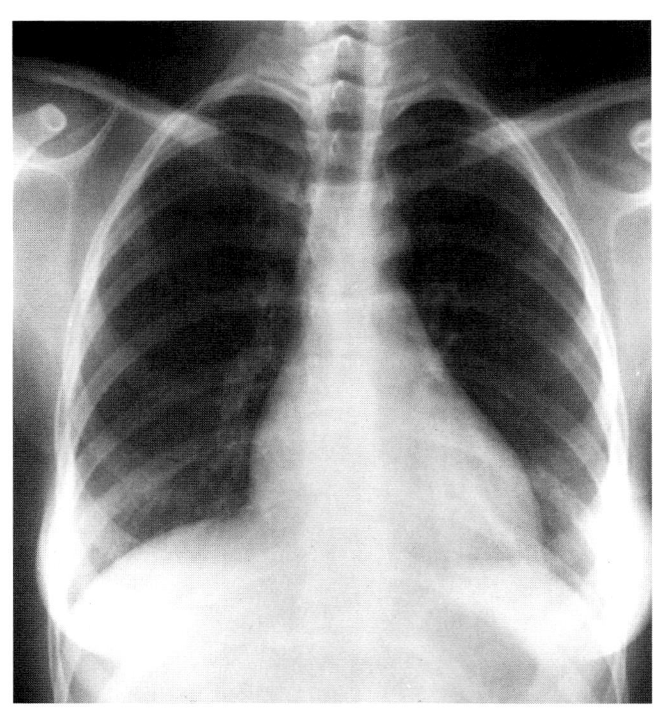

Figure 10.6A. This film appears to show cardiomegaly, but it is only a normal heart at expiration.

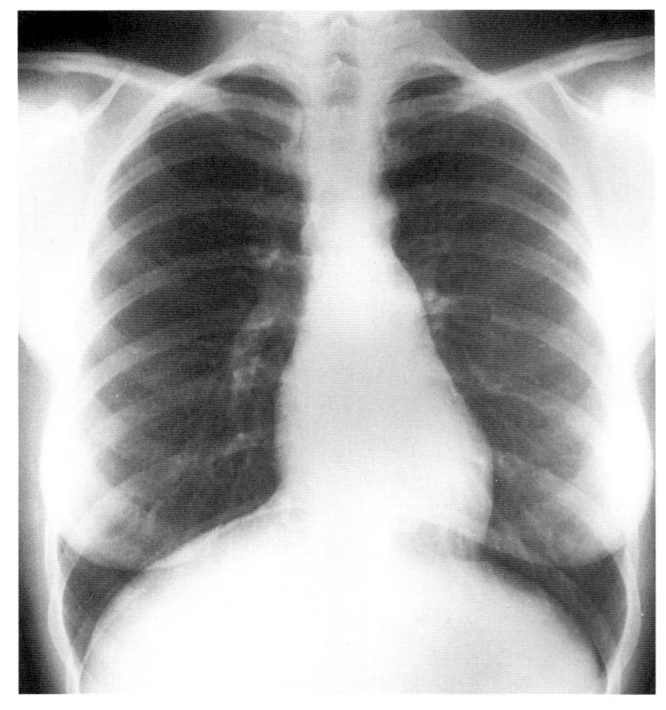

Figure 10.6B. The same heart at inspiration.

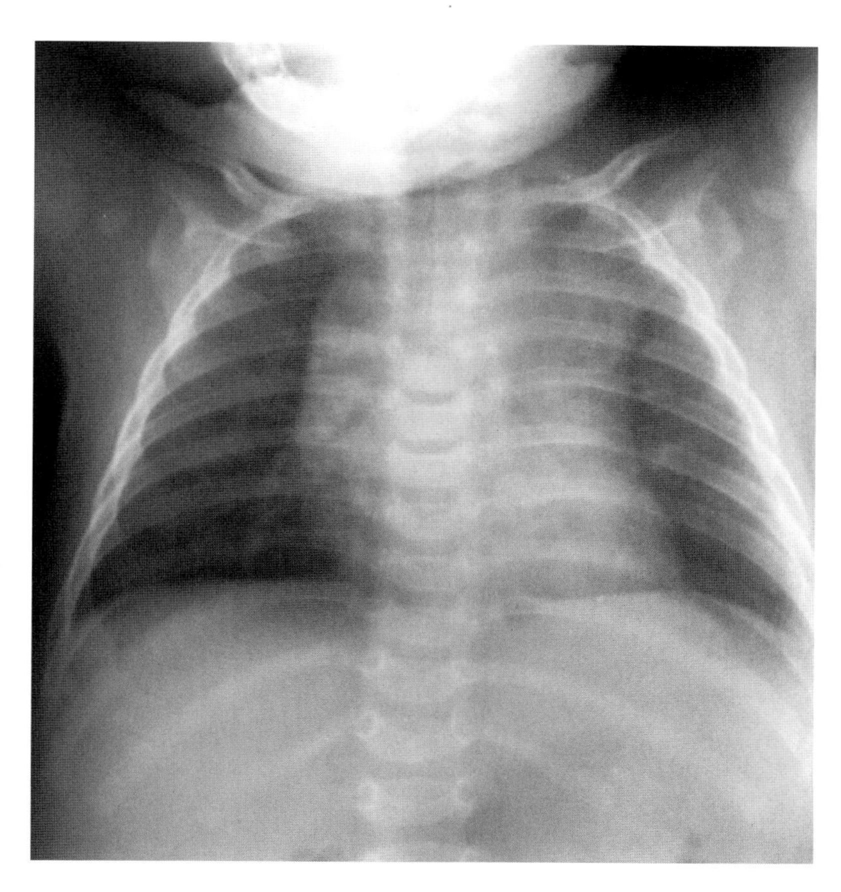

Figure 10.7. AP chest film of an infant. The cardiac shadow looks large for several reasons. Since this 6-month-old child could not be instructed to take a deep breath, the film was not obtained at full inspiration. Also, it was taken with the AP technique, the child lying on his back on the x-ray table. The fullness of the upper mediastinum shown extending down over the upper right and left heart borders is nothing more than the normal thymus in a child of this age.

the end of systole. The shadow is slightly different at the extremes of the cardiac cycle, but the difference is not usually enough to matter in a rough estimate such as the cardiothoracic ratio, at least in adults.

It is vital that you develop an immense degree of caution with regard to making pronouncements about apparent enlargement of the *infant* heart as you see it on chest films. This is particularly true in the infant under 1 year of age (Figure 10.7). Because an infant cannot be requested to take a deep breath and because of the basic difference in proportion of abdominal size to thoracic size, the normal diaphragmatic level in an infant is higher than in an adult. Infants are usually filmed AP and supine. They wriggle and are hard to immobilize, which produces rotation. They have not yet developed the proportion of lung size to heart size characteristic of adults and present already in older children. Remember that the thymus overlying the heart may also mimic cardiomegaly. Always beware of x-ray appearances suggesting cardiac enlargement in patients under a year old, unless you have supporting clinical evidence.

But always remember that *overdistension of the lungs* for any reason compresses the heart and mediastinal structures from both sides and narrows their PA shadow. In the dyspneic patient with a low diaphragm and in the emphysematous patient, therefore, the heart size as measured on the PA chest film may be *deceptively small* and not inform you at all reliably about the cardiac status. In patients with chronic emphysema the heart is often found at autopsy to be enlarged by weight as a result of right ventricular hypertrophy (cor pulmonale), although no cardiac enlargement had ever been noted radiologically.

Noncardiac disease may mask true cardiac enlargement. If you think back through the earlier chapters, you will have no difficulty appreciating the degree to which mediastinal or pulmonary disease may render the dimensions of the heart unobtainable. Any density that obscures one cardiac profile makes it futile to try to estimate heart size. Thus neither the size nor the shape of the heart can be studied from plain chest films in the patient who has a massive pleural effusion, consolidation in the anterior part of either lung, or a large anterior mediastinal mass.

True mediastinal shift is usually the result of some important change in intrathoracic dynamics (collapse of an entire lung, for example) and may alter the position of the heart so much that measurements are meaningless.

Examples of Apparent Abnormality in Heart Size and Difficulties in Measuring

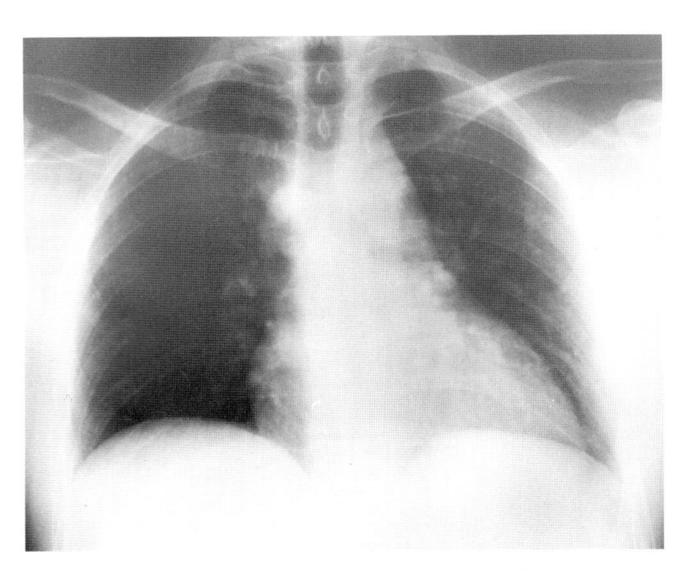

Figure 10.8A. This chest film made AP recumbent simulates cardiac enlargement.

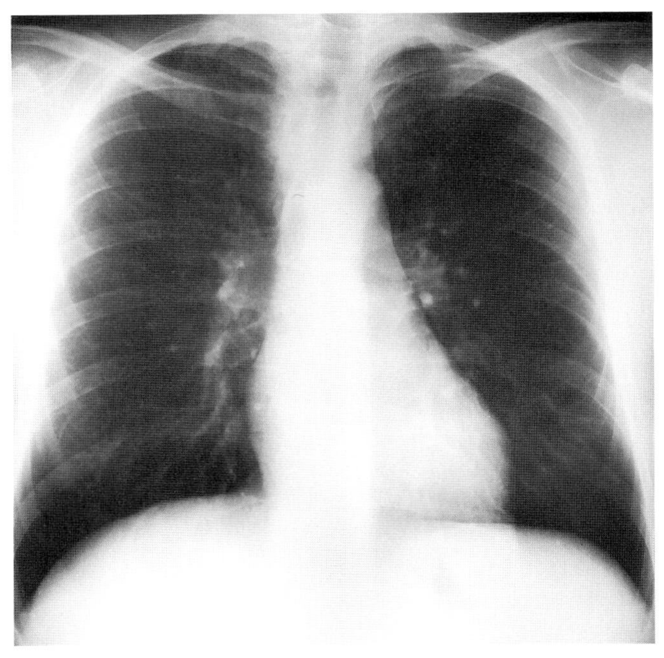

Figure 10.8B. PA film of the same patient standing.

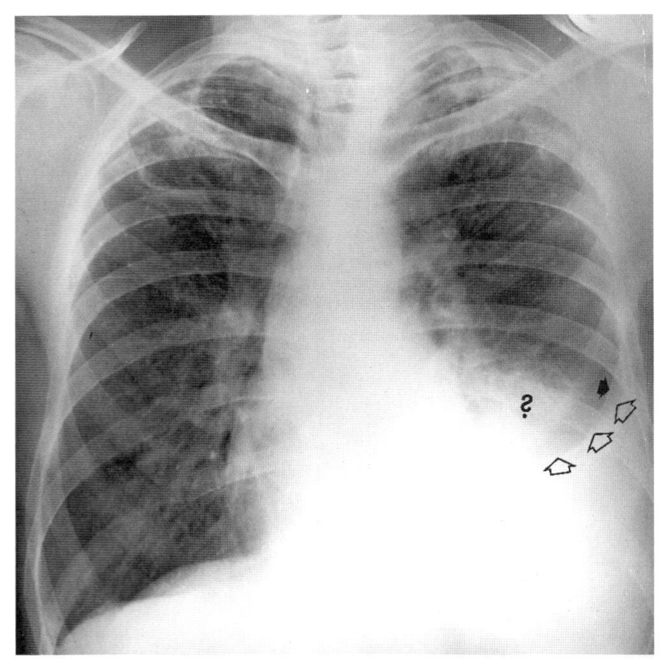

Figure 10.9. Here, because of a combination of pleural and pulmonary pathology, heart size cannot be estimated. (The *open arrows* mark a fluid line of a left pleural effusion crossing a rib margin, *solid black arrow.* The *question mark* indicates probable lingula consolidation, further obscuring the left heart border in this patient with known upper lobe tuberculosis.)

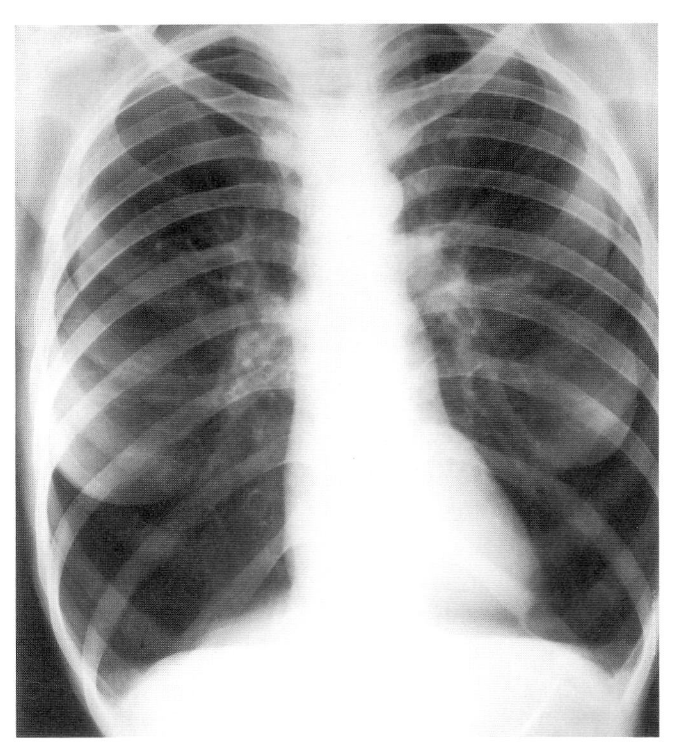

Figure 10.10. The heart does not appear enlarged, but cor pulmonale is present in this patient with emphysema. The low level of the diaphragm and hyperinflated lungs may make cardiac pathology difficult to detect.

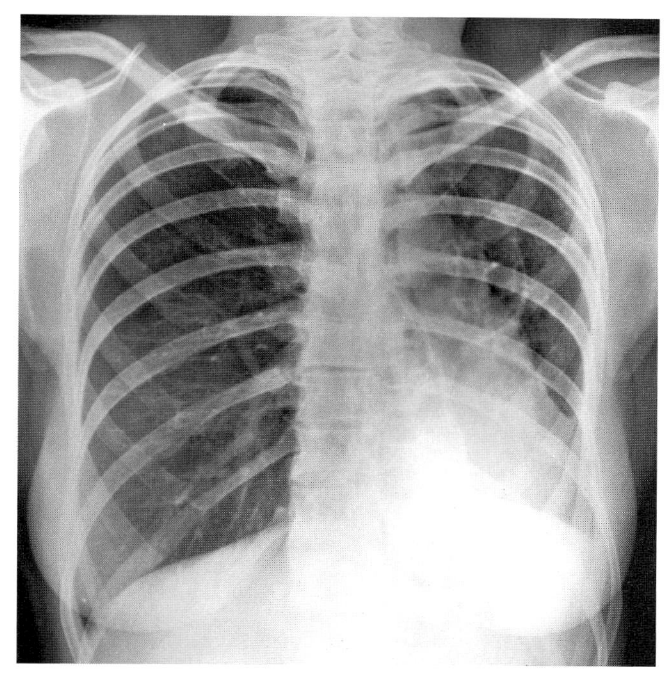

Figure 10.11. The heart is not well seen in collapse of the left lower lobe (or of the whole left lung).

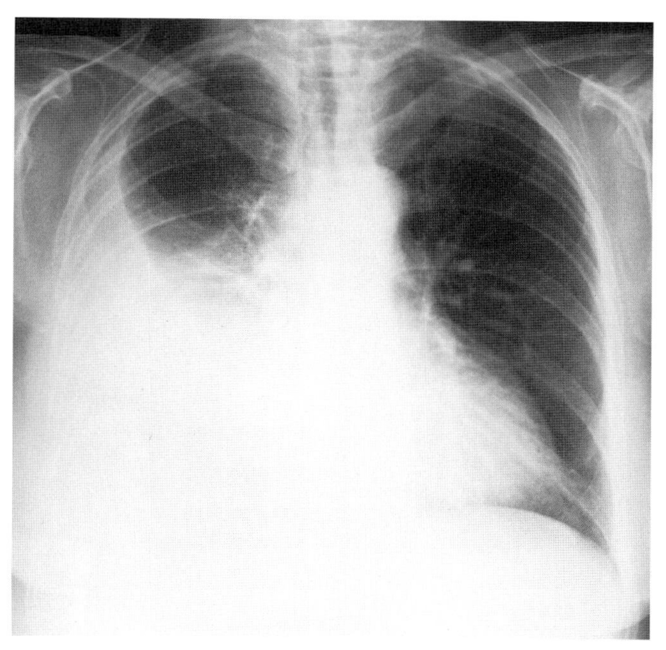

Figure 10.12. The heart size cannot be estimated in massive pleural effusion.

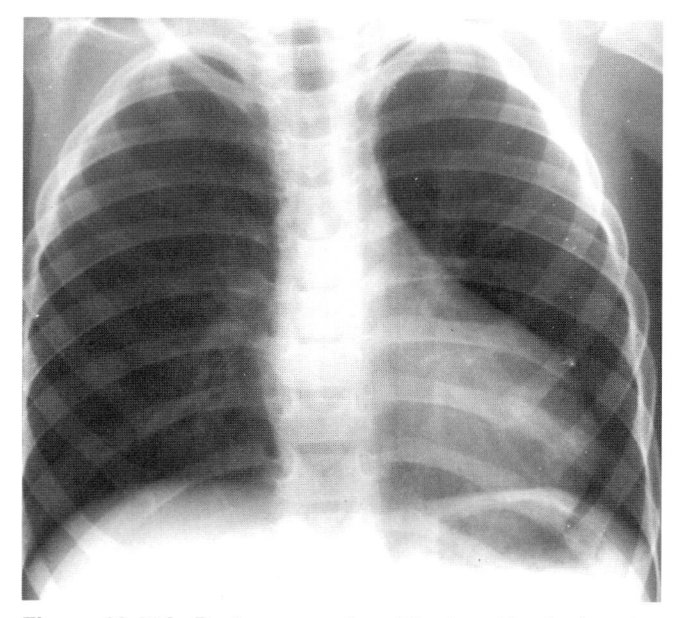

Figure 10.13A. Pectus excavatum. The heart is displaced slightly to the left.

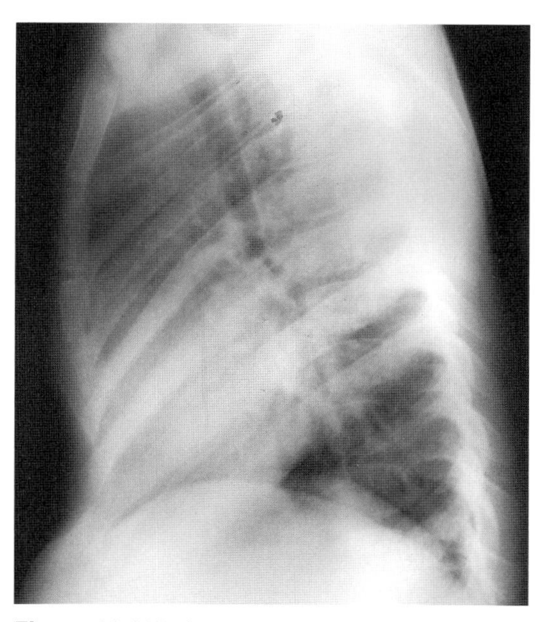

Figure 10.13B. Lateral view of the same patient shows depression of the sternum at the level of the heart in a young man whose sternal growth plates have not yet fused. The AP dimension of the chest at heart level is decreased and the heart is displaced posteriorly (posterior margin behind the inferior vena cava).

Interpretation of the Measurably Enlarged Heart Shadow

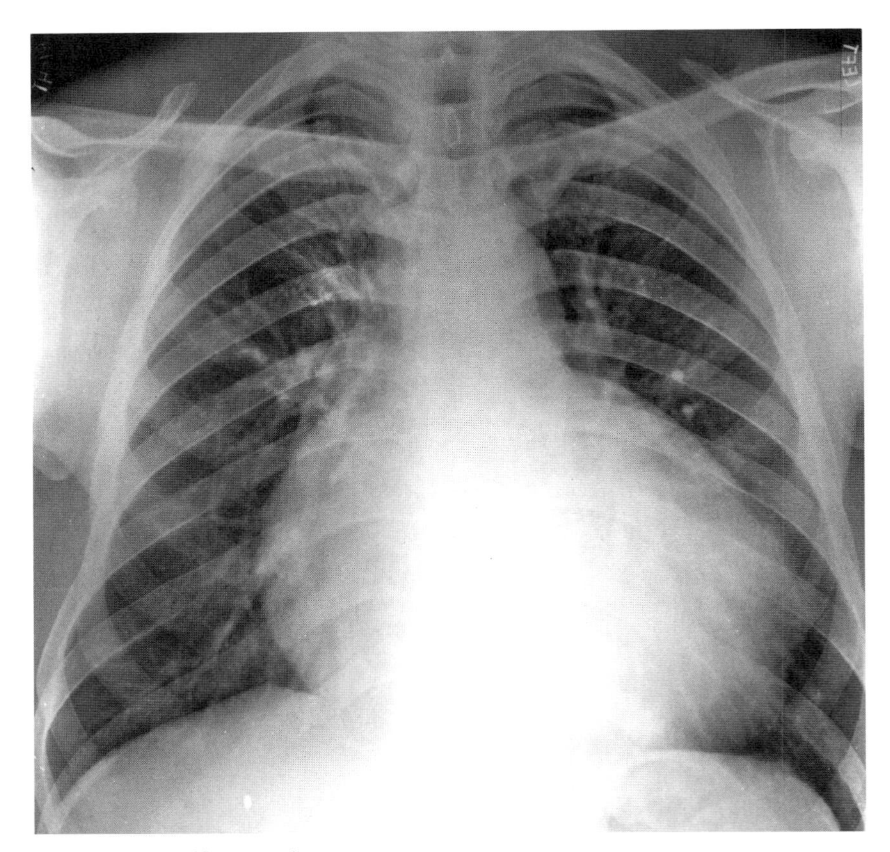

Figure 10.14. (See text.)

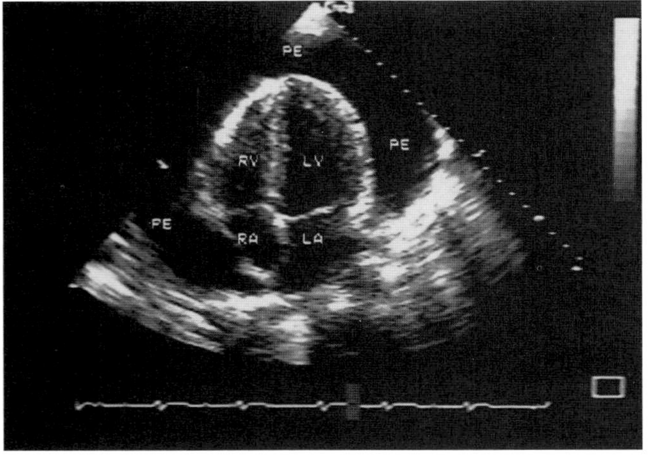

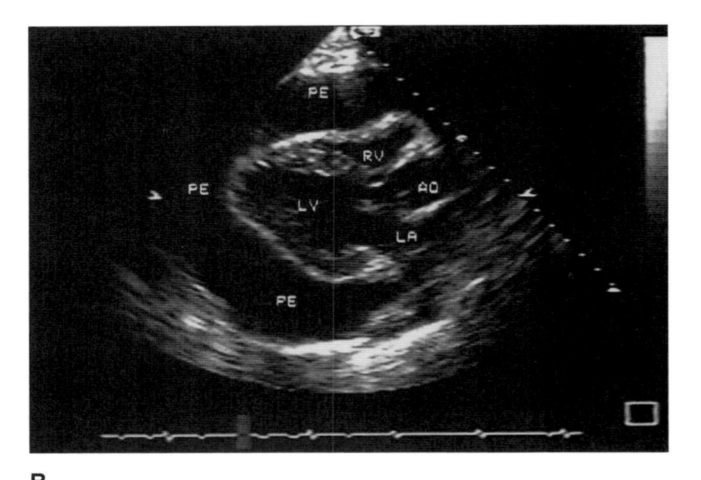

A B

Figure 10.15. Echocardiogram showing the presence of a large pericardial effusion *(PE)* around the heart. A: Four-chamber view depicting a ring of anechoic (no echo) fluid around the right ventricle *(RV),* left ventricle *(LV),* right atrium *(RA),* and left atrium *(LA).* B: Long axis view showing the aortic valve between the left ventricle *(LV)* and ascending aorta *(AO);* again visible is the large pericardial effusion *(PE)* surrounding the heart.

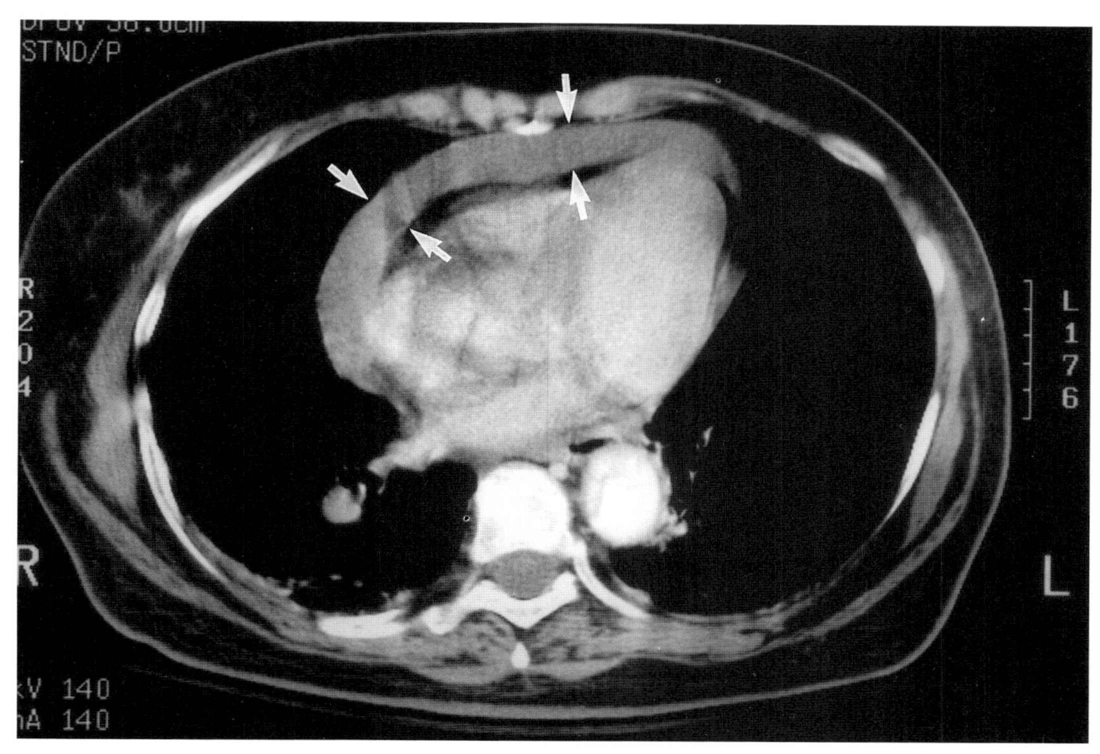

Figure 10.16. CT scan of a patient with a pericardial fluid collection. Note the thick ring of fluid density *(arrows)* around the heart. The black curve behind the pericardial fluid represents epicardial fat between the pericardial space and the myocardium. Intravenous contrast material opacifies the cardiac chambers and the ascending aorta. Compare the normal chest CT scans in Chapter 3.

Consider now a chest film in which, after checking out all potentially misleading factors, you find that the measured heart shadow exceeds its allowed 50 percent of the transthoracic diameter. How can you distinguish between the shadows cast by cardiac hypertrophy, cardiac dilatation, and pericardial effusion around the heart?

Figure 10.14 gives you an example. A patient with acute rheumatic fever and pancarditis shows obvious enlargement of the heart shadow. You know from your study of pathology that there may well be valvular involvement, myocardial damage, and pericarditis with effusion. The heart disease is properly termed pancarditis; and dilatation of the chambers due to poorly functioning valves and an inflamed, inefficient myocardium, as well as the presence of pericardial fluid, could all be contributing to the production of such a large shadow. In fact, all were present.

As you look at PA and lateral chest films on any patient having an enlarged heart, there are a number of findings that will help you. You may be able to recognize a predominantly enlarged *left* ventricle from the extension to the left in the PA view and posteriorly in the lateral. *Right* ventricular enlargement will show no posterior extension on the lateral film but will show anterior fullness filling in the lower part of the anterior clear space. Remember that plain chest films may show ventricular enlargement but do not differentiate hypertrophy from dilatation.

If the heart is decompensating, it will tend to shapelessness and extend to both right and left in the PA view, suggesting either failure or pericardial effusion. Effusion is usually documented by *echocardiography,* a sonographic record of the reflection of sound waves from the wall of the heart and from the pericardium, separated by a layer of fluid (Figure 10.15). Pericardial effusions may also be detected by CT (Figure 10.16) and MR imaging. In clinical practice a review of the patient's old films is probably the best way to assess the development of cardiac enlargement, in and out of failure. Sudden shapeless increase in size should suggest pericardial effusion to you.

Enlargement of the Left or Right Ventricle—Help from the Lateral Film

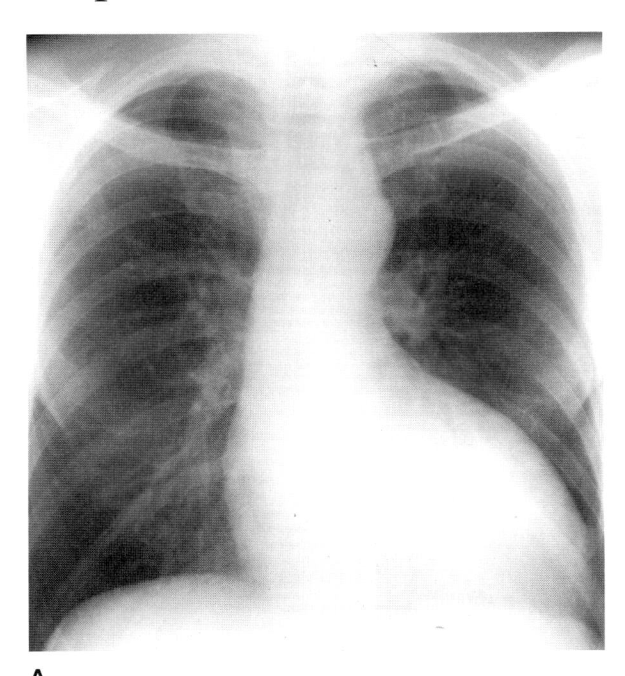

A

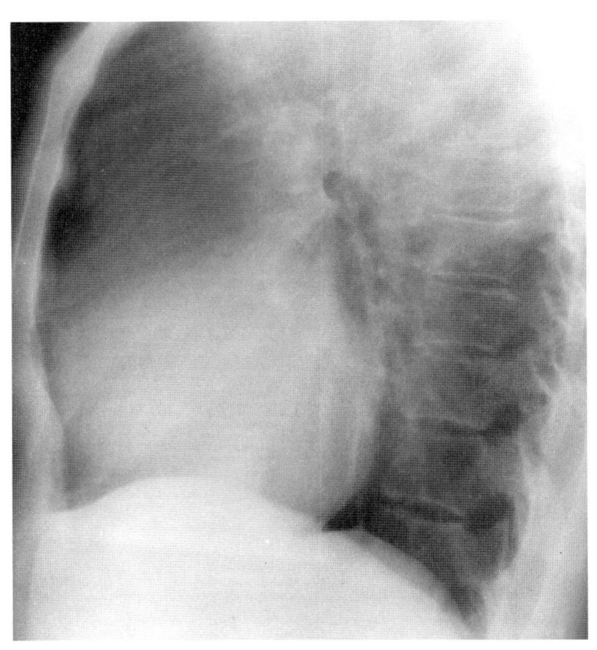

B

Figure 10.17. A: Left ventricular enlargement with characteristic shape. B: Lateral view, same patient. Note the extension posteriorly of the left ventricle.

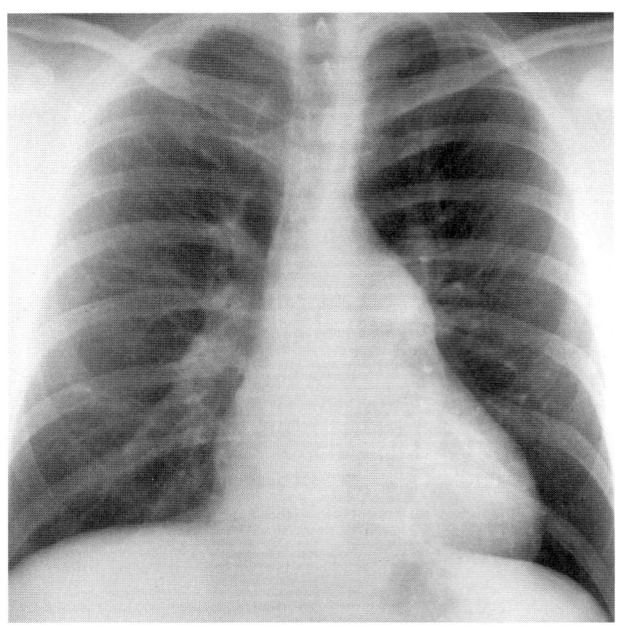

A

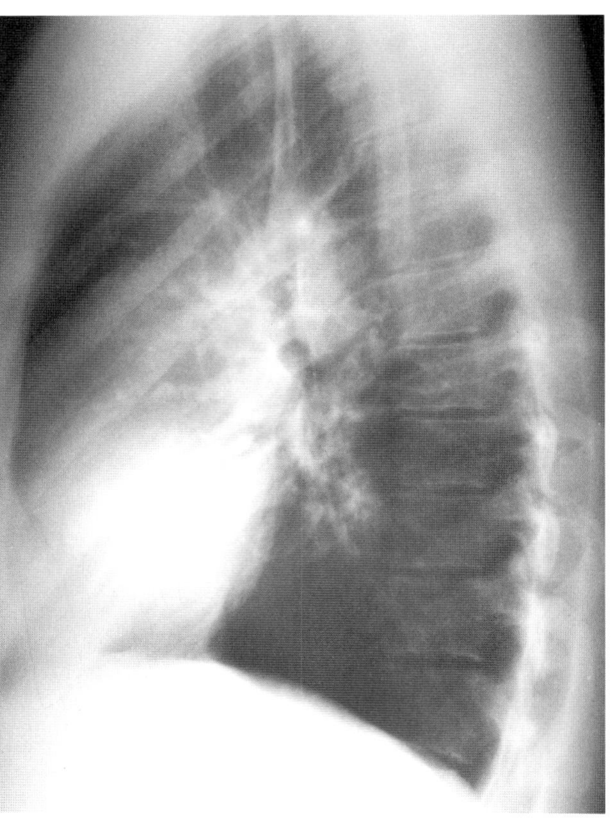

B

Figure 10.18. Patient with right ventricular enlargement. A: The PA view shows only generalized cardiac enlargement. There is a straightened left heart border and enlargement of the pulmonary artery. B: The lateral film shows the enlarged right ventricle rising up anteriorly, filling in the anterior clear space. Note that the heart is flat posteriorly.

The Heart in Failure

The appearance of the hilar and pulmonary vessels is an excellent indicator of the physiological state of the heart, failing or healthy. In the failing heart, in addition to observing the increasing size and shapelessness of the cardiac shadow, you should look for evidence of pulmonary venous engorgement. The vessels are seen to extend farther than normal into the lung field. The normally thin-walled and hence indistinguishable bronchi become "framed" in the interstitial fluid accumulating around them. When seen end-on they appear as white rings (Figure 10.21A). This is often called peribronchial cuffing and can be observed to decrease as the patient improves under therapy and the lung interstitium is cleared of water (Figure 10.21B). Pleural effusion in cardiac failure may be bilateral or unilateral, and is more frequent on the right.

The lungs appear hazy and less radiolucent than normal because of retained water, and soon Kerley's B-lines appear. These are short, horizontal white linear densities very close to the peripheral margin of the lung. They have been proven to represent the thickened, edematous interlobular septa, and may ultimately remain present in the patient who has been repeatedly in and out of failure. Kerley's B-lines may also be seen in other conditions that thicken the interlobular septa, including lymphangitic spread of malignancies within the lung parenchyma (Figure 10.20A).

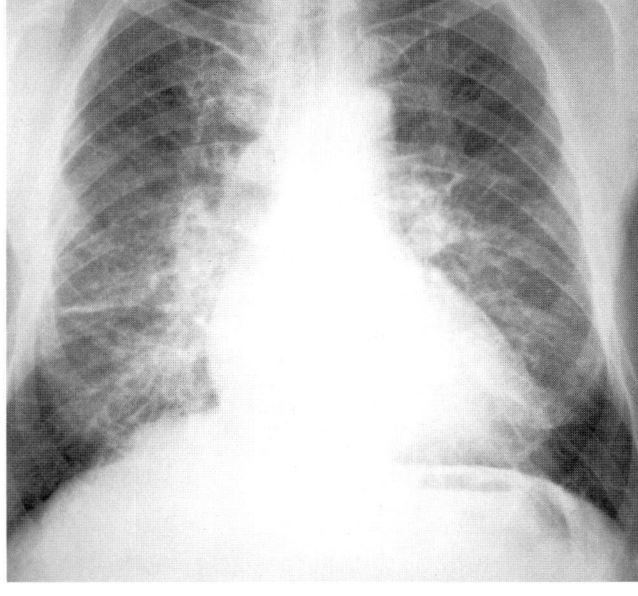

Figure 10.19. Moderate congestive failure. Note the general increase in vascular markings, engorged hila, Kerley's B-lines, and fluid in the horizontal fissure in this patient with primarily interstitial pulmonary edema.

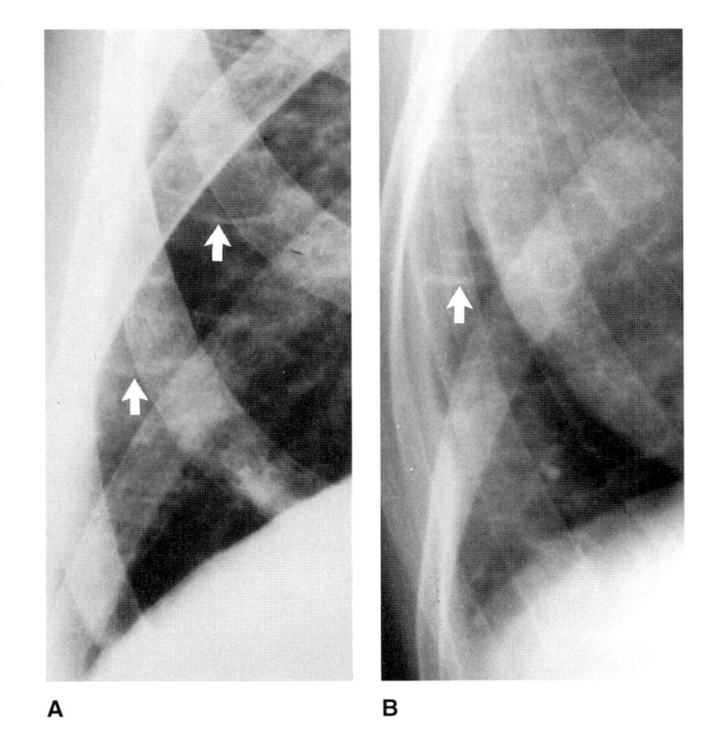

Figure 10.20. A: Kerley's B-lines *(arrows)* represent thickened interlobular septa seen tangentially close to the chest wall. This patient had lymphatic spread of carcinoma, not cardiac failure. B: Kerley's B-lines *(arrow)* in a patient with mitral valvular disease and a history of repeated episodes of congestive failure.

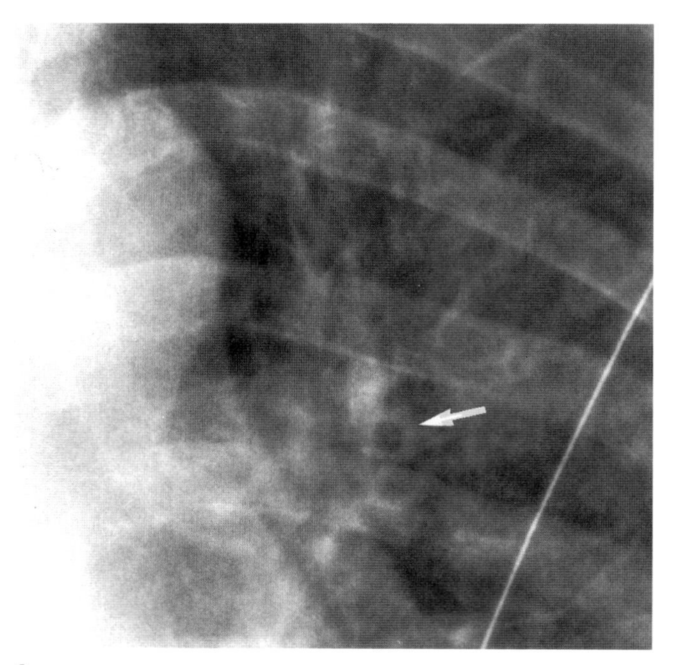

A

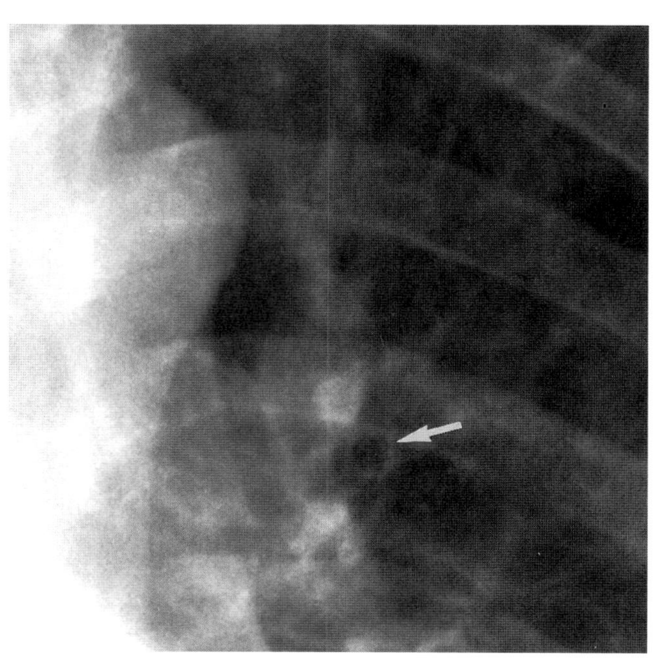

B

Figure 10.21. A: Detail of the area around the left hilum in a patient with congestive failure. Note the increased vascularity with interstitial fluid accumulating about all hilar structures, blurring the outline of the vascular trunks and producing "peribronchial cuffing" *(arrow)*. B: Clearing after therapy; note the much thinner ring of density around the same bronchus.

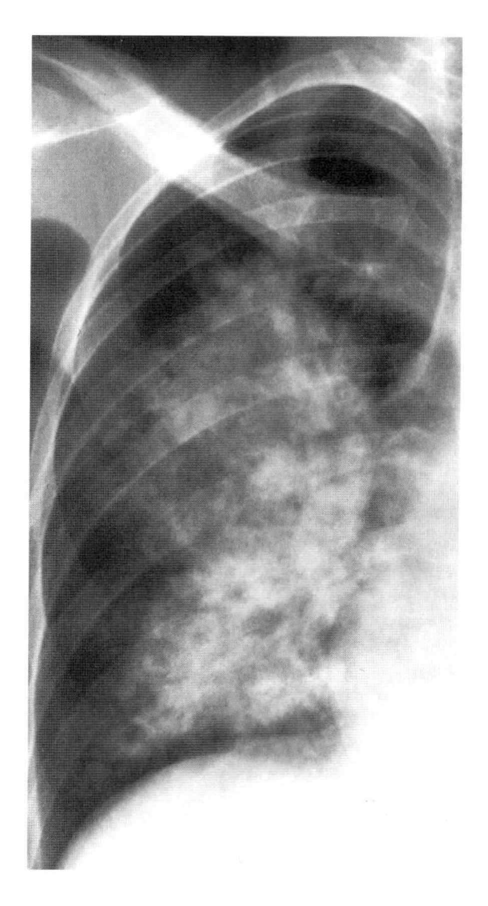

Figure 10.22. Alveolar (air-space) pulmonary edema in a perihilar distribution.

Often, rapid accumulation of interstitial fluid spills over into the alveoli and causes the development of alveolar (air-space) *pulmonary edema,* as shown in Figure 10.22, a film of a patient with a massive myocardial infarction whose left ventricle failed very rapidly. Pulmonary edema also occurs in noncardiac conditions (fluid overload, renal failure, heroin overdose, and inhalation injury or burns); in these instances the chest film will show findings of pulmonary edema but a heart normal in size.

Pulmonary edema may be unequal bilaterally, but in general it produces the so-called bat-wing appearance you are familiar with, symmetrical about both hila. It may appear rapidly after sudden left ventricular failure or it may be superimposed on the more gradual roentgen findings of cardiac failure. Note that the vessels of the hilum, bathed in interstitial fluid as they are, disappear in Figure 10.22, together with the superimposed shadows of innumerable fluid-filled alveoli. Compare the hilar vessels in Figure 10.19, in which they are becoming indistinct. Similar early loss of the interface outlines of the hilar vessels is seen in the patient in Figure 10.21A; in the B film, after good response to therapy, they are seen again emerging from their fluid bed. A similar response is seen in Figure 10.23B.

Serial chest films provide an excellent means of following the progress of a cardiac patient through an episode of failure, and you will find that they tally well with other clinical signs available to you. If possible, the patient should be sent to the radiology department because the films obtained there are superior to those taken with portable equipment.

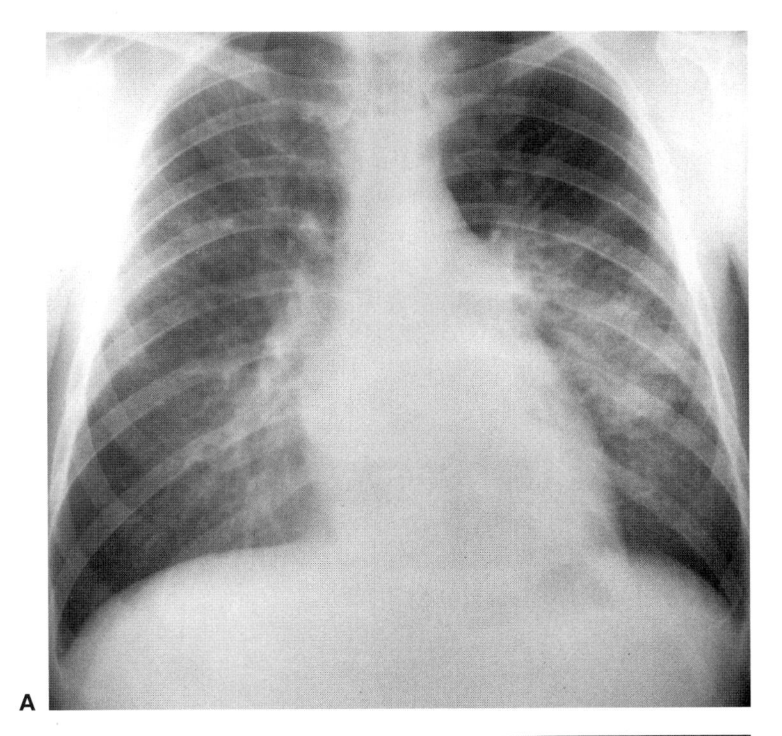

A

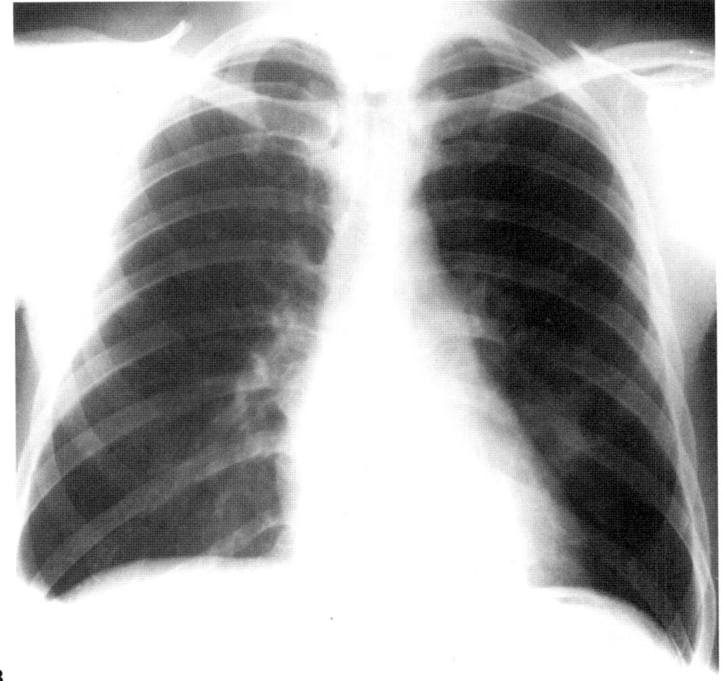

B

Figure 10.23. A: Interstitial and air-space pulmonary edema during an episode of congestive heart failure. B: Clearing five days later.

Variations in Pulmonary Blood Flow

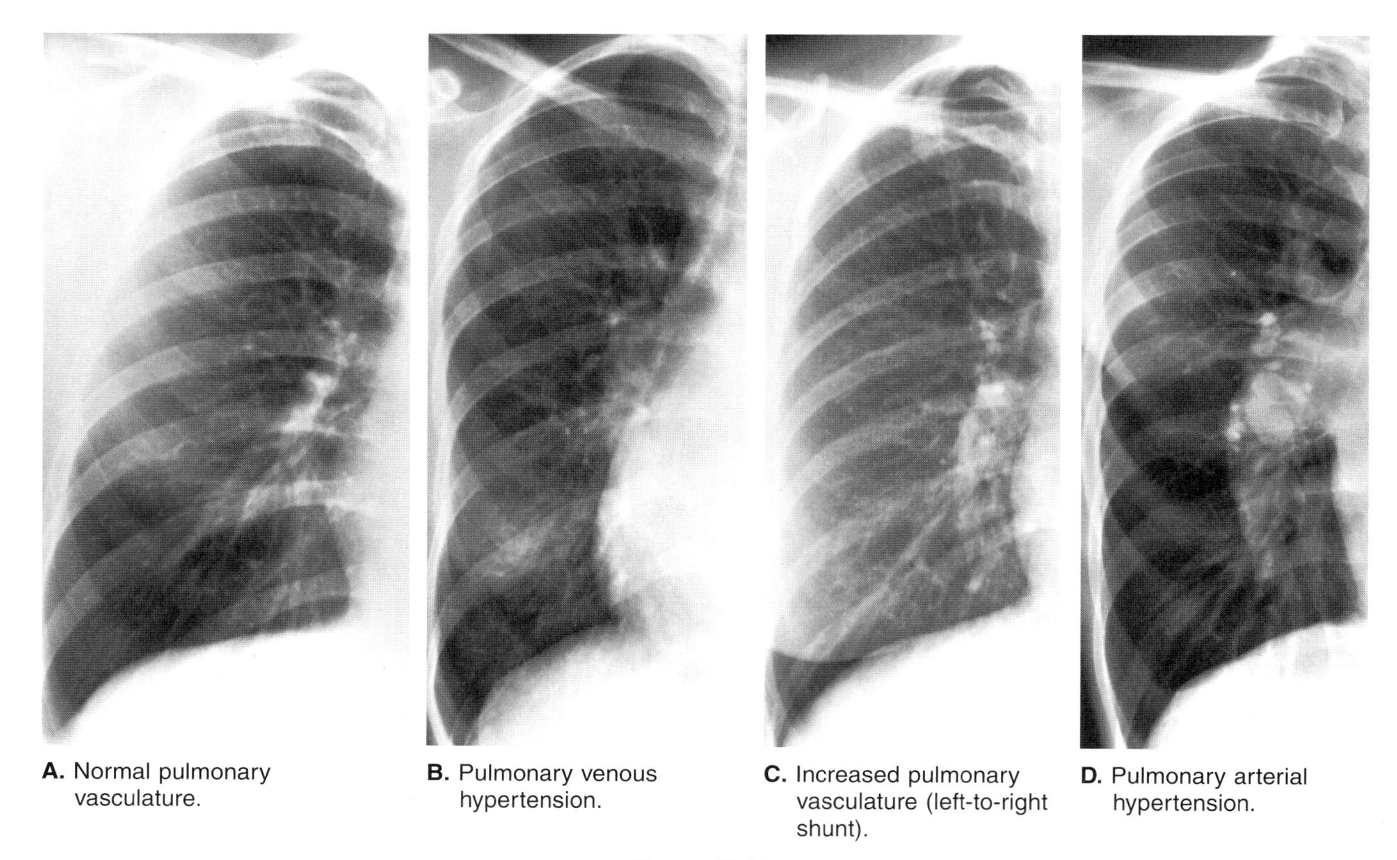

A. Normal pulmonary vasculature.

B. Pulmonary venous hypertension.

C. Increased pulmonary vasculature (left-to-right shunt).

D. Pulmonary arterial hypertension.

Figure 10.24

1. Normal Pulmonary Vasculature

Pulmonary blood flow in the upright normal patient is much greater at the bases of the lung than it is at the apex. This is apparent on normal chest films and is partly due to the fact that the lung is pyramidal in shape, with more lung tissue and more vessels superimposed at the base nearer the diaphragm (Figure 10.24A).

2. Pulmonary Venous Hypertension

The pattern of pulmonary venous hypertension is seen in patients with elevated pulmonary venous pressure, usually caused by left ventricular failure or by obstruction of the left atrial outflow, as in mitral stenosis. This pattern is identified on chest films by increased prominence and thickening of the upper lobe vessels, decreased prominence of the lower lobe vessels, and haziness of the hilar vessels (Figure 10.24B). Note that the lower lobe vessels are barely visible compared with the other three patterns. In congestive failure the pulmonary venous hypertension pattern is seen *underlying* varying degrees of interstitial and alveolar pulmonary edema.

3. Increased Pulmonary Vasculature

Increased pulmonary blood flow may be seen in both cardiac and noncardiac conditions. In patients with congenital heart disease and a left-to-right shunt, the pulmonary circulation is constantly being overloaded with blood returned to the right heart from the left heart. This is commonly seen in interatrial or interventricular septal defect, and, to a lesser degree, in patent ductus arteriosus. An example of a noncardiac cause of increased pulmonary blood flow is an arteriovenous fistula or malformation elsewhere in the body. The chest film (Figure 10.24C) shows increase in the caliber and prominence of both upper and lower lobe blood vessels centrally, in the midlung and at the lung periphery.

4. Pulmonary Arterial Hypertension

Pulmonary arterial hypertension is caused by conditions that *decrease* the flow of blood through the pulmonary capillary bed. A common cause is emphysema. Decrease in the volume of the peripheral vascular bed of the lungs may be idiopathic or may result from showers of pulmonary emboli, vasoconstrictive states, and long-standing intracardiac shunts. You have seen an example of the pruned-tree appearance of such a constricted arterial bed in Figure 5.17, a wedge arteriogram. As you would expect, the lung field in such patients shows the decreased vasculature you see exemplified in Figure 10.24D. Note that the hilar trunks are enormously dilated in response to the constricted arterial bed.

Cardiac Calcification

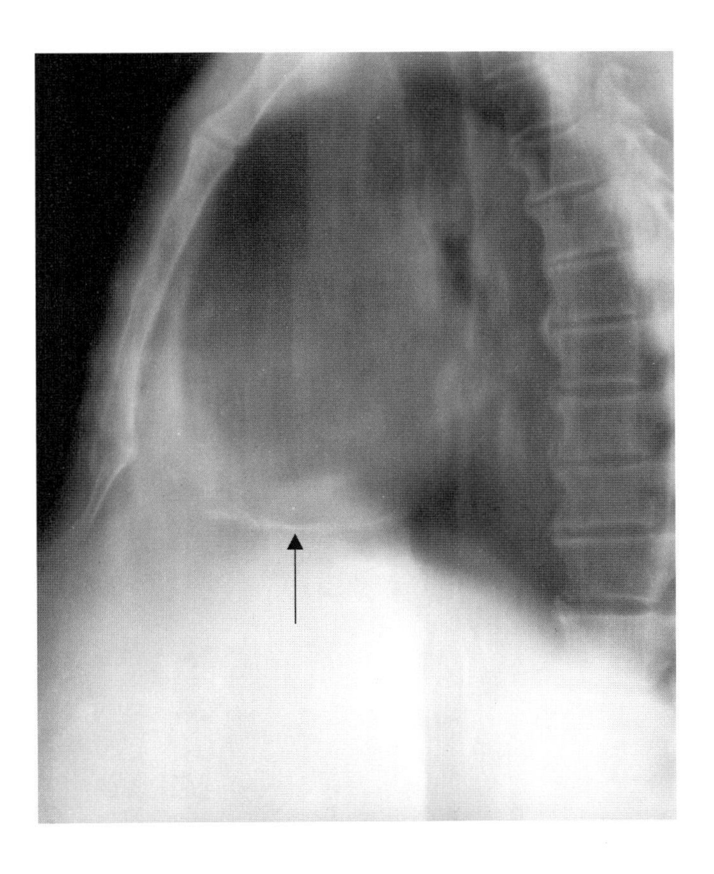

Various parts of the heart may calcify. These include valve leaflets, valve rings, coronary arteries, ventricular wall aneurysms, and the pericardium itself. Thrombi in the left atrium and cardiac neoplasms also calcify. Large areas of calcification can often be seen on plain films, but most calcifications are small and may be identified in the moving heart only by fluoroscopy or ultrasound. Figure 10.25 shows a shell of pericardial calcification around the heart in a patient with constrictive pericarditis. Cardiac fluoroscopy has been superseded today by echocardiography (cardiac ultrasound), which offers more detailed information about the interior of the moving heart without the use of ionizing radiation. With this procedure you can observe ventricular wall motion, movement of the valve leaflets, and blood flow.

Figure 10.25. Lateral conventional tomogram showing pericardial calcification *(arrow)* along the inferior margin of the heart in a patient with constrictive pericarditis.

The Anatomy of the Heart Surface

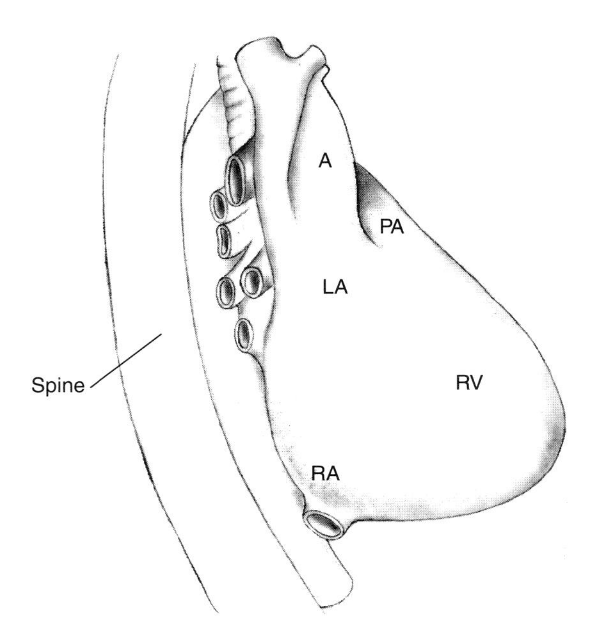

Figure 10.26A. Orientation of the heart for a right anterior oblique view. Figures 10.27 and 10.28 *(below)* show you how the patient is oriented with the right anterolateral surface of the chest closest to the film cassette.

Right and Left Anterior Oblique Views

The right and left anterior oblique films of the chest are still used, so you need to understand them. Justify these two projections with the help of the diagrams and drawings on this page spread. Turn to the angiocardiograms (next page spread) in order to realize how the opaque-filled masses of the various chambers project in the frontal view, first during the period when the opaque bolus is passing through the right chambers and then after it returns from the lungs and opacifies the left chambers.

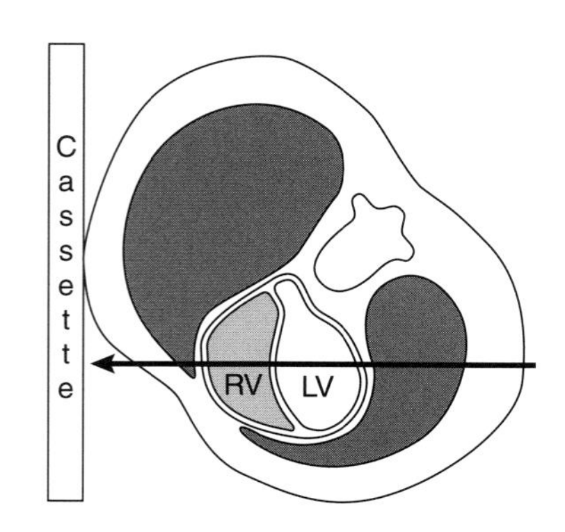

Figure 10.27

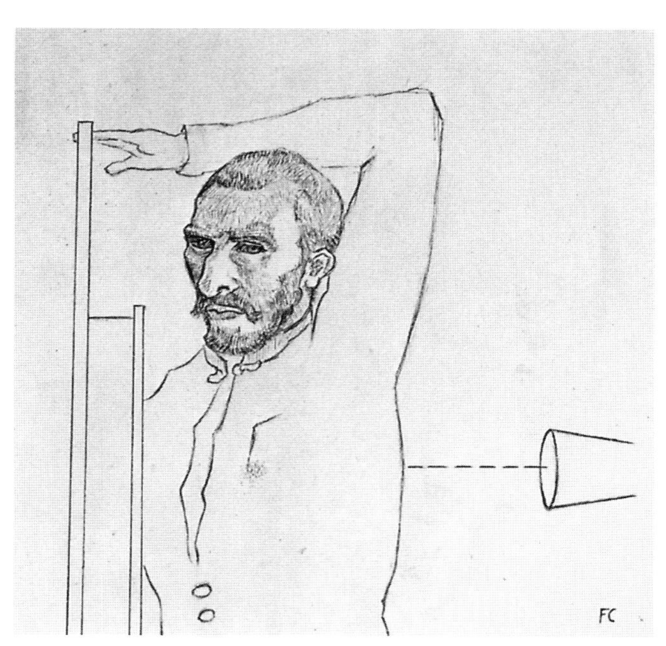

Figure 10.28

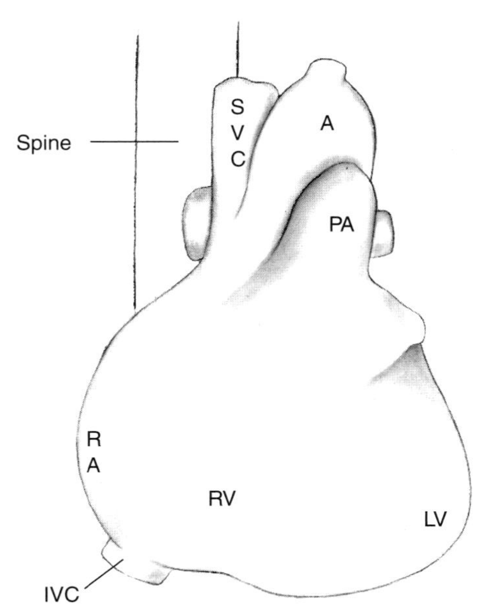

Figure 10.26B. The surface of the heart as it would appear on a routine frontal chest film.

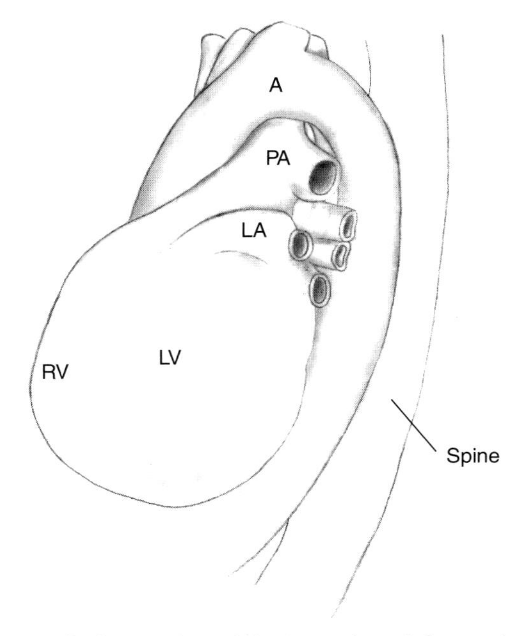

Figure 10.26C. Orientation of the heart for a left anterior oblique view. Figures 10.29 and 10.30 *(below)* show you how the patient is oriented with the left anterolateral surface of the chest closest to the cassette.

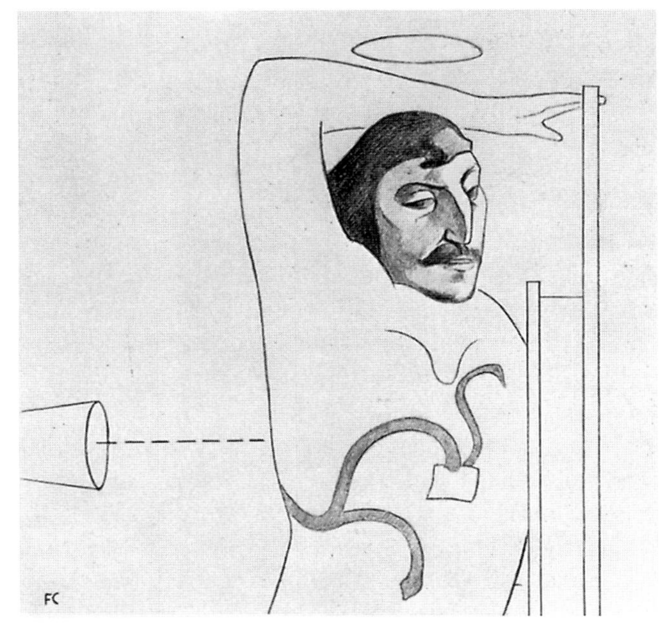

Figure 10.29

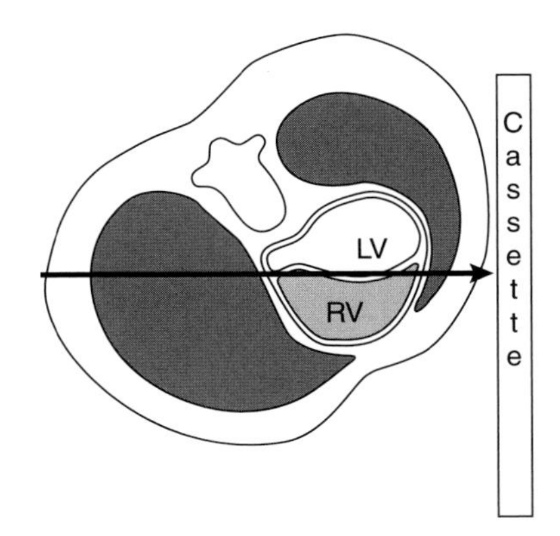

Figure 10.30

Identifying Right and Left Anterior Oblique Views

On the normal right anterior oblique film, the heart projects to the left of the spine (your right) and looks triangular with a flat posterior surface.

On the left anterior oblique film, the heart projects to the right of the spine and is much more bulbous in shape, with a rounded posterior surface. On both of the oblique views in Figure 10.31 (three views of the same patient), the esophagus has been opacified with barium.

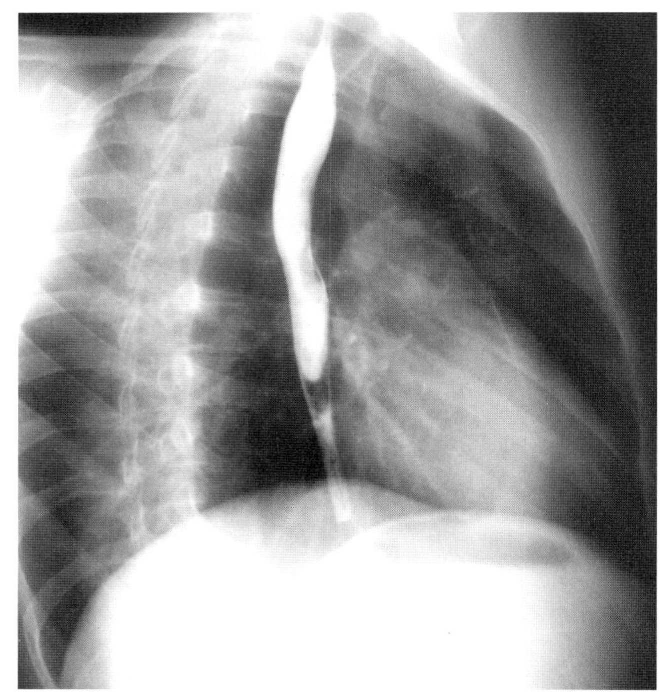

Figure 10.31A. Right anterior oblique.

The Anatomy of the Heart Interior

To help you understand the interior anatomy of the heart, study the angiocardiogram shown in Figures 10.32A and 10.32B, in which the cardiac chambers have been opacified with contrast material. For this procedure a percutaneously inserted venous catheter from a femoral vein approach was advanced until its tip was in the right atrium. A bolus of contrast material was then injected, during which rapid serial films were made, in the frontal projection. The time when contrast medium opacifies the right side of the heart is called the *dextrophase,* and an image obtained then is called a *dextrocardiogram* (Figure 10.32A). The *arrow* indicates the branch of the pulmonary artery supplying the left lower lobe. Note that the right atrium is thin walled and makes up most of the right heart border. The indentations in the outline of the opacified right ventricle cavity are muscle trabeculae. Did you remember that the right ventricle is more trabeculated than the left? Compare the two on these images. If you look carefully, you can locate the site of the pulmonic valve.

The *levocardiogram* was filmed after the bolus of contrast medium had passed through the pulmonary circulation and returned to the heart *(levophase),* where it opacified the left chambers. The *arrow* points to the pulmonary vein draining the left lower lobe. Note that it enters directly into the left atrium. At this point in the cycle the ascending aorta is only faintly opacified. Notice that the left ventricle constitutes the left heart border. You may have already guessed that the left ventricle and aortic valve could also have been opacified by catheterization via an *artery* (such as a femoral artery), with retrograde passage of the angiography catheter all the way up the aorta, around the aortic arch, across the aortic valve, and into the left ventricle cavity. With contrast injection and filming, excellent details of left ventricular anatomy and function can be recorded. In the postmyocardial infarction patient this technique may show areas of diminished ventricular wall motion *(hypokinesis),* absent wall motion *(akinesis),* paradoxical wall motion *(dyskinesis),* or even a left ventricular aneurysm.

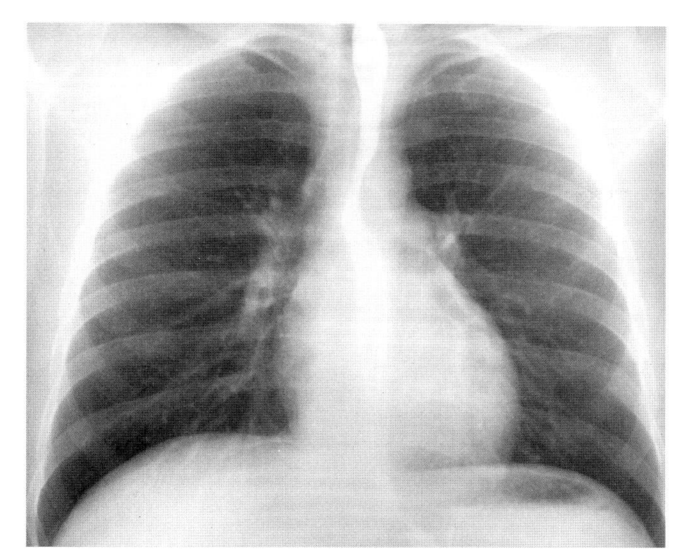

Figure 10.31B. Posteroanterior view.

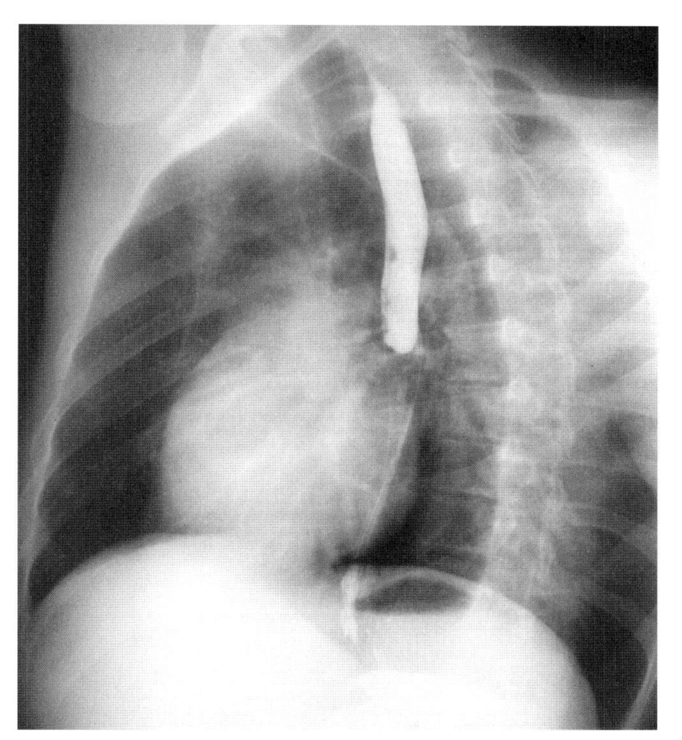

Figure 10.31C. Left anterior oblique.

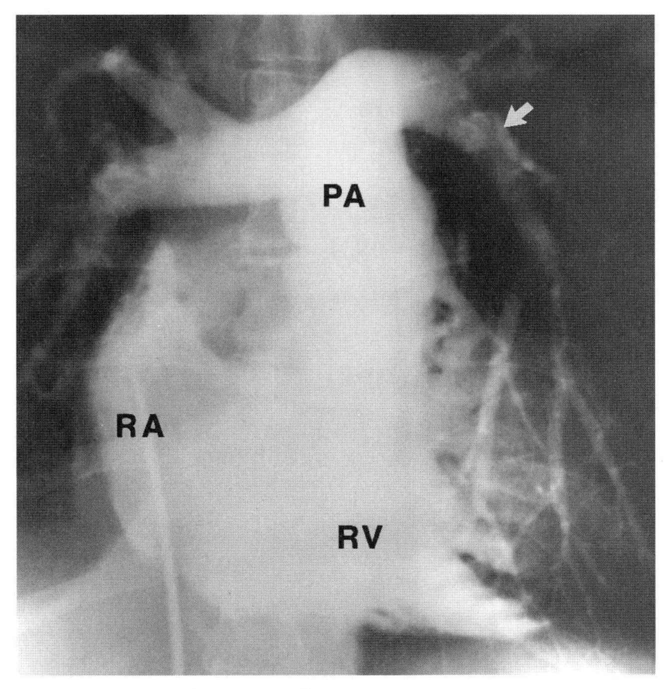

Figure 10.32A. Dextrocardiogram.

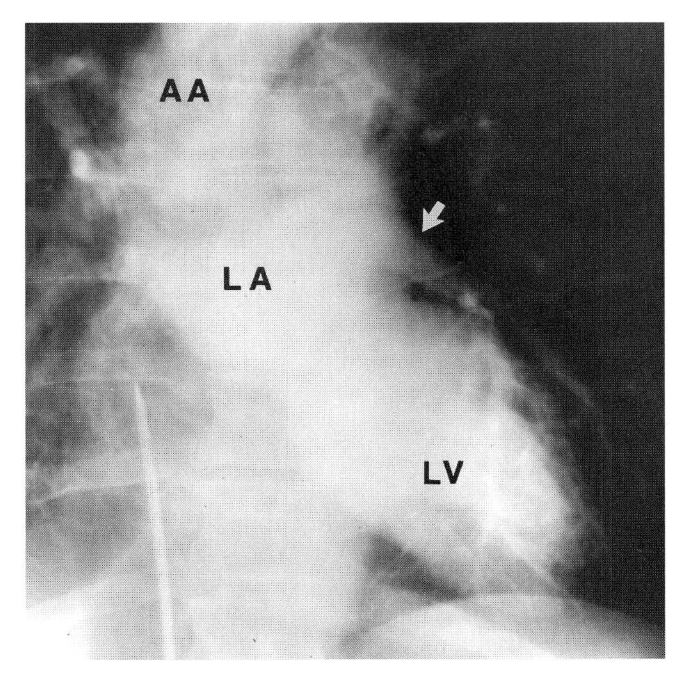

Figure 10.32B. Levocardiogram.

Coronary Arteriography

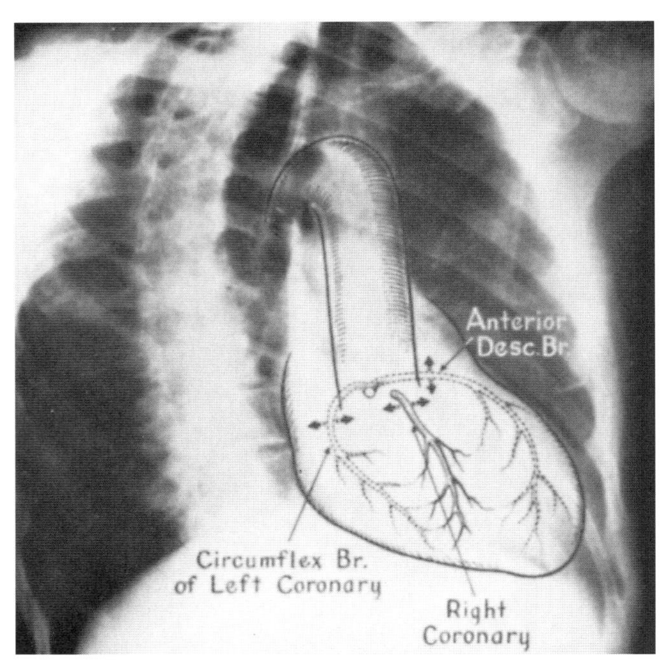

Figure 10.33A. Coronary artery anatomy, right anterior oblique projection.

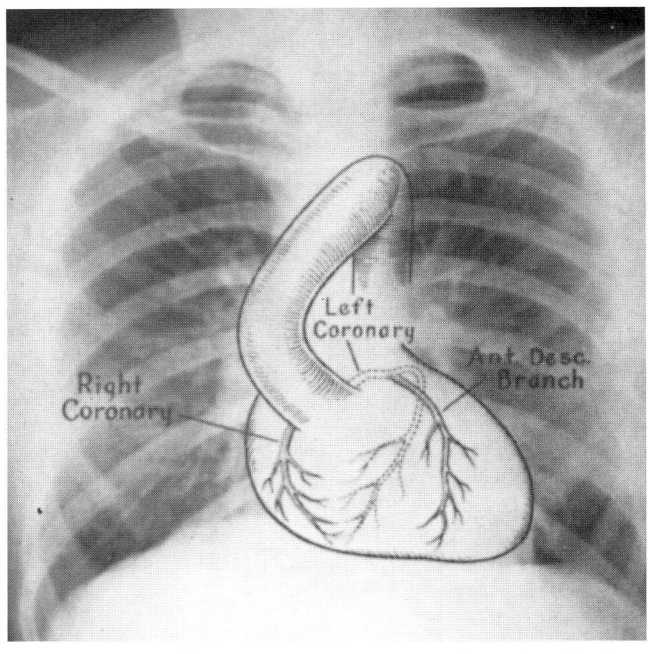

Figure 10.33B. Coronary artery anatomy, frontal projection.

Coronary arteriography is carried out commonly in patients with symptoms of ischemic heart disease and may be recommended to confirm an anatomic cause for angina, to evaluate asymptomatic patients with abnormal exercise tolerance tests, to evaluate patients before cardiac surgery, to evaluate patients after coronary artery bypass graft surgery, and to determine whether a patient who has suffered a myocardial infarction is a candidate for interventional therapy, such as balloon angioplasty of a coronary artery stenosis. Coronary arteriography involves selective catheterization of both coronary arteries under fluoroscopic control with a flexible angiographic catheter inserted through a femoral or a brachial artery. Hand injection of contrast material combined with filming produces images of the coronary arteries in excellent detail. Films are made in standard projections and varied as required to best show the course of the arteries and their branches. Stenoses and occlusions caused by atherosclerotic disease are identified precisely by this procedure. A 75 percent reduction in cross-sectional area is required to cause a significant reduction in blood flow. Remember that a 50 percent reduction in diameter corresponds to a 75 percent reduction in cross-sectional area. Collateral blood flow usually develops when the stenosis is greater than 85 percent.

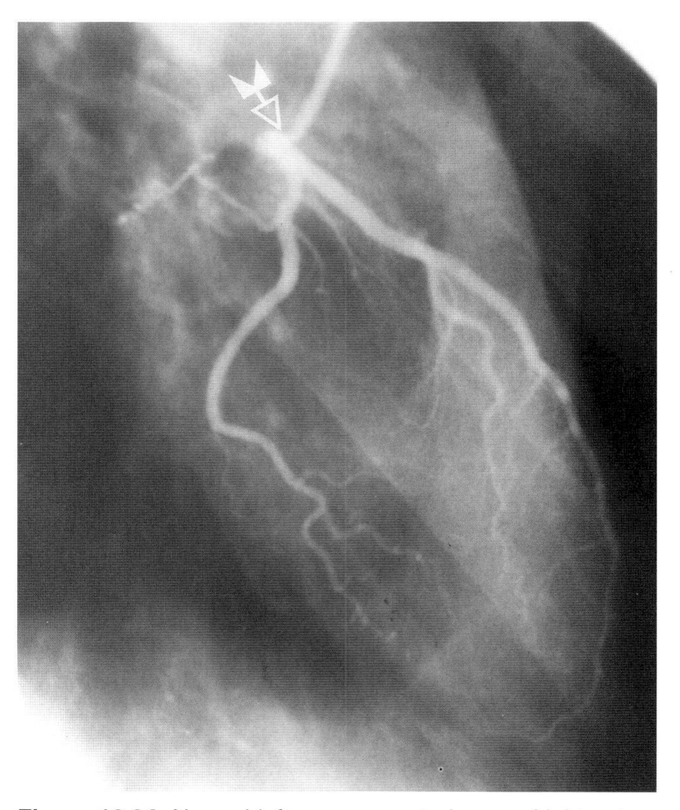

Figure 10.34. Normal left coronary arteriogram (right anterior oblique view). In Figures 10.34–10.37 an *open arrow* indicates the catheter tip in the origin of the artery.

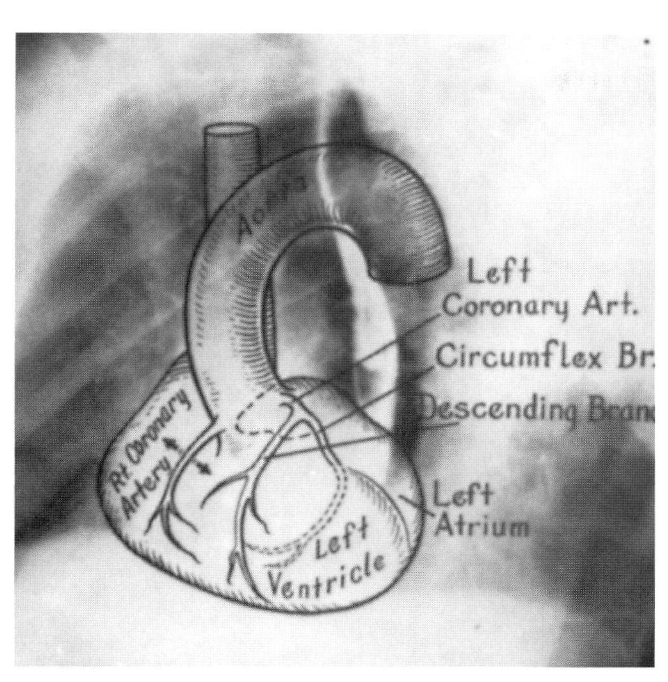

Figure 10.33C. Coronary artery anatomy, left anterior oblique projection.

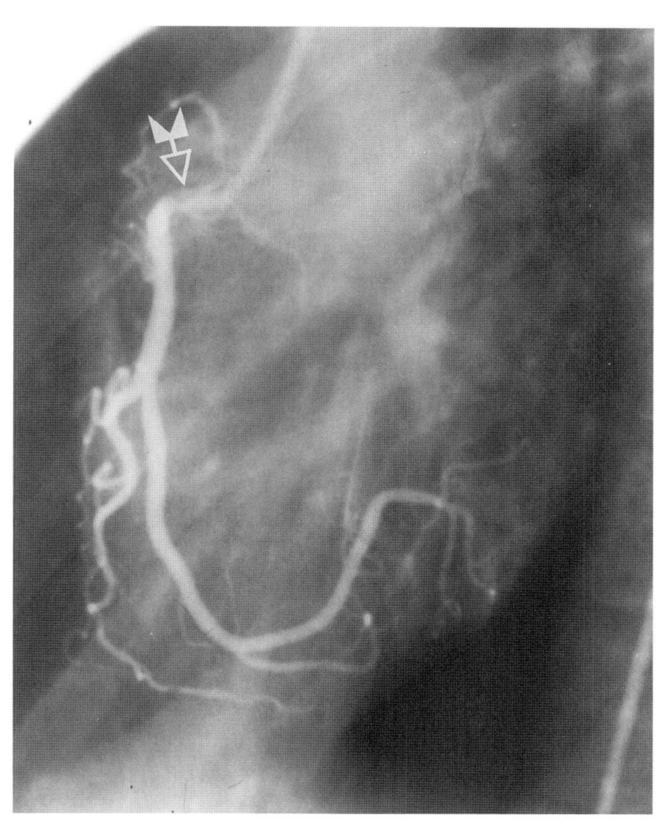

Figure 10.36. Normal right coronary arteriogram (left anterior oblique view).

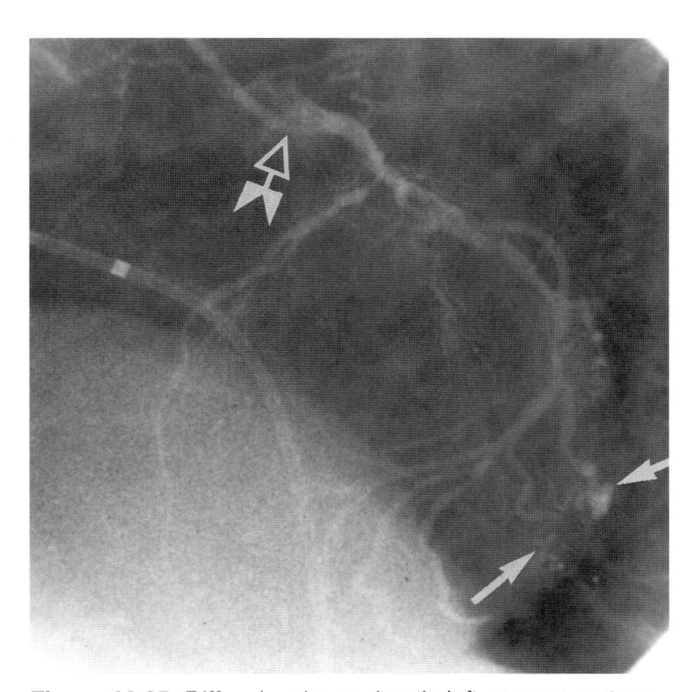

Figure 10.35. Diffusely atherosclerotic left coronary artery. Note the luminal narrowing of the entire arterial tree, multiple sites of stenosis, and the formation of collateral branches (*solid arrows* to tortuous collaterals).

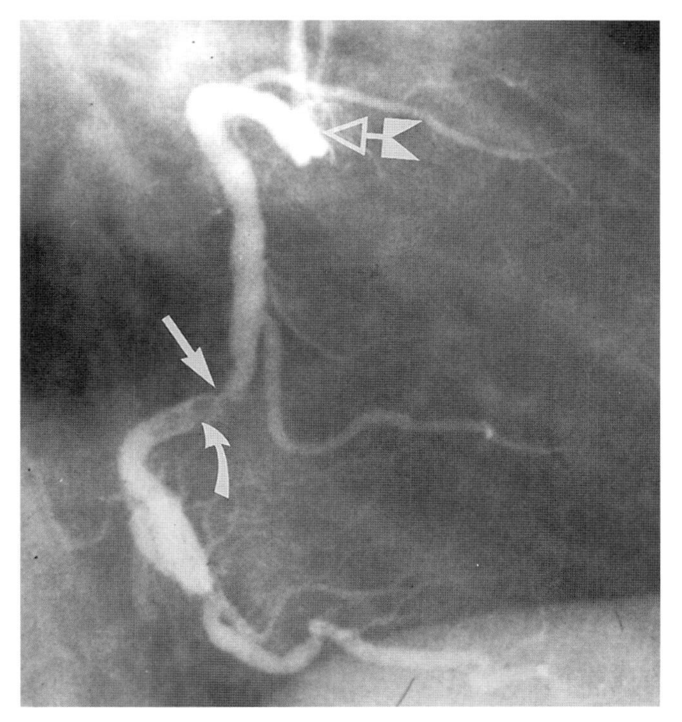

Figure 10.37. Atherosclerotic right coronary artery in a patient with an acute myocardial infarction. Note the tight stenosis *(solid straight arrow)* with intraluminal thrombus *(solid curved arrow)* just distal to the stenosis.

Classic Changes in Shape with Chamber Enlargement

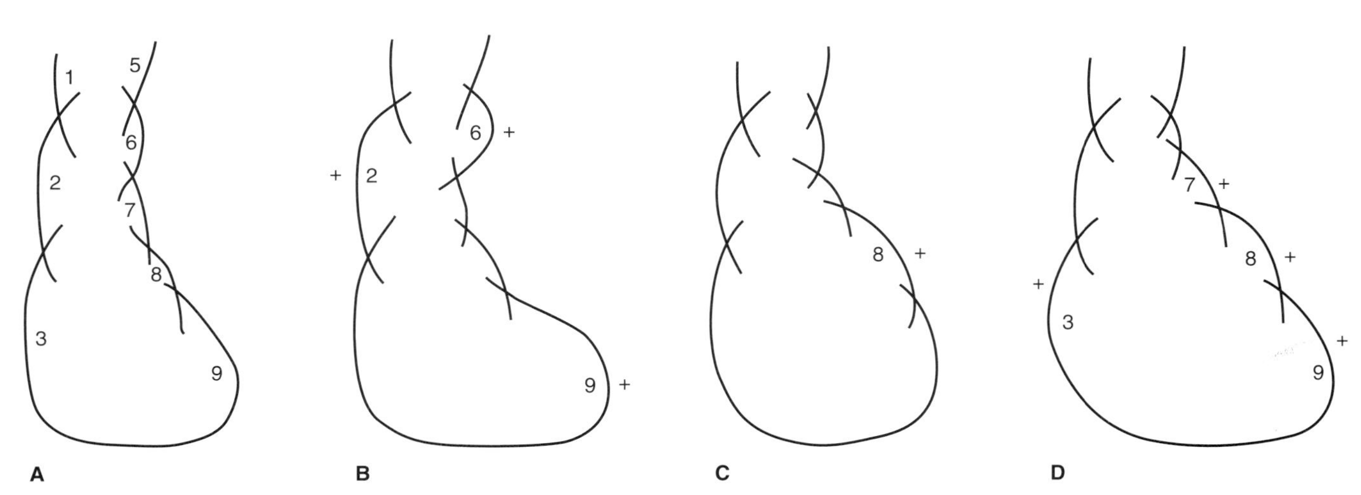

Figure 10.38. Changes in the shape of the heart with specific chamber enlargements, expressed in terms of alteration of the nine intersecting arcs responsible for the PA profile of the heart and great vessels (compare Figure 9.2). Note: the normally concave slope between arcs 6 and 9 is often called the cardiac waistline.

As you looked at the shadows of the heart in Figure 10.33 on the preceding page spread, you were accepting the fact that the heart size was normal and also the fact that the *shape* of the heart was normal.

Try now to imagine what would happen to the size and shape of the heart with increasing degrees of *left ventricular enlargement.* Predict the change in shape of the heart that you might expect in long-standing resistance to outflow through the aortic valve—as in aortic stenosis, coarctation of the aorta, or systemic hypertension. You can appreciate that the shadow of the left ventricle will project farther to the left on the PA view (Figure 10.38B) and that it will extend farther posteriorly in the lateral view. The posterior surface of the heart (as seen in the lateral view) should normally clear the anterior surface of the vertebral column, but in advanced cardiac disease with left ventricular enlargement, especially in decompensation, the posterior surface of the heart is often seen to overlap the spine to some extent. Left ventricular enlargement is often associated with aortic stenosis and chronic hypertension, both of which may cause enlargement of the aorta (Figure 10.38B, *2* and *6*). With aortic valve stenosis this may be due to poststenotic dilatation.

Now imagine that the *left atrium* enlarges, as it would in a patient with mitral valvular disease with stenosis and insufficiency. What would you expect that to do to the heart shadow on the PA view? Remember that the left atrium lies posteriorly just underneath the carina, higher than the major chambers. When it enlarges it will produce fullness across the heart at about the level of the waistline (arc 8), producing the shape you see in Figure 10.38C. In the left oblique view the normally open aortic window between the ascending and descending aorta will be filled in by the enlarging atrium. A well-penetrated PA film will show the air-containing carina to be splayed, and usually the widening of the subcarinal angle will be due primarily to elevation of the left main bronchus as it rides over the expanding left atrium.

In the right anterior oblique view, the displacement of the esophagus posteriorly will be evident too, as you see in Figure 10.40. The patient swallowed thick barium, which outlined the esophagus as it lies against the posterior surface of the heart. Normally the esophagus lies just anterior to the spine in the midline, bisecting the chest in the lateral view.

In the PA view, as the left atrium enlarges in mitral disease, it first fills in the left cardiac waistline so that the left heart border becomes convex instead of concave, and then the atrium extends to the right so that its margin is visible along the *right* heart border, above the profile of the right atrium and overlapping it—the "double shadow" so frequently referred to as a classic sign in left atrial enlargement (Figure 10.41).

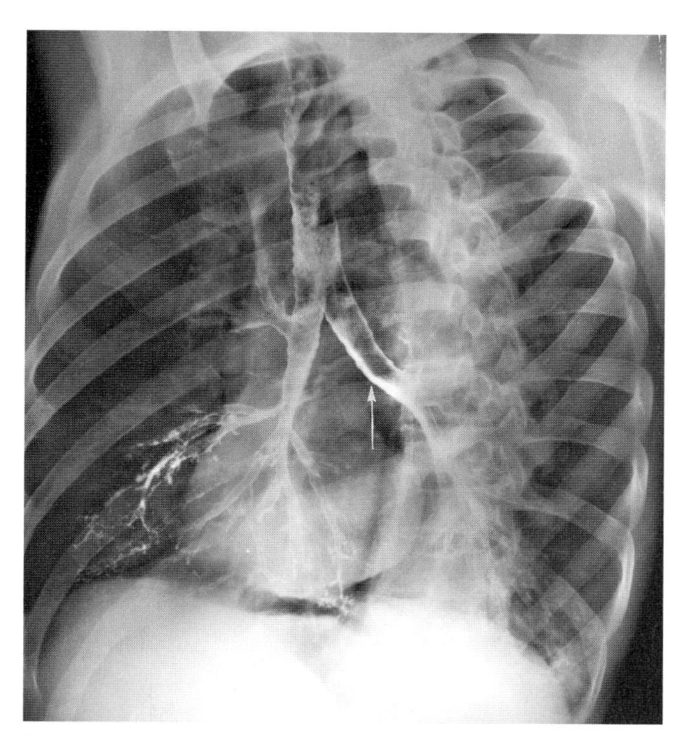

Figure 10.39. Left anterior oblique view showing normal subcarinal angle during a bronchogram. The carina sits atop the left atrium; when that chamber dilates, the left main bronchus *(arrow)* will be lifted, so that the subcarinal angle is increased.

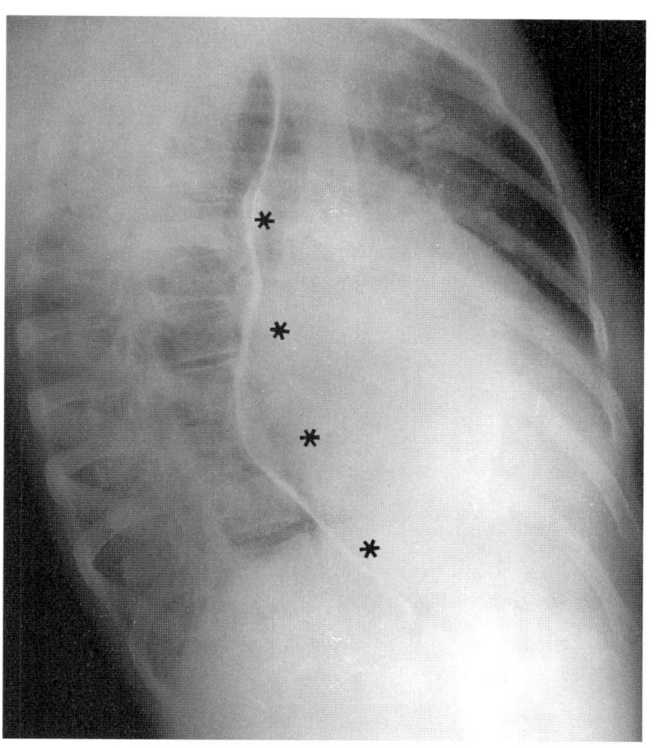

Figure 10.40. Right anterior oblique view of a patient with mitral valvular disease and insufficiency, showing posterior displacement of the barium-filled esophagus by the dilated left atrium. The *asterisks* indicate the normal course of the esophagus.

Observe in Figure 10.41 that in addition to the enlargement of the left atrium there is also extension of the left ventricle to the left, as in Figure 10.38D, showing left ventricular enlargement in mitral disease of long standing. This may result from work overload on the left ventricle, which has to overcome the inefficient backward flow of some blood into the left atrium through the poorly closing mitral valve. Enlargement of the left ventricle may also be seen in patients with combined aortic and mitral valvular disease in whom the resistance and insufficiency of a deformed aortic valve adds to the work the left ventricle must do.

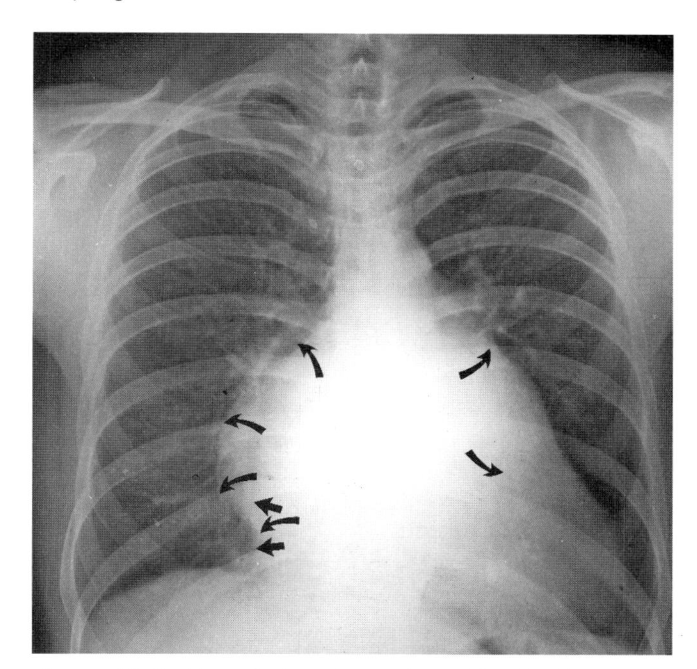

Figure 10.41. Frontal view of a patient with mitral insufficiency and a huge left atrium (outlined by *curved arrows),* which projects both to the left and to the right. The profile of the right atrium is indicated by two *straight arrows.* This creates the double shadow along the right heart border seen in advanced mitral insufficiency.

Problems

Try analyzing the four hearts on these pages before reading the text.

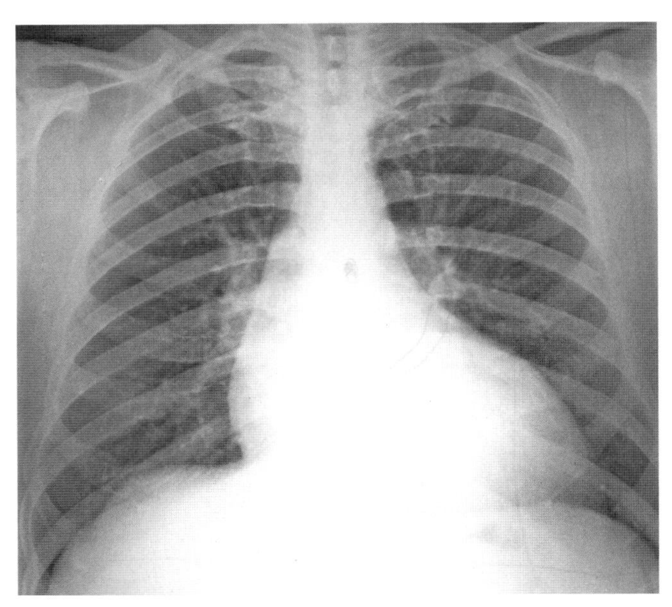

Figure 10.42

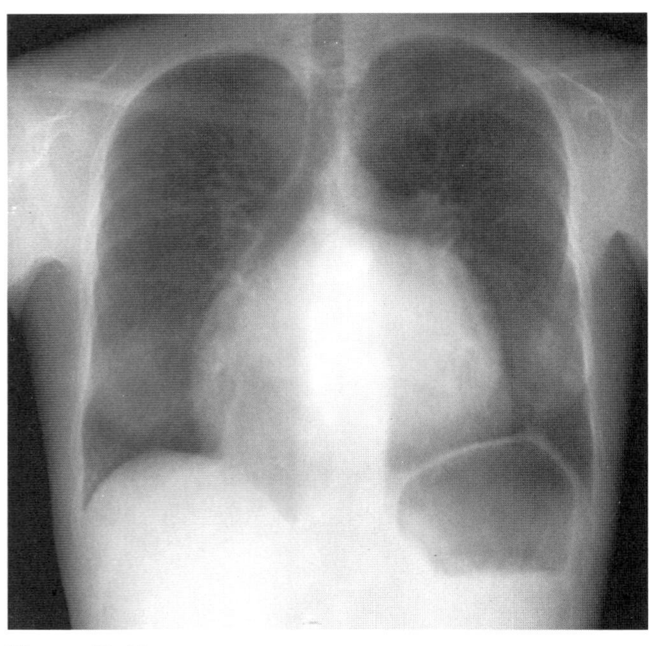

Figure 10.43

You begin now to have a feeling for the basic differences in shape between the heart shadow with predominantly left ventricular enlargement and the heart in which the left atrium is dilated. They are the most important specific chamber enlargements for you to be able to recognize. When, from the PA view, you suspect either, you should try to confirm your impression by examining the obliques and the lateral views.

The patient in Figure 10.42 has measurable cardiac enlargement and a shape that must suggest to you left ventricular enlargement. Note the concave left waistline and extension of the apex to the left, which would be confirmed in the lateral view with posterior projection of the left ventricle. This patient had classic murmurs of aortic stenosis and insufficiency, probably of rheumatic origin. Note the flat shadow of the aortic arch.

Change in the shape of the heart resulting from enlargement of the *right* chambers is much more difficult to recognize. The right atrium is immensely dilated in Ebstein's malformation, as you would anticipate. We do not reproduce an example here, because that entity is exceedingly rare. The right ventricle enlarges in cor pulmonale and in pulmonic stenosis. When it does (as you have already seen in Figure 10.18), the PA film may show the heart to be deceptively normal or show the normal left ventricle displaced to the left. (Since the right ventricle is located anteriorly, no part of it is seen in profile in the PA view.) When you examine the lateral film, however, you will be struck by the filling in of the lower part of the anterior clear space and by the flat posterior surface of the heart, unlike the rounded posterior projection of left ventricular enlargement. Of course, in today's practice of medicine and radiology you are not dependent solely on the data available from plain chest films. Ultrasound, CT, and MR imaging have added new dimensions of accuracy in appraising the relative size of the chambers of the heart, the thickness of the cardiac wall, and the function and structure of the valves, as you will see in the next few pages.

The patient in Figure 10.43 is seen to have a clearly measurable enlargement of the heart in this PA film, made with an overexposed technique so that you can see the air in the carina and the marked increase of the subcarinal angle with elevation of the left main bronchus. The patient had the murmurs of both mitral stenosis and mitral insufficiency and a markedly dilated left atrium. This is a classic example of a mitral heart.

Straightening of the left heart border may be a normal finding and does not always signify increased left atrial size. Remember too that filling in of the normally concave waistline may be due to fullness that is either posterior (as in left atrial dilatation) or anterior (as in any condition causing dilatation of the main pulmonary artery, such as poststenotic dilatation in pulmonic stenosis, or dilatation due to patent ductus arteriosus).

The patient in Figure 10.44 also had mitral stenosis and insufficiency. She has measurable cardiac enlargement in this single view, elevation of the left main bronchus seen just above the left eighth rib, and a double shadow along the right heart border, all classic signs of left atrial enlargement in mitral disease, although less marked than in the patient in Figure 10.43. The patient in Figure 10.45 was short of breath because of an incipient attack of asthma—of which she had a long history. She has slight straightening of the left border, to be sure, but has no cardiac enlargement and no clinical signs of cardiac disease. The straightened left heart border here is probably due to slight fullness of the main pulmonary artery, which may be a normal finding, especially in young women.

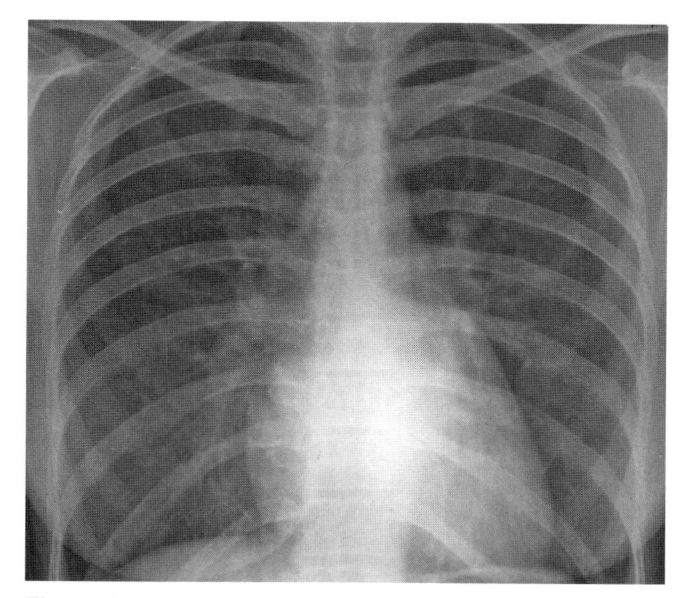

Figure 10.44

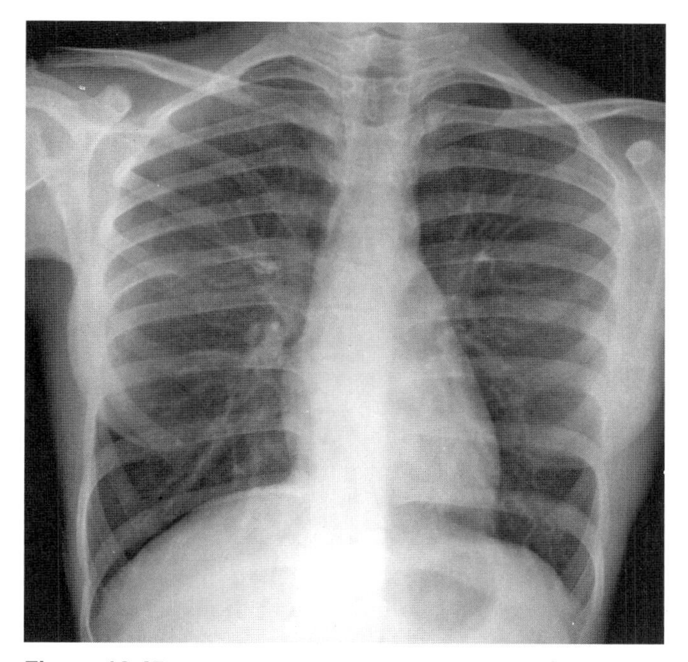

Figure 10.45

Nuclear Cardiac Imaging

When the heart is enlarged, it is difficult if not impossible to differentiate on plain films between cardiac hypertrophy and chamber dilatation. Of course, the two conditions can be distinguished by imaging techniques that depict ventricular wall thickness, such as angiocardiography, CT, and MR imaging. The past two decades have seen an explosion of techniques for imaging the heart. Only coronary arteriography, as you have seen in Figures 10.35 and 10.37, can document precisely and visually the location of atherosclerotic plaques in the coronary circulation that produce stenoses and occlusions. It is an invasive procedure, however, requiring arterial catheterization and injection of contrast material.

Several noninvasive radioisotope techniques are currently being used for the diagnosis of ischemic cardiac disease. One involves the use of *radioactive thallium* to evaluate coronary artery perfusion. Thallium, like potassium, accumulates in well-perfused, well-oxygenated muscle cells within a few minutes after intravenous injection. A normal heart shows a uniform distribution of radioactive thallium throughout the myocardium, whereas an ischemic heart shows areas of decreased thallium activity, or "cold spots."

Electrocardiogram-gated myocardial blood pool studies permit examination of cardiac wall motion. Radioactive agents that label a patient's blood pool, such as technetium-labeled red blood cells or human serum albumin, are injected into a patient, and the heart is imaged under a gamma camera that is coupled with a computer. The image information can be manipulated and analyzed to determine features of ventricular function such as stroke volume and ejection fraction *(radionuclide ventriculography)*. Blood within the chambers of the heart can be imaged either with a complex of computed frames or by cine technique, to show the size and shape of the cardiac chambers, the position of the great vessels, the thickness of the ventricular walls, filling defects in the chambers, wall motion, and any dyskinetic segments.

In Figures 10.46A and B you see illustrations of normal and abnormal thallium perfusion scans. When a patient is at rest, perfusion to the myocardium distal to a stenosis is generally normal or nearly normal. When the heart is under stress, the increased demands placed on the myocardium are reflected by a demand for increased oxygen and increased perfusion. In this circumstance the decreased coronary vascular reserve distal to a stenosis is best demonstrated. Consequently, thallium scanning is performed during exercise, usually on a treadmill. For both scans shown here thallium was injected at peak exercise; scanning with a gamma camera was performed shortly thereafter. In the normal patient (Figure 10.46A) thallium is shown readily concentrated in well-functioning myocardium so that the image obtained in the left anterior oblique projection appears as a doughnut-shaped ring, the ventricular myocardium. In Figure 10.46B, though, there is a defect *(arrow)* in the septum and the inferior wall where no thallium has yet entered the muscle cells, because that portion of myocardium is underperfused and poorly oxygenated. A later coronary arteriogram showed a tight stenosis in the left coronary artery.

Thallium stress imaging has proven to be a highly sensitive and specific test for the detection of coronary artery disease, in fact, more sensitive than exercise electrocardiography. Nevertheless, thallium stress imaging should not be considered as a replacement for stress ECG (electrocardiogram) examinations in patients with suspected coronary disease; rather, it should be reserved for those patients in whom the ECG is nondiagnostic. Thallium stress imaging is considerably more expensive than an ECG.

A relatively new radioisotope imaging technique called SPECT (single-photon emission computed tomography) produces tomographic radioisotope images of the heart. Figure 10.47 illustrates a normal scan obtained after injection of radioactive thallium. In the top row are tomographic short axis images. You can identify a series of tomographic cuts across the left cavity, beginning at the apex, and extending back toward the base of the heart. The second row is a tomographic series in the vertical long axis, and the bottom row is a tomographic series in the horizontal long axis. The diagrams to the left should assist you with the orientation.

More commonly, thallium imaging is done by a gamma camera, which produces planar views. Rather than being rotated around the patient, like a SPECT scanner, the gamma camera is placed against the patient to produce three different radioisotope views of the heart: anterior, 45-degree, and 70-degree left anterior oblique. For each position both initial and delayed images are obtained. You have already seen a photograph of a gamma camera in Chapter 2.

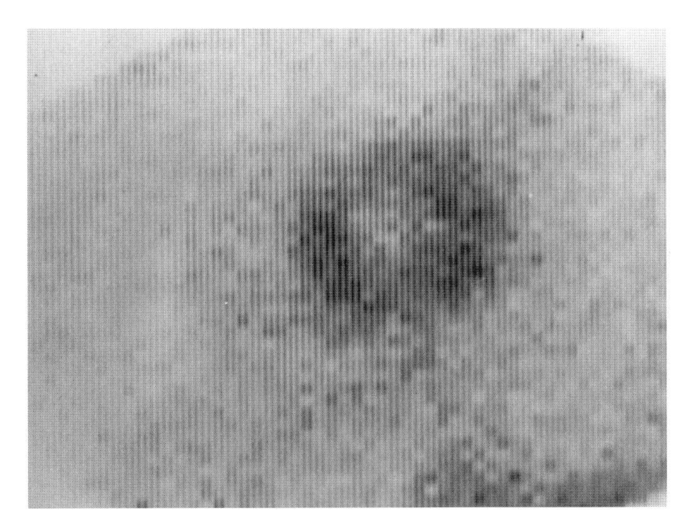

Figure 10.46A. Normal thallium perfusion scan during exercise. Note the uniform distribution of the isotope.

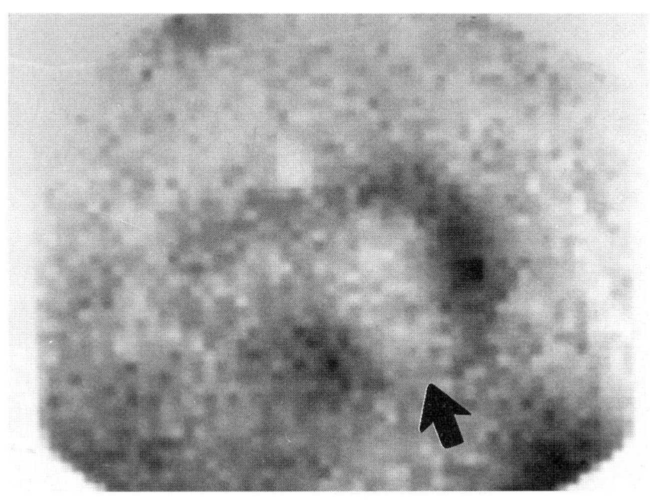

Figure 10.46B. Thallium perfusion scan during exercise in a patient with coronary artery stenosis. The *arrow* points to an ischemic area having decreased thallium activity.

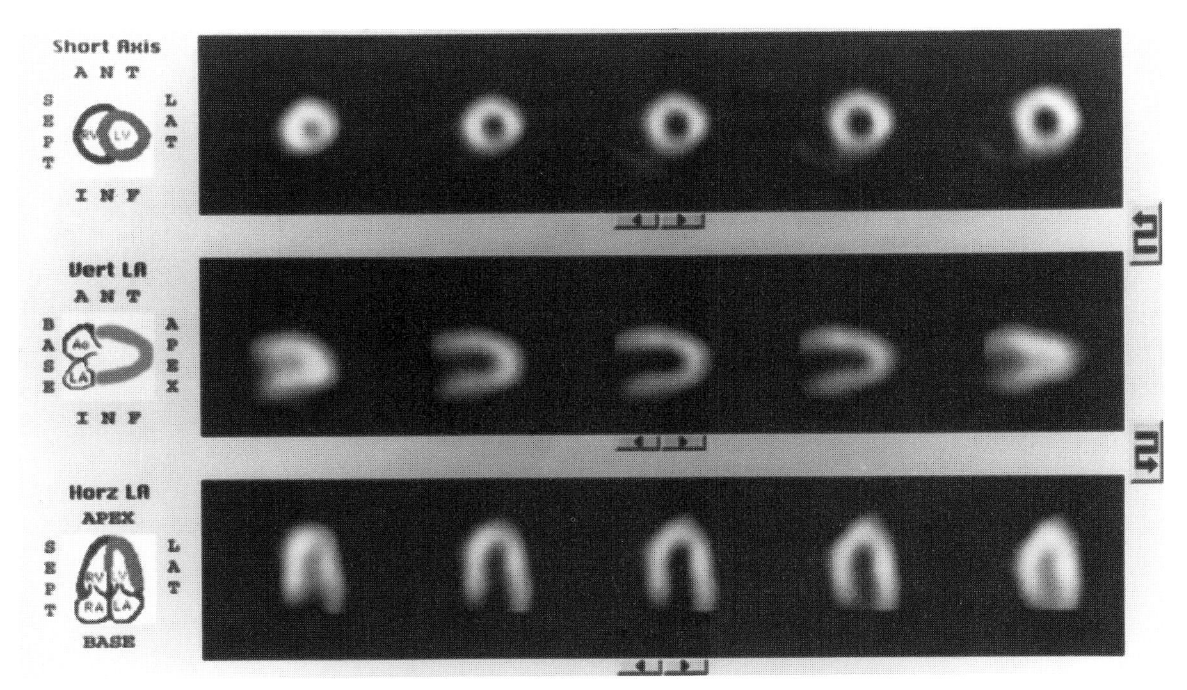

Figure 10.47. Normal thallium perfusion scan obtained with the SPECT technique, which produces a series of tomographic images in various imaging planes. *Top row, short axis view:* myocardium appears to have a doughnut shape. *Middle row, vertical long axis view:* myocardium appears as a horizontal U. *Bottom row, horizontal long axis view:* myocardium appears as an inverted U. (See text.)

Figure 10.48A shows a normal scan and Figure 10.48B an abnormal one, with a dilated left ventricle, increased pulmonary thallium activity, and a defect in the inferior wall *(arrow)*. For each examination there are six images. The left column shows the initial views, the right column the delayed views. The anterior views are in the top row, 45-degree views in the middle row, and 70-degree views in the bottom row. Note that the inferior perfusion abnormality in the abnormal patient (B) does not change on the delayed views.

A recent advance in nuclear cardiology has been the development of technetium-labeled compounds for use in assessing myocardial perfusion. Technetium is more readily available than thallium, and these compounds produce better-quality images because of their optimal gamma energy and higher count rates. In addition, these compounds can be used to perform first pass ventriculography during injection. One of the most popular compounds is technetium methoxy-isobutyl-isonitryl (Sestamibi). Sestamibi, like thallium, is distributed to the myocardium proportional to blood flow.

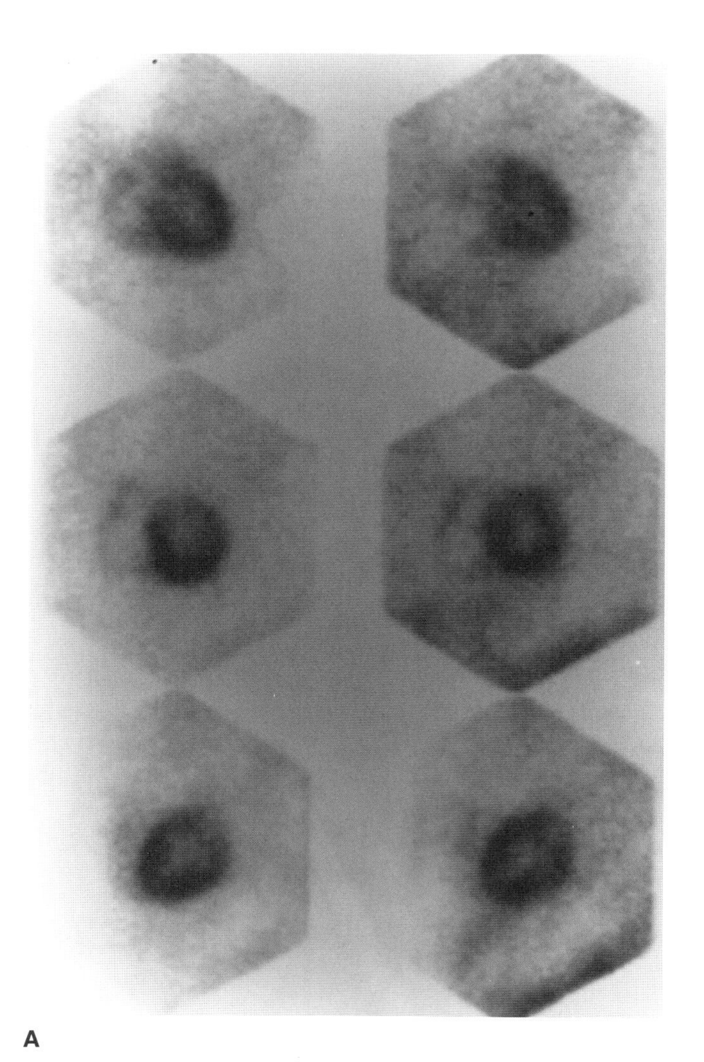

A

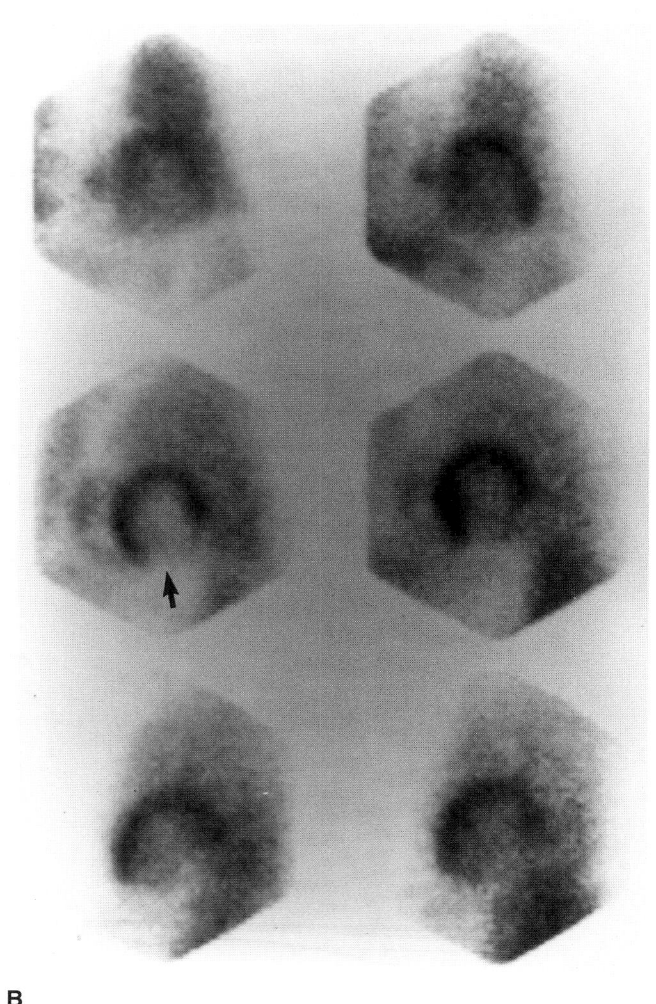

B

Figure 10.48. Normal (A) and abnormal (B) thallium perfusion scans obtained with a gamma camera imaging in anterior, 45-degree, and 70-degree left anterior oblique projections. *Left column:* initial views; *right column:* delayed views. The *arrow* indicates a perfusion defect in the inferior wall. (See text.)

MR Images of the Heart—Coronal Plane

To help you understand the three-dimensional anatomy of the four cardiac chambers and their associated great vessels, we suggest that you review the labeled anatomic structures on a series of MR scans. These coronal images have been slightly obliqued to best show the relevant anatomy. Figure 10.49 is the most anterior section; Figure 10.54 is the most posterior. (Compare the radiographs of coronal slices of a cadaver in Figures 2.14 through 2.17). Note the thicker muscular wall of the left ventricle, as compared to the right. The pulmonary artery can be seen arising from the right ventricle and the aorta from the left. The ventricles are more anteriorly positioned than the atria. Note the superior and inferior cava draining into the right atrium. The pulmonary veins entering the left atrium are not well seen.

Remember that moving blood in cardiac chambers and blood vessels appears black at MR on these T1-weighted images. With other MR imaging parameters, moving blood may appear white. MR images are acquired as tomographic slices through any selected imaging plane; consequently, unlike CT, MR scans can yield primary coronal, sagittal, or oblique slices of the heart, as well as conventional axial slices. One popular oblique angle of slicing is parallel to the long axis of the heart (Figure 10.55). Electrocardiographic gating may be used during MR scanning to obtain slice-specific images that match various phases of the cardiac cycle (Figure 10.56). Thus MR cardiac imaging permits not only anatomic but also functional examination of the heart. You may observe not only chamber wall thickness but wall motion as well. Ejection fractions may also be determined, along with other physiological findings.

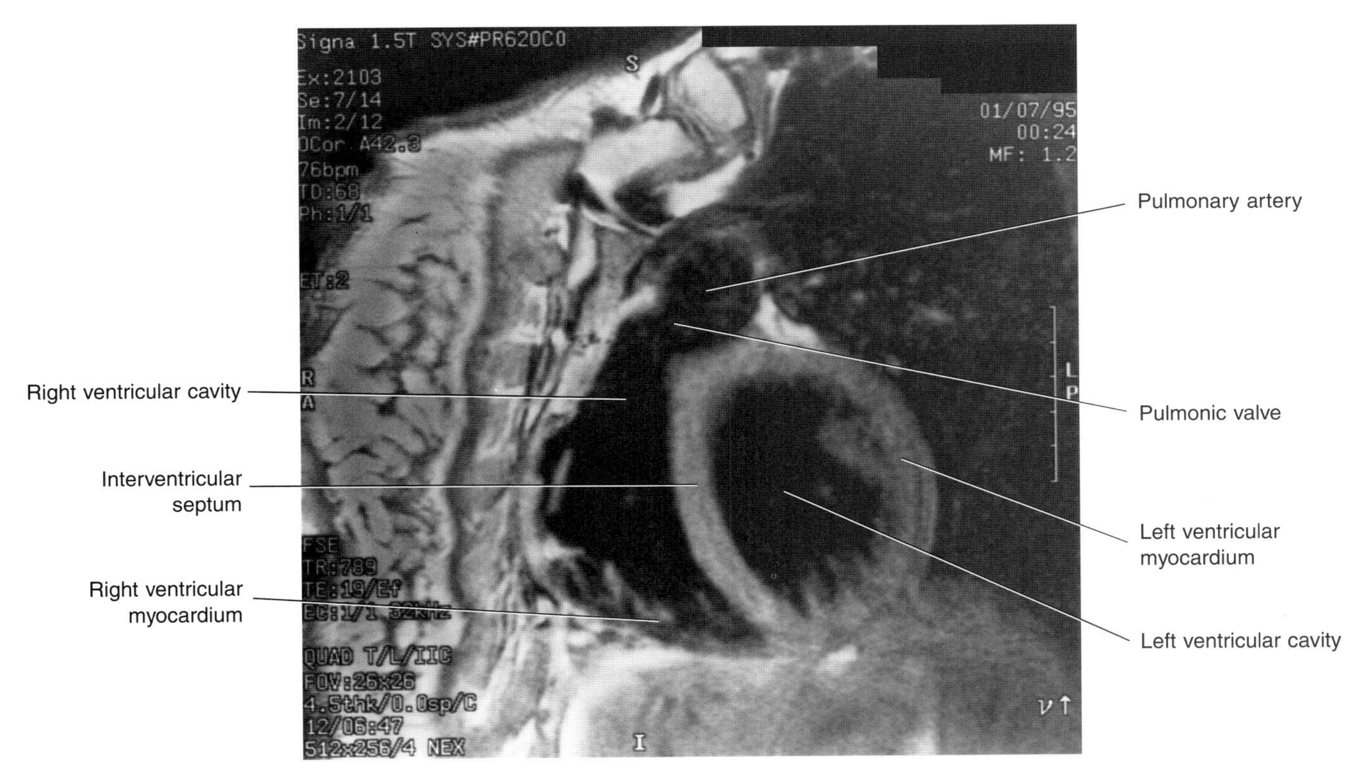

Figure 10.49. Oblique coronal MR scan, level of the pulmonic valve.

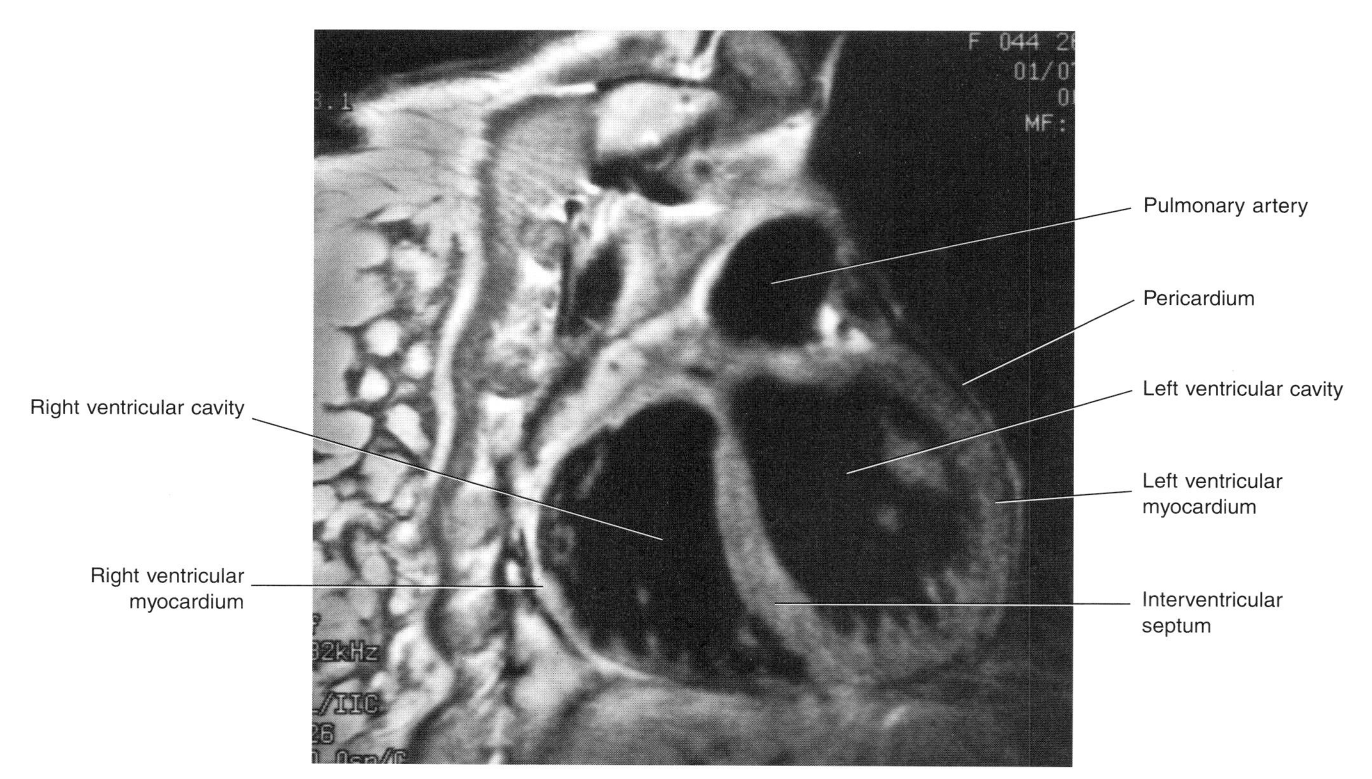

Figure 10.50. Oblique coronal MR scan, level of the midventricles.

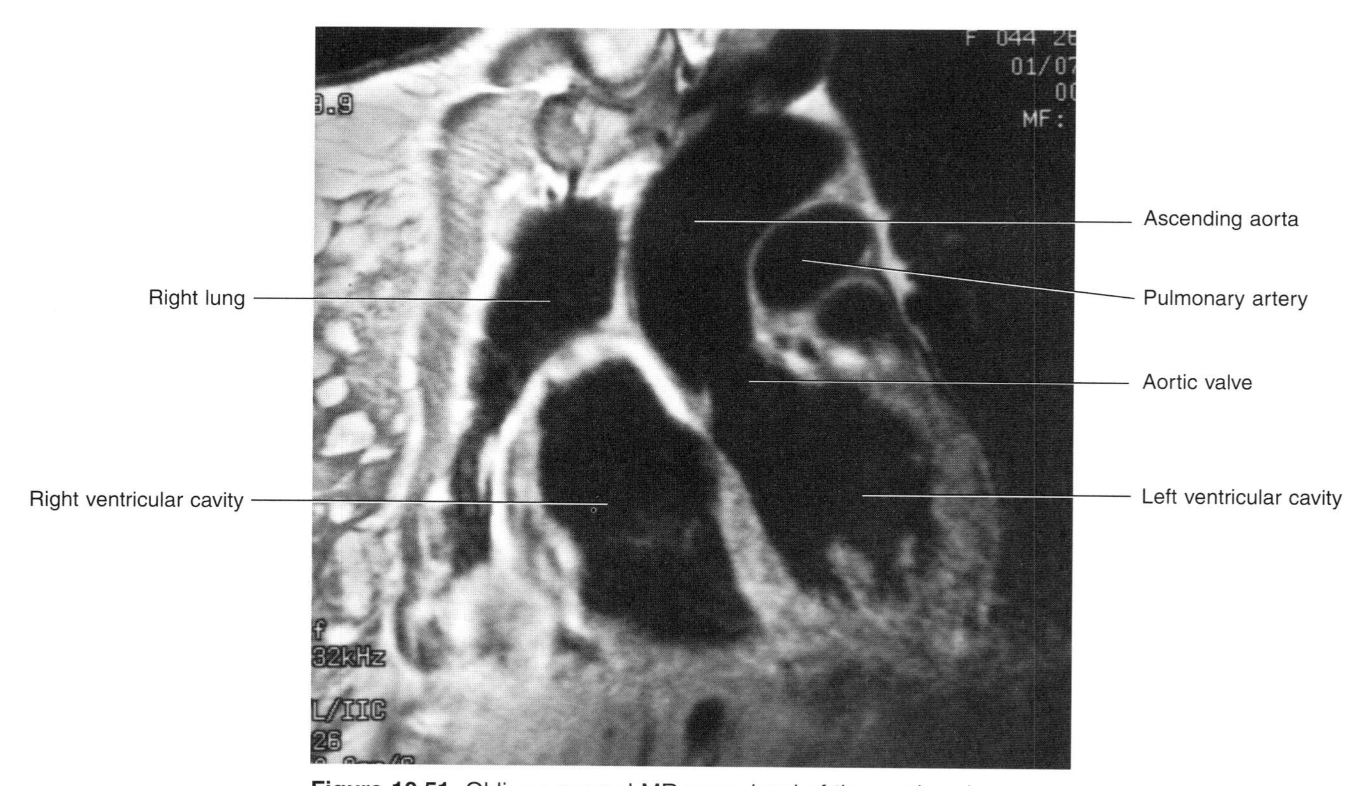

Figure 10.51. Oblique coronal MR scan, level of the aortic valve.

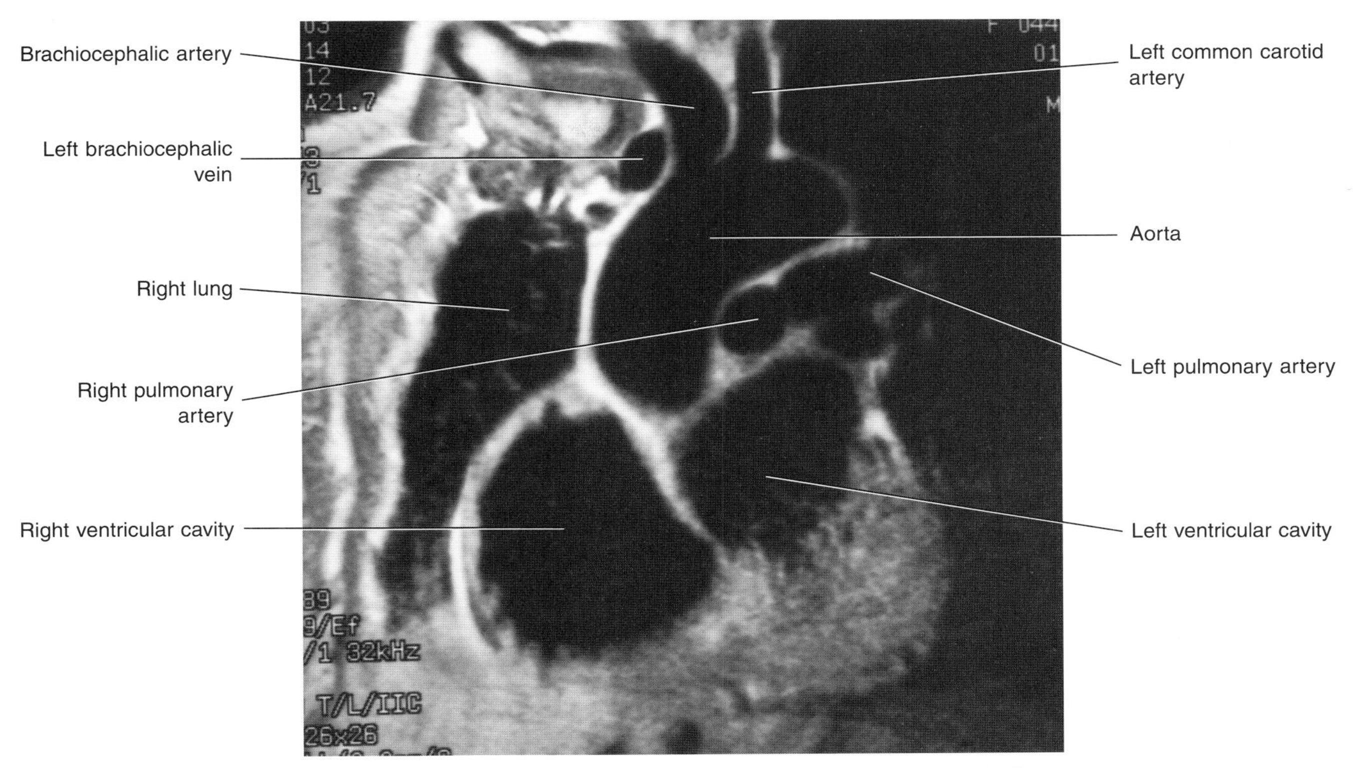

Figure 10.52. Oblique coronal MR scan, level of the brachiocephalic and left common carotid artery branches of the aorta.

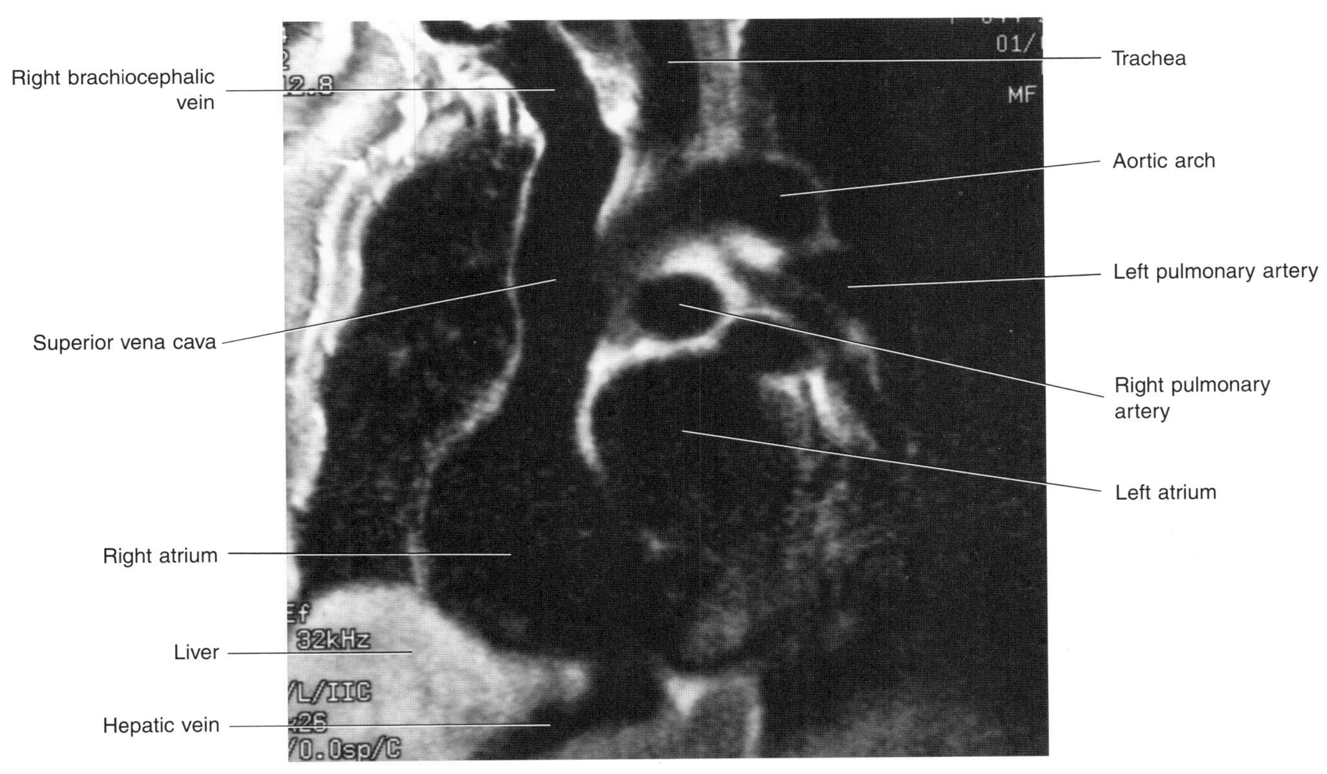

Figure 10.53. Oblique coronal MR scan, level of the superior vena cava and right atrium.

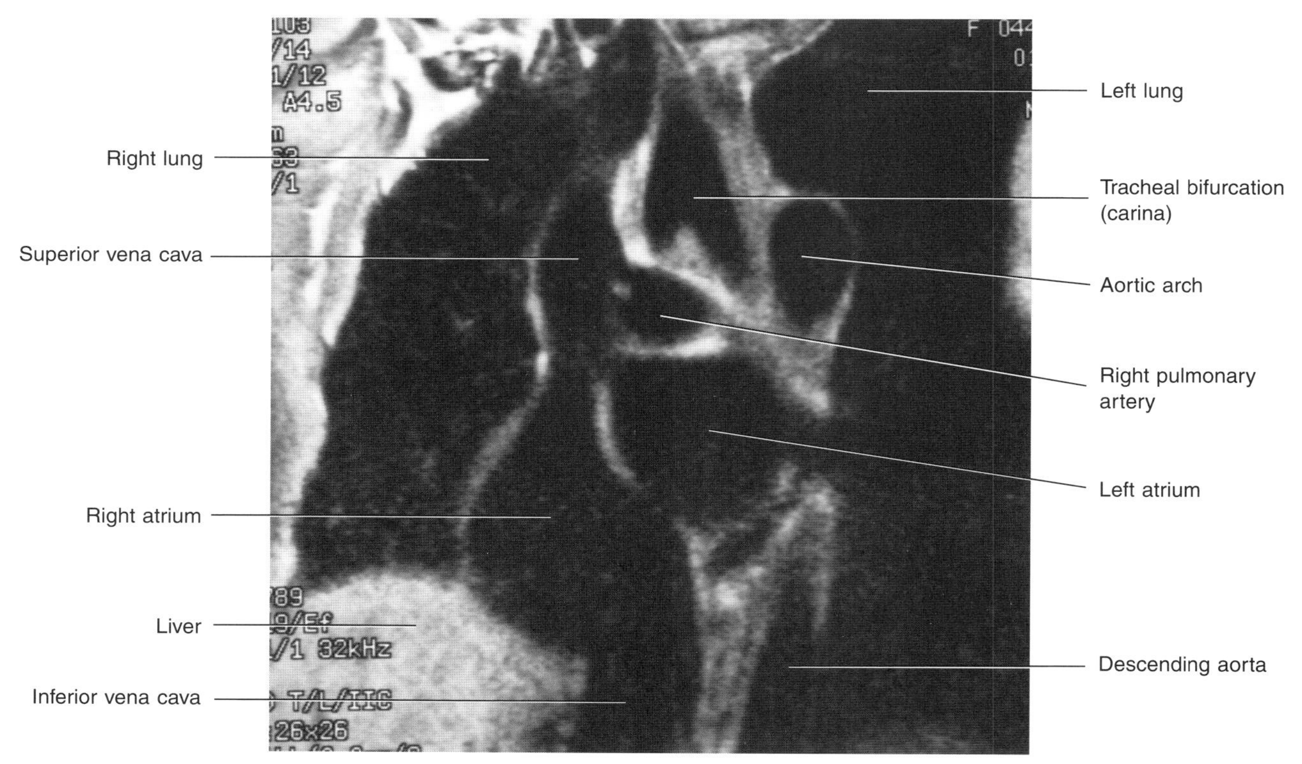

Figure 10.54. Oblique coronal MR scan, level of the left atrium and carina.

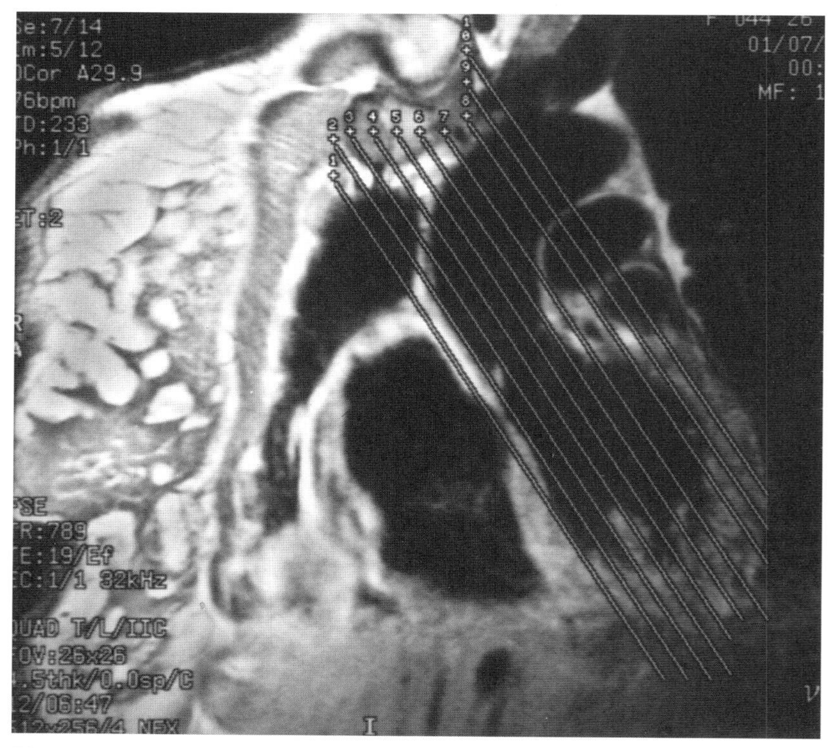

Figure 10.55A. Oblique coronal MR scan showing the orientation of long axis slices of the heart.

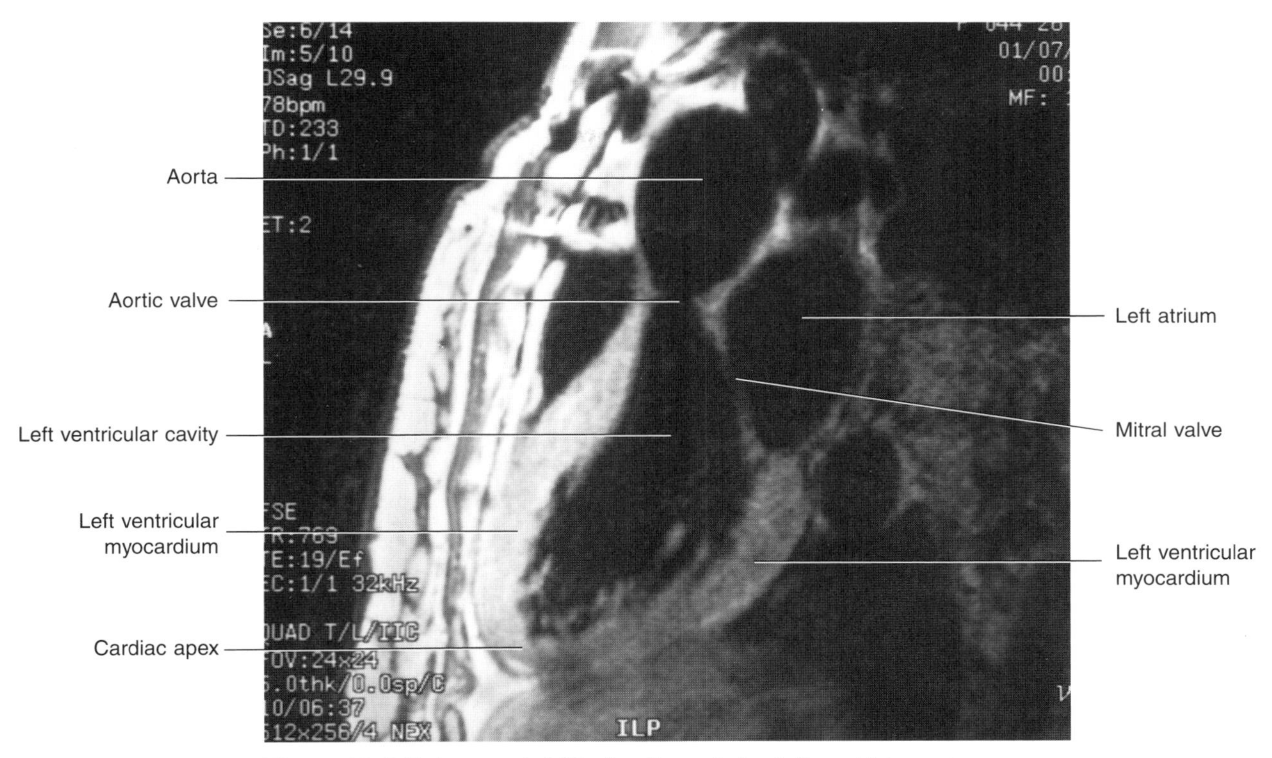

Figure 10.55B. Long axis MR slice through the left ventricle and aortic valve (slice 5 from Figure 10.55A).

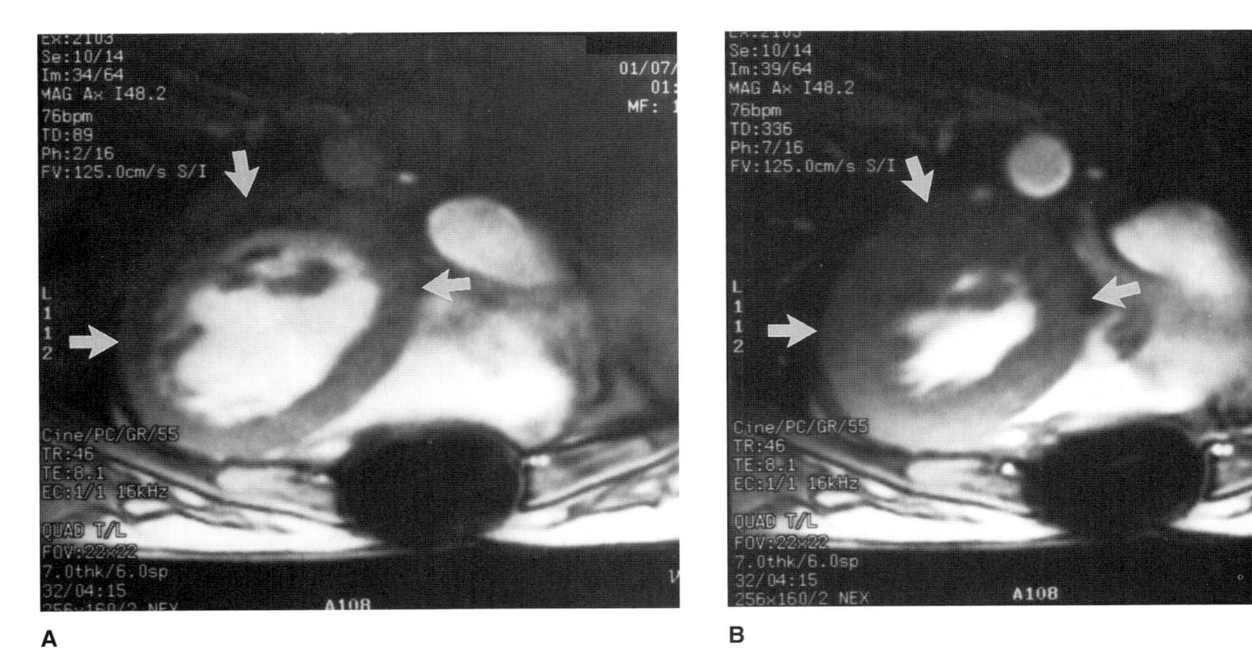

Figure 10.56. Gated MR scans showing left ventricle activity *(arrows).* A: Diastole. B: Systole. These scans were obtained with a gradient echo technique that gives blood within the chambers an intense white MR signal.

CT Images of the Heart—Axial Plane

This series of axial CT scans shows the anatomy of the heart and great vessels from the division of the pulmonary artery at level 1 down through the top of the right hemidiaphragm. These images were obtained by following an intravenous bolus of contrast material with an ultra-fast CT scanner that decreases artifacts caused by cardiac motion. The study was performed on a middle-aged man following coronary artery bypass graft (CABG) surgery to determine graft patency. As you are no doubt aware, this procedure involves placement of vein grafts from the root of the aorta down alongside the great vessels and heart into the ventricular myocardium to supply blood to ischemic areas of myocardium.

In level 1 (Figure 10.57) note the anterior location of the ascending aorta and the posterior location of the descending aorta. Metallic clips can be identified in levels 1 and 2 at sites of bypass graft surgery. The highest cardiac chamber, the left atrium, is seen in level 3. Note the patent grafts (opacified with contrast) in levels 4 through 8.

In level 5 the right atrium, right ventricle, and a portion of the left ventricle are first seen. This scan is located at the level of the aortic valve. The leaflets of the mitral valve are well shown in level 6 between the left atrium and the left ventricle. The tricuspid valve can be identified in level 7 between the right atrium and the right ventricle. Note the calcified atherosclerotic plaques in the descending aorta. The left atrium can no longer be identified, since level 7 is below it. Level 8 best shows the interventricular septum between the right and left ventricular cavities. Once again, note the thick wall of the left ventricular myocardium (as compared with the right wall) and the bundles of papillary muscle extending into the left ventricular cavity.

In clinical practice, CT can identify pericardial thickening and fluid collections, pericardial calcifications, and intracardiac masses and thrombi, as well as central pulmonary emboli and thoracic aortic aneurysms and dissections.

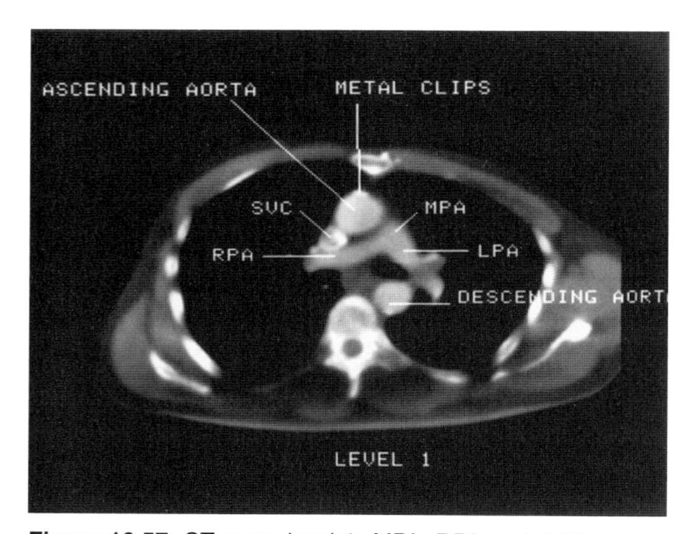

Figure 10.57. CT scan, level 1. *MPA, RPA,* and *LPA* indicate the main, right, and left pulmonary arteries.

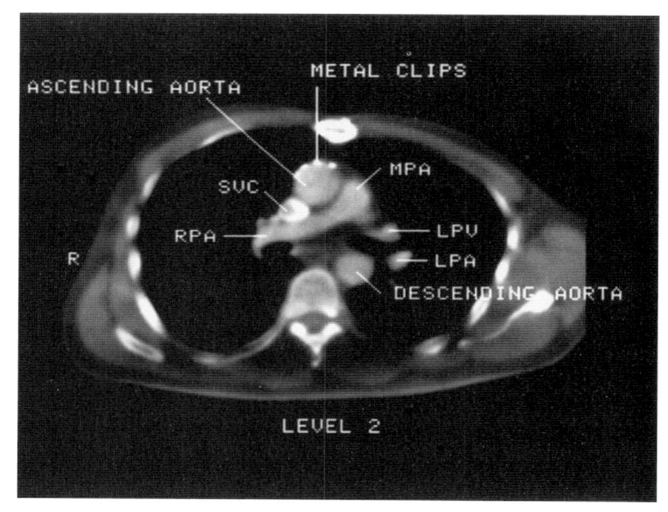

Figure 10.58. CT scan, level 2.

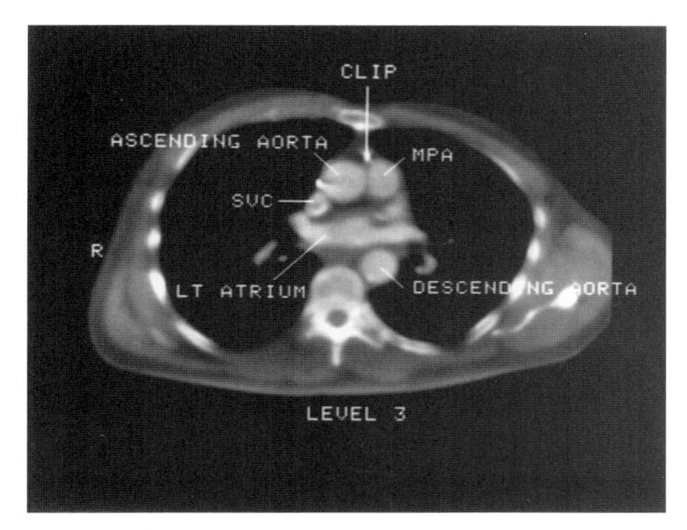

Figure 10.59. CT scan, level 3.

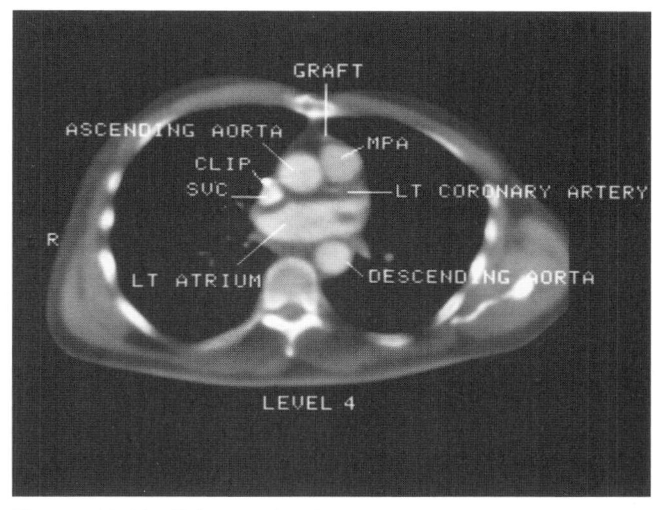

Figure 10.60. CT scan, level 4.

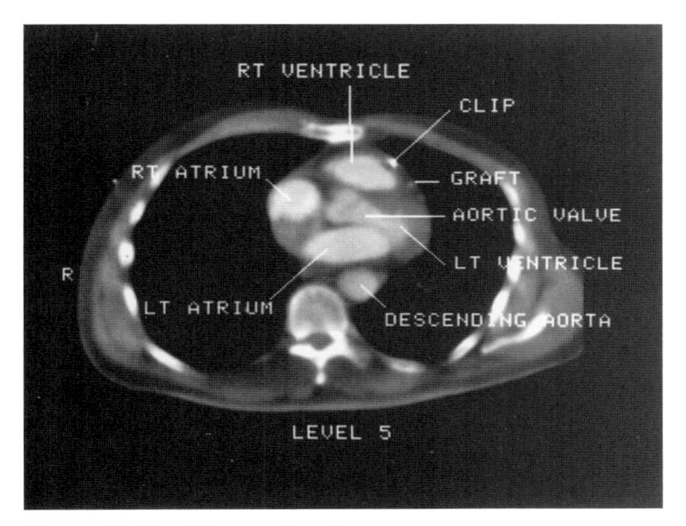

Figure 10.61. CT scan, level 5.

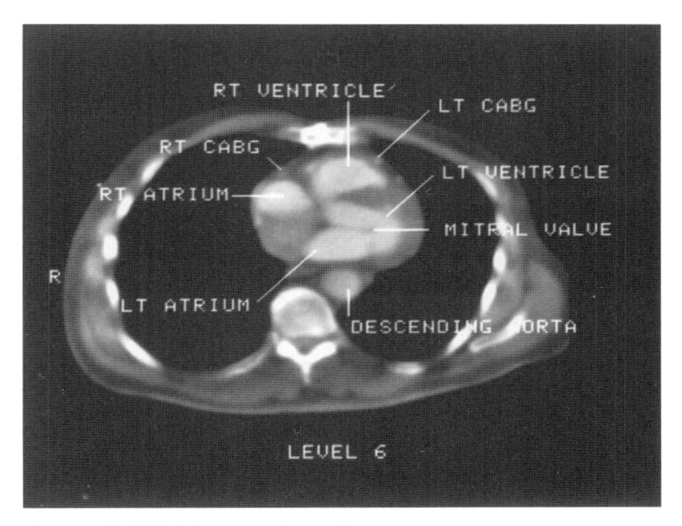

Figure 10.62. CT scan, level 6.

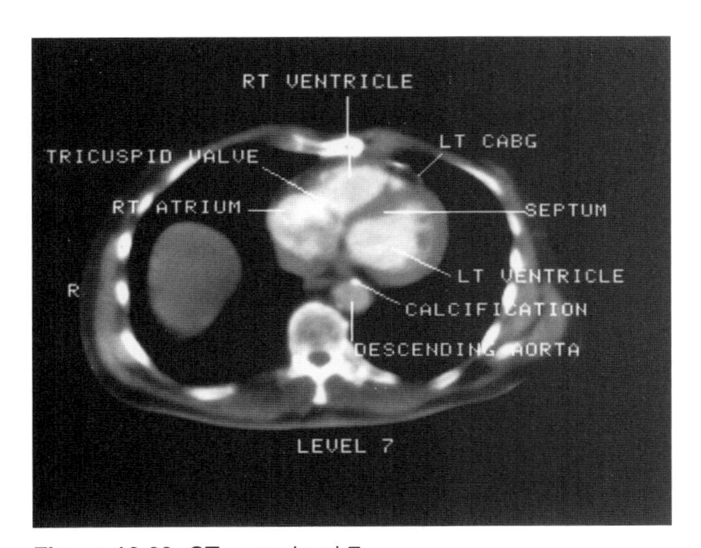

Figure 10.63. CT scan, level 7.

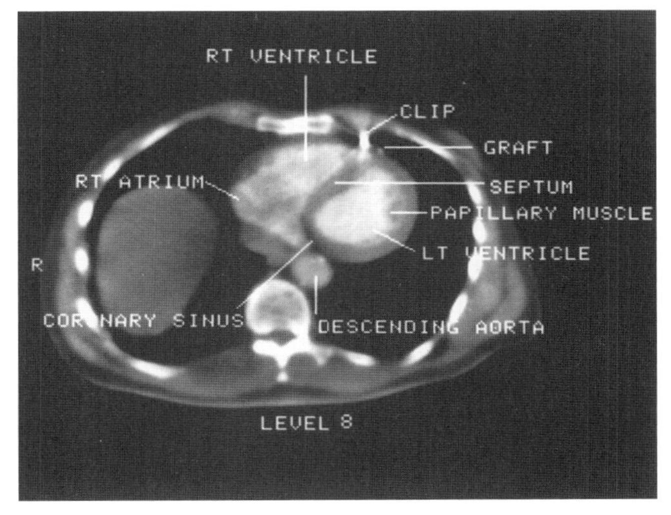

Figure 10.64. CT scan, level 8.

11 How to Study the Abdomen

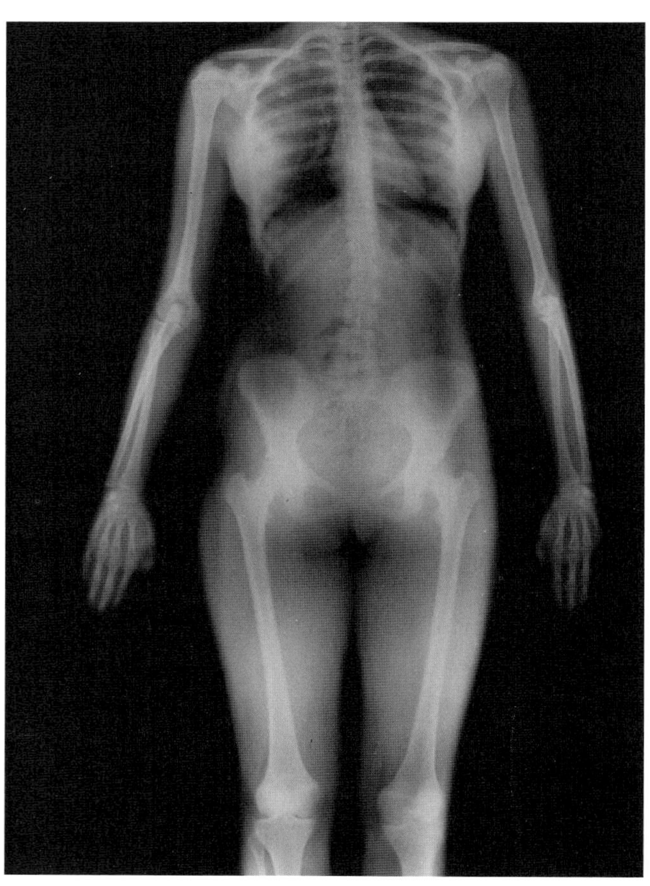

Figure 11.1. The abdomen and chest filmed together for comparison of radiodensities.

Radiographic study of the abdomen is perhaps a little more difficult and a little more subtle than that of the chest, but it is equally interesting from the standpoint of the opportunity it affords you to discover how much radiological information you can obtain through reasoning. Material you learn by appreciating the logic of its appearance is easily retained because you can reason it out again if you forget it. In this chapter we will address first the way in which you should examine plain film radiographs of the abdomen. More about abdominal CT and abdominal ultrasound follows.

The Plain Film Radiograph

The wide differences in radiodensity of the chest structures provide profiles and margins that are easy to see and to interpret as you first begin to look at x-ray films. In the abdomen, however, organ masses and great vessels may merge into a confluent gray shadow so that their borders and profiles vanish. *Only when some structure of differing density lies against one you wish to know about can you see its boundary*—often only a small segment of that boundary—from which you may be able to deduce something about the size and shape of the organ in question.

The striking radiolucency of air within the bowel should occur to you at once as providing the sort of boundaries and outline segments you need. You will use this information constantly in assaying from plain films the size and shape of organs and masses within the abdomen.

Thus, for example, all air-containing structures may be swept into the left side of the abdomen by a grossly enlarged liver, whose mass is seen as a large gray shadow but whose *margin* is often only visible to you outlined by air in the colon (Figure 11.3). The stomach, when filled with fluid, lies against the spleen and blends with its shadow invisibly, giving no information about its size; but if the stomach is inflated with air, it may be seen to be clearly indented from the left and displaced medially by an enlarged spleen. All air-containing structures may be displaced upward out of the pelvis by a large ovarian cyst or a distended bladder. Individual large-bowel and small-bowel loops, like a circle of dark beads, will outline the upper surface.

Variable as the content of air in the gut certainly is, it will still prove extremely useful to you in tagging abnormalities in the size and shape of other organs, and you will soon form a visual baseline with regard to the amount and location of air-in-gut that you can expect to see. There is normally at least a little air in the stomach and a fair amount distributed throughout the colon. In the healthy, ambulatory adult the small bowel usually contains little or no air, but normal

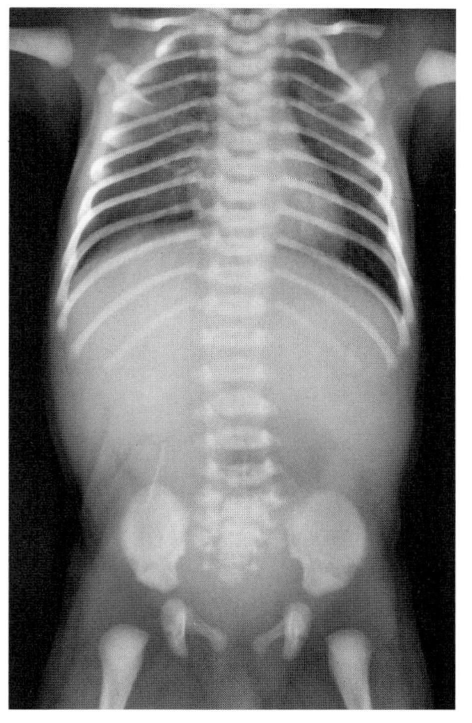

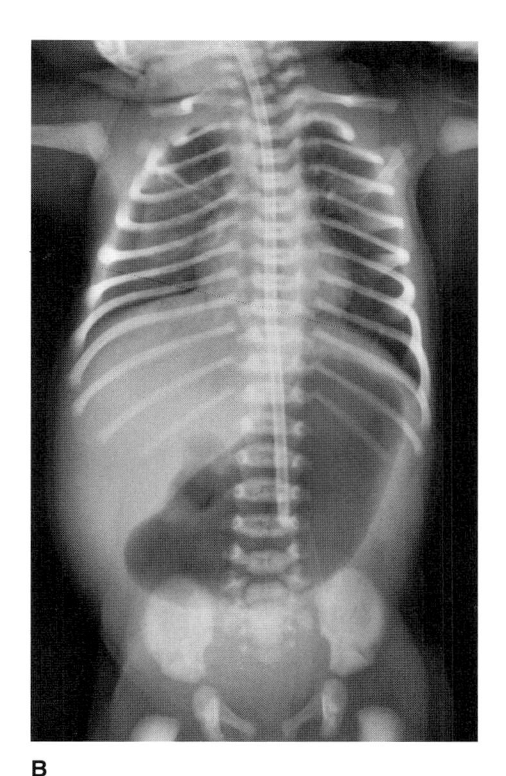

Figure 11.2. A: Airless abdomen in a 4-day-old girl who had been vomiting since birth. All the organs blend together as one confluent gray shadow. B: Same patient, stomach inflated with air. Note the density of the nasogastric tube. The right side of the stomach lies against the liver and outlines its margin.

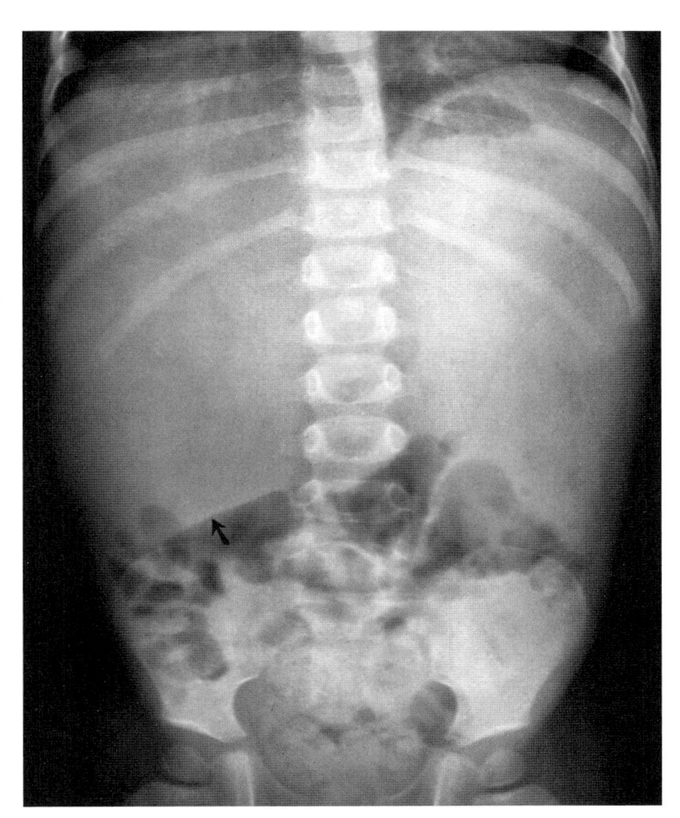

Figure 11.3. The margins of the liver *(arrow)* and spleen are outlined by air in the bowel. You know the film was made with the patient standing because of the fluid level in the fundus of the stomach. The liver and spleen in this small boy were immensely enlarged by lymphoma.

infants and bedridden adults frequently show considerable amounts of small-bowel air without any abdominal pathology to account for it.

By definition a "plain film" is a film made without any artificially introduced contrast substance. For this reason you should call it a *plain film* rather than a "flat plate," an obsolete term that derives from a time when images were made on large glass plates instead of x-ray film. The so-called KUB (kidney, ureter, bladder) is another name for an abdominal plain film. Although we will in this chapter discuss first the plain film of the abdomen, you will find that you will better anticipate the location and appearance of the air-outlined bowel *after* you have seen all parts of the gastrointestinal tract filled with barium. *In barium studies the lumen of the gastrointestinal tract is rendered visible, but in the plain film you have to depend on transient air content alone for information,* and you need to remember that many parts of the bowel are ordinarily invisible because they contain fluid, food, or feces, or are collapsed. Sometimes parts of the colon are outlined by their content of semisolid feces with which bubbles of air have been mixed. This casts a distinctive speckled shadow and may be just as useful as air-filled gut in indicating the position of neighboring structures or in locating parts of the colon itself. Such speckled fecal shadows always identify the colon and are not seen in the small bowel.

Identifying Parts of the Gastrointestinal Tract from Intraluminal Barium and Then from Air Content

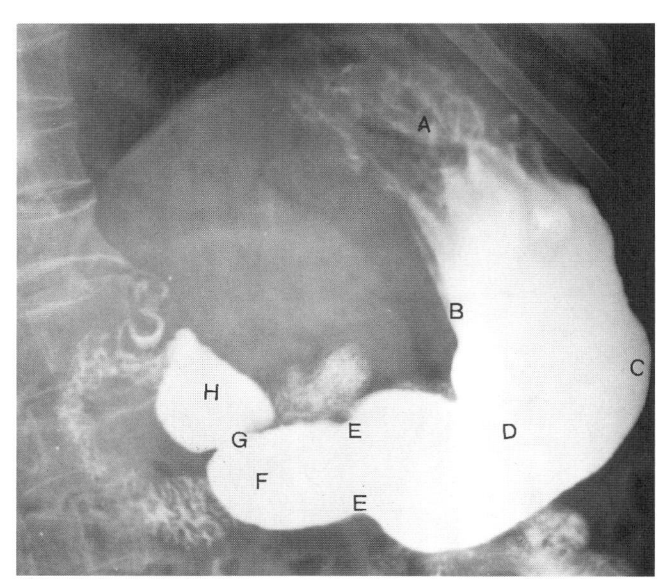

Figure 11.4. The stomach is visible here because it is filled with barium sulfate contrast material. *A,* fundus; *B,* lesser curvature; *C,* greater curvature; *D,* body; *E–E,* indentation of a peristaltic wave; *F,* pyloric antrum; *G,* pyloric canal; *H,* first portion of the duodenum (the duodenal cap or bulb).

Barium casts of various parts of the gastrointestinal tract produce white shadows on the film, the margins of which distinctly reproduce the character of the mucosal pattern. The rugae of the stomach are quite different from the valvulae conniventes (plicae circulares) of the small intestine and from the smoother, more widely spaced haustral indentations of the colon.

The distribution through the abdomen of air in the bowel is determined to some extent by the degree of fixation of several structures. The stomach may vary considerably in size, but it is fixed to the diaphragm and to the duodenum, which is partly retroperitoneal. The small bowel enjoys the liberty of its ample mesentery, folded into the midabdomen. The transverse colon varies widely in position, hanging from its mesocolon; the ascending and descending portions of the large bowel are relatively fixed laterally in the anterior compartment of the retroperitoneum. Within this degree of latitude you will learn to identify different parts of the air-filled bowel by their locations as well as by their distinctive mucosal patterns.

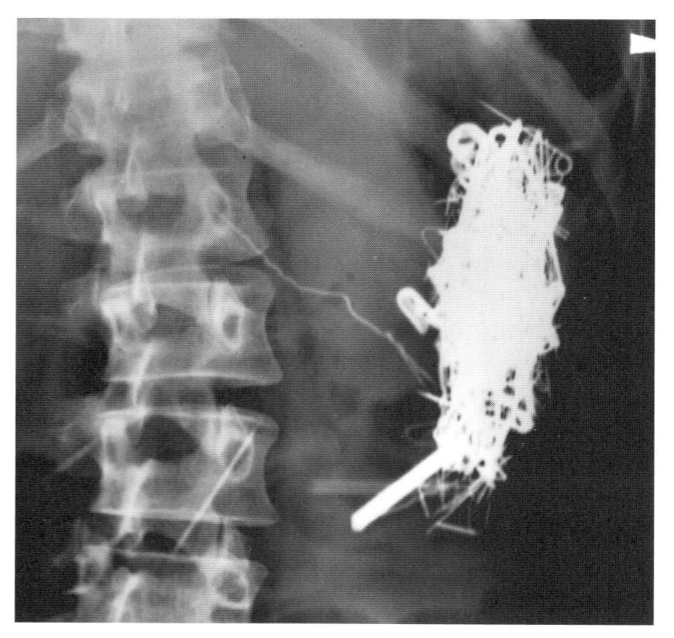

A

B

Figure 11.5. A psychiatric patient claimed to have "swallowed a pin" and complained of abdominal pain. The radiograph (A) shows an overlapping tangle of many metallic objects. At gastrotomy 287 metallic and glass objects were removed (B).

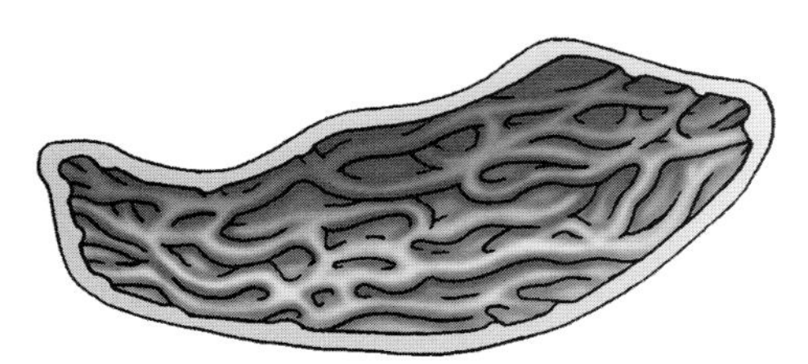

A. Mucosal folds of stomach

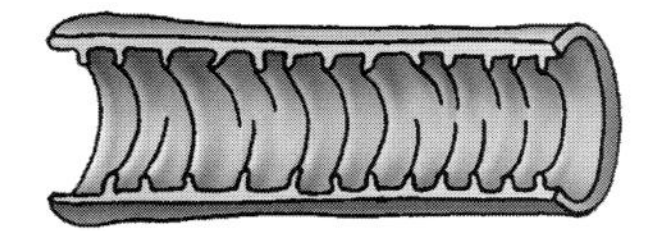

B. Valvulae conniventes of jejunum

C. Valvulae conniventes of ileum

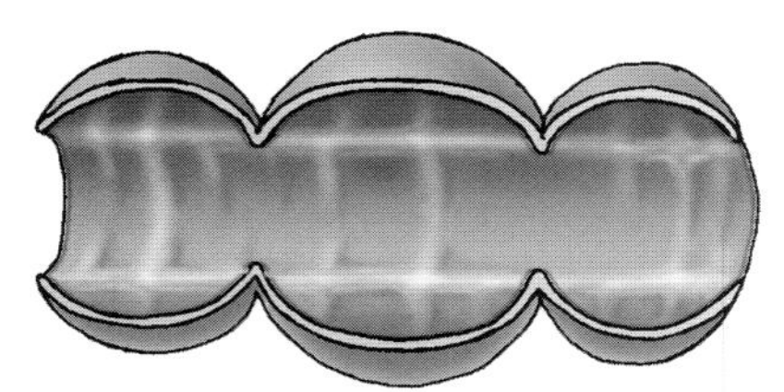

D. Haustral indentations of colon

Figure 11.6. The mucosal lining of segments of the gastrointestinal tract differ enough anatomically to be identified from the appearance of their respective air shadows. Compare the pattern of the mucosal folds of the stomach (A) and the valvulae conniventes of the jejunum (B) and ileum (C) with the haustral indentations of the colon (D). Remember that hollow organs *full of barium* (as in Figure 11.4) look quite different from the same organs *filled with air* or lightly coated with a smear of barium.

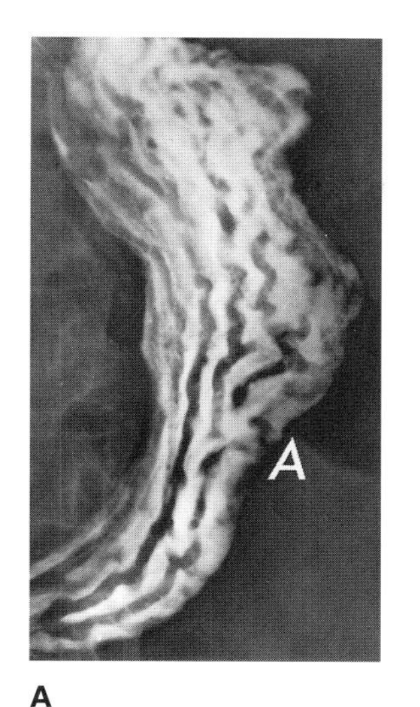

A

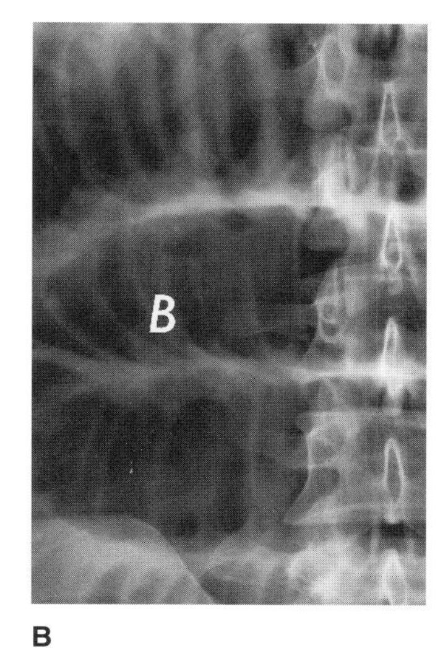

B

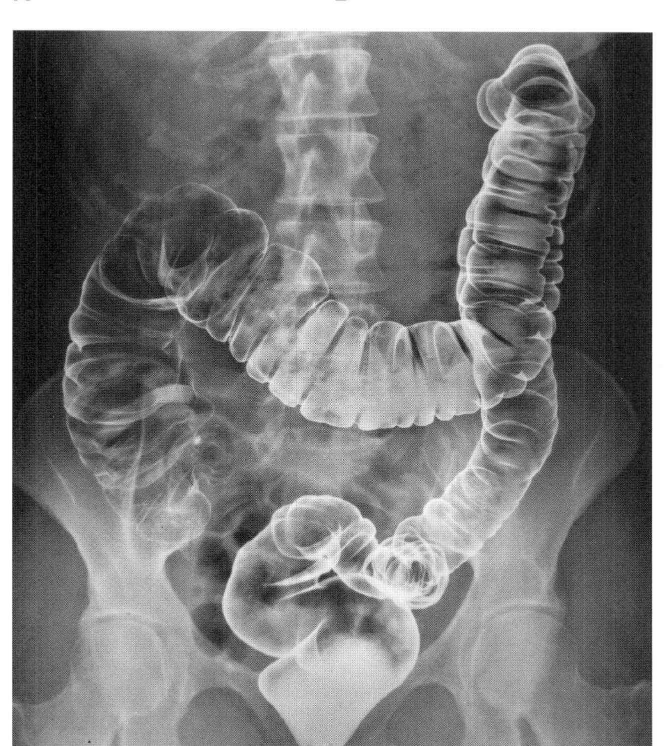

C

Figure 11.7. Clinical examples to help you. A: The rugae of the stomach are seen as black wavy shadows, opaque barium filling the valleys between them. B: The distinctive plical folds inside the jejunum appear as transverse ridges with air between them (here seen inside the distended small bowel of a patient with intestinal obstruction). C: The normal colon coated inside with barium and distended with air.

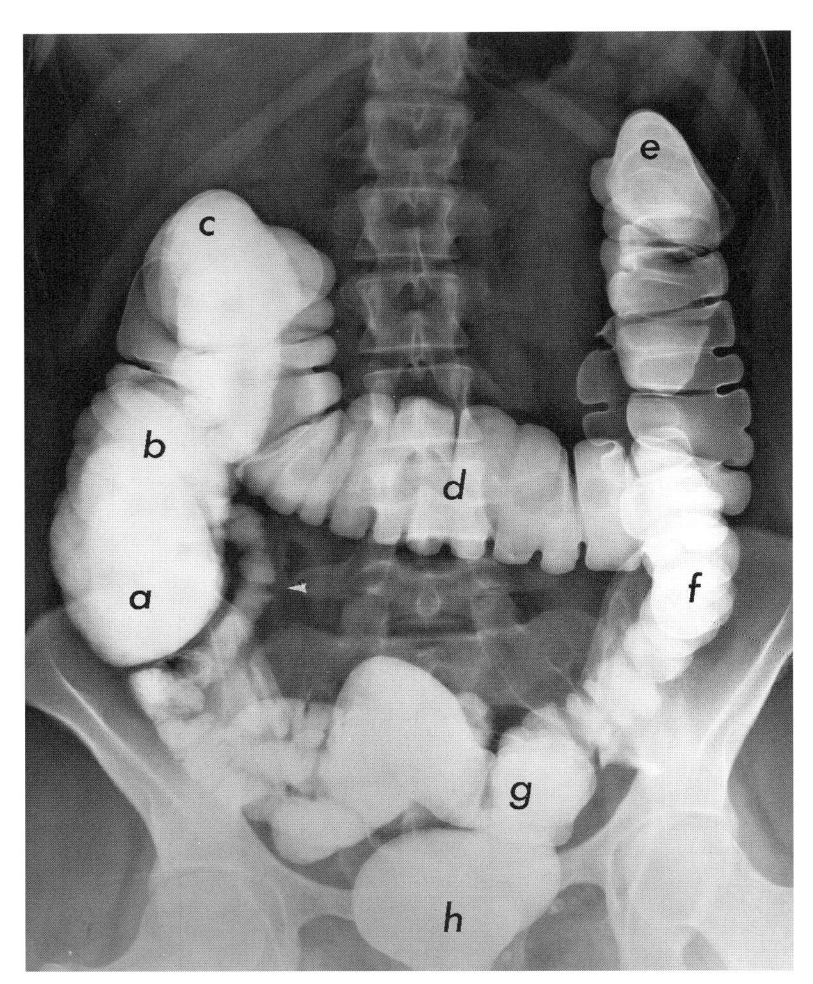

Figure 11.8. Normal colon anatomy: *a,* cecum; *b,* ascending colon; *c,* hepatic (right) flexure; *d,* transverse colon; *e,* splenic (left) flexure; *f,* descending colon; *g,* sigmoid colon; and *h,* rectum. Note the overlap at flexures in the frontal view. When the patient is turned, they are seen unrolled. The *small arrowhead* points to the terminal ileum, which often fills during a barium enema examination.

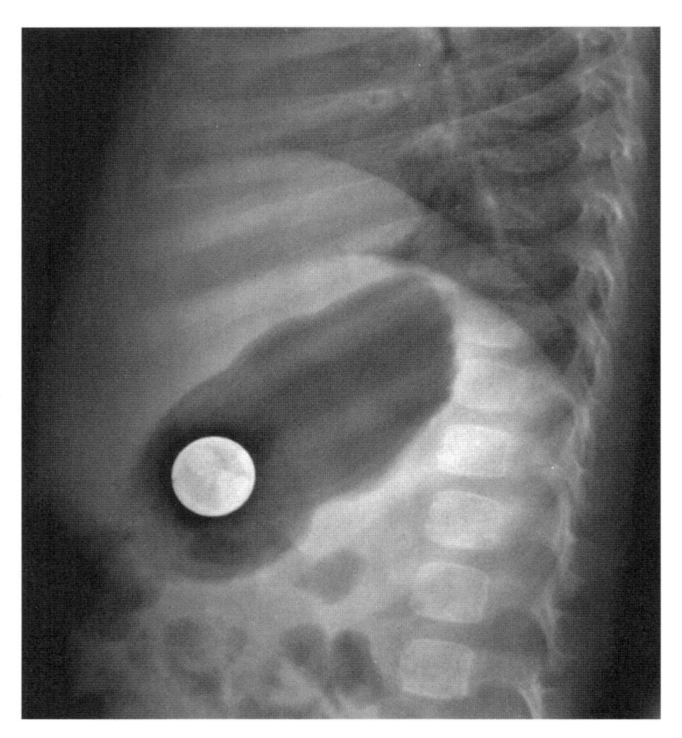

Figure 11.9. Lateral view of an air-filled stomach containing a penny. (Is it a Lincoln-head or an Indian-head penny?)

Identifying Fat Planes, Tangentially Viewed, as Normal Markers

Fat distributions within the abdomen also help you to make certain decisions about the structures they invest. The wide apron of the omentum will *not* help you, because it is distributed across the abdomen and never seen in tangent. The perirenal envelope of fat, however, provides a tangential radiolucent layer outlining the kidney mass with a dark line where more x-rays reach the film to blacken it.

In precisely the same way, the fatty layer next to the peritoneum in the abdominal wall is seen where the sagittally directed ray of a supine plain film strikes it and indicates the lateral limit of the peritoneal cavity. At each side where the fat layer turns posteriorly toward the patient's back, the beam catches it tangentially and the dark line produced on the film is called the flank stripe.

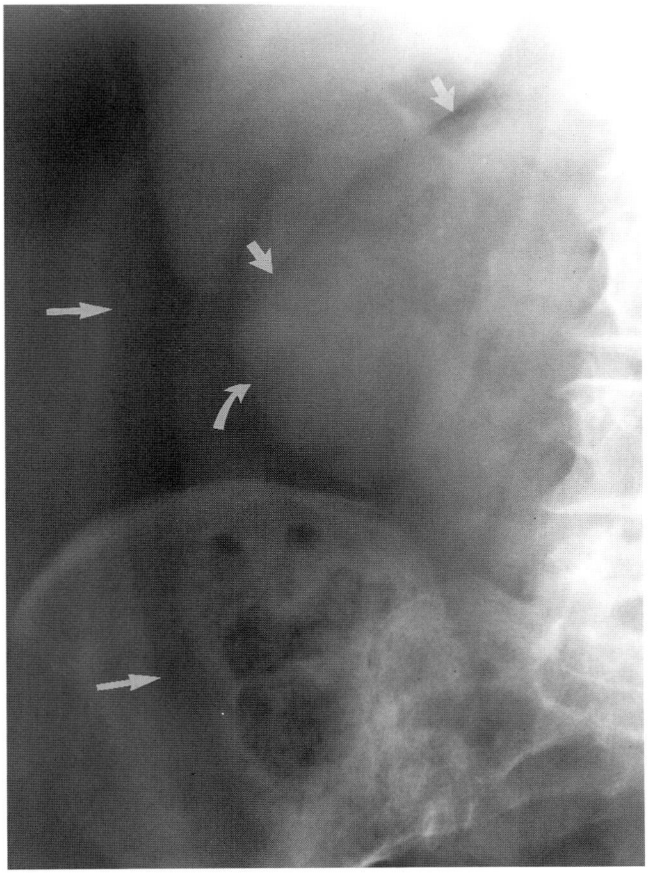

The flank stripe disappears when the flank itself becomes edematous. This is perfectly logical: fluid infiltrating the fat renders it as dense to the x-ray beam as the muscle that adjoins it. With inflammation near the flank (as in appendiceal abscess, for example), the flank stripe on that side may disappear, while the opposite one remains normal. In exactly the same fashion, perirenal inflammation may erase the perirenal fat.

Concentrated accumulations of fat, such as may be present within a dermoid cyst, produce localized, round radiolucent shadows on the film, visible because of their juxtaposition with surrounding structures of greater density.

As you will see later, it is their surrounding radiolucent fat that renders abdominal organs and masses so readily identifiable in cross section by computed tomography and magnetic-resonance imaging. Obese patients are harder to examine clinically and often have suboptimal plain films, but their abdominal organs are usually better shown on CT and MR scans than the abdominal organs of thin patients.

Figure 11.10A. Plain film of the abdomen showing the right flank stripe *(longer arrows)*. Close against it lies the ascending colon—indicated, as it so often is on the plain film, by the characteristic speckled shadow of feces mixed with air. The inferior margin of the liver is clear *(shorter arrows)* and the *curved arrow* indicates the right kidney, which is surrounded by perirenal fat.

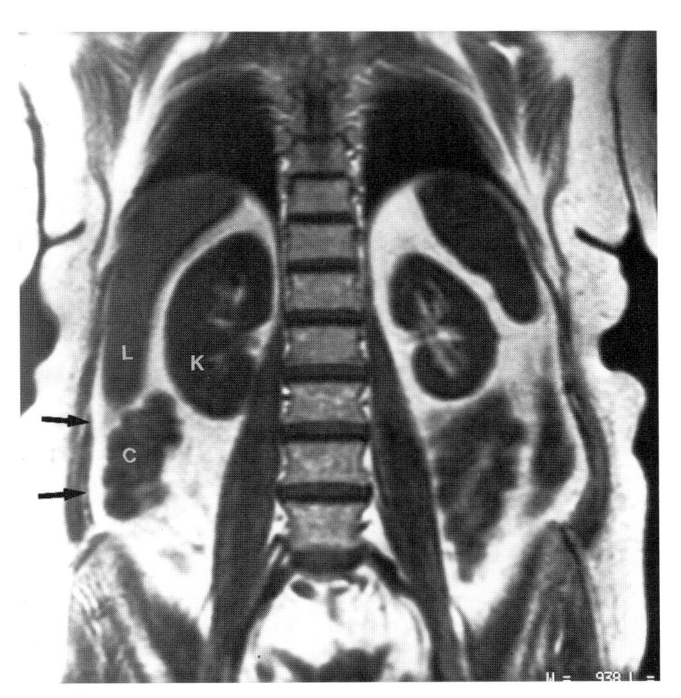

Figure 11.10B. Coronal MR scan of the posterior abdomen of an obese patient. The *black arrows* indicate the fat density of the right flank stripe, as well as fat surrounding the liver *(L)*, ascending colon *(C)* and right kidney *(K)*; you no doubt remember that fat with its high signal will appear white at many common MR settings.

A

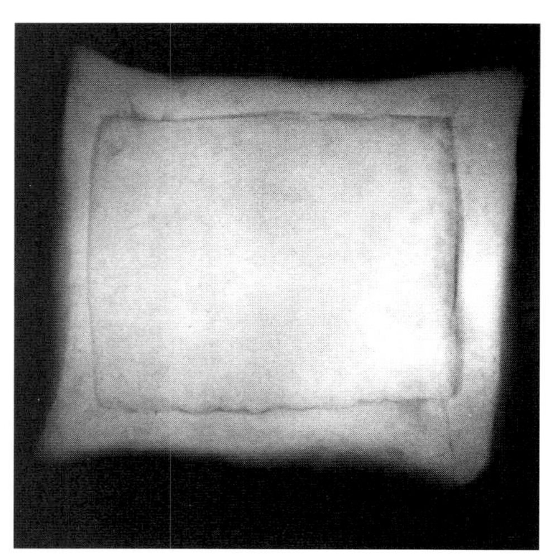

B

Figure 11.11. A: Clay tablets from the ancient Sumerian city of Ur. Business transactions were recorded on such tablets, and because the tablets were fragile and the records precious, an outer envelope of clay was added, bearing the same information in duplicate. B: The envelope of air between the inner and outer layers of baked clay is well shown in this radiograph of the intact clay tablet. The parallel between this and any fat-encased or gas-encased anatomic structure of greater density is obvious.

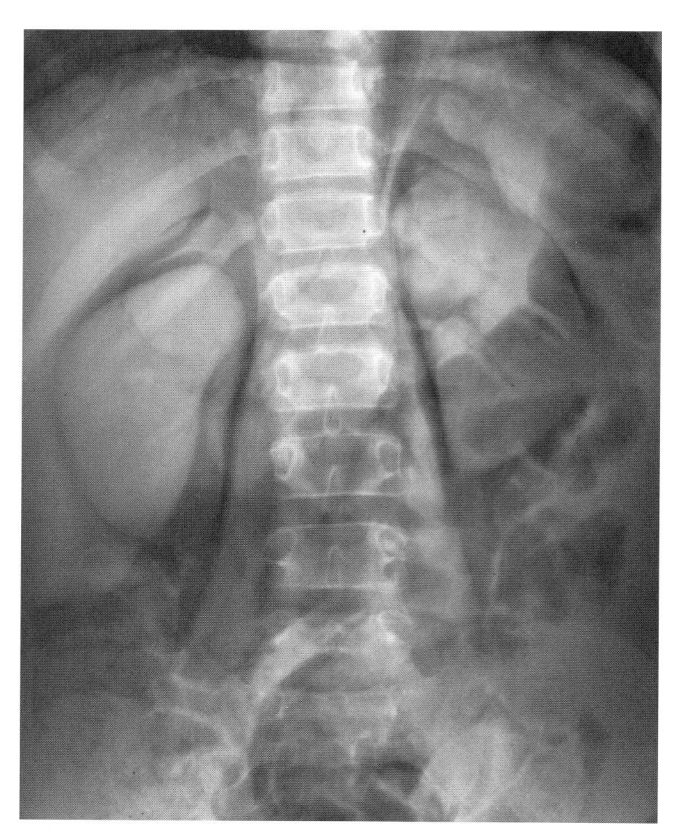

Figure 11.12. A gas injected into the retroperitoneal space (a procedure no longer performed today) has outlined the kidneys even more dramatically than does the fatty envelope you usually depend on to locate them on a plain film. Compare with Figure 11.13.

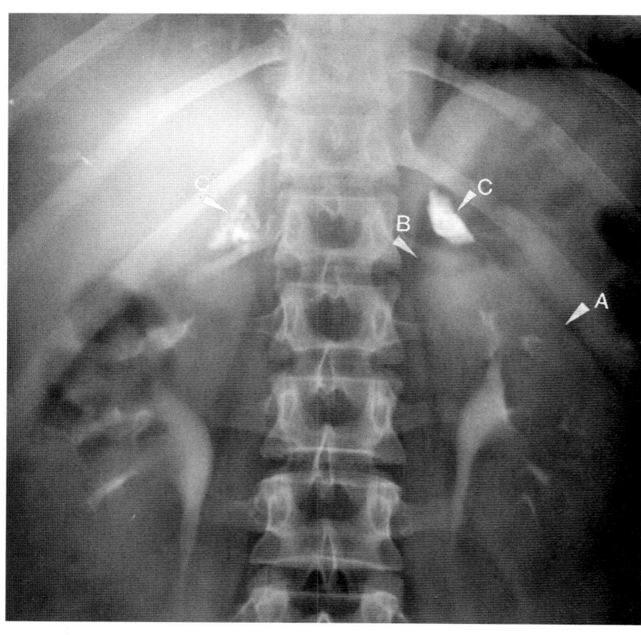

Figure 11.13. In this film *A* is the margin of the renal parenchyma, *B* the upper pole, and *C* the calcified adrenal gland. This patient had Addison's disease. The renal calyces and the pelvis are seen because a renally excreted contrast material has been intravenously injected (the intravenous urogram).

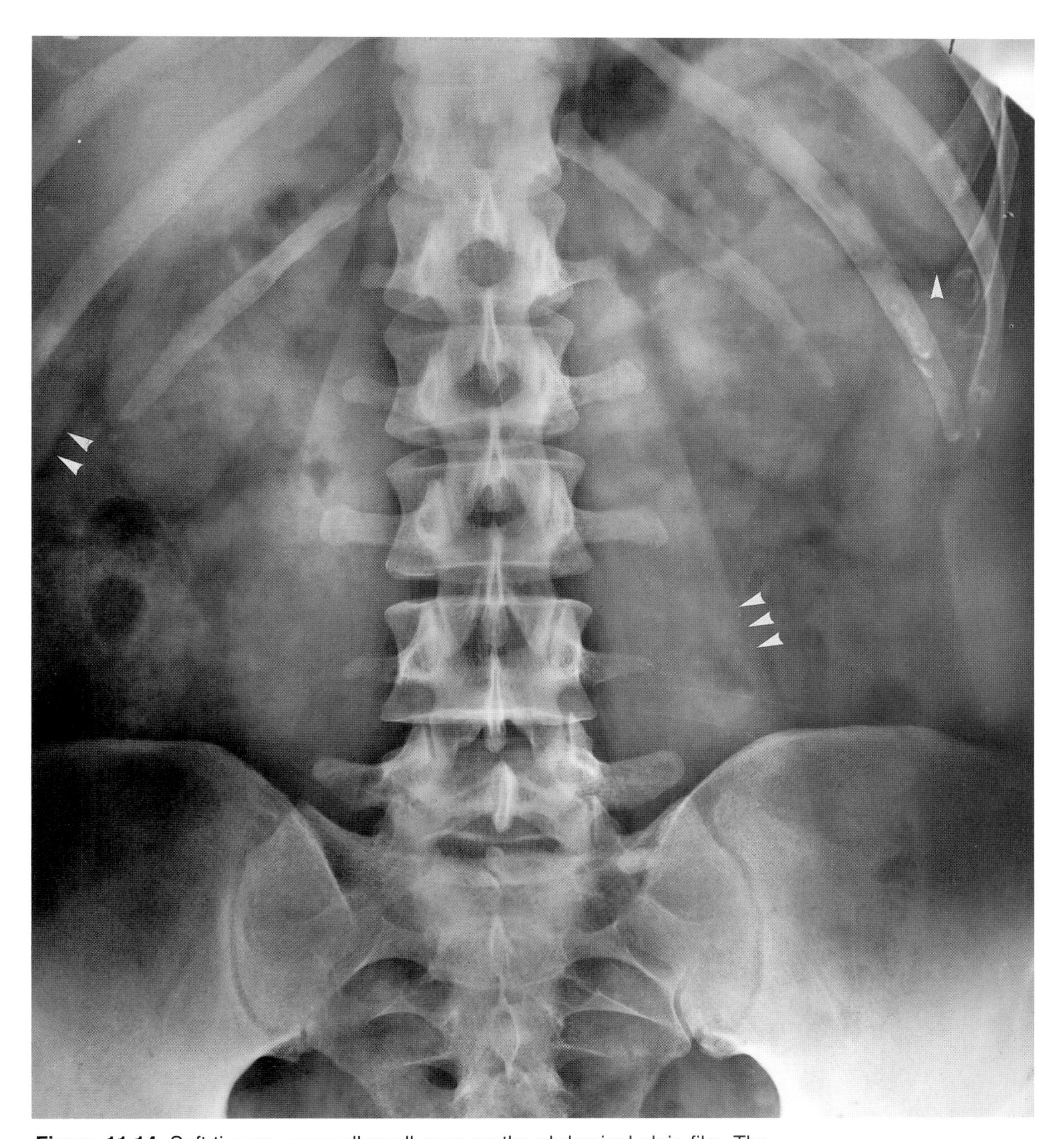

Figure 11.14. Soft tissues unusually well seen on the abdominal plain film. The *single arrowhead* indicates the tip of the spleen. The *double arrowhead* marks the lower margin of the liver, which you can follow obliquely upward across the shadow of the kidney. The *triple arrowhead* indicates the left psoas margin. The psoas shadows are generally symmetrical and are seen sharply because of fat surrounding the psoas sheath. Here the lower part of the right psoas muscle is obscured by something of equal density lying against it. Note the dark air in the stomach overlying the upper pole of the left kidney, and the haustrated, air-filled transverse colon superimposed on the middle of the left kidney. The entire outline of the right kidney is well seen. Note the left flank stripe.

Identifying Various Kinds of Abnormal Densities in the Abdomen

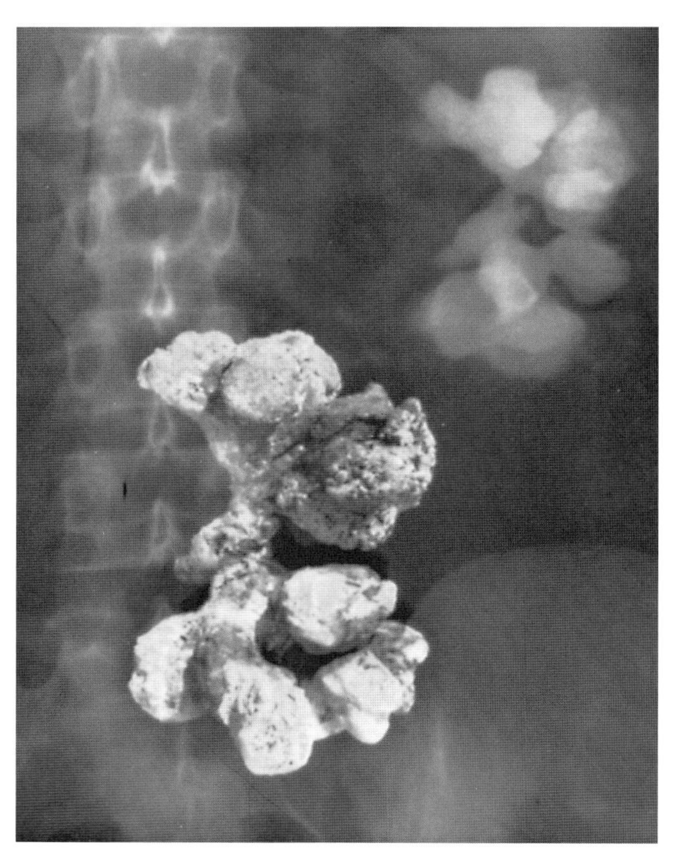

Figure 11.15. Stag-horn calculus in the left kidney. The photograph of the surgical specimen has been superimposed on the abdominal plain film and photographed with both transmitted and direct illumination.

Abnormal radiodensities can be provided by any area of calcification sufficiently large to absorb some of the beam. For example, phleboliths (calcified thrombi in the veins) are seen as dense white nuggets. Gallstones calcify much less commonly than kidney stones, but both have a characteristic location and often show a distinctive radiographic structure in their shadows as well. Calcified gallstones are frequently laminated and faceted. Gallstones, more often than kidney stones, form over a long period of time in a pool of fluid of slowly changing metabolic composition—hence the lamination seen in the radiograph. They are also more often multiple and are made to rub against one another with the contractions of the gallbladder—hence the faceting. The very characteristic stag-horn renal calculi (Figure 11.15) fill up the entire renal pelvis and calyces, closely resembling the shadow of opaque fluid you see on a urogram.

Calcification of the capsule of any organ resembles the radiograph of an eggshell, more dense peripherally where it is caught tangentially by the ray (Figure 11.18). Calcification in the wall of a hollow organ looks very similar, and you will often see this type of calcification in the aorta of older patients. Sometimes it indicates the presence of an aneurysm of that great artery (Figure 11.20). Plaques of calcium caught by the ray in tangent will provide an interrupted white outline, as you would expect. An artery of smaller caliber, when its wall becomes calcified, shows linear white margins like the rose stem in the first chapter, and they will be serpiginous and parallel if the artery describes a tortuous course (Figure 11.19). You will recognize them as characteristic of a cylinder of dense material about a more radiolucent core, x-rayed from the side.

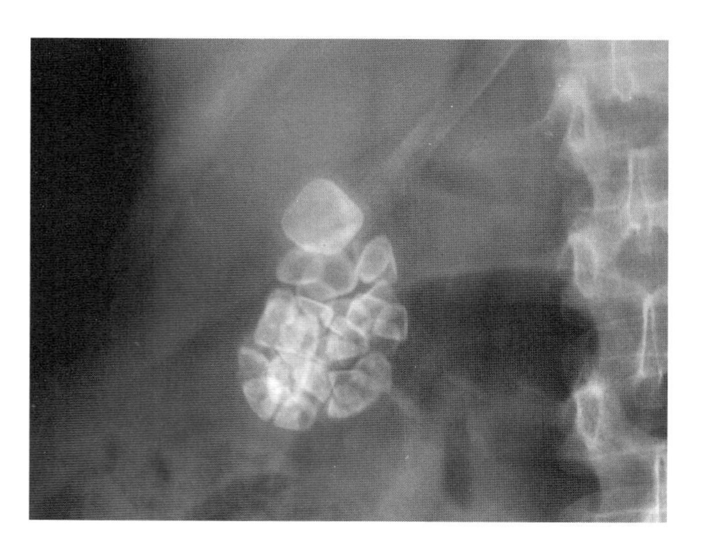

Figure 11.16. A cluster of faceted calculi in the gallbladder. Note that they have formed in such a way that their outer surfaces contain more calcium than their central portions. Now look closely. The large, square, uppermost calculus shows a new layer of lesser density. This illustrates the development of lamination in large gallstones that have existed for a long time.

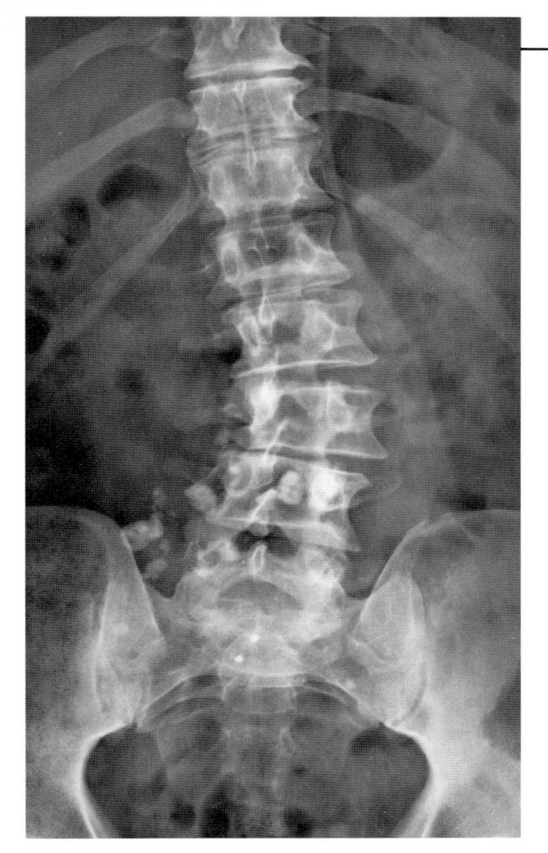

Figure 11.17. A cluster of calcified mesenteric lymph nodes overlies the course of the right ureter near the upper border of the right sacroiliac joint. This patient was placed as straight as possible in the supine position. Note the obliquity produced in the midlumbar spine by his degree of scoliosis. Only the left psoas muscle can be seen.

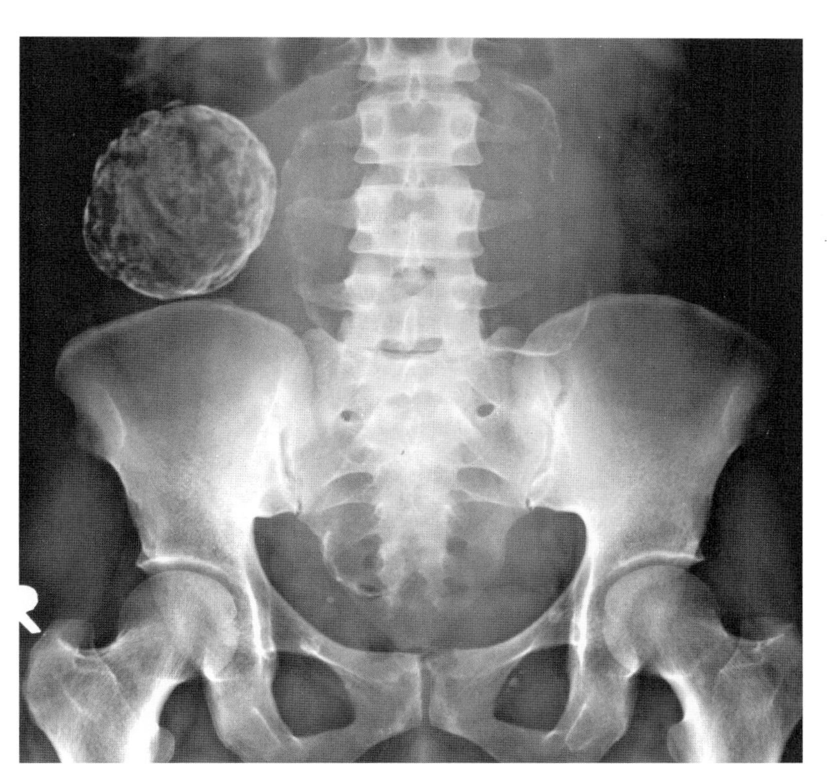

Figure 11.18. Multiple enormous uterine leiomyomas (fibroid tumors) in a 42-year-old woman complaining of constipation and dysuria. The patient had lived all her life in a remote rural area and had never consulted a physician before. The specimen removed at surgery showed many intramural, pedunculated, and submucosal fibroids with varying degrees of calcification. The only visualized air-filled loops of bowel are displaced into the upper abdomen by the large uterine tumors.

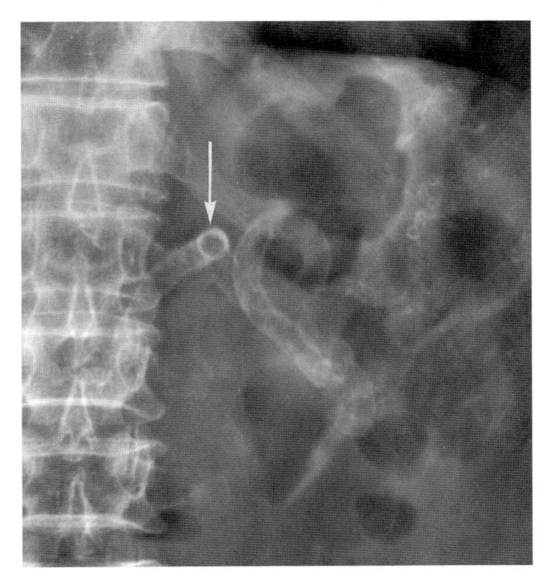

Figure 11.19. Calcified tortuous splenic artery. Note the parallel winding white lines. The *arrow* indicates a segment passing AP, hence filmed end-on and appearing as a white ring.

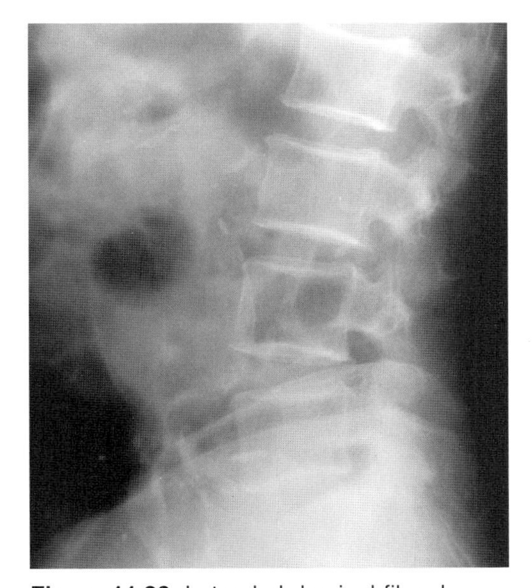

Figure 11.20. Lateral abdominal film showing a large egg-shaped aortic aneurysm just in front of the spine with a calcified wall both anteriorly and posteriorly. You may need to look closely to find the anterior wall.

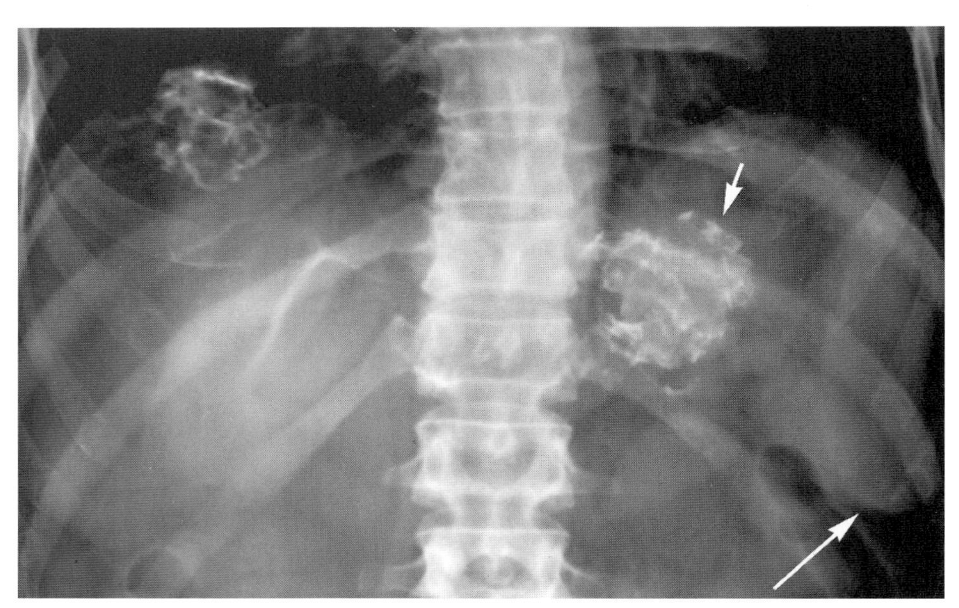

Figure 11.21. Calcified amebic abscesses in the liver. The *long arrow* indicates the tip of the spleen, which is not enlarged. The *short arrow* points to a large calcified abscess in the left lobe of the liver.

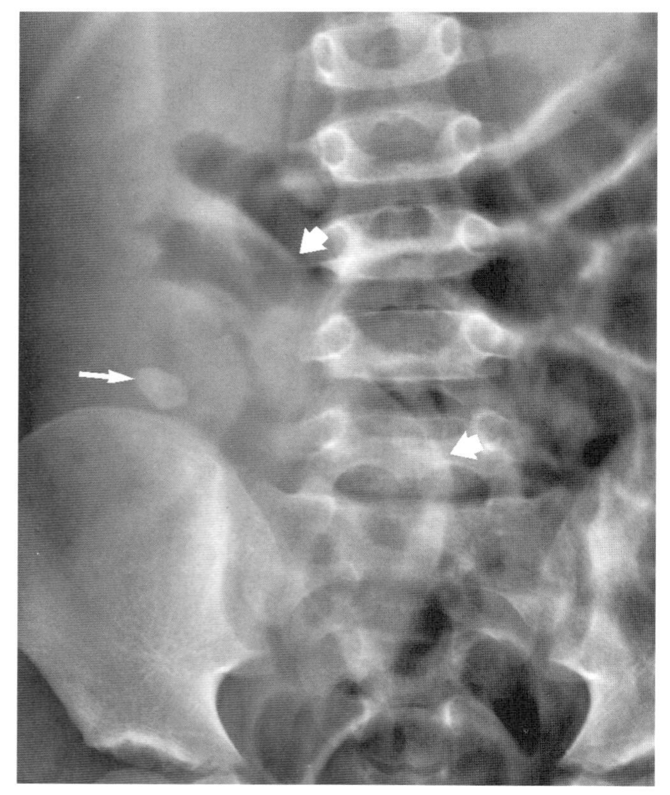

Figure 11.22. Abdomen of a 5-year-old boy. Air-filled bowel is seen displaced away from the right flank by a large soft tissue mass *(wide arrows)* in which there is an oval dense calcification *(thin arrow)*. The calcification proved to be an appendolith within a large appendiceal abscess. Note scoliosis concave toward the abscess.

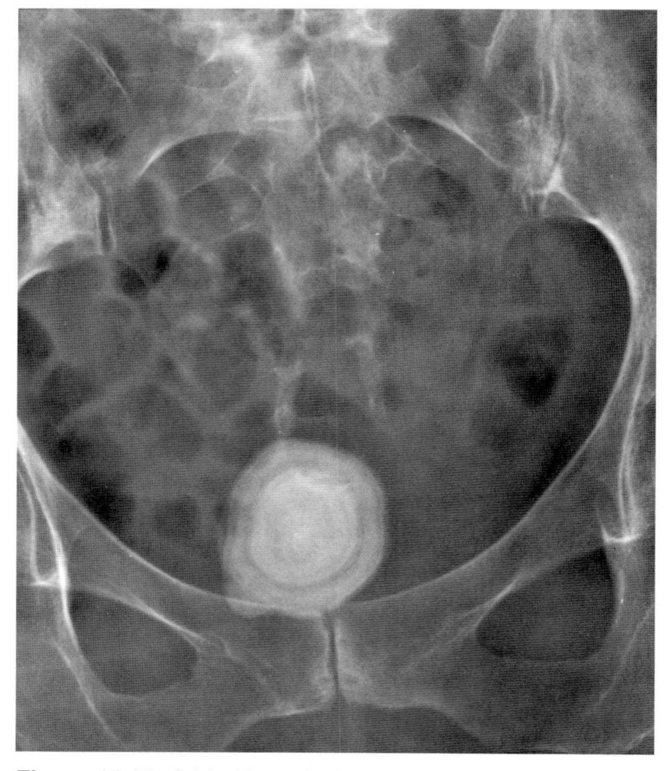

Figure 11.23. A bladder calculus present for some time was finally removed via the suprapubic route. Note the lamination.

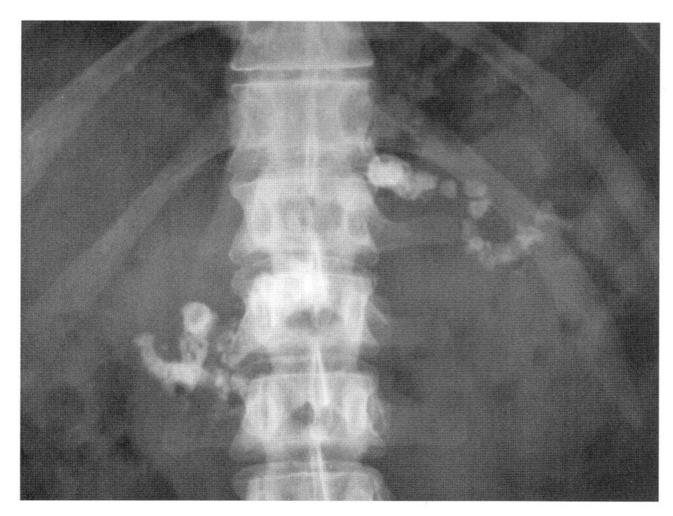

A

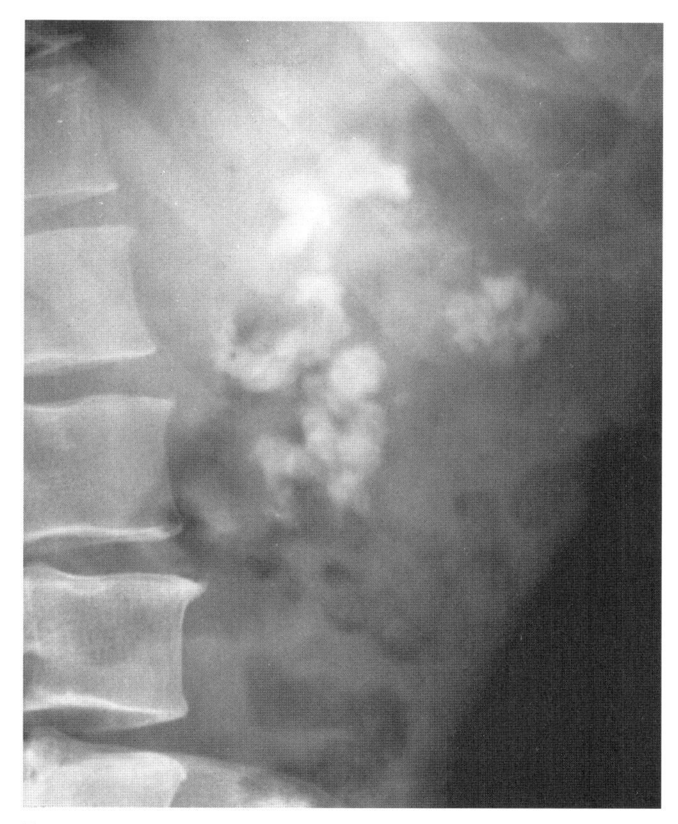

B

Figure 11.24A and B. AP and lateral films of an alcoholic patient with chronic pancreatitis. Note the irregular calcifications in the pancreas, seen overlying the gastric air shadow on the AP projection and anterior to the spine on the lateral projection.

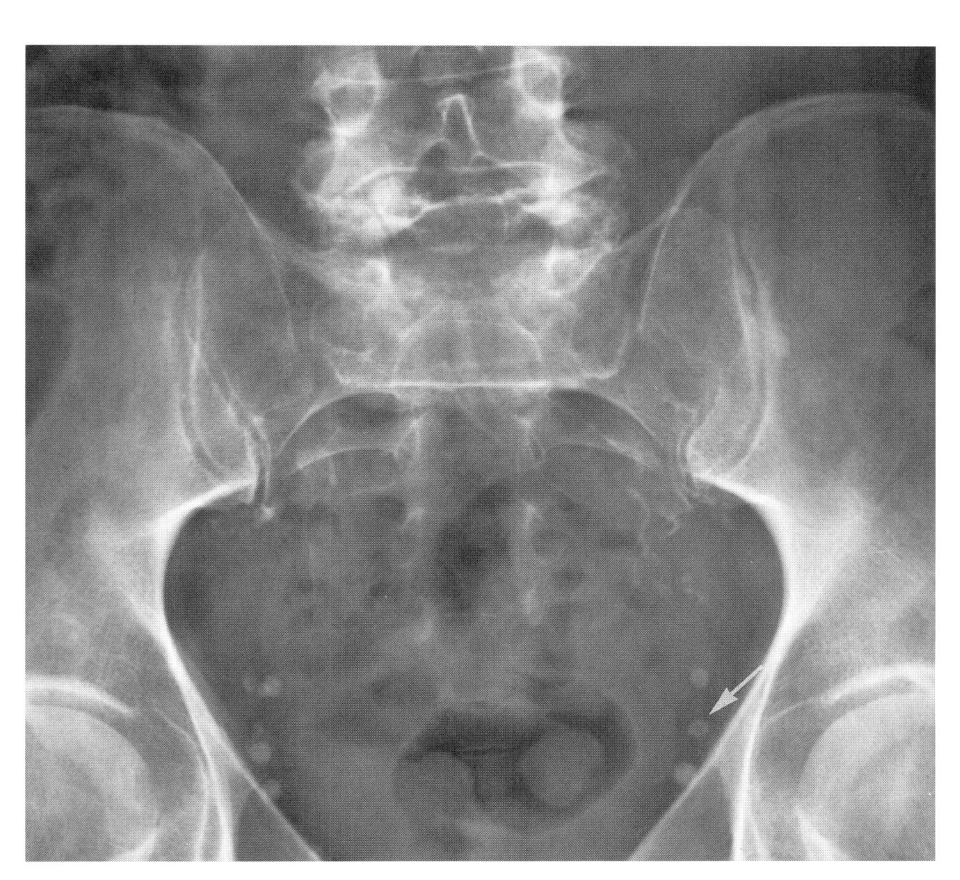

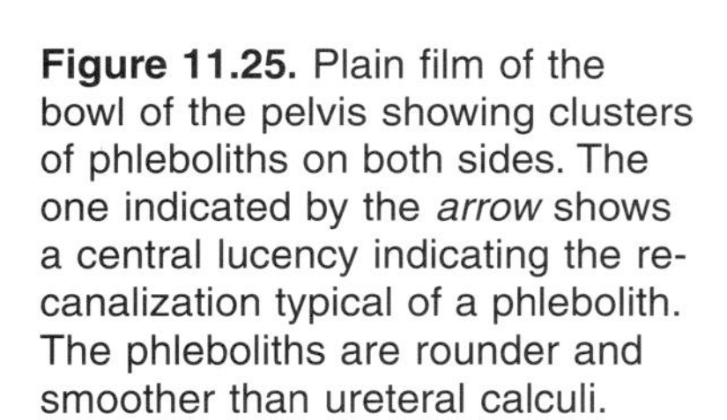

Figure 11.25. Plain film of the bowl of the pelvis showing clusters of phleboliths on both sides. The one indicated by the *arrow* shows a central lucency indicating the recanalization typical of a phlebolith. The phleboliths are rounder and smoother than ureteral calculi.

Systematic Study of the Plain Film

The plain film of the abdomen is important because it is simple to obtain, involves no discomfort for the patient, and can be immensely informative. So much can be learned, in fact, from plain films of the abdomen in so many different conditions that every physician should be familiar with them.

There are many subtleties in the interpretation of abdominal plain films, to be sure, and it is easy to feel, when you first begin to look at them, that you are missing important and obvious findings. At your stage of learning, an orderly manner of approaching the analysis of an abdominal plain film is strongly recommended. We suggest that you make a practice of *looking first at the bones* on a plain film of the abdomen (vertebral column, lower ribs, pelvis), excluding from your mind's eye all other structures. (If you do not look at the bones first, you will almost certainly forget them later.)

Then examine carefully the *soft tissues* of a series of smaller areas, including the left upper quadrant, right upper quadrant, both flanks, midabdomen, and pelvis, in that order. In each soft-tissue zone you should *check border indicators, organ masses, and fat lines, looking for calcification and for any shift in position or change in shape of the structures you see and identify.*

Then check out the gastrointestinal tract, accounting for all parts of it in order, ascertaining whether they are distended with gas or contain only a small amount, and recognizing some parts of the colon by their content of solid or semisolid feces.

Finally, decide whether there are any gray soft-tissue shadows or radiolucencies not yet accounted for in your survey, and if there are, try to reconcile them with the patient's history and physical examination.

Begin with the spine . . .

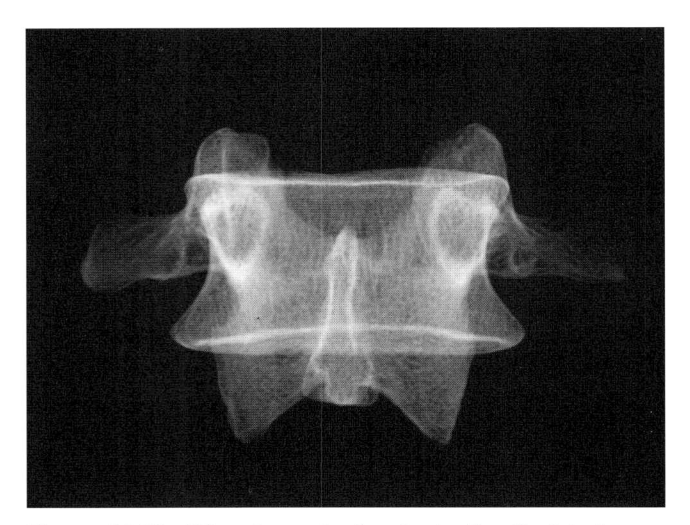

Figure 11.26. AP radiograph of a single disarticulated vertebra.

The shadows of the *lumbar vertebrae* may appear very confusing when you first look at them, but they will be easy to comprehend and remember once you have analyzed them part by part. To begin with, the box-like body of a vertebra, extending anteriorly, would have a very simple radiological structure if it could be seen by itself and not superimposed on the complex posterior articulating processes. Because it is literally a flat cylindrical box of dense compact bone filled with spongy bone, you would expect it to radiograph with a shell of tangentially seen—and therefore denser—bone outlining it, and an interior of many superimposed slender white trabeculae with dark marrow spaces between them.

Add to this box outline, now in the AP projection, the two pedicles, cylinders of compact bone extending straight backward on either side of the spinal canal. Because they too are filled with spongy bone, they will radiograph as cylinders seen end-on and, as you can predict, will appear on the film as two white circles. Now as you look at the vertebrae on any plain film of the abdomen, you can account for the two "eyes" that you see superimposed on the upper part of each vertebral body.

The centrally located white teardrop is the compact bone investing the spinous process, also filled with spongy bone. The pairs of superior and inferior articular processes are also to be seen as wings of bone extending upward and downward from each vertebral

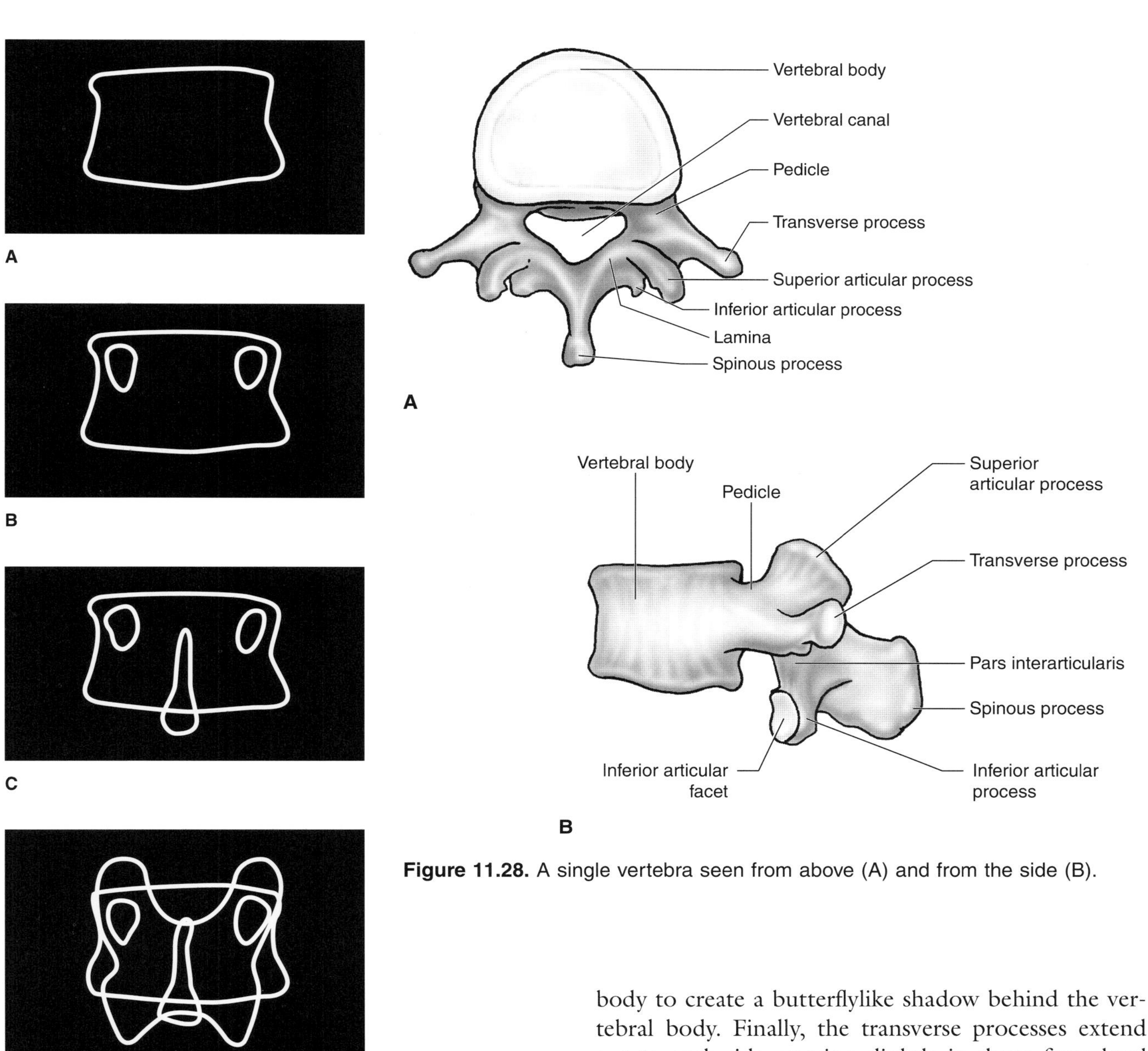

Figure 11.28. A single vertebra seen from above (A) and from the side (B).

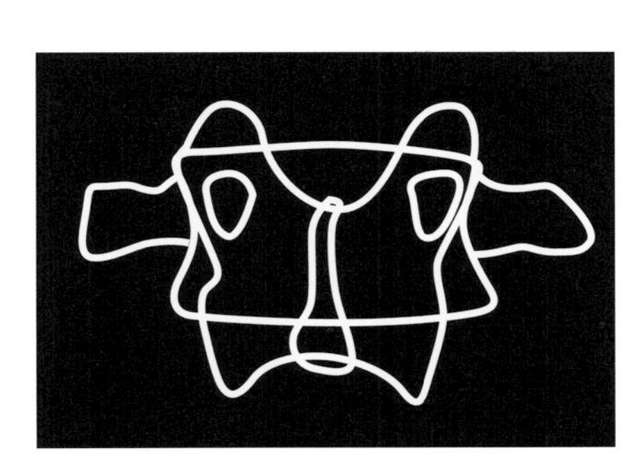

Figure 11.27

body to create a butterflylike shadow behind the vertebral body. Finally, the transverse processes extend out to each side, varying slightly in shape from level to level.

If you always think first of the outline of the vertebral body as you look at the spine on an abdominal film, and then add the posterior structures one by one, you will not be confused by the jumble of overlapping bony parts. A pathological process that destroys any part of these bony structures will cause the disappearance of a shadow you are now expecting to see and can usually trace because of its symmetry with the same structure on the other side, or above or below at a different level.

The bilateral symmetry of the posterior structures superimposed upon the vertebral body will be useful to you in another way, because it tells you that the ray passed sagittally through the patient. Routinely, plain

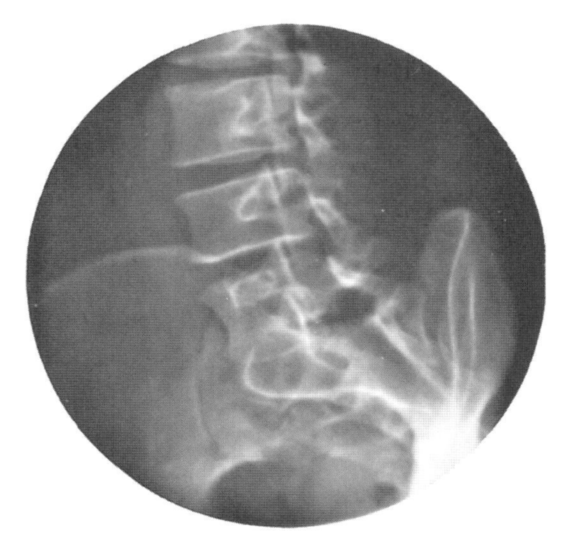

Figure 11.29. Oblique view of the lower lumbar vertebrae.

films of the abdomen are made AP with the patient supine, but some barium studies are made PA with the patient prone. (Why?) Many types of special procedures are performed in conventional degrees of obliquity, and you will find that the appearance of the vertebrae indicates the obliquity of the ray.

Thus if you see the boxy bodies of the vertebrae cleanly separated from their posterior structures, you will know that you are looking at a *lateral* film; but if the bodies and posterior structures are precisely superimposed and bilaterally symmetrical, you are looking at a film made with a sagittal ray. *Oblique films* will show you the vertebrae about as you see them in Figure 11.29, and note that now you can see through the obliquely directed posterior articulations.

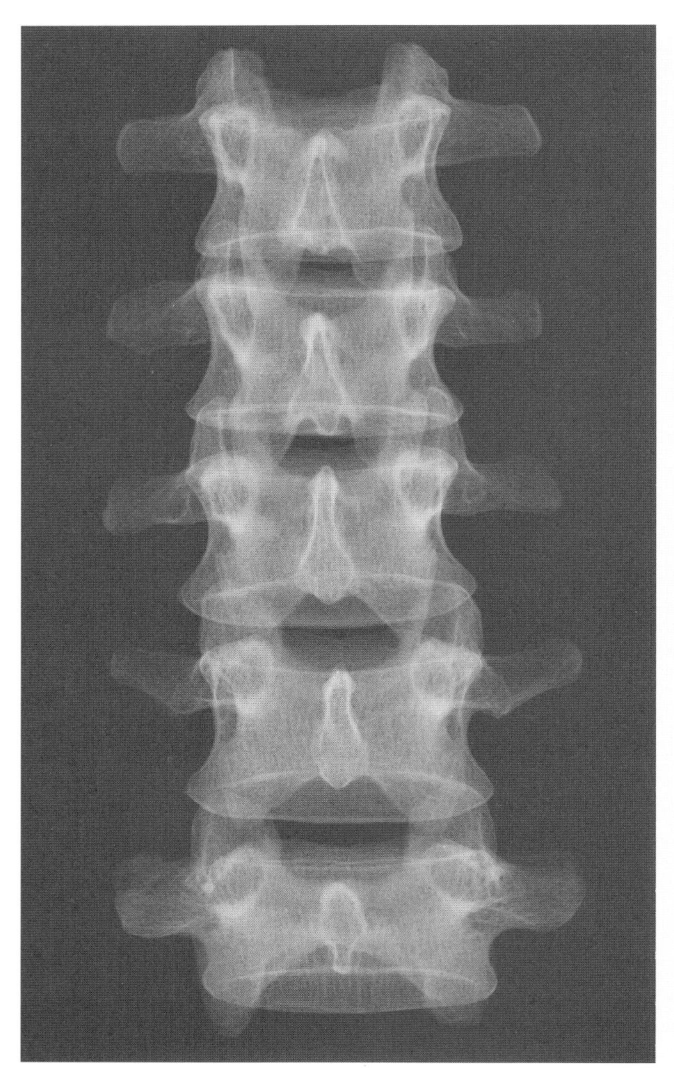

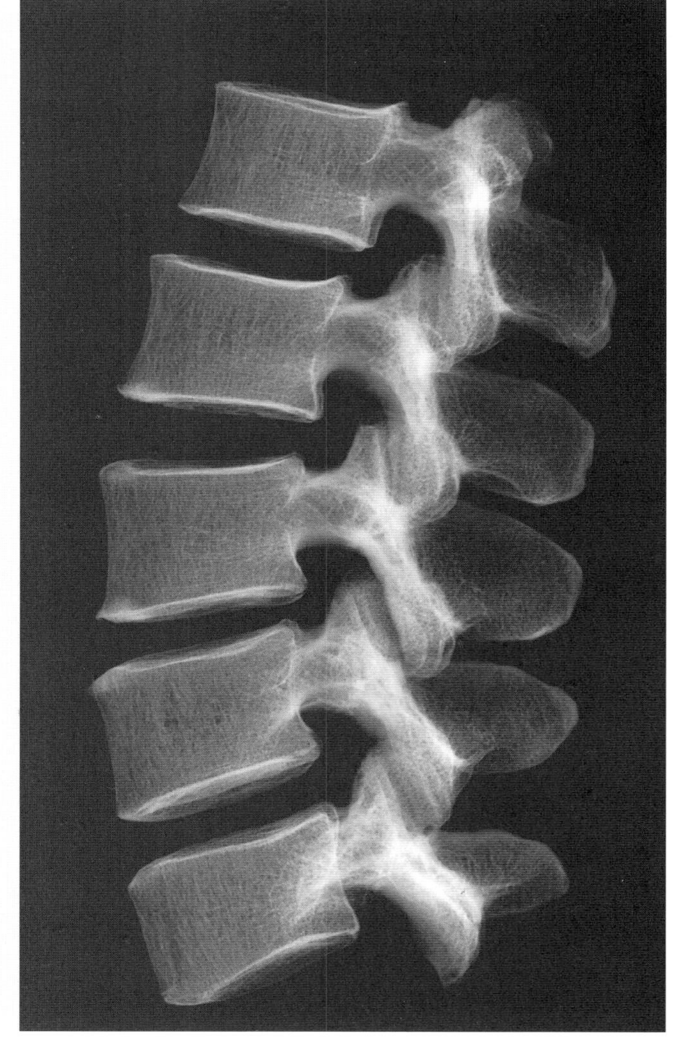

Figure 11.30. AP and lateral radiographs of a disarticulated specimen of lumbar vertebrae. Without turning back to the previous page, identify the various parts.

. . . then study the ribs . . .

All the abdominal films you see will have been made with a Bucky grid, and you will observe that you can see the ribs below the diaphragm much better than you do in chest films.

Calcifications in the costal cartilages, which normally are radiolucent and invisible, may cause some confusion when they are seen superimposed on intra-abdominal calcifications within the gallbladder, kidney, or adrenal gland. Costochondral calcifications can usually be distinguished by tracing the expected course of the rib cartilages anteriorly.

. . . and the pelvis and upper femora . . .

The shape of the shadows of the bones of the *pelvis* differs when the ray passes through PA and when it passes through AP, as you can anticipate if you think of the flared and tilted wings of the ilium. These are "flattened out against the film," appearing round and wide on a supine plain film but narrow and more vertical on films made with the patient prone. They look this way because the ray that passes through the patient PA is much more nearly tangential to the surface of the iliac wings, so that they are more approximately filmed on edge.

The ray is usually centered on the umbilicus in making a plain film of the abdomen, so that roughly half the air-in-gut shadows are below this point. You will therefore expect to see (superimposed on the bones of the pelvis and sacrum) the air in the cecum, sigmoid, and rectum, as well as in the small-bowel loops when they do contain air. A loop of air-containing bowel overlying the iliac wing on a supine plain film is often very difficult to differentiate from a round area of bone destruction, and the procedure is to examine several films of the area: small-bowel air changes in shape and location from film to film, but an area of bone destruction remains in exactly the same relation to the margins of the bone in which it is present (Figure 11.31).

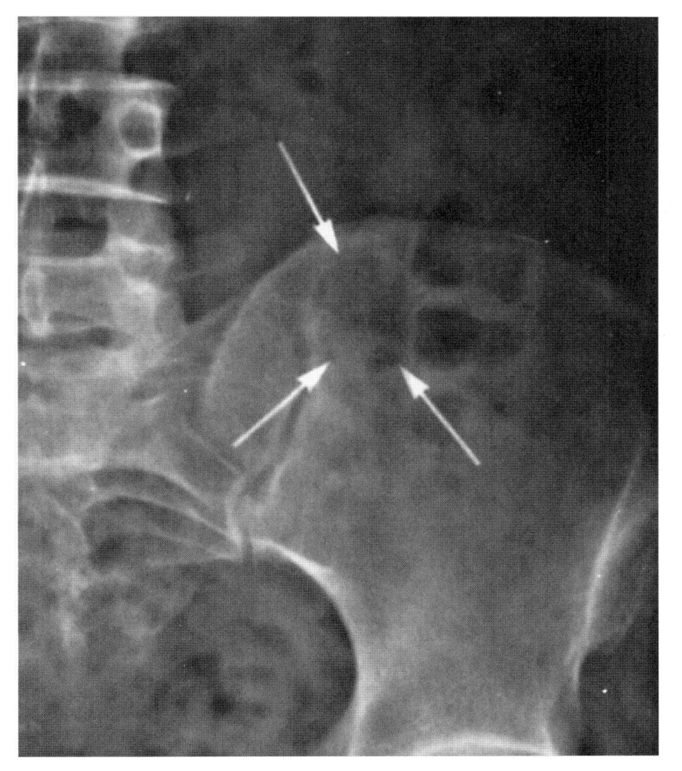

Figure 11.31. The radiolucent area of bone destruction indicated by the *arrows* did not change from film to film in relation to the margin of the sacroiliac joint. The darker air shadows lateral to it, representing gas in bowel, did change.

Note in Figure 11.32 that you see through the cartilaginous part of the anterior portion of the sacroiliac joint and through the symphysis pubis. Identify the spines and tuberosities of the ischia, and note that the hip joint is "seen" because it is bounded on both sides by the cortical bone of the acetabulum and femoral head, seen in tangent. Later, when you have read Chapter 15, you will look at the bones on a plain film of the abdomen with a more precise eye for abnormalities, but for now leave them and go on to a study of the series of soft-tissue zones.

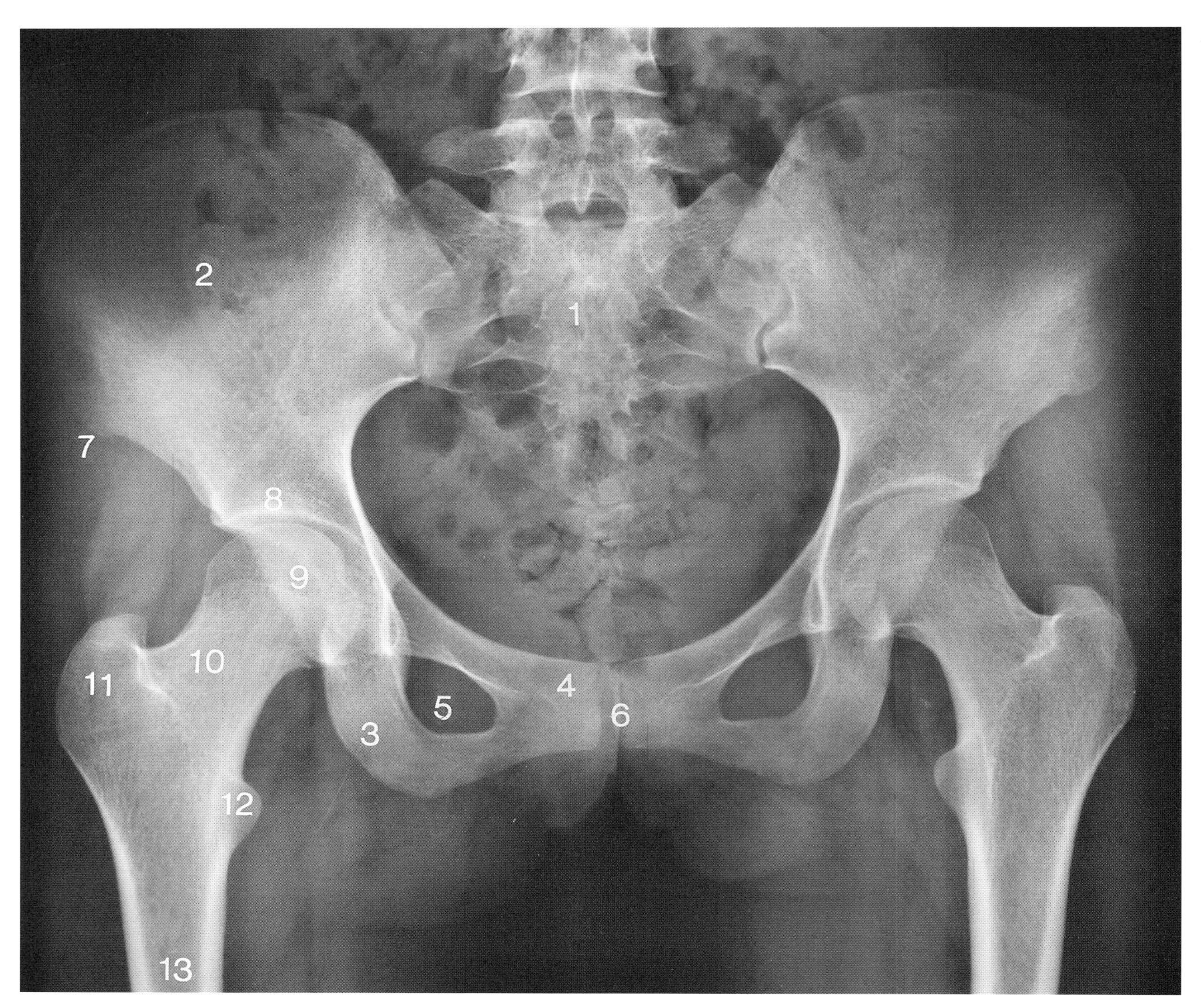

Figure 11.32. AP radiograph of the pelvis and upper femurs.
(1) Sacrum
(2) Ilium
(3) Ischium
(4) Pubis
(5) Obturator foramen
(6) Symphysis pubis
(7) Anterior inferior iliac spine
(8) Acetabulum
(9) Femoral head
(10) Femoral neck
(11) Greater trochanter
(12) Lesser trochanter
(13) Femoral shaft

. . . then search the upper quadrants, flanks, and midabdomen for organ masses and calcifications . . .

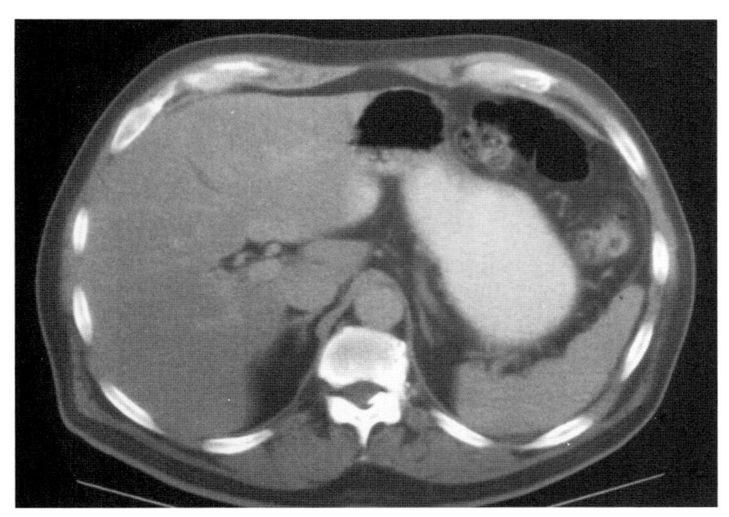

Figure 11.33. CT scan through the liver and spleen showing normal-sized organs. The gastric lumen has been opacified with oral contrast material.

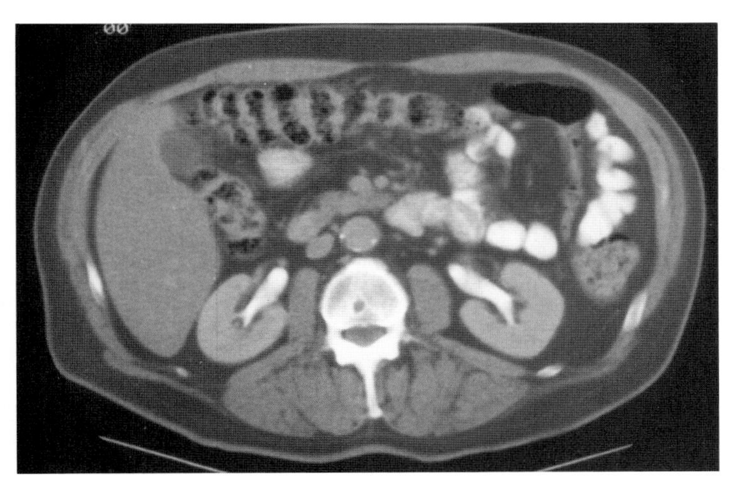

Figure 11.34. CT scan at a lower level through the kidneys shows how the right lobe of the liver decreases in size inferiorly. Oral contrast material can be seen in the bowel and intravenous contrast material has opacified the renal collecting systems (renal pelves and calyces).

Abnormality in size or shape of an organ is frequently evident from the plain film alone, but there are some differences in the degree of accuracy with which size may be judged by x-ray. The radiological shadow of the liver is very misleading as an index to its size, for example, and it must be grossly enlarged before you can assume hepatomegaly from the plain film. Partly because of its shape and partly because of variation in the tilted position of the liver within the abdomen, pronouncements with regard to liver enlargement based on the plain film are risky. You will find that as an assay of liver size, old-fashioned palpation is a more reliable method. If you are still in doubt after physical examination of the patient, request CT or MR scanning, both of which provide an accurate measurement of liver size.

The *liver*, so much larger than the spleen, normally tends to depress the organs in the right upper quadrant. Its margin may be seen either as the inferior limit of a gray mass or as a boundary outlined by air in the transverse colon and hepatic flexure of the colon. The hepatic flexure is usually lower than the splenic flexure, but may occasionally overlap a part of the liver shadow. CT, MR, and ultrasound techniques are becoming increasingly practicable for routine use in studying the liver, and they also can easily localize metastatic and other lesions within the liver parenchyma.

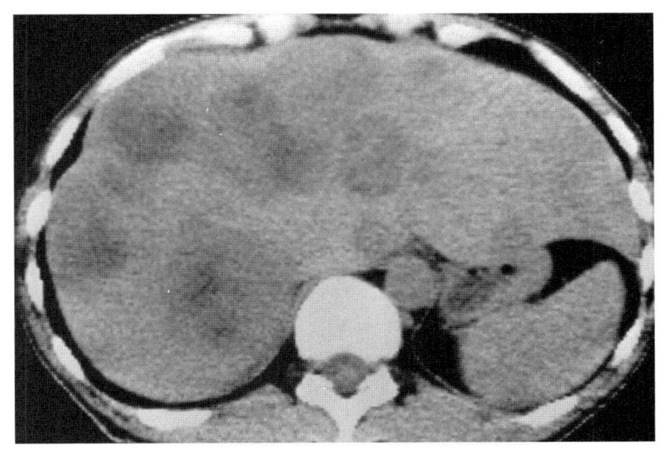

Figure 11.35. CT scan of a grossly enlarged liver, with numerous metastases from carcinoma of the prostate seen as multiple hepatic masses of decreased density (low CT attenuation, compared with the normal hepatic parenchyma).

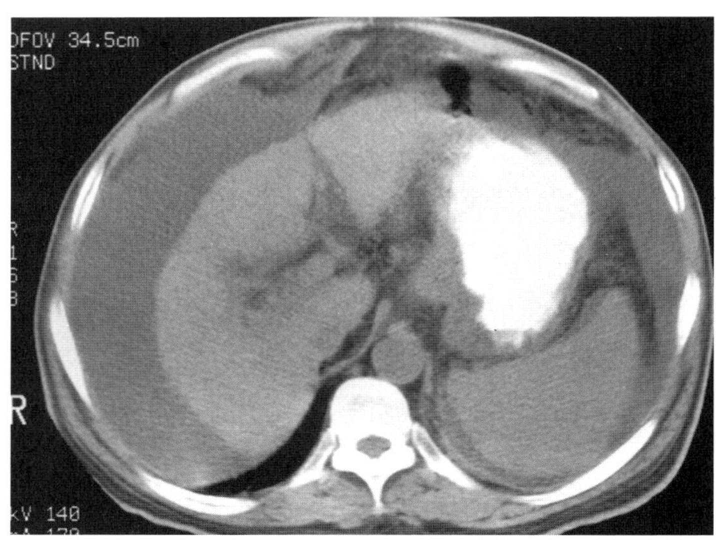

Figure 11.36. CT scan of an alcoholic patient with cirrhosis, splenomegaly, and ascites. Note the shrunken appearance of the liver and the large volume of ascites in the peritoneal cavity surrounding the liver, stomach, and spleen.

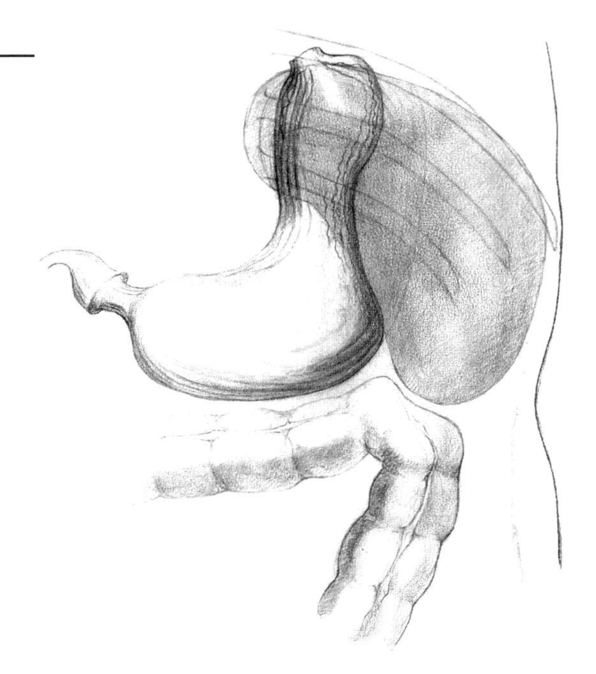

Figure 11.37. The normal stomach indented by an enlarged spleen, which depresses the splenic flexure of the colon. It is shown as it would look if you could see all three structures.

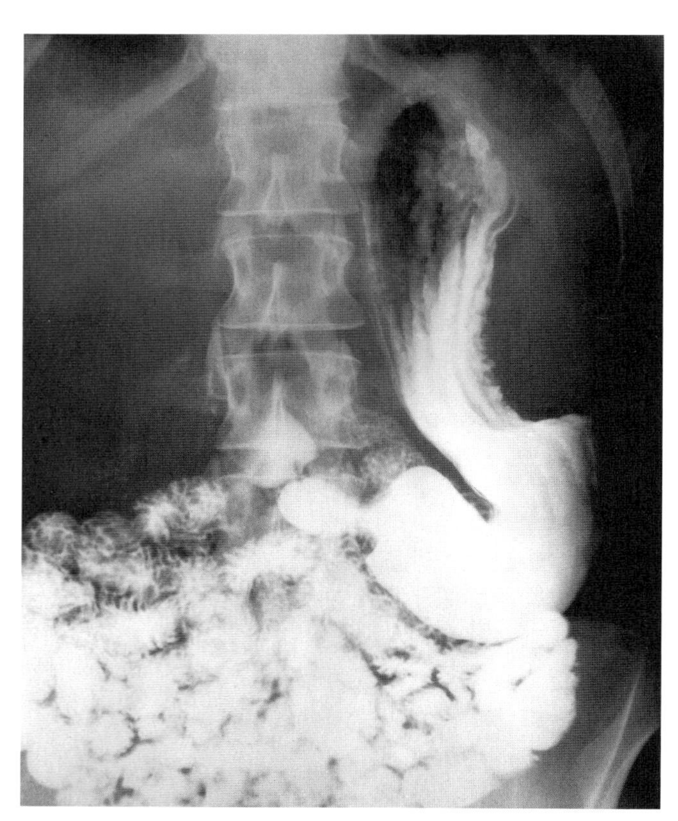

Figure 11.38. The normal stomach filled with barium and indented by an enlarged spleen.

The *spleen* may cast a shadow on the plain film that is unquestionably increased in size. You should learn to place a good deal of reliance on a radiological impression of splenomegaly. A very large spleen is not difficult to recognize and may reach well below the iliac crest and across the midline. A plain film suggesting splenomegaly may be confirmed today with CT and other imaging methods, some of which may also identify pathological conditions such as splenic trauma, neoplasm, abscess, or cyst.

The *splenic flexure* of the colon is quite variable in position. It may be indented by the tip of the spleen, overlap it partially, or extend over it as high as the diaphragm. It should not be hard for you to identify in such cases, for it will have the characteristic smooth haustral indentations of the colon as opposed to the crinkled margin of the stomach shadow.

The *stomach* itself is almost never difficult to identify. In the supine film, the air in the stomach rises into the more anteriorly placed body of the stomach, outlining the heavy rugal folds of gastric mucosa. In the prone film, whatever air is present rises into the posteriorly placed fundus, so that on a prone plain film made with a sagittal ray the air bubble of the stomach appears as a round dark shadow with a wrinkled margin nearer the diaphragm. These differences, in fact, will also help you to decide whether a film has been made supine or prone. The same principles apply to barium studies of the stomach, as you will see later.

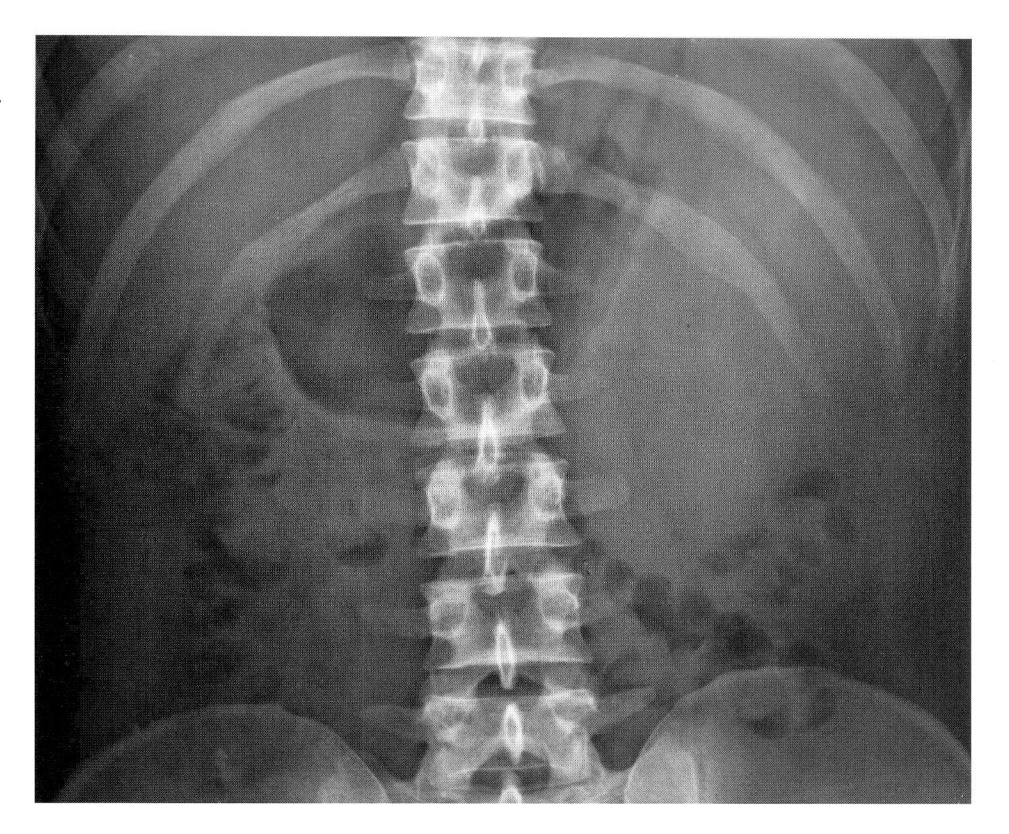

Figure 11.39. A ruptured spleen surrounded by a large perisplenic hematoma that is displacing the normal stomach to the right and extending down the left flank. Note the coarse edematous rugal folds of the stomach, which probably shared in the trauma. Was this film made supine or prone?

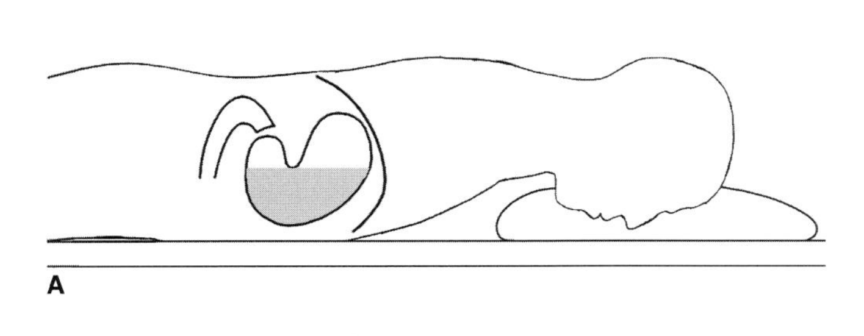

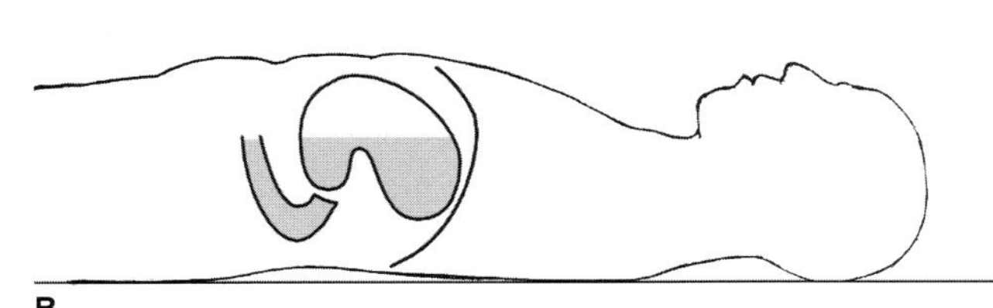

Figure 11.40. Explanation of the difference in the stomach air bubble on plain films made with the patient prone *(above)* and supine *(below).* On this page spread Figures 11.38 and 11.41 were made prone with the fundus filled with air. Figure 11.39 was made supine.

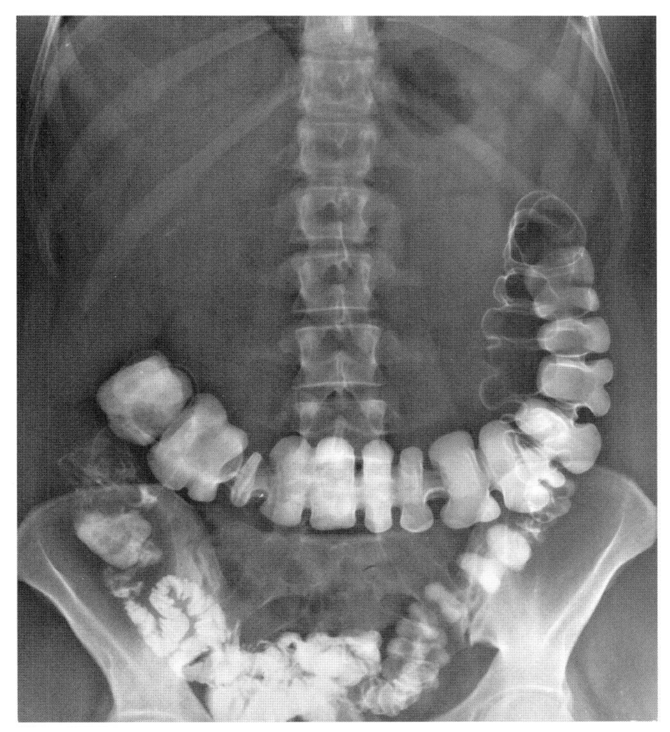

Figure 11.41. Identify the splenic and hepatic flexures and the tips of the liver and spleen. Neither the liver nor the spleen was felt manually to be enlarged.

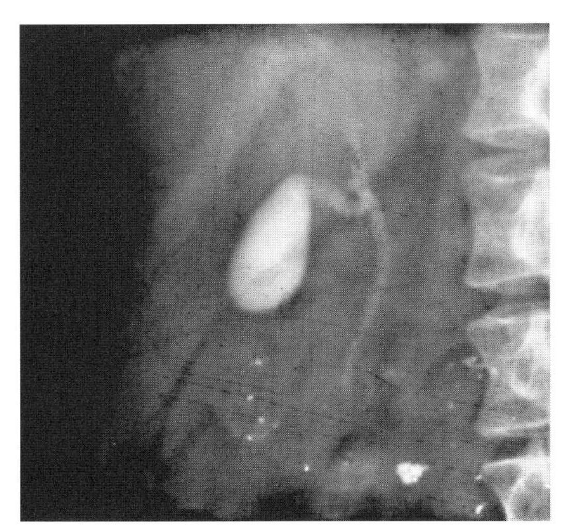

Figure 11.42. The normal gallbladder filled with physiologically concentrated opaque material (oral cholecystogram, or OCG). A fatty meal has just been given and the gallbladder is contracting, so that the cystic duct, common bile duct, and, by reflux, a part of the common hepatic duct are seen containing contrast material. Scattered flecks of dense white are leftover (unabsorbed) opaque material in the bowel.

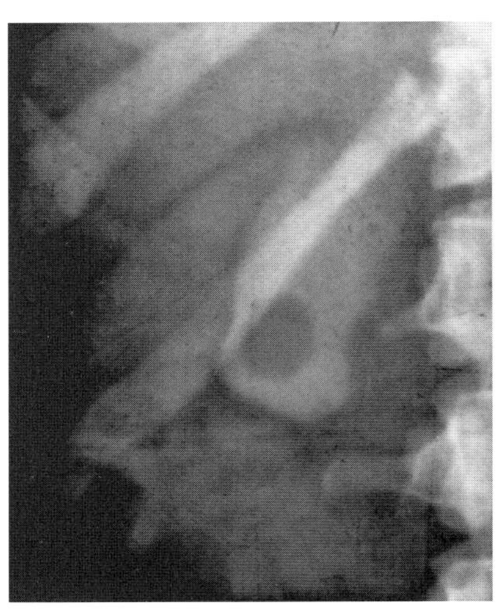

Figure 11.43. Cholecystogram showing a concentration of opaque material and a radiolucent stone composed of low-density material. The stone was invisible on the plain film.

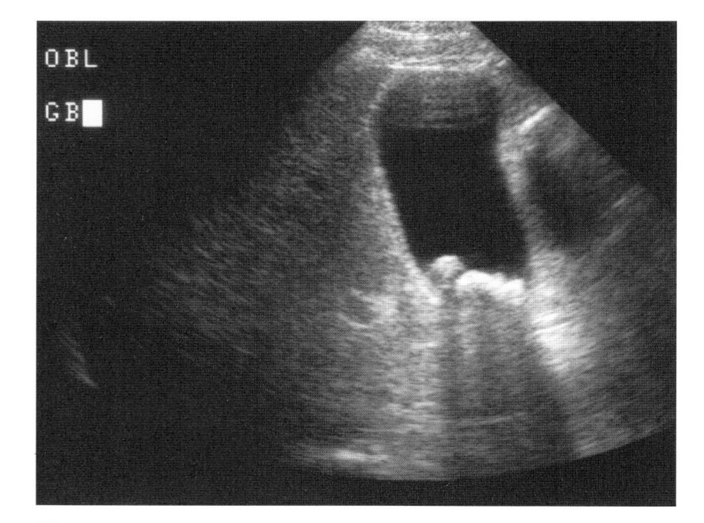

Figure 11.44. Ultrasound showing a fluid-filled gallbladder, on the dependent posterior wall of which rest multiple stones. The sound beam enters through the anterior abdominal wall of this supine patient *(top of scan).* No sound waves (echoes) are reflected from the bile, so it appears solid black (anechoic). Echoes are reflected by the stones, hence their white (echogenic) appearance. Note the distinct acoustic "shadows" extending posteriorly from the stones as areas that reflect little sound, because most of the ultrasound beam has been stopped and reflected by the stones.

The gallbladder is usually not seen on abdominal plain films, although its shadow may occasionally be seen as a rounded shadow superimposed on the liver margin and the right kidney. It had for many years been successfully *rendered* visible by administering, by mouth, radiopaque compounds that are excreted by the liver. These materials do not concentrate sufficiently within the liver cells to visualize them, but the normal gallbladder, by extracting water and concentrating bile, also concentrates the excreted opaque substance so that contrast visualization of the gallbladder is possible. This is called *oral cholecystography* (OCG).

Only about 15 percent of gallbladder stones are shown to be calcified on plain films. *The rest are radiolucent and thus invisible on a plain film,* an important fact for you to remember. During cholecystography, in gallbladders that are still capable of concentrating the opaque substance excreted by the liver, such common radiolucent stones appear as "filling defects," radiolucent relative to the opacified bile. Unfortunately, poor concentration in chronic cholecystitis has in the past rendered the visualization of such stones difficult or impossible.

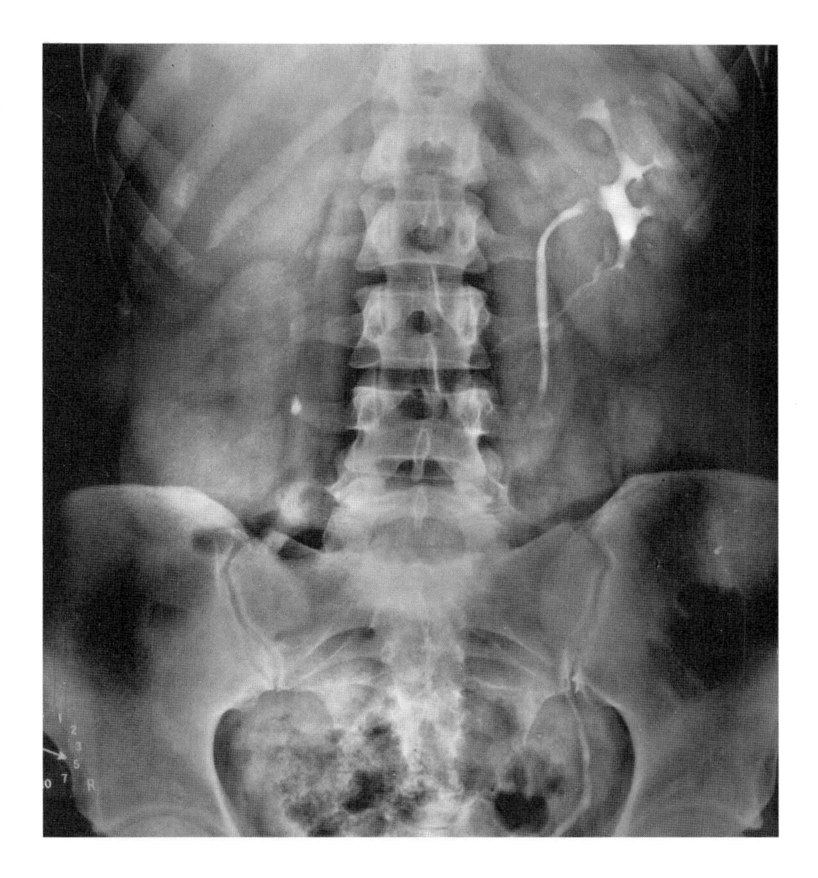

Figure 11.45. Intravenous urogram of a patient who presented with right flank pain, hematuria, and a history of passing stones. The right kidney parenchyma is denser than that on the left because the fluid-and-contrast flow through this kidney is delayed by the obstructing dense calculus in the upper ureter. The right kidney is also ptotic. Note the discontinuous visualization of contrast material in the left ureter. This is normal, since the ureter is swept by waves of peristalsis.

Today the investigation of the gallbladder and biliary tree by ultrasound and CT is so successful that cholecystography is rarely used. Ultrasound of the gallbladder and biliary tree is discussed later and in Chapter 14.

Farther posterior on either side of the spine are situated the *kidneys,* the left a little higher than the right in most patients and retroperitoneal. The upper poles of both are tilted toward the midline, against the psoas muscles. You should trace their outlines as completely as possible. You will have some difficulty outlining the kidneys on a good many plain films because they are nearly always partly obscured by varying amounts of gas and stool in the bowel. If you see no kidney shadow at all on a film of good quality, without too much overlying gas, that may mean that the kidney is absent, very small, or has little or no perirenal fat, for example.

The kidneys, close to the film in the supine patient and outlined by fat, are much better described by the plain film evidence than they can ever be on physical examination, and you will discover that a difference in size of the two kidneys, for example, or a bulge in the kidney outline, is repeatedly brought to your attention by the radiologist. A difference in length may have significance as an indication of disparity in function. Normally the kidney length should be 3.7 times the height of the second lumbar vertebra (L2) in a growing child; in the adult the kidney should measure not less than 9 centimeters and not more than 15 centimeters in length. A parenchymal renal cyst or tumor enlarges the kidney by ballooning out the pole in which it is present, pushing the perirenal fat before it. Ptosis is common and easily recognized when you find the kidney farther down the psoas shadow than is normal. Any rotation of the kidney about its transverse axis alters the shadow it casts.

The kidneys are studied specifically by the use of renally excreted radiopaque contrast materials (the intravenous urogram) or by injecting similar contrast substances through catheters placed in the ureters during cystoscopy (the retrograde pyelogram). Radiographic study of the kidneys has been revolutionized by ultrasound, CT, angiography, and other special techniques that will be discussed in more detail in Chapter 14. Your concern here is with morphology: learning to interpret changes in the size, shape, and position of organs that you can identify on the plain film.

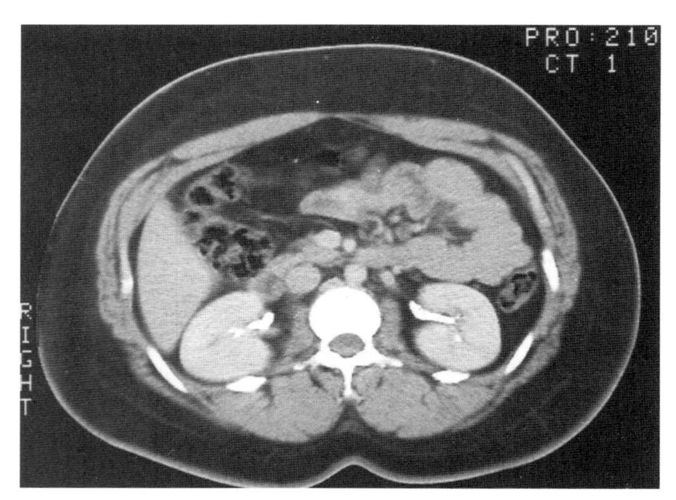

Figure 11.46. CT through the midkidneys (in a different patient). Intravenous contrast has been given. Note the enhanced density of the renal parenchyma and the concentrated contrast material in the renal pelves leaving the anteromedially directed hila of the kidneys. They are, of course, lying in the retroperitoneal compartment surrounded by lucent perirenal fat. Note the psoas muscle masses against the vertebral body and the aorta just anterior to it.

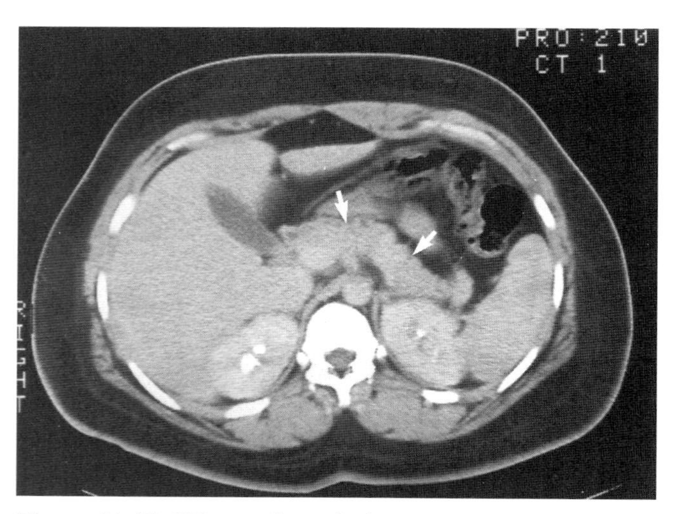

Figure 11.47. CT scan through the pancreas *(arrows).* Note that the kidneys are sectioned through their upper-pole calyces above the indentation of parenchyma at the hilum of the kidney. (Compare with those in Figure 11.46.) The gallbladder is seen as a relatively lucent area against the liver.

The *pancreas,* its head encircled by the loop of the duodenum, lies below and behind the antrum and body of the stomach, curving over the bodies of the upper lumbar vertebrae as they project forward into the abdomen. The tail of the pancreas is directed toward the hilum of the spleen. It is of course retroperitoneal and not visible on plain films, unless it contains calcifications and stones scattered through it to outline and mark its position (as in Figure 11.24). It may be indirectly bounded by other organs that normally lie close to it, and a large mass in the head of the pancreas, for example, may expand the duodenal loop and displace forward the pyloric antrum of the stomach. The pancreas may be imaged with ultrasound, CT, endoscopic retrograde cholangiopancreatography (ERCP), angiography, or MR, as you will learn later.

Your special survey of the upper *midabdomen,* then, includes some structures that you have "looked for" already, since only an arbitrary division can separate those structures that lie in the upper right and left quadrants from those that, like the pancreas, lie partly also in the midabdomen. Think of the midabdominal structures three-dimensionally, beginning with those most anteriorly placed. The *body of the stomach,* with its J-shaped streak of air in the supine patient, lies anteriorly against the abdominal wall and just above the curve of the air-containing *transverse colon.* The pyloric antrum and duodenal bulb (or cap) turn and point posteriorly, so that they are best seen in the lateral view. The bulb and descending limb of the *duodenal loop,* which is partly retroperitoneal, turn downward, passing around the head of the pancreas, to the left and upward again toward the ligament of Treitz to join the jejunum. The duodenum is generally fluid filled and invisibly merged on the plain film with the shadows of other solid or fluid-containing structures near it, although you will see the duodenal loop regularly on barium studies (covered in Chapter 13) and from it construe the position of the always invisible pancreas.

. . . and finally examine the flanks and lower abdomen.

Examine the *flanks* next, on both sides of the abdomen. The flank stripes may be obscured by the intense black of this area, which is often "burned out" at exposures calculated to penetrate the bones of the pelvis and spine. Even so, the flank stripes may often be seen well if you place the film against a bright light, kept handy by radiologists for the illumination of dark areas on films. The flank stripes are usually symmetrical bilaterally in thin patients who have been positioned carefully. The dark haustrated colon should be seen lying close against the flank stripe. Free peritoneal fluid or blood may make the flank bulge out and will separate the colon appreciably from the flank stripe. In the presence of inflammation nearby, the flank stripe on that side will be seen to be smudged because of edema of the fat, rendering it indistinguishable from other water-dense soft tissues, such as muscle.

Study the *lower midabdomen* next; follow the known course of the ureters along the psoas shadows down to the bladder, looking for any shadow that could represent a calculus. The ureters are invisible on plain films, but their shadows outlined with contrast material on urograms will help you learn the variations in their position. Many plain films show small, round, calcium-dense shadows just inside the brim of the bony pelvis. These are calcified thrombi in the pelvic veins, or phleboliths, and generally lie closer to the margin of the bowl of the pelvis than any part of the ureter. Their position, then, will help you distinguish them from calculi, and you will also be aided by the fact that calculi may be any shape at all and often have jagged points in their shadow profiles, whereas phleboliths are invariably smooth and round and sometimes show a central radiolucency, like beads ready for stringing (see Figure 11.25).

The lower midabdomen is a common location for calcified mesenteric nodes, and they will not usually be mistaken for ureteral stones because they tend to look like clusters of small concretions and are somewhat denser than most stones. They vary widely in position from film to film as a rule, because of the mobility of the small-bowel mesentery. They often overlap the bony structures of the vertebral column and sacrum.

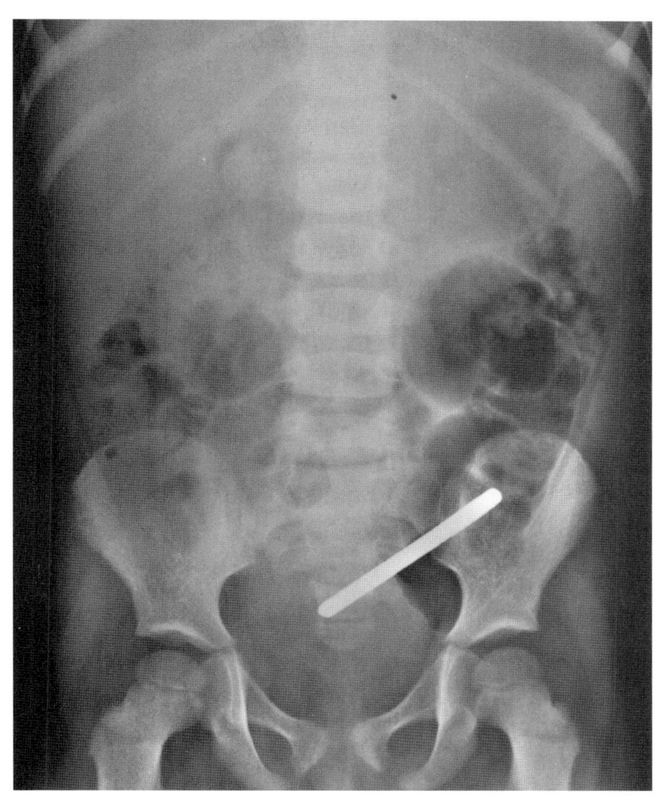

Figure 11.48. Plain film of a child admitted with fever, constipation, periumbilical pain, and an elevated white count. The clinical working diagnosis was appendicitis. At surgery the appendix was found to be retrocecal and perforated. There was inflammation and edema of the tissues of the right gutter and right flank. Note the well-seen left and the absent right flank stripe. (The white bar is an artifact.)

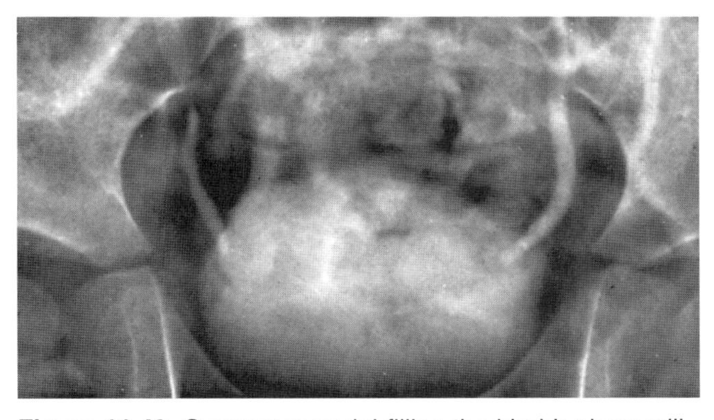

Figure 11.49. Contrast material filling the bladder here will help you recognize the shadow cast by the urine-filled bladder on plain films. Note the location of the ureters as they enter the bladder.

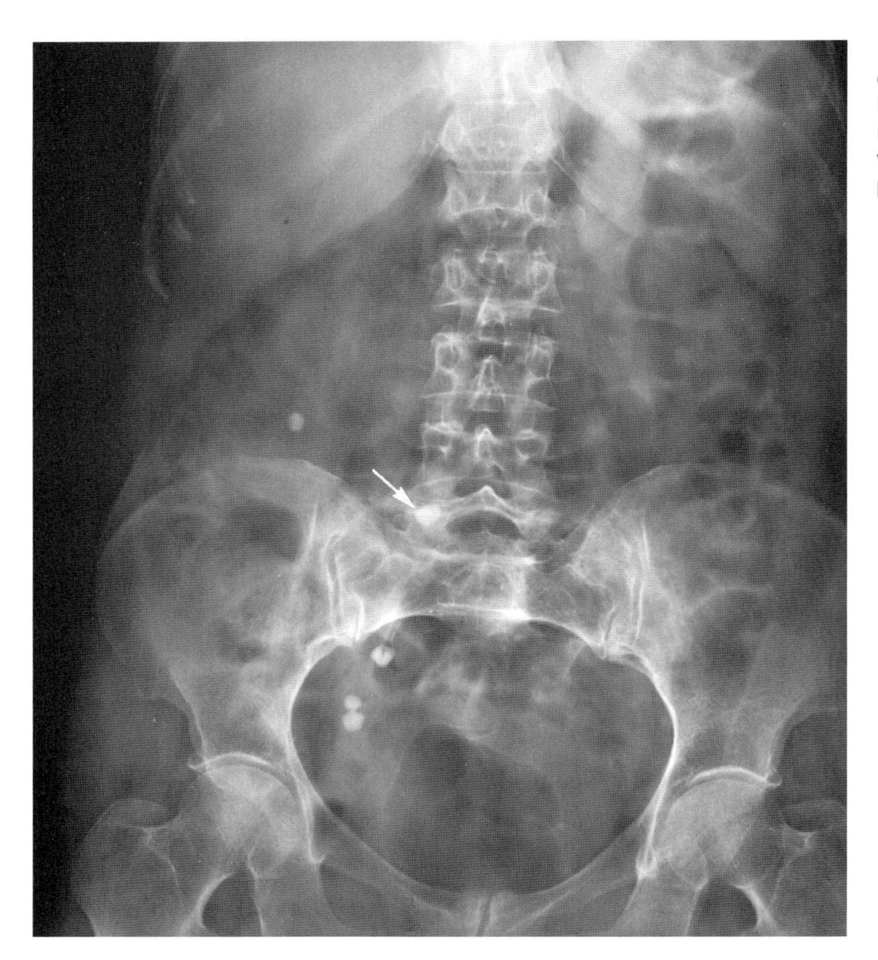

Figure 11.50. This plain film shows several calculi in the lower right ureter (*arrow* to the highest one) and a calcified mesenteric node farther lateral above the top of the iliac wing. Note the air-filled rectum superimposed on the bladder.

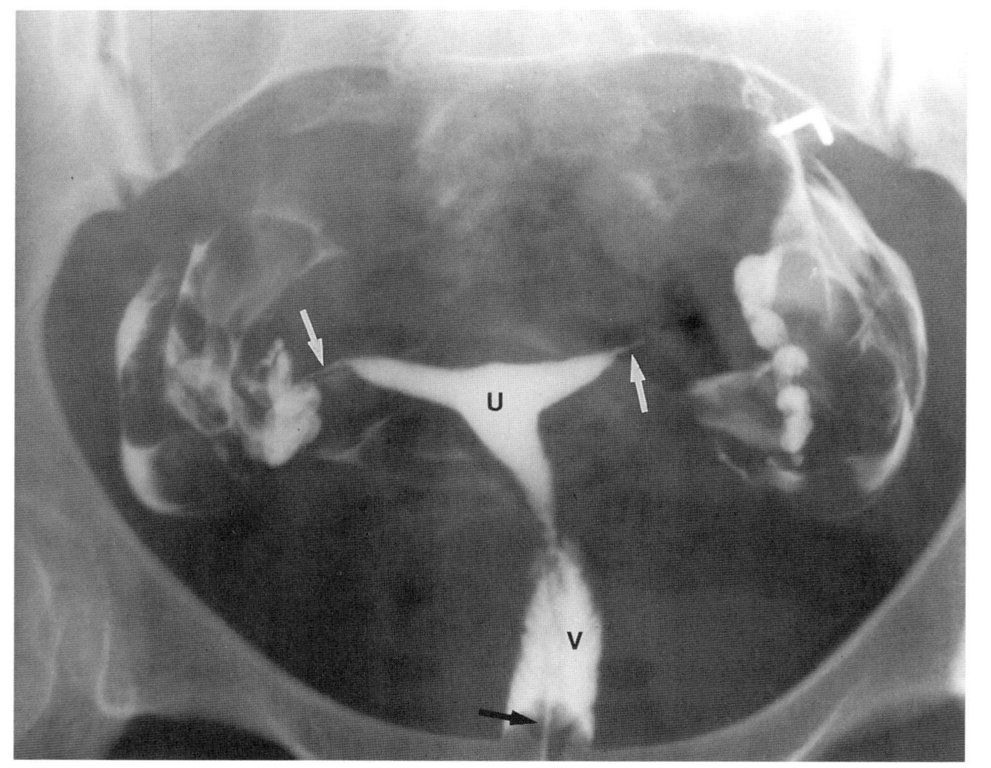

Figure 11.51. A hysterosalpingogram performed as part of a fertility workup. The uterus *(U)* and uterine (fallopian) tubes *(white arrows)* have been filled with contrast material injected via a cannula *(black arrow)* inserted into the uterine os. There is retrograde opacification of the vagina *(V).* Normally there is spill of opaque material from the uterine tubes into the peritoneal cavity, as shown here; this finding indicates patency of the uterine tubes. Intraperitoneal contrast material can be seen surrounding loops of bowel. The ovaries are not seen.

The many loops of small bowel are contained for the most part in the lower midabdomen and pelvis. In the normal ambulatory adult, as we have said, they contain fluid and little or no air; but since you will be seeing plain films mostly on patients sick enough to be hospitalized, you will become accustomed to seeing some air-outlined loops of small bowel overlying the lower lumbar vertebrae and pelvis.

The soft tissues within the bowl of the pelvis, finally, include the urinary bladder and lower ureters, the sigmoid colon and rectum, the uterus and adnexae in women, and the prostate and seminal vesicles in men. The bladder, when it contains a moderate amount of urine, is commonly visible on the plain film as a somewhat flattened oval shadow within the pelvis. When it is greatly distended, it may rise up to the umbilicus as a uniformly gray, rounded shadow, not infrequently mistaken for a pathological mass. The rectum is generally visible superimposed on the shadow of the bladder and outlined with contained air or feces. The uterus, adnexae, seminal vesicles, and prostate are not visible on plain films but can be imaged by ultrasound, CT, and MR.

CT of the Abdomen

As you discovered in Chapter 2, a computed tomogram provides you with a range of *density* values for a particular slice of your patient, which you will learn to study with the regional cross-sectional anatomy in mind. An awareness of the relative attenuation of different tissues and organs, and their interfaces with fat planes in particular, helps you as you look at CT scans. As you do so, you will be better able to understand anatomic relationships and proportions by knowing that those on CT scans are *living* dimensions rather than the necessarily distorted ones in the desiccated cadaver. There are also physiological implications in some of the observations you will make in serial CT scans. For example, in scans made across the upper abdomen the inferior vena cava can be seen to distend to twice its normal size during the Valsalva maneuver (in which the patient bears down against a closed glottis).

CT scans can be performed in the supine or prone position or with the patient lying on his side (decubitus). Usually, however, the patient is supine, because he is most comfortable and most relaxed in that position and can more readily keep still. Always remember: you are looking *up* at any CT scan from the patient's feet.

Contrast media are frequently used during abdominal CT scanning for two basic reasons:

1. The GI tract lumen can be opacified and thus fluid-filled loops of bowel can be distinguished from neighboring normal structures (and abnormal masses) by having the patient swallow dilute opaque material.
2. Intravenous administration of contrast material produces a temporary increase in the density of arteries, of all capillary-perfused parenchyma, and, finally, of peripheral veins during CT scanning. This is called *enhancement* and is extremely useful. Dilated low-density branching channels within the homogeneous parenchyma of the liver, for example, on an initial scan without contrast material could be either fluid-filled bile ducts or blood vessels. If the radiologist enhances the vascular bed with contrast medium, and if the observed channels are unchanged on a subsequent scan, they are bile ducts, not vessels. Intravenous contrast material is excreted renally, and hence the renal collecting systems and ureters will be opacified with the contrast material.

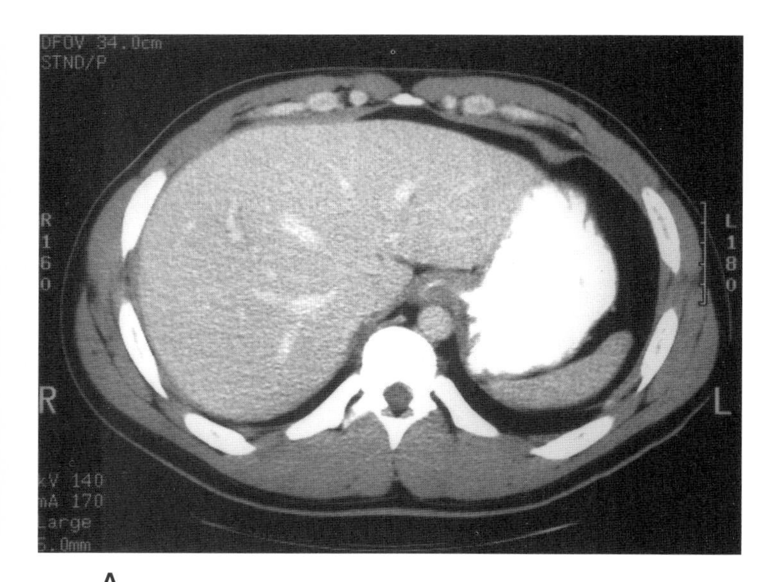

A

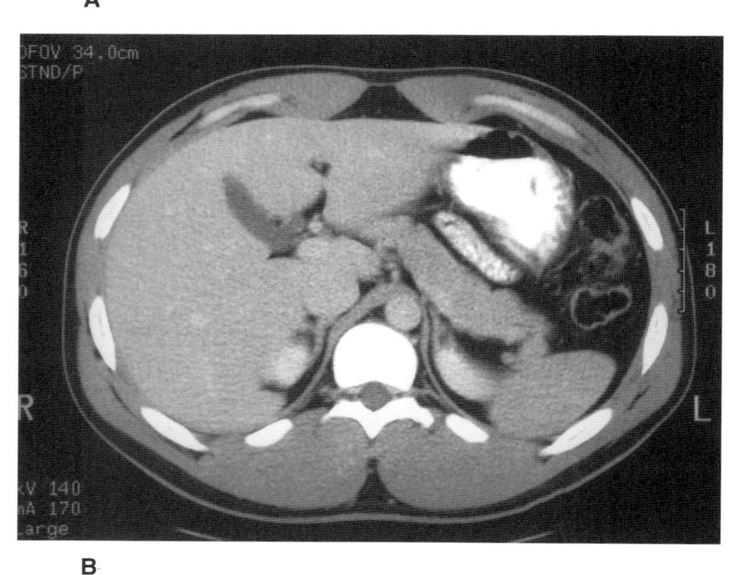

B

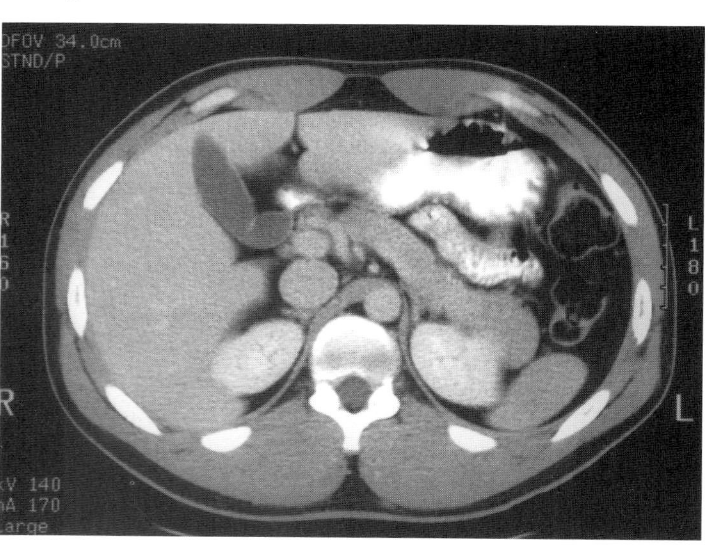

C

Computed tomography carried out in the abdomen is usually made transaxially with contiguous slices 0.5 to 1.0 centimeter thick, although each CT scanning study must be tailored to the particular patient's problem.

Identify the liver in Figure 11.52A. Note that as you look up from the patient's feet, the liver is on your left and anterior. Ignore everything else and follow the change in shape and size of the liver down through all six sections of Figure 11.52. Then note how the dense vertebra changes shape depending on whether the cut is through the transverse processes in F or above or below them. The vertebral canal is visible, as is the cord within it.

Locate the aorta just anterior to the body of the vertebra and slightly to the left of the midline. Because this series of scans was carried out after intravenous injection of contrast material, perfused structures and vessels are slightly whiter (enhanced). Note the branching white vessels in A in the homogeneous parenchyma of the liver. The intravenous contrast material used is excreted by the kidneys, so you see the renal parenchyma and collecting systems as whiter than other organs.

The stomach and small bowel are opacified by oral contrast material that has not yet passed into the colon. Note that the wall of the stomach is less dense than its fluid content, and as you look from scan to scan you cross-section first the body and then the pyloric antrum of the stomach; in E and F you can see contrast material in the duodenum.

Figure 11.52. Normal CT scans of the upper abdomen.

The black, low-density areas on either side of the vertebra and surrounding the kidneys represent retroperitoneal fat. Follow those areas downward from section to section and study the kidneys, noting the initial appearance of their upper poles in B and the doughnut-shaped parenchyma around the upper calyces in D before the hilum has been reached in F. The two kidneys are at nearly the same level in this patient; normally the left kidney is higher. The two hila are sectioned in F, where the kidney is seen as a crescent of parenchyma curving around the medially and anteriorly directed hila.

Now find the inferior vena cava in C, just anterior to the vertebra a little to the right of the midline. If you follow it downward you can see it in F receiving the left renal vein, which normally crosses in front of the aorta from left to right at this level. Locate the tail and body of the pancreas in B. The dragon-shaped mass of the pancreas is seen first in section B, its tail extending toward the hilum of the spleen. (You can check your findings by referring to the labeled CT scans of other levels of this patient that appear at the end of Chapter 3.)

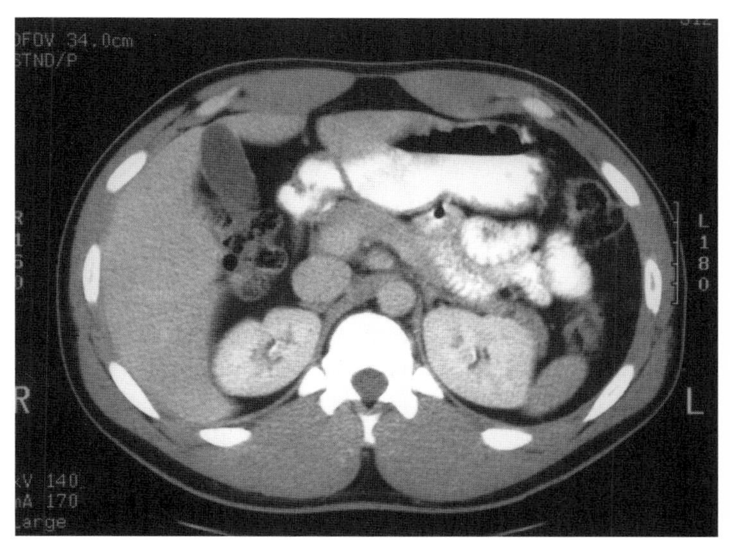

D

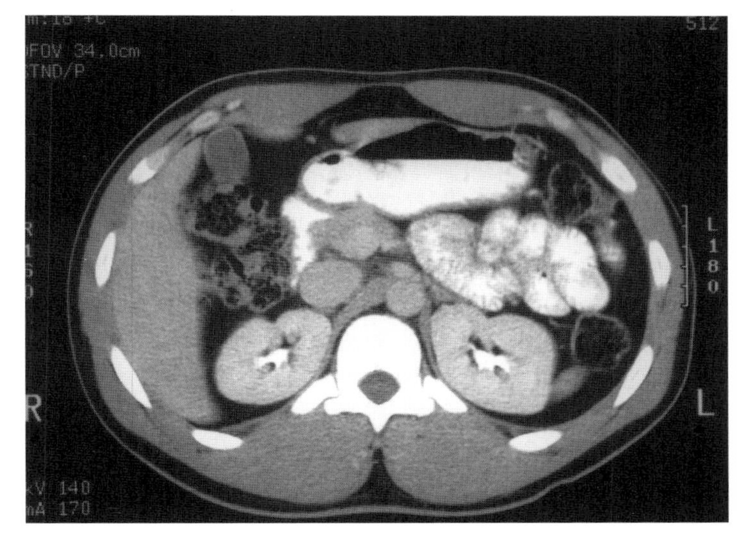

E

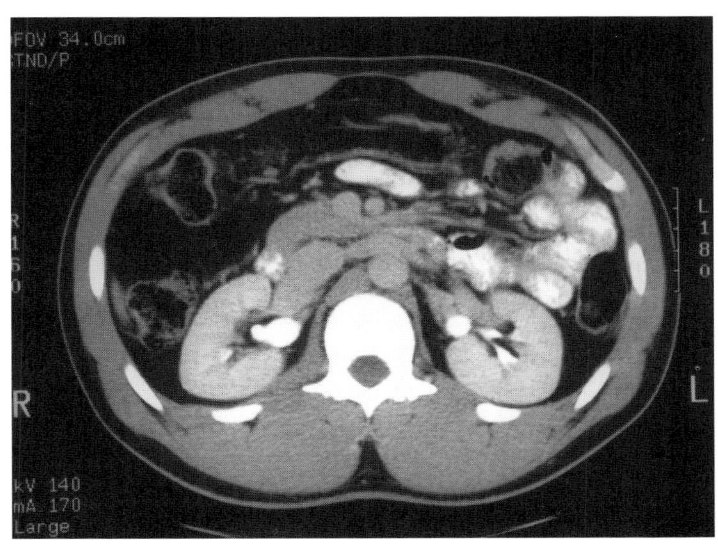

F

Now here is another series of normal abdominal CT scans for you to practice your skills on, studying all four sections and identifying serially the liver, spleen, stomach, duodenum, kidneys, aorta, inferior vena cava, and pancreas. Compare each structure with the same one in the patient in Figure 11.52.

By observing the density of the kidney collecting systems and stomach contents, decide whether intravenous or oral contrast medium was given to this patient. Can you distinguish the head of the pancreas from the descending second portion of the duodenum in C? Where is the gallbladder? Find the transverse colon in D. Multiple loops of small bowel containing contrast material are seen in D. Has the contrast material reached the colon, or does the haustrated colon contain only fecal material? The superior mesenteric artery arises from the aorta just above the point where the inferior vena cava receives the left renal vein, which crosses to the right in front of the aorta. Find the superior mesenteric artery in C.

In this patient you can recognize the adrenal glands embedded in fat in sections A and B. The left adrenal lies to the left of the aorta and posterior to the tail of the pancreas and splenic artery in this patient. The right adrenal is always seen just posterior to the inferior vena cava, as in A.

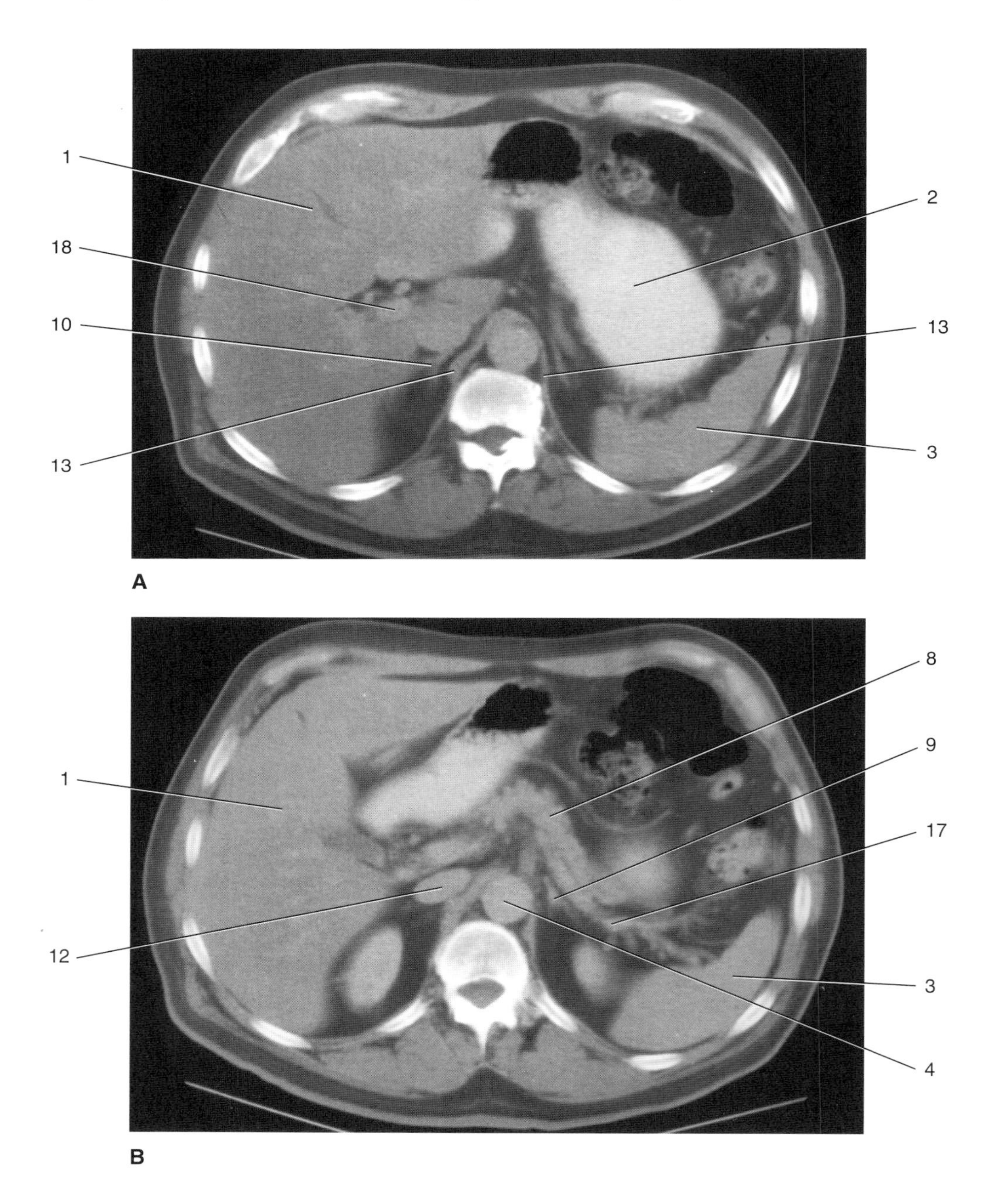

Figure 11.53. A series of upper abdominal CT scans on another patient. Use them as an exercise to test your progress. You can check the list below if you are unsure about the identity of any structure.

(1) Liver
(2) Stomach with water-soluble oral contrast medium
(3) Spleen
(4) Aorta
(5) Kidneys
(6) Duodenum
(7) Head of pancreas
(8) Body and tail of pancreas
(9) Left adrenal
(10) Right adrenal
(11) Superior mesenteric artery
(12) Inferior vena cava
(13) Diaphragmatic crura
(14) Transverse colon
(15) Ascending colon
(16) Descending colon
(17) Splenic artery and vein
(18) Portal vein
(19) Gallbladder
(20) Small bowel

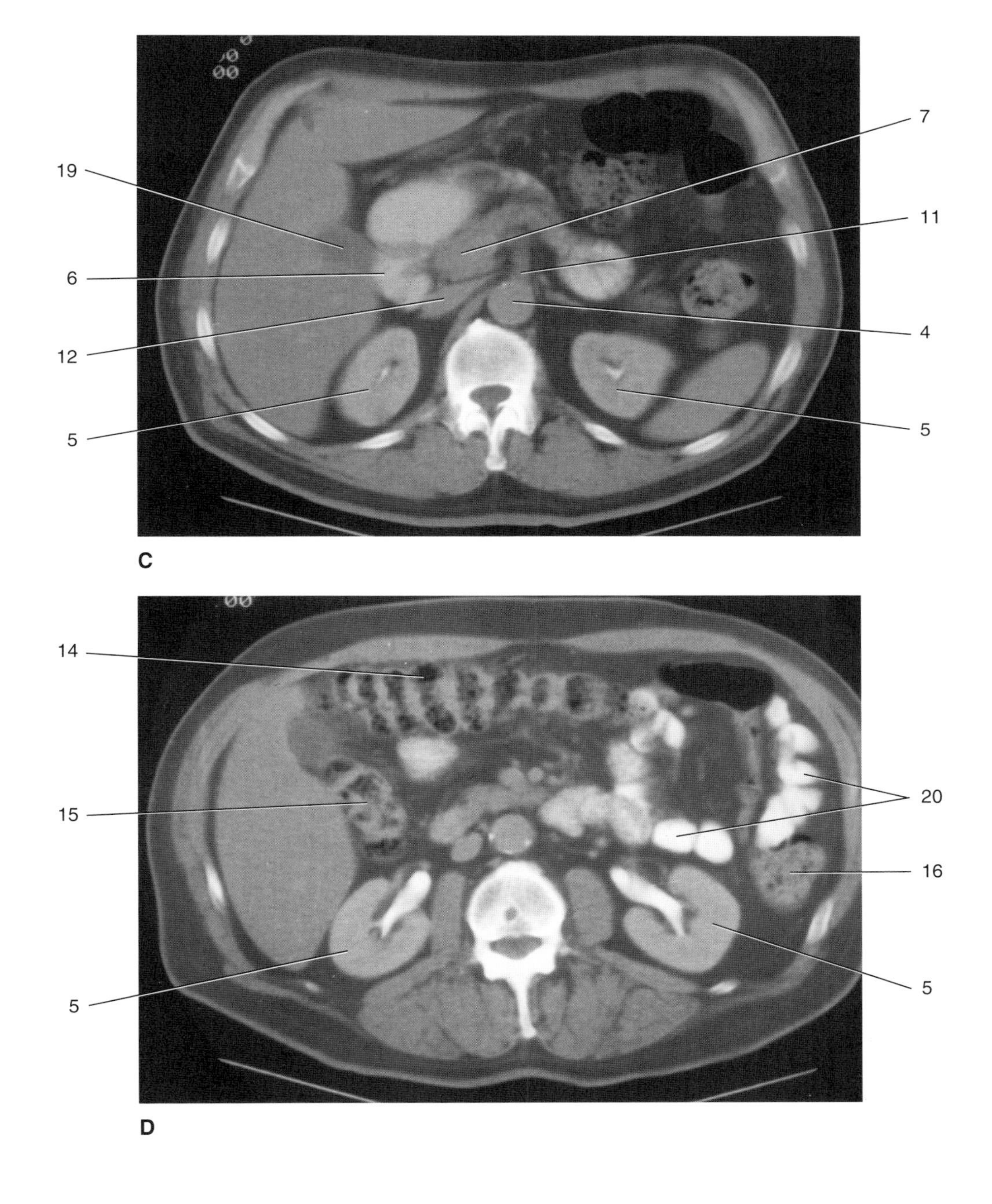

Ultrasound of the Abdomen

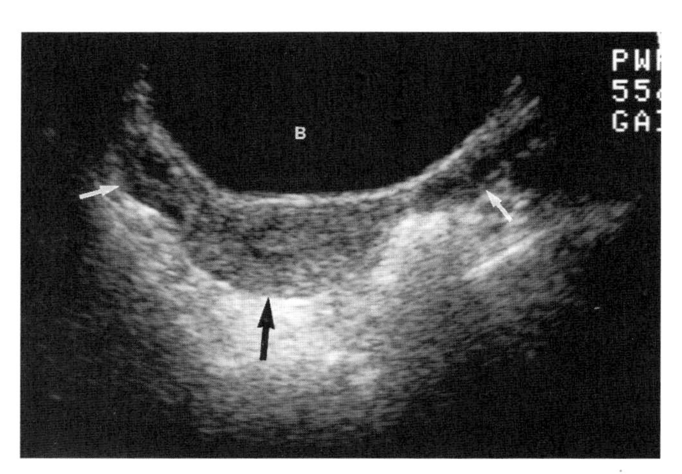

Figure 11.54. Normal transabdominal ultrasound in the transverse plane of the female pelvis. *B* is the bladder. The *black arrow* indicates the uterus and the *white arrows* point to the adnexa.

Now that you have addressed yourself to the study of plain films and to the intellectual discipline of understanding serial CT scans of the abdomen, you have certainly acquired an even more expert feeling for considering body structures always in three dimensions. Even if you were blind, had therefore never seen an *apple* or known that such a shape existed, you would be able intellectually to reconstruct its three-dimensional form precisely from assembled slices that you had been able to touch-examine with your fingers. Now you must address in further detail the information available through *ultrasound* (or ultrasonography).

It is a curious fact that high-frequency sound waves beyond the range of the human ear were produced experimentally some 15 years before Roentgen discovered the x-ray. No attempt was made to develop ultrasonography for medical use until after the stimulus of its use during World War II in the underwater detection of submarines. Since the 1950s, however, ultrasound has earned an important place in medical diagnosis, and physicians find they employ it on some of their patients practically every day.

The fact that sound waves are reflected by many solid substances but transmitted readily by fluids meant that a beam of sound could be projected through water toward a submarine, and its reflection from the surface of the hull timed so that the presence of the submarine and its distance from the beam's origin were known.

Translated to medical use, this means that a beam of sound waves projected into the body from the surface of the skin will be transmitted forward by *echolucent* (fluid) substances, a part of the beam being reflected back when it encounters an interface with a substance or structure of different *acoustic impedance,* such as a gallstone bathed in bile in the gallbladder. The gallstone would be echogenic.

The time needed for the signal to return to the transducer *locates* the depth of the interface from which it was reflected. The rest of the initial beam continues on into and through the encountered organ or mass, reflecting more and more of the beam from other interfaces at measurable distances, until finally the last of the beam is absorbed.

Whenever the beam encounters a fluid-filled structure inside the body, the sound is transmitted through with negligible absorption until it reaches the interface of the far wall of the cystic structure, from which it is reflected. *Sound is transmitted well through any fluid, but poorly or not at all through bone, air, and barium.*

The beam of sound is produced in pulses or periodic bursts, very brief in duration, and the same transducer then "listens" for the returning echoes until the next burst of outgoing sound.

Returning echoes are electronically converted into a video image on a monitor, echoes appearing as white dots on a black background. The result is a *picture of a wedge-shaped slice of the patient.* Because of the variety of degrees of obliquity of image plane used, it is not easy to reconcile such pictures with your intellectual awareness of cross-sectional anatomy in standard planes. In fact, the ultrasonographer must identify the anatomic segment and image plane on each scan for the benefit of the viewer.

Unlike the important tissue injury of ionizing radiation, which has to be taken into consideration in electing to carry out any radiographic study (including CT), pulsed diagnostic ultrasound has no injurious effect in the range used for medical technology. This makes ultrasound a particularly valuable tool, especially in imaging pelvic structures in obstetrical and gynecological practice.

A patient being prepared for conventional pelvic ultrasound is instructed not to void for some time beforehand, a full urinary bladder providing a sonolucent mass (acoustic window to transmit sound waves) in the lower abdomen, which displaces undesirable

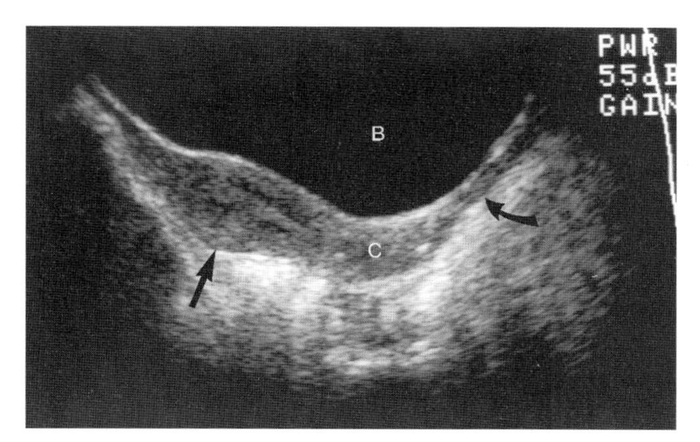

Figure 11.55. Normal transabdominal ultrasound in the sagittal plane of the female pelvis. *B* is the urinary bladder; *C,* the cervix. The *straight black arrow* indicates the uterus and the *curved black arrow* the collapsed vagina.

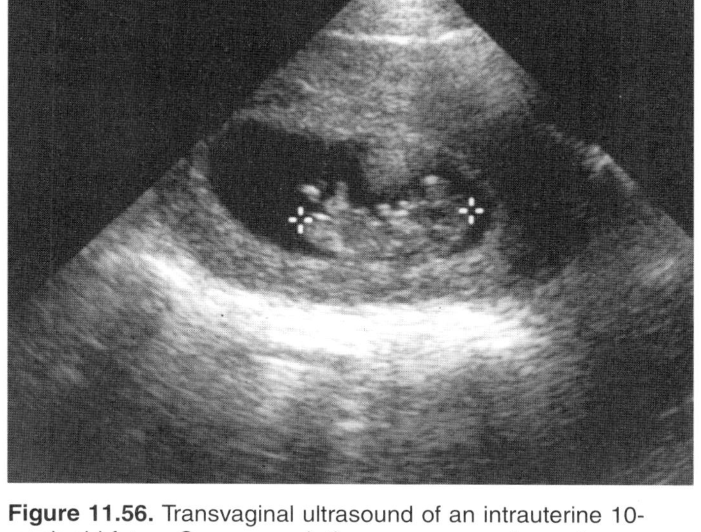

Figure 11.56. Transvaginal ultrasound of an intrauterine 10-week-old fetus. *Cursors* mark the crown-rump dimension.

air-containing loops of bowel up out of the pelvis so that a good beam of sound transmitted through the bladder will encounter the pelvic organs posterior to it. The ultrasound technologist applies the ultrasound transducer to the lower anterior abdominal wall, making good skin contact with a special gel applied between the transducer and the patient's skin. The transducer is then moved over the skin surface, tilted and rotated until the technologist finds the structures to be included in the examination, in this case, the uterus and adnexa. By rotating the transducer 90 degrees, the technologist can change the imaging plane from transverse to sagittal. Pelvic ultrasound can also be performed with a special transducer that is placed directly within the vagina. This technique is called *transvaginal ultrasound,* and it yields greater detail of the uterus and adnexa than the conventional technique of *transabdominal ultrasound.* Transvaginal examinations do not require a full bladder because the sound waves are emitted directly within the pelvis; no acoustic window is necessary for imaging.

Figure 11.54 is a *transverse* pelvic ultrasound, obtained just above the pubic symphysis. The beam of sound has entered through the anterior abdominal wall *(top of figure).* You are looking up at the slice from the patient's feet. The large, uniformly black structure *(B)* is the urine-filled, echolucent (or anechoic) bladder. Posterior to it is an oval mass *(black arrow)* representing the normal uterus. The *white arrows* point to the adnexa, representing the ovaries and uterine tubes. Remember: *echogenic* structures appear white and *echolucent* structures (fluid collections) appear black. Fatty tissues generally are very echogenic, and you can see their white echoes surrounding the bladder and uterus. Note the many dots representing echoes within the muscular wall of the uterus and the slitlike uterine cavity with its darker ring of endometrium.

Figure 11.55 shows the uterus *(straight black arrow)* in a midline *sagittal* ultrasound. The patient's head is to your left. Again observe the echogenic posterior wall of the bladder *(B)* and the echogenic uterine cavity. The cervix *(C)* and collapsed vagina *(curved black arrow)* are well shown.

Notice that the uterus in Figure 11.55 is empty. Compare it with the uterus in the transvaginal ultrasound in Figure 11.56, which contains a 10-week-old fetus. The two cross-shaped artifacts are ultrasound markers called *cursors,* which are positioned by the ultrasonographer to measure distances. After the cursors are positioned, the ultrasound computer calculates the distance between them. In this instance the crown-rump dimension measures 39.5 millimeters, normal for a fetus of this age. The echogenic fetus is surrounded by sonolucent (black) amniotic fluid, contained between the muscular walls of the uterus. Obstetrical and gynecological uses of ultrasound will be discussed further in Chapter 16.

Ultrasound has many other useful applications in the abdomen, although certain parts of the body are not amenable to sonographic study because of their barriers to ultrasound transmission (bone and air).

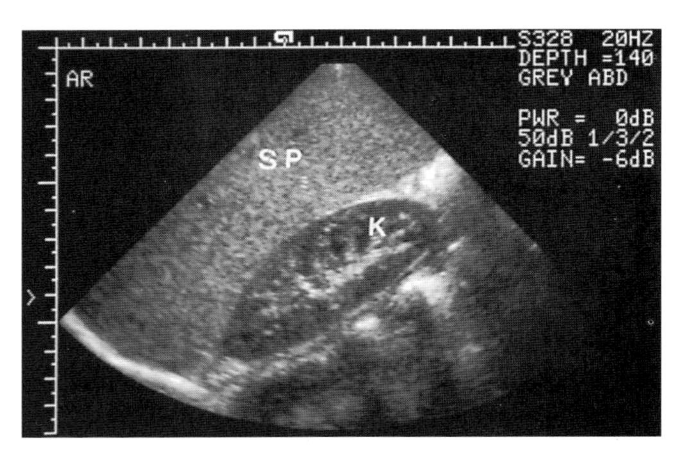

Figure 11.57. Normal longitudinal ultrasound of the left kidney. *SP* is the spleen, *K* the kidney.

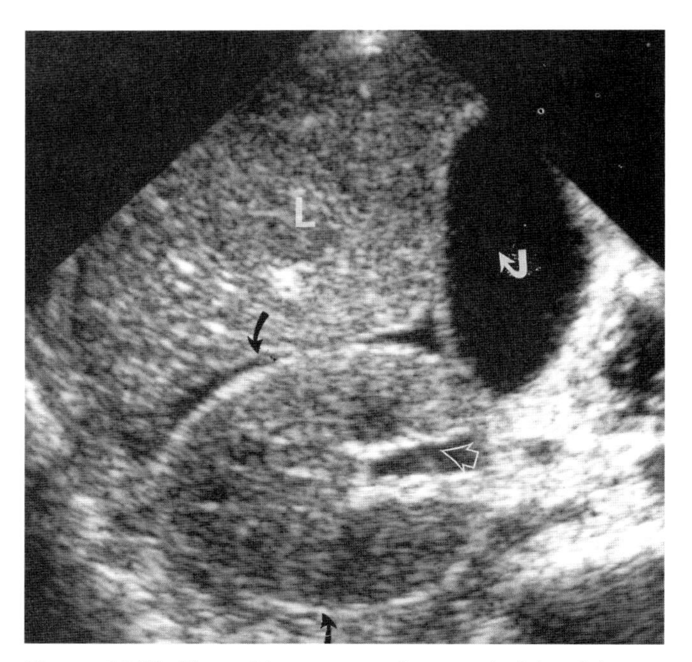

Figure 11.58. Normal transverse ultrasound of the right kidney.

For example, the adult brain cannot be imaged by ultrasound because the bony skull completely surrounds it. And it is not possible to image pulmonary nodules and other lung masses with ultrasound because of the failure of air in the surrounding normal lung to transmit sound waves. The mediastinum, however, is routinely imaged. You have already seen examples of cardiac ultrasound (echocardiography) in Chapter 10.

In the neck, ultrasound is frequently used to diagnose carotid artery stenoses and occlusions in patients with suspected cerebrovascular disease. Vascular ultrasound will be discussed in Chapter 17. Thyroid cysts can be differentiated from thyroid tumors with ultrasound, and even parathyroid masses may be discernible.

The most common use of diagnostic ultrasound is in the abdomen and pelvis where, as in Figure 11.57, a *sagittal* scan, you see the left kidney lying posterior to an enlarged spleen. Note the central echoes produced by peripelvic fat in the kidney. The kidney parenchyma is partially surrounded by echogenic (white) perirenal fat. Note the homogeneous echotexture of the spleen. The minute lucencies are splenic blood vessels.

Figure 11.58 is a magnified *transverse* ultrasound of the right kidney *(between black arrows)*. The liver is seen anteriorly *(L)*, and the gallbladder, a sonolucent fluid-filled oval mass, is just medial to it *(bent white arrow)*. The dark echolucent quadrangle indenting the kidney from its medial surface *(open arrow)* is the renal pelvis. Renal ultrasound is used to diagnose hydronephrosis, renal cysts, and neoplasms.

Figure 11.59 is a gallbladder ultrasound of a patient with a large number of calculi. The echogenic stones produced shadows behind them, because the beam does not penetrate the stones. Acute cholecystitis and bile duct obstruction may also be recognized with the help of ultrasonography. In fact, ultrasound is usually the initial imaging procedure in patients who present with new jaundice.

Figure 11.60 is a sagittal scan of the liver. The patient's head is to your left, and the echogenic arc *(arrow)* represents the diaphragm closely applied against the surface of the liver. The lung above the diaphragm is not well seen, but an echolucent pleural effusion would be. Ultrasound is often used to identify small pleural effusions and to localize them for thoracentesis. Under ultrasound guidance thoracentesis needles can be placed directly into small effusions. Anteriorly (at the top of the scan) you can see a relatively sonolucent round neoplasm in the liver parenchyma, with cursors to measure its size. The other echolucent areas are, as you would expect, hepatic blood vessels and bile ducts. Ultrasound of the liver easily identifies metastases as well as solitary primary tumors, liver cysts, and other conditions. You will see in Chapter 14 that ultrasound can be helpful in the diagnosis of a wide variety of other abdominal conditions.

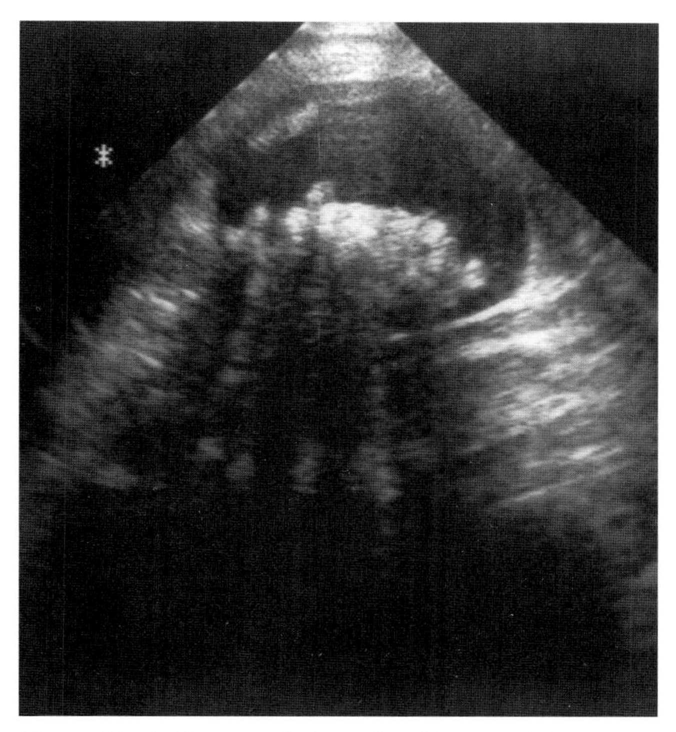

Figure 11.59. Ultrasound of a gallbladder containing multiple echogenic calculi. Note the acoustic shadows beyond the stones, areas from which no echoes are returned.

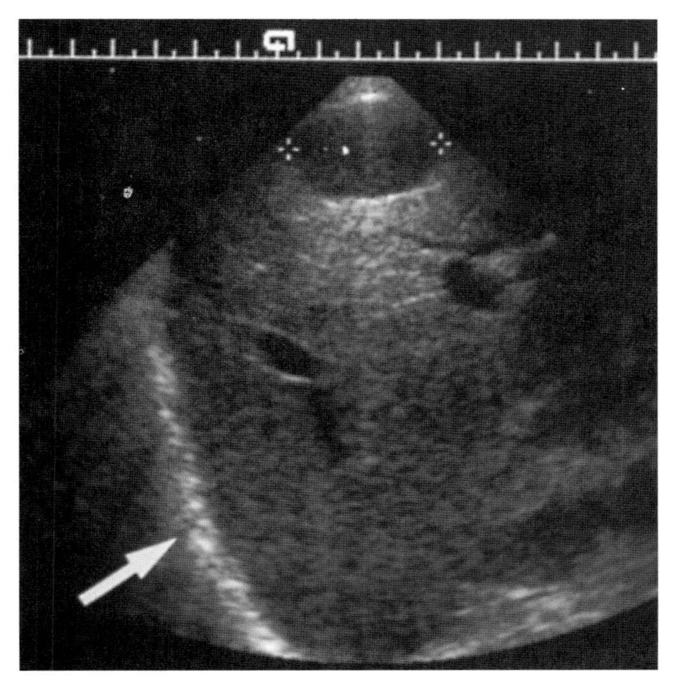

Figure 11.60. Ultrasound of the liver and diaphragm (see text).

12 Bowel Gas Patterns, Free Fluid, and Free Air

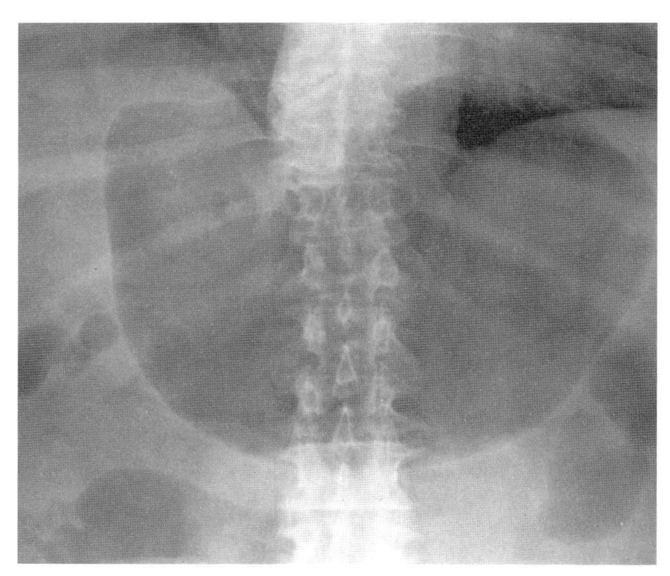

Figure 12.1. The stomach distended with air in agonal diabetic coma.

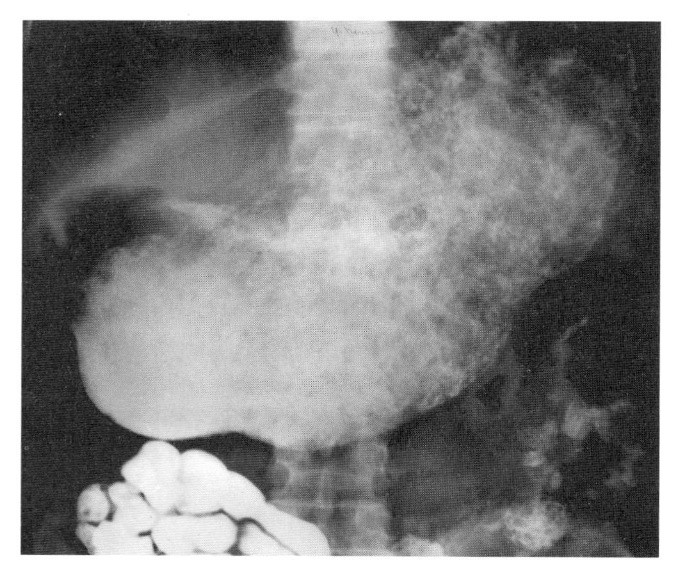

Figure 12.2. The stomach distended with food and barium mixture in a patient with gastric outlet obstruction, the result of scarring from long-standing duodenal ulcer disease. The film was made 4 hours after the patient swallowed the barium, some of which is moving into the small bowel.

Today perhaps the greatest usefulness of the abdominal plain film is in recognizing mechanical bowel obstruction and paralytic (adynamic) ileus, the subjects of this chapter.

The Distended Stomach

After you have studied the bones and the soft-tissue zones and profiles and decided whether there seems to be evidence of organ enlargement or displacement, you should look at the plain film as a whole, focusing on the gas distribution and content. Normally the air in the stomach ranges from the small, round, wrinkled fundal bubble seen on the prone film to a few oblique streaks of air in the body of the stomach, which you see in supine films. You seldom see the entire stomach filled with swallowed air. In Figure 12.1 you see the stomach of a patient in diabetic coma with agonal dilatation.

In pyloric obstruction the stomach may be grossly dilated, of course, and it often contains a large amount of food and retained secretions. On the plain film such a stomach appears as an ill-defined density extending across the upper abdomen, and if the radiologist tries to study it fluoroscopically, swallowed barium appears to sink into a bog (Figure 12.2). Attempted barium study of such a stomach is futile until the stomach has been emptied, for only a clean mucosa can be examined with barium. The stomach is often seen distended with air in paralytic ileus, in diabetic coma, and in air swallowing.

The Distended Colon

Abdominal plain films of hospitalized patients often show a certain amount of air in the small bowel, particularly the ileum, even though there is no clinical evidence to suggest the presence of either adynamic (paralytic) ileus or mechanical obstruction. You will find that the amount of air in the intestine is increased in plain films made after any kind of painful interventional procedure. Unfortunately, this increased air usually overlies the kidneys, and the intersecting lines produced by the folded walls of air-filled bowel confuse the details of the shadows of the kidneys and psoas muscles.

Truly distended loops of small bowel may approach and even exceed the caliber of the normal colon. When they are filled with air, their distinctive mucosal markings usually identify them; but when they are filled with fluid, they cast vague, confluent, or sausage-shaped gray shadows across the midabdomen. Bubbles of air are often superimposed upon these shadows in the supine film.

The colon, particularly its distal half, usually contains some air. You may see air outlining solid fecal material within the lumen. The cecum and ascending colon often contain semisolid feces (Figure 12.3), which produce a characteristic speckled shadow.

With moderate obstructive distension of the colon, the haustral indentations become shallower but are still visible. *More of the colon than is usual is continuously outlined with air* (Figure 12.4). Thus when a tumor obstructs at the level of the sigmoid, air may be seen outlining and distending all the colon proximal to that point (Figures 12.3 and 12.4). With a tumor obstructing at the midtransverse colon, the proximal half of the transverse colon, hepatic flexure, ascending colon, and cecum are distended with air even though the left half of the colon is empty, cleared, invisible. The cecum may eventually balloon to enormous proportions in obstruction of the distal colon. When that happens, all haustral indentations are lost, and the cecum appears as a huge air-filled structure occupying the right side of the abdomen. Still-incomplete obstruction to the forward flow of gas and feces may be temporarily overcome by retrograde instillation of fluid (cleansing enemas), as you see in the patient in Figures 12.3 and 12.4. The initial plain film on such patients will show accumulation of feces and

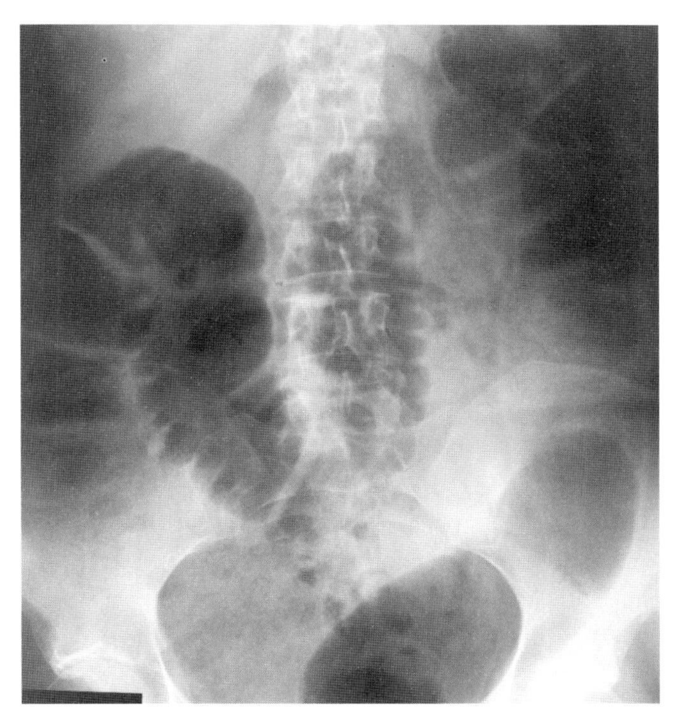

Figure 12.3. Large-bowel obstruction in carcinoma of the sigmoid. Note the retained feces.

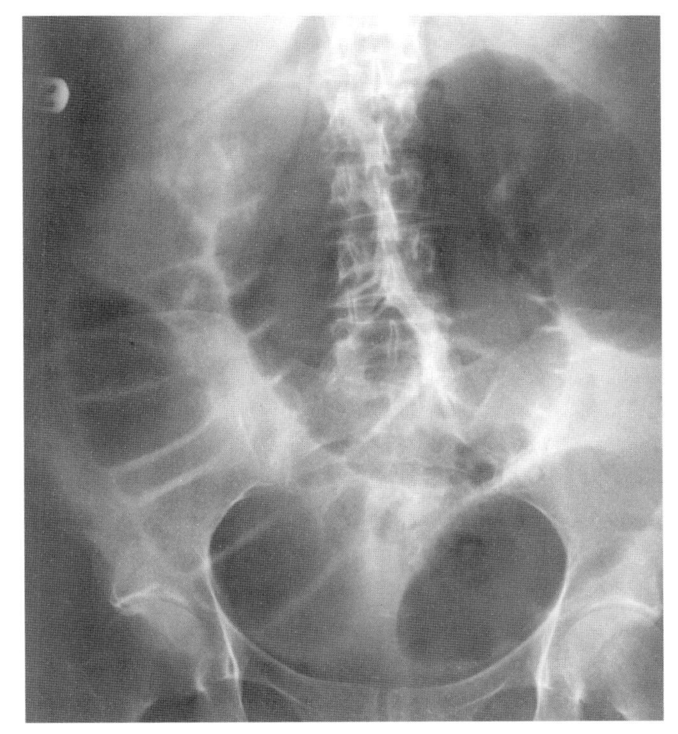

Figure 12.4. The same patient after cleansing enemas; the clean colon is distended with air. The ileocecal valve is competent (no air in the small bowel).

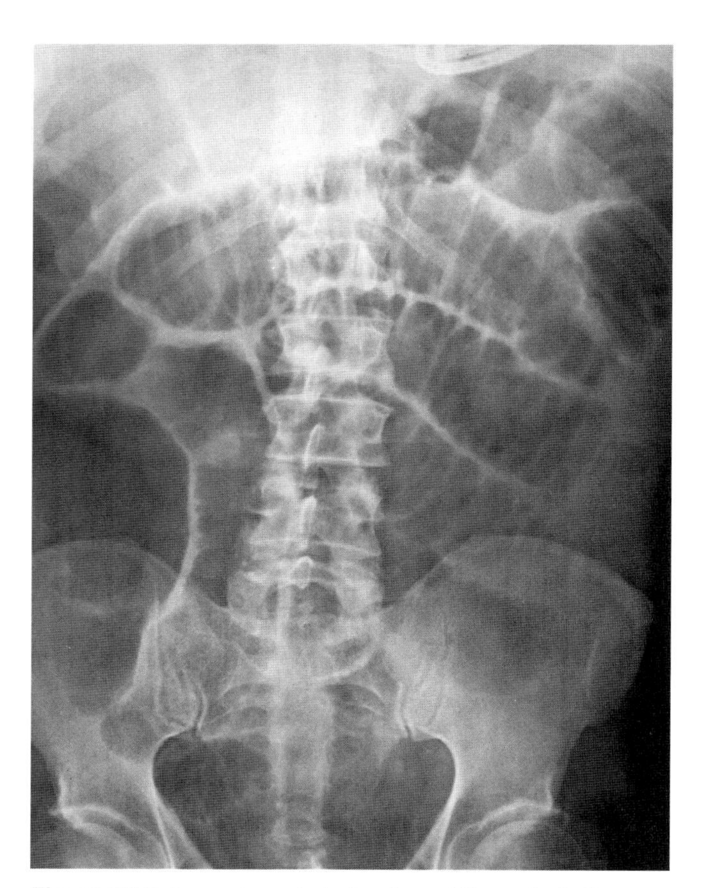

Figure 12.5. Large-bowel obstruction at the splenic flexure with decompression backward through an incompetent ileocecal valve. Note that the descending colon, sigmoid, and rectum are empty.

gas above the tumor, which has narrowed the lumen. After the cleansing enemas, continuously air-filled distended colon is seen.

An important point in the diagnosis of all types of mechanical obstruction is that *the compensatory increase in peristalsis which develops is carried beyond the point of obstruction and results in the clearing of air from the bowel distal to that point* (that is, from that portion of the bowel which *can* be cleared). Thus in obstruction at the splenic flexure, you would expect eventually to find the descending colon and sigmoid completely empty, and therefore invisible (Figure 12.5). It is thus possible to make a strong presumptive diagnosis of large-bowel obstruction from a single plain film, with the next step a barium enema and demonstration of the lesion from its distal side.

In low large-bowel obstruction, air gradually fills most of the colon and, if the patient is fortunate, the ileocecal valve will become *incompetent,* allowing the colon to decompress backward into the small bowel (Figure 12.5).

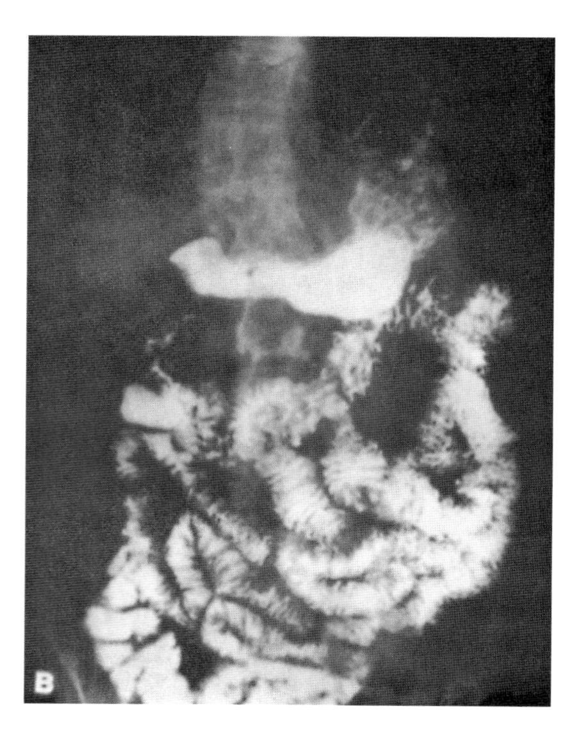

The Distended Small Bowel

Figure 12.6 shows you the caliber of the small bowel filled with barium in a normal person and can be compared with Figure 12.7, in which obstructed and distended small bowel is visualized with barium and air. The *arrow* indicates the point of obstruction, beyond which no barium has passed. Barium given by mouth in a patient with *colonic* obstruction could cause a dangerous barium impaction. Colonic obstruction should be ruled out by barium enema or colonoscopy in any patient whose symptoms and plain film suggest it. This is not a problem with small-bowel

Figure 12.6 (left). Normal small bowel. Barium was given by mouth and is present in the stomach. The upper loops of jejunum have a more "feathery" appearance and a slightly larger caliber than the lower loops of ileum.

obstruction because the small-bowel contents remain liquid proximal to the site of obstruction. The study of an obstructing colonic lesion by barium enema from its *distal* side is an entirely different matter, since the barium is readily evacuated. Sometimes in colonic obstruction a water-soluble contrast substance is used instead of barium if colonoscopy is planned.

Figure 12.7 shows you the caliber of moderately distended loops of jejunum with their characteristic cross striations representing the valvulae conniventes. You may think that they resemble the haustral indentations of the colon—and they do, superficially at least. They differ in their periodicity, however, being more numerous than haustral indentations and more narrowly spaced even when the small bowel is distended. They also cross the lumen from one side to the other, as opposed to haustral indentations, which indent but do not usually cross the colon and often are not precisely opposite the indentation on the other side. In addition, small-bowel loops tend to line up in rows, three and four parallel loops of bowel appearing close beside each other (Figures 12.7 and 12.8). The colon, when it distends, almost never gives this "arranged" appearance.

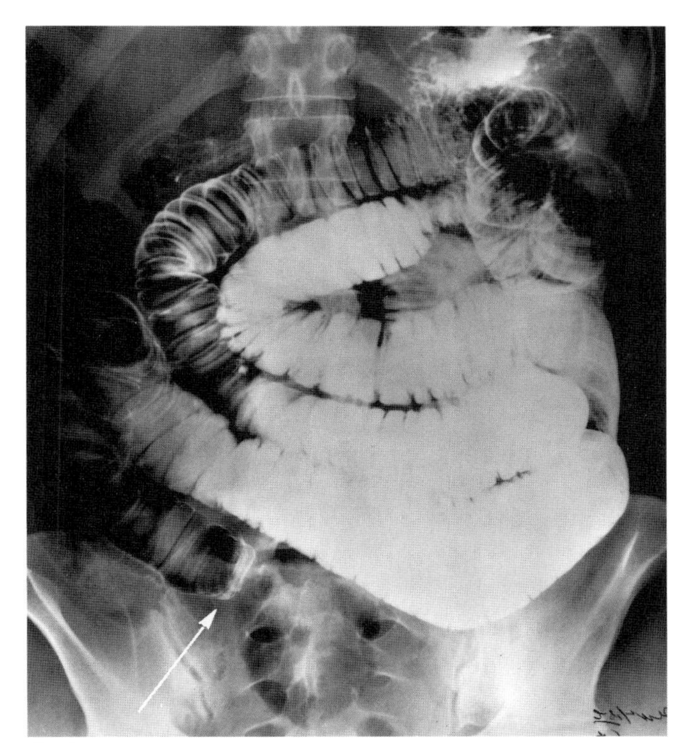

Figure 12.7. Distended small bowel above an obstructing lesion *(arrow)*. Note the cleared (invisible) colon.

In mechanical small-bowel obstruction precisely the same principle applies that we described for the large bowel: clearing of all bowel beyond the point of obstruction so that it is empty of gas, collapsed, and invisible (Figure 12.8). If you make a practice of looking for the colon as soon as you recognize distended small bowel, someday you will find yourself looking at an unknown plain film on which you can find no haustrated air shadows and hence no part of the colon. You will then realize that you must be looking at the radiographic findings in mechanical small-bowel obstruction, the *entire* colon having been swept clear of gas and feces.

In paralytic, or adynamic, ileus, by contrast, both large and small bowel will be seen distended with air, since peristalsis is generally decreased. This is a far less distinctive radiological picture than that for mechanical obstruction, and makes for indecision and a sense of confusion in trying to interpret plain films in patients with abdominal symptoms. Still, once you have learned to recognize the picture of *mechanical*

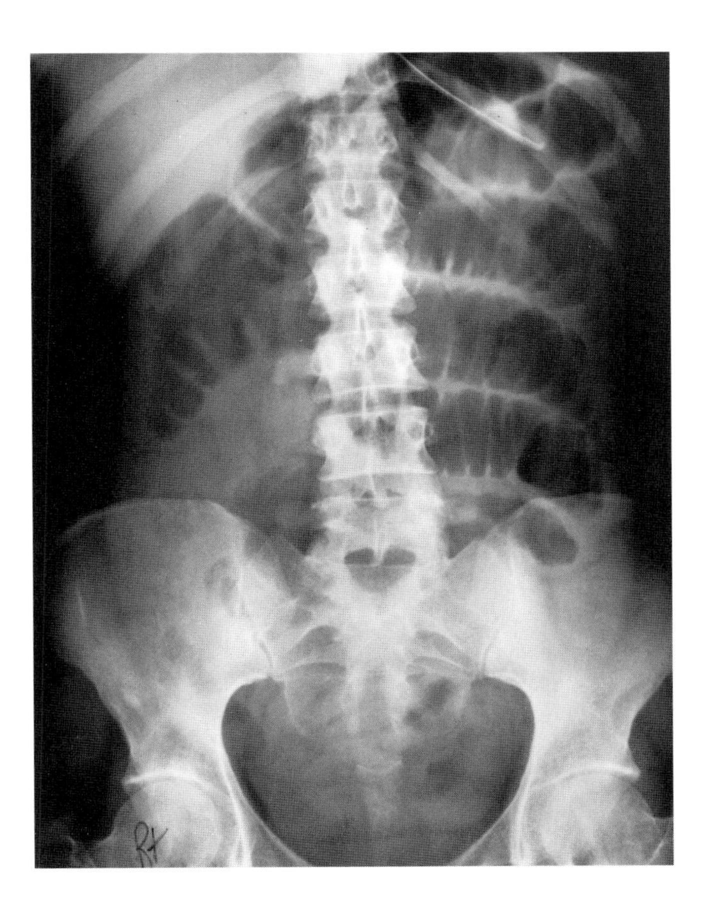

Figure 12.8 (right). Small-bowel obstruction in a patient with inflammatory bowel disease and stricture of the terminal ileum. Note the clearing of the colon here.

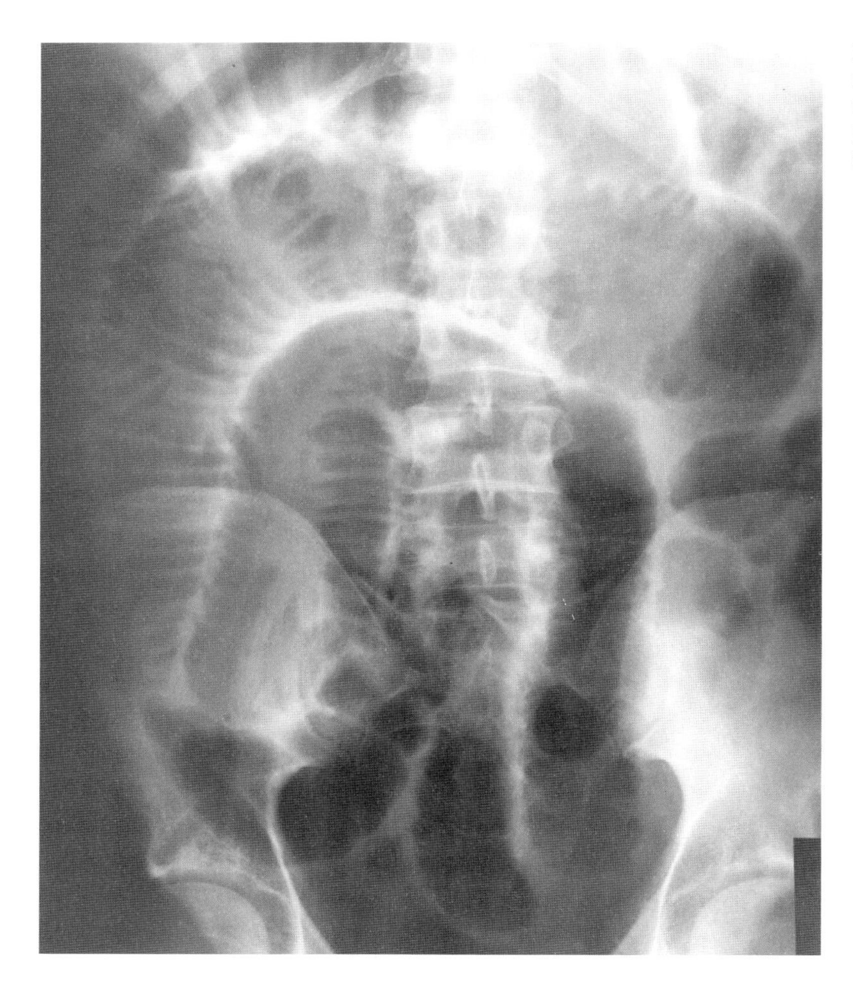

Figure 12.9. Small-bowel obstruction, with many loops of small bowel distended with air and almost complete clearing of the colon. At laparotomy an adhesive band was lysed and the patient recovered.

obstruction when it is definitive, you will feel somewhat more comfortable about studying the equivocal findings so often seen in plain films.

Time is the important factor we often forget to consider in looking at a single film. It is all too easy, when you are worried about a patient and in quest of diagnostic help, to forget that a single examination is a point on a curve and nothing more. How long has the obstruction in the midtransverse colon been present? Has it been a complete obstruction *long enough* to allow the bowel beyond that point to become cleared of air? If not, then the presence of air in both large and small bowel may be hard to distinguish radiographically from paralytic ileus. *Serial films in patients with abdominal problems are often very informative,* indicating the developing change more clearly than any other investigative procedure.

In Figure 12.9 loops of distended small bowel are easy to identify, but no air is seen in the colon. This patient had *mechanical small-bowel obstruction* from an adhesive band in the right lower quadrant.

The patient in Figure 12.10 had paralytic ileus. You can identify air in the stomach, small bowel, and large bowel. Note, however, that air does not truly distend the bowel; there is a varying amount of air in the colon and small bowel, which are *discontinuously air containing.* Parts of the colon have deep haustral indentations with little resemblance to the blown-up, distended appearance seen in Figure 12.3. Contrast the *continuously distended* and elongated colon you would expect in any obstructed flexible tubular structure.

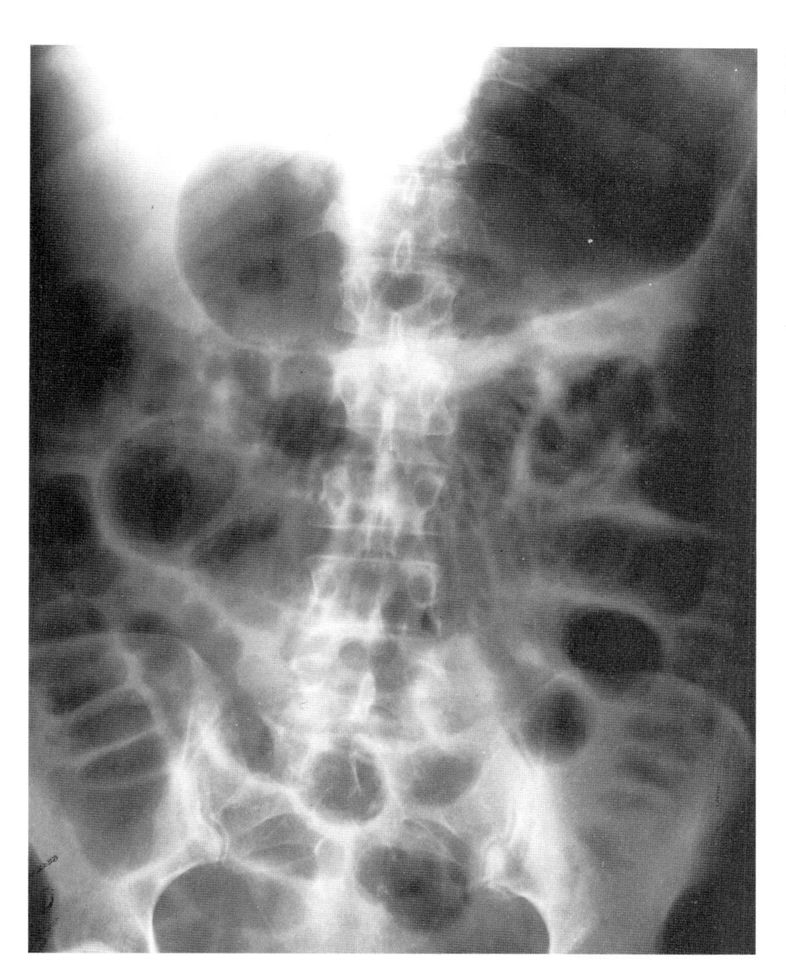

Figure 12.10. Paralytic ileus in a patient after an automobile accident. (See text.) The picture did not change on serial films made at 1 and 2 hours.

Differentiating Large-Bowel and Small-Bowel Mechanical Obstruction from Paralytic Ileus

I. *Too much air in either colon or small bowel, but none in the other*
—is either:
A. Small-bowel obstruction old enough to have allowed the colon to clear
or
B. Large-bowel obstruction with a competent (tight) ileocecal valve

II. *Too much air in both parts of the bowel*
—is one of the following:
A. Paralytic (adynamic) ileus
B. Large-bowel obstruction with an incompetent ileocecal valve, allowing the distended colon to decompress backward into the small bowel
C. Small-bowel obstruction that is—
1. Early (colon has not had time to clear)
or
2. Intermittent (loop of small bowel caught from time to time in a hernia or behind an adhesion)

Large or Small Bowel? Mechanical Obstruction or Paralytic Ileus? Determine what Figures 12.11 and 12.12 show before you read on.

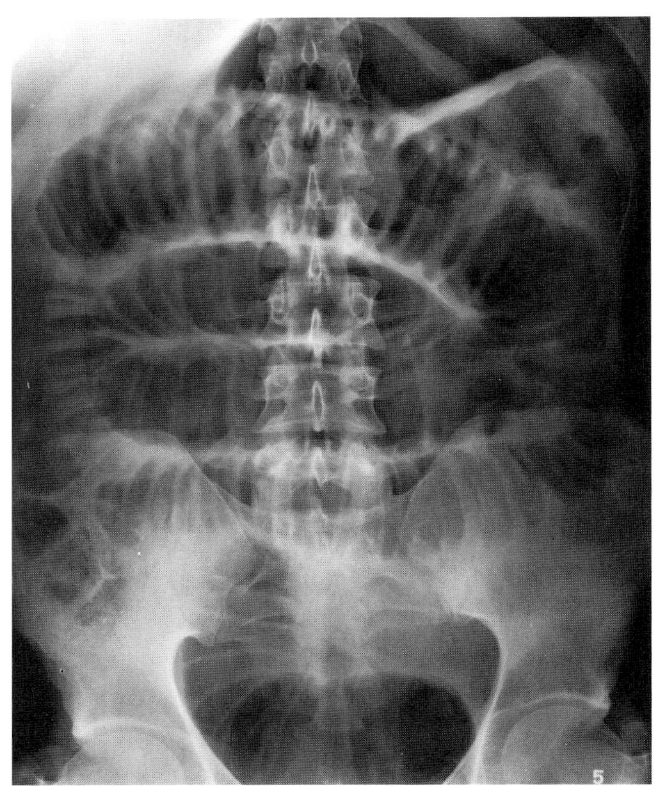

Figure 12.11 ***(Unknown 12.1)***

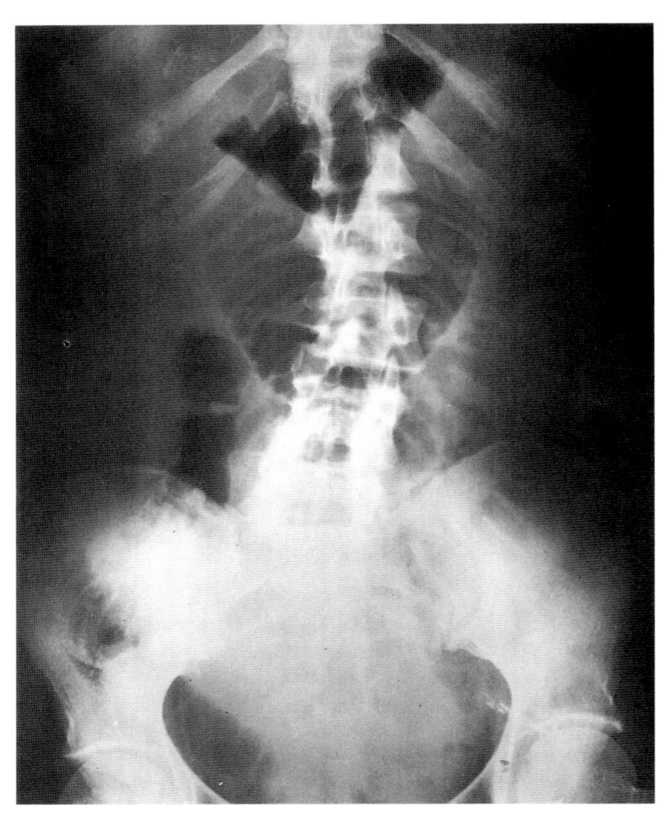

Figure 12.12 ***(Unknown 12.2)***

When a patient is admitted with abdominal pain and the initial plain film shows both the large and the small bowel to be distended, the findings are equivocal, as we have said, in that they may represent either paralytic ileus or an early or intermittent mechanical obstruction. The activity of the bowel sounds, repeatedly observed over a period of time, may or may not clarify the issue. *Serial films,* however, may show changes in the radiological findings that provide helpful clues.

The small intestine always contains fluid, and when obstructed or paralyzed it accumulates additional fluid and air. On the erect plain film (Figure 12.13) the air-fluid interfaces inside distended loops of bowel appear as fluid levels, varying in length according to their size and the relative quantities of air and fluid within them. Loops entirely filled with fluid cast ill-defined gray shadows; loops with little fluid and a great deal of air appear on the supine film like those in Figure 12.11. Loops three-quarters full of fluid and containing a relatively small amount of air may be very deceptive, since in the erect and decubitus films they show short fluid levels (Figure 12.14) and in the supine plain film rather unimpressive bubbles of air superimposed on the indefinite gray of the fluid (Figure 12.12).

Note that Figure 12.14 is the decubitus film on the woman in Figure 12.12. You can appreciate the fact that such a patient may not have a very startling initial plain film, and yet be sicker and in a more advanced stage of obstruction than the patient in Figure 12.11. Always remember that a plain film with obstructed, fluid-filled loops with relatively little air may indicate that your patient is dehydrated and in marked electrolyte imbalance. The patient in Figures 12.12 and 12.14 had late small-bowel obstruction from an adhesion near the ileocecal valve and markedly abnormal electrolytes.

Except in rare cases the use of erect films of the abdomen has not proven to be of significant value in differentiating mechanical bowel obstruction from paralytic ileus. It is true that the width of the fluid levels gives you an idea of the caliber of a distended loop. The relative height of the levels, however, has proven unreliable in making the distinction between mechanical obstruction and paralytic ileus. Even normal people with no abdominal symptoms at all may have fluid levels on erect films, since swallowed air and intestinal juices are always present.

A vital diagnosis that must not be overlooked in any patient with acute abdominal complaints is perforation of the bowel with free peritoneal air. This may be a complication of bowel obstruction. The diagnosis of a small amount of free air cannot be made from the supine plain film because the bubble of free air floats anteriorly under the abdominal wall and may look just like a loop of bowel (if it is seen at all). On the erect film it is easy to identify free air interposed between the liver and the dome of the diaphragm. But patients with this symptom, who may have a bowel perforation, ought not to be asked to stand; they may fall and injure themselves. On the left-side-down decubitus film (made AP with a horizontal beam) it is easy to identify free air between the right lobe of the liver and the lateral portion of the diaphragm. Further discussion of free air is to follow, but remember that the diagnosis of intestinal obstruction as well as free air is a radiological one.

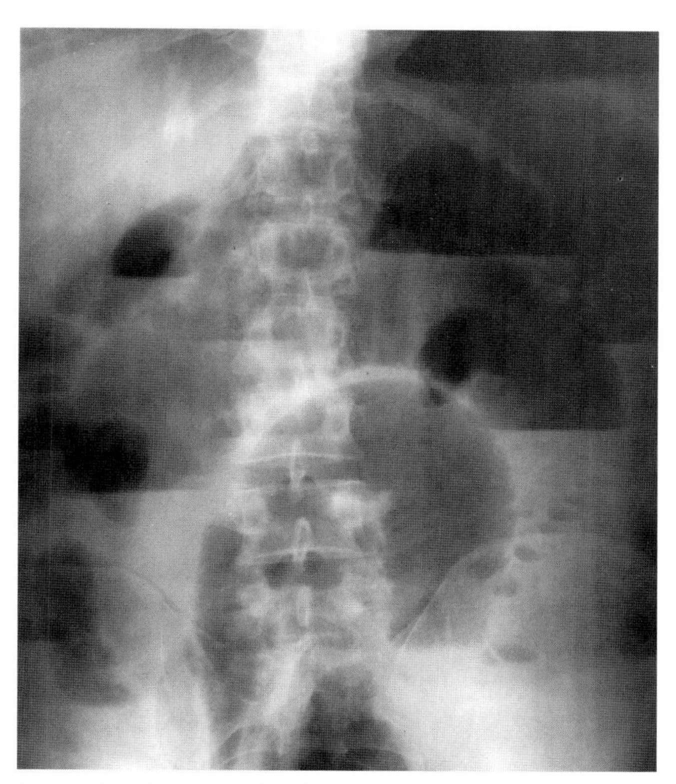

Figure 12.13. Air-fluid levels in obstructed small-bowel loops rarely help in differentiating obstruction from paralytic ileus. Levels are present in both conditions (and sometimes even in normal patients) on erect films.

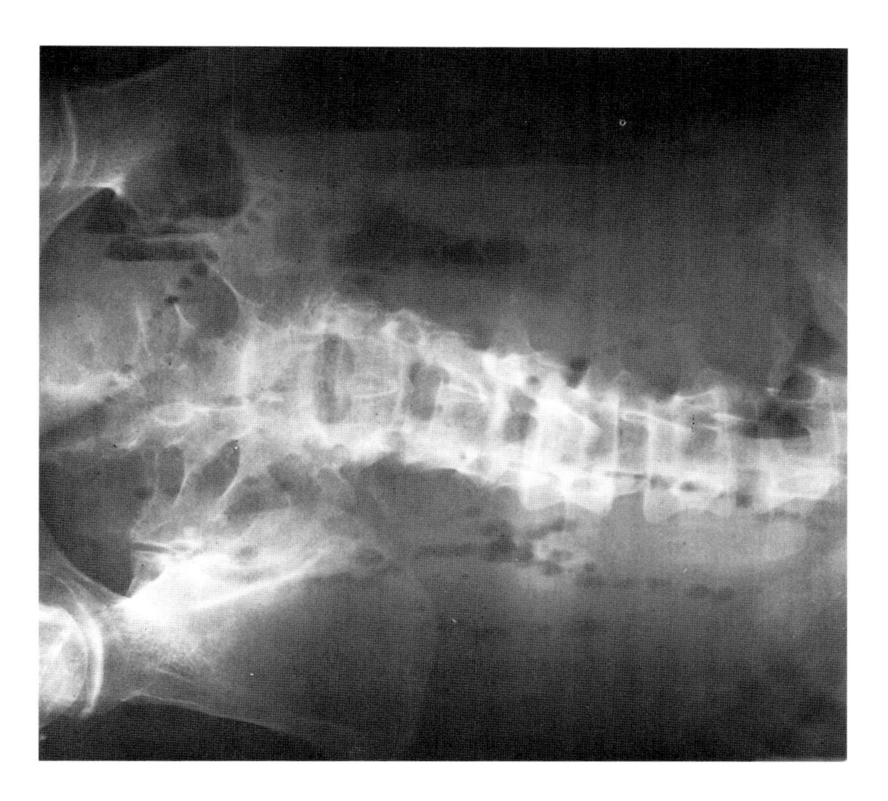

Figure 12.14. Left lateral decubitus film on the patient in Figure 12.12. Note the small bubbles of air caught against the valvulae conniventes of the obstructed small bowel, which is almost completely filled with fluid.

Free Peritoneal Fluid

Now direct your attention to the *general density* of the plain film. To state the oversimplified extremes first, as usual: large amounts of *free air* in the peritoneal space increase the radiolucency of the abdomen, just as you would expect, and the film looks darker. Large amounts of *free fluid* add to the radiodensity of the abdomen and the film appears grayer than usual. These statements are true for the conventional exposure techniques used for radiography of the abdomen. Remember too that if intestinal loops are filled with fluid, the effect is that of adding more thickness to the patient, and for the same exposure factors such a film is also gray and indistinct.

When there is a small amount of fluid free in the peritoneal space, it gravitates to the most dependent part of the peritoneal cavity, which in the supine patient is the bowl of the pelvis, as you see in the diagrams in Figures 12.17 and 12.18. Such relatively *small amounts of free fluid* probably often go unobserved on plain films, because we are more or less accustomed to seeing the pelvis filled with the density of a distended bladder. But small amounts of peritoneal fluid can easily be detected by CT and ultrasound.

Larger quantities of peritoneal fluid spill over into the abdominal cavity, flowing up the flanks on either side (Figure 12.19). Fluid collected in the flank displaces the colon medially away from the flank stripe (Figure 12.21), and with even greater accumulations air-filled loops of bowel float up under the anterior abdominal wall. They are seen on the supine plain film as a cluster of radiolucent shadows in the central abdomen surrounded by the uniform gray of the peritoneal fluid (Figure 12.15).

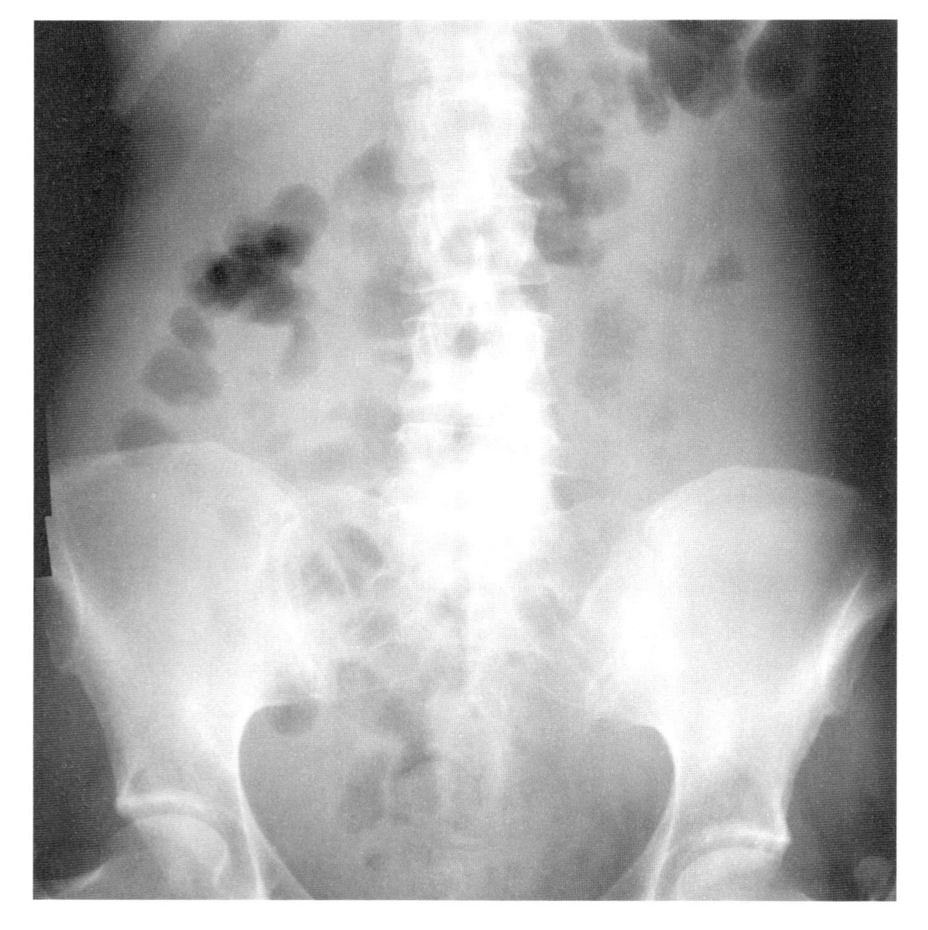

Figure 12.15. Ascites. Note the generalized opacity from free fluid. Since the patient is supine, loops of air containing small and large bowel float centrally under the anterior abdominal wall.

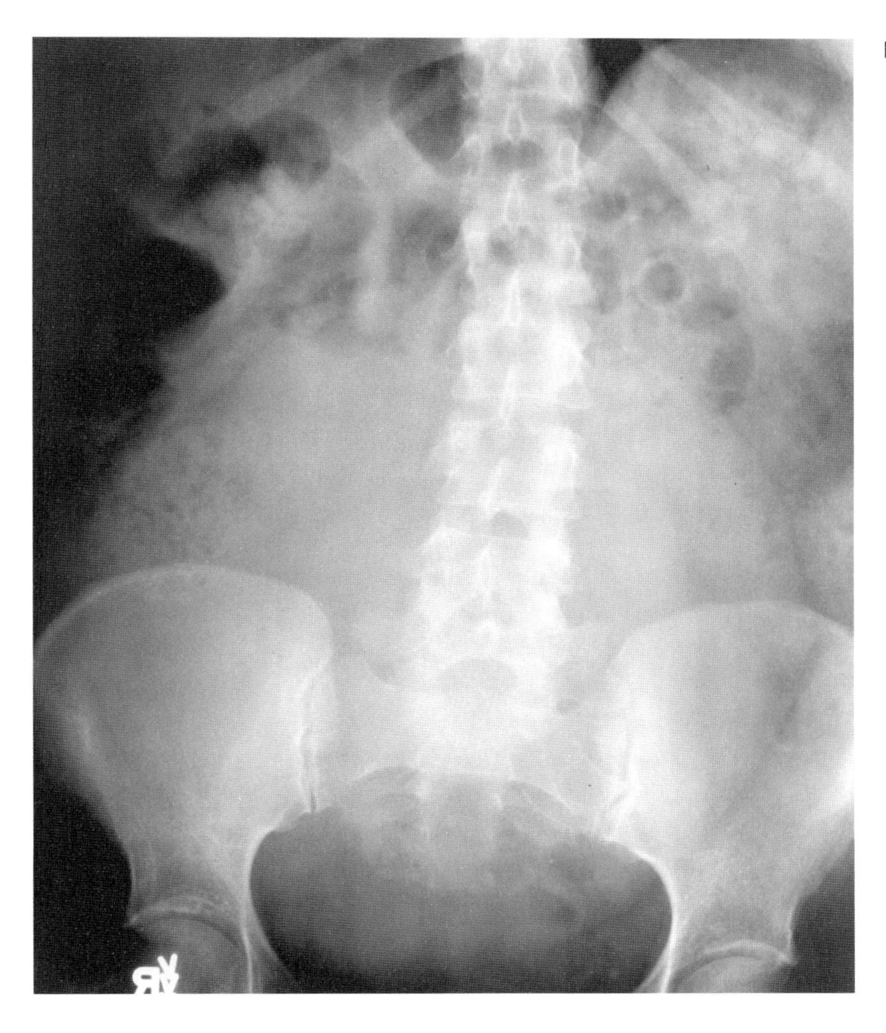

Figure 12.16. Is this ascites?

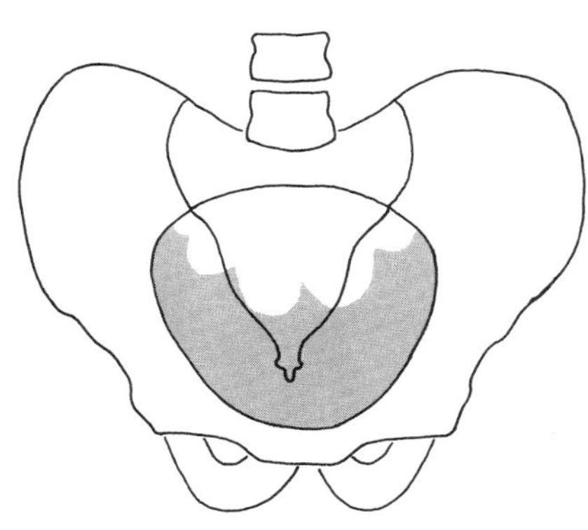

Figure 12.17. The half-moon of free fluid in the pelvis often engulfs the urinary bladder. The scalloped upper border results from loops of ileum dipping into fluid.

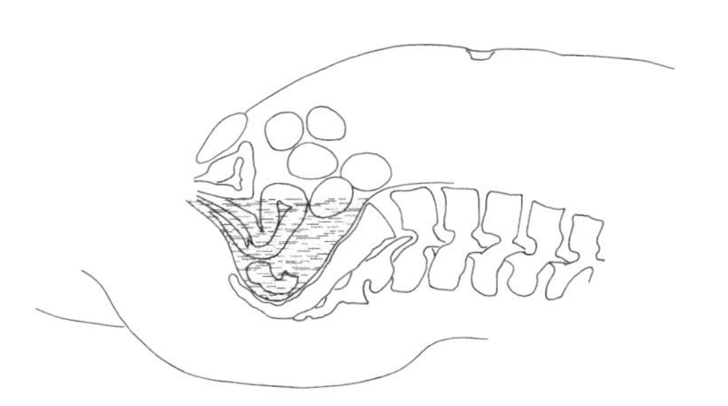

Figure 12.18. Lateral drawing of a supine patient with free peritoneal fluid accumulating in the most dependent part of the peritoneal cavity.

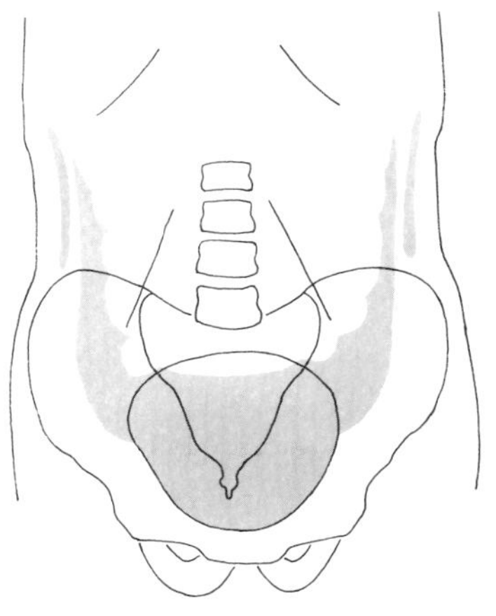

Figure 12.19. Increasing amounts of fluid flow into the flanks, displacing the colon medially, as in Figure 12.21.

In sum, a large amount of free fluid (ascites, blood, bile, etc.) is easy to recognize on plain films, but CT and ultrasound can show the presence of smaller amounts of fluid. In the CT scan in Figure 12.20, note the large volume of peritoneal fluid bathing the liver in a patient with malignant ascites. Compare the CT scan in Figure 12.22, a patient with bile peritonitis, due to a postoperative bile leak after biliary surgery, with the normal CT scan in Figure 12.23, taken at approximately the same level. A large amount of fluid is seen within the peritoneal cavity in Figure 12.22, with central migration of the bowel loops and mesentery. In the normal patient only bowel loops and fat are seen within the peritoneal cavity. Also note that

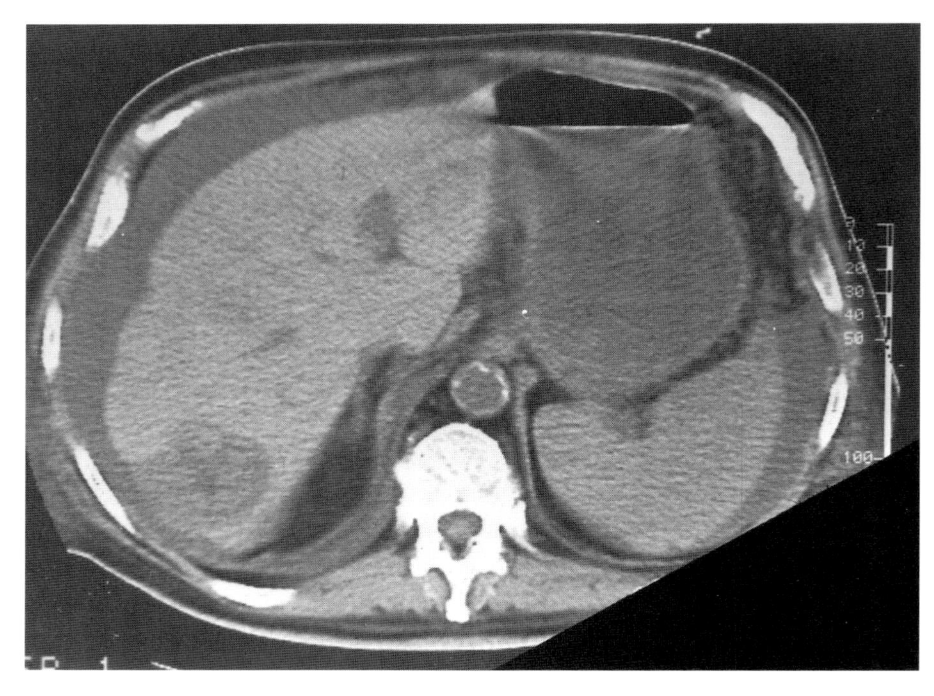

Figure 12.20. CT scan showing liver metastases from carcinoma of the colon. Note the ascitic fluid around the liver and spleen. The peritoneum was seeded with metastases.

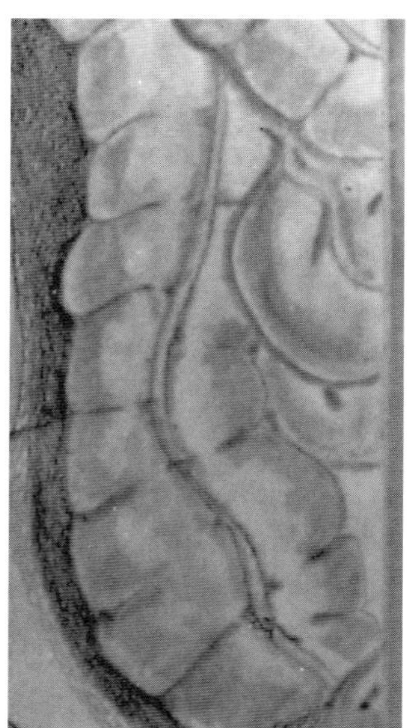

A

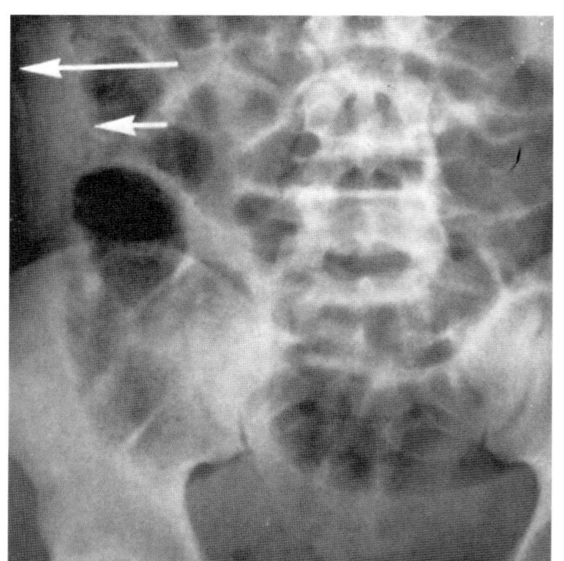

B

Figure 12.21. Liver laceration with hemoperitoneum after an automobile accident. Note in A and in the supine plain film (B) the interposition of fluid (blood in this case) between the flank stripe *(longer arrow)* and the medially displaced air in the colon *(shorter arrow).*

the volume of peritoneal fluid produces bulging flanks, not seen in the normal scan.

Look again at Figure 12.16. It could not show ascites, since the air-containing bowel is displaced up and laterally toward the flanks by a round central mass, which proved to be an enormous uterine myoma.

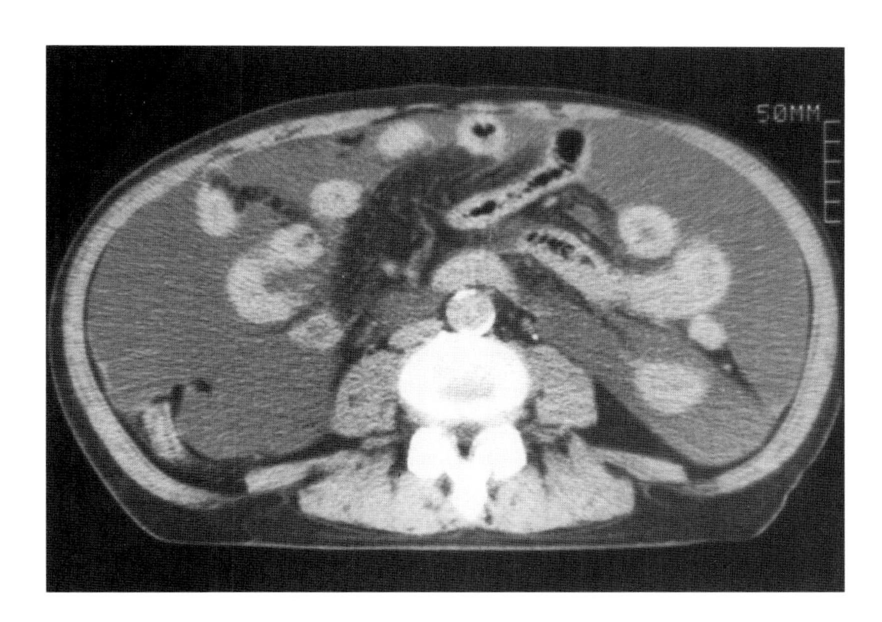

Figure 12.22. CT scan of a patient with a large peritoneal fluid collection, showing central migration of the bowel loops and bulging of the flanks. Compare Figure 12.23.

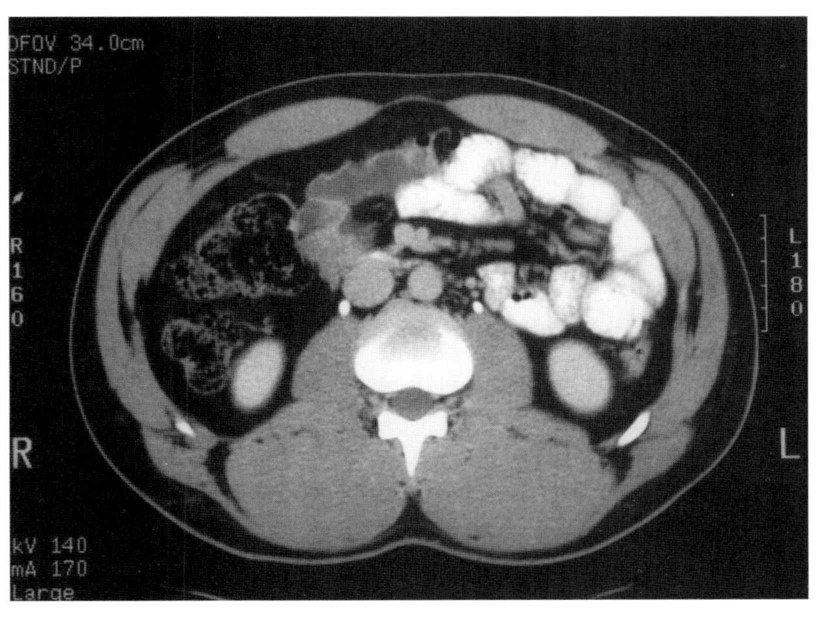

Figure 12.23. Normal CT scan for comparison with Figure 12.22.

Free Peritoneal Air

A large amount of free peritoneal air strikingly outlines the organ masses of the liver and spleen, including their lateral and superior (diaphragmatic) surfaces (Figure 12.24). The volume of free air is so large in Figures 12.24 and 12.25 that detection can be made on supine films. In Figure 12.25 the identification of free air is made on the basis of seeing both sides (mucosal and serosal surfaces) of the bowel wall. In a normal person air should only be seen margining the bowel lumen (mucosal surface). Small amounts of free peritoneal air may require upright or decubitus films or even CT for their detection.

Small amounts of free air may be quite as important to detect as larger amounts, for they occur most frequently with subtle or early perforation of the bowel and could be missed on a routine plain film examination. A patient with these symptoms, if well enough to stand, will show crescents of radiolucent air interposed between the diaphragm and the liver, and such a finding is occasionally first detected on chest films (Figure 12.26).

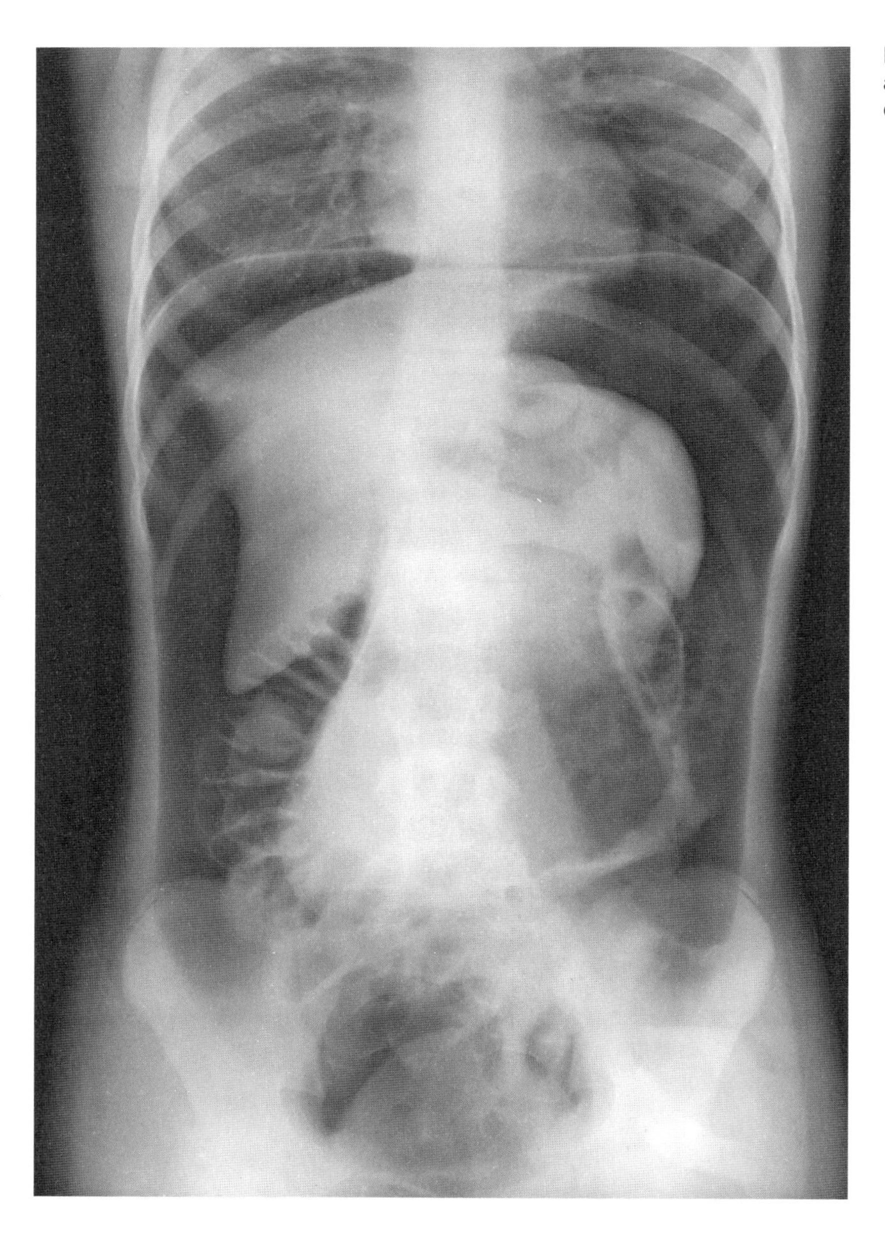

Figure 12.24. Free peritoneal air in large amounts after traumatic rupture of the duodenum during an assault.

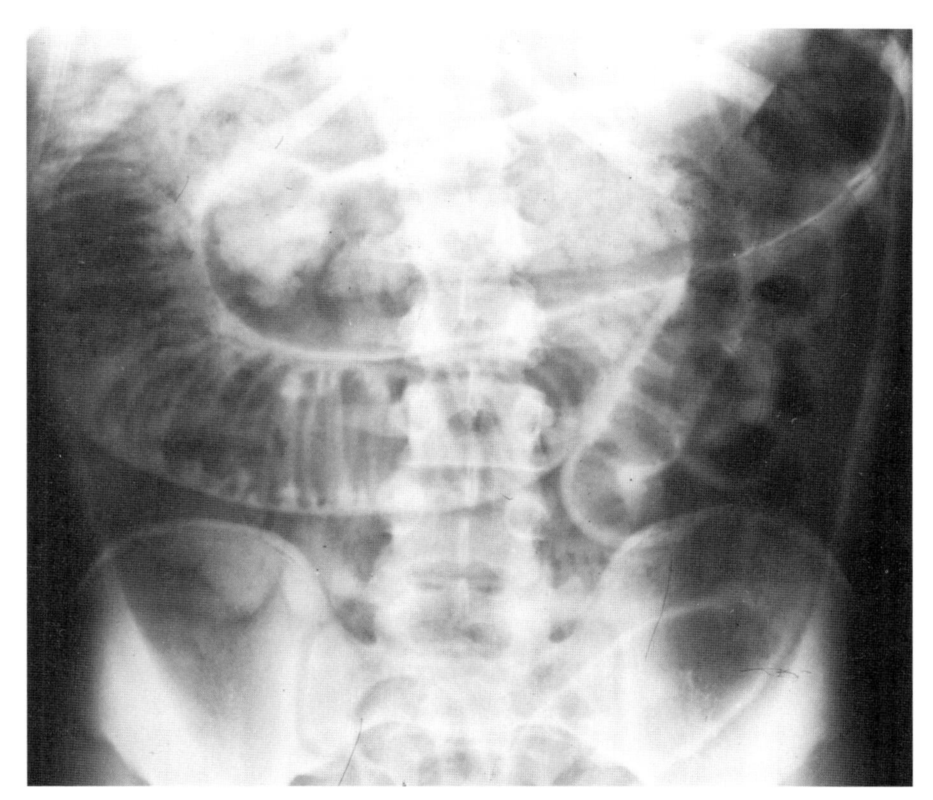

Figure 12.25. Free peritoneal air appreciated on the supine plain film. Both the mucosal and serosal surfaces of the bowel wall are seen outlined by air.

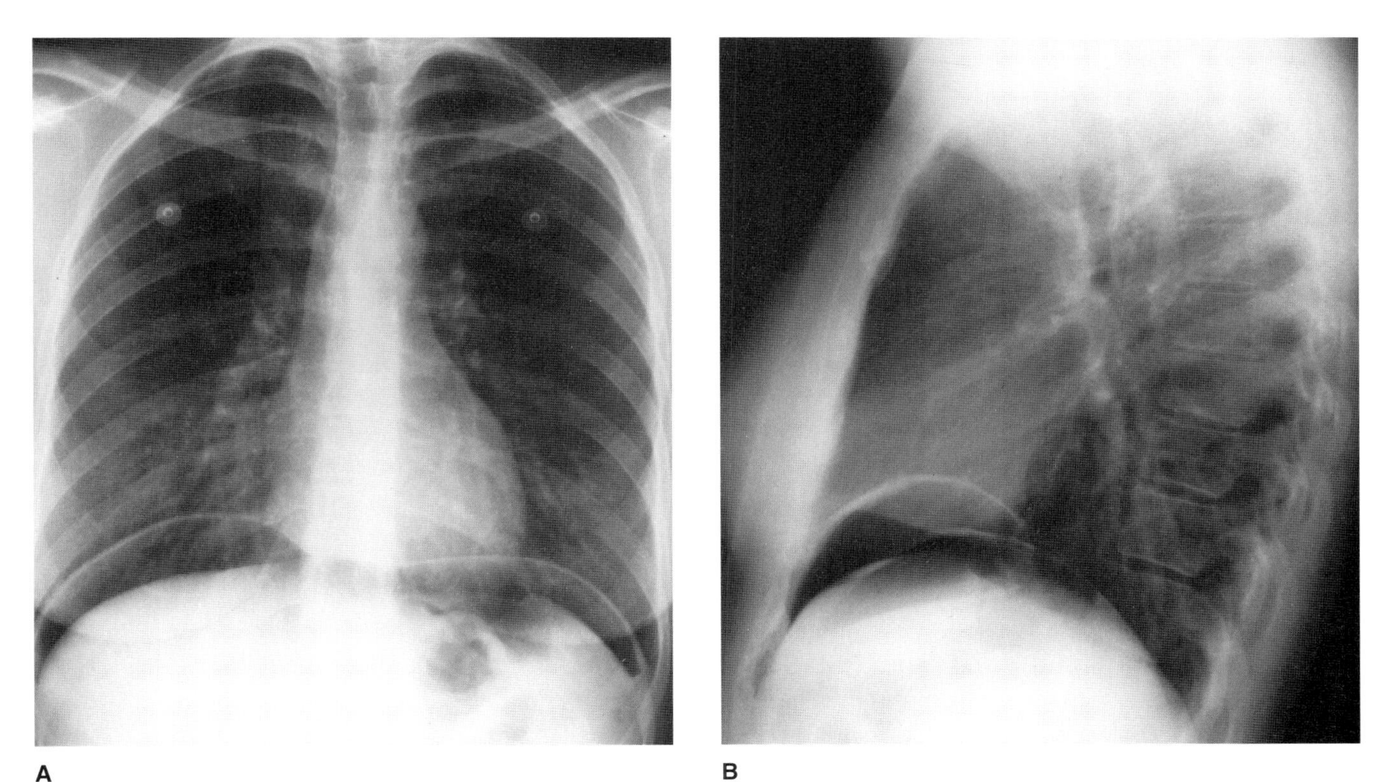

A B

Figure 12.26. PA (A) and lateral (B) chest films showing a smaller amount of free air under the diaphragm than in the patient in Figure 12.25.

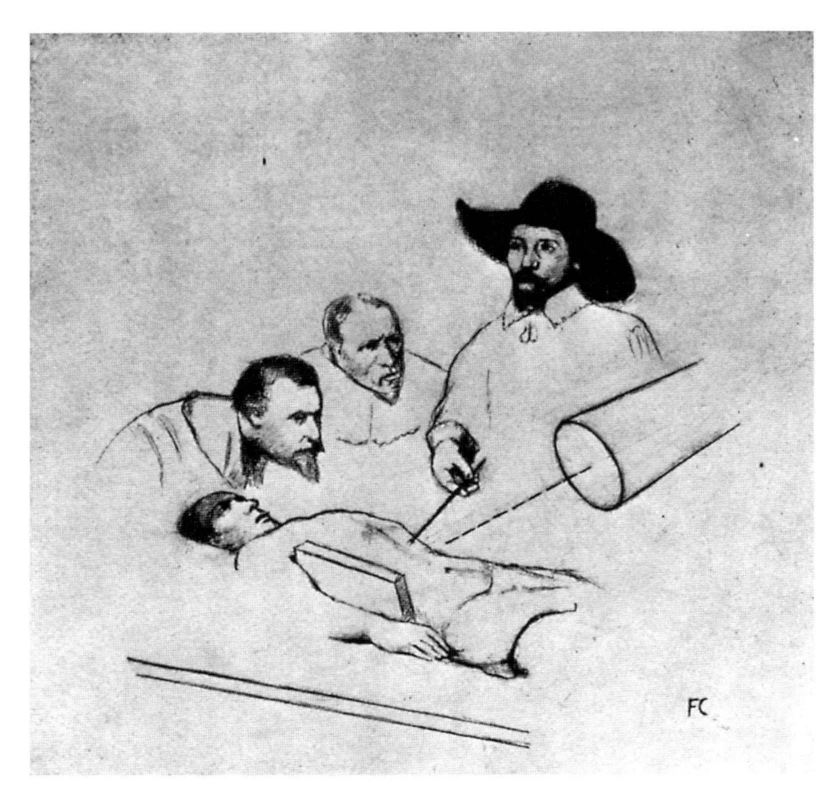

Figure 12.27. Lateral abdominal radiography with the patient supine (cross-table lateral). Dr. Tulp indicates the use of this projection to two eager residents.

The patient with a perforated viscus is often too ill to stand, however, and in any event ought not to be disturbed any more than is absolutely vital. For this reason the search for free air is much more often carried out by placing the patient on his left side and radiographing him anteroposteriorly with a horizontal beam. The resulting film, a left lateral decubitus film, will show even a small amount of free air lateral to the liver (Figure 12.29). *It is impossible to see small amounts of free air on a supine plain film.*

The horizontal beam can also be used in a patient too ill even to be turned on his side, as in Figure 12.27. The patient lies supine and is radiographed from side to side, a cassette being placed vertically against his flank. In this way air free against the underside of the anterior abdominal wall can be seen. Figure 12.28 shows you such a horizontal-beam lateral film in a patient with a large amount of free air.

Remember that on any of the films made with the horizontal beam (erect, left lateral decubitus, or cross-table lateral), the free air is apparent because it lies against some structure not normally outlined with air (undersurface of the diaphragm, lateral side of the liver, anterior abdominal wall). Never forget that even a moderate amount of free air against the anterior abdominal wall cannot be recognized on the ordinary supine plain film made with a vertical beam, because the bubble of air looks just like another loop of bowel. *Never request only a supine plain film to "rule out free air."*

CT can identify a smaller amount of free air in the peritoneal cavity than can plain films. You should always search for free air on the abdominal CT scans of patients whose clinical condition suggests it. Figure 12.30 is the CT scan of an elderly woman injured in a fall. The free air resulted from a rupture of her transverse colon. The finding of free air in an abdominal trauma patient usually indicates the need for immediate laparotomy to identify and repair a bowel injury. At CT free air is best seen in the least dependent portion of the peritoneal cavity, usually in the anterior upper abdomen overlying the liver.

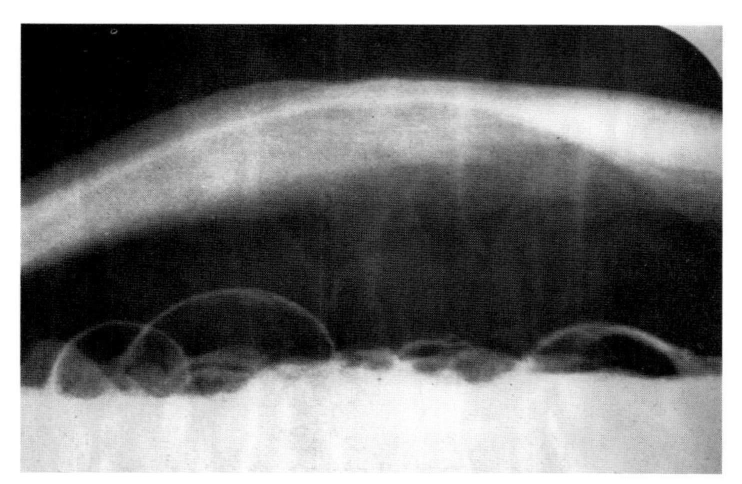

Figure 12.28. Horizontal-beam lateral radiograph made with the patient supine, as in Figure 12.27. The abundant quantity of free air under the abdominal wall outlines the serosal surface of the air-filled loops of the bowel.

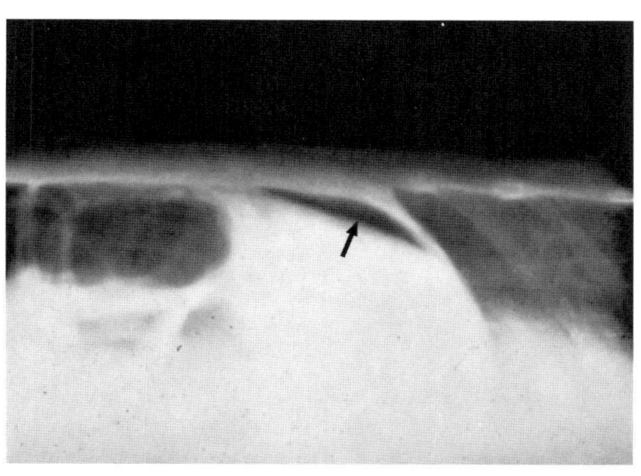

Figure 12.29. Left lateral decubitus film, showing free air *(arrow)* lateral to the liver in a patient suspected of a perforated viscus, in this case a duodenal ulcer.

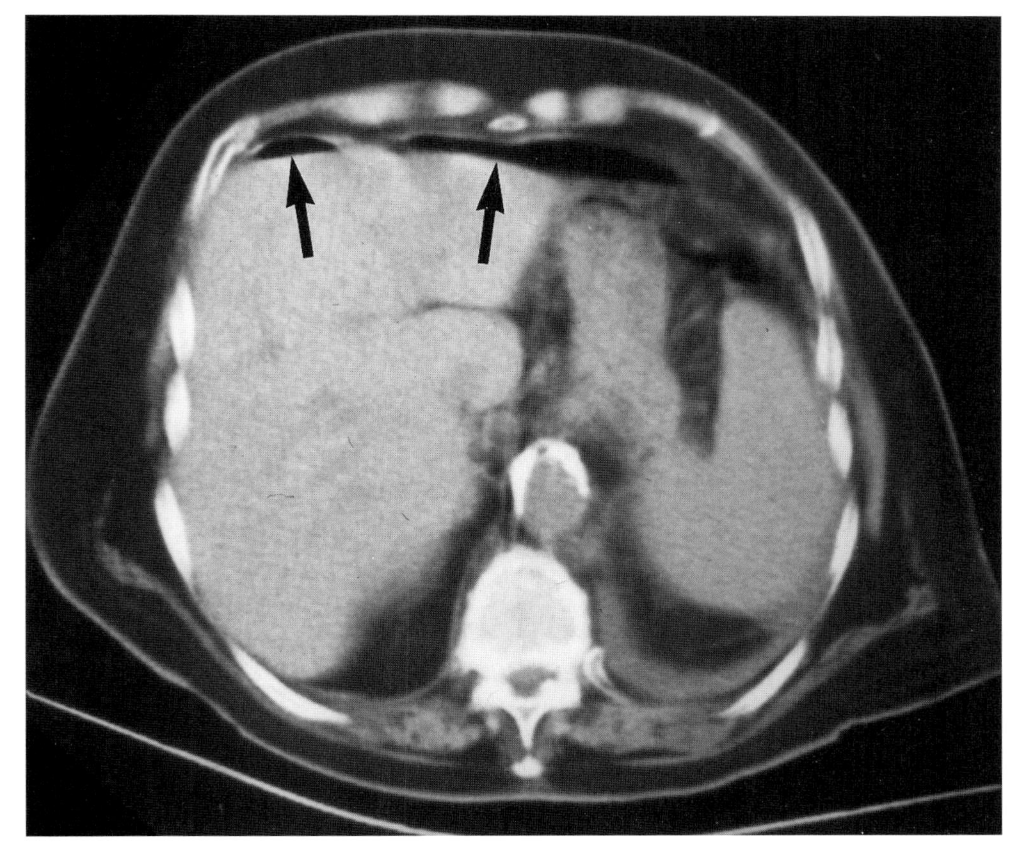

Figure 12.30. CT scan showing a small amount of free air *(arrows)* over the liver. As you would expect in a supine patient in a CT scanner, the free air migrated anteriorly, to the least dependent portion of the peritoneal cavity. This elderly patient had suffered a colon rupture due to blunt abdominal trauma sustained in a fall.

13 Contrast Study and CT of the Gastrointestinal Tract

Many abnormalities of the gastrointestinal tract can be readily identified by barium examination. In other cases, CT provides more precise information; and in some conditions, only CT can provide images that allow an accurate diagnosis. We will begin with images of barium and associated studies, and then go on to consider CT.

Principles of Barium Work

Barium study of the GI tract involves the interpretation of shadows not only of whole barium sulfate fluid casts of hollow structures but also of the far more complex shadows of thin films of opaque substance caught against the mucosal irregularities of the inner surface of the bowel. Such *mucosal relief studies,* as they are often called, are carried out with minimal amounts of opaque material, manipulated and spread over the surface of the clean mucosa during fluoroscopic study, *spot films* being obtained at frequent intervals by either mechanically substituting a small cassette for the fluoroscopic screen or producing hard copy digital images from the fluoroscopic screen itself.

In certain cases, and when considered desirable, the radiologist inflates the stomach or colon with air, a procedure termed an *air-contrast* or *double contrast study,* which may provide additional information, especially about superficial lesions. The technical expertise required by these procedures and the judgment and experience needed for interpreting correctly the observed shadows constitutes one of the most sophisticated skills of the radiologist.

Nevertheless, the conclusions the radiologist draws are basically no less logical than everything else you have been learning to appreciate about the field. In order to comprehend the reliability of the evidence offered by roentgen data obtained from barium studies, you should understand some of the fundamental implications of the various kinds of roentgen observations based on such procedures.

To that end, examine the hypothetical drawings in Figure 13.2. The gastrointestinal tract is essentially a tube, and the roentgen principles for examining it vary only in degrees even in the stomach, cecum, and rectum, where its tubular structure has been modified by nature. The simple tube in *A,* filled with an opaque substance and radiographed, would produce a shadow like the one you see in *a,* smooth bordered and uniformly dense. A polyp protruding into its lumen upon a stalk, like that in *B,* would produce a barium cast-shadow like the one in *b.* A solid tumor growing in its wall like that in *C,* and protruding into the lumen as a sessile growth, would produce a shadow like *c.*

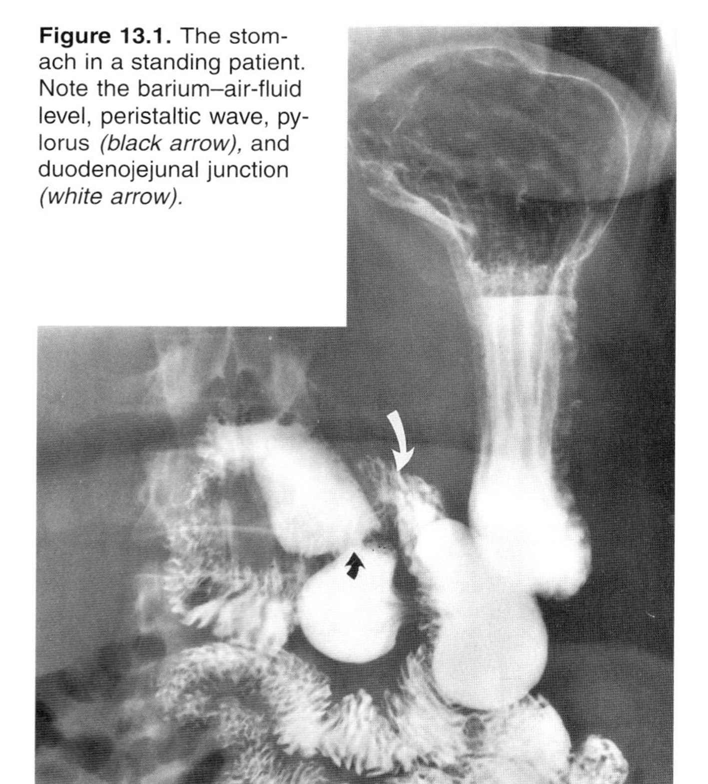

Figure 13.1. The stomach in a standing patient. Note the barium–air-fluid level, peristaltic wave, pylorus *(black arrow),* and duodenojejunal junction *(white arrow).*

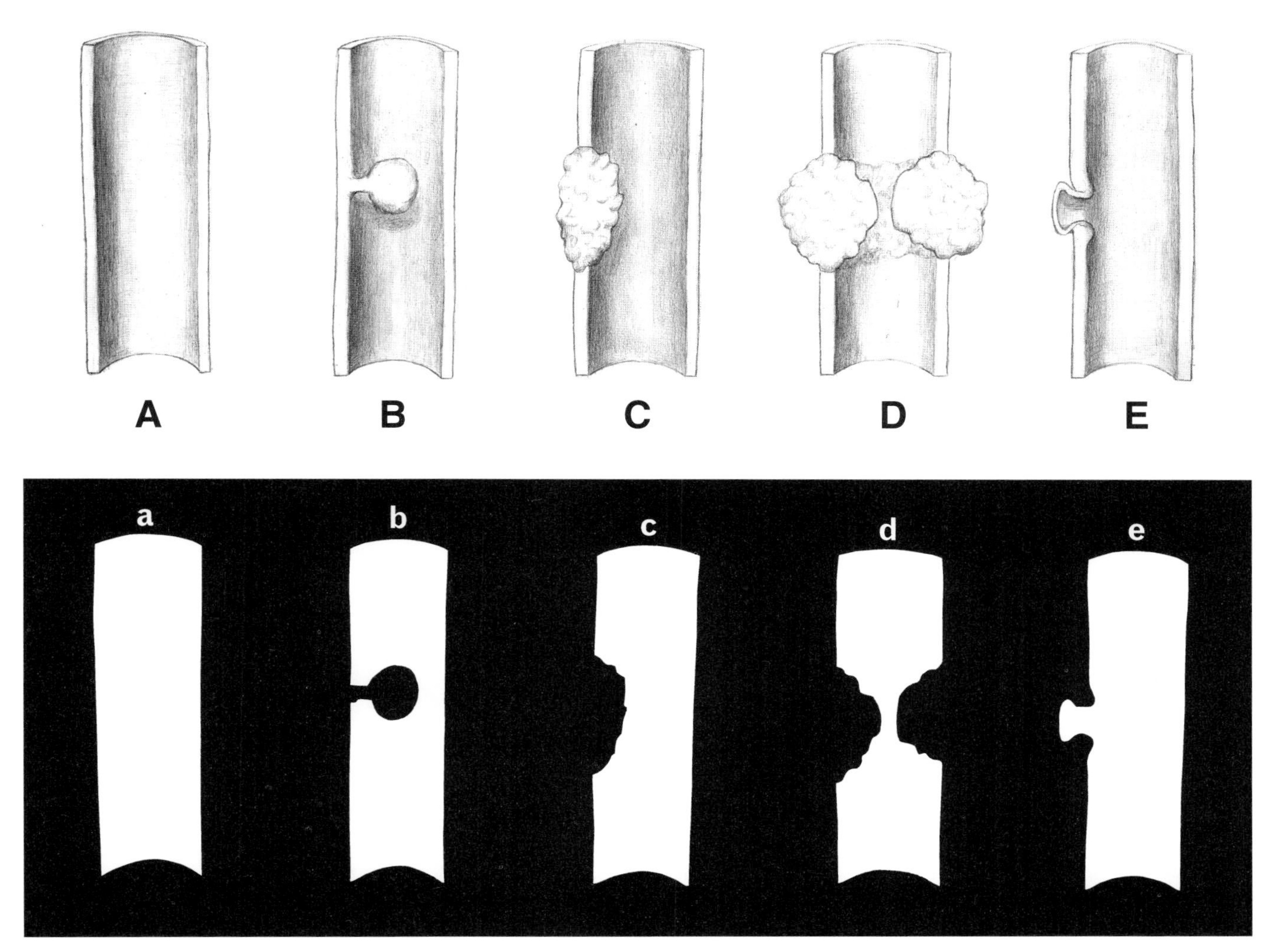

Figure 13.2

Both of these alterations in the original normal tubular shadow are what are called *filling defects.* A part of the expected luminal shadow has been subtracted, because barium has been displaced by radiolucent soft tissue.

The growth you see in *D* has entirely encircled the tubular structure being examined, so that a constriction of the lumen is produced. This is often called an *annular defect.* Whenever you see a barium shadow like the one in *d,* you ought to *reconstruct mentally* the rigid annular lesion that has produced it, representing tumor or other soft tissue wherever the barium has been displaced. The abrupt and often angular change in the shadow where the normal luminal wall meets the margin of a tumor is frequently and aptly referred to as a *shelf* or *shoulder,* and you should interpret its consistent appearance on film after film in the same location as reliable evidence of rigidity of some sort in the otherwise distensible wall of the bowel. *E* and *e* represent the appearance of a benign ulcer crater in the wall of the bowel that extends the luminal shadow and is surrounded by a *collar* of regenerating mucosa, seen in tangent about the neck of the crater.

You will find that a filling defect and its shelflike margin are often so precisely the same from film to film in a series made during the barium study that they may be superimposed on each other over a bright light. If on two or more such films you can bring into perfect register the margin of a filling defect suspected of representing a malignant tumor, the probabilities that it *is* a tumor are greatly enhanced. If, of course, two such films do not superimpose, it may mean either that the area in question is not rigid, and therefore changes slightly, or that the two films were made in different projections.

Normal Variation versus a Constant Filling Defect

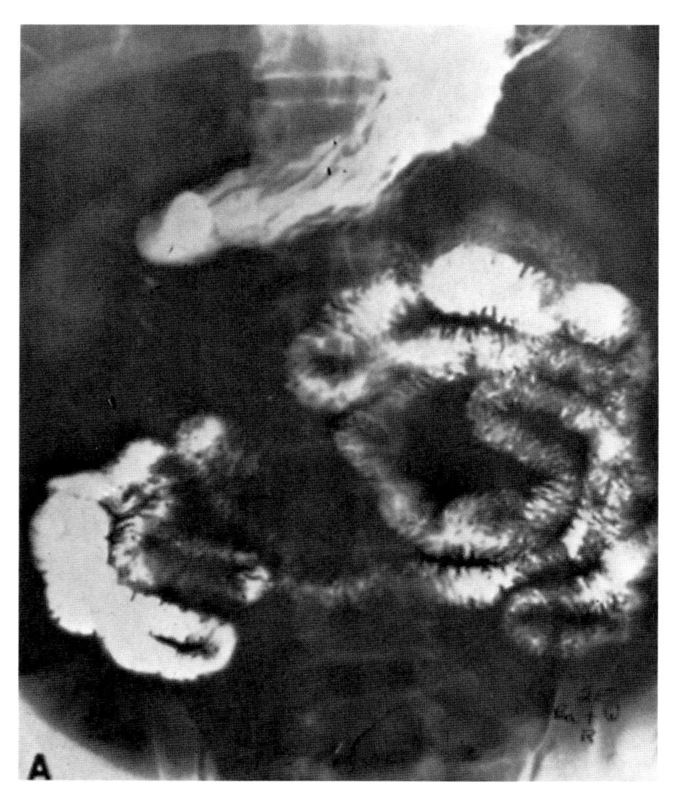

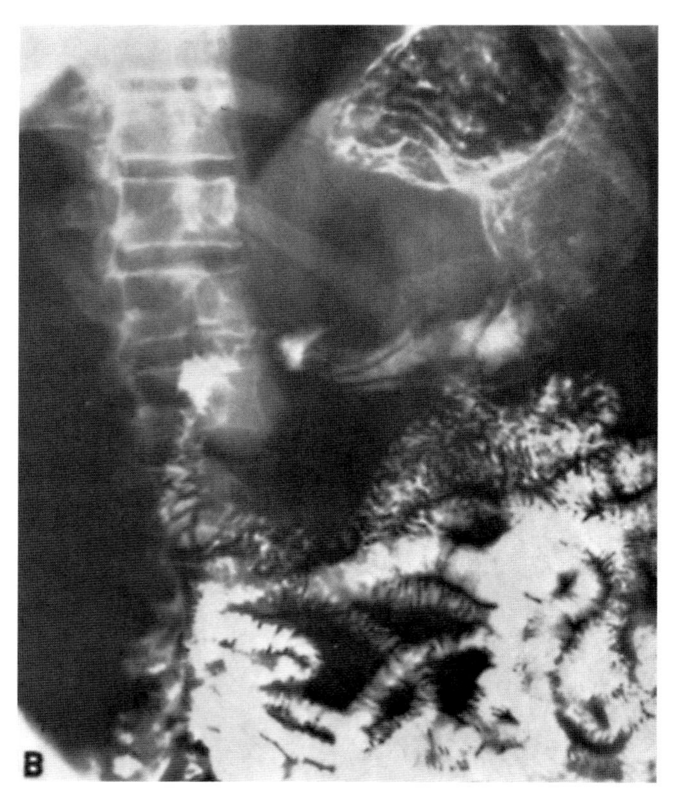

Figure 13.3A-C. Normal upper gastrointestinal series, large overhead films only. Films made in other projections and spot films made during fluoroscopy would complete the series.

The variation of barium shadows within the gastrointestinal tract is well illustrated by these films made during an *upper GI series* with small-bowel follow-through (Figure 13.3). A, made prone half an hour after 10 ounces of barium suspension had been swallowed, shows the stomach partly emptied and the jejunum and upper ileum filled. B, made 15 minutes later in an oblique position, shows most of the small bowel filled, and C, made an hour after the administration of barium, shows the right colon filling. The transit time for barium from the stomach to the right colon varies, but is about 1.5 hours in a fasting normal patient. (Later on, the presence of barium in the hepatic flexure and transverse colon would interfere with refilming of the stomach and duodenum.) The gallbladder is opacified from orally administered gallbladder contrast material.

You may be inclined to reject the possibility of demonstrating any structure consistently in so changeable a barium pattern, but look carefully at Figure 13.4, two spot films of the stomach and duodenum of a patient who had guaiac-positive stools. She had no hemorrhoids, negative findings on colonoscopy and barium enema study, was not anemic, and enjoyed excellent health. What could have caused her gastrointestinal bleeding?

If up to now barium studies have tended to confuse you, if you have wondered how any firm conclusions at all can be derived from them, this is the time to tell you that the radiologist arrives at reliable interpretations largely by demonstrating a finding *repeatedly.* No finding present only on a single film is worth very much, and the resident in training in radiology soon finds that positive diagnostic observations made

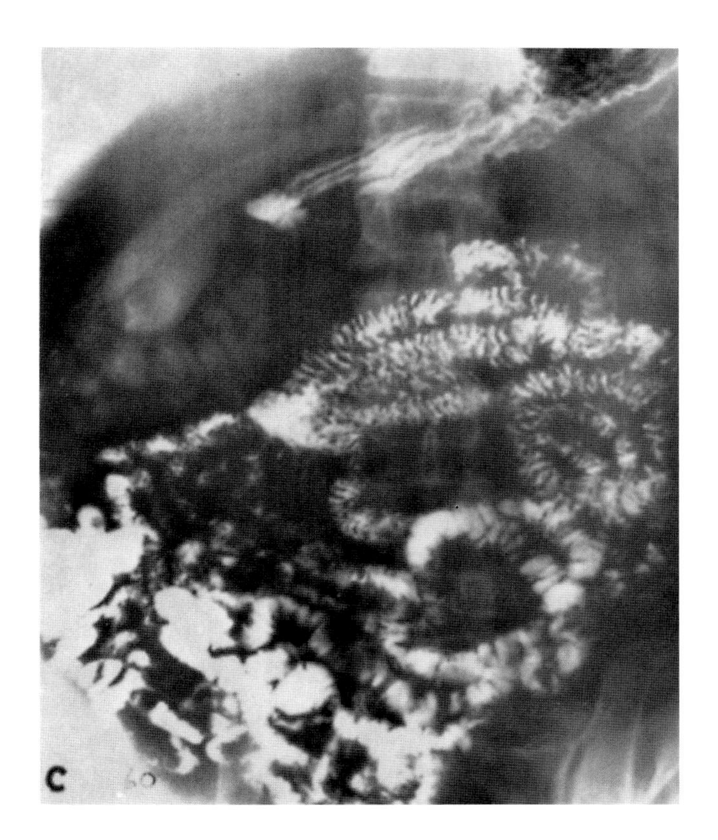

from barium studies must be *consistently demonstrable* if they are to be believed. So variable and shifting are the shadows presented by opaque substances within the constantly changing gastrointestinal tract that only those consistently present should be taken seriously. A transient finding seen on only one film may represent an area of spasm, a peristaltic wave, or food. This is no less true for other spheres of roentgen investigation, to be sure—nor for that matter for other branches of medicine. Any single positive test, always negative thereafter, is unlikely to weigh much in the balance of evidence, and the principle is no different for the more complex investigative procedures.

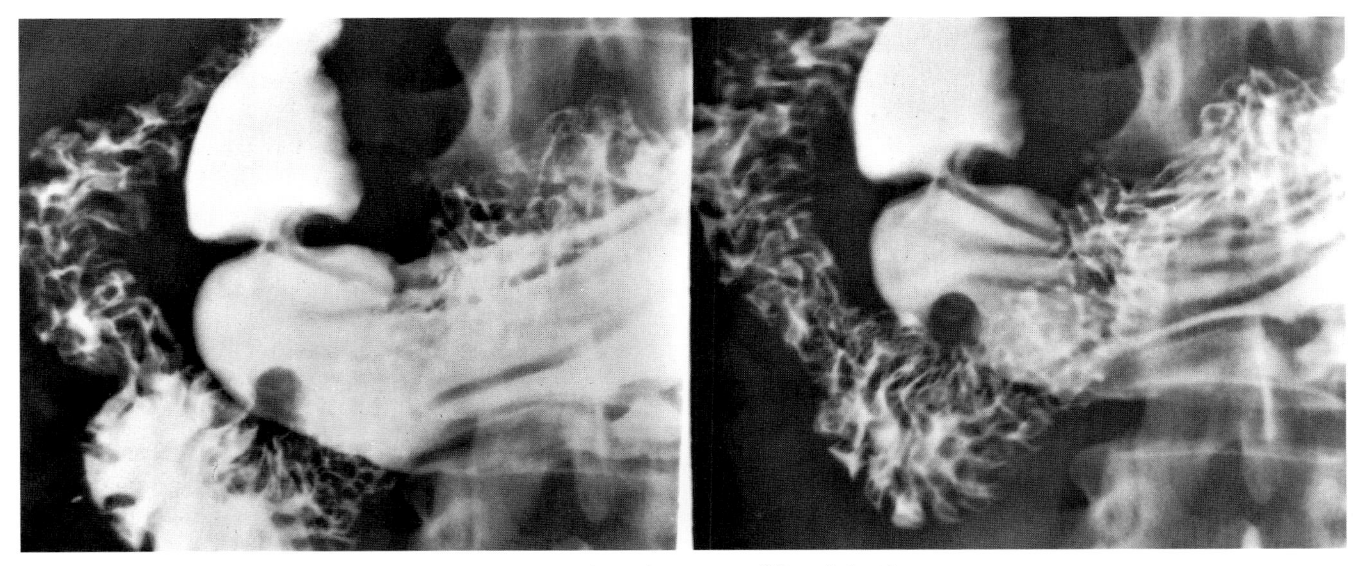

Figure 13.4 ***(Unknown 13.1).*** Can you spot the consistently present filling defect?

The Components of the Upper GI Series

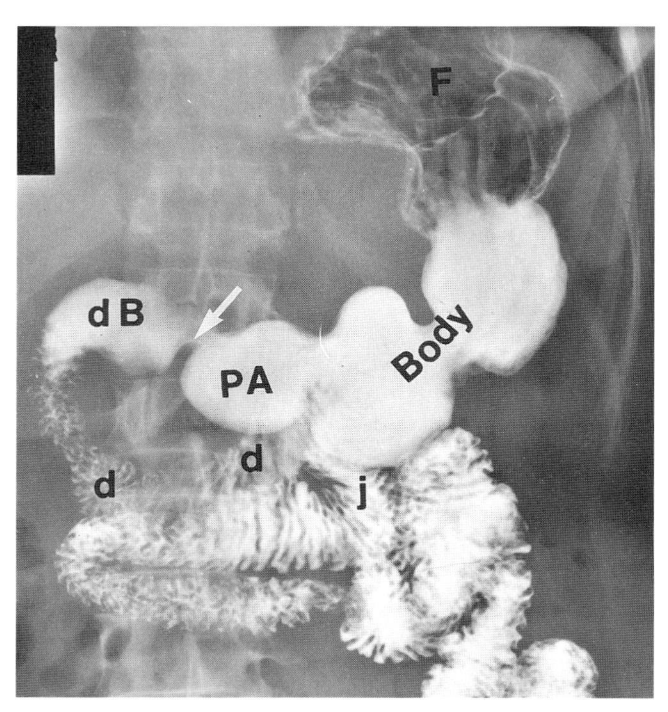

Figure 13.5. Prone film of a barium-filled stomach with fundus *(F), body,* pyloric antrum *(PA),* pyloric canal *(arrow),* duodenal bulb *(dB),* C-loop of duodenum (*d*'s), *and* jejunum *(j)* identified.

The student who looks at radiographs of the barium-filled stomach for the first time often has some difficulty identifying the various parts of the stomach, in particular the pyloric canal, and this difficulty generally stems from the fact that each of the conventional views of the stomach is made in a different projection with the patient differently positioned.

The radiologist usually begins with the patient standing, and examines the esophagus and stomach with a small amount of barium to study the mucosal relief pattern. Then the radiologist tilts the power-driven fluoroscopy table into the horizontal position and arranges the patient prone and turned slightly up on his right side. The patient drinks more barium (through a straw) and is turned into the lateral and then into the supine position. Routine spot films are made at intervals and whenever the radiologist sees anything on the screen that merits recording. Next, large films are made by an x-ray technologist in a series of specified projections. These generally include one made straight prone, one prone but turned slightly to the right (the right anterior oblique), one made in a straight lateral projection with the patient on his right side, and one made supine with the patient rolled to the left slightly so as to fill the antrum of the stomach with air. The radiologist may elect to study the stomach by *air-barium double contrast* for certain suspected conditions. This is accomplished by giving the patient effervescent materials with the barium.

Study the series of normal stomachs in various positions on this and the following pages. Note the varying shape of the stomach and of the duodenal bulb, as well as the deep indentations in both curvatures (peristaltic waves, which progress when seen at fluoroscopy). You can figure out the position of the patient and the projection of the film by noting (1) the appearance of the spine and (2) the presence of barium or air in the fundus of the stomach. If the spine appears *symmetrical* (vertebral pedicles on either side of the spinous process), then the film was made either AP or PA with the beam in the sagittal plane (frontal view). If the *bodies of the vertebrae* appear completely separated from the complex posterior structures, you are looking at a *lateral view* and can see the anterior and posterior walls of the stomach. If the spine is seen with the complex posterior structures overlapping the posterior third of the body of the vertebra, you are looking at an oblique film. You can easily decide whether it is a prone oblique or a supine oblique film, because in any prone film the fundus contains air, whereas in any supine film barium flows back into the more posterior fundus. Decide for each of the following films whether the patient was filmed prone, lateral, or supine, and whether the film is frontal, lateral, or oblique. The pylorus is located for you in each study by an *arrow*. You will find the answers at the bottom of page 267.

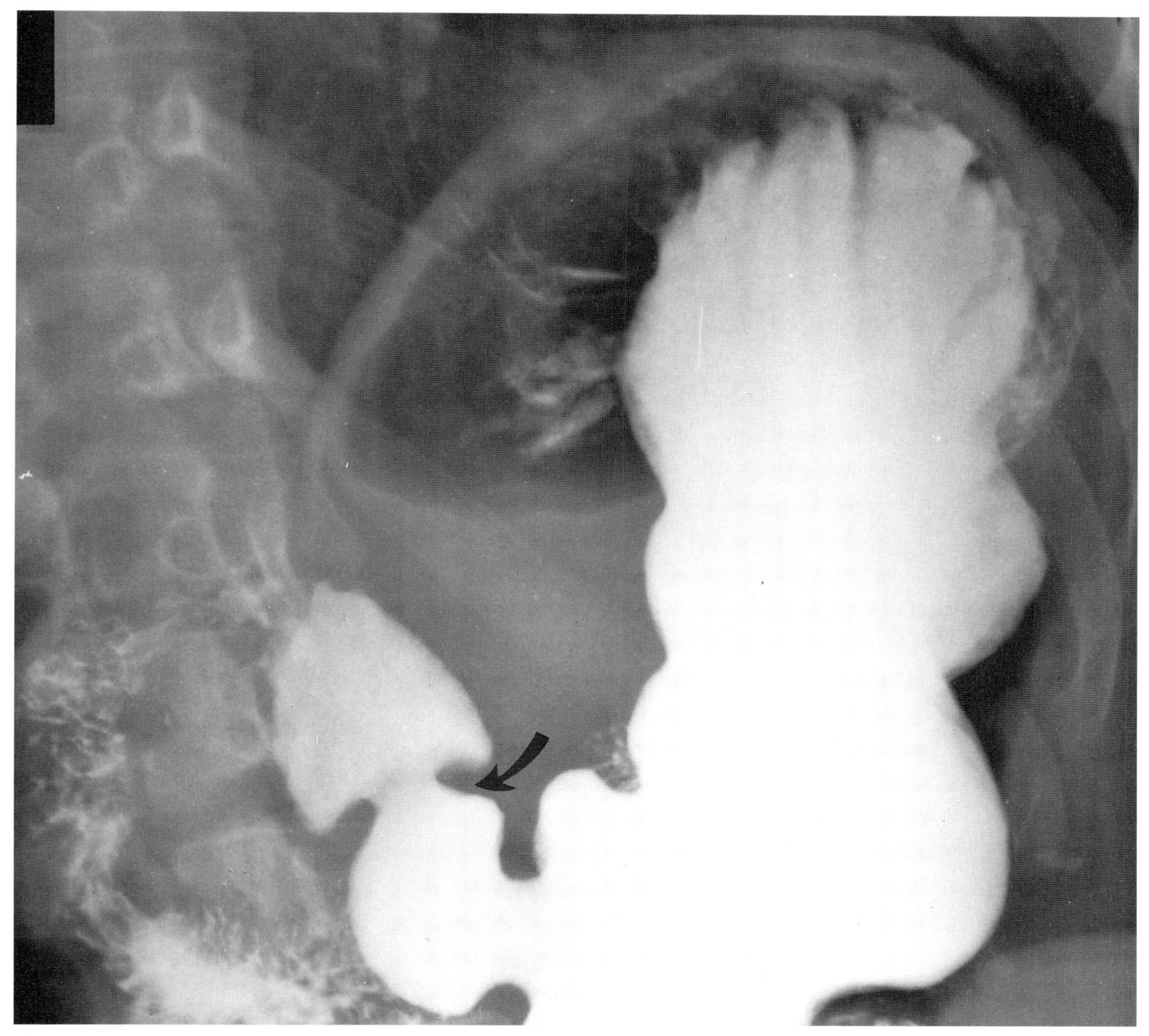

Figure 13.6

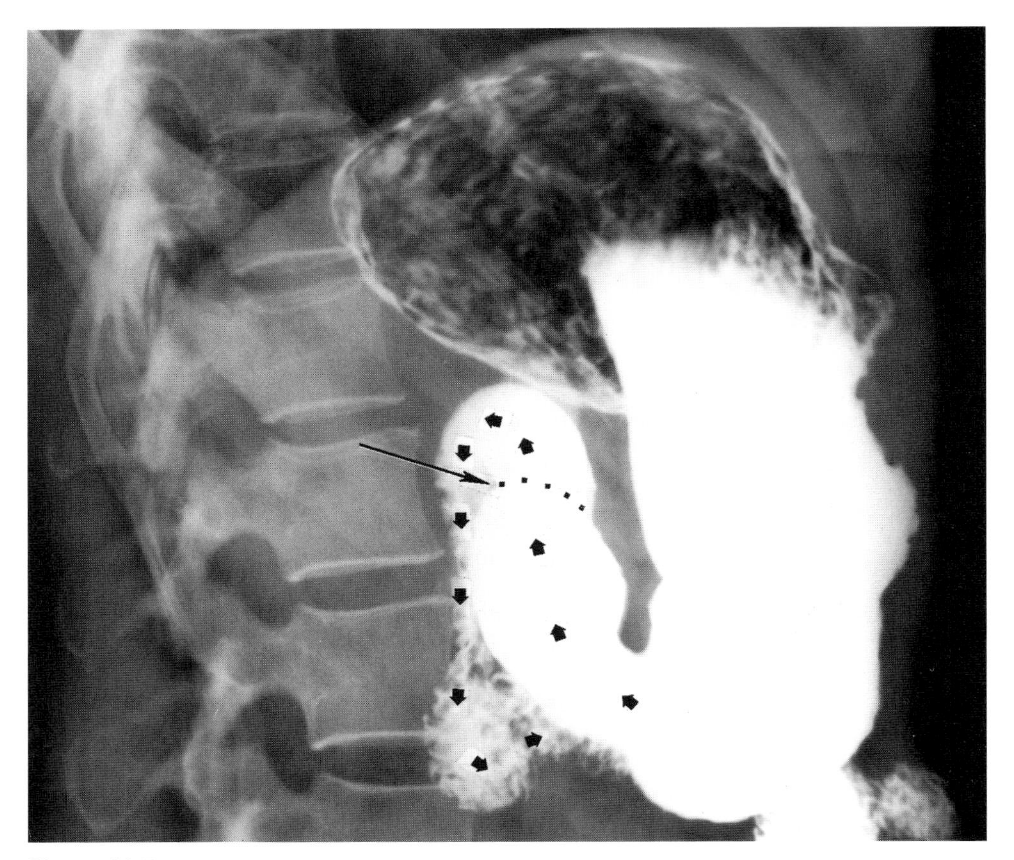

Figure 13.7

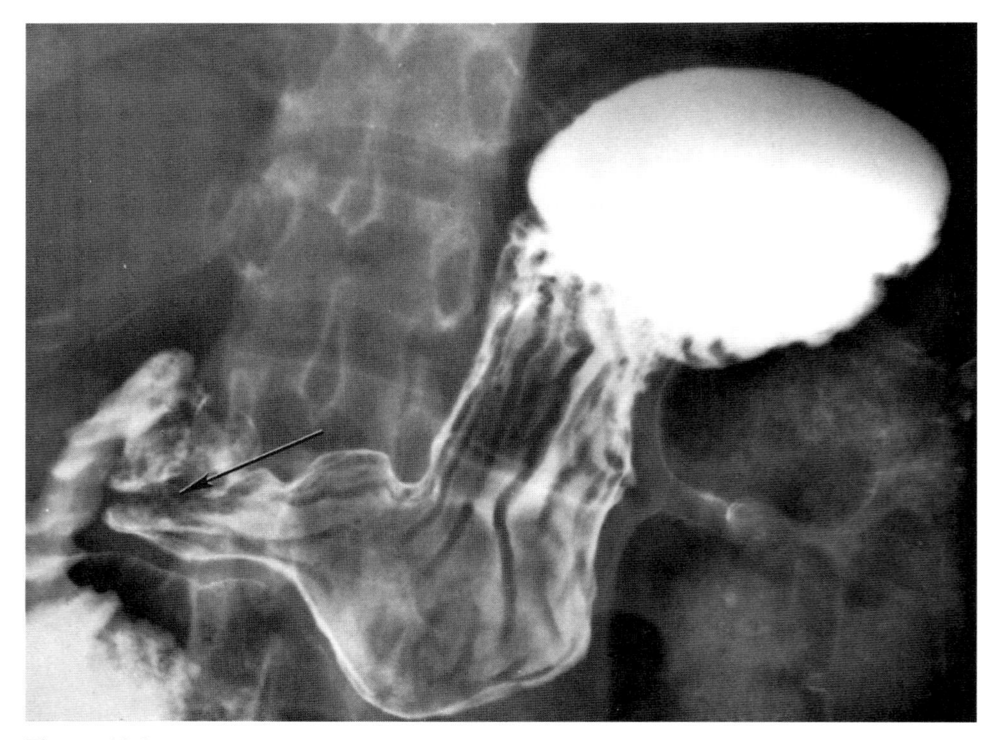

Figure 13.8

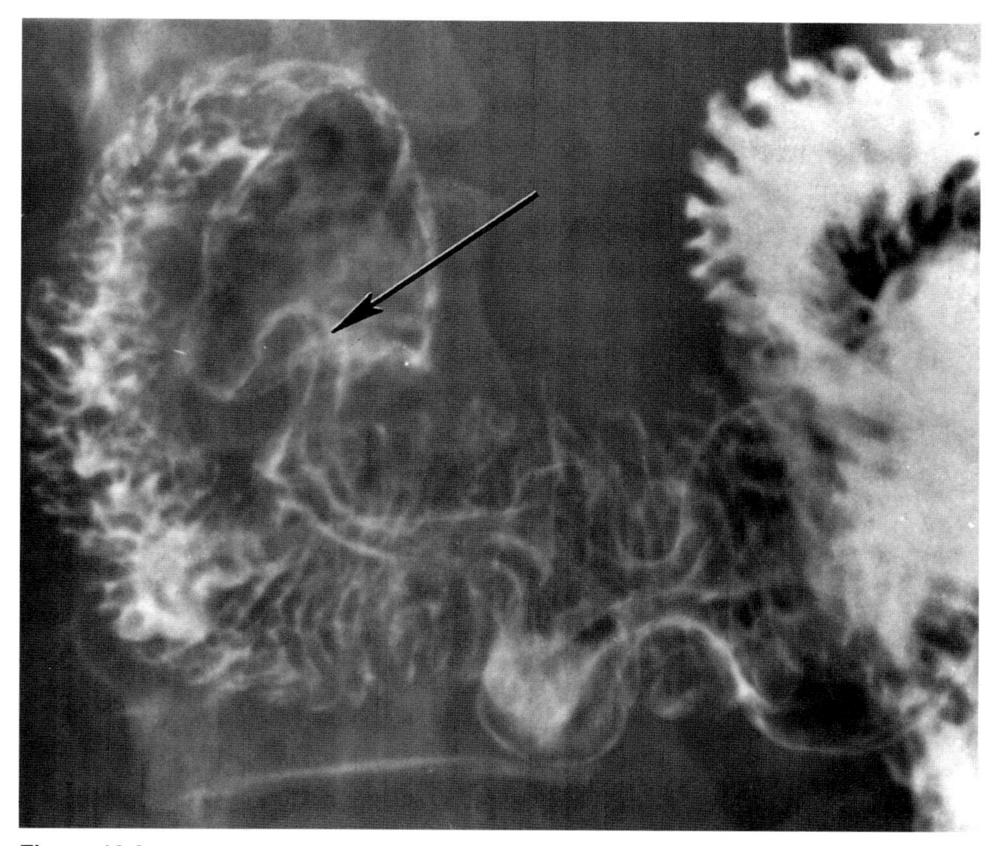

Figure 13.9

Answers

Figure 13.5 is a frontal prone film. Note the air in the fundus.

Figure 13.6 is a right prone oblique film with air in the fundus (the highest part of the stomach in the prone patient). The pyloric canal *(arrow)* and the normal, nondeformed duodenal bulb are seen.

Figure 13.7 is a right lateral film. The spine is clearly seen in the true lateral projection; the patient is lying on his right side, and air is filling the fundus. Note that ulcer craters on the posterior gastric wall might be well seen in this view. The *long arrow* indicates the site of the pyloric canal; the *short arrows* indicate the direction of fluid flow.

Figure 13.8 is a supine film. Most of the barium has flowed back into the now-dependent fundus, and air is seen inflating the barium-coated body and pyloric antrum. The *arrow* indicates the pyloric canal.

Figure 13.9 is a detail view of a supine oblique. The patient is rolled onto his left side, and air is flowing through to inflate the duodenal bulb. Again, the *arrow* indicates the pyloric canal.

Rigidity of the Wall

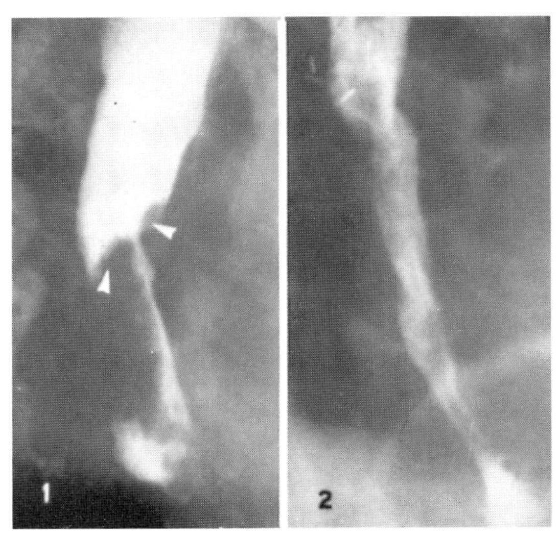

Figure 13.10. A patient with dysphagia. *Left (1):* Carcinoma of the esophagus narrowed the lumen to a tunnel a few millimeters wide and 10 centimeters in length, a rigid segment that never changed either at fluoroscopy or on the other films. The *arrows* indicate the tumor shelf. *Right (2):* Appearance after radiation therapy.

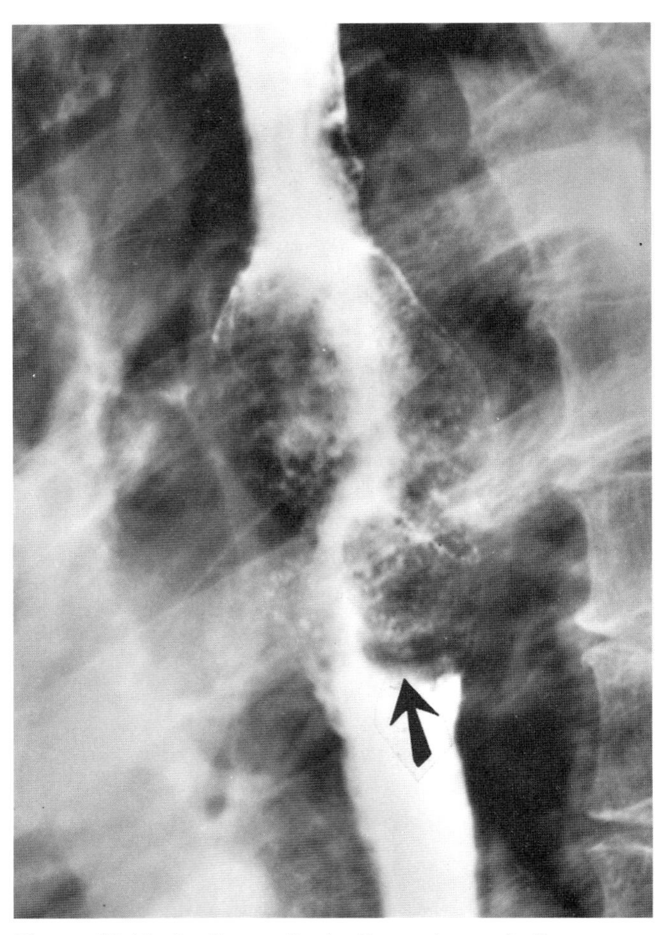

Figure 13.11. Another patient with carcinoma in the midesophagus. Note the irregular rigid lumen. The *arrow* indicates the shelf of tumor.

In studying the gastrointestinal tract, the radiologist looks for a mucosal relief pattern that seems to be within normal limits of variation. The radiologist first searches specifically for *ulcer crater niches* and *filling defects*, and then fills the succeeding parts of the bowel with barium, testing their distensibility and looking for areas of *rigidity*, which may indicate even without any apparent ulceration that the wall is invaded by new growth or scarred by inflammation.

The recognition of an area of rigidity in the bowel wall is more difficult in many ways than the recognition of a crater, because the wall of the bowel varies so much from part to part normally and because early infiltration with sheets of tumor cells does not render the wall entirely rigid but rather limits its elasticity, much in the way that a sheet of rubber changes with age. If you can imagine a remarkably distensible organ like the stomach, into the wall of which has been set a piece of rubber that has lost some of its elasticity, you will have a fair idea of the behavior that can be expected from such a segment under the fluoroscope. Barium pushed against it will fail to produce quite the prompt bulging expected. Barium pushed upward into normal bowel will show a marginal pattern of wrinkles or folds as the flexible wall is mechanically displaced; the rigid or infiltrated segment will fold sluggishly and less deeply. Peristalsis too will be altered, and if you watch the normal passage of ringlike constrictions along the organ, you will see that they are resisted by the suspicious segment, which indents less readily with the passage of the wave. Flexibility of the wall can be limited by the infiltration of tumor cells, by edema, or by postinflammatory changes. Decisions with regard to the flexibility of gut wall, then, are based on how the stomach or intestine is seen to distend with barium, to respond to manipulation, and to contract physiologically.

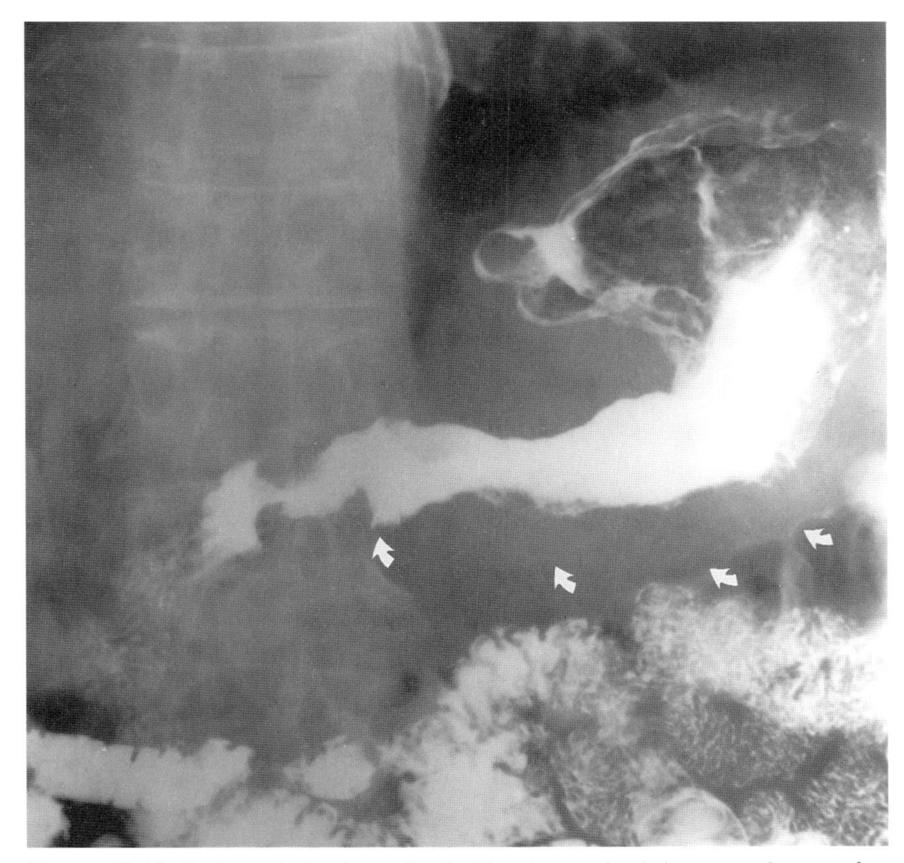

Figure 13.12. Barium study of a patient with advanced scirrhous carcinoma of the stomach (linitis plastica). The *arrows* indicate the outer margin of the thick tumor-invaded gastric wall.

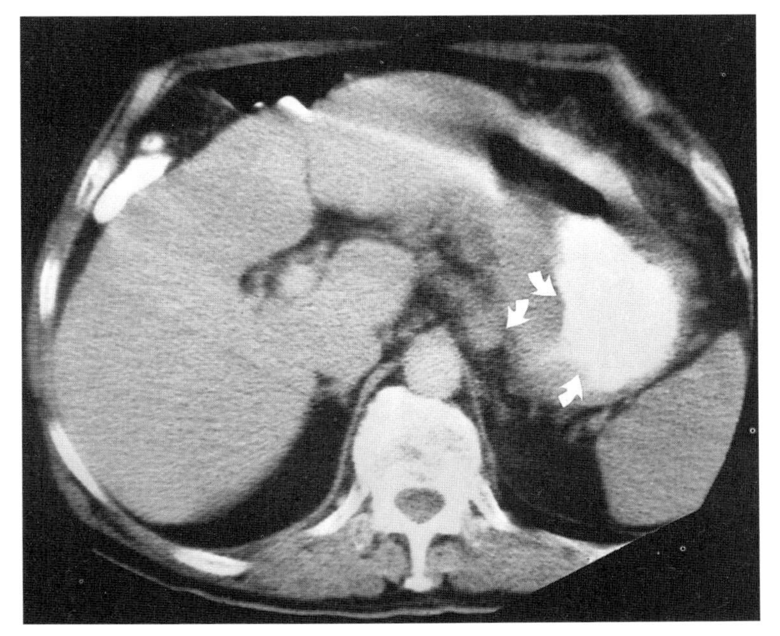

Figure 13.13. The same patient at CT. The *arrows* indicate the thick tumor-invaded wall of the stomach, which is displacing barium within the gastric lumen.

Filling Defects and Intraluminal Masses in the Stomach and Small Bowel

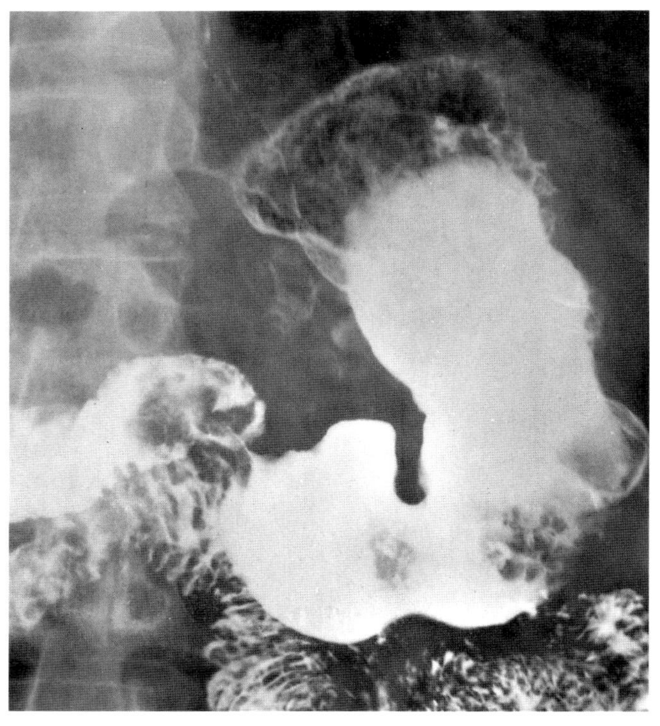

A

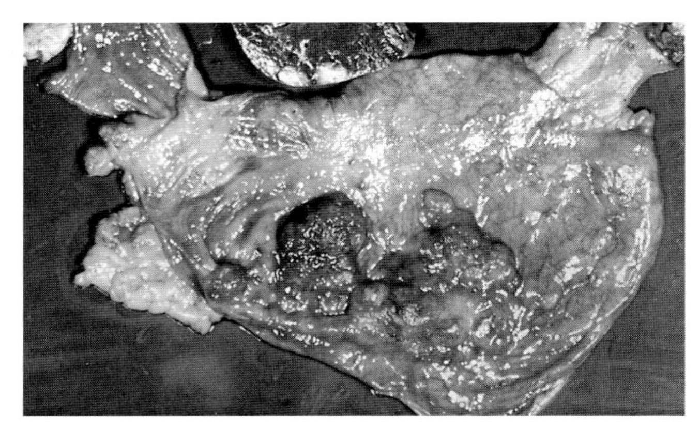

B

Figure 13.14. A *(left):* Filling defects along the greater curvature in a patient with "anemia." There is also a defect inside the duodenal bulb. The patient refused treatment. B *(above):* The open stomach at postmortem. Polypoid lesions along the greater curvature proved to be adenocarcinoma. There was also a larger carcinomatous polyp found prolapsed into the duodenal bulb.

Intraluminal masses, which you see illustrated here, may take many forms. Polyps range in size from a millimeter to several centimeters (Figure 13.14). Polypoid tumors may fill the lumen of the bowel (Figure 13.15). Barium passing between the tumor and the normally distensible wall of the bowel will outline the tumor, showing the normal mucosal markings stretched over the tumor. These can often be seen as a double moulage, the one distinguishable from the other, like those in Figure 13.17, where the normal mucosa of the posterior or anterior wall is seen streaming across the coated lesser curvature tumor.

Intraluminal masses can sometimes be seen to be free floating in the barium, unattached to any wall. Occasionally, matted intraluminal masses are formed of foreign substances like hair or vegetable fibers, becoming too large to be passed and eventually causing symptoms. These are called bezoars (Figure 13.16) and are similar to the hairballs animals vomit. They can usually be differentiated from intraluminal soft-tissue masses because barium mixes with the matted bezoar, giving an appearance quite different from tumor coated on the outside with barium.

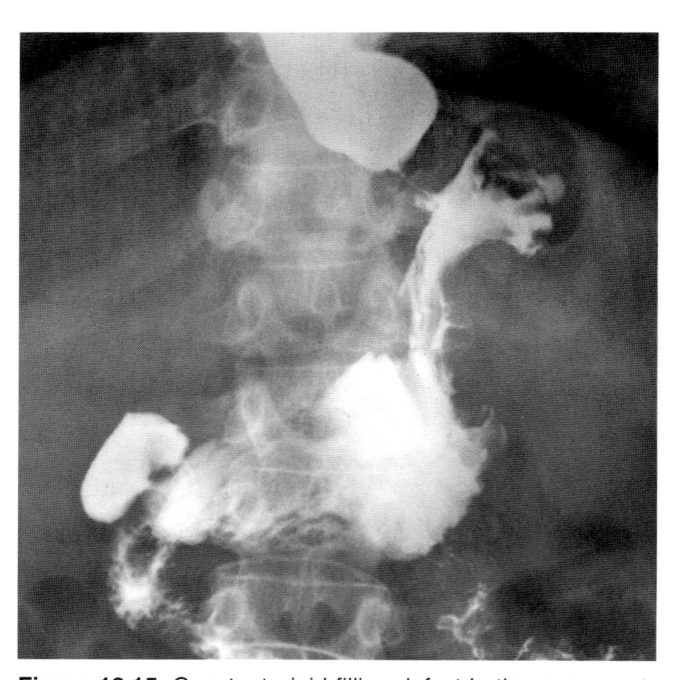

Figure 13.15. Constant, rigid filling defect in the upper part of the stomach. Note the infiltration of the cardia; the esophagus does not empty. This tumor proved to be an adenocarcinoma, originating in the stomach.

Figure 13.16. Bezoar in the stomach and duodenum, composed of matted hair, in a 16-year-old who was known to chew her hair. A: The bezoar is easy to see when surrounded by barium.
B: Can you outline the same bezoar on the plain film as a gastric mass, where it is partially composed of mottled air and surrounded by curvilinear segments of air?

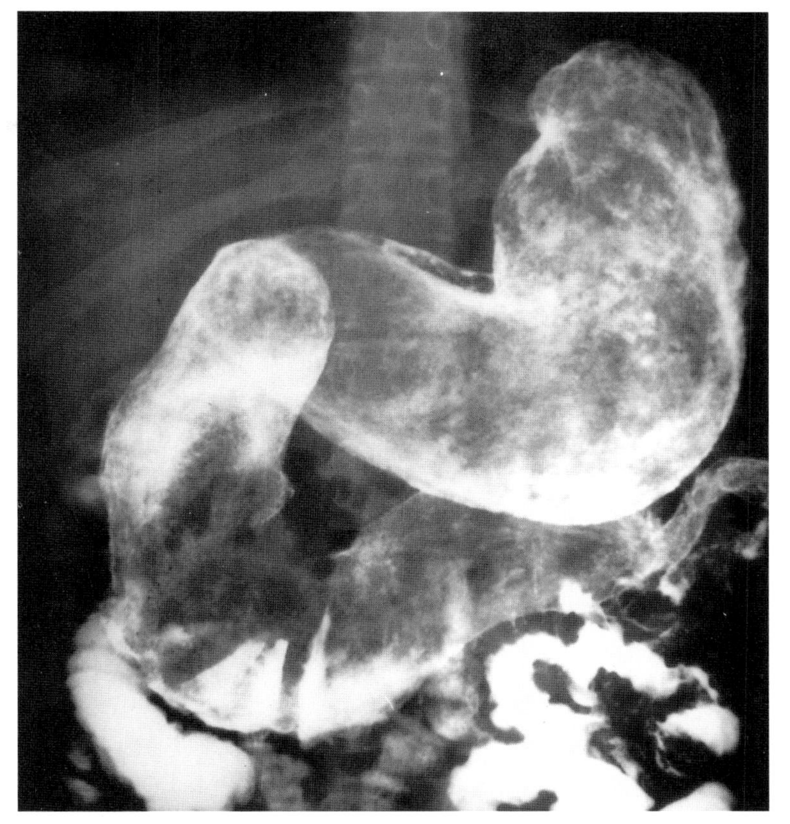

A

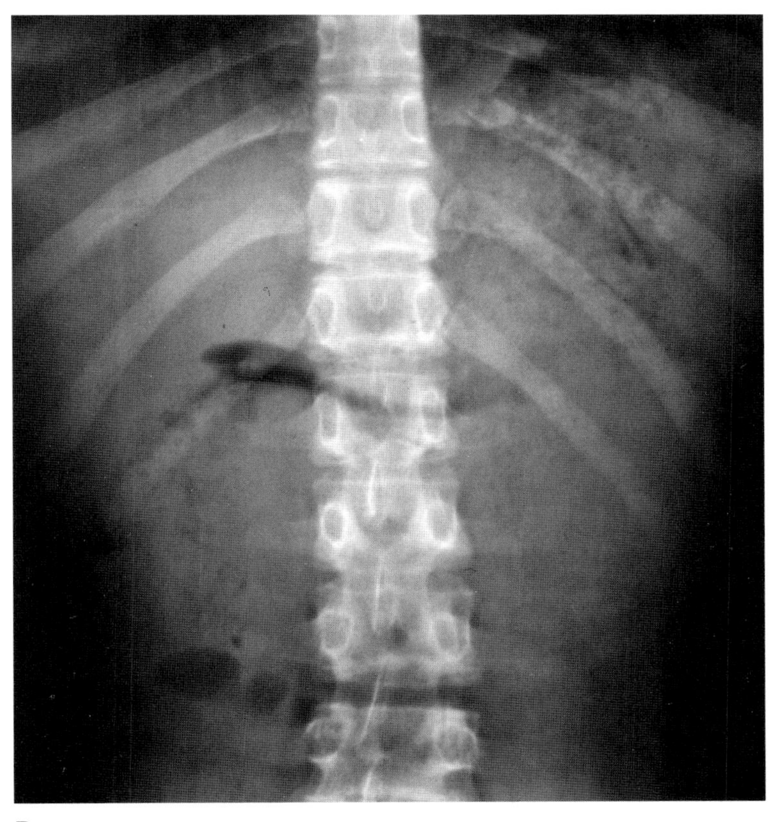

B

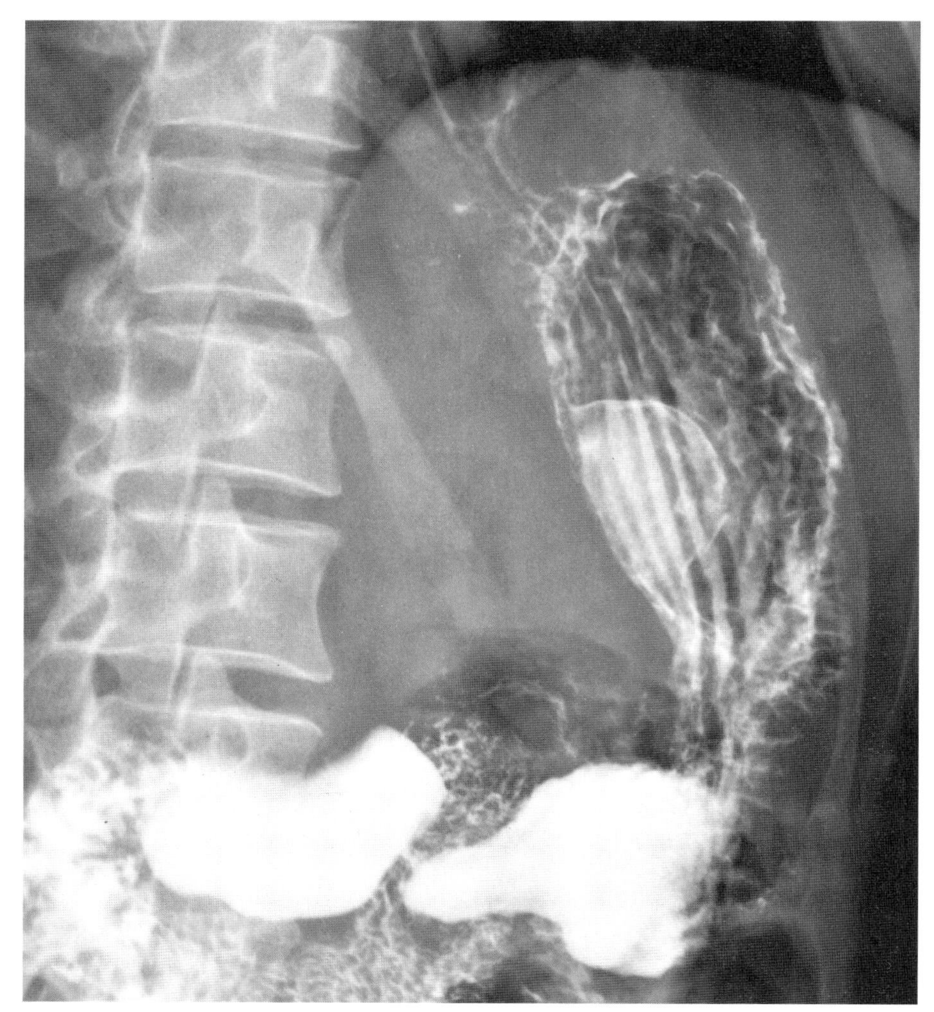

Figure 13.17. Benign tumor projecting into the lumen of the stomach from its sessile base high on the lesser curvature. The normal rugal folds either behind it or in front of it are seen outlined with barium.

Figure 13.18 shows another sort of intraluminal mass. You can see now why it is vital to have the patient fast overnight before a gastrointestinal examination with barium: the mucosa must be perfectly clean and the lumen clear of food and feces.

From the teaching you have had about *endoscopy* you may wonder if it is really necessary for you to learn much about barium studies. As a matter of fact, endoscopy and radiographic examination of the gastrointestinal tract are *complementary*. Certain patients may benefit from a combination of these studies; others may need only one or the other. For example, a young person suspected of having duodenal ulcer disease is usually diagnosed accurately with an upper GI series, which is quicker and more comfortable for the patient than endoscopy. But when acute gastritis is suspected, the patient is better studied by endoscopy because the subtle mucosal changes cannot be seen with barium. A barium enema would be the appropriate study for a patient with suspected low large-bowel mechanical obstruction, whereas the superficial lesions of early ulcerative colitis may be easier to detect by sigmoidoscopy or colonoscopy. Remember too that the small bowel cannot be visualized endoscopically and *must* be studied with oral barium and a small-bowel follow-through, or with CT. As compared to oral contrast examinations, which only cast an image of the small-bowel lumen, CT shows the bowel wall itself and can be used to diagnose conditions that cause small-bowel wall thickening, such as Crohn's disease.

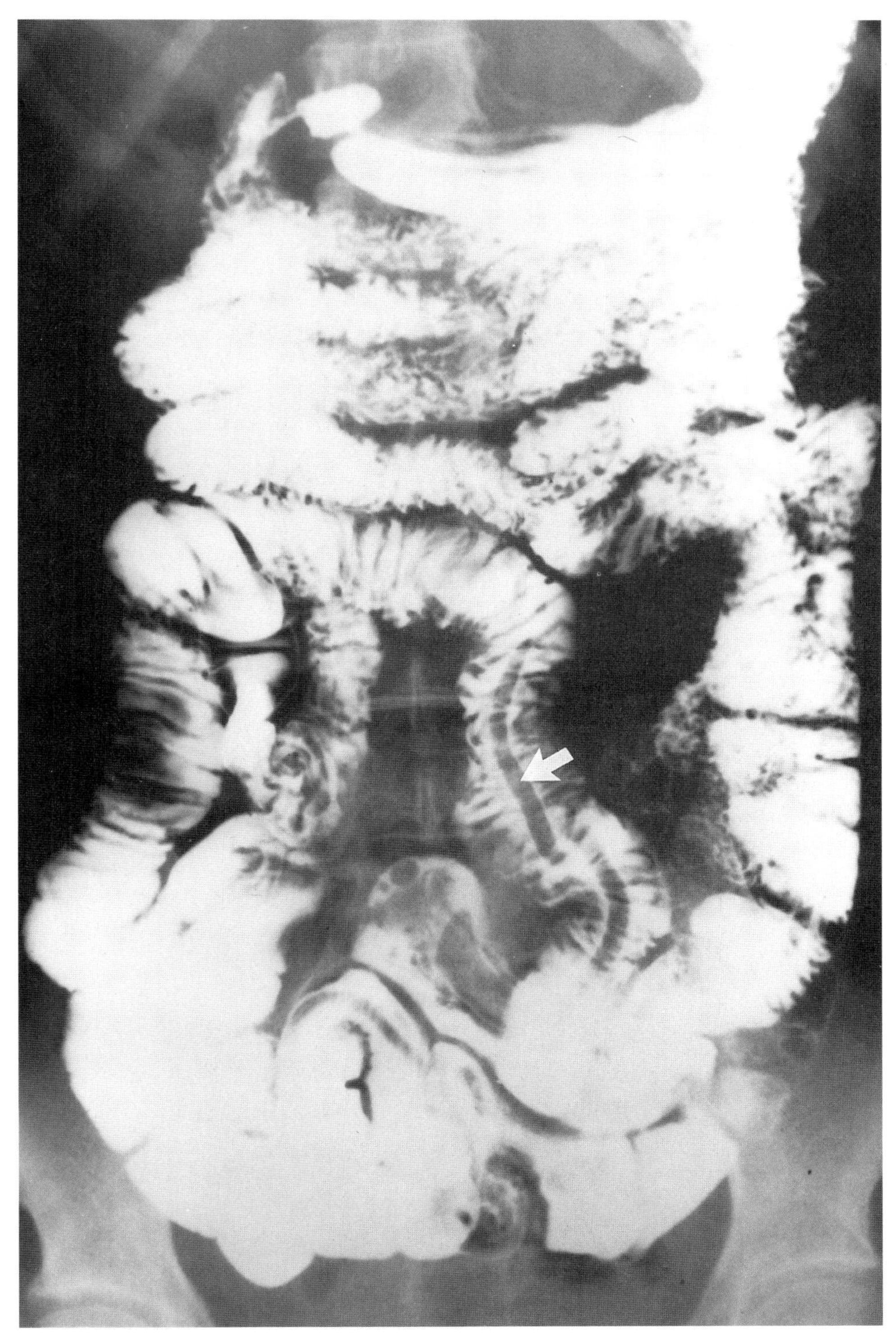

Figure 13.18. Ascaris infestation of the small bowel. The worms are seen as long intraluminal filling defects *(arrow)* in the small bowel.

Gastric Ulcer

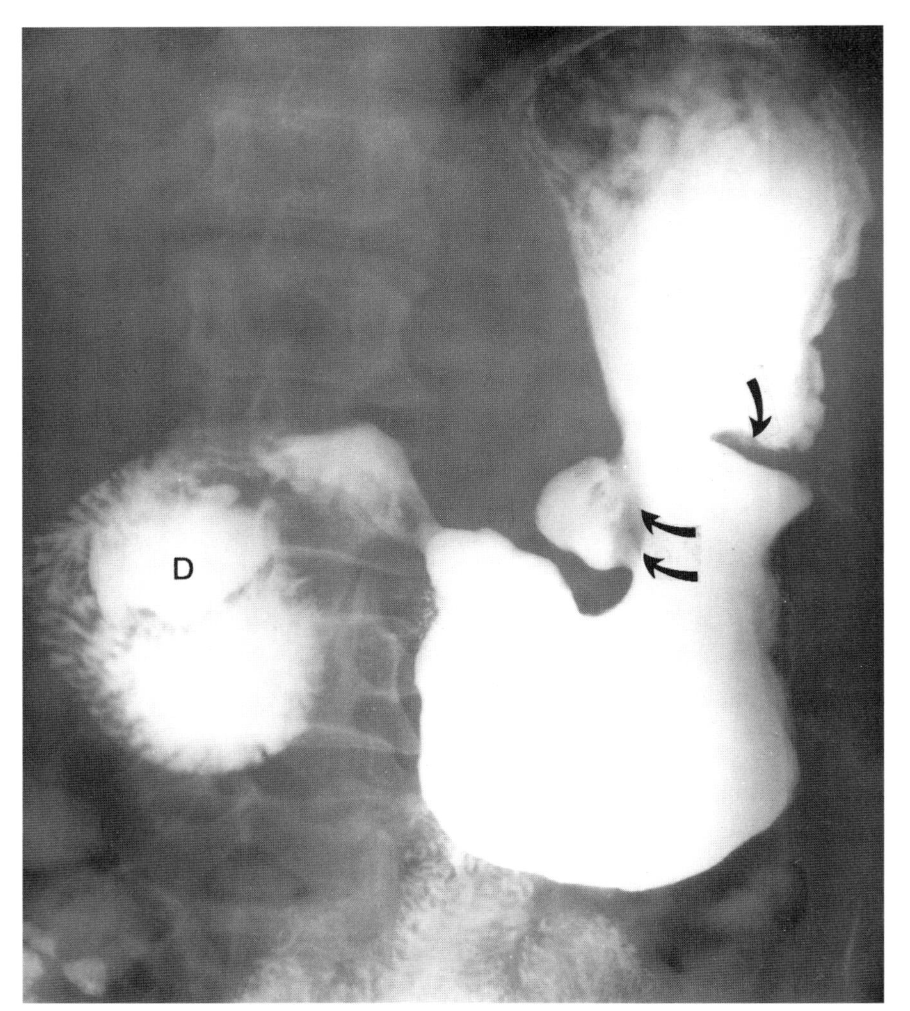

Figure 13.19. Large benign ulcer crater projecting from the midlesser curvature. Note the radiolucent collar of granulation tissue (Hampton's line, *twin arrows*) across the base of the ulcer. The upper *single arrow* indicates a partial constriction of the lumen by an area of spasm along the greater curvature in response to the large active ulcer. *D* is a diverticulum, arising from the medial wall of the descending limb of the duodenal loop.

In the last of the five hypothetical tubular structures in Figure 13.2 an ulceration has occurred in the wall of the tube, forming a small additional hollow space into which the opaque substance can flow. In radiological parlance this type of shadow is called a *niche* or *crater*, and you will find both terms used to refer to the projecting shadow of a barium-filled ulceration in the esophagus, stomach, or duodenum. Only when seen in profile, of course, will it be a projection from the normal margin of the bowel wall.

When a barium-filled crater is seen en face, it appears as a spot of white more dense than the surrounding shadow because it is frequently encircled by a rolled-up margin of granulation tissue. Ulcer craters that are filled with blood clot or food particles at the time of examination with barium may not be visualized at all. As they heal, they fill in from the sides, becoming sharp and thorn-shaped in profile and finally disappearing altogether.

Because the tubular gastrointestinal tract is flexible, fills and empties in response to waves of peristalsis, and has opposing walls coated with barium, there are myriad small angular barium shadows in most of the films you examine. To find among them one that can with confidence be labeled a niche or crater requires that it have certain characteristics. In the first place, it is deeper than most of the valleys between the folds of mucosa. Therefore its shadow will be *denser*, because it contains a slightly greater thickness of barium. Because it is an ulcer it has no mucosal pattern, and because it is generally surrounded by inflammatory reaction it is less flexible than the rest of the gastric wall. Accordingly, the shadow of the niche or crater is *constant in shape and size*. It is *consistently demonstrable* in the same place from film to film, and all these characteristics enable the radiologist to find and identify it over the course of study.

Numerous other details help to interpret the findings. For example, when radiologists observe that the nearby mucosal folds in the stomach *converge toward a demonstrable ulcer niche or crater*, they may report that the ulcer is almost unquestionably benign, because in differentiating the two types of ulcer in the stomach, the *convergence of folds* has proven to be the most reliable indication of benignity.

You will hear much about the differentiation of benign and malignant ulcers in the stomach, and you will also find that a benign ulcer of long standing may be so embedded in scar tissue, so rigid, and so reluctant to heal on medical management that it is believed to be malignant by the referring physician, radiologist, and surgeon; and only the pathologist, using microscopic evidence, can establish the fact that it is benign. Of course, most patients with ulceration of this type should have endoscopic biopsy during the workup. The majority of benign gastric ulcers, however, *are* convincingly benign at upper GI series. Keep in mind that:

1. Most gastric ulcers are benign (90 percent).
2. The size and location of a gastric ulcer crater are *not* an index of malignancy. Convergence of folds, the shape of the crater, its projection beyond the stomach lumen, and its propensity to heal *are* reliable characteristics of benign ulcer craters.
3. *Every* "benign" *gastric* ulcer must be followed by x-ray until it has disappeared because failure to heal may be a sign of *carcinoma in situ*. This does not apply at all to duodenal ulcers, which are uniformly benign and do not require either endoscopy or reexamination.

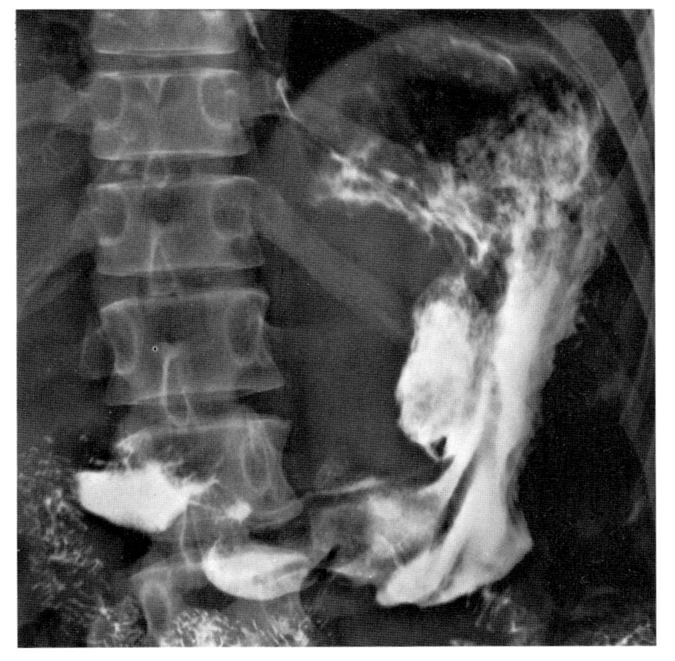

A

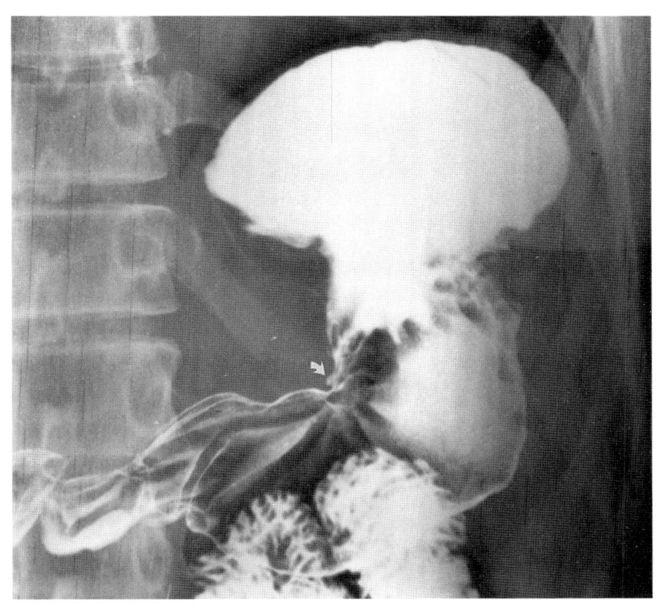

B

Figure 13.20. Natural history of healing in a large benign gastric ulcer. In B, a film made 6 weeks later than the one in A, the healing thorn-shaped crater *(arrow)* might easily be missed. The convergence of folds toward the small remaining crater reinforces its probable benign character.

The Clearly Malignant Ulcer

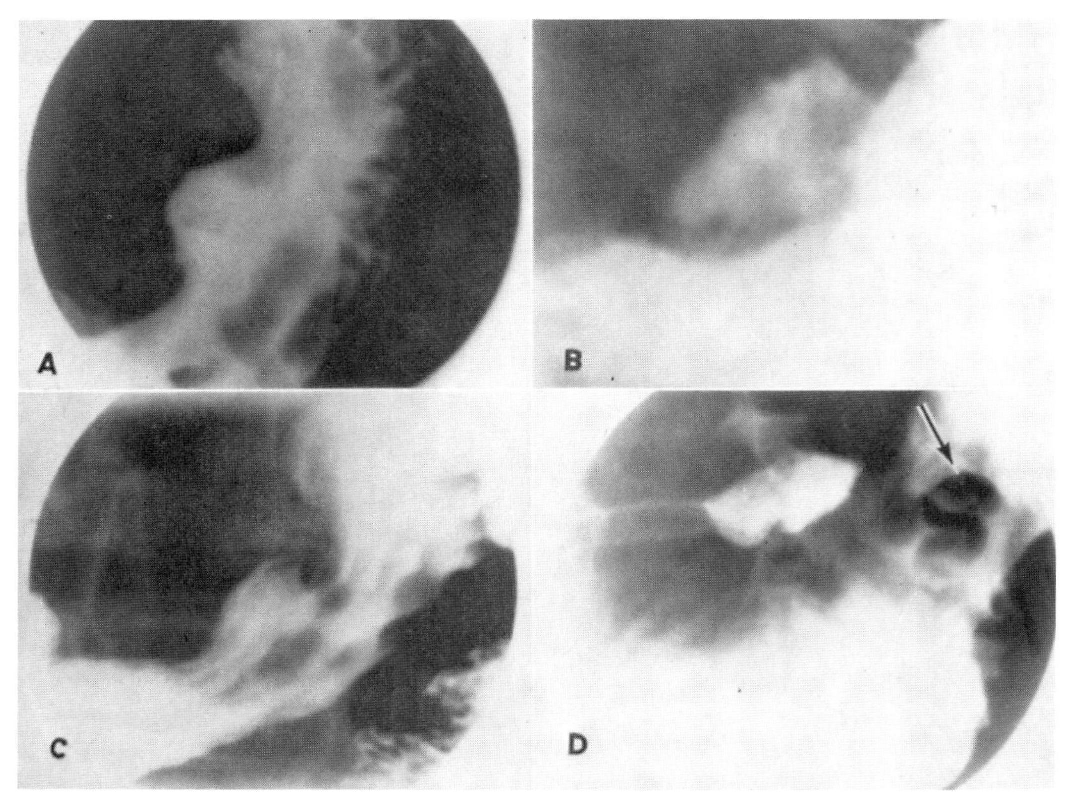

Figure 13.21. Four proven malignant lesser curvature ulcers. The ulcer is *within a mass,* and there is marked distortion and infiltration of the surrounding mucosa *(D, arrow).* All four would probably be labeled malignant by the radiologist, but should certainly have endoscopic biopsy.

Any radiologist would recognize the ulcer craters above as malignant. They represent ulcerations within a mass. Convergence of folds up to the margin of the crater is not present, and folds are irregularly interrupted by the surrounding tumor mass. Of course, these patients should be endoscoped and biopsied so that the surgeon can confirm malignancy and plan optimal management.

Duodenal Ulcer

What is true about the identification of stomach ulcers is true about duodenal ulcer craters, in that they are consistently demonstrable collections of barium. In the duodenum, however, the problem is somewhat different, since instead of a wide sac the structure to be examined is a narrow tube with a bulb or ampulla at its commencement just distal to the pyloric canal. Although the crater itself is demonstrated in much the same way as it is in the stomach, still more important and informative in the long run are the changes due to scar tissue formation in this characteristically recurrent condition.

The most common location by far for the crater is in the center of the posterior wall of the bulb. Duodenal ulcer disease has a recurrent course; and after several episodes of ulceration and healing, permanent strands of scar tissue may develop, which constrict the lumen of the duodenal bulb and limit its free distensibility. These limiting bands of scar tissue produce distinctive changes in the shape of the shadow of the barium-filled bulb, so that its cavity seems to be divided into several cavities bulging outward from the central point at which the ulcer crater has been present or may still be seen. This appearance has been called the *cloverleaf deformity* of the duodenal bulb or cap (Figure 13.22).

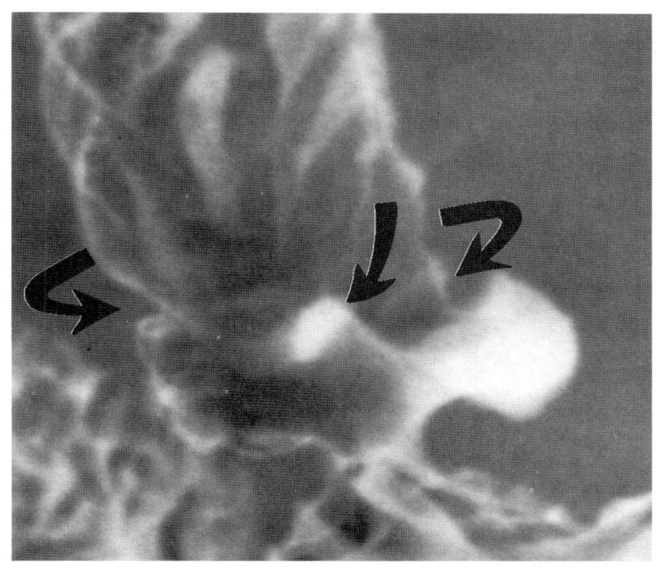

Figure 13.22. Classic location of a duodenal ulcer *(center arrow)* on the posterior wall, midcap. Note the indentation on the greater and lesser curvature sides of the cap *(bent arrows),* and the beginning of scarring that will produce a typical cloverleaf deformity as the proximal and distal portions of the cap distend in response to constriction about the ulcer (prestenotic and poststenotic dilatation).

This is only one of the scarring patterns in long-standing duodenal ulcer disease and is by no means present in all of the advanced cases you will see. Another common pattern of scarring is the gradual development of a stenosed apex of the cap, eventually producing a high degree of obstruction. Still other patterns of scarring may flatten one side of the bulb asymmetrically.

Several chapters could be written on the subject of duodenal ulcer disease, and you will gradually become familiar with the problems of diagnosing this condition. One very important point to remember is that in its early episodes a duodenal ulcer crater is quite easy to demonstrate with barium. After scar formation has become fairly well advanced, however, the crater itself becomes more and more difficult to visualize with each successive attack, and at length the radiologist will find it almost impossible to demonstrate the crater in spite of unquestioned reactivation suggested by the patient's symptoms. Of course, the permanent deformity does not change.

Once the diagnosis has been made, the clinician does well to be guided by the patient's symptoms alone. Only with a significant change in the patient's long-standing symptoms or with increasing evidence of obstruction need reexamination be carried out. Remember that the diagnosis can be made from either a crater or typical scarring, and that the patient needs to have the usually recurrent nature of the disease explained to him so that he will not seek reexamination with barium during each episode.

Examples of Duodenal Ulcer Disease

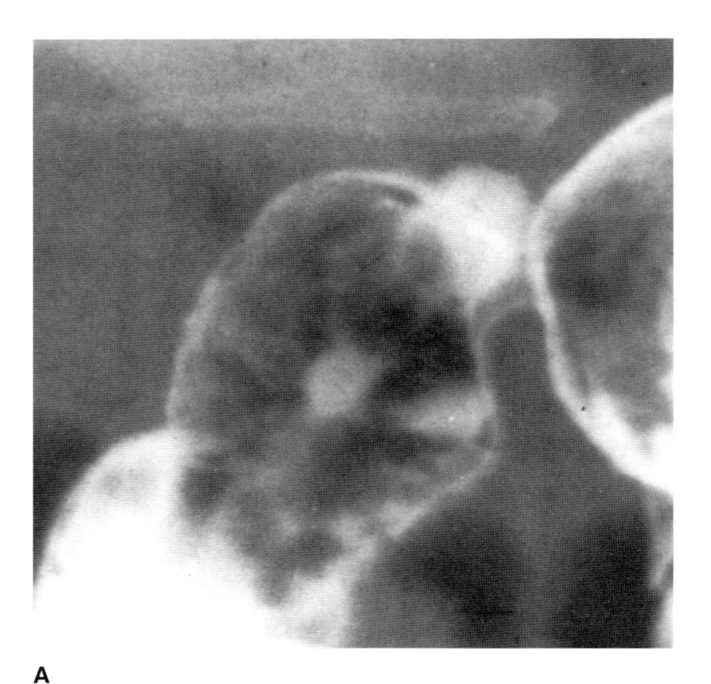

A

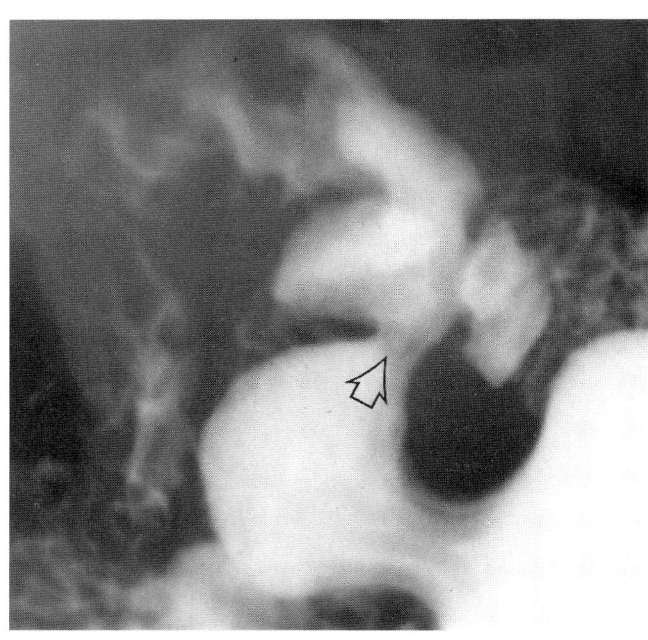

B

Figure 13.23. A: Duodenal ulcer crater in the classic location on the posterior wall in a patient radiographed supine with air filling the cap and, as yet, no deformity. B: Deformed cloverleaf cap. The *arrow* indicates the pyloric canal.

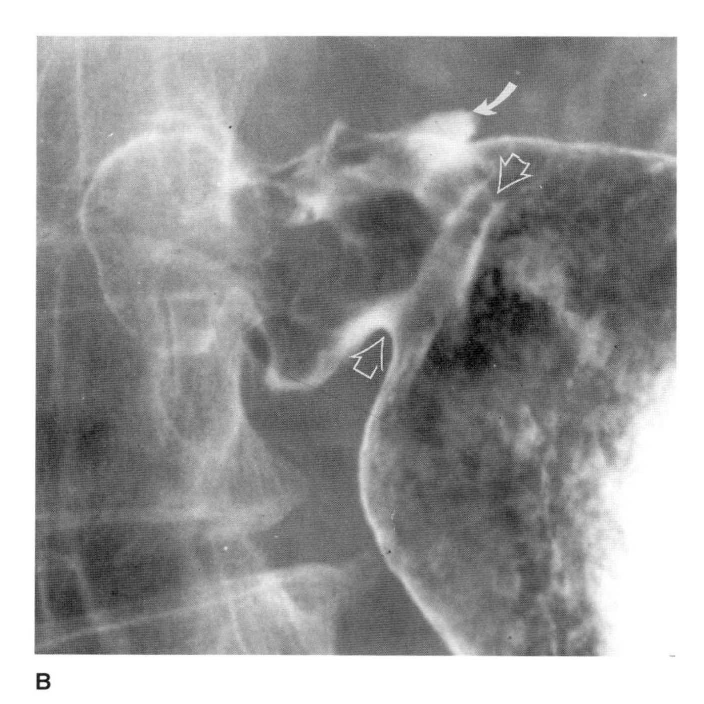

B

A

Figure 13.24. A: Cloverleaf deformity of the cap in duodenal ulcer disease, probably active, but a crater is not demonstrated. The *arrow* indicates the pyloric canal. B: Duodenal ulcer disease in a patient whose crater is located eccentrically just beyond the pyloric canal on the lesser curvature side of the cap *(curved arrow).* The wide, patulous pyloric canal is indicated by *open arrows.* Scarring has flattened the cap and resulted in the eccentric canal.

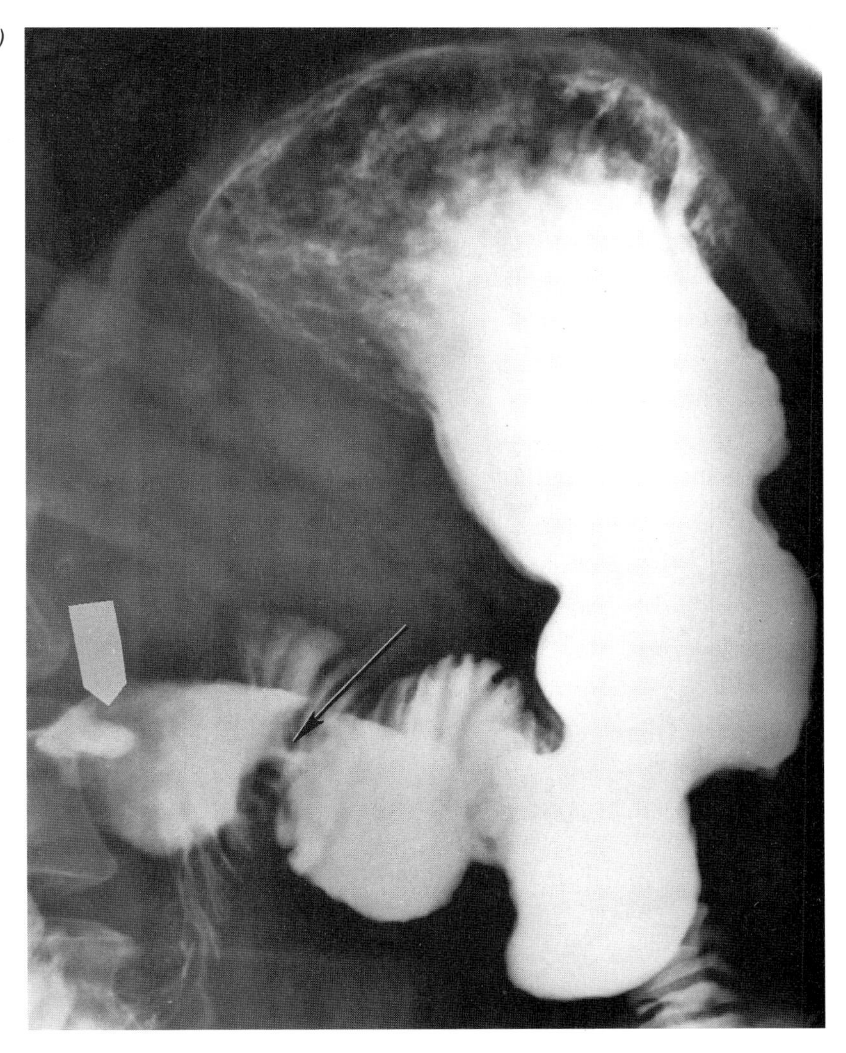

Figure 13.25. A large ulcer crater *(white arrow)* at the apex of the cap. Scarring produced by craters in this location is likely to lead to eventual obstruction. The *black arrow* indicates the pyloric canal.

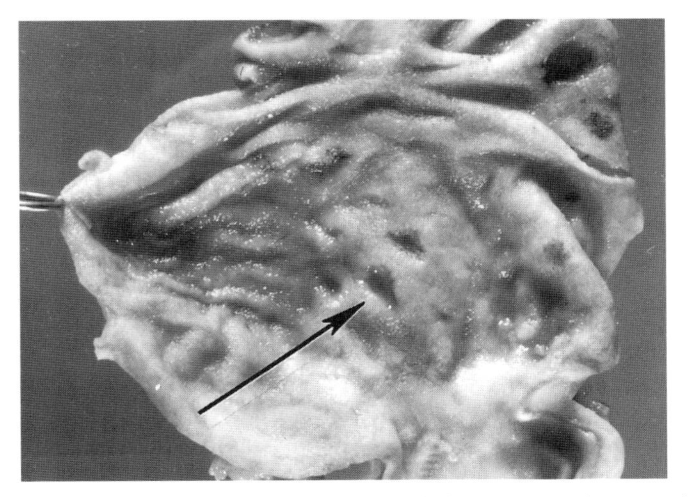

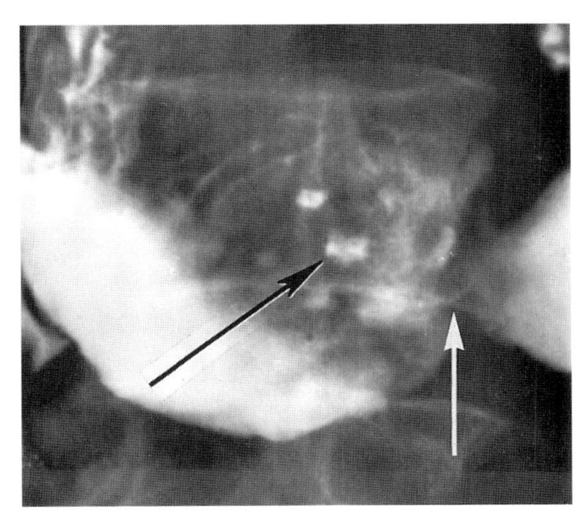

Figures 13.26 ***(left)*** **and 13.27** ***(right).*** Surgical specimen and radiograph of a patient with multiple ulcer craters on the posterior wall of the duodenal bulb. The black arrow indicates the large central crater, around which the others are arranged in a circle. The *white arrow* indicates the pyloric canal.

The Barium Enema

The cecum and rectum, like the stomach, are expanded sections of the bowel. They are difficult to examine and present special problems for the radiologist. The colon can only be examined properly after it has been thoroughly cleansed. This usually requires catharsis, although patients in whom catharsis is contraindicated may be studied after two days on a low-residue diet followed by two days on a liquid diet and cleansing enemas. The cecum should not be considered as seen in its entirety until there is retrograde filling of either the appendix or the terminal ileum. This is a vital point in patients with unexplained anemia, in whom carcinoma of the cecum must be ruled out.

Carcinoma of the rectum should be diagnosed by the clinician on physical examination and proctosigmoidoscopy. Because of the great distensibility of the rectal ampulla, the radiologist knows that it is easy to miss completely a sizable carcinoma in this location, which may well be obscured by the barium surrounding and concealing it.

During the conventional single-column barium enema patients are examined as the barium is being instilled by gravity into the rectum, sigmoid, descending colon, transverse and ascending colon, and cecum. In the supine position the flexures are studied in various degrees of obliquity. When the colon is filled, multiple overhead films are obtained. Then the patient

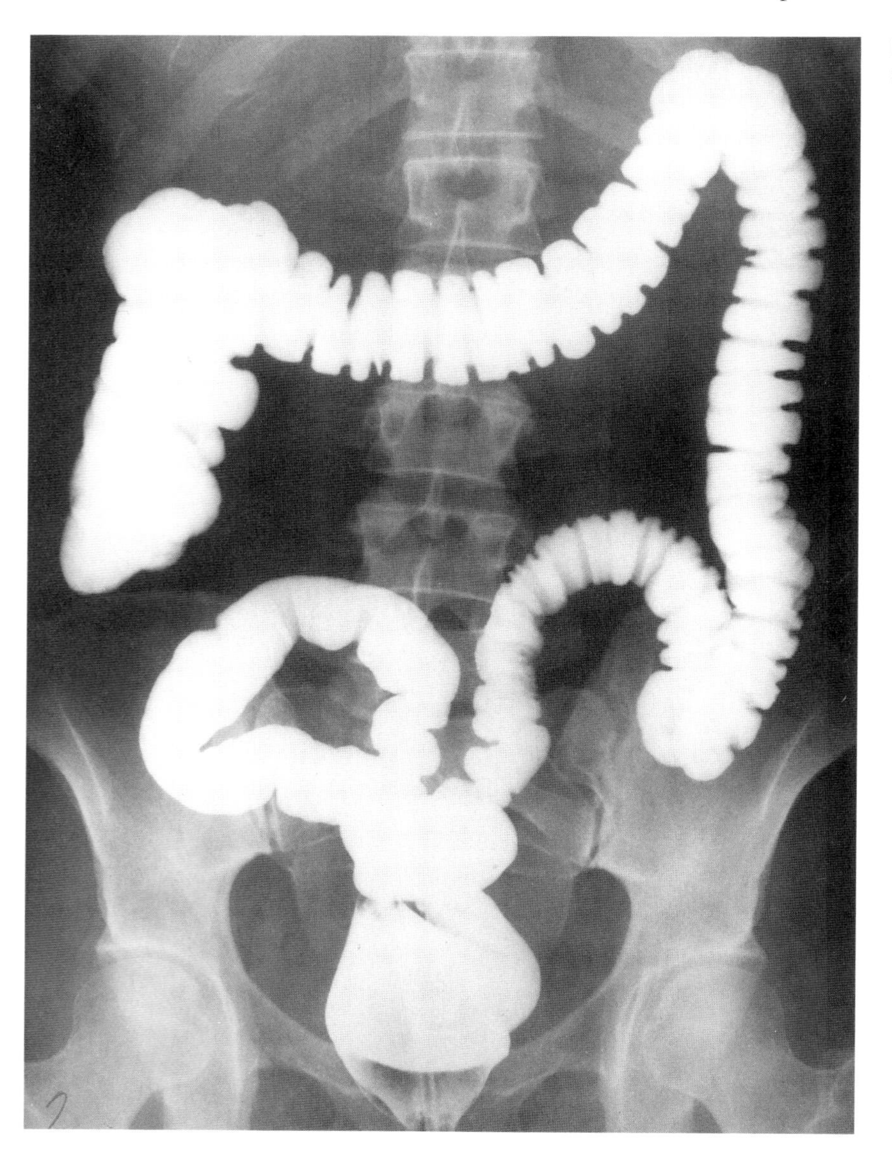

Figure 13.28. Normal barium enema without reflux into the terminal ileum.

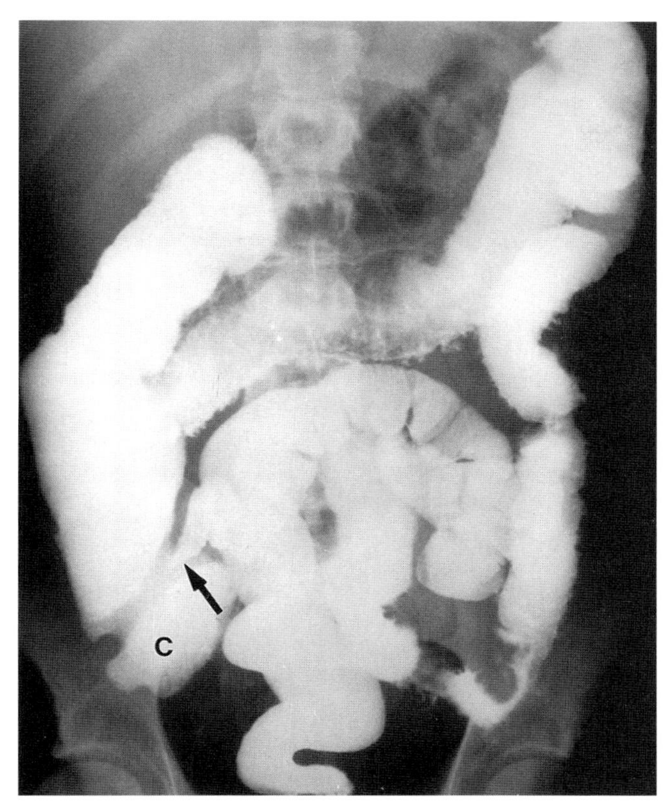

Figure 13.29. Barium enema in ulcerative colitis with reflux into the terminal ileum. Opacified loops of small bowel can be seen centrally within the frame of ascending, transverse, and descending colon. The *arrow* points to the ileocecal valve; *C* indicates the cecum. Compare this with the normal barium enema shown in Figure 13.28, and note the loss of haustral markings and the shaggy ulcerated appearance of the mucosal surface.

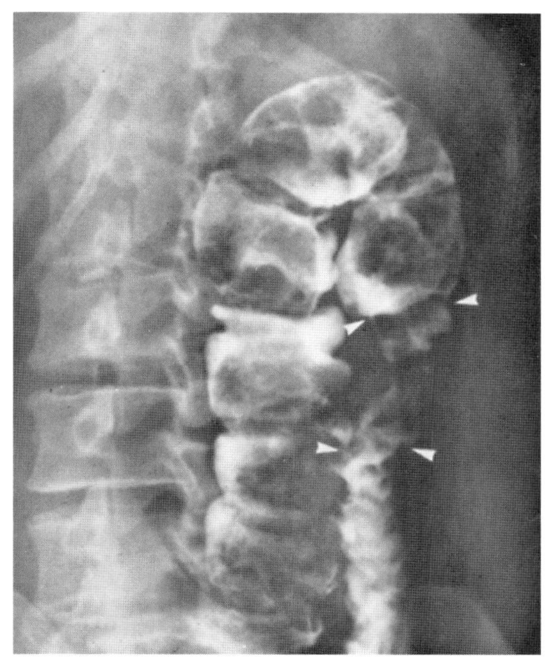

Figure 13.30. An annular area of narrowing in the descending colon *(arrows)* just distal to the splenic flexure was constant on all films obtained during a barium enema. Note the evidence of low-grade obstruction; the transverse colon and splenic flexure are dilated and contain scybala (fecal boluses) outlined by barium. The descending colon below the lesion shows the pattern expected after evacuation. Colon carcinoma was found at surgery.

is allowed to evacuate the barium, after which a prone film is made to show the emptied large bowel and its mucosal relief pattern.

The radiologist, searching for intraluminal tumor masses in the colon, may allow the patient to evacuate most of the barium and then insufflate the colon with air so that a *postevacuation air-contrast study* can be filmed. Or the radiologist may elect to use the *double contrast method* from the start of the examination, instilling a small amount of barium and then air as needed. A double contrast study is especially useful in identifying small polyps and mucosal lesions. In planning the workup of your patient you may wish to discuss with the radiologist the advisability of the barium enema or colonoscopy or both, and in what order.

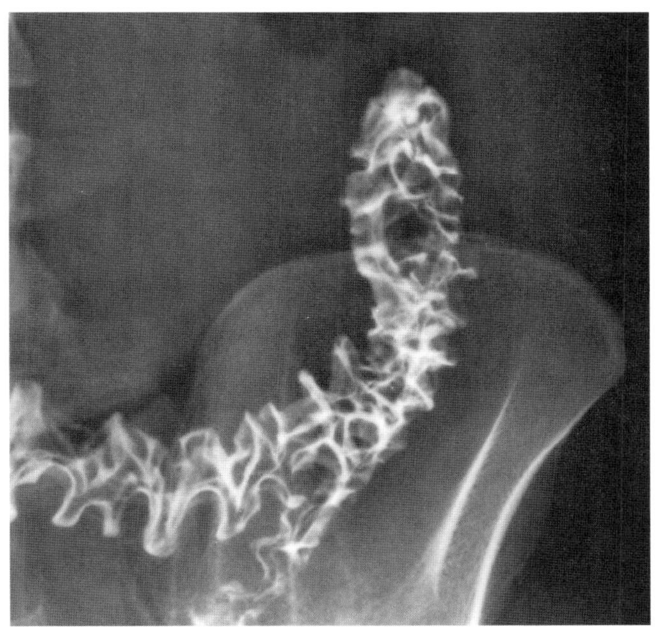

Figure 13.31. Normal empty splenic flexure after evacuation of barium enema.

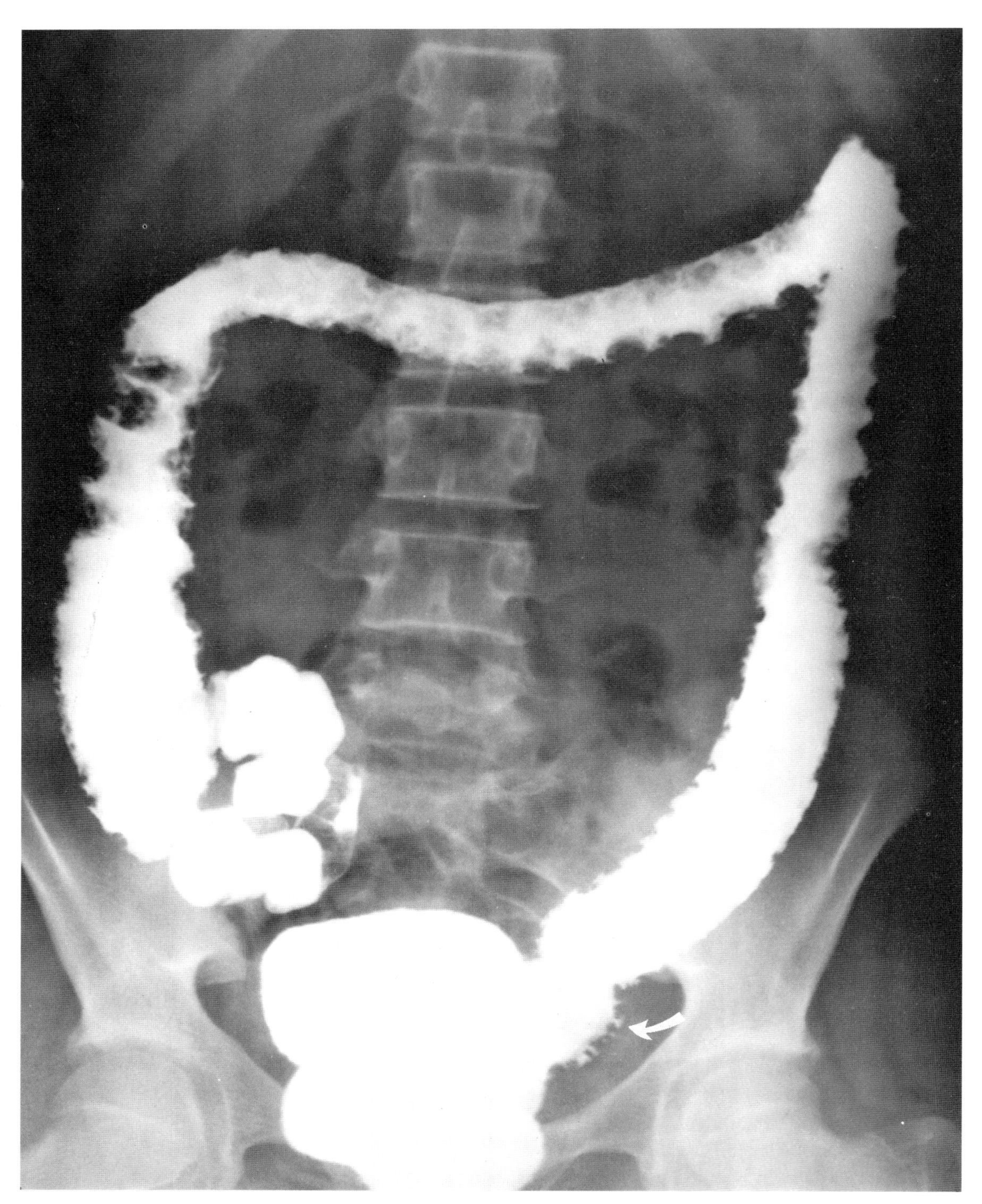

Figure 13.32. Ulcerative colitis in a young boy. Here ulceration is seen from cecum to rectum. The *arrow* indicates the classic "collar button" shape often taken by these ulcers. Note also that there is shortening of the colon (straightened flexures), loss of normal haustral markings, nodular indentations (pseudopolyps), reflux into the terminal ileum, and that the cecum is much smaller than normal because of chronic scarring.

Filling Defects and Intraluminal Masses in the Colon

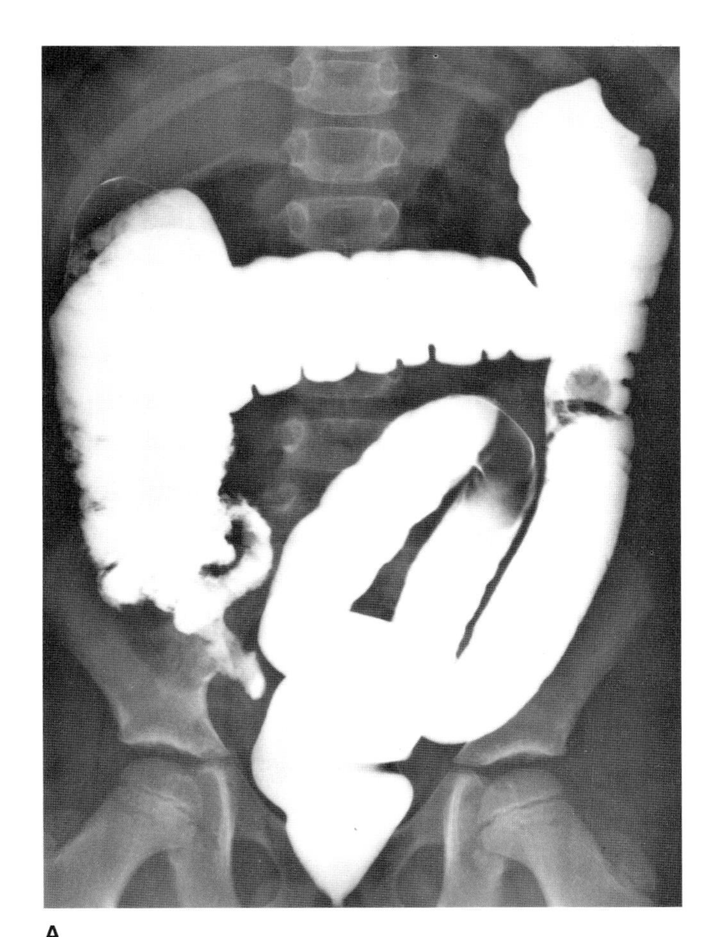

A

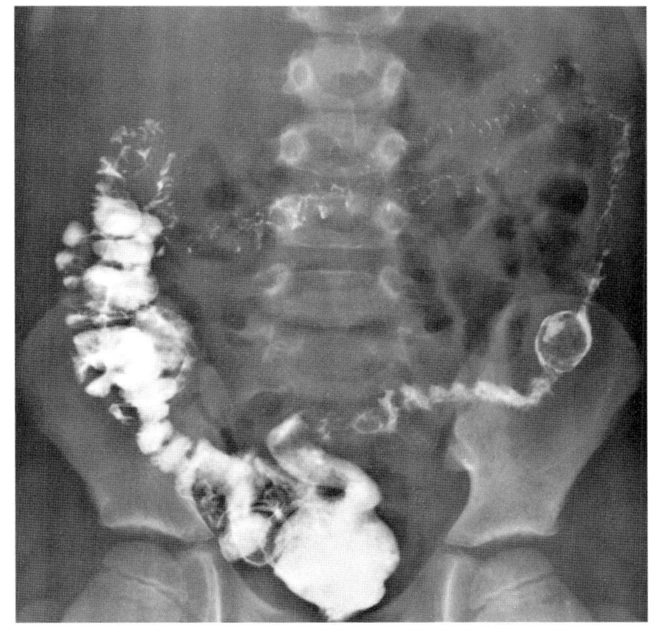

B

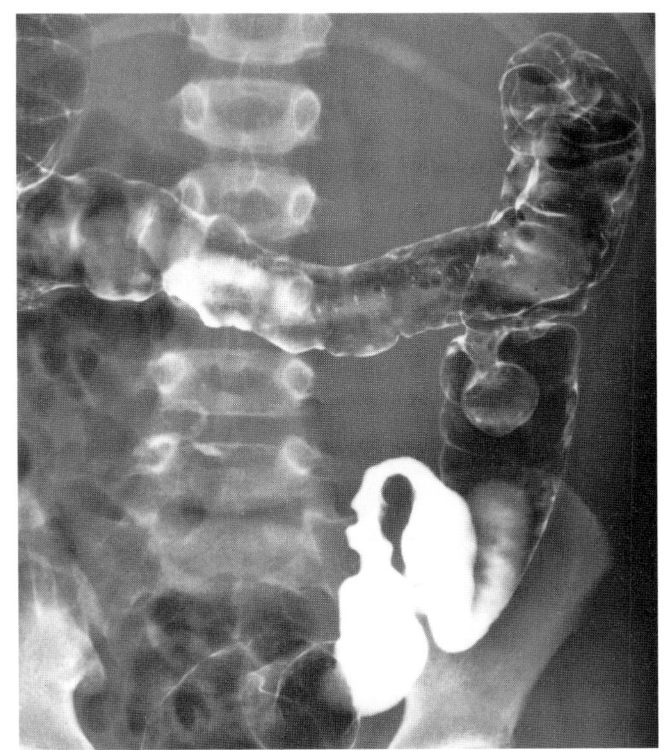

C

Figure 13.33. Polyp in the descending colon in a boy with bloody stools and crampy left abdominal pain. Note that the polyp is seen demonstrated in three different ways, always in the same location. In A it is seen as a radiolucent filling defect in the opaque barium column. In B, after evacuation, the polyp is seen because it prevents collapse of the barium-coated walls of the colon, which are seen above and below it. In C the polyp, coated with opaque material, is seen outlined by air following air insufflation. Notice how clearly you can see the stalk of the polyp.

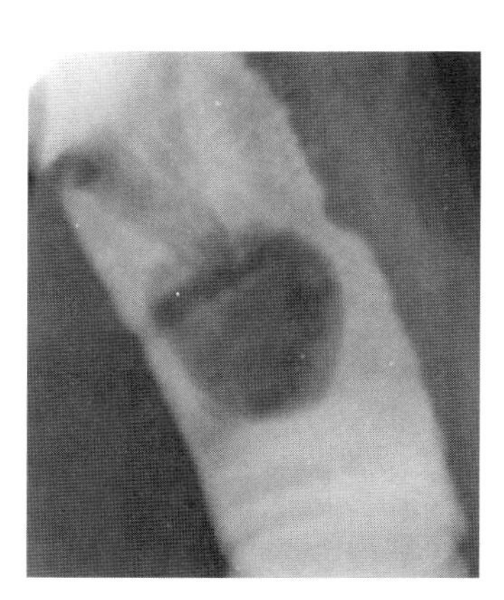

Figure 13.34. Polyps within the colon may be demonstrated as radiolucent filling defects displacing the contrast substance. Note the stalk, which is well seen in this patient.

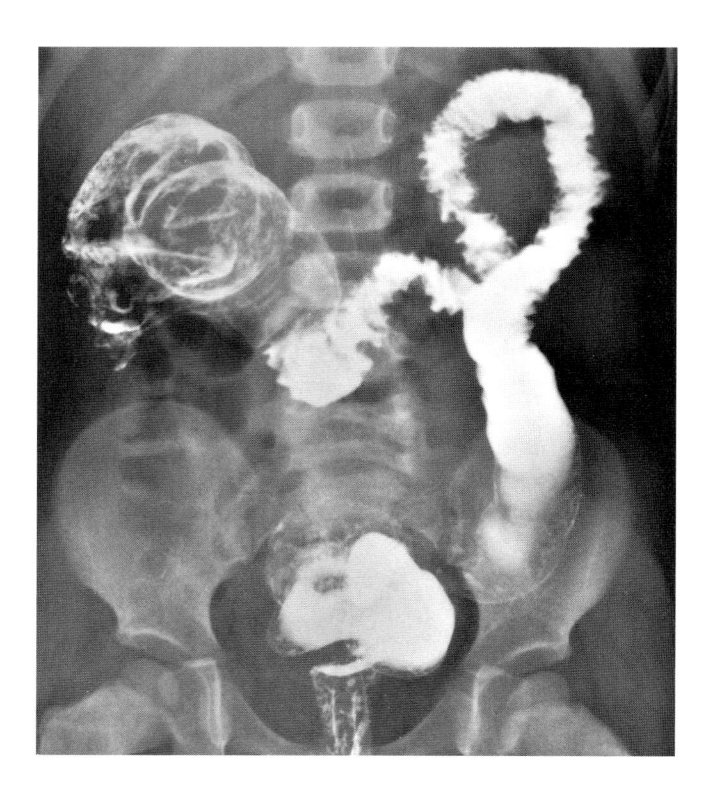

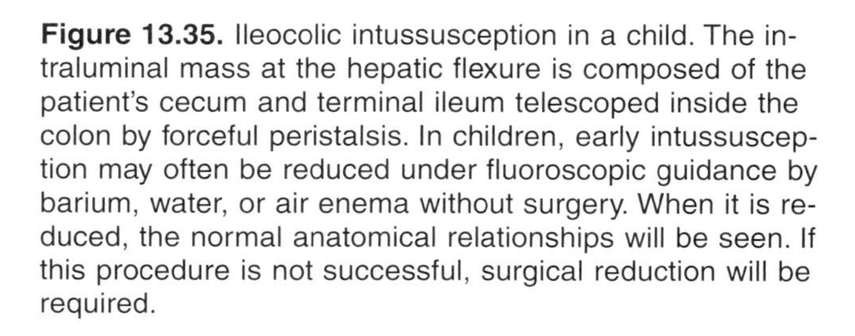

Figure 13.35. Ileocolic intussusception in a child. The intraluminal mass at the hepatic flexure is composed of the patient's cecum and terminal ileum telescoped inside the colon by forceful peristalsis. In children, early intussusception may often be reduced under fluoroscopic guidance by barium, water, or air enema without surgery. When it is reduced, the normal anatomical relationships will be seen. If this procedure is not successful, surgical reduction will be required.

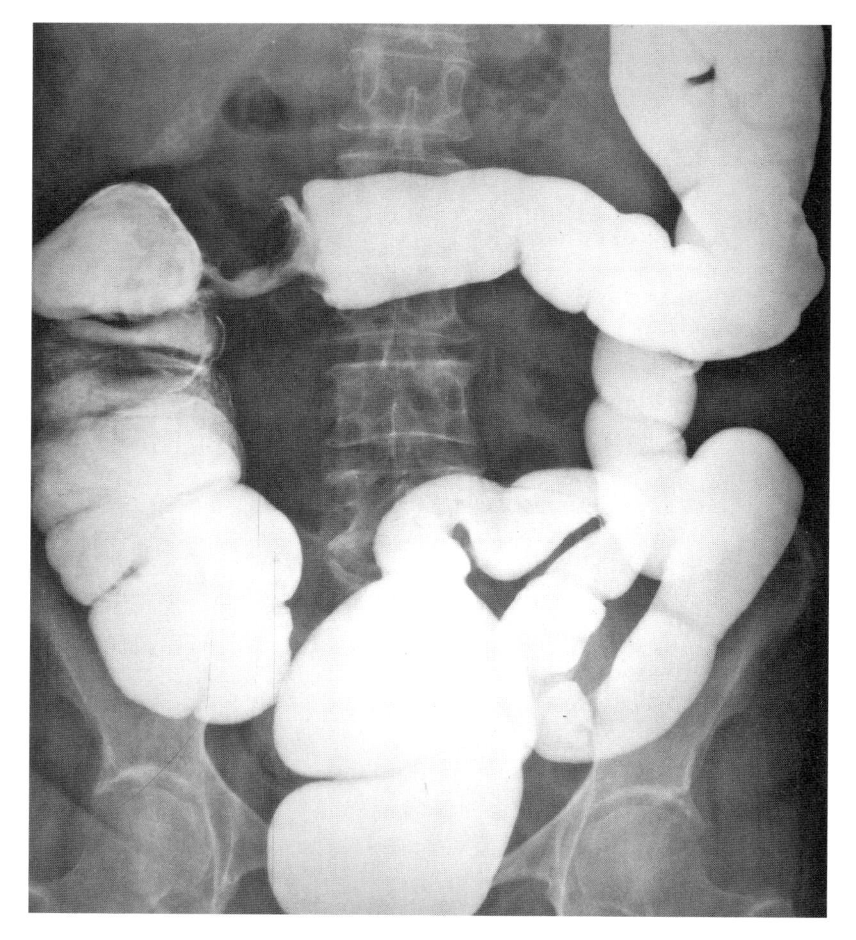

Figure 13.36. The constant annular lesion in the proximal transverse colon, just beyond the hepatic flexure, proved to be a carcinoma, as expected.

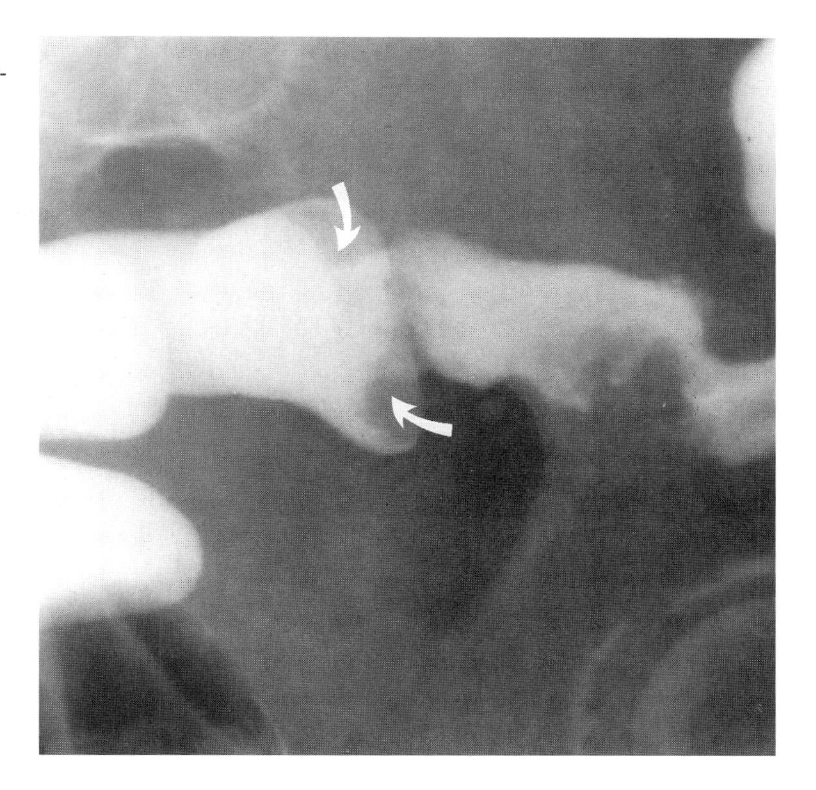

Figure 13.37. Carcinoma of the midsigmoid colon. The normal colon suddenly showed an annular constriction that was rigid and identical in all films. The *arrows* indicate the tumor shelf.

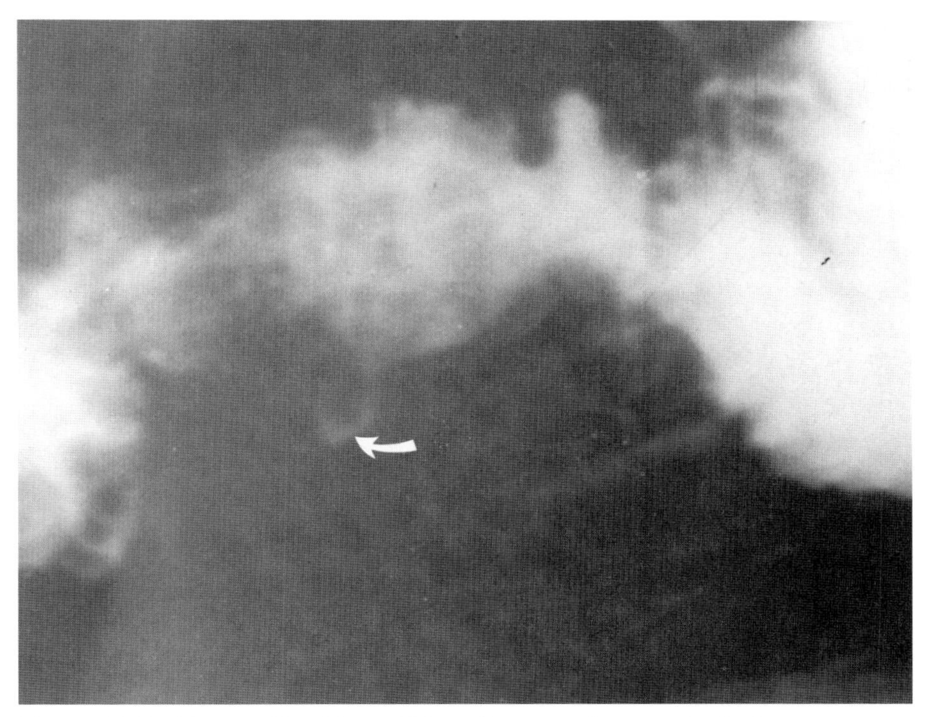

Figure 13.38. Diverticulitis of the midsigmoid colon. The *arrow* indicates barium extravasation into the middle of an intramural abscess on the inferior wall of the colon.

The Sigmoid Colon

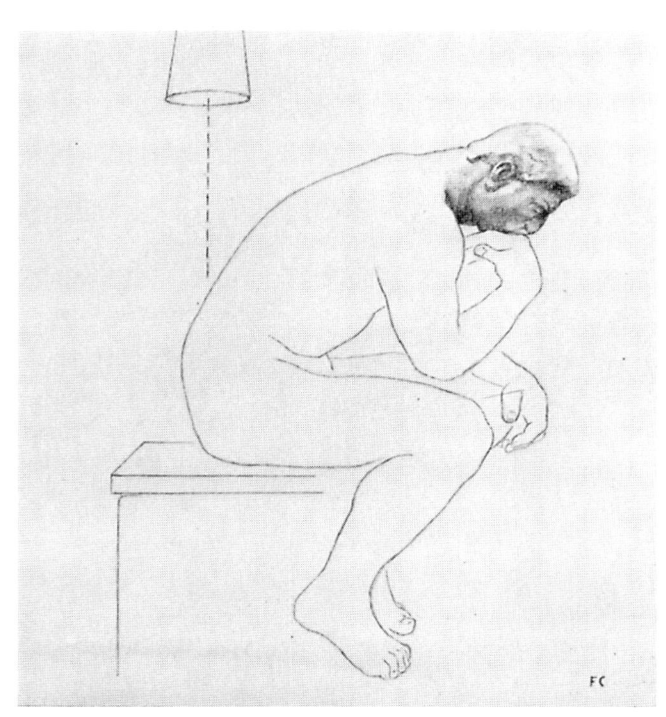

Figure 13.39. A special position in which the patient may be examined in order to unroll the sigmoid colon and reveal abnormalities concealed by its usual redundancy.

The rectosigmoid colon, because of its great redundancy and overlap in the pelvis, is very difficult to "unroll" and therefore to visualize in every part. A number of ingenious maneuvers have been designed to help locate malignant lesions in the sigmoid. Patients have been examined head down at a sharp incline on a table, so that the loops of bowel are pulled up out of the pelvis by their own heavy barium content and the sigmoid is straightened. Patients are routinely examined in oblique projections and laterally, and many radiologists use a view in which the central ray is directed obliquely caudad in the sagittal plane in a prone patient. The patient may also be examined sitting up, the ray directed downward through the back, as in Figure 13.39. The cooperation of the patient makes a great deal of difference in the success of these procedures. For this reason it ought to be a rule with physicians to explain to the patient beforehand approximately what is going to be done.

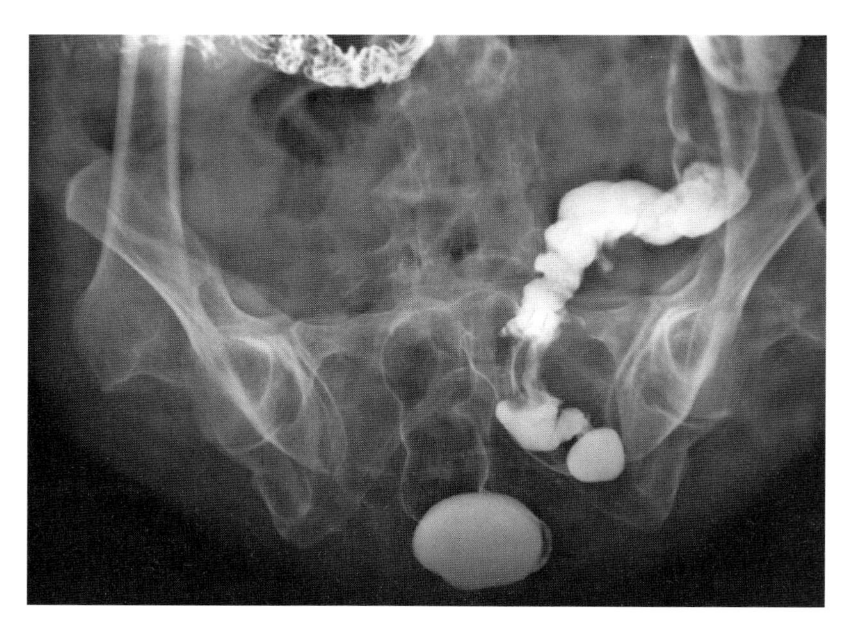

Figure 13.40. Annular carcinoma in a patient filmed as in Figure 13.39. The lesion was not seen on routine barium enema films.

Contraindications to Barium Studies

Some barium contrast studies are virtually harmless; others have well-recognized contraindications. So many of these studies incur little risk and minimal discomfort and have become so routinely a part of the diagnostic plan, however, that perhaps too few physicians could list half a dozen conditions in which a gastrointestinal series or a barium enema, for example, should *not* be carried out, or in which it would be better to postpone the procedure and the preparation for it. Barium sulfate in water suspension is itself inert, and none of it is absorbed during its passage through the gastrointestinal tract; but almost any such study involves the taking of numerous films and perhaps repeated fluoroscopic inspections. The process is fatiguing for the patient, particularly for one who is acutely ill. The anxiety of the patient is always a vital part of the hazard of the procedure. That anxiety can be allayed to an appreciable extent by intelligent preparation by the referring physician, who should take the time to explain in advance a little about the procedure.

The patient with symptoms suggesting any kind of large-bowel obstruction should *not* be given barium by mouth, because it becomes dehydrated in the colon. Furthermore, the preparation of a patient for a barium enema requires thorough cleansing of the bowel, usually by catharsis and cleansing enemas, and such measures are sometimes contraindicated clinically in the debilitated, elderly, dehydrated patient, or in the patient with electrolyte imbalance. For a patient whose presenting complaint is diarrhea, the usual cathartic agents should be decreased in volume or replaced by milder agents.

In sum, then, the complex barium contrast study should not be requisitioned without consideration of the entire clinical problem, nor should it be undertaken when the patient is not in a reasonably safe condition to undergo it. You would do well to observe one example of each of the major barium procedures for yourself, so that you will understand not only what it may require of your patient in terms of energy, stamina, and patience, but also the degree to which the patient's cooperation may be required for the success of the procedure. A patient who is paralyzed will not be able to stand for certain parts of a gastrointestinal examination that are usually carried out in that position. A patient who speaks no English will be particularly difficult to examine; he must hold his breath on command during the exposure of films, and if he does not understand and continues to breathe the films obtained will often be valueless. Obviously, this kind of difficulty may be prevented by your discussing the procedure with the radiologist before it is carried out. *If intelligent consideration for the patient is the primary concern of both radiologist and referring physician, undesirable developments resulting from any sort of procedure will be kept to a minimum.* (Needless to say, this applies no more to radiology than to any other branch of medicine.)

CT of the Gastrointestinal Tract

Barium examinations provide images of casts of the interior of the stomach, duodenum, small bowel, and large bowel. The existence of pathological processes of the gastrointestinal tract is implied radiologically by the demonstration of alterations of these barium "casts." Conditions that alter the bowel lumen can be identified by barium examination, but those that do not will not be revealed. Conditions that alter the appearance of the mucosa only, such as early gastritis or colitis, may not affect the barium column and require endoscopy for diagnosis. Conditions that alter the thickness of the bowel wall, or that produce changes in the soft tissues adjacent to the bowel, can be identified by CT.

Study the gastrointestinal tract in the four normal CT scans in Figure 13.41. The patient has swallowed oral contrast material that opacifies the stomach, duodenum, and proximal small bowel. For CT, a very dilute solution of water-soluble iodinated contrast media or a very dilute barium suspension is used for oral

Figure 13.41. Normal CT scans for review of the gastrointestinal tract. Study the appearance of the gastric wall in A and B, the duodenum in B and C, the small bowel (both contrast opacified and nonopacified loops) in B-D, and the colon in A-D.

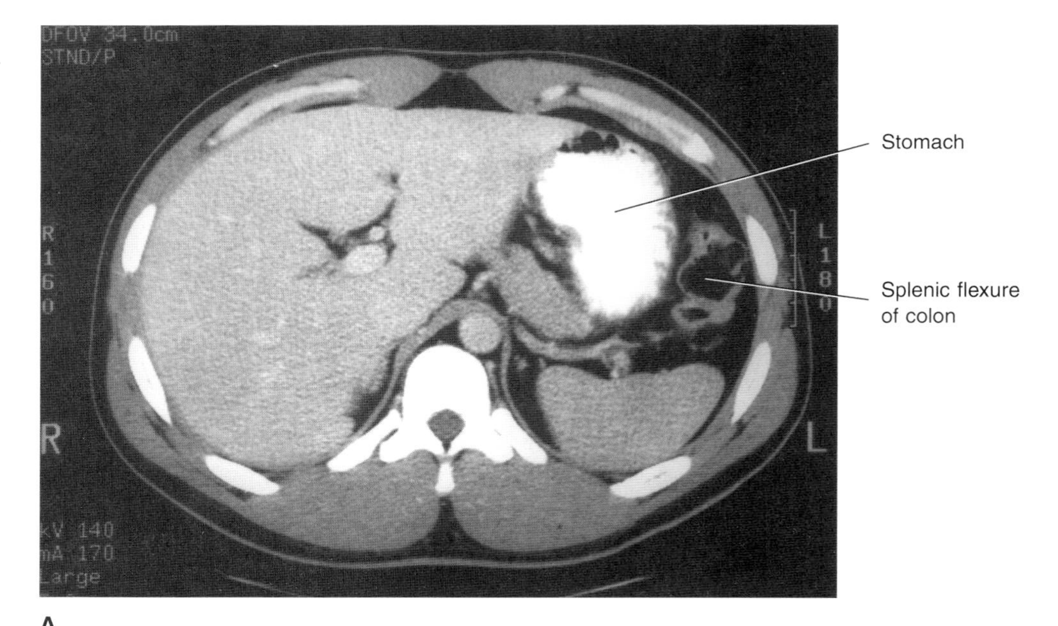

A

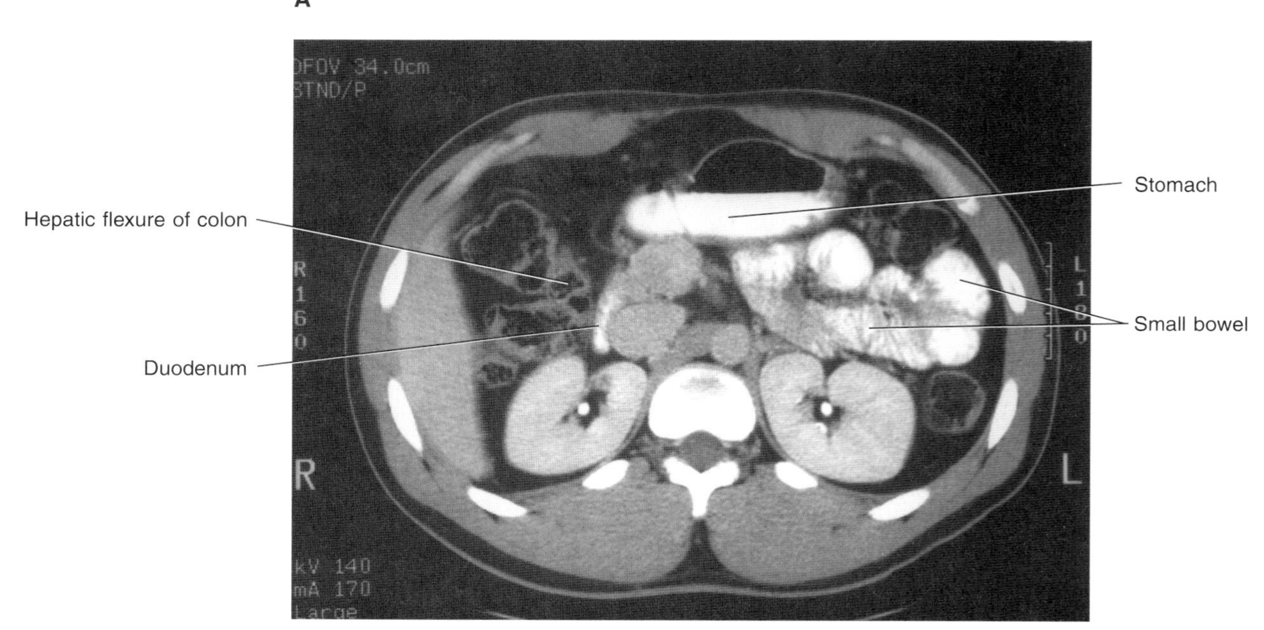

B

contrast. No contrast is seen in the distal small bowel or colon. Note the bowel wall thickness, the characteristic patterns of the differing segments, and the sharply defined contours of the serosal surfaces, abutting the low-density mesenteric fat. As you will see shortly, these serosal margins are disrupted and, because of inflammatory and neoplastic processes of the bowel that extend into the mesenteric fat and other surrounding structures, they may disappear at CT. Note the gastric rugal folds in A and B; the stomach is not fully distended. The duodenum is not distended either; its characteristic mucosal folds are seen in B and C. Small-bowel valvulae conniventes are well highlighted by contrast material in B and C. Loops of more distal small bowel containing only water-dense intestinal contents can be seen in D. In all four scans you can identify segments of the colon containing air and varying amounts of fecal material. Haustral markings in the transverse colon are well shown in C.

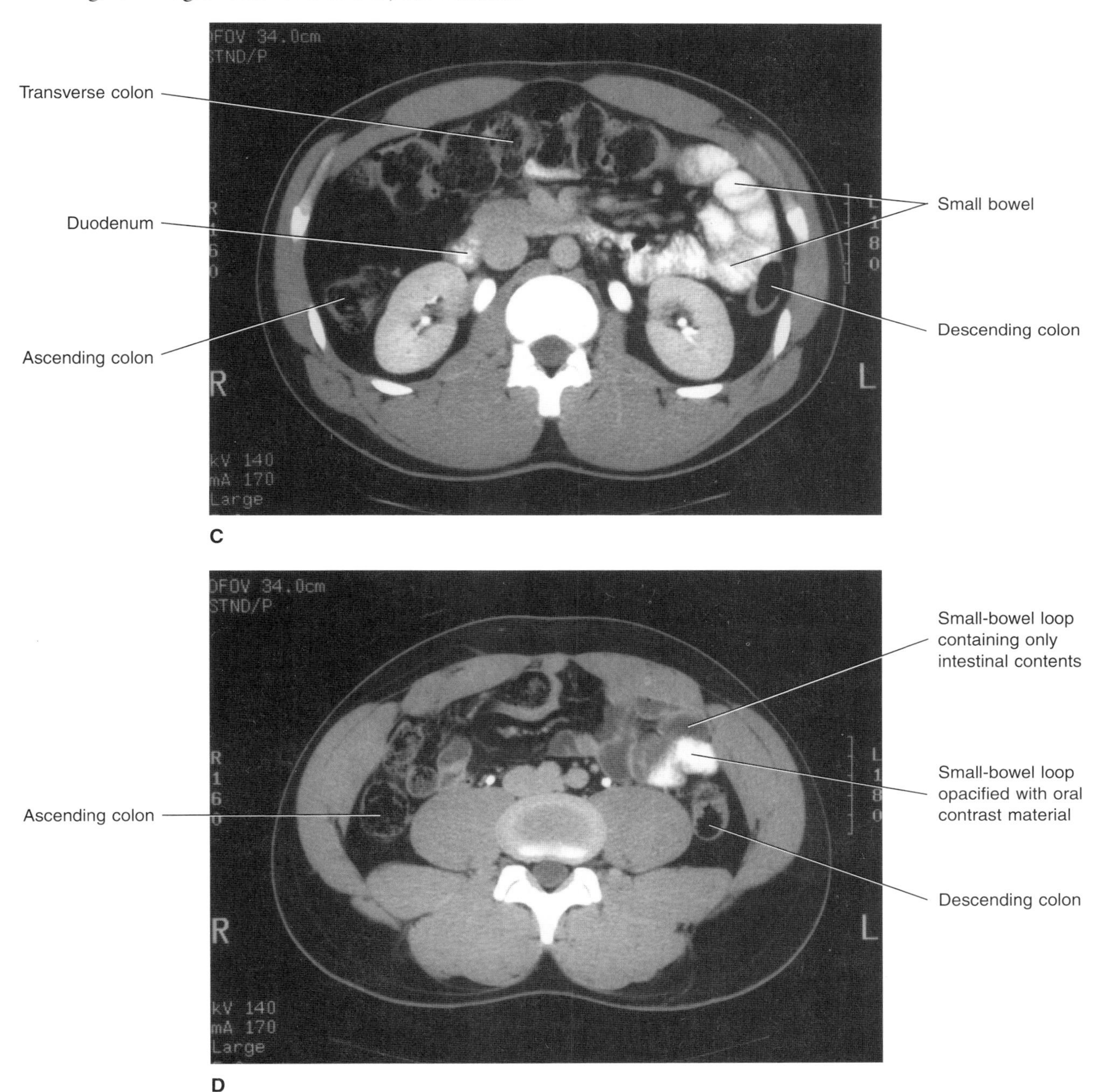

CT of the Thickened Bowel Wall

Diseases that thicken the bowel wall are well shown by CT. The patient in Figure 13.42 has Crohn's disease. He presents with right lower quadrant pain, fever, and an elevated white count. Note the marked thickening of the diseased small-bowel wall compared with the normal caliber wall of the descending colon. Marked inflammatory change is visible in the surrounding mesenteric fat, changing the black fat density to gray soft-tissue density. Note that portions of the serosal margins are obscured by the inflammatory change. CT can also show complications of bowel diseases such as Crohn's disease, including bowel perforation, abscess formation, fistula formation, and strictures causing mechanical small-bowel obstruction. B, a lower-level image, shows a large abscess resulting from a perforation of the terminal ileum; this abscess was the cause of the patient's abdominal pain and fever. Other conditions that may thicken the bowel wall which are well shown by CT include lymphoma, amyloidosis, Whipple's disease, and infections such as giardiasis, tuberculosis, and cryptosporidiosis (in immunosuppressed patients).

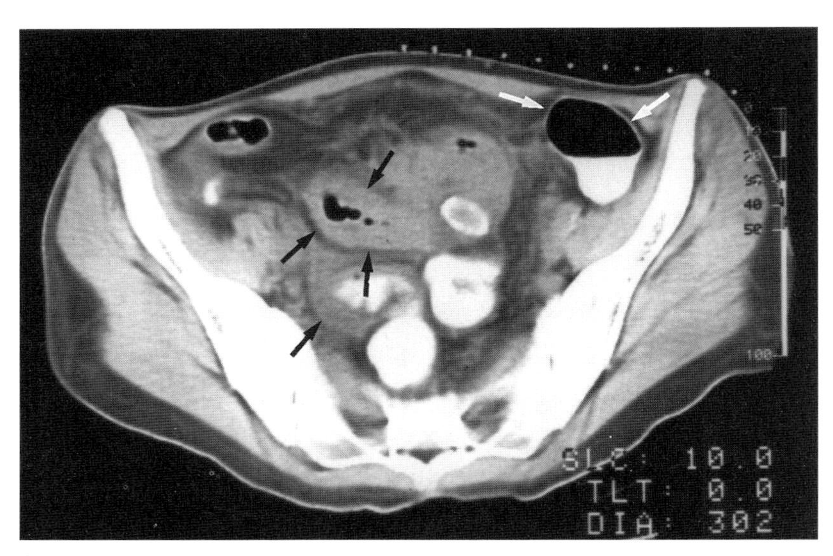

A

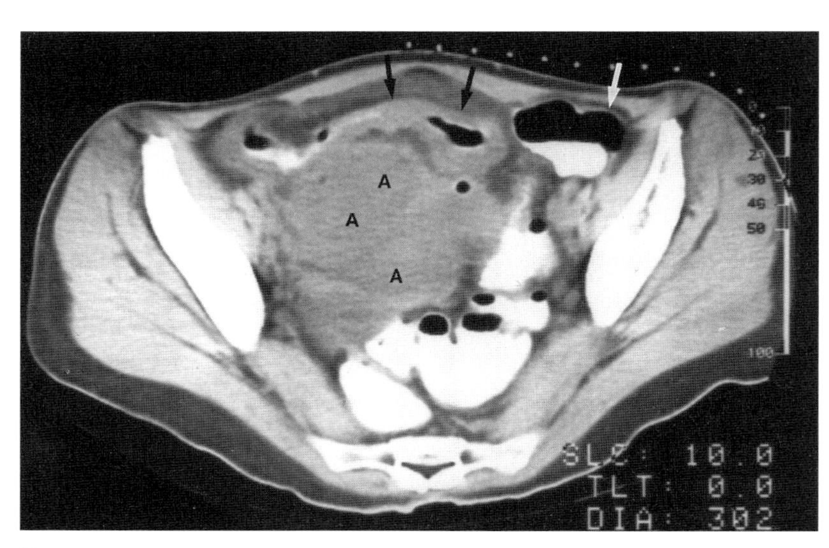

B

Figure 13.42A and B. CT slices of a patient with Crohn's disease showing bowel wall thickening in diseased loops of small bowel *(black arrows);* compare the normal caliber of the wall of the descending colon *(white arrows).* This patient suffered a common complication of Crohn's disease: bowel perforation with abscess formation (*A*'s).

CT of Diverticula Disease

Diverticulosis is a common large-bowel condition resulting from the herniation of portions of the mucosal and submucosal layers of the colonic wall through the muscular layer, producing multiple, small, thinly walled outpouchings, or diverticula. The condition is related to the low-fiber diet of the industrial world, and the incidence increases with age. The sigmoid colon is the part of the colon most commonly involved. Figure 13.43 is the CT scan of a patient with extensive, asymptotic diverticulosis of the sigmoid colon. The diverticula appear as multiple, air-filled outpouchings; the lumen of the sigmoid colon is filled with fecal material.

Diverticulosis may remain asymptomatic, or two complications may occur, bleeding and diverticulitis. Bleeding occurs when branches of vasa recta underlying the thin mucosal wall of the diverticula are abraded. Painless lower gastrointestinal hemorrhage results, the diagnosis and treatment of which will be discussed in Chapter 19.

Diverticulitis occurs when fecal material becomes trapped within a diverticulum, resulting in perforation of the diverticulum and abscess formation. The abscess may be confined to the wall of the colon or may extend outside the wall as a pericolic abscess. The clinical signs include a painful mass, often in the left lower quadrant, localized peritoneal inflammation, and an elevated white count. Most diverticula abscesses are quickly walled off and localized, although free perforation of pus and free air into the peritoneal cavity may occur, leading to peritonitis. In addition, sinus cavities may form, with fistula to adjacent structures such as the bladder.

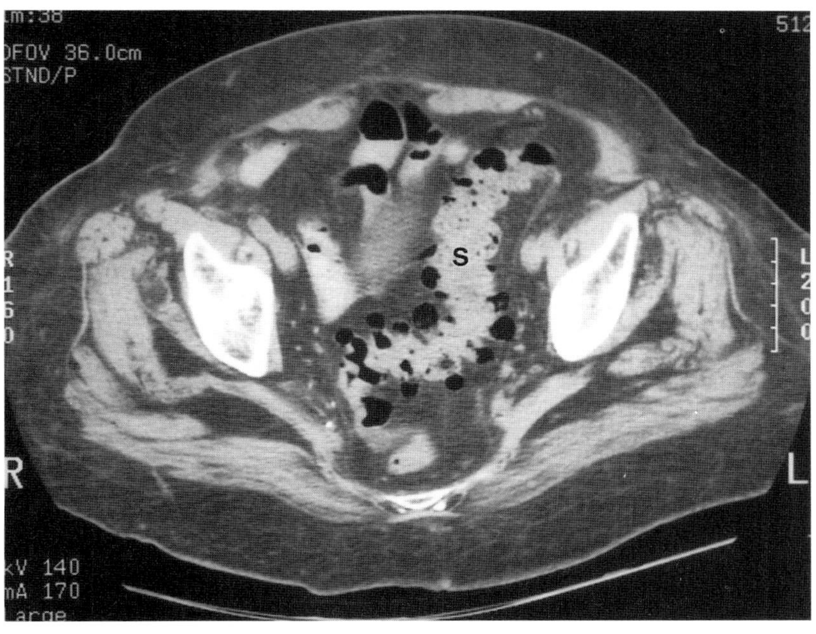

Figure 13.43. Diverticulosis of the sigmoid colon *(S)*. The diverticula appear as air-filled outpouchings.

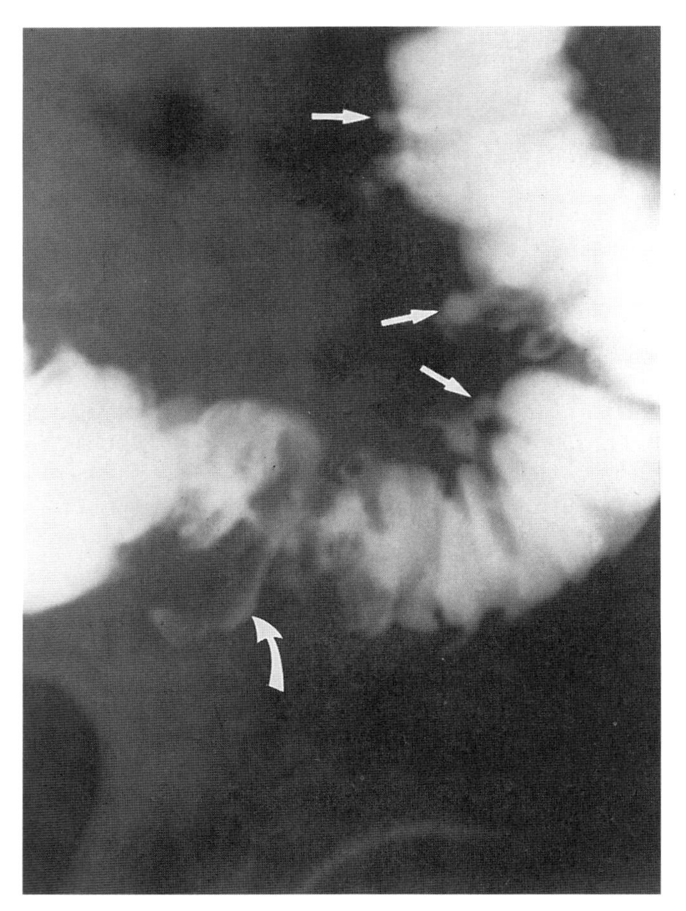

Figure 13.44. Barium enema examination of diverticulitis with a small intramural perforation *(curved arrow)*. The *straight arrows* indicate asymptomatic diverticula.

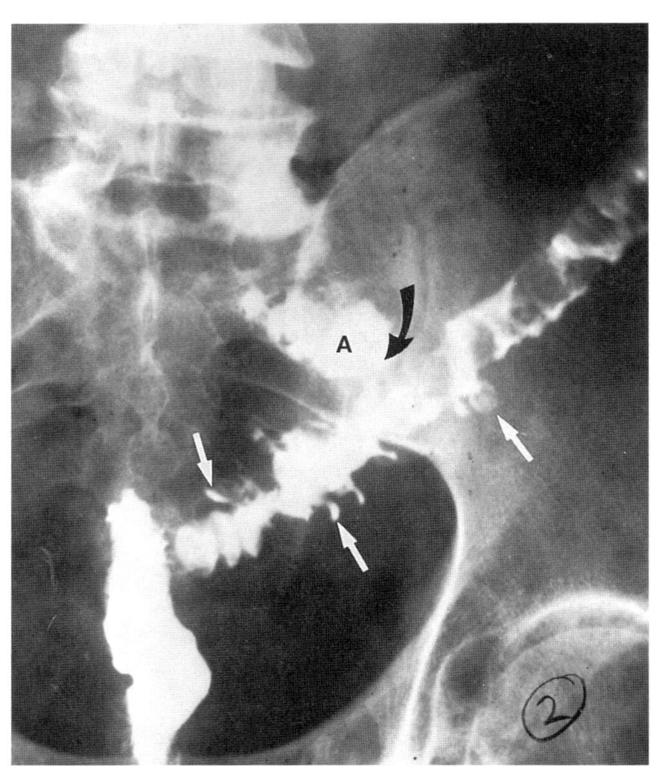

Figure 13.45. Barium enema examination of diverticulitis with a larger perforation *(curved arrow)* and pericolic abscess *(A)* formation. The *straight arrows* indicate asymptomatic diverticula.

In the past, diverticulitis was diagnosed by barium enema examination. Figures 13.44 and 13.45 show two patients with different-sized perforations. In Figure 13.44 the perforation is small with an abscess that is confined to the bowel wall (intramural abscess). In Figure 13.45 the perforation is more extensive, showing barium extravasation into a large pericolic abscess within the peritoneal cavity.

Today clinically suspected diverticulitis is diagnosed with CT, which better demonstrates extension of the process outside the colon. Figure 13.46 is the CT scan of an elderly man with diverticulitis of the sigmoid colon complicated by a paracolic abscess and fistula formation to the bladder. In A you can identify the thickened and inflamed wall of sigmoid colon *(white arrows)* and an adjacent abscess *(A)* containing pus and air. Compare the wall thickness of the diseased segment of sigmoid with the uninvolved segment *(black arrows)*, where haustral markings can be identified and the normal thin colonic wall is barely perceptible. In B, note that the bladder *(B)* contains air and fecal material floating in urine; gas and feces had passed into the bladder through a fistulous tract that had developed. This patient also had pneumonia.

In Figure 13.47 we see the CT scan of another patient with diverticulitis of the sigmoid colon *(S)* with free perforation of free air *(white arrows)* and pus *(P)* into the peritoneal cavity. CT has provided a great advance in bowel imaging by permitting visualization of the bowel wall and of structures adjacent to the bowel.

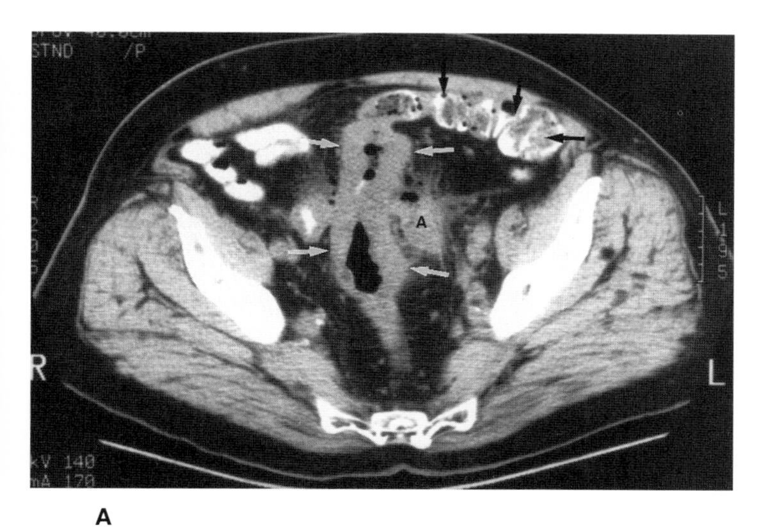

A

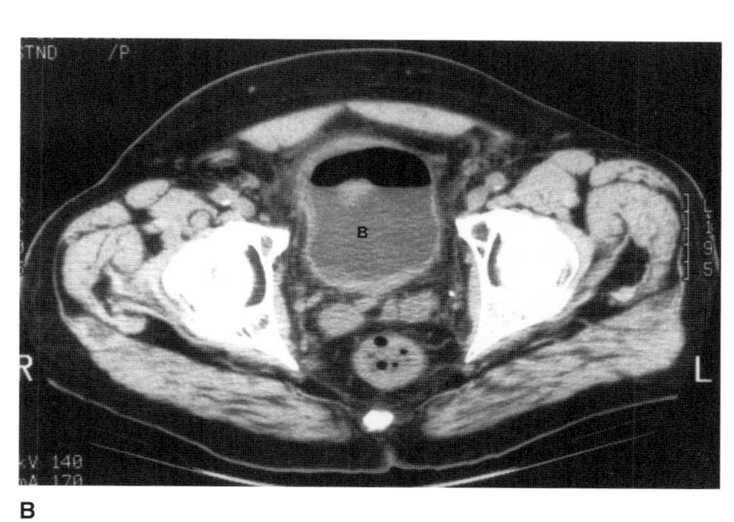

B

Figure 13.46. CT scan of sigmoid diverticulitis with a paracolic abscess and bladder fistula (see text).

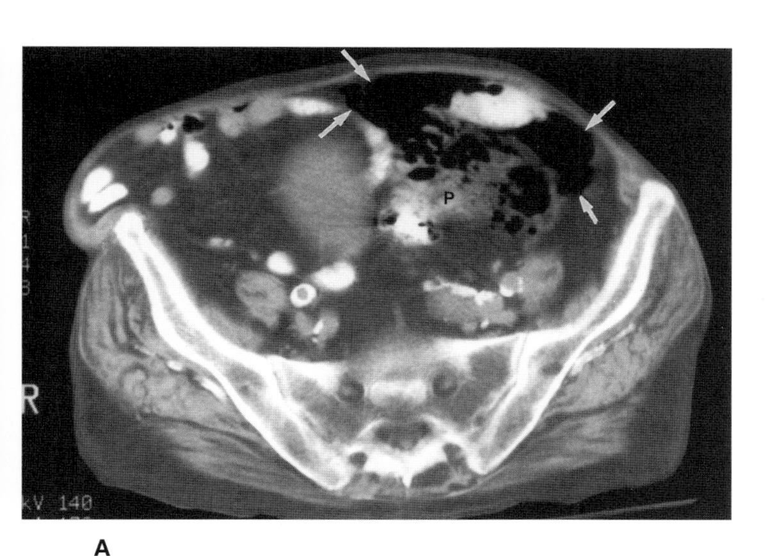

A

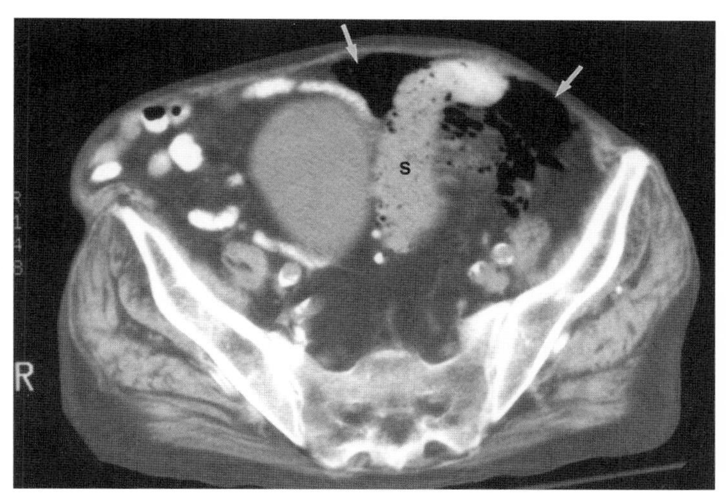

B

Figure 13.47. CT scan of sigmoid diverticulitis with free perforation of free air and pus into the peritoneal cavity (see text).

CT of Appendicitis

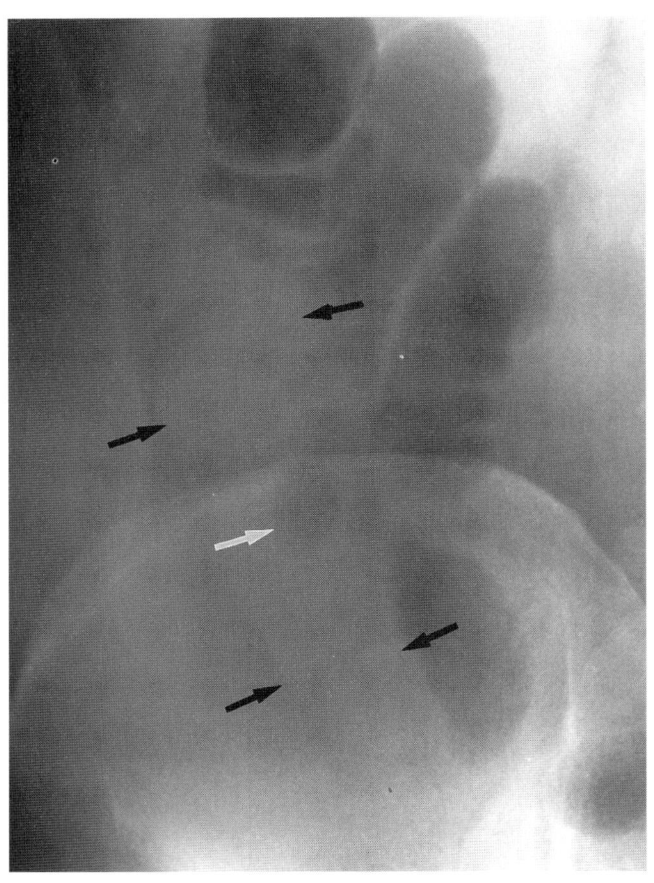

Figure 13.48. Plain film detail of appendicitis. Air-containing loops of bowel are displaced from the flank stripe by a sausage-shaped soft-tissue mass *(black arrows)* that is an appendiceal abscess. A small air collection *(white arrow)* is seen within the abscess.

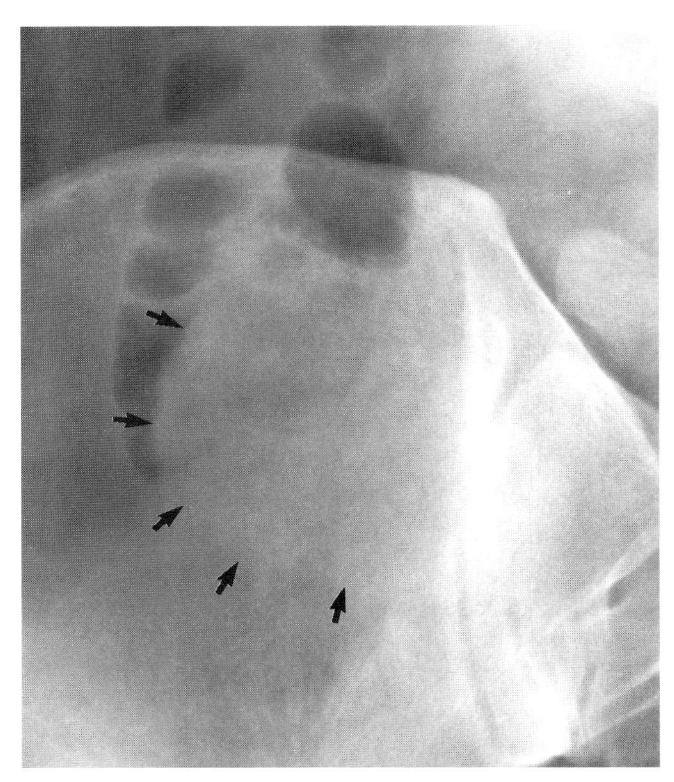

Figure 13.49. Plain film detail of appendicitis. The *arrows* indicate a soft-tissue mass extending into the cecum that is an appendiceal abscess.

Acute appendicitis is one of the most common causes of an acute abdomen. In the past, when patients presented with typical right lower quadrant pain, rebound tenderness, and elevated white count, no diagnostic imaging was performed before surgical management. But when the presentation was atypical and the diagnosis was in doubt, a radiological workup was of great assistance. Today, diagnostic imaging is recommended for all patients with suspected appendicitis.

Acute appendicitis occurs when the appendiceal lumen becomes obstructed and continued mucosal secretions cause dilatation of the appendix with increased intraluminal pressure. This impedes the appendiceal blood flow, leading to mucosal ulceration. Any superimposed bacterial infection results in gangrene and appendiceal perforation with abscess formation. Plain films, CT, and ultrasound may all demonstrate these pathological changes in the appendix and surrounding tissues, although CT shows them best.

An appendiceal calculus (appendolith) is seen on plain films in about 14 percent of patients with acute appendicitis. You have already seen an example in Figure 11.22, and you may wish to turn back to look at it again. The patient's appendix had ruptured, resulting in a large periappendiceal abscess. Two other patients with appendiceal abscesses are shown in Figures 13.48 and 13.49. In the first example the appendiceal abscess separates loops of bowel from the right flank stripe; in the second, the abscess indents the cecum. Barium enema examination in appendicitis is usually nonspecific, although if the appendix does not fill with barium, you may conclude that the appendiceal lumen is obstructed with appendicitis. You may also see a mass of inflammation indenting the cecum.

Ultrasound may be helpful in diagnosing appendicitis by demonstrating an enlarged appendix with or without an appendolith. Normally the appendix at ultrasound measures less than 6 millimeters when observed with a graded compression technique; an appendix larger than that usually indicates appendicitis. However, the diagnostic accuracy of the procedure depends on the skill of the sonographer and the clinical condition of the patient.

Because of its very high degree of accuracy, computed tomography has become the most popular method for imaging appendicitis. CT can determine not only whether appendicitis is present but also whether perforation and abscess formation have occurred. Figure 13.50 shows a patient with uncomplicated appendicitis. In A you can see the thickened and inflamed wall of the cecum, narrowing the cecal lumen *(C)*. The arrow points to an appendolith, at the junction of the appendix and cecum. In B, a slightly lower-level scan, you can identify the dilated appendix *(arrow)* and see surrounding inflammatory change in the pericolic fat. There is no evidence of perforation.

Figure 13.51 shows the CT examination of a patient with perforated appendicitis and abscess formation. In A, the cecum *(C)* is markedly distorted, with a thickened and inflamed wall. Just behind the cecum is an abscess *(A)* containing lower-density pus and collections of gas (shown as small black bubbles). Note the extensive inflammatory change in the surrounding mesenteric fat. The appendix is not well seen. In B, a scan taken at a slightly lower level, you can see more of the abscess *(A)* and surrounding inflammatory change. From the CT scan it is easy to understand why this condition produces right lower quadrant pain and tenderness to palpation. In the United States today, one out of five appendectomies removes a normal appendix. The routine use of CT in all patients with suspected appendicitis should eliminate unnecessary surgery.

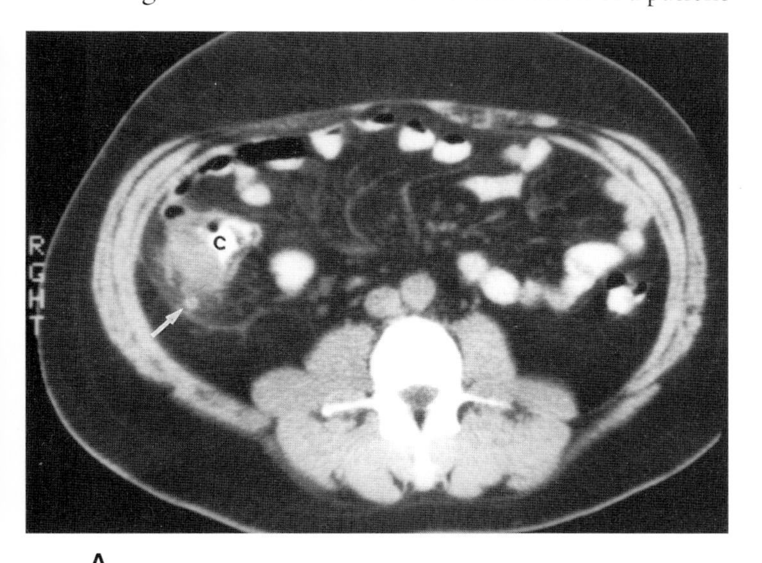

A

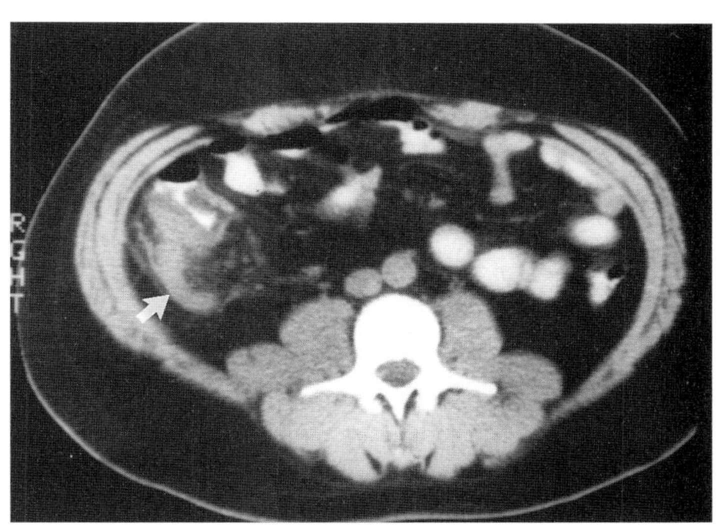

B

Figure 13.50. CT scan of uncomplicated appendicitis (see text).

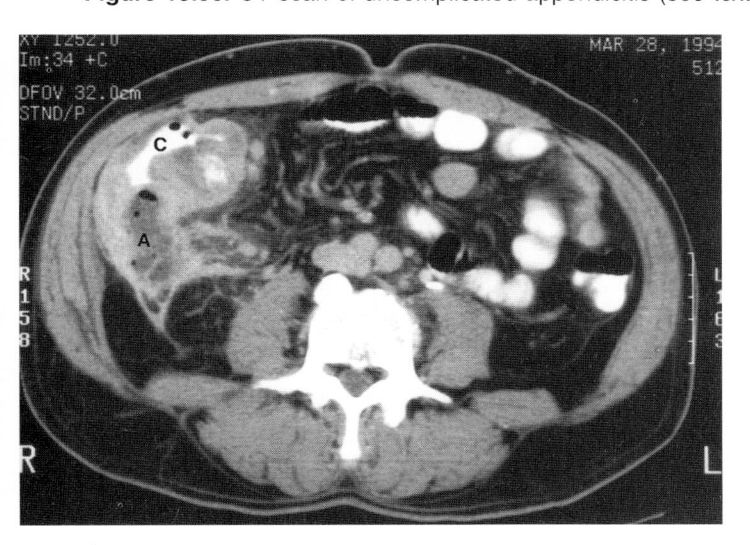

A

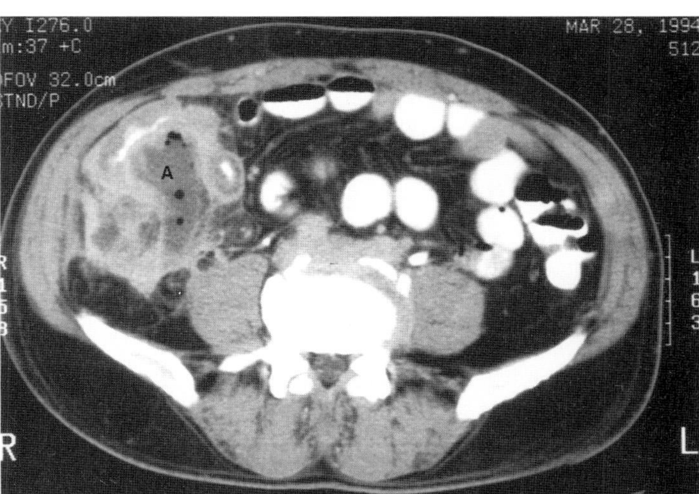

B

Figure 13.51. CT scan of perforated appendicitis with abscess formation (see text).

CT of Bowel Obstruction

In patients with suspected bowel obstruction, CT may be helpful in identifying the cause in cases when the plain films show only signs of mechanical small-bowel obstruction. The CT examination may demonstrate a neoplasm, intussusception, incarcerated hernia, or adhesions.

The patient illustrated in Figure 13.52 presented with clinical signs of bowel obstruction. His supine (A) and upright (B) plain films showed several markedly dilated loops of proximal small bowel *(arrows)* in the left upper quadrant, with air-fluid levels. In the background, a frame of normal caliber colon can be identified. These two films are consistent with a mechanical obstruction of the proximal small bowel. But what is the etiology?

The CT scan identified the cause, intussusception (small bowel intussuscepted within small bowel), the leading point (intussusceptum) being a small-bowel tumor. The loop within which the leading point is herniated is called the intussuscepiens. In C, a CT scan at the level of the midkidneys, you can identify several markedly dilated loops of small bowel *(S)*. Note that these loops contain mostly fluid, with only a small amount of air, rising anteriorly within the loops. In D, a slightly lower-level scan, you can see that a proximal segment of the small bowel (intussusceptum, indicated by the *black arrow*) has herniated within a slightly more distal loop (intussuscepiens, indicated by the *white arrow*). The fat-dense structure represents mesenteric fat with accompanying blood vessels that have been drawn within the intussusception. In E, an even lower-level scan, you can see that the leading point of the intussusception is a small-bowel tumor *(T)* seen within the intussuscepiens *(white arrow)*. At surgery the tumor proved to be a small-bowel melanoma. The patient was given oral contrast material before surgery, and F is a plain film that shows the transition site between the obstructed small bowel and the intussusception *(arrows)*.

Intussusception is common in children and usually ileocolic, with the terminal ileum herniating into the cecum and ascending colon. In most cases no cause can be found. By contrast, intussusception in adults is usually associated with a neoplasm or other abnormality as the lead point, and may involve other segments of the bowel.

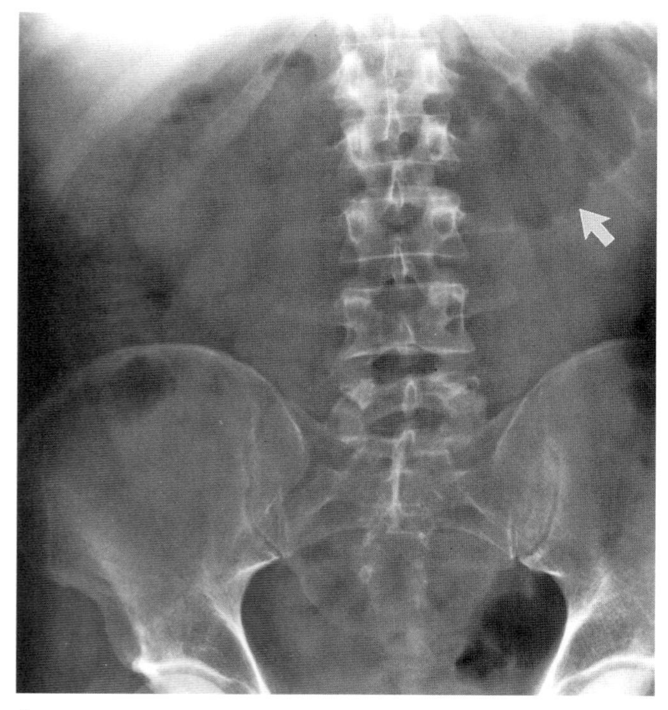

A

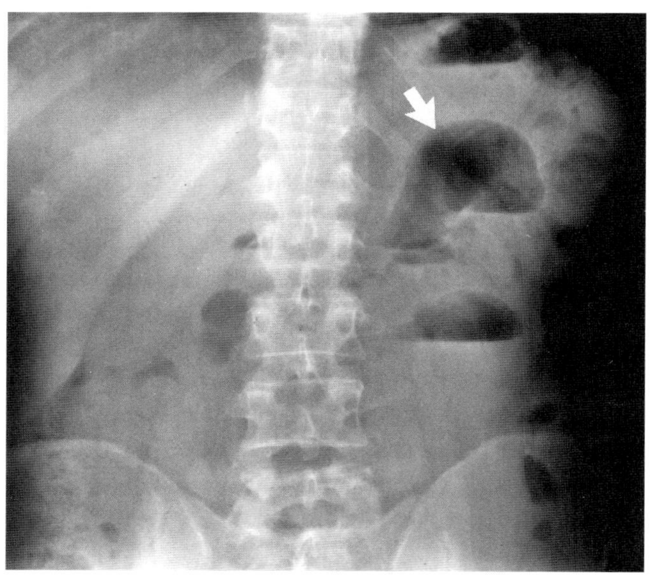

B

Figure 13.52. Small-bowel obstruction caused by intussusception (see text).

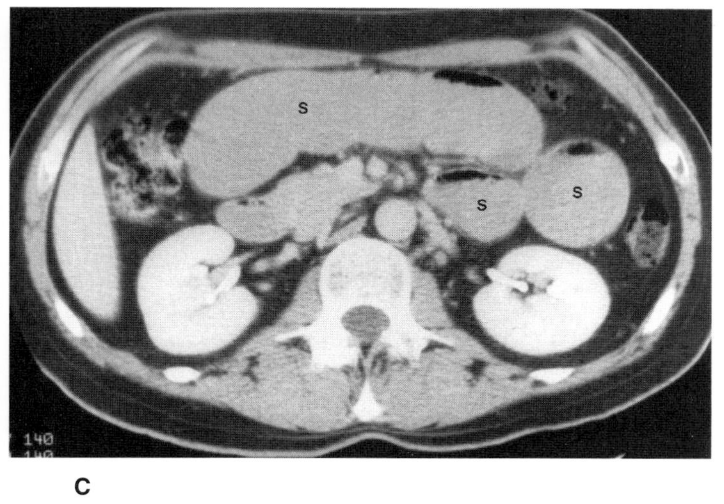

C

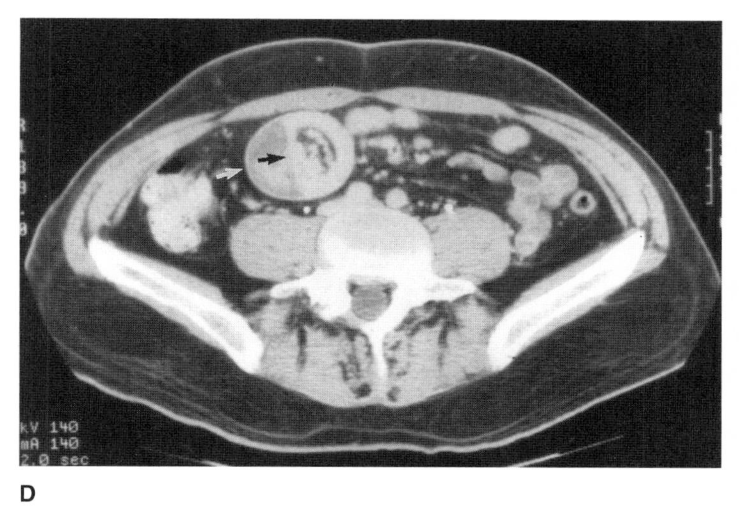

D

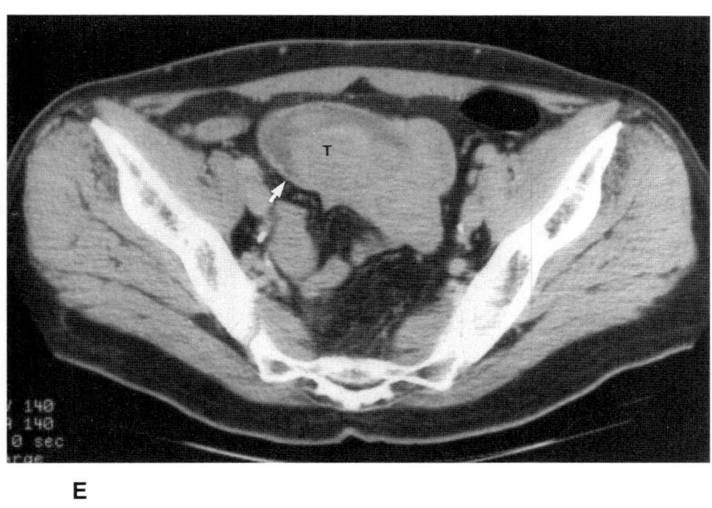

E

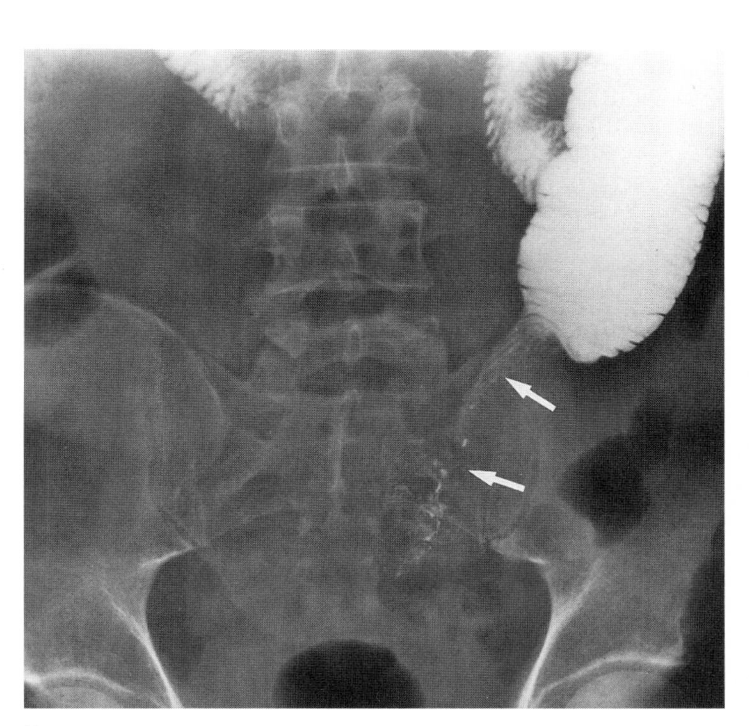

F

CT of Bowel Ischemia

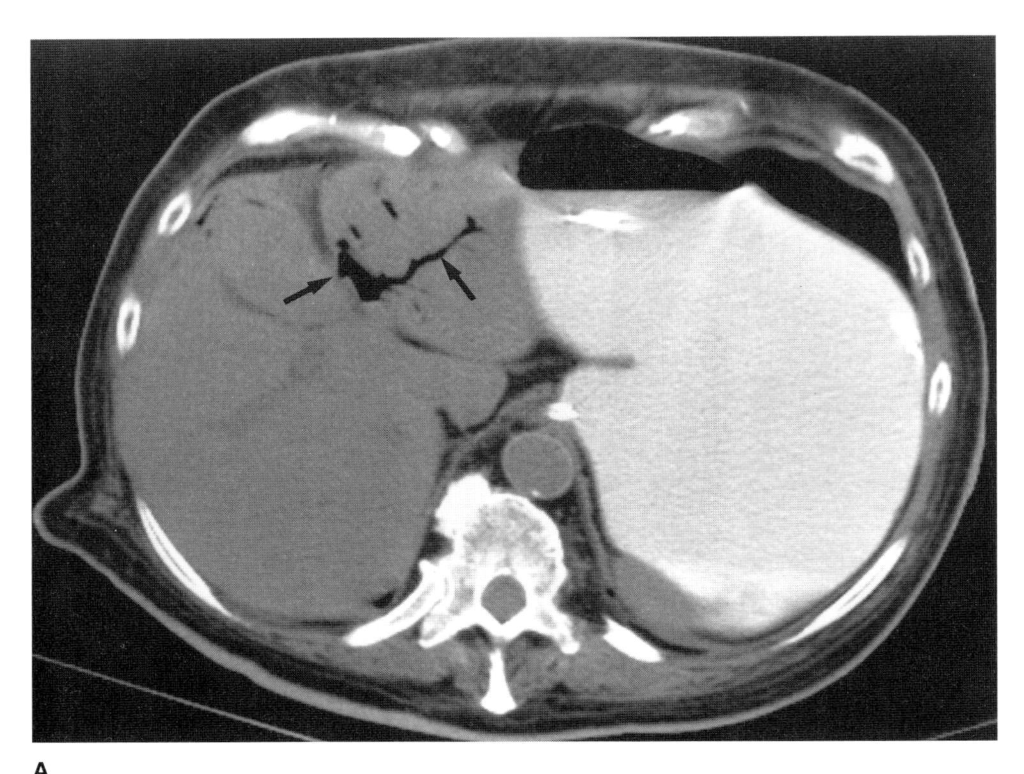

A

Figure 13.53. CT scan of an ischemic bowel (see text).

Interruption of blood flow to the bowel can result in bowel ischemia and necrosis, a diagnosis that may be difficult to establish, but that is imperative to confirm for optimum patient management. Patients usually present with diffuse abdominal pain, and may or may not have peritoneal signs on physical examination. The diagnosis may be suggested by concurrent conditions that cause bowel ischemia, such as atrial fibrillation or a new arrhythmia that may have produced emboli to the superior mesenteric or other bowel arteries.

CT can identify bowel infarction by showing the presence of gas within the bowel wall and portal venous system. Figure 13.53 shows an example, an elderly woman with atrial fibrillation and new abdominal pain. In C, the lowest-level scan, you can identify collections of gas *(arrows)* within the walls of the involved bowel segments. Note that the gas collections do not migrate anteriorly to the less dependent portions of the bowel because they are trapped within the wall rather than being able to move freely within the bowel lumen. In B, gas is seen within the mesenteric vein *(black arrow)* draining the bowel. The superior mesenteric artery is the small white dot adjacent to the vein. In A, you can identify gas within the less dependent, anterior portal venous branches *(arrows)* of the liver.

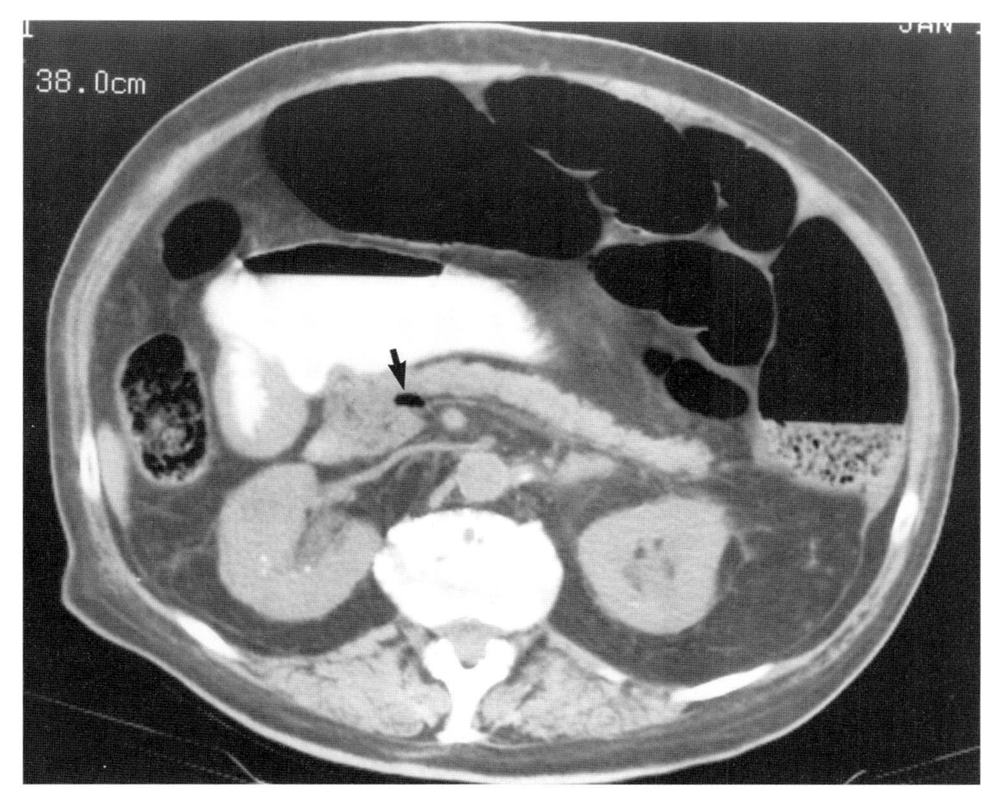

B

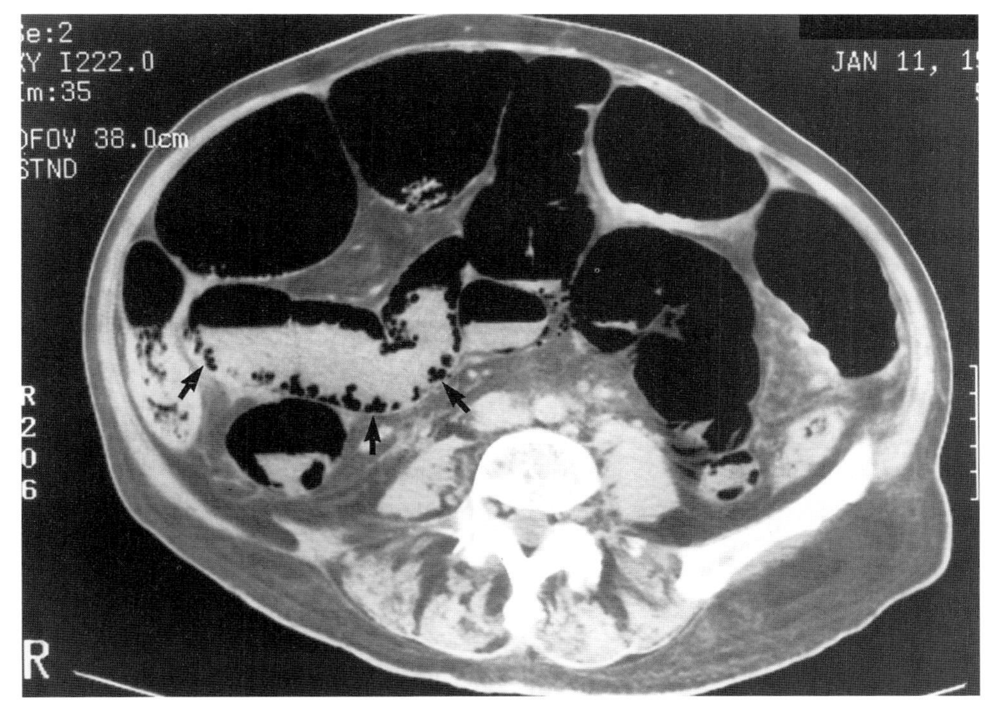

C

14 The Abdominal Organs

So far you have learned about the plain film appearance of the abdomen and the imaging of the gastrointestinal tract. In this chapter you will study techniques for imaging the remaining abdominal organs. We will proceed organ by organ, discussing the most common disease conditions requiring diagnostic imaging.

Today such a variety of techniques are available for imaging the abdomen that it is often difficult for a young physician to decide which is most appropriate for a given clinical problem. We hope that this chapter will give you a substantial grounding from which to work. Of course, our comments apply to current recommendations; with the passage of time imaging technologies will change. Certain examinations may increase or decrease in popularity, not only because of their diagnostic accuracy, but also because of their availability, safety, costs, and patients' acceptance of them. The last factor you should pay great attention to in your practice. You may, for instance, be surprised to hear that a patient requiring postoperative imaging of an abdominal aortic aneurysm graft prefers a catheter arteriogram to an MR arteriogram (the latter requiring no vascular catheterization or contrast material injection). But you will understand when you learn that this patient has severe claustrophobia and was unable to cooperate for the completion of an MR scan on a previous occasion.

Always listen to your patients! In addition to choosing an alternative procedure, you may on your own, or in consultation with the radiologist, determine ways of making procedures more comfortable for and acceptable to your patients, for example by providing sedation for an MR scan or by inviting a family member or friend to accompany the patient into the scanner suite to provide moral support and encouragement.

In the following pages you will observe a great variety of techniques for imaging the abdominal organs. If after completing the chapter you feel somewhat lacking in confidence in choosing the most appropriate methods of abdominal imaging, remember that consultation with the radiologist is always possible and will prove useful to you in selecting imaging examinations for your patients. The radiologist is often able to suggest a test to answer your clinical question that may be less expensive and easier for your patient than the alternatives.

The Liver

The liver can be imaged with a variety of techniques, including plain films, ultrasound, radioisotope scanning, CT, MR, and angiography. You have already seen the liver shadow on plain films. But plain films have a number of limitations. The determination of hepatomegaly on plain films, for instance, is generally unreliable and can be made with certainty only when the liver is massively enlarged. Plain films may show liver calcifications and air within the biliary tree or abscesses; they do not show soft-tissue abnormalities such as fluid-filled cysts, liver tumors, and dilated bile ducts.

Ultrasound can easily diagnose dilated bile ducts in patients with obstructive jaundice, as well as cysts, abscesses, and tumors, which appear as focal areas of decreased or increased echogenicity. But the detail is less than that shown by CT, the accuracy of diagnosis varies with the skill of the operator, and some anatomic portions of the liver are difficult to evaluate with ultrasound.

A radioisotope liver scan is performed after an intravenous injection of a radioisotope-labeled sulfur colloid (usually technetium). The colloid particles are phagocytized by the reticuloendothelial (Kupffer's) cells of the liver and spleen, and then imaged with a gamma camera. Today this procedure has been virtually replaced by ultrasound, CT, and MR scanning. A normal radioisotope liver scan is shown in Figure 14.1. Cysts, abscesses, and tumors would displace the normal reticuloendothelial system, appearing as "cold spots" (focal areas of diminished or absent radioactivity) on isotope liver scans.

At present, CT with intravenous contrast material is the procedure of choice for imaging the liver parenchyma. It can identify a wide range of hepatic pathology with excellent detail and is the primary method of examining patients with suspected liver metastases. MR is particularly helpful in differentiating common benign hepatic hemangiomas from primary liver tumors and metastases; hemangiomas may appear as metastases on CT scans. MR is also valuable in the evaluation of primary liver tumors and liver metastases. Today hepatic angiography is reserved primarily for those patients who need detailed evaluation of their hepatic vasculature, including the portal venous system. Such vascular road-mapping is often required before liver surgery.

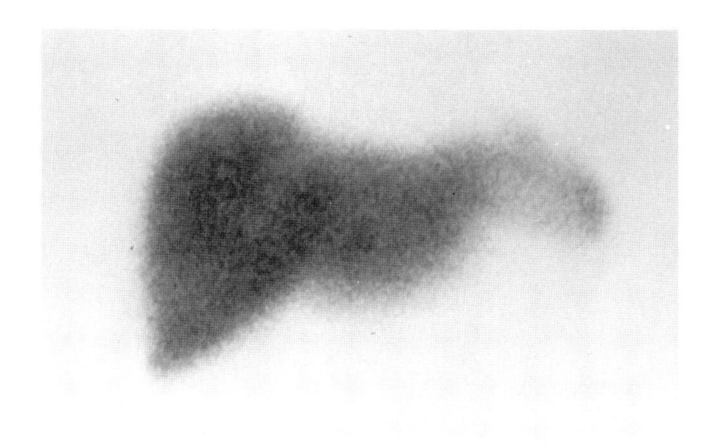

Figure 14.1. Normal radioisotope liver scan.

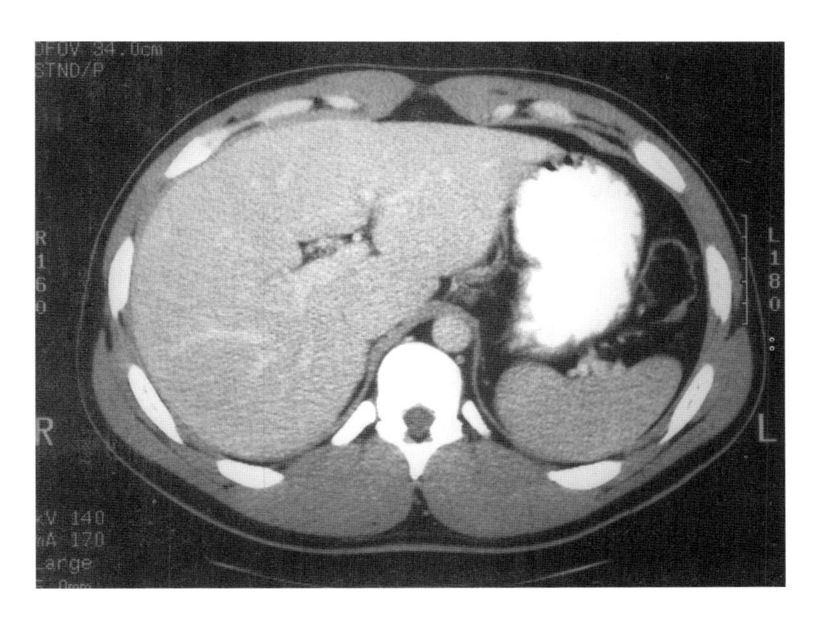

Figure 14.2. Normal CT scan of the liver. The portal vein and other vascular structures are opacified with intravenous contrast material. Oral contrast material is seen in the stomach.

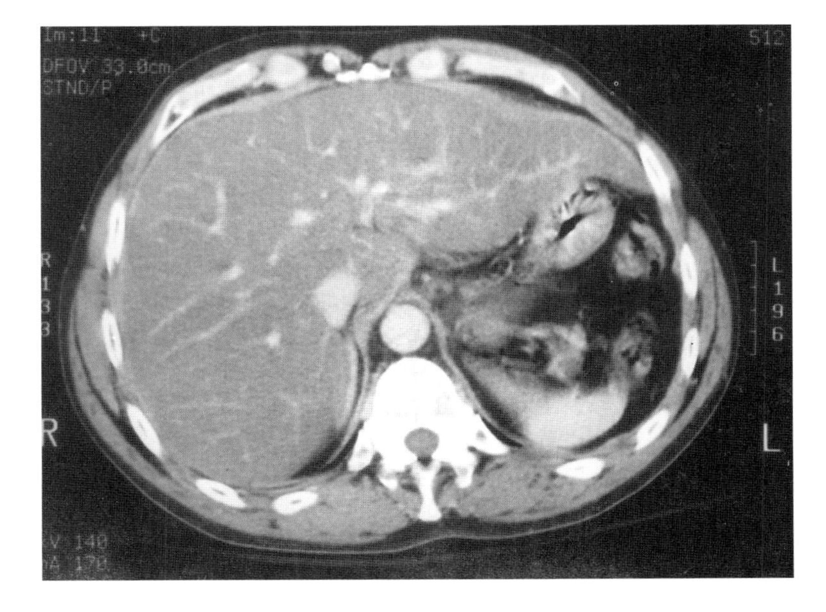

Figure 14.3. CT scan of a fatty liver in an alcoholic patient. Compare the CT attenuation (density) of the abnormal liver with the liver in Figure 14.2. The fatty liver is less dense (darker) than the chest wall musculature seen around the ribs; in the normal patient the liver is denser (whiter) than the chest wall musculature. Note that the hepatic blood vessels are much better seen in the fatty liver, silhouetted against the less dense hepatic parenchyma.

Liver Metastases

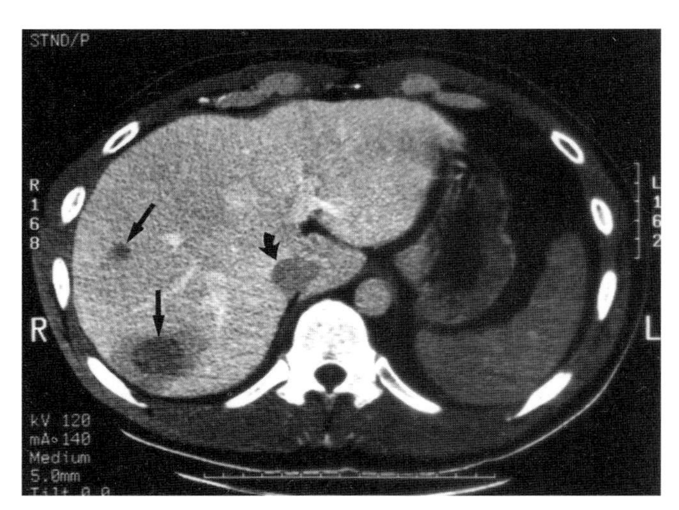

Figure 14.4. CT scan of liver metastases. Two low-attenuation metastatic deposits *(straight arrows)* are well shown against the contrast opacified hepatic parenchyma. The normal inferior vena cava *(curved arrow)* is not yet opacified as brightly as the adjacent aorta.

You may wish to examine the liver for possible metastases in a patient with a recently diagnosed primary cancer that is known to spread to the liver. The most common arise from the lung, breast, colon, rectum, stomach, and pancreas. Approximately 30 percent of patients with liver metastases have normal liver chemistries at the time of initial staging; therefore, imaging the liver should not be neglected if the laboratory values are normal. In addition, you may wish to image the liver in a patient with known metastatic disease to evaluate the response to treatment with chemotherapy or radiation therapy.

CT with intravenous contrast material is presently the most accurate, readily available technique for identifying liver metastases. Both large and small lesions can be recognized and their locations pinpointed. The entire liver can be evaluated in a single examination. The remainder of the abdomen is also shown, so that extrahepatic nodal and other metastatic sites can be seen as well. If the origin of the primary tumor is abdominal or pelvic, and it is unresectable, then it also can be evaluated in the same abdominal CT scan. In addition, new computer programs can calculate the volumes of liver metastases and compare them on serial scans to document either growth over time or decrease in size in response to chemotherapy.

Metastatic tumors are generally of lower CT density than the surrounding contrast-enhanced hepatic parenchyma (Figures 14.4 to 14.6). They are less well

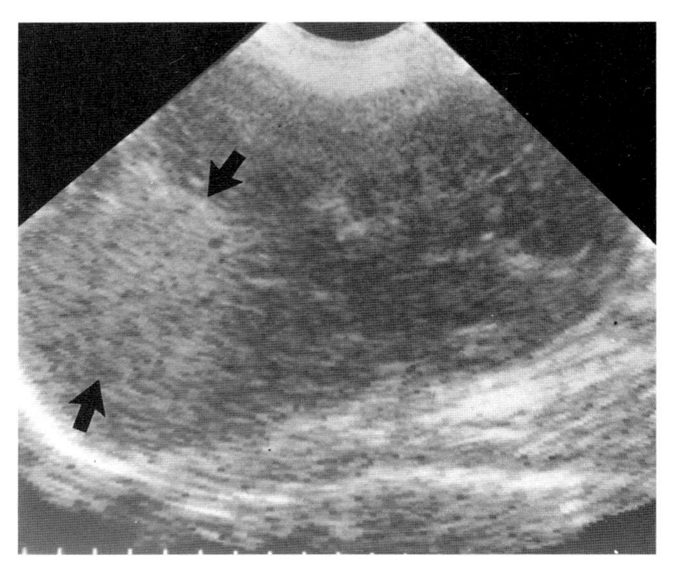

A

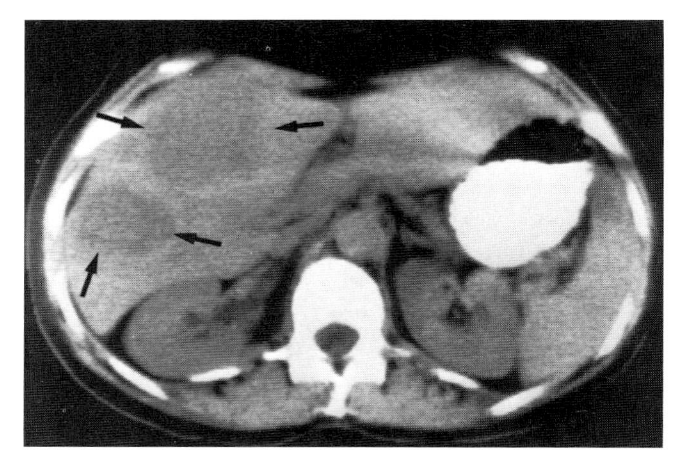

B

Figure 14.5. A: Ultrasound of liver metastases from colon cancer. This image was obtained in the sagittal plane; the patient's head is to your left, the feet to your right. The very echogenic right diaphragm is seen as a white curve in the lower left portion of the scan. The *arrows* indicate a metastatic deposit appearing as an echogenic mass. B: CT scan of the same patient showing large, low-attenuation metastases *(arrows).*

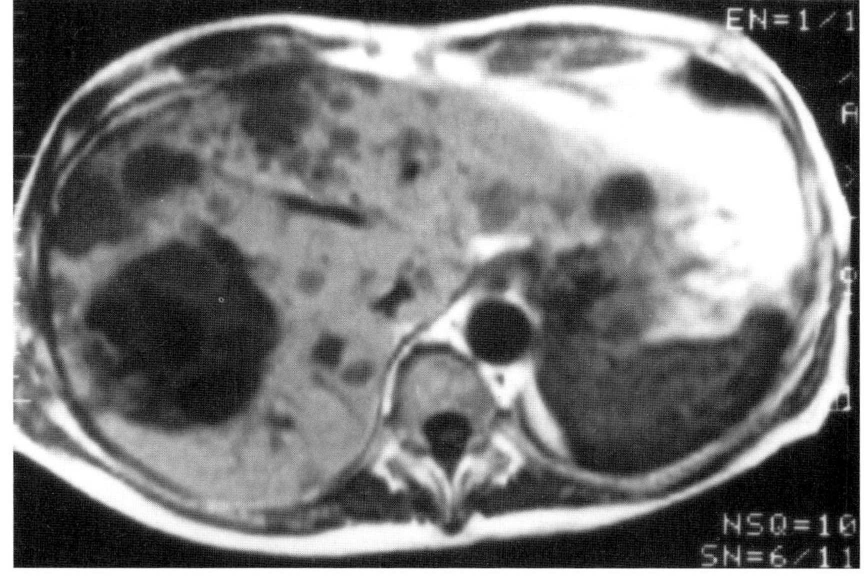

A

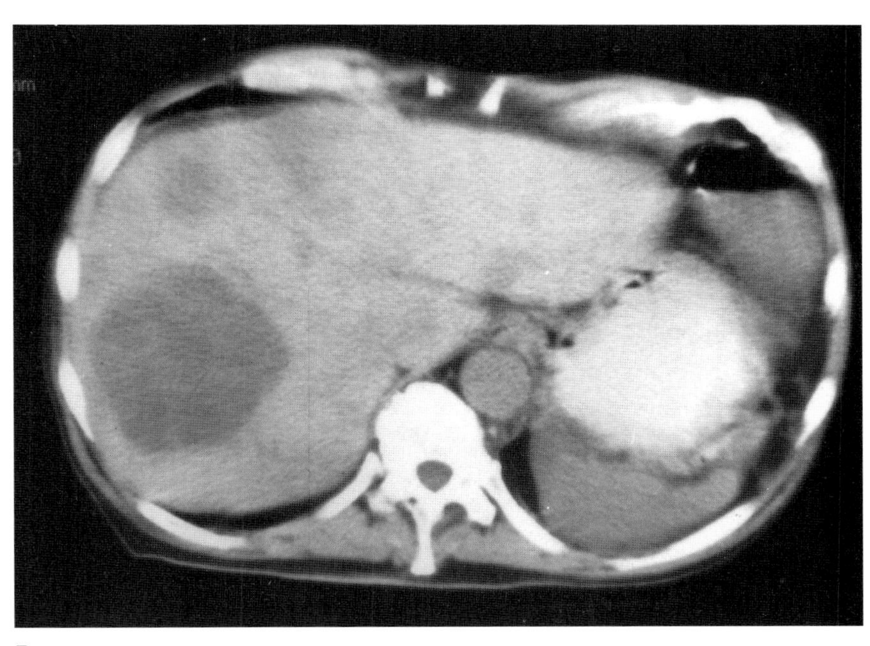

B

Figure 14.6. A: MR scan of multiple liver metastases from a malignant pancreatic islet cell tumor. The metastatic deposits have decreased signal strength (appear blacker) compared with the normal liver parenchyma on this T1-weighted image. B: Comparison CT scan of the same patient at a similar level.

circumscribed than liver cysts, and a measure of their CT attenuation would show it to be much higher than that of cysts (you will remember that the density of cysts is similar to that of water). Also, CT can usually determine by location and size whether a single metastatic tumor is amenable to surgical resection.

Although ultrasound can show large metastases in portions of the liver that are more easily examined by this technique, it is not the procedure of choice for determining the number and volume of liver metastases. But ultrasound may show *unsuspected* metastases in patients in whom an ultrasound examination was performed for some other reason. Generally metastases are echogenic as compared with normal liver parenchyma (Figure 14.5) and are not echolucent like liver cysts. Small metastases, especially those less than 1–2 centimeters in diameter, may very well be overlooked by ultrasound.

MR is an excellent technique for imaging abnormalities in soft-tissue structures, and consequently liver metastases are well shown on MR scans (Figure 14.6). Because of MR's ability to differentiate varying types of soft tissues, liver metastases may be easier to see with MR than with CT.

Primary Tumors of the Liver

Patients with primary tumors of the liver often present with right upper quadrant pain; on physical examination a mass or hepatomegaly may be detected. Liver chemistries may be normal or abnormal. Benign tumors such as hepatic adenoma and focal nodular hyperplasia are more common in young and middle-aged women who have been taking birth control pills or hormonal replacement therapy. Hepatocellular carcinoma, or hepatoma, is more common in cirrhotic patients.

At ultrasound, primary tumors will appear as echogenic masses, distinguishing them from anechoic cysts. CT with intravenous contrast material is an excellent technique for imaging liver tumors. Most will appear as solid masses having less CT density than the surrounding liver parenchyma. Hypervascular tumors, however, may show increased CT density. Liver tumors tend to have ill-defined margins and sometimes necrotic centers and calcification. CT accurately localizes tumors within the liver, usually making it possible to determine whether resection is possible or not. Preoperatively, hepatic angiography is also performed to delineate the major hepatic arteries and veins. If certain major vascular structures, such as the portal vein or inferior vena cava are encased, resection may not be possible.

Figure 14.7 shows the CT examination and the hepatic arteriogram of a 36-year-old woman with a benign hepatic adenoma of the left lobe of the liver. It is *hypervascular* on both examinations. On the CT examination the mass has *increased CT attenuation* as compared with the surrounding hepatic parenchyma. Compare the bizarre hypervascular arterial pattern within the tumor with the normal arteries in the right liver lobe. This pattern is typical of "tumor vasculature."

Figure 14.8 shows the liver scan, CT, and arteriogram of a 58-year-old man with a hepatoma of the left lobe. Note the large, round cold area *(M)* between the right lobe of the liver and the spleen on the radioisotope liver scan. Hepatomas are *hypovascular* at CT; the mass has *decreased CT attenuation* compared with the surrounding parenchyma. And, as predicted by the CT scan, the mass is shown to be *hypovascular* on the late phase of a celiac arteriogram; compare its density with the normal vascular density of the right lobe of the liver beyond the right margin of the tumor.

The most common benign liver neoplasm is the cavernous hemangioma, found in up to 7 percent of the general population. In 10 percent of cases multiple lesions are present. Most are less than 5 centimeters in diameter and cause no symptoms whatsoever; they are, for the most part, considered to be incidental findings. But they can mimic primary malignant liver neoplasms or liver metastases on ultrasound and CT scans. Consequently, correct recognition of a cavernous hemangioma is mandatory, to differentiate this benign neoplasm from more serious conditions. In a patient undergoing staging examinations for a cancer

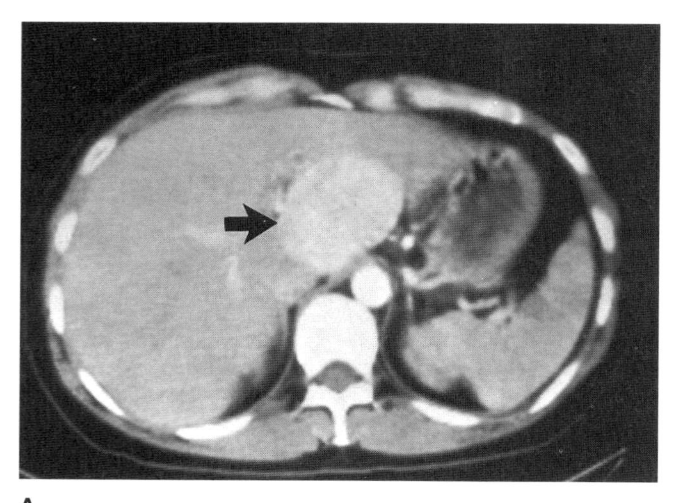

A

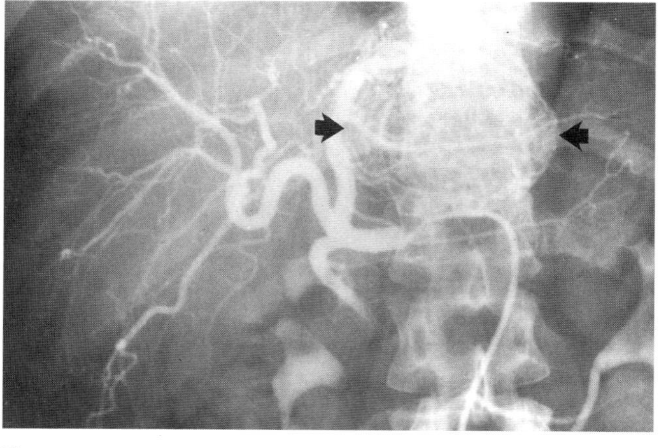

B

Figure 14.7. Benign hepatic adenoma in a 36-year-old woman. A: CT scan shows a hypervascular left lobe of liver mass *(arrow)*. B: Hepatic arteriogram confirms a hypervascular mass *(arrows)* with tumor vasculature.

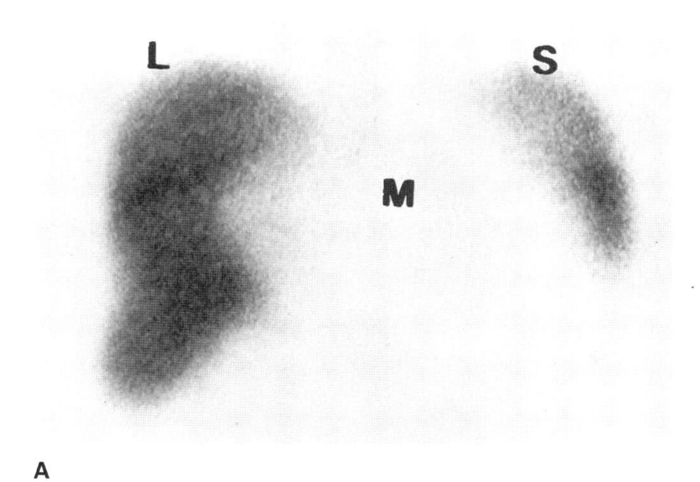

A

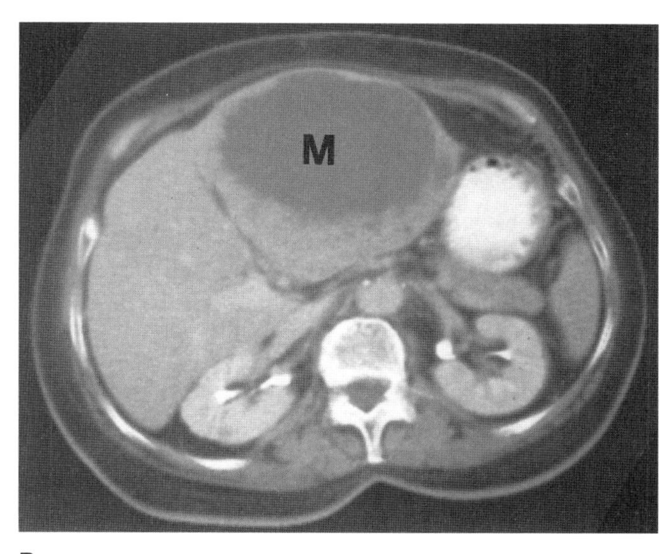

B

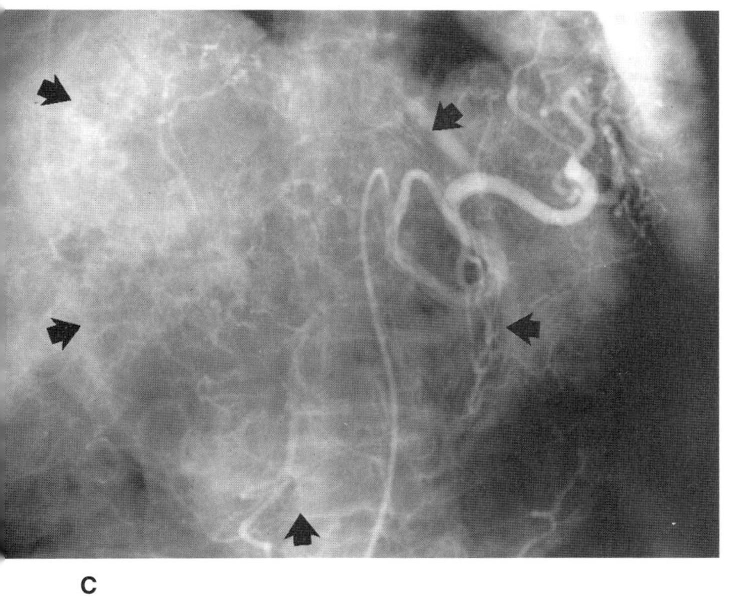

C

Figure 14.8. Hepatoma of the left lobe of the liver. A: Radioisotope scan shows a large area of absent uptake *(M)* between the right lobe of the liver *(L)* and the spleen *(S)*. B: CT scan demonstrates a large, low-attenuation mass *(M)* in the left lobe. C: Celiac arteriogram confirms a large hypovascular mass *(arrows)* in the left lobe of the liver.

that may metastasize to the liver (such as colon cancer), a cavernous hemangioma could be confused with a liver metastasis. Fortunately, there are techniques for identifying cavernous hemangiomas, the most common being MR scanning, where they display a characteristic pattern. These lesions are hypo-intense or iso-intense on a T1-weighted image, but brighten markedly with T2-weighting (Figure 14.9).

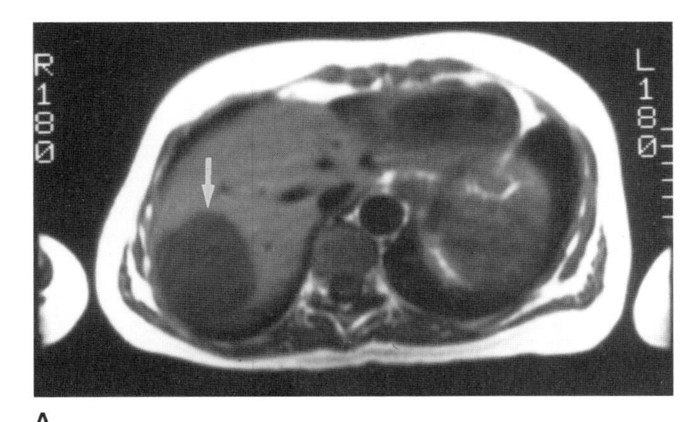

A

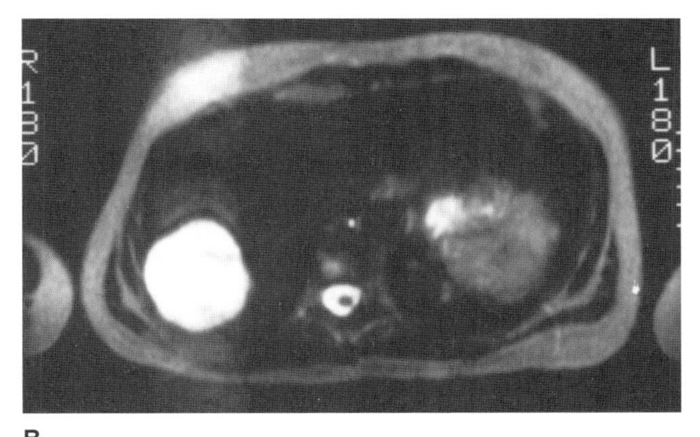

B

Figure 14.9. MR scan of a benign cavernous hemangioma of the liver. A: T1-weighted image showing a hypo-intense (decreased MR signal) mass *(arrow)* in the posterior right lobe of the liver that could represent a primary or metastatic liver tumor, or a benign hemangioma. B: T2-weighted image at the same level shows that the mass is very hyperintense (markedly increased MR signal), characteristic of a benign hemangioma.

Hepatic Cysts and Abscesses

Two nonneoplastic liver masses that may be easily diagnosed with cross-sectional imaging techniques are hepatic cysts and abscesses. Patients with either condition may present with right upper quadrant pain, a mass, and hepatomegaly. Patients with abscesses, having signs of fever and septicemia, usually appear sicker than those with cysts. Liver abscesses are usually caused by *Escherichia coli, Staphylococcus aureus, Streptococcus,* or anaerobic bacteria. The diagnosis is usually confirmed by percutaneous aspiration, performed by the radiologist, using CT or ultrasound guidance.

At ultrasound a hepatic cyst appears as a sharply defined round mass with a thin wall that is echo-free compared with the normal echogenic liver parenchyma. Echoes may be seen within a cyst if a complication has occurred, such as bleeding into the cyst, or if the cyst is complex, as in echinococcal disease. A liver abscess may be echo-free or echogenic depending on the density and consistency of the fluid within the abscess. Its outline may or may not be spherical and its wall may be thicker than a cyst. Often abscesses are multiloculated.

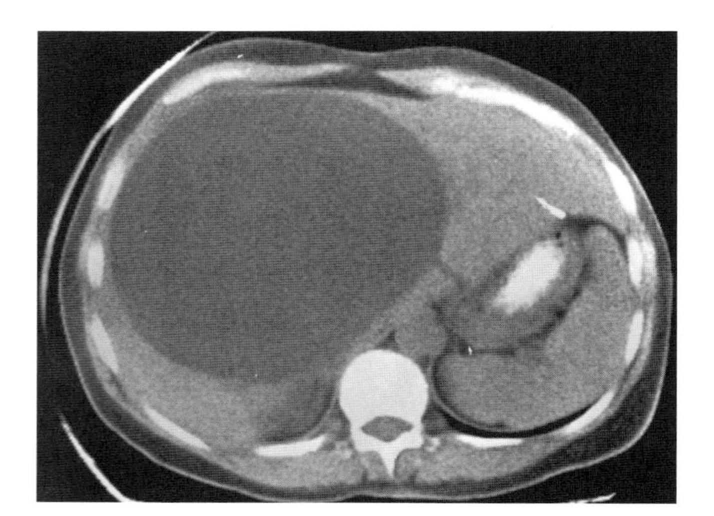

Figure 14.10. Enormous fluid-filled liver cyst.

Figure 14.11 ***(below).*** Large pyogenic liver abscess. The *arrow* points to a dot of gas produced by the Gram-negative organism.

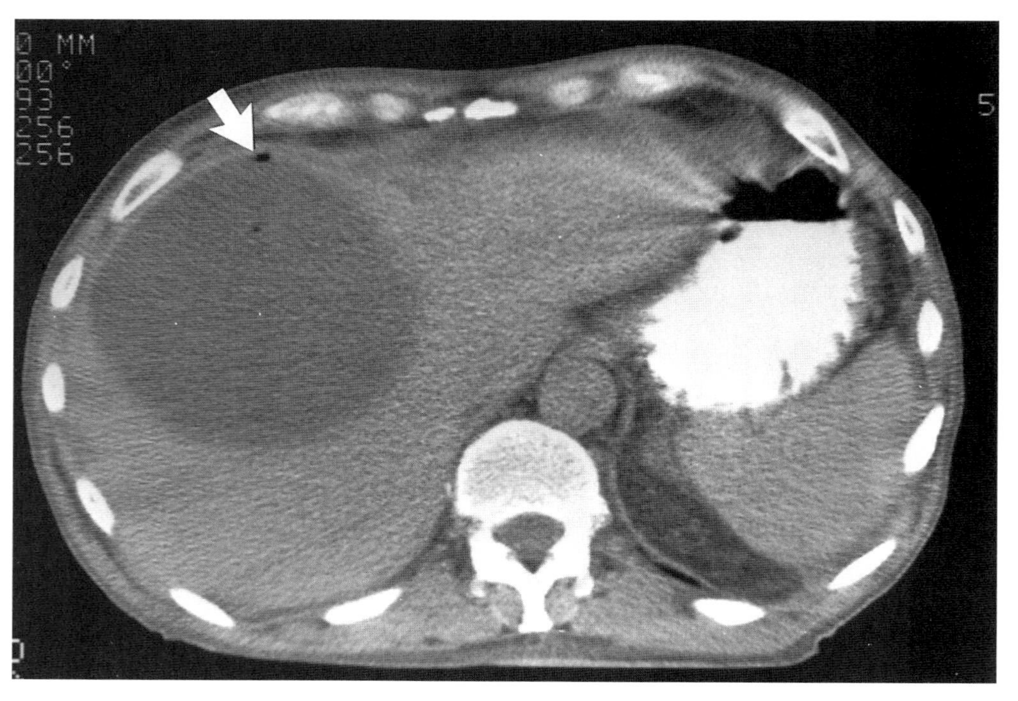

An ideal imaging technique for diagnosing liver cysts and abscesses is CT. Both appear less dense at CT than the normal hepatic parenchyma. Cysts look similar to their appearance in other organs, with a sharply defined margin, thin wall, and spherical shape (Figure 14.10). Abscesses may look similar to cysts, but they can be clearly identified if the CT scan also shows collections of gas (Figure 14.11), produced by gas-forming organisms. Hepatic abscesses may also vary from a spherical shape and may have thick irregular walls.

Multiple liver cysts may be seen in conditions such as polycystic disease, associated with multiple cysts of the kidneys and pancreas. At first glance (Figure 14.12), multiple cysts look very much like liver metastases, but in fact their CT density is much less (darker compared with the normal liver) because they are water dense rather than tissue dense. If any doubt exists, the CT scanner can measure the density of the cysts themselves to prove that they are water dense (approximately 0.0 Hounsfield unit). Not to be confused with multiple cysts are multiloculated cysts, such as might be seen with echinococcal cysts (Figure 14.13). The multiloculated appearance results from the presence of daughter cysts within the parent cyst.

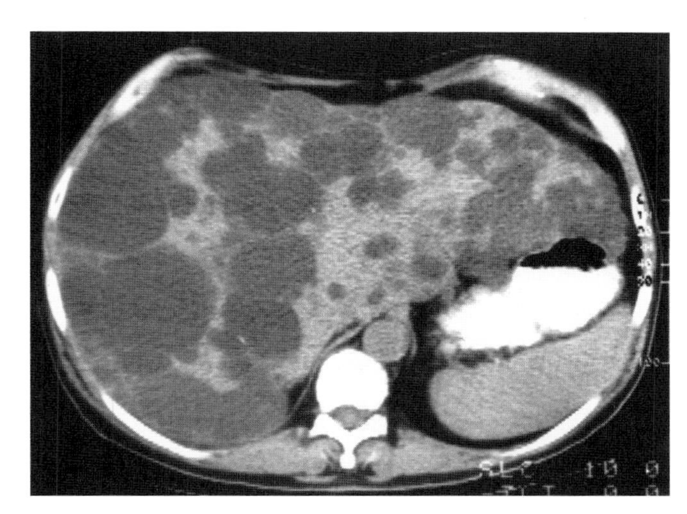

Figure 14.12. Multiple liver cysts in an adult with polycystic disease.

Figure 14.13 ***(below)*****.** Multiloculated echinococcal liver cyst in the posterior right lobe. Calcium in some portions of the cyst wall produces a white border.

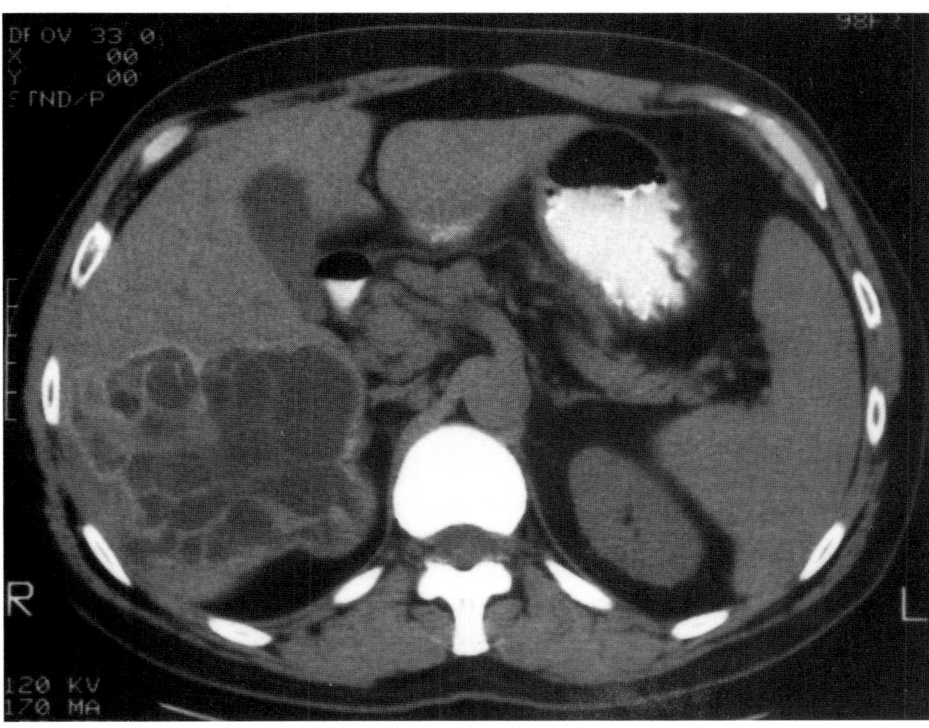

Liver Trauma

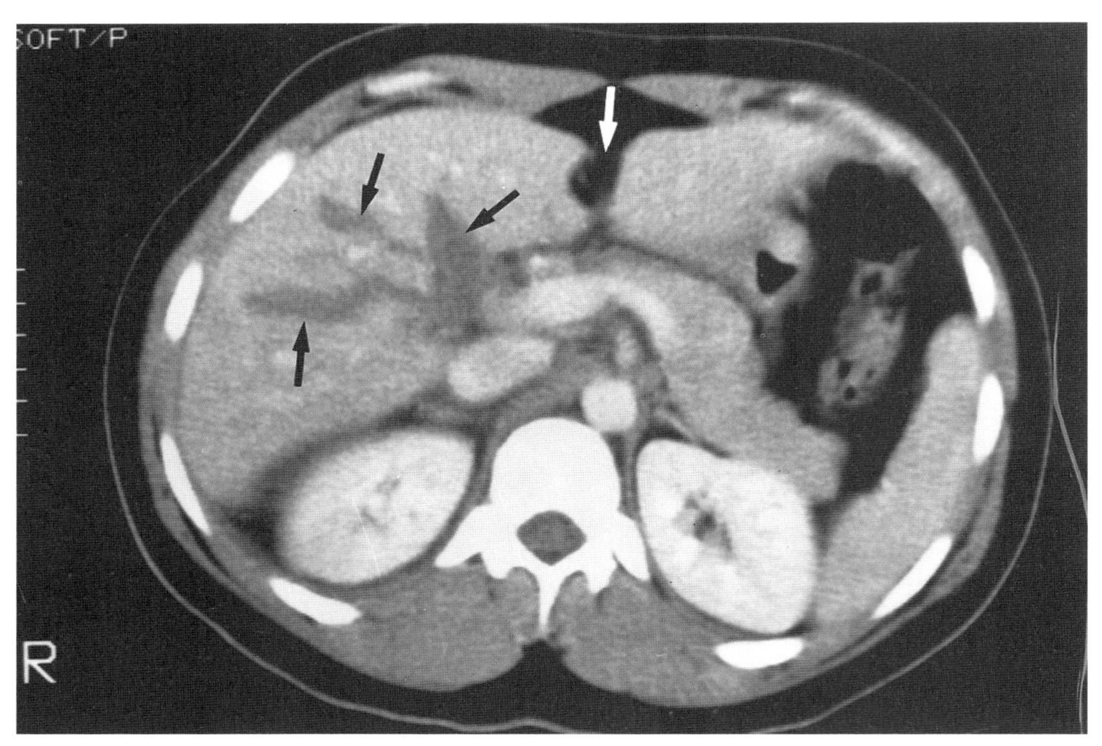

Figure 14.14. Palmate liver laceration after blunt trauma. The *black arrows* indicate three irregularly margined, blood-filled liver lacerations (note the CT density within the lacerations). The *white arrow* indicates a normal liver fissure for the ligamentum teres, which is filled with fat (same black CT density as other fatty structures). No hemoperitoneum was identified on the remainder of the scan and the patient made a good recovery with conservative (nonsurgical) treatment.

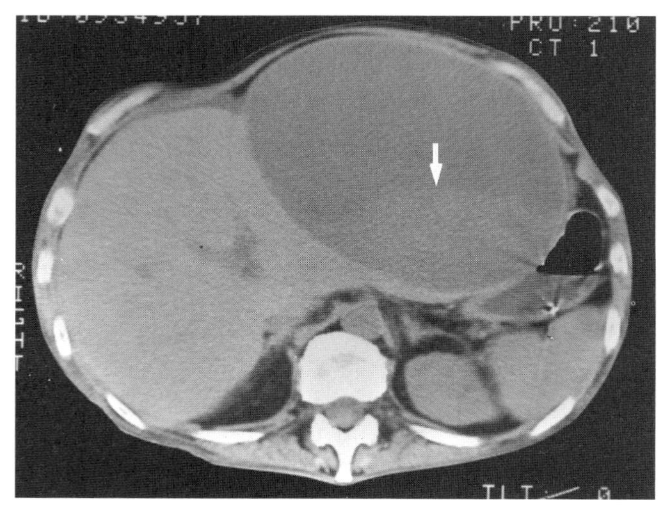

Figure 14.15. Large subcapsular hematoma in the left lobe of the liver. Note the mound-shaped hematocrit level *(arrow)* that has formed within the hematoma.

The liver may be injured by either blunt or penetrating trauma. The former is common in motor vehicle accidents and falls; the latter in stab wounds, gunshot injuries, and hemorrhagic complications of liver biopsy. For any patient who has suffered blunt or penetrating abdominal trauma, which could affect the liver or any other abdominal organ, the ideal imaging procedure is CT with intravenous and oral contrast media. In fact, CT has revolutionized the diagnostic workup of the abdominal trauma patient because it is fast, accurate, readily available in nearly all major trauma centers, and can evaluate *all* the organs in the peritoneal and retroperitoneal compartments in one quick imaging examination. Keep in mind that to be examined by CT abdominal trauma patients must be sufficiently stable to endure a trip to the CT scanner (plan at least 30 minutes: 10 minutes for travel in each direction and 10 minutes for scanning). *An unstable trauma patient should not be taken to the radiology department for a diagnostic procedure.* Unstable abdominal trauma patients can be examined in the trauma center with an emergency portable ultrasound examination, which can quickly and accurately detect the presence of peritoneal fluid (hemoperitoneum), or they can be examined with peritoneal lavage. If they become stable after fluid resuscitation, they can be taken to CT.

CT can detect hepatic lacerations, hematomas, and vascular injuries. CT can also ascertain whether hemoperitoneum accompanies the liver injury, a major factor in determining whether the injury will require surgical management. Intrahepatic lacerations that remain within an intact liver capsule and do not bleed into the peritoneal cavity can usually be treated conservatively, without surgery. Major lacerations with significant hemoperitoneum may require immediate surgical repair. When liver trauma patients are treated conservatively, follow-up CT exams can be performed to assess healing.

Hepatic lacerations are often linear and single, but they may be multiple in a branching palmate or stellate configuration. Lacerations are differentiated from normal hepatic fissures by their irregular and jagged edges, their location (where fissures are not expected to be present), and the fact that they contain blood-dense (30–45 Hounsfield units) material *(hepatic hematomas)* as compared with fissures, which contain fat (Figure 14.14). *Hepatic hematomas* are simply large collections of blood in hepatic lacerations. A *subcapsular hematoma* (Figure 14.15) appears as a crescent-shaped or oval mass immediately below the liver capsule and indenting the normally expected liver margin. Although subcapsular hematomas may occur as a consequence of blunt trauma, they often result from a penetrating injury, including liver biopsy.

Liver lacerations and hematomas are made easier to identify at CT by the injection of intravenous contrast material, which enhances (makes whiter) the normal liver parenchyma to better show hematoma (which appears darker because the blood does not enhance with contrast medium). Liver lacerations may occur in combination with other injuries. Those resulting from a blow to the right trunk may be associated with right rib fractures, a right pneumothorax, right lung contusion, and a right renal injury. Those resulting from an anterior midline (epigastric) blow—such as that sustained by an unrestrained driver struck with the steering wheel in a car accident—may have associated with them sternal fracture, cardiac or pericardial injury, bowel injury, or pancreatic fracture (Figure 14.16). In a few

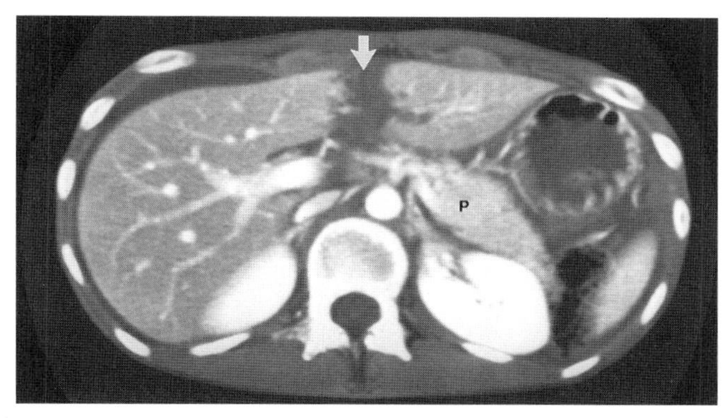

A. Slice through the liver shows the blood-filled liver laceration *(arrow)*. The body and tail of the pancreas *(P)* are separated from the head of the pancreas, seen in B.

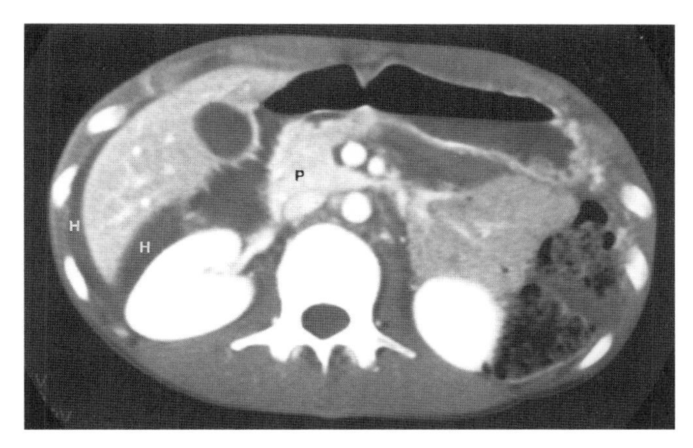

B. Hemoperitoneum (*H's)* is seen around the liver. The head of the pancreas *(P)* is disconnected from the remainder of the pancreas.

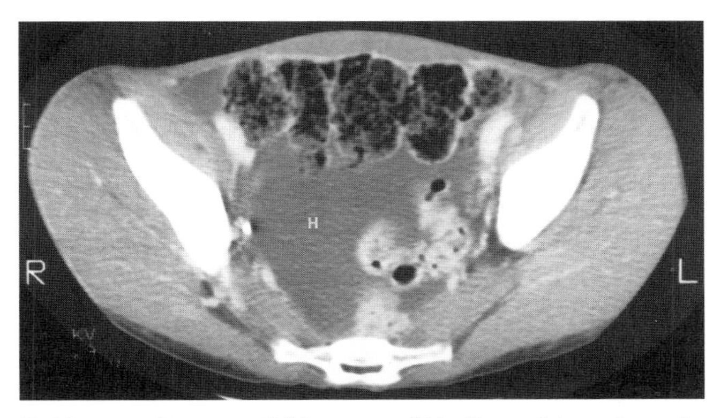

C. Hemoperitoneum *(H)* is seen within the pelvis, adjacent to loops of bowel.

Figure 14.16. CT scan of vertical laceration of the left lobe of the liver and pancreatic fracture in an unrestrained driver who was struck in the epigastrium with the steering wheel in a deceleration car crash. The liver and pancreas were compressed against the spine and fractured.

cases right truncal trauma may result in a rupture of the right hemidiaphragm with herniation of portions of the liver into the right chest (Figure 14.17).

When monitoring abdominal trauma CT scans, radiologists measure the CT attenuation of many of the abnormalities they see. Not only do they measure the attenuation of material filling apparent hepatic lacerations to ascertain that it is indeed blood, but they nearly always measure the attenuation of any peritoneal fluid seen. This is most important because all fluids seen in the peritoneal cavity of trauma patients are not blood. Those that measure 30 to 45 Hounsfield units or greater do indeed represent blood, but those that have the density of water (0–5 Hounsfield units) may represent urine (due to bladder rupture), bile (gallbladder rupture), or intestinal contents (bowel rupture).

Figure 14.17. Rupture of the right hemidiaphragm with herniation of the liver into the lower right chest, after right chest and abdomen trauma in a construction accident.

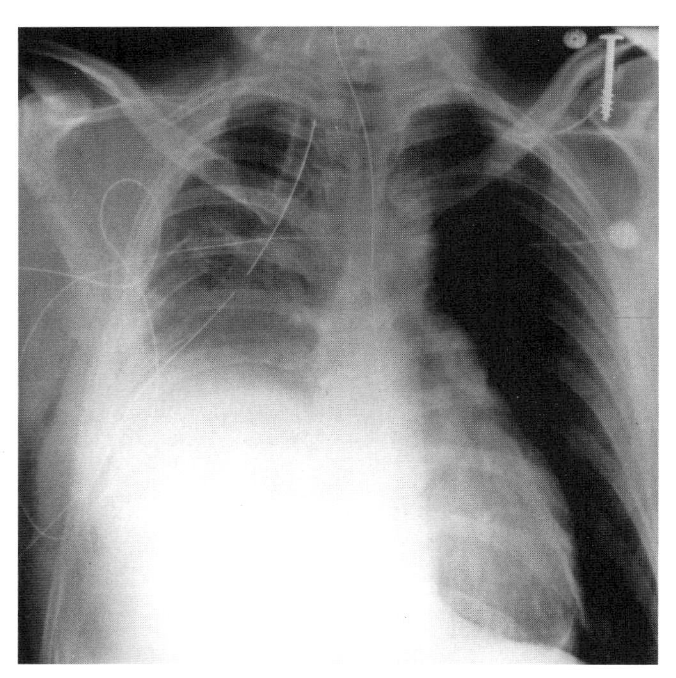

A. Portable chest film showing multiple right rib fractures and what appears to be elevation of the right hemidiaphragm; compare the normally positioned left hemidiaphragm.

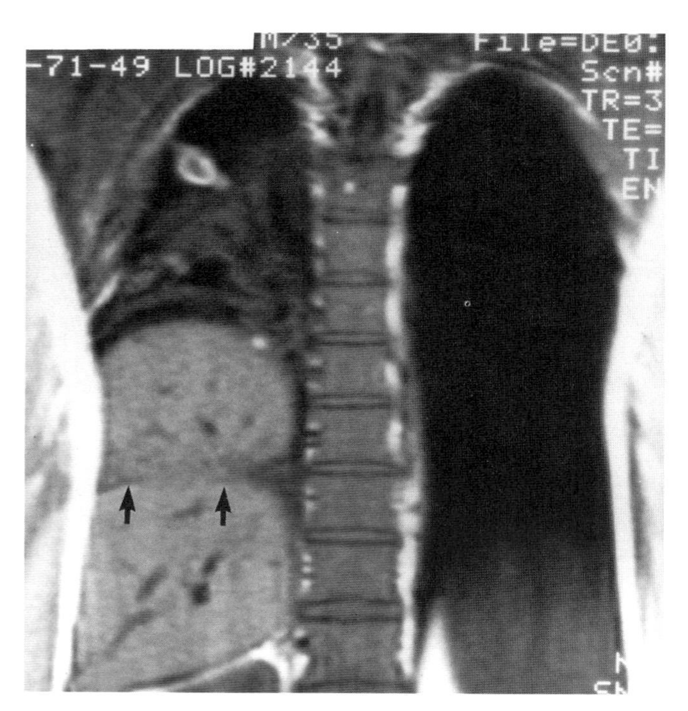

B. Coronal MR scan showing herniation of the liver through a large diaphragm laceration (the *arrows* indicate traumatic opening in the diaphragm).

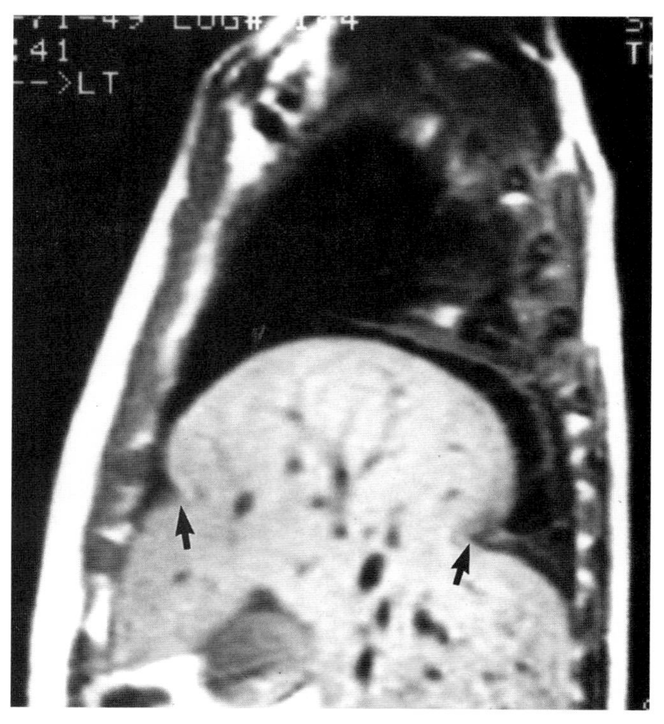

C. Sagittal MR scan showing liver herniation through a large diaphragm laceration *(arrows).*

Cirrhosis, Splenomegaly, and Ascites

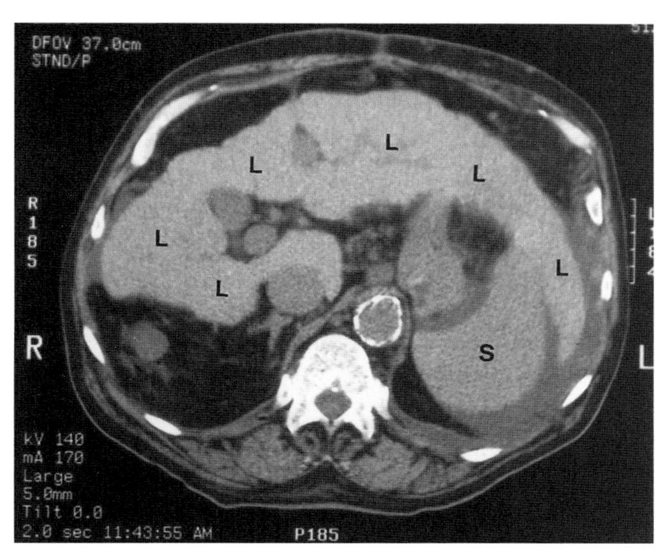

A

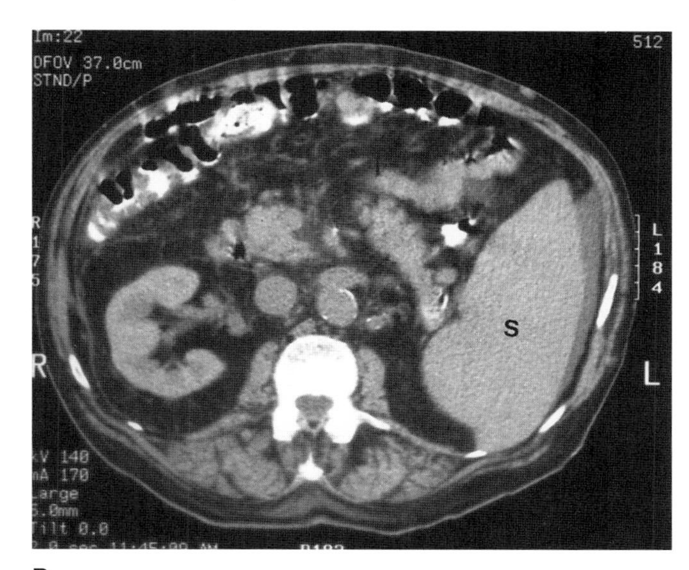

B

Figure 14.18. CT scan of an alcoholic patient showing a cirrhotic liver, splenomegaly, and ascites. A: Scan through the cirrhotic liver (*L*'s) shows it to be small and scarred. The left lobe of the liver extends laterally, in front of the spleen. A small amount of ascitic fluid can be seen around the spleen *(S)*. B: A slightly lower scan shows the full cross section of the enlarged spleen *(S)*, which extends downward into the midabdomen.

Diffuse parenchymal diseases of the liver, such as acute hepatitis, may produce only minimal abnormalities on cross-sectional imaging. With hepatitis the radiologist may detect only diffuse increased echogenicity at ultrasound and only increased liver size with decreased liver attenuation at CT. In a few hepatitis patients, associated thickening of the gallbladder wall may be shown by either examination. But hepatitis can be reliably diagnosed only on the basis of history, physical examination, laboratory tests, and liver biopsy.

Posthepatitic cirrhosis, however, and cirrhosis resulting from chronic alcoholism may be detected when these processes alter liver size, contour, or density—alterations in density ranging from fatty infiltration to advanced fibrosis. These changes are best imaged with CT. Normally the liver has a slightly greater CT attenuation than the spleen. This ratio is reversed with fatty liver infiltration, which may be diffuse or focal. You have already seen an example of a fatty liver in Figure 14.3 and you should turn back to look at it again. Notice how well the hepatic blood vessels are shown against the fatty infiltrated and consequently less dense liver. Later, in advanced cirrhosis, the liver will appear smaller in size and irregular in shape because of fibrous scarring, segmental atrophy, and regenerating nodules. The CT scan may also show splenomegaly (from portal hypertension, of course) and ascites. Note the large spleen, small scarred nodular liver, and peritoneal fluid (ascites) in the CT scan of the alcoholic patient shown in Figure 14.18.

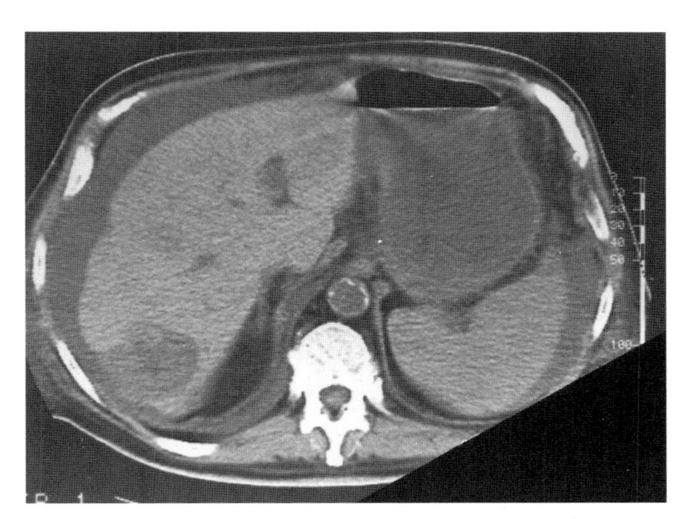

Figure 14.19. Malignant ascites from metastatic colon cancer. Low-density liver metastases are also evident.

The patient in Figure 14.19 also has ascites (in this case malignant ascites from metastatic colon cancer), but the spleen is normal in size. Did you notice the liver metastases?

Splenomegaly can usually be detected by physical examination. When the diagnosis is uncertain, spleen size may be estimated from plain films of the abdomen or CT. In earlier chapters you have seen the plain film findings of splenomegaly in several patients. In addition to confirming splenic enlargement, CT may show the etiology—whether it is a splenic tumor, abscess, cyst, or other process. The presence of gas confirms the diagnosis of a suspected splenic abscess in Figure 14.20, the CT scan of a patient who presented with fever, chills, and left upper quadrant pain.

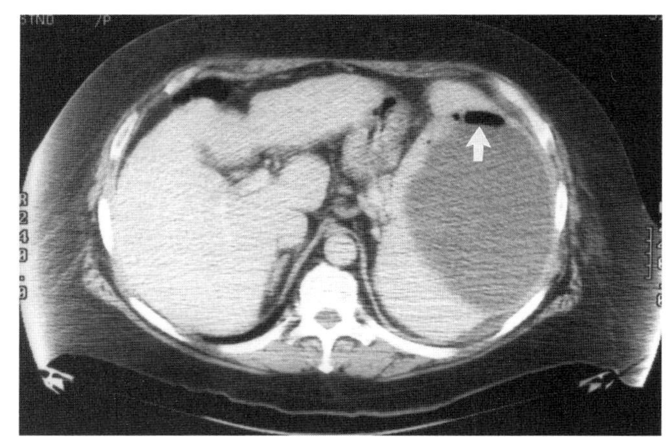

Figure 14.20. CT scan of a pyogenic splenic abscess. The *arrow* indicates a small collection of gas within the abscess.

Splenic Trauma

The spleen is the organ most frequently injured during blunt abdominal trauma. Splenic injuries are commonly seen after motor vehicle accidents, falls, and assaults, especially when left lower rib fractures are present. CT can diagnose all types of splenic injuries quickly and accurately, including splenic lacerations,

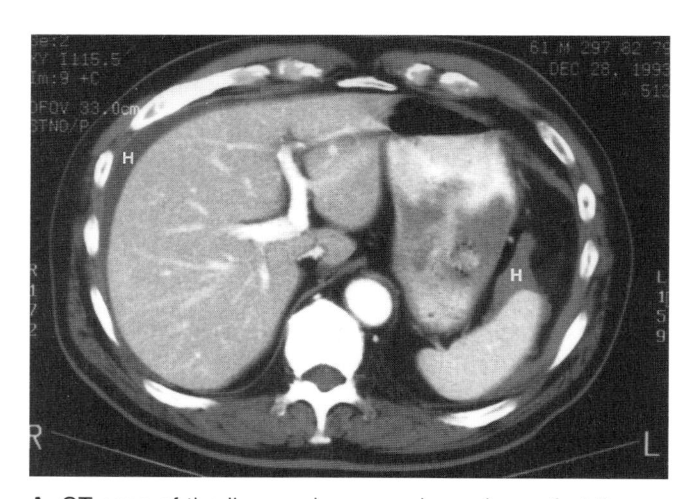

A. CT scan of the liver and upper spleen shows that the hemoperitoneum *(H)* adjacent to the spleen is denser than the hemoperitoneum *(H)* adjacent to the liver. No spleen injury is seen at this level scan.

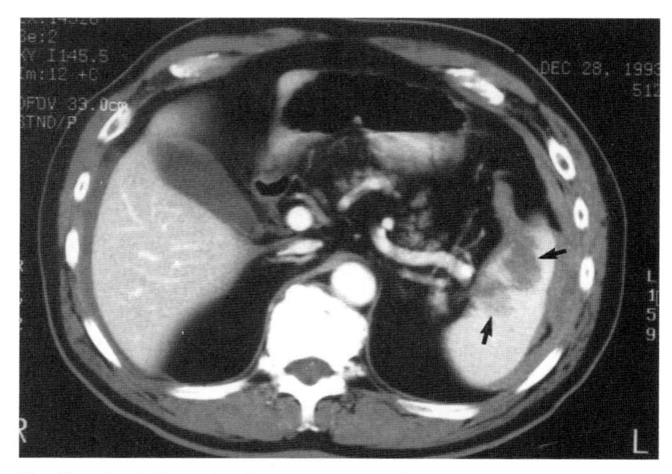

B. Blood within splenic lacerations *(arrows)* is less dense than the opacified splenic parenchyma.

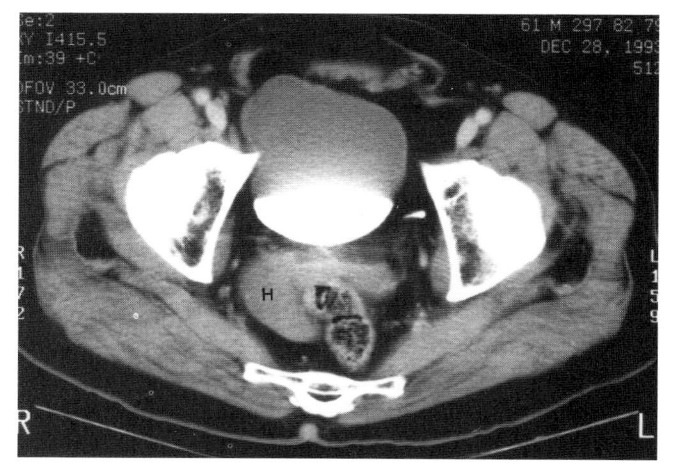

C. Hemoperitoneum *(H)* is seen in the pelvis, between the opacified bladder and the stool-containing rectum.

Figure 14.21. Splenic lacerations and hemoperitoneum in a 61-year-old man injured in a car accident.

fractures, and subcapsular hematomas. CT can also determine whether hemoperitoneum is present. But in cases of spleen trauma the decision to operate or treat conservatively is determined more by the clinical course of the patient and the patient's age than by the findings on the CT scan. Children and young adults do well on conservative therapy for even severely injured spleens, whereas older patients, especially those over 55, require surgery for even minor splenic lacerations. Today physicians appreciate the immunological contributions of the spleen and try to avoid splenectomy whenever possible, using follow-up CT scans to monitor the course of patients who are managed without surgery or whose spleens are repaired rather than resected. Ultrasound can also detect splenic injuries and hemoperitoneum, but with less accuracy than CT. Still, you may wish to consider a portable ultrasound examination for any unstable patients with suspected splenic trauma.

Like liver lacerations, splenic lacerations appear as irregularly margined, blood-containing spaces within the contrast-opacified splenic parenchyma (Figure 14.21). If the splenic capsule is torn, hemoperitoneum can be identified around the spleen and in other portions of the peritoneal cavity. Note particularly that the hemoperitoneum adjacent to the spleen in Figure 14.21A is denser than the more remote hemoperitoneum adjacent to the liver, because dense blood clots surround the spleen while thinner blood appears farther away from the injury. This important CT sign (Figure 14.22) assists the radiologist in identifying the site of bleeding in abdominal trauma patients. The densest blood in the peritoneal cavity is usually closest to the site of the injury.

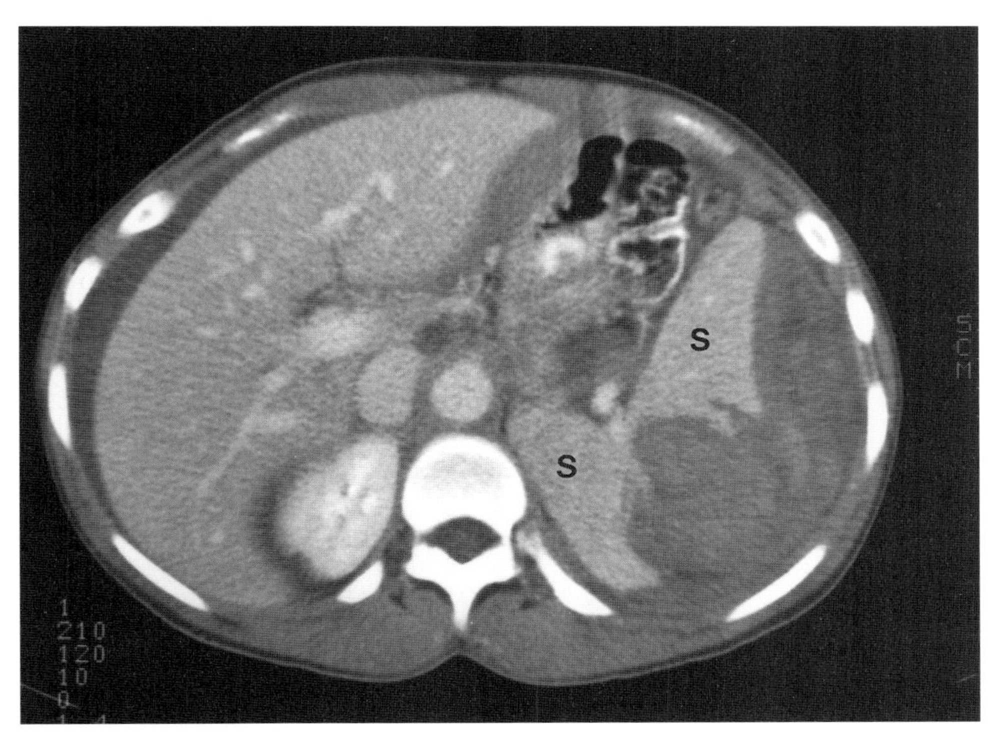

Figure 14.22. CT scan of a fractured spleen. A large laceration separates the two halves of the spleen (*S*'s). Laminated blood clots of varying CT densities are evident within the laceration and surrounding the spleen laterally. The distant free blood adjacent to the liver is less dense than the perisplenic blood clots so well shown here.

Cholelithiasis and Cholecystitis

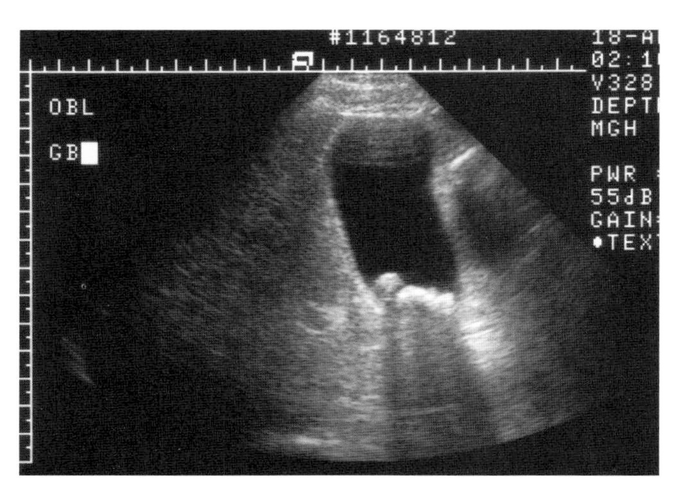

Figure 14.23. Gallbladder ultrasound showing echogenic stones and acoustic shadows behind the stones.

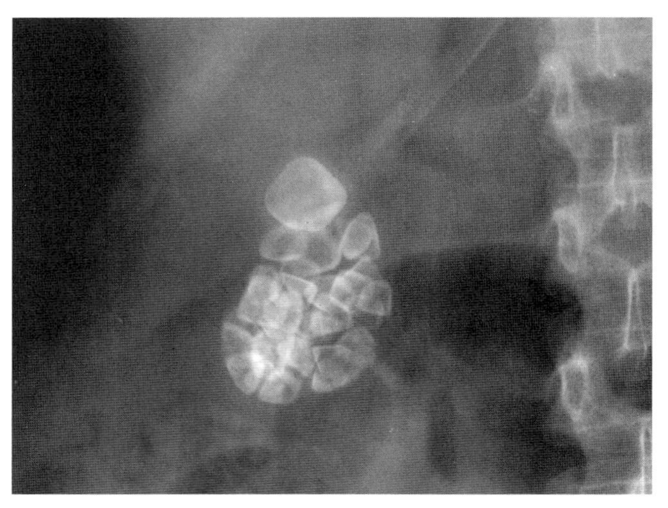

Figure 14.25. Detail of a plain film of the right abdomen showing a cluster of calcified and faceted gallbladder stones.

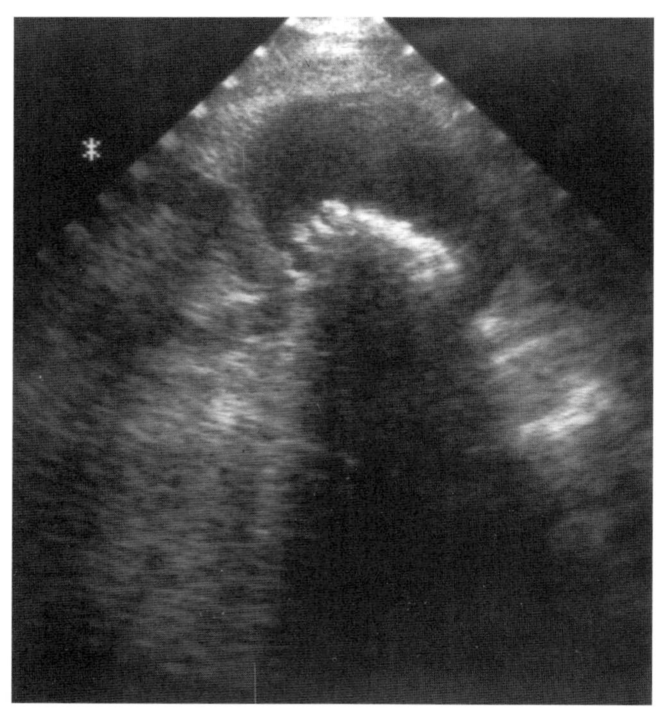

Figure 14.24. Gallbladder ultrasound in a patient with a larger number of stones than the patient in Figure 14.23.

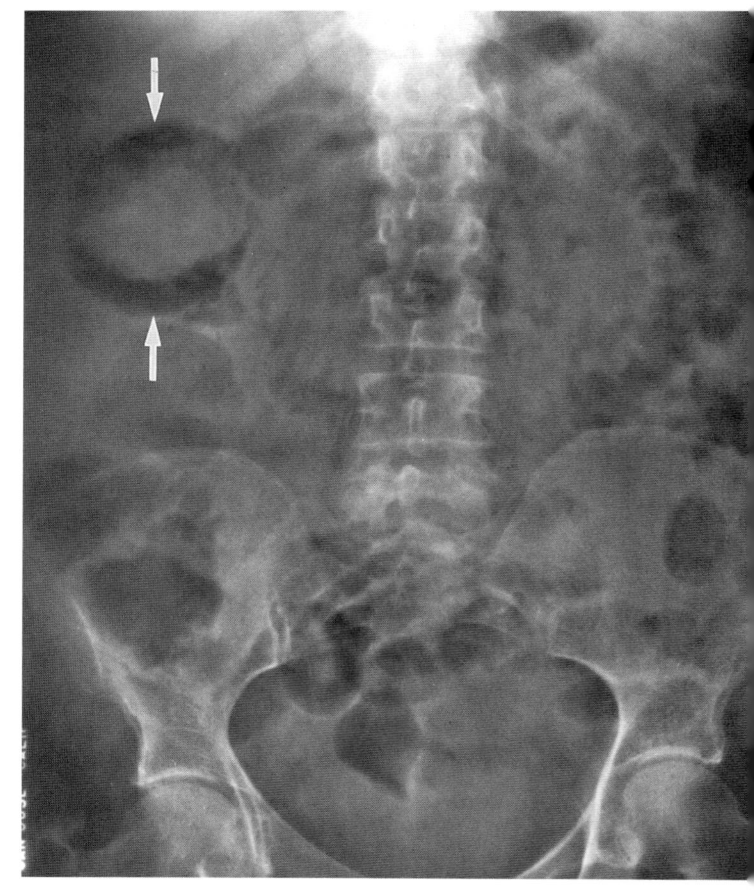

Figure 14.26. Plain film of the abdomen showing emphysematous cholecystitis. The *arrows* point to the circular collection of air in the gallbladder wall.

Ultrasound is the imaging procedure of choice for evaluating gallbladder disease. It is highly accurate, comfortable for patients, convenient, cost-effective, relatively inexpensive, and has no associated ionizing radiation. It is the best test for detecting cholelithiasis, identifying 95 percent of all stones. At ultrasound, gallbladder stones appear as rounded echogenic foci that are associated with *acoustic shadows* (Figures 14.23 and 14.24). The sound waves that hit the stones are reflected by them and consequently cast acoustic shadows, whereas adjacent sound waves that do not hit the stones pass through the gallbladder uninterrupted. Stones and their associated shadows can be shown to move within the gallbladder as the patient changes position. Obese patients and those with a significant amount of bowel gas (which interferes with the transmission of sound waves) may have suboptimal ultrasound examinations.

Approximately 10 percent of gallbladder stones are radiopaque and may be visualized on plain films (Figure 14.25). Plain films may also show calcification of the gallbladder wall and air within the gallbladder wall or lumen. The presence of gallbladder air indicates that the patient has *emphysematous cholecystitis* (Figure 14.26), a manifestation of acute cholecystitis resulting from infection of the gallbladder with a gas-forming organism, usually *E. coli* or *Clostridium perfringens*. CT can also show calcified gallbladder stones (Figure 14.27), as well as calcification of the gallbladder wall (Figure 14.28) and emphysematous cholecystitis. Gallbladder wall calcification, or *porcelain gallbladder*, is usually a manifestation of chronic cholecystitis in an obstructed and chronically inflamed gallbladder.

Another type of gallbladder examination, rarely used today but often used before the advent of ultrasound, is the oral cholecystogram (Figure 14.29). In this test, contrast material, ingested by the patient the night before the examination, is concentrated in the gallbladder; stones appear as filling defects within the pool of

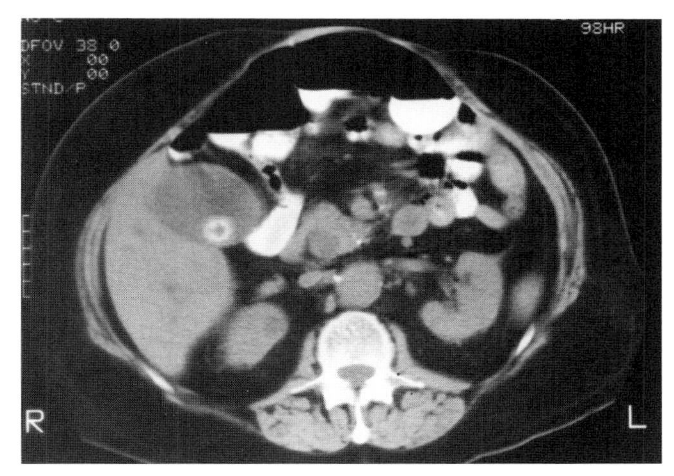

Figure 14.27. CT scan showing a single calcified stone within the gallbladder.

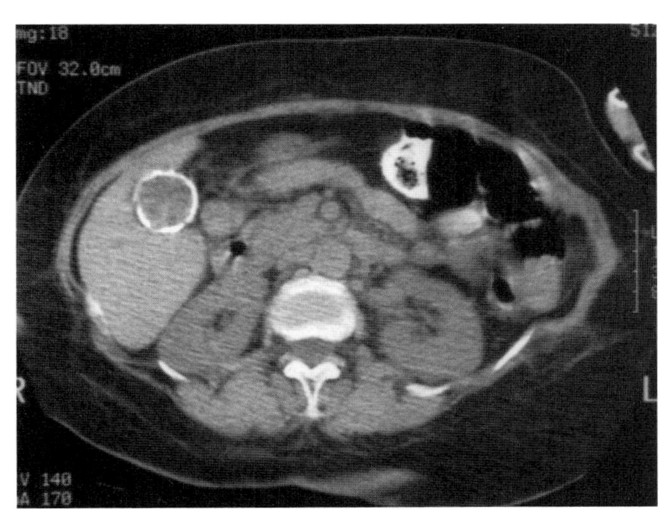

Figure 14.28. CT scan showing a porcelain gallbladder, with dystrophic calcification of the gallbladder wall, in a patient with chronic cholecystitis.

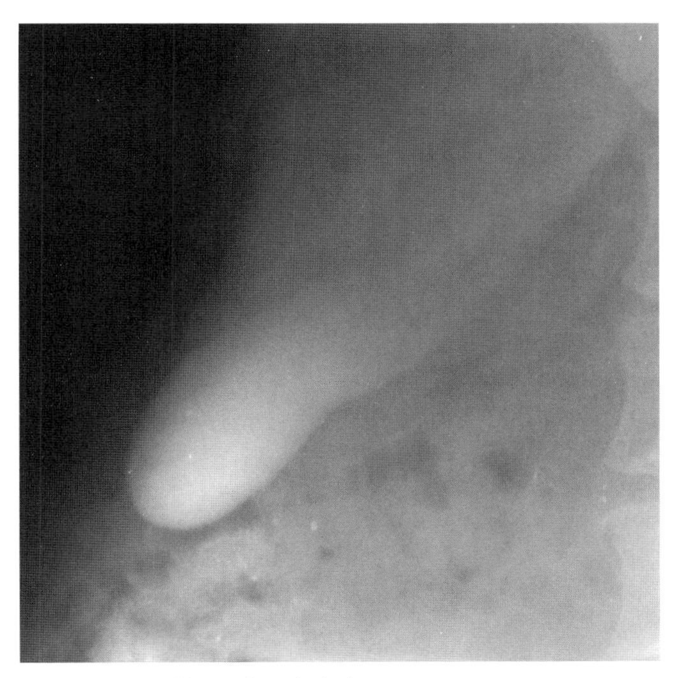

Figure 14.29. Normal oral cholecystogram.

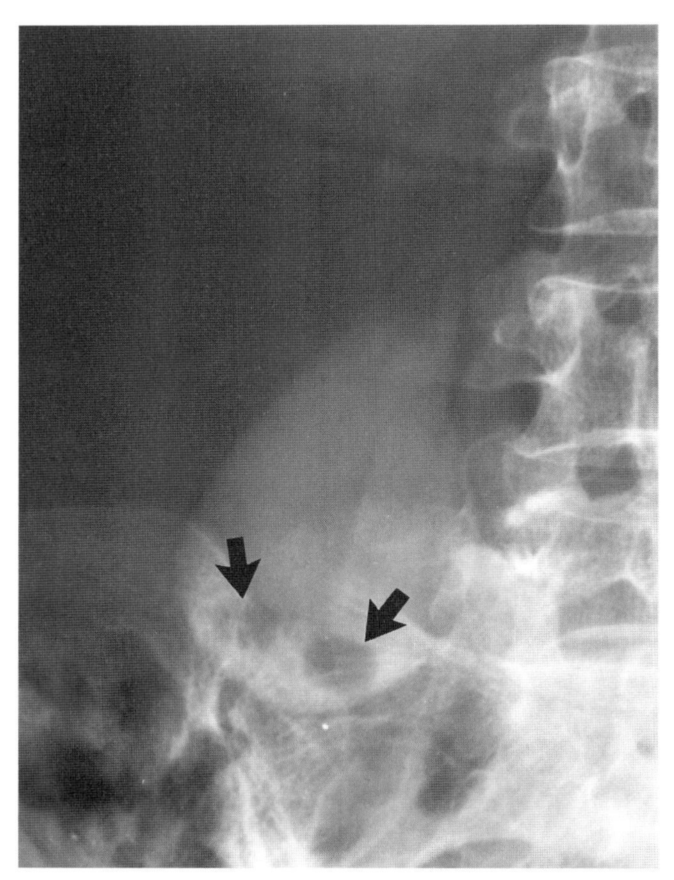

Figure 14.30. Abnormal oral cholecystogram in a patient with two lucent stones *(arrows).*

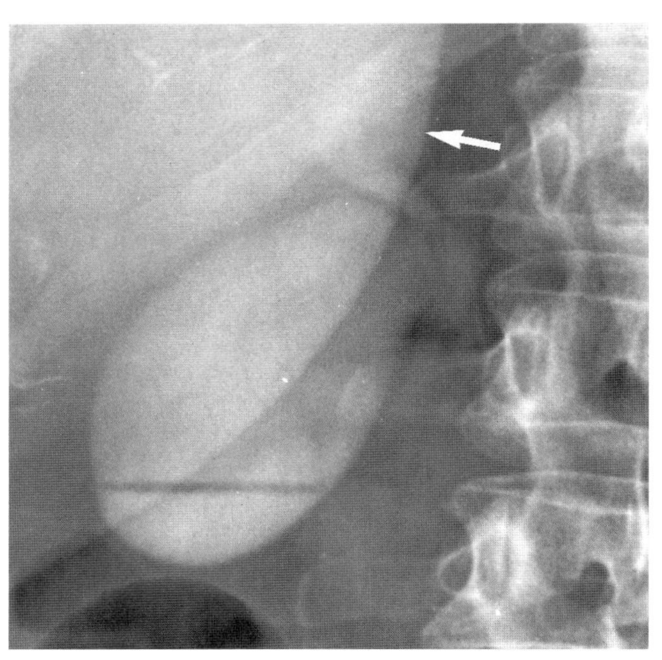

Figure 14.31. This upright spot film obtained during oral cholecystography shows a row of radiolucent stones layering (floating in a layer of similar-density bile) within the gallbladder. The *arrow* points to a right breast shadow. Supine films may miss small stones; consequently, the radiologist always obtains upright films.

contrast material on radiographs taken the next day (Figures 14.30 and 14.31). This test misses at least 10 percent of stones and can be performed only in patients capable of concentrating the opaque material, which requires a bilirubin level below 2 milligrams per 100 milliliters. A variety of other conditions can also interfere with opacification of the gallbladder during this examination.

So far, you have had the opportunity to see the gallbladder on many CT scans, and you may wonder why CT is not routinely used for diagnosing cholelithiasis. The reason is that many gallbladder stones are *isodense* (have the same CT attenuation as) the surrounding fluid within the gallbladder and therefore are *invisible* on a CT scan. In fact, CT can detect only 80–85 percent of gallbladder stones.

Acute cholecystitis, or acute inflammation of the gallbladder, is a serious condition caused in 90 percent of cases by obstruction of the cystic duct by gallstones. When this condition occurs without stones, it is called *acalculous cholecystitis,* the inflammation being caused by ischemia or bacterial infection from an adjacent source. In patients whom you suspect of having acute cholecystitis, two examinations are helpful for diagnosis, ultrasound and a radioisotope test called *cholescintigraphy.* Ultrasound may reveal gallbladder stones, an ultrasonic Murphy's sign (tenderness under the ultrasound transducer when pressure is applied directly over the gallbladder), thickening of the gallbladder wall due to edema, and the presence of pericholecystic fluid (Figure 14.32).

Cholescintigraphy images the gallbladder with intravenously injected hepatobiliary iminodiacetic acid (HIDA or related compounds) conjoiners. These are excreted by the hepatocytes into the biliary system, from which they flow into the bowel. Normally there is reflux of the isotope up the cystic duct into the gallbladder (Figure 14.33). Because the cystic duct is nearly always occluded in acute cholecystitis, there is generally no visualization of the gallbladder, even on delayed images (Figure 14.34).

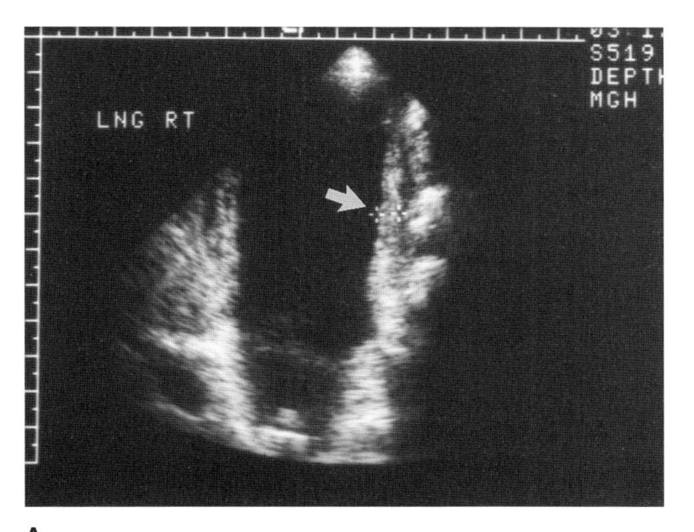

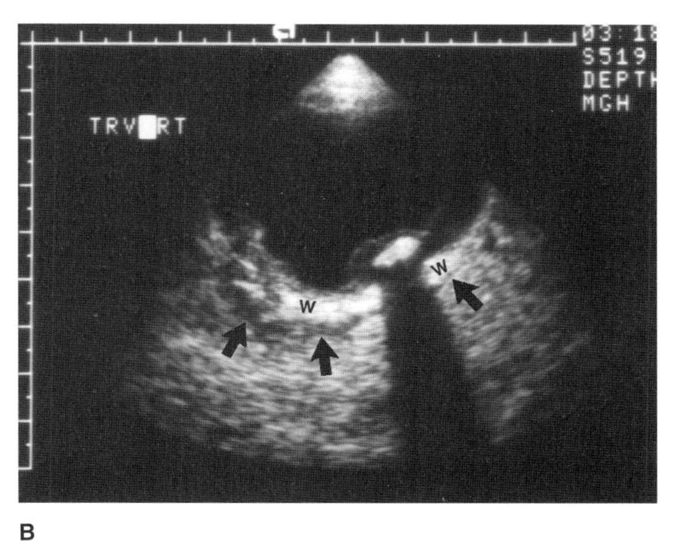

A B

Figure 14.32. Two views of a gallbladder ultrasound examination showing evidence of acute cholecystitis. A: The longitudinal scan shows the cursors *(arrow)* that measure the gallbladder wall, which at 6 millimeters is thickened; the upper limit of normal is 3 millimeters. B: The transverse view shows a gallbladder stone with a shadow and echolucent pericholecystic fluid *(arrows)* just outside the thickened gallbladder wall (*W*'s). This patient also had a pronounced ultrasonic Murphy's sign.

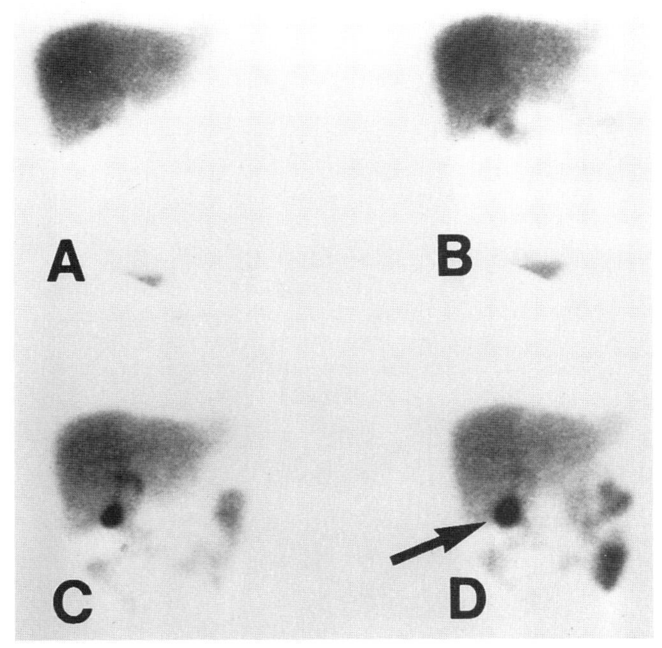

Figure 14.33. Normal HIDA scan showing the isotope only in the liver initially *(A)*, then in the biliary tree (*B* and *C*), with later concentration of the isotope in the gallbladder *(arrow)* in *D*, with evidence of flow of the isotope into loops of the bowel.

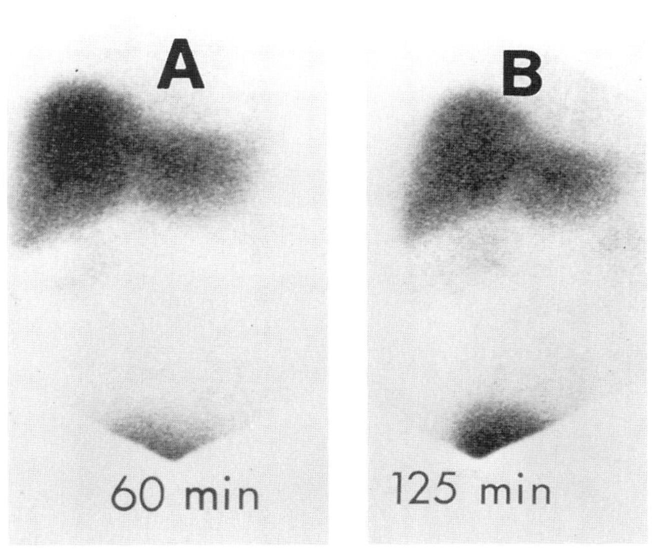

Figure 14.34. Abnormal HIDA scan showing no isotope in the gallbladder or bladder on films delayed as long as 60 minutes *(A)* and 125 minutes *(B)*. This patient proved to have obstruction of the cystic duct due to acute cholecystitis.

Although CT may not be the procedure of choice when acute cholecystitis is *expected*, a CT scan performed to determine the cause of "abdominal pain" may make possible a diagnosis of acute cholecystitis when this condition is *not expected*. The findings at CT include gallbladder stones, a thickened and distended gallbladder wall, and pericholecystic fluid (Figure 14.35). With gas-forming organisms air may be seen within the gallbladder lumen (Figure 14.36) or wall.

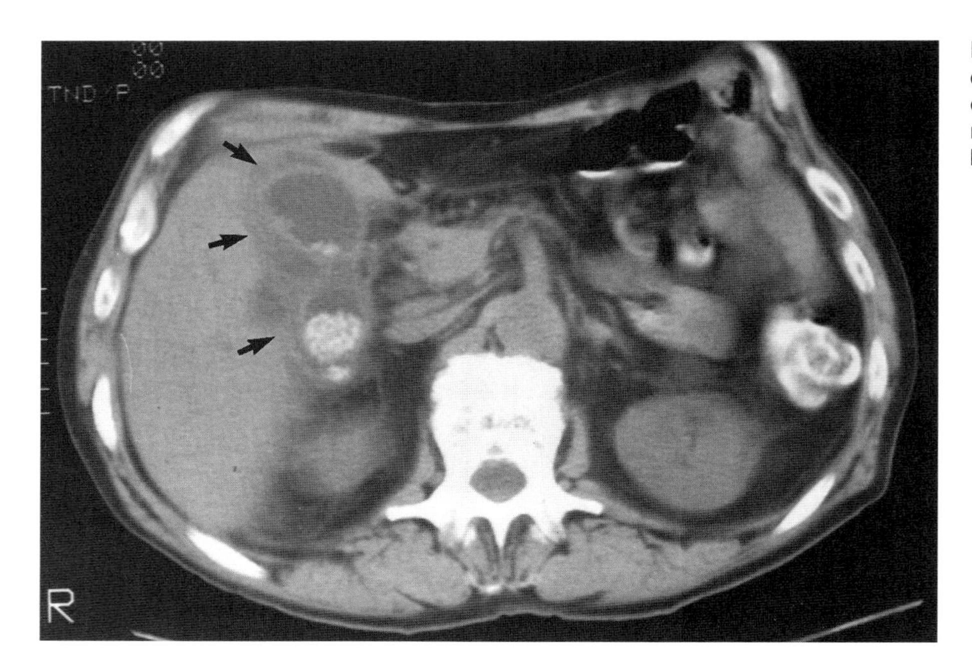

Figure 14.35. CT scan of acute cholecystitis. The redundant gallbladder has a thickened wall, contains multiple stones, and is surrounded by pericholecystic fluid *(arrows).*

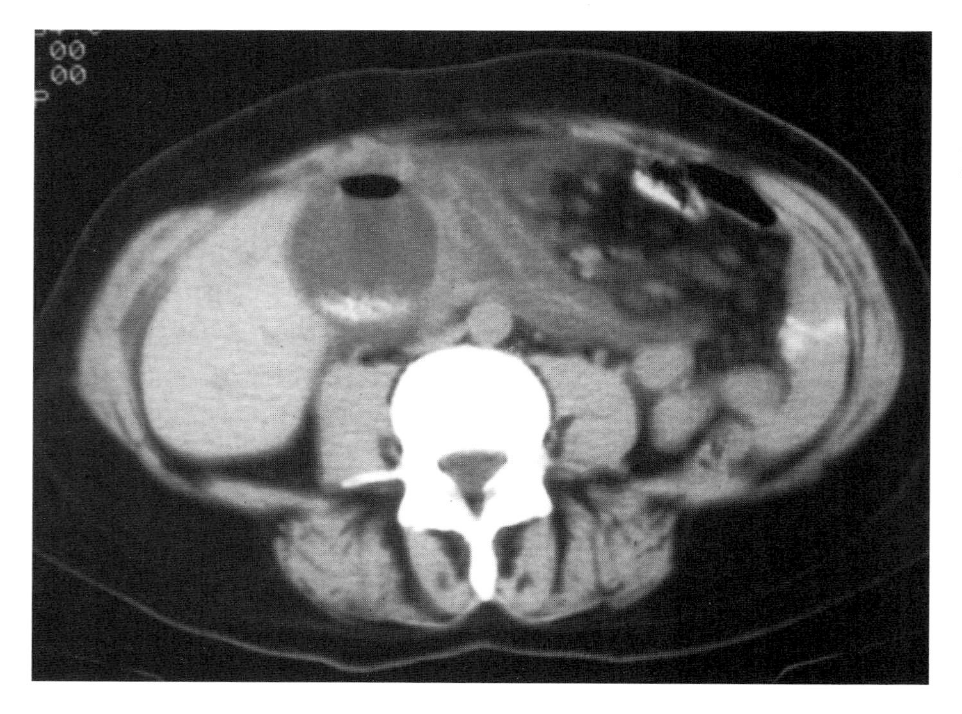

Figure 14.36. CT scan of acute emphysematous cholecystitis. The gallbladder contains air, fluid, and stones, and is surrounded, especially medially, by pericholecystic fluid and inflammation.

Obstruction of the Biliary Tree

A drawing of the biliary tree and pancreas is provided for you in Figure 14.37. At the porta hepatis, biliary trunks from the right and left liver lobes join to form the hepatic duct. The short hepatic duct is joined by the cystic duct draining the gallbladder to form the common bile duct. This structure drains into the bowel at the duodenal papilla (papilla of Vater), which has a common orifice for the termination of both the common bile duct and the greater (main) pancreatic duct. The accessory pancreatic duct drains into the duodenum through a separate orifice.

In Figure 14.38 the biliary tree has been opacified by injection of contrast material through a surgically placed T-tube, after cholecystectomy and common bile duct exploration. The anatomy of the biliary system is fully displayed and residual common duct stones can easily be identified. Note the round filling defects (stones) in the distal common duct. The majority of patients whose biliary systems you may wish to study will not have had surgery and T-tube placement, and therefore will not offer such easy access for the administration of contrast medium; you will have to employ another imaging technique, usually ultrasound.

One of the most frequent reasons for imaging the biliary system is to determine whether it is dilated or not in patients presenting with new jaundice. Dilatation of the common bile duct and intrahepatic biliary tree is most easily and quickly detected by ultrasound. If the biliary system *is not dilated* at ultrasound, then you should consider a hepatocellular disease such as hepatitis to account for the patient's jaundice. If the biliary system *is dilated,* you should consider causes for mechanical obstruction of the biliary tree, such as a common duct stone or pancreatic carcinoma. Less common causes include bile duct stricture and tumor.

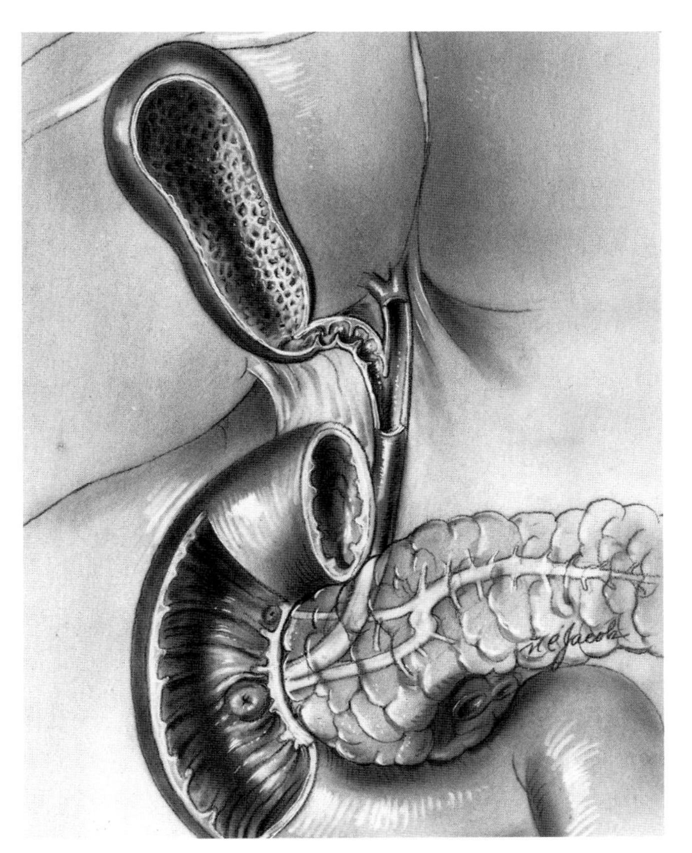

Figure 14.37. The biliary tree, duodenum, and pancreas. The gallbladder has been reflected upward.

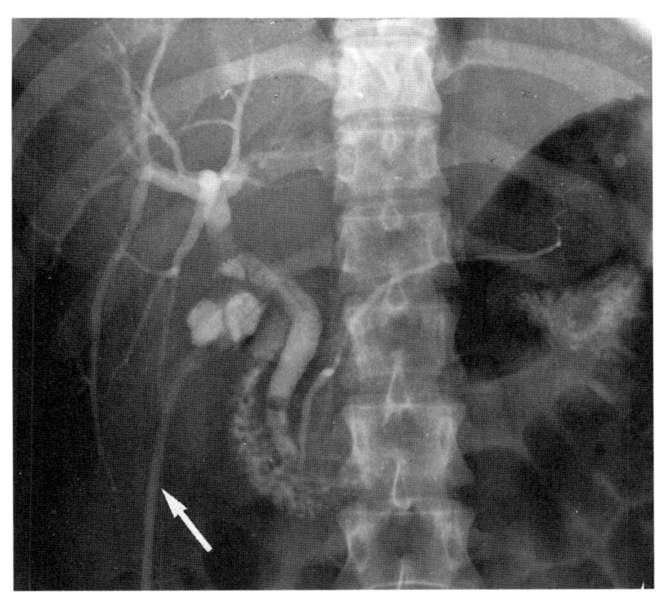

Figure 14.38. Intraoperative T-tube cholangiogram. This procedure was performed immediately after cholecystectomy and common duct exploration to evaluate the patency of the biliary tree and search for any retained common duct stones. The *arrow* points to the T-tube, which terminates in the common bile duct and through which contrast material was injected. The intrahepatic biliary tree is opacified and contrast material refluxes into the pancreatic duct, behind the dark, air-filled stomach. Also there is opacification of the duodenal lumen, where the common bile duct empties. The round radiolucencies in the distal common bile duct represent residual common duct stones, which were removed after detection by this examination.

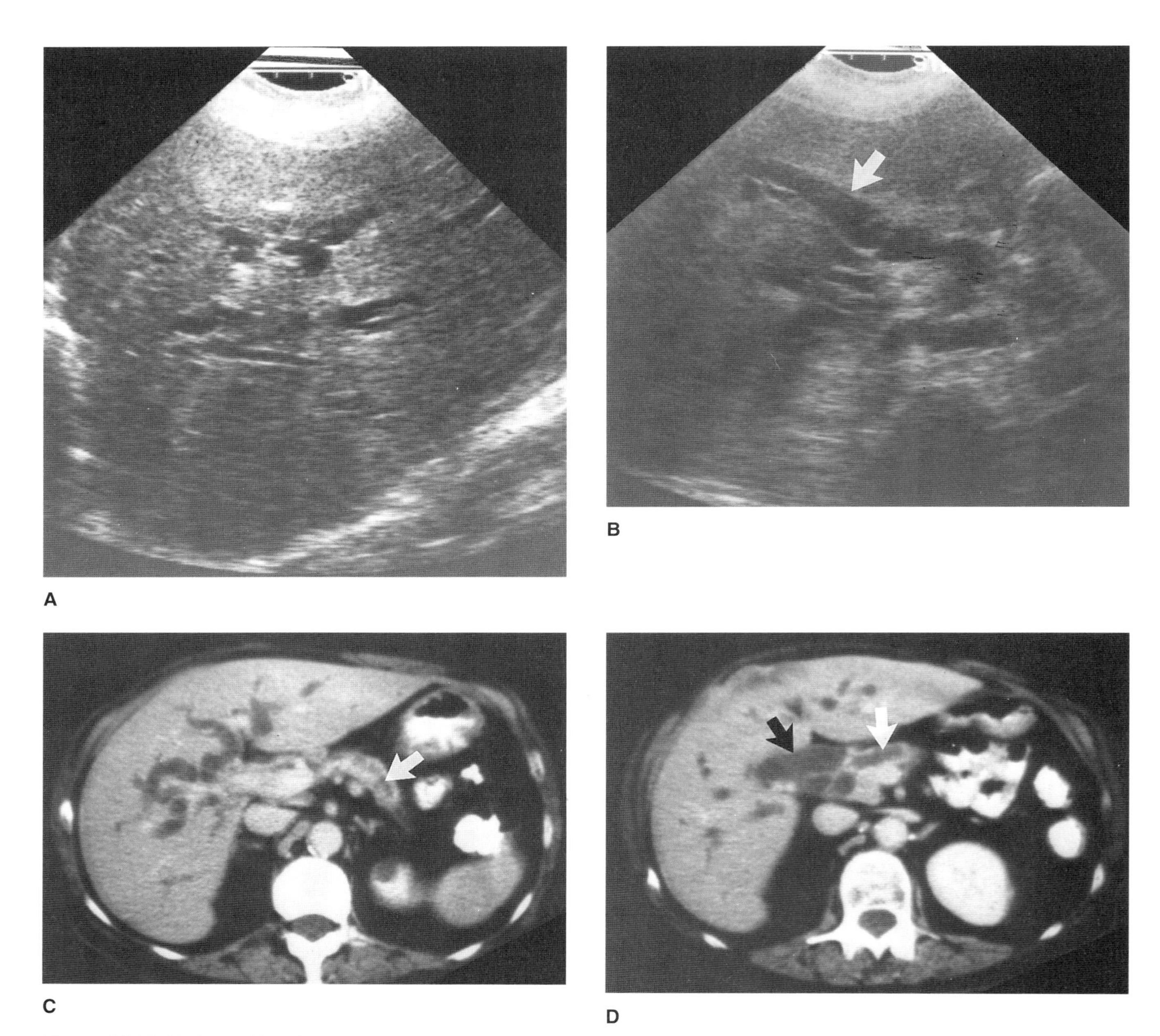

Figure 14.39. Patient with painless jaundice due to obstruction of the common bile duct by a small tumor at the duodenal ampulla (ampulla of Vater). A: Parasagittal liver ultrasound, showing dilatation of the intrahepatic biliary radicals. The echogenic, white curving diaphragm is located to your left. B: Transverse liver ultrasound, showing the dilated common bile duct *(arrow)*. C: CT scan, showing dilatation of the intrahepatic biliary radicals. The pancreatic duct is also dilated *(arrow)*. D: Lower-level CT scan, showing dilatation of the common bile duct *(black arrow)* and the pancreatic duct *(white arrow)*.

In addition to recognizing biliary dilatation, ultrasound may also show the obstructing common duct stone or pancreatic mass. Nevertheless, the fine detail of ultrasound imaging is limited, and bowel gas and body habitus may make ultrasound evaluation of the common bile duct difficult. Consequently, if the ducts are dilated but the level of obstruction is uncertain, additional imaging procedures such as CT, percutaneous transhepatic cholangiography (PTC), and endoscopic retrograde cholangiopancreatography (ERCP) may be required.

The patient illustrated in Figure 14.39 is a 52-year-old woman who presented in an emergency center with new, painless jaundice. Two images from her in-

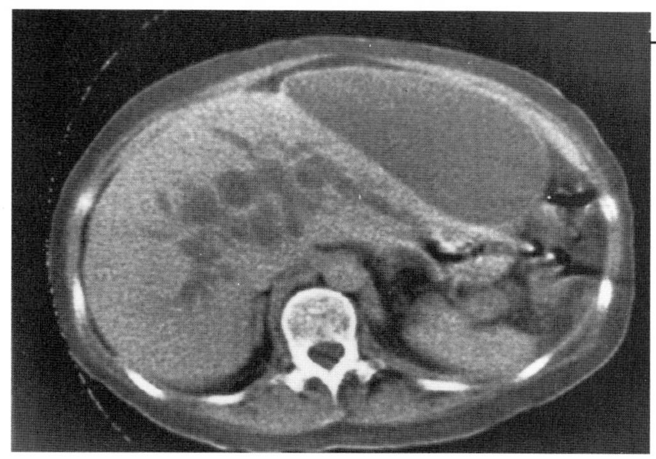

A

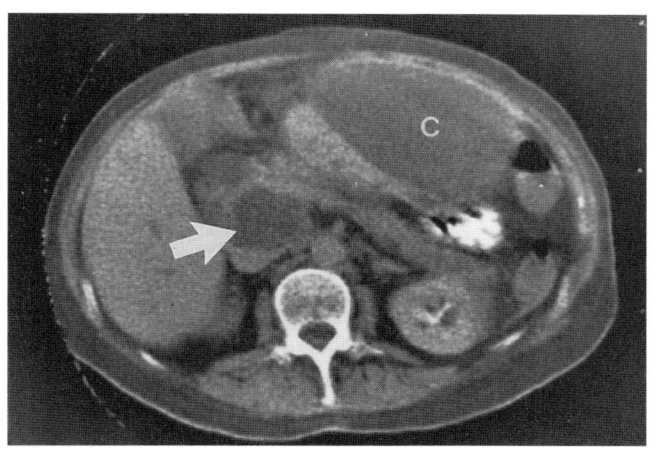

B

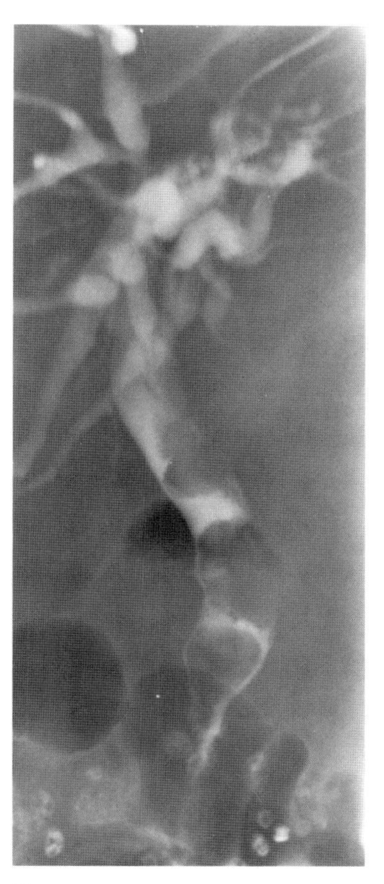

C

Figure 14.40. A and B: CT scans of another patient with obstructive jaundice. Note the dilated intrahepatic biliary radicals, dilated common bile duct *(arrow),* and pancreatic pseudocyst *(C).* C: Percutaneous transhepatic cholangiogram of the same patient, showing the large stones within the common bile duct that caused the biliary obstruction.

itial ultrasound examination showed dilatation of the intrahepatic biliary radicals (A) as well as dilatation of the common bile duct itself (B). Dilated intrahepatic ducts appear as branching, tubular structures at ultrasound, coursing parallel to the routing of the portal venous system. The common duct diameter can be measured at ultrasound, and extrahepatic biliary obstruction can be diagnosed when the common duct measures more than 4 to 5 millimeters. This patient's common bile duct measured 1.2 centimeters in diameter. CT confirmed dilatation of the intrahepatic ducts and the common bile duct (C and D) and showed a dilated pancreatic duct. The patient proved to have a small tumor at the ampulla of Vater.

Percutaneous transhepatic cholangiography is carried out by injection of a water-soluble contrast material through a fine needle introduced directly into the liver through the skin. The dilated biliary tree is shown in great detail and the site and cause of biliary obstruction are usually readily apparent.

In Figures 14.40A and B you see CT studies performed on a 65-year-old woman who had been hospitalized for evaluation of jaundice and a history of bouts of abdominal pain over the previous 20 years. Ultrasound (not illustrated here) indicated dilated bile ducts as well as the presence of pancreatic pseudocysts. A CT examination was requested for further evaluation. Note the dilated biliary radicals in Figure 14.40A and the dilated common bile duct in Figure 14.40B. The patient's long history of pain suggested either cholecystitis with the passage of gallstones or pancreatitis or both. Her development of obstructive jaundice was worrisome and could have been caused by stones in the common bile duct or by the development of carcinoma in the head of the pancreas.

The next step in this patient was, logically, a percutaneous transhepatic cholangiogram, which you see in Figure 14.40C. Several large filling defects are visible in the dilated common bile duct. At surgery several large stones were removed successfully and the pancreatic pseudocyst arising from the tail of the pancreas was drained externally.

The Pancreas

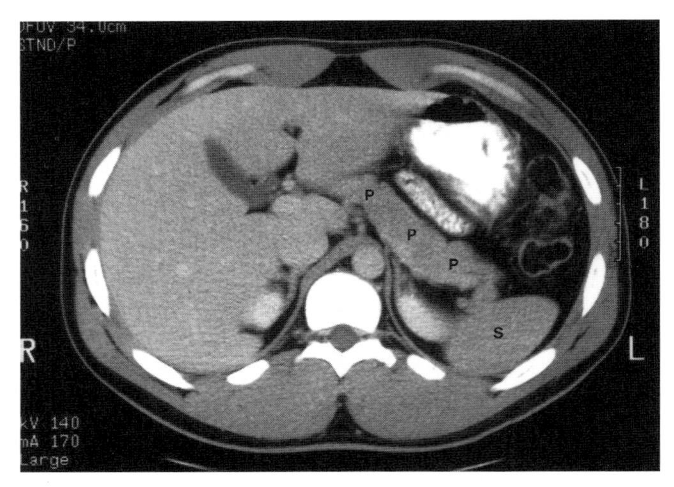

A

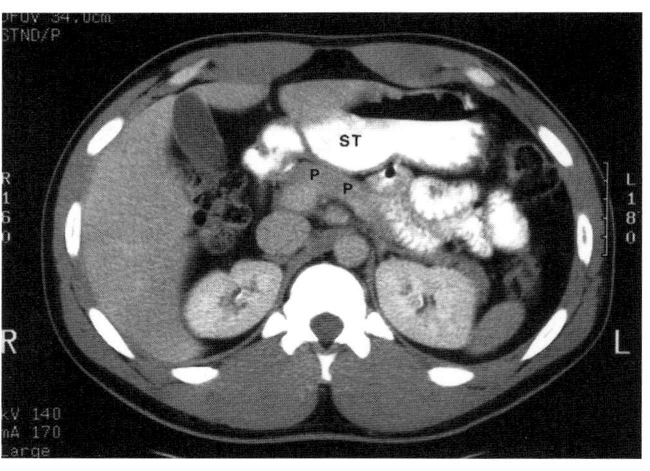

B

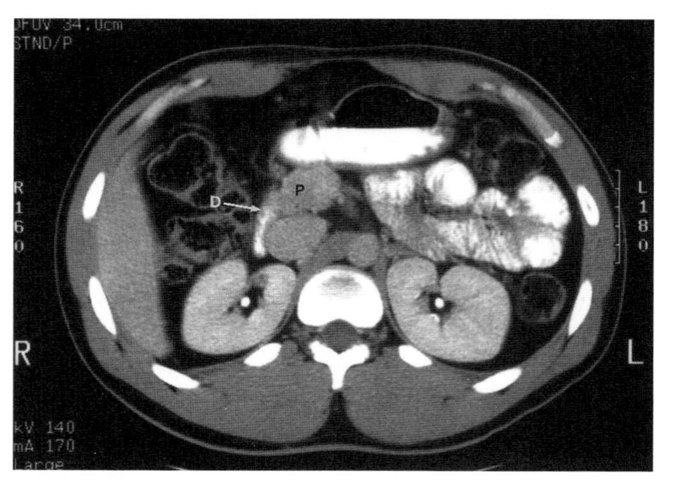

C

Before the advent of ultrasound and CT the pancreas was considered to be one of the "concealed" organs of the abdomen, imaged only indirectly. Today, with the newer cross-sectional imaging methods we can visualize pancreatic tissue directly (Figure 14.41), and the pancreatic duct itself can be opacified by ERCP (Figure 14.42).

The pancreas is not visible on plain films, but pancreatic calcifications may be evident in some patients and are usually diagnostic of chronic pancreatitis (see Figure 11.24). When large pancreatic masses are present, they may be detected indirectly at upper gastrointestinal series by anterior displacement of the barium-filled stomach and widening of the duodenal loop, which wraps around the head of the pancreas.

Arteriography does not opacify the pancreatic parenchyma as brightly as the parenchyma of other organs such as the liver, spleen, and kidneys. Consequently, the angiographic diagnosis of pancreatic cancer is based on the demonstration of encasement of, or tumor vessels associated with, the major blood vessels adjacent to or within the pancreas.

Although better seen by CT, the pancreatic parenchyma can be visualized directly with ultrasound in most patients. Normally pancreatic tissue is more echogenic than the adjacent liver, and alterations in echogenicity may be important signs of pancreatic dis-

Figure 14.41. Normal CT scans of the pancreas. The pancreas is oriented along a diagonal course across the upper abdomen, with the pancreatic head and neck lower in position than the pancreatic body and the tail. A: Upper-level scan showing the pancreatic body and tail (*P*'s) pointing toward the spleen *(S)*. B: Slightly lower-level scan showing the pancreatic neck posterior to the stomach *(ST)*. C: Even lower-level scan showing the pancreatic head in close proximity to the contrast-opacified duodenum *(D, arrow)*.

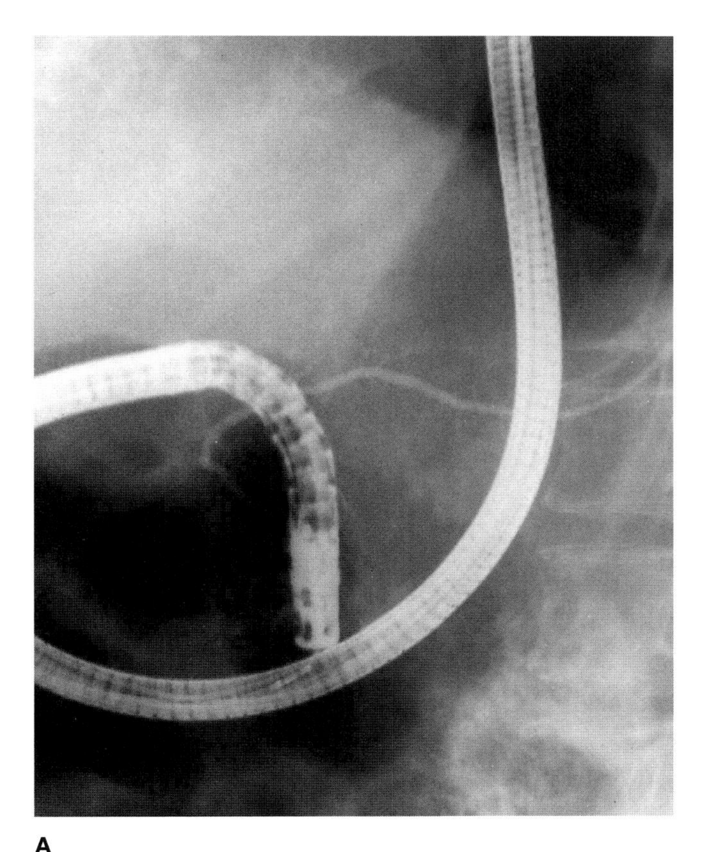

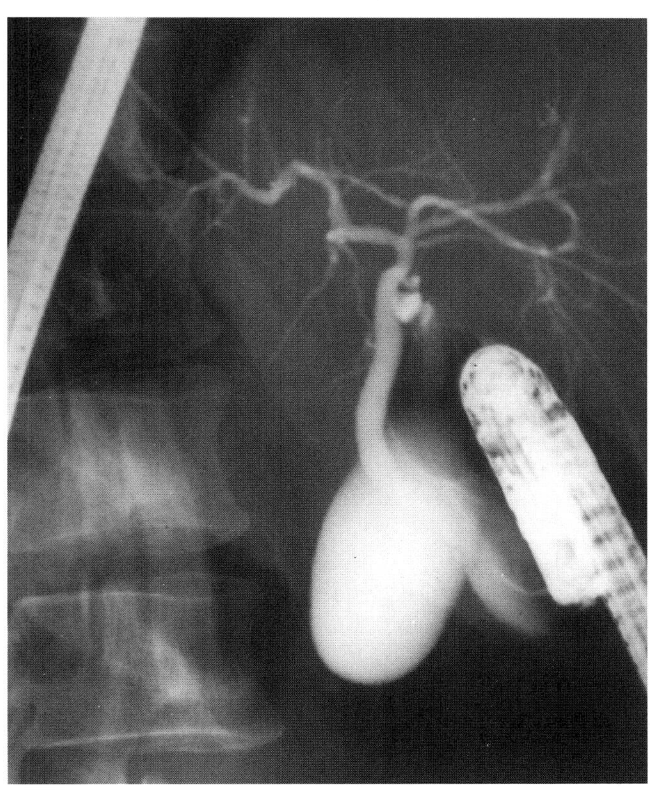

A B

Figure 14.42. Endoscopic retrograde cholangiopancreatography (ERCP). The pancreatic duct (in A) and the common bile duct (in B) have been individually cannulated and injected with contrast material.

ease. Ultrasound may show pancreatic masses, pseudocysts, pancreatic duct dilatation, and evidence of pancreatitis.

Pancreatic anatomy can be clearly delineated by CT, which shows the pancreas arcing anteriorly over the spine, its head adjacent to the duodenum and its tail extending toward the spleen. The pancreas may look smooth or lobulated. To differentiate pancreatic tissue from the adjacent blood vessels and duodenum, both oral and intravenous contrast materials are used.

ERCP (Figure 14.42) involves cannulation, under fluoroscopic guidance, of the common bile and pancreatic ducts via the ampulla of Vater during upper endoscopy. Generally this requires the combined efforts of the endoscopist and the radiologist. Contrast material injected into the lower common bile duct can show pathology at this level when the common bile duct is totally occluded from above. When the ultrasound or angiographic diagnosis of suspected pancreatic disease is uncertain, ERCP with contrast injection into the pancreatic duct is often conclusive. The course of the pancreatic duct is affected by any mass lesion within the pancreas.

In your clinical practice you will refer patients for pancreatic imaging who are suspected of having pancreatic neoplasms (carcinoma and islet cell tumors), pancreatitis, pancreatic cystic lesions (pseudocysts), and pancreatic trauma.

Pancreatic Tumors

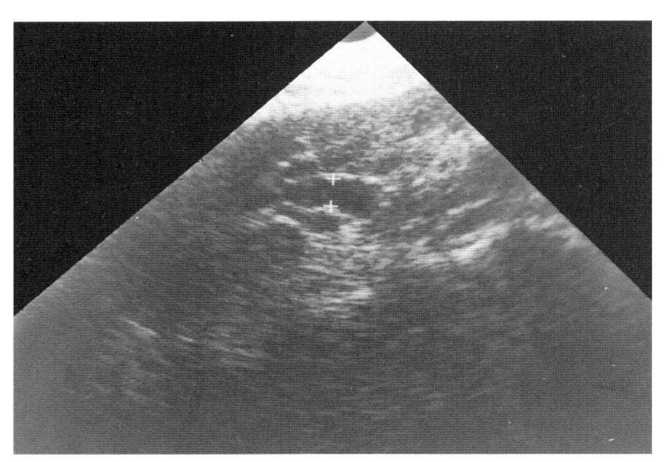

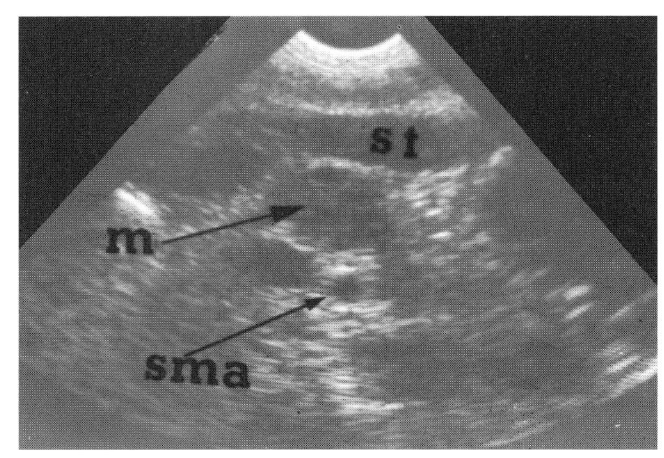

A B

Figure 14.43. Obstructive jaundice caused by a small pancreatic carcinoma. A: The initial ultrasound survey showed that the common bile duct was dilated to a diameter of 9 millimeters (measured by ultrasound cursors, the two white crosses placed on the lateral margins of the duct). B: Transverse ultrasound showed a mass *(m)* between the stomach *(st)* and superior mesenteric artery *(sma).* This would be the location of the pancreatic head. C: CT scan through the same level confirmed a mass *(m)* between the stomach *(st)* and the superior mesenteric artery *(sma).* An *arrow* points to the dilated common bile duct *(cd).* The gallbladder *(gb)* is also indicated. D: Endoscopic retrograde cholangiopancreatography identified a marked stenosis *(arrow)* of the common bile duct, which was encased by pancreatic tumor. Note the marked dilatation of the common duct proximal to the stenosis.

Carcinoma of the pancreas may be a difficult diagnosis to make. The symptoms of abdominal pain, weight loss, and early satiety are often nonspecific, and the small size of some tumors may make them difficult to image. When pancreatic cancer is clinically suspected, the diagnostic procedure of choice is CT. If the patient is jaundiced, ultrasound is usually performed first to determine if biliary obstruction is present; ultrasound may even show the pancreatic mass (Figure 14.43).

At ultrasound, pancreatic cancers appear as masses or bulges, more echogenic than the remainder of the organ (Figure 14.43B). At CT, pancreatic cancers also appear as masses disturbing the normal organ contours (Figure 14.43C). Although the tissue density of pancreatic carcinoma is similar to that of the normal pancreatic parenchyma, CT can show more subtle alterations in pancreatic shape and size than ultrasound. CT may also show secondary findings better, such as pancreatic duct dilatation (as you have previously seen in Figure 14.39C and D). It may also show signs of local and distant abdominal metastases.

Under CT guidance (which involves following the advancing needle tip with serial CT scans), biopsy of pancreatic masses by percutaneous needle aspiration may be obtained in order to provide a tissue diagnosis. This technique allows confirmation of cancer when the diagnosis is in doubt. In addition, tumors deemed unresectable by CT may be biopsied so that a patient can be started on appropriate radiation or chemotherapy, without the further stress of a laparotomy biopsy.

When there is any doubt about the CT identification of a pancreatic mass, ERCP is usually performed as the next imaging procedure. It may confirm a mass suspected at CT by showing encasement of, or oc-

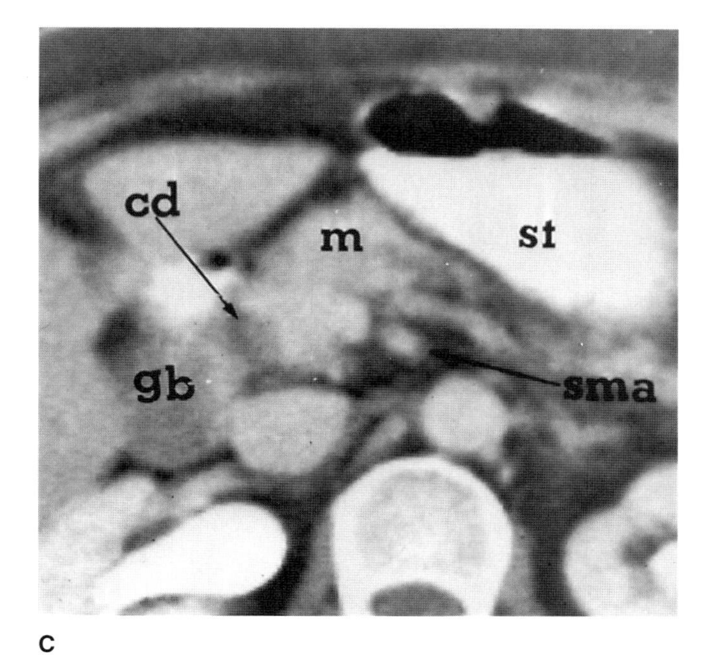

C

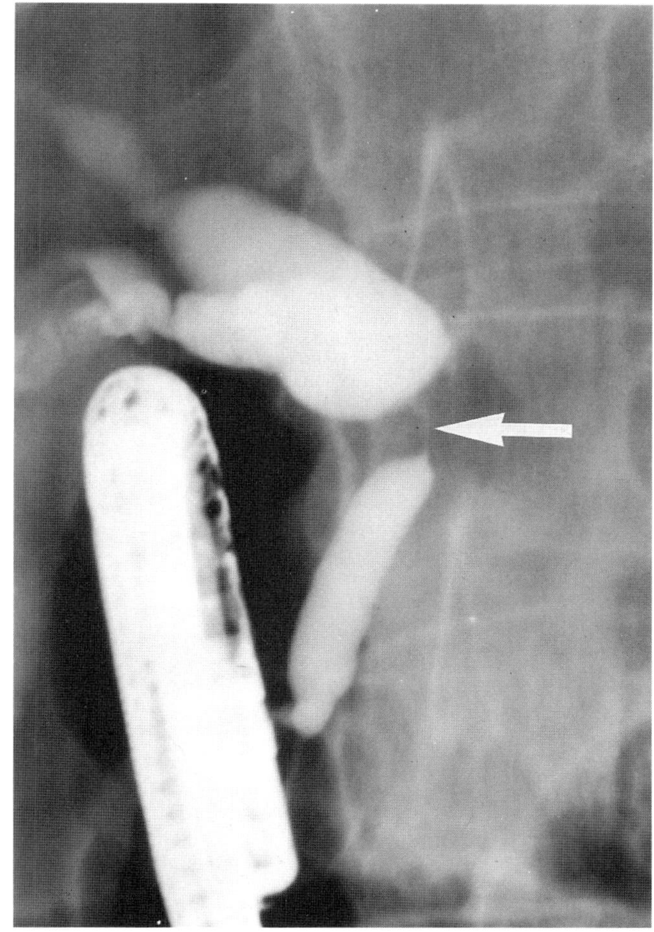

D

clusion of, either the pancreatic duct or the common bile duct. The outcome depends of course on whether the mass is located in the head, body, or tail of the pancreas. (In Figure 14.43D the ERCP examination shows that the common bile duct is encased by the mass detected in the head of the pancreas at CT.)

Angiography is used less than it once was for the diagnosis of pancreatic carcinoma, although it is often performed for preoperative staging after a pancreatic tumor has been detected by CT. If a pancreatic cancer encases a major blood vessel such as the portal vein, it is usually unresectable. Since blood vessel encasement is generally better shown by angiography than by CT, many surgeons refer their patients for preoperative angiography.

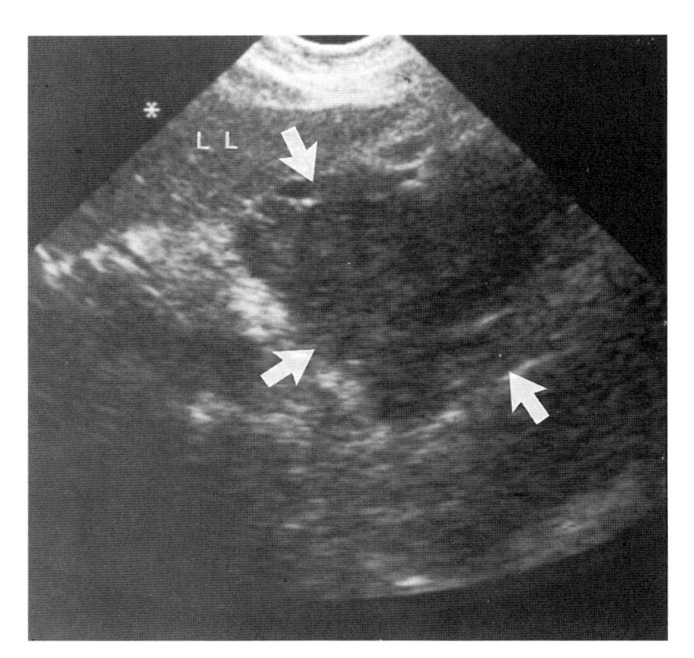

A

Figure 14.44. Islet cell tumor (insulinoma) of the pancreas. A: Transverse ultrasound shows a large mass *(arrows)* in the body of the pancreas behind the left lobe *(LL)* of the liver. The mass is less echogenic than the liver or surrounding fat. B: CT confirms a large mass *(m)* in the body of the pancreas, which indents the contrast-filled stomach posteriorly *(arrow)*. C: Selective splenic arteriogram in the arterial phase shows that the mass is hypervascular *(arrows);* this would be more typical of an islet cell pancreatic tumor than a pancreatic carcinoma. To the left of the mass is the normal splenic parenchyma. D: Venous phase of the splenic arteriogram shows marked collateral circulation through veins (varices) of the lesser and greater curvature of the stomach (*c*'s) because of splenic vein occlusion by the tumor. The portal vein *(pv)* remains patent. This patient has *prehepatic portal hypertension* and could bleed from gastric varices.

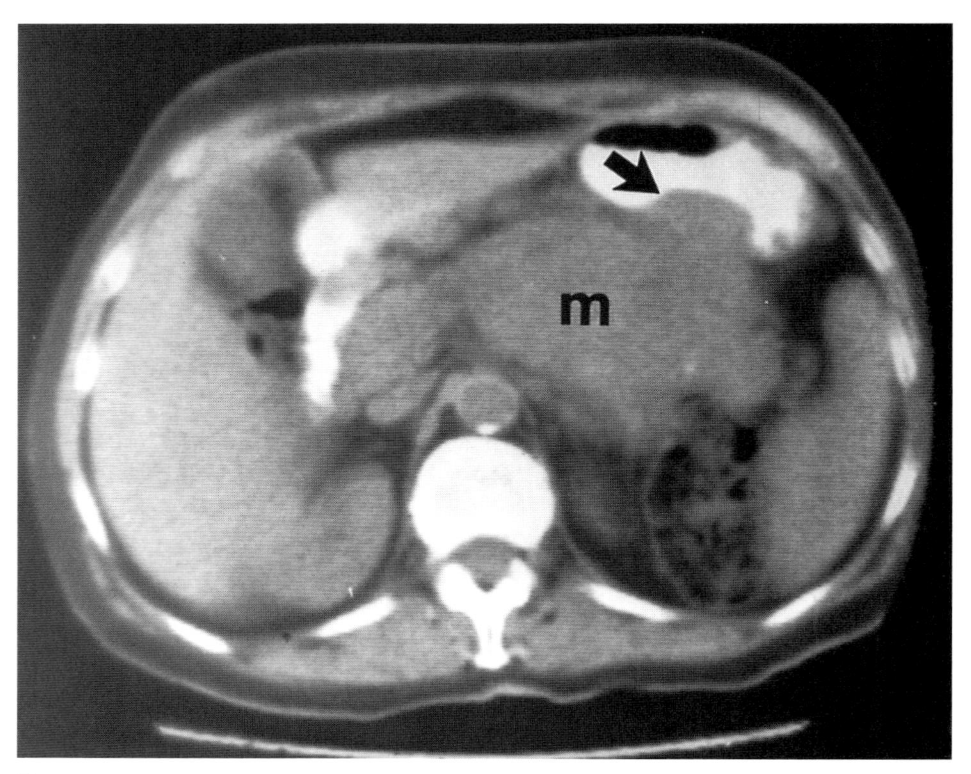

B

At angiography the vascularity of the tumor may also be better appreciated than with CT. Pancreatic carcinoma is nearly always *hypovascular,* whereas islet cell tumors, such as insulinomas (Figure 14.44), are usually *hypervascular.* This may not be such an important differentiation to make by imaging, because most islet cell tumors usually present with symptoms and signs of their associated hormonal syndromes. You will generally know that your patient has an islet cell tumor—insulinoma, gastrinoma, or glucagonoma—by clinical and laboratory findings, even before you refer your patient for CT.

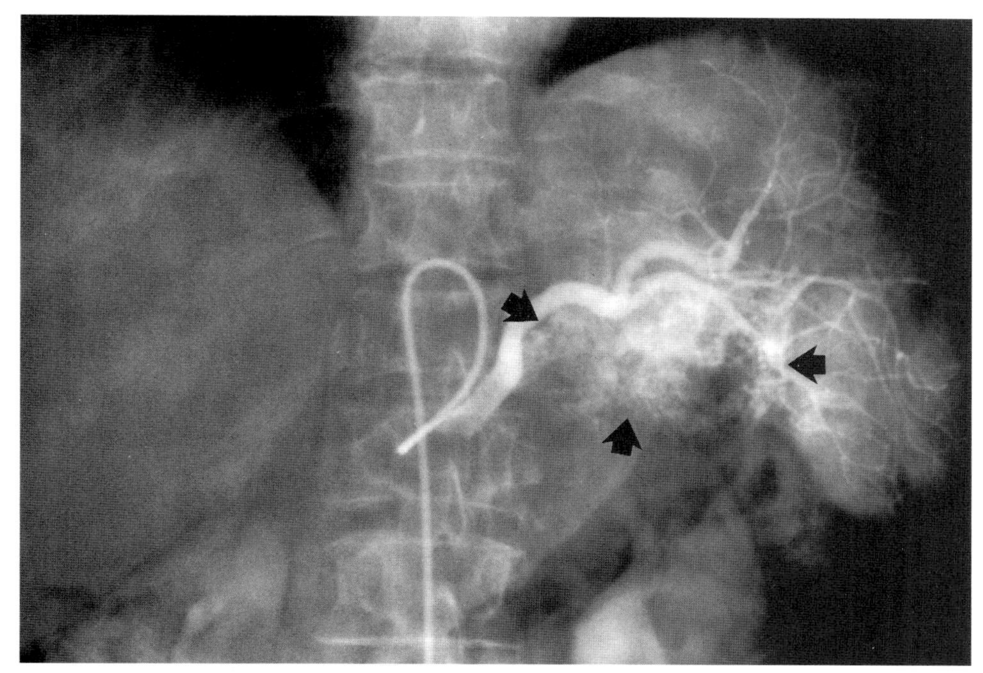

C

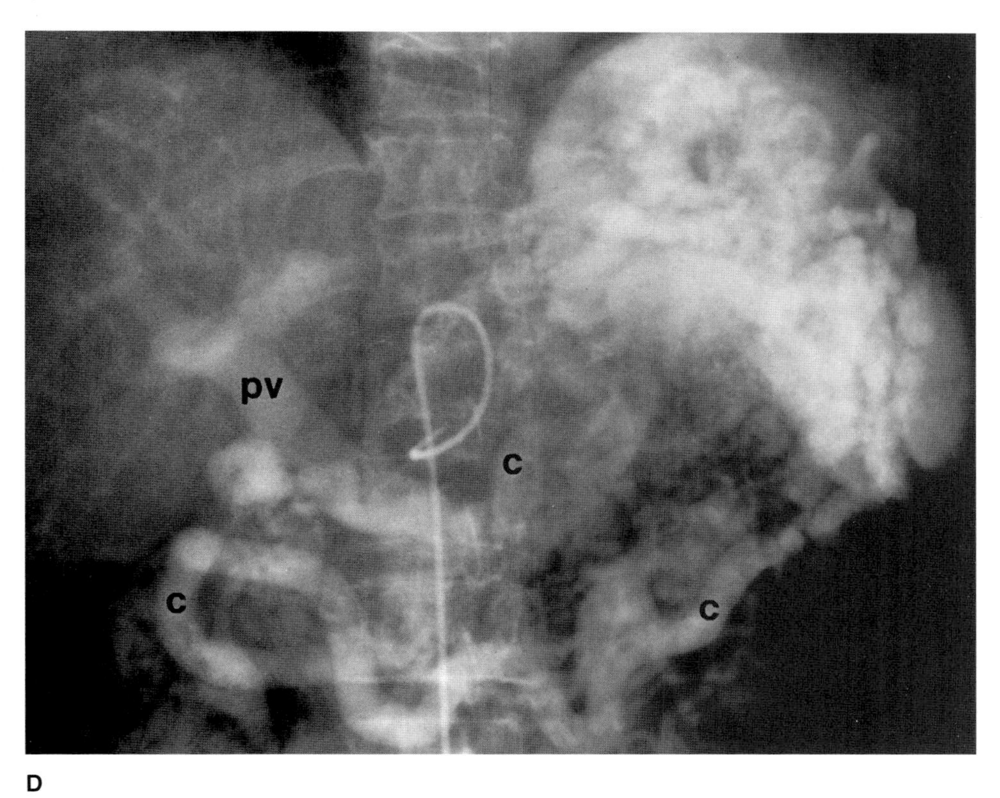

D

Pancreatitis and Pancreatic Abscesses

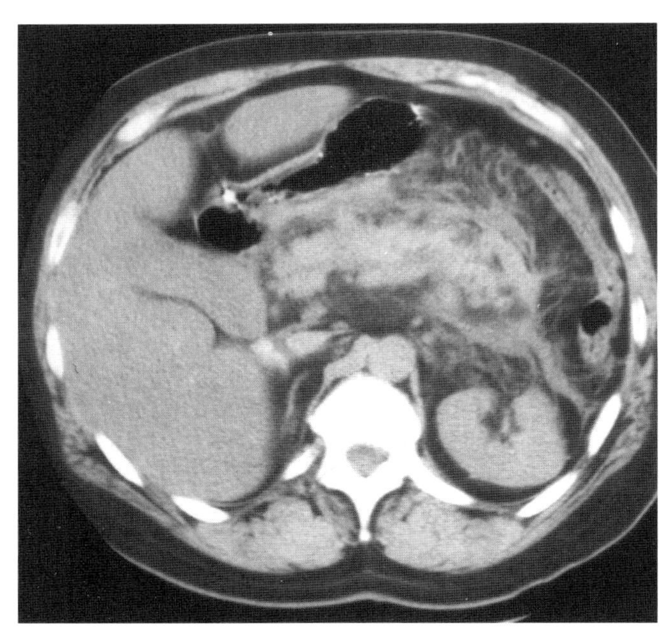

Figure 14.45. CT scan of moderate pancreatitis showing enlargement and irregularity of the pancreas with inflammatory stranding within the surrounding soft tissues. Compare Figure 14.41.

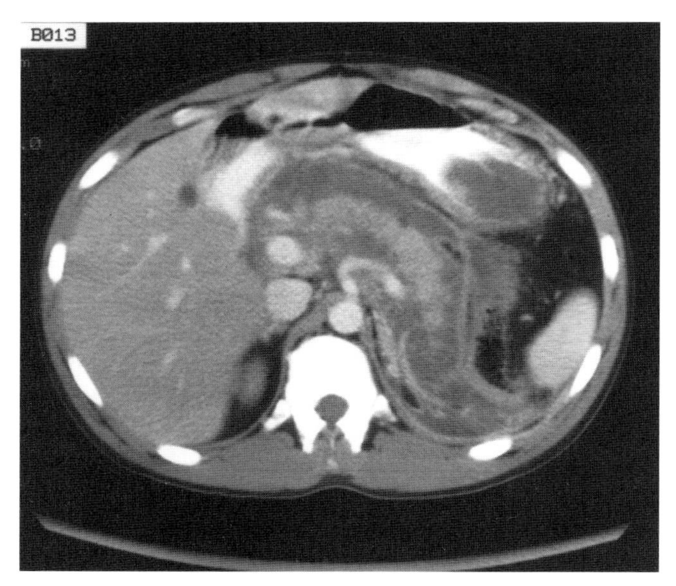

Figure 14.46. CT scan of severe pancreatitis showing the pancreas entirely surrounded by inflammatory fluid.

Making a diagnosis of acute pancreatitis may not require any imaging at all. The diagnosis may be obvious from the clinical findings of epigastric pain and tenderness, nausea, vomiting, and elevation of serum amylase. The causes include alcoholism, obstruction of pancreatic duct drainage (by a distal common bile duct stone, for example), trauma, recent surgery, and certain therapeutic drugs. Imaging is recommended when the diagnosis is in doubt or when complications of pancreatitis, such as pseudocyst and abscess formation, are suspected.

Acute pancreatitis is nearly always associated with edema and enlargement of the pancreas. This is best demonstrated with CT (Figure 14.45), which usually also shows irregularity of the pancreatic outline and inflammatory stranding in the surrounding soft tissues. In mild cases, however, the CT examination may be entirely normal, or it may show only mild enlargement. In severe cases peripancreatic fluid may be visible (Figure 14.46), as well as evidence of pancreatic necrosis (Figure 14.47) and phlegmon, consisting of a large inflammatory mass of pancreatic and peripancreatic tissues. The ultrasound findings of acute pancreatitis also include pancreatic enlargement and an overall decrease in the echogenicity of the organ. Pseudocysts are easily detected with ultrasound.

A dreaded complication of acute pancreatitis is pancreatic abscess formation, which is suspected clinically when the patient also presents with fever and septicemia. The pathognomonic finding is the presence of gas bubbles in the pancreatic bed. These bubbles are best shown by CT (Figure 14.48) when there is only a small amount of gas, but they may also be seen on a plain abdominal film when copious gas is present.

Pancreatic calcifications may also be identified on plain films, providing evidence of previous pancreatitis, and gas may be seen in the pancreatic bed. Free air in the peritoneal space may also be seen, indicating a perforated viscus, such as a perforated peptic ulcer, which may mimic the clinical signs of pancreatitis. CT can also show pancreatic calcifications from chronic pancreatitis, as well as complications from this condition such as pseudocyst formation (Figure 14.49).

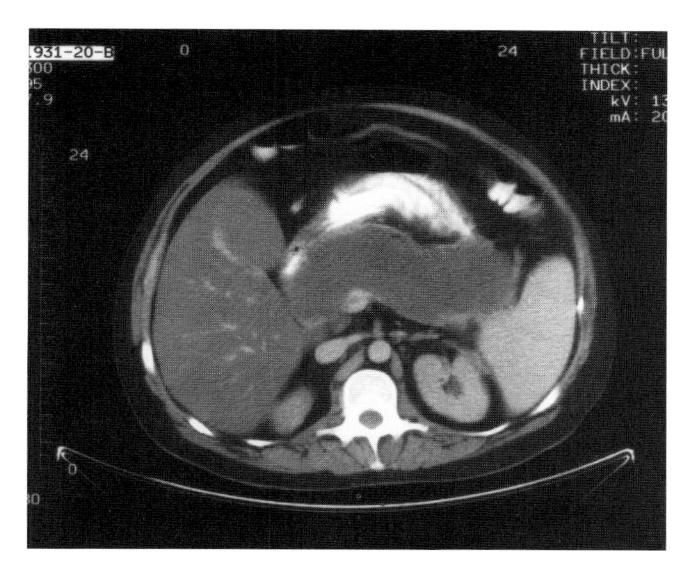

Figure 14.47. CT scan of severe pancreatitis with necrosis of the pancreas and replacement with fluid.

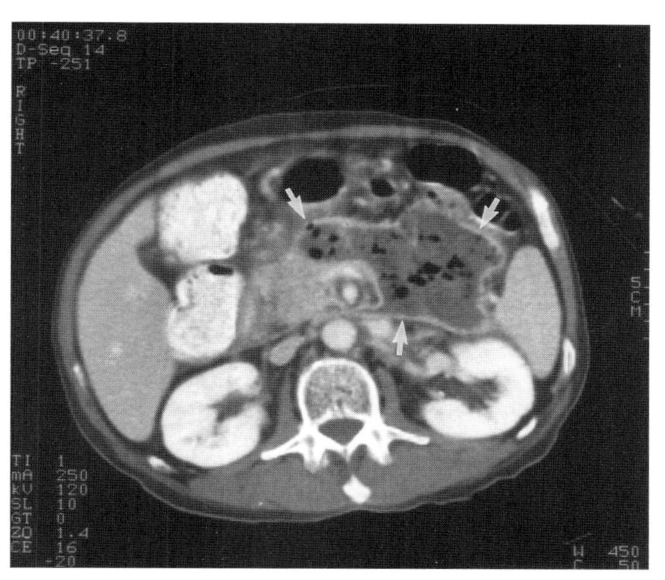

Figure 14.48. CT scan of a pancreatic abscess. The body and tail of the pancreas have been replaced by a thinly walled abscess *(arrows)* containing pus and gas.

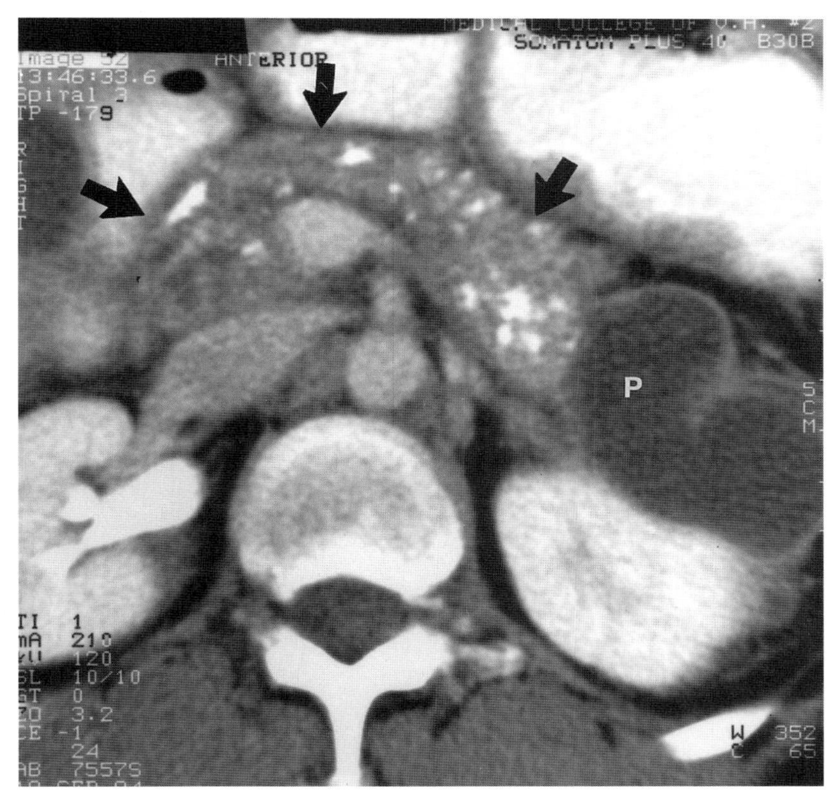

Figure 14.49. Detail of a CT scan of chronic pancreatitis with pseudocyst formation. Note the bright white calcifications in the neck and body of the pancreas *(arrows)*. A fluid-filled pseudocyst *(P)* replaces the tail of the pancreas.

Pancreatic Trauma

With blunt abdominal trauma, especially anterior midline blows, the pancreas may be compressed against the spine, producing pancreatic lacerations, hematomas, and transections. These may be associated with bleeding within the anterior pararenal space (anterior compartment of the retroperitoneum), where the pancreas is located. The diagnosis should be considered in patients with anterior abdominal trauma, especially when laboratory tests indicate an elevation in serum amylase. As with all abdominal traumas, the imaging procedure of choice is a CT scan (Figure 14.50). If CT shows a laceration, the patient should also be examined with ERCP to visualize the pancreatic duct. If the duct is intact, the patient can generally be managed conservatively; if the ERCP shows disruption of the duct, providing evidence that the duct is unable to drain the distal pancreas, surgery is indicated, usually a distal pancreatectomy.

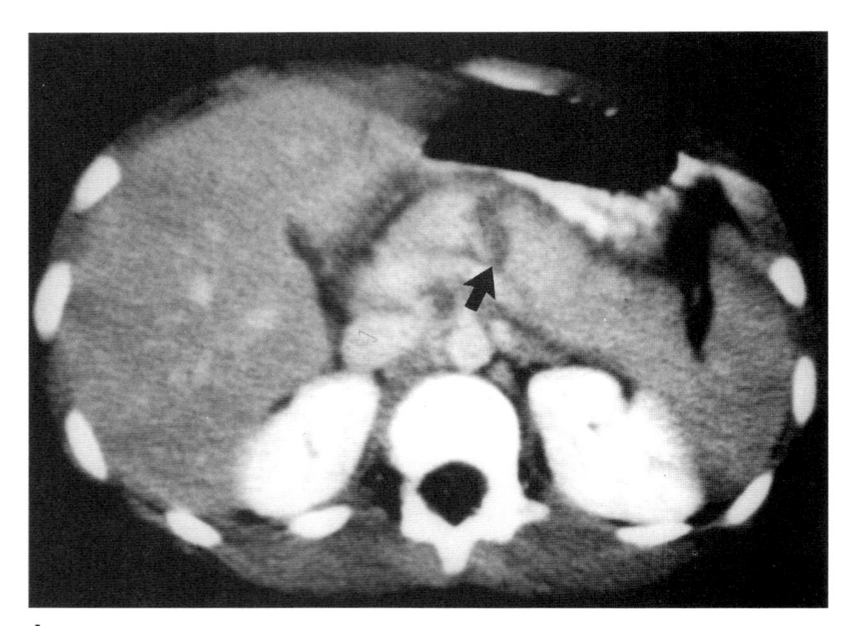

A

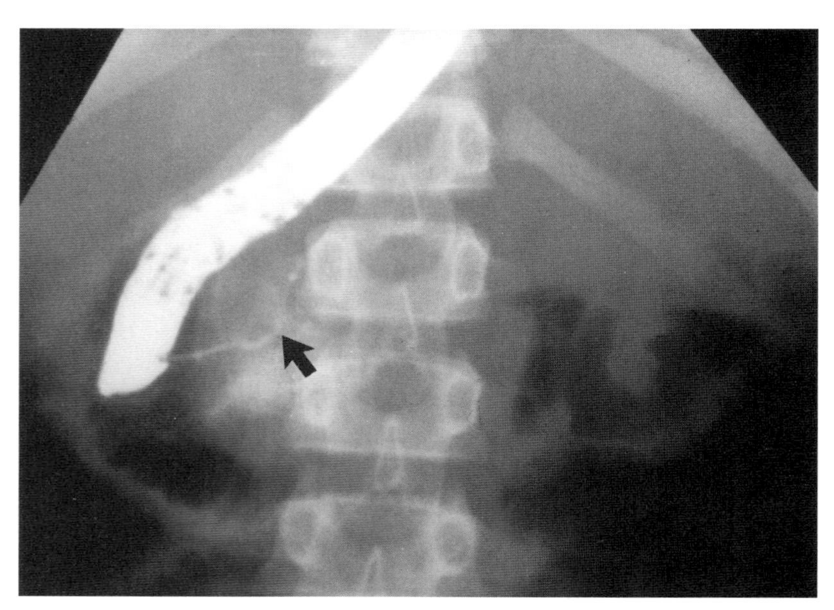

B

Figure 14.50. Pancreatic fracture in a child who suffered anterior abdominal compression in a sledding accident. A: CT scan showing that the pancreas was compressed against the spine, producing a vertical fracture *(arrow)*. B: ERCP showing that the pancreatic duct is interrupted at the fracture line *(arrow);* consequently, pancreatic juices produced by the tail of the pancreas could not be drained. The patient underwent a distal pancreatectomy. Compare the patient in Figure 14.16, page 309, who had both a pancreatic fracture and a liver laceration.

The Urinary Tract

The traditional technique for imaging the urinary tract is the intravenous urogram (IVU), referred to in some hospitals as the intravenous pyelogram, or IVP. Although ultrasound and CT can provide better diagnostic images of the kidneys, IVU was until very recently the first step in the imaging workup of suspected stone disease and hematuria. For this examination an intravenous infusion of iodinated contrast material is followed by a series of x-ray films that show the opacified kidneys, ureters, and bladder. The intravenous urogram should be considered a morphological rather than a functional examination, because you cannot equate the density of contrast material seen on the x-ray films with overall kidney function. Kidney function is a highly complex matter comprising many interrelated processes: glomerular filtration, tubular excretion, and water reabsorption. The density of contrast substance in the draining structures and parenchyma of the kidney must be interpreted in the light of any variation in the *fluid flow through the kidney* as well as any fluctuations in renal physiology.

The contrast materials most widely used are excreted by the kidney almost entirely in the glomerular filtrate. Less than a minute after the intravenous infusion is started, enough opaque substance is present in the glomeruli and tubules of the renal parenchyma to give an appreciable whitening to the kidney shadow on an abdominal film. This has been called the *nephrogram phase* of urography, and is the proper time to observe the size and shape of the kidneys. The rule in studying the morphology of the kidneys on plain films and nephrograms is that *the normal length of the kidney is 3.7 times the height of the second lumbar vertebra.* This measures approximately 9 to 15 centimeters in adults.

The calyces, renal pelves, ureters, and bladder are viewed in sequence following the nephrogram phase. These draining structures (the collecting system and bladder) may be seen to fill, at the earliest, within minutes of the nephrogram phase; the filling increases to a peak and then gradually fades. The entire length of the ureters is not normally seen filled on any single

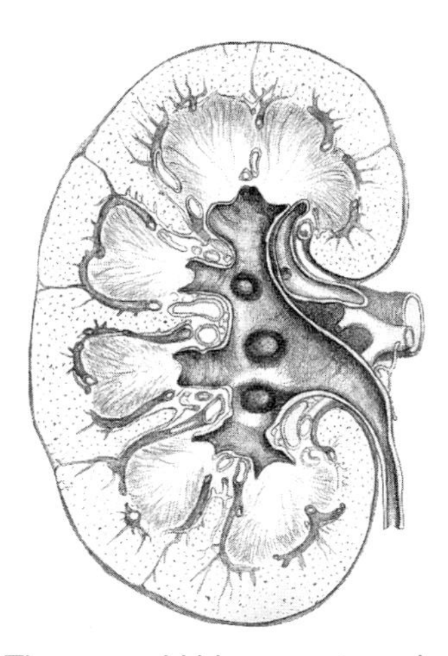

Figure 14.51. The normal kidney anatomy in the coronal plane.

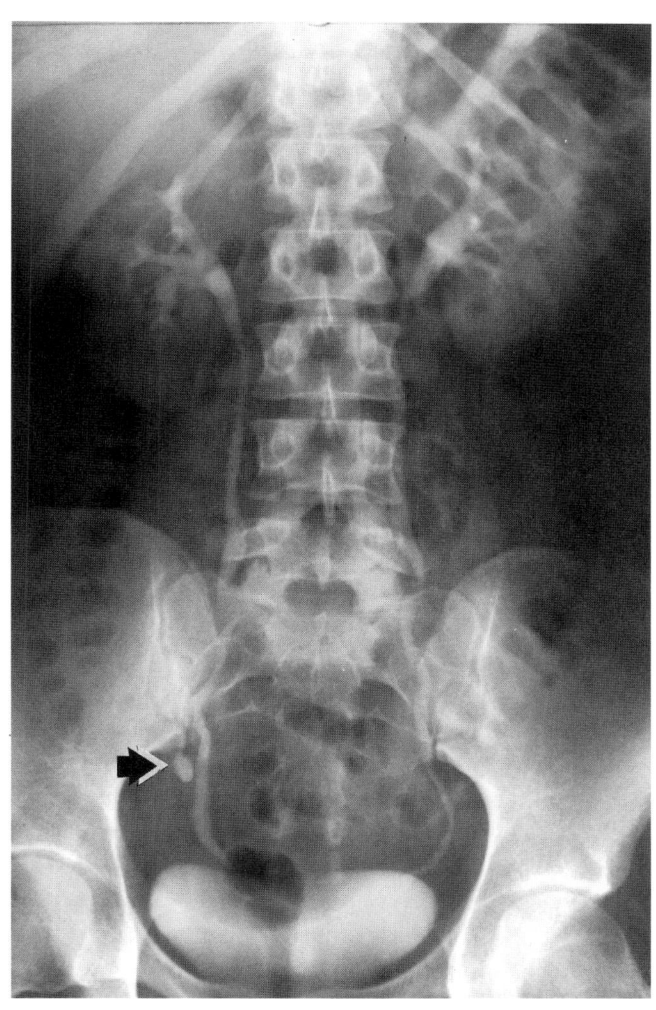

Figure 14.52. Almost normal intravenous urogram. The distal right ureter is displaced slightly medially within the bowl of the bony pelvis by what later proved to be a calcified right ovarian teratoma *(arrow).* The intrarenal collecting systems, left ureter, and bladder are normal.

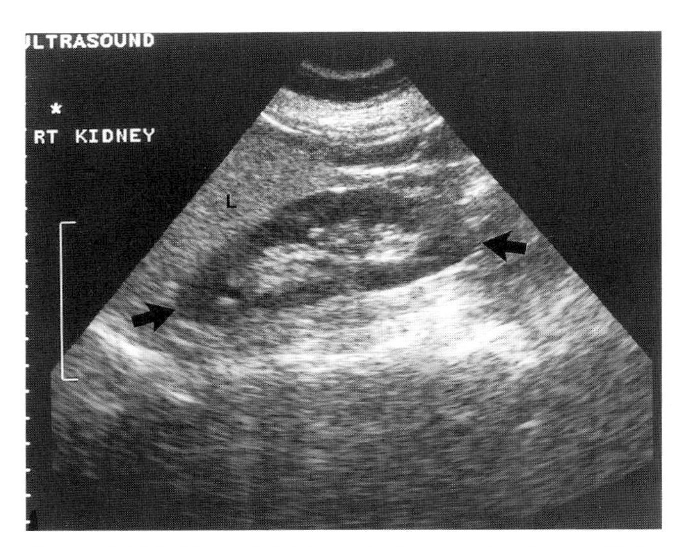

Figure 14.53. Normal right renal ultrasound in the sagittal plane. The *arrows* indicate the upper and lower poles. The echogenicity of the renal cortex is normally less than that of the liver *(L)*. Note the normal echogenic sinus fat within the central portion of the kidney.

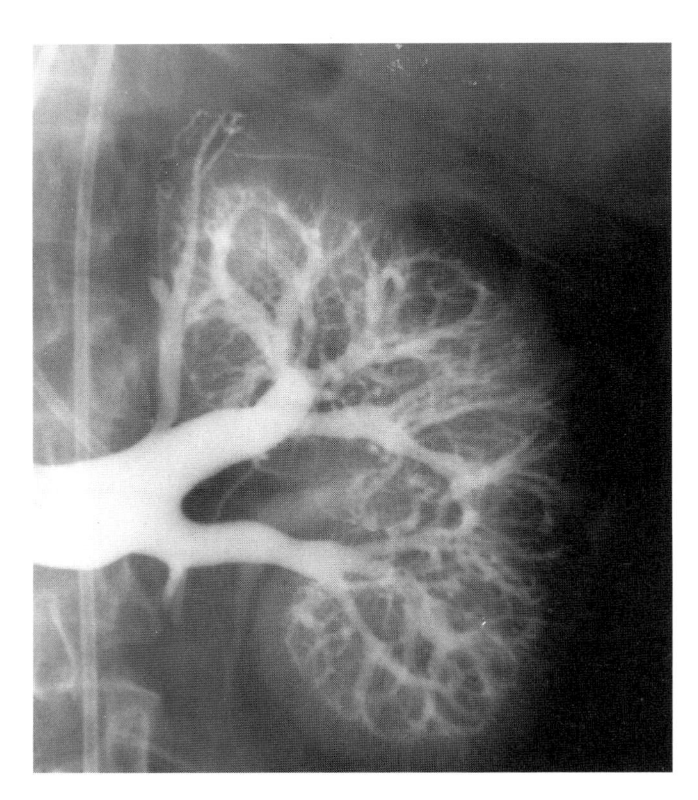

Figure 14.54 ***(right).*** Normal left renal venogram.

film, since they are constantly being swept by peristaltic waves. After 20 to 30 minutes the collecting system will be seen too faintly for further study and all the visible opaque will have accumulated in the urinary bladder. Of course the kidneys continue to excrete the remaining opaque until the bloodstream is cleared.

The density of the opaque material seen in the collecting system will be decreased if there is some degree of ureteral obstruction. There will be a delay before the contrast substance appears, and the filming sequence will have to be lengthened accordingly.

The *conventional intravenous urogram* commences with a preliminary plain film, which the radiologist always examines before injecting contrast material, to screen for calcification or stones that might be obscured by the contrast material. (Ninety percent of kidney stones are radiopaque and dense enough to be seen on the plain film.) Then films are made at intervals; the radiologist reviews each one until satisfied that the examination has been completed for that patient. Thus every intravenous urogram is essentially a custom-tailored study. If the abnormality revealed on the first films suggests the need, additional special views may be obtained while the contrast material is still present. Even tomography of the kidneys (nephrotomography) may be performed.

Pyelography (either *retrograde,* in which visualization of the urinary collecting system is achieved via a cystoscope, ureteral catheterization, and retrograde injection of contrast medium; or *antegrade,* via percutaneous puncture of the collecting system for contrast injection) can be performed when the intrarenal collecting system (calices and renal pelvis) and ureter cannot be opacified by intravenous techniques. This may occur when high-grade obstruction interferes with opacification of the collecting system, or when renal failure is present and the patient is unable to concentrate the intravenous contrast material for opacification of the collecting systems. Retrograde or antegrade pyelography shows only the collecting systems (renal pelvis and associated structures); it does not yield information about the parenchyma of the kidney itself.

Ultrasound is the imaging technique of choice for screening patients with suspected hydronephrosis. It can quickly and accurately identify a dilated renal col-

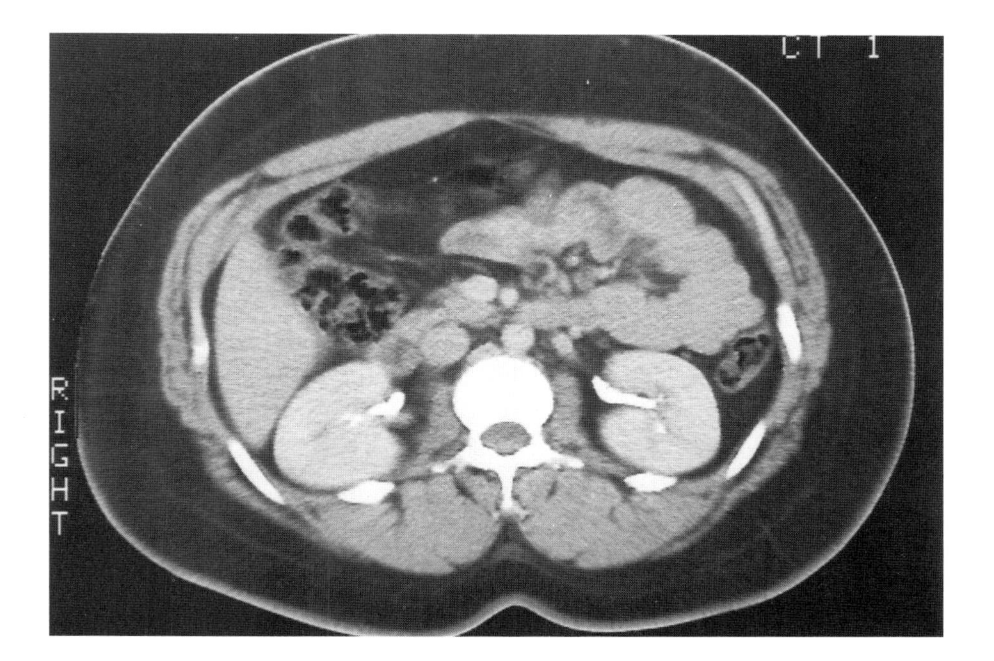

Figure 14.55. Normal CT scan through the midkidneys obtained with IV contrast material. The renal pelves and calices are brightly opacified.

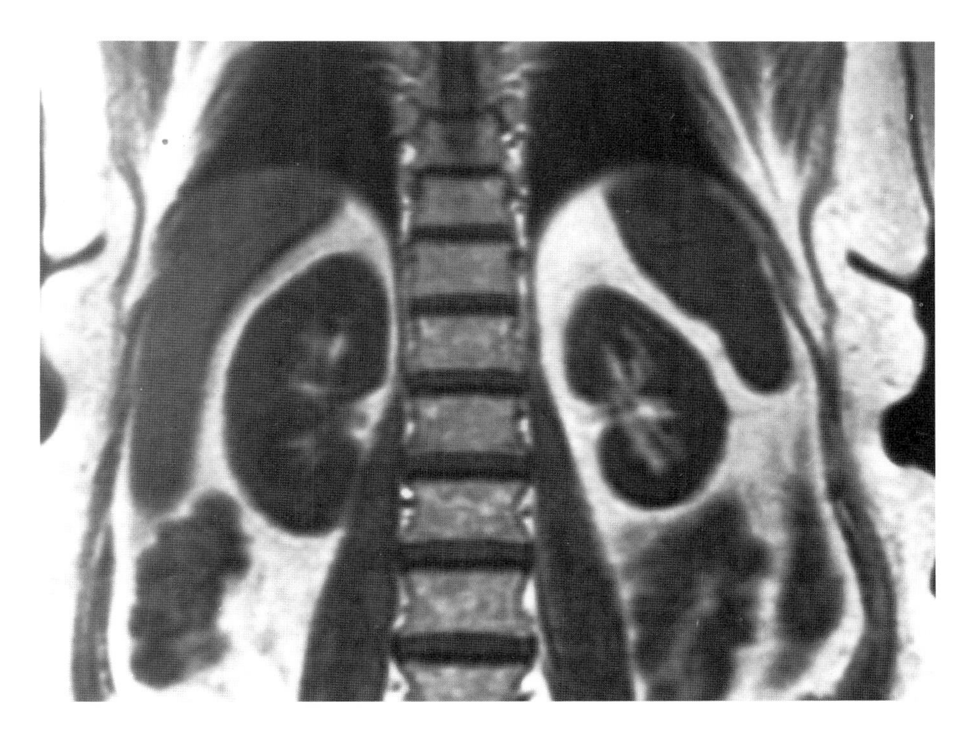

Figure 14.56. Normal coronal MR scan through the midkidneys. The renal sinus fat and surrounding perirenal fat appear white due to the strong MR signal of fat on this T1-weighted image.

lecting system without intravenous contrast injection or radiation of the patient. Furthermore, ultrasound can do it in a patient in renal failure who would be unable to concentrate contrast material for a urographic demonstration of hydronephrosis. In patients with renal masses ultrasound can easily differentiate anechoic benign cysts from echogenic renal tumors. CT is also an excellent imaging technique for the urinary tract; it can show renal tumors with great detail and stage the spread of malignant tumors.

Obstructive Uropathy

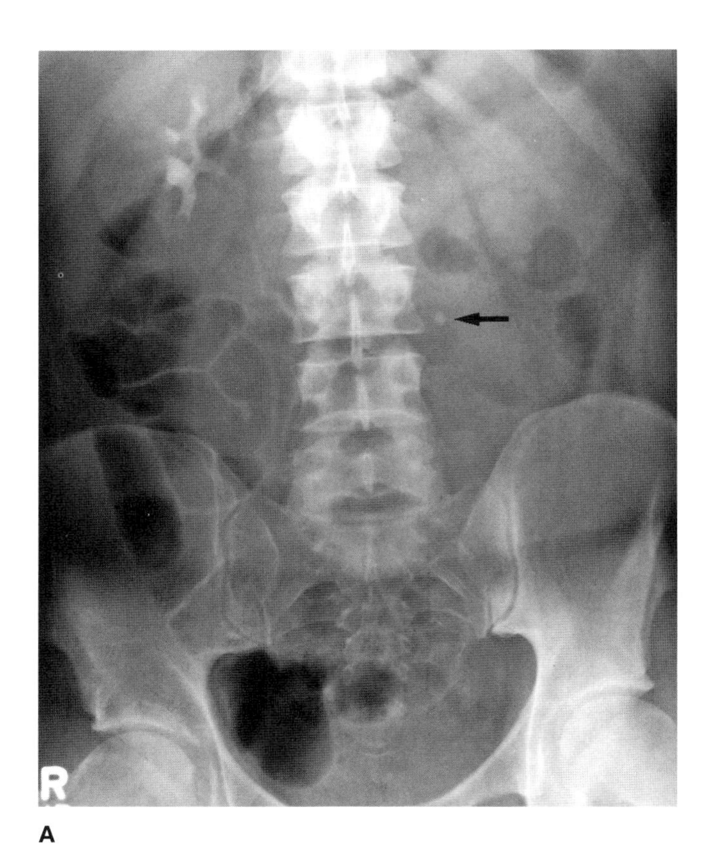

A

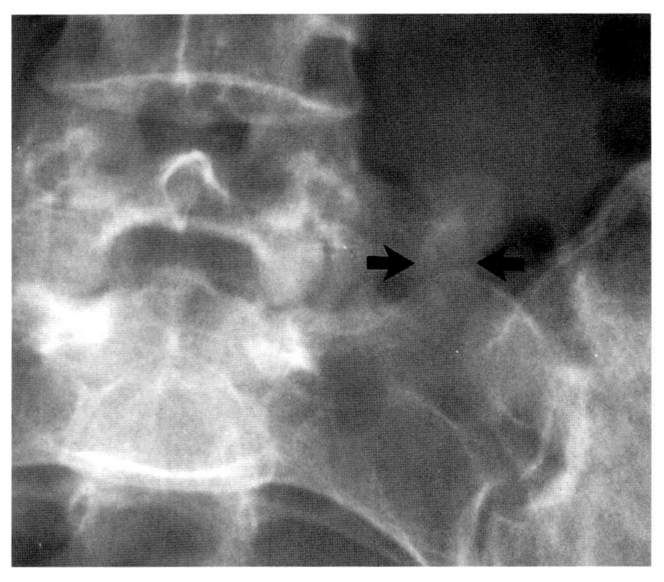

A

Figure 14.58 ***(above and on facing page).*** Chronically obstructing left ureteral stone that has produced left hydronephrosis. A *(above):* Plain film shows a calcified stone *(arrows).* B *(facing page):* Urogram shows left hydronephrosis; the *arrow* indicates the site of the obstruction.

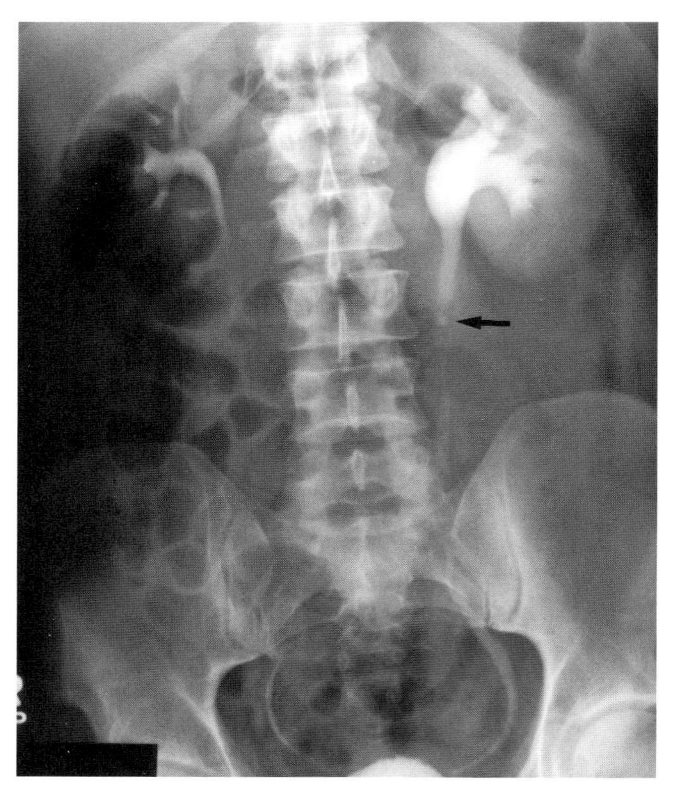

B

Figure 14.57 ***(left, above and below).*** Intravenous urogram of an acute, partially obstructing left ureteral stone. A: Early film showing delay in the appearance of contrast material in the left intrarenal collecting system, as compared with the right one, where the calyces are opacified (this finding indicates left-sided obstruction). The *arrow* points to a calculus overlying the course of the left ureter. B: Delayed film showing that the calculus *(arrow)* is indeed within the left ureter. Note the dilatation of the left intrarenal collecting system (which is partially obstructed) as compared with the right. The left renal parenchyma is hyperdense compared with the right; this finding is another sign of obstructive uropathy.

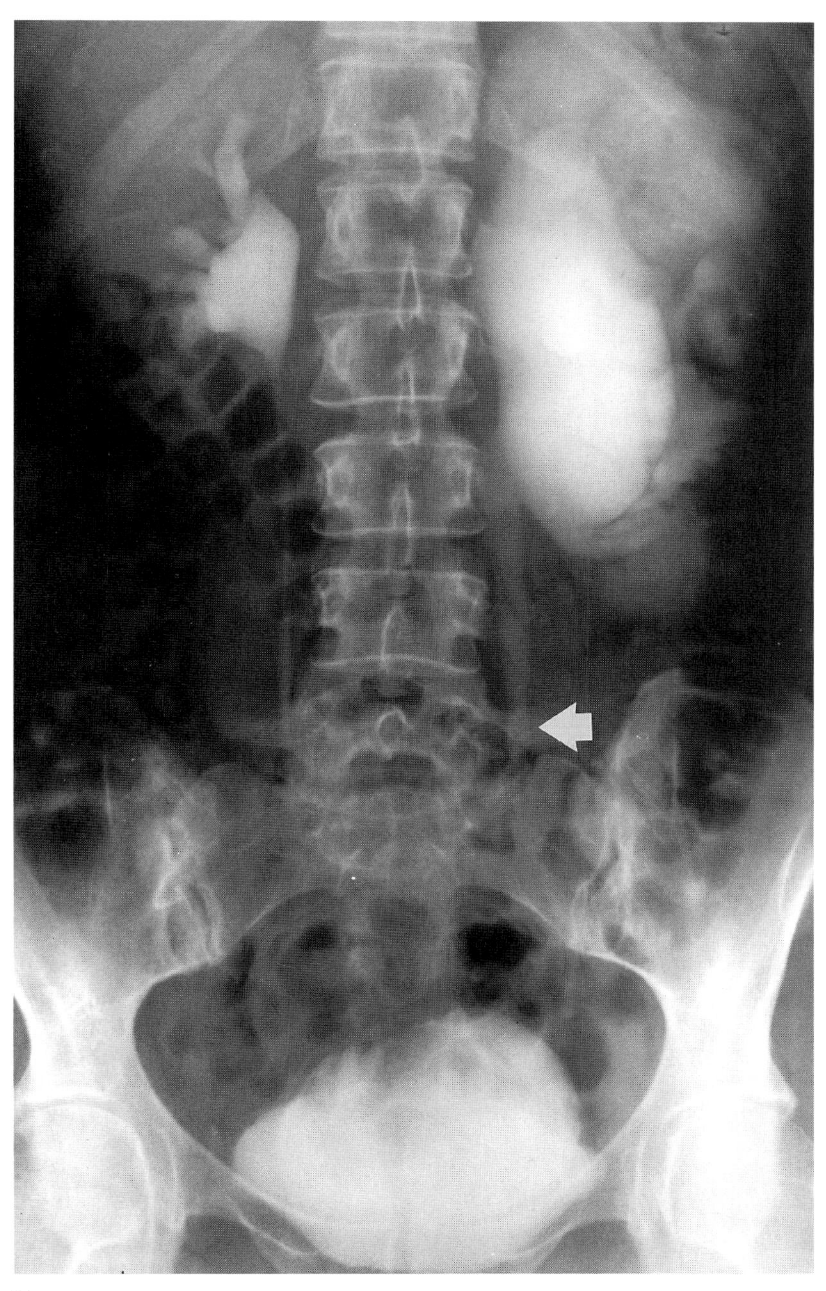

B

Urine flow out of the kidney may be prevented by numerous conditions, but among the commonest is an obstructing *ureteral calculus.* Since 90 percent of renal calculi are radiopaque, the chance of recognizing the presence of a calculus on the initial plain film of an intravenous urogram is excellent, and the study for a stone should begin there. With one ureter obstructed by a stone and unopacified urine backed up above it, a delay in the appearance of contrast medium in that kidney is to be expected; sometimes several hours are needed before the nephrogram finally appears. This phenomenon has been called the late white kidney of acute renal obstruction. Ultimately, opacification of the ureter down to the point of the obstruction will occur as excreted contrast material mixes with retained urine. The degree of dilatation of the intrarenal collecting system and ureter proximal to the site of obstruction (degree of hydronephrosis) will depend on the degree of obstruction (partial versus nearly complete) and the length of time that obstruction has been present (Figures 14.57 and 14.58).

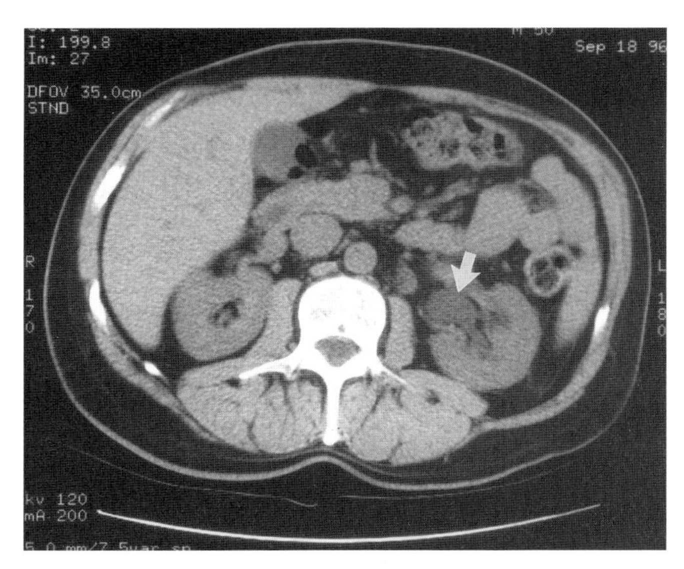

A. Axial CT scan through the mid left kidney shows a dilated (hydronephrotic) left intrarenal collecting system *(white arrow).*

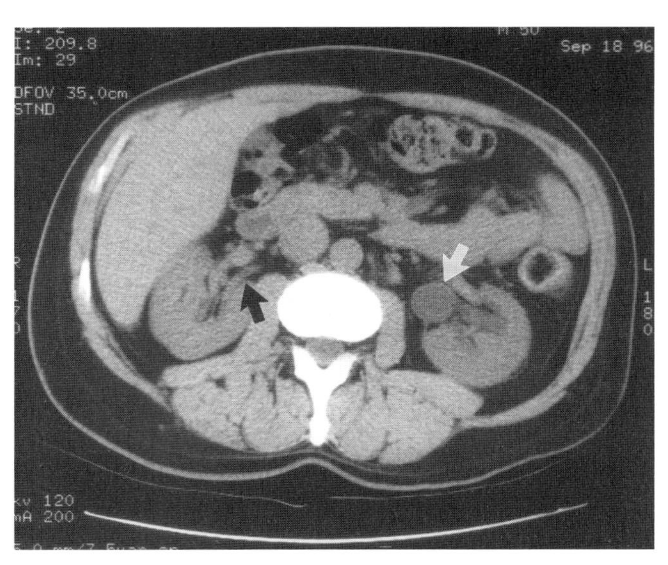

B. Axial CT scan, at a slightly lower level than A, shows a dilated left renal pelvis *(white arrow).* Compare the normal, nondilated right renal pelvis *(black arrow).*

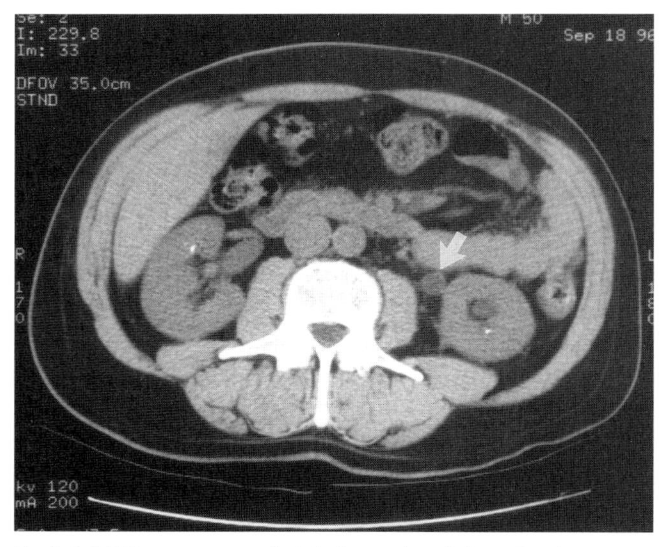

C. Axial CT scan, at a slightly lower level than B, shows dilatation of the left ureter *(white arrow).* Tiny calculi are visible in both kidneys.

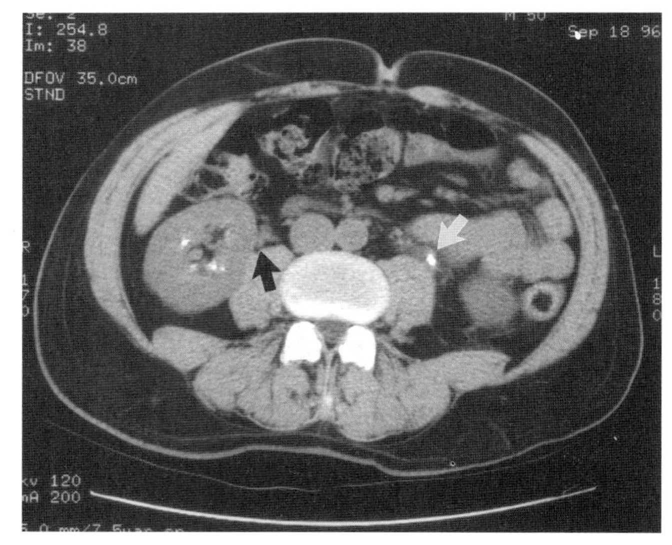

D. Axial CT scan, at a slightly lower level than C, shows the cause of obstruction, a ureteral stone *(white arrow).* Note the normal small caliber of the right ureter *(black arrow).* Multiple calculi are shown in the right kidney.

Figure 14.59. Noncontrast helical CT of obstructing left ureteral stones in a patient with multiple renal calculi.

A new technique for diagnosing an obstructing ureteral stone is noncontrast helical CT (Figure 14.59). No intravenous contrast material is required, and the procedure can be performed in a small fraction of the time needed for an intravenous urogram. Helical CT of the urinary tract can be completed with only 90 seconds of scanning, whereas an intravenous urogram may require a series of x-ray films taken over 30 to 60 minutes to diagnose and localize an obstructing stone. Patients suffering recurrent stone disease are delighted with the speed of CT and pleased not to have an intravenous injection of contrast material.

Another benefit of helical CT is that it can identify other causes of flank pain when no stone is present. When CT fails to show obstructive uropathy, the radiologist reviews the scan for other conditions that may mimic the pain of an obstructing ureteral stone, such as a leaking abdominal aortic aneurysm, diverticulitis, or pyelonephritis.

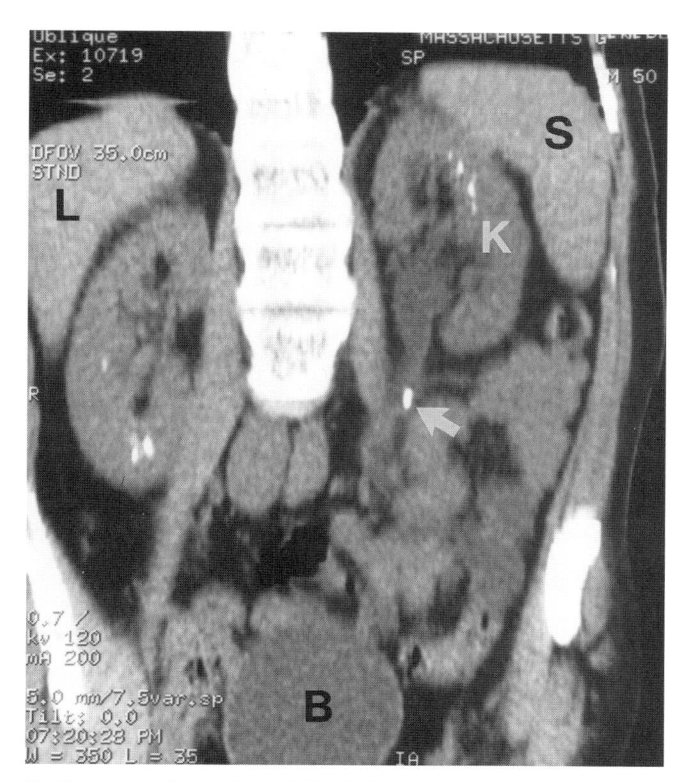

E. Coronal reformation of the helical axial scans shows a stone in the proximal ureter *(arrow).* Note the ureteral dilatation between the left kidney *(K)* and the stone. *L* indicates the liver, *S* the spleen, and *B* the bladder. Again multiple calculi are visible in both kidneys.

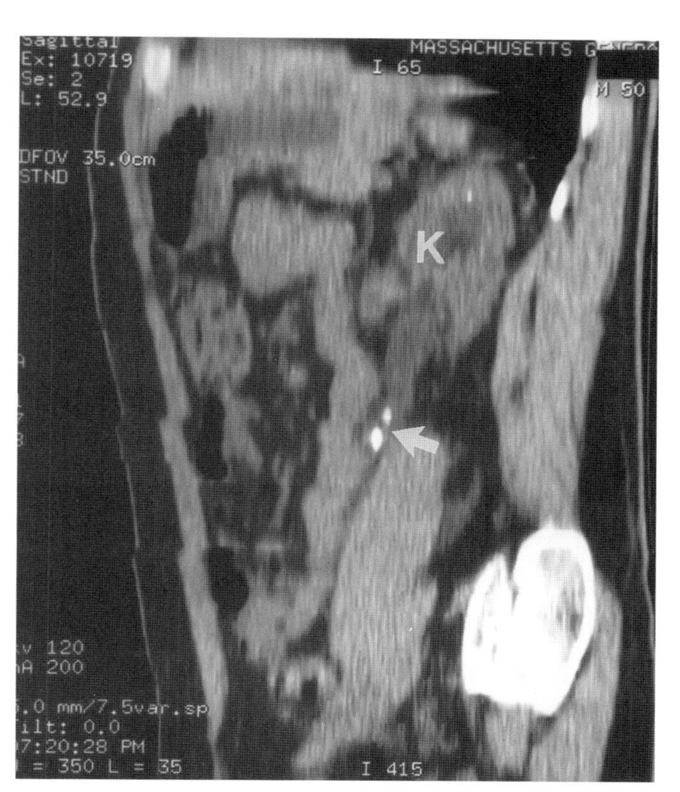

F. Sagittal reformation through the left kidney *(K)* shows two left ureteral stones *(arrow).*

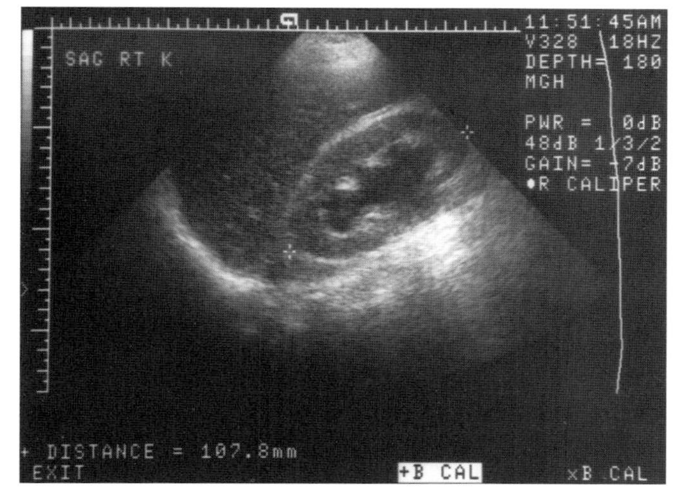

Figure 14.60A. Parasagittal ultrasound showing the liver and right kidney of a patient with right hydronephrosis. Compare the dark, echolucent, fluid-filled center of the kidney with the normal right kidney in Figure 14.60B.

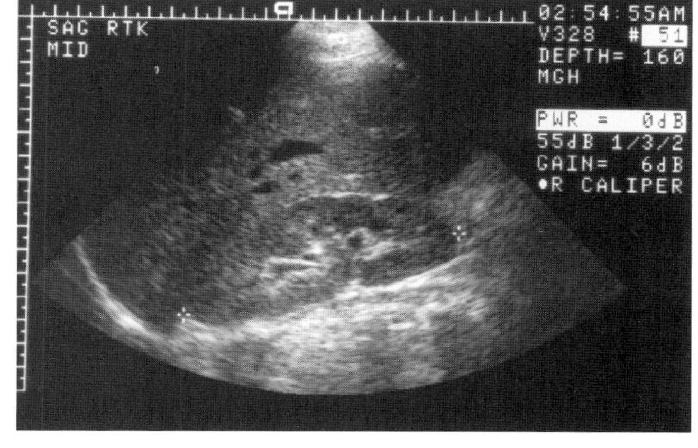

Figure 14.60B. Parasagittal ultrasound of a normal right kidney. Note the echogenic (white) kidney center; the echoes are produced by normal, central, renal sinus fat. Small echolucent areas represent blood vessels and normal portions of the renal collecting system.

An obstructed kidney can also be readily diagnosed by ultrasound if significant hydronephrosis has occurred. The fluid-filled dilated calyces of the hydronephrotic kidney produce an echo-free center for that kidney at ultrasound (Figure 14.60A), as compared with a normal kidney (Figure 14.60B), which shows an echogenic center without a dilated collecting system.

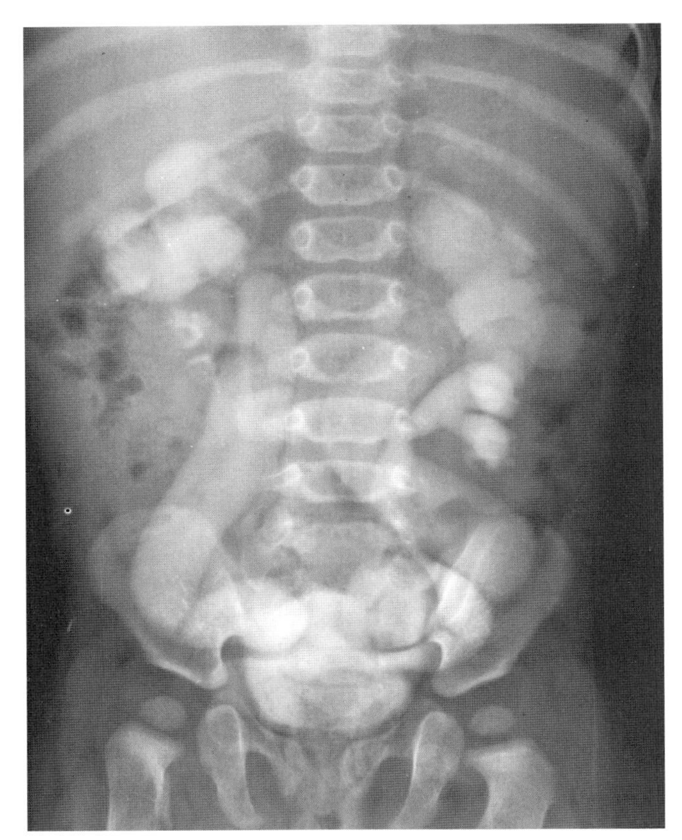

Figure 14.61. Advanced bilateral hydronephrosis in a child with congenitally defective drainage of the lower urinary tract.

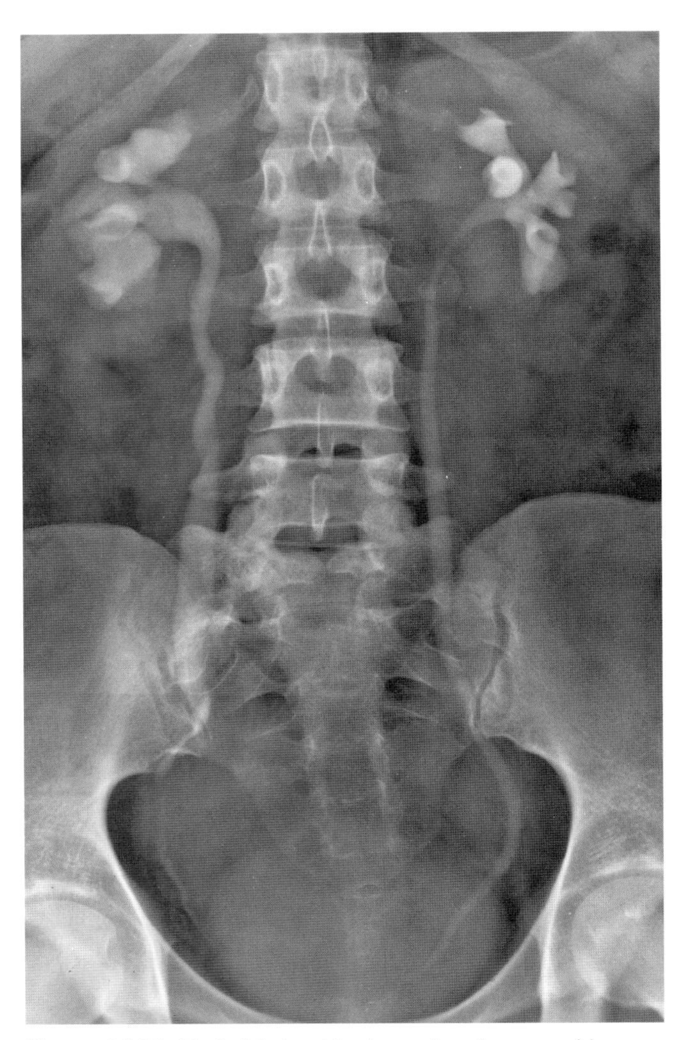

Figure 14.62. Early bilateral hydronephrosis caused by external compression from a pelvic mass.

It is surprising that any function remains in a chronically obstructed kidney; in advanced hydronephrosis the kidney has been converted into a thin-walled sac with only a slender rim of renal parenchyma remaining. Figure 14.61 is a late urogram film that shows bilateral obstruction in an infant with a congenital obstruction in both uretero-vesicle junctions. Note the marked dilatation of the ureters. A lesser degree of obstruction is evident in Figure 14.62.

In sum, then, *in studying plain films and intravenous urograms* you must carefully assess the size and shape of the two kidneys, the outline of parenchyma for each and its homogeneity, and the appearance of the collecting structures and their rate of filling and emptying. *Failure of the kidney to visualize* at urography should make you ask whether there is evidence for:

1. No blood getting into the kidney (renal artery occluded);
2. No blood getting out (renal vein thrombosis);
3. Blocked drainage (such as malignancy obstructing the ureter); or
4. Destruction of the nephron system.

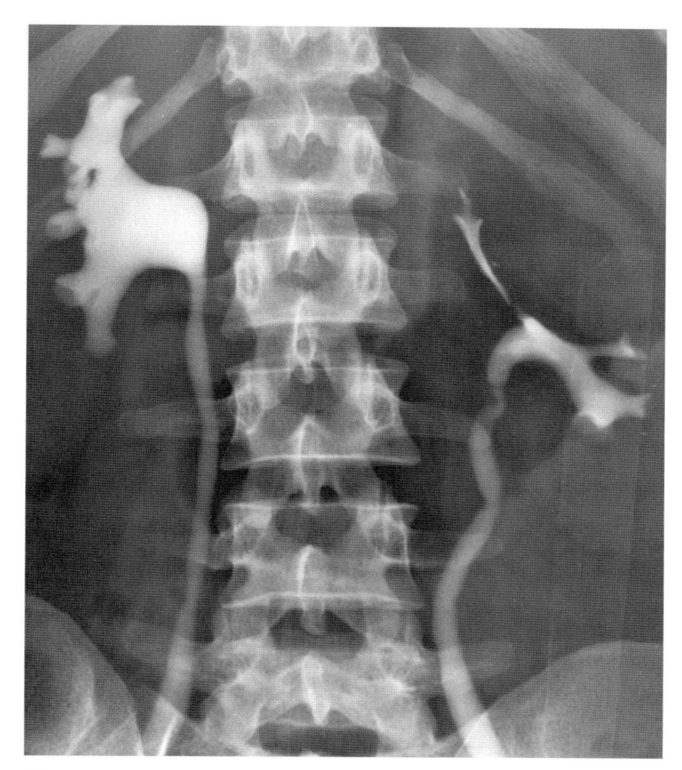

Figure 14.63. Retrograde pyelogram after injection catheters have been withdrawn. Note the stretching of the upper and middle calyces in the left kidney around a "mass," which proved at ultrasound to be a cyst.

Cystic Disease of the Kidneys

Because the kidneys secrete fluid, they are subject to the development of retention cysts of various kinds. Although these do not fill with contrast material because they are walled-off collections of fluid, their presence in the kidneys distorts the parenchyma and the draining structures on urography in a characteristic fashion (Figure 14.63). Even more precise is the recognition of cystic renal masses by either ultrasound (Figure 14.64) or CT (Figures 14.65 and 14.66). A

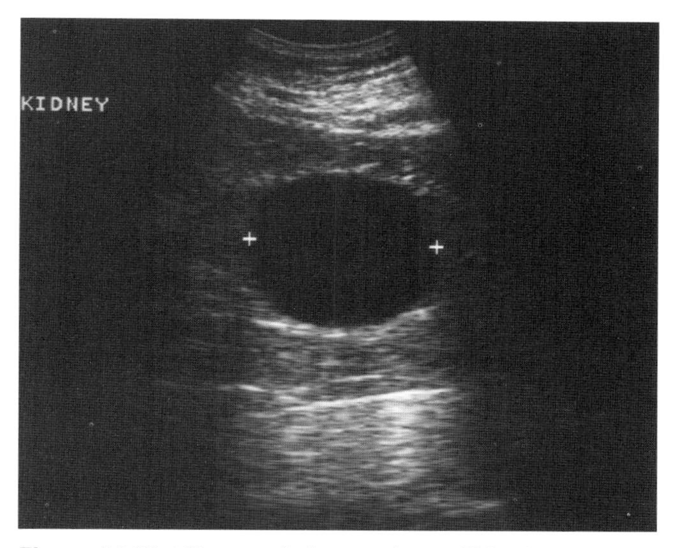

Figure 14.64. Ultrasound of a renal cyst. This simple cyst is anechoic, and has thin walls and *through transmission* (more echoes arise from behind the cyst than from behind portions of the kidney adjacent to the cyst, because the sound waves pass more efficiently through the cyst fluid than through adjacent renal parenchymal tissue).

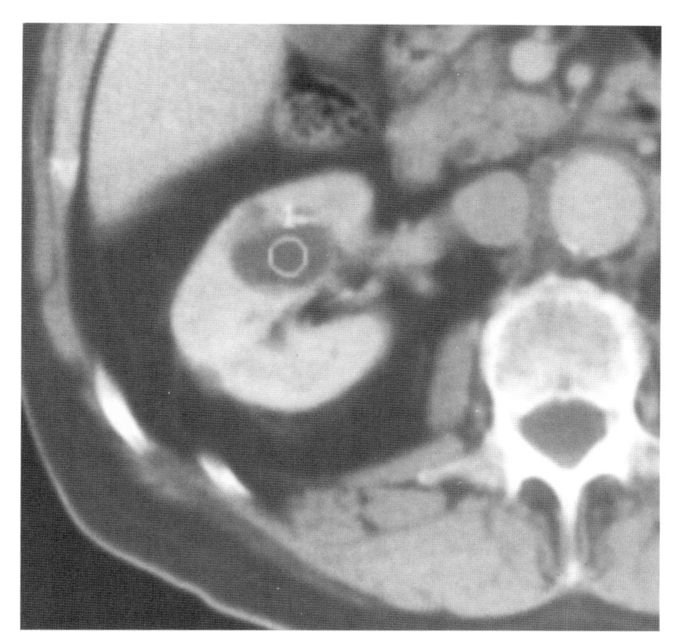

Figure 14.65. Detail of a CT scan of a right renal cyst. It appears as a round, sharply defined area of diminished CT attenuation (darker), as compared with the renal parenchyma. The white circle overlying the cyst represents a cursor positioned for computer sampling of CT density. This sample measured 3 Hounsfield units, a value which is similar to that of water and is typical of a benign cyst.

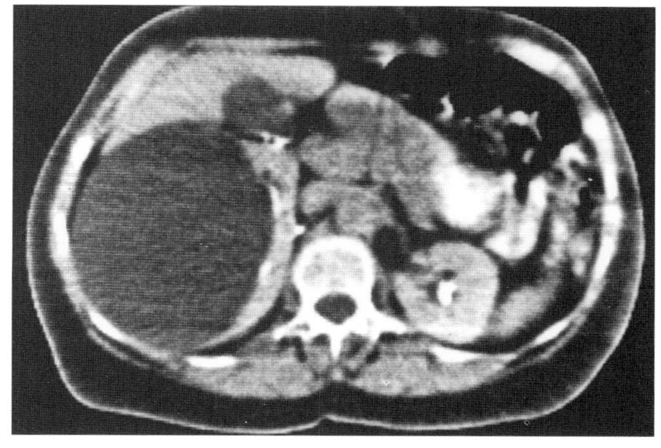

Figure 14.66. CT scan of a patient with an enormous right renal cyst. Note the splaying of the right renal parenchyma around the cyst.

renal mass discovered incidentally, by plain films or by urography, should be studied at once with ultrasound to determine whether it is cystic or solid, because if solid it may represent a malignant tumor.

In a patient with no hematuria, determination by ultrasound that a renal mass is a *simple cyst*—by demonstration of a spherical, echo-free fluid collection with a thin surrounding wall, and good through transmission of sound waves—is usually sufficient evidence to terminate the workup. But if ultrasound shows that the mass is a *complicated cyst*, indicated by the presence of echoes within the cyst and a thick wall, a CT scan with or without cyst puncture should be performed to rule out a cystic malignancy. Cyst puncture (Figure 14.67) can be carried out for confirmation; obtaining a clear aspirate with no abnormal cytology proves the diagnosis of a benign cyst. The patient with a cystic mass and hematuria must also have additional diagnostic workup with CT or cyst aspiration.

If the initial ultrasound exam shows that the incidentally discovered renal mass on plain films or intravenous urography is *solid*, the patient must be evaluated for a renal tumor. This workup will be discussed shortly.

Polycystic kidneys (Figure 14.68) have so classic and pathognomonic an appearance on both ultrasound and CT that those procedures are used for screening in families of patients with polycystic disease. Not only may cysts of the kidneys be seen, but also cysts of the liver and pancreas.

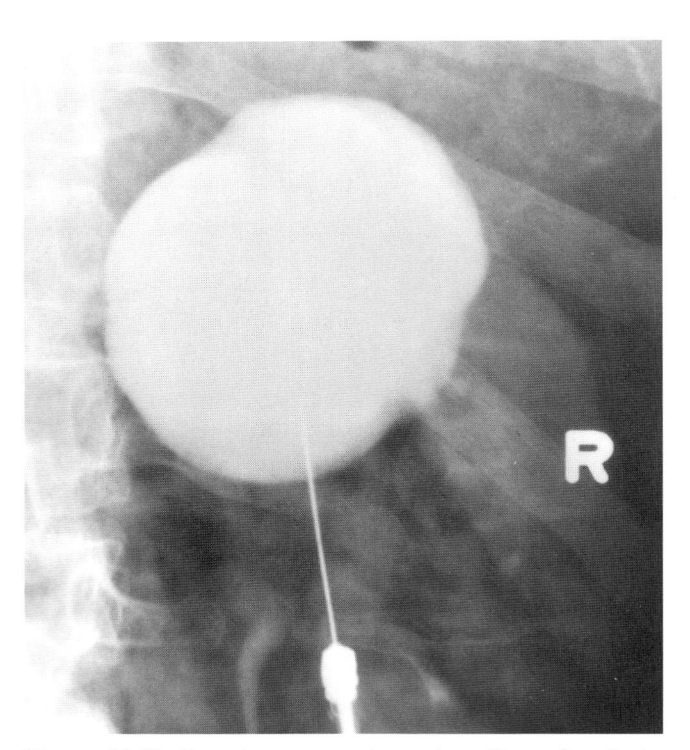

Figure 14.67. Percutaneous cyst puncture. To confirm that an ultrasound-diagnosed cyst was indeed benign in a patient with microscopic hematuria, a percutaneous needle aspiration was performed. A spot film (the patient is lying prone, and therefore the film is correctly marked with an *R*) shows the cyst cavity opacified with water-soluble contrast materials. Analysis of the aspirated cyst fluid showed no evidence of blood or malignant cells. Note that cyst puncture may be performed under fluoroscopic, ultrasound, or CT guidance.

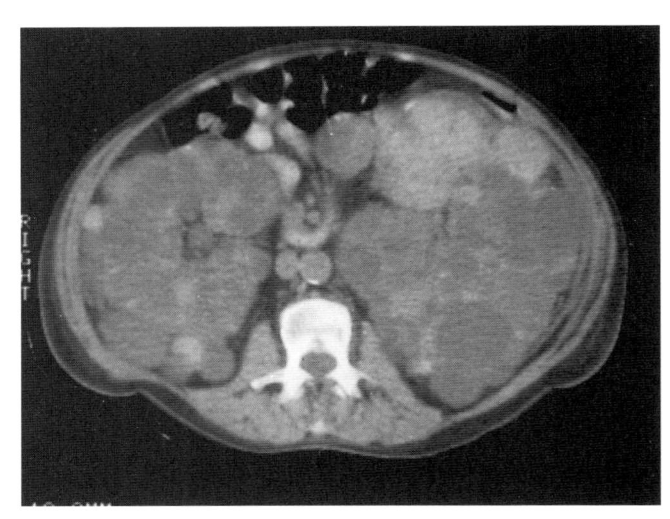

Figure 14.68. CT scan of a patient with polycystic kidneys.

Urinary Tract Infection

You should not routinely request intravenous urography on adult patients with urinary tract infections. Of course, the indications for urography in all patients with flank pain suggesting stone and those with hematuria are perfectly clear, but in the adult patient with only urinary tract infection the examination is often futile. Children with urinary tract infections, however, need diagnostic imaging at the time of initial workup because they may have a congenitally obstructing lesion or vesicoureteral reflux.

Even with *acute pyelonephritis,* caused by ascending infections with Gram-negative organisms such as *E. coli,* urography may not show any abnormality at all. But a urogram may show diminished renal function, swelling of the kidney, and loss of the renal outline due to inflammation in the perirenal fat (Figure 14.69). CT with intravenous contrast material can usually demonstrate subtle changes in the renal parenchyma, including patchy segments of decreased density and a striated nephrogram, although CT is not necessary to make the diagnosis. But both CT and ultrasound can be helpful in showing any possible complications, such as renal or perirenal abscess formation.

Most kidneys chronically infected from childhood do show important morphological changes, demonstrable either by urography with tomograms or by CT. The renal alterations consist of a decrease in size of the affected kidney with localized thinning of the parenchyma at sites where calyceal blunting is associated with overlying cortical scarring (Figure 14.70). This condition, called chronic pyelonephritis, is usually related to vesicoureteral reflux of infected urine. When you examine a plain film or urogram, always estimate the size and shape of the kidney and trace its normal smooth outline.

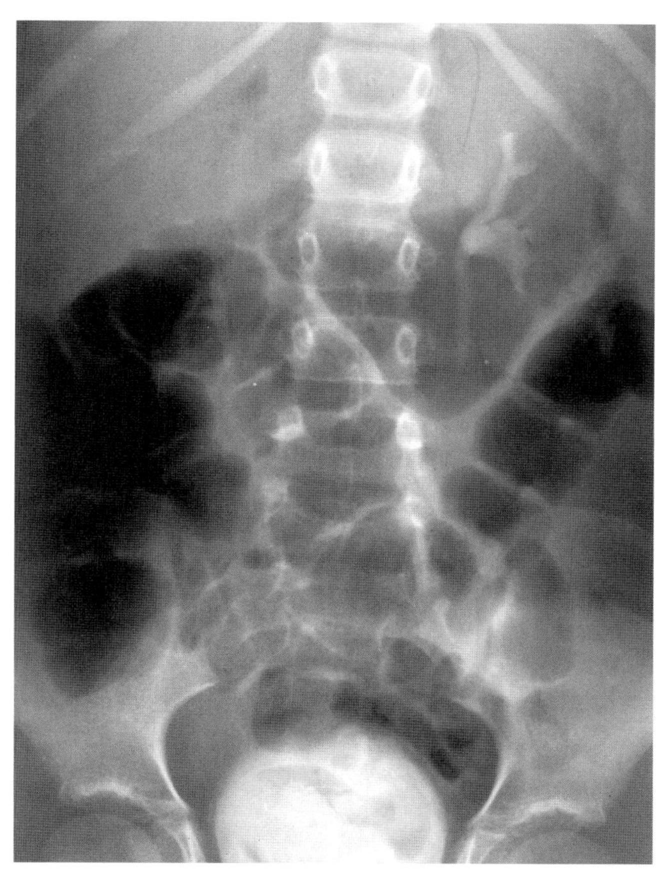

Figure 14.69. Intravenous urogram of *acute* right-sided pyelonephritis in a child. Little excreted contrast material appears within the right intrarenal collecting system, in contrast to the amount seen in the left kidney. The right kidney is swollen, resulting in poor visualization of the right renal outline and psoas margin.

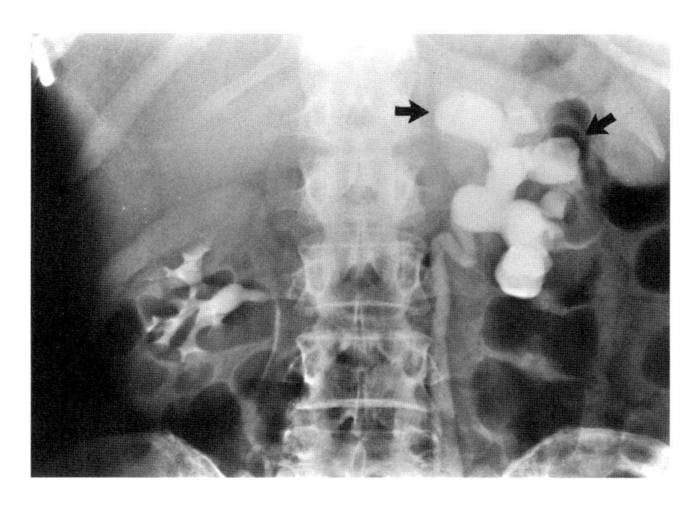

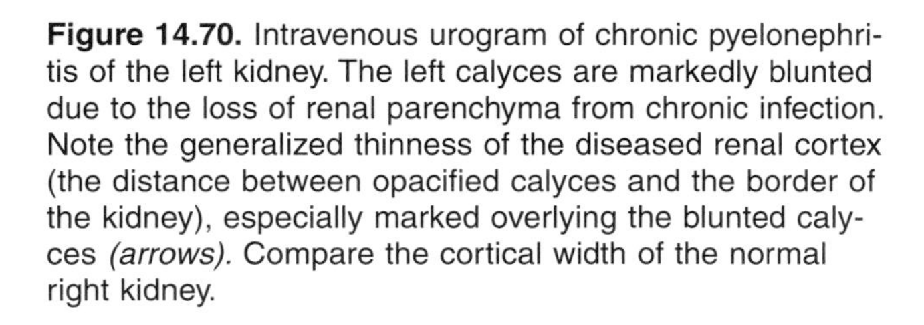

Figure 14.70. Intravenous urogram of chronic pyelonephritis of the left kidney. The left calyces are markedly blunted due to the loss of renal parenchyma from chronic infection. Note the generalized thinness of the diseased renal cortex (the distance between opacified calyces and the border of the kidney), especially marked overlying the blunted calyces *(arrows).* Compare the cortical width of the normal right kidney.

Renal Tumors

When a renal mass is discovered by clinical examination, plain films, or intravenous urography, and ultrasound indicates that the mass is solid, the patient should be referred for computed tomography. CT with intravenous contrast material can characterize tumors in more detail than ultrasound. It can better delineate the extent of the tumor, and show the degree of vascularity, the presence or absence of a necrotic center, and the presence or absence of local invasion of adjacent structures such as the renal vein and inferior vena cava. If the mass shown at urography is large and irregular (which suggests it is not a benign cyst) or shows invasion of the intrarenal collecting system, or if hematuria is present, the patient may be referred directly to CT, bypassing ultrasound.

In masses that seem cystic at ultrasound but have thick walls (complicated cysts), you must suspect a cystic tumor or a tumor that is partially necrotic or has central hemorrhage. CT may show conclusive evidence of a tumor; if not, refer the patient for percutaneous needle aspiration to provide a specific tissue diagnosis. Tumors are also known to arise occasionally in the wall of a benign cyst, and when they do, they may cause hemorrhage and pain. At ultrasound you may see a cystic structure with a thick wall and internal echoes (blood mixed with cystic fluid); at CT you may see a thick-walled cyst, perhaps with a localized tumor mass in one portion of the wall, and blood-attenuation material within the cyst cavity. Percutaneous aspiration biopsy can confirm the diagnosis.

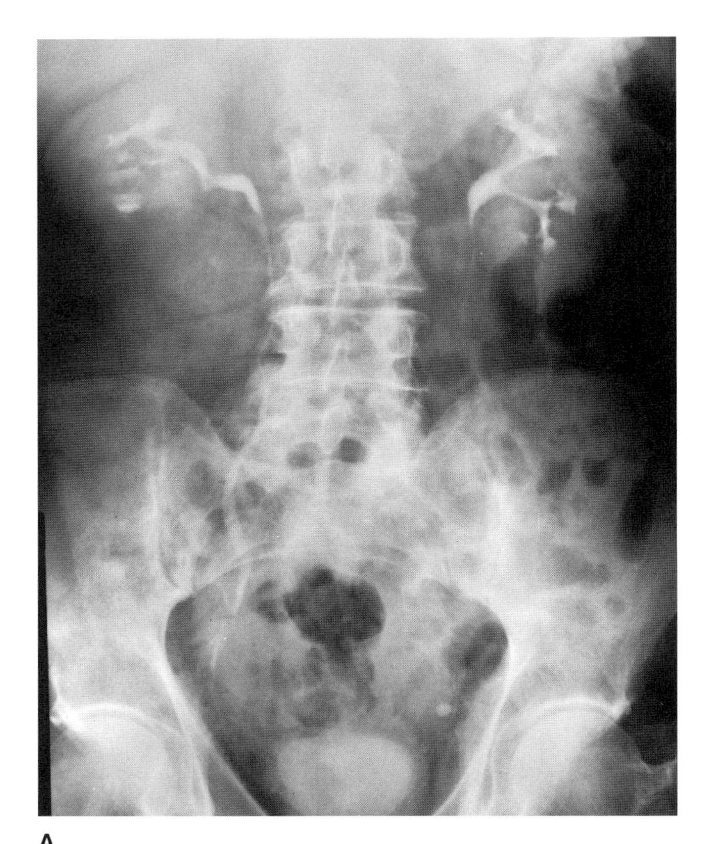

A

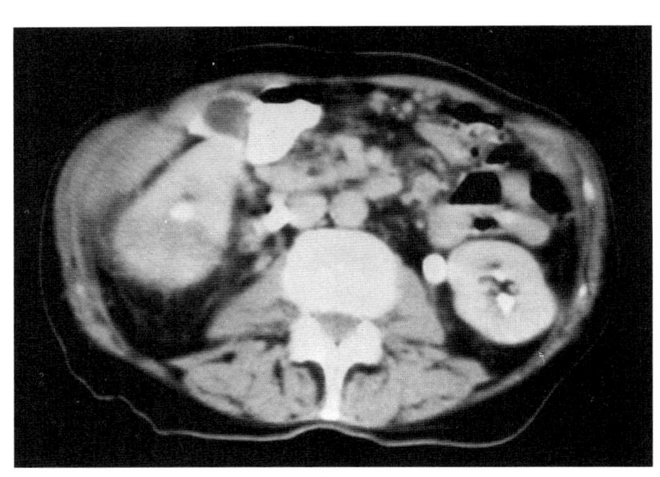

B

C

Figure 14.71. Renal cell carcinoma. A: Intravenous urogram shows a lower pole right renal mass that displaces the proximal right ureter medially. B: CT scan through the midkidneys reveals a low-attenuation mass posteriorly in the right kidney. C: Slightly lower-level CT scan shows a large right renal tumor with a low-density necrotic center.

The patient illustrated in Figure 14.71 is a 62-year-old man with right flank pain and microscopic hematuria. His urogram (A) shows a large right lower pole renal mass that produces a bulge in the medial kidney margin, splaying the collecting structures away from it, and displaces the ureter medially. Because of the hematuria, the patient was referred directly to CT and a large lower pole solid tumor was identified (B and C). It has a low CT attenuation and the irregularly margined center typical of central tumor necrosis; this finding is common with renal cell carcinoma (hypernephroma), which this mass proved to be.

Renal arteriography or venography frequently is indicated to determine whether the tumor has invaded the renal veins and the inferior vena cava. A case illustrating invasion of the latter will be presented later, on pp. 494–495. In addition, angiography can show the vascularity of the tumor itself. The arterial and venous phases of the selective right renal arteriogram of another patient with renal cell carcinoma are shown in Figure 14.72. They depict a markedly hypervascular tumor that does not invade the right renal vein.

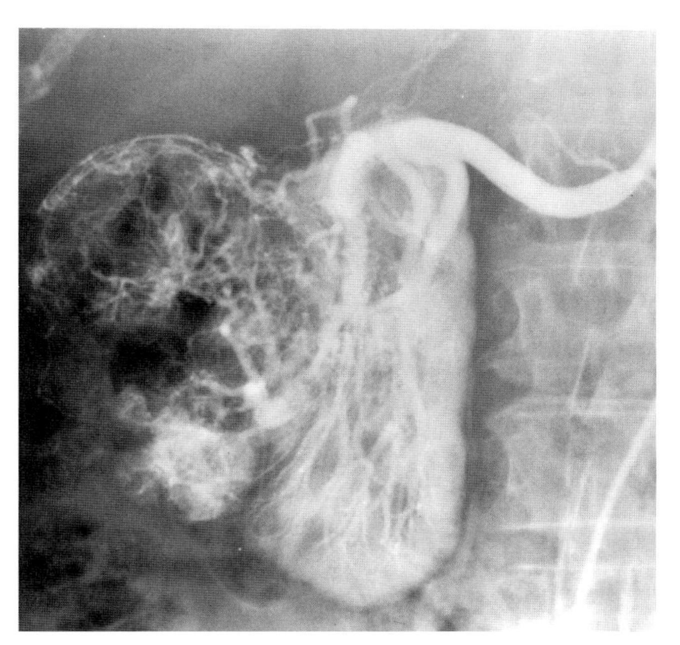

A

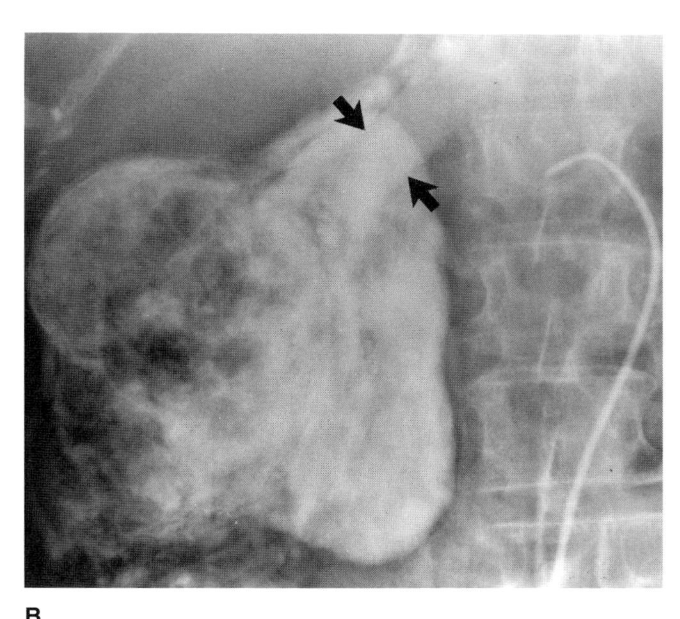

B

Figure 14.72. Arteriography of another patient with renal cell carcinoma.
A: Arterial phase of a right renal arteriogram shows a large vascular tumor.
B: Venous phase shows no evidence of tumor invasion of the right renal vein *(arrows)*.

Intravenous Contrast Materials

Iodinated radiographic contrast materials injected intravascularly for urography, CT, and angiography may occasionally produce undesirable reactions, and you should be aware that intravenous administration of contrast agents may result in such episodes while the patient is being examined.

Most commonly these reactions occur with the older, *high-osmolar (ionic)* contrast agents. Reactions may be quite mild: consisting of nausea, vomiting, hives, sweating, anxiety, or bronchospasm. These reactions occur in less than 5 percent of all such contrast examinations. They may or may not require treatment by the radiologist (such as injection of an antihistamine for hives), and usually do not require premature termination of the radiological exam.

Reactions to these older contrast agents may also, very rarely, be severe, consisting of laryngeal edema, hypotension, bradycardia, shock, seizures, and anaphylactoid reactions. Fatalities (generally due to cardiovascular collapse in anaphylactoid reactions) are reported to occur once in 40,000 to once in 140,000 cases. (The incidence of all types of reaction is similar to the incidence of reaction to penicillin.) For this reason no such studies should be carried out without the ready availability of an emergency cardiopulmonary support system and the presence of a physician (usually the radiologist) who knows how to treat contrast reactions. Nearly all are treatable.

Today newer types of contrast agents are available for intravascular use that can diminish the likelihood of all types of contrast reactions by at least 50 percent. These are the *low-osmolar (nonionic)* agents, sometimes called LOCA. Death is so uncommon with LOCA that the incidence has not been yet determined. Unfortunately, these agents are exceedingly expensive, costing many times more than ionic agents. But LOCA should always be used in patients who have a history of a prior contrast reaction. Their use in your practice should be individually determined for each patient.

Renal Trauma

Renal trauma may be suspected in patients with hematuria or flank pain occurring after any kind of injury. The trauma may be slight or catastrophic, and varying degrees of trauma between these extremes may or may not require surgical repair. Consequently, it is important not only to make a diagnosis of renal trauma, but to stage the degree of renal injury. This can be best accomplished with CT (Figures 14.73 and 14.74).

In trauma patients an abdominal CT scan is superior to an intravenous urogram; CT can recognize a smaller degree of injury than can urography, and CT can evaluate the remainder of the abdominal organs in the same examination. When trauma has occurred, CT can identify renal contusions, small cortical lacerations, and subcapsular hematomas, which do not require surgery, and differentiate them from larger lacerations with contrast extravasation, shattered kidneys, and renal artery injuries, which usually do require emergency surgical intervention. In addition, CT is faster than intravenous urography, requires less patient manipulation, and uses less radiation. CT's ability to image adjacent abdominal structures is especially valuable because patients with kidney trauma frequently have sustained injuries to other organs. Right renal injuries are often associated with liver injuries and left renal injuries with spleen trauma. Injuries to either kidney may be associated with lumbar spine injuries, often fractures of the transverse processes.

Emergency CT is available in nearly all major trauma centers. If, however, you have to evaluate a patient with suspected renal trauma where CT is unavailable, intravenous urography may help by showing findings indicative of acute renal trauma. As you might expect in patients with traumatic renal artery injury and occlusion (renal pedicle injury), the kidney on the affected side does not opacify with contrast material on the urogram. The urogram may not show any signs of a renal contusion or a small cortical laceration. But when a renal cortical laceration has extended into the collecting system, the urogram may show extravasation of contrast medium around the kidney, within the surrounding perirenal space, and in the laceration itself. With perirenal hemorrhage, the renal outline becomes less apparent and the psoas margin on that side may disappear.

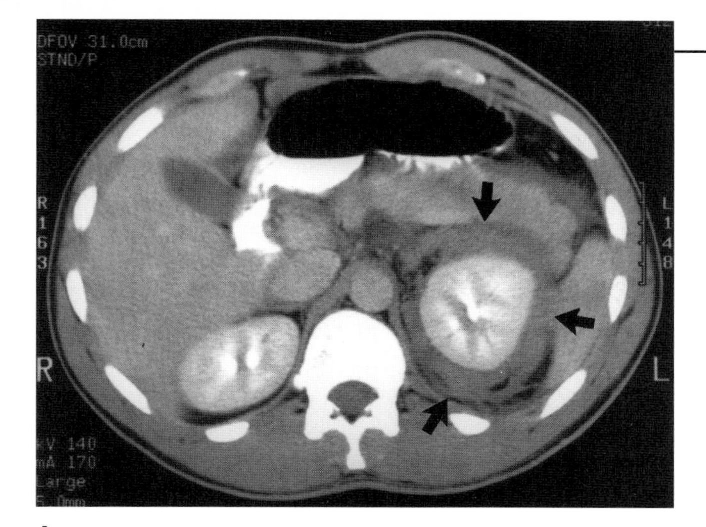

A

B

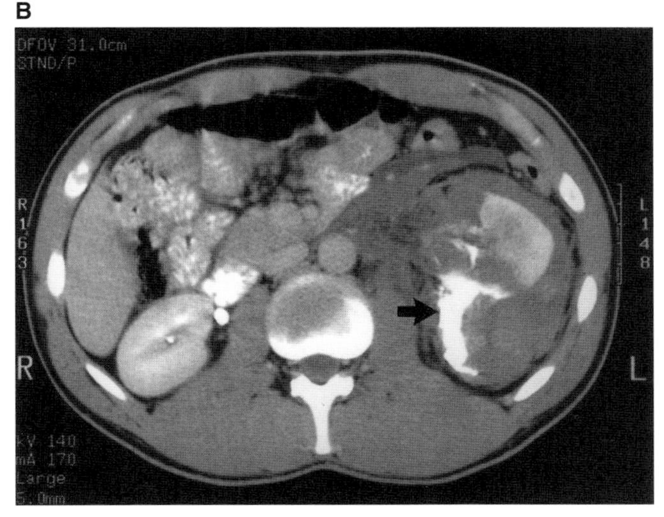

C

Figure 14.73. CT scan of a fractured left kidney. A: Slice at the level of the upper kidney poles shows no laceration, but reveals blood in the left perirenal space *(arrows)*, surrounding the left kidney. B: Slightly lower-level slice shows lacerations of the left renal parenchyma and more perirenal hemorrhage. C: Even lower-level image shows a fractured portion of the left renal parenchyma laterally and extravasation of intravenous contrast material *(arrow)* from the extension of the fracture line into the renal collecting system.

Injuries of the ureters and bladder may be shown, by either computed tomography or intravenous urography, as extravasation of opacified urine at sites of ureter or bladder laceration. Ureteral lacerations extravasate within the retroperitoneum; bladder lacerations (which, of course, result in bladder rupture) may extravasate within the peritoneal cavity *(intraperitoneal bladder rupture)* or extraperitoneal soft tissues *(extraperitoneal bladder rupture)*.

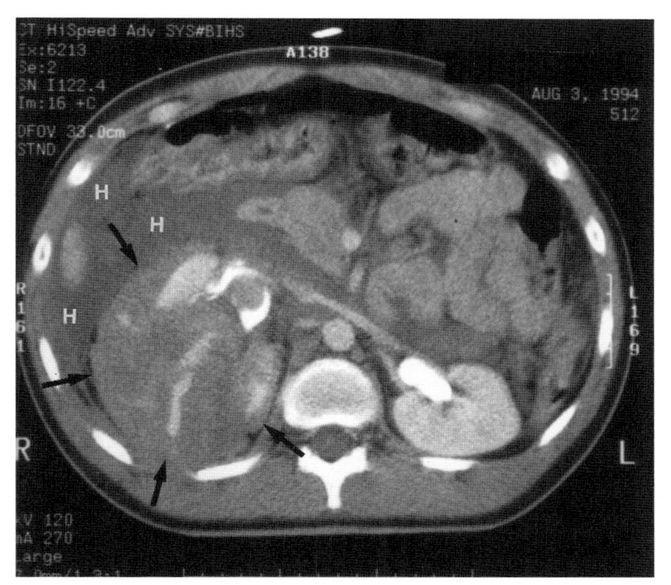

A

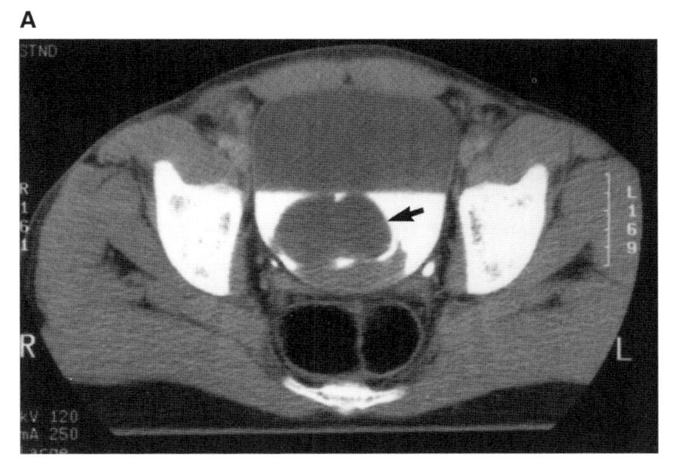

B

Figure 14.74. CT scan of a shattered right kidney with hemoperitoneum from an associated liver laceration. A: Slice through the midkidneys shows a shattered right kidney with contrast extravasation and a large collection of blood in the perirenal space *(arrows)*. Renal trauma does not usually produce hemoperitoneum (*H*'s); but when hemoperitoneum is seen with renal trauma, an additional injury of an intraperitoneal organ must be suspected. B: Slice through the pelvis shows blood clots in the bladder *(arrow)* in this patient with gross hematuria.

The Urinary Bladder

The urinary bladder is routinely opacified and filmed at intravenous urography, and you have seen several examples already. Conditions, such as bladder cancer and bladder stones, that produce filling defects in the bladder lumen can thus be readily detected at intravenous urography. In addition, masses adjacent to the bladder that displace it can also be identified. Figure 14.75 shows a portion of the intravenous urogram of an elderly man with benign prostatic hypertrophy; the bladder floor is elevated by the enlarged prostate gland.

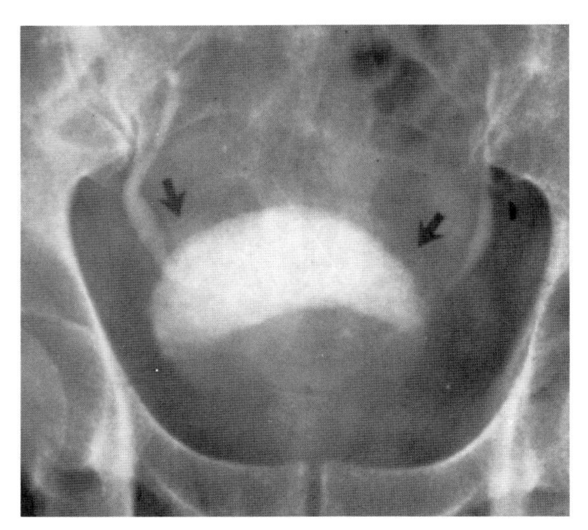

Figure 14.75. Coned-down view of the bladder in a patient with prostatic hypertrophy. The mass elevating the floor of the bladder represents an enlarged prostate. The *arrows* indicate the thickened bladder wall, resulting from chronic bladder outlet obstruction.

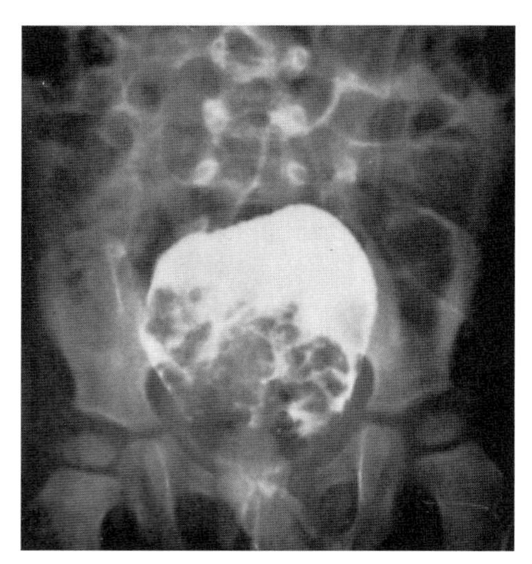

Figure 14.76. Retrograde cystogram showing multiple radiolucencies within the bladder, outlined by contrast material. These represent a nodular sarcoma of the bladder in this pediatric patient.

Improved opacification of the bladder can be obtained by direct injection of contrast material into the bladder through a bladder catheter (a Foley catheter, for example) inserted via the urethra. This procedure is called a *retrograde cystogram*. Figure 14.76 shows an example in a patient with a large nodular bladder cancer. Retrograde cystography may also be indicated when bladder opacification is required in a patient with renal failure whose kidneys are unable to concentrate contrast material, or in a patient whose bladder does not opacify because of bilateral ureteral obstruction.

Bladder injuries may be shown by either intravenous urography, CT, or retrograde cystography. But none of these examinations demonstrates injuries of the male urethra, which require examination by a retrograde urethrogram. Urethral injuries should be suspected in trauma patients with fractures of the anterior pelvis or a bloody urethra discharge. Retrograde urethrography is performed with a small-caliber injection catheter under fluoroscopic control. In the unstable trauma patient it may need to be done portably in the emergency department. This examination will be discussed in Chapter 16. If the urethra is patent, a retrograde cystogram can be obtained. Figure 14.77A shows a cystogram of a multiple-trauma patient with a bladder rupture. Note the teardrop shape of the bladder, caused by compression of the lateral bladder walls by bilateral pelvic hematomas. You can see extravasation of urine and contrast material into the pelvic soft tissues. A CT examination (Figures 14.77B through E) was performed after cystography for evaluation of the abdominal organs. No other injuries were found, although the CT scan confirmed the presence of intraperitoneal rupture, showing extravasated contrast material and urine throughout the peritoneal cavity. Differentiating between intraperitoneal and extraperitoneal bladder rupture is important in planning patient management because intraperitoneal bladder rupture always requires surgical intervention and extraperitoneal rupture usually does not.

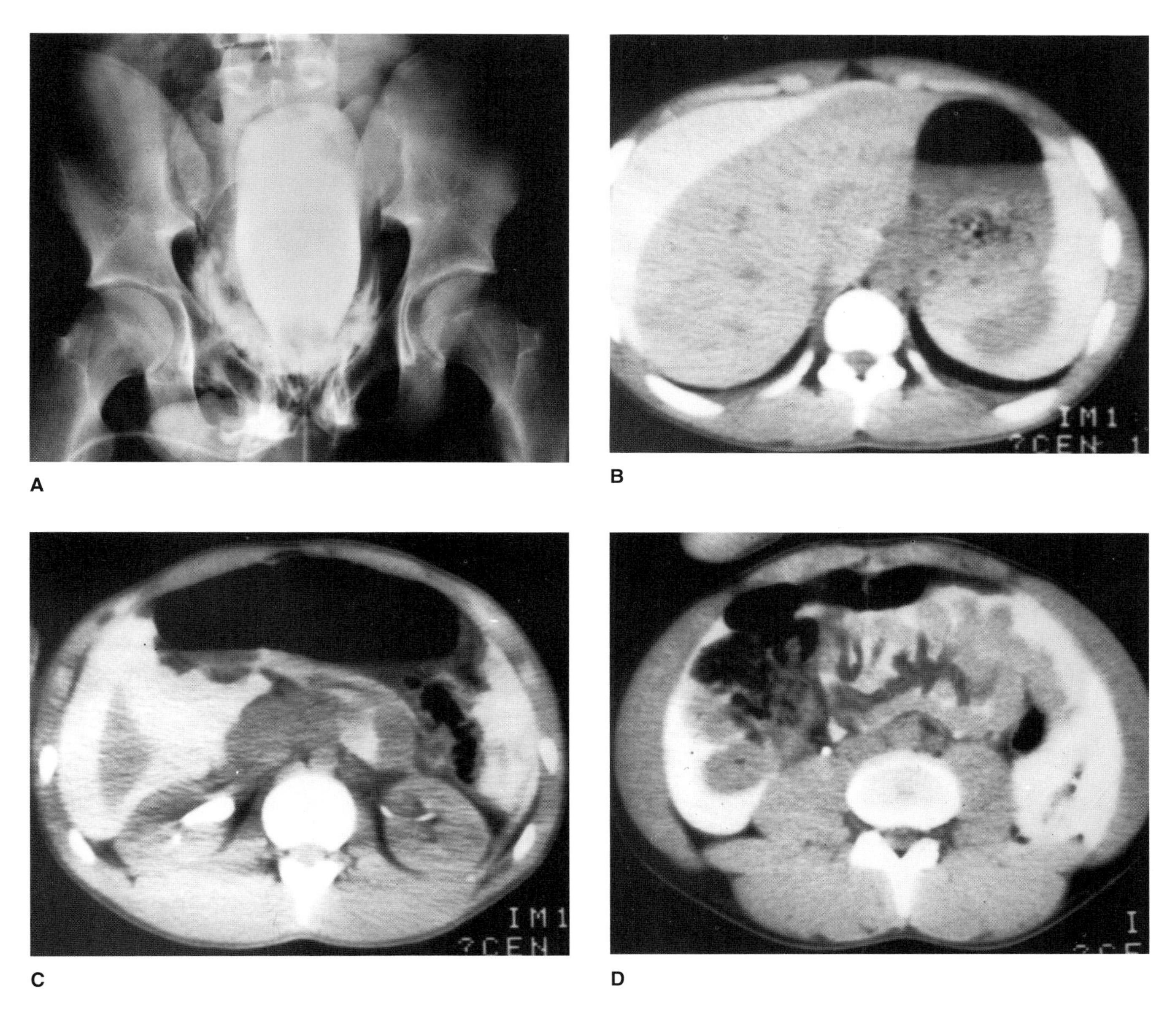

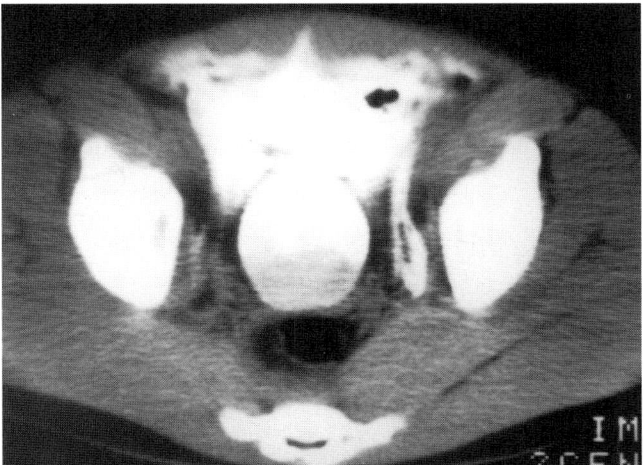

E

Figure 14.77. Multiple-trauma patient with a bladder rupture. A: Cystogram film shows a teardrop-shaped bladder and extravasation of contrast medium. B: CT scan of the upper abdomen after administration of intravenous contrast medium shows opacified fluid (urine opacified with contrast medium leaking into the peritoneal space via the bladder rupture) in the peritoneal space, surrounding the liver, stomach, and spleen. C: CT scan through the midkidneys shows opacified fluid surrounding the tip of the liver and in the left paracolic gutter. D: Lower-level CT scan shows opacified urine in both paracolic gutters. E: CT scan through the pelvis shows contrast medium extravasation into the anterior pelvic soft tissues and peritoneal cavity.

You should be aware of a very important fact about the imaging of suspected bladder rupture in trauma patients. If the bladder is not full and distended on the urogram or CT scan, there may not be sufficient intravesicular pressure to extravasate opacified urine through a small laceration to demonstrate the injury. Therefore if the radiologist sees that the bladder is not full on the urogram or CT, the patient will be given a retrograde injection of contrast material, as was the patient in Figure 14.77, and re-imaged. In the trauma patient without suspected urethral injury, the contrast material is administered through a Foley catheter that terminates in the bladder itself. Such re-imaging by conventional films is called a retrograde cystogram; when CT is used it is called a CT cystogram.

Various imaging techniques are useful in diagnosing bladder cancer and lesions. Intravenous urography may not reveal a small bladder cancer because it shows only the lumen of the bladder, not the thickness of the bladder wall. Cystoscopy, however, shows the bladder mucosa well, and whenever a small bladder cancer is suspected but not seen at urography, cystoscopy is usually performed. Ultrasound, CT, and magnetic-resonance imaging have been helpful in the diagnosis of bladder lesions that affect the bladder wall and surrounding tissues more than the lumen. For example, in the staging of bladder cancer, CT can show the extent of involvement of the bladder wall and the presence or absence of invasion of the surrounding pelvic structures.

The Adrenal Glands

CT, MRI, and ultrasound have significantly advanced the imaging diagnosis of adrenal disease. Before the advent of these techniques, patients were subjected to lengthy and uncomfortable angiographic procedures to determine whether adrenal masses or enlargement existed. Although ultrasound can readily detect large adrenal masses, patients with suspected adrenal disease are usually referred for a CT examination because its finer anatomic resolution can identify even small adrenal abnormalities. In fact, CT is the procedure of choice and appropriate first imaging examination to perform in patients with suspected adrenal disease. In recent years MR scanning has become popular for adrenal imaging because it provides better soft-tissue discrimination than CT.

At CT the normal left adrenal gland is triangular in shape, whereas the right is crescent shaped. You have seen normal examples already in Chapter 3. The indications for adrenal CT include paroxysmal hypertension suggestive of pheochromocytoma, clinical findings of Cushing's disease or other adrenal cortex hypersecretion syndrome, and a new flank or abdominal mass that could represent an adrenal carcinoma or neuroblastoma. Finally, the adrenal glands are routinely examined with CT in patients undergoing staging of

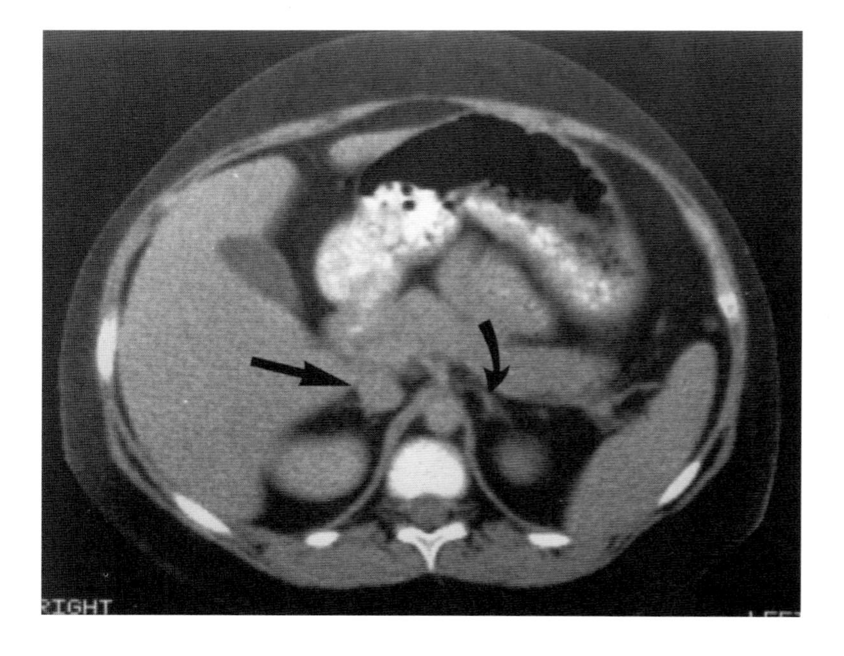

Figure 14.78. Enlargement of the right adrenal gland *(straight arrow)* due to the presence of an adrenocortical tumor, an aldosteronoma. The *curved arrow* points to the left adrenal gland, which is normal in size.

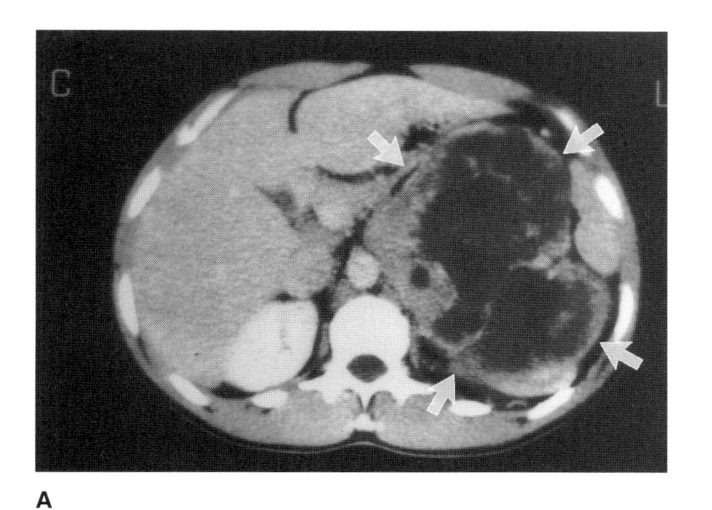

A

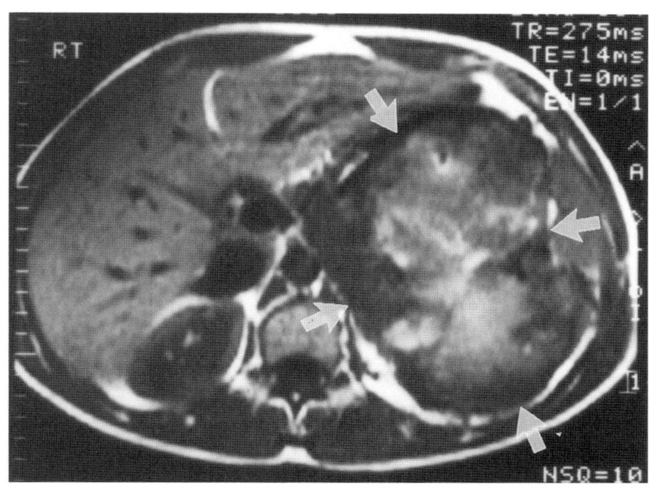

B

Figure 14.79. CT and MR scans of a patient with a large left adrenal pheochromocytoma *(arrows).* A: CT scan. B: MR scan.

Figure 14.80. CT scan of an enlarged right adrenal gland *(arrows),* caused by a metastatic deposit from a bronchogenic carcinoma of the lung.

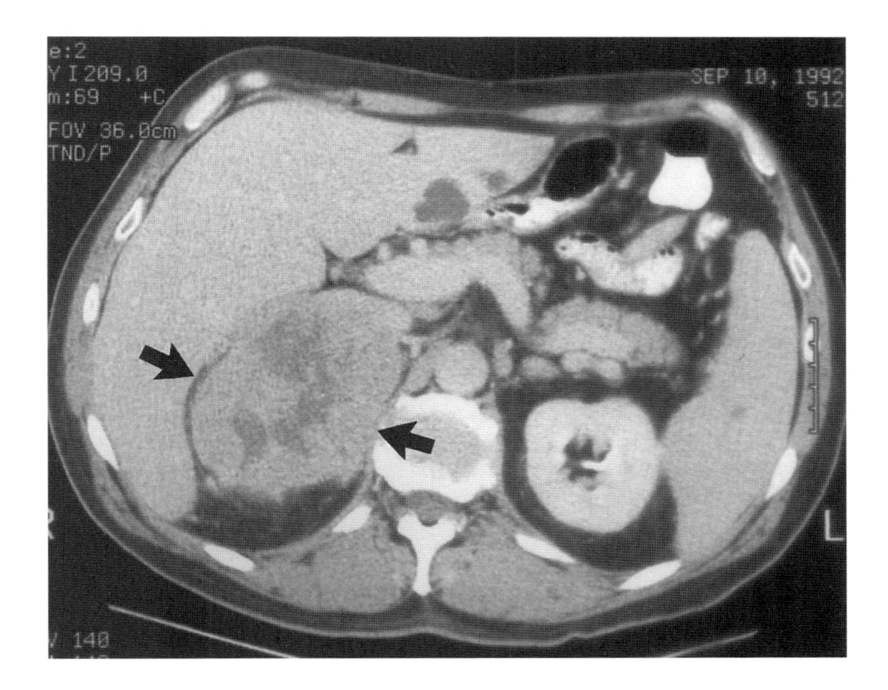

lung and breast cancers, which frequently metastasize to the adrenal glands. In fact, adrenal metastases are exceedingly common; in autopsy series they are found in 27 percent of patients with malignant disease. Other sources of adrenal metastases include melanomas and primary tumors of the gastrointestinal tract or kidneys. It may be impossible to differentiate benign nonfunctioning adenomas of the adrenal glands from actual metastases at CT and MR; both may show only an enlarged or nodular gland. Percutaneous biopsy may be very helpful in this situation.

Bilateral enlargement of the adrenal glands is seen at CT with adrenal hyperplasia. Unilateral enlargement usually signifies a benign or malignant adrenal mass, such as a pheochromocytoma or cortical adenoma. Metastatic disease, however, may cause enlargement of one or both adrenal glands. Consequently, it is imperative to correlate the CT examination with the clinical findings. The patient in Figure 14.78 had an aldosteronoma, the one in Figure 14.79 a large pheochromocytoma, and the patient in Figure 14.80 metastases from a bronchogenic carcinoma.

Problems

Unknown 14.1 (Figure 14.81)

This middle-aged man noted increasing prominence of his abdomen after recovering from a laparotomy one year ago. Study the two slices from his CT scan, which was performed with oral contrast material but no intravenous contrast material. Can you find the cause? You may wish to compare these images with the normal abdominal CT scans in Chapter 3.

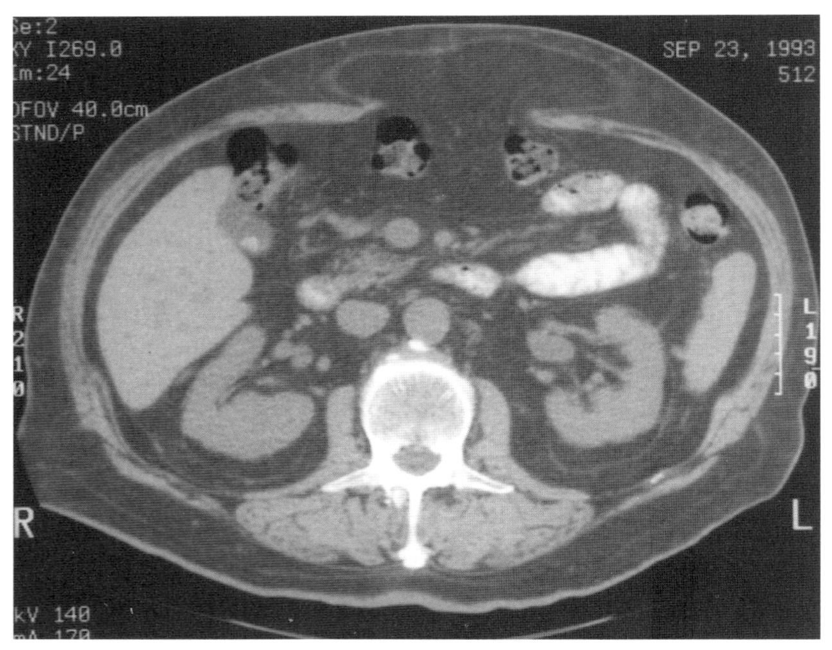

A

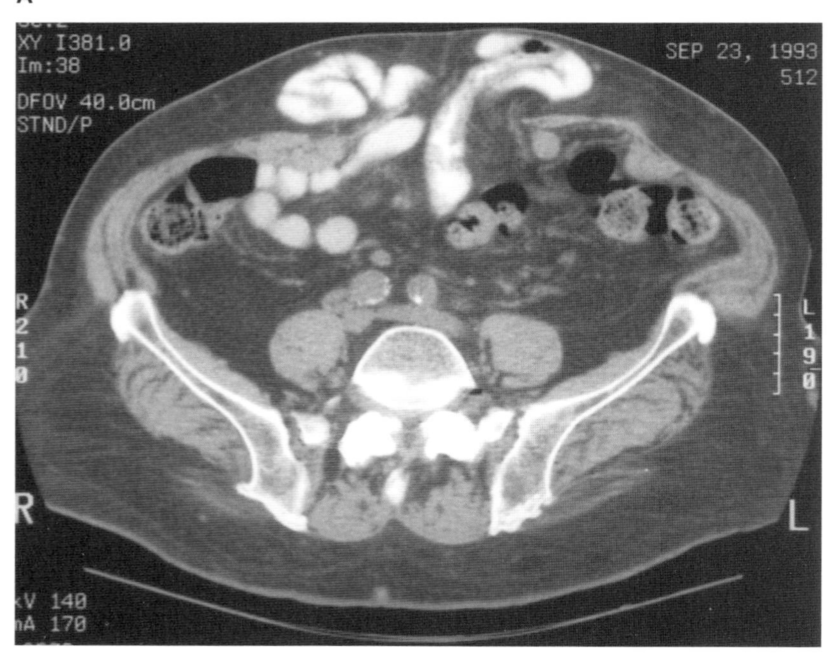

B

Figure 14.81 ***(Unknown 14.1)***

Unknown 14.2 (Figure 14.82)

Analyze this CT scan, which was obtained with intravenous contrast material in a middle-aged man with chronic alcoholism now complaining of progressive right abdominal pain. Compare the liver and spleen. What are your observations and conclusions?

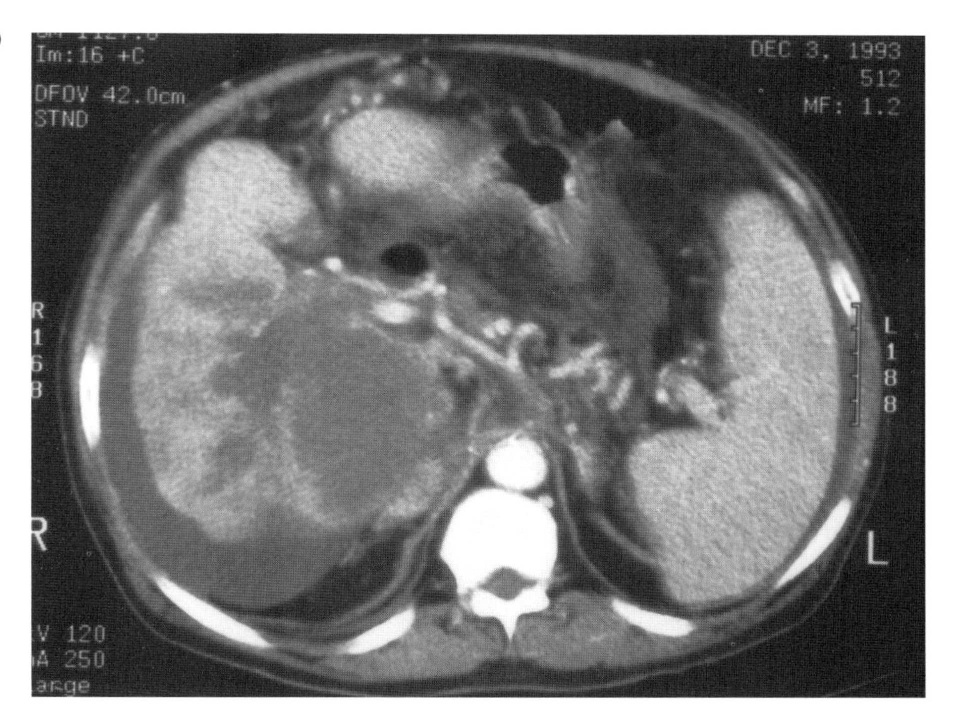

Figure 14.82 ***(Unknown 14.2)***

Unknown 14.3 (Figure 14.83)

This young man suffered multiple trauma in a motor vehicle accident. He is hypotensive. What do you see?

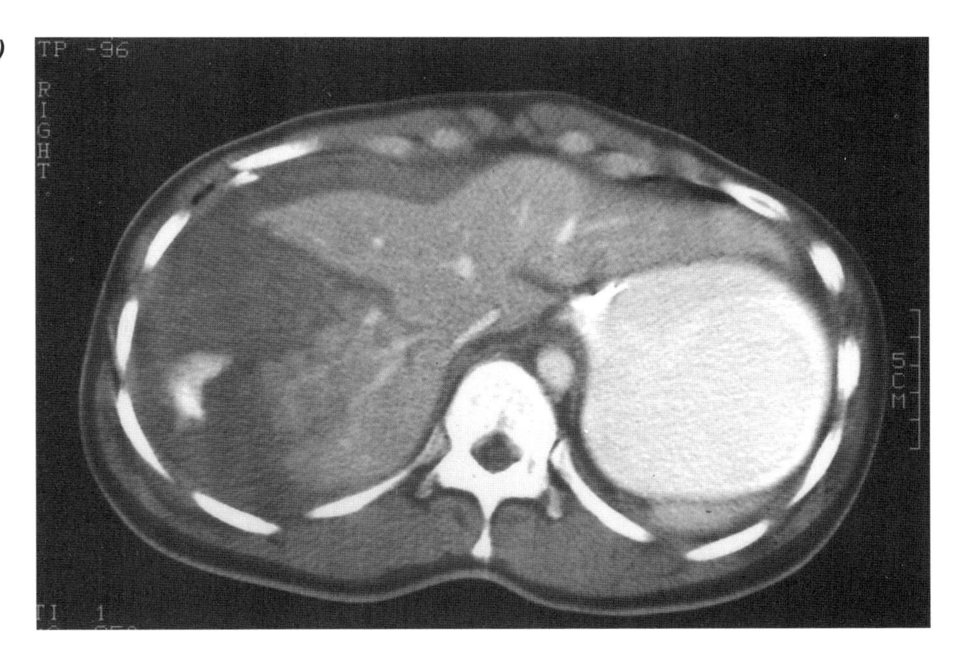

Figure 14.83 ***(Unknown 14.3)***

15 The Musculoskeletal System

We will begin our study of the musculoskeletal system with the bones, which constitute the major component of musculoskeletal imaging, and at appropriate points we will address muscle and other soft-tissue imaging.

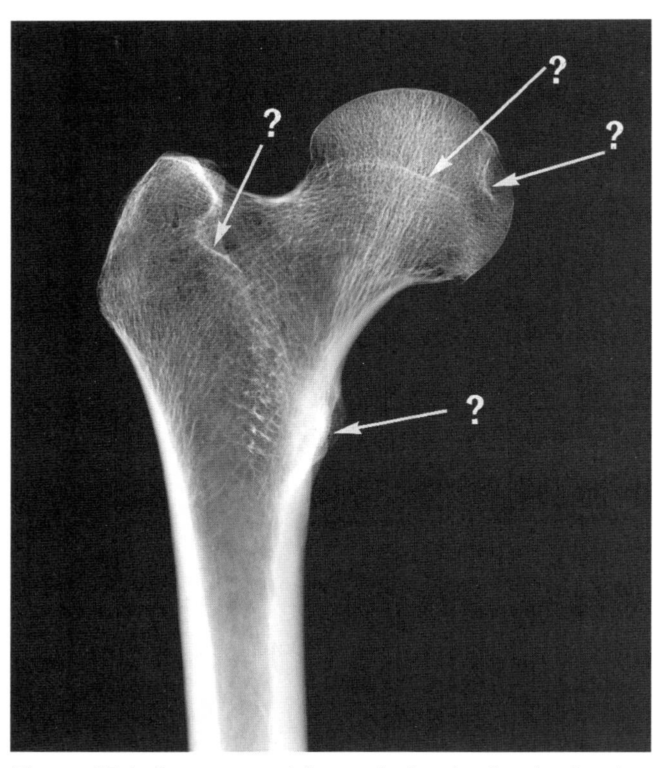

Figure 15.1. Can you explain precisely why the details of the radiographic shadows marked with arrows should have been produced by the anatomic structures they represent? (See text.)

How to Study Radiographs of Bones

Bones are much more interesting than you may have realized during your first year in medical school. The skeleton probably did not seem fascinating during the tedium of memorizing the origins and insertions of muscles, and perhaps the remembrance of that tiresome and difficult task clouds the subject of bone for you, even at a time when great advances are being made in the diagnosis of bone disease.

But you cannot afford to neglect the bones—their function and change, their growth and mature microscopic structure in health and disease. You must be able to imagine what is going on in the bony skeleton of the immobilized patient with a healing fracture of the pelvis, not just at the site of the fracture but throughout the body as a result of enforced inactivity. You must be able to predict the degree to which invasion of your patients' bones by metastatic tumor will alter their blood chemistry. And you must understand the metabolic reasons for the formation of kidney stones in hyperparathyroid patients in whom there is a marked increase in the rate of bone breakdown and a significant increase in serum calcium and calcium excretion.

You must not think of the bony skeleton as a static and principally structural fabric. It is constantly changing throughout life, and since the functional lifetime of any microscopic plate of spongy bone or any osteone (Haversian) segment of compact bone is only about 7 years, there is continuous bone breakdown and replacement. The rate of turnover varies somewhat by age. Bone production is more rapid than bone breakdown in children because their skeletons are increasing in size constantly. In the elderly the rate of bone replacement lags and the bones gradually become thinner and more fragile.

Although these changes are all appreciable to the trained eye and intellect by means of a variety of imaging procedures, the degree to which they may be noted on radiographs is an important part of your training in radiology and medicine. Begin by examining the two radiographs on this page spread.

You will have identified the fovea capitis in Figure 15.1, but did you account for the white streak that you see in the depth of that hollow in the femoral head as the cortical bone, there caught tangentially by the ray? Similarly, the white line crossing the neck of the femur just where the epiphyseal line used to be is the tangentially viewed plate of somewhat more densely crowded trabecular plates that developed there as the growth plate fused when growth was complete. Another *arrow* indicates the tangentially viewed ridge of bone of the intertrochanteric line on the posterior surface of the femur, and it is being projected superimposed on all the spongy bone and compact cortical bone anterior to it. You must think of radiographs of bones as summation shadowgrams, just as you do when looking at chest films. Finally, the fourth *arrow* points to the lesser trochanter, the surface of which is compact bone seen tangentially. Thus you can carry forward to your detailed examination of bone films what you learned in the first chapter about the curving sheets of the rose petal.

The legs of the mummy in Figure 15.2 were broken postmortem by an ancient Egyptian funeral director to make the body fit into a burial case the embalmer had on hand. The lower femurs have been removed, and the arms also are missing.

You can tell the mummy is that of a child because there are as-yet-unfused epiphyses present at the femoral heads and proximal ends of the tibias. The secondary centers of ossification at the ends of long bones appear on radiographs when mineralized bone forms at the center of the cartilaginous anlage of those epiphyses. The growth and development of the epiphyses and the ultimate fusion to their growth plates have been radiographically documented; both first appearance and fusion occur at predictable ages, and thus the age of unidentified skeletons can easily be determined.

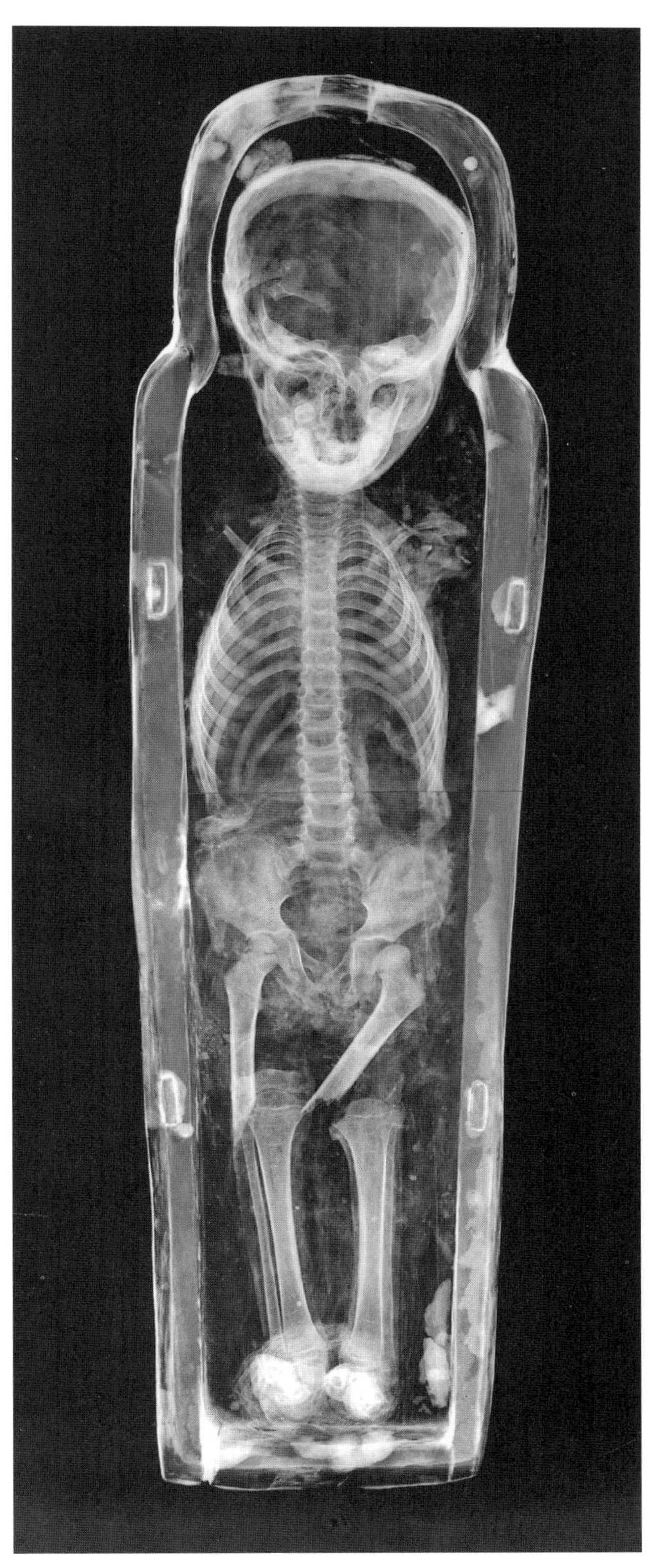

Figure 15.2. Radiograph of the mummy of a child. What can you determine about the remains? How do you know it is a child? How could experts determine the child's age at death? (See text.)

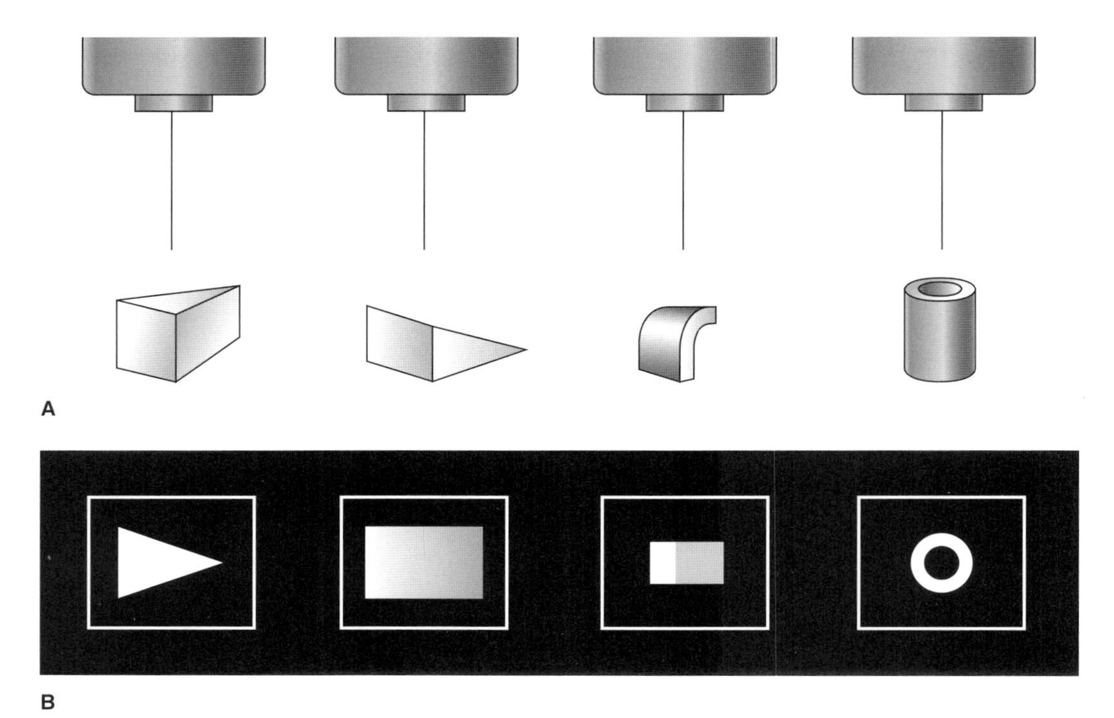

Figure 15.3. (See text.)

Examine the several hypothetical examples of details of bony structure (A) and their approximate radiographic images (B) in Figure 15.3. By a simple rule of summation the wedge casts a different shadow according to the direction in which the ray traverses it. The curved sheet of bone obeys the same principles as the curved rose petal and leaf in Chapter 1. The cylinder seen end-on becomes a dense circle in the radiograph, and if it were radiographed from the side it would produce two parallel lines of tangentially projected cortex, just as you see in the lower part of Figure 15.4C. Realize too that in Figure 15.4C at the level of the lesser trochanter you are seeing the x-ray shadows of the tangentially viewed lateral and medial compact bone on either side and, between them, two layers of compact bone viewed en face—the anterior and posterior cortex, superimposed on the network of spongy bone inside.

You will have noticed on the films you have just seen of the upper femur that the cortical bone is very thick in the shaft or diaphysis, and that it thins out around the upper femur, where the bone is filled with trabeculae of spongy bone. These arch in parallel struts, which closely follow the lines of stress developed within this part of the bone in humans, who stand erect and carry the weight of their trunks upon the two femoral heads. The arrangement of these struts to bear weight differs slightly from that of similar spongy bone in quadrupeds, and it is interesting to reflect how those trabeculae must have adapted themselves when our distant ancestors first stood to walk on their hind legs.

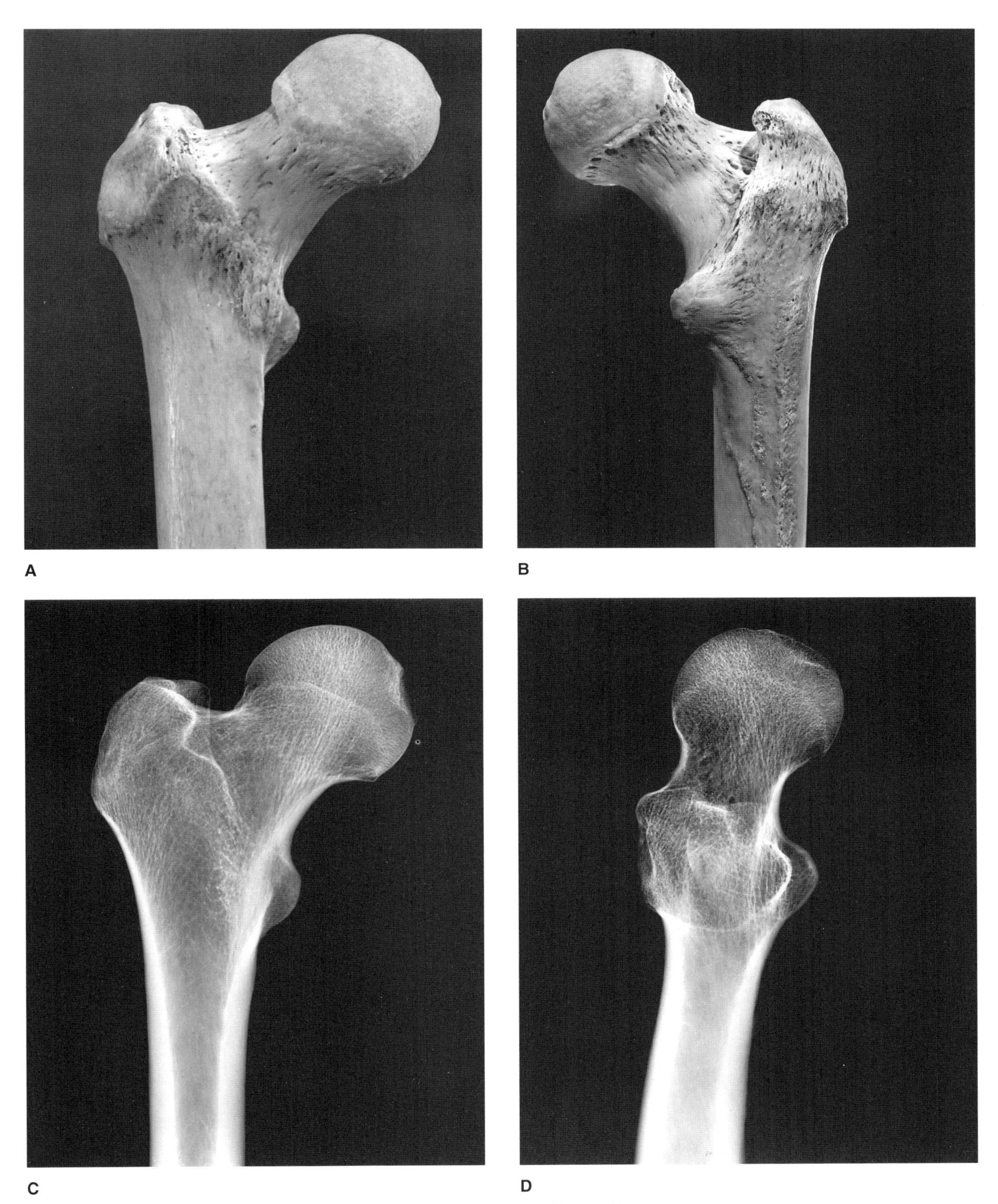

Figure 15.4. Anterior and posterior photographs of the upper femur (A and B) to help you account for details in the AP and lateral radiographs in C and D.

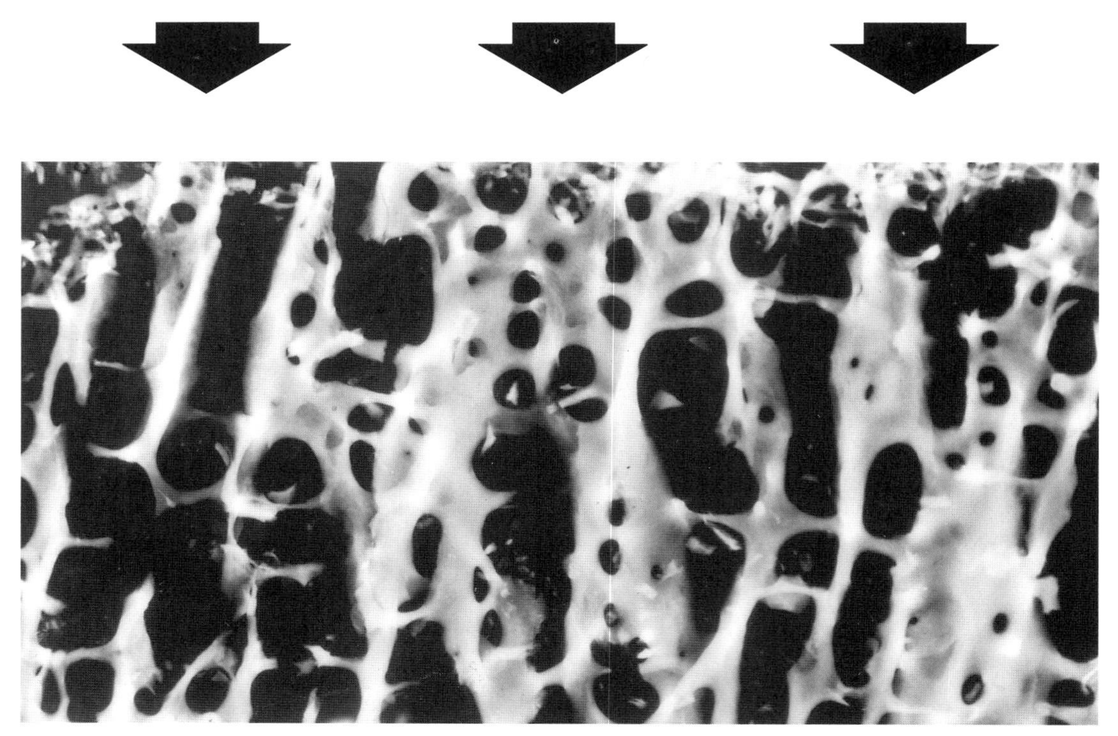

Figure 15.5. Photomicrograph showing trabecular arrangement developed with weight bearing *(arrows).*

Figure 15.5 is a photomicrograph of the trabeculae in a coronal 3-millimeter slice of bone from the upper tibia showing adaptation to weight bearing. Note that the vertical struts are heavier and thicker, and that they are joined by much lighter and thinner ones called secondary trabeculae, which reinforce them structurally. In areas having spongy bone that is not so directly weight bearing, such as the central parts of the vertebral bodies or the scapula, for example, the size and thickness of the trabeculae are more uniform, the marrow spaces between being enclosed in a bony sponge that does not show any thicker struts. Apply these observations to the lateral radiograph of the foot in Figure 15.6, and identify the stress vectors in the talus and calcaneus.

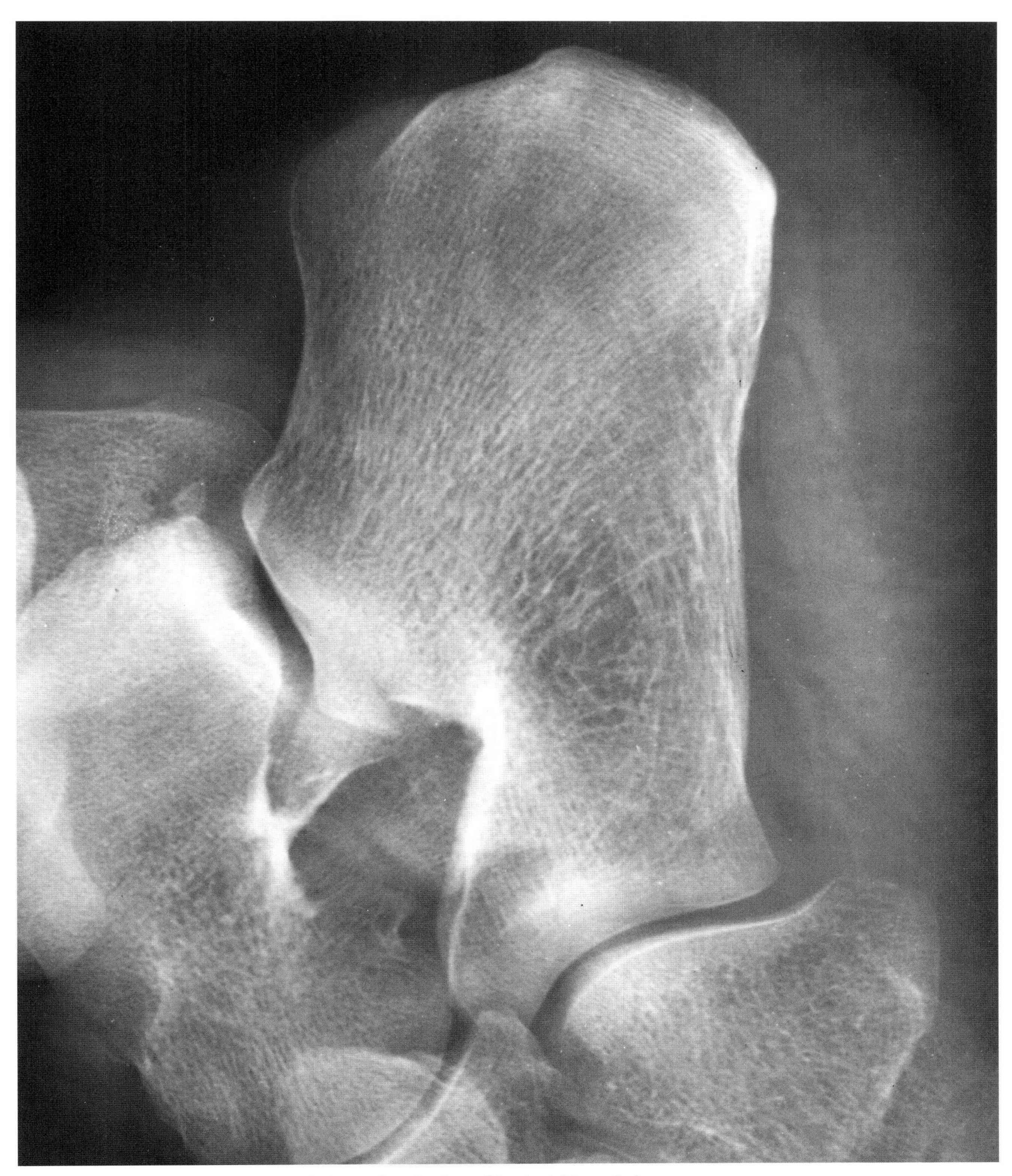

Figure 15.6. Magnification x-ray study. Lateral projection of the foot (the sole is parallel to the side margin of the page). Note the trabeculae, which arch backward and downward from the calcaneotalar joint toward the weight-bearing point under the heel.

Requesting Films of Bones

Figure 15.7. Mother Whistler has her foot x-rayed.

The various projections in which films of the bones are made are largely a matter of convention based upon which ones produce the most information. You will easily become accustomed to the AP and lateral views, which are routine—two views made at 90 degrees to one another being essential in almost all bone filming. Additional views have been designed to show to best advantage lesions in particular portions of the bony skeleton. Some of these special views are referred to by the structure being examined (such as the *scaphoid view* of the wrist), others by the name of the individual who first described the view and recommended it (*Neer's view* of the shoulder, for example).

You may be relieved to know that you do not have to request particular views; you only have to indicate the *bony area of interest to be filmed* on the requisition, followed by the word *films* (for instance, cervical spine films, right knee films, or left thumb films). Each radiology department has determined which views are to be routinely taken by its radiology technologists for each anatomic area. You should also indicate the location of any pain and describe any relevant clinical findings on the requisition form. Thus you would have requested "right foot films" for Mother Whistler, since the pain was located in the midfoot, where you would

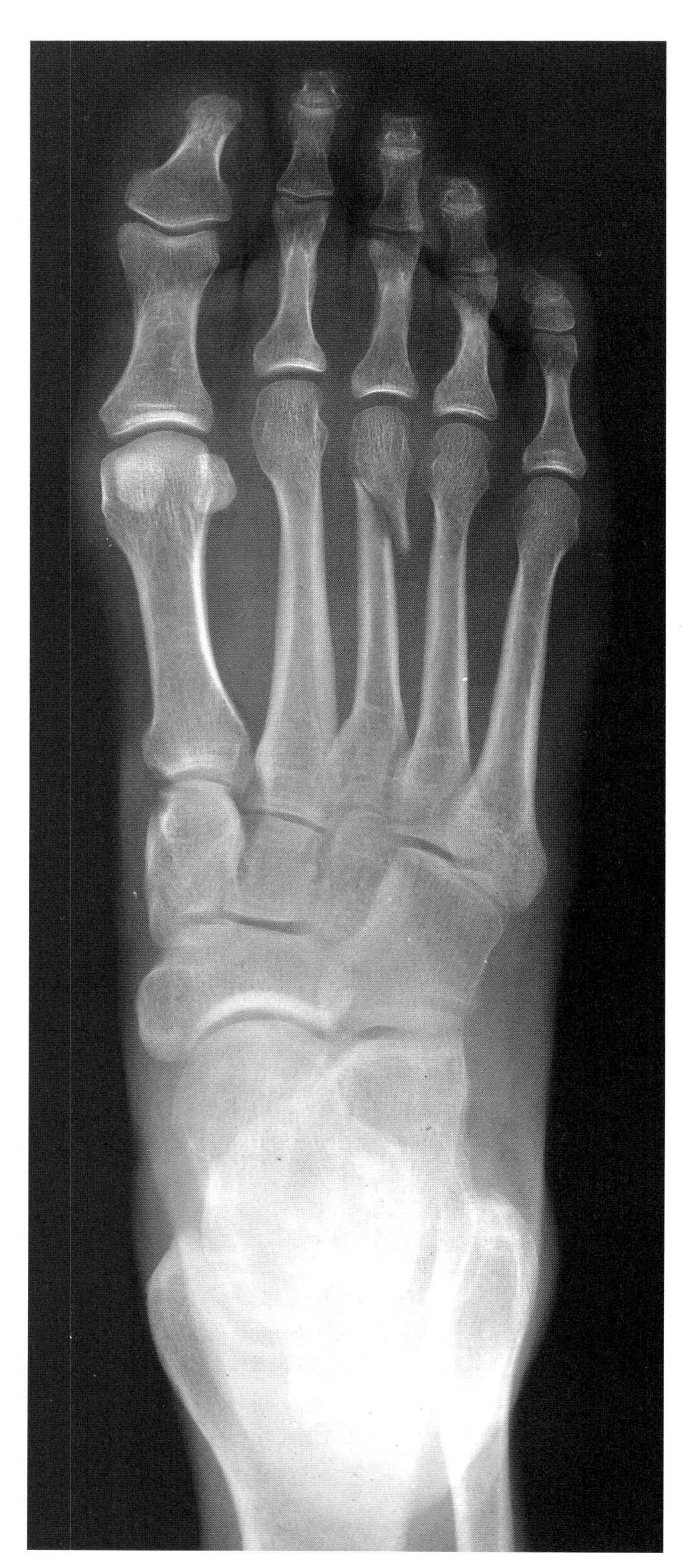

A

Figure 15.8A *(above)* and B *(facing page).* Two different views of the right foot. How would you have to rearrange Mother Whistler's comfortable position to obtain B?

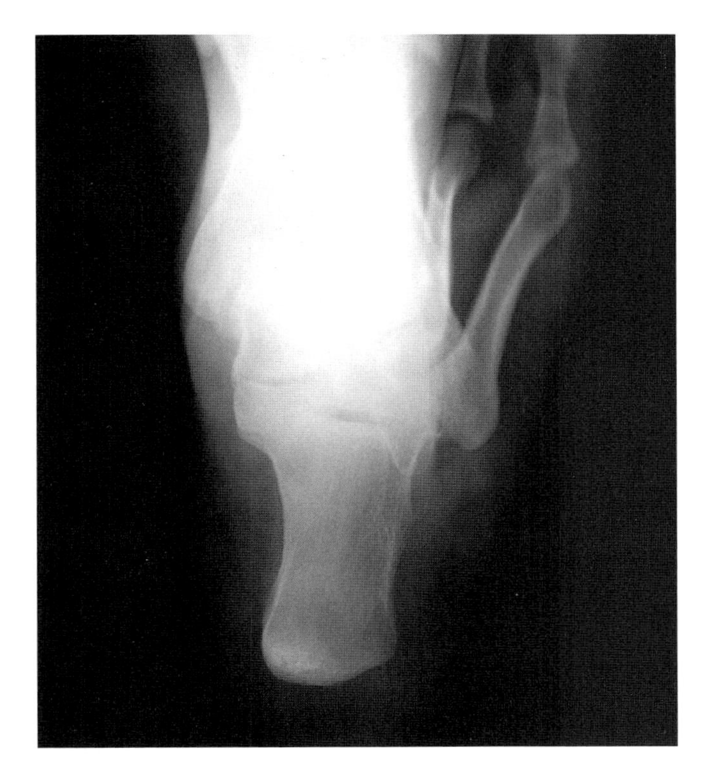

B

have discovered the fractured third metatarsal on x-ray. If she had complained of pain farther up, you might have requested "films of the foot and ankle," and then an additional routine ankle series would also have been obtained. Note that in Figure 15.8A the lower parts of the tibia and fibula are superimposed on the proximal part of the foot. Therefore, a radiographic examination of the foot is inadequate for diagnosis of possible fracture of the ankle. Precise location of the pain and point tenderness often helps you to decide whether you need films of the foot or the ankle or both. It also helps the radiologist determine whether the correct film series has been requested.

X-rays of injured parts of bone are usually obtained after fractures have been reduced and set in plaster casts, to determine the position and alignment of fracture fragments. With the superimposition of the cast material, of course, much less detail of the bones can be seen.

You will have to develop a familiarity with the details of the anatomy of bones on radiographs, anticipating that when bones are unavoidably superimposed (as are the femoral head and the acetabulum in Figure 15.9) you must account for every detail in the x-ray image by subtracting intellectually the parts that belong to one bone from the superimposed images of parts that belong to the other.

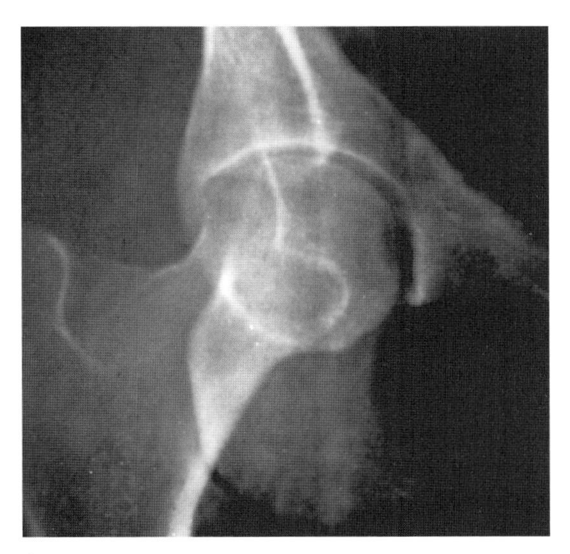

A

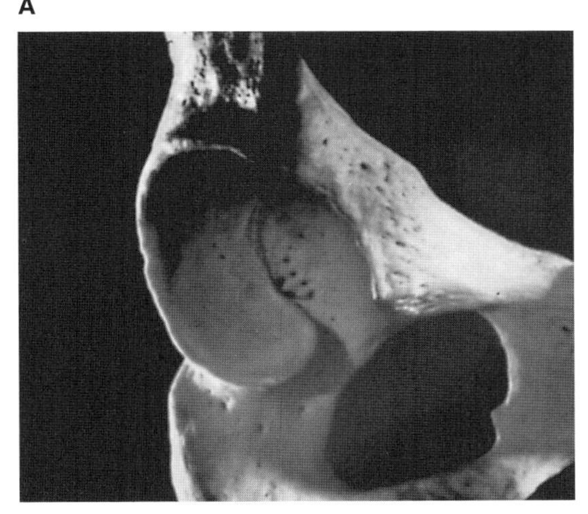

B

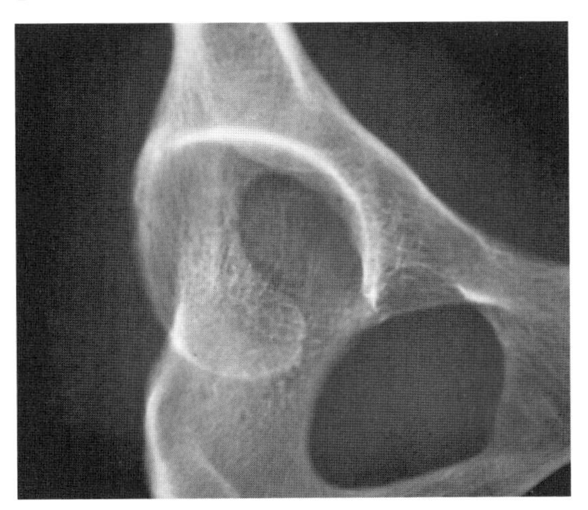

C

Figure 15.9. The shadow of the acetabulum superimposed on that of the upper femur. A: Radiograph of a patient. B: Photograph of a specimen from another patient; the femur has been removed. C: Radiograph of the acetabulum of the specimen in B.

Fractures

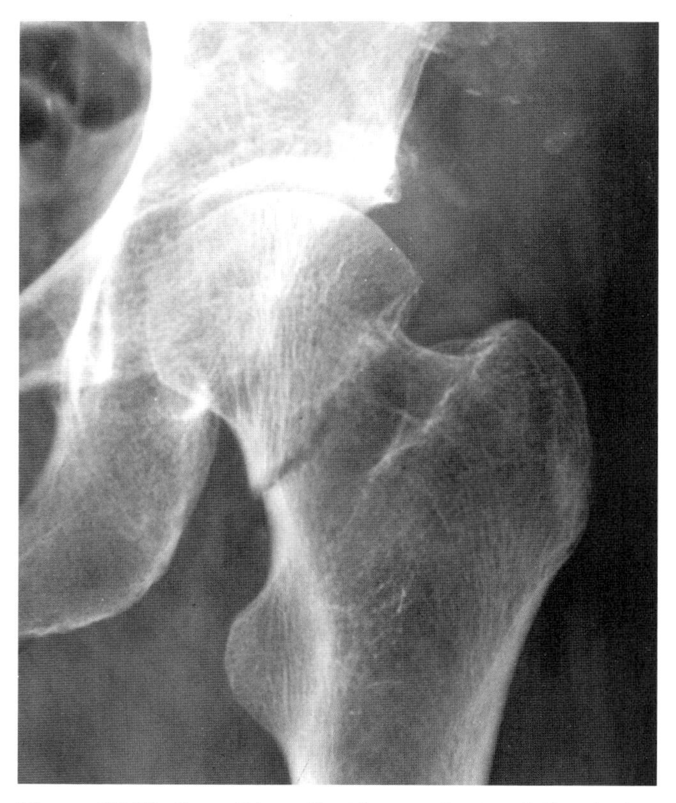

Figure 15.10. Does this patient have a fracture? See text.

Figure 15.11. Child's hip showing growth plates *(arrows),* to be distinguished from fractures. The ossification of the epiphysis for the femoral head appears before 8 months and fuses at about 18 years. The one for the greater trochanter appears at around 2 years and fuses to its metaphysis at about 16 years.

Fractures of bones are common and many different types are recognized. They may be *transverse, spiral, oblique, simple, comminuted* (when there are several fragments and intersecting fracture lines), *impacted,* or *pathological* (in the presence of underlying disease). Fractures that are open to the skin surface are called *compound fractures.* A fracture that extends through both cortices is a *complete fracture,* whereas one that extends only through one cortex is an *incomplete fracture.* The latter is seen most commonly in children.

An *avulsion fracture* occurs when a fragment of bone is traumatically pulled off by a tendon or ligament, at an origin or insertion site. Such fractures are often small, and they may be overlooked on the x-ray if a careful search for them is not made. Generally, though, there is excellent clinical correlation with avulsion fractures; that is, if the fracture site is palpable, the physician can almost always detect overlying point tenderness. When you review radiographs you should scrutinize sites of ligament and tendon insertions for possible fractures.

Fractures that extend into joint spaces are called *intra-articular* fractures, and they are usually accompanied by bleeding into the joint space. Such collections of joint fluid are easily seen in the knee, ankle, and elbow joints, and their presence, when detected radiographically, should alert you to search for an underlying fracture. These bloody effusions may displace anatomically normal fat planes surrounding joints, causing *displaced fat-pad* signs of trauma.

When abnormal stress is placed on a bone, a *stress fracture* may result. Two types of stress fractures are recognized. *Fatigue fractures* occur when abnormal stresses are placed on normal bone. Often the stress is of a repetitive nature, such as the repeated stress that causes the "march" fractures that occur in the metatarsals of new military recruits. *Insufficiency fractures* occur when normal stresses are placed on abnormal bone; an example is the vertebral compression fractures that occur with minimal trauma in elderly osteoporotic women.

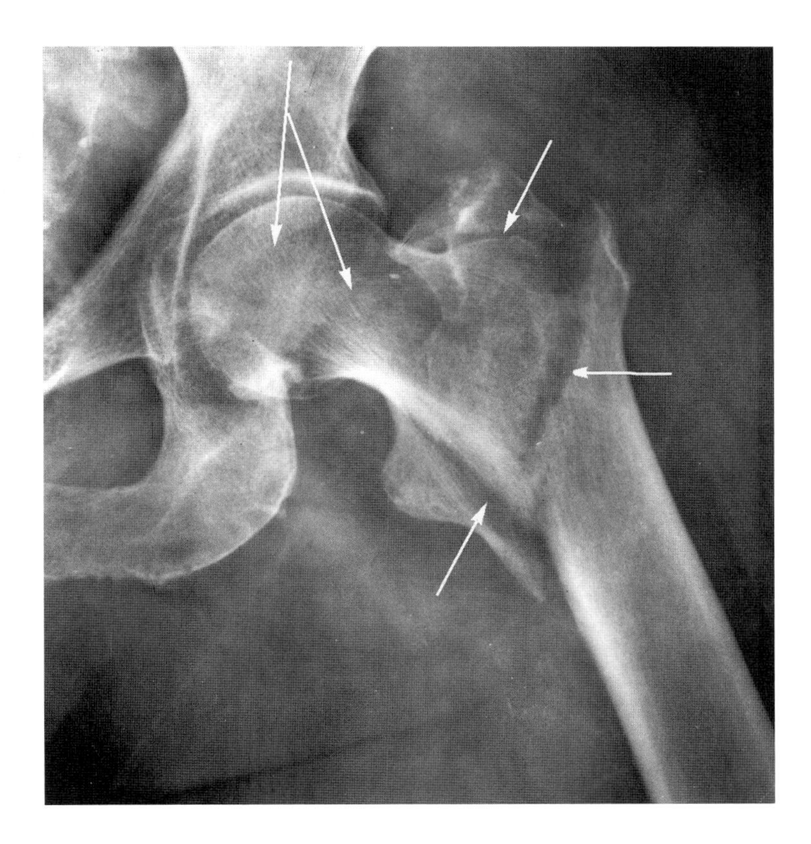

Figure 15.12. Comminuted intertrochanteric fracture of the left femur. The *long arrows* indicate the overlapping margins of the anterior and posterior rims of the acetabulum. The *short arrows* indicate three communicating fracture planes.

Radiographic examinations for skeletal trauma must include a minimum of two views taken at 90 degrees to one another, because a fracture line, especially when the fracture fragments are only minimally displaced, may be invisible on only one projection. More complex areas of bony anatomy, such as occur at joints, often require additional oblique or other projections. Again, as the referring physician, you do not need to memorize the required views for various bony parts; you only need to request "ankle films," "wrist films," or "cervical spine films," for example.

In many cases it is wise to radiograph the entire bone you believe to be fractured. This is especially important when you are dealing with the paired bones of the lower leg and the forearm. Obvious fracture of the lower tibia is quite often accompanied by a subtle fracture of the fibula near the knee, and you should request "tibia-fibula films" that include both knee and ankle in such patients. Of course, if you have examined the patient carefully, you will have observed point tenderness in both locations.

You should have no difficulty in recognizing the simple fracture of the neck of the femur in Figure 15.10. Fractures appear on radiographs as dark streaks across the bone where the continuity of both cortical (compact) bone and spongy bone is interrupted. Hemorrhage and soft tissue, torn and injured in the area, are often interposed to some extent between the fractured fragments. If the fracture is comminuted, several fragments and several separate but communicating fracture planes must be present (as in Figure 15.12). Sometimes one of these planes of fracture is not obvious on the films obtained because it is oblique to the ray. In Figure 15.12 three planes of fracture may be seen and one more supposed—that for the greater trochanter.

In examining films for fracture you should look closely everywhere for interruption of the normal line of the periosteal surface of the cortex, because it may be the only indication of fracture if there is no separation of fragments. Remember that spongy bone can also be fractured, and close examination may show discontinuity of major trabeculae with only slight separation of the plane of the undisplaced fracture. Depending on the direction and character of the trauma, spongy bone may be impacted. If so, the trabeculae will have become enmeshed so that innumerable fragments of bone across the plane of the fracture lie closer together than normally and produce an abnormally dense white area across the bone where the *impacted* plane of fracture is seen tangentially.

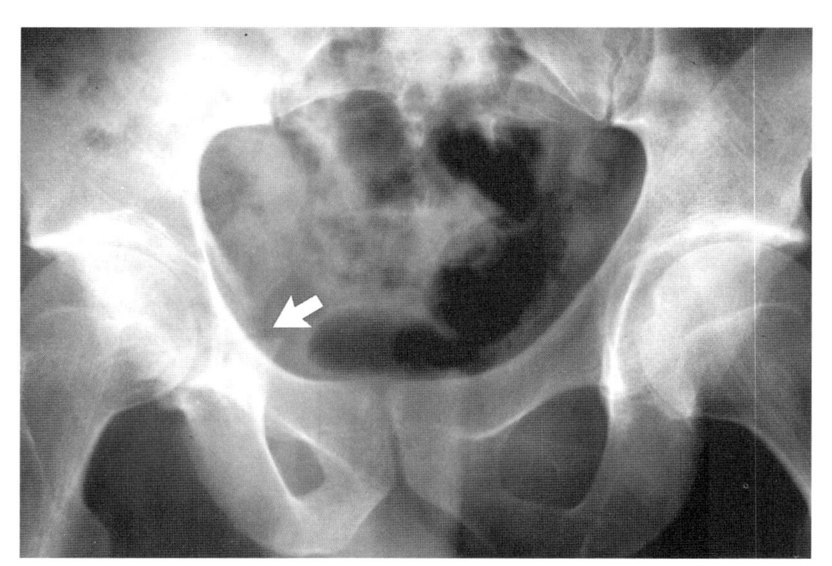

A

Figure 15.13. A: Radiograph of a pelvic fracture at the level of the right acetabulum *(arrow).* B: CT scan of the fracture shows additional fractures of the posterior rim of the acetabulum that were not obvious on the plain film. Note the two displaced fracture fragments posterior to the femoral head. If the posterior rim of the acetabulum is not reconstructed surgically with fixation of these fragments, the patient will suffer recurrent posterior dislocation of the femoral head.

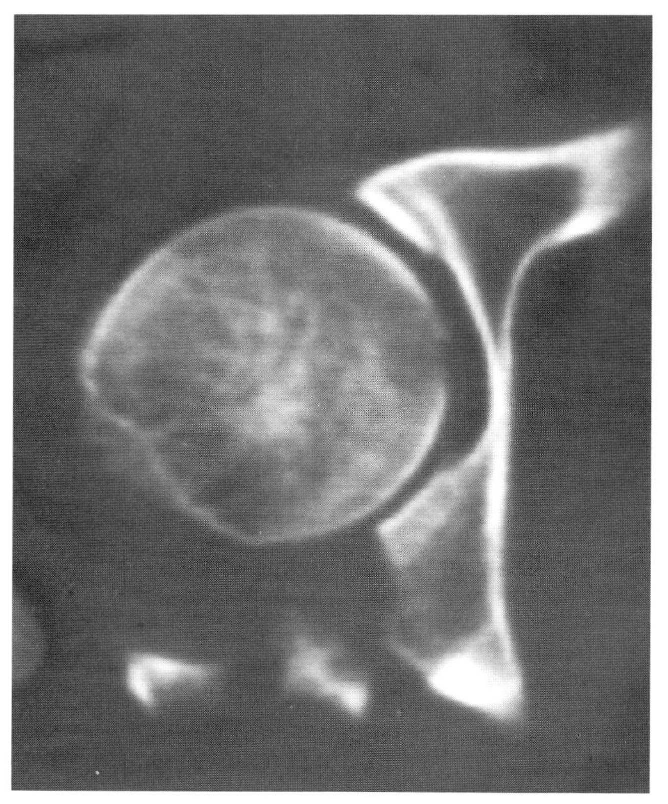

B

You will not mistake the similar white boundary of an epiphyseal growth plate like that in Figure 15.11 for a fracture, once you have studied a few of them. The linear dark area crossing the bone is the tangentially viewed cartilaginous growth plate, and its margin shows not an abrupt interruption of cortex, but a smoothly curved one at an expected location.

You should know that in some patients with an acute fracture the initial plain film examination may be negative when the fracture is nondisplaced. Certain structures are notorious for having initially normal radiographs when fractured; they include the scaphoid bone of the wrist, the hip (proximal femur), and the spine. Usually, repeat films on patients with such injuries become positive within a few days as slight displacement occurs along the fracture line or bony reabsorption adjacent to the fracture line makes the fracture apparent radiographically. But you should never overlook *the possibility of an acute fracture;* whenever the clinical examination suggests fracture (severe pain, hematoma, disruption of function) and the plain films are negative, the patient must be examined with another imaging procedure, usually CT (sometimes, but less frequently, a bone scan).

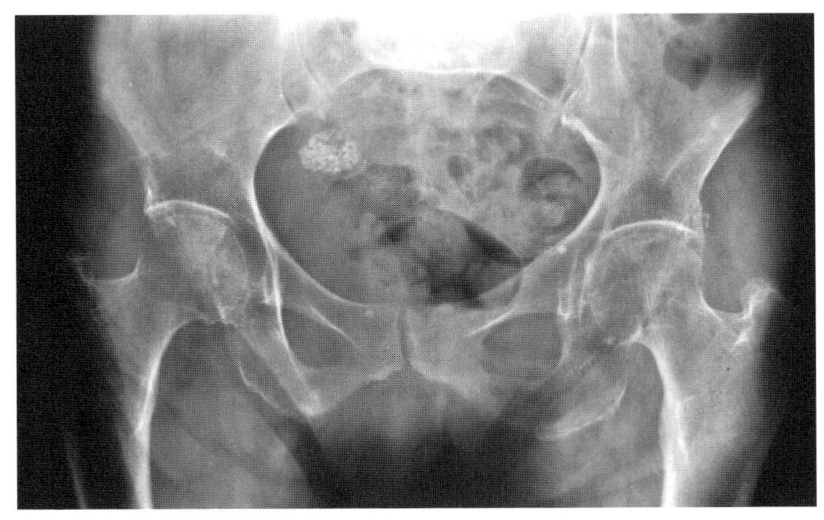

A

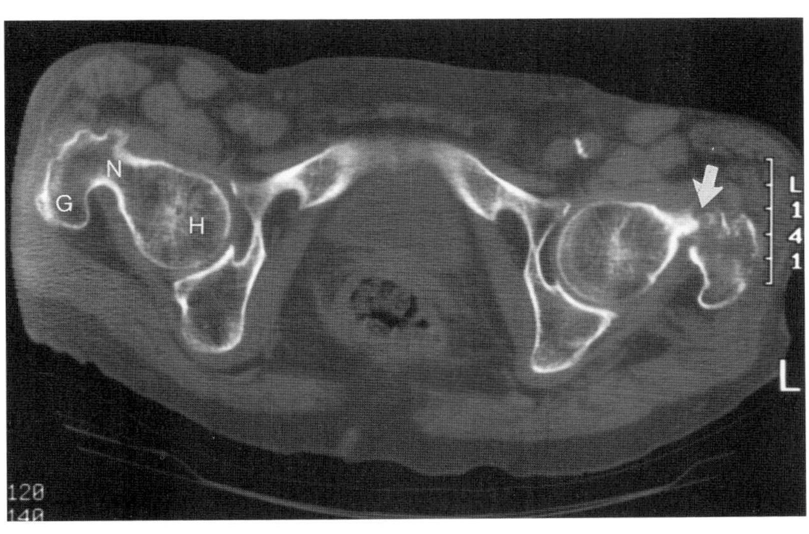

B

Figure 15.14. A: Nondisplaced fracture of the left hip in an elderly woman shows almost no abnormality on the initial plain film examination. B: Her CT scan shows an acute fracture *(arrow)* of the left femoral neck. Note the breaks in the left femoral neck cortical bone, and compare the normal right hip. *H* is the femoral head, *N* the femoral neck, and *G* the greater trochanter. When your trauma patients' levels of pain and clinical story are compatible with fracture but the initial plain films appear normal, consider further study with a CT scan, which can determine whether or not a nondisplaced fracture is present.

CT can show the fracture lines better than plain films and can identify even the subtle cortical abnormalities of nondisplaced fractures. A bone scan would show increased uptake of isotope at the site where the nondisplaced fracture was located. But CT is preferable to the bone scan because CT can show the fracture lines in greater detail and the position and orientation of fracture fragments better than any imaging procedure (Figure 15.13). CT is especially helpful in showing fractures in portions of the skeleton where the bony anatomy is complex, such as the spine, pelvis and hips, face, shoulder girdle, and foot (Figure 15.14). In addition, it can be performed quickly, without uncomfortable positioning of the patient, and with less radiation than conventional tomograms. After the axial slices are obtained, the CT scanner computer can produce sagittal and coronal reformations, as well as three-dimensional models of the injured area, that display the position and orientation of the major fracture fragments so that you view them as if you were looking at the patient's injury with the soft tissues removed (Figure 15.15).

MR imaging does not show fracture lines or fragments as well as CT because there is no MR signal from cortical bone. But MR can show bony injuries, such as *bone bruises,* that do not involve cortical bone disruption but produce hemorrhage and edema within the marrow space (Figure 15.16).

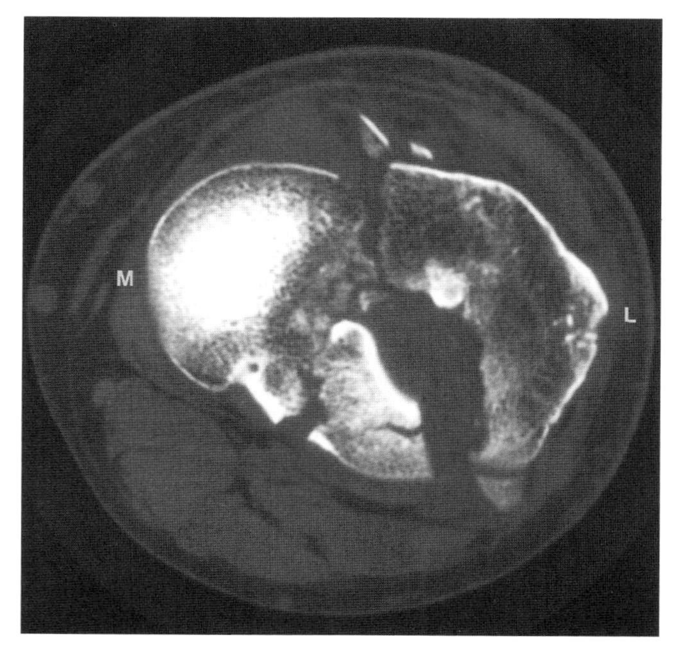

A

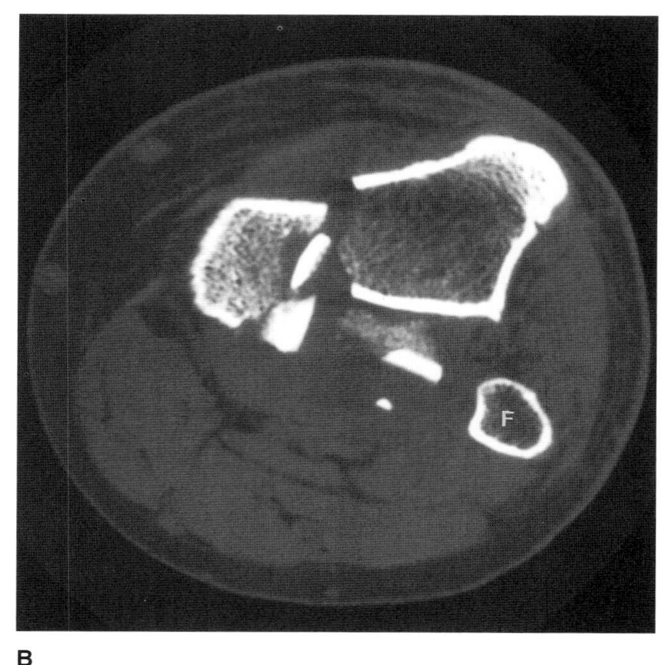

B

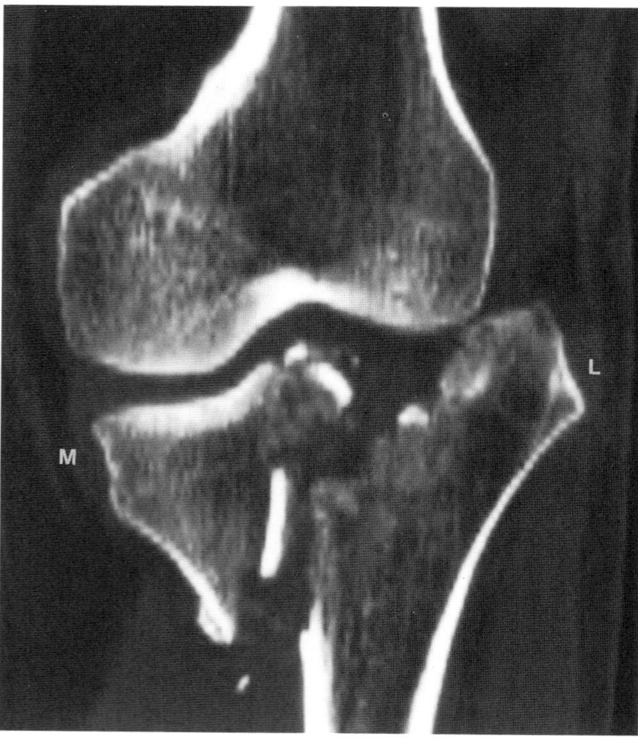

C

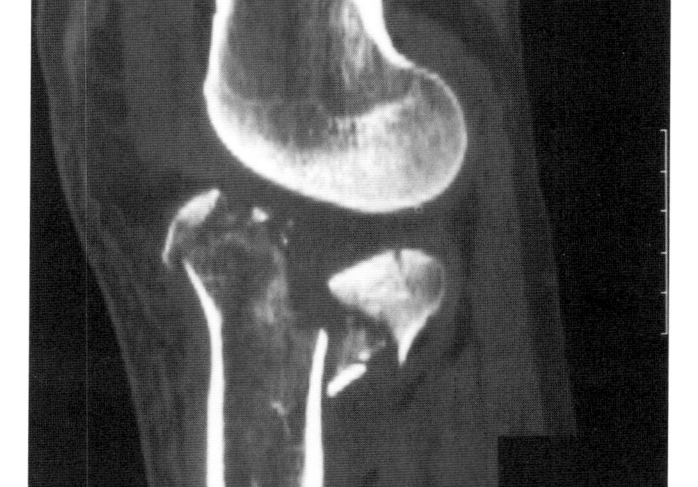

D

Figure 15.15. CT scan of a left tibial plateau fracture with coronal, sagittal, and three-dimensional reformations. A: Axial slice of the upper tibia shows extensive fracturing of the tibial plateau; the medial plateau *(M)* is to your left and the lateral *(L)* to your right. B: Axial slice several centimeters lower than that in A shows a comminuted vertical fracture line extending down the tibia. The fibula *(F)*, now seen, is intact. C: Computer reformation in the coronal plane (the CT computer stacked all the axial slices to allow computer generation of slices in the coronal and other planes) shows a major comminuted oblique fracture line extending below the medial tibial plateau, which has been displaced inferiorly and medially. The femur is intact. D: Sagittal reformation through the medial aspect of the joint shows that portions of the medial tibial plateau have also been displaced posteriorly. E *(top of facing page):* Three-dimensional reformation filmed from the anterior aspect shows the relationship between the depressed medial tibial plateau and other bony components of the left knee joint. Find the patella, femur, fibula, and smaller fracture fragments. F *(top of facing page):* Three-dimensional reformation filmed from the posterior aspect. Note the medial and lateral condyles of the femur and the excellent delineation of the medial tibial plateau depression.

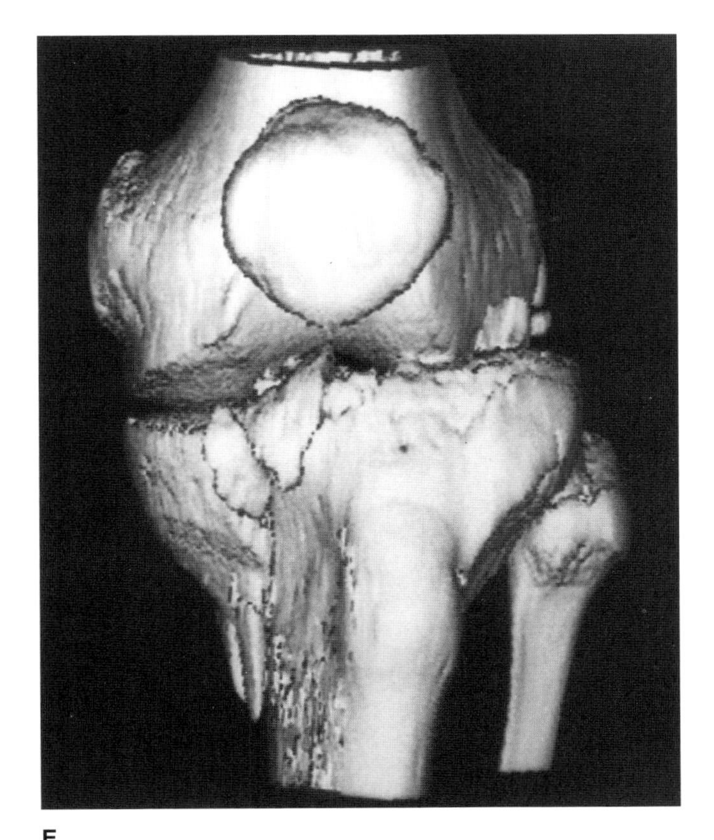

E

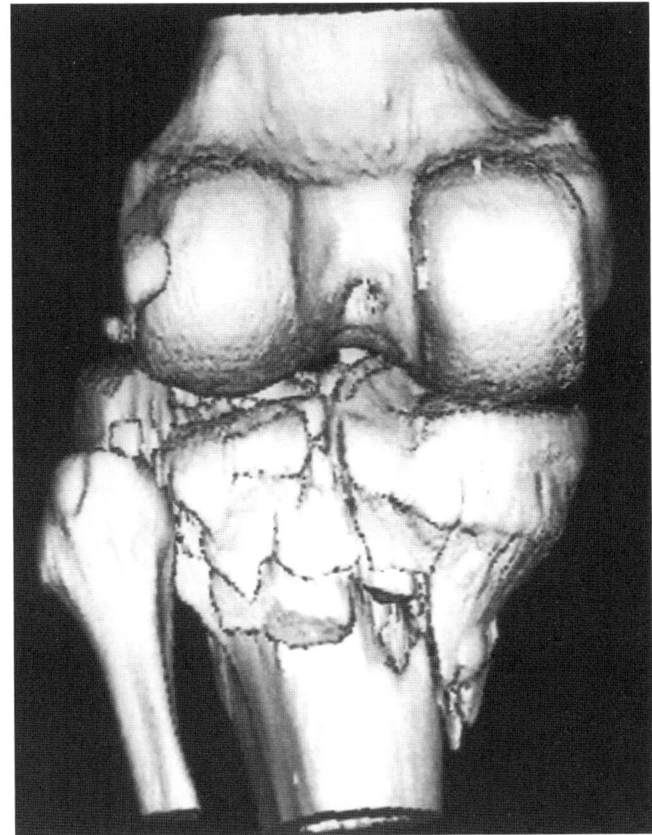

F

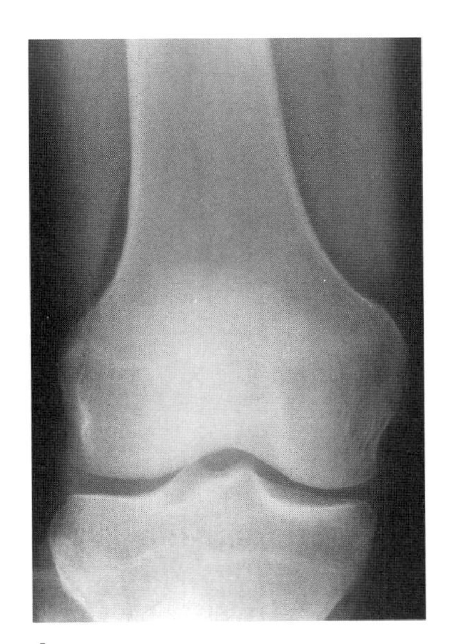

A

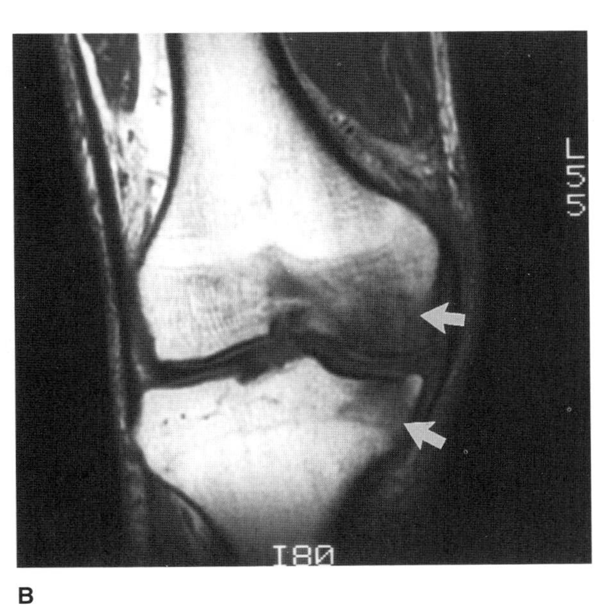

B

Figure 15.16. Images of a bone bruise in a young man who injured his right knee in a ski accident. The plain films were entirely normal, as you can see in the AP view shown in A. The MR scan (B) shows bone bruises *(arrows)* of the medial femoral condyle and medial tibial plateau. Normal medullary bone has a strong MR signal (appears bright white) on a T1-weighted image due to the presence of marrow fat. When a bone is bruised, the fat is replaced in the injured areas with hemorrhage and edema, which decrease the MR signal. With more severe trauma this patient could have suffered a medial tibial plateau fracture or a fracture of the medial femoral condyle.

Fracture Clinic

The illustrations on the next few pages constitute an exercise in fracture diagnosis, framed for you as unknowns. Not every film shows a fracture, as would be the case if you had seen these patients at random in the emergency department. But all these patients *had* been injured.

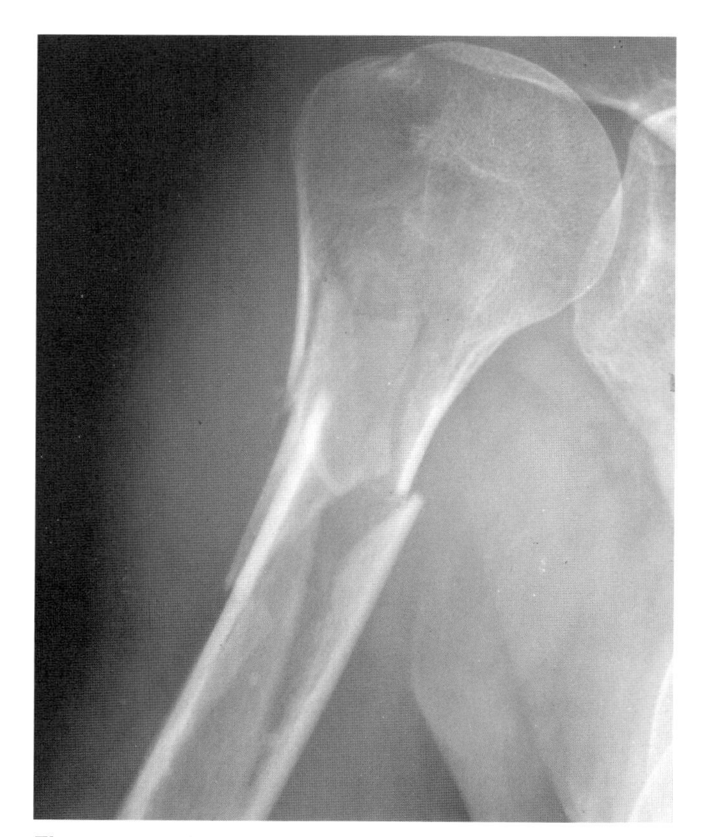

Figure 15.17 ***(Unknown 15.1)***

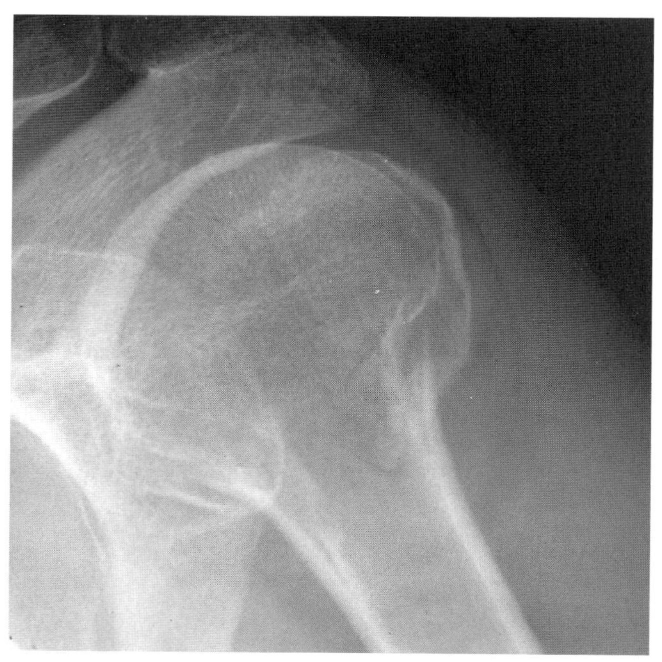

Figure 15.18 ***(Unknown 15.2)***

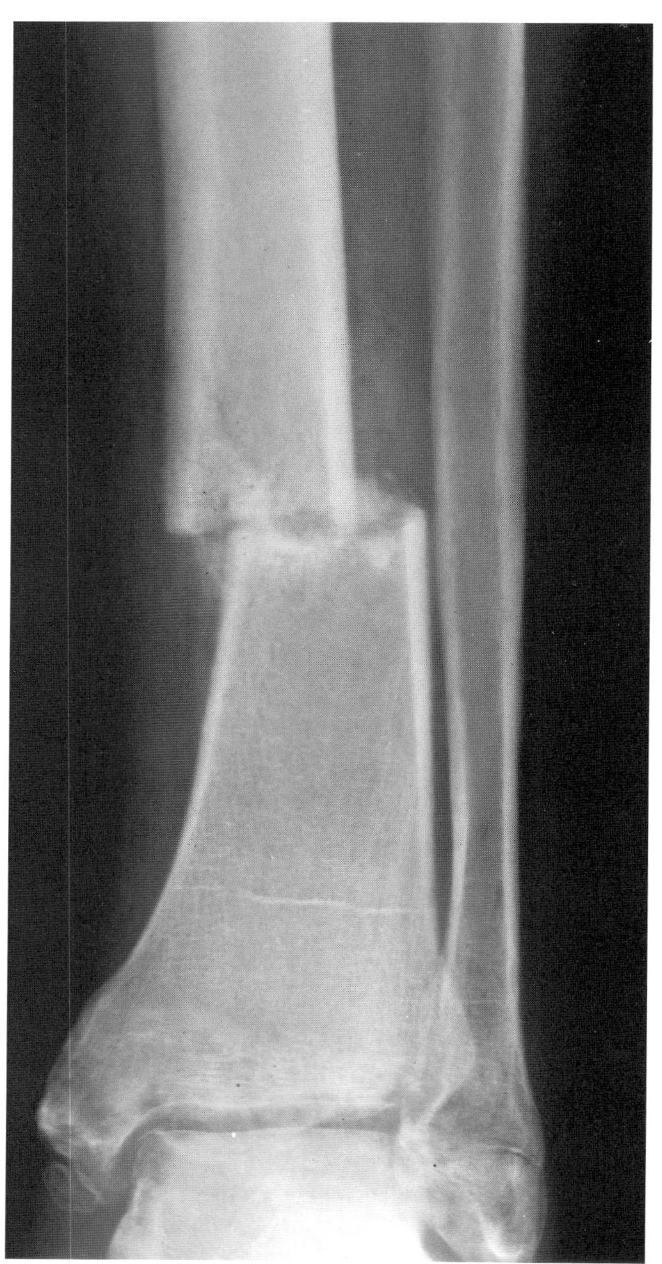

Figure 15.19 ***(Unknown 15.3)***

The roentgen findings you are looking for are the following:

1. Breaks in the continuity of cortex
2. Radiolucent fracture lines
3. Overlap of both cortical bone and spongy bone creating an abnormally white zone
4. Unexplained fragments of bone, even in the absence of a visible fracture
5. Denser areas where impaction of bone has occurred, seen in two views
6. Flocculent density in soft tissues adjoining bone in healing fractures (callus), visible only after it calcifies

Remember that in describing these films to the consultant orthopedic surgeon, for example, you should not only state the location and type of fracture, but also comment on the alignment of the fracture fragments and any overriding that is present. Open or compound fractures may not be apparent at all radiographically unless there is obvious lucent air in the soft tissues.

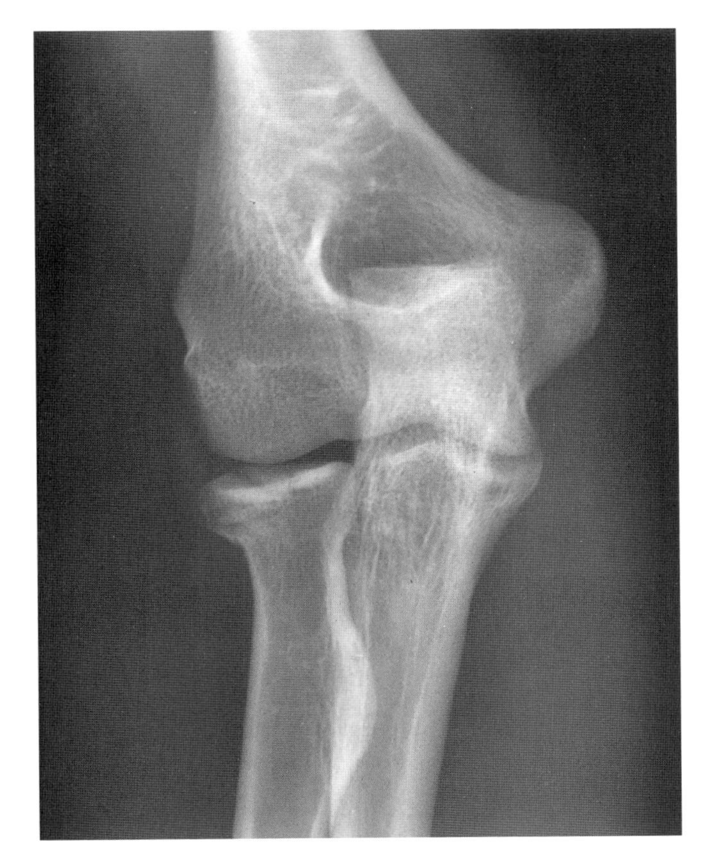

A

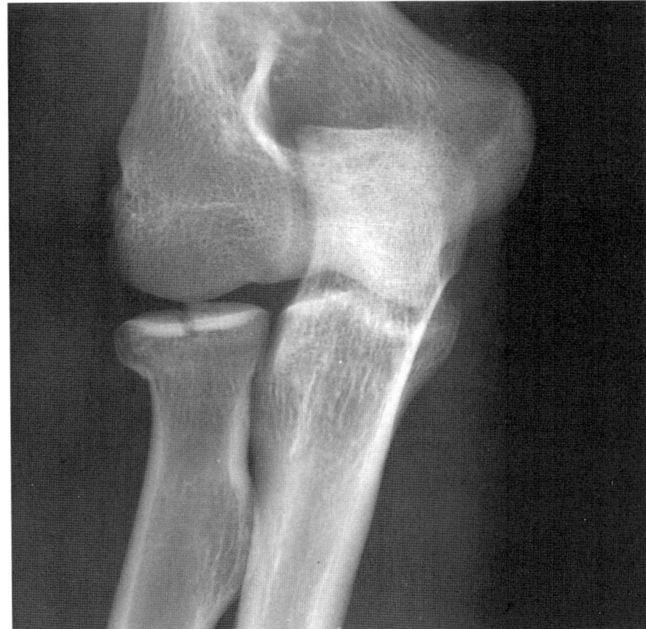

B

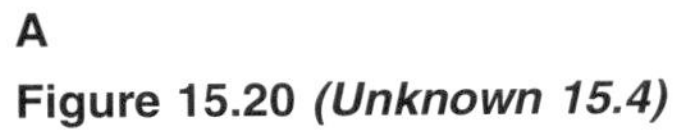

Figure 15.20 ***(Unknown 15.4)***

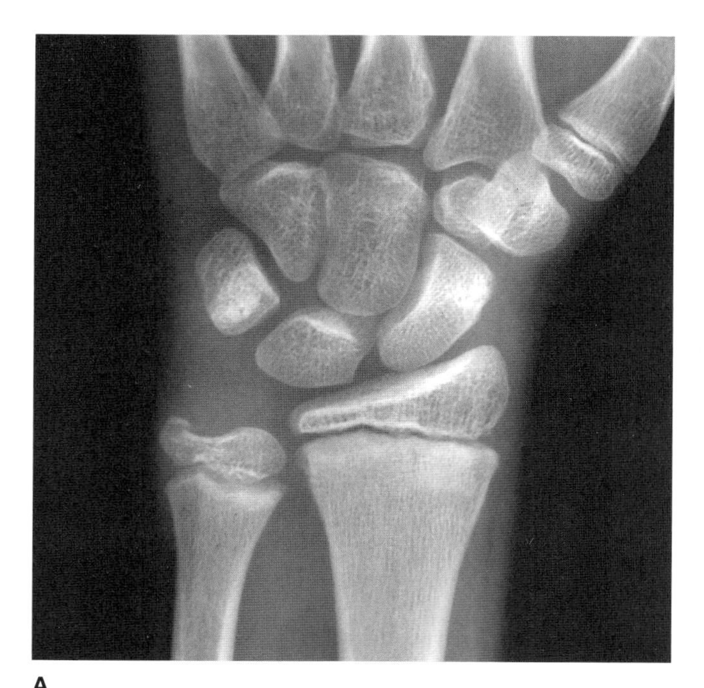

A

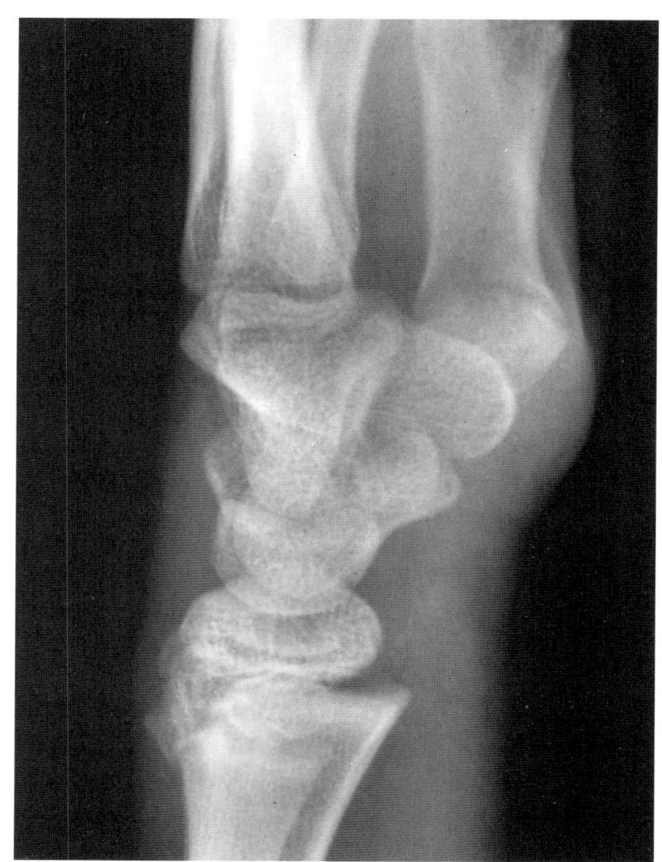

B

Figure 15.21 *(Unknown 15.5)*

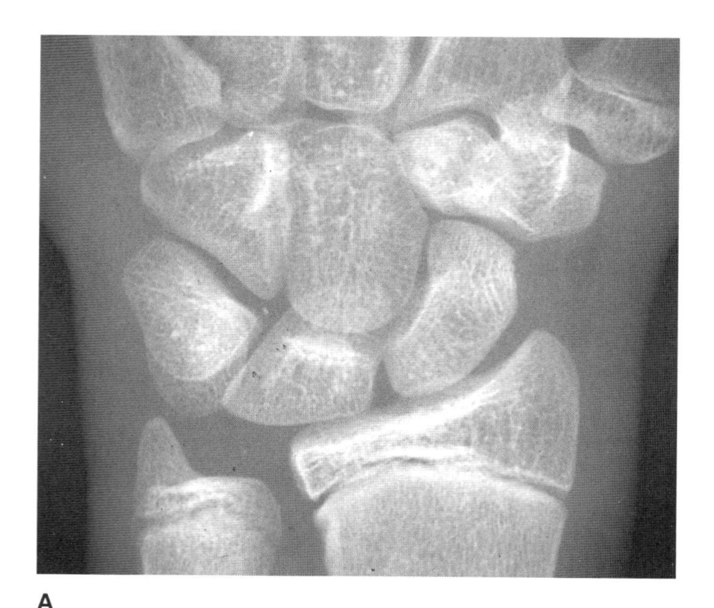

A

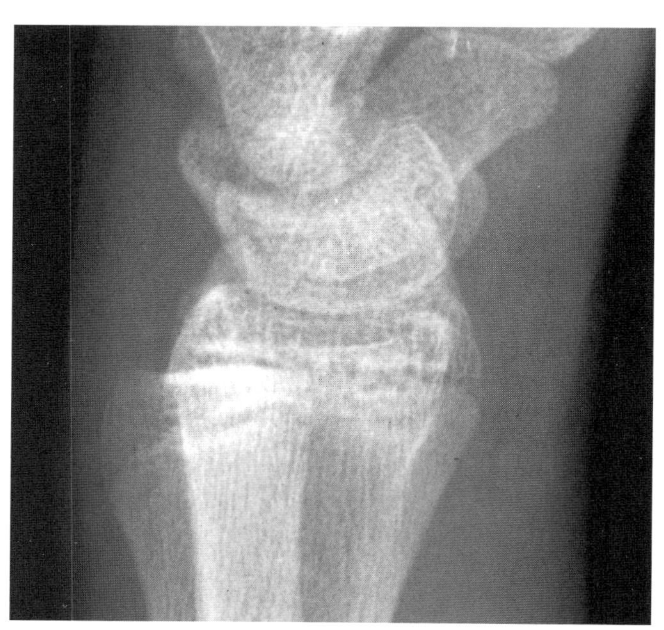

B

Figure 15.22 *(Unknown 15.6)*

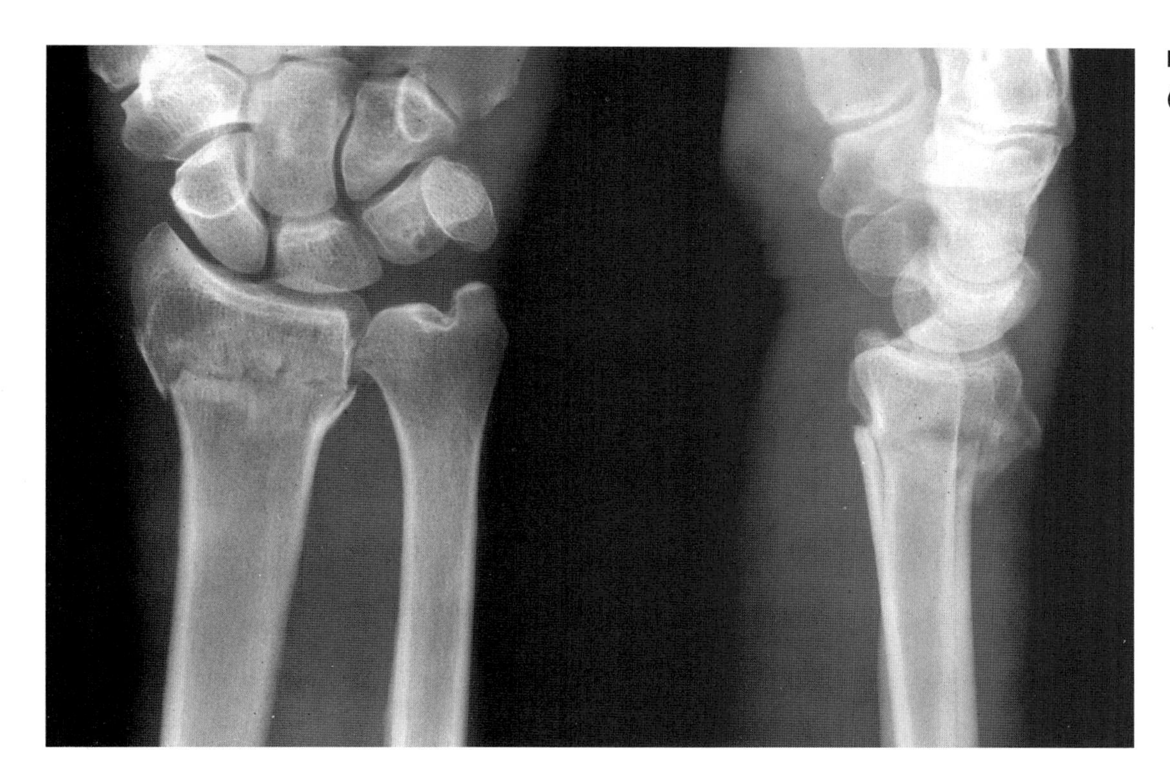

Figure 15.23
(Unknown 15.7)

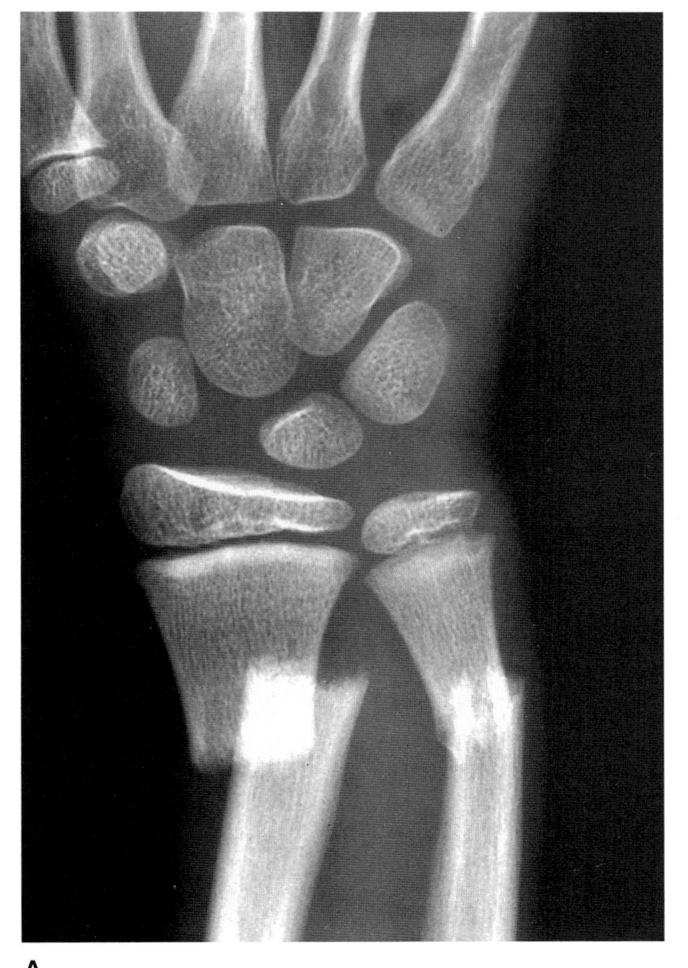

A

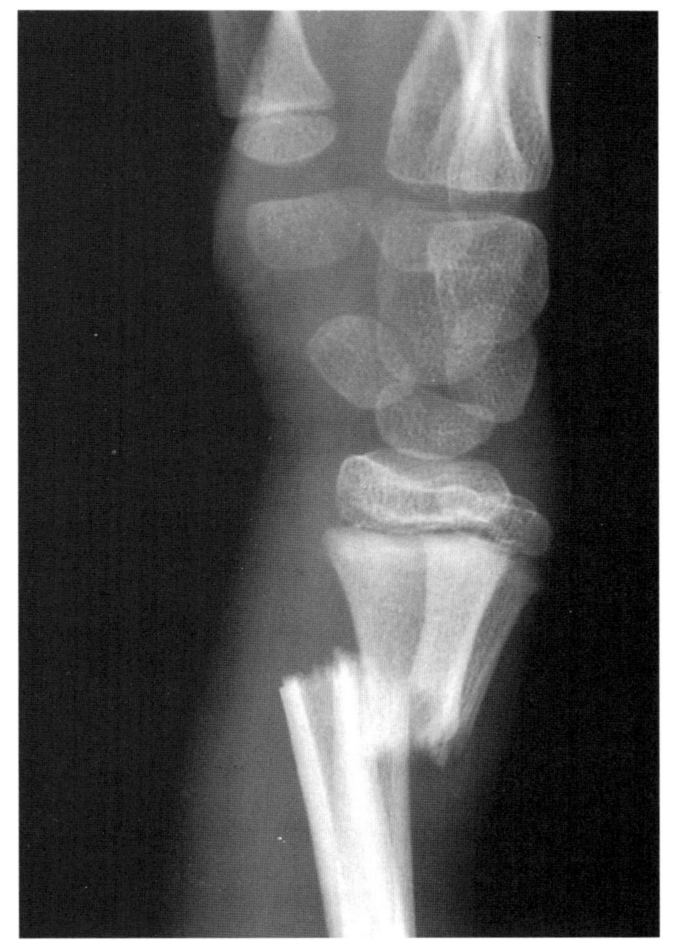

B

Figure 15.24 *(Unknown 15.8)*

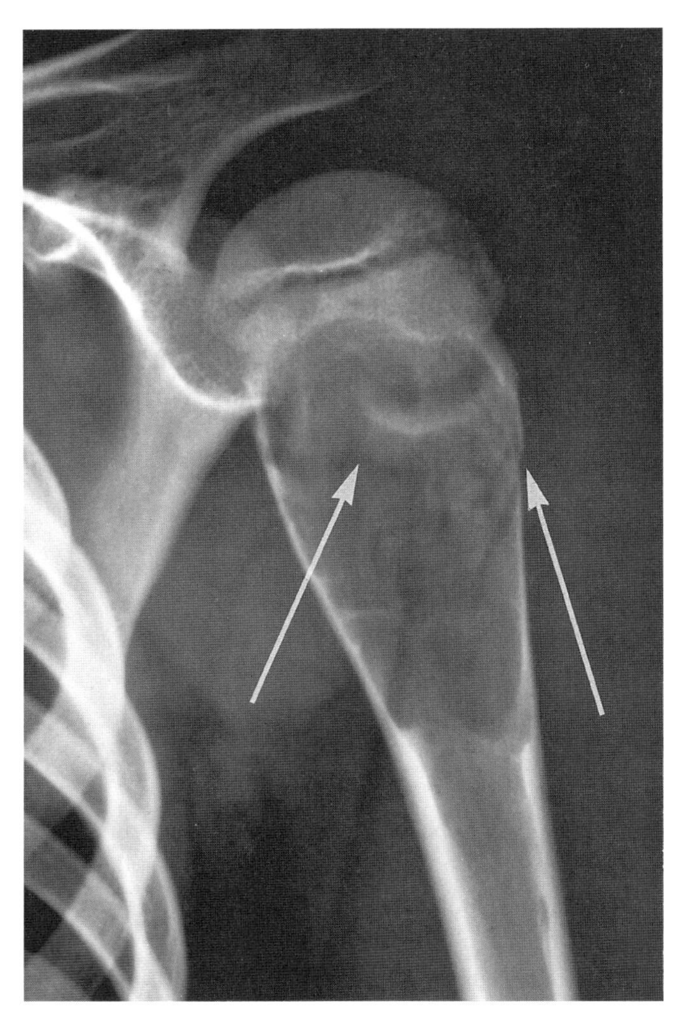

Figure 15.25. Pathological fracture through a large bone cyst in a classic location in the metaphysis of a growing bone.

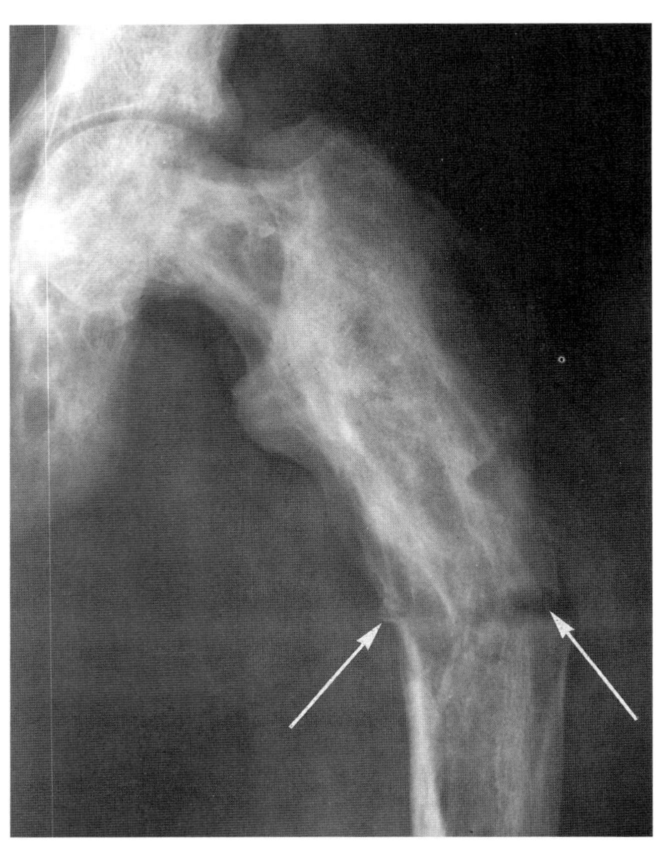

Figure 15.26. Pathological transverse fracture in the upper femoral shaft of a patient with Paget's disease.

When a fracture appears to have occurred through bone that was already abnormal and may therefore have been unusually fragile, it is called a *pathological fracture*. Figures 15.25 and 15.26 are examples.

In Figure 15.25 a cylindrical cuff of bony cortex has been eroded by pressure from within. You observe the thinning best where you see cortex in tangent medially and laterally, although you know it has occurred anteriorly and posteriorly as well. The bone looks expanded, subperiosteal new bone having been added as the endosteal layers were removed by pressure. Across the thinned segment of bone a jagged fracture has occurred and is seen on the lateral margin as a distinct interruption of cortex. This is an example of a *bone cyst* in a child. It was fractured by a very trivial trauma—throwing a ball. The fact that the bone looks expanded is an indication of its benign nature, since there has been time for such change to occur. Malignant tumors are aggressively destructive and usually produce new bone in the soft tissues.

Figure 15.26 is the upper femur of a patient with Paget's disease. A transverse fracture has occurred across the shaft several centimeters below the lesser trochanter. Although it characteristically thickens bone, Paget's disease also weakens its structure, and the bone withstands stress less well than normal tubular long bone does. It is also more usual for a normal upper femur to fracture irregularly in ragged points and with comminution; this bone has fractured transversely. Paget's disease typically produces enlargement of the bone, thickening of the cortex, and grotesque disarrangement of the trabecular pattern.

Dislocations and Subluxations

Fractures are often accompanied by *dislocation* of joints, but dislocations may occur without fractures. A partial dislocation is called a *subluxation*. Figure 15.27 shows you an anterior dislocation of the shoulder and its appearance after reduction. Nearly all shoulder dislocations are anterior, with the humeral head displaced anteriorly, medially, and inferiorly, making them easy to diagnose radiographically. The rare posterior dislocation may be missed on AP films because the displacement is much less, and often only special views can establish its presence. Dislocations and subluxations can occur at nearly every joint in the body, including all the joints on the extremities. A useful sign for detecting dislocation is overlap of bone at a joint instead of visualization of the clear joint space. On the AP foot film in Figure 15.28 note the overlap at the first metatarsophalangeal (MP) joint and compare with the other four MP joints. The dislocation is more apparent on the lateral view.

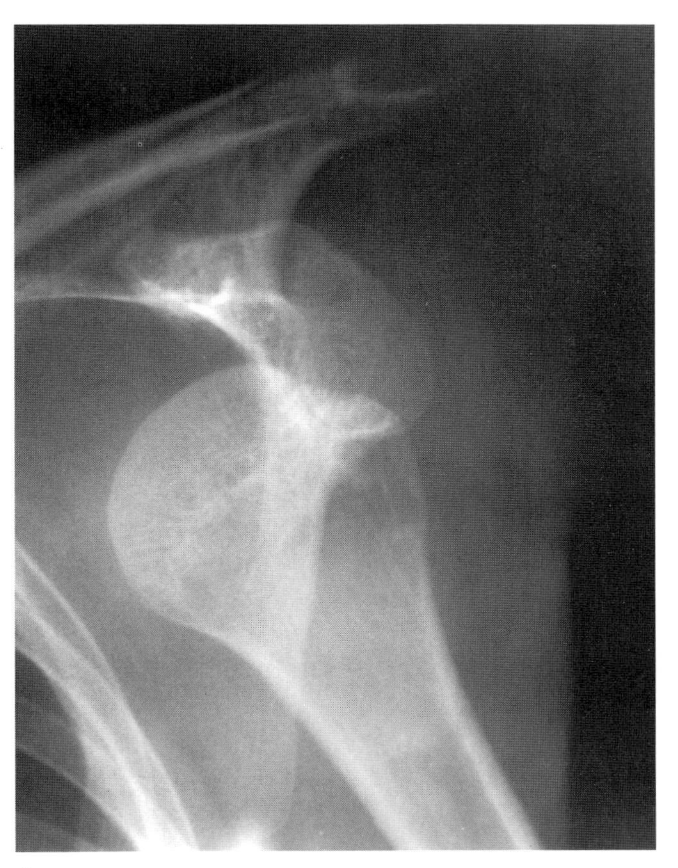

A

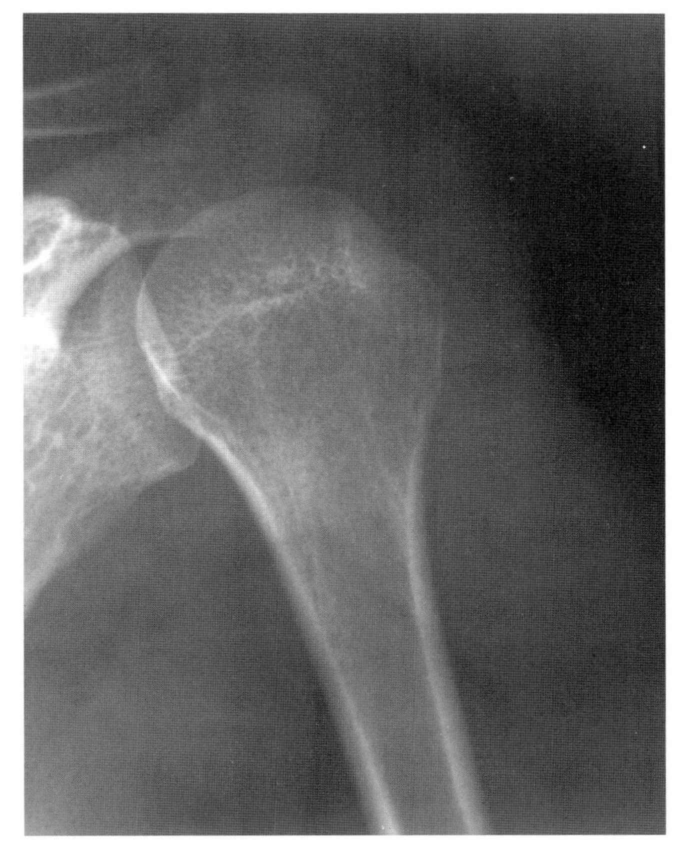

B

Figure 15.27. Anterior dislocation of the shoulder before (A) and after (B) reduction.

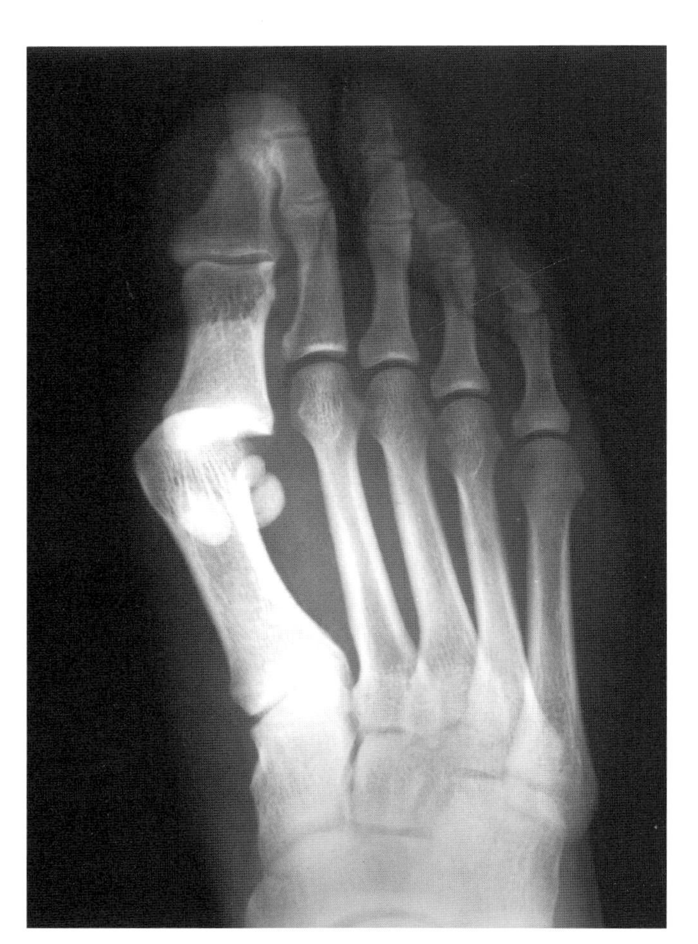

A

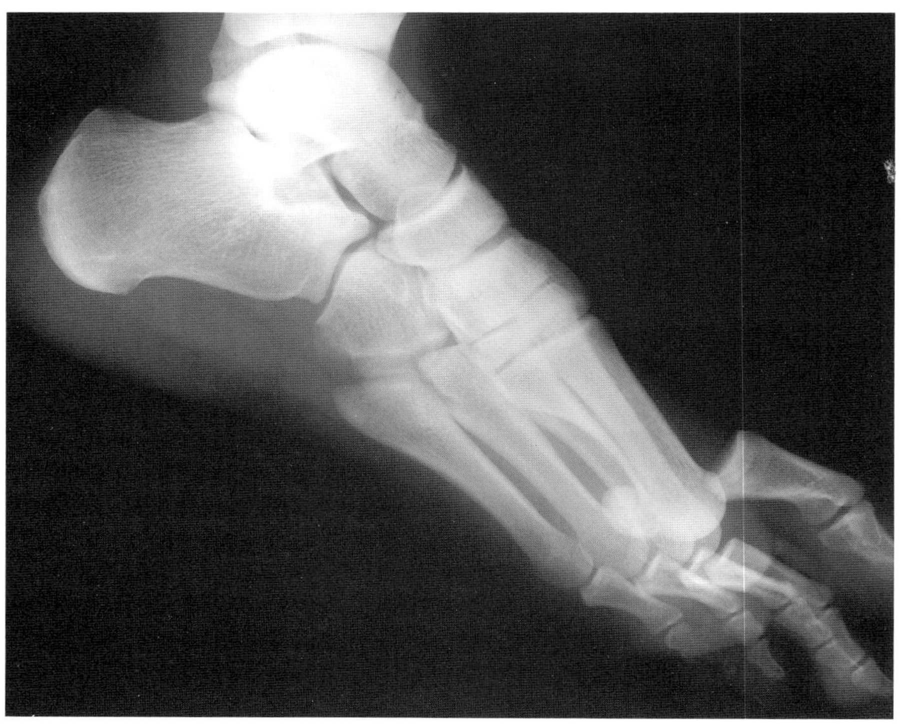

B

Figure 15.28. Dislocation of the first metatarsophalangeal joint of the right foot. A: AP view shows overlap of bone at the first metatarsophalangeal joint rather than a clear joint space. B: Lateral view shows that the first toe has been dislocated dorsally. The articular surfaces of the first metatarsal and proximal phalanx of the first toe are no longer in contact. This dislocation was successfully reduced.

Osteomyelitis

Osteomyelitis can occur in almost any bony location and in patients of any age. Most frequently this condition involves the long bones near the metaphysis, and it is usually blood borne. It is seen today in drug addicts, the most common causative organism being *Staphylococcus aureus.* The changes that are seen in osteomyelitis (bone destruction and periosteal reaction) unfortunately are not likely to appear until two weeks after the start of the process. When detectable on plain films, the bone destruction appears as a lytic lesion (Figure 15.29), which may or may not be expansile, which may or may not have a sclerotic margin, and which may or may not have an associated periosteal reaction. Usually there is also swelling and loss of fat planes in the adjacent soft tissues.

Because plain films do not show infection until late in the process, patients clinically suspected of having osteomyelitis should have an early bone scan; it will indicate the presence of acutely reactive bone turnover within 48 hours so that therapy can be started. The subtle later bony changes of osteomyelitis can be better seen with CT than with plain films, and the marrow involvement can be better shown with MR. Remember that any new lytic bone lesion may represent osteomyelitis.

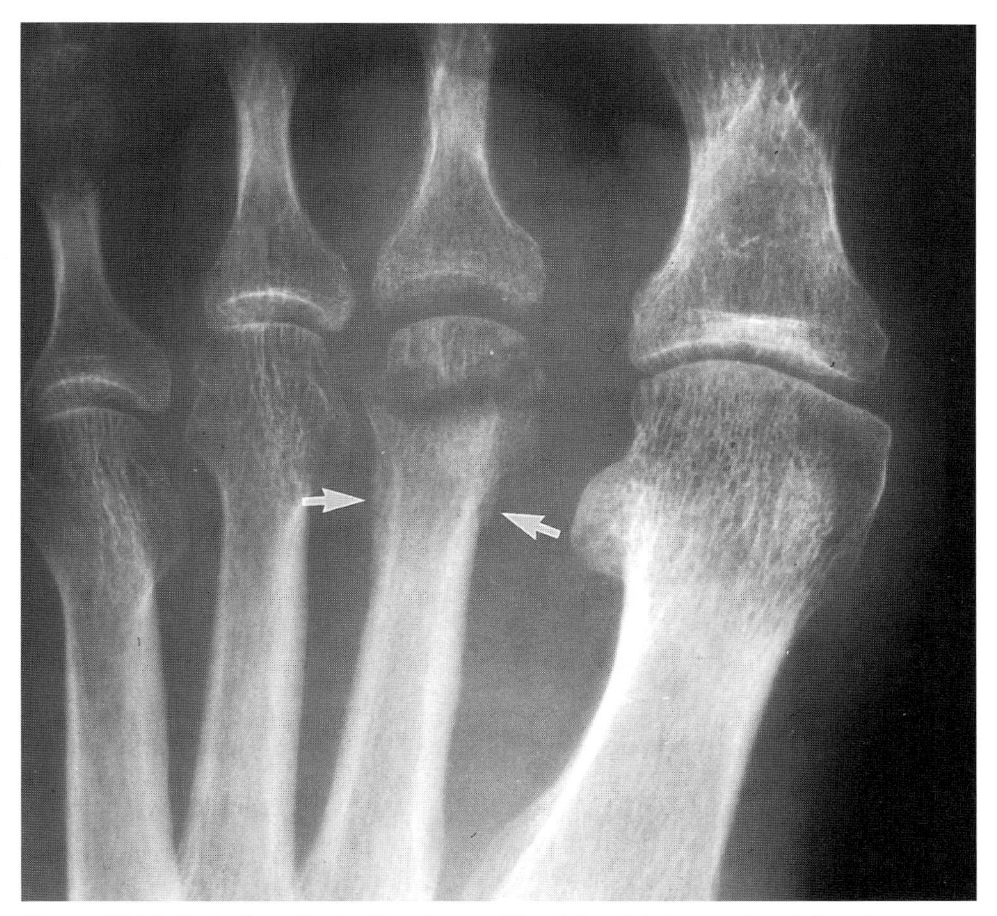

Figure 15.29. Diabetic patient with osteomyelitis of the distal second metatarsal. Note the zone of lytic destruction transversing the metatarsal head. The *arrows* indicate periosteal reaction with associated periosteal thickening and calcification. Compare the similar portions of the third and fourth metatarsals. The widening of the second metatarsophalangeal joint is due to the presence of pus in the joint space (septic arthritis).

Arthritis

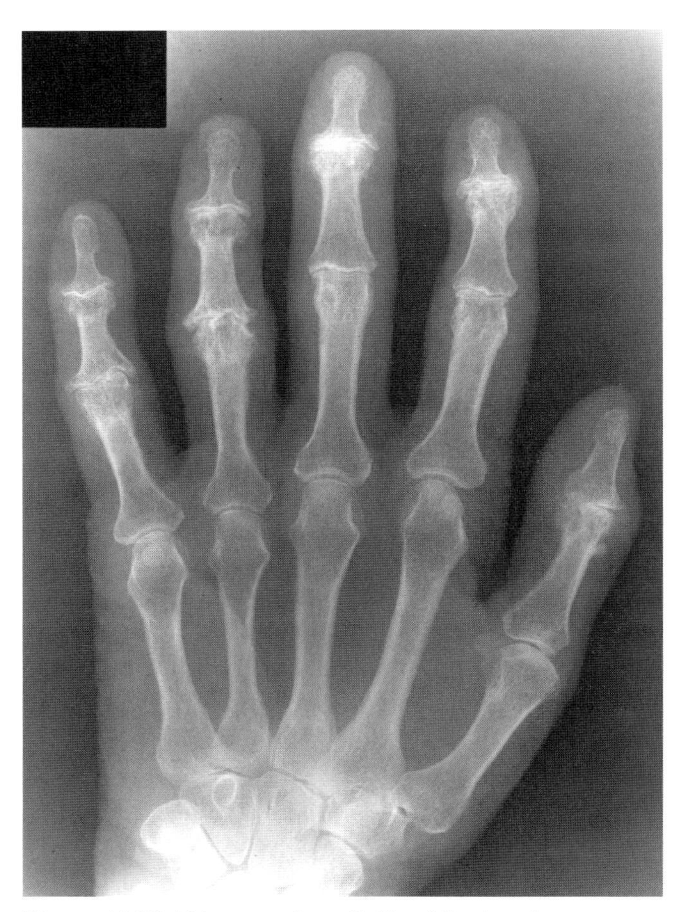

Figure 15.30. Primary osteoarthritis of the hand.

When a patient presents with symptoms of arthritis, the goal of the consulting physician is to confirm the diagnosis and to determine the type of arthritis in order to initiate an appropriate management plan. Many forms of arthritis have characteristic clinical manifestations and laboratory findings, such as the distal interphalangeal joint involvement of osteoarthritis in elderly women, and the hyperuricemia of gout. Radiology, however, may play an important role in the differential diagnosis when the clinical findings are not clear. In addition, radiological evaluation is helpful in staging the degree and extent of involvement for optimal treatment planning. Here we will review three common types of arthritis: osteoarthritis, rheumatoid arthritis, and gout. Remember that your future patients may have other less common forms of arthritis that will yield different clinical and radiological findings.

Osteoarthritis (*degenerative arthritis,* DJD) is the most common form of arthritis. It is prevalent in the elderly but may be seen in younger patients following repeated trauma (professional baseball pitchers, for example, often have early osteoarthritis of the elbow). The characteristic radiological findings are narrowing of the joint space, sclerosis appearing as productive bone of increased density on both sides of the involved joint, osteophyte formation margining the joint spaces, and small cysts in the bone near the joint resulting from fractures of the joint cartilage and penetration of the joint fluid into the juxta-articular bone. The radiological findings may vary depending upon the patient's level of bone mineralization. The degree of sclerosis and formation of osteophytes may be diminished in elderly patients with osteoporosis.

Two types of osteoarthritis are recognized: primary and secondary. Secondary osteoarthritis is generally what has been referred to as DJD, and is trauma related. It is particularly common in the spine, hands, and weight-bearing joints such as the hip and knee. When it is advanced in the hip and knee, treatment may require a prosthetic joint replacement.

Primary osteoarthritis is a congenital form of arthritis that commonly involves the hands in middle-aged women. It too is quite characteristic radiologically, affecting the distal interphalangeal joints most commonly, with the proximal interphalangeal joints next most frequently showing changes. These changes are symmetric bilaterally.

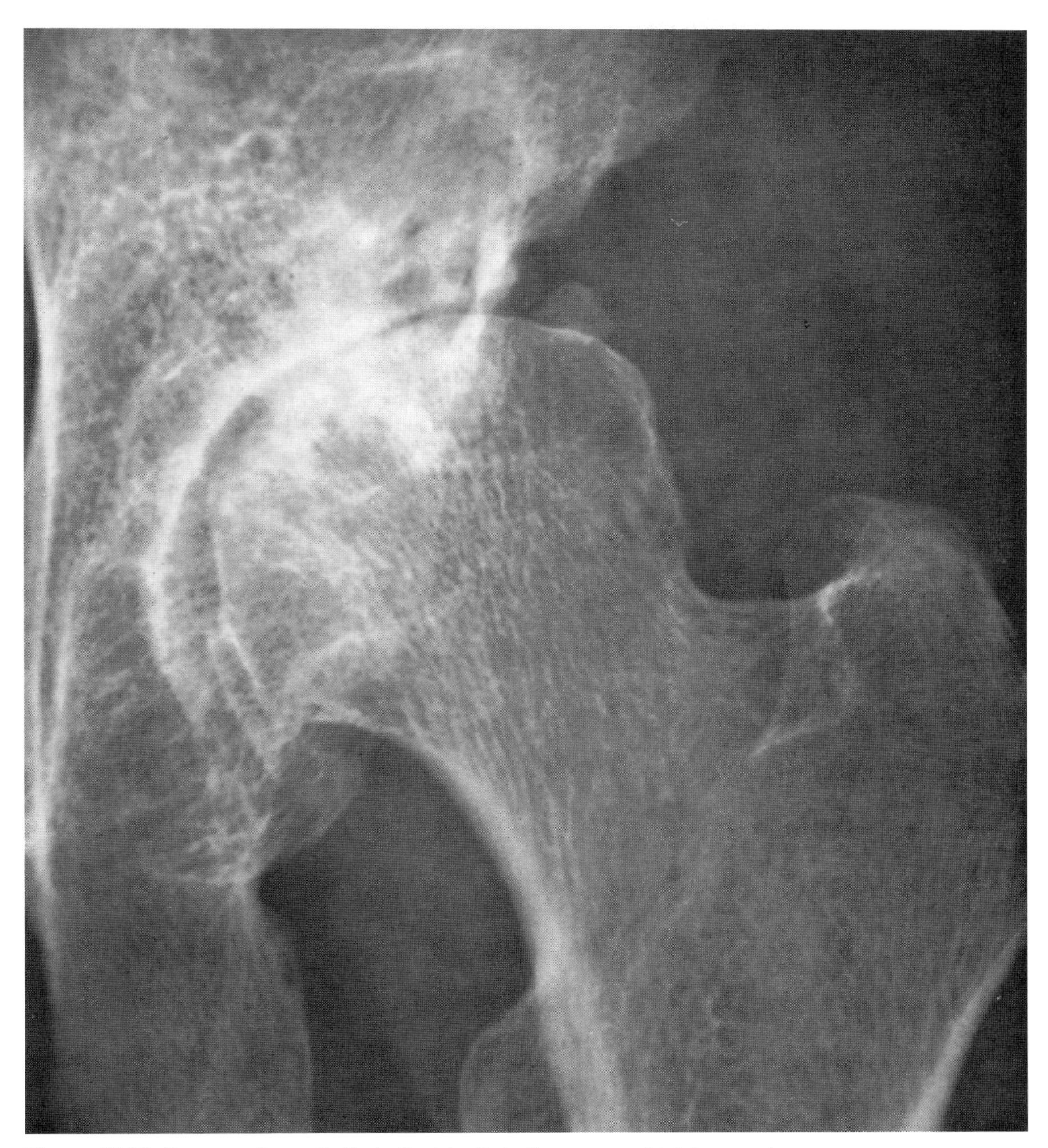

Figure 15.31. Degenerative arthritis in the hip. Note the narrowed joint space in the weight-bearing area, where the cartilage must be very thin and deteriorated. There is reactive sclerosis of bone on both sides of the joint, with lucent cysts forming in the juxta-articular bone.

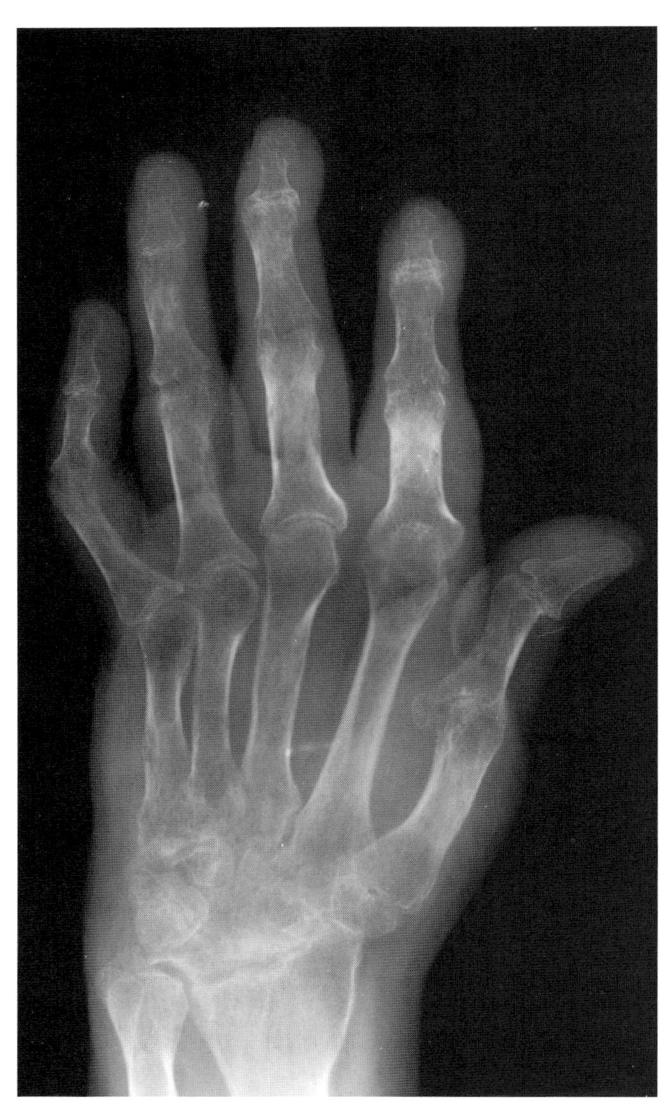

Figure 15.32. Rheumatoid arthritis of the hand.

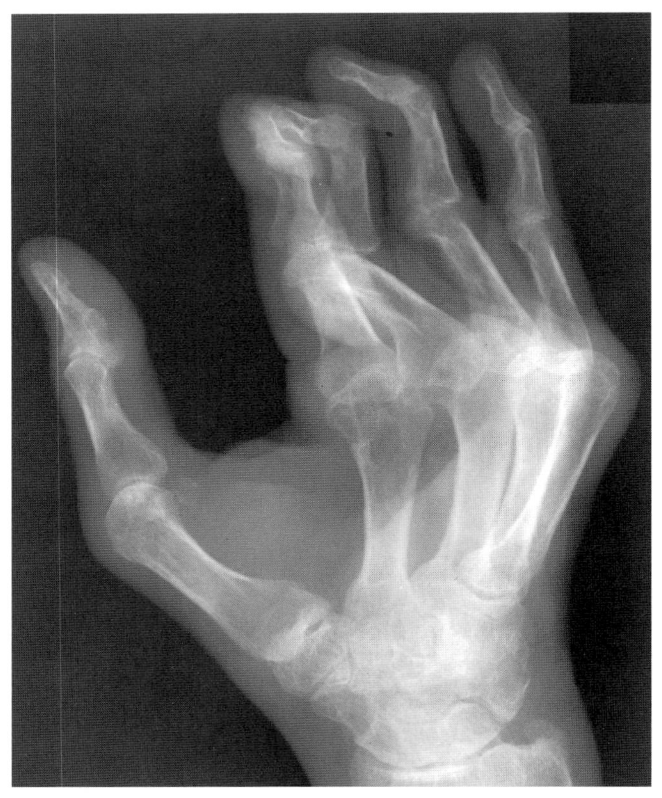

Figure 15.33. Late rheumatoid arthritis with subluxation of the metacarpophalangeal joints.

Classically, *rheumatoid arthritis* occurs in women from 24 to 45 years of age. It is a connective tissue disorder of unknown cause that can affect any of the synovial joints. Patients complain of stiffness, swelling, and pain in the joints, especially the hands, although other joints are commonly involved. The classic radiological findings are soft-tissue swelling, osteoporosis, narrowing of the joint space, and marginal erosions. Rheumatoid arthritis is usually bilateral and symmetrical, and x-rays usually show soft-tissue swelling about the metacarpophalangeal and proximal interphalangeal joints. The juxta-articular erosions of bone are produced by hypertrophied synovium. Ultimately, there will be destruction and even subluxation of the joints, resulting in joint deformity. The disease affects the carpal joints, and the wrist is classically seen to be fused in advanced cases. Late in rheumatoid disease some of the findings of osteoarthritis may be superimposed on the radiographic findings of rheumatoid arthritis, an important point to remember.

Gout is a metabolic disorder associated with hyperuricemia in which monosodium urate crystals are deposited in many tissues, including synovium, bone, soft tissues, and joint cartilage. It should not be thought of as a joint disease. It is at least ten times as common in men as it is in women and is seen most commonly in the hands and feet, especially in the first metatarsophalangeal joint, in which extremely painful soft-tissue swelling accompanies the joint involvement—classic podagra.

The radiological signs of gout are quite specific, but it may take 4–6 years for the x-ray manifestations of the disease to become evident. Most patients will have been diagnosed by clinical and laboratory findings, and successfully treated, before any destructive joint changes occur. Consequently, you may see many normal plain film examinations in your practice of patients who do actually have gout. When radiological findings are present, they typically show large erosions of bone, often with sclerotic, overhanging, hooklike margins where soft-tissue nodules (the tophis) adjoin the erosions. The distribution of joint involvement is random; no particular *group* of joints is affected, and many may be completely normal in appearance. The tophi may or may not calcify. Remember that both rheumatoid arthritis and gout may involve bone erosions. In rheumatoid arthritis the erosions are margined by osteoporotic bone; in gout they are margined by dense bone and even bony sclerosis.

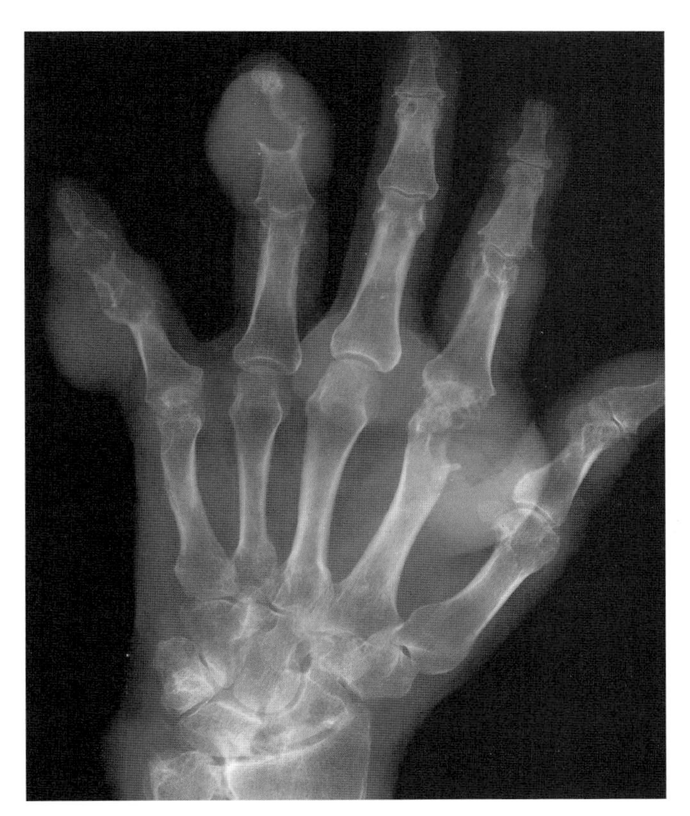

Figure 15.34. Gout of the hand. Note the classic tophi in the soft tissues, scattered rather than regular joint destruction, endosseous tophous inclusions, and hooklike bone spicules remaining near the larger tophi.

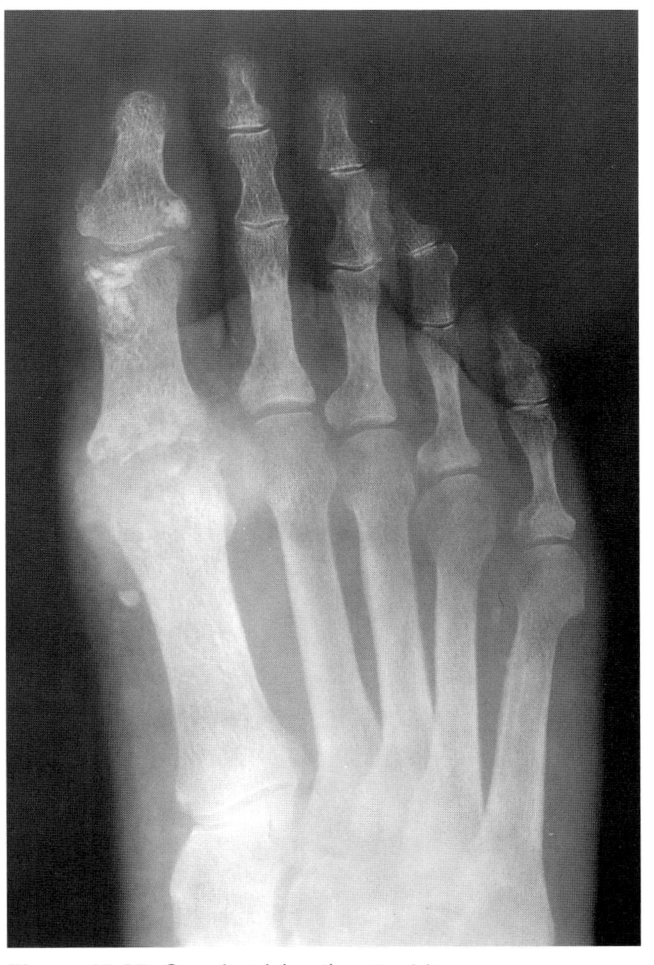

Figure 15.35. Gout involving the great toe (classic podagra).

Osteonecrosis

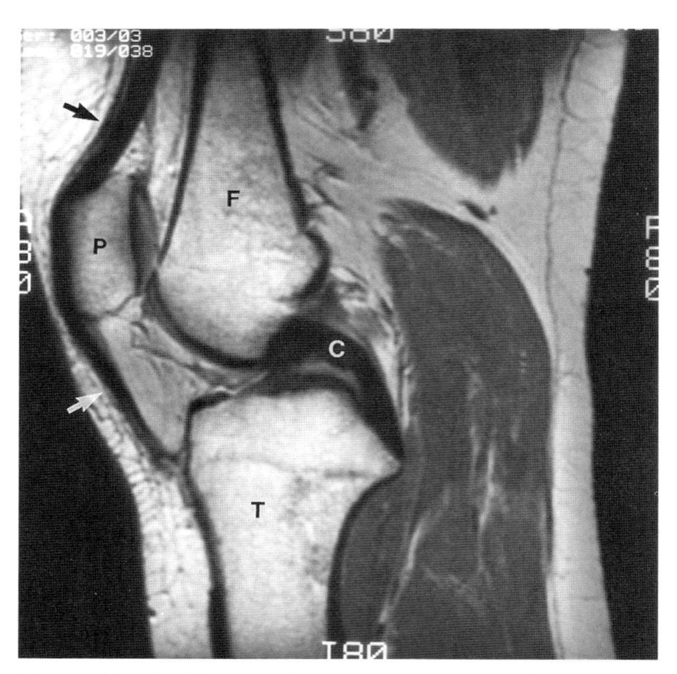

Figure 15.36. MR scan of a normal knee. On this T1-weighted image in the sagittal plane, fat in the subcutaneous fascia (beneath the skin and surrounding the muscles and tendons) and in the bone marrow has a strong MR signal (appears white). Cortical (compact) bone and tendons have little or no MR signal and appear black. Muscles with an intermediate MR signal appear gray. Note the patella *(P),* femur *(F),* tibia *(T),* posterior cruciate ligament *(C),* quadriceps tendon *(black arrow),* and patella tendon *(white arrow).* The soleus and gastrocnemius muscles are seen behind the knee joint and tibia; the hamstrings can be seen higher up, behind the femur.

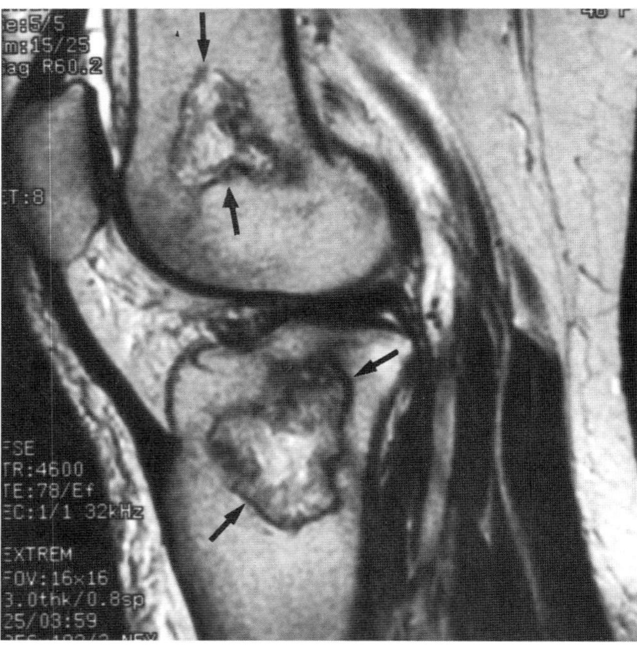

Figure 15.37. MR scan showing osteonecrosis of the knee (after steroid treatment for an organ transplant) involving the marrow of the distal femur and proximal tibia. Note the areas of abnormal marrow signal *(arrows)* characteristic of this condition.

Ostoenecrosis (avascular necrosis, aseptic necrosis), or ischemic necrosis of bone, results from interference with the blood supply to a bone or an involved portion of a bone, which can be caused by trauma, hemoglobinopathies, steroids, and a wide variety of systemic conditions. The changes are not immediate but occur through a progressive series of ischemic injury. The bone structure is initially unaltered, and consequently the early x-ray appearance on plain films is normal because the bone matrix has not yet changed even though cell death has occurred. Later, patchy radiolucencies are seen on plain films, with areas of sclerosis and bone collapse, involving especially the articular surface.

Osteonecrosis is most commonly seen in the epiphyseal marrow cavities of long bones, especially the femoral heads, but can involve almost any joint. Today, one of the most common causes is high-dose steroid therapy after an organ transplant or for treatment of asthma, arthritis, or spinal cord injury. An MR scan is the diagnostic imaging procedure of choice for the early detection of osteonecrosis. After the initial ischemic insult and the death of fat cells, an abnormal marrow signal can be seen. Compare the marrow of the osteonecrotic patient in Figure 15.37 with that of the normal knee shown in the MR scan in Figure 15.36. Always consider an MR scan for patients on steroid therapy who have bone pain and normal bone films. The MR scan can show evidence of osteonecrosis even when bone films and the radioisotope bone scan are normal (Figure 15.38). Later the plain films may show changes in the areas of MR abnormality.

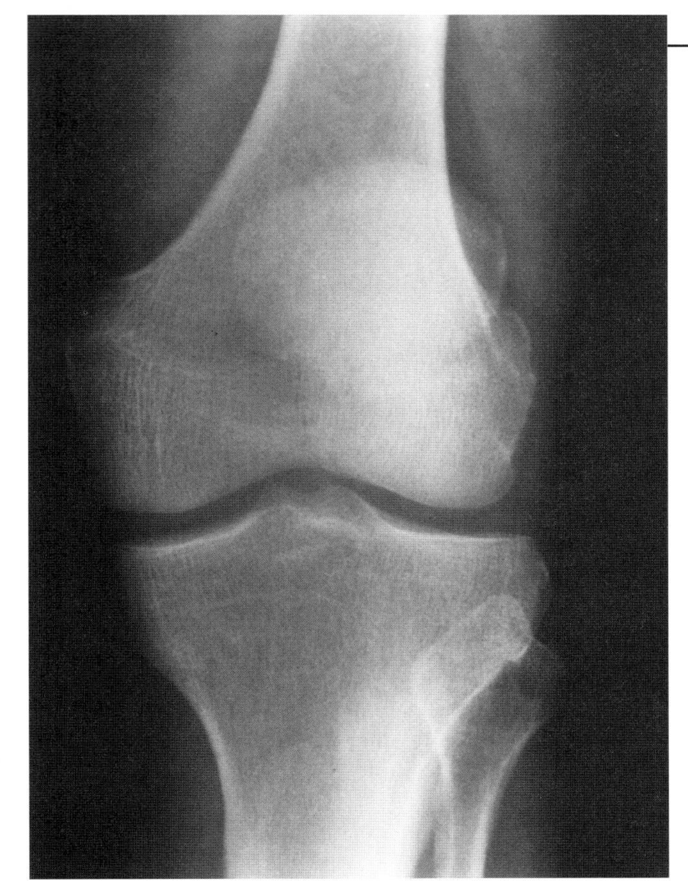

A

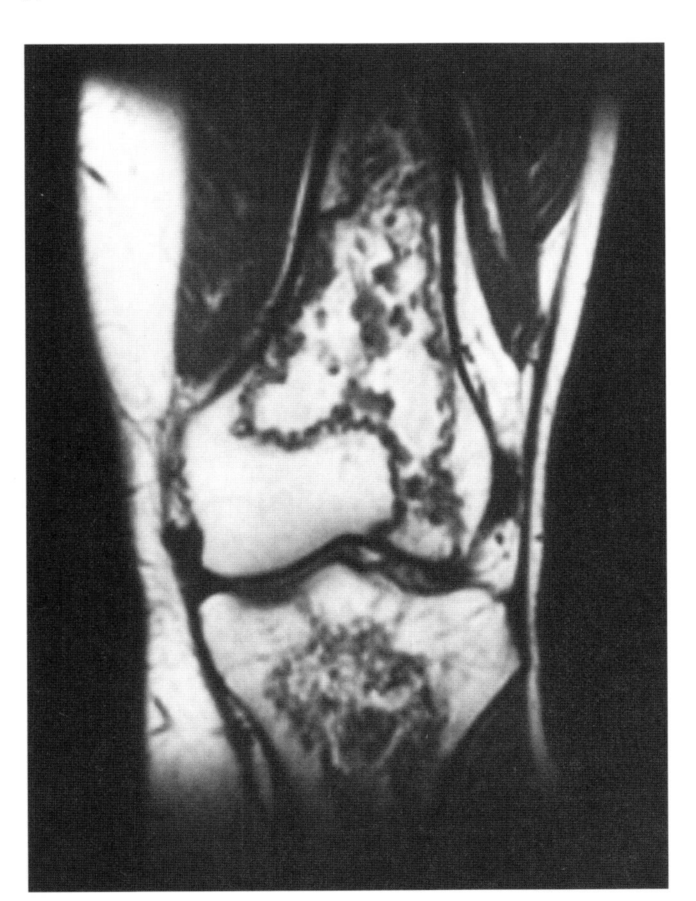

B

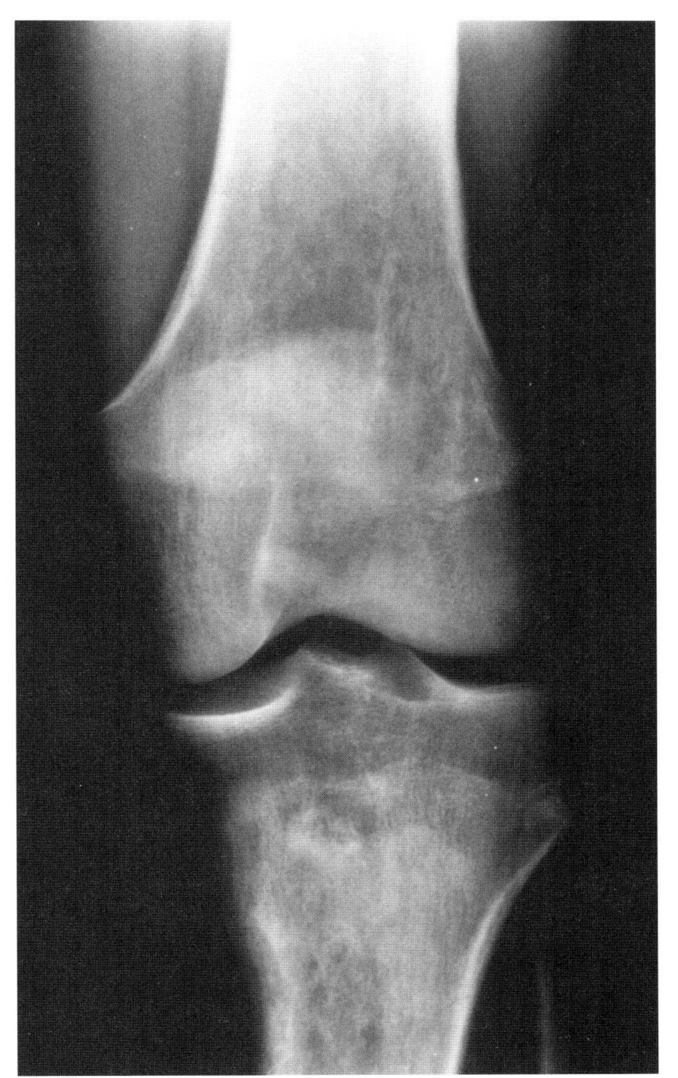

C

Figure 15.38. Another patient with osteonecrosis of the knee occurring after steroid treatment for Crohn's disease. At the time of initial symptoms the plain film examination (A) was entirely normal, but the MR scan (B) showed extensive osteonecrosis in the femur and tibia. Two years later changes caused by osteonecrosis were apparent on the plain film (C).

Microscopic Bone Structure and Maintenance

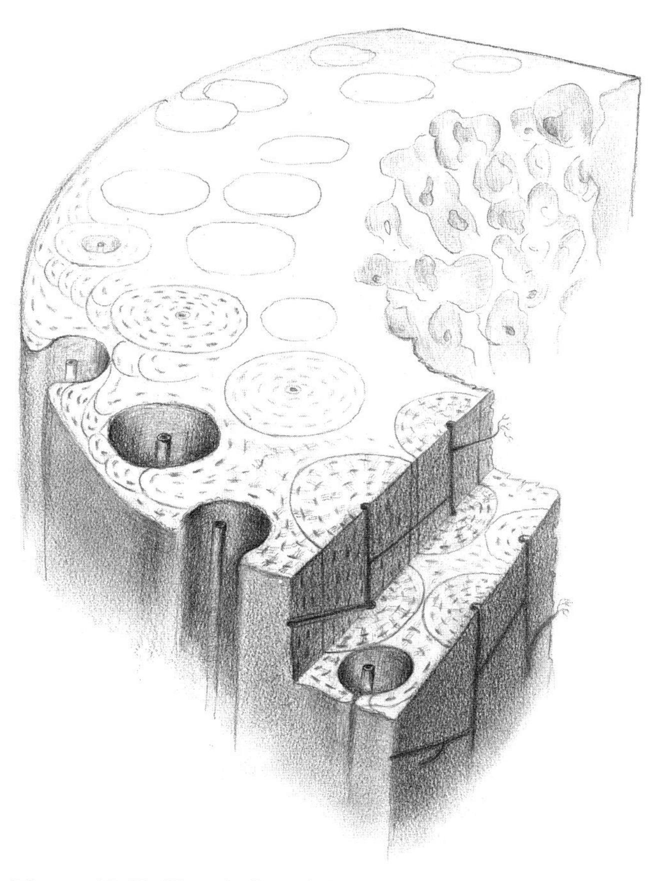

Figure 15.39. Knowledge of the structure of compact bone is important to an understanding of bone disease. This drawing shows both horizontal and vertical cut faces of the femoral cortex. The diagram has been simplified for clarity; of course there are many more osteones in the full cortical thickness (see Figure 15.40).

If you focus on a cross section of the full thickness of the cortex of the femoral shaft you were looking at in Figure 15.1, you will see the structure diagrammed above in Figure 15.39. The cross-cut face reveals numerous sectioned *osteones* (or Haversian systems) with their central vessels. The vertical cut enables you to recall that the basic functioning adult bone unit is a cylinder of very minute dimensions. These cylinders, the osteones, are connected with one another via their branching central arteries and compose, in effect, a "breccia," units mortared together to form a mass. They are of a high order phylogenetically and do not exist in the bones of many lower animals. Structurally they produce a type of bone (and there are several types, remember) that is of excellent resilience and beautifully designed for adaptation to changing needs.

Such a composite of arterially connected units begins to be laid down in the bones of a human infant. This replaces a far less well designed type of immature bone, phylogenetically much earlier in type and resembling a woven fabric rather than a masonry wall. Even in the infant, osteones are concentrated in regions of particular stress, such as important tendon insertions. Eventually in the adult, most compact bone is composed of osteones mortared together by a lamellar bone matrix, as you can see in Figure 15.40, a microradiograph of a thin-ground cross section of the shaft of a long bone, which might have been sliced off the face of the diagrammed bone wedge in Figure 15.39.

The osteones in such microradiographs are seen as rings of varying density around dark holes that once contained arteries. Osteones vary in radiodensity because they are of different ages and therefore contain somewhat different amounts of mineral apatite. The mineral is precipitated very rapidly into the organic collagen bone matrix when it is first laid down, and after that more slowly, over several years.

The useful life expectancy of an individual osteone is around 7 years, at the end of which it is removed by erosion from within until an empty cylinder exists where once it was traversed by the central artery and lined by sheets of mesenchymal cells. These cells differentiate into osteoblasts and lay down concentric new layers of bone matrix, one within another, until the central artery is again surrounded by a new osteone. The osteoblasts become engulfed in the bone matrix they elaborate (after which they are called osteocytes), continuing to function via minute canaliculi that radiate from their surfaces in all directions like the spines of a burr. These communicate with the canaliculi of other osteocytes nearby, so that the bone is able to be perfused with fluid and electrolytes, functioning throughout life as an organ no less important than the liver or kidney.

Because the osteoblasts become engulfed in concentric cylindrical layers of matrix, *in cross section* they appear to be arranged in concentric circles around the artery, just as you see them in the diagram and in the microradiograph. You would *expect* the cells to be radiolucent compared with the mineralized matrix around them and could predict that in the microra-

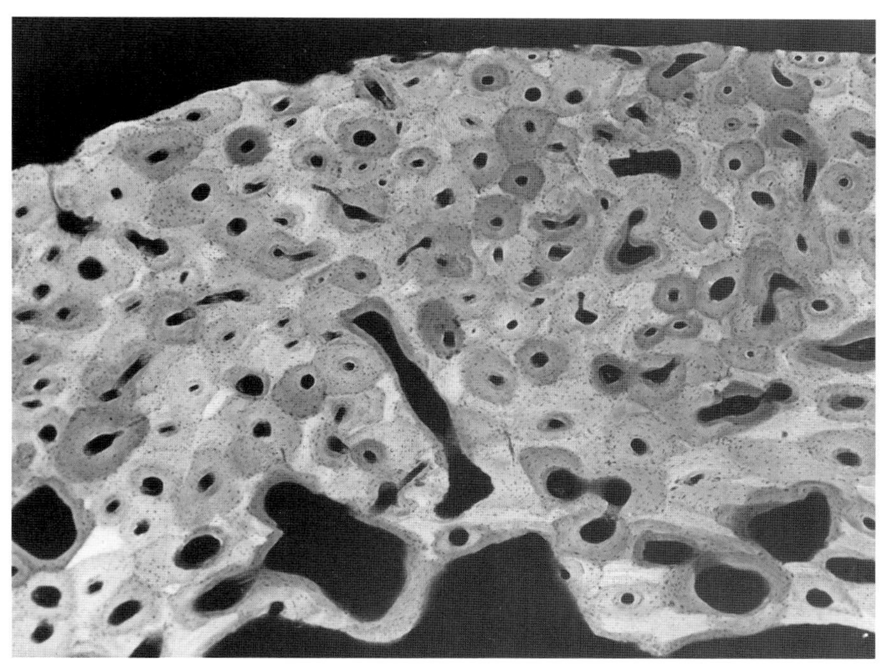

Figure 15.40. Microradiograph of a transverse section of the normal femoral cortex. The specimen was not prepared by sectioning decalcified bone but was sawed off and then ground down to a thickness of a few microns in a fresh state, retaining its normal calcium content. The radiograph was made with the specimen in close contact with photographic film.

diograph they would appear as minute black dots. Note that some of the osteone circles in the microradiograph are very dark; these are the younger ones, less completely mineralized than their white, denser, older neighbors.

If the patient in Figure 15.40 had recently been given an injection of tagged (artificially rendered radioactive) calcium, the younger osteones that you have just identified would now contain much larger amounts of that calcium load, since they would be mineralizing at a more rapid rate than their seniors. An *autoradiograph*, made by placing a fine-grained photographic film in close contact with a section of bone like the one x-rayed to produce Figure 15.40, would show darker spots in the precise locations of the younger osteones, because radioactivity of the calcium isotope produces silver precipitation in the film. When microradiographs, autoradiographs, photographs made with polarized light, and special stain studies of the same bone section are matched in register, there emerges an invaluable means of studying bone pathophysiology at the microscopic level.

By means of *serial* studies of these types, the rate at which osteones form and are mineralized, removed, and restored can be recorded for healthy as well as abnormal bone. The gross appearance of a bone is so suggestive of permanence and durability that it is difficult to accept intellectually the degree to which bones are being constantly changed and remodeled throughout life, and the extent to which they reflect and share in virtually every disease condition. In learning to comprehend these changes, you will find it essential to think clearly of the "flow" prevailing normally in the bones at various periods of life. The infant and young child grow in many ways and at different speeds from year to year, and the growth of their bones has been charted and documented extensively. Most of us fail to comprehend fully, however, that bone laid down in one site this year *may* begin next month to be removed, to accommodate developing changes.

Take any given tendon insertion site as an example. During childhood growth spurts, when the long bones are increasing in length very rapidly, an important muscle tendon inserted at one point soon functions at a disadvantage unless it is moved again close to the joint it subtends. The adductor muscles, arising from the pubis and inserting into the femur posteriorly along its entire length, powerfully adduct the thigh.

The adductor longus inserts into the linea aspera on the posterior surface of the bone about midshaft, a location at which little or no change is occurring, while the adductor magnus inserts farther down into a more limited area by means of a heavy aponeurosis and just above the margin of the epiphyseal growth plate. Here new bone is forming and extending the length of the femur at a very rapid rate indeed. If the tendon insertion of the adductor magnus remained attached at one point, it would soon be inserting farther up the femur and would adduct the thigh much less efficiently.

The concentration of osteones along the linea aspera in early childhood and at all tendon sites allows for a mechanism of adaptation. Through destruction and reconstruction of osteones at slightly different locations, heavy cortex is maintained where the tendon fibers insert. Thus the bones, like every other tissue, adapt and change with growth until maturity. Through the prime years many individual osteones doubtless manage to live out their 7-odd years of usefulness in the same location, but others are removed before that time in order to accommodate to changes pertinent to the habits or activity or health of the individual.

The Development of Metabolic Bone Disease

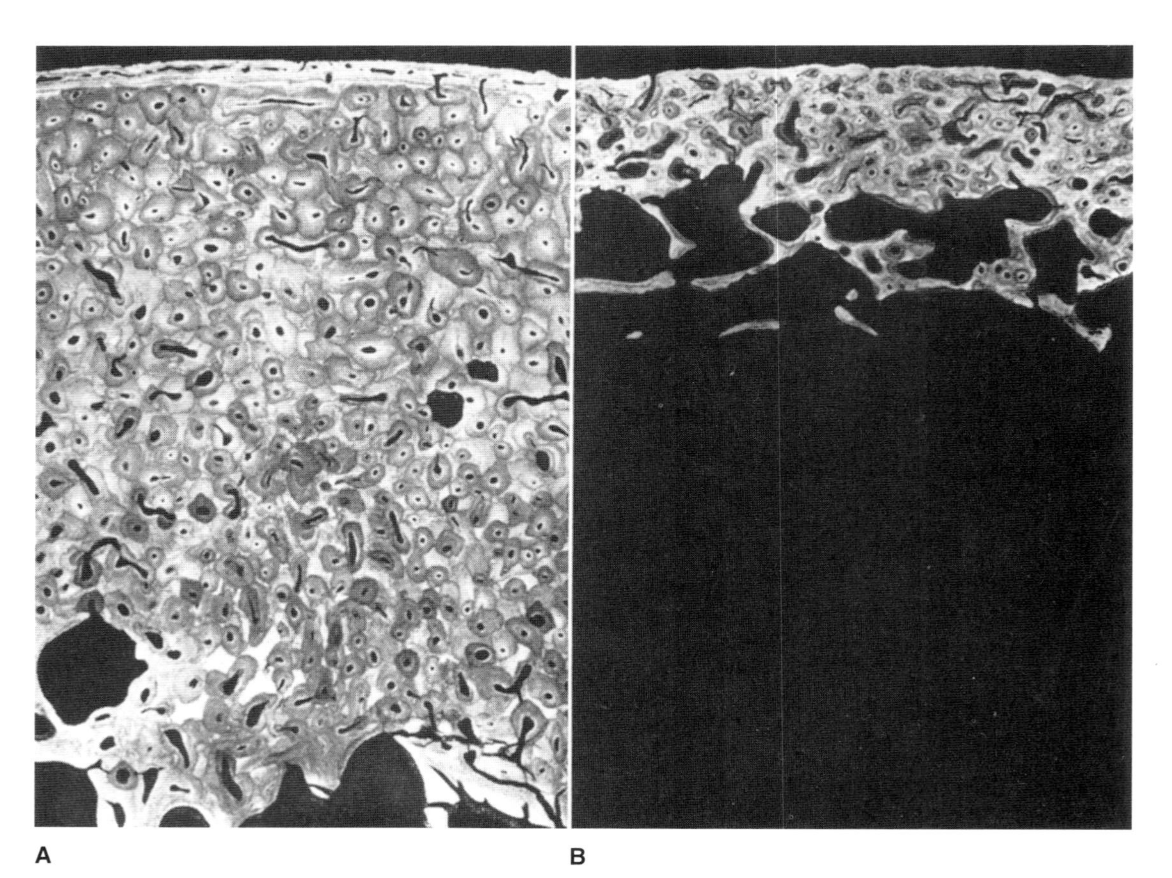

Figure 15.41. Microradiographs of a normal cortex (A) and the cortex in an elderly woman (B) taken during autopsy from precisely the same location on the lateral surface of the femur. The periosteal surface is at the top and the medullary surface below. (See text.)

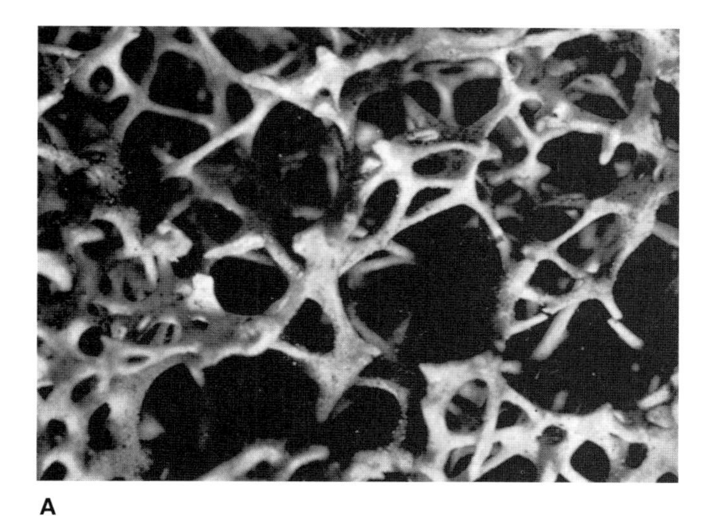
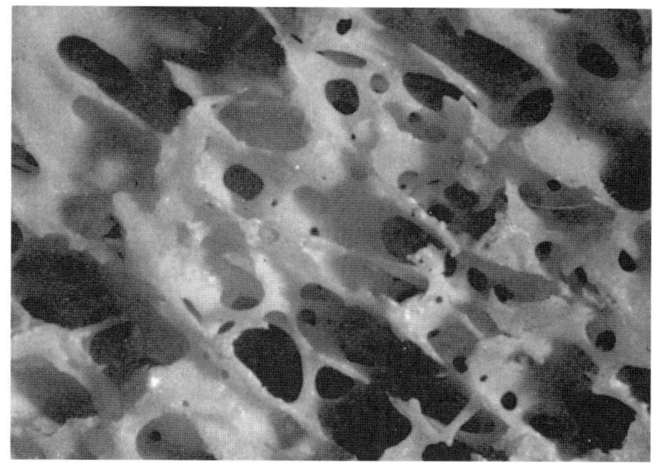

A B

Figure 15.42. Cancellous bone, porotic (A) and sclerotic (B). Normal bone would be somewhere in between. Osteoporotic spongy bone should be thought of as composed of connecting threads of bone, whereas normal bone is composed of intersecting plates of bone.

With the advent of postmaturity and the waning years, the process of bone replacement flags. The normal stimuli to the maintenance of healthy bone begin to diminish. Activity decreases. The appetite declines and the body gets inadequate supplies of the proteins, vitamins, and minerals essential to proper bone building. Hormonal stimuli to bone maintenance gradually abate with advancing years in both men and women, although these changes, occurring earlier in women, have time to produce in them the atrophy of bone known as *postmenopausal osteoporosis,* which might best be thought of as the *net decrease in bone mass prevalent in old age.* In Figure 15.41 you can judge for yourself the lengths to which this decrease in bone mass may be carried. The cortex of the elderly woman in B has gradually been decreased to one-fourth the normally maintained cortex in A, taken from exactly the same location on the femur. As osteones fail to be replaced, the juxtamedullary part of the cortex is gradually converted into a mesh of thin bone segments, which are then removed altogether.

Cancellous bone within the medullary canal also shares in this attritional process. The high-magnification photographs in Figure 15.42 will give you an unforgettable concept of net decrease and increase in bone mass of *cancellous bone,* which is composed not of osteones but of sheets of lamellar bone, laid down or removed by the surface activity of osteoblasts and osteoclasts.

When you consider that bone mass may be decreased either by failure to replace it when it is removed in the course of normal bone maintenance (osteoporosis) or by some extraordinary process of bone destruction, you will understand that the *radiographs* of bones in both conditions look similar: both a thin cortex and thinner, fewer plates of spongy bone make bones appear more radiolucent.

The acceleration of bone destruction in hyperparathyroidism contrasts with the much more gradual reduction in bone mass occurring in osteoporosis. A distinctive (in fact, pathognomonic) x-ray finding is seen in hyperparathyroidism: *subperiosteal erosion of bone.* This is often best appreciated in bones of the hands, where a very thin piece of bone, the phalanx, can be studied in tangent. Of course, the process of accelerated bone destruction is going on throughout the bony skeleton and may also be seen in large bones when they are examined closely.

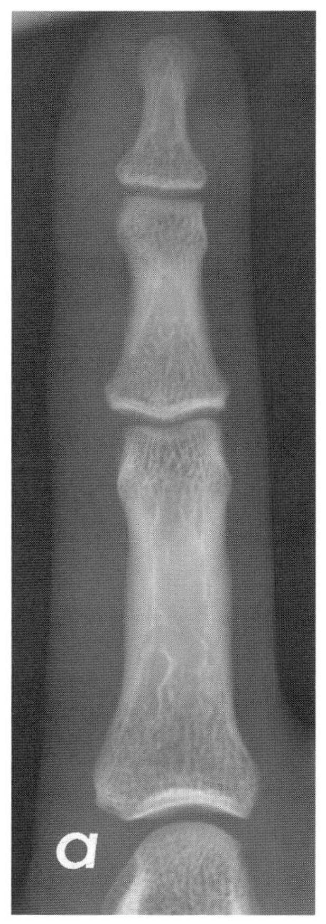

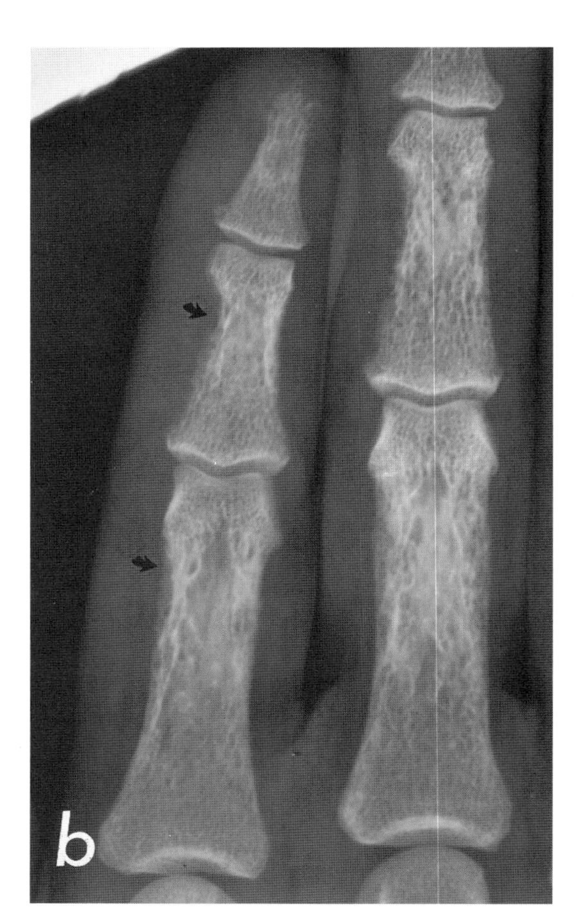

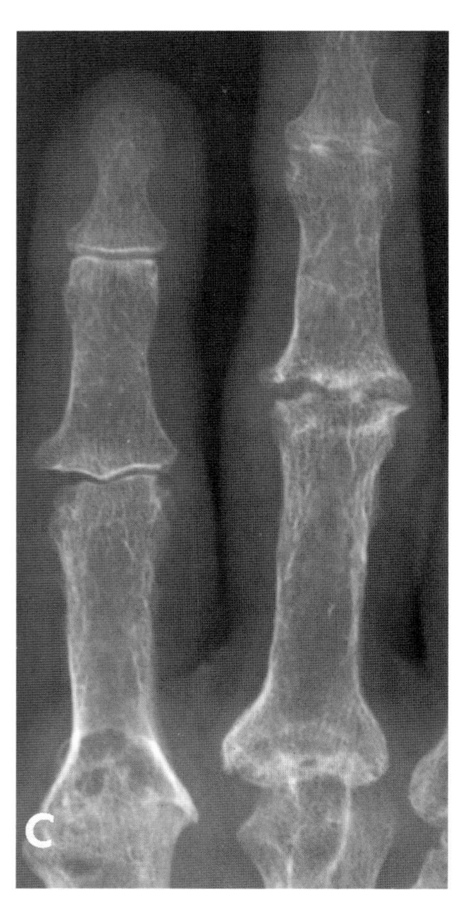

Figure 15.43. (See text.)

Variations in the appearance of cortical (compact) and cancellous (medullary) bone that can be recognized from radiographs may be gauged from the six studies of fingers in Figures 15.43 and 15.44.

The finger in *a* is from a normal young man. Compare the relative thickness of compact bone in midshaft in the proximal phalanx, as well as the size of the individual trabeculae and the marrow space intervals between them, with the abnormal bones in the other five radiographs.

In *b,* you can recognize the bone destruction *(black arrows)* occurring subperiosteally in hyperparathyroidism, which we were just discussing. Note the erosion of the terminal tuft (tuberosity) of the distal phalanx as well as the appearance of destruction in the subperiosteal compact bone, effectively thinning the cortex.

In *c,* the fingers of an elderly man whose activity had been limited for a prolonged period of time by generalized rheumatoid arthritis, the destructive joint changes, especially marked in the metacarpophalangeal joints, are the finding which first strikes you. But look at the cortex and note how thinned and osteoporotic it is even in midshaft, well away from the joints. In any severe generalized illness in which activity is sharply reduced, the gradual development of osteoporosis is inevitable, because the normal stimulus of stress, weight bearing, and muscle pull is so much decreased. In the case of very sudden and almost total interruption of activity, as in the patient who is put into traction in bed after a fracture of the femur, disuse osteoporosis occurs; this may become appreciable radiographically, but only after about half the bone is lost. Thus in many painful bone conditions the typical radiographic picture has *superimposed on it* a degree of disuse osteoporosis, with thinned cortex and trabeculae.

In *d,* sharply margined areas of bone destruction are seen scattered throughout the bones, pressure erosion

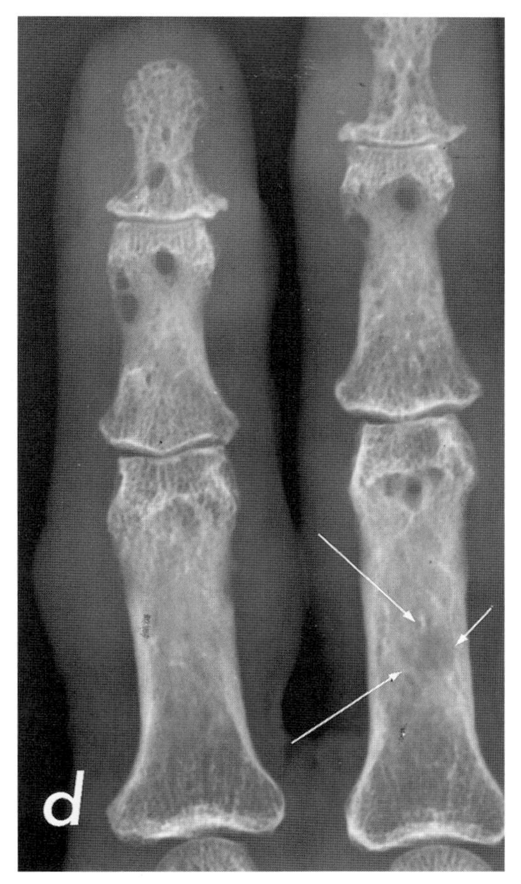

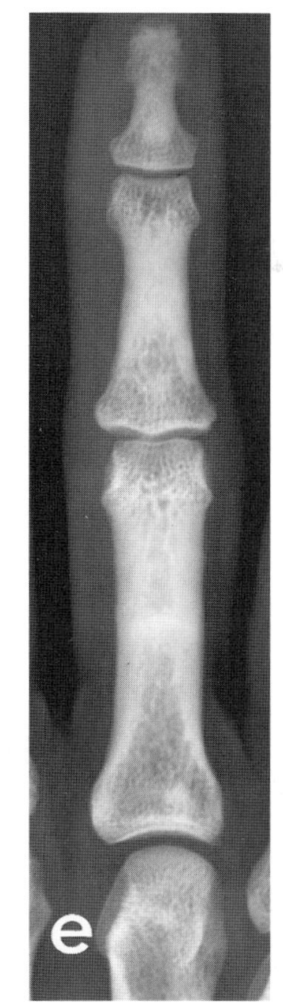

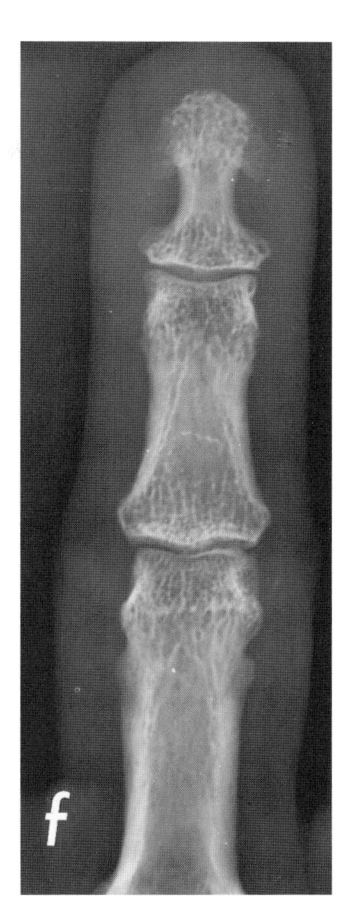

Figure 15.44. (See text.)

from granulomatous foci in a patient with sarcoidosis. Note that where these areas involve principally the cortical bone, they are easier to recognize. When the destruction is mainly of trabecular bone *(white arrows)*, a good deal more bone must be missing before that loss is apparent radiographically. The reason is of course that the areas of loss are masked by normal bone in front of and behind the destroyed area. In a lateral radiograph of the spine, areas of destruction in the body of the vertebra up to 1 centimeter in diameter may be invisible even in retrospect when their presence has been established by autopsy. It is for this reason that today more sophisticated imaging procedures are used to screen for early evidence of the spread of metastases to bone. Although early metastases can be identified by CT, MR, and a radioisotope bone scan, the last is the procedure of choice because the isotope bone scan can image every bone in the body in one examination. This would not be possible with CT and MR. Radioisotope bone scans are exceedingly sensitive and can show evidence of bony metastases long before they are apparent on plain films.

The patient in *e* had osteopetrosis, or marble bones, a rare, inherited fault in which destruction of bone is impaired and old bone accumulates. The cortex is abnormally thick and the marrow cavities may be obliterated, so that patients suffer from failing hematopoiesis. The net bone mass is strikingly increased.

In *f*, the form is faulty. This finger would be recognized by experts as having the typical alteration in form seen in acromegaly. Compare the widely flanged tuft of the terminal phalanx with the others on these two pages. The bones are broad and the bases of the phalanges splayed. Note the distinctive abundance of soft tissues. Roentgen findings of this sort may suggest a diagnosis not suspected clinically, or confirm one that is.

Osteoporosis of the Spine

The spine poses special problems in radiological diagnosis because it is very complex in structure, with very many overlapping bony parts of diverse shapes. CT is, as we have said, of immense help in pinpointing obscure sites of fracture, or in confirming (or excluding) clinically suspected disease when plain bone films have shown no abnormality. CT is also used to measure the degree of osteoporosis, and serial CT scans can evaluate the patient's response to therapy.

In Figure 15.45 you see lateral films of the thoracic and lumbar spine of an older woman with pronounced

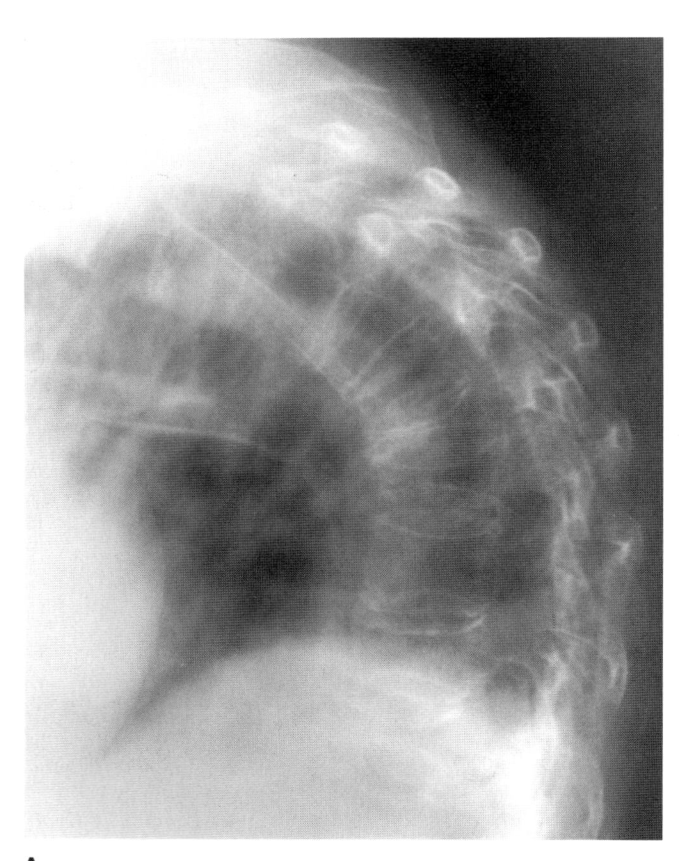

A

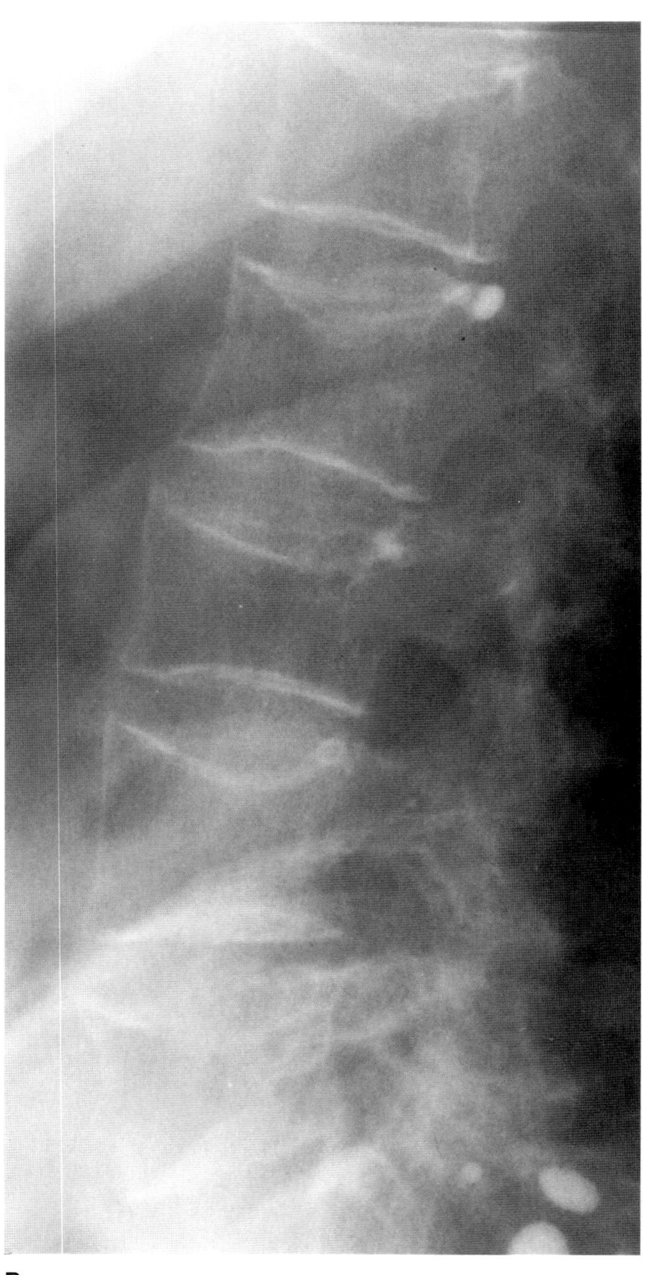

B

Figure 15.45. Osteoporosis of the thoracic and lumbar spine of an elderly woman. Note the extreme kyphosis in the thoracic region at the site of several adjacent compression fractures.

osteoporosis. She has a marked kyphosis of the thoracic spine, produced over the years by gradual loss of bone mass, with thinning of both cortical and spongy bone complicated by multiple compression fractures of vertebral bodies. Each of these individual compression fractures was no doubt symptomatic at the time of fracture as an episode of back pain, lasting several days to weeks, occurring after strenuous activity or mild or moderate trauma. If this patient presented to you with new upper back pain and you questioned whether a new compression fracture might be the cause, could you tell from this film? No, you could not, because new and old vertebral compression fractures in patients with osteoporosis may be indistinguishable on a plain film examination. But you could retrieve any prior spine films or even a prior lateral chest film, and compare the individual vertebrae to differentiate between an old or new fracture. Another alternative would be to order a radioisotope bone scan, which would show increased uptake of isotope with a new compression fracture but no increased uptake with old fractures.

In this woman's lumbar vertebrae, there is no anterior wedging, but there is marked thinning of cortex and radiolucency of the bodies of the vertebrae, largely the result of the loss of spongy bone within them. Compare them with the normal lumbar vertebrae in Figure 15.46; those appear much more dense, with a thicker surface investment of compact bone and denser central spongy bone. Observe also that the normal vertebrae have flat parallel vertebral body endplates on either side of the lucent intervertebral disc spaces. Now compare these spaces with the lozenge-shaped disc spaces in Figure 15.45B. There, because of decreased bone mass and repetitive trauma, the discs have herniated with the endplates (actually endplate compression fractures) into the vertebral bodies, producing a shape for the vertebral bodies that is roughly a biconcave disc. Such vertebrae have been called "fish" vertebrae because many of the larger fishes normally have vertebral bodies so shaped.

Of course, the deformity of vertebrae produced in this way is not unique to osteoporosis, for it can occur whenever there is a decrease in bone mass—as, for example, in Cushing's disease, hyperparathyroidism, or prolonged steroid therapy. Always remember as you examine this kind of change in vertebral body shape that the metabolic process by which it is produced varies from the slow failure of replacement of bone in osteoporosis to the rapid destruction of bone in hyperparathyroidism.

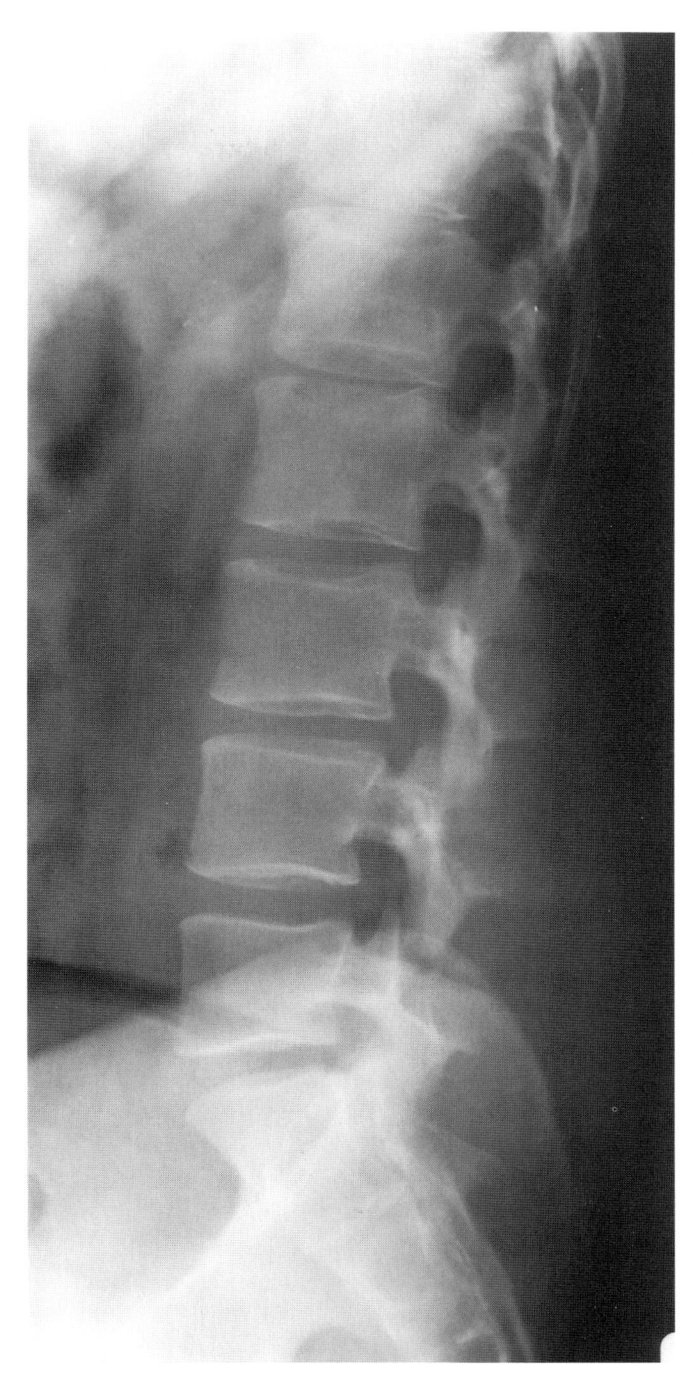

Figure 15.46. Normal lumbar spine, for comparison with preceding figures and those to follow.

Spine Fractures

The patient in Figure 15.47 suffered neck trauma in a motor vehicle collision. At the accident scene he was placed in a cervical immobilization collar by the emergency medical technicians, moved onto a trauma board, and transported to the emergency department. Whenever an injured patient is suspected of having a cervical spine fracture, enormous care in handling and management must be undertaken to prevent a cervical cord injury because the patient may have a cervical spine fracture that is unstable. It would be a terrible tragedy to injure a spinal cord *after* the accident by careless manipulation of the patient. Consequently, all emergency departments, in cooperation with their radiology departments, maintain strict protocols regarding the imaging of suspected cervical spine injuries. The key features are the following. The cervical spine immobilization collar is *not removed* in the radiology area until an unstable spine fracture has been ruled out, by carefully obtained and monitored plain films. If a fracture is detected on plain films, an emergency CT examination is performed for further evaluation. CT may show more fracture lines and fragments than the plain films, and the CT scan can determine whether there was any compromise of the neural canal by fracture fragments. Furthermore, any patient with cervical spine trauma *and neurological signs and symptoms* should also be examined with an emergency MR scan to rule out a spinal cord injury; the MR scan will show any signs of cord contusion, laceration, or hematoma that may require emergency medical or surgical management. The early detection of cord contusion and treatment with high-dose steroids have markedly improved the prognosis of patients with this injury. Always remember that *in the spine trauma patient MR should be performed whenever neurological signs and symptoms are present* even though the plain films are normal, because a soft-tissue injury or hematoma of the spinal cord may occur without spine fracture.

The first plain film obtained is a cross-table lateral film in the immobilization collar. In the patient in Figure 15.47 it showed a fracture of the pars interarticularis of the second cervical vertebra (C2), consistent with a hangman's fracture, a not uncommon fracture after a hyperextension injury. After the lateral view, the patient should be examined with four more plain films, an AP view, oblique views, and a coned-down film of the odontoid process. All of these can be taken with the immobilization collar in place, or the radiologist may choose to remove the collar and

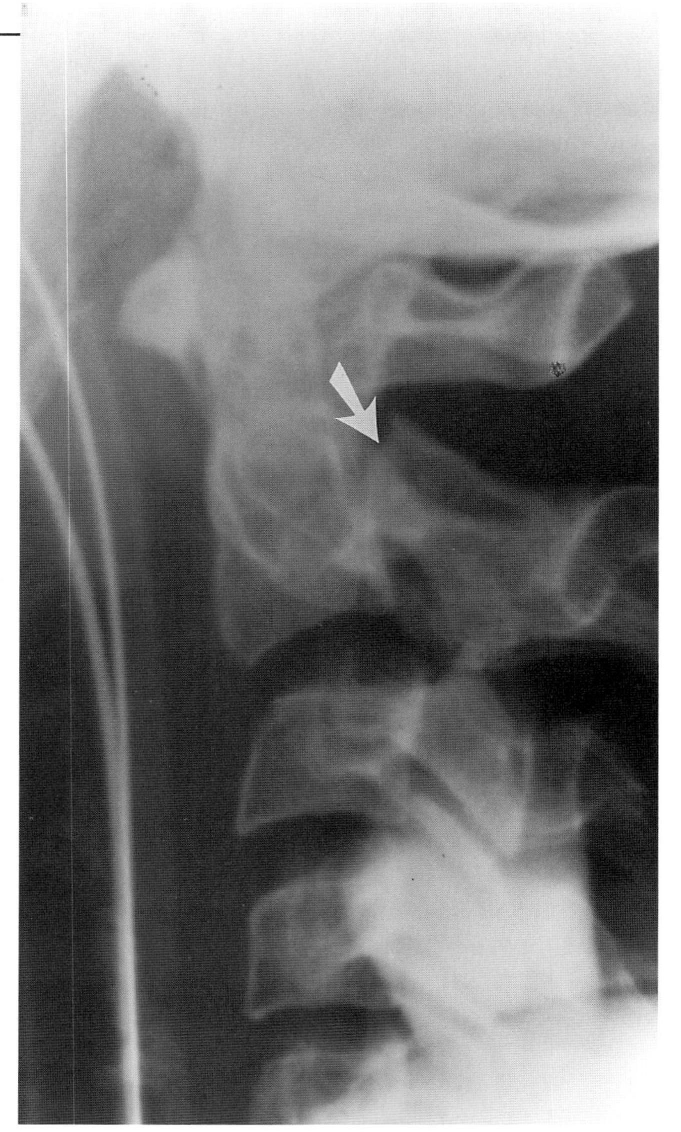

A

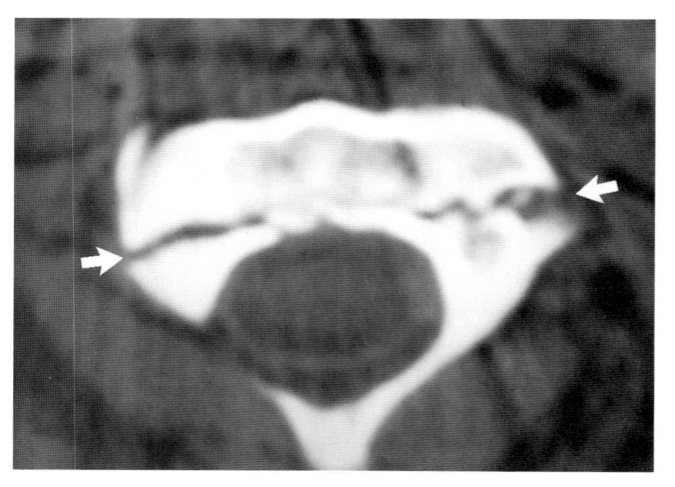

B

Figure 15.47. Hangman's fracture (see text). A: Cross-table lateral film. B: CT scan clearly delineating the location and extent of the fracture. Note that the vertebral canal containing the spinal cord and related structures is not impinged upon by fracture fragments. Compare Figure 15.48.

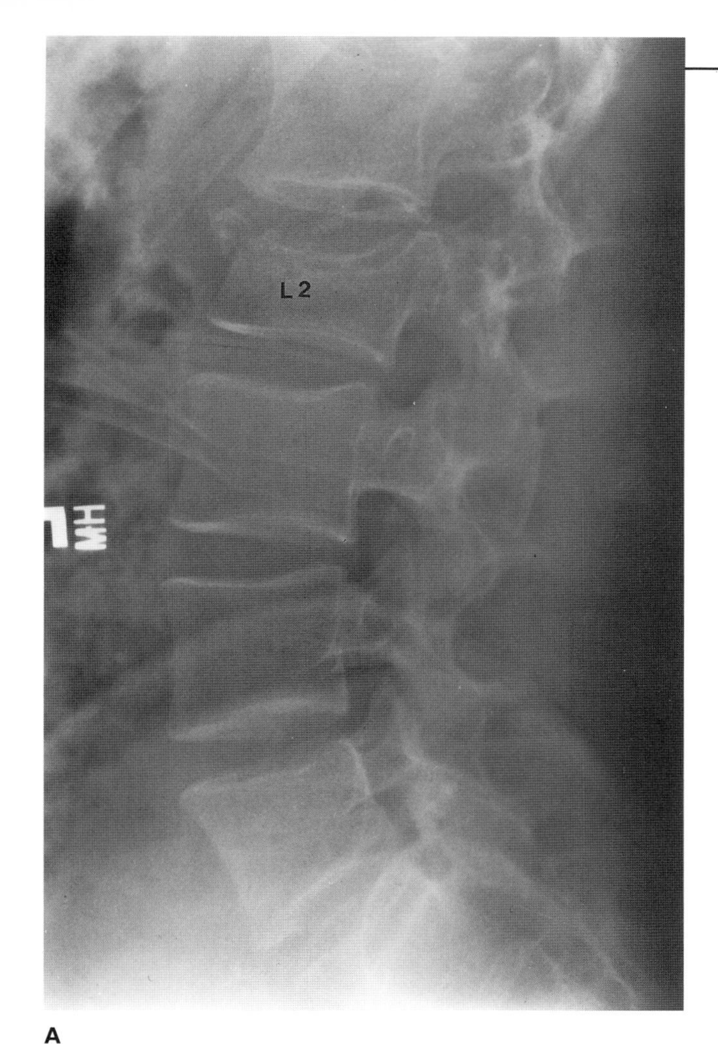

A

B

C

Figure 15.48. Fracture of the L2 vertebral body. A: Lateral plain film shows a compression fracture. B: CT scan reveals a much more serious *burst* fracture, with a large fracture fragment retropulsed into the vertebral canal, impinging upon the spinal cord. C: Lower-level CT slice of the same vertebra, taken below the level of the retropulsed fragment, shows the vertebral canal with a more normal anterior-posterior diameter.

obtain the oblique views with the patient turning his or her own neck, if no fracture was seen on the lateral, AP, and odontoid films. The collar was not removed, of course, for the patient in Figure 15.47; after his plain film series he was taken directly to CT. The hangman's fracture was confirmed, no fragments were identified in the neural canal, and no other fractures were seen in the adjacent vertebra. Today all acute spine fractures shown on plain films should be examined with CT for exact fracture staging and evaluation of the vertebral canal. CT is also indicated when the clinical presentation, such as severe focal spine pain, suggests fracture and the plain films are normal. CT in this circumstance may show a nondisplaced fracture that is not apparent on the original plain films.

Fractures of the spine are frequently obvious on x-ray films, but on further examination they often prove to be more complex than first thought. In the patient in Figure 15.48, the obvious (and painful) fracture of the spine resulting in compression of L2 was suspected clinically and proven with plain radiographs. At CT this fracture was shown to be a *burst fracture* rather than a simple *compression fracture*. Burst fractures are comminuted fractures of vertebral bodies with *retropulsed* fragments in the neural canal (see Figure 15.48B). These fragments may later produce clinical symptoms and signs even when the patient does not initially have a neurological deficit. The identification of a burst fracture is paramount, because its management is very different from the management of a compression fracture. Compression fractures are usually treated conservatively, whereas burst fractures usually require surgical decompression of the neural canal fragments in order to prevent neurological injury.

Osteomyelitis of the Spine

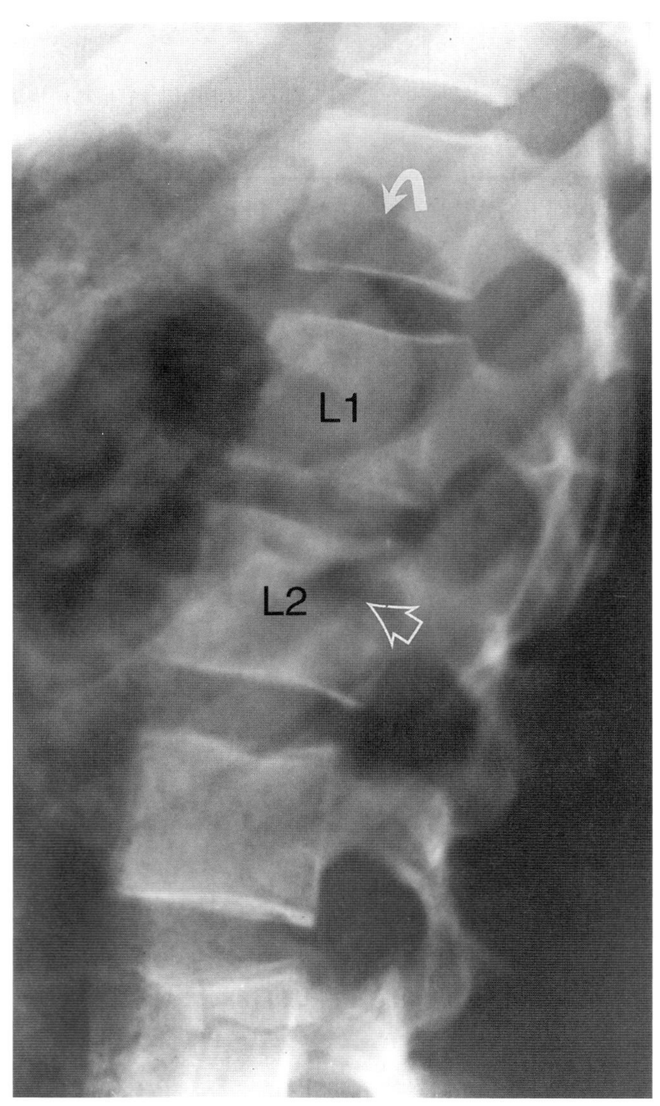

A

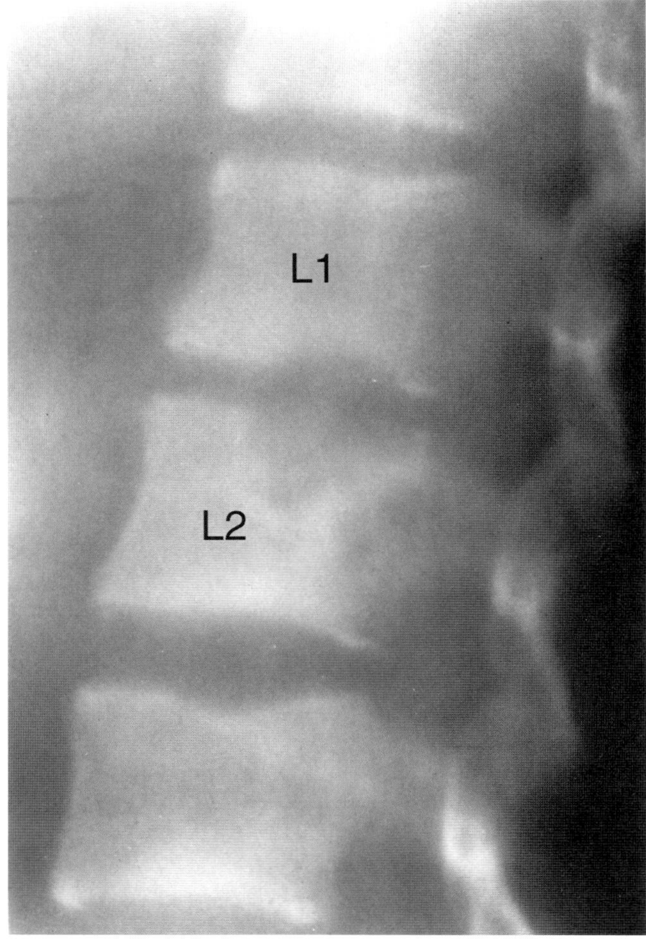

B

Figure 15.49. Osteomyelitis of the spine (see text). A: Lateral plain film. B: Conventional lateral tomogram. C: Lateral MR scan. D: Posterior view bone scan.

The young woman in the four illustrations in Figure 15.49 presented with fever and pain in her back. Plain films of the spine centering across the L1–L2 area were suspicious for bone destruction in the posterior part of the body of L2. There is definite narrowing of the L1–L2 interspace. In Figure 15.49A the *bent arrow* indicates air in the bowel overlying the body of the twelfth thoracic vertebra (T12), but the *open arrow* points to an area of questionable lucency, which is confirmed on the conventional tomogram in B, where areas of bone destruction are clearly seen on both sides of the disc space, typical of the bone destruction in osteomyelitis.

In the lateral MR scan, C, you can determine that there is no involvement of the spinal canal (no abscess). The cord is white and tapers off at the level of L2. Cerebrospinal fluid in the dural sac is black on this MR image. The findings of osteomyelitis are readily apparent at MR. The normal intervertebral discs appear as white, high-signal structures between vertebral bodies. This disc signal is lost at L1–L2, where infection is present. Extending into the L1 and L2 vertebral

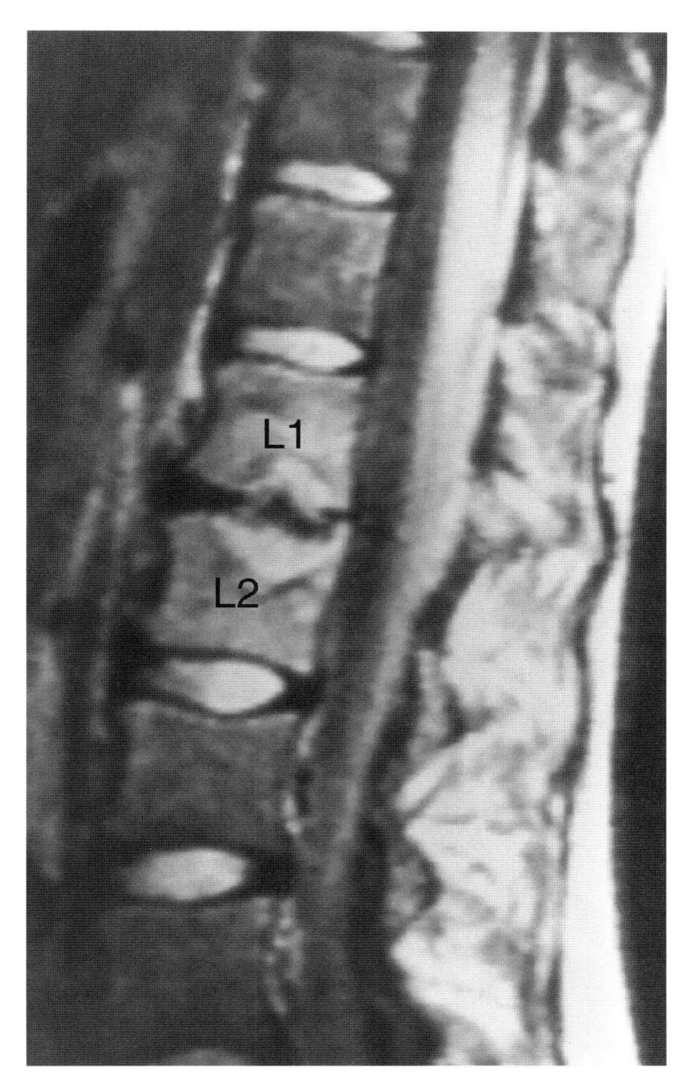

C

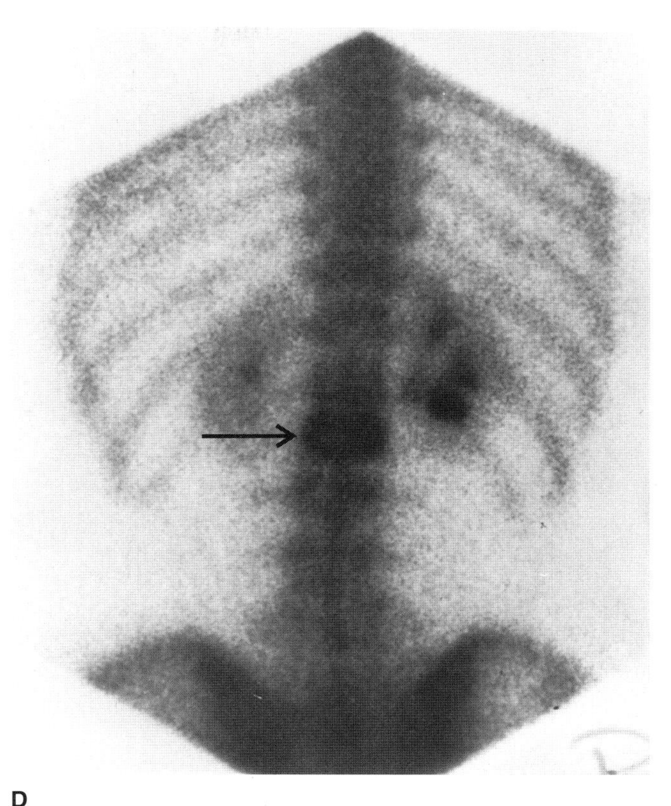

D

bodies are high-signal collections of pus and inflammation that can be differentiated from the normal bone marrow, which looks darker. Note that in D, the bone scan on this patient, much more radioactivity emanates from the area in question *(arrow),* and to a lesser extent from L1, than from the other vertebrae. On further questioning, the patient informed her physician that she had had her ears pierced 2 weeks earlier. Presumably, infection had hematogenously spread to her spine from one of the ear-piercing sites.

Metastatic Bone Tumors

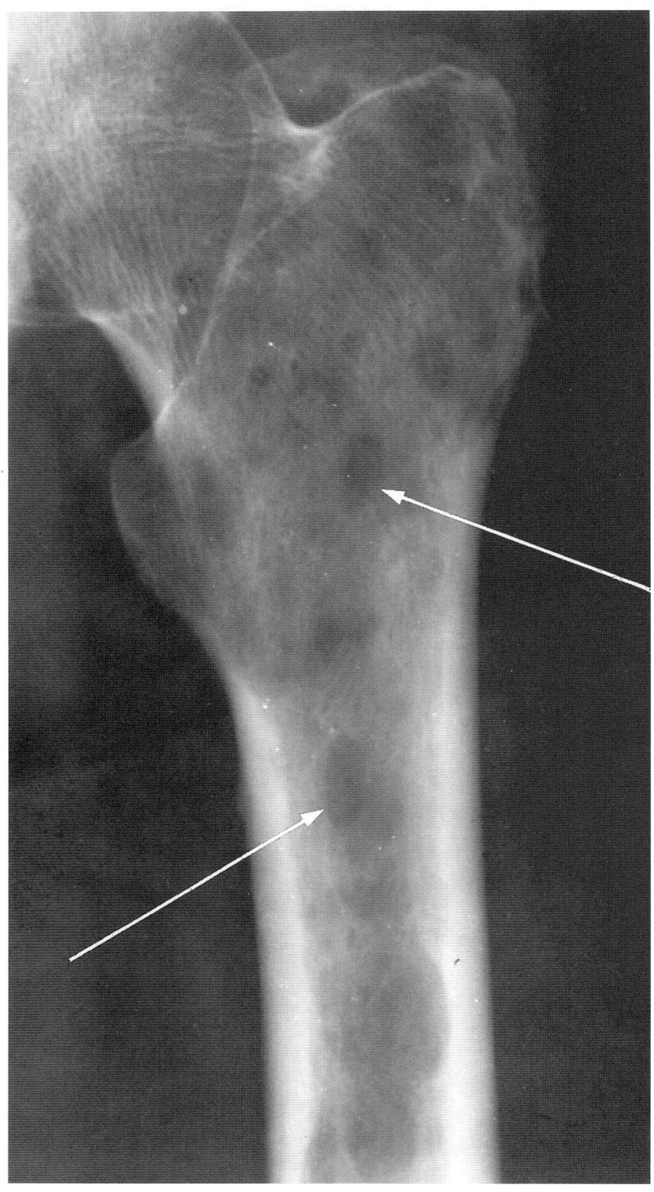

Figure 15.50. Osteolytic metastases. Lytic areas of destruction by growing tumor are the rule from *kidney, lung,* and *thyroid* carcinoma. Carcinoma of the breast may be lytic or blastic, or mixed.

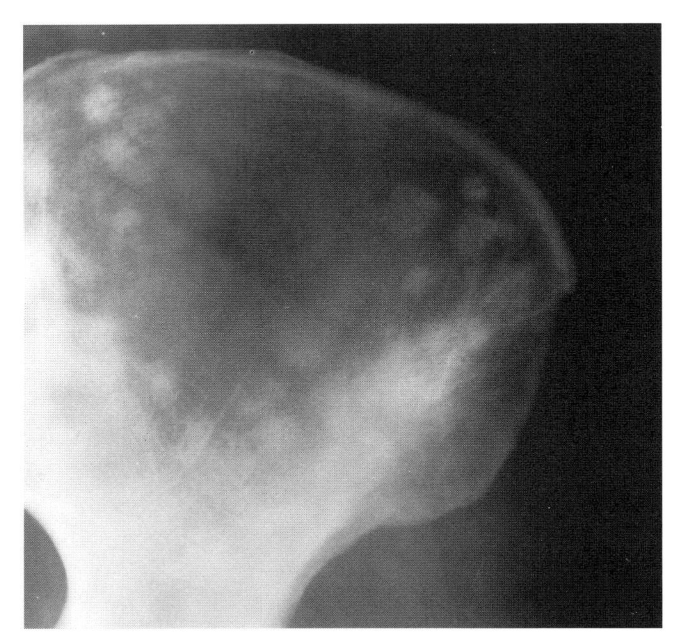

Figure 15.51. Osteoblastic metastases from carcinoma of the *prostate.* New bone formation locally in areas of spread to bone may be spotty, as here, or diffuse, as in Figure 15.52.

During your career you will probably see many more patients with metastatic bone tumors than primary bone tumors, for the latter are very rare. You may not actually be responsible for making the diagnosis of either, which is in part radiological and in part a tissue diagnosis. But you must be familiar with the role of imaging in making that diagnosis.

Almost any kind of malignant tumor may metastasize to bone, but the most common originate from five organs: breast, kidney, lung, prostate, and thyroid. They produce either *lytic* (lucent) or *blastic* (opaque) areas scattered through normal bone on routine bone films. Those that are characteristically lytic are most commonly from carcinoma of the lung, thyroid, and kidney; metastases from carcinoma of the breast are usually lytic (Figure 15.50), although they may turn blastic with therapy. Blastic spread to bone in men is most commonly seen with carcinoma of the prostate. It can be either spotty, as you see in Figure 15.51, or diffuse, as you see in Figure 15.52.

Bony metastases may be detected on a radioisotope bone scan before they are large enough to produce sufficient alteration of normal bone to be seen on plain films. For that reason it is common practice today to procure bone scans in patients known to have malignancy elsewhere who have bone pain, or in whom

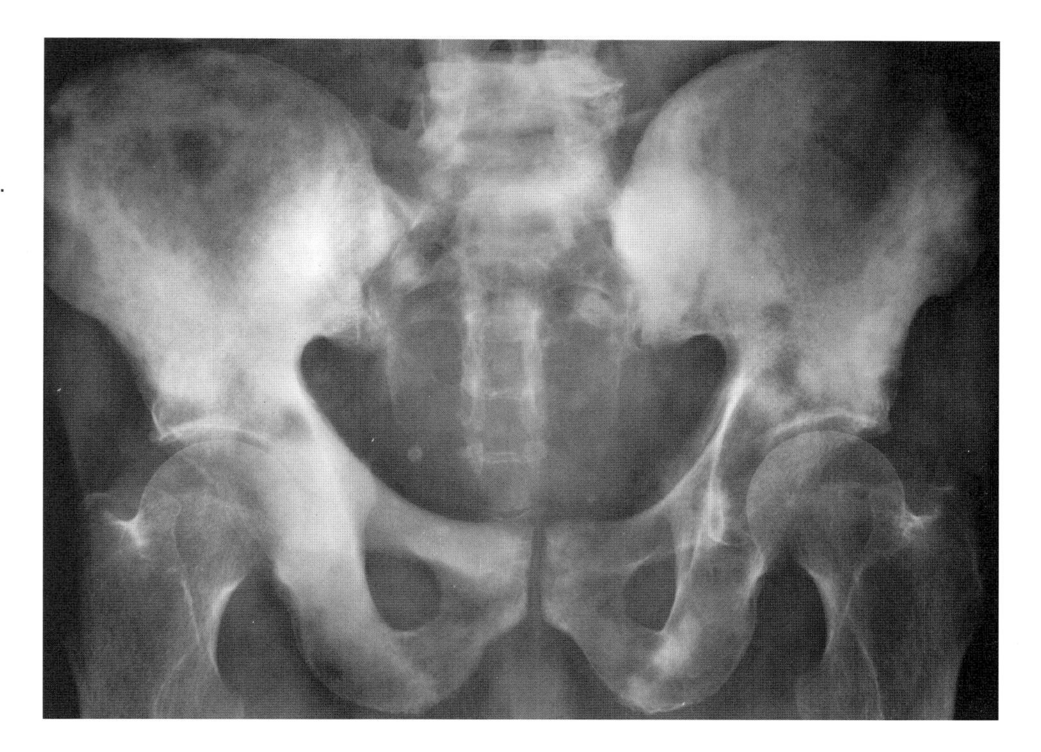

Figure 15.52. The pelvis in a patient known to have carcinoma of the prostate. Note that the bone is not enlarged as it is in the patient with Paget's disease (Figure 15.53).

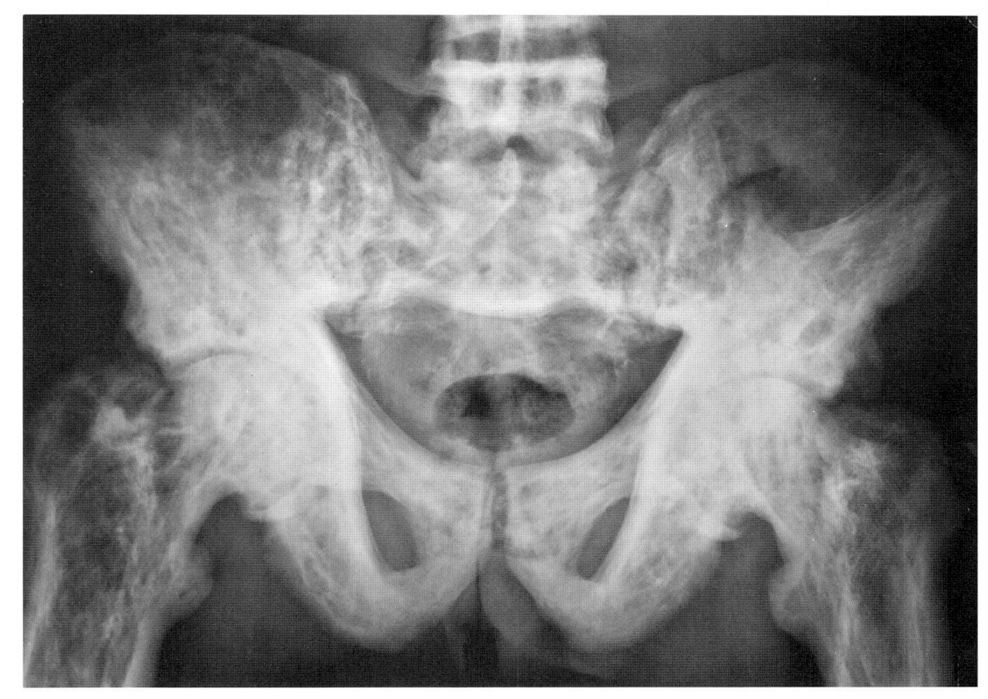

Figure 15.53. The pelvis in Paget's disease. Note the characteristic linear streaking of abnormal disarranged trabeculae and the enlargement of bone due to subperiosteal new bone formation. It is important to distinguish this condition from the increased density in prostatic carcinoma metastases, because the two diseases are apt to be seen in the same group of patients (elderly men).

therapy might be altered if you knew there was already spread of the disease to the bones. Generally, if there are "hot spots" in asymmetrical areas on the bone scans in such patients, the radiologist assumes that they represent metastases—although the bone scan is not specific for tumor, and an intense isotope signal indicates only augmented bone turnover, which can result from trauma, arthritis, infection, and Paget's disease, for example, as well as a tumor. The patients in Figures 15.52 and 15.53 would both have had widely positive bone scans, although only the first had a tumor.

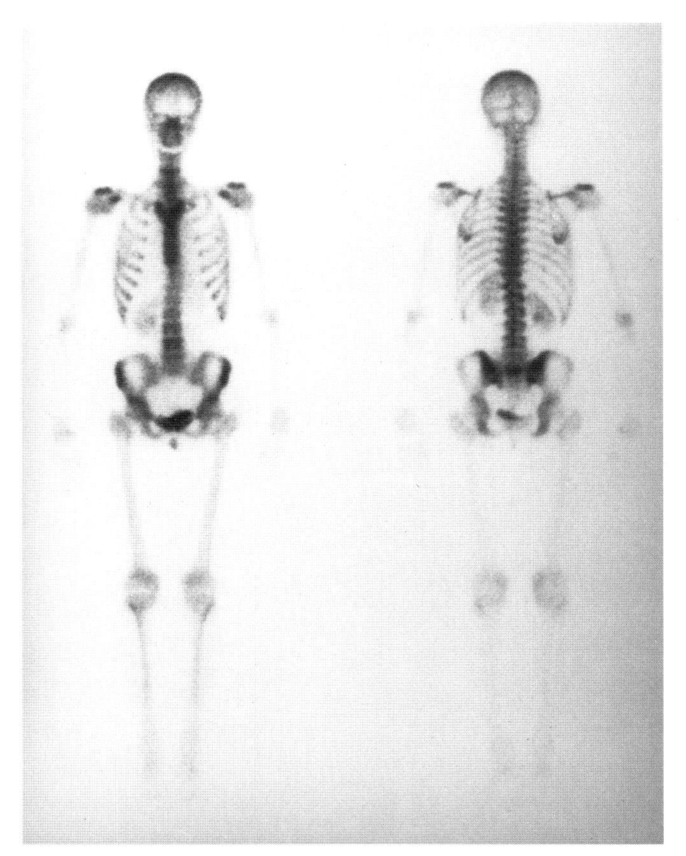

Figure 15.54. Normal adult bone scan: anterior *(left)* and posterior *(right)* views.

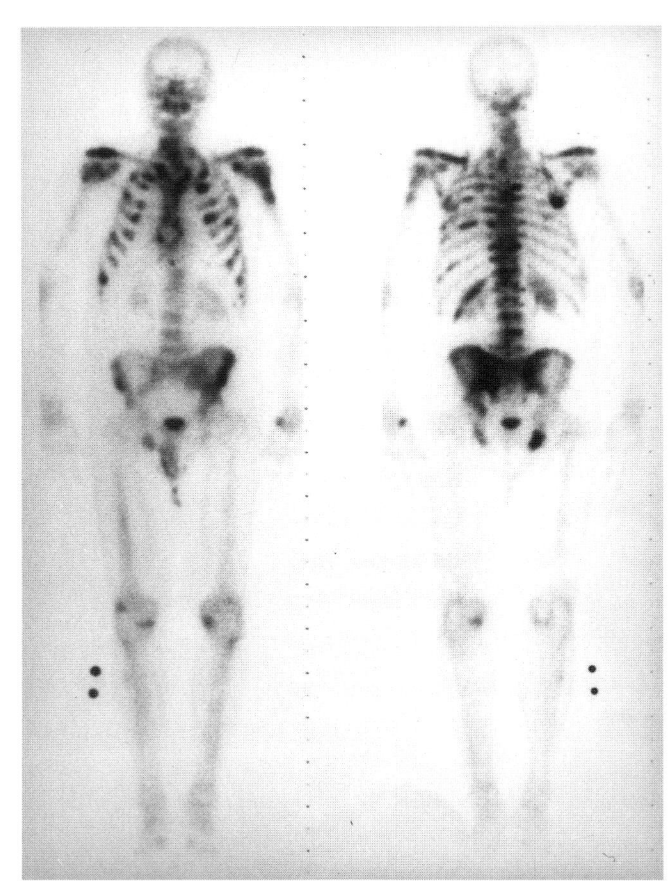

Figure 15.55. Abnormal bone scan in a woman with metastatic breast cancer: anterior *(left)* and posterior *(right)* views.

Compare the normal adult radioisotope bone scan in Figure 15.54 with the abnormal bone scan of a woman with metastatic breast cancer in Figure 15.55. Note the multiple *asymmetric* sites of increased isotope activity in the ribs, spine, pelvis, and shoulders, representing bony metastases. Now compare both of these scans with the scan of a child in Figure 15.56; the *symmetric* sites of increased activity at the joints are normal for a growing child.

In another patient with breast cancer and back pain a lateral lumbar spine film (Figure 15.57A) showed a pathological compression fracture (compression of the vertebral body plus increased bone density consistent with a blastic metastasis) of L2 *(arrow)*, but normal-appearing L1 and L3. It is difficult to be sure of abnormality in L4, which is partly overlapped on the density of the iliac wings. Pathological fractures may occur in bones affected by primary and metastatic malignancies, osteoporosis, osteonecrosis, and benign conditions that weaken bone, such as bone cysts. The

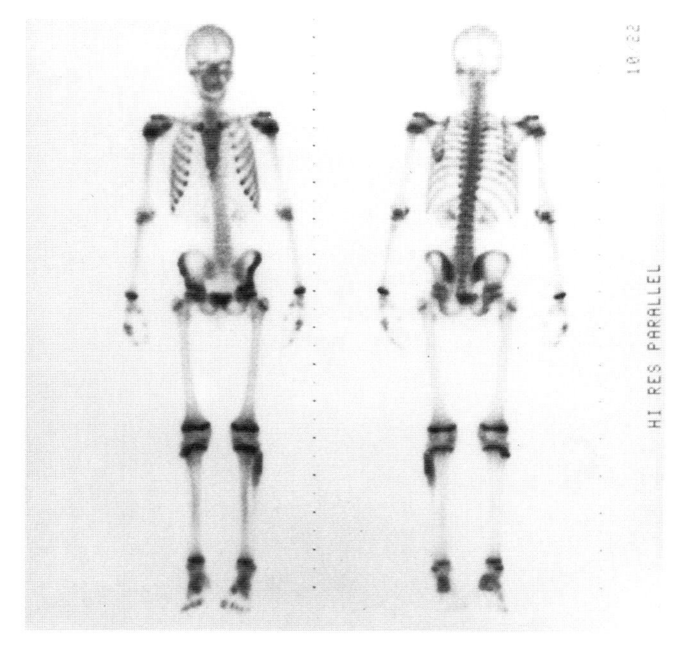

Figure 15.56. Abnormal bone scan: a child with a bone tumor in the proximal left fibula.

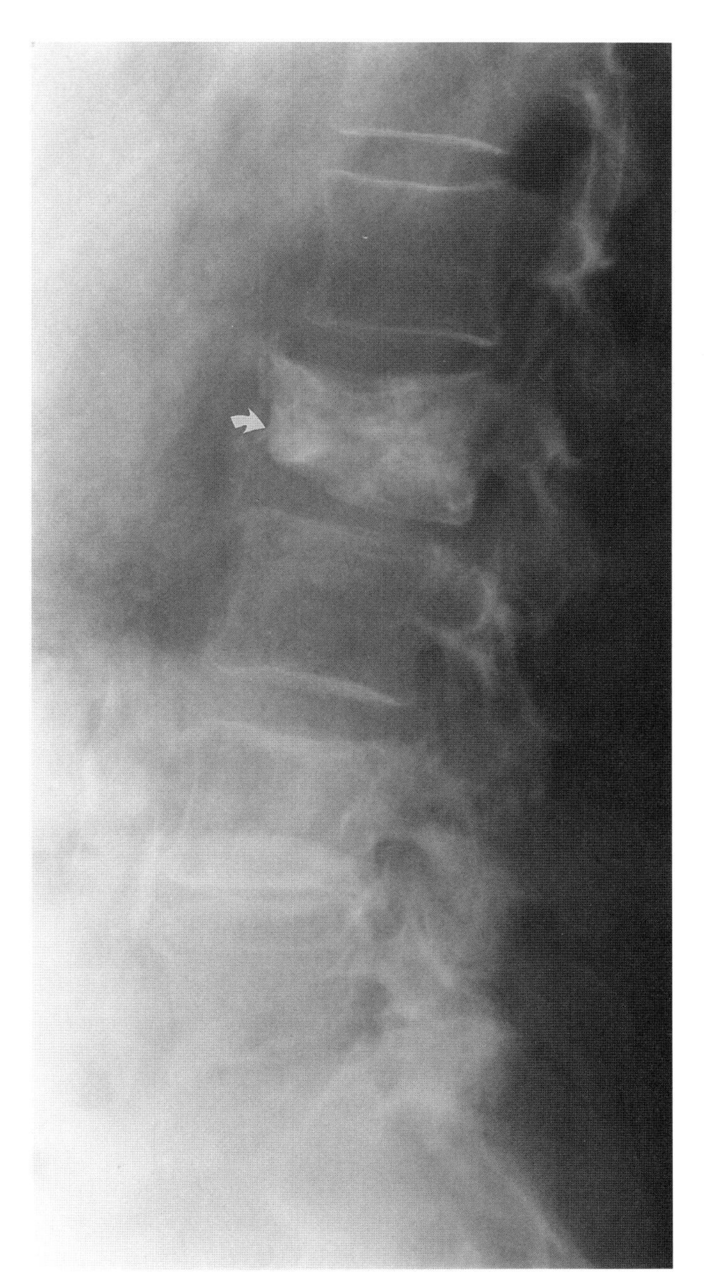

A

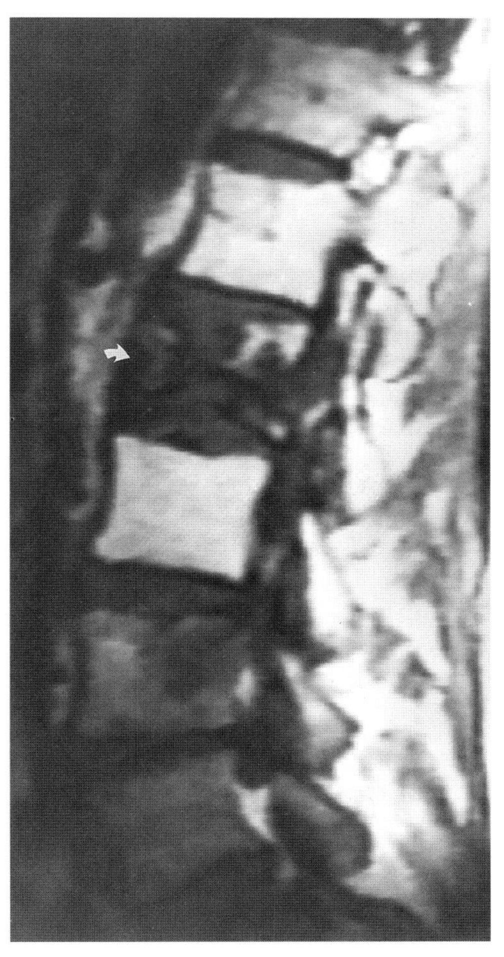

B

Figure 15.57. A: Lateral lumbar spine film of a patient with breast cancer and back pain (see text). B: Sagittal MR spine scan of the same patient (see text).

plain films of the fracture may show evidence of its pathological etiology by the appearance of the underlying bone itself.

Figure 15.57B shows an MR image of the same segment of spine. The second lumbar vertebra (the *arrow* indicates L2) gives off almost no signal, because its fatty marrow has been replaced by tumor, which gives no signal. In contrast, the bodies of L1 and L3 on either side of it give an intense fatty bone marrow signal (nearly homogenous white signal throughout the vertebral bodies). But L4 and L5 are also involved: the decreased marrow signal of these vertebrae is consistent with metastatic spread.

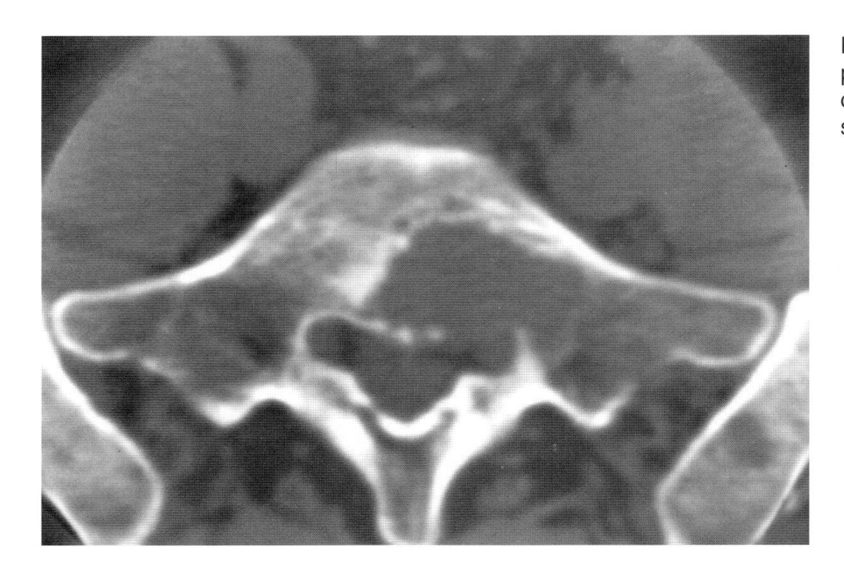

Figure 15.58. Breast cancer in a different patient. Plain films showed no spread of the cancer but the CT scan revealed it had metastasized to the sacrum.

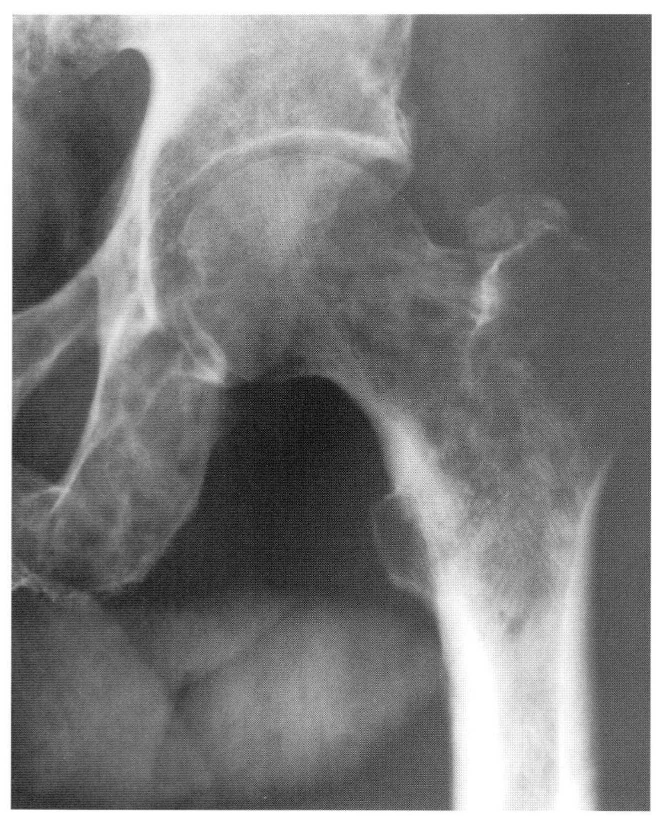

Figure 15.59. Multiple myeloma.

Figure 15.58 is another patient with known breast carcinoma who had lower back pain in the sacral area but whose bone films did not show a definite lesion because of overlying gas in the bowel. This CT scan across the sacrum shows a large area of destruction due to metastatic disease. Like the bone scan, CT and MR may show bony metastases before they would appear on plain films; but since a radioisotope bone scan images the *entire* skeleton in one examination, it is the ideal initial survey study. CT and MR may be helpful when only a limited portion of the skeletal system needs investigation. Although the bone scan is an extremely *sensitive* exam for bony metastases, it is not very *specific,* and may show "hot spots" due to other bony abnormalities in a patient with no metastases. In this situation CT, MR, and plain films may be called upon to determine the nature of a hot spot.

The scattered lytic lesions in Figure 15.59 may look somewhat like those you have been seeing, but they are not from kidney, thyroid, lung, or breast cancer; they are the late "punched-out" areas of bone destruction in multiple myeloma. The patient was known to have myeloma and had this film made when he complained of hip pain. Such lesions occur rather late in the disease and often are preceded by years of bone pain in which plain bone films are entirely negative, showing only diffuse decrease of bone mass with loss of both cortical and spongy bone but no localized destruction. A diagnosis of osteoporosis is frequently offered on the basis of these films before the localized areas of destruction ultimately develop.

The man in Figure 15.60A was being studied because of an 8-week fever of unknown origin. He was 46 years old and also had anemia, but his urinalysis was normal. He complained of right shoulder pain. The plain film shows a lytic lesion in the right acromion *(arrows)*. A radioisotope bone scan showed this to be the only such area of abnormal isotope activity. CT study through the level of the kidneys (Figure 15.60B) shows the source of the metastasis, a primary tumor of the left kidney. Renal cell carcinomas are often locally asymptomatic but usually are associated with anemia and fever; their first clinical sign may be that of symptomatic metastases.

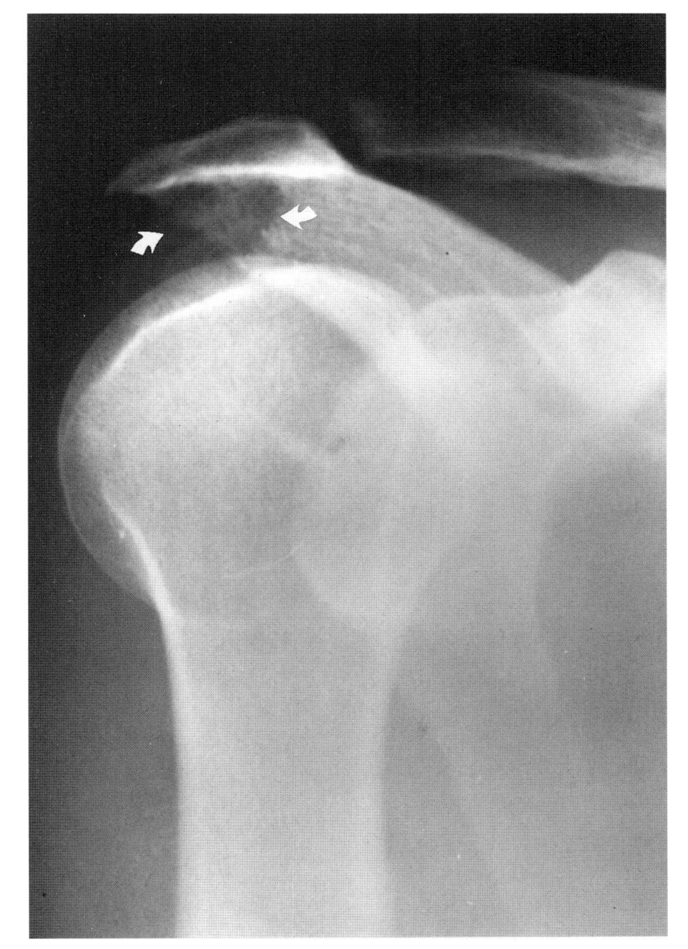

A

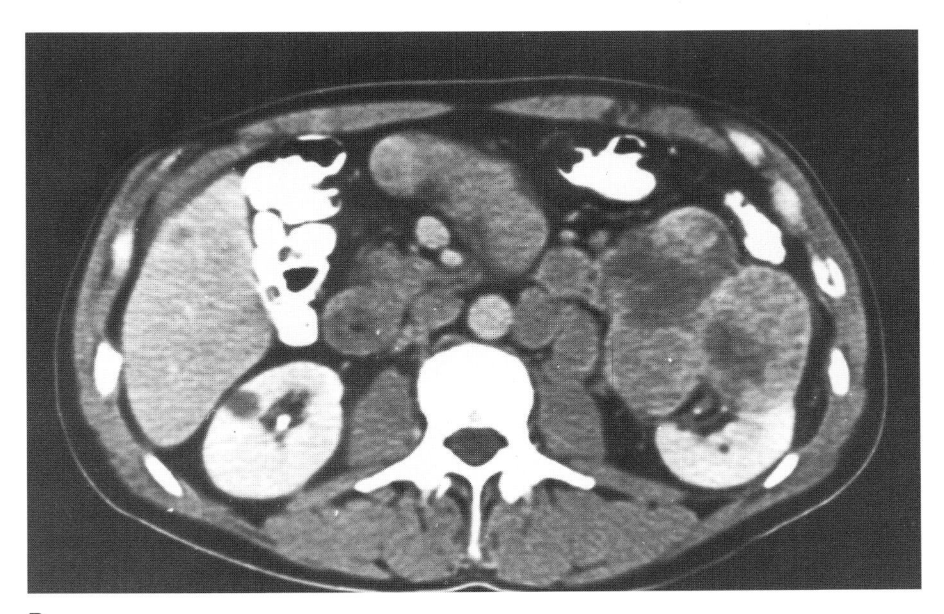

B

Figure 15.60. A: Solitary lytic lesion *(arrows)* in the acromion in a man with pain in the shoulder, fever, and anemia. B: CT scan through the kidneys, same patient.

Primary Bone Tumors

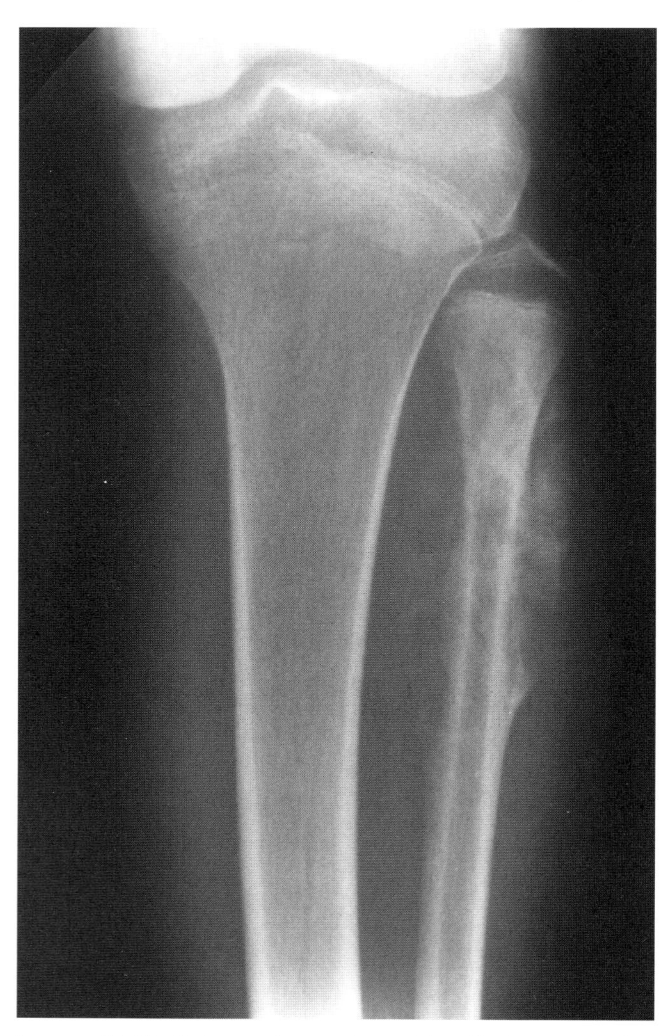

Figure 15.61. Osteogenic sarcoma of the fibula (see text).

Primary malignancies of bone fortunately are rare. Osteogenic sarcoma characteristically is seen in young males, 10 to 25 years old. Ewing's sarcoma is usually seen in younger children, and chondrosarcoma rarely appears before the age of 40 and usually in older patients. There are several types of osteosarcoma, depending on the predominant malignant tissue (osteogenic sarcoma versus fibrosarcoma). In Figure 15.61 you see the proximal fibula in a teenager with osteosarcoma having the classic densely ossified soft tissue and intraosseous mass, thickening of the cortex from periosteal reaction, and raylike extension outward into the soft tissues.

MR and CT have proved valuable in the estimation of soft-tissue involvement with primary bone tumors. Figure 15.62 shows the coronal MR scan and axial CT scan of a 17-year-old with a distal femur osteosarcoma. On the MR scan, you can clearly see soft-tissue extension *(white arrows)* as well as extension into the bone marrow *(black arrows)*. The axial CT scan also shows extensive soft-tissue involvement *(arrows)* as well as raylike calcifications like those seen on the plain film in Figure 15.61.

The radiographic identification of the various types of bone tumors is based upon the age of the patient, the particular bone involved, the location within that bone, the character of its border with normal bone, the presence or absence of soft-tissue involvement, and the presence or absence of periosteal reaction. Consider, for example, the giant cell tumor. This tumor occurs only in a patient with closed epiphyses (compare the osteosarcoma in Figure 15.61). Nearly every giant cell tumor is epiphyseal and abuts the articular surface of the involved bone. This lesion is usually eccentrically located in the bone and has a sharply defined border that is not sclerotic.

Figure 15.62. Osteogenic sarcoma of the femur (see text). A: Coronal MR scan. B: Axial CT scan.

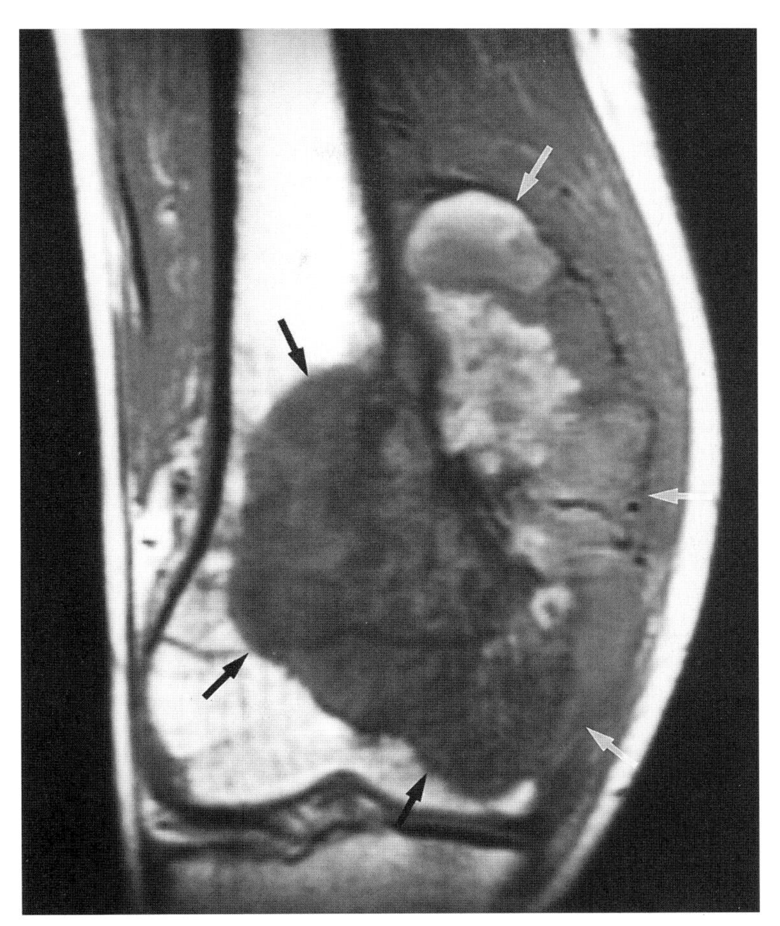

A

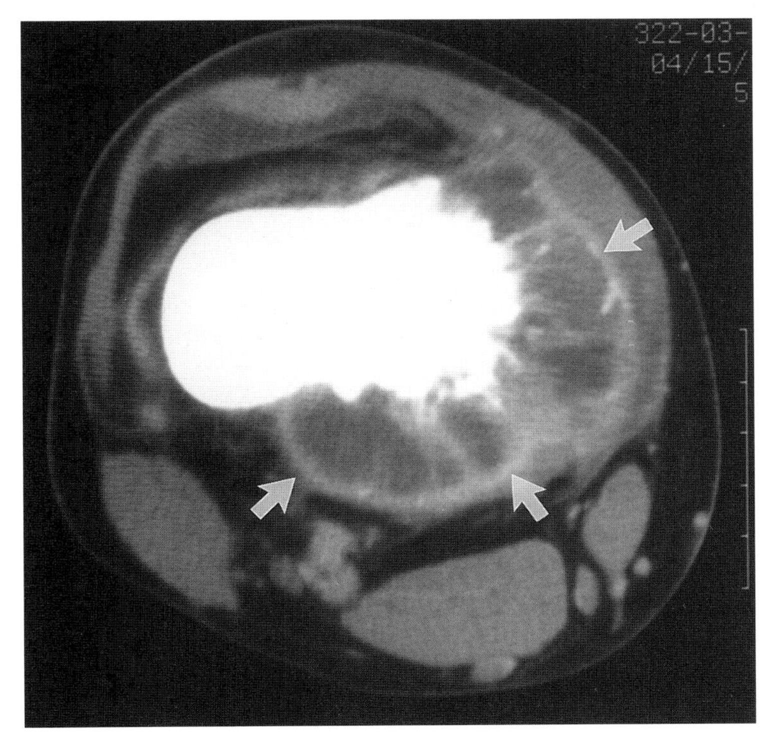

B

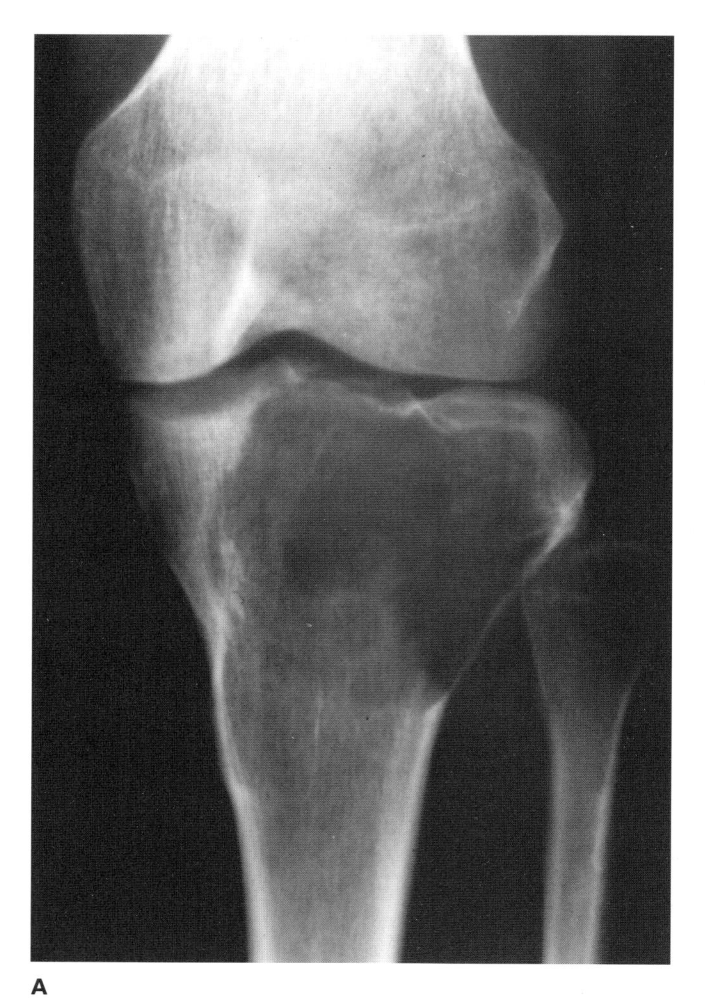

A

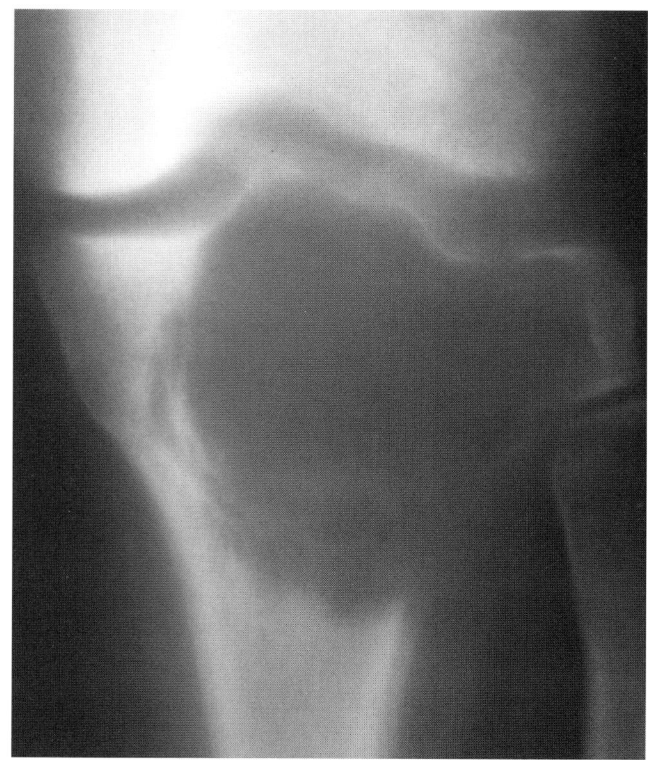

B

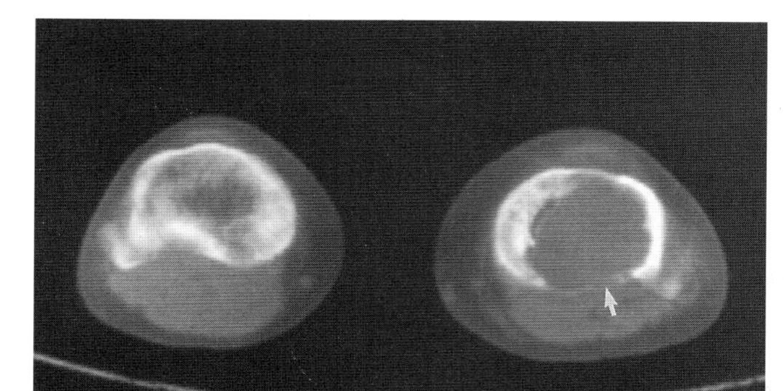

C

D

Figure 15.63. Giant cell tumor of the left tibia (see text). A: Plain film. B: Conventional tomogram. C: Axial CT scan at the level of the tibial plateau. D: Slightly lower-level CT scan through both tibias and fibulas. The *arrows* in C and D indicate sites where cortical bone has been destroyed by the growing tumor.

Now study the plain film and tomographic film in Figure 15.63, which depict a middle-aged man with new, nontraumatic knee pain, and determine whether the tibial lesion could be a giant cell tumor. Definitely yes! Its radiographic appearance fits all the criteria for a giant cell tumor, and that is exactly what it proved to be. The CT examination showed the reason for his new knee pain, extension of the tumor through the cortex of the tibia *(arrows)*.

Many kinds of benign bone tumors may also occur. Enchondromas are growing, expanding, cartilaginous tumors, common in the metaphyses of long bones and in the hands and ribs. Osteochondromas (often called exostoses) are discovered at any age, are most common in the knee area, and may be multiple or solitary. Osteoid osteomas usually present in young males with intense bone pain, which responds dramatically to aspirin. They may not be tumors, in fact; they may instead be inflammatory in origin and show a lucent area centrally with a surrounding shell of bony density. They must be differentiated from localized bone abscesses.

The 24-year-old man in Figure 15.64A had a palpable but otherwise asymptomatic mass over his left scapula. The plain film shows an osseous mass associated with the scapula, which might be either benign (an osteochondroma) or malignant (an osteosarcoma). Fortunately, the CT scan (Figure 15.64B) clearly shows it to be benign in character, an osteochondroma, with only normal muscles around it and no soft-tissue tumor mass. Figure 15.65 is another smaller osteochondroma of the scapula for you to compare.

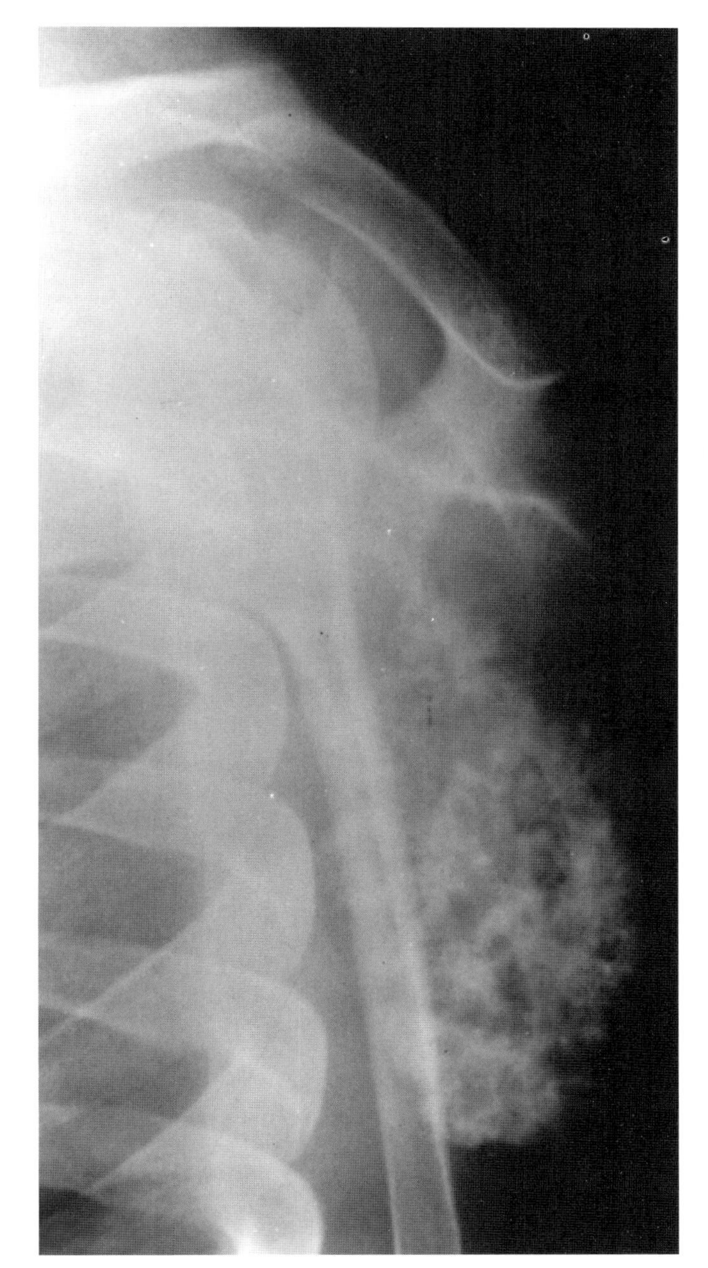

A

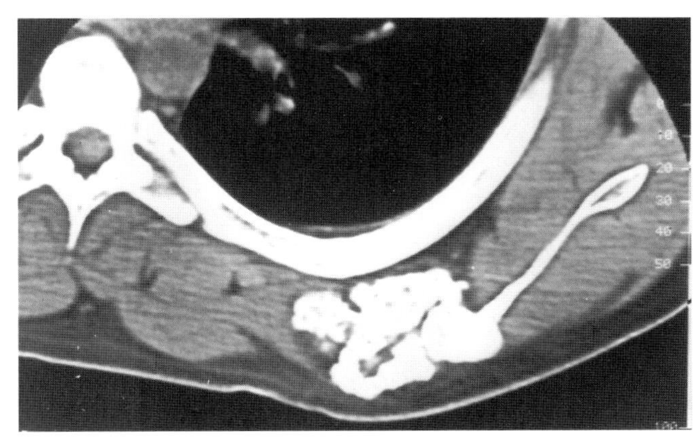

B

Figure 15.64. Osteochondroma of the scapula. A: Plain film showing an osseous mass. B: CT scan, showing the typical appearance of a benign tumor.

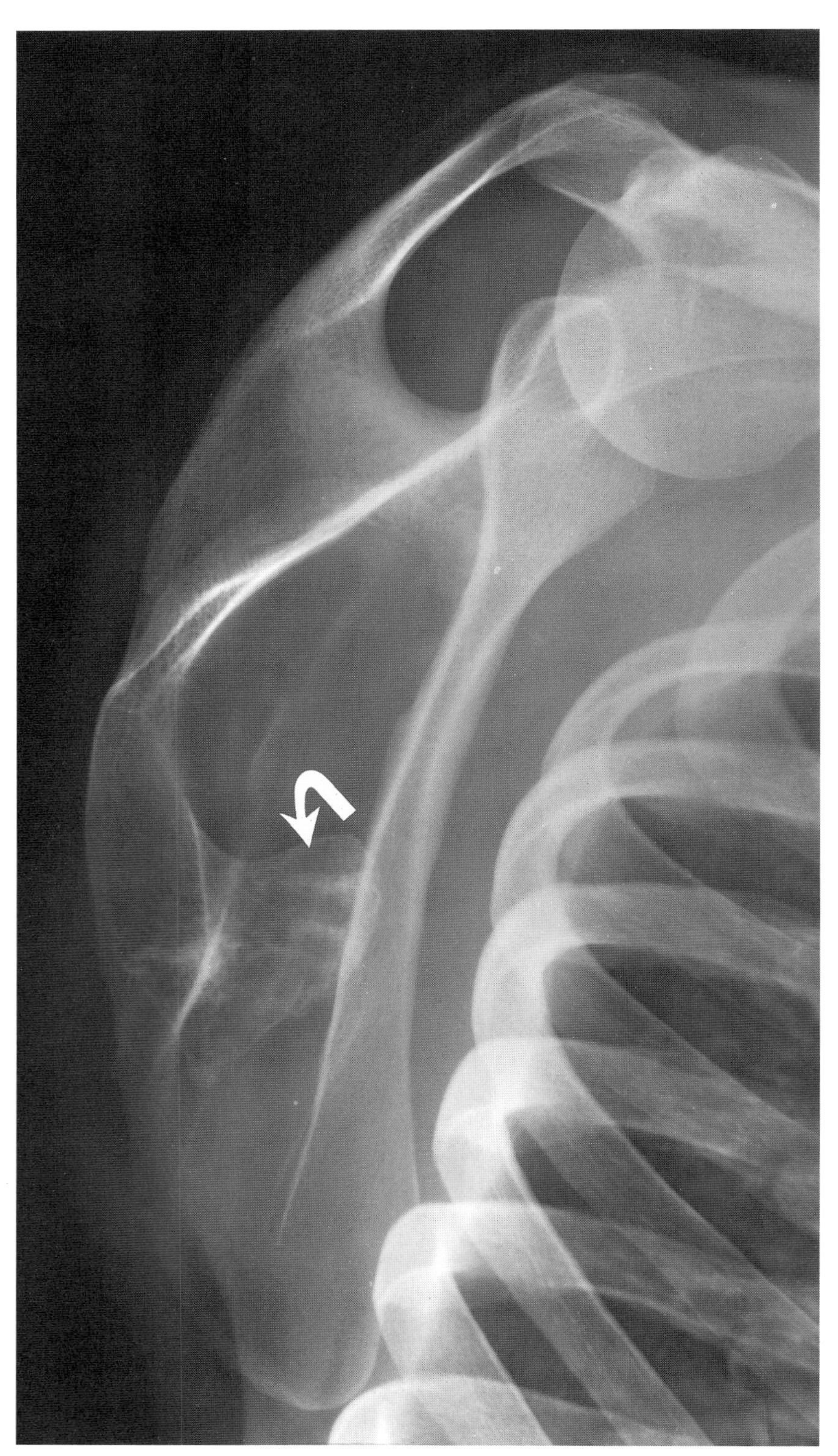

Figure 15.65. Another patient with osteochondroma *(arrow)* of the scapula.

Musculoskeletal MR Imaging

Although physicians have been imaging bones for over a century, we have only been able to directly image the soft-tissue components of the musculoskeletal system for the past two decades, first with CT and later with MR. Of these two cross-sectional imaging methods, the superior one for soft-tissue musculoskeletal imaging is MR because of its ability to produce sectional images in almost any imaging plane (coronal, sagittal, axial, and oblique planes), and its superior tissue differentiation, which allows individual depiction of tendons, ligaments, blood vessels, nerves, hyaline cartilage, and fibrocartilage. Compact bone, fibrocartilage, fasciae, ligaments, and tendons have low signal strength on most MR pulse sequences and appear dark. Review these structures on the normal T1-weighted knee MR scan in Figure 15.66. Muscles have an intermediary signal intensity. Fat has a high signal intensity (white appearance) on T1-weighted images and a low signal intensity on T2-weighted images, whereas fluid (joint effusion) has a high signal intensity on T2-weighted images and a low one on T1-weighted images.

So far you have seen examples of MR imaging in the diagnosis of several varieties of bone disease, including trauma, osteonecrosis, osteomyelitis, and malignancy. In the field of musculoskeletal imaging, MR is the best method of imaging joints, because it visualizes both bones and soft-tissue structures in the same examination. MR is routinely used to evaluate abnormalities of the hip, knee, ankle, foot, shoulder, elbow, wrist, hand, and temporomandibular joints.

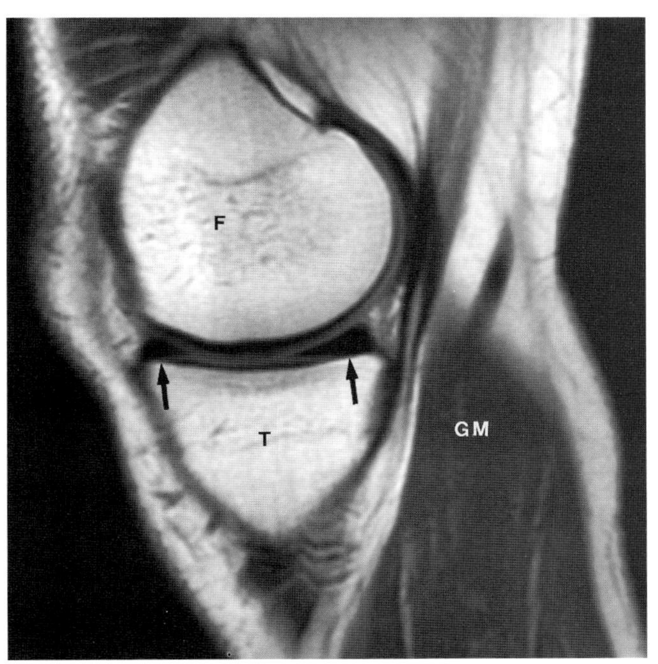

A

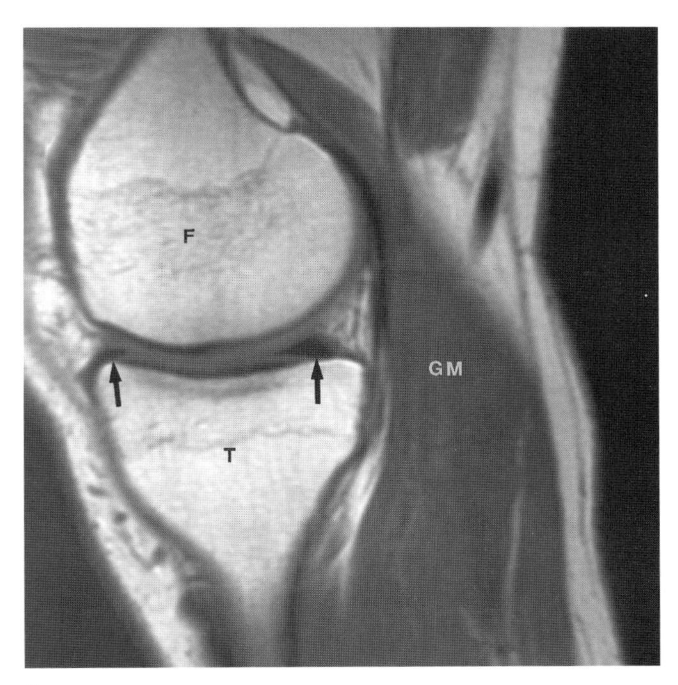

B

Figure 15.66A and B. Normal MR scan of the knee in two adjacent sagittal planes of the medial aspect of the joint, showing the C-shaped medial meniscus as low-signal (appear black) triangles *(arrows).* Bone marrow within the femur *(F)* and tibia *(T)* have a high signal (appear white) because of the presence of fat. Fat is also seen in the subcutaneous fascia surrounding the bones and muscles. The gastrocnemius muscle *(GM)* appears dark gray because muscle has an intermediate-strength MR signal.

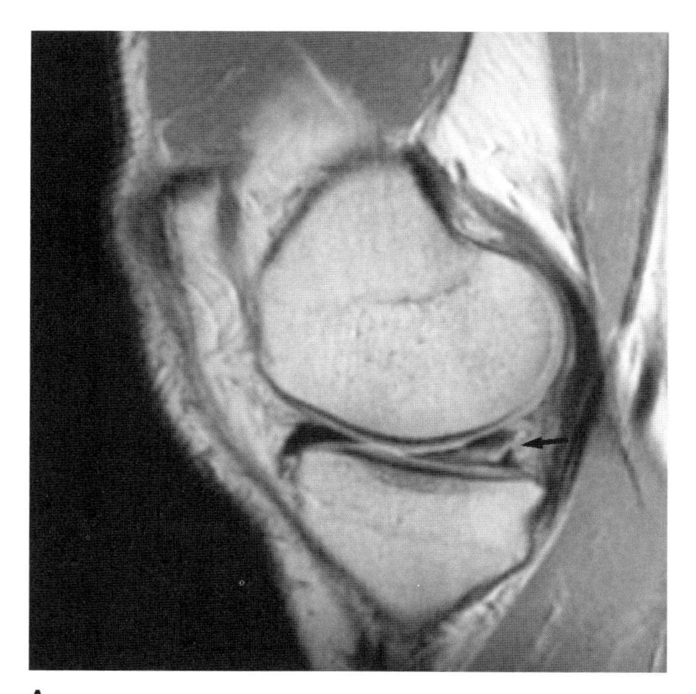

A

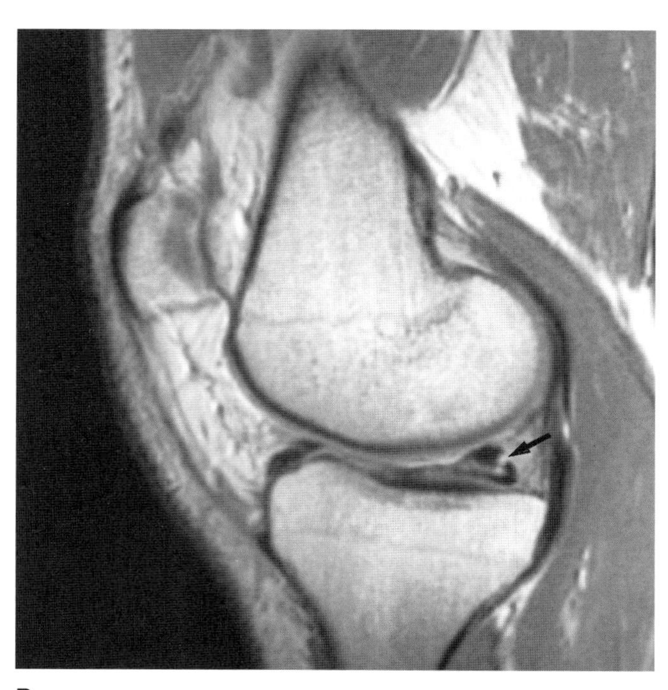

B

Figure 15.67A and B. MR scan of the knee showing a tear of the lateral meniscus *(arrow).* Note the increased MR signal at the site of the injury.

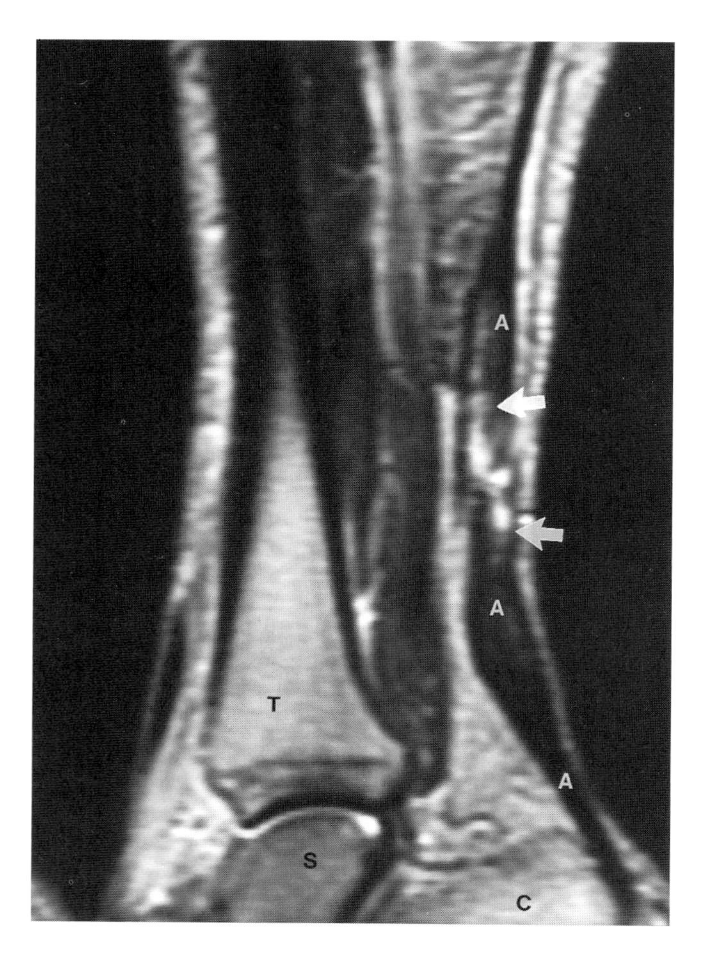

In the knee, for example, MR can provide excellent images of several important soft-tissue structures that may be injured by trauma, including the medial and lateral menisci, the anterior and posterior cruciate ligaments, the medial and lateral collateral ligaments, and the articular cartilages. The menisci appear as triangular-shaped structures of low signal intensity; injuries and tears increase the MR signal (Figure 15.67). Before the advent of MR, contrast injection into the knee joint space was required to visualize the menisci; called arthrography, this procedure, unlike MR, was uncomfortable for patients.

Figure 15.68. Sagittal MR scan of the lower leg showing a rupture *(arrows)* of the Achilles tendon (*A*'s). Normal tendons are black at MR; when injuries are present with hemorrhage and edema, an increased signal will be detected. Find the tibia *(T)*, talus *(S)*, and calcaneus *(C)*.

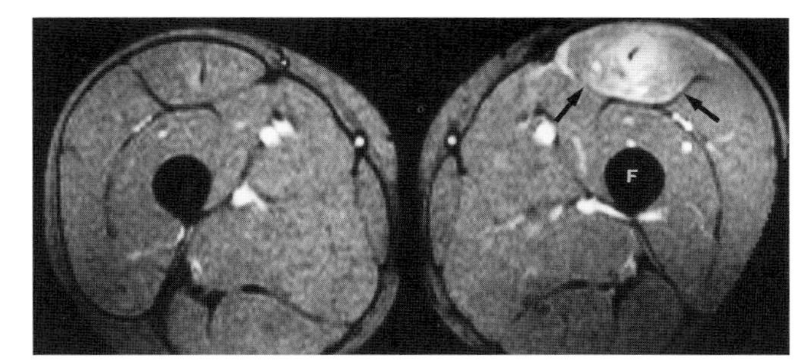

A

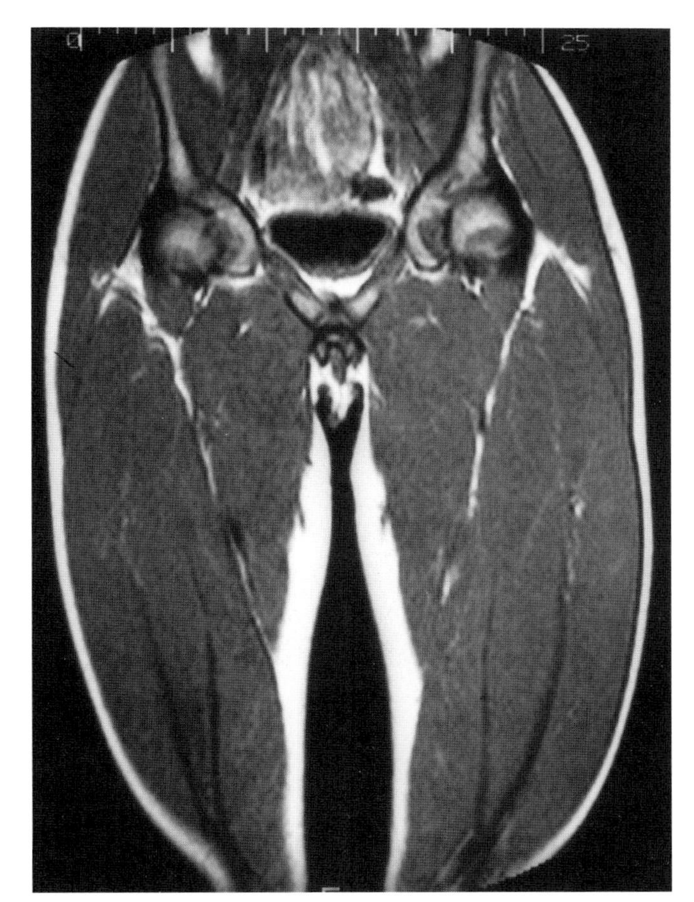

C

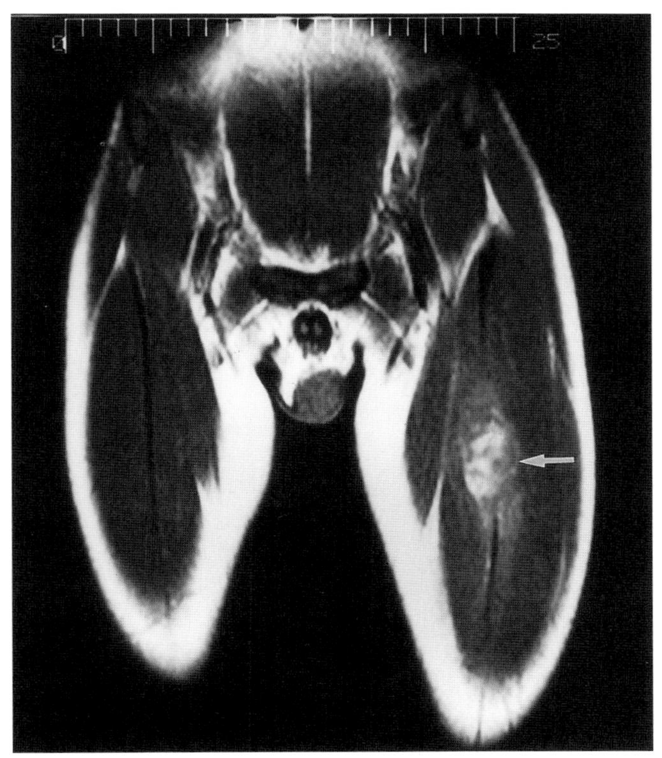

B

Figure 15.69. MR examination of a left rectus femoris muscle hemorrhage resulting from a jumping injury. A: Axial scan showing enlargement and increased signal in the left rectus femoris muscle *(arrows). F* identifies the left femur. B and C: Coronal slices of the quadriceps femoris muscles. An increased signal (*arrow* in B) can be seen in the left rectus femoris muscle. Compare the left and right sides on both the axial and the coronal images.

MR scanning also beautifully shows rotator cuff tears of the shoulder. Other tendon injuries, such as biceps tendon rupture or Achilles tendon rupture, are easily diagnosed with MR (Figure 15.68).

MR shows the muscles well, and can be used to diagnose muscle injuries and muscle tumors. Figure 15.69 depicts the MR examination of a young man with a left rectus femoris hematoma resulting from a jumping injury. Note the enlargement of the left rectus femoris muscle and the high-signal blood within the injured muscle.

16 Men, Women, and Children

So far you have learned mostly about the radiology of conditions that affect both sexes. In this chapter you will study the imaging of conditions that affect only men, only women, or only children. In the women's section you will learn about breast, gynecological, and obstetrical imaging and in the men's section, the imaging of conditions affecting the male genitourinary tract. The children's section is designed to remind you that medically and radiologically children are not just small adults; they have imaging issues peculiar to them and often suffer from age-related disorders. Of course, we cannot provide an encyclopedic review of all the conditions affecting men, women, and children in this chapter, but we will cover many of the common disorders you will be seeing in your future practices.

General Issues in Women's Imaging

Here we will cover breast imaging and imaging of the female reproductive organs. You should be familiar with both screening and problem-solving mammography and the techniques available for imaging common obstetrical and gynecological conditions.

The Female Breast

Breast cancer is the leading cause of *nonpreventable cancer death* in women; lung cancer is, in fact, the most common cause of cancer death in women, but most lung cancers can be prevented. In the United States each year breast cancer is diagnosed in more than 180,000 women, and every year over 46,000 women die from this condition. The risk of developing breast cancer increases steadily with age, especially after a woman is over 40. It has been estimated that 1 woman in 8 will develop breast cancer during her lifetime and that 1 in 33 will die from breast cancer. In addition to age, certain other factors are associated with an increased risk of developing breast cancer: a family history of breast cancer, early menarche, late menopause, nulliparity, and a late first full-term pregnancy.

X-ray mammography is the radiographic examination used to detect breast cancer. Because mammography can detect cancers *before they can be detected by palpation*, it has become a very valuable screening examination; in fact, mammography is the primary tool for the early detection of breast cancer. Early detection is extremely important because women whose breast cancers are detected at the *nonpalpable stage*, when the tumor is small, have the best chance for survival. Cancers first detected by self-examination or by the physician by palpation are usually at a later stage of growth.

Routine breast cancer screening should be performed with both mammography and a breast clinical examination. Although mammography can detect cancers smaller than those that can be palpated, mammography may not reveal some palpable cancers because they are located in portions of the breast that are difficult to image or because their presence is radiologically masked by cystic or other changes in breast tissue. Always remember that (1) *mammography can detect the presence of early breast cancer at the nonpalpable stage*, but (2) *mammography cannot rule out breast cancer in a patient with a palpable mass or other breast abnormality on clinical examination.* Mammography can rule in cancer, but not rule it out. A negative mammogram should never delay the evaluation of symptoms reported by the patient or suspicious findings discovered on physical examination, such as a breast lump, skin changes, pain, or spontaneous nipple discharge.

Currently controversy exists over which age groups of women should be routinely examined with *screening mammography*. There is general agreement that all women over 50 years of age should be examined annually with screening mammography, and in this group mammography can be expected to reduce cancer deaths by 30 percent. If a cancer is present in a woman over 50, mammography has a 90 percent

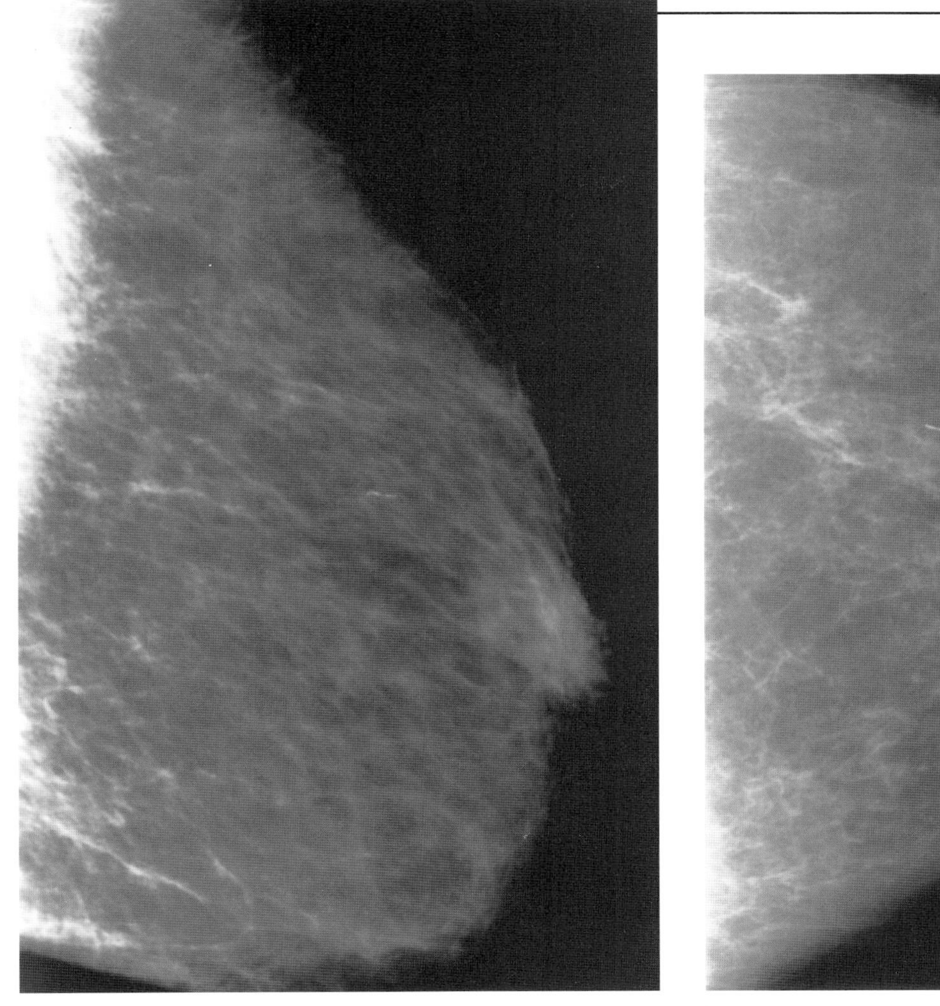

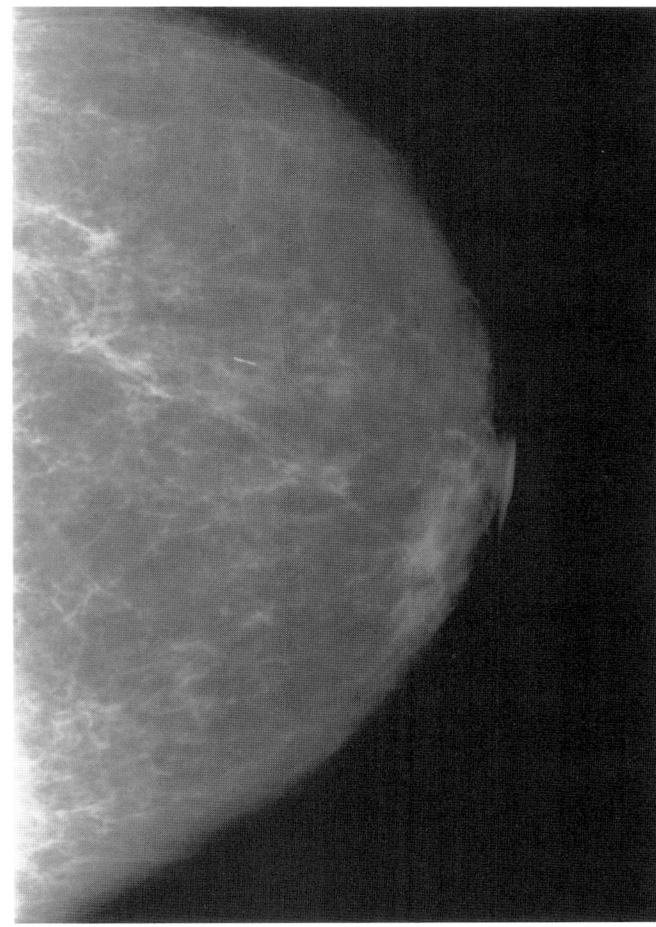

A B

Figure 16.1. Normal mammogram. A: Mediolateral oblique (MLO) view. B: Craniocaudal (CC) view.

chance of detecting it. Nearly every insurance plan covers the cost of screening mammography for women over 50; you should know that currently no other medical imaging method is covered by insurers for screening purposes. The data are not so clear for premenopausal women from 40 to 49 years of age. Although the American Cancer Society recommends screening mammography every 1 to 2 years for this group, there is not universal agreement on the benefits of such screening.

Screening mammography is performed on asymptomatic women to detect unsuspected breast cancer at an early stage. By comparison, *diagnostic mammography* (problem-solving mammography) is performed to evaluate abnormal findings in the breasts such as palpable masses, nipple discharge, nipple retraction, or skin changes. Mammography requires special equipment and specially trained technologists and radiologists. All physicians should verify that the mammography center where they wish to refer patients has been approved by the American College of Radiology; only then can they be assured of high-quality examinations. Low-quality mammography reduces the sensitivity of the screening examination.

A mammogram consists of two views of each breast: a mediolateral oblique (MLO) view (Figure 16.1A) and a craniocaudal (CC) view (Figure 16.1B). The breasts must be vigorously compressed to spread out the breast tissue (Figure 16.2). *You should warn your patients that although mammography should not be painful, it is usually very uncomfortable.* If the radiologist is attempting to resolve a particular breast problem, such as a palpable mass, other views, including spot compression and magnification views, may be obtained.

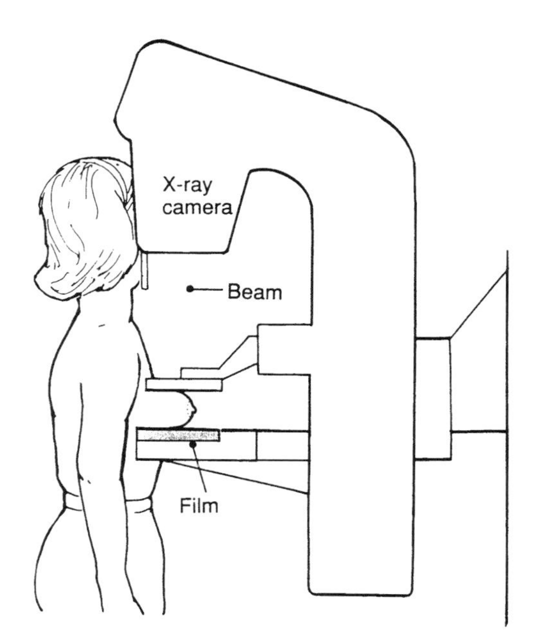

Figure 16.2. Patient positioned for a mammogram.

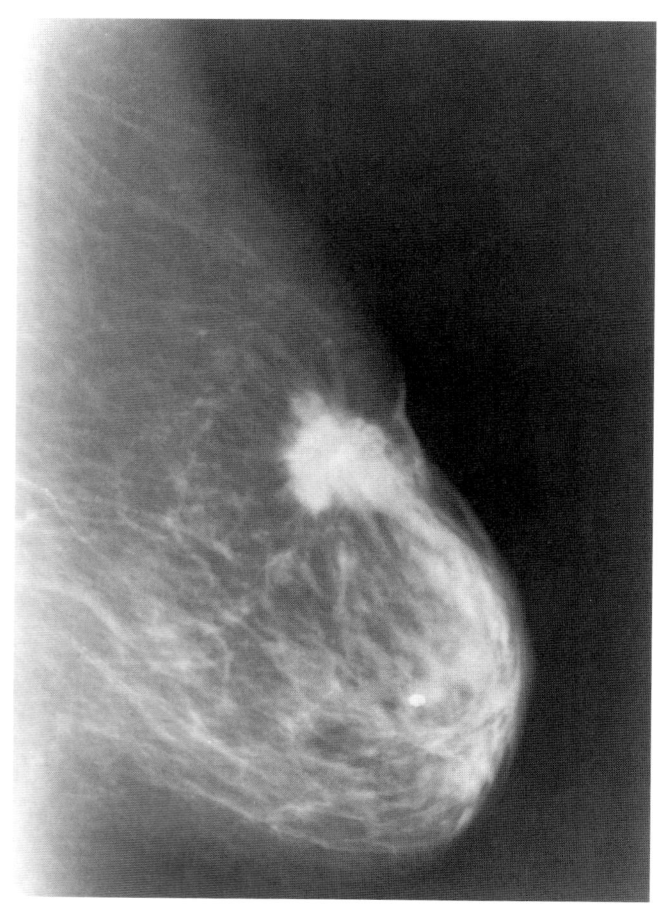

Figure 16.3. Mammogram of a large breast cancer, appearing as a mass with spiculated margins and overlying skin retraction.

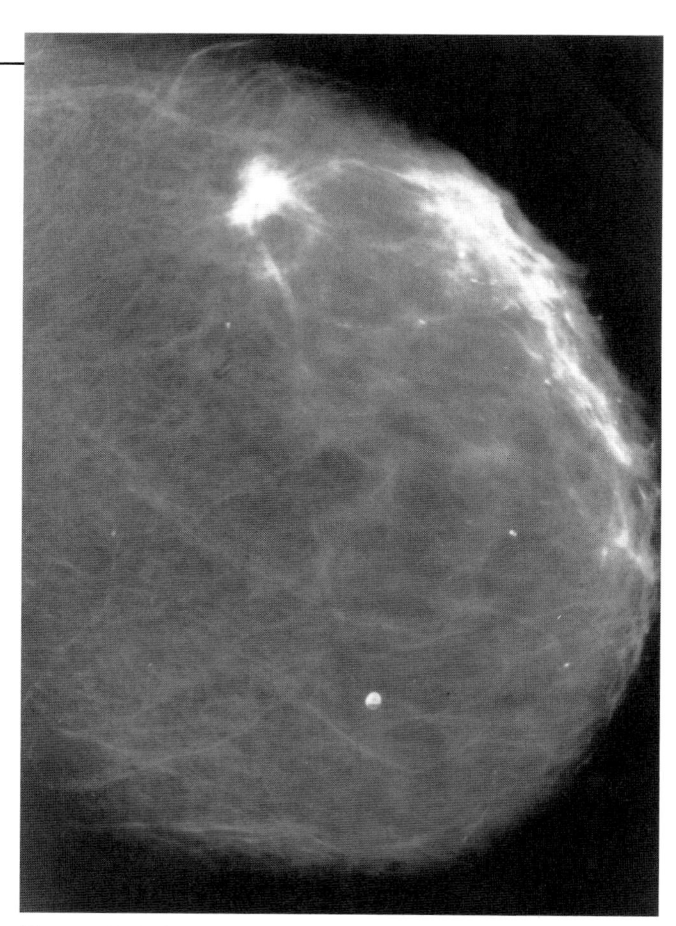

Figure 16.4. Mammogram of a breast cancer that appears as a mass with spiculated margins that also contains some tiny microcalcifications, which can be seen at the periphery of the mass. Another indication of malignancy is infiltration of the overlying skin.

On mammograms cancers typically appear as stellate masses with spiculated margins (Figure 16.3). If the cancer is close to the skin there may be skin retraction associated with it. Breast cancers may also appear as clusters of microcalcifications, with or without an associated stellate mass (Figure 16.4). Sometimes a cancer may be indicated only by asymmetry of the breast tissue. Other signs of breast cancer include skin thickening, nipple retraction, and venous engorgement. Mammograms are difficult to interpret, and only experts should render final interpretations. You may be surprised to learn that in many radiology departments screening mammograms are *double read* (read independently by two different radiologists); such double readings have been shown to increase the detection rate of cancers on screening mammograms.

Benign abnormalities may also be seen on mammograms. Normal calcifications may be seen in the skin, in the breast arteries, and in the walls of benign cysts. Benign fibroadenomas may contain dense popcorn-like calcifications (Figure 16.5). Fibroadenomas may

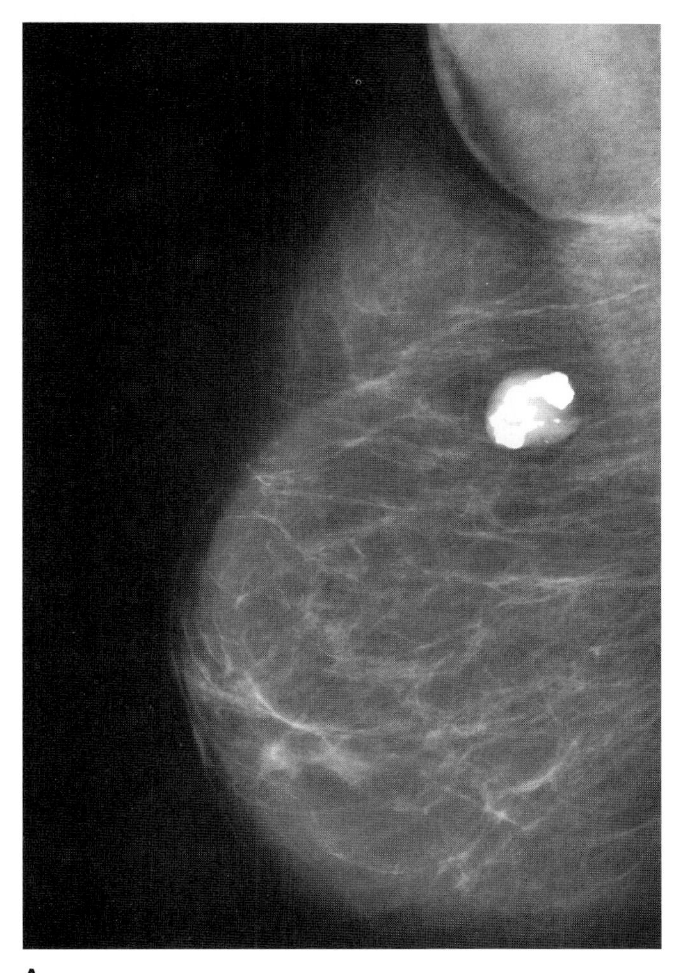

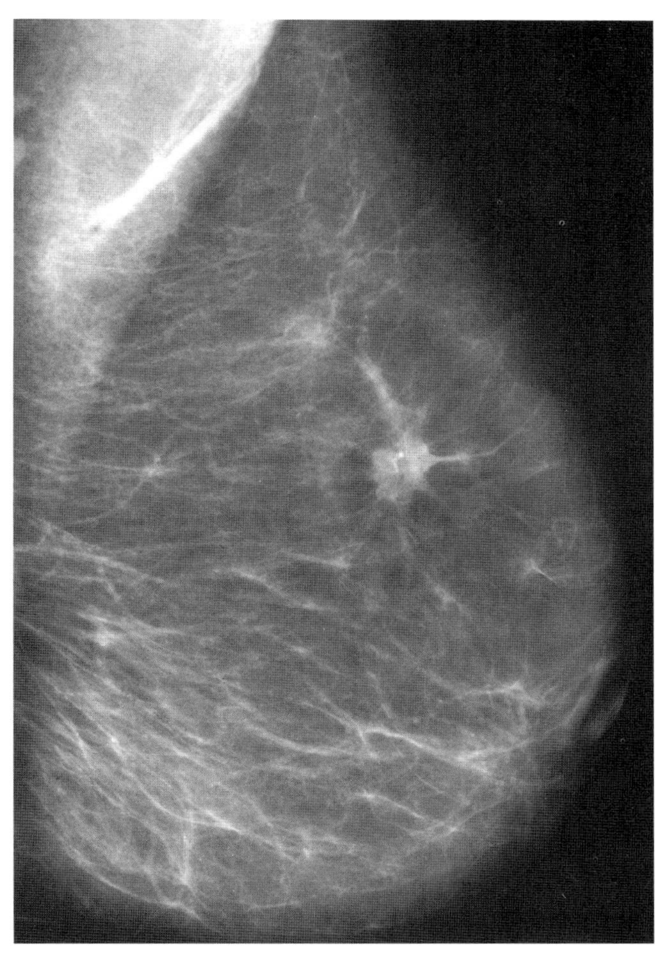

Figure 16.5. Mammograms of both breasts of a woman who had discovered a palpable mass in her right breast on self-examination. A: The right breast mass is clearly a benign fibroadenoma containing typical dense popcorn-like calcifications. B: The left breast, however, shows an unsuspected breast cancer that could not be detected on physical examination.

also appear as noncalcified masses, as do intramammary lymph nodes and cysts. As you might guess, breast cysts are well circumscribed at mammography and are shown to be anechoic at breast ultrasound. A *needle aspiration* can also determine whether a mass is cystic or solid.

When a nonpalpable suspicious lesion is seen on a mammogram, it can be localized by the radiologist for biopsy and resection with mammographic or ultrasound guidance. The tip of a needle with a wire hook is placed directly into the radiographic lesion. With this *needle localization* technique the lesion can be exactly localized by the breast surgeon for evaluation or surgical management.

CT and ultrasound may be indicated in some patients for breast imaging. CT can show very posterior breast masses, close to the chest wall, that may be

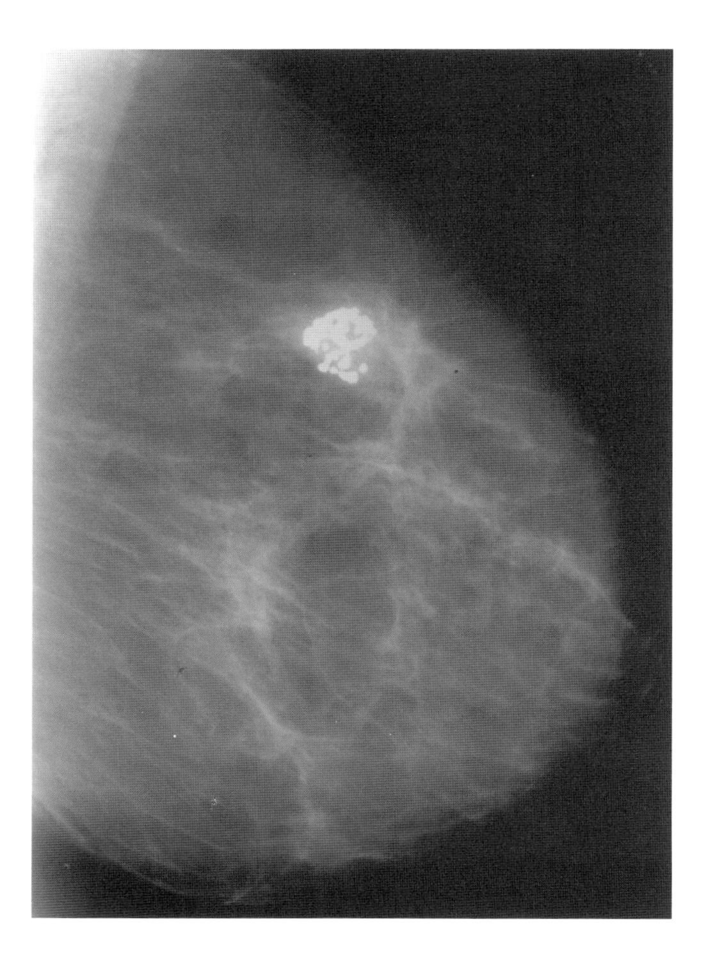

difficult to image by mammograms. Ultrasound's ability to distinguish between cystic and solid lesions is particularly valuable in breast imaging. Current investigations suggest that MR may be helpful in the future in the evaluation of patients with abnormal mammograms.

Finally you should be aware of some disturbing facts. Despite strong evidence that the early detection of breast cancer by screening mammography could cut breast cancer deaths by one-third, it was recently estimated by the National Cancer Institute that only 30 percent of women in their fifties and 36 percent of those 60 and older had had a mammogram in the past year. Furthermore, 40 percent of women over 50 had never had a mammogram! The underuse of mammography results in a needless loss of women to breast cancer. Primary care physicians should refer women for routine screening mammography so that breast cancer can be diagnosed and treated early, when the chances of cure are high.

Figure 16.6. Mammogram of another patient with a benign fibroadenoma containing the typical large, coarse, and irregular calcifications seen in benign lesions.

The Female Pelvis

The female pelvic cavity contains the rectum, sigmoid colon and terminal loops of ileum posteriorly, and the urinary bladder, uterus, vagina, ovaries, and uterine (fallopian) tubes anteriorly. Review the anatomic drawings in Figures 16.7 and 16.8. The uterus is a pear-shaped organ with thick muscular walls that varies in size with age. In young nulliparous adults it averages 8 centimeters in length and 5 centimeters in width. The uterine cavity is shaped like an inverted triangle; the upper two points are directed toward the uterine tubes and the lower point at the cervical canal, which connects the uterine cavity to the vagina. Normally the uterus is bent forward (anteverted), forming an angle of 90 degrees with the vagina. The uterine tubes, which are normally 10 centimeters long, connect the uterine cavity with the peritoneal cavity near the two ovaries. The uterine tubes course through the upper border of the broad ligament. Each ovary is attached to the broad ligament laterally, lying against a depression in the lateral wall of the pelvis, called the ovarian fossa. The ovaries measure approximately 4 by 2 centimeters, and also vary in size with age, shrinking after menopause. The vagina is about 8 centimeters long and has anterior and posterior walls that are normally in apposition.

Several imaging techniques are used to evaluate the female pelvis. The primary method is ultrasound, and you have already seen examples of pelvic scans in Chapter 11. The sonographer can perform pelvic ultrasound *transabdominally,* using the patient's full bladder as an acoustic window, or *transvaginally,* using a specially designed ultrasound transducer that is placed

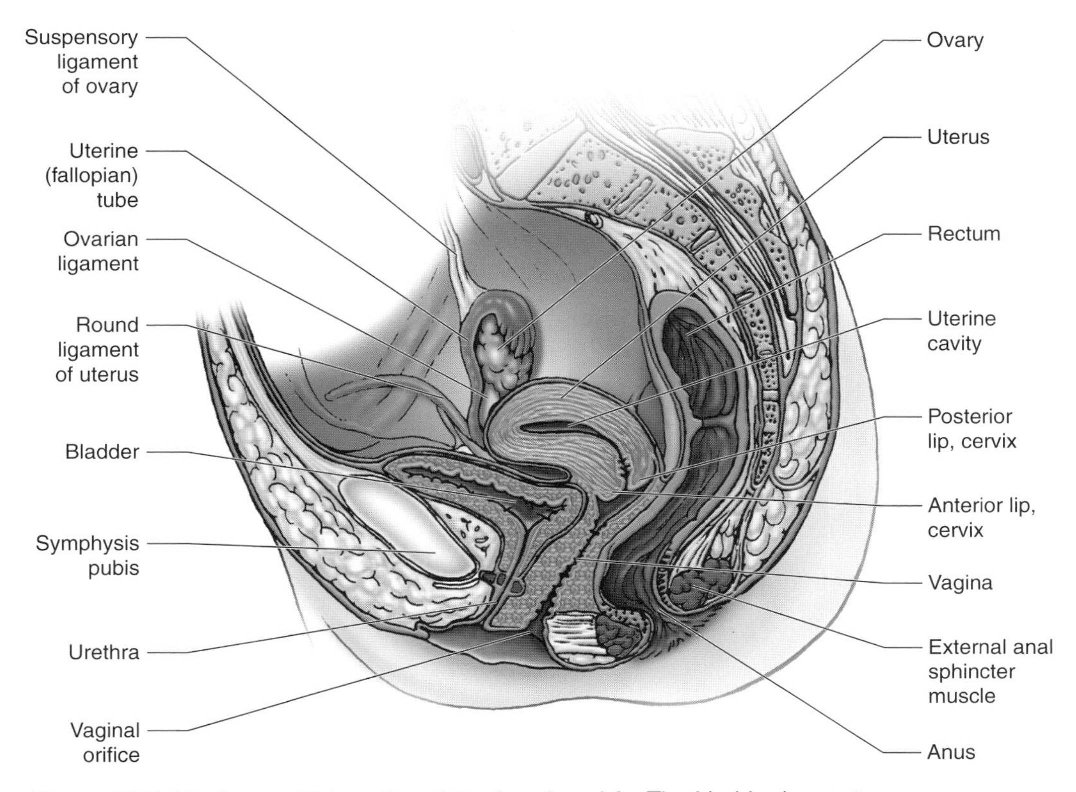

Figure 16.7. Median sagittal section of the female pelvis. The bladder is empty; the uterus and vagina are slightly dilated.

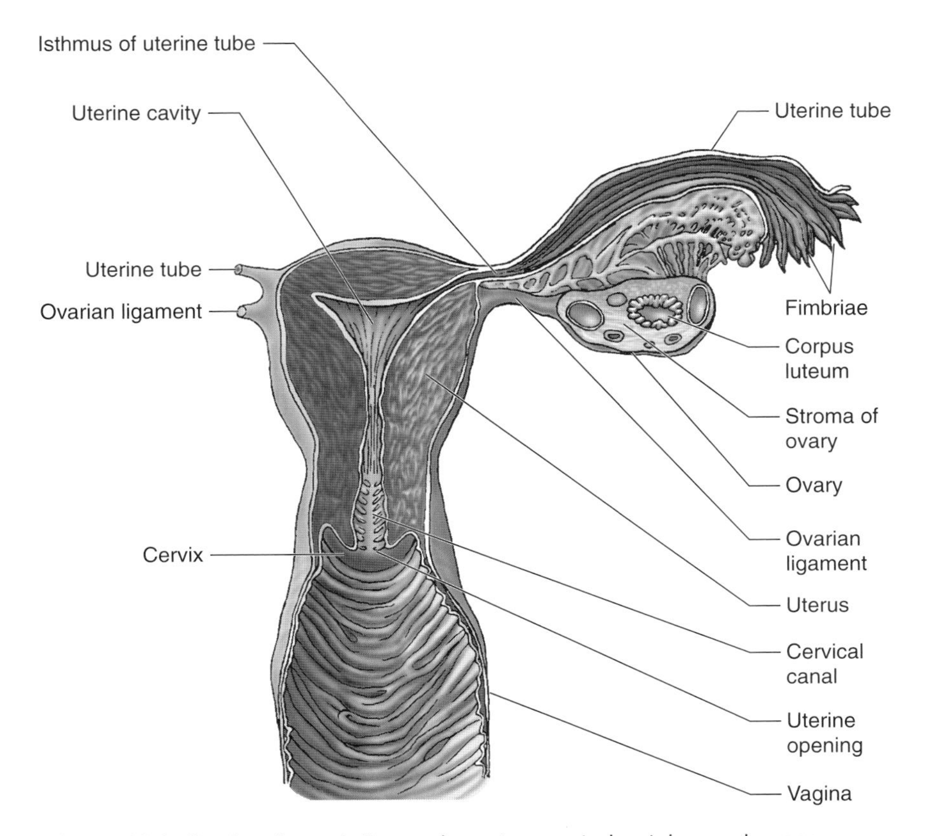

Figure 16.8. Section through the vagina, uterus, uterine tube, and ovary.

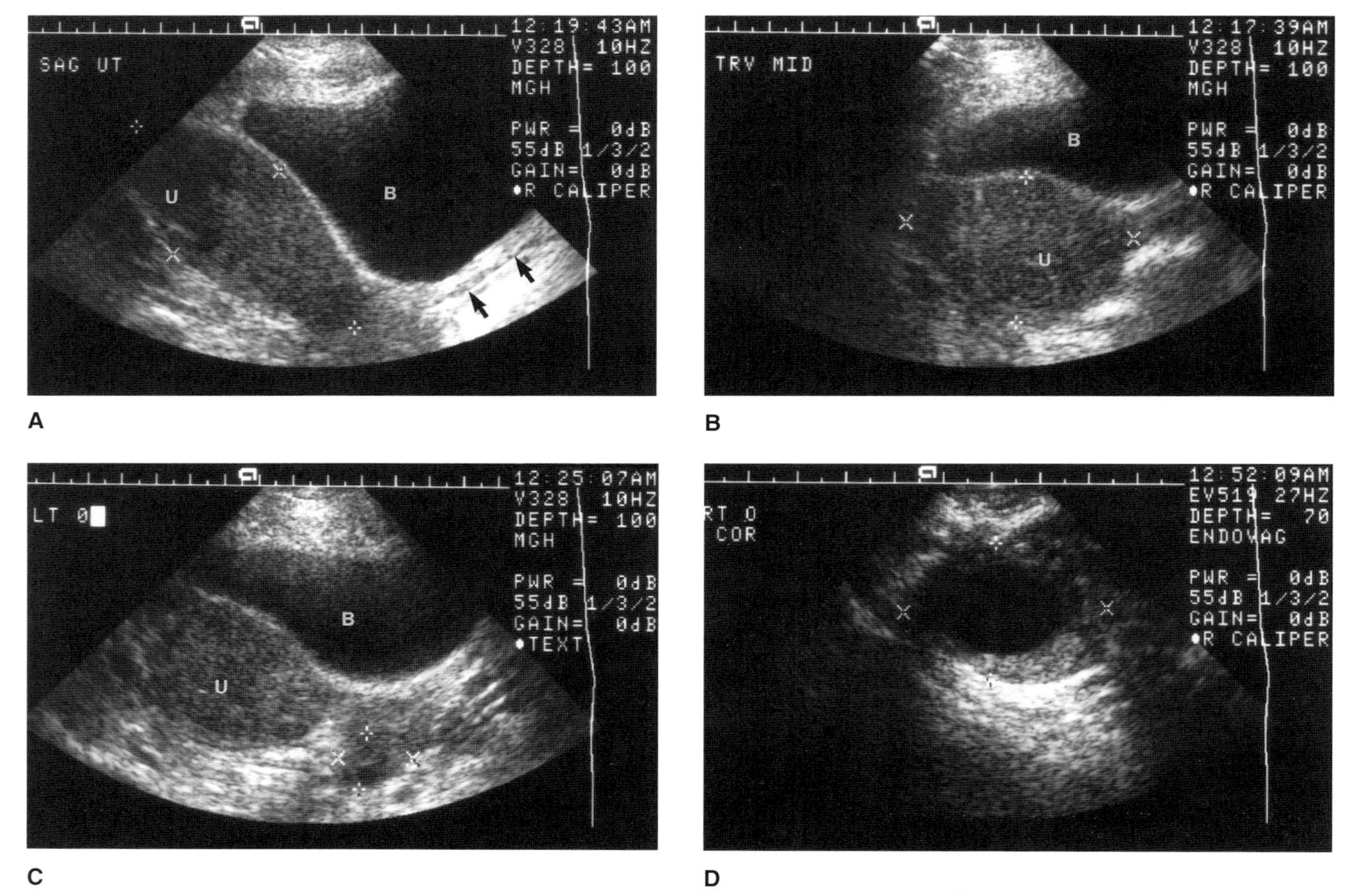

Figure 16.9. Normal ultrasound examination of the female pelvis (see text). A: Midline sagittal transabdominal scan shows a hypoechoic uterus *(U)* marked by cursors, a full bladder *(B)*, and the vagina *(arrows)*. B: Transverse transabdominal scan in the midline shows the uterus *(U)* posterior to the bladder *(B)*. C: Sagittal transabdominal scan to the left of the midline shows the left ovary (marked by cursors), which appears to contain a small anechoic cyst. Portions of the uterus *(U)* and bladder *(B)* are again seen. D: Transvaginal scan of the left ovary of the same patient shows markedly improved visualization of the ovary (marked by cursors), which contains an anechoic physiological cyst.

inside the patient's vagina for scanning. Transvaginal ultrasound improves the visualization of small structures and is especially valuable in obstetrical imaging to depict first-trimester development and diagnose ectopic pregnancy.

Figures 16.9A and 16.9B are midline sagittal and transverse ultrasound scans of the pelvis obtained by transabdominal scanning of a patient with a full bladder. The uterus is hypoechoic and has a uniform echotexture. The ultrasound appearance of the endometrial lining of the uterine cavity varies in accordance with the phase of the menstrual cycle. The vagina usually appears as a hypoechoic linear structure with an echogenic lumen. The ovaries (Figure 16.9C) change in appearance with age and during various phases of the menstrual cycle as physiological cysts appear, increase in size, and become atretic. The ovaries are best seen on a transvaginal scan (Figure 16.9D).

CT and MR can produce excellent images of the female pelvis (Figures 16.10 and 16.11), and compared with ultrasound, can provide better delineation of structures surrounding the uterus and adnexa. Consequently, CT and MR are particularly helpful in the staging of pelvic malignancies.

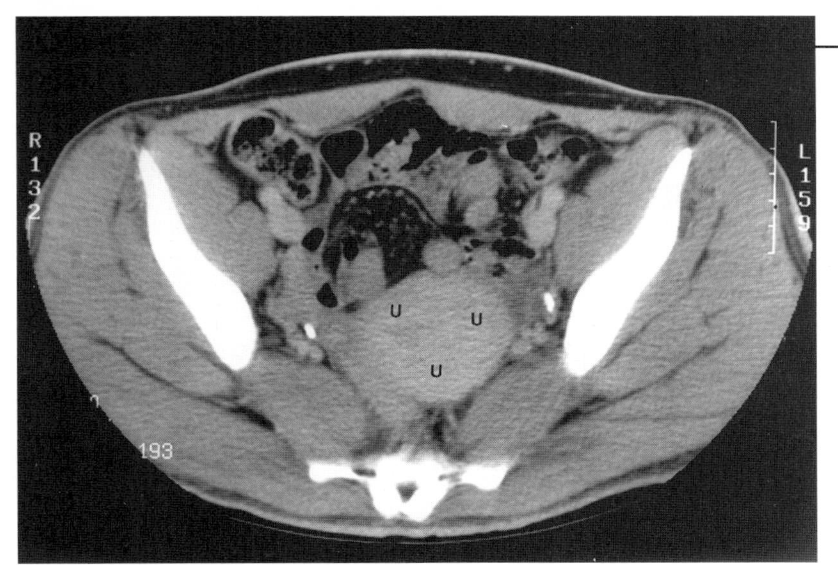

A

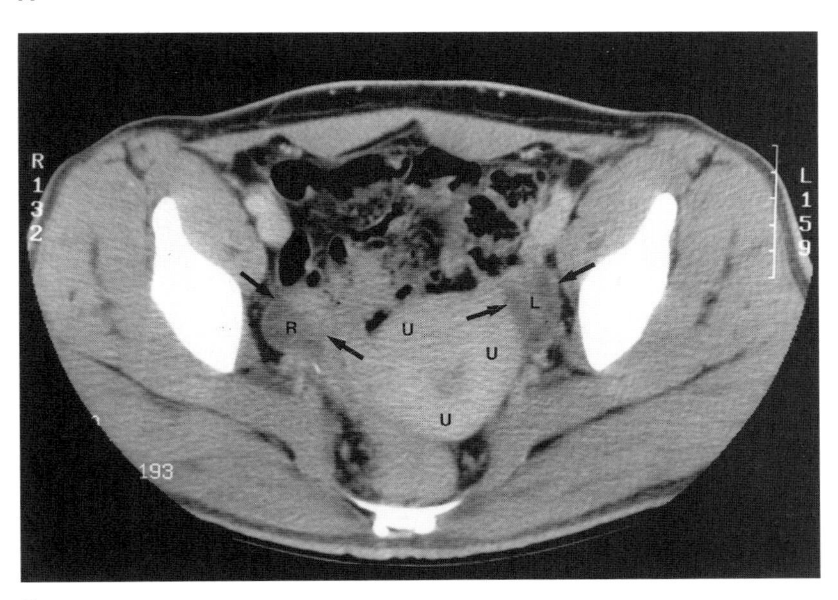

B

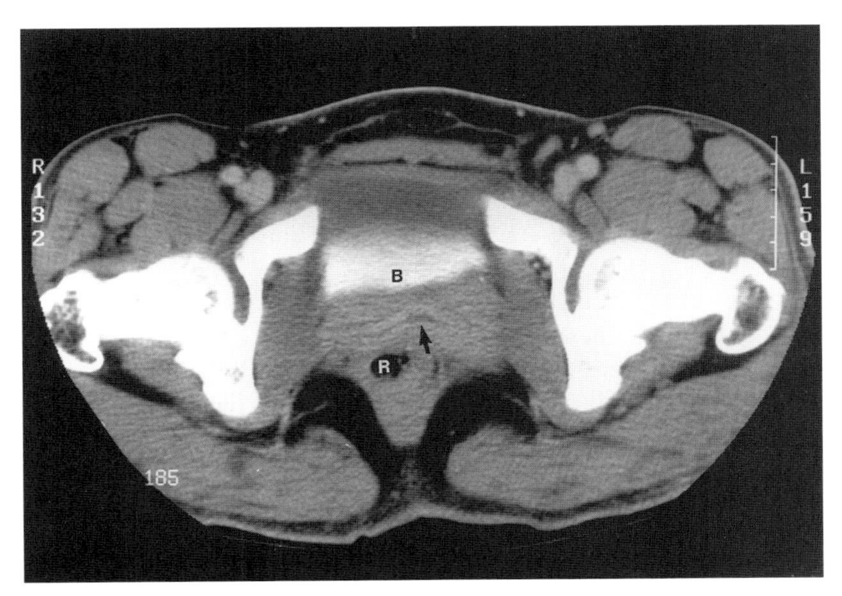

C

Figure 16.10. Normal CT scans of the female pelvis. A: Upper-level scan shows the dome of the muscular uterus (*U*'s). B: Slightly lower-level scan shows the mid-uterus (*U*'s) and its almost central, less dense, uterine cavity. The right *(R)* and left *(L)* ovaries *(arrows)* are shown adjacent to the uterus. C: Lowest-level scan, taken below the uterus, shows a collapsed vagina *(arrow)* between the contrast-opacified bladder *(B)* and the air-containing rectum *(R).*

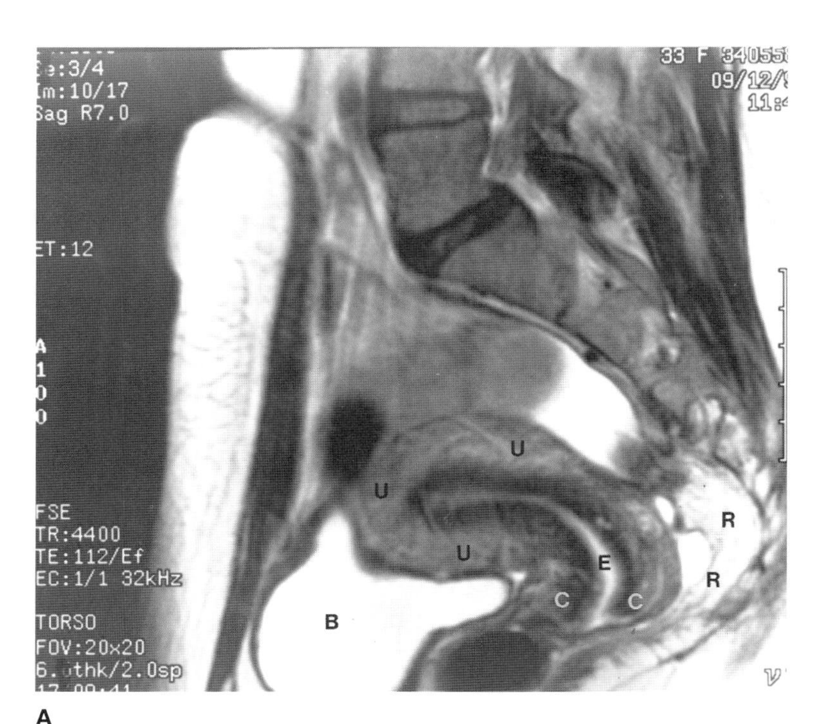

A

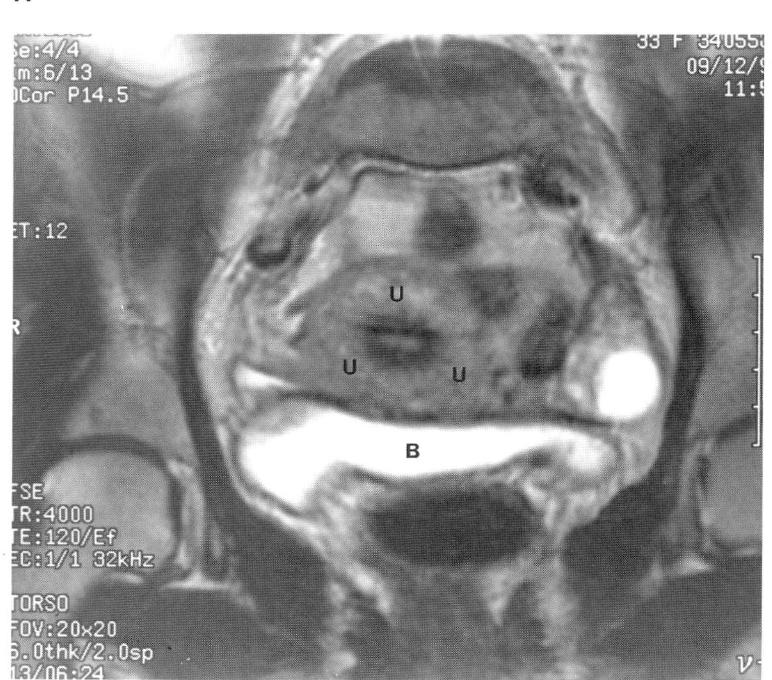

B

Figure 16.11. Normal MR scans of the female pelvis. A: Midline T2-weighted sagittal scan showing the anteroflexed uterus (*U*'s) and cervix (*C*'s) between the bladder *(B)* and rectum (*R*'s). Note the bright white signal of the endometrial *(E)* lining of the uterine cavity. B: Coronal scan of the same patient shows the uterus (*U*'s) superior to the nondistended bladder *(B)*.

Another way of imaging the female pelvic organs is *hysterosalpingography,* which you have already seen demonstrated in Figure 11.51. To perform this procedure, the cervix is cannulated and a contrast material is injected to opacify the uterine cavity and tubes. Examination of a normal uterus will show the contrast material flowing freely from the uterine tubes into the peritoneal cavity (Figure 16.12). Hysterosalpingography is used to diagnose congenital anomalies of the uterus such as a septated uterus or bicornuate uterus, which may also be diagnosed by MRI and, less specifically, by ultrasound. Hysterosalpingography is also requested in the evaluation of infertility; failure to demonstrate free flow of contrast material through a uterine tube into the peritoneal cavity indicates an obstructed tube (Figure 16.13).

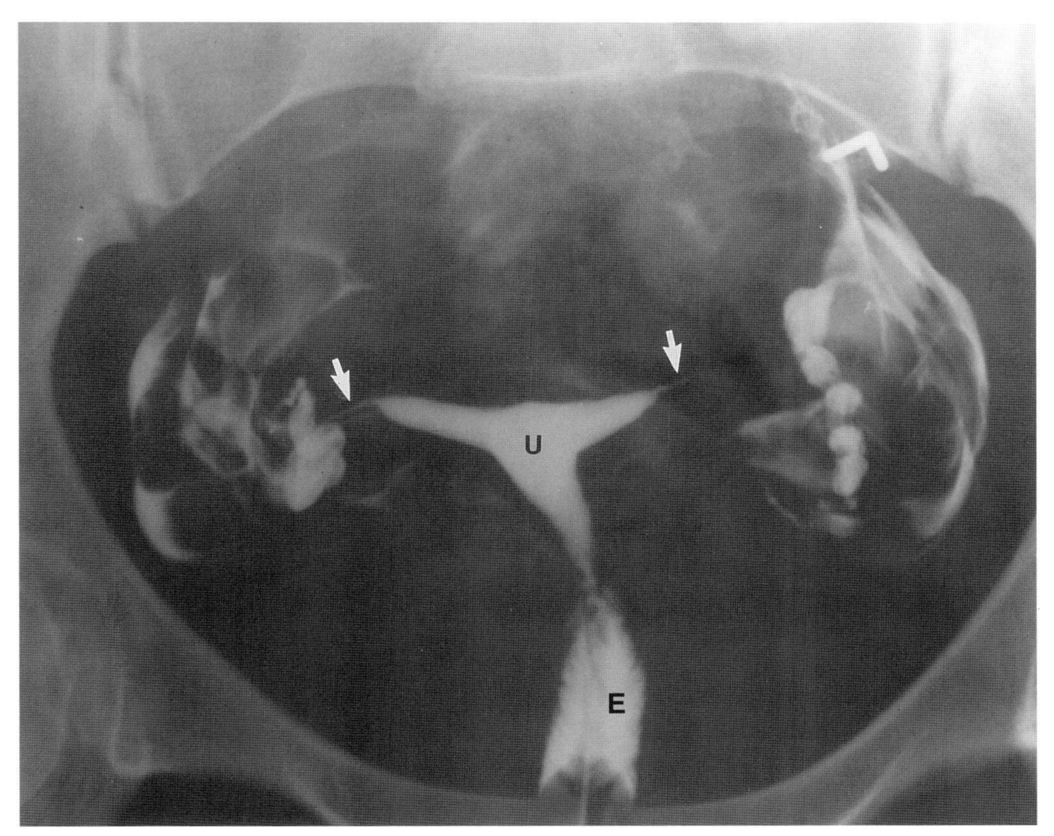

Figure 16.12. Normal hysterosalpingogram. Contrast material has been injected into a catheter placed within the cervix. The uterine cavity *(U)* is opacified as well as the uterine tubes *(arrows).* There is retrograde flow of contrast material into the endocervical canal *(E);* the injection catheter can be seen passing through the vagina. The free flow of contrast material into the peritoneal cavity shows that the patient's uterine tubes are patent. The ovaries are not seen.

Figure 16.13. Abnormal hysterosalpingogram. Because of scarring from prior inflammation, both uterine tubes are occluded, which prevents flow of the contrast material into the peritoneal cavity. The tubes are also dilated, indicating the presence of hydrosalpinx. The *arrow* points to the injection catheter.

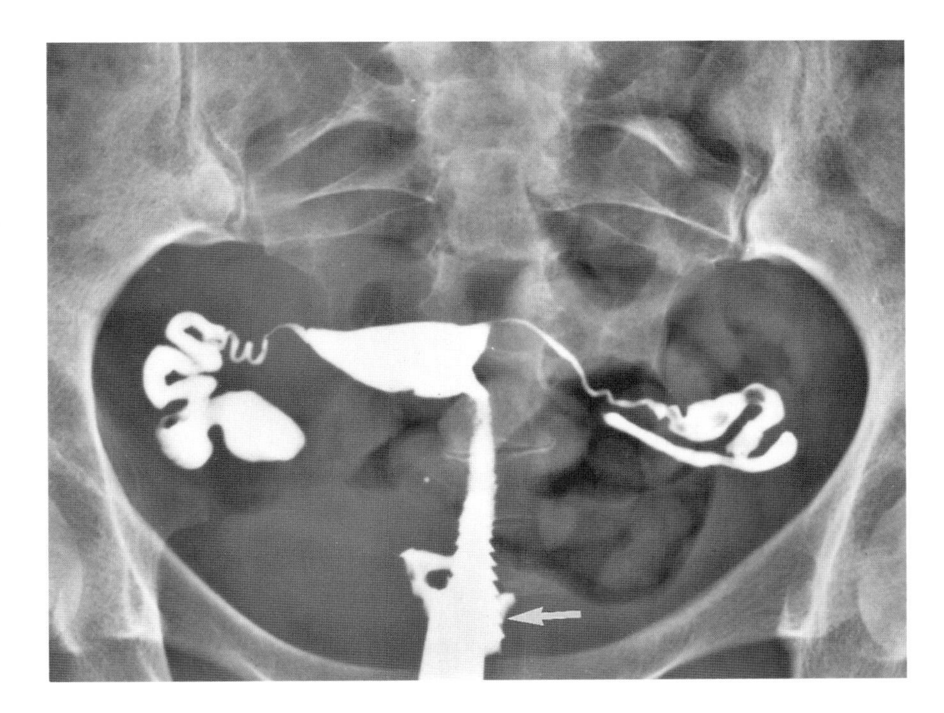

Gynecological Conditions

A wide variety of benign pelvic conditions can be diagnosed with ultrasound, including ovarian cysts, pelvic inflammatory disease, endometriosis, and benign tumors of the uterus (leiomyomas) and ovaries (cystadenomas, cystic teratomas). Ovarian masses may be cystic, solid, or complex. Functional or physiological cysts in premenopausal women are the most common adnexal masses. Ovarian cysts may vary in size from 5 millimeters to 8–10 centimeters in diameter. Usually they are asymptomatic and regress spontaneously, and are discovered as incidental findings on pelvic ultrasound examinations (Figure 16.14). But acute symptoms may be associated with hemorrhage into ovarian cysts and torsion of large cysts (Figure 16.15).

Solid ovarian masses may represent benign tumors, such as teratomas or stromal tumors, or ovarian cancer; most ovarian neoplasms, such as cystadenomas, are epithelial in origin and are cystic. Figure 16.16 shows the ultrasound examination of a woman with bilateral ovarian masses representing bilateral endometriomas. Both ovaries are slightly enlarged, the right measuring 68 millimeters in length and the left 48 millimeters. In Figure 16.17 you see a cystic tumor of the right ovary compared with a benign ovarian cyst in another patient. Ultrasound, with the assistance of CT and MR, can characterize these lesions by showing calcification, fat density, and other tissue features that can help identify the exact nature of these masses. Figure 16.18 shows the CT scan of a child with a cystic teratoma (dermoid) of the right ovary containing teeth, fat, and hair. And Figure 16.19 shows CT

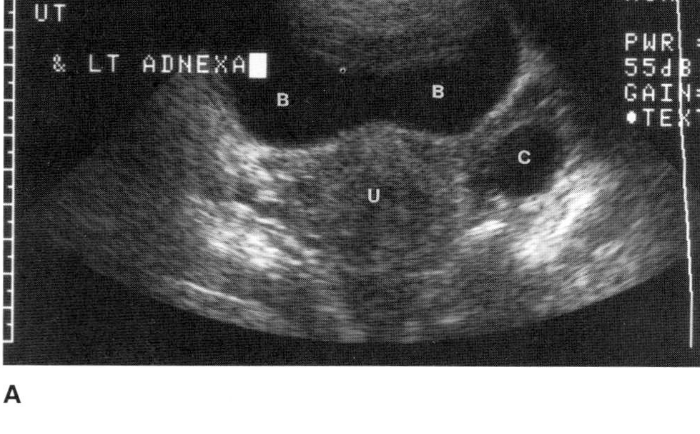

A

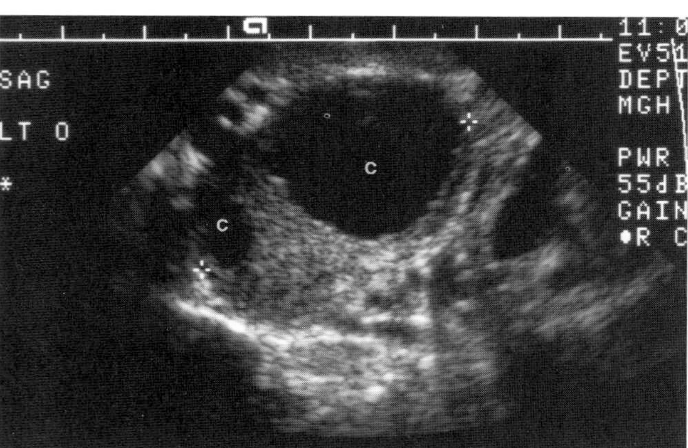

B

Figure 16.14. Ultrasound of an asymptomatic left ovarian cyst. A: Transabdominal scan in the transverse plane shows a small cyst *(C)* in the left ovary. The bladder (*B*'s) and uterus are shown medially. B: Transvaginal scan of the left ovary (margins marked with cursors) shows two cysts (*C*'s); only the larger one was apparent on the transabdominal scan.

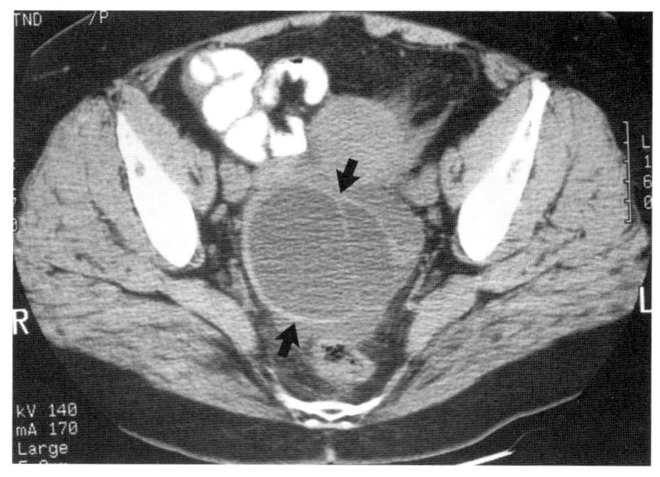

Figure 16.15. CT scan of a torsed right ovarian cyst. The patient underwent a CT examination for the acute onset of severe lower abdominal pain. A large cystic structure *(arrows)* was identified in the pelvis. At surgery a torsed ovarian cyst was identified and resected. The CT scan made the diagnosis of a cyst, probably ovarian, but could not detect the torsion. It was the combination of the patient's clinical presentation and the CT finding that led to the correct preoperative diagnosis.

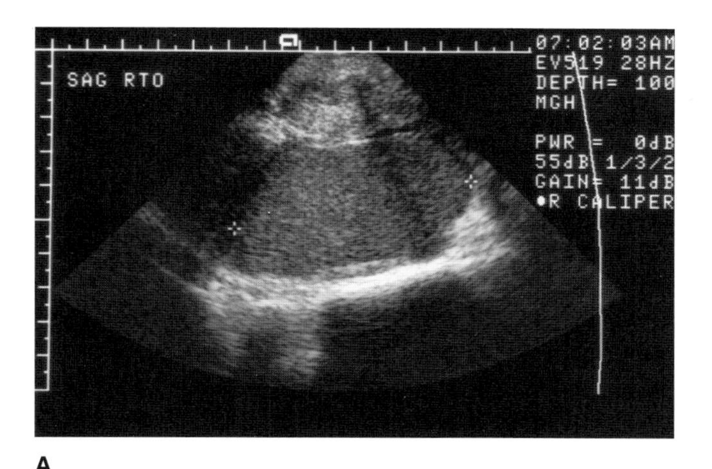

A

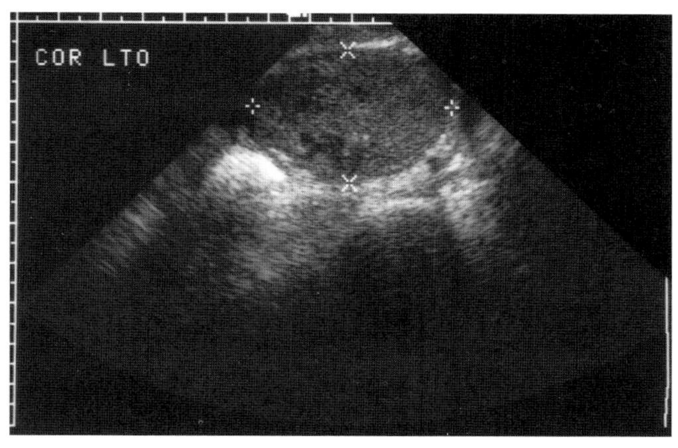

B

Figure 16.16. Bilateral ovarian endometriomas (see text). A: Sagittal transvaginal ultrasound of the right ovary. B: Coronal transvaginal ultrasound of the left ovary.

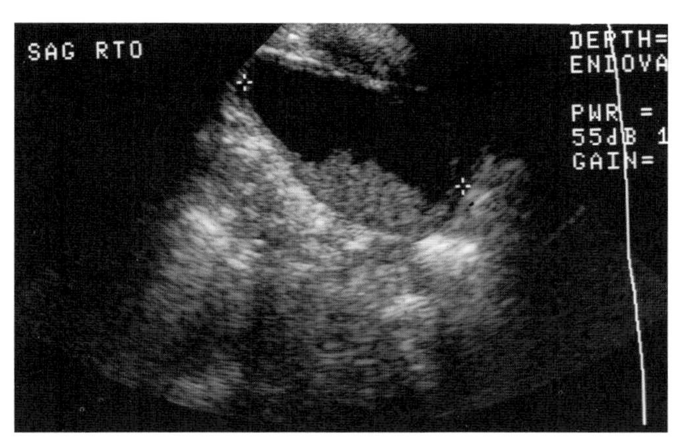

A

Figure 16.17. Comparison of a cystic ovarian tumor (A, *above*) with a benign ovarian cyst (B, *at right*) in another patient. The cystic tumor has enlarged the right ovary to nearly twice its normal size, measuring 56 by 73 millimeters; the cystic space within the tumor is irregularly margined and not spherical in shape, as is the benign cyst.

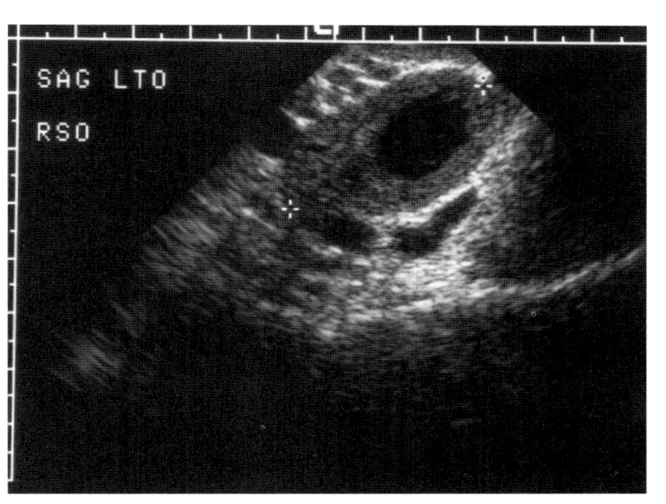

B

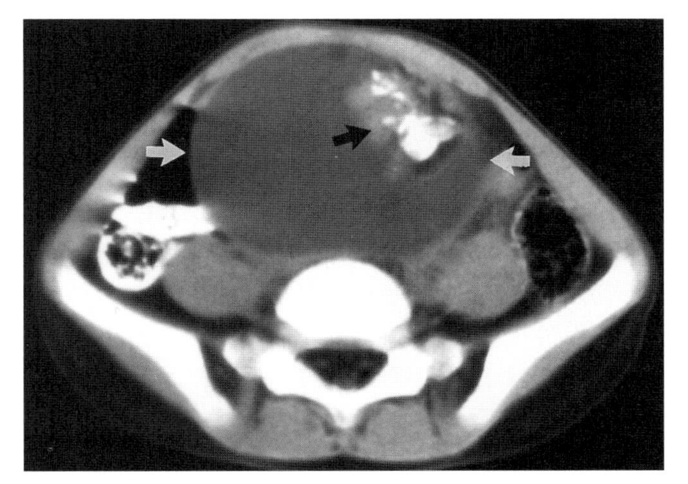

Figure 16.18. CT scan of a child with a large cystic teratoma of the right ovary. The *white arrows* indicate the extent of the cystic mass. The *black arrow* points to the dermoid plug, containing teeth, fat, and hair.

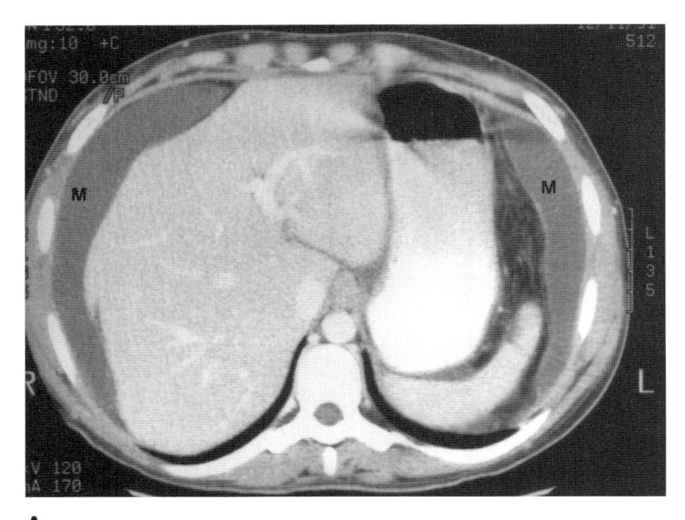

A

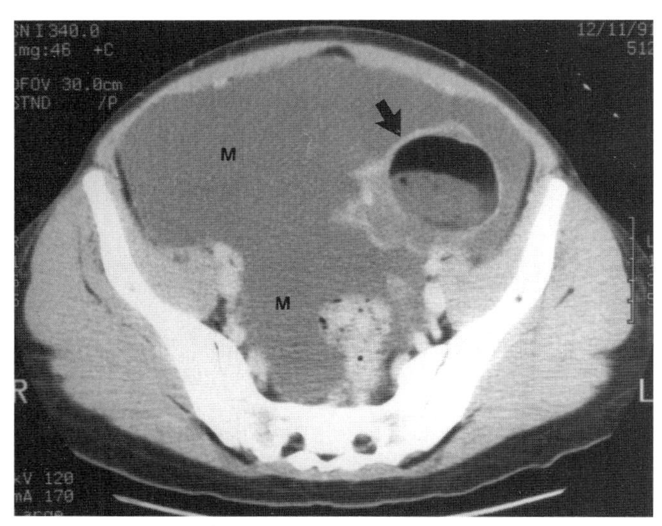

B

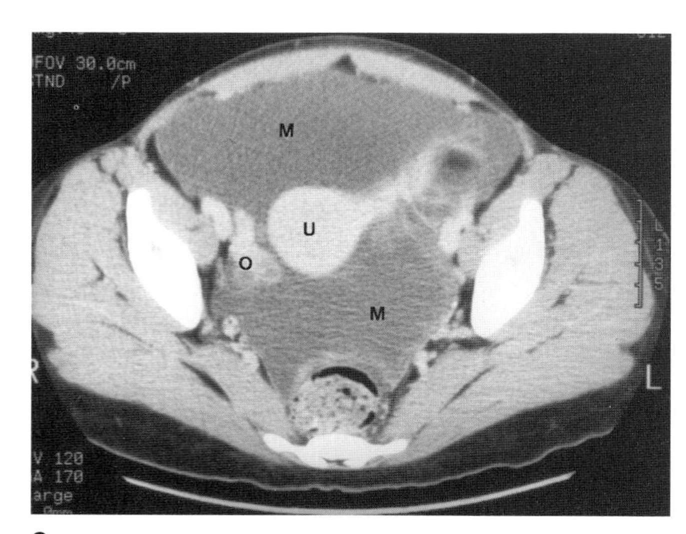

C

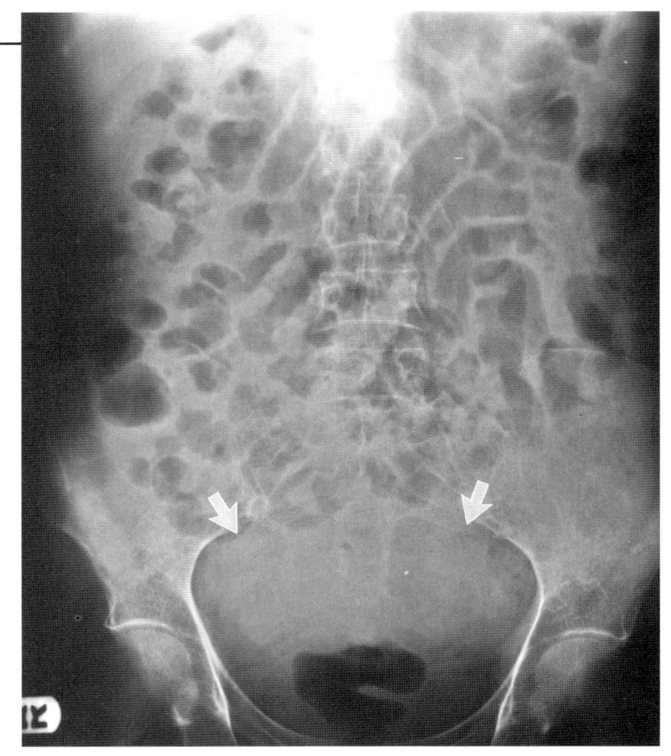

A

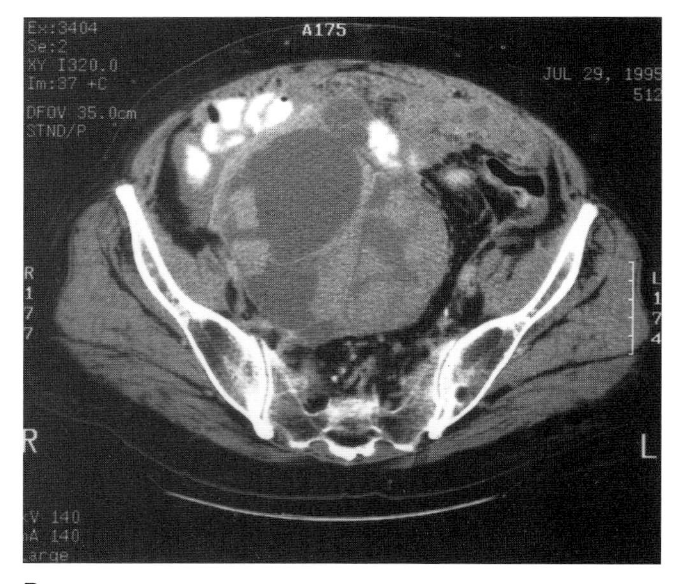

B

Figure 16.20. Large tumor of the right ovary. A: Plain film shows the air-filled loops of bowel lifted out of the pelvis by a large mass *(arrows).* B: CT scan reveals a large, complex cystic ovarian tumor.

Figure 16.19A, B, and C (left column). CT scans of a patient with a mucinous cystadenoma of the left ovary with rupture and development of pseudomyxoma peritonei. Peritoneal spread of mucin-secreting cells from the left ovarian tumor (*arrow* in B) has filled the peritoneal cavity with mucin *(M),* clearly seen in A. The lowest-level scan (C) shows a normal uterus *(U)* and right ovary *(O).*

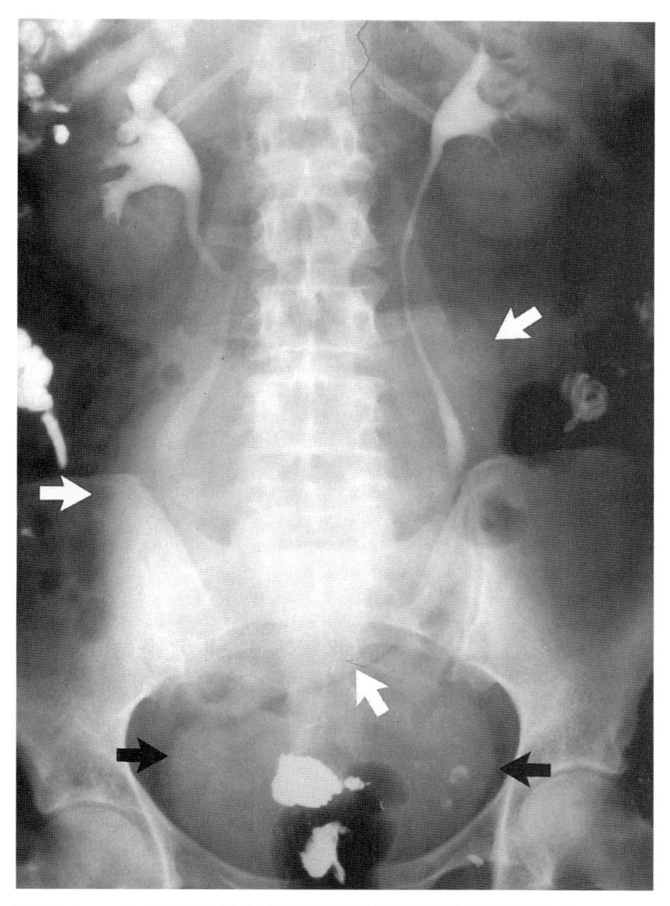

scans of a patient with a mucinous cystadenoma with rupture and peritoneal spread of mucin. Ovarian carcinoma usually spreads by peritoneal seeding, with tumor nodules implanting on loops of bowel, the mesentery, and the omentum; this spread is best shown with CT, which is recommended for staging and follow-up after treatment. Large ovarian tumors (Figure 16.20) may be detected on plain films.

The most common benign tumors of the uterus are leiomyomas (fibroids) and those caused by adenomyosis (ingrowth of endometrial tissues within the uterine muscle, or myometrium). Leiomyomas are so common that they occur in 40 percent of women over 35. You have already seen the plain film of a patient with enormous calcified uterine leiomyomas in Chapter 11 (Figure 11.18). Figure 16.21 shows the intravenous urogram of another patient with an enormous noncalcified uterine leiomyoma. Adenomyosis is less common and occurs mostly in women over 40. Both

Figure 16.21 *(left)*. Intravenous urogram of a woman with an enormous noncalcified uterine leiomyoma *(white arrows)*. A thin, lucent, fat line separates the leiomyoma from the opacified urinary bladder *(black arrows)*. You no doubt also noted the residual barium contrast material scattered through the large bowel.

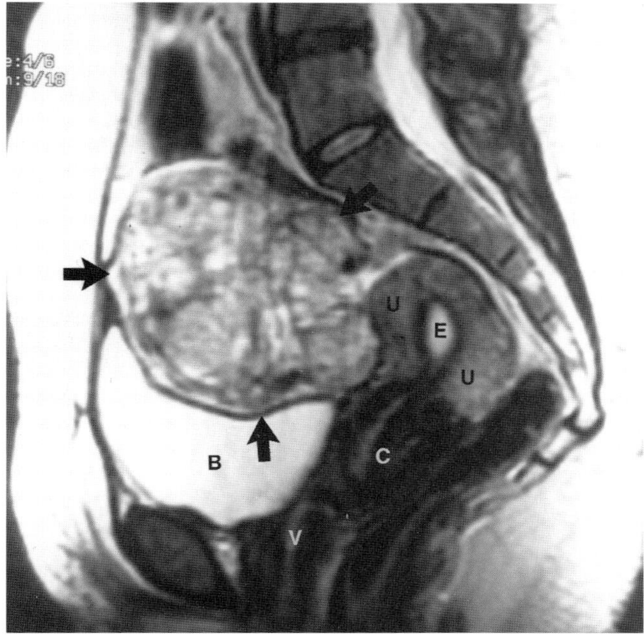

A

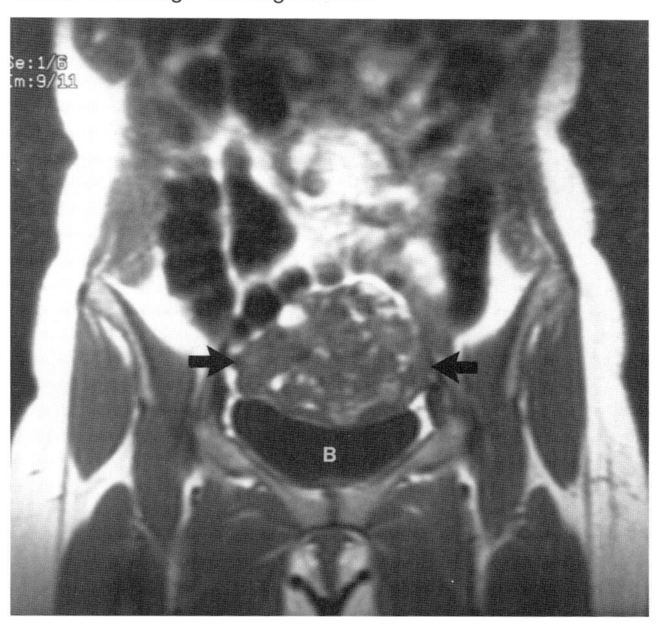

B

Figure 16.22. Pelvic MR scans of a woman with a large right ovarian teratoma. A: Lateral T2-weighted image shows a large tumor *(arrows)* on top of the bladder *(B)*, displacing the uterus *(U)* posteriorly. *E* indicates the endometrium, *C* the cervix, and *V* the region of the vagina. B: Coronal T1-weighted image again shows the tumor *(arrows)* on top of the bladder *(B)*.

conditions can cause pelvic pain and hypermenorrhea. Ultrasound is the most commonly used screening examination, but MR is superior at differentiating the two conditions. The correct diagnosis is very important because the treatment of the two is radically different: leiomyomas can be resected surgically without depriving the patient of her uterus and her fertility; symptomatic adenomyosis requires hysterectomy.

Cervical cancer is the most common gynecological malignancy, and also the most common malignancy in women under 50. Local spread is common, into the vagina, uterus, bladder, and rectum. CT and MR have been most successful in staging this cancer, assessing tumor volume and local and distant extension. Endometrial carcinoma occurs in older women, presenting with postmenopausal bleeding. Ultrasound can detect the neoplasm within the uterine cavity; MR is superior at showing the depth of invasion into the myometrium and the extent of the involvement of surrounding structures.

Pelvic inflammatory disease (PID), usually caused by infection with gonococcus or chlamydia, produces a spectrum of conditions from endometritis to salpingitis with hydrosalpinx. When the involvement is early or minimal, an ultrasound exam may be normal. Later, when hydrosalpinx is present, ultrasound will show an array of complex cystic masses representing the dilated, fluid-filled uterine tubes (Figure 16.23).

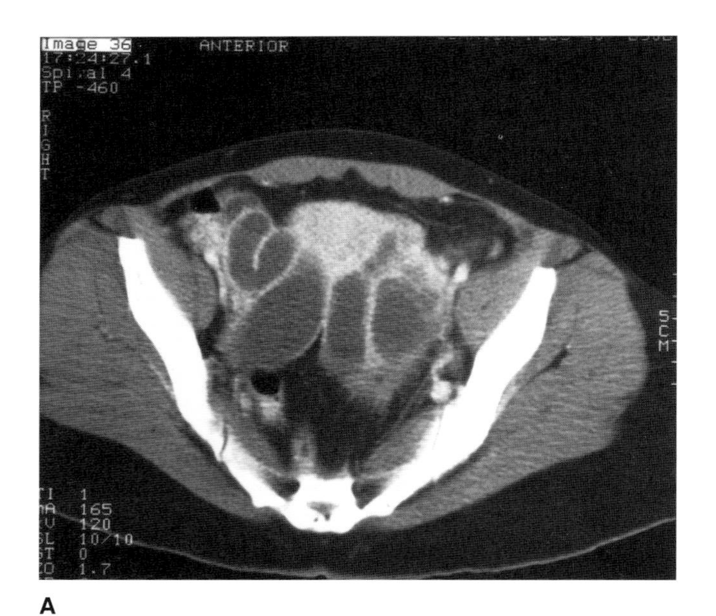

A

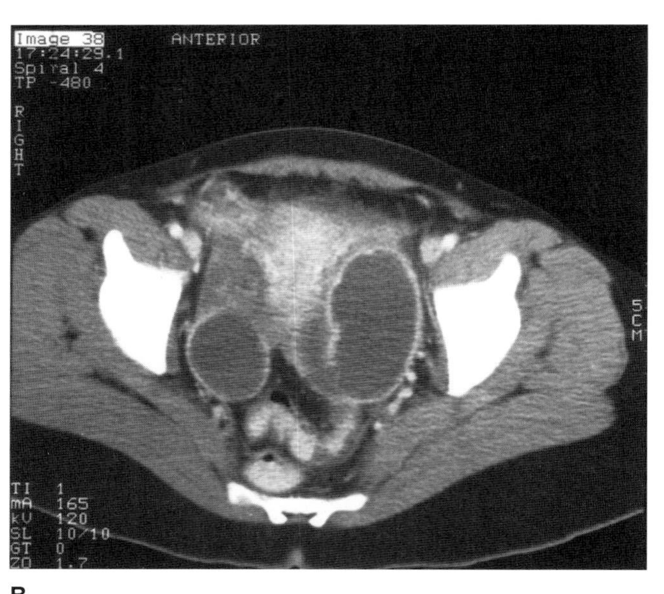

B

Figure 16.23. CT scans of a young woman with hydrosalpinx resulting from pelvic inflammatory disease. The large fluid-filled tubular structures in the pelvis represent markedly dilated and obstructed uterine tubes.

Obstetrical Imaging

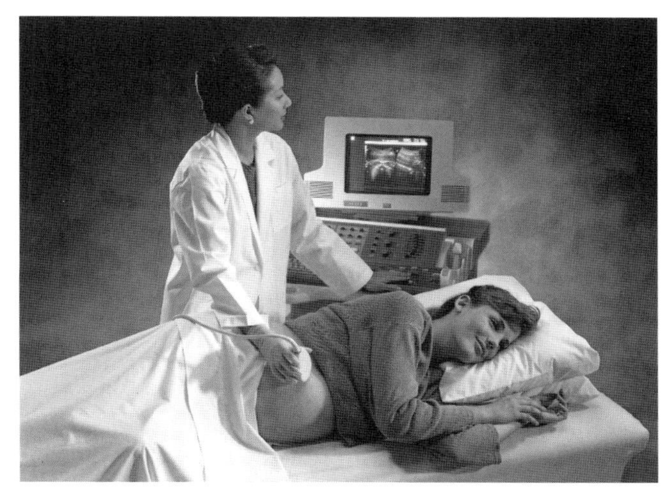

Figure 16.24. Obstetrical ultrasound examination.

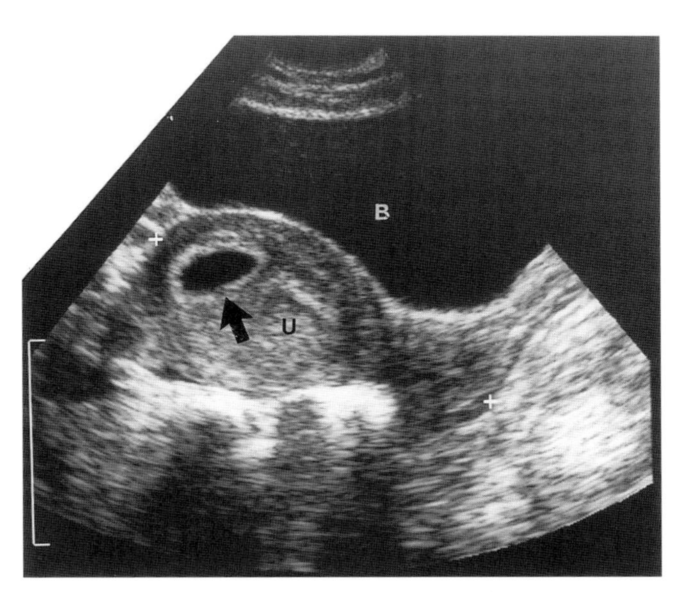

Figure 16.25. Transabdominal ultrasound of the normal gestational sac at 6 weeks. The gestational sac appears as a fluid collection within the uterus *(U)* surrounded by an echogenic rim *(arrow)*. The anechoic, fluid-filled bladder *(B)* is used as an acoustic window.

Ultrasound during pregnancy (Figure 16.24) can accurately date the pregnancy, detect multiple pregnancies, monitor fetal growth, and assess fetal well-being. With ultrasound's real-time motion images you can observe fetal cardiac motion and fetal movements; a decrease or absence of these activities would, of course, indicate an endangered or dead fetus. With the excellent resolution of transvaginal ultrasound it is possible to detect fetal heart motion as early as 5 weeks after gestation.

Ultrasound can quickly and accurately detect first-trimester problems such as ectopic pregnancy (implantation of the pregnancy outside the uterine cavity) and spontaneous abortion (termination of the pregnancy before 20 weeks' duration by natural causes), and third-trimester difficulties such as placenta previa (placenta covering the cervical os) and placental abruption (premature separation of a normally positioned placenta). Transvaginal scanning is usually necessary during the first trimester, but abdominal scanning suffices later in pregnancy. Ultrasound can also detect a wide range of fetal abnormalities involving the head (anencephaly), spine (spina bifida), chest (diaphragmatic hernia), heart (congenital heart disease), abdomen (fetal hydronephrosis or bowel obstruction), and skeleton (achondroplasia).

Ultrasound imaging is indicated in pregnant women who develop vaginal bleeding, pelvic pain, or premature labor, or who suffer trauma. Controversy exists over whether to use a screening ultrasound examination to assess all pregnancies—even normal and asymptomatic pregnancies. Some obstetricians and medical centers examine every pregnant woman; others scan only patients with a worrisome history or clinical examination.

When routine ultrasound assessment of pregnancy is performed, the scanning protocol varies with the stage of pregnancy. First-trimester scanning documents the appearance and location of the gestational sac, identifies the embryo, measures its crown-rump length, and verifies fetal cardiac activity. Second- and third-trimester scanning evaluates fetal well-being, estimates the volume of amniotic fluid, determines the fetal position and placental location, and determines several important fetal measurements, such as head size (biparietal diameter and head circumference), abdominal circumference, and femur length. Various parts of the fetal anatomy are observed for possible abnormalities, including the cerebral ventricles, heart chambers, kidneys, stomach, bladder, spine, and umbilical cord.

During the first trimester a small gestational sac can be seen as early as 4.5 weeks with transvaginal scanning and 5 weeks with transabdominal scanning. The sac appears as an intrauterine fluid collection surrounded by a rim of moderate echoes (Figure 16.25).

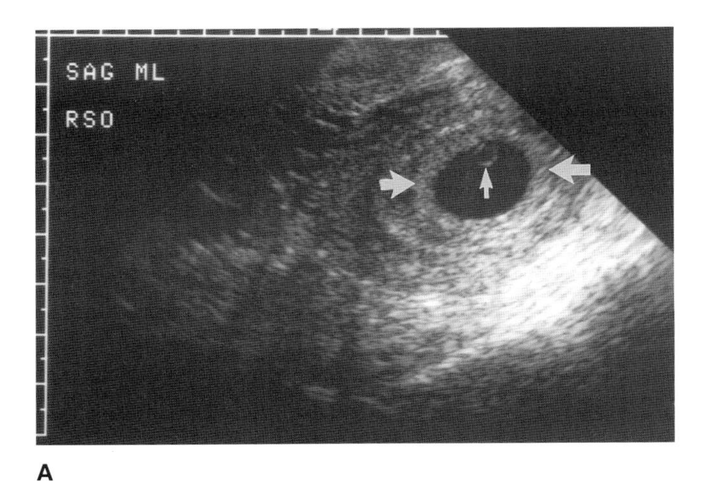

A

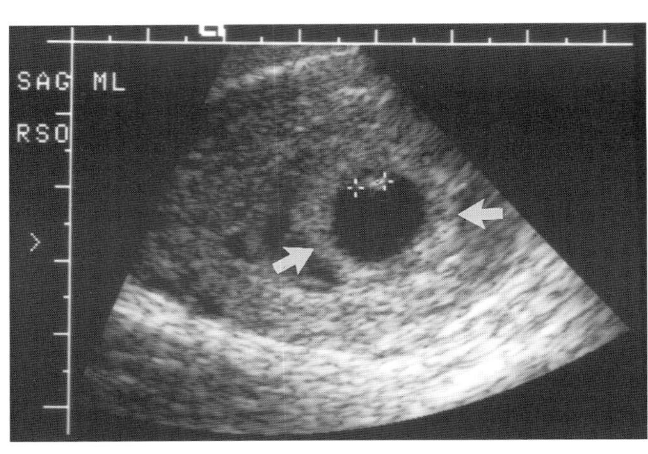

B

Figure 16.26. Ultrasound of a gestational sac at 5½ weeks. A: Transvaginal scan shows a yolk sac *(small arrow)* within the gestational sac *(large arrows).* B: Transvaginal scan also shows a tiny (5.3 millimeters long) embryo (marked by cursors) within the sac *(arrows).* Embryonic cardiac activity was detected.

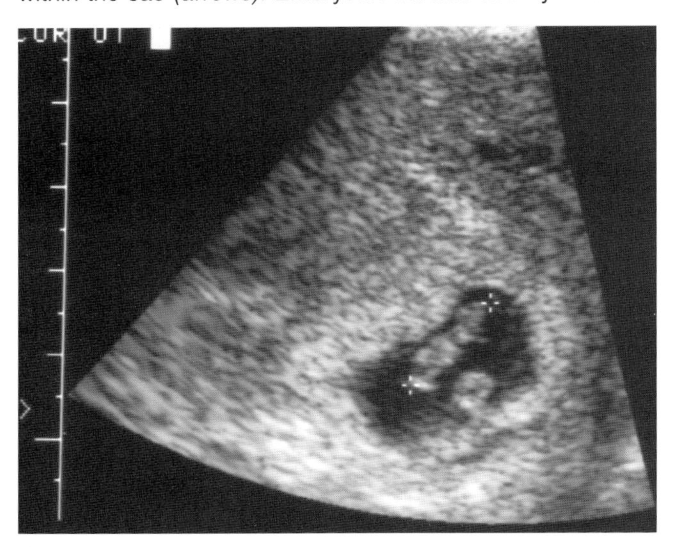

A

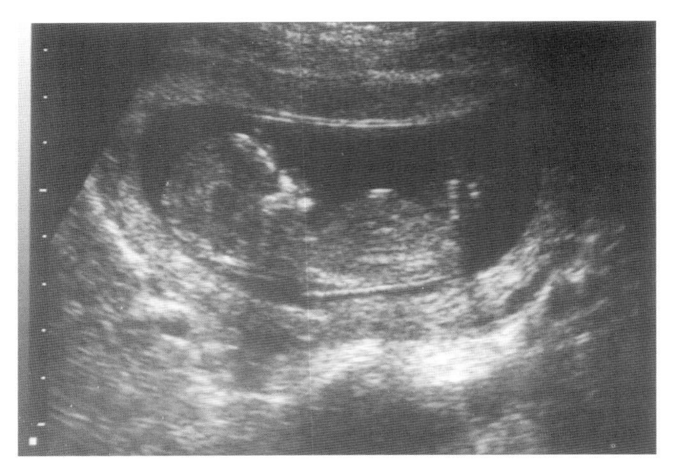

B

Figure 16.27. Transvaginal ultrasound scans of a 7-week-old fetus (A, *at left*) and a 12-week-old fetus (B).

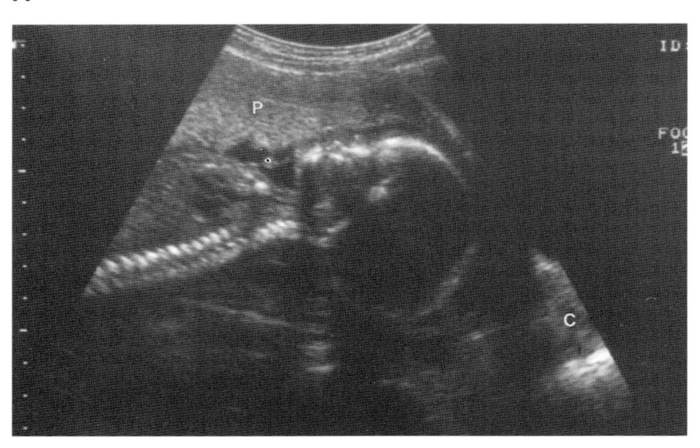

Figure 16.28. Longitudinal transabdominal ultrasound at 27 weeks menstrual age. The fetus is in a cephalic, face-up presentation. Note the relationship between the fetal head and cervix *(C).* Identify the head, face, and spine. The placenta *(P)* can be seen anterior to the fetus.

The yolk sac is the first object seen within the gestational sac, where it appears as a small rounded structure (Figure 16.26A). The embryo can first be seen in close proximity to the yolk sac when the crown-rump length is greater than 5 millimeters, and embryonic cardiac activity can then be observed (Figure 16.26B). Cardiac activity is one of the most important indicators of fetal life and well-being. For comparison, look at the 7-week-old fetus and the 12-week-old fetus in Figure 16.27. As the fetus enlarges, scanning in the second and third trimesters is performed transabdominally (Figure 16.28).

Ectopic Pregnancy

Ectopic pregnancy is one of the leading causes of maternal death during pregnancy, occurring in 1–2 percent of all pregnancies and accounting for 15 percent of maternal deaths. The incidence of ectopic pregnancy has been steadily increasing because more and more women are contracting pelvic inflammatory disease or undergoing in vitro fertilization, both of which increase the likelihood of ectopic implant. Fortunately, the ability to make a rapid and accurate early diagnosis with ultrasound has decreased the mortality of this condition. You should consider ectopic pregnancy in your first-trimester patients who present with pain, abnormal vaginal bleeding, and a palpable adnexal mass. As you might expect, the findings at transvaginal ultrasound include the absence of a gestational sac, or embryo, within the uterine cavity. If echogenic fluid is detected in the cul-de-sac, it usually represents blood and indicates that the ectopic pregnancy has ruptured (Figure 16.29). In about one-third of cases the actual ectopic sac (Figure 16.30), or embryo, may be identified, often located in a uterine tube, but sometimes within the ovary, abdominal cavity, or cervix.

Two other facts about ectopic pregnancy are important for you to know. First, a ruptured ectopic pregnancy can occur in a woman *who does not realize she is pregnant.* You should therefore consider the possibility in all women of child-bearing age whose clinical signs are suspicious for possible ectopic pregnancy, and you should immediately request a β-hCG (human chorionic gonadotrophin) level. If that level is elevated, pregnancy is confirmed, and emergency transvaginal ultrasound must be performed to rule out an ectopic pregnancy. Second, simultaneous intrauterine and ectopic pregnancies, although rare, do occur, and their incidence is increasing (as is the incidence of multiple pregnancies) in women undergoing fertility treatment with ovulation-inducing drugs. Consequently the ultrasonographer will continue to search for an ectopic gestational sac, or embryo, even when a normal intrauterine pregnancy has been identified at the beginning of the ultrasound examination.

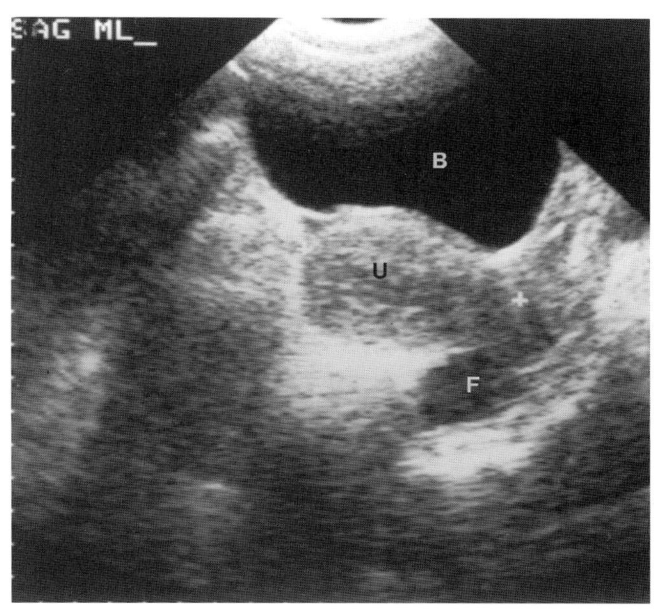

Figure 16.29. Transabdominal ultrasound of a ruptured ectopic pregnancy. In this patient with an elevated β-hCG (human chorionic gonadotrophin) level, ultrasound would normally show an intrauterine gestational sac. But no sac is seen within the uterus *(U),* and free fluid *(F),* representing blood, is seen behind the uterus. The white cursor overlies the cervix, and the bladder *(B)* is seen anteriorly.

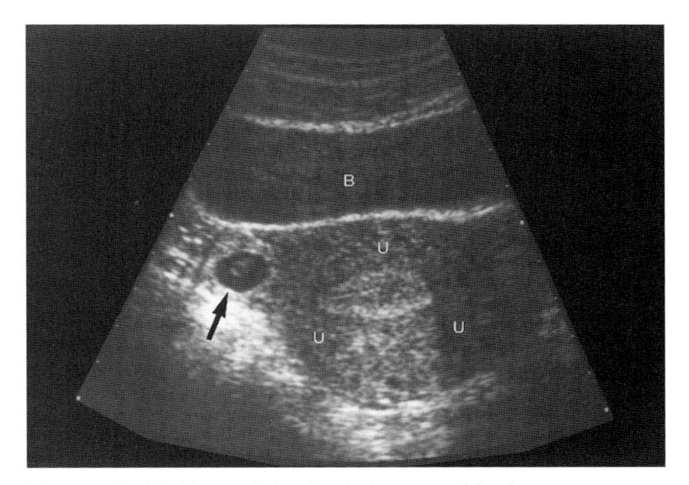

Figure 16.30. Transabdominal ultrasound in the transverse, or coronal, plane showing an ectopic gestational sac *(arrow)* within the right adnexa. No rupture had occurred. Note the tiny yolk sac within the gestational sac. Only echogenic endometrium is seen within the cavity of the uterus (*U*'s). The echolucent bladder *(B)* is shown anteriorly.

Placenta Previa

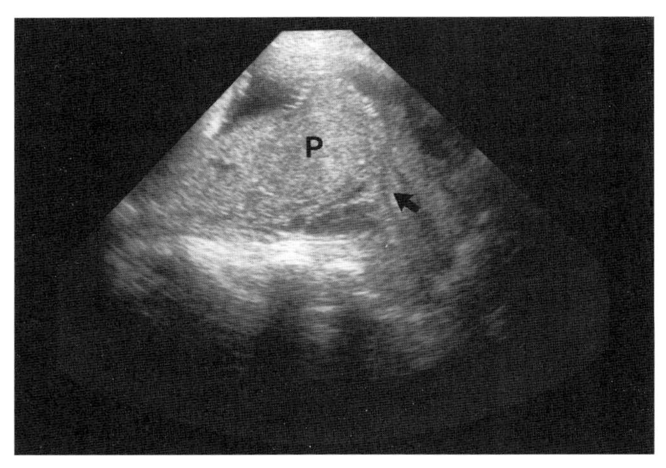

Figure 16.31. Ultrasound of placenta previa. The placenta *(P)* is low, covering the region of the cervical os *(arrow)*.

Painless vaginal bleeding in the third trimester may be an indication of placenta previa, a condition in which the placenta covers the internal os of the cervix. Patients with this symptom should have an ultrasound examination to determine the position of the internal os in relation to the placenta, which can be identified by its shape and granular echotexture. In a *complete previa* the placenta covers the entire cervical os (Figure 16.31); in a *partial previa* the placental margin extends to but not across the internal os. Knowledge of placenta previa will assist the obstetrician in planning the appropriate management of the pregnancy.

Placental Abruption

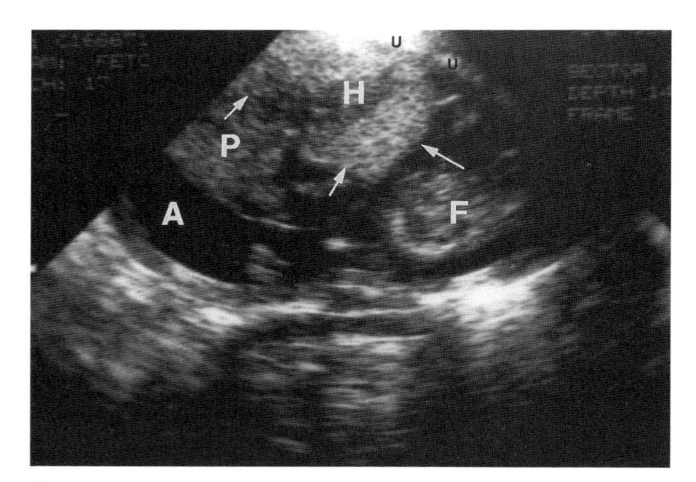

Figure 16.32. Ultrasound of placental abruption. The placenta *(P)* is separated from the wall of the uterus (*U*'s) by an echogenic hematoma *(H, arrows)*. Anechoic amniotic fluid *(A)* and the fetus *(F)* are also shown within the uterine cavity.

Because the placenta contains many blood vessels, hemorrhages are one of the most common placental complications. When the hemorrhage is retroplacental (between the placenta and the uterine wall), placental abruption may occur, separating the placenta from the uterus. Abruption can produce pain, vaginal bleeding, and hypovolemic shock. If abruption has occurred, an ultrasound examination will show an echogenic collection of blood in the retroplacental area (Figure 16.32). Placental abruption is a major cause of fetal mortality and accounts for up to 15–20 percent of perinatal deaths; maternal morbidity and mortality may also, unfortunately, occur with this condition.

General Issues in Men's Imaging

For optimum management of your male patients you should be familiar with the techniques for imaging the scrotum and its contents, the prostate gland, and the urethra. Testicular tumors may occur in younger men and prostatic cancer in older ones. The most common indication for imaging the male urethra is trauma.

The Scrotum

Before the advent of ultrasound and cross-sectional imaging, the scrotum and its contents could be examined only by palpation and transillumination. Today a wide range of intratesticular and extratesticular conditions can be diagnosed with scrotal ultrasound, which is the procedure of choice for scrotal imaging. Ultrasound can identify testicular tumors, testicular trauma, testicular torsion, epididymitis (infection of the epididymis), orchitis (infection of the testis), hernias involving the scrotal sac, and fluid collections such as hydroceles, hematoceles, and abscesses. Although MRI can also diagnose many of these same scrotal conditions, ultrasound is generally faster, easier to obtain, and less expensive.

The scrotum (Figure 16.33) is a cutaneous pouch that contains the testes, the epididymides, and the lower ends of the spermatic cords . Each *testis* is mobile and ovoid, measuring 4–5 centimeters in length and 2.5–3.0 centimeters in width; the testes are suspended from the spermatic cords. A septum divides the scrotum into a separate compartment for each testis. The lobules of the testes are drained by seminiferous tubules that open into a network of channels, the rete testis; small tubules connect the rete testis to the *epididymis* (Figure 16.34). The epididymis lies posterior to the testis; it has an expanded head at the upper pole of the testis, a body, and a tail that is directed downward, and then continues as the vas

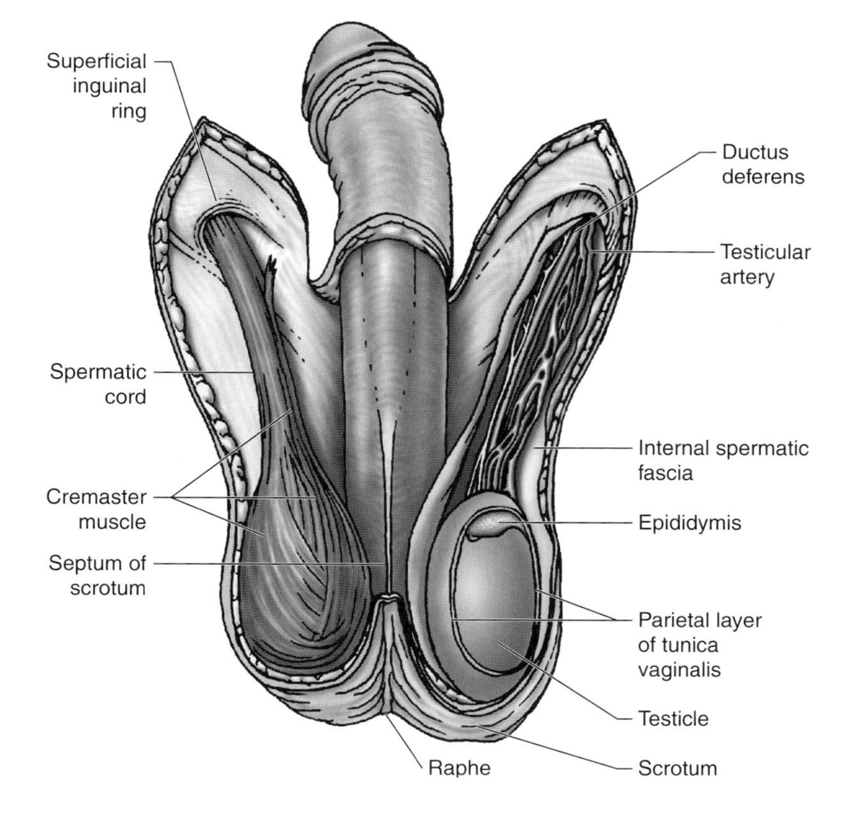

Figure 16.33. The scrotum. The penis has been turned upward, and the anterior wall of the scrotum has been removed. On the right side, the spermatic cord, the internal spermatic fascia, and the cremaster muscles are displayed; on the left side, the internal spermatic fascia has been divided by a longitudinal incision passing along the front of the cord and the testicle, and a portion of the parietal layer of the tunica vaginalis has been removed to display the testicle and a portion of the head of the epididymis covered by the visceral layer of the tunica vaginalis.

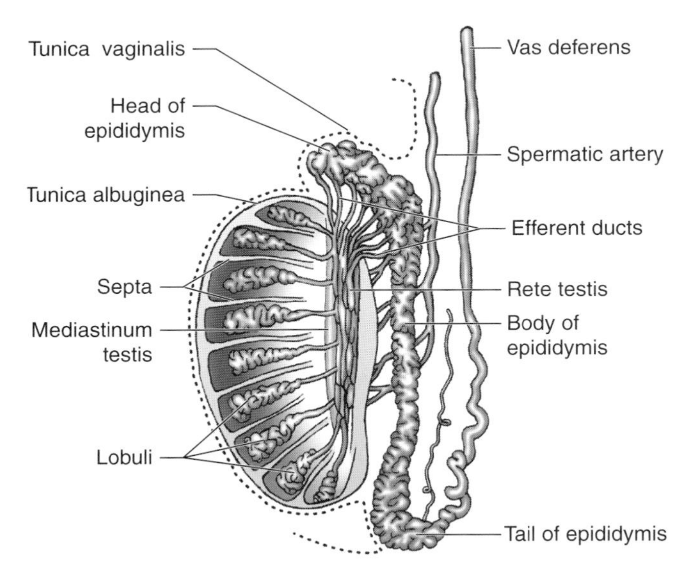

Figure 16.34. Vertical section of the testis showing the arrangement of the ducts.

deferens. A serous cavity, the tunica vaginalis, surrounds the anterior, medial, and lateral surfaces of the testes; fluid collections such as hydroceles may form in this cavity.

A scrotal ultrasound examination will no doubt prove to be a most unusual experience for your male patients. You should always carefully explain the procedure to your patients to reassure them. Two techniques are used. The scrotum may be draped over a towel for support while the ultrasonographer scans with a high-resolution transducer in direct contact with the scrotum. Or the ultrasonographer may hold the scrotum in one gloved hand to permit testicular and epididymal palpation correlation with the real-time ultrasound images, while scanning with a transducer held in the opposite hand. Scrotal ultrasound is usually not painful, but may be when the patient has an acute inflammatory condition such as epididymitis.

Normally the testes and epididymides display a fine, homogeneous echotexture at ultrasound (Figure 16.35). The ultrasonographer obtains images in both the transverse and the longitudinal planes by rotating the transducer while scanning. The testes appear round on transverse (coronal) images (Figures 16.35A and B) and ovoid on longitudinal (sagittal) images (Figures 16.35D, E, and F). The expanded head of the epididymis makes it easier to find that organ. You see a transverse scan of the head of the epididymis in Figure 16.35C and a longitudinal scan in Figure

Figure 16.35. Normal testicular ultrasound examination.

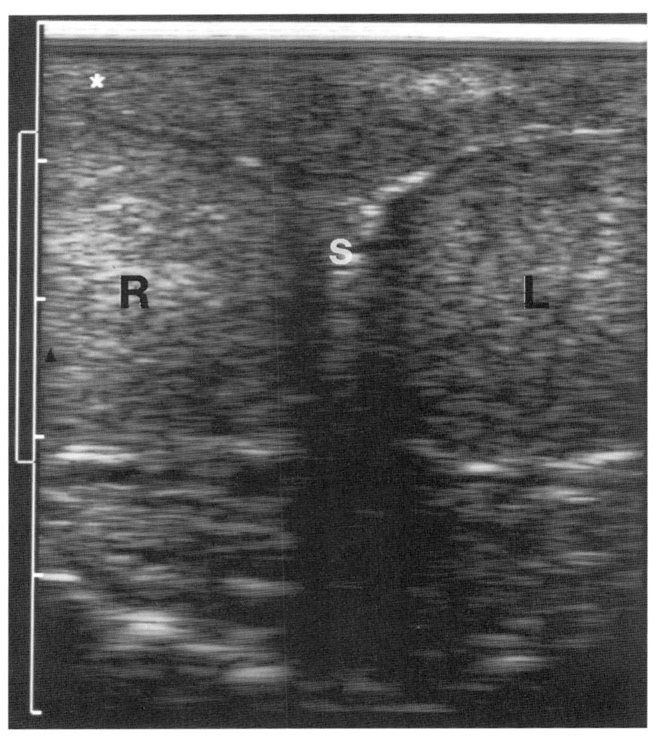

A. Transverse (coronal) scan of both testes. Note the homogeneous echotexture of the right *(R)* and left *(L)* testes, which are separated by the echogenic scrotal septum *(S)*.

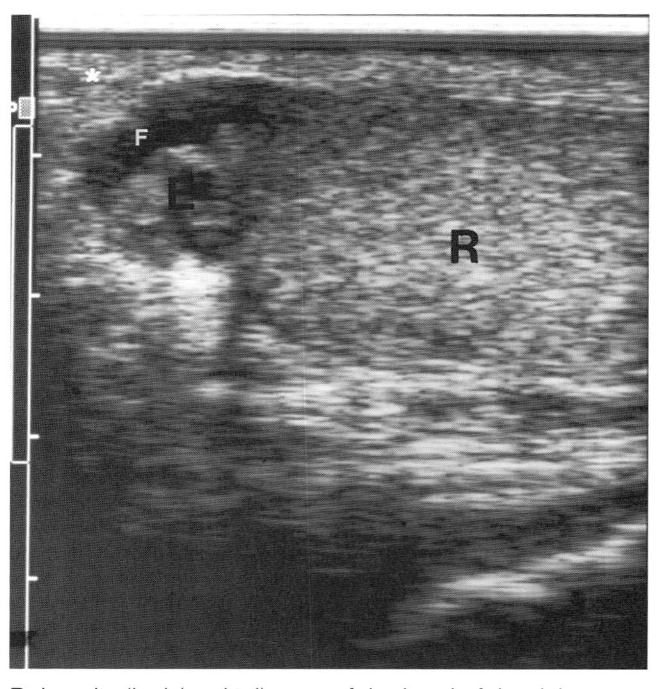

D. Longitudinal (sagittal) scan of the head of the right epididymis *(E)* and upper pole of the right testis *(R)*. Again physiological fluid *(F)* is seen around the epididymis.

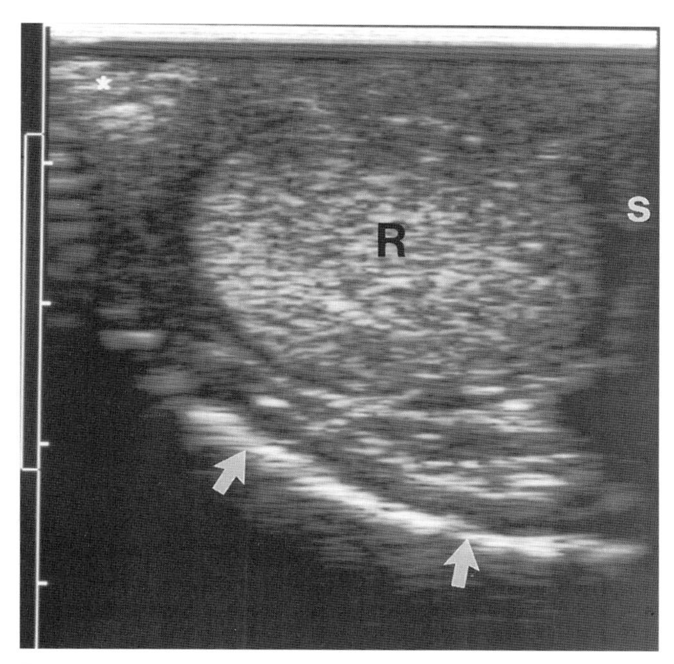

B. Transverse (coronal) scan of the right *(R)* testis. The septum *(S)* is shown just medial to the testis. The *arrows* indicate the echogenic and echolucent stripes produced by the various layers of the scrotal wall.

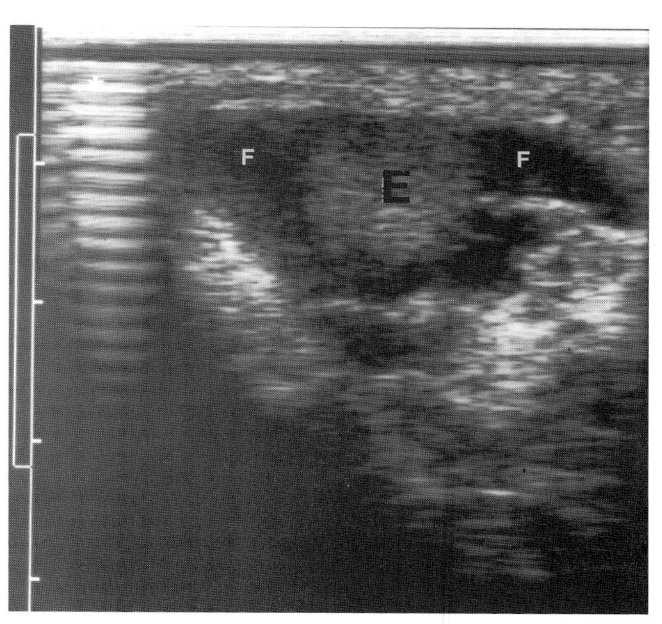

C. Transverse (coronal) scan of the head of the right epididymis *(E)*. Echolucent physiologic fluid *(F)* is seen around the epididymis.

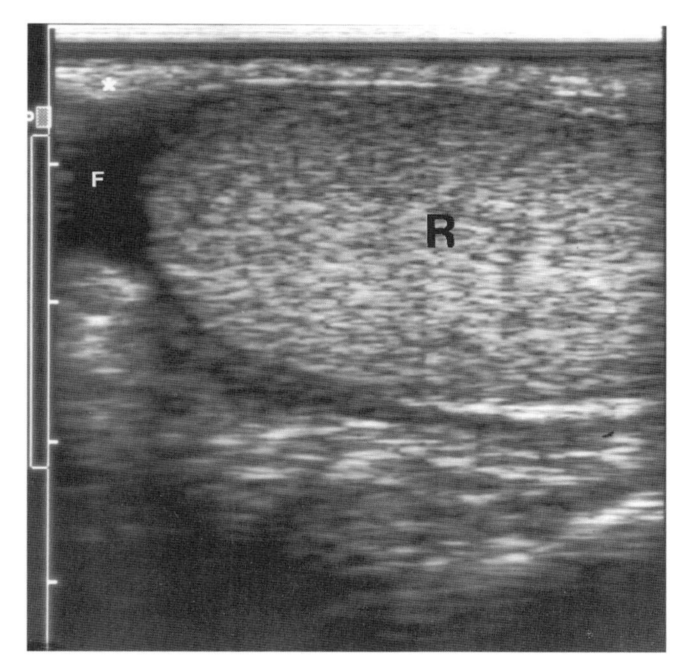

E. Longitudinal scan of the body of the right testis *(R)* with physiological fluid *(F)*.

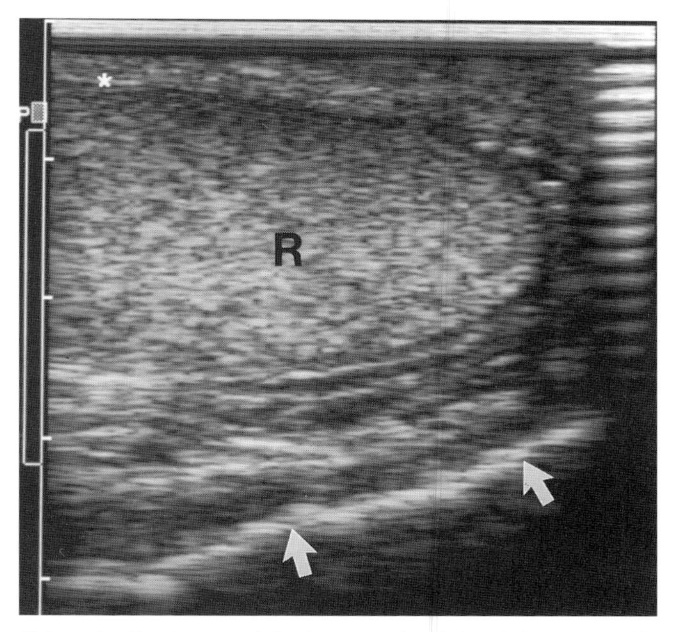

F. Longitudinal scan of the lower pole of the right testis *(R)*. The *arrows* point to the various layers of the scrotal wall.

16.35D; in both a small amount of echolucent physiological fluid is seen around the epididymis. The various layers of the scrotum appear as echogenic stripes around the testes on both transverse (Figure 16.35B) and longitudinal (Figure 16.35F) scans.

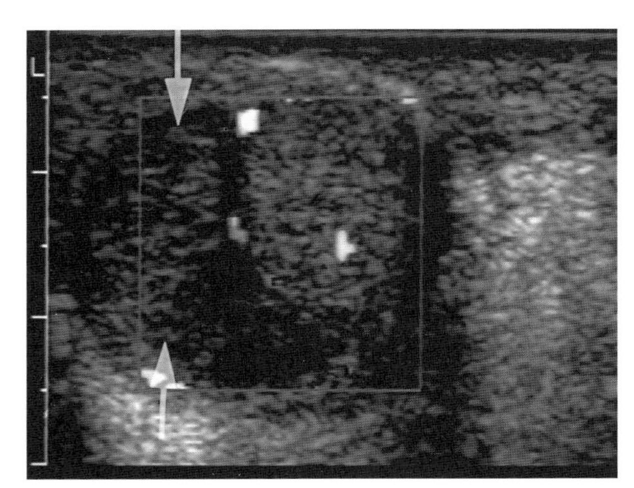

Figure 16.36. Ultrasound of epididymitis. The epididymis *(arrows)* is markedly enlarged and hypoechoic compared with the normal epididymis seen adjacent to it. Color Doppler ultrasound would show increased blood flow.

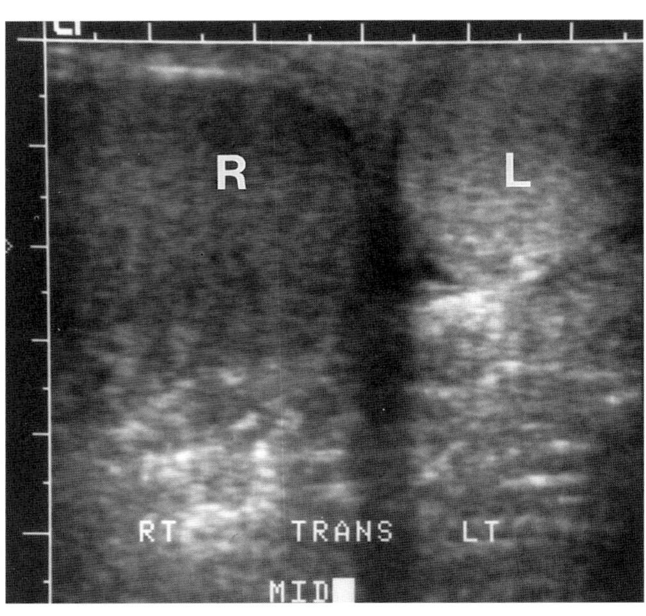

Figure 16.37. Transverse (coronal) ultrasound of orchitis of the right testis. The right testis *(R)* is enlarged, showing decreased testicular echogenicity. Compare the size and echopattern with the normal left testis *(L).* Also compare this scan with the normal transverse scan in Figure 16.35A.

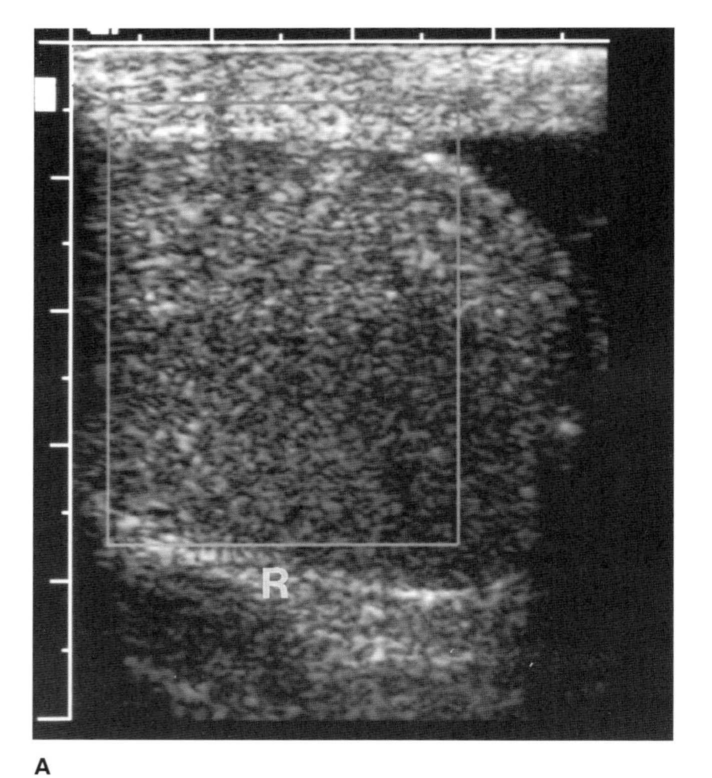

A

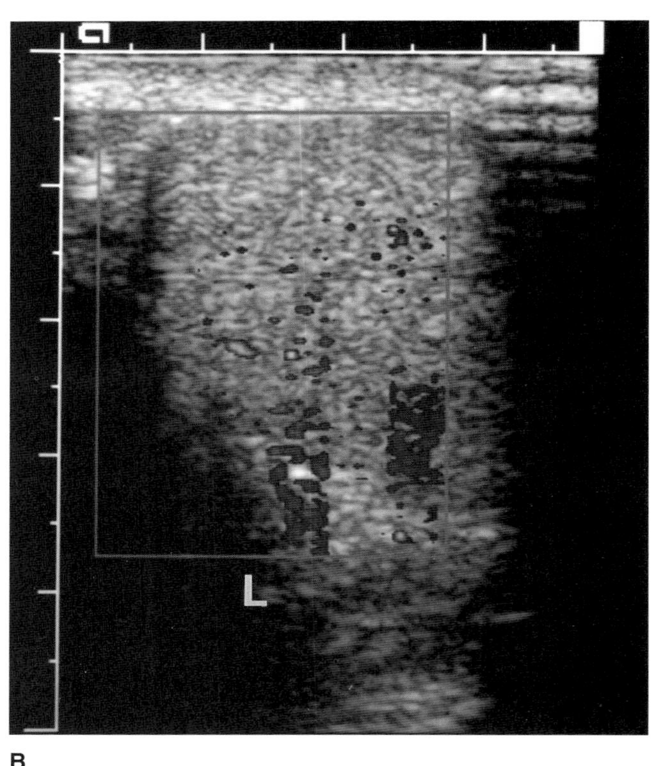

B

Figure 16.38. Ultrasound of testicular torsion. A: Doppler mode scan (area within rectangle) of the right kidney shows no Doppler signals, because there is no right testicular blood flow. B: Normal Doppler signals (course echoes overlying the normal testicular echotexture) are seen on the left, representing normal blood flow. Of course, these Doppler signals would be easier to see if photographed in color, as you can see in Figure 17.12H.

Patients should be referred for diagnostic scrotal ultrasound who present with scrotal pain, scrotal enlargement, a palpable scrotal mass, or scrotal trauma. Acute scrotal pain is most frequently caused by inflammatory conditions such as epididymitis and orchitis (usually from retrograde infection from the urinary tract). But testicular torsion, which is an acute surgical emergency, can also cause acute scrotal pain. Ultrasound can differentiate between inflammations and torsion. With *epididymitis* the epididymis appears enlarged and hypoechoic at ultrasound (Figure 16.36), as does the testis with *orchitis* (Figure 16.37). Color Doppler ultrasound, discussed in Chapter 17, would show *normal to increased blood flow* to the affected organs. With *testicular torsion* (Figure 16.38), which produces obstruction of blood flow in the spermatic cord and accompanying testicular artery resulting in testicular ischemia, color Doppler ultrasound of the affected testis would show *decreased or no blood flow* (see Figure 17.12H). Undiagnosed and untreated torsion leads to testicular infarction (Figure 16.39), shown at ultrasound as a heterogeneous and hypoechoic testicular echotexture.

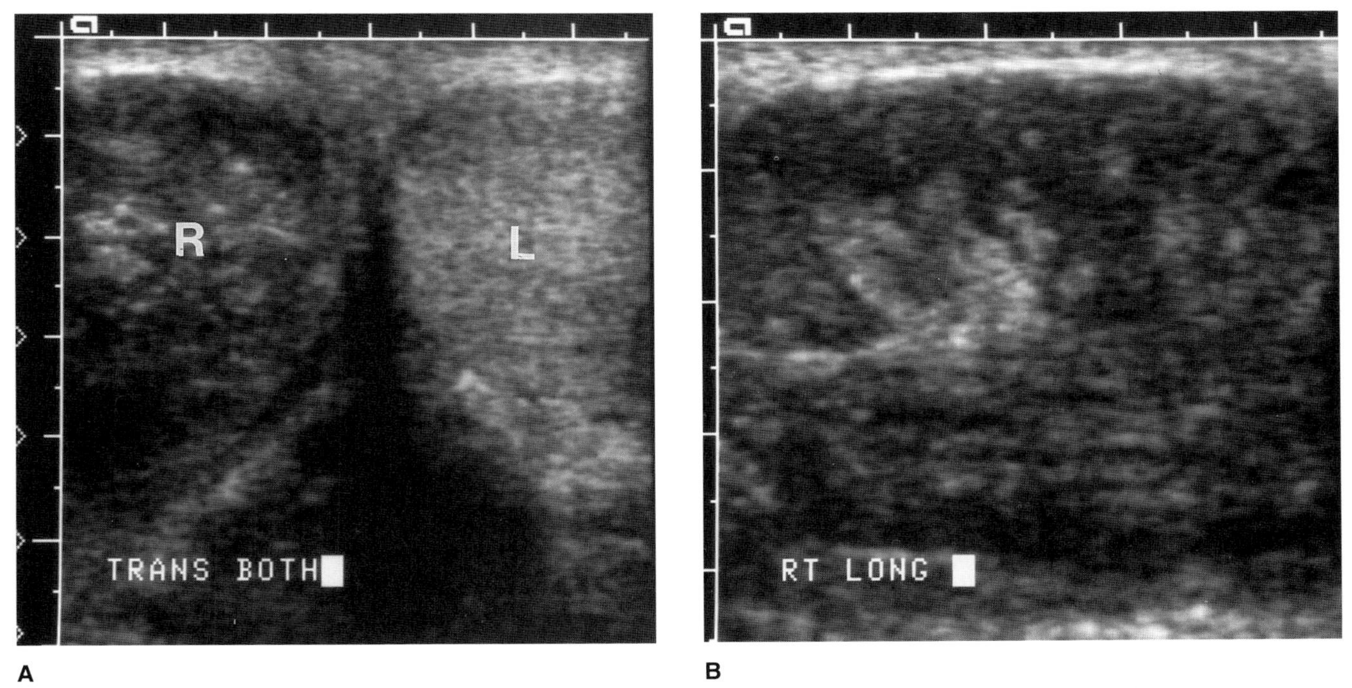

Figure 16.39. Ultrasound of infarction of the right testis. A: Transverse (coronal) scan shows that the echopattern of the right testis *(R)* is heterogeneous and hypoechoic, compared with the normal left testis *(L)*. B: Longitudinal (sagittal) scan of the infarcted right testis; compare the normal longitudinal scan of the right testis in Figure 16.35E.

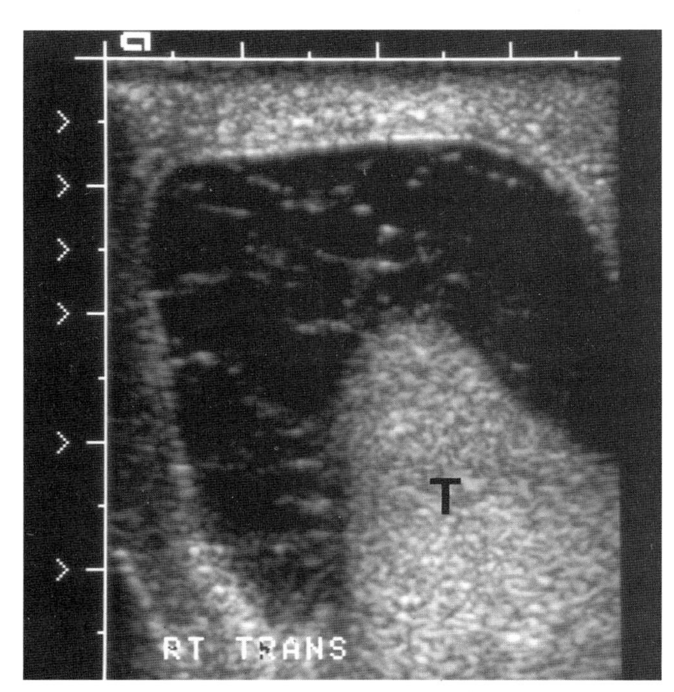

Figure 16.40. Ultrasound of a pyocele. A fluid collection surrounds the testis *(T),* which has many echogenic septations. Compare Figure 16.41.

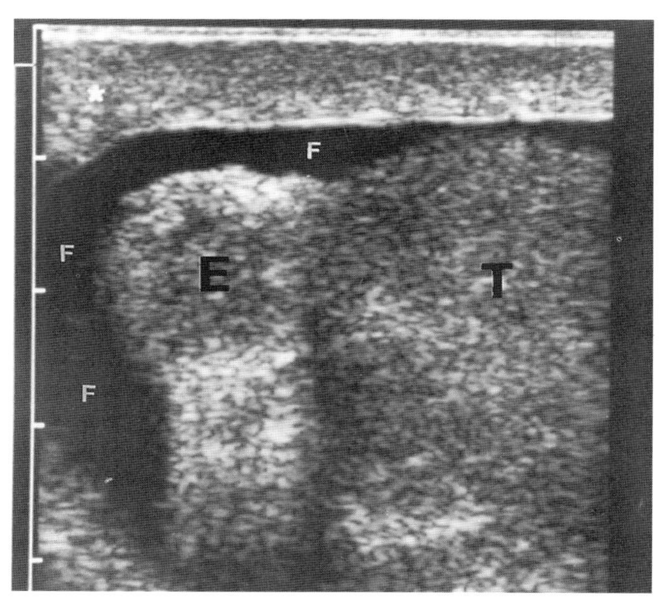

Figure 16.41. Ultrasound of a hydrocele. Anechoic fluid *(F)* surrounds the testis *(T)* and epididymis *(E).*

Acute scrotal enlargement may be caused by extratesticular fluid collections in the space between the visceral and parietal layers of the tunica vaginalis; they include hydroceles (serous fluid), hematoceles (blood), and pyoceles (pus). At ultrasound they appear as fluid collections surrounding the testis; hematoceles and pyoceles appear as echogenic or complex fluid collections (Figure 16.40), while hydroceles (Figure 16.41) are anechoic, yielding only pure fluid transmission of sound. Scrotal enlargement can also be caused by scrotal hernias; at ultrasound you can actually watch the peristalsis of the loop of bowel herniated within the scrotum.

New scrotal masses may represent malignant or benign tumors. Ultrasound can differentiate between intratesticular masses (Figure 16.42), which are generally malignant (seminoma, embryonal cell carcinoma, teratoma, choriocarcinoma), and extratesticular masses, which are generally benign. Most malignant testicular tumors are more hypoechoic than the adjacent normal testicular echotexture. Areas of hemorrhage, calcification, and necrosis may vary the echopattern within these tumors. A benign testicular cyst or abscess may present as a scrotal mass, as may a varicocele, a collection of dilated and tortuous veins associated with the epididymis and vas deferens.

Scrotal trauma may cause laceration and fracture of the testis and testicular rupture (Figure 16.43), which usually requires emergency surgical repair. Scrotal trauma may also cause only an extratesticular hematoma (hematocele), which can be treated conservatively. Patients with both testicular rupture and extratesticular hematoma may present with similar scrotal pain and swelling after trauma; ultrasound can differentiate between intra- and extratesticular injuries and promptly identify patients who need emergency surgical management.

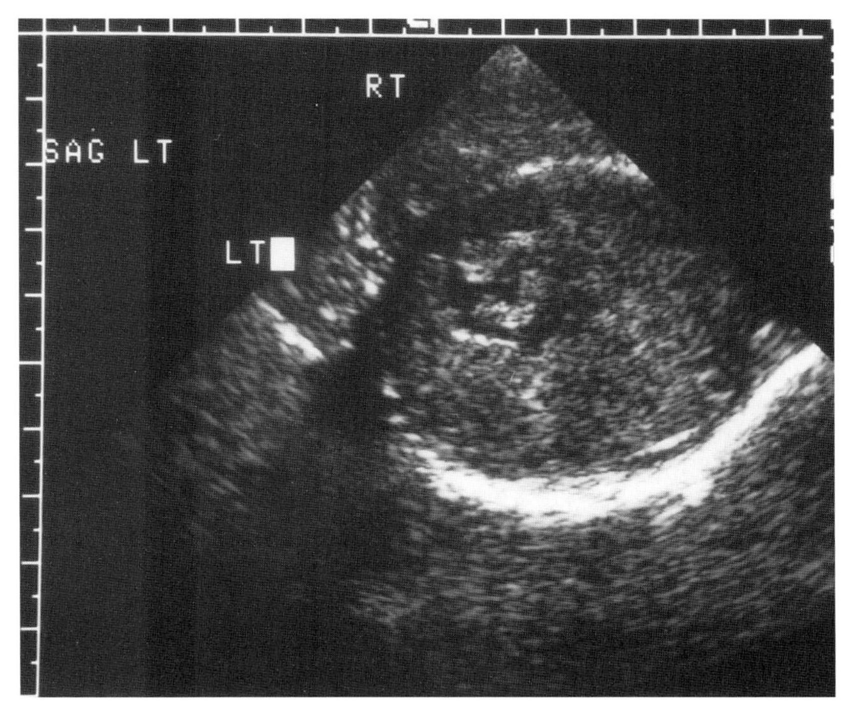

Figure 16.42. Ultrasound of a testicular tumor. Compare the normal scan, Figure 16.35. This testis contains an inhomogeneous mass with areas of increased and decreased echogenicity.

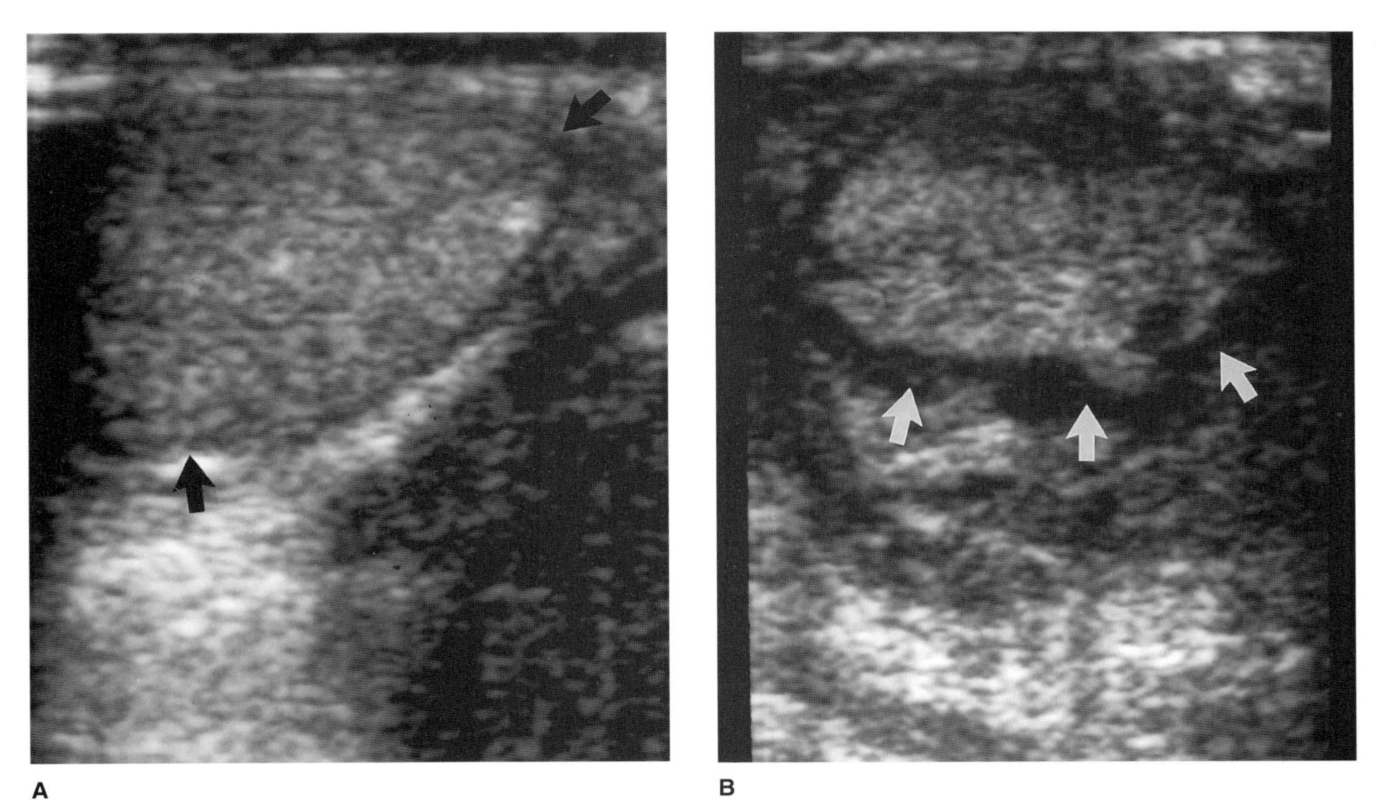

Figure 16.43. Ultrasound of a young man with scrotal trauma. A: The right testis *(black arrows)* is normal. B: Note the irregular fracture line *(white arrows)* coursing across the ruptured left testis. Testicular parenchyma can be seen both above and below the fracture line.

The Prostate

The prostate gland (Figure 16.44), located just below the bladder, and surrounding the first or prostatic segment of the urethra, is not visible on plain films. But when it is enlarged, the prostate may be detected on an intravenous urogram as an elevation or indentation of the base of the contrast-filled bladder (Figures 14.75 and 16.45A). After prostatic enlargement has reached a critical size, radiographic signs of bladder outlet obstruction (prostatism) will be obvious, and the postvoid residual bladder volume will be revealed as abnormally large on a repeat film taken after the patient has voided. Ultrasound examination of the bladder when filled and after voiding is also useful for evaluating postvoid residual volume. Prostatic hypertrophy can also be identified by CT (Figure 16.45).

Prostatic enlargement may be caused by benign prostatic hypertrophy (BPH) or prostate cancer, and it is therefore essential to determine the cause of prostatic enlargement. Furthermore, small prostatic cancers may occur in unenlarged prostate glands, and it is valuable to have a technique for imaging the texture of normal as well as enlarged glands. That technique is transrectal ultrasound.

Although the prostate gland lies deep within the soft tissues of the pelvis, it can be approached ultrasonographically with a high-frequency transrectal probe. Transrectal ultrasound provides excellent definition of the internal glandular structure of the prostate, showing either normal size and echotexture or prostatic enlargement and abnormal echotexture.

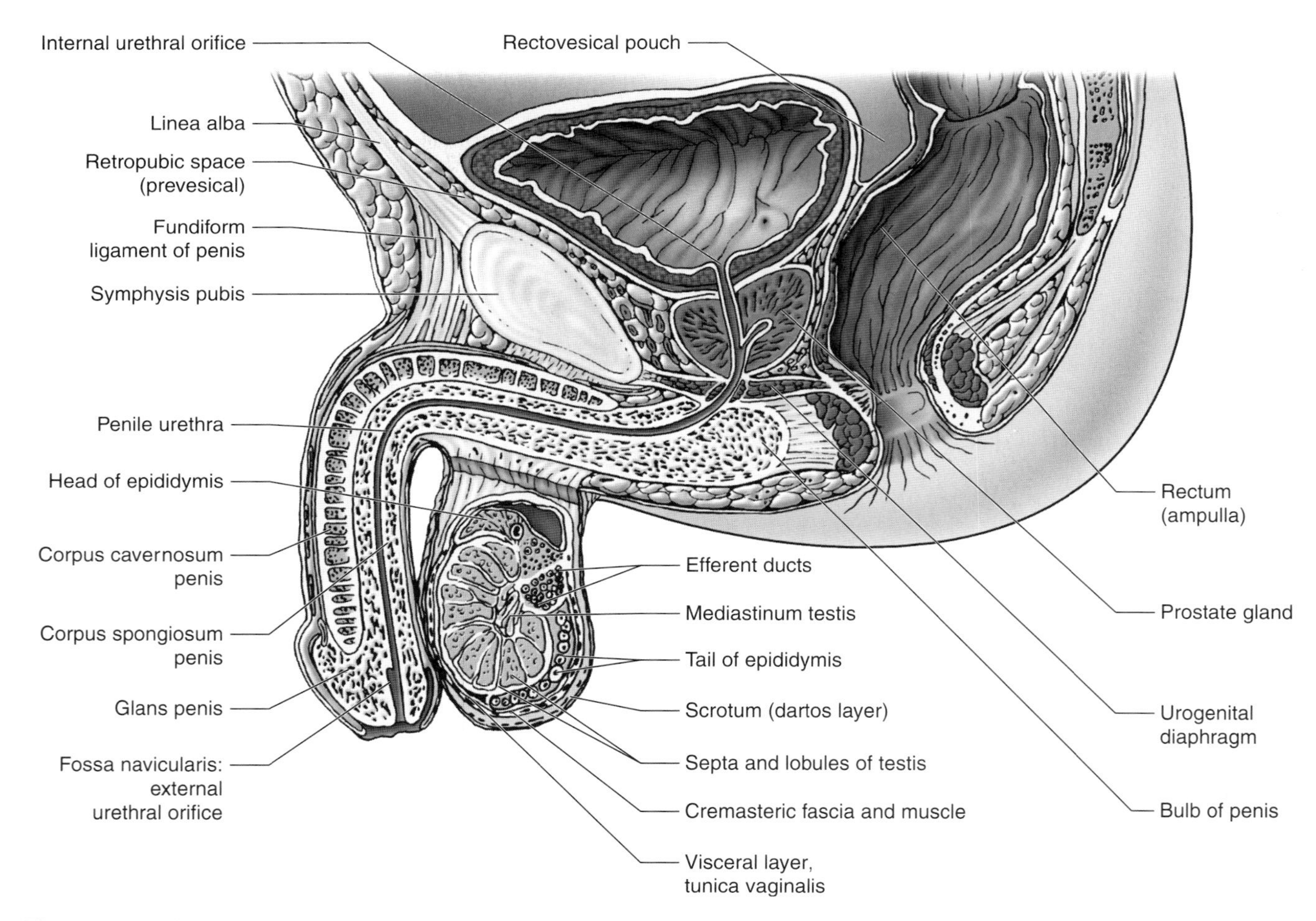

Figure 16.44. Median sagittal section of the male pelvis.

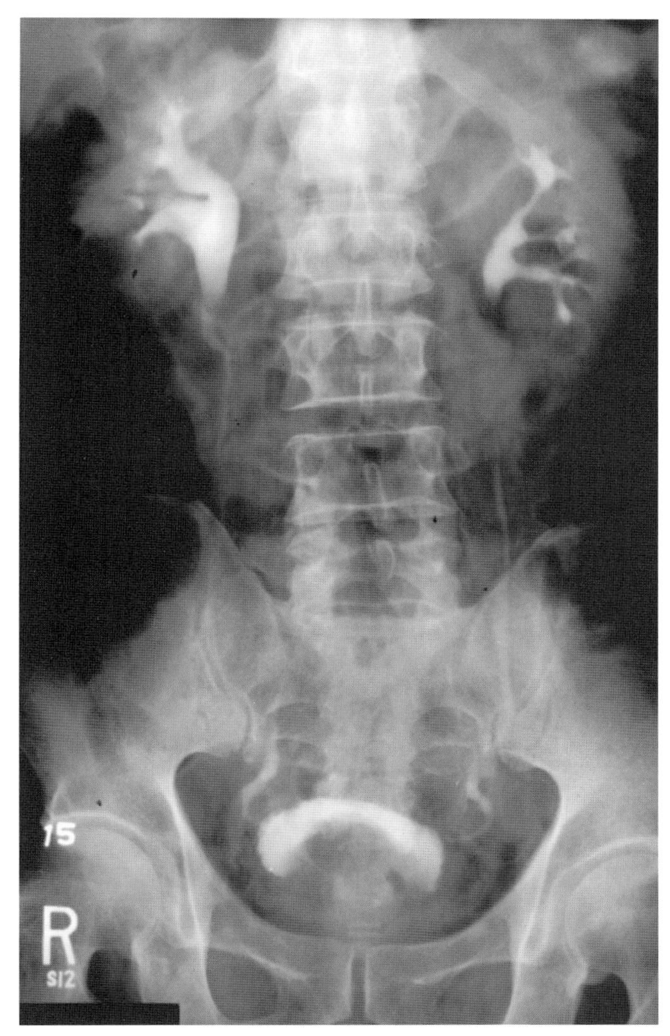

A

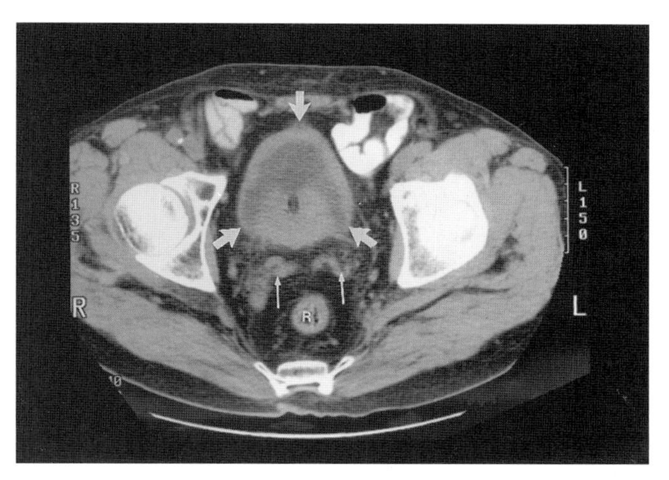

B

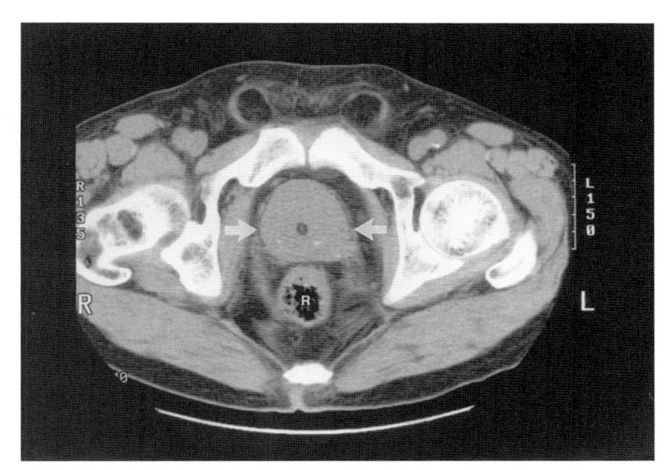

C

Figure 16.45. Intravenous urogram and CT scans of a patient with prostatic hypertrophy. A: Urogram shows elevation of the base of the bladder by the mass of the enlarged prostate. B and C: CT scans at two different levels within the pelvis show enlargement of the prostate gland *(large arrows),* anterior to the seminal vesicles *(small arrows)* and rectum *(R).* Did you note the central prostatic urethra within the prostate?

Anatomically the prostate is divided into four glandular parts (Figure 16.46): the peripheral zone, transitional zone, central zone, and periurethral glandular area. A nonglandular region on the anterior surface of the prostate is called the anterior fibromuscular stroma. On a normal transrectal ultrasound, the gland can be separated into two parts, the *peripheral zone* and the *central, or inner, gland,* which includes the transitional zone, central zone, and periurethral area (Figure 16.47). The seminal vesicles are also well shown by transrectal ultrasound.

Ultrasound-guided prostatic biopsy, done with a needle-guided biopsy system that attaches to the side of the ultrasound probe, permits very accurate tissue diagnosis of ultrasonically shown prostatic abnormalities. The procedure can be performed on outpatients, often directly after a diagnostic examination.

Benign prostatic enlargement is common in older men, and the size of the gland does not always correlate with the symptoms of prostatism. The size of the gland can, however, actually be measured by ultrasonic volumetric techniques. The most common appearance of BPH at ultrasound is enlargement of the inner gland, compared with the peripheral zone. The inner gland may also appear hypoechoic, although the echopattern may vary. Calcifications, if present, appear as hyperechoic nodules.

Prostatic cancer is the third leading cause of cancer deaths in men. Today a laboratory test is available for detecting prostatic cancer by measuring the level of prostatic-specific antigen (PSA), which is elevated by almost all prostatic cancers and by certain benign conditions. Since all glandular tissues of the prostate produce PSA, it is present in normal, hyperplastic, inflammatory, and neoplastic conditions. PSA is prostate-specific but not disease-specific. Cancer, however, elevates PSA much higher than does BPH. Generally a PSA level higher than 4 nanograms per milliliter indicates the patient should be referred for transrectal ultrasound. Patients may also be referred for ultrasound because of an abnormal digital rectal examination during physical examination.

The ultrasound appearance of prostatic cancer depends upon the size of the cancer and the nature of the surrounding prostatic tissue. Cancers usually appear as hypoechoic masses, as shown in Figure 16.48.

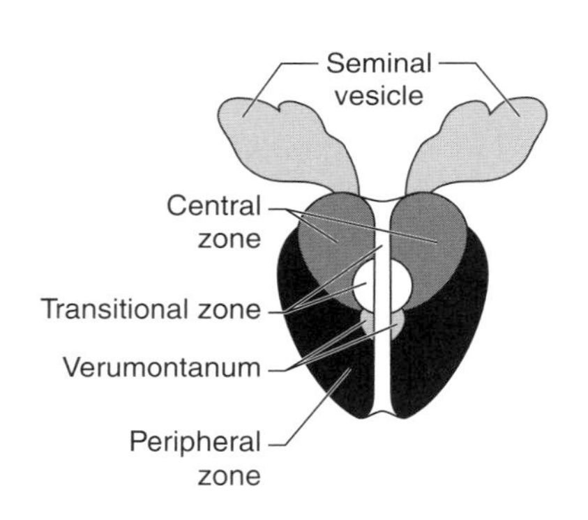

A. Coronal section.

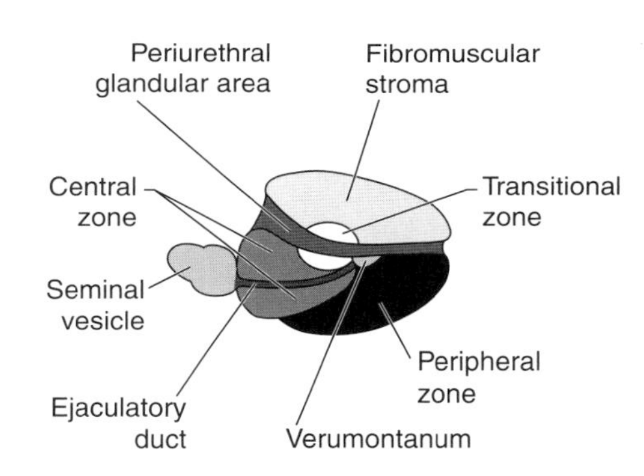

B. Sagittal midline section.

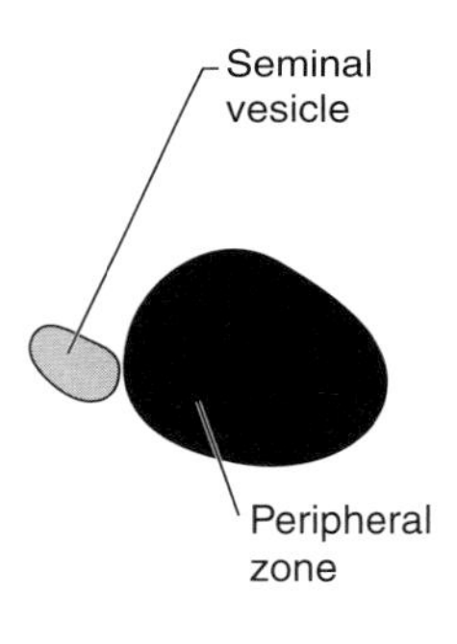

C. Sagittal section.

Figure 16.46. Anatomy of the prostate gland.

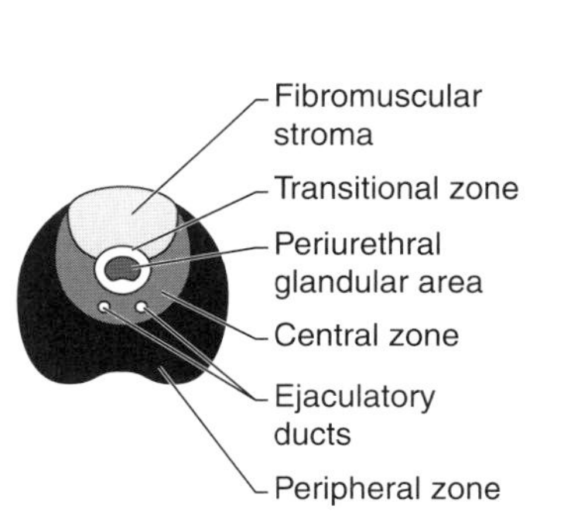

D. Axial section.

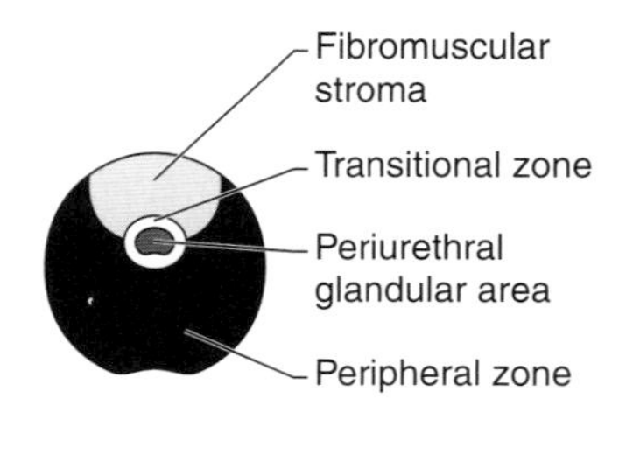

E. Axial section.

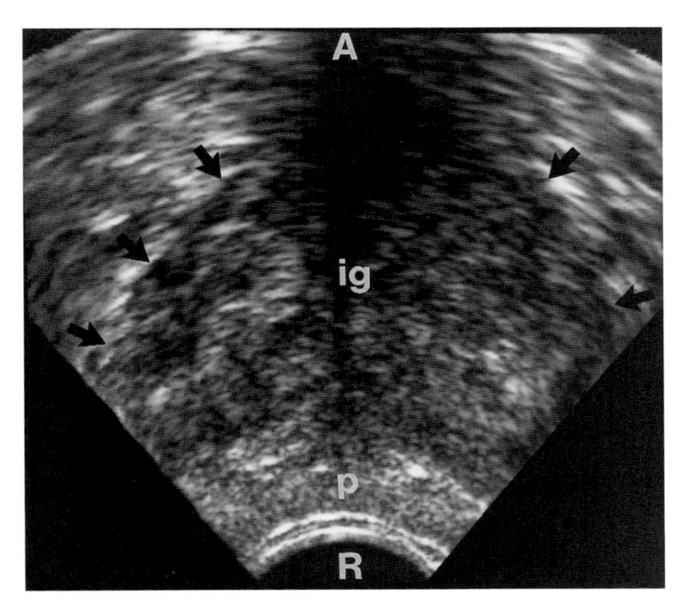

Figure 16.47. Normal prostate ultrasound in the axial plane. The ultrasound transducer is located within the rectum *(R)* and therefore the anterior *(A)* aspect of the patient is located at the top of the image. The *arrows* outline the margins of the prostate gland; *p* indicates the peripheral zone and *ig* the inner gland of the prostate.

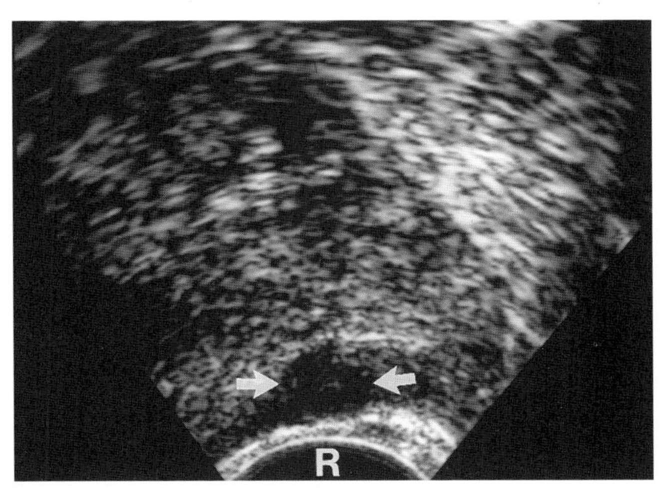

Figure 16.48. Ultrasound of a typical small prostate cancer. The *arrows* point to a small hypoechoic mass in the peripheral zone of the prostate gland, just anterior to the rectum *(R)*. Compare Figure 16.47. Tumors in this location may or may not be palpable on rectal examination.

Infrequently they may appear hyperechoic or even isoechoic. The isoechoic lesions, as you should know, are impossible to detect because they cannot be differentiated from the surrounding normal tissue, although secondary signs of asymmetric enlargement may indicate their existence. About 70 percent of cancers arise in the peripheral zone, and these are the ones most easily detected by ultrasound. Ultrasound is also useful in staging prostatic cancers because it can demonstrate local invasion of the gland capsule or seminal vesicles. The presence or absence of invasion of other pelvic structures is best shown by MR.

The Male Urethra

The male urethra is usually imaged by *retrograde urethrography*, commonly called *RUG*. To perform this procedure, water-soluble contrast material is injected into the distal urethra via a small balloon catheter that occludes the meatal orifice. The contrast material flows backward, opacifying the urethra for x-ray filming, and ultimately flowing into the bladder. An alternative technique is *voiding cystourethrography*. To perform this procedure the radiologist first fills the bladder with contrast material via a bladder catheter, then removes the catheter and has the patient urinate on the fluoroscopy table so that films can be made of the passage of contrast material through the urethra.

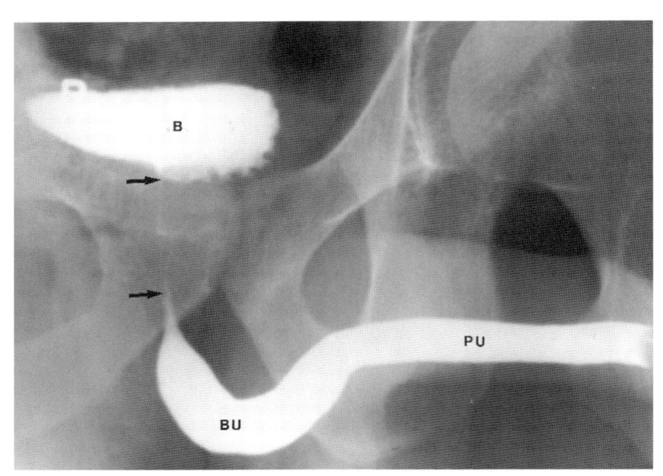

Figure 16.49. Normal retrograde urethrogram. Contrast material was injected through the urethra to opacify the urethra and bladder. The *posterior urethra* consists of the *prostatic urethra (between the arrows)* within the prostate gland and the short *membraneous urethra* within the 1-centimeter-thick urogenital diaphragm. The *anterior urethra* extends from the urogenital diaphragm to the external urethral meatus and consists of the *bulbous urethra (BU)* until the penile-scrotal junction, and then the *penile urethra (PU)* until the meatus.

On a normal RUG (Figure 16.49) you can identify the anterior and posterior portions of the urethra, which are divided by the urogenital diaphragm (Figure 16.44). The *posterior urethra* consists of the prostatic urethra (the portion of the urethra that passes through the prostate gland) and the membranous urethra (the portion that runs through the urogenital diaphragm). The *anterior urethra* consists of the bulbous urethra (short, running from the urogenital diaphragm to the penoscrotal junction) and the penile urethra (longer, extending to the urethra meatus).

Urethrography can diagnose urethral strictures, urethral diverticula, and acute urethral trauma. Strictures may develop after trauma (including traumatic catheterizations), surgery (such as prostatectomy), and infections (usually gonorrhea). One of the most common indications for urethrography is suspected acute urethral trauma. In the multiple-trauma patient the usual clinical sign is gross bleeding from the urethral meatus, as contrasted with hematuria, which in the trauma patient is usually a sign of renal, ureteral, or bladder trauma. Injury to the posterior urethra is particularly common in patients with pelvic fractures involving the anterior, pubic bones. At urethrography (Figure 16.50) you will see extravasation of contrast material at sites of urethral lacerations and transections.

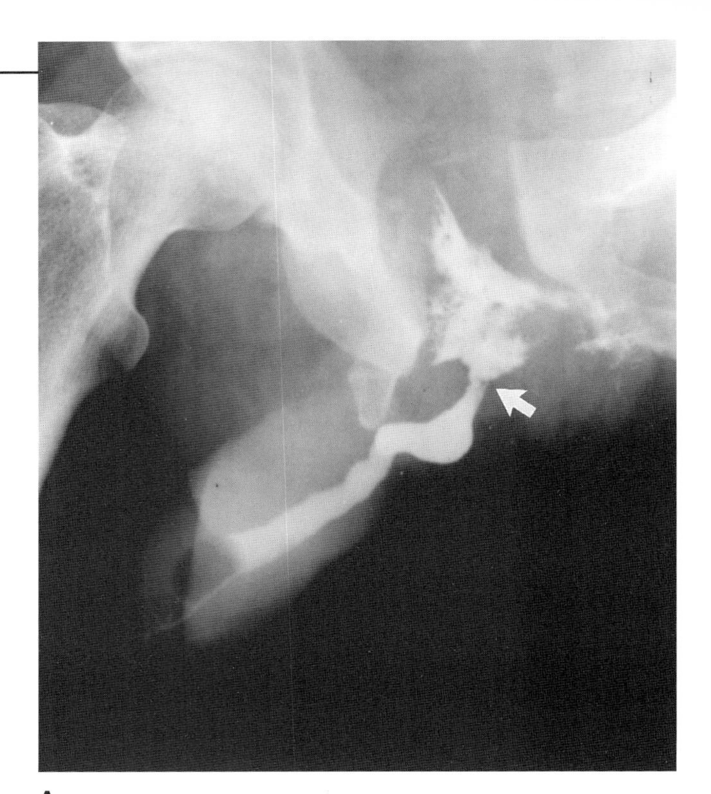

A

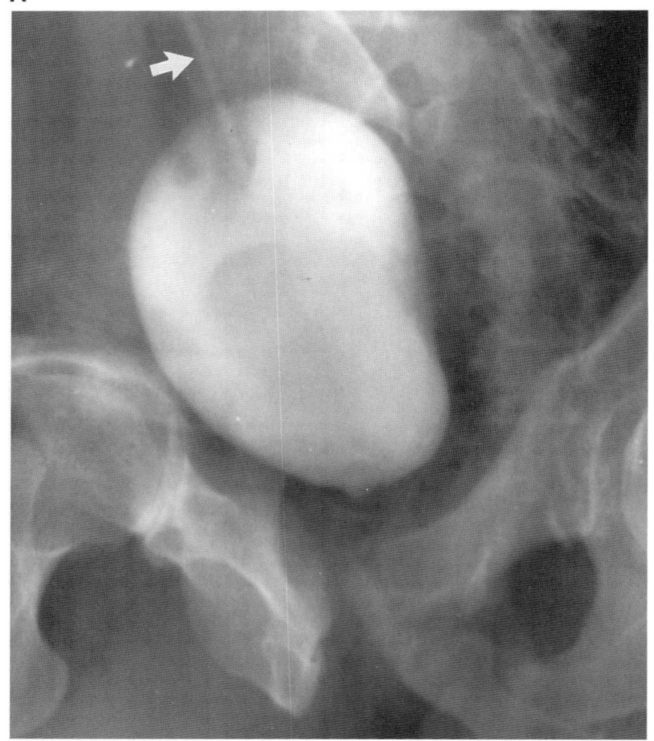

B

Figure 16.50. Abnormal retrograde urethrogram and cystogram of a trauma patient with pelvic fractures. A: Retrograde injection of contrast material reveals a complete transection of the urethra from the bladder at their junction *(arrow).* Contrast material extravasates only into the pelvic soft tissues; there is no filling of the bladder. B: Cystography, performed through a suprapubically inserted Foley catheter *(arrow),* shows the bladder filling but no flow of contrast material into the urethra.

General Issues in Children's Imaging

The goal of pediatric imaging is to obtain a high-quality diagnostic examination with the least amount of radiation. Consequently, the routine views recommended for specific imaging examinations are often less extensive than those obtained in adults, and radiation shielding is used whenever possible to cover parts of the body that do not need to be exposed.

Restraints or sedation may be required, especially for longer examinations. As a general rule, children from ages 1 day to 1 year do not move very much, and may require only minimal restraining with velcro straps. And most children over 3 years of age are completely cooperative. But children from 1 to 3 years old may provide an imaging challenge to the radiology department, requiring immobilization restraints for short procedures such as plain films and conscious sedation for CT, MR, and radioisotope scanning. When sedation is necessary, it is usually arranged by the radiology department. The type of sedation depends upon the length of the procedure and whether or not it is painful. In most cases sedation is accomplished by agents that are administered orally, rectally, intravenously, or intramuscularly. Only occasionally is general anesthesia required.

Another challenge for radiologists is to get children to drink agents for contrast examinations such as barium for gastrointestinal studies and water-soluble contrast agents (often mixed with juice for children) for CT. Again the problem is less severe in infants and older children. For children under 1 to 2 years of age, achieving cooperation may be as simple as withholding the last feeding and then giving the hungry child a bottle filled with the contrast agent. Children over 3 to 4 years old can usually be persuaded to cooperate. With children 1 to 3 years old who will not cooperate, it is sometimes necessary to insert a nasogastric tube to administer the contrast material.

It is important for you to remember these imaging difficulties as you manage pediatric patients. Radiologists may invite your patient's parents to be present during certain imaging procedures to comfort the child, assuring better cooperation and a less stressful experience.

In the rest of this chapter we will discuss a few of the important conditions that commonly affect children, particularly those involving the chest, abdomen, and bones. For further information on the many other pediatric conditions, including the vast number of congenital and developmental anomalies, you should consult a textbook on pediatric radiology.

Pediatric Chest Conditions

Many of the adult chest conditions you have studied so far, including air-space and interstitial pneumonias, atalectasis and collapse, pneumothorax and pleural effusion, occur also in children. Aspirated endobronchial foreign bodies are, in fact, more common in children than in adults. The most common child-specific chest conditions you will encounter are croup and epiglottitis, viral pneumonia, bronchiolitis, and cystic fibrosis.

Croup and Epiglottitis

Croup produces acute airway obstruction in young children; it is caused by infection with influenza and parainfluenza viruses. The peak incidence is between the ages of 6 months and 3 years. Although the infection may involve the entire airway, producing an extensive laryngotracheobronchitis, the critical site is immediately below the larynx, where edema narrows the subglottic trachea. Compare the normal subglottic trachea in Figure 16.51 with the trachea of a child with croup in Figure 16.52. On the AP view of the neck in Figure 16.52, the narrowed subglottic trachea has an extended inverted-V appearance that is characteristic of croup.

Epiglottitis is usually caused by *Hemophilus influenza* bacteria and is a much more dangerous condition than croup. The epiglottis and surrounding structures become edematous and enlarged, producing inspiratory stridor, restlessness, and even dysphagia. Acute epiglottitis can be a life-threatening condition, and the

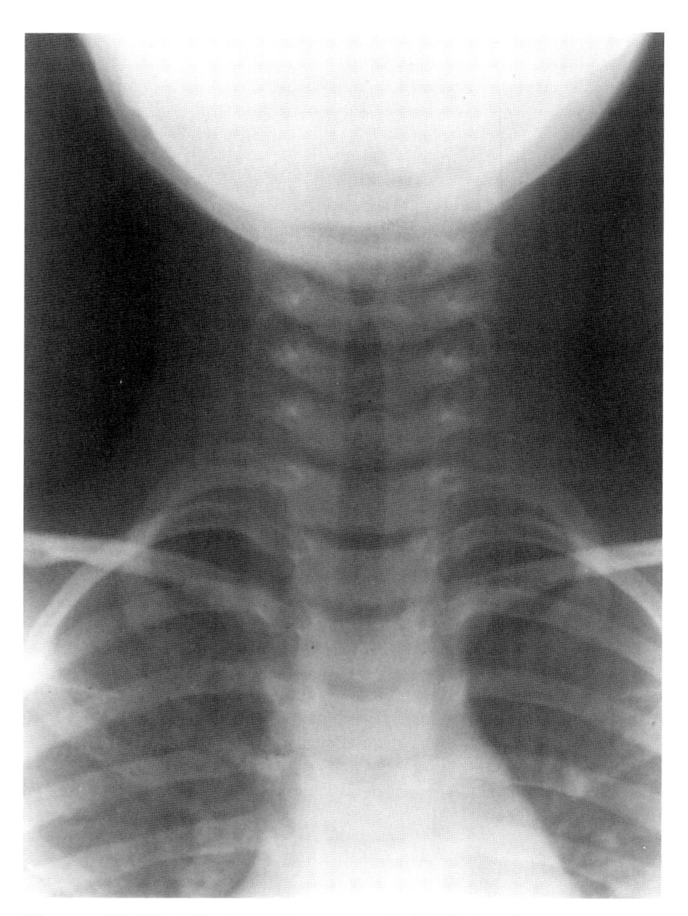

Figure 16.51. AP view of a normal child's trachea.

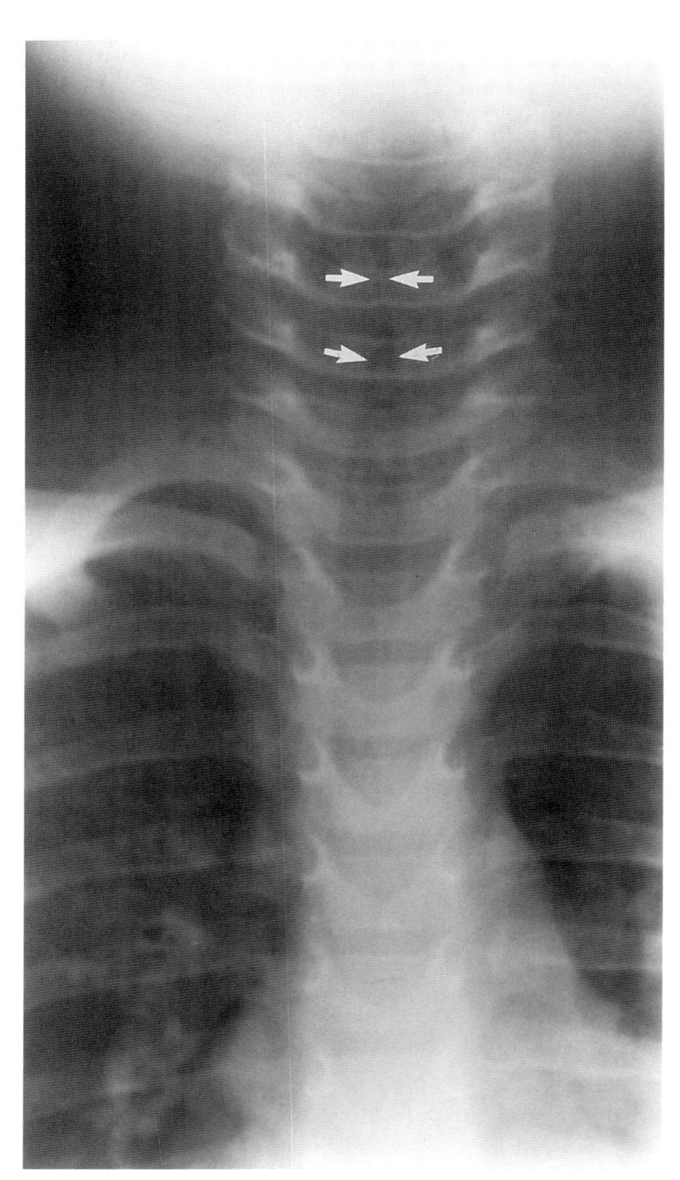

Figure 16.52. AP view of the trachea of a child with croup. Edema narrows the subglottic trachea *(arrows)* producing respiratory distress.

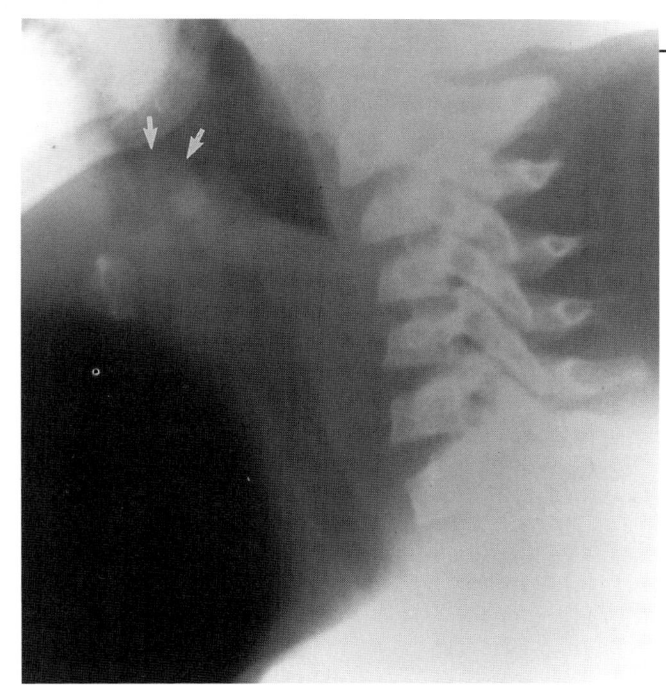

Figure 16.53. Child with epiglottitis. The epiglottis *(arrows)* is thickened and enlarged, with edema of the surrounding laryngeal soft tissues.

child should be closely monitored if radiological confirmation is required. Films should be taken with the patient upright so as not to make breathing more difficult. A single lateral soft-tissue film of the neck will show marked enlargement of the epiglottis and thickening of the surrounding tissues (Figure 16.53). Compare the normal thin epiglottis in Figure 16.54. Before reading on, decide what is bothering the child in Figure 16.54. Of course, this child has swallowed a quarter, which has lodged in the cervical esophagus. Note its position on the lateral film behind the air-containing trachea.

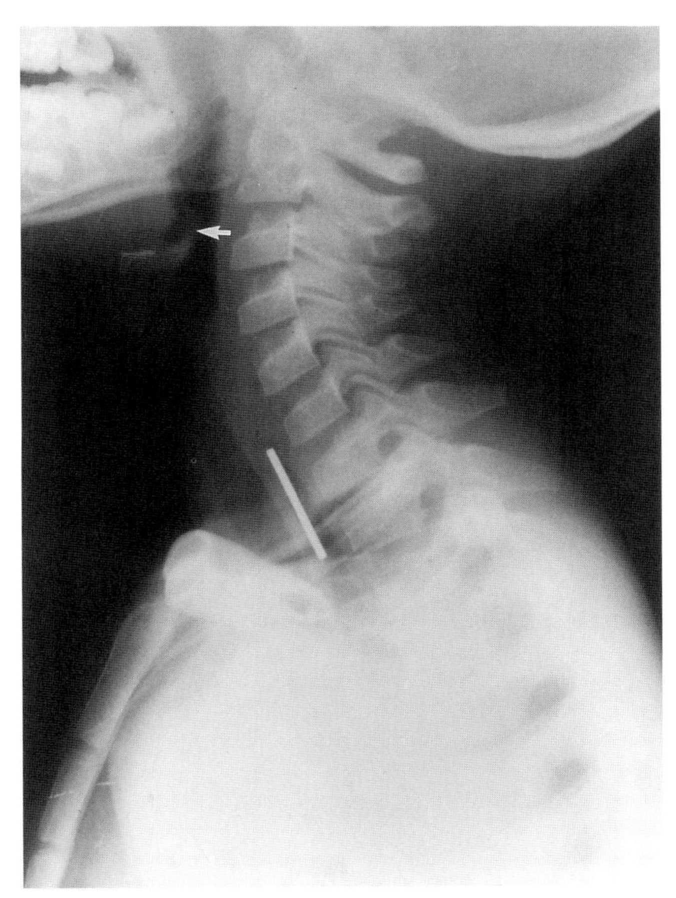

A. Lateral view.

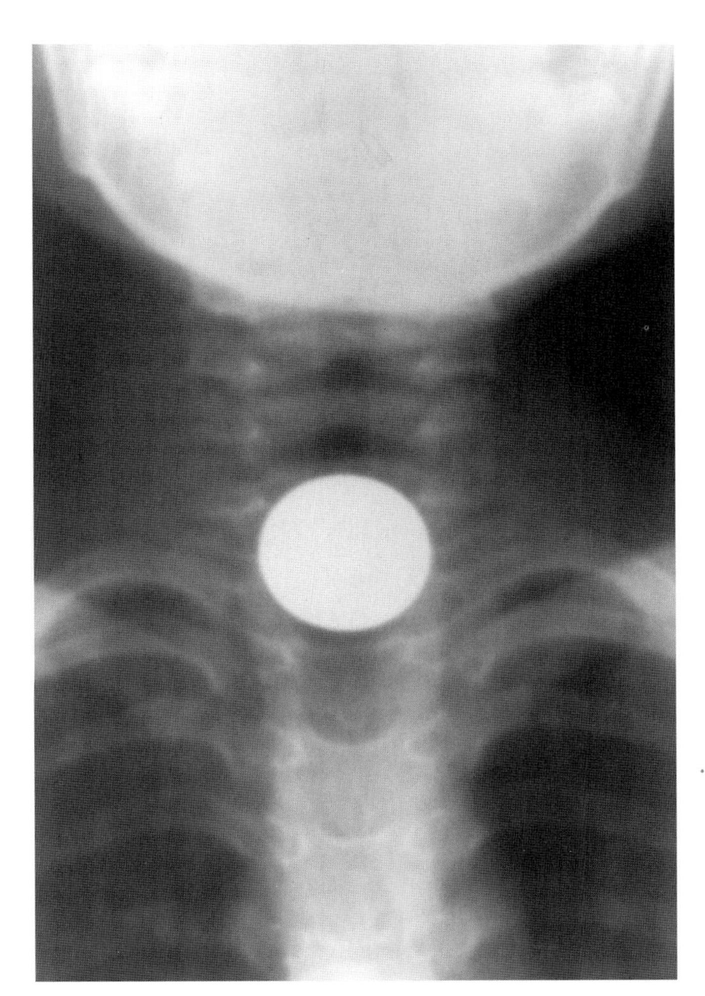

B. AP view.

Figure 16.54. This child has a normal epiglottis (*arrow,* lateral view) but his films are not normal. What is wrong?

Pneumonia, Bronchiolitis, and Bronchitis

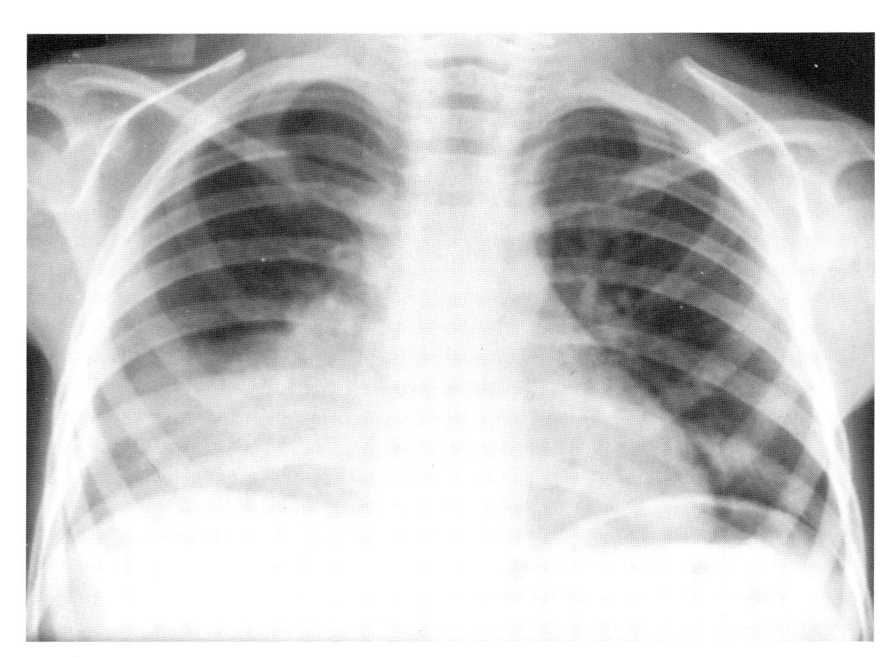

A. AP view.

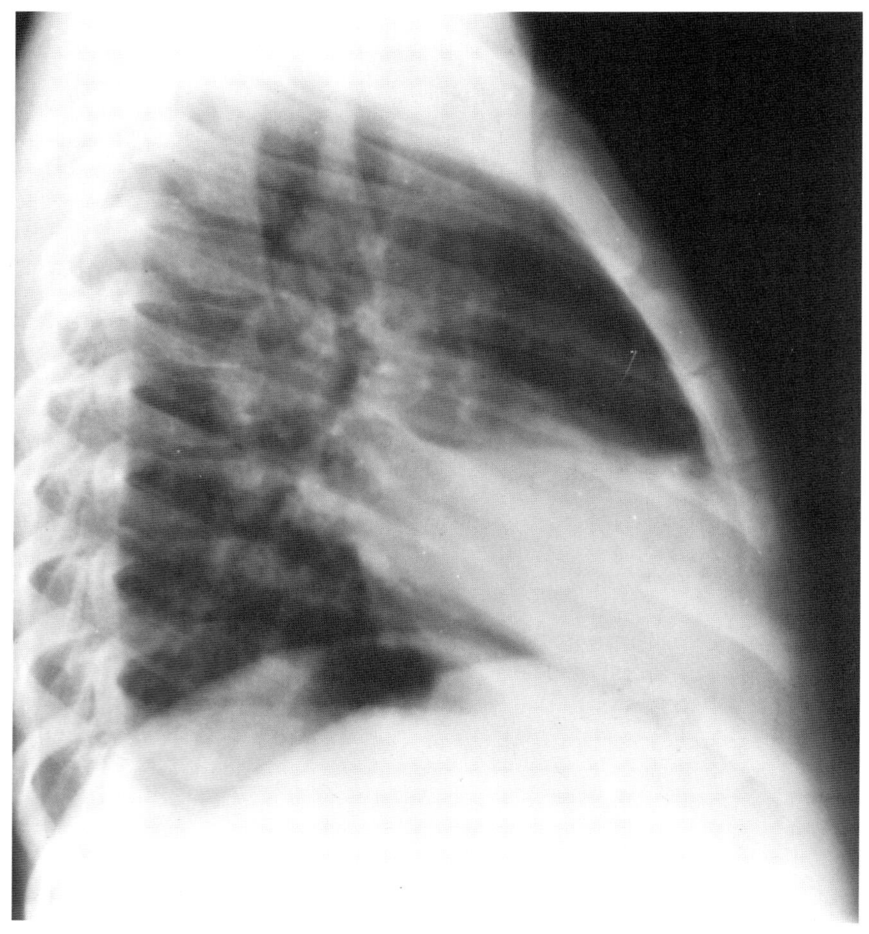

B. Lateral view.

Figure 16.55. Right middle lobe bacterial (pneumococcus) pneumonia in a child.

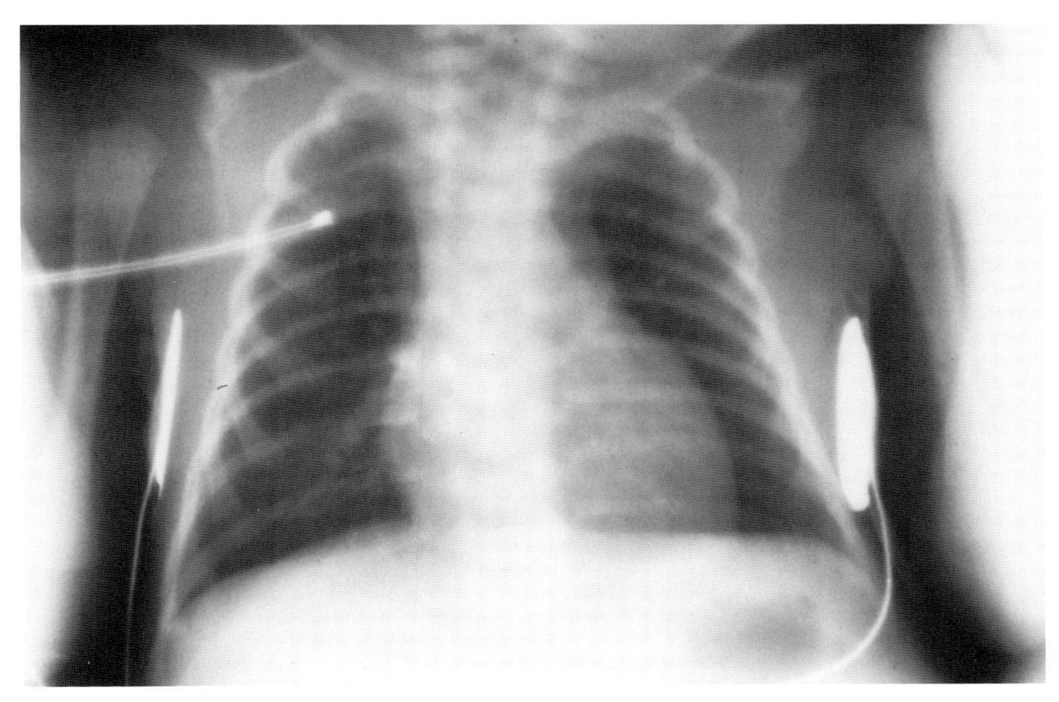

A. PA view.

Figure 16.56. PA and lateral chest films of an infant with bronchiolitis. Note the marked hyperinflation of both lungs, especially apparent on the lateral view. Central bronchial wall thickening and increased linear markings are other characteristic signs of this condition.

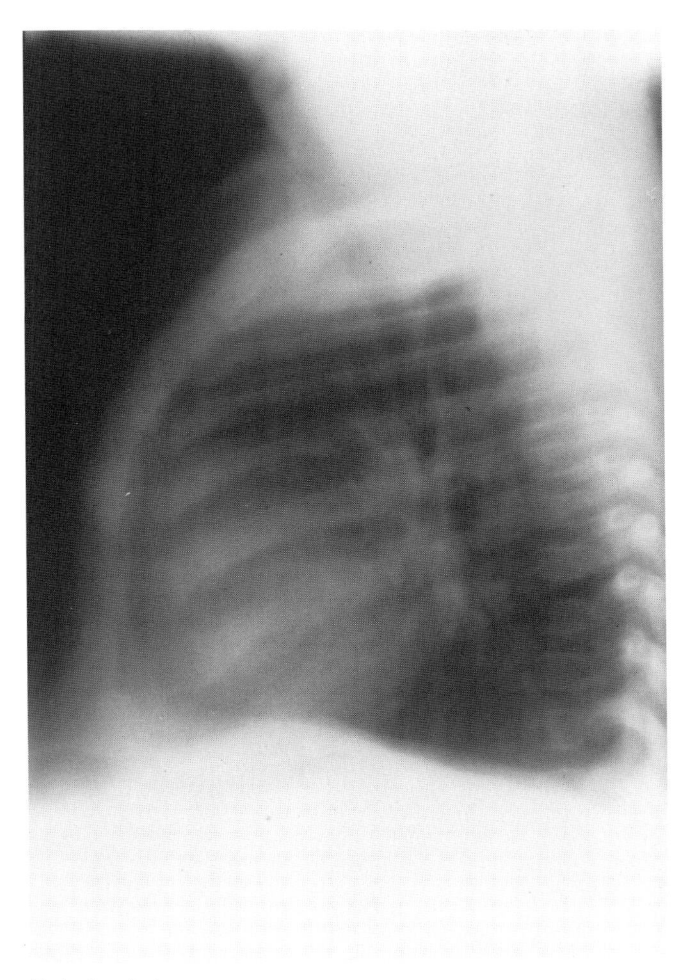

B. Lateral view.

Pneumonia in children may be produced by the same agents that produce pneumonia in adults (Figure 16.55). In addition, viruses (such as respiratory syncytial virus, parainfluenza virus, adenovirus) are a common cause of pneumonia in children under 5 years of age. These viruses produce an infection of the respiratory mucosal membranes with edema and inflammatory infiltration of the bronchioles (bronchiolitis) and bronchi (bronchitis); the air spaces are usually spared. Chest films (Figure 16.56) show thickening of the bronchial wall, opacification of the surrounding bronchi (peribronchial cuffing), hyperaeration, and increased linear lung markings. Bronchiolitis usually occurs in infants less than a year old; bronchitis affects older infants and children.

Cystic Fibrosis

Cystic fibrosis is an inherited disease of exocrine gland dysfunction associated with progressive pulmonary and gastrointestinal disease. The pulmonary manifestations present in late infancy and include chronic cough, recurrent pulmonary infections, and obstructive pulmonary disease. The late stages are associated with pulmonary arterial hypertension and respiratory insufficiency.

In infants and young children, the chest film may be entirely normal, the diagnosis of cystic fibrosis having been made clinically. In older children (Figure 16.57) the films may show hyperaeration, peribronchial cuffing, increased linear markings, and dilated bronchi (bronchiectasis). A variety of pulmonary complications may occur that chest films can identify, such as pneumonia, lung abscess, pneumothorax, atalectasis, and lung collapse.

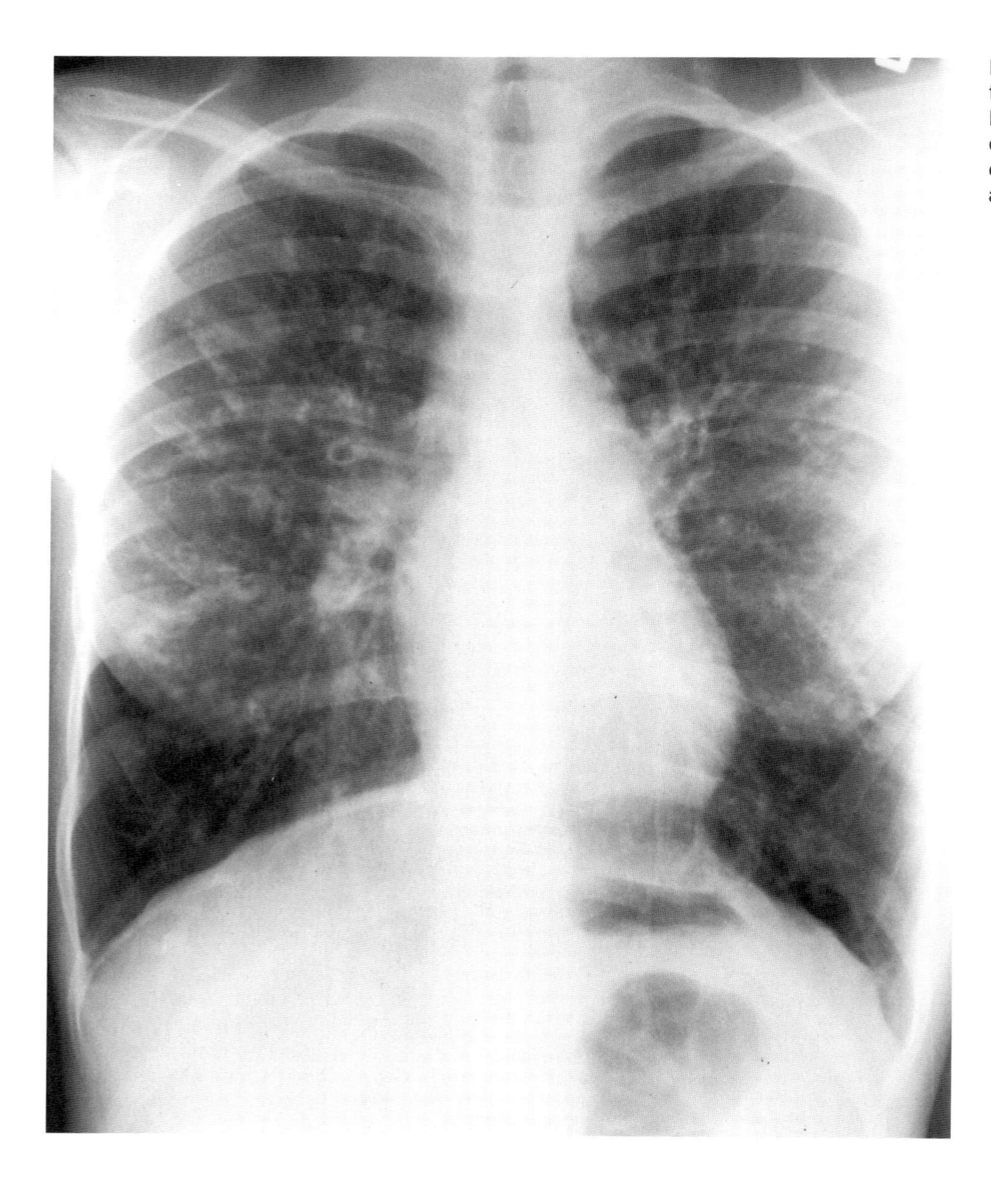

Figure 16.57. Plain film of a teenager with cystic fibrosis. Note the low hemidiaphragms due to hyperinflation, the thickened bronchial walls, and the abnormal linear lung markings.

Pediatric Abdominal Conditions

Children may suffer from some of the same abdominal conditions that affect adults. But the causes of an acute abdomen in pediatric patients are usually childhood specific, and the incidence of these conditions almost always varies with the age of the child. Neonates with an acute abdomen often suffer from conditions of a congenital or developmental nature, such as intestinal atresia and meconium ileus, or conditions resulting from prematurity, such as necrotizing enterocolitis. Hirschsprung's disease usually presents sometime between birth and the first 6 weeks of life, but may present later, after the age of 5 years. During the first 3 months of life, infants may present with inguinal hernia and hypertrophic pyloric stenosis. In children 6 months to 2 years of age, a common cause of an acute abdomen is ileocolic intussusception; in children over 2 years old appendicitis is often responsible. Below we describe pediatric abdominal conditions that have dramatic radiological manifestations and discuss abdominal masses in children.

Hypertrophic Pyloric Stenosis

Hypertrophic pyloric stenosis (HPS), a common condition in infants, causes pyloric obstruction because of hypertrophy of the circular musculature of the pylorus. The etiology is unknown, and male children are affected four times as frequently as females. The presenting symptom of vomiting that progresses from simple regurgitation to projectile vomiting after each feeding usually occurs during the second to sixth week of life. On physical examination a palpable mass may be detected in the area of the pylorus. Radiological examination is not routinely requested; many infants are scheduled for surgical repair on the basis of a convincing history and a palpable pyloric mass. When the clinical diagnosis is uncertain, it may be confirmed with either an upper gastrointestinal series (UGI) or abdominal ultrasound. Plain films of the abdomen may show a dilated stomach with little distal bowel gas. UGI series (Figure 16.58) will show a delay in gastric emptying and an abnormal pylorus. The pyloric canal will appear aperistaltic, nondistensible, and elongated, with thickening of the surrounding pyloric musculature. Ultrasound will show a soft-tissue mass (hypertrophied muscle) surrounding the pyloric canal.

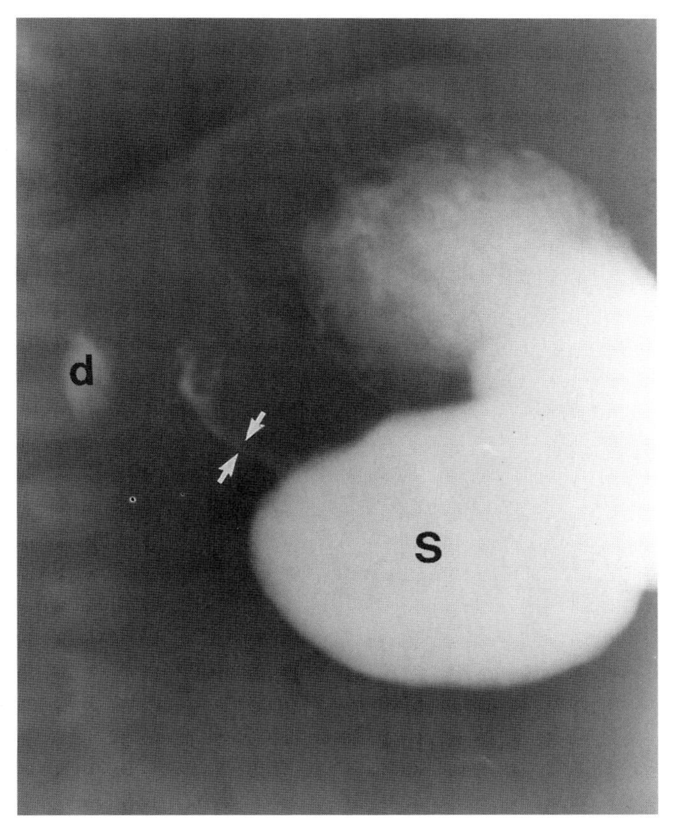

Figure 16.58. Upper gastrointestinal series of a child with projectile vomiting caused by hypertrophic pyloric stenosis. The stomach *(S)* is filled with barium and gastric emptying is delayed; only a tiny volume of liquid barium has passed into the duodenum *(d)*. The pyloric channel *(arrows)* is markedly narrowed and elongated due to hypertrophy of the surrounding pyloric musculature.

Ileocolic Intussusception

In intussusception a segment of bowel invaginates into a segment of bowel just distal to it, producing bowel obstruction. This process (Figure 16.59) may occur in the small bowel (ileoileal intussusception), in the colon (colocolic intussusception), or at the junction of the small and large bowel (ileocolic). The proximal bowel that has become invaginated is called the *intussusceptum,* and the surrounding bowel that receives the intussusceptum is called the *intussuscepiens.* In adults, intussusception nearly always indicates an un-

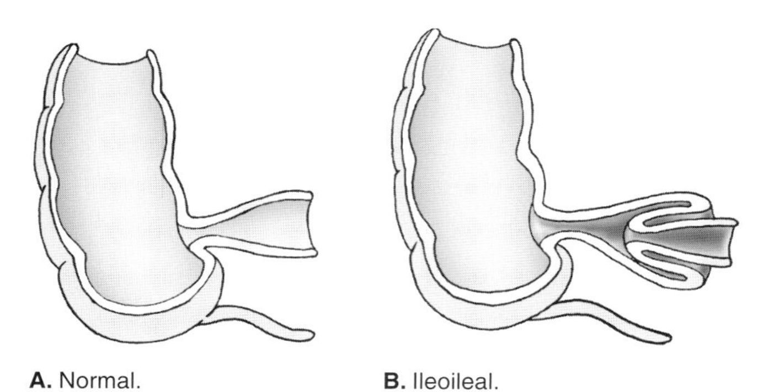

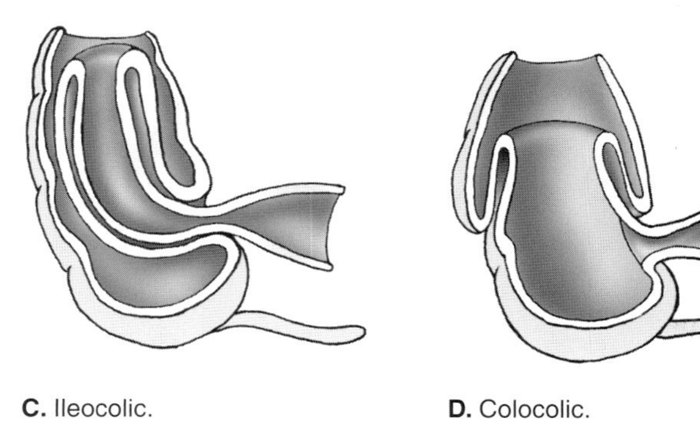

A. Normal. **B.** Ileoileal. **C.** Ileocolic. **D.** Colocolic.

Figure 16.59. Various types of intussusception.

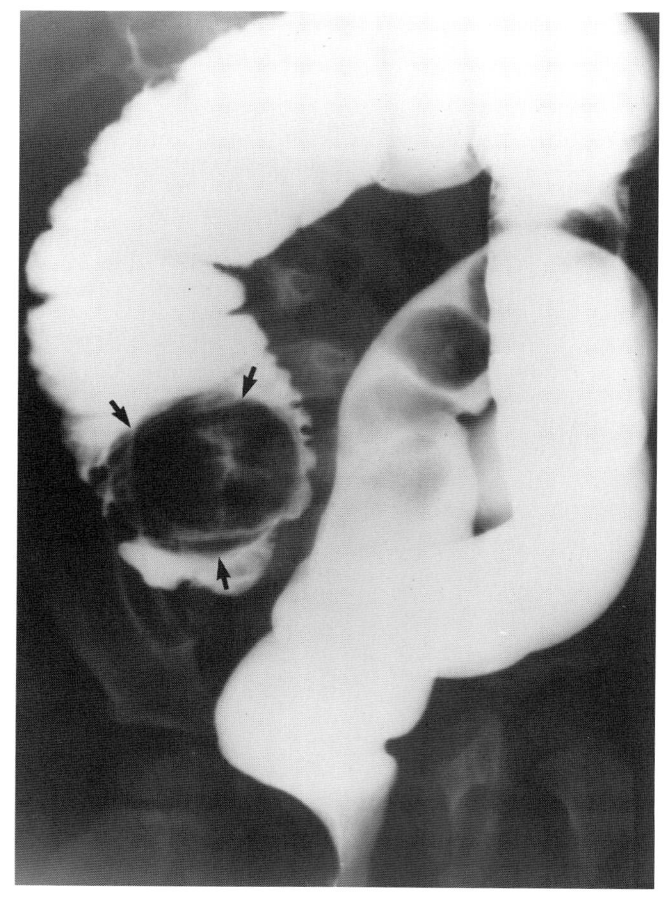

Figure 16.60. Barium enema of ileocolic intussusception in the right colon. The retrograde flow of barium is halted by the intussuscepted terminal ileum *(arrows).*

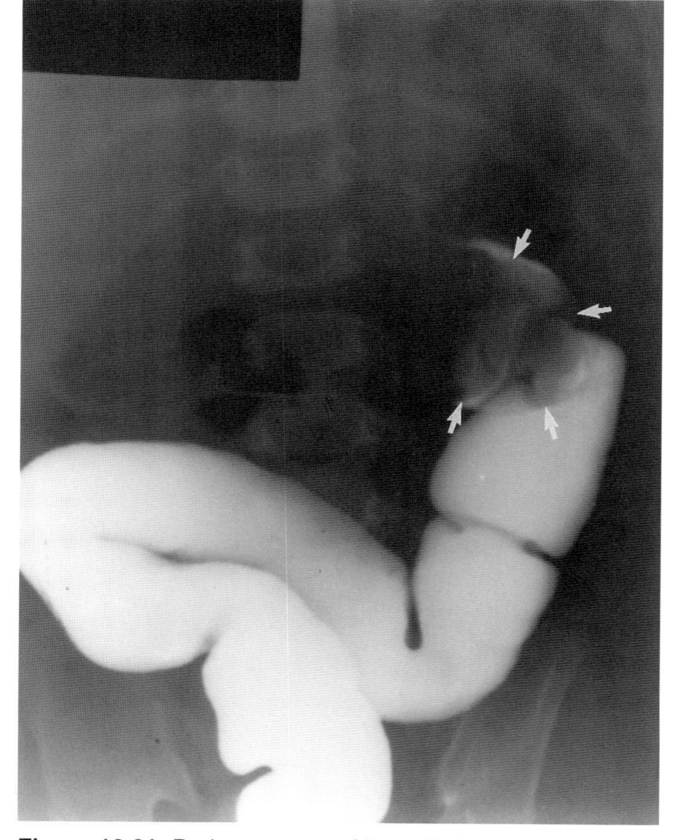

Figure 16.61. Barium enema of ileocolic intussusception into the left colon. The intussuscepting terminal ileum *(arrows)* has passed all the way through the right and transverse segments of colon and now terminates in the proximal left colon.

derlying bowel abnormality, usually a bowel tumor that has been swept along by peristalsis to become the leading point of the intussusceptum. In children, intussusception occurs without any mass as the lead point and is usually ileocolic. The lead point in most cases is hypertrophy of lymphoid tissue in the terminal ileum.

A patient with intussusception usually presents with abdominal pain, vomiting, bleeding from the rectum, and sometimes a palpable abdominal mass. A supine plain film of the abdomen may be normal, but if the child is positioned to fill the right colon with air (as in a left lateral decubitus view or a prone view), a mass (the intussusception) may be seen, surrounded by air, in the right abdomen. A definitive diagnosis can be made by a barium enema examination (Figures 16.60 and 16.61), which will show obstruction of the colon by a convex filling defect. The barium enema examination can be used to reduce the intussusception by hydrostatic pressure. When reduction is complete, the radiologist will see retrograde flow of contrast material into the small bowel.

Hirschsprung's Disease

Hirschsprung's disease is caused by an absence of ganglion cells in the distal colon, which results in a functional obstruction of the bowel. Affected children present with intestinal obstruction or intermittent diarrhea and constipation. Plain films, as you would expect, show dilated large and small bowel, suggesting low bowel obstruction. The radiological diagnosis can be confirmed with a barium enema examination (Figure 16.62), which will on the lateral radiograph of the rectum show a transitional zone (aganglionic segment) with disordered and irregular contractions; immediately proximal to this zone the colon will be dilated. Delayed films will show little evacuation of the colon 24 or 48 hours after the administration of barium.

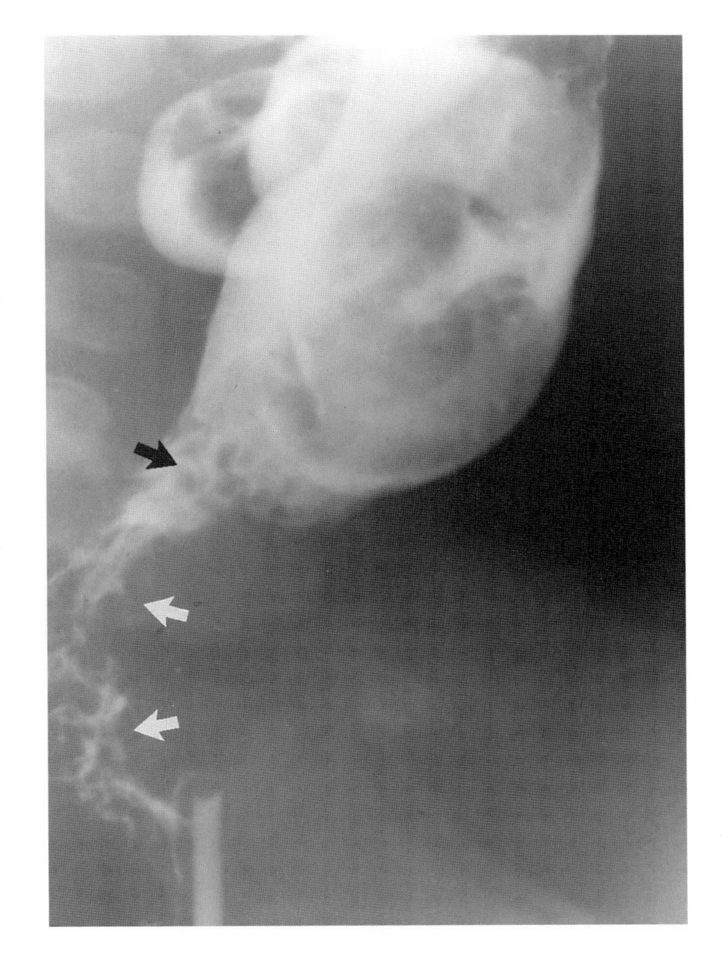

Figure 16.62. Barium enema of a patient with Hirschsprung's disease. The coned-down view of the distal colon shows the narrowed, aganglionic rectum *(white arrows)* and the transitional zone *(black arrow),* above which the colon is dilated. The tubular structure in the distal rectum is the rectal tube through which the barium contrast material was injected.

Abdominal Masses in Infants and Children

A not infrequent indication for pediatric abdominal imaging is the diagnostic evaluation of a clinically detected abdominal mass. Although finding an abdominal mass in their child is alarming for new parents, more than half of these masses in neonates are renal in origin, most are benign, and usually the prognosis is excellent. Hydronephrosis is the single most common cause of a neonatal abdominal mass, responsible for 25 percent of cases. It is usually a result of congenital or developmental anomalies of the urinary tract that cause obstruction, such as ureteropelvic junction stenoses (junction of the renal pelvis and ureter), posterior urethral valves, and ectopic ureteroceles. Other causes of neonatal masses include a variety of cystic, developmental, and neoplastic conditions of the genitourinary tract and gastrointestinal tract.

In older infants and children, the majority of abdominal masses are also renal in origin and most are benign, although significantly more tumors are malignant than is the case for neonates. In 22 percent of this age group the renal mass is a Wilms' tumor; hydronephrosis accounts for another 20 percent of masses. The other common tumor in this age group is neuroblastoma, responsible for 21 percent of abdominal masses.

As you might have guessed, ultrasound is the recommended initial imaging examination for pediatric abdominal masses. Ultrasound can accurately detect hydronephrosis, identify the organ of origin of other masses, and determine whether these are cystic or solid. When hydronephrosis is detected, the child is usually referred for urographic study (intravenous urogram, voiding cystourethrogram) to determine the cause of the obstruction. When ultrasound detects a solid mass (Figure 16.63), the patient is usually referred for a CT scan, which can more accurately characterize the mass's extent and show its effects on adjacent organs.

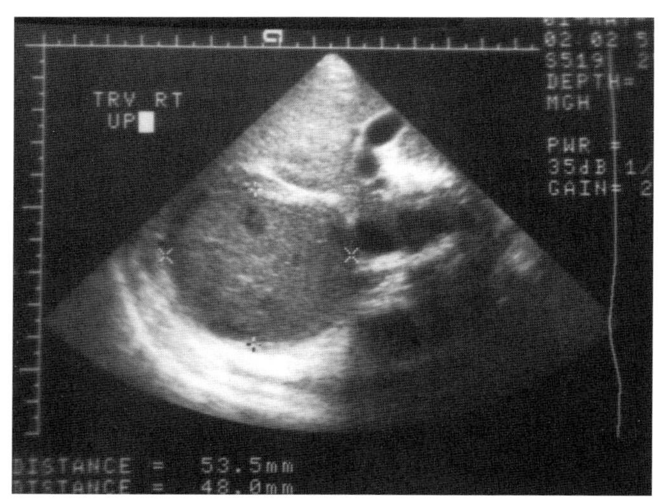

A

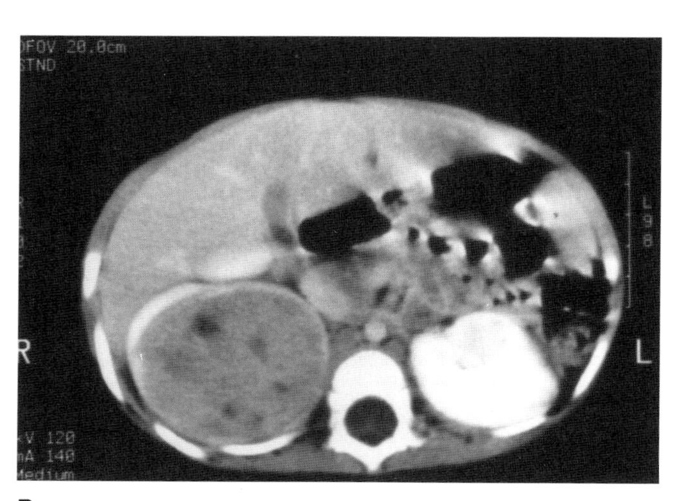

B

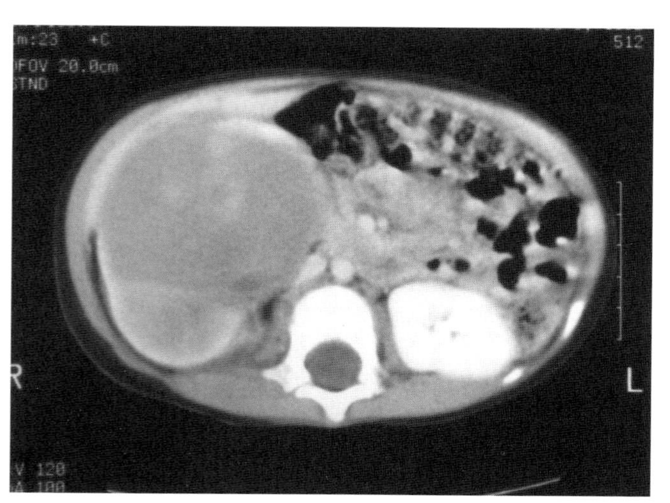

C

Figure 16.63. Young child with a palpable mass in the right abdomen. Ultrasound (A) showed that the mass (cursors) was solid (echogenic). For further evaluation, CT was performed with intravenous contrast material. The mass (B and C) was shown to be a large solid tumor of the right kidney. The opacified right renal parenchyma is stretched around the mass as a thin white rim. The left kidney is normal. As you no doubt guessed, this neoplasm proved to be a Wilms' tumor.

Normal Pediatric Bones

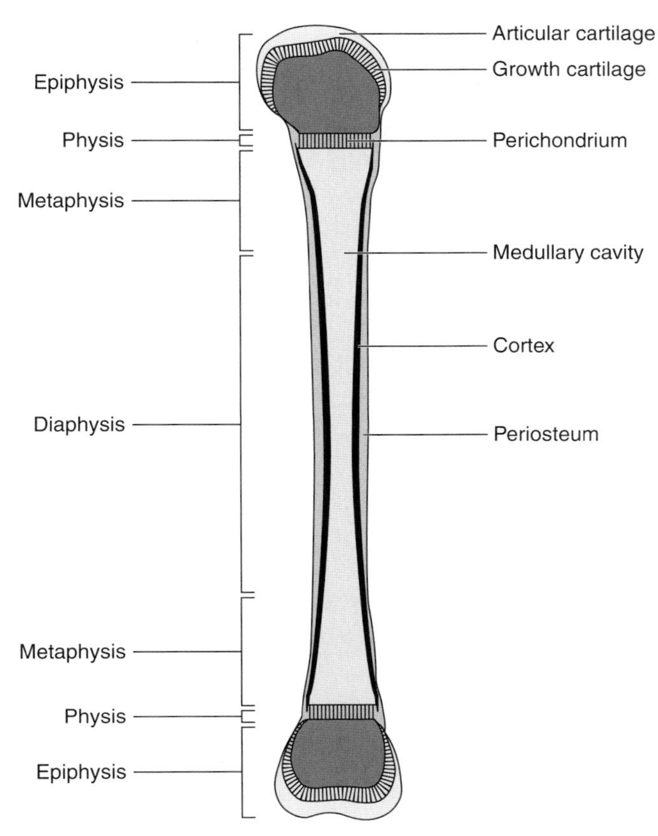

Figure 16.64. Drawing of a normal pediatric long bone.

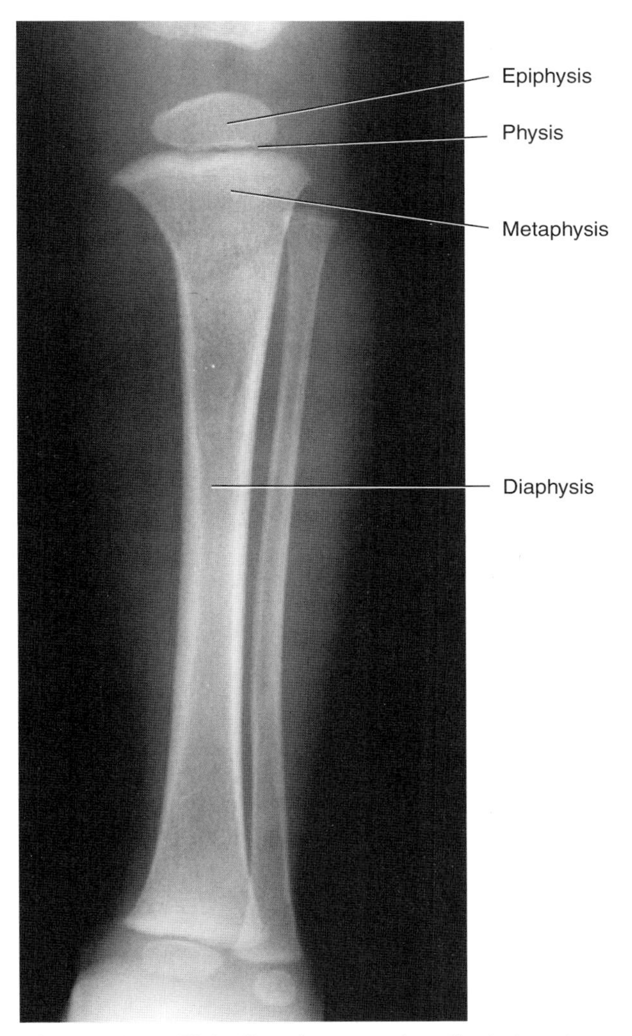

Figure 16.65. Plain film of a normal pediatric long bone.

You must be familiar with the appearance of normal growing bones before you can begin to appreciate the changes caused by conditions that affect the pediatric skeleton. A normal growing tubular bone (Figure 16.64) is composed of two ossification centers *(epiphyses)*, which are separated from the shaft (*diaphysis* and *metaphysis*) by the growth plates *(physes)*.

Normal bone development involves growth in length, increase in shaft diameter, and epiphyseal enlargement. Longitudinal growth occurs at the physes. The shaft diameter is increased by the deposition of periosteal membranous new bone along the shaft. In very young children the epiphyses are not visible radiographically because they are formed in cartilage; after they ossify the epiphyses enlarge. As the bones continue to grow, the physes gradually diminish in width and finally as the physes close completely, longitudinal bone growth stops. Since the epiphyses of different bones make their radiographic appearance at different but known times, and grow at different but known rates, it is possible to determine skeletal maturation (bone age) radiographically. Note the appearance of the epiphyses and growth plates in the pediatric radiograph in Figure 16.65.

Fractures in Children

Pediatric fractures differ from adult fractures. Since young bones are more pliable than older ones, fractures may occur in children that result in bending of the bone *(acute plastic bowing fracture),* buckling of the cortex of the bone *(torus fracture),* or bending on the concave side of the bone combined with an incomplete fracture on the convex side *(greenstick fracture).* These are depicted in Figure 16.66, along with a complete fracture *(complete diaphyseal fracture),* which may also occur in children. The child in Figure 16.67 suffered a torus fracture *(white arrow)* of the radius and a greenstick fracture *(black arrow)* of the ulna. Note that the radial fracture is bowed on the convex side and buckled on the concave side. The ulna fracture is bowed on the concave side with an incomplete fracture on the convex side.

Epiphyseal plate (physeal) fractures are also common in children and are produced by the same forces that can cause a dislocation in an adult. In children, the epiphyseal complex is composed of the epiphysis (radiopaque because of bone density), the cartilaginous growth plate (radiolucent because of cartilage den-

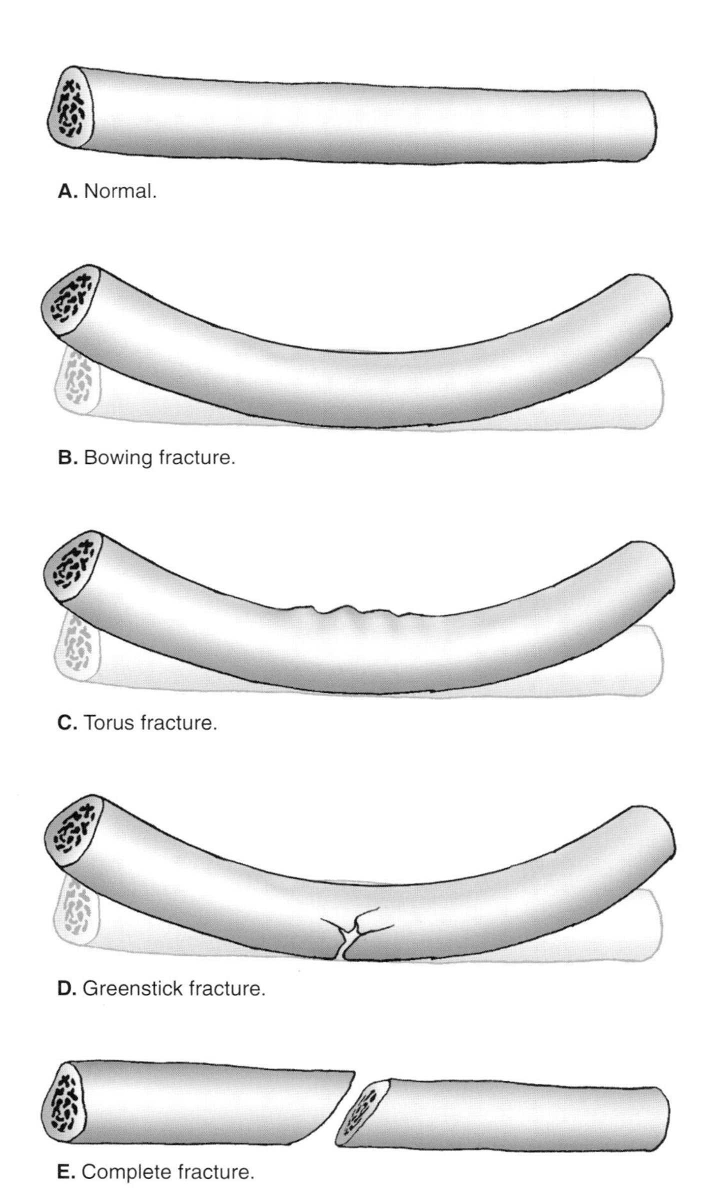

A. Normal.

B. Bowing fracture.

C. Torus fracture.

D. Greenstick fracture.

E. Complete fracture.

Figure 16.66. Drawing of types of pediatric fractures.

Figure 16.67. Torus and greenstick fractures (see text).

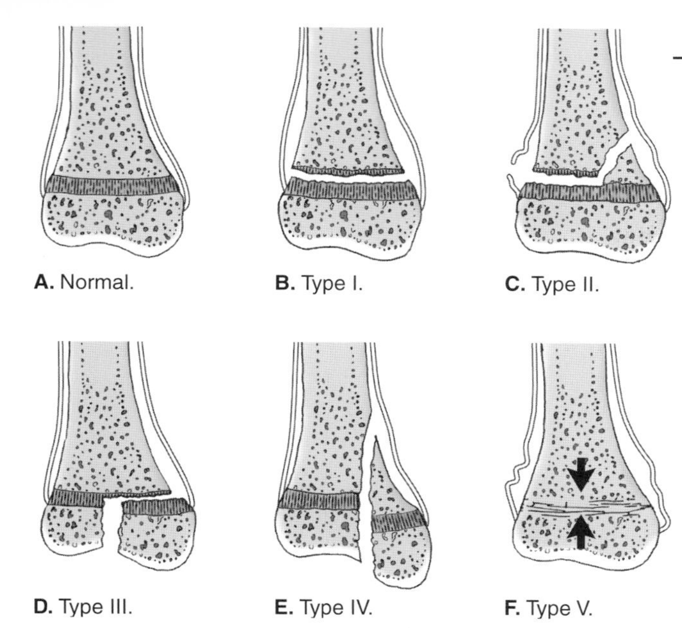

Figure 16.68. The Salter-Harris classification of epiphyseal fractures.

sity), and the metaphysis (radiopaque because of bone density). Drs. Robert Salter and R. I. Harris have described five types of epiphyseal fractures, categorized in the *Salter-Harris classification* system (Figure 16.68). *Type II* fractures are the most common in children, and you will see them frequently if you treat cases of pediatric trauma. A Type II injury involves a fracture of the growth plate that extends into the metaphysis, producing a small metaphyseal fragment (Figure 16.69). Another important injury is the *Type I* fracture. Only 6 percent of epiphyseal fractures are Type I, but you must be aware of them because patients with this injury *may have normal x-rays*. Whenever a patient presents with pain, limited motion, and swelling and point tenderness over the growth plate, but has normal x-rays, you should suspect a Type I fracture.

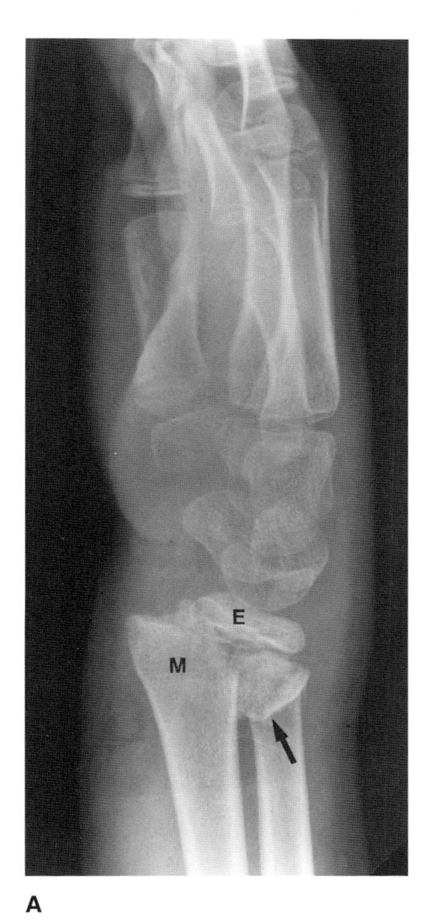

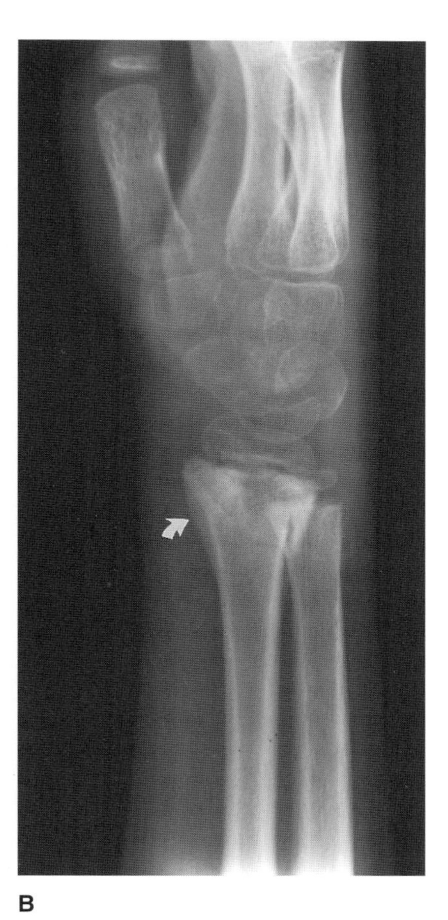

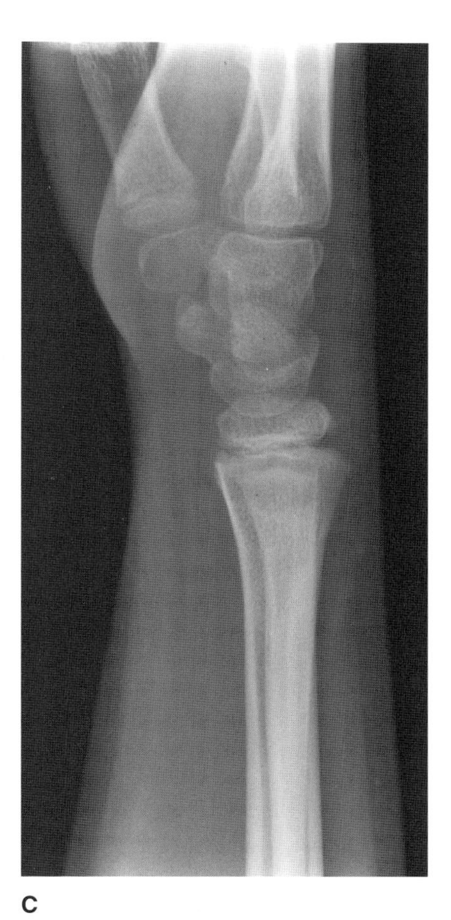

Figure 16.69. Plain films of a Salter-Harris Type II fracture of the distal radius of a child who fell on his outstretched hand. A: Lateral view shows a fracture of the radial growth plate with posterior displacement of the radial epiphysis *(E)* compared with the radial metaphysis *(M)*. The *arrow* indicates the metaphyseal fracture fragment characteristic of the Type II fracture. B: Film taken 2 months later, after reduction of the fracture and healing. Note the callus *(arrow),* an excellent sign of healing, and the disuse osteoporosis of the hand resulting from immobilization. C: Film taken 2 years later shows almost no sign of the fracture and good bone mineralization.

Child Abuse

The radiological manifestations of child abuse include healing fractures of various ages, fractures at the edges of metaphyses, metaphyseal and epiphyseal injuries, posterior rib fractures, and compression fractures of vertebral bodies (Figure 16.70). On CT and MR scans of the heads of abused children you may find bilateral subdural hematomas, global cerebral edema, punctuate hemorrhages, and brain atrophy from prior injuries (Figure 16.71). Abdominal CT scans may show evidence of visceral injuries, such as hepatic lacerations and duodenal hematoma.

Imaging examinations are extremely important whenever child abuse is suspected because they may supply evidence of *multiple episodes of abuse* (such as old fractures or cerebral atrophy), while the clinical examination and history may provide evidence of only the most recent injury.

The imaging examination you should request when you suspect child abuse is a *skeletal survey* consisting of AP and lateral views of the skull, AP views of the chest, abdomen, and pelvis, and AP views of the long bones of the extremities, including the hands and feet. For an infant or small child these can all be included on only a few x-ray films. A bone scan, because of its increased sensitivity, may show more fractures than can be seen on the skeletal survey.

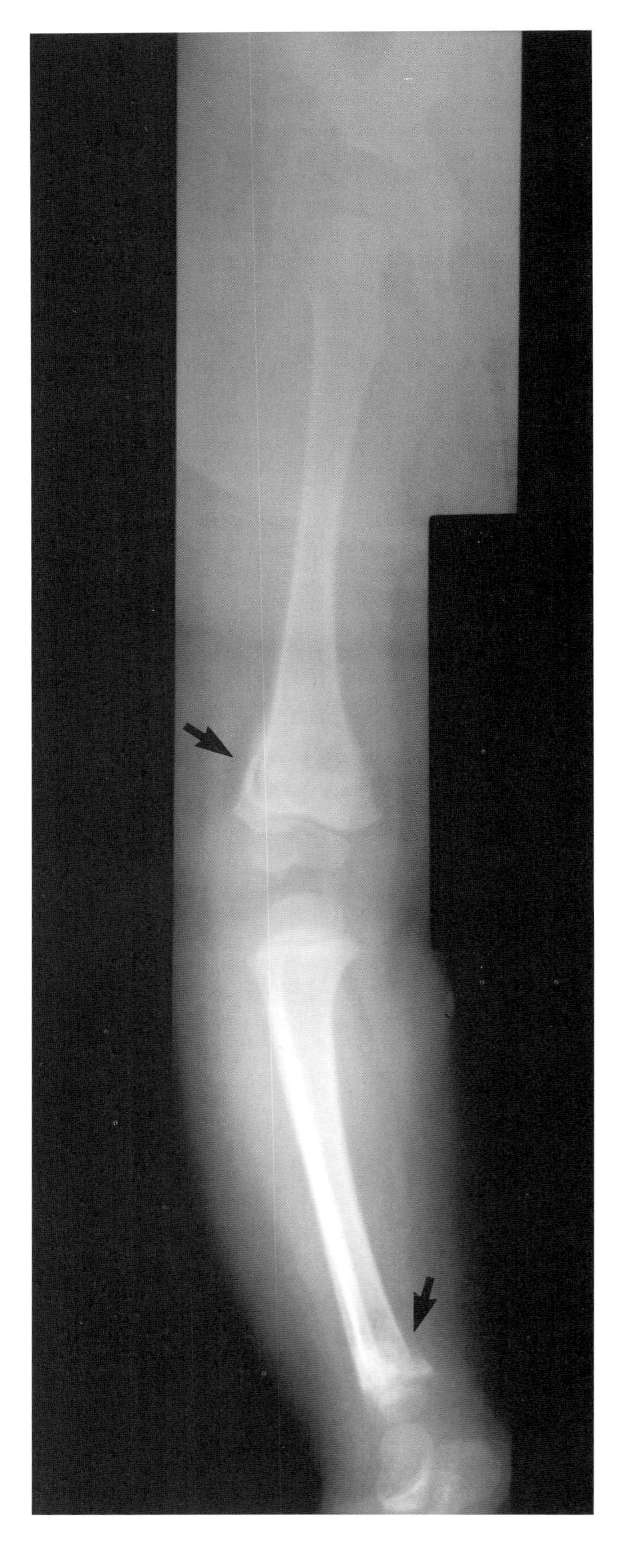

Figure 16.70A. Plain film of a child abuse victim showing multiple fractures *(arrows)* of the right leg in varying stages of healing. Periosteal reaction, a sign that the fracture is several weeks old, is well shown about the distal femur fracture.

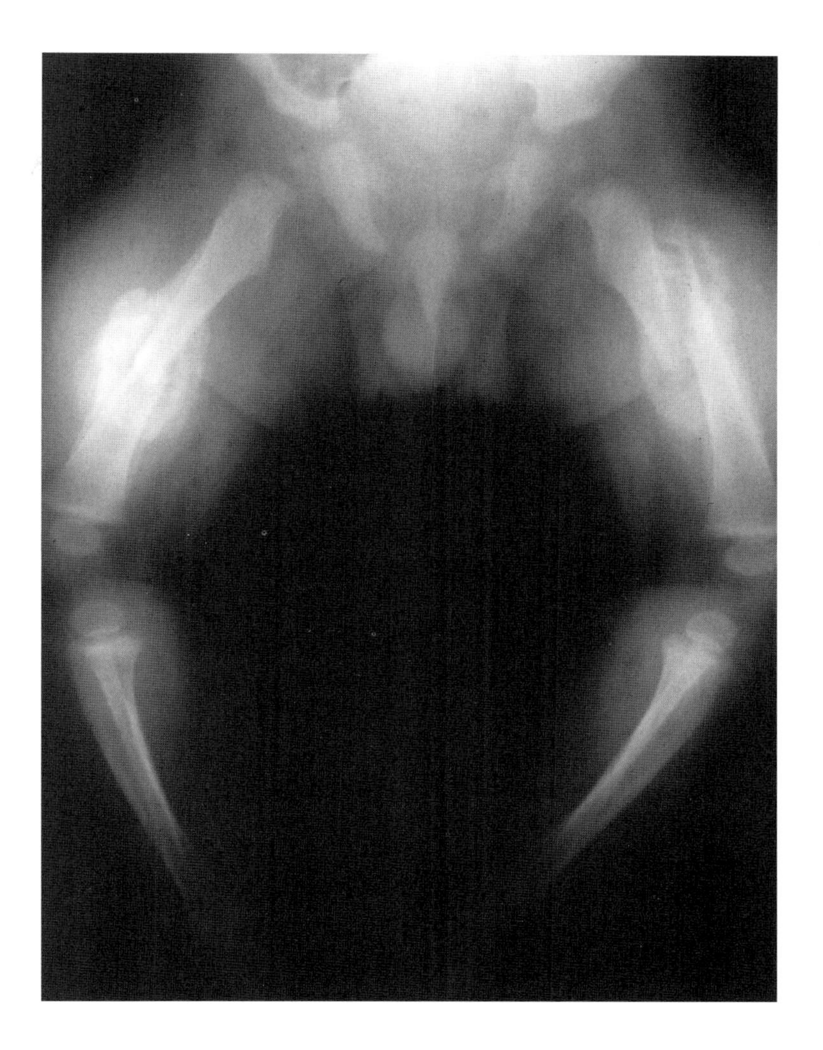

Figure 16.70B. Plain film of another child abuse victim, an infant only a few months old. Note the bilateral femur fractures with exuberant callus formation.

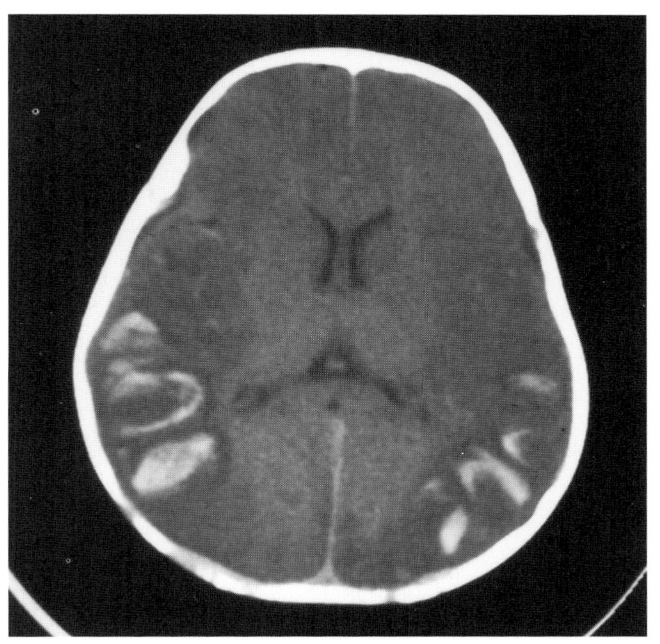

Figure 16.71. CT scan of the head of an abused child, showing bilateral cerebral hemorrhage.

Pediatric Cranial Ultrasound

You may well have thought that cranial ultrasound is impossible, because the brain is surrounded by the bony calvaria and ultrasound waves do not pass through bone. It is impossible in adults, of course, but it is possible to image the brain of neonates and infants with ultrasound through the still-open fontanels (Figure 16.72). Many features of normal anatomy, including the brain, ventricles, and blood vessels, can be visualized in both the coronal and the sagittal planes. This procedure can be performed quickly and with portable equipment, even in the newborn nursery.

One of the most frequent reasons for cranial ultrasound is the confirmation of neonatal intracranial hemorrhage. This should be considered in any neonate who develops bulging fontanels, has an increase in head size, shows a change in mental status, or has a falling hematocrit. Intracranial bleeding may result from birth trauma, but the most common cause is prematurity and hypoxia. Intracranial hemorrhage is a common cause of neonatal death, especially in premature babies. In most cases the bleeding occurs within the brain parenchyma or ventricles, and is easy to detect with ultrasound (Figure 16.73). Cranial ultrasound can also determine ventricular size in patients suspected of having hydrocephalus and detect fluid collections, cystic lesions, and solid parenchymal masses.

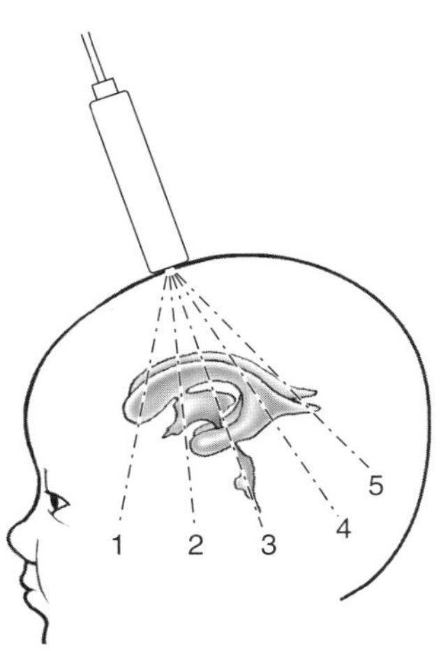

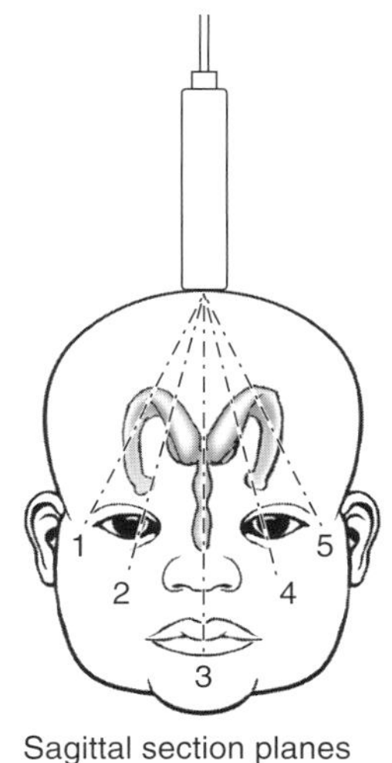

A

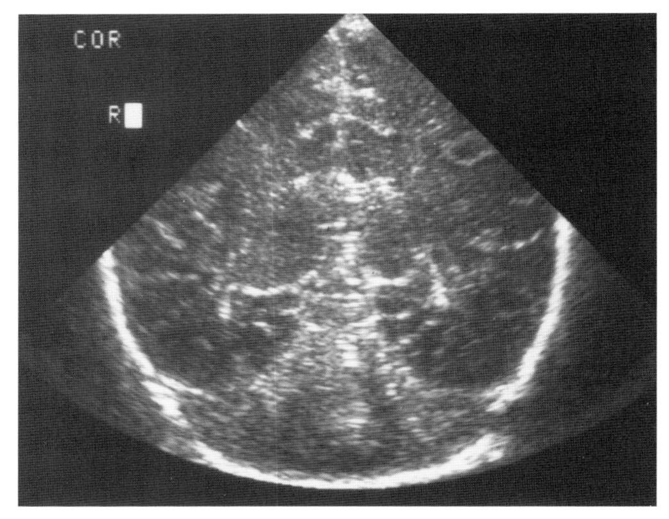

B

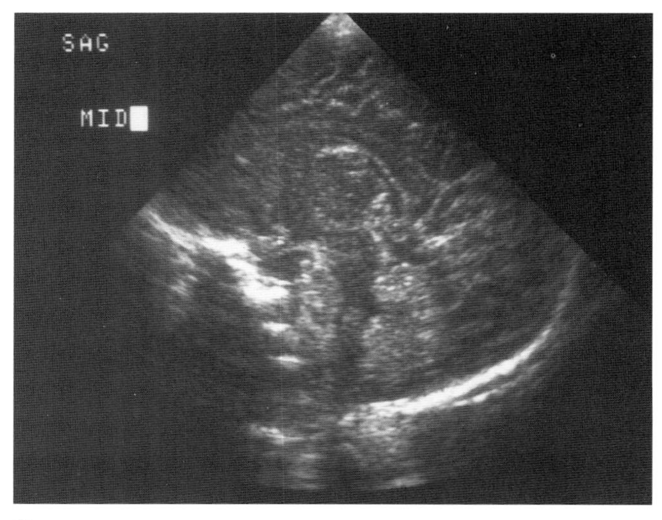

C

Figure 16.72. Normal neonatal intracranial ultrasound examination. A: Drawing showing that transfontanel ultrasound examinations include multiple images taken in the coronal and sagittal planes. B: Normal coronal scan. C: Normal sagittal scan. The ventricles are barely visible. Compare Figure 16.73.

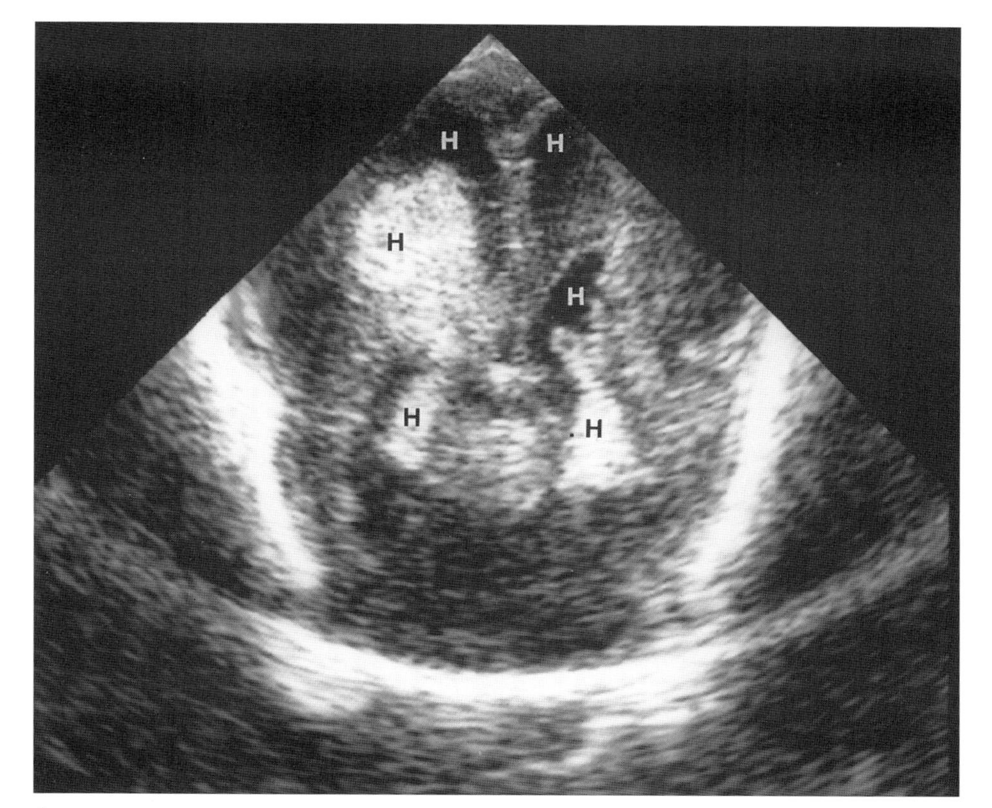

A

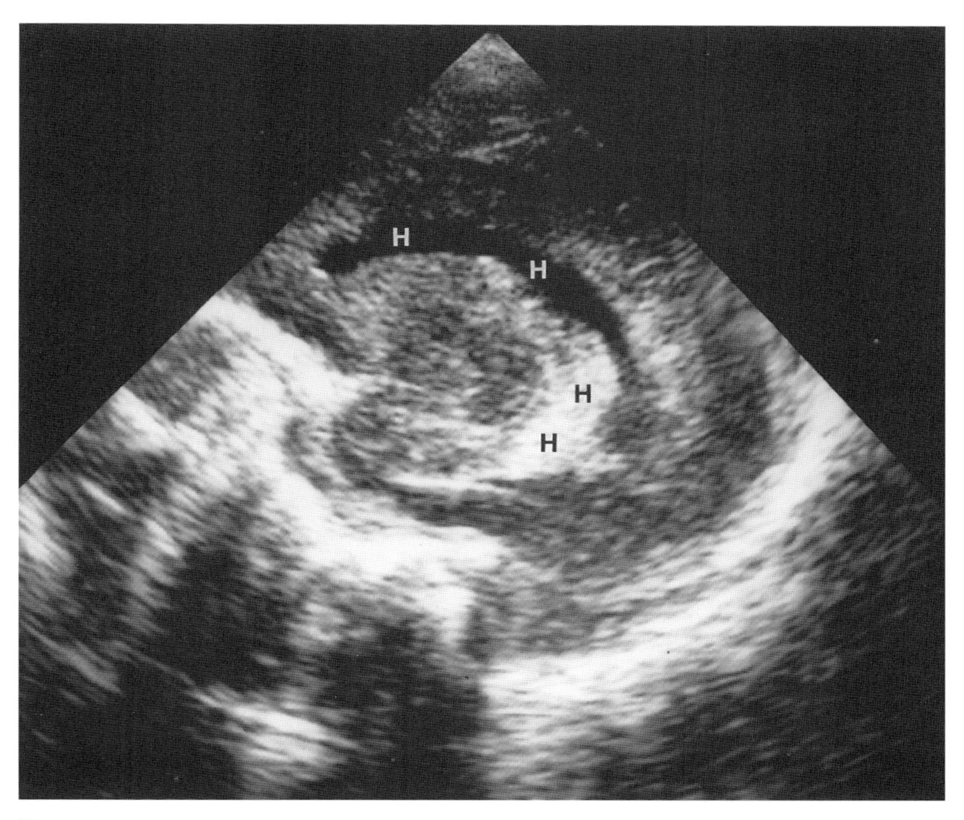

B

Figure 16.73. Ultrasound of neonatal intracranial hemorrhage in a premature intensive care unit baby. The coronal (A) and sagittal (B) scans show dilatation of the lateral ventricles which are filled with echogenic (black *H*'s) and echolucent blood (white *H*'s). The dependent blood in the ventricles is more echogenic due to the hematocrit effect.

Problems

Unknown 16.1 (Figure 16.74)

This elderly man complains of scrotal swelling. Can you determine the cause from this plain film?

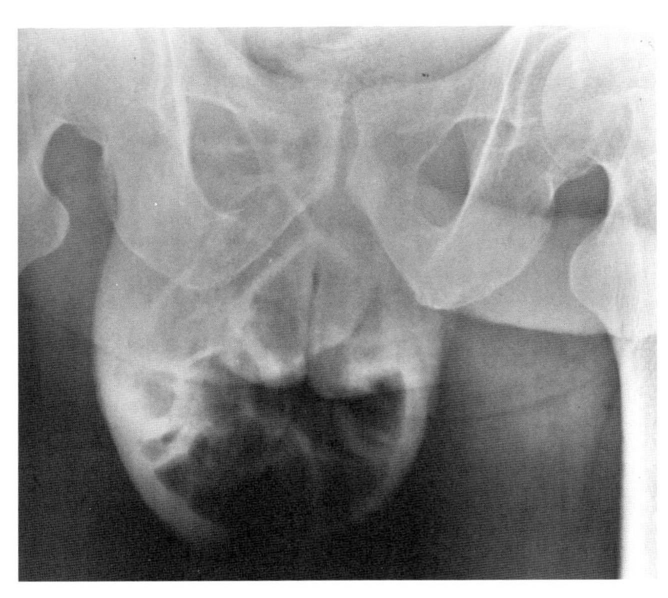

Figure 16.74 ***(Unknown 16.1)***

Unknown 16.2 (Figure 16.75)

Can you determine what echogenic structure is located within the uterus (*U*'s) of this 30-year-old woman?

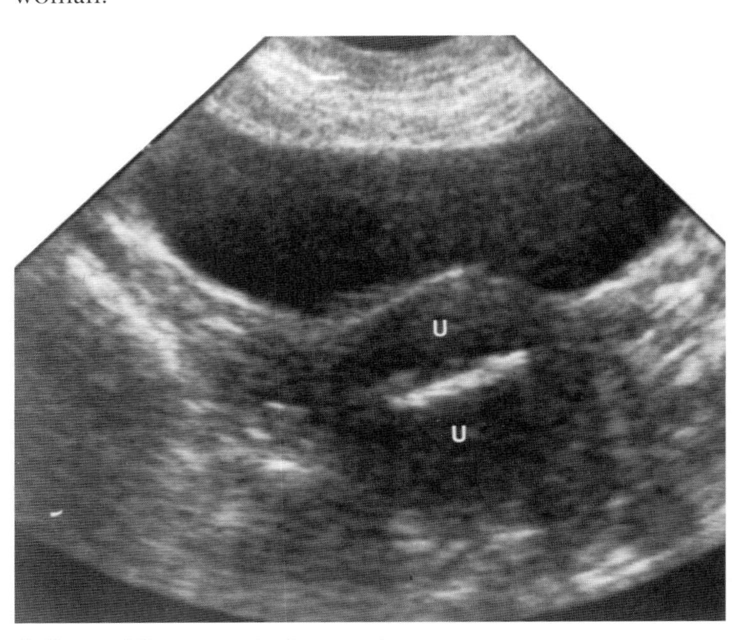

A. Coronal (transverse) ultrasound.

Figure 16.75 ***(Unknown 16.2)***

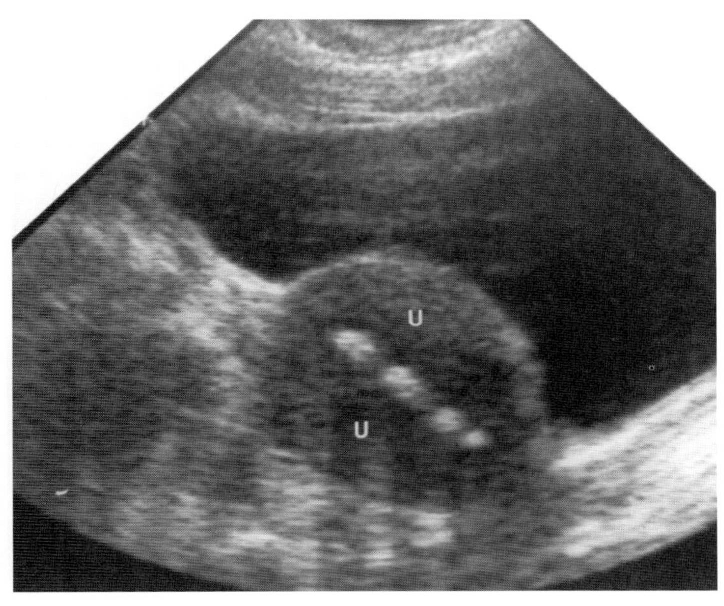

B. Sagittal ultrasound.

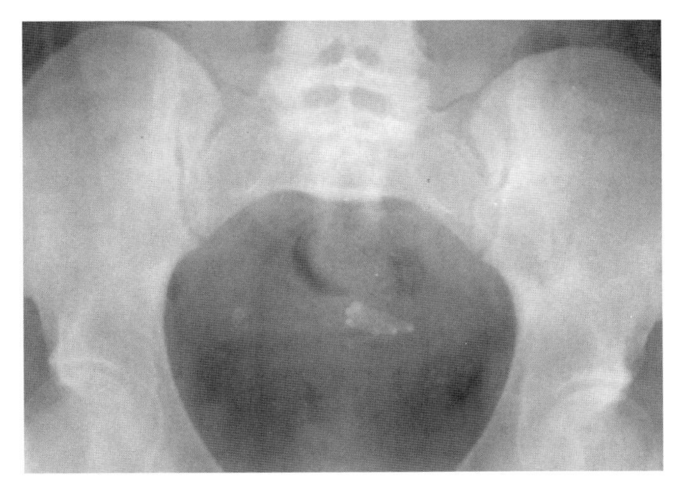

C. Plain film.

Unknown 16.3 (Figure 16.76)

Study the pelvic CT scan of this 78-year-old woman who presented with weight loss, pelvic pain, and vaginal bleeding. The *arrows* indicate the margins of her uterus. What are your conclusions?

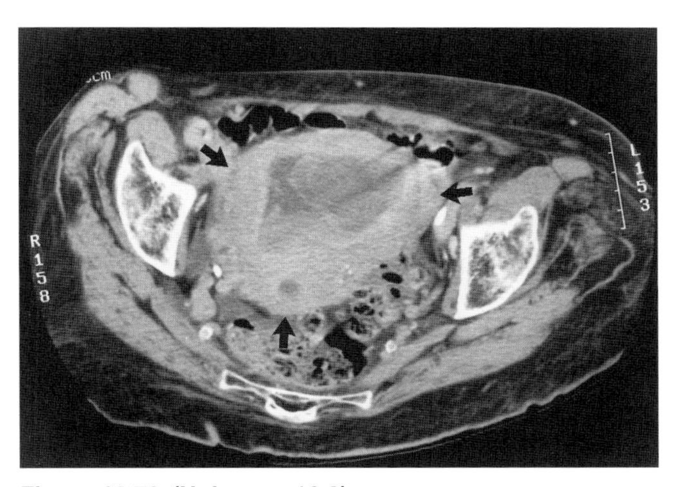

Figure 16.76 ***(Unknown 16.3)***

17 The Vascular System

Traditionally, vascular imaging required arterial or venous catheterization or needle puncture with injection of contrast material and filming with conventional x-ray techniques. An *angiogram* is an imaging examination of either an artery or a vein: an *arteriogram* is an arterial study (contrast injection into an artery) and a *venogram* is a venous study (contrast injection into a vein). Similarly, a *lymphangiogram* is a study filmed after injection of a contrast substance into the lymphatic system.

In recent years several new techniques have become available that permit vascular imaging without arterial or venous catheterization. They include ultrasound (especially color *Doppler ultrasound*), magnetic-resonance imaging, *magnetic-resonance angiography (MRA)*, computed tomography, and *computed tomography angiography (CTA)*. You have already seen many examples of blood vessel imaging by some of these techniques in previous chapters. The advantages of these newer "noninvasive" techniques are decreased patient morbidity and discomfort and reduced expense. A catheter arteriogram usually costs two to three times as much as a CT or MR scan. An additional benefit of the ultrasound and MR techniques is that they do not require the administration of vascular contrast materials. Consequently, these techniques are often recommended for patients who are at risk for injection of vascular contrast material because they have a history of reaction to contrast material or because their renal function is impaired.

For many clinical conditions the noninvasive ultrasound, CT, and MR techniques provide enough diagnostic information and detail for optimal patient management. A typical example is diagnosis of deep venous thrombosis of the lower extremity by vascular ultrasound. But several other clinical conditions, such as cerebral aneurysm of the circle of Willis, still require traditional angiography with arterial or venous catheterization and contrast injection to obtain the detail required for a definitive diagnosis and appropriate treatment planning.

In this chapter we will first survey the various imaging techniques for arteries and veins (lymphangiography, a specialized procedure used much less often, will be covered at the end of the chapter). Then we will discuss several common and important vascular disorders. Pulmonary arteriography and the imaging of pulmonary embolism have already been covered in Chapters 5 and 7. Cerebrovascular disease and stroke will be discussed in Chapter 18. Interventional vascular procedures such as angioplasty and therapeutic embolization will be covered in Chapter 19.

Conventional Arteriography

In *conventional arteriography* the tip of a small-caliber vascular catheter is advanced under fluoroscopic guidance into the artery under investigation, a selected volume of vascular contrast material is injected, and a series of x-ray images is obtained of that vascular bed as the contrast material sequentially opacifies the arteries *(arterial phase)*, capillaries *(capillary phase)*, and veins *(venous phase)* (Figure 17.1). The images in Figure 17.1 show examples of normal blood vessels in a normal anatomic structure.

Arteriography of a tumor will show it to be *hypervascular*, when it is associated with more blood vessels or has brighter contrast opacification than the normal organ, or *hypovascular*, when it is associated with fewer blood vessels or has less contrast opacification than the

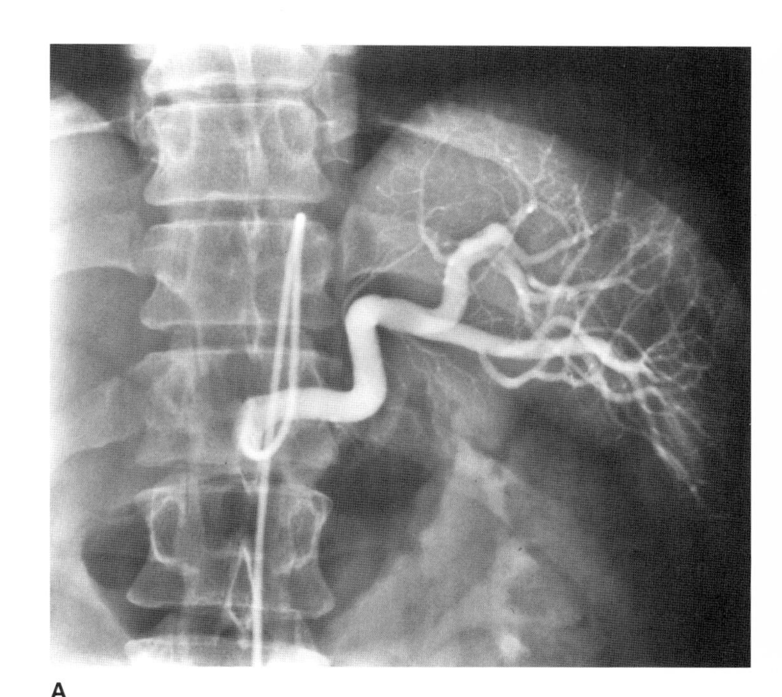

A

B

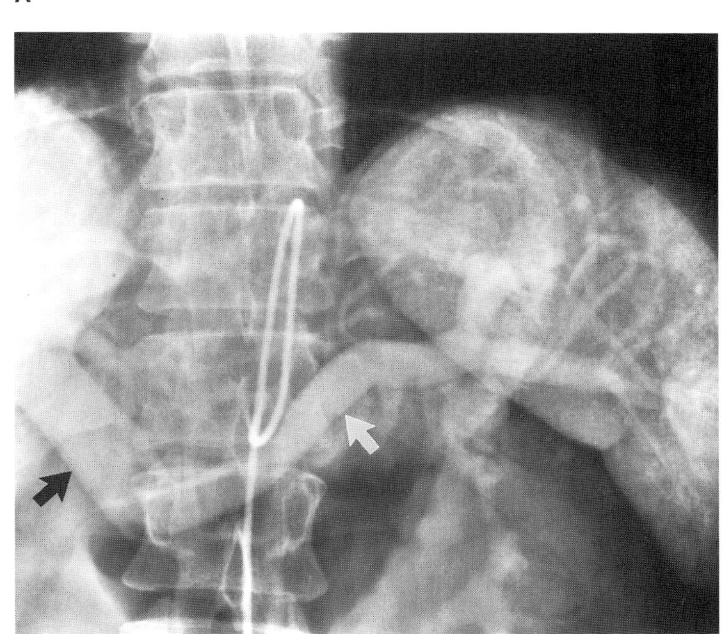

C

Figure 17.1. Selective splenic arteriogram showing examples of (A) *arterial,* (B) *capillary,* and (C) *venous* phases. These three films were selected from the series of radiographs obtained during and following injection of vascular contrast material into the arterial catheter. The *arterial phase* shows excellent contrast opacification of the splenic artery and its branches; the splenic parenchyma is only minimally opacified. The *capillary phase* best shows the splenic parenchyma. The *venous phase* shows less contrast opacification of the splenic parenchyma and superior opacification of the splenic vein *(white arrow),* which drains into the portal vein *(black arrow).*

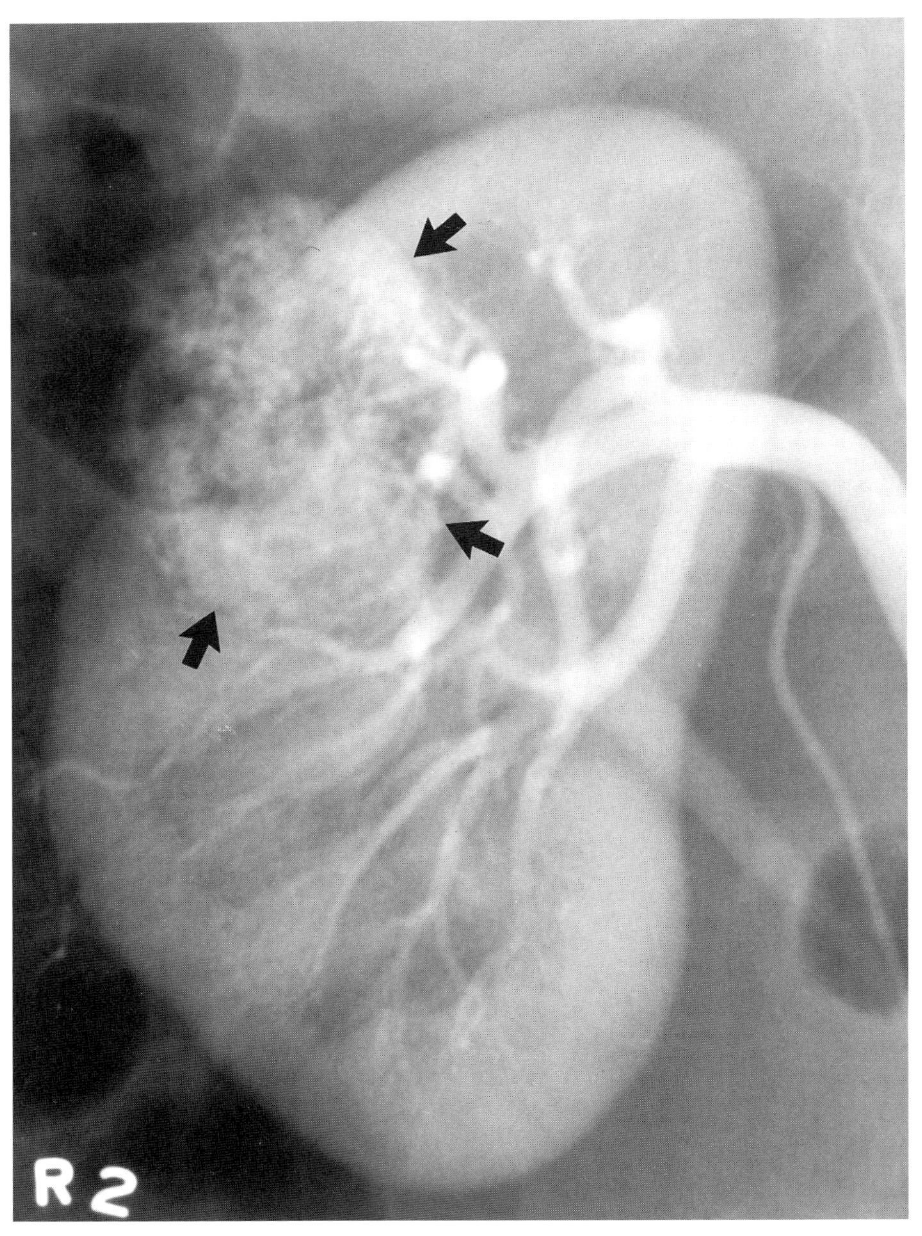

A

normal organ. Figure 17.2 shows examples of both these types of tumor. Tumor vessels may be associated with bizarre branching patterns and abnormal-looking vessels. Tumors that have the same vascularity as the surrounding uninvolved tissue are termed normovascular.

A specially designed procedure room, or angiographic suite, illustrated earlier (Figure 2.23, page 26), is required for conventional angiography. This room is furnished with fluoroscopic equipment, a mechanical contrast injector, a rapid film changer, and patient monitoring devices.

Patients are awake for arteriographic procedures, but they are usually sedated for their comfort. They should be well hydrated before the examination so that they can best tolerate the load of renally-excreted contrast material (usually the same type of contrast material injected for intravenous urography) needed for the procedure. The arterial catheter is usually inserted through a femoral artery; following removal of the catheter and compression of the puncture site by the radiologist, the patient is required to lie flat with the involved leg kept straight for several hours.

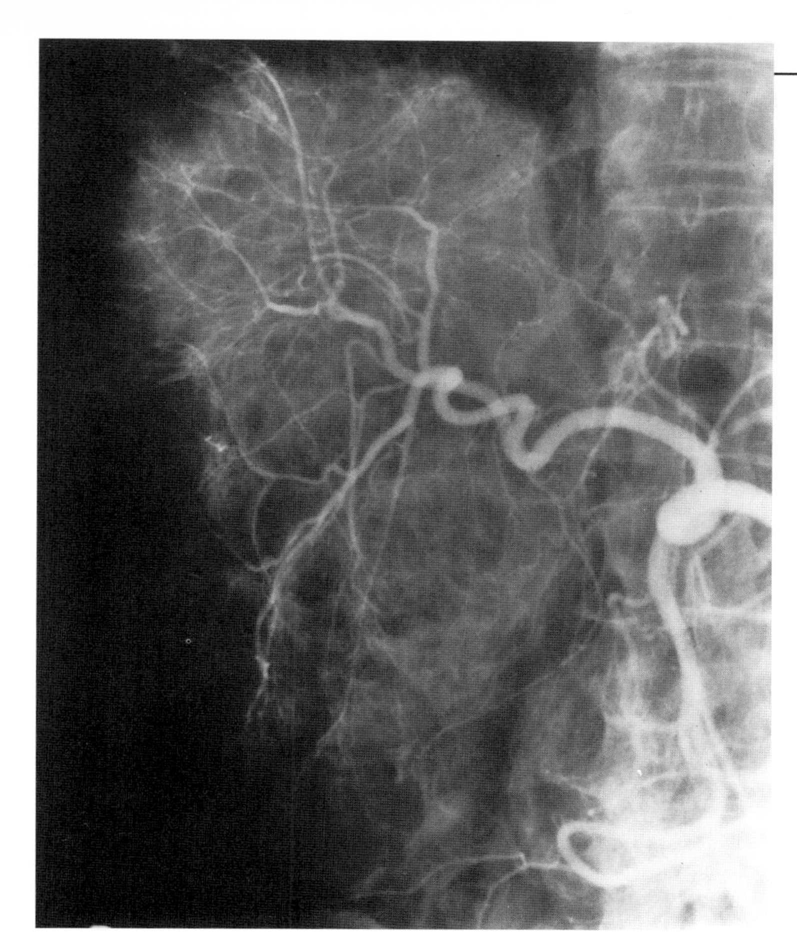

B

Figure 17.2. Hypervascular and hypovascular tumors. A *(facing page):* Right renal arteriogram showing a *hypervascular* renal cell carcinoma *(arrows)* of the right kidney. Compare the abnormal vascularity of the small tumor with the normal adjacent renal vasculature. B and C: Arterial and capillary phases of a hepatic arteriogram showing large *hypovascular* metastases *(arrows)* from metastatic colon cancer. These lesions are less vascular than the surrounding normal hepatic parenchyma, which is compressed around the lesions.

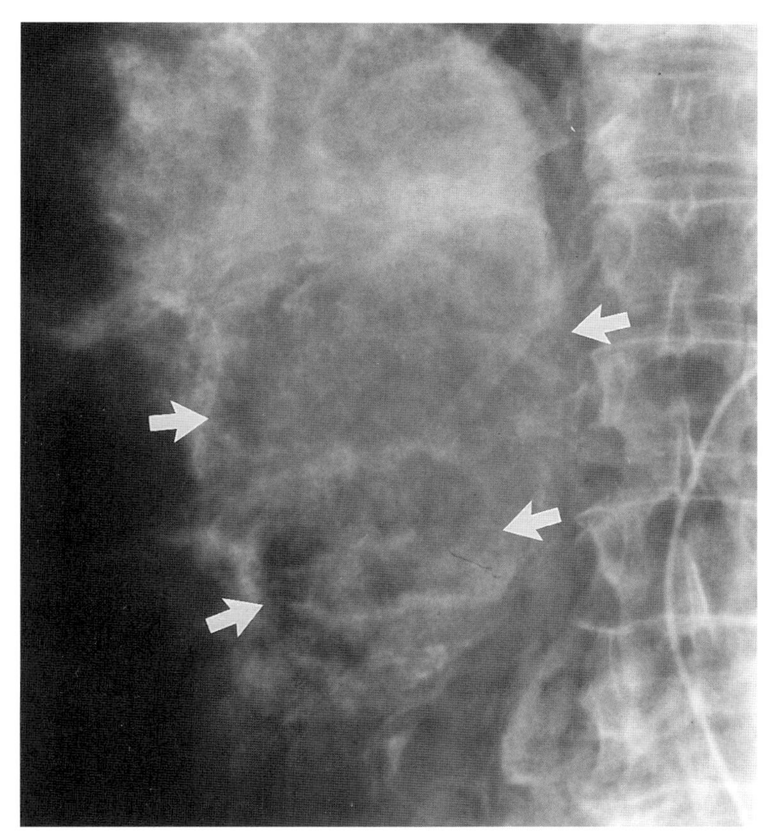

C

Digital Subtraction Angiography

A newer way of obtaining angiographic images is *digital subtraction angiography (DSA)*, which is very similar to conventional arteriography (or venography) except that instead of exposing a series of x-ray films with a rapid film changer, the radiologist obtains serial images electronically and stores them digitally in a computer. When the radiologist reviews the images on a CRT monitor, the initial image of the field without contrast material in the blood vessels is electronically subtracted from the contrast material images; the resultant digital subtraction vascular images appear to show black blood vessels on a blank background (Figures 17.3 and 17.4). One advantage of DSA is that blood vessels appear denser because of the improved contrast resolution; note in the hepatic arteriogram in Figure 17.5A how well the hypervascular liver metastases stand out against the subtracted (almost blank) background. DSA is also faster than conventional filming and usually requires less contrast material. But DSA gives less detail (less spatial resolution) than does conventional "film" angiography, and can produce subtraction artifacts generated by patient motion or bowel motion during the imaging sequence.

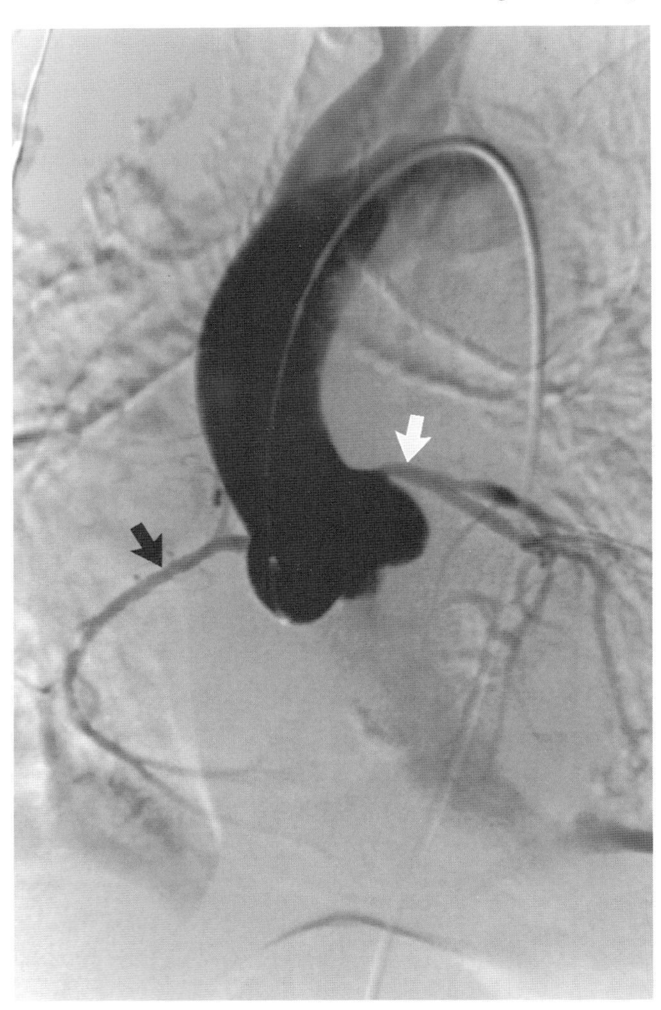

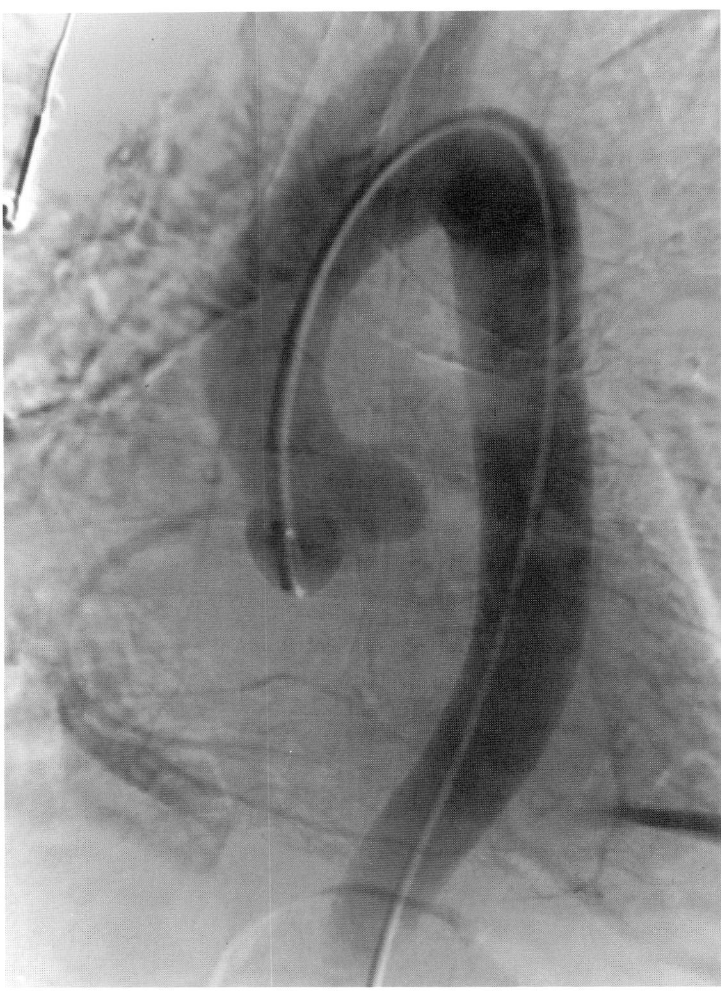

A B

Figure 17.3. Digital subtraction arteriogram of the thoracic aorta (DSA aortogram). A: Early film showing maximum contrast opacification of the ascending aorta, sinuses of Valsalva, right coronary artery *(black arrow),* and left coronary arteries *(white arrow).* The brachiocephalic branches to the head and arms are faintly opacified. B: Slightly later image with contrast opacification of the aortic arch and descending aorta. The opacification of the ascending aorta is diminished as the contrast material is "washed out" by flowing blood.

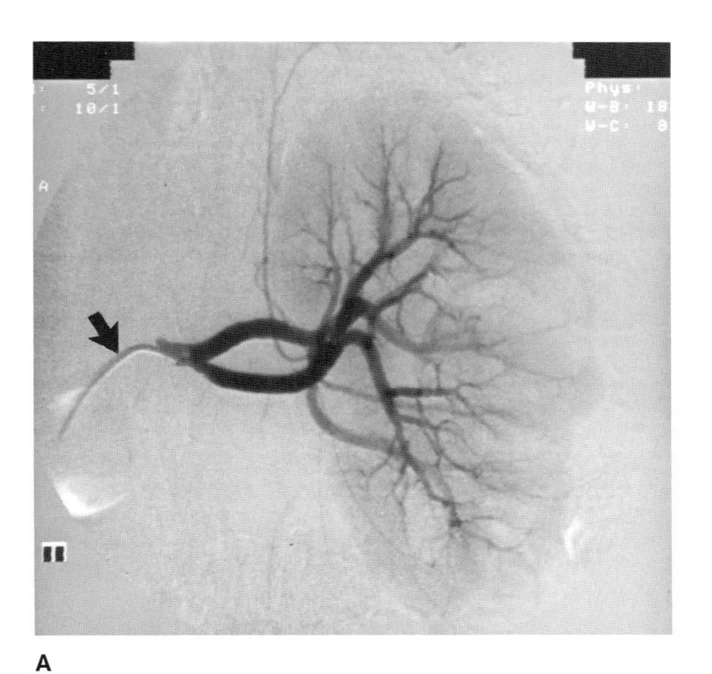

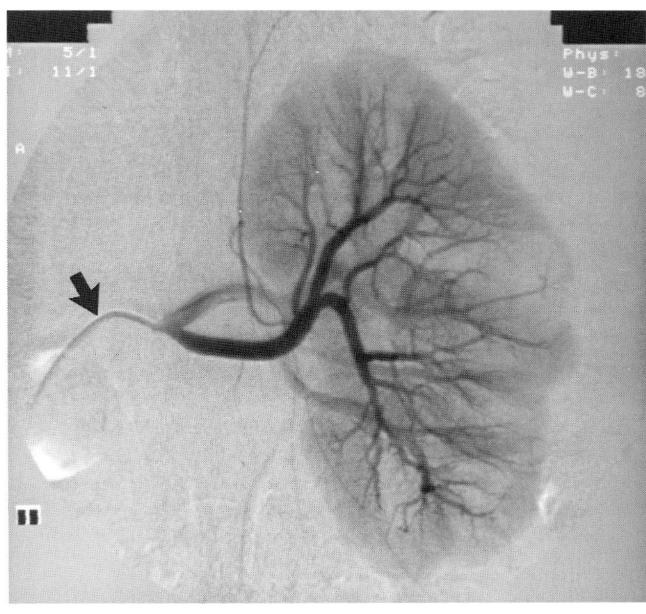

A B

Figure 17.4. Digital subtraction arteriogram of the left renal artery. A: Early image with dense opacification of the proximal renal artery branches. B: Later image during the capillary phase, showing early opacification of the renal parenchyma. The *arrows* point to the curved-tipped angiographic catheter, which was positioned in the proximal left renal artery for selective contrast injection.

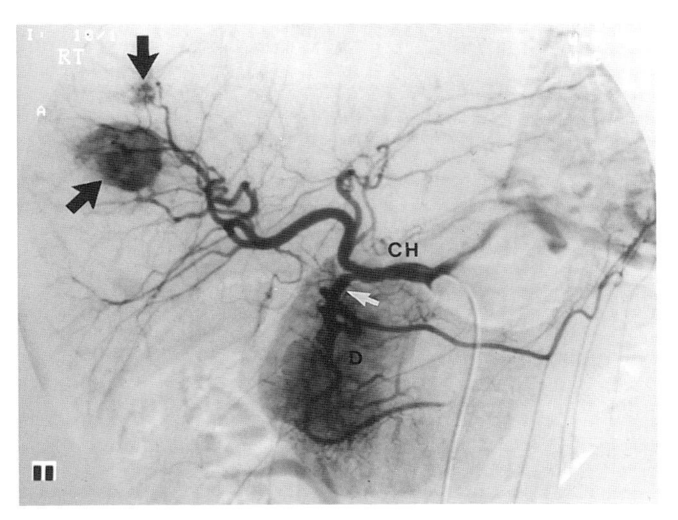

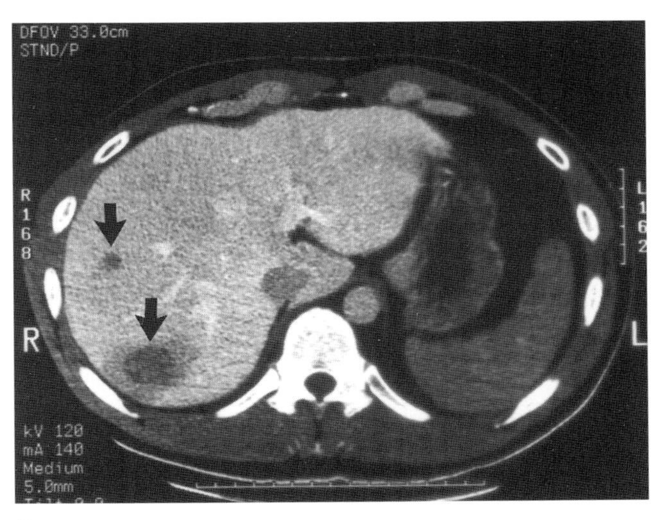

A B

Figure 17.5. Patient with liver metastases. A: Digital subtraction arteriogram of the common hepatic artery *(CH)* showing two hypervascular liver metastases *(black arrows).* The large hypervascular structure just below the common hepatic artery is the normal duodenum *(D).* It is supplied by the gastroduodenal branch *(white arrow)* and usually demonstrates a bright hypervascular "blush" at arteriography. B: CT scan showing the relationship of the metastases *(arrows)* to the remainder of the liver.

Conventional Venography

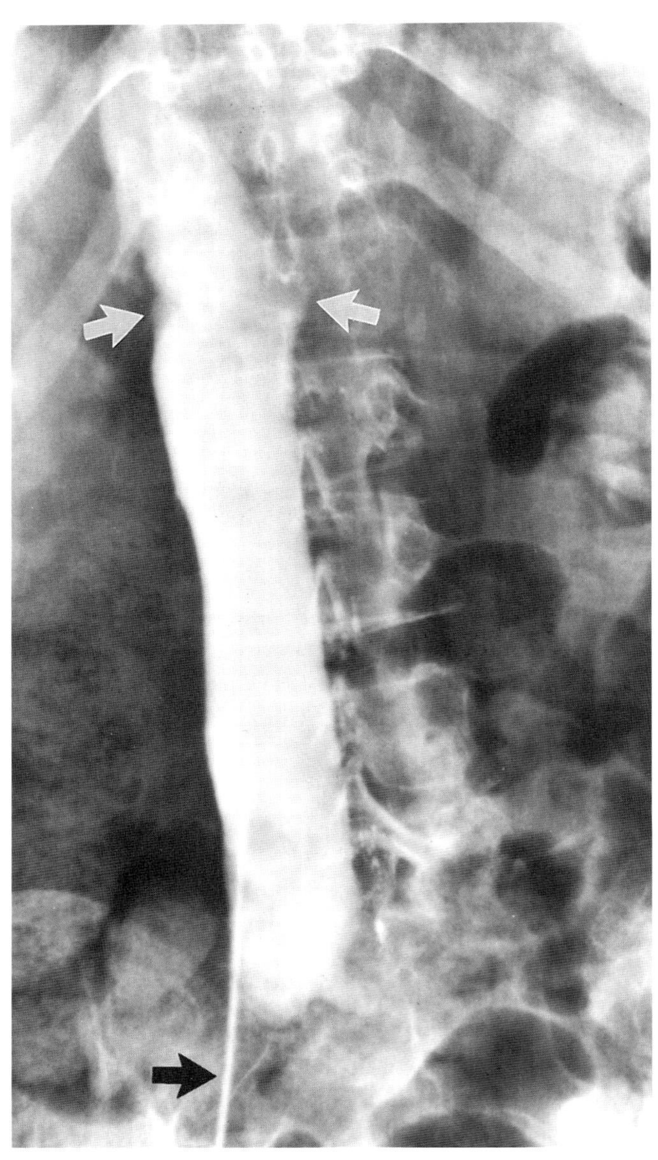

Figure 17.6. Normal inferior vena cavagram. The *black arrow* points to the angiographic catheter, which was placed in the IVC through a right femoral vein. The *white arrows* indicate the level of the renal veins where unopacified blood from the kidneys is entering the IVC, producing flow defects in the column of opaque contrast. Above the renal veins the IVC narrows as it traverses the liver to reach the right atrium.

Conventional venography may require venous catheterization for the imaging of large vessels, such as the inferior vena cava (IVC) (Figure 17.6), or for selective contrast injection, as is required for a renal venogram (Figure 17.7). The veins of the extremities can be opacified with only a venipuncture of a distal peripheral vein. The hand-injected contrast material will be transported passively by venous blood flow within the extremity, where it will opacify the deep venous system. For a leg venogram (Figure 17.8) contrast material is injected through a venipuncture of a small vein on the dorsum of the foot. Venography can be filmed either with a series of conventional x-ray films or electronically with digital subtraction.

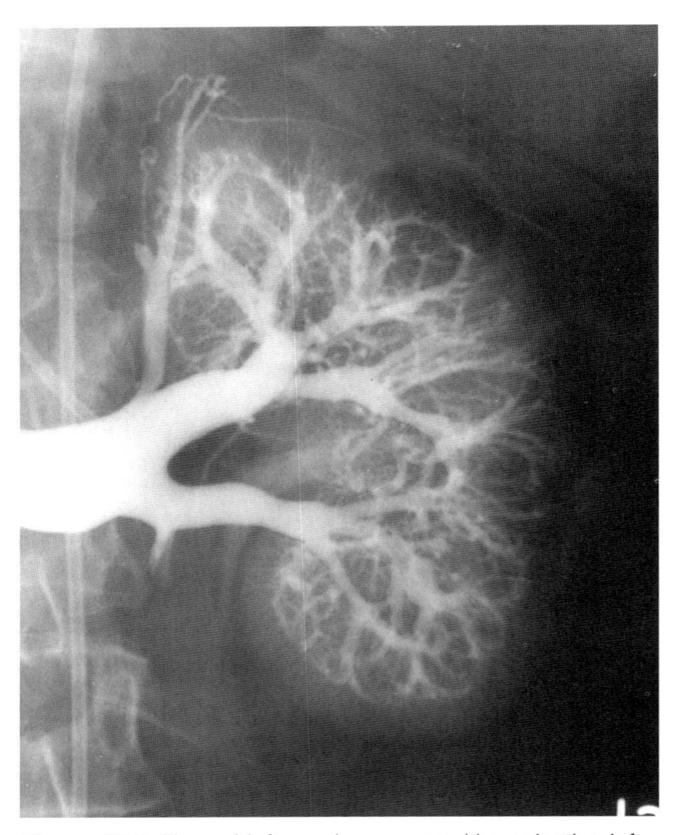

Figure 17.7. Normal left renal venogram (the selective left renal vein catheter is not visible on this film because the dense contrast material within the renal vein obscures it).

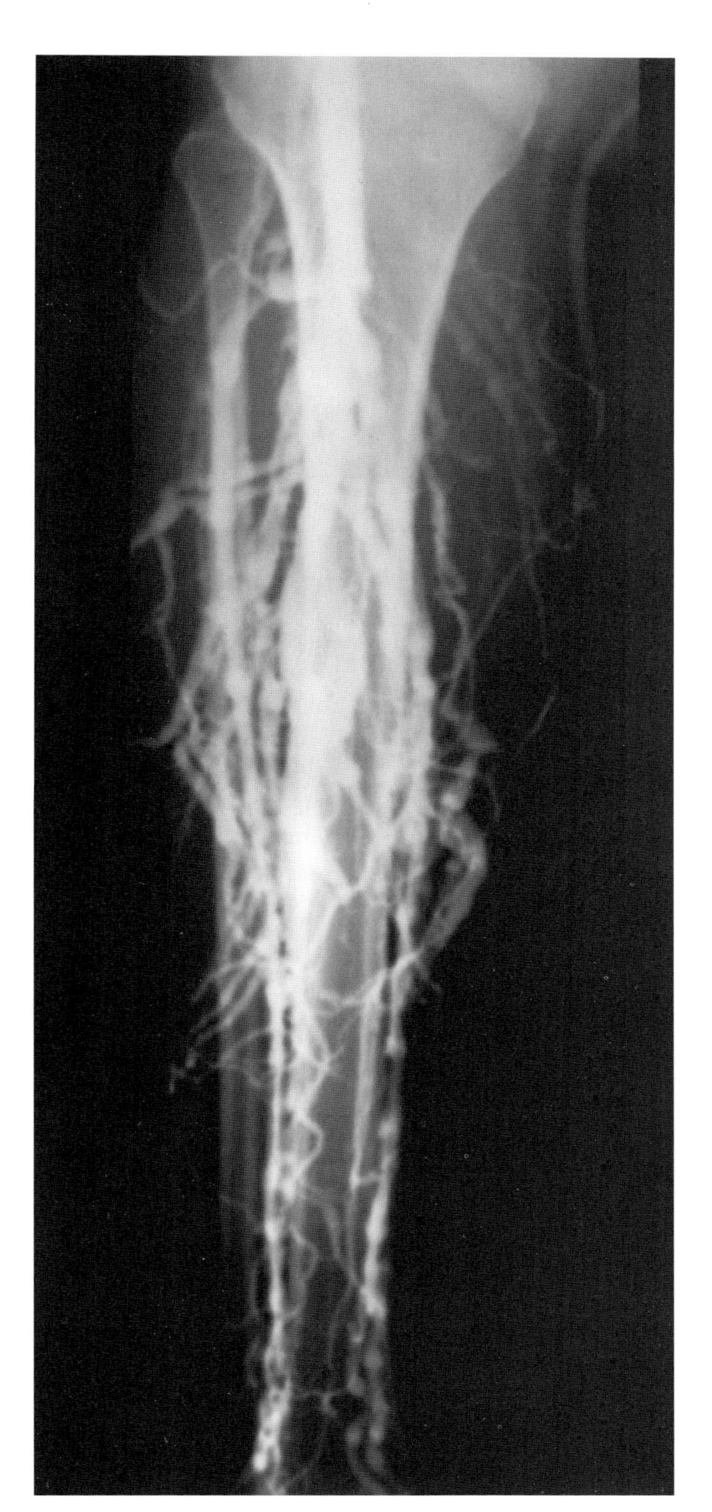

Figure 17.8. Film from a right leg venogram showing opacification of the deep veins of the lower leg. Contrast material was injected by venipuncture in the right foot.

Ultrasound and Color Doppler Ultrasound

You have already learned that *clear fluid collections* produce no echoes at ultrasound and are thus anechoic. You have seen such collections in benign renal cysts, which are very easy to recognize on ultrasound examinations as black round spaces within the echopattern of the surrounding renal parenchymal tissue. Collections of *not-completely clear liquids* are also quite recognizable at ultrasound but are usually hypoechoic, because some echoes are produced when the ultrasound waves encounter debris, red blood cells, or pus within the fluid. You have seen hypoechoic fluid within the gallbladder. Ultrasound also shows the blood within arteries and veins as fluid collections, making vascular structures visible. It is possible to use ultrasound to diagnose a wide range of vascular conditions. The blood may appear anechoic or hypoechoic or a mixture of the two, depending on the ultrasound settings selected. Ultrasound can visualize blood vessels of varying sizes and in many locations in the body.

Since vascular ultrasound examinations do not require the injection of contrast material, they are well tolerated by and comfortable for most patients. In Figure 17.9 you see a patient positioned for a vascular ultrasound examination of the neck. A normal carotid

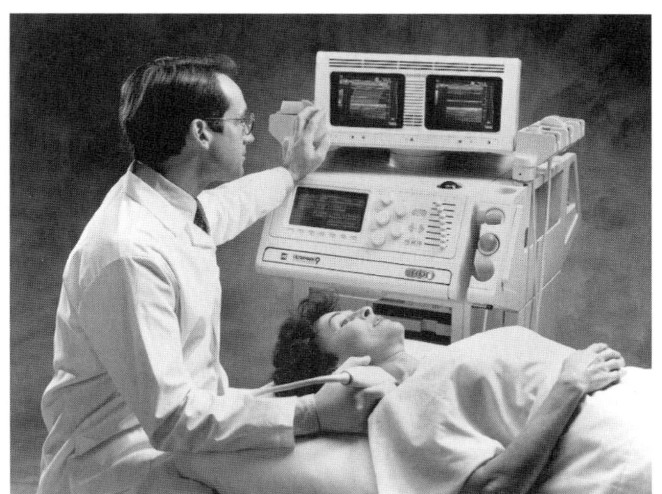

Figure 17.9. Patient positioned for a carotid artery ultrasound examination.

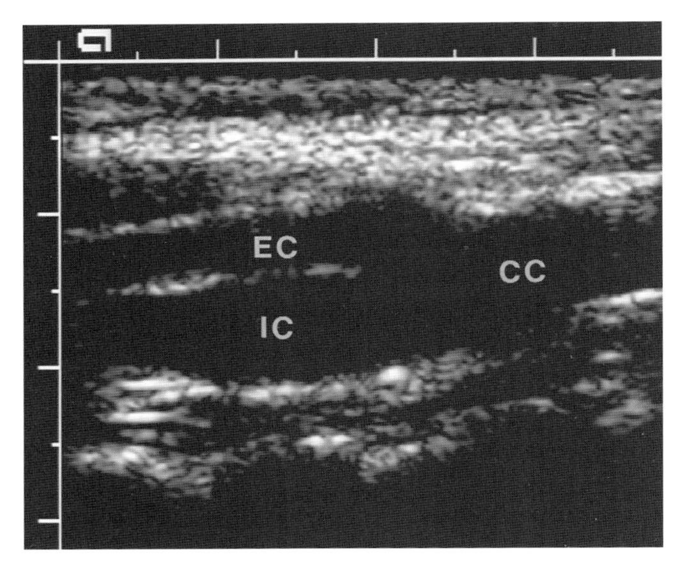

Figure 17.10. Ultrasound of a normal carotid artery bifurcation. The patient's head is to your left, the feet to your right. Blood within the vascular system appears black (anechoic). This image shows the common carotid *(CC)* artery bifurcation into the internal carotid *(IC)* and external carotid *(EC)* artery branches.

artery bifurcation would appear as shown in Figure 17.10; the arterial blood is anechoic and no stenosis is present. Arteries and veins in other parts of the body have similar ultrasound characteristics, as Figure 17.11 shows.

An important advance in vascular ultrasound has been the development of color Doppler ultrasound, which depicts the flow of blood within arteries and veins. Study the color Doppler ultrasound images in Figure 17.12. The ultrasound computer arbitrarily colors flow in one direction red and flow in the opposite direction blue. Lack of a color signal indicates the absence of blood flow in the vessels being imaged. Color Doppler ultrasound can determine whether blood flow in an artery, vein, or other vascular structure is present or absent, normal or diminished. For this reason color Doppler ultrasound has become an invaluable tool in a variety of clinical situations. It can determine whether vascular grafts have remained patent or not, demonstrate the significance of atherosclerotic arterial stenoses, and show whether or not a structure has been deprived of its blood supply (see Figure 17.12H). Tracings of the character of pulsatile flow can also be recorded for arteries and veins. Color Doppler ultrasound is very commonly used today to evaluate carotid artery disease in stroke patients and peripheral vascular disease in patients with atherosclerosis, and to monitor blood flow within the fetus and umbilical cord.

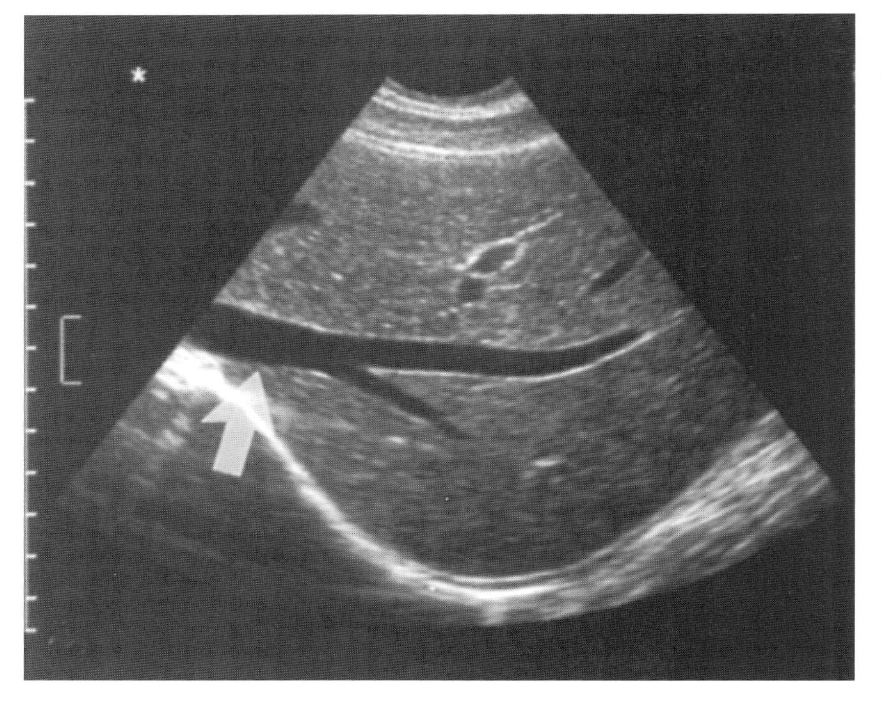

Figure 17.11. Abdominal ultrasound showing a normal hepatic vein branch *(arrow)* within the liver.

Figure 17.12. Color Doppler Ultrasound Exhibit

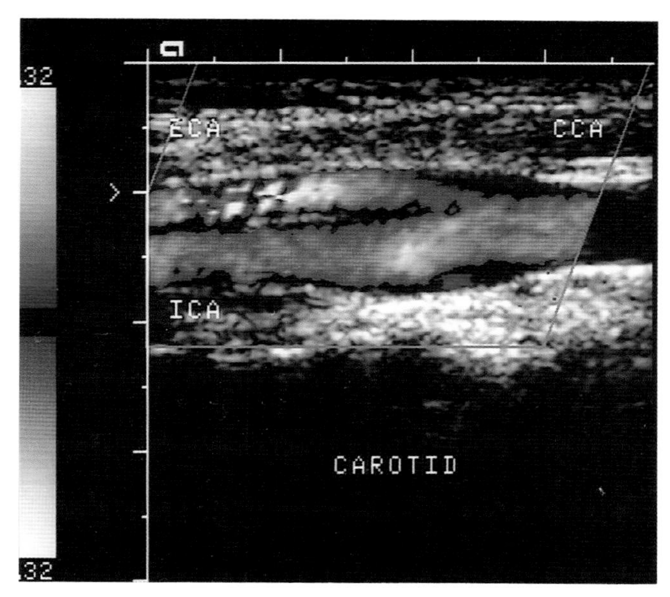

A. Scan of a normal carotid artery in a longitudinal plane showing the common carotid artery *(CCA)* bifurcating into the internal carotid artery *(ICA)* and the external carotid artery *(ECA).*

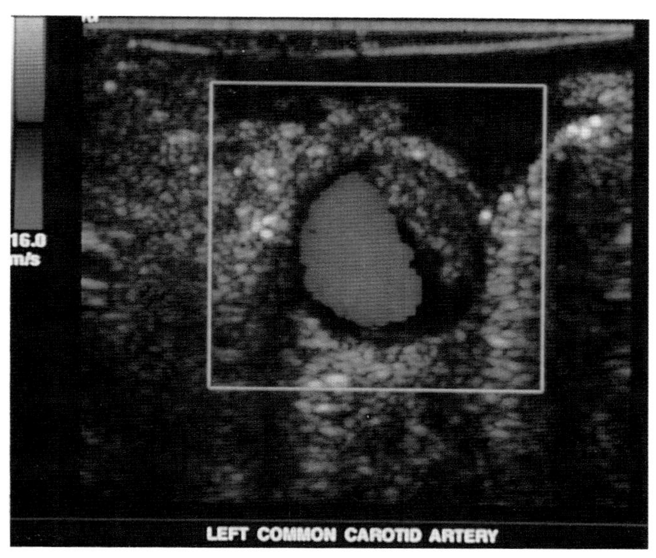

B. Cross-sectional scan of the left common carotid artery showing a large atherosclerotic plaque narrowing the lumen *(red)* by nearly 50 percent.

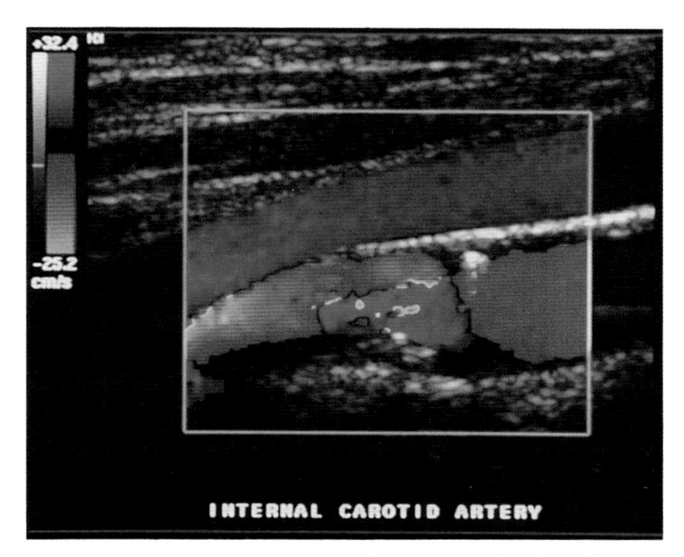

C. Longitudinal scan showing a focal stenosis of the internal carotid artery *(red)* by an atherosclerotic plaque. The larger blue area above it represents blood flow in the jugular vein; it appears deep blue because the direction of the flow is opposite to the flow in the carotid artery.

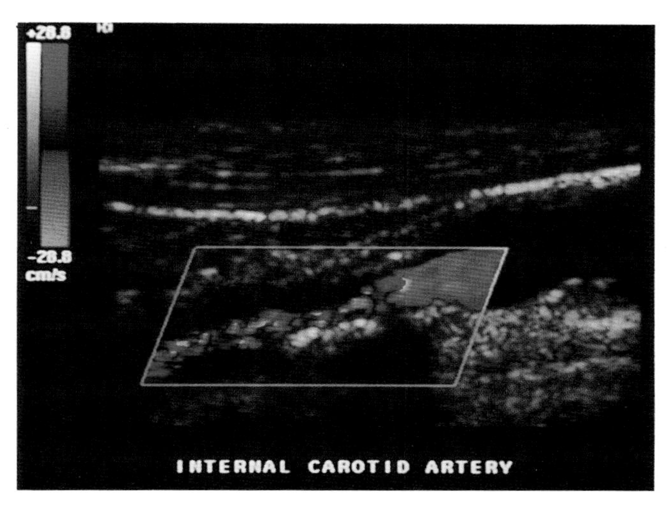

D. Longitudinal scan showing a long atherosclerotic stenosis of an internal carotid artery.

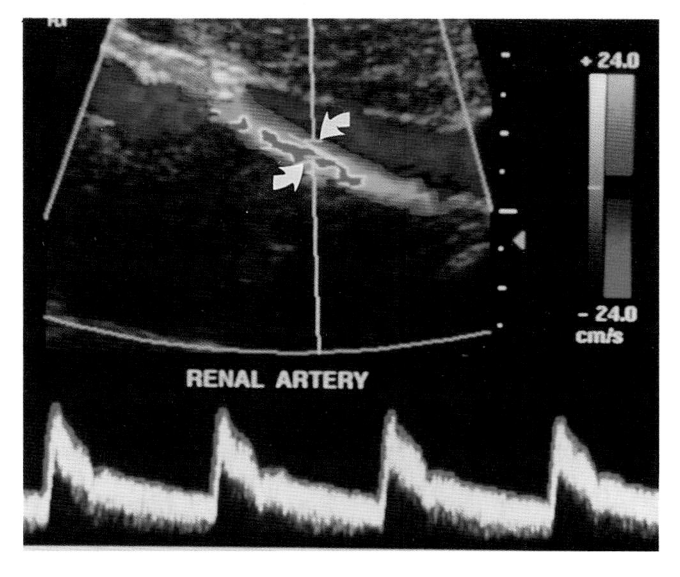

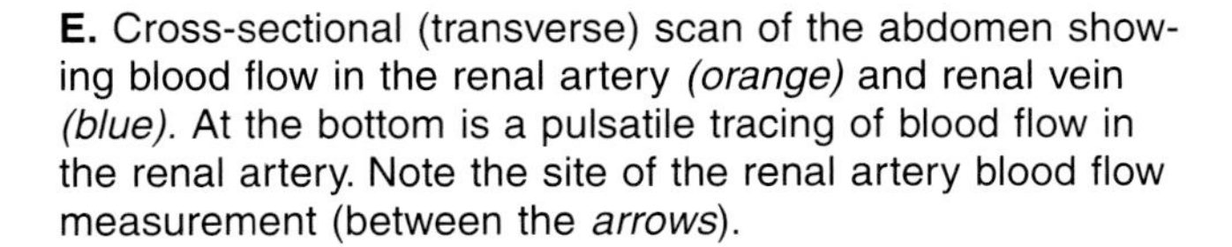

E. Cross-sectional (transverse) scan of the abdomen showing blood flow in the renal artery *(orange)* and renal vein *(blue).* At the bottom is a pulsatile tracing of blood flow in the renal artery. Note the site of the renal artery blood flow measurement (between the *arrows*).

Figure 17.12. Color Doppler Ultrasound Exhibit *(cont.)*

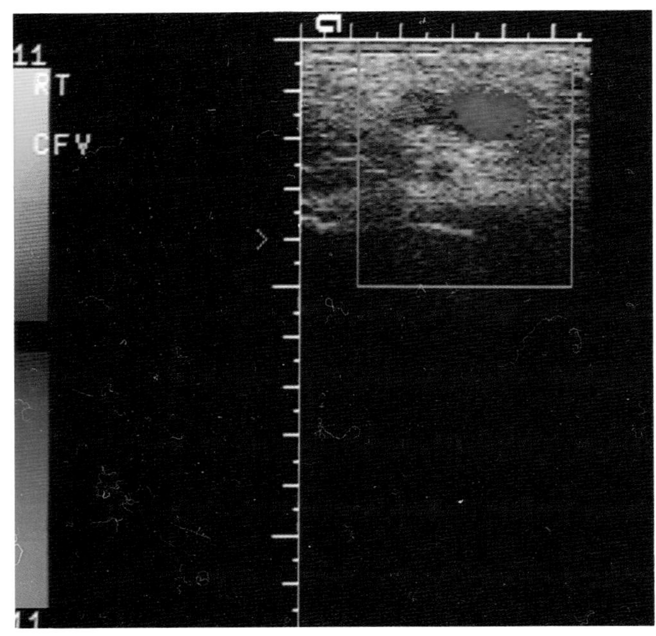

F. Normal scan of the femoral area showing blood flow in the femoral artery *(red)* and femoral vein *(blue).*

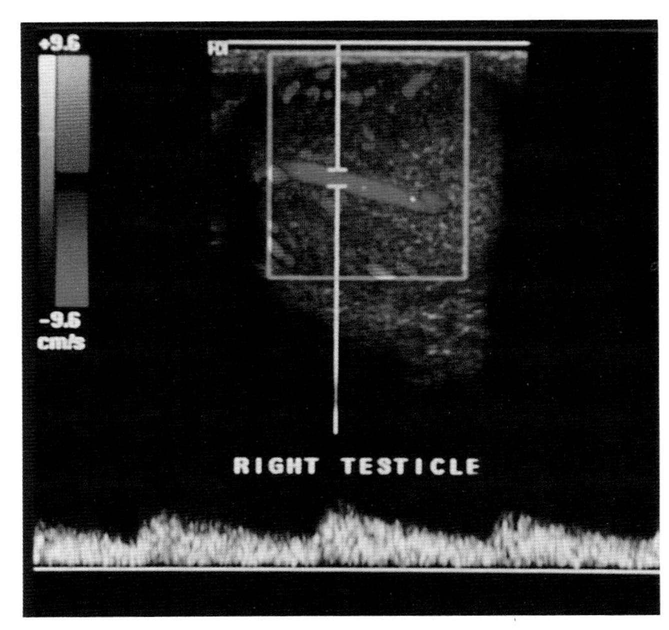

G. Normal testicular scan showing blood flow within the testicle *(multiple color signals).*

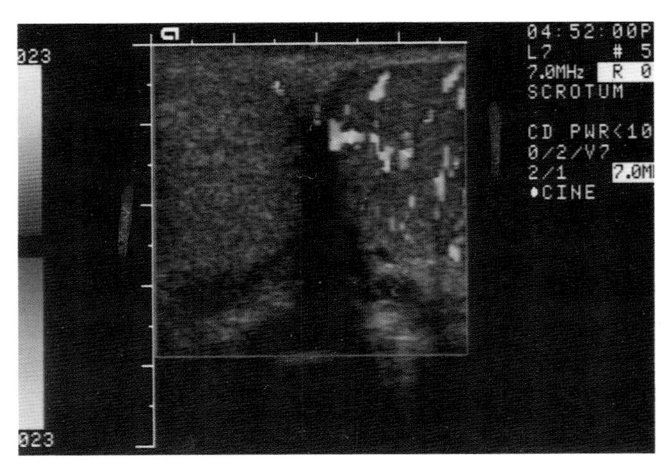

H. Coronal scan of a child with right testicular torsion. No blood flow can be detected within the right testicle *(no color signals);* multiple color signals are seen in the normal left testicle.

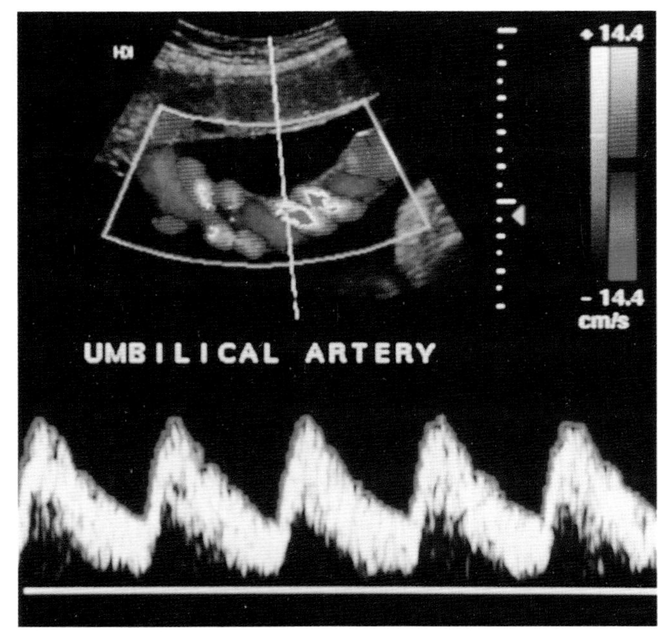

I. In utero scan of an umbilical cord showing blood flow in the umbilical arteries *(red)* and veins *(blue).*

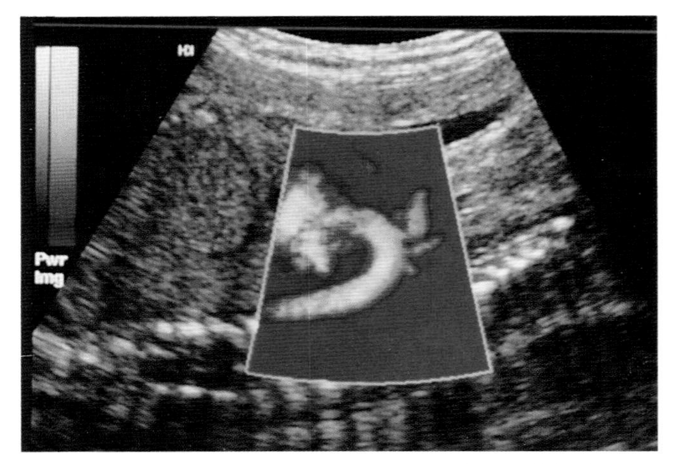

J. In utero scan of a fetus showing blood flow within the heart and aortic arch.

MR Angiography

As you have learned previously, moving blood produces no signal on a conventional spin-echo MR scan and the lumina of vessels consequently appear black (owing to the absence of white MR signals) (Figure 17.13). Other MR sequences can turn flowing blood into a strong signal, viewed as "white blood." Three-dimensional software programs can convert a stack of contiguous MR slices of white blood vessels into a three-dimensional angiographic model. This procedure is called magnetic-resonance angiography (MRA). The computer can also rotate the 3D vascular model so that individual vascular structures can be viewed from different directions and separated from overlapping neighboring vessels.

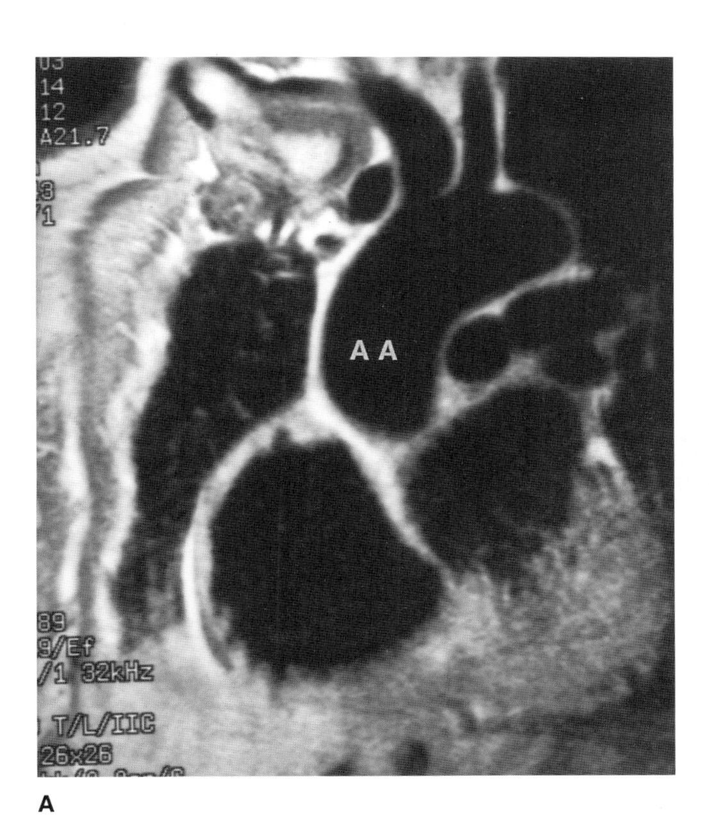

A

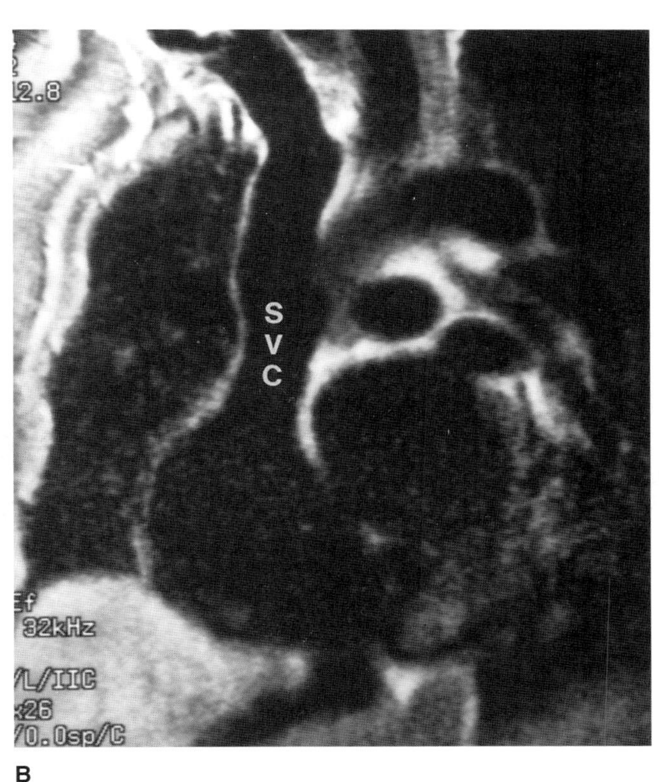

B

Figure 17.13. Cardiac MR scan showing "black" blood (absent MR signal) in the blood vessels and cardiac chambers. A: Scan at the level of the ascending aorta *(AA)*. B: Scan at the level of the superior vena cava *(SVC)*.

MRA has proven most valuable for neurovascular imaging of the blood vessels of the neck and brain. Routinely today, MRA is used to diagnose carotid artery stenoses (Figure 17.14), cerebral aneurysms, and cerebral arteriovenous malformations. MRA often does not require any exogenous contrast material and in such cases the procedure can be repeated without any risks to the patient. When contrast material is required, a gadolinium-based MR agent is used. Consequently, MR is a good way to image the blood vessels of patients in renal failure, who cannot be given the iodinated vascular contrast material that is required for conventional angiography. MRA is also useful in evaluating aortic abnormalities, such as dissections and aneurysms, and atherosclerotic peripheral vascular disease in the lower extremities. Because ventilatory motion and peristalsis degrade MRA images, MRA is less useful in imaging the abdomen and chest.

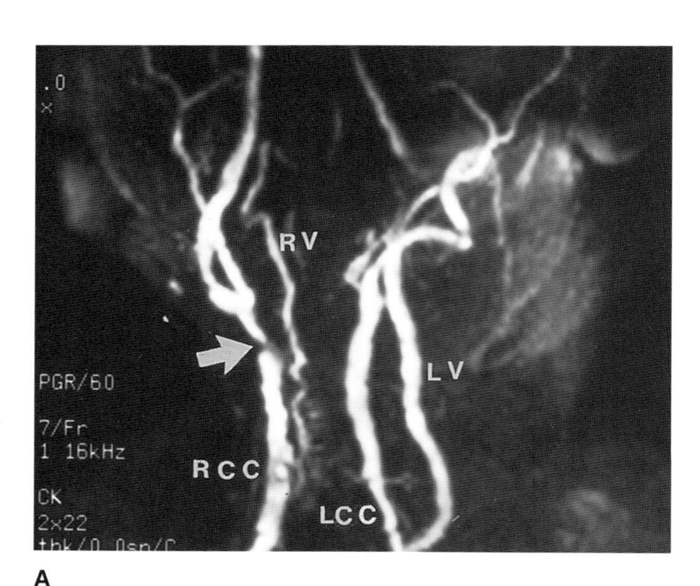

A

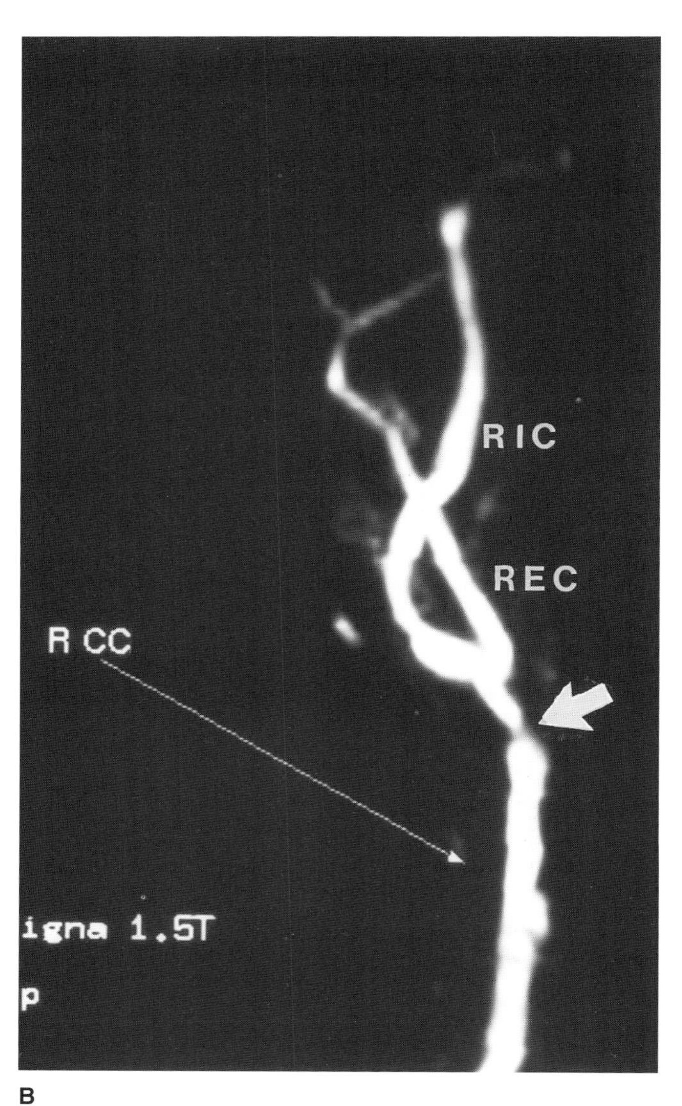

B

Figure 17.14. Magnetic-resonance angiography showing a moderate stenosis *(thick white arrow)* of the right common carotid artery proximal to the bifurcation. A: Anterior MRA image of all 4 neck arteries to the head: *RCC* (right common carotid artery), *LCC* (left common carotid artery), *RV* (right vertebral artery), and *LV* (left vertebral artery). The left vertebral artery is dominant and much larger than the right vertebral artery. B: Isolated image of the right common carotid artery *(RCC)*. The right internal carotid artery *(RIC)* and external carotid artery *(REC)* branches are labeled. The intracranial branches are not shown here.

CT Angiography

Blood vessels are well shown in cross section by CT, especially when the procedure is performed with intravenous contrast material (Figure 17.15). You have seen many examples already in previous chapters. A newer technique for displaying blood vessels imaged by CT is *computed tomography angiography,* or *CTA,* which employs the same computer reconstruction principles as MRA. To perform CTA, the radiologist rapidly scans a region of anatomy (preferably with a fast spiral or helical scanner) to obtain very thin, contiguous CT slices while the patient is injected with an infusion of contrast material to opacify the vascular system. On the resulting CT slices the blood vessels are contrast-opacified and appear white. For CTA the CT computer is instructed to make a three-dimensional model of all the *white* structures. The result is a 3D model of the blood vessels, and also of the bones, because bones and contrast-opacified blood vessels have overlapping CT attenuations. If bones are included on the contributory CT slices and appear on the final 3D model, they can be removed by giving additional directions to the CT computer, so that only the vascular model remains (Figure 17.16).

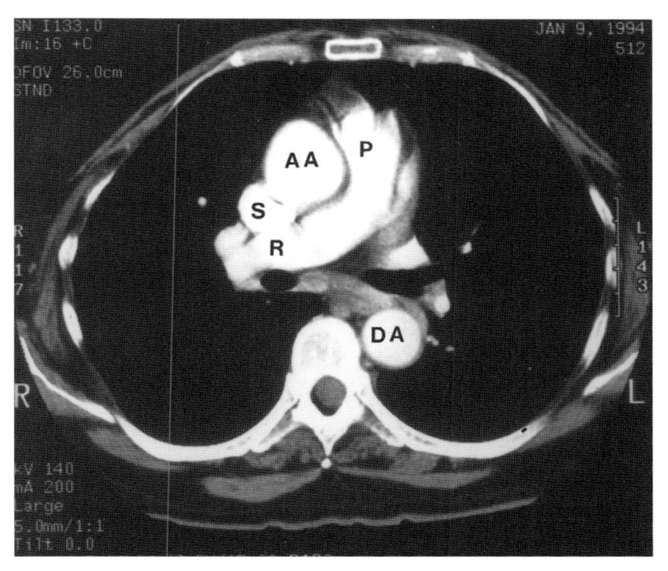

Figure 17.15. Axial CT scan of the chest, performed with intravenous contrast material, at a level just below the aortic arch. *AA,* ascending aorta; *DA,* descending aorta; *S,* superior vena cava; *P,* main pulmonary artery; *R,* right pulmonary artery.

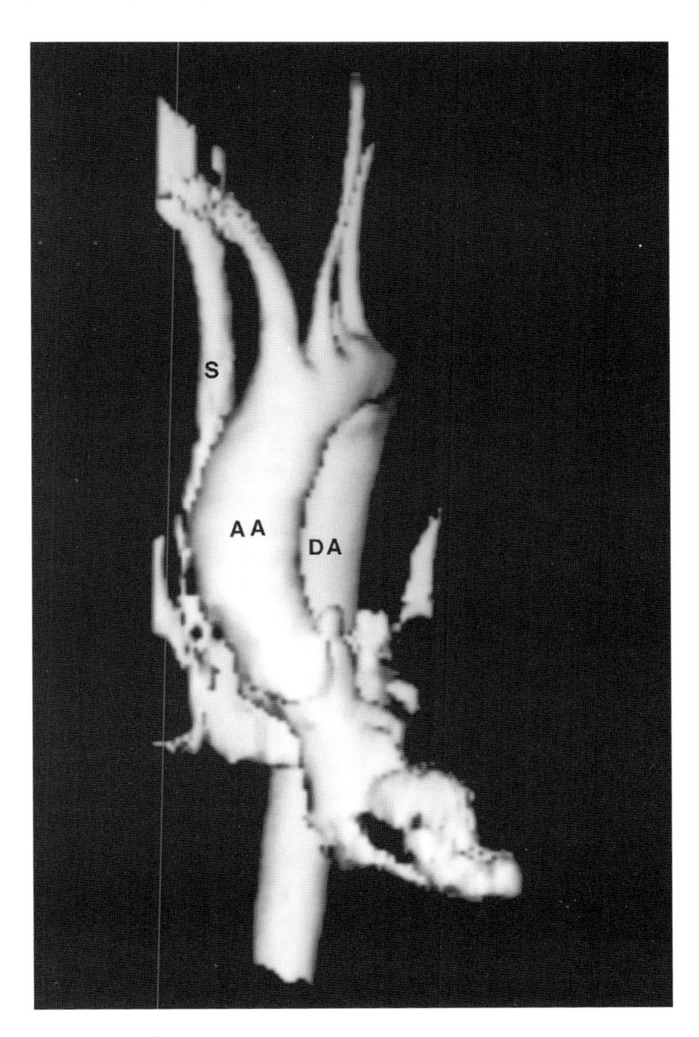

Figure 17.16. Computed tomography angiography of the chest. *AA,* ascending aorta; *DA,* descending aorta; *S,* superior vena cava.

Arterial Anatomy

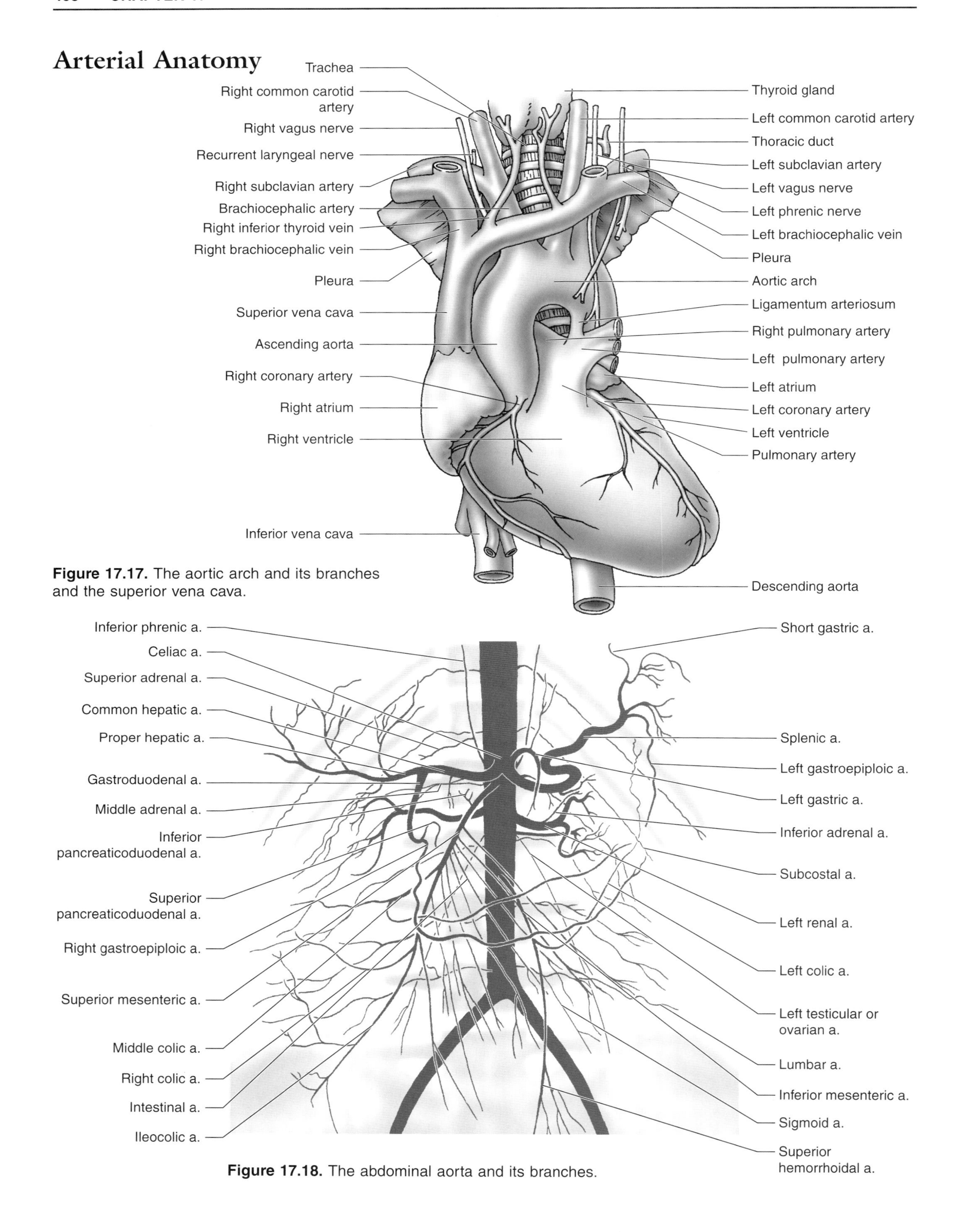

Figure 17.17. The aortic arch and its branches and the superior vena cava.

Figure 17.18. The abdominal aorta and its branches.

Before studying abnormalities of the arterial system, it will be helpful for you to review basic arterial anatomy. (We will consider venous anatomy later.) You may wish to compare the vascular anatomy drawings on pages 468–471 with the vascular images you have already seen as well as with the normal CT images in Chapter 3 and the MR images in Chapter 10.

The *thoracic aorta* (Figure 17.17) arises from the aortic valve, curves upward *(ascending aorta)* to form an *arch,* and then descends downward *(descending aorta)* through the diaphragm to become the abdominal aorta. The left and right coronary arteries arise from dilatations just above the aortic valve leaflets called the sinuses of Valsalva. Three aortic branches supplying the arms and head arise from the arch: the brachiocephalic artery (innominate artery), the left common carotid artery, and the left subclavian artery.

The *abdominal aorta* (Figure 17.18) begins at the diaphragm and extends caudally to its bifurcation into the right and left common iliac arteries. The internal iliac arteries give off several branches supplying the pelvis (Figure 17.19); the external iliac arteries continue on, eventually becoming the common femoral arteries supplying the lower extremities.

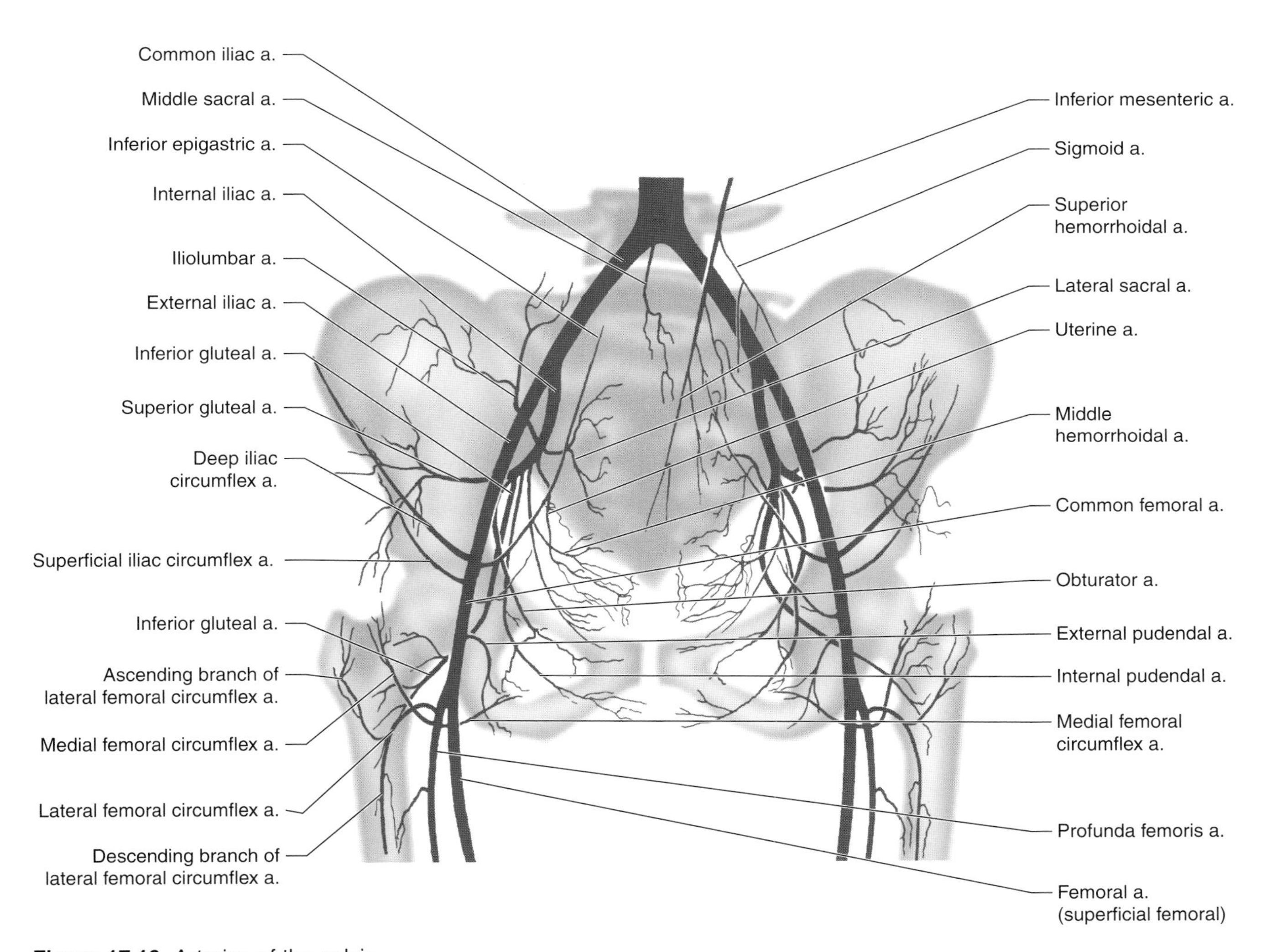

Figure 17.19. Arteries of the pelvis.

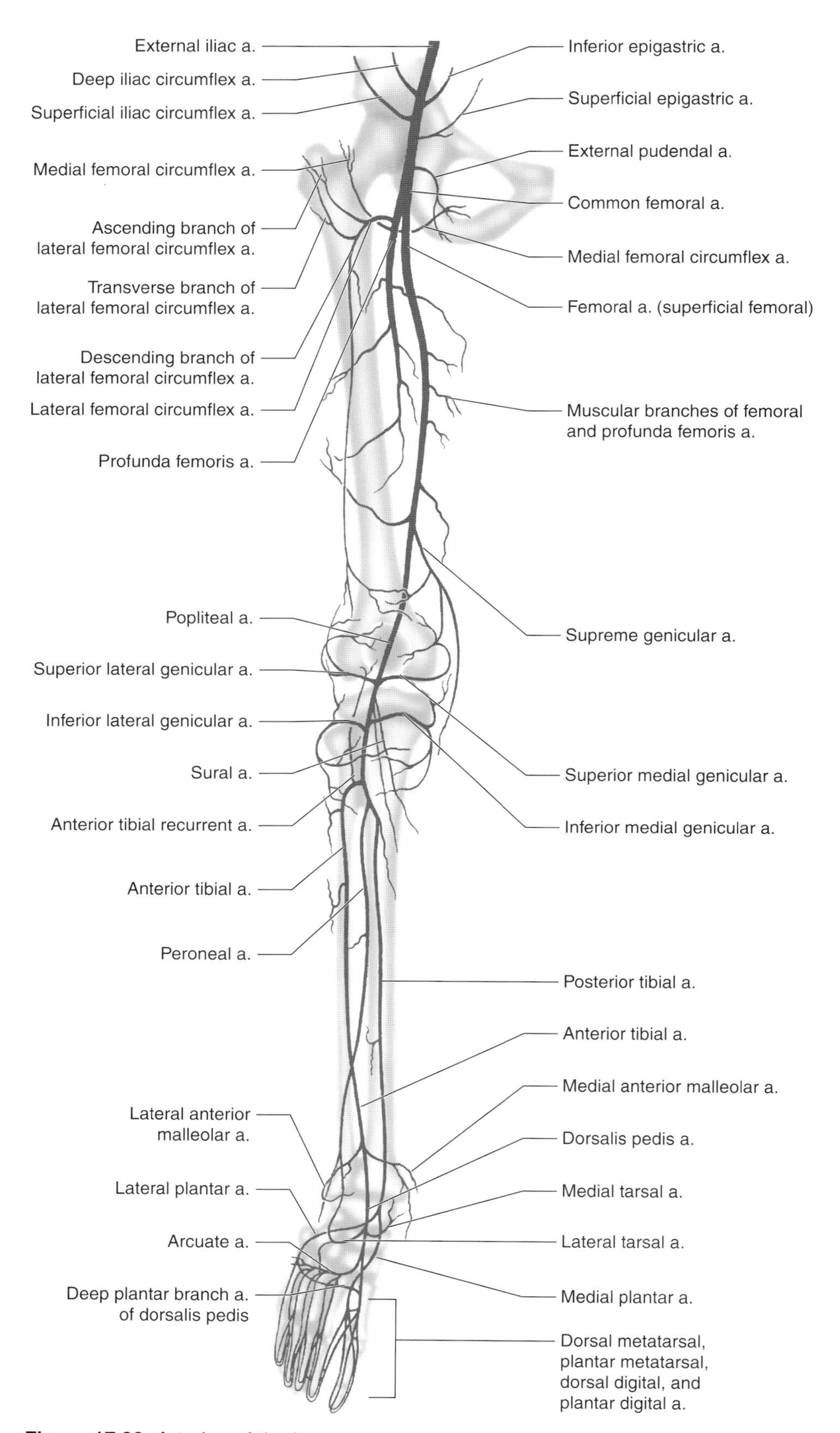

Figure 17.20. Arteries of the leg.

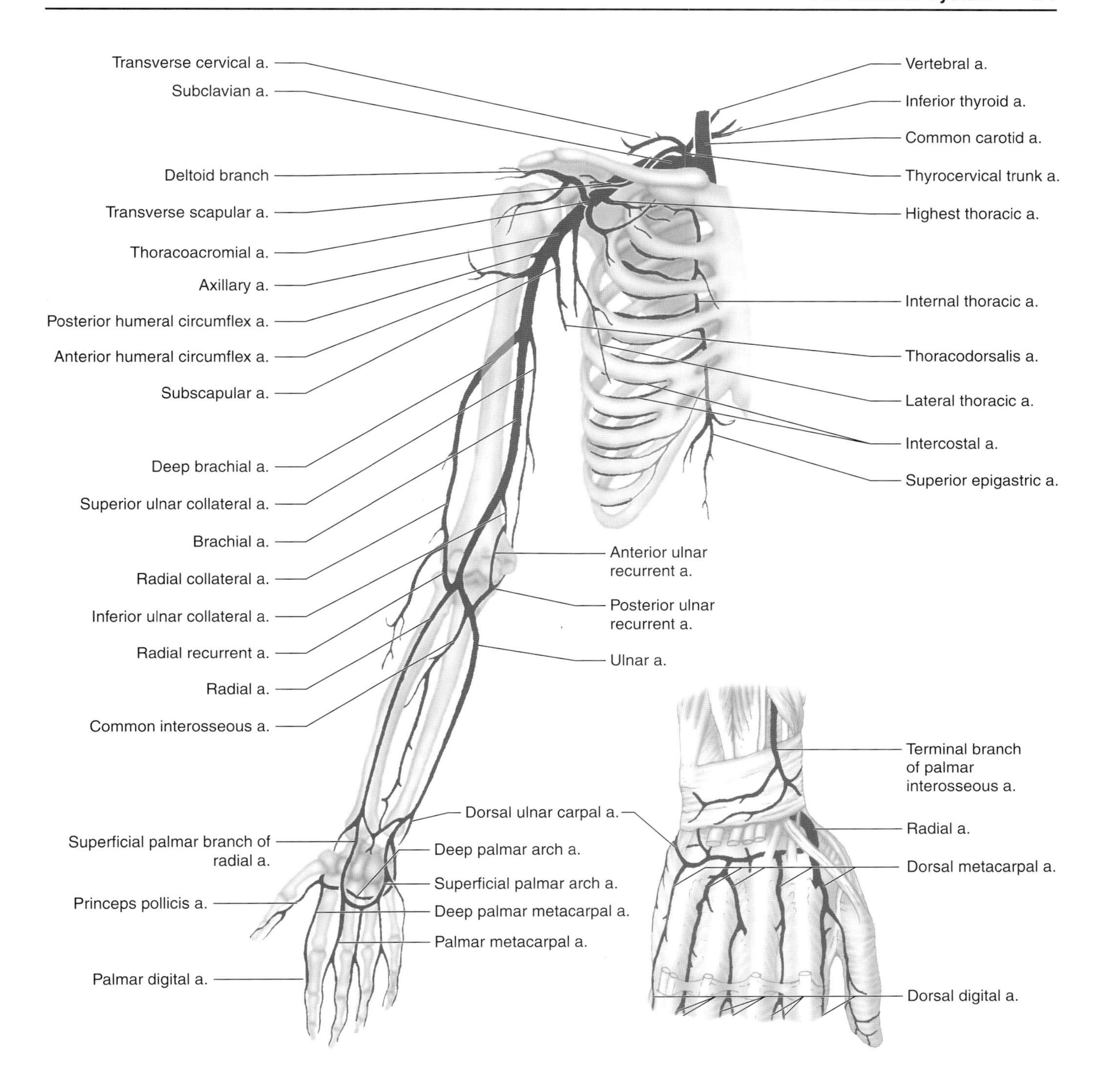

Figure 17.21. Arteries of the arm.

The arteries to the legs are shown in Figure 17.20 and the arteries to the arm in Figure 17.21. You do not need to memorize all these blood vessels right now, but you may wish to use these illustrations as a reference when observing angiographic procedures and reviewing angiographic images.

Aortic Aneurysm

Most aortic aneurysms involve the abdominal aorta and are caused by atherosclerosis. Aortic wall ischemia and weakening leads to progressive dilatation of the aortic wall and lumen and then to formation of an aneurysm. As the aortic lumen dilates, the linear blood flow becomes turbulent and the developing aneurysm may become lined with thrombus. Atherosclerotic abdominal aortic aneurysms are usually *fusiform* (Figure 17.22), although occasionally they are *saccular;* most saccular aneurysms, however, are caused by trauma, surgery, or infection. The major complications of an abdominal aortic aneurysm are leakage and rupture, causing massive hemorrhage and death. Atherosclerotic aneurysms may also occur in the thoracic aorta, and aneurysms of all types may involve almost any artery in the body.

An abdominal aortic aneurysm (AAA) is clinically suspected when a midline, palpable, and pulsatile abdominal mass is discovered on physical examination. It may be an asymptomatic incidental finding on a routine examination, or it may be palpated in a patient who seeks medical help for the acute onset of lower back or abdominal pain. This is an important differentiation, because it will alter the initial imaging workup. Asymptomatic aneurysms are usually not emergencies, and the clinically suspected aneurysm can quickly be confirmed with an ultrasound examination. You have already learned that some aneurysms calcify and therefore can be identified on plain films; however, most do not. Fortunately, ultrasound not only can detect nearly all abdominal aortic aneurysms, but can also indicate their size accurately and demonstrate the presence or absence of thrombus within the lumen.

Figure 17.23 shows an example of a patient, a 69-year-old man, with an asymptomatic large pulsatile abdominal mass. Note that ultrasound (A and B) can visualize the aneurysm in both axial (transverse) and sagittal (longitudinal) planes, and can clearly differentiate thrombus within the aneurysm from the echo-free lumen containing flowing blood. The aortic branches, however, are not well seen on ultrasound images, and any free blood in the retroperitoneum that may result from a leak cannot be as accurately identified by ultrasound as by CT. In most individuals

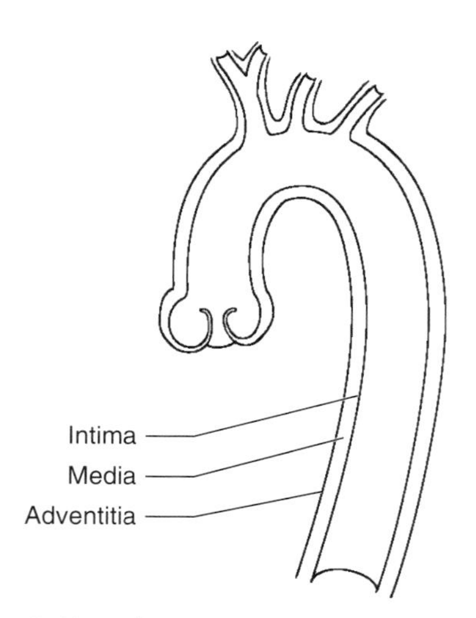

A. Normal aorta.

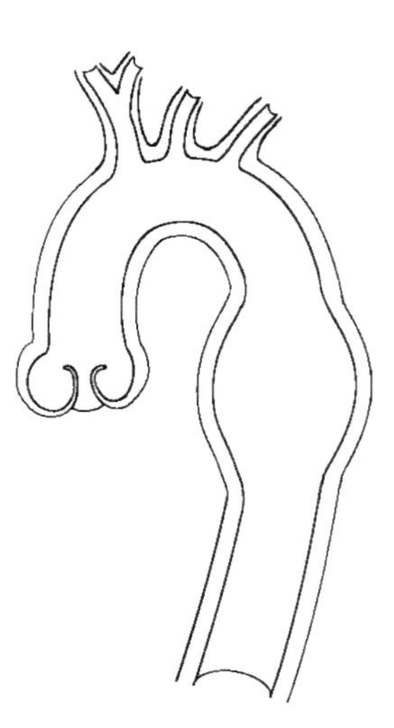

B. Fusiform aneurysm.

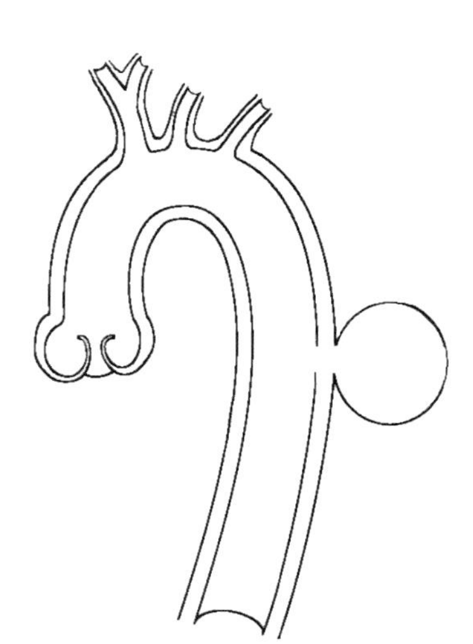

C. Saccular aneurysm.

Figure 17.22. Normal aorta (A), fusiform aortic aneurysm (B), and saccular aortic aneurysm (C). A fusiform aneurysm is characterized by a spindle-shaped dilatation of the aorta that includes dilatation of all three layers of the aortic wall (intima, media, adventitia). A saccular aneurysm is often eccentric, with a neck that is smaller than the maximum width of the aneurysm. Saccular aneurysms may not be lined with all three layers of the arterial wall. In the one shown in C, the intima and medial have been injured and only a layer of adventitia invests the aneurysm.

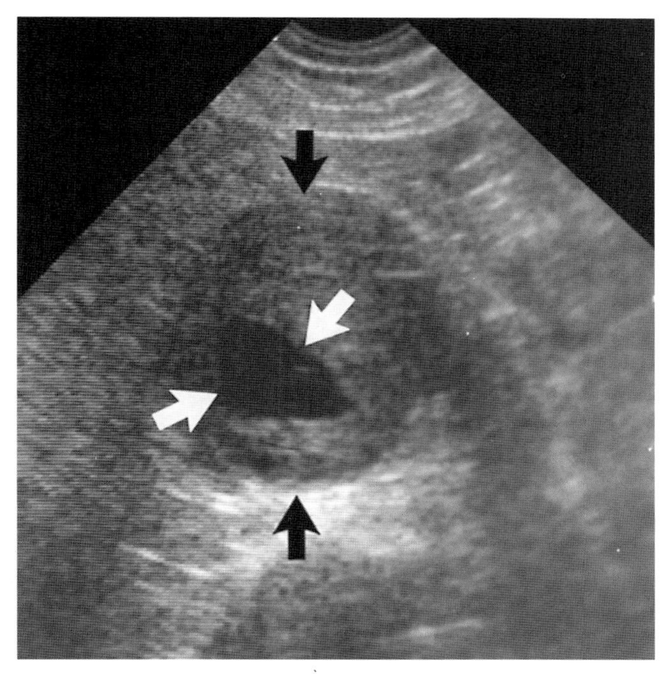

A

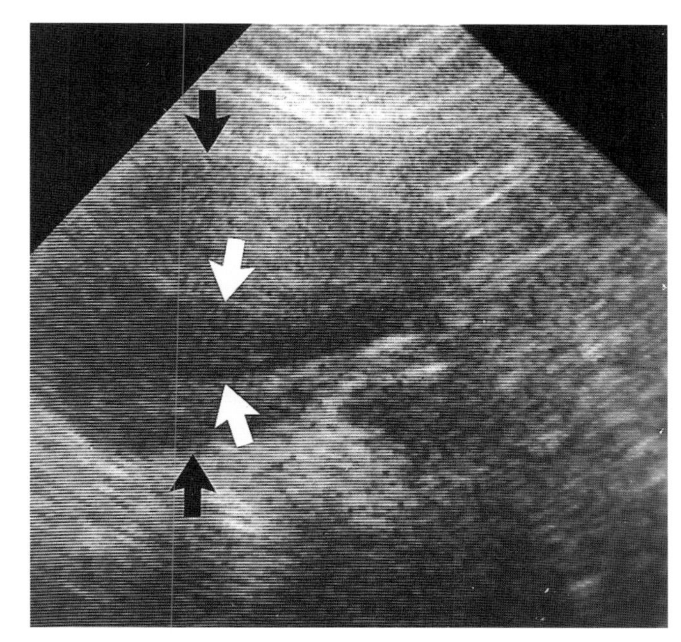

B

Figure 17.23. Large asymptomatic abdominal aortic aneurysm. A: Transverse ultrasound shows a large aneurysm in the retroperitoneum, just anterior to the spine. The *black arrows* indicate the overall size of the aneurysm; the *white arrows* indicate the diameter of the lumen (echolucent), containing free-flowing blood. The echogenic material between the two pairs of arrows represents thrombus lining the interior of the aneurysm. B: Midline sagittal ultrasound; the patient's head is to your left and his feet to your right. Again, the *black arrows* indicate the overall size of the aneurysm and the *white arrows* the lumen. C: Lateral aortogram best relates the location of the aneurysm to the origins of the major aortic branches. Note that the aneurysm begins well below the superior mesenteric artery *(sma),* which courses anteriorly, and just below the renal artery branches *(black arrows),* which course posteriorly. The celiac artery can be seen just above the superior mesenteric artery.

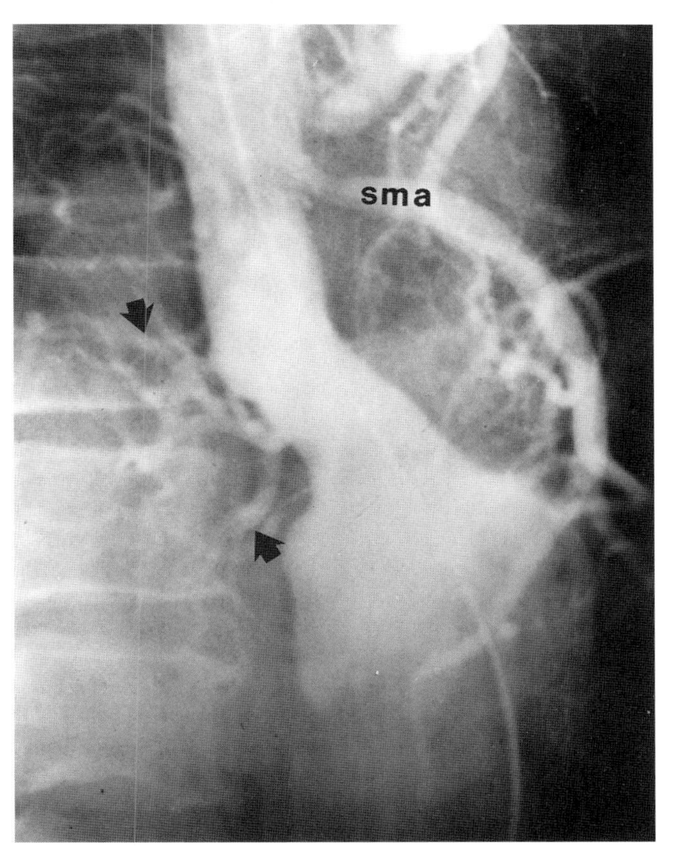

C

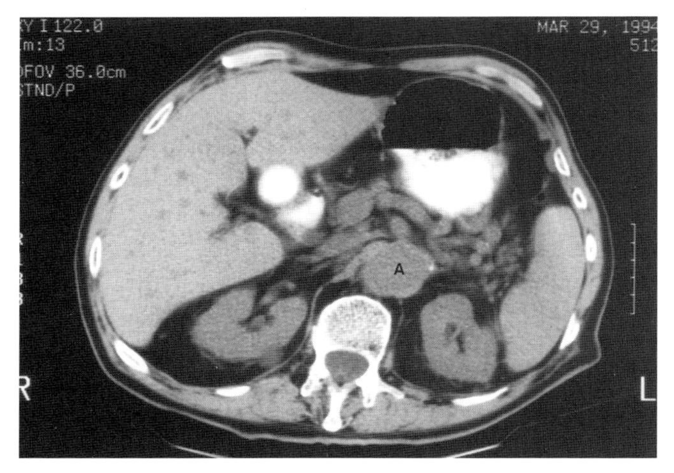

A

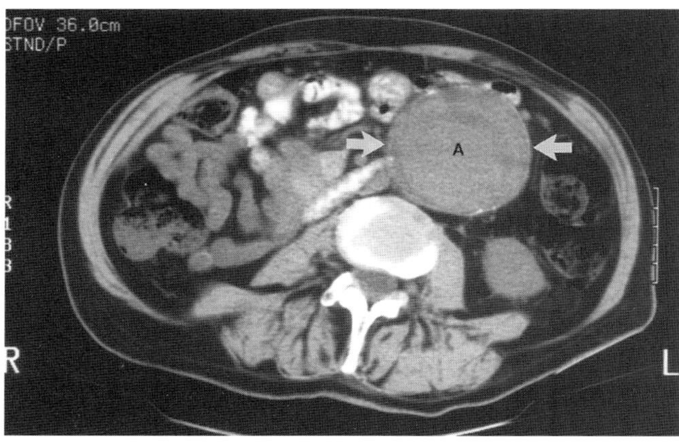

B

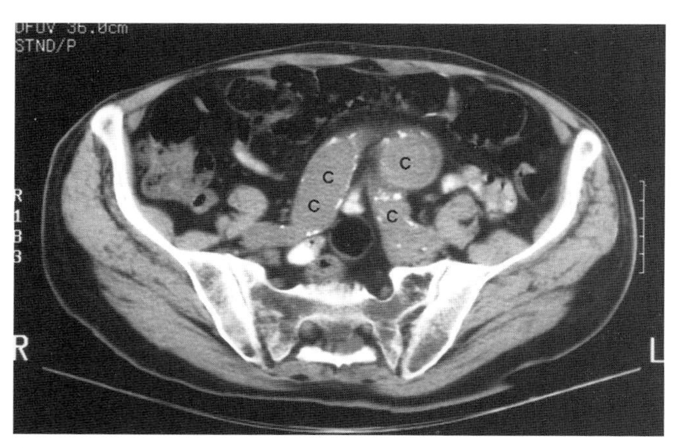

C

Figure 17.24. CT scans of an asymptomatic abdominal aortic aneurysm. A: Scan at the level of the renal arteries shows a normal caliber aorta *(A).* B: Scan several centimeters lower reveals a large aortic aneurysm *(A, arrows).* C: Scan just below the aortic bifurcation shows aneurysmal dilatation of both common iliac arteries (*C*'s). It can be concluded from this CT examination that the AAA begins below the renal arteries and extends beyond the aortic bifurcation to involve the iliac arteries.

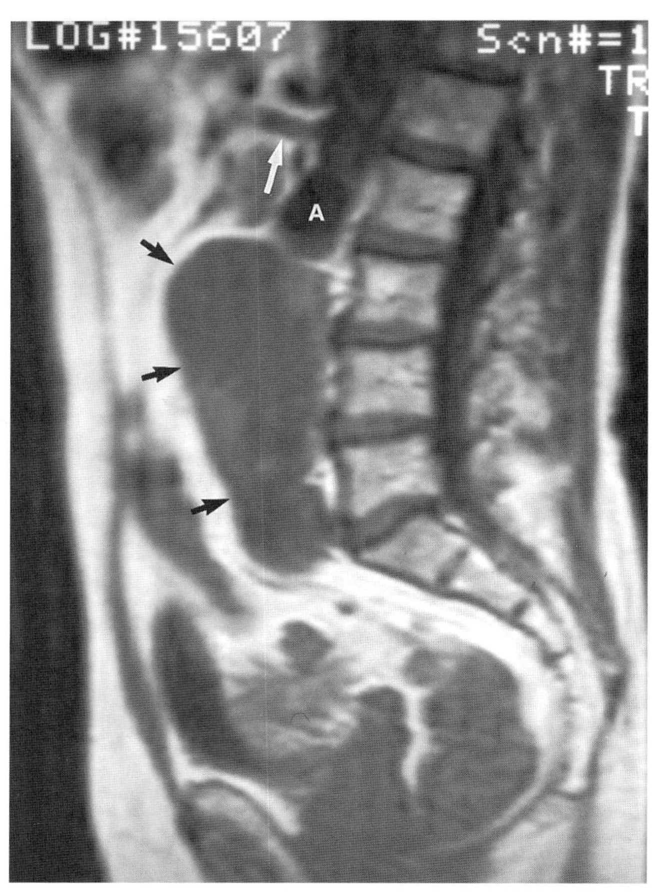

A

Figure 17.25. MR scans of an asymptomatic abdominal aortic aneurysm. A *(above):* Lateral scan shows a large abdominal aortic aneurysm *(black arrows)* anterior to the spine. Note the normal caliber aorta *(A)* above the aneurysm, its superior mesenteric artery branch *(white arrow)* coursing anteriorly. B *(top of facing page):* Coronal scan shows that the aneurysm *(black arrows)* extends beyond the aortic bifurcation into the common iliac arteries (*C*'s), which are shown to be dilated.

an abdominal aortic diameter at ultrasound larger than 3 centimeters is consistent with an aneurysm. Asymptomatic abdominal aortic aneurysms can also be detected by CT (Figure 17.24) and MR (Figure 17.25).

A patient with a symptomatic or tender aneurysm should be examined with an emergency CT scan. The man shown in Figure 17.26 presented in an emergency center with acute back and right flank pain. On physical examination a tender, pulsatile abdominal mass was discovered. CT not only confirmed the presence of the aneurysm but showed free blood in the retroperitoneum consistent with an acute leak. The patient was taken immediately to the operating room, where the leaking aneurysm was successfully treated with a bypass graft.

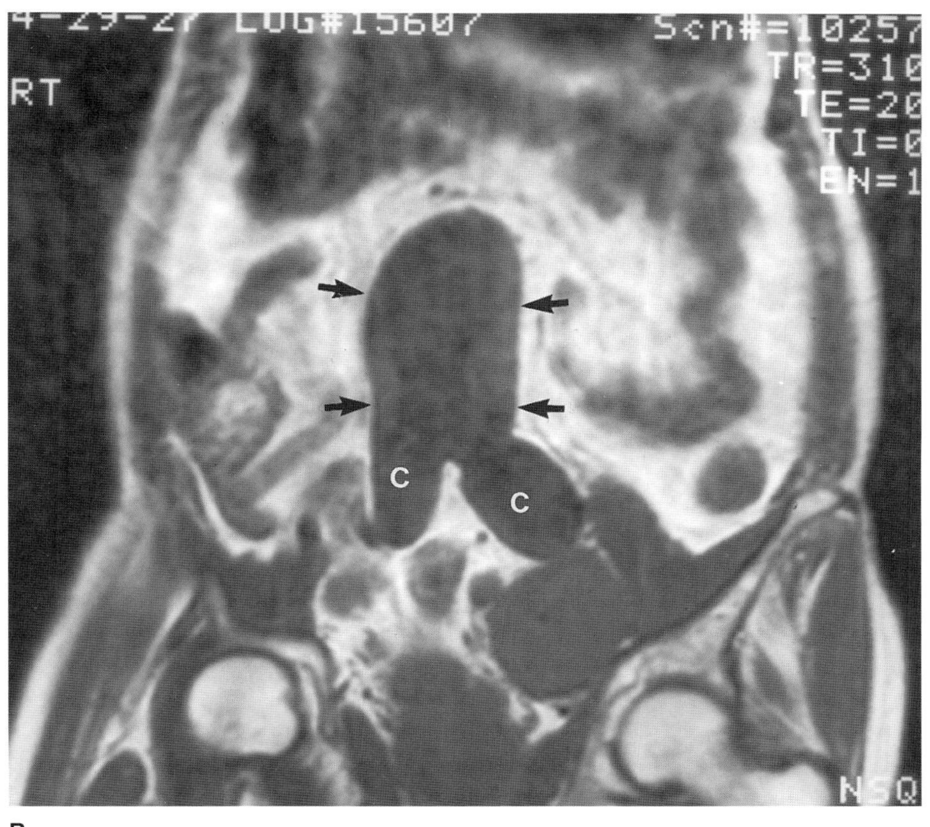

B

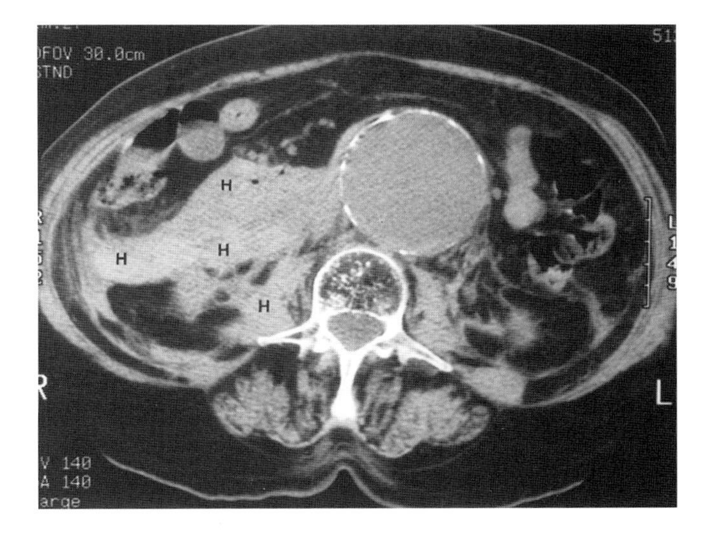

Figure 17.26. CT scan of a leaking abdominal aortic aneurysm. This patient presented with sudden onset of severe back and right flank pain; a pulsatile midline abdominal mass was detected on physical examination. The CT scan showed a large calcified aneurysm with hemorrhage (*H*'s) into the retroperitoneal and peritoneal cavities.

Remember that *a patient with a clinically obvious aneurysmal leak or rupture and unstable vital signs should not be referred to CT, but should be taken directly to the operating room.* Emergency CT should be reserved for stable patients with pain and a clinically suspected aneurysm in whom the diagnosis of a leak is uncertain. This group includes patients with suspected aneurysms whose acute back, flank, or abdominal pain could be caused by a urinary tract stone, diverticulitis, a ruptured lumbar disc, or several other conditions. CT can prevent many patients from undergoing unnecessary emergency laparotomies by demonstrating that their aneurysms are not leaking and can be repaired by elective surgery.

For patients whose aneurysms are scheduled for elective repair, an imaging procedure is generally performed to show the relationship of the aneurysm to the branches of the abdominal aorta. The technique

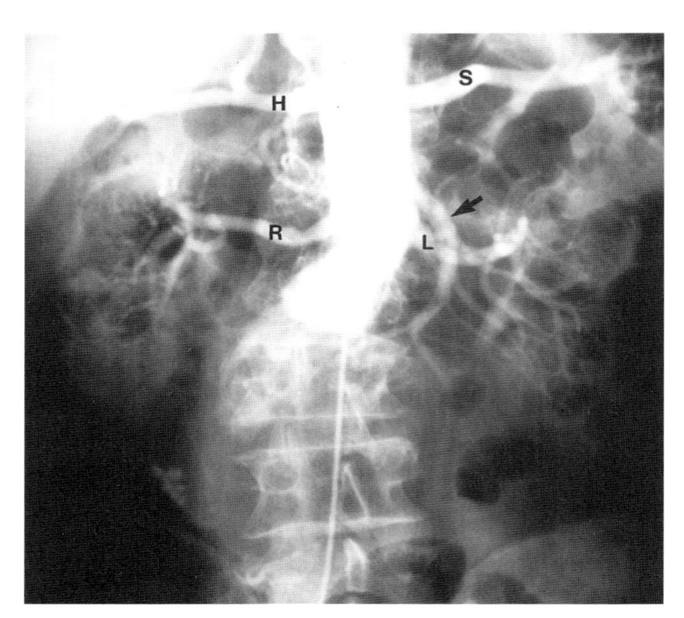

A

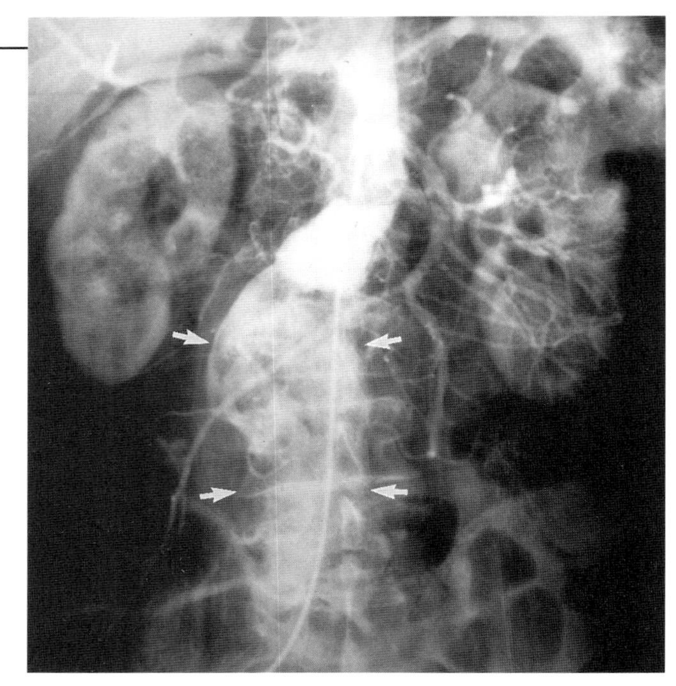

B

has traditionally been conventional catheter arteriography (Figure 17.27), but CT angiography (Figure 17.28) or an MR scan can also accomplish the task.

Have you noticed already that arteriography shows only the lumen of the aneurysm and cannot determine its overall size? The arteriogram opacifies only the space indicated by the *white arrows* on the ultrasound exam in Figure 17.23A and B. Note also that angiography does not show free blood in the retroperitoneum when a leak is present; CT is the appropriate procedure to perform if a leak is suspected.

Another type of aneurysm, the mycotic aneurysm, is caused by the growth of microorganisms within the arterial wall, resulting in weakening of the artery and aneurysm formation. Mycotic aneurysms are nearly always saccular. Patients often present with a fever of unknown origin (FUO). Common among the predisposing factors are septicemia associated with intravenous drug abuse, bacterial endocarditis, and an immunocompromised state. Figure 17.29 illustrates the imaging of a patient with an FUO due to a mycotic aneurysm. Initially the patient was examined with a technetium-labeled immunoglobulin-G (Tc-IGG) radioisotope scan to assist in identifying the site of infection. This isotope concentrates at sites of infection and inflammation. When such a site was seen overlying the abdominal aorta, an abdominal CT examination was requested; it showed a mass adjacent to the aorta. An arteriogram revealed the mass was a saccular aneurysm, which proved to be mycotic in origin on pathological examination of the surgical specimen.

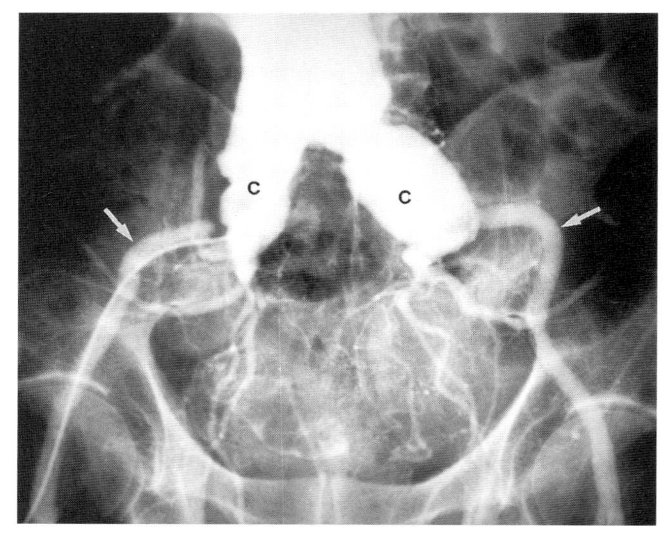

C

Figure 17.27. Conventional arteriography of an asymptomatic abdominal aortic aneurysm. A: Early-phase abdominal aortogram shows opacification of the normal caliber upper abdominal aorta. Find the common hepatic *(H)* and splenic *(S)* branches of the celiac artery, the superior mesenteric artery *(arrow),* and the right *(R)* and left *(L)* renal arteries. The aneurysm, located below the renal arteries, is just starting to fill with contrast material. Did you note the injection catheter inserted from below coursing within the aneurysm? B: Later-phase abdominal aortogram shows better opacification of the aneurysm lumen *(arrows).* Remember that angiography opacifies only the lumen of an aneurysm; it cannot determine its outside diameter, as can ultrasound, CT, or MR. C: Pelvic arteriogram with contrast injection within the aneurysm lumen (catheter pulled down), just above the aortic bifurcation. The aneurysm extends to involve both common iliac arteries (*C*'s), but the external iliac arteries *(arrows)* are of normal caliber. Does this aneurysm look familiar? It is the one shown in Figure 17.25.

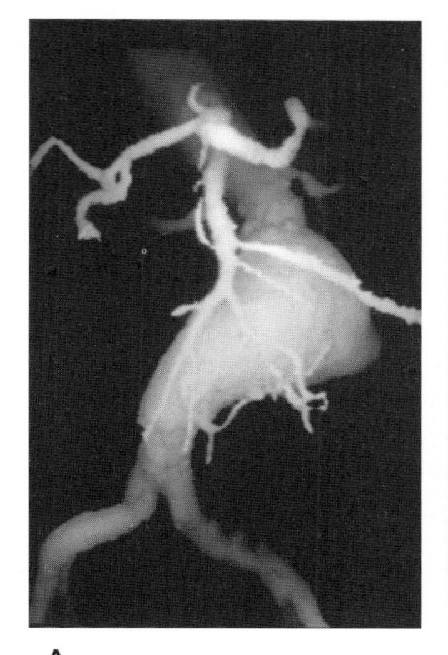

A

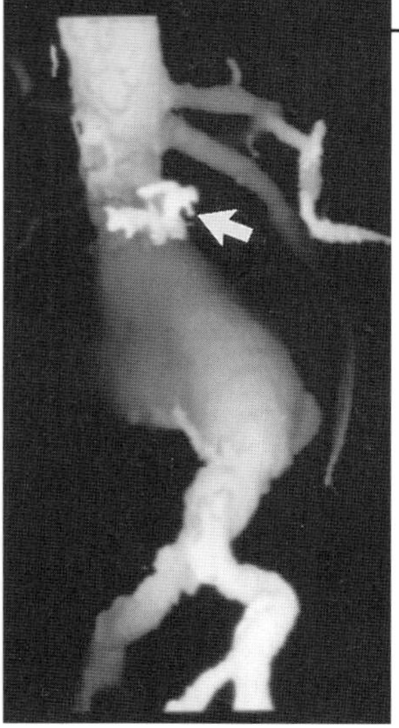

B

Figure 17.28 *(left).* CTA (CT aortogram) of an asymptomatic abdominal aortic aneurysm. A: Frontal view shows that the aneurysm begins below the celiac, superior mesenteric, and renal branches. Identify them. Note also that the aneurysm does not extend across the aortic bifurcation into the common iliac arteries. B: Lateral view (anterior to your right) shows the celiac and superior mesenteric arteries coursing anteriorly above the aneurysm; the right renal artery *(arrow)* is extending toward the viewer.

Figure 17.29 *(below).* Mycotic aneurysm of the abdominal aorta (see text). A: Tc-IGG radioisotope scan shows abnormal uptake *(arrow)* overlying the abdominal aorta. Normal liver *(L)* activity is seen above it. B: CT scan at the level of the renal arteries shows a normal aorta *(A).* C: CT scan at a slightly lower level shows a mass *(arrows)* contiguous with the aorta *(A).* D: Abdominal aortogram reveals that the mass is a saccular aneurysm *(arrow).*

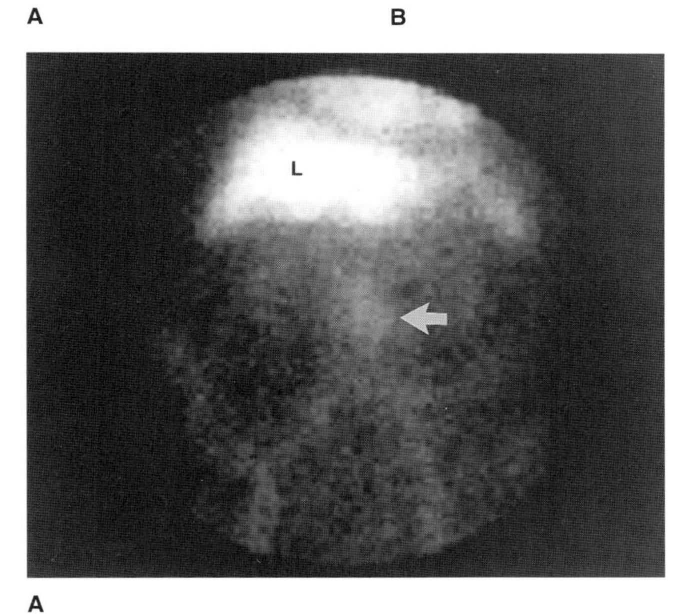

A

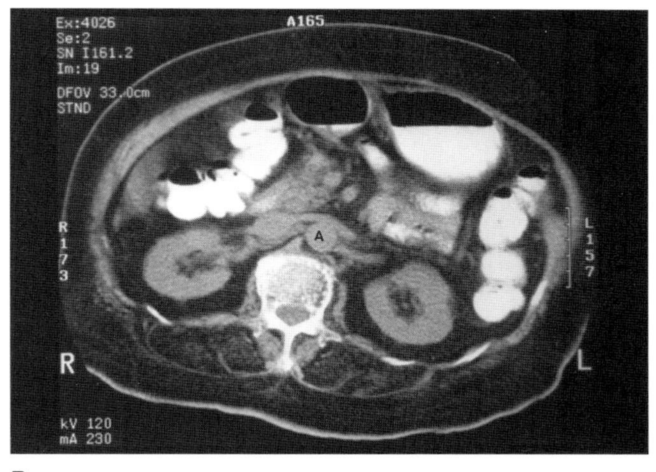

B

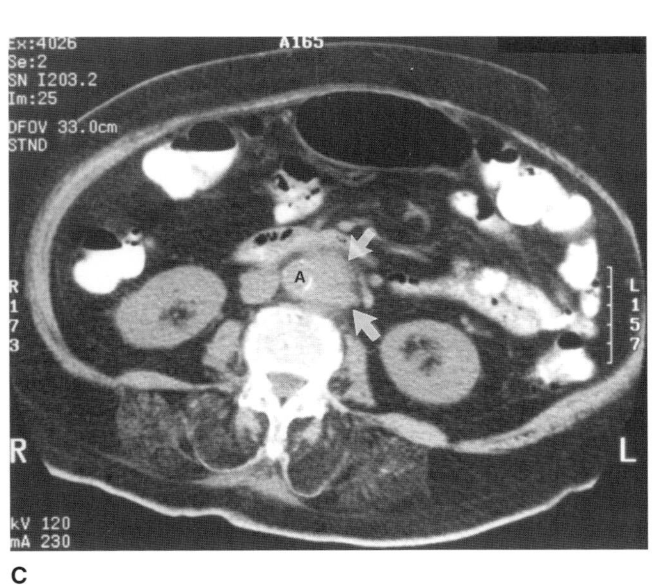

C

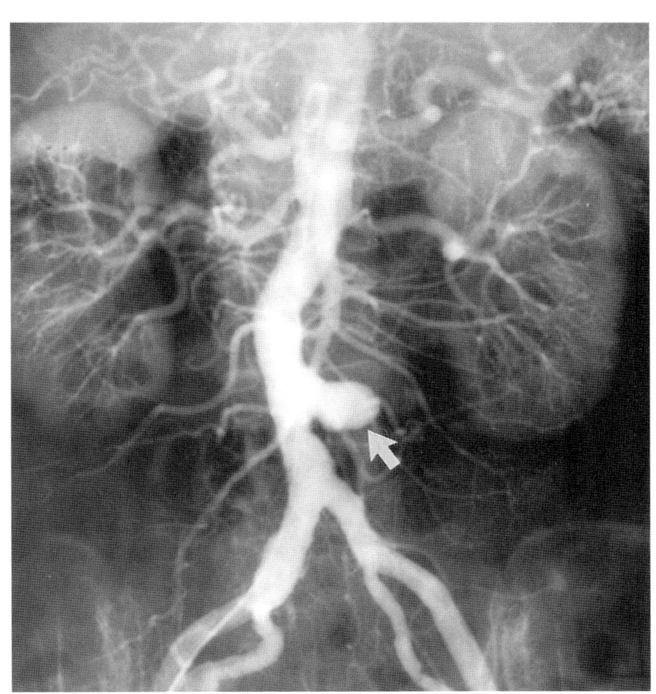

D

Aortic Dissection

The wall of the aorta, like that of all other arteries, is composed of three layers: a thin, smooth inner layer called the *intima*, a thicker muscular layer known as the *media*, and an outer layer called the *adventitia*. *Aortic dissection* is a separation of the layers of the aortic wall, usually through the outer third of the medial layer. In this process a new lumen *(false lumen)* is formed; blood may flow through this false lumen or it may thrombose, resulting in a *dissecting hematoma*. Patients with aortic dissection classically present with the sudden onset of "tearing" chest pain, radiating to the back. The clinical condition that most frequently must be differentiated from aortic dissection is acute myocardial infarction.

Dissections are categorized according to their extent and site of origin (Figure 17.30). One classification system was developed by Dr. Michael De Bakey in 1964. In his system, *Type I* dissections involve both the ascending and the descending aorta, *Type II* dissections only the ascending aorta, and *Type III* dissections only the descending aorta. Types I and II originate just above the aortic valve; Type III begins just distal to the left subclavian artery. Types I and III may extend distally into the abdominal aorta. According to a second classification system, Types I and II are combined to become *Type A* (involves the ascending aorta and is treated surgically); Type III becomes *Type B* (involves only the descending aorta and is usually treated medically).

Since a dissection extends through an aorta, individual aortic branches may become totally occluded or compromised (because of lower blood flow through the false lumen), which results in ischemia of the structures supplied by those branches. Each patient's clinical presentation will vary according to the aortic branches affected. If the dissection involves the coronary arteries, for example, myocardial infarction may occur. Involvement of the carotid arteries may result

Type I Type II Type III

Type A Type B

Figure 17.30. Types of aortic dissection (see text).

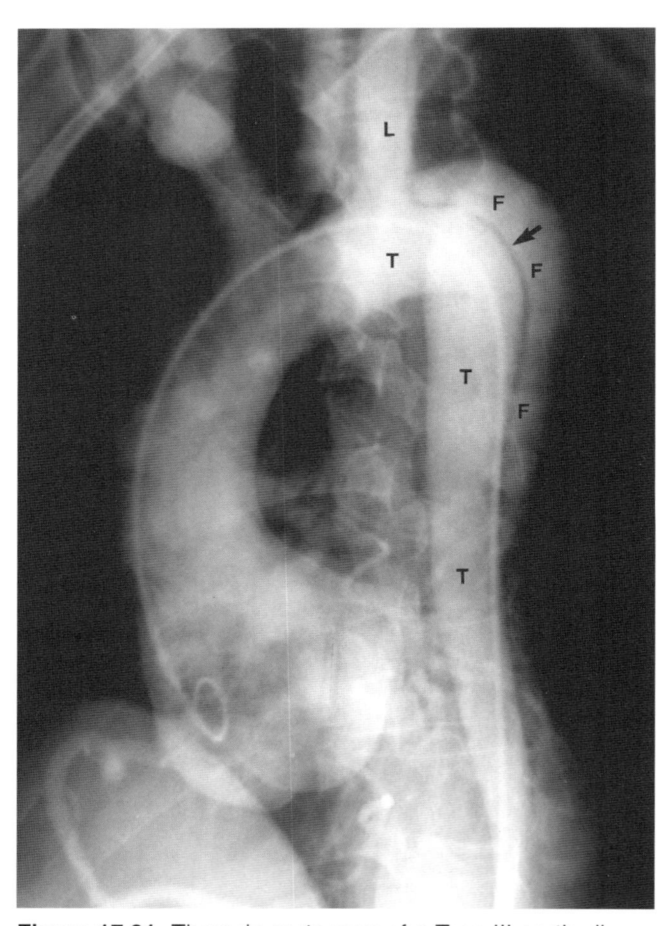

Figure 17.31. Thoracic aortogram of a Type III aortic dissection that begins at the left subclavian *(L)* artery origin. A radiolucent intimal flap *(arrow)* can be seen separating the true lumen (*T*'s) from the false lumen (*F*'s). Since both lumens are opacified with contrast material, free-flowing blood must be moving within them. A white (radiopaque) contrast injection catheter can be seen within the true lumen.

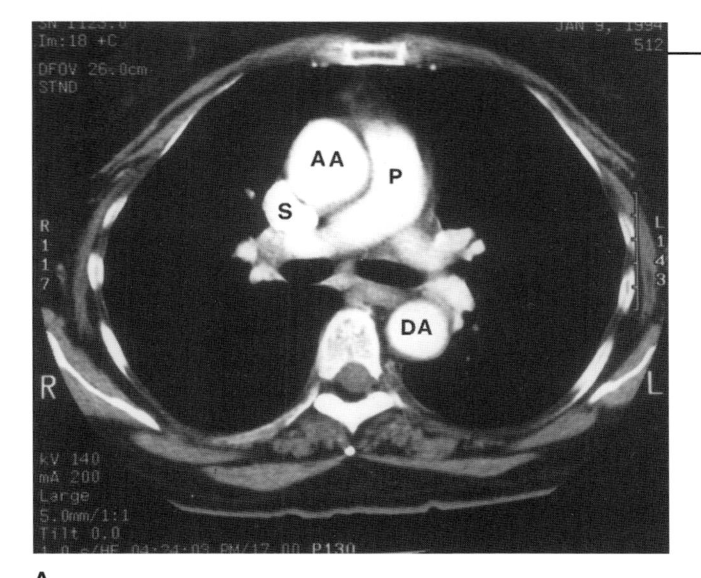

A

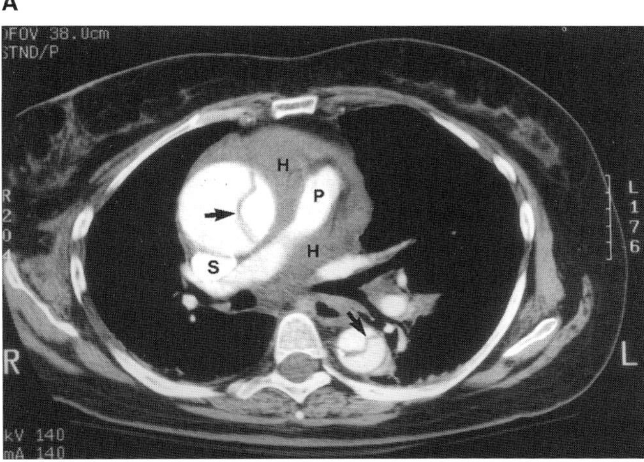

B

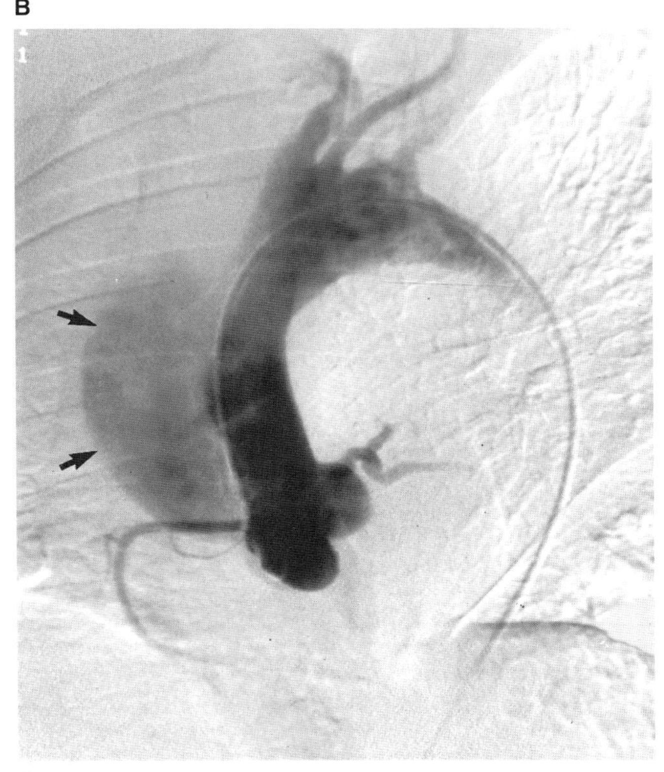

C

in a stroke, while involvement of the superior mesenteric artery may result in bowel ischemia. A catastrophic complication is leak or rupture of the thinly lined false lumen.

Aortic dissection can be diagnosed by conventional catheter arteriography, CT, MR, or an esophageal ultrasound examination. The ultrasound test, transesophageal echocardiography (TEE), uses a new type of ultrasound transducer that is inserted orally and advanced into the esophagus for scanning. The choice of procedure depends, of course, on the individual clinical situation.

Conventional catheter arteriography of the aorta (aortography), the traditional imaging technique, provides an excellent depiction of the extent of dissection with the aorta and the presence or absence of involvement of the aortic branches. Aortography offers a detailed preoperative road map for surgical planning. The most characteristic finding is the demonstration of a longitudinal *intimal flap* (Figure 17.31) separating the true aortic lumen from the false lumen. The true lumen is surrounded by intima, whereas the false lumen was established by creation of a new lumen within the media; consequently, the two are separated by a layer of intima (the intimal flap).

CT is faster and easier to perform than conventional aortography and can quickly confirm the presence of aortic dissection. It is less expensive than arteriography and no arterial catheterization is required. Compare the CT scan of the patient with a Type I dissection in Figure 17.32 with the normal scan and the digital subtraction aortogram of the dissection.

Figure 17.32. Normal CT scan and CT scan and aortogram of a Type I aortic dissection. A: Normal CT scan (slice taken a few centimeters below the aortic arch) of patient with no aortic dissection for comparison: *AA,* ascending aorta; *DA,* descending aorta; *S,* superior vena cava; *P,* pulmonary artery. B: CT scan at the same level of a patient with a Type I dissection. Note the intimal flap *(arrows)* in both the ascending aorta and the descending aorta separating the true lumen and false lumen. The ascending aorta is markedly dilated because there is aneurysmal dilatation of the false lumen of the ascending aorta with bleeding (*H*'s, hemorrhage) into the mediastinum. The superior vena cava *(S)* and pulmonary artery *(P)* are displaced posteriorly by the enlarged ascending aorta and mediastinal hematoma. C: Early-phase digital subtraction aortogram of the same patient. The injection catheter can be seen within the somewhat compressed true lumen of the ascending aorta, which is supplying both coronary arteries. A cloud of contrast material is seen flowing into the dilated false lumen *(arrows).*

Since CT requires the use of intravenous contrast material and MR does not, MR (Figure 17.33) should be considered for patients with compromised renal function. But it is important to remember that MR is difficult to perform on an acutely ill patient who requires careful monitoring. Within the bore of the magnet patients cannot be as closely observed as with CT, and an MR scan takes more time to complete than a CT scan. Transesophageal echocardiography, the newest imaging technique for visualizing aortic dissection, can (like other ultrasound examinations) be performed portably and can therefore offer a means of diagnosing aortic dissection in patients who are not sufficiently stable to be transported to the radiology department. In choosing the best procedure to diagnose aortic dissection in your future patients you would do well to consult with the radiologist.

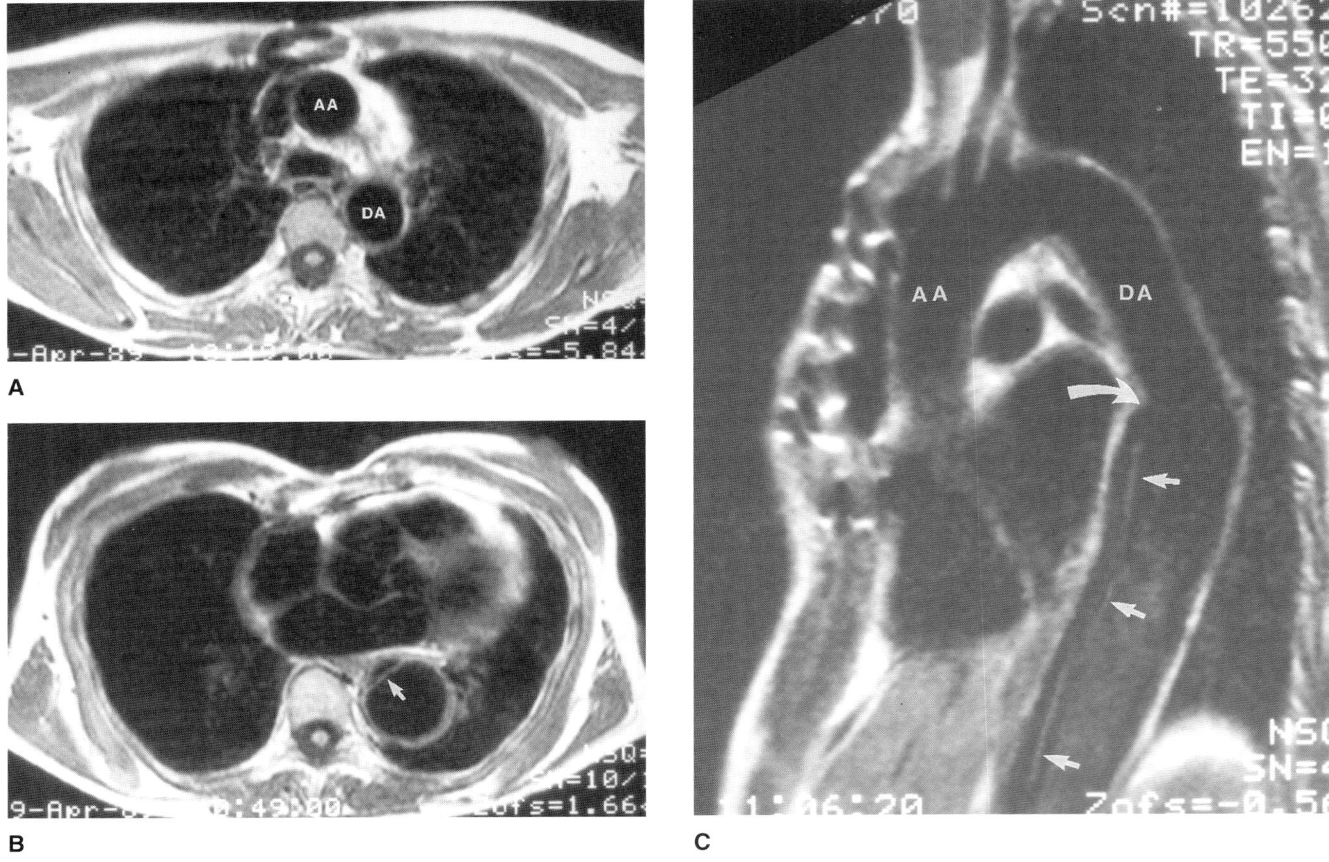

Figure 17.33. MR scans of a Type III aortic dissection. A: Axial scan of the upper chest taken just below the aortic arch shows a normal ascending aorta *(AA)* and normal descending aorta *(DA)*. B: Axial scan of the lower thorax shows a dissection of the descending aorta; the *arrow* points to the intimal flap. C: Oblique sagittal scan shows that the dissection begins *(curved arrow)* in the mid-descending aorta and extends distally into the abdominal aorta. The *straight arrows* point to the intimal flap. A normal ascending aorta *(AA)* and upper descending aorta *(DA)* are seen proximal to the dissection.

Traumatic Aortic Injury

Traumatic aortic injury is a life-threatening condition resulting from sudden deceleration trauma, usually sustained in a high-speed motor vehicle accident or, less commonly, in a fall. The aorta may be lacerated or completely transected. The laceration may involve only the intimal and medial layers, leaving the adventitia intact, or extend through all three aortic wall layers. The injuries typically occur at two sites: (1) the aortic root and (2) a location just distal to the origin of the left subclavian artery, near the attachment of the ligamentum arteriosum. Aortic root injuries are more common, but since they usually cause death at the scene of the accident, the second type of injury is seen more commonly in patients who survive to reach an emergency center.

Traumatic aortic injury should be considered in any patient who has suffered a deceleration accident. A portable chest film is usually the first imaging procedure performed on a trauma patient, and it may show signs of aortic injury (Figure 17.34). The most suggestive are an abnormal aortic arch (fuzzy arch, double arch, lumpy arch), abnormal superior mediastinum (widened by mediastinal hemorrhage), a left apical cap (accumulation of blood in the extrapleural space overlying the lung), the presence of a left mediastinal stripe (hematoma) above the aortic arch, and widening of the right paratracheal area (hematoma extending to the right side of the mediastinum).

Patients suspected of having a traumatic aortic injury because of their history and a suspicious chest film are usually referred for an emergency CT scan or aortogram. Arteriography is considered to be the most accurate technique for confirming this injury or eliminating it from consideration. CT may overlook the tiny intimal disruptions that may be the only signs of injury in some patients. Aortography will show the intimal disruption associated with this injury as well as any false aneurysm formation that may occur when the intima and media are lacerated and the adventitia is ballooned out in an attempt to confine the bleeding.

Despite its tendency to overlook tiny intimal disruptions, CT is helpful in screening trauma patients for aortic injury when chest films show a widened or otherwise abnormal mediastinum. CT can determine whether the widening is due to the presence of mediastinal hemorrhage or some other cause, such as mediastinal fat deposition in an obese patient (Figure 17.35). If mediastinal hemorrhage is present (as it is in Figure 17.32B, though there it is not caused by

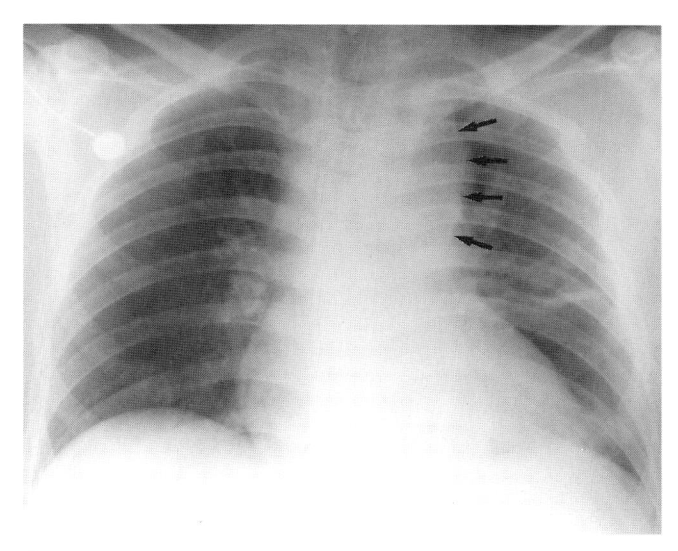

A

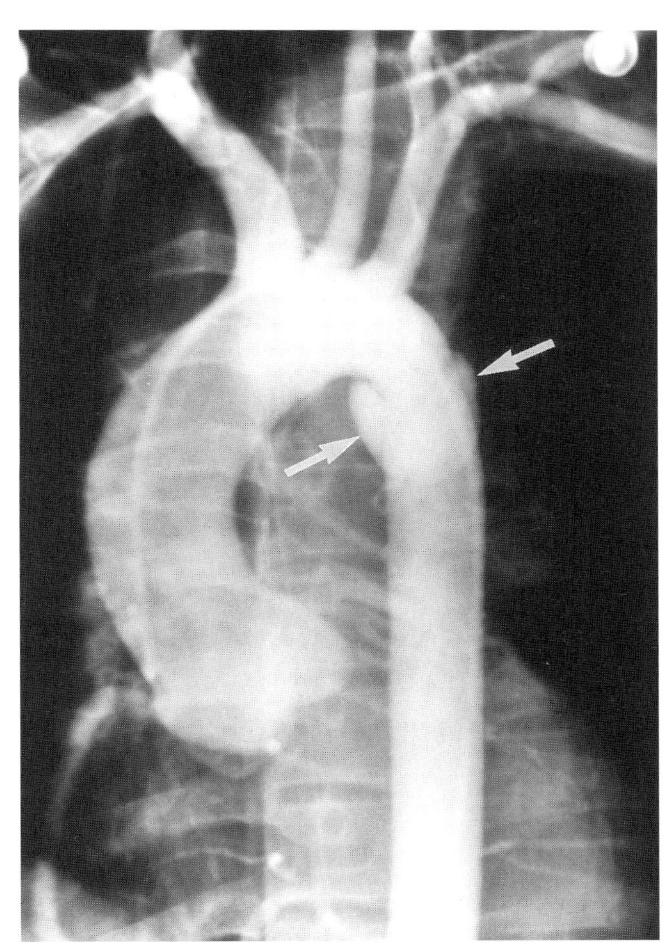

B

Figure 17.34. Chest film and aortogram of a young man with a traumatic aortic injury resulting from a deceleration automobile accident. A: The initial portable chest film shows widening of the upper left mediastinum and enlargement and irregularity of the aortic arch *(arrows).* B: The aortogram shows a saccular false aneurysm *(arrows)* at the site of the aortic laceration.

trauma), the patient must be referred for an emergency aortogram. If no mediastinal hemorrhage is present, the patient has not suffered an aortic injury and does not need an aortogram. Today, faster CT scans that eliminate aortic wall pulsatile motion permit more detailed depiction of vascular structures. Consequently, direct visualization of aortic injuries, including intimal flaps and false aneurysms, is now possible.

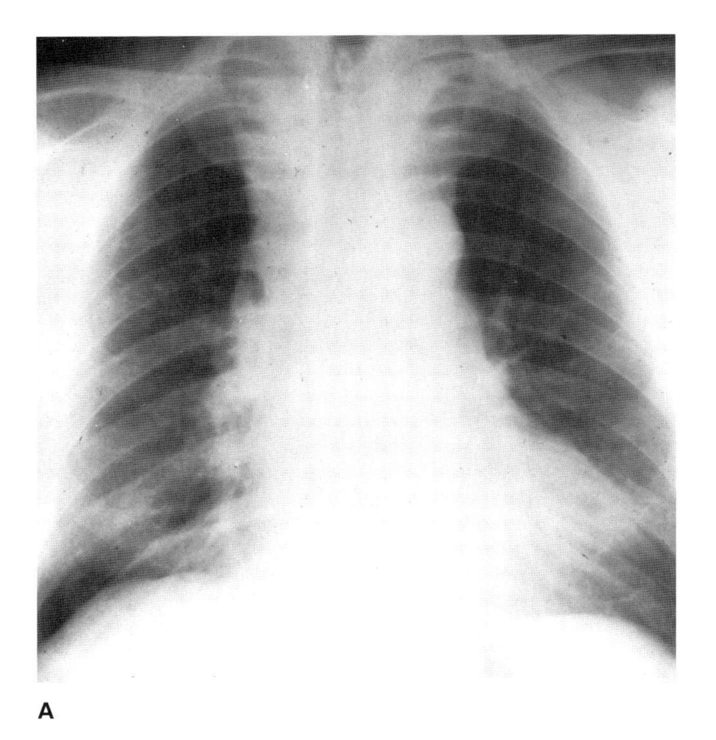

A

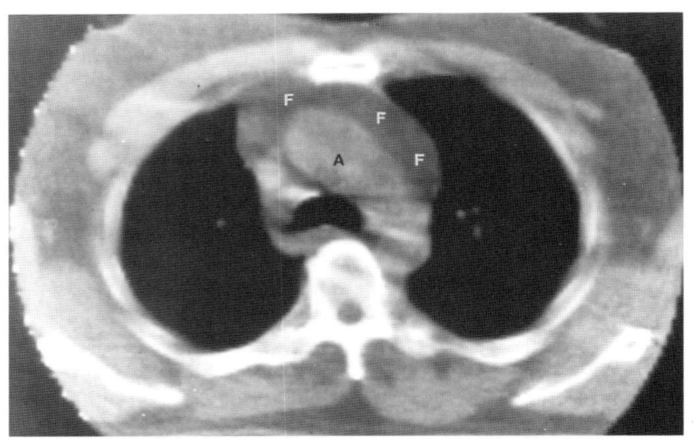

B

Figure 17.35. The use of CT for confirming or ruling out suspected aortic injury. A: Portable chest film of an obese trauma patient with a wide mediastinum. B: CT scan shows that the widening is due to the presence of mediastinal fat (*F*'s). The aorta (*A*) appears normal. There is no sign of mediastinal hemorrhage and therefore this patient will not need an aortogram.

Atherosclerotic Arterial Occlusive Disease

Atherosclerotic disease is the primary cause of occlusive disease of the systemic arterial system. When this disease involves the blood supply to the lower extremities, the clinical manifestations range from *intermittent claudication* (pain caused by exercise-induced muscle ischemia) to *rest pain,* including coolness and numbness of the involved extremity and perhaps even ischemic ulcers and gangrene. The sites of arterial stenosis and occlusion causing ischemia may be anywhere from the aorta down to the pedal arteries. Patients whose symptoms warrant treatment are usually investigated by an arteriographic examination commonly called an *aortogram and runoff arteriogram.*

To perform this procedure, an angiographic catheter is inserted percutaneously, usually into a femoral artery, and advanced proximally into the aorta. Vascular contrast material is injected while films are obtained of the arterial tree of the abdomen, pelvis, and legs. Figure 17.36 shows a normal examination.

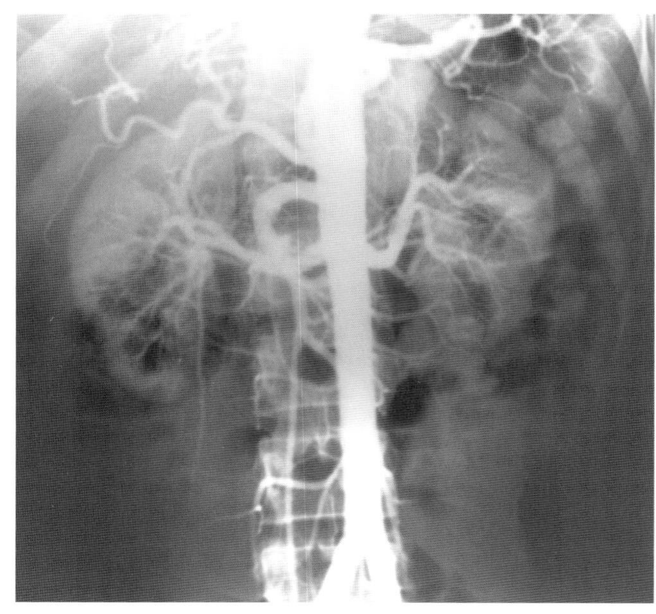

A

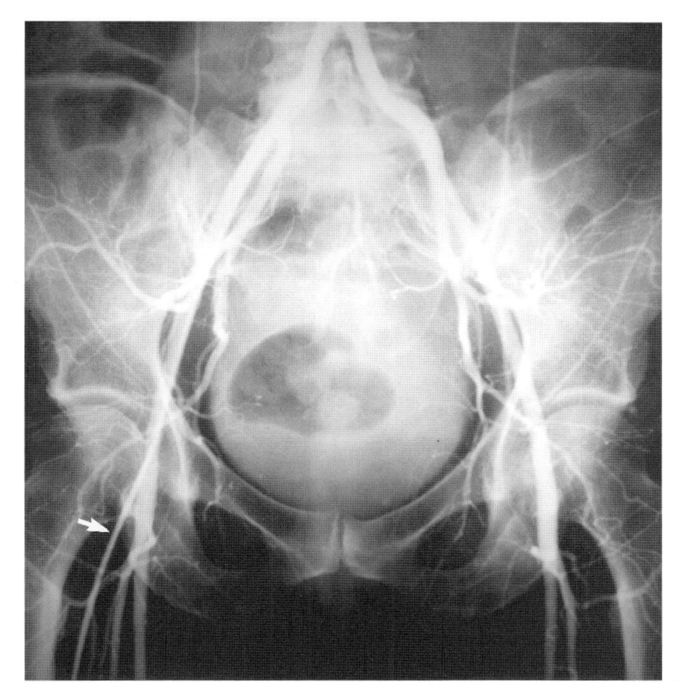

B

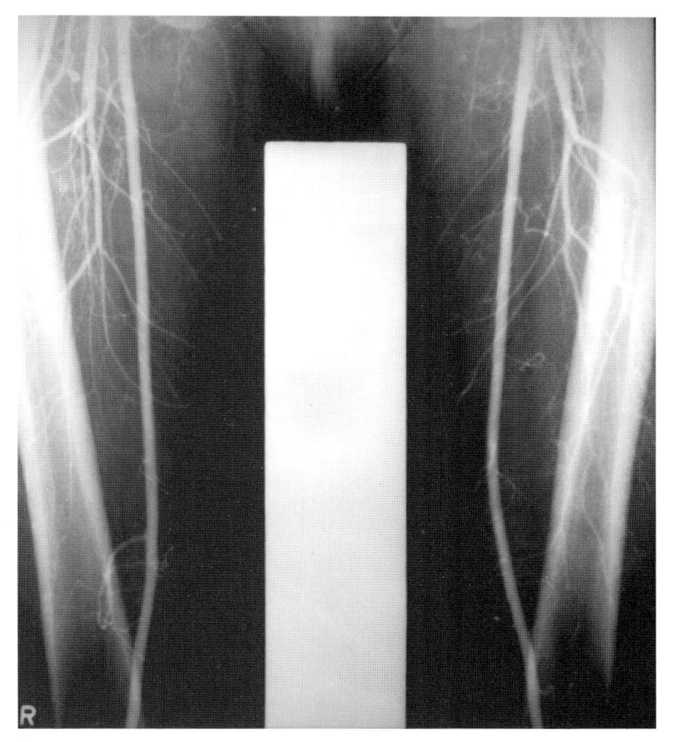

C

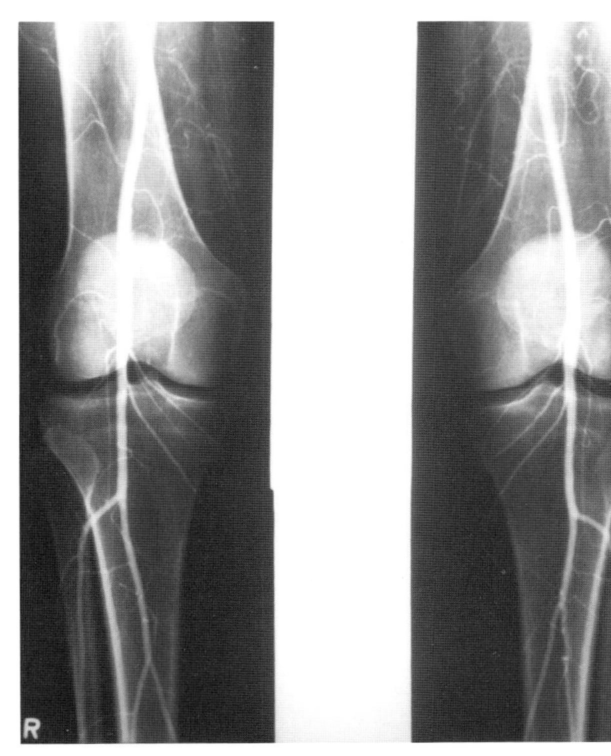

D

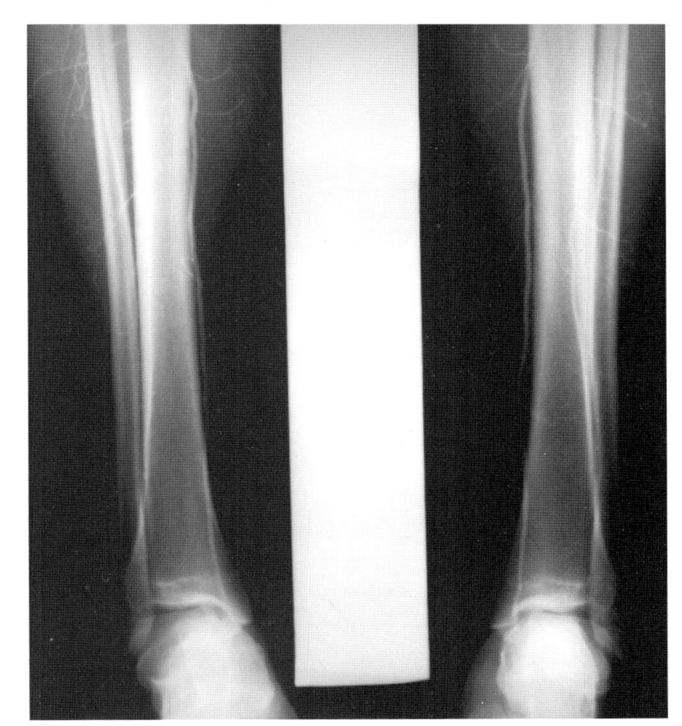

E

Figure 17.36. Normal abdominal aortogram and runoff arteriogram of the pelvis and legs. Review the major arterial anatomy by comparing these films with the anatomic diagrams in Figures 17.18 to 17.20. A: Abdomen. Identify the aortic branches. B: Pelvis. Identify the common iliac arteries, the internal and external iliac arteries, the common femoral arteries—site where the right injection catheter *(arrow)* was inserted—and the profunda femoris and femoral (superficial femoral) arteries. C: Upper legs. Identify the profunda femoris and femoral (superficial femoral) arteries. D: Knees. Identify the popliteal arteries and the three branches to the lower legs. E: Lower legs. The anterior tibial arteries are seen laterally over the upper fibulas; the posterior tibial arteries are shown medially. The peroneal arteries are faintly seen between the anterior and posterior tibial arteries.

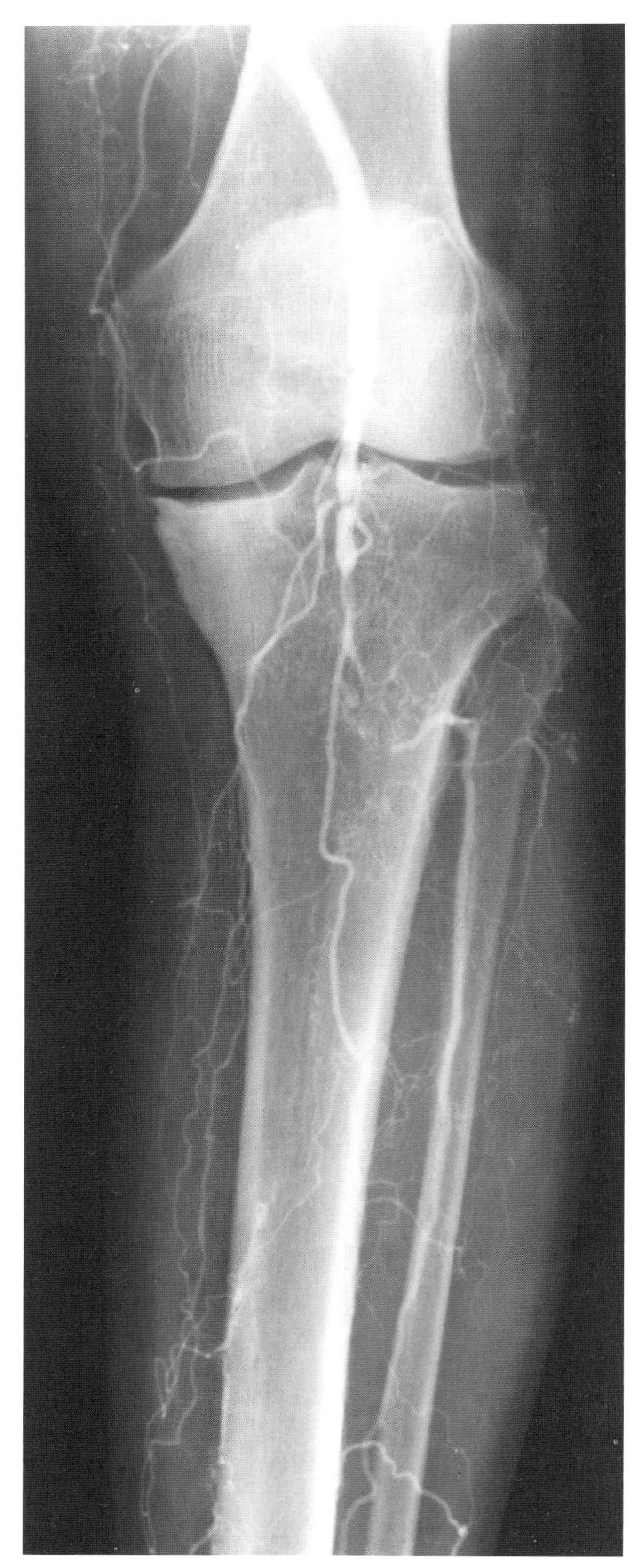

A

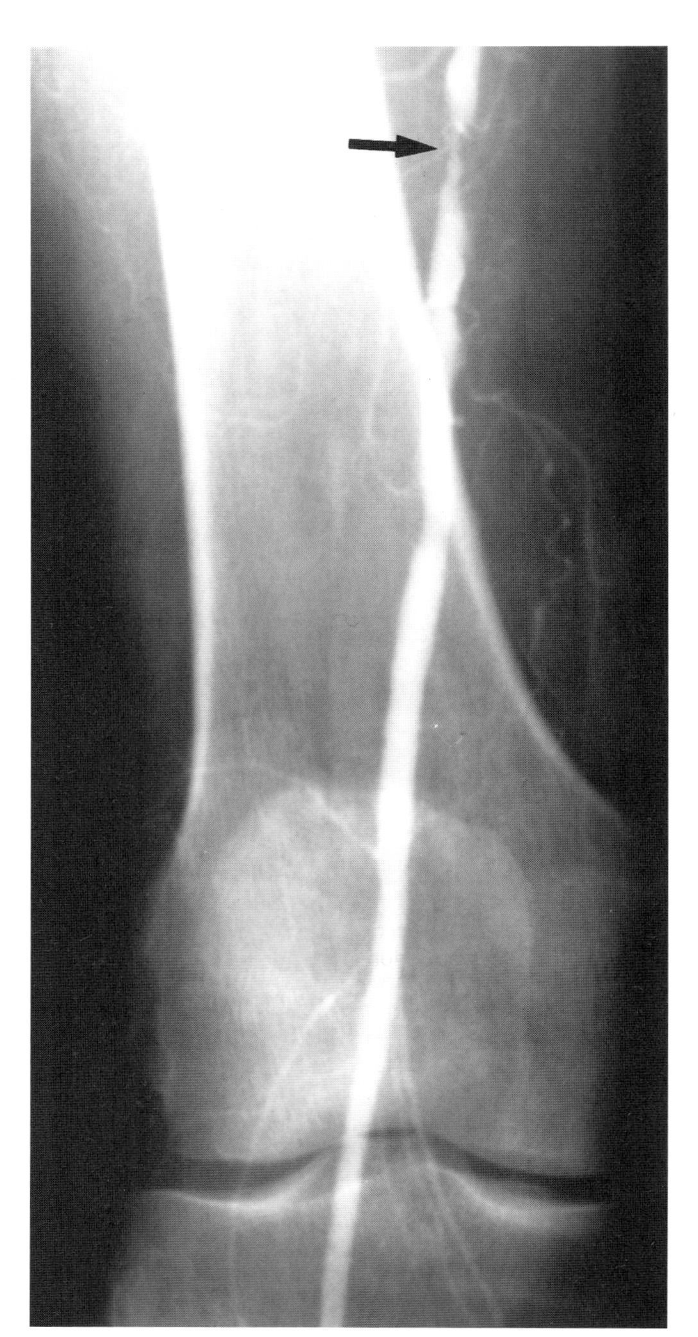

B

In abnormal cases the runoff arteriogram may demonstrate one or more arterial stenoses or occlusions (Figure 17.37). The radiologist and vascular surgeon can then determine which occlusions or stenoses need to be treated, based on the clinical history, physical findings, the measurements of blood pressures and blood flow (taken in the vascular laboratory), and the arteriogram. Patients with short arterial stenoses or occlusions may be candidates for *angioplasty,* which is angiographic dilatation of the stenosis or occlusion with a balloon catheter, a procedure we will discuss in Chapter 19. Patients with involvement of the abdominal aorta and long occlusions in the pelvic and extremity arteries are usually treated with bypass graft surgery.

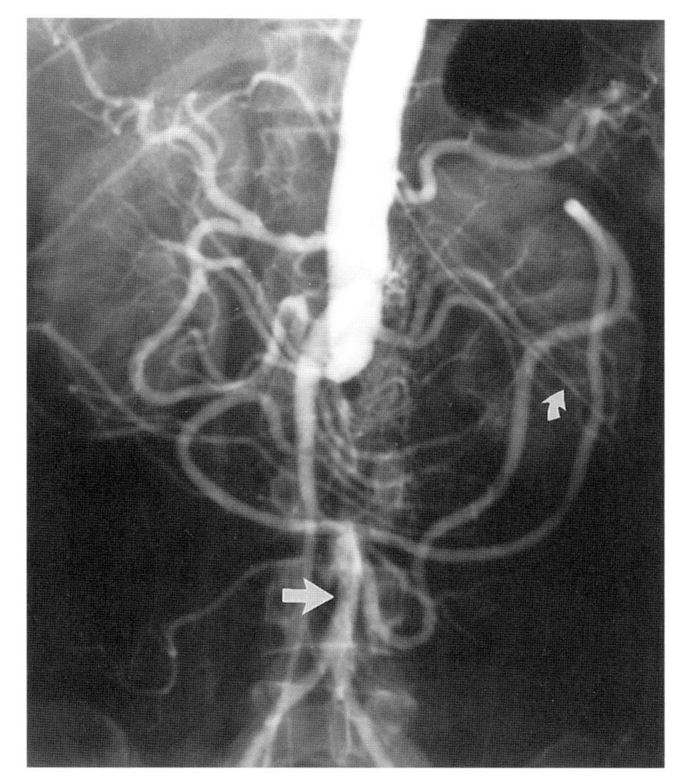

C

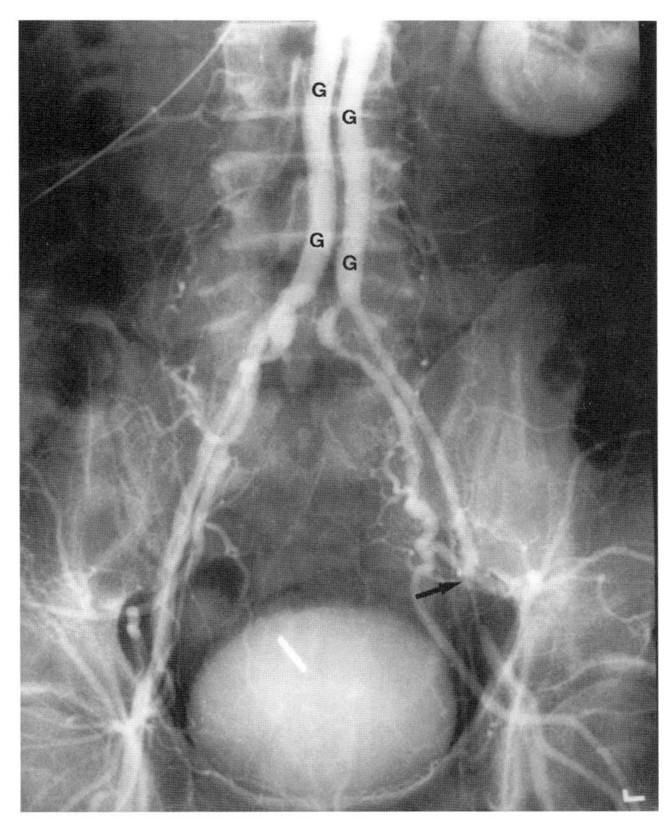

D

Figure 17.37. Examples of atherosclerotic arterial stenoses and occlusions in different patients. A: Occlusion of the popliteal artery with complete occlusion of the posterior tibial and peroneal arteries and partial occlusion of the anterior tibial artery. This unfortunate patient is not a candidate for either bypass surgery or angioplasty because there are no major patent distal arteries. B: Short tight stenosis *(arrow)* of the distal femoral artery. This patient is an excellent candidate for angioplasty and you will see the results in Figure 19.2 C: Total occlusion of the infrarenal portion of the abdominal aorta. Note the reconstitution of the aorta distal to the occlusion *(arrow)* by intestinal collateral circulation. This patient is a candidate for aorto-iliac bypass surgery. Obviously the catheter was not inserted by the femoral artery because the patient had no palpable femoral pulses. Under these circumstances arteriography can be done by catheterization through an axillary artery or by translumbar catheterization (used in this patient). The curved *arrow* points to the catheter, which under fluoroscopic guidance was inserted percutaneously into the aorta over a trocar. D: Short occlusion *(arrow)* of the left common femoral artery; this patient could be treated by a bypass graft or angioplasty. Did you notice that this patient has already had bypass surgery? Observe the unusual junction of the aorta and iliac arteries. The aortic bifurcation appears high because an aorto-iliac bypass graft (*G*'s) has replaced an aortic occlusion similar to the one shown in C. Note the corrugated pattern of the graft.

MR angiography is showing promise in the demonstration of lower extremity arterial supply (Figure 17.38) and may decrease the need for catheter arteriograms in many future patients with atherosclerotic arterial occlusive disease.

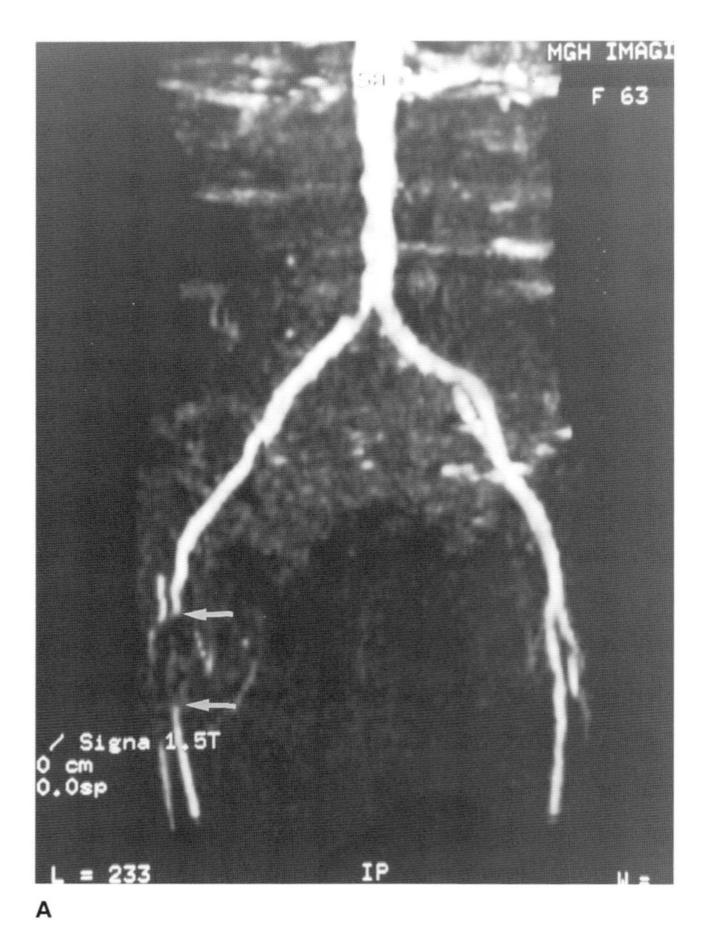

A

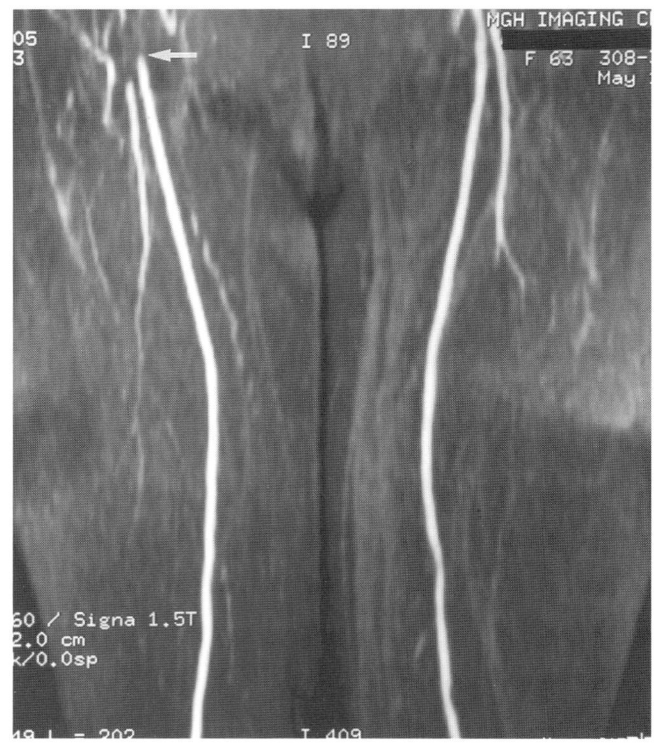

B

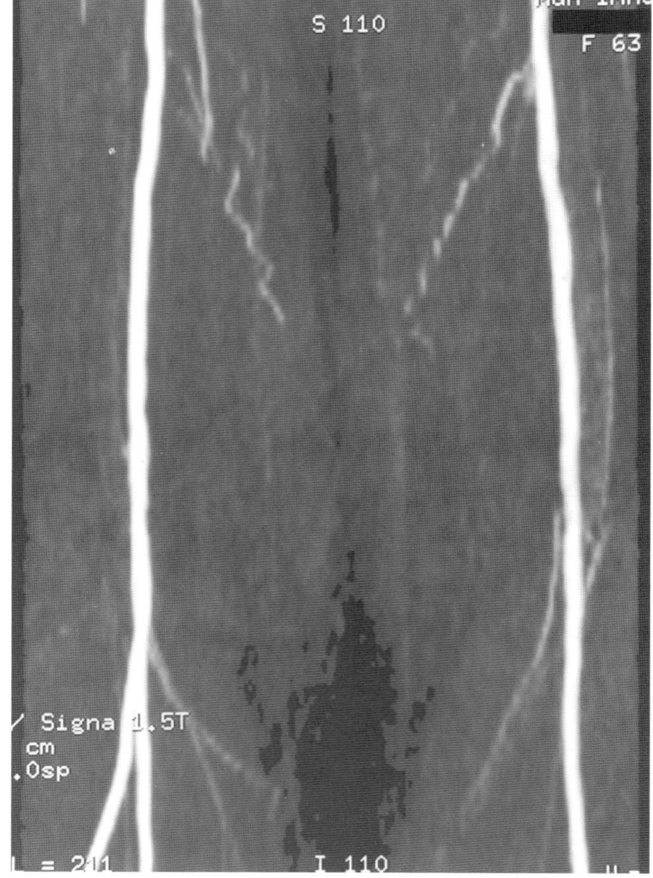

C

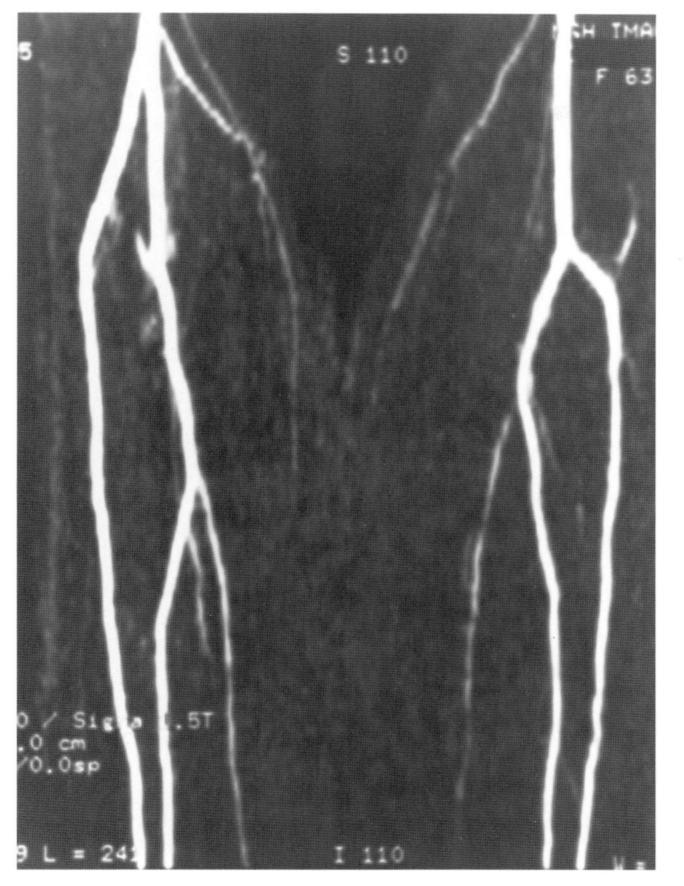

D

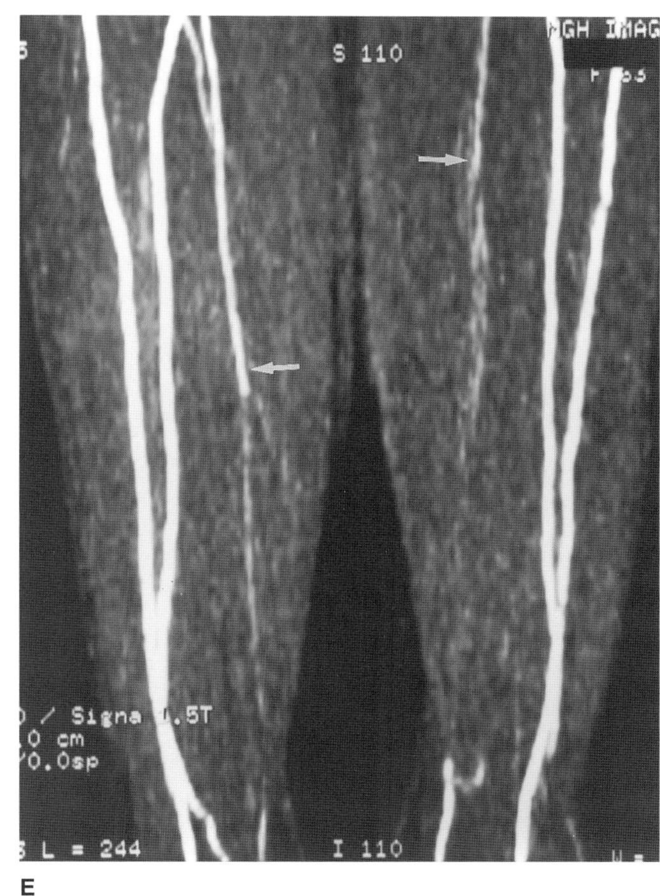

E

Figure 17.38. MR runoff arteriogram of an elderly man with right leg claudication. A: Pelvis. Note that the proximal right superficial femoral and profunda femoris arteries are occluded *(between the arrows).* Irregularities in the outlines of the aorta and iliac arteries are caused by the presence of atherosclerotic plaques. B: Upper legs. Right superficial femoral artery occlusion again seen *(arrow).* C: Knees. The popliteal arteries are patent. D and E: *Lower legs.* Both anterior tibial arteries (*arrows* in E) are diseased, with long stenoses and occlusions.

Renovascular Hypertension

There are several ways in which the arterial supply to the kidney can be compromised. Acutely, this can be caused by an embolus to a renal artery, the patient presenting with acute flank pain. More often the cause is chronic, with progressive atherosclerotic disease producing a renal artery stenosis or occlusion. Gradually, atherosclerotic plaques narrow the renal artery lumen, eventually occluding it entirely (Figure 17.39). Affected patients usually have atherosclerotic disease elsewhere, especially in the abdominal aorta. Another condition causing renal artery narrowing and ischemia is fibromuscular disease, which occurs only in young patients (Figure 17.40).

Renal ischemia with resulting increased production of *renin* by the kidney distal to a renal artery stenosis or occlusion has been demonstrated to be the cause of hypertension in a small percentage of the hypertensive population. Since this is a treatable cause, it is an important one to diagnose. Renin, you will remember, is the proteolytic enzyme produced by the renal juxtaglomerular cells in response to decreased blood pressure in renal arterioles; renin acts on angiotensinogen to form angiotensin-I, which is then converted into angiotensin-II, a potent vasopressor compound.

Renovascular hypertension should be considered in young patients with hypertension (who may have fibromuscular disease) as well as in patients with hypertension of new onset (new embolus or new thrombosis), hypertension and an abdominal bruit (bruit caused by the stenosis), hypertension and renal failure

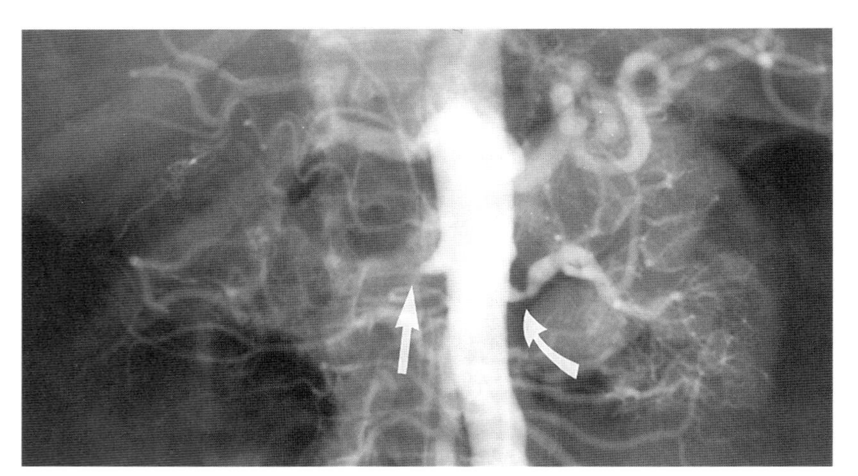

A

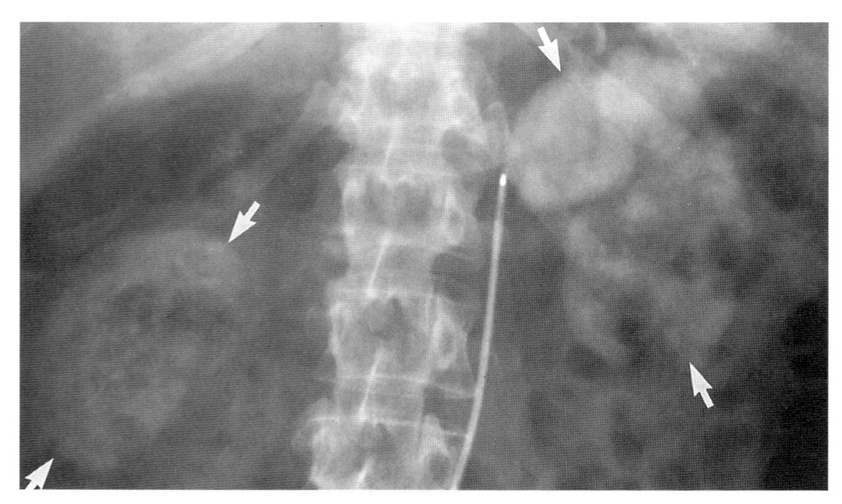

B

Figure 17.39. Aortogram of a patient with renovascular hypertension. A: Early-phase film, showing a total occlusion of the right renal artery *(straight arrow)* and a stenosis *(curved arrow)* in the proximal left renal artery. B: Later-phase film, showing a small, poorly opacified right kidney, compared with the left kidney. Venous sampling of the right renal vein demonstrated a significantly elevated renin level.

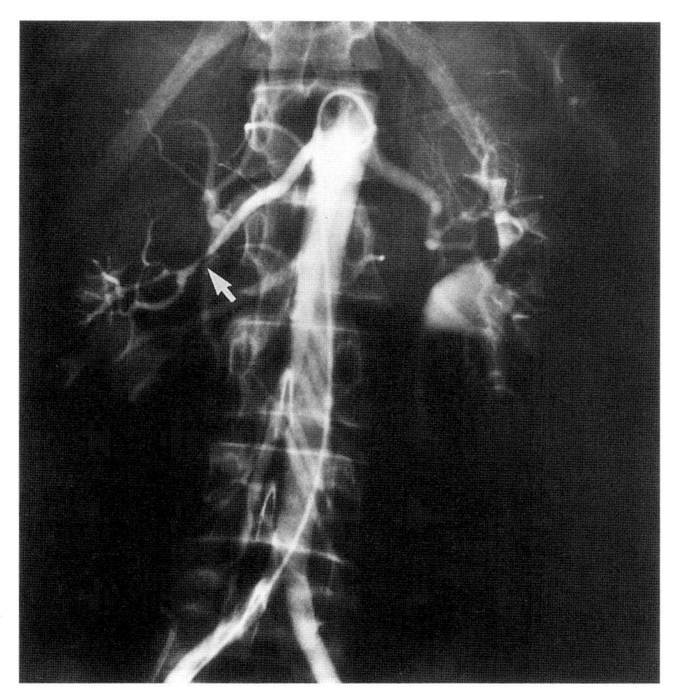

A

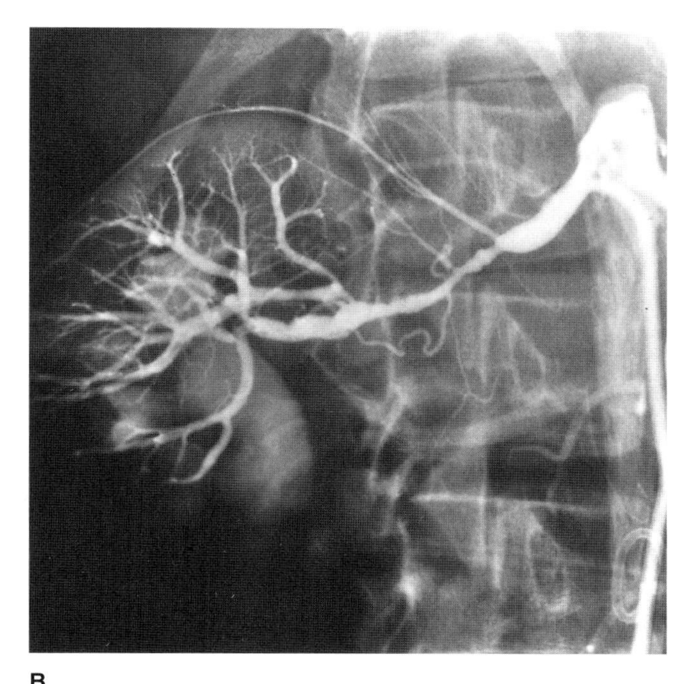

B

Figure 17.40. Young woman with right renal artery stenosis caused by fibromuscular disease. A: Aortogram shows a right renal artery stenosis *(arrow).* B: Better visualization of the right renal artery is obtained with a (selective right renal arteriogram, which shows a long, smoothly margined, undulating stenosis characteristic of hyperplasia of the arterial wall. Note the absence of atherosclerotic plaques in the abdominal aorta and its branches. Did you notice that the patient has a second right renal artery (a common normal variation)? Look back at the aortogram; the second accessory (lower) renal artery supplies the lower pole of the right kidney.

(diminished renal function when the ischemia is bilateral), and hypertension with an elevated peripheral renin level.

When renovascular hypertension is suspected clinically, the patient should be referred for diagnostic imaging. Although intravenous urography may show the delayed appearance of contrast material and decreased kidney size in cases of renal ischemia, it is such an inadequate screening procedure for renovascular hypertension that it is not recommended today. Instead, the renal arteries should be visualized directly by conventional arteriography, MR angiography, CT angiography, or vascular ultrasound. If a renal artery stenosis or occlusion is found at arteriography, selective catheter sampling of blood from both renal veins may be performed to confirm the diagnosis. The samples should show a significant elevation in the renal vein renin level from the kidney with the arterial stenosis or occlusion. It is important to take such samples, because many patients with essential hypertension may coincidentally have a renal artery stenosis. Patients with true renovascular hypertension generally benefit from either balloon angioplasty performed by the radiologist or surgical repair with a renal artery bypass graft.

The limited role of radiological imaging in the workup of the hypertensive patient is important for you to understand, because most hypertensive patients you will see in a lifetime of practice will have essential hypertension. You should suspect *coarctation of the aorta* when you find unequal limb blood pressures on physical examination, and confirm (or disconfirm) the diagnosis with conventional arteriography, CTA, or MRA. You should also suspect endocrine-related adrenal hypertension caused by conditions such as pheochromocytoma or Cushing's syndrome from the clinical presentation and laboratory investigation; CT, as you saw in Chapter 14, can provide fast and accurate confirmation of a clinically suspected adrenal mass.

Venous Anatomy

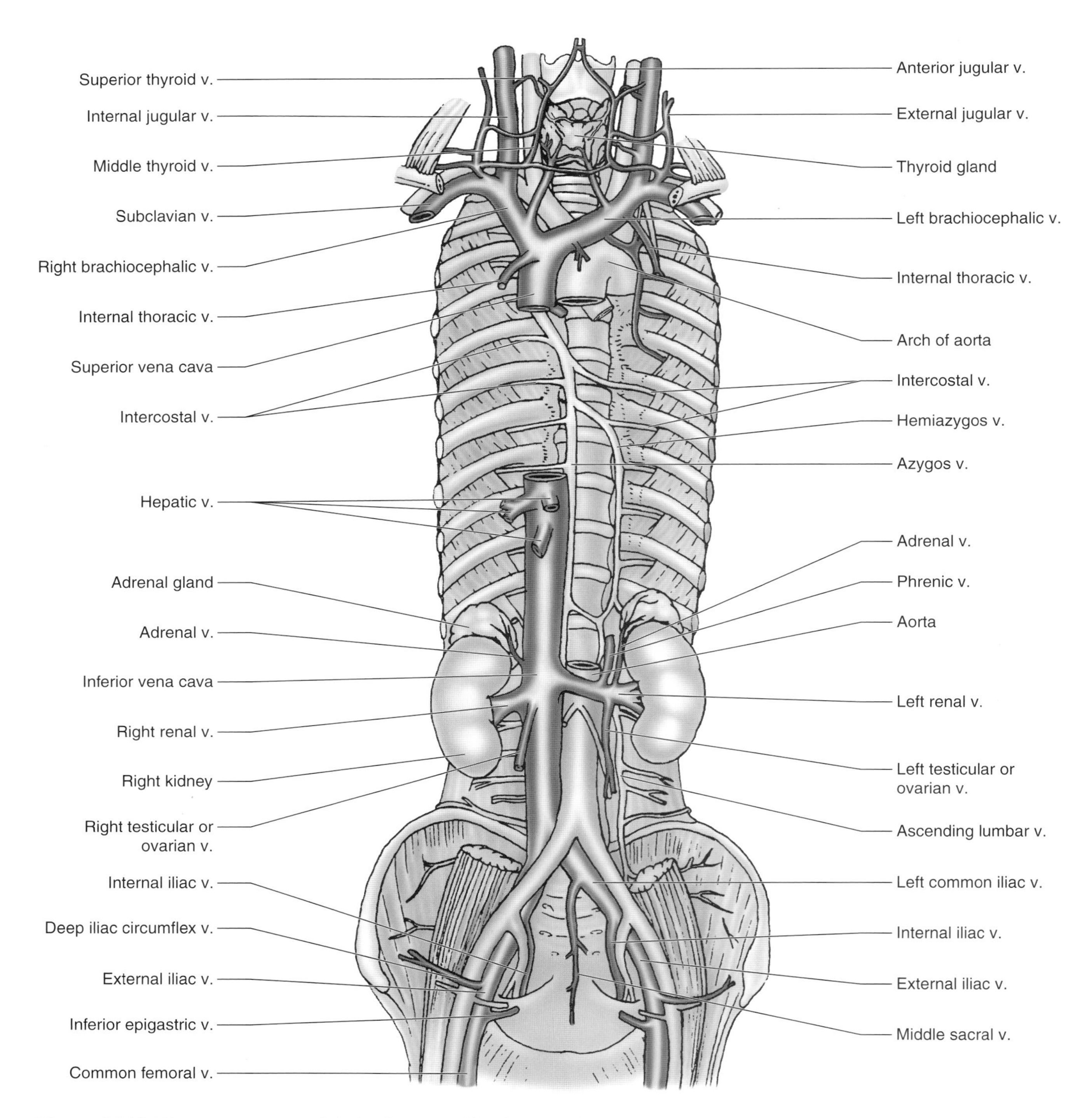

Figure 17.41. The vena cavae and their tributaries. The heart and many other organs have been removed.

Before turning to abnormalities of the venous system, it will be useful for you to refresh your memory of venous anatomy, especially the structure of the vena cavae.

The superior vena cava (SVC) extends down from the junction of the right and left brachiocephalic veins to join the right atrium. Figure 17.41 depicts the anatomy of both vena cavae and their tributaries with the heart and several other organs removed. Turn back to Figure 17.17 to review the anatomy of the SVC with the heart intact. You have already seen the SVC identified on chest CT scans in this chapter and in earlier chapters on the chest.

The inferior vena cava (IVC) is formed from the junction of the right and left common iliac veins. It ascends in the retroperitoneum to the right of the aorta and drains into the right atrium. You are already familiar with its appearance at CT from the chapters on the abdomen. The veins of the upper arms drain predominately through the *superficial venous system,* consisting mostly of cephalic and basilic veins (Figure 17.42). The veins of the legs (Figure 17.43, on the next page) are drained primarily by the *deep venous system,* consisting of paired peroneal, anterior, and posterior tibial veins, which combine at the knee to become the popliteal vein, which continues as the femoral and iliac veins to drain into the inferior vena cava. The major superficial veins in the legs are the lesser and greater saphenous veins.

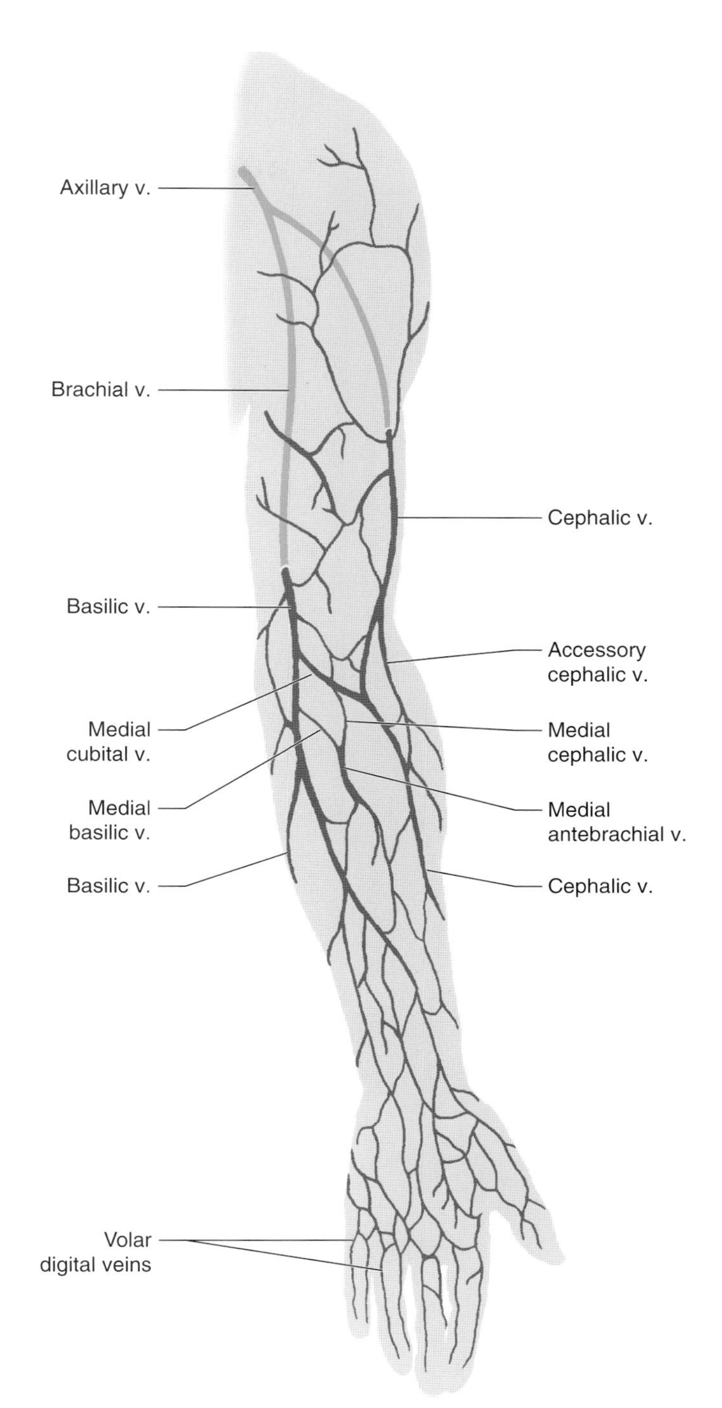

Figure 17.42. Veins of the arm.

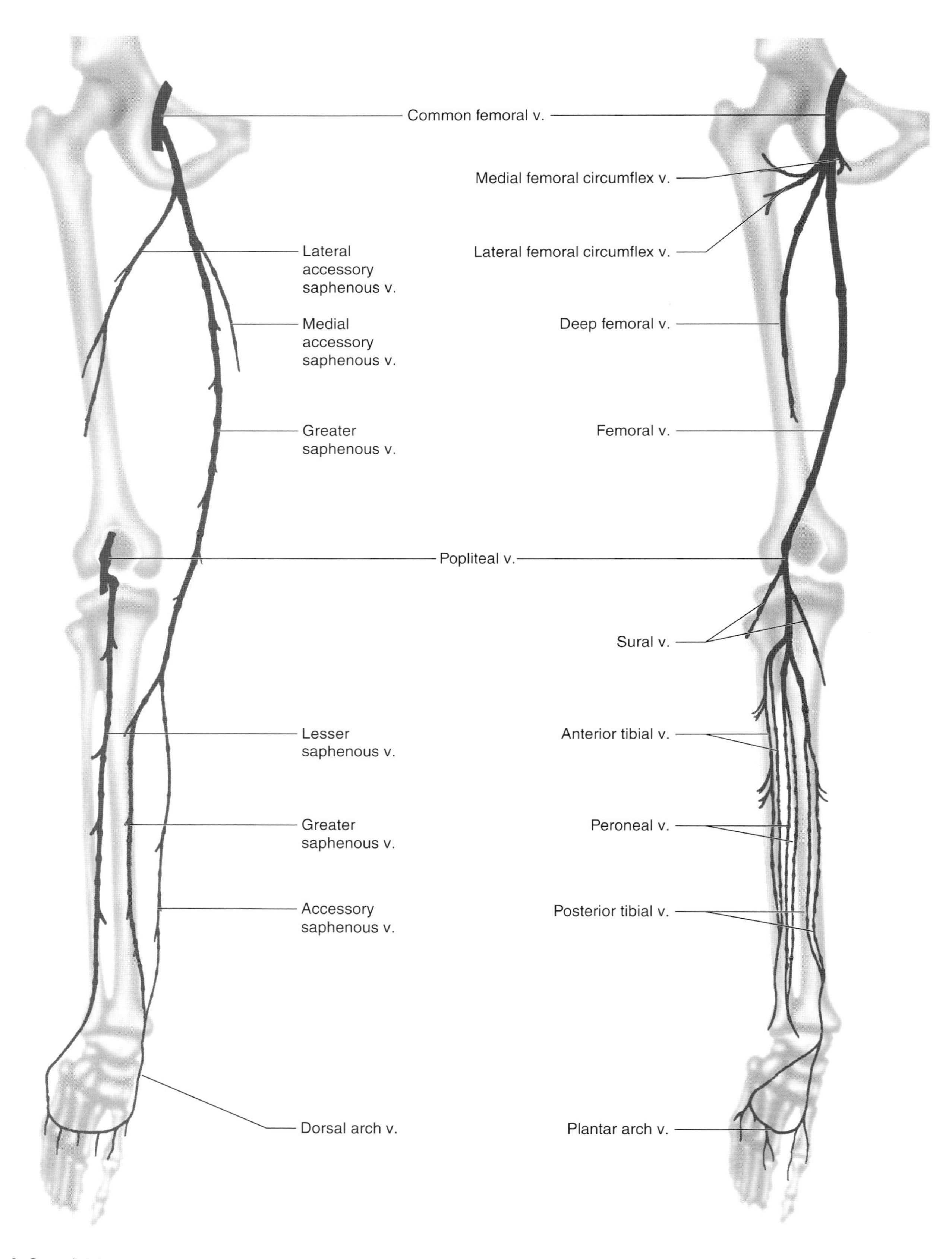

Figure 17.43. Superficial and deep veins of the leg.

Obstruction of the Superior Vena Cava

Obstruction of the superior vena cava may be caused by external compression or intraluminal thrombosis. The result of obstruction is venous hypertension of the head and arms, characterized by progressive dilatation of the veins of the face, neck, and arms accompanied by cyanosis and edema. Malignant tumors of the chest, such as bronchogenic lung cancer, are the most common cause, producing either external compression of the SVC or invasion of it. SVC compression may also be caused by nonmalignant processes, such as granulomatous diseases. Thrombosis of the SVC is often the result of the long-term presence of an indwelling venous catheter. When SVC obstruction is suspected, the diagnosis can be confirmed by venography (Figure 17.44). In some patients, superior vena cavography requires only a hand injection of contrast material through venipuncture in the arm; other patients require venous catheterization. CT and MR are particularly helpful in showing such causes of obstruction as malignant tumors.

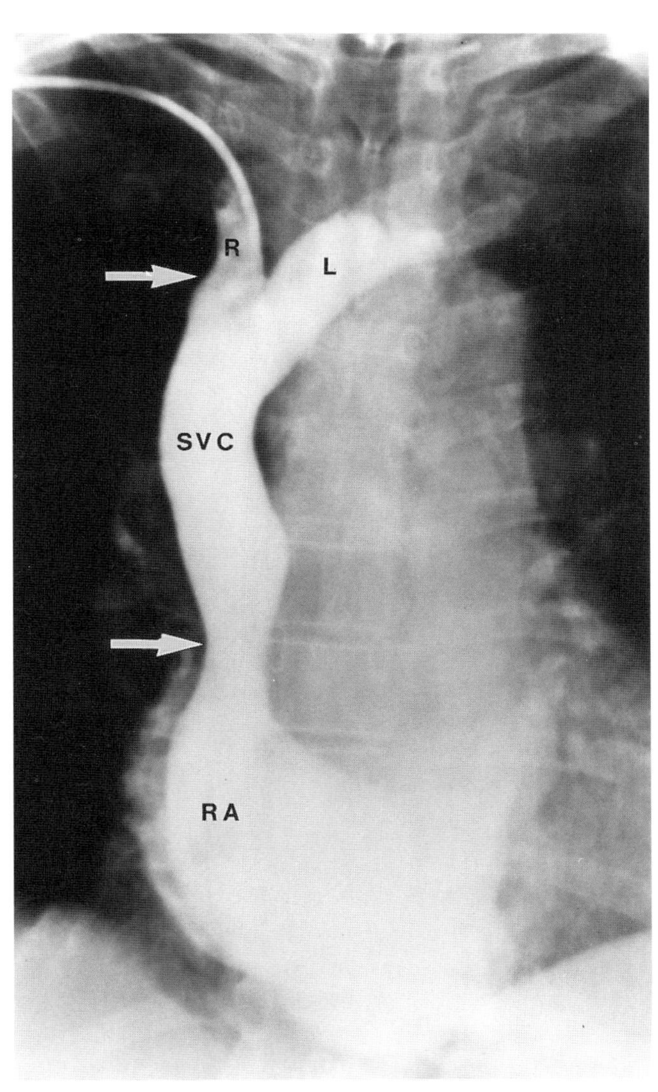

A

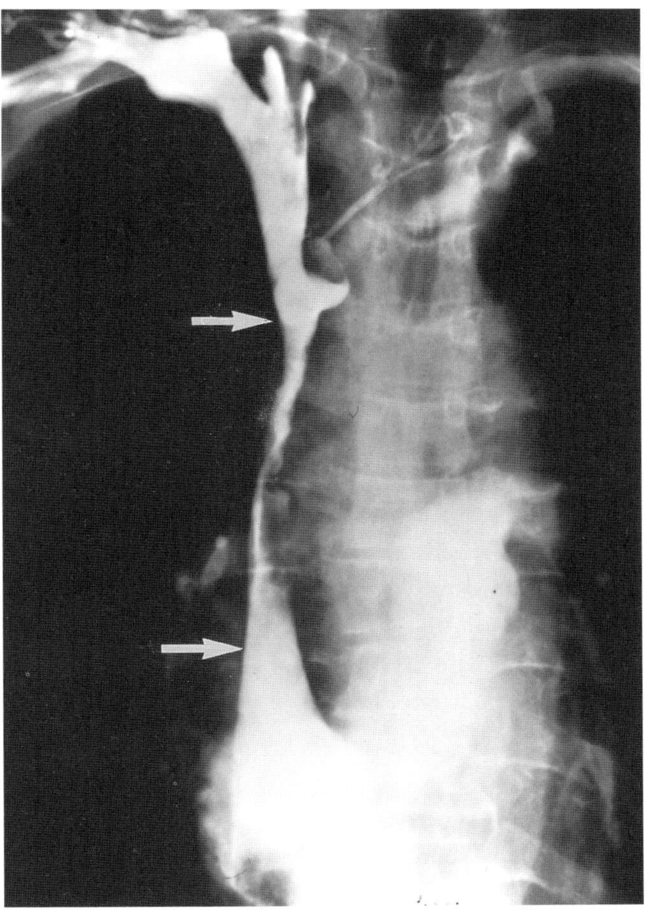

B

Figure 17.44. A: Normal superior vena cavagram: *SVC,* superior vena cava *(arrows); RA,* right atrium of the heart; *R* and *L,* right and left brachiocephalic veins. B: A patient with superior vena cava syndrome due to encasement of the cava *(arrows)* by metastatic bronchogenic carcinoma of the lung. Although superior vena cavography can be performed by injecting contrast material via an arm venipuncture, in both these examinations contrast material was injected through a right arm venous catheter. Can you find the catheters?

Disorders of the Inferior Vena Cava

The inferior vena cava may be imaged with a variety of techniques, including ultrasound, CT, MR, and inferior vena cavography. By now you are very familiar with the IVC's teardrop shape at CT, located in the prevertebral retroperitoneum, just to the right of the aorta. To perform an inferior vena cavagram, an angiographic catheter is inserted percutaneously via a femoral vein puncture and advanced proximally into a common iliac vein. A bolus of angiographic contrast material is then injected while x-ray filming or digital imaging of the abdomen is carried out. The contrast material ascends passively within the inferior vena cava, carried by the flowing blood. The veins draining *into* the inferior vena cava (renal veins, adrenal veins, hepatic veins) are not opacified, as are the branches of the aorta on an aortogram, because contrast is not flowing into those veins. Rather, they are transferring unopacified blood and depositing it in the cava, which can be imaged as "wash-out" on cavagrams at sites where these branches enter the inferior vena cava (Figure 17.45).

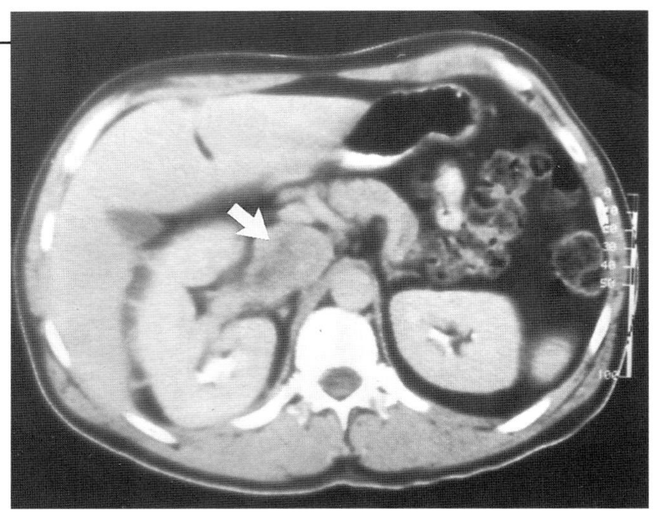

A

Figure 17.46. Renal cell carcinoma invading the inferior vena cava. A: CT scan showing a low-attenuation tumor within the right renal vein extending into the inferior vena cava *(arrow).* B: Slightly lower-level CT scan showing the maximal diameter of the renal tumor with central tumor necrosis. C: Inferior vena cavagram showing invasion of the cava with a tumor mass extending from the right renal vein.

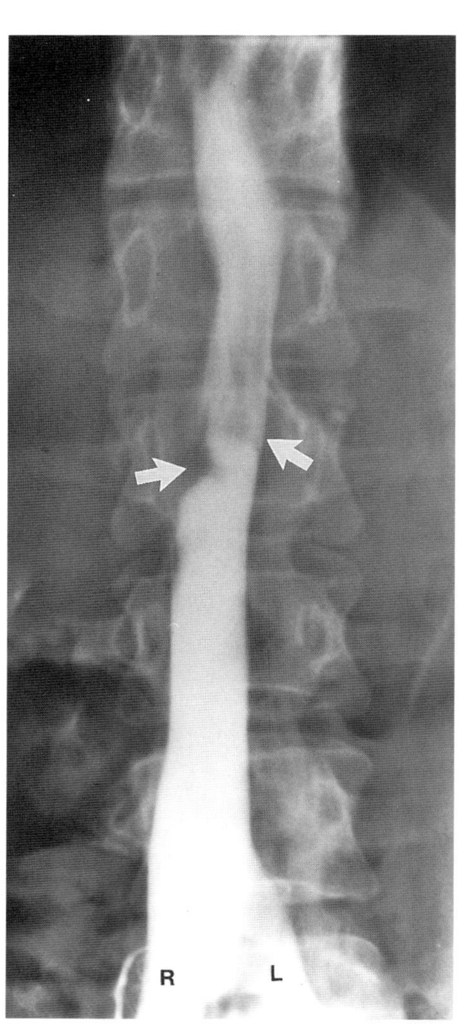

Figure 17.45. Normal inferior vena cavagram. The contrast material was injected into a right femoral vein catheter, the tip of which (not included on the film) was positioned in the right *(R)* common iliac vein. Some reflux of contrast material is seen flowing down into the left *(L)* common iliac vein. The *arrows* indicate the sites of the right and left renal veins where unopacified blood from the kidneys has diluted the contrast material flowing in the cava, producing a "wash-out" pattern in the upper half of the cava.

The two most common indications for imaging the inferior vena cava are suspected tumor invasion from an adjacent neoplasm and suspected caval thrombosis from extension of lower extremity or pelvic vein thrombosis. Although both of these caval conditions may be diagnosed by ultrasound, CT, and MR, inferior vena cavography is often performed because it shows the cava in greatest detail.

A tumor that frequently invades the inferior vena cava is renal cell carcinoma. Figure 17.46 illustrates such a case, a 62-year-old man with right flank pain and hematuria. He also complained of the recent onset of ankle swelling. His intravenous urogram showed a right renal mass, and a CT scan was performed for further evaluation. The CT examination, shown in Figures 17.46A and B, confirmed a tumor in the lower pole of the right kidney that was growing into the inferior vena cava *(arrow).* Note the low-attenuation areas within the mass, typical of central tumor necrosis. The patient's inferior vena cavagram, Figure

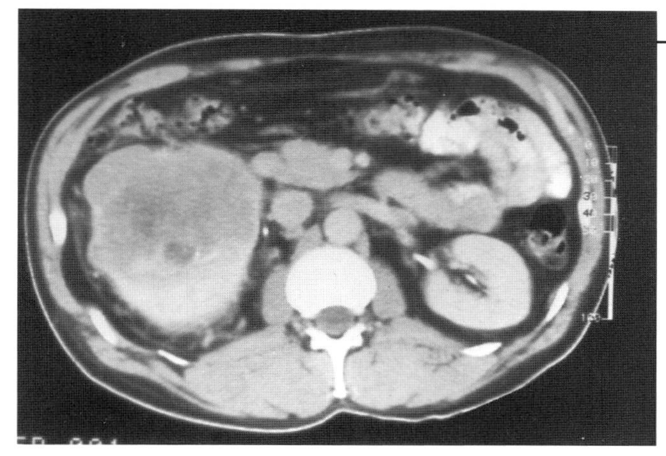

B

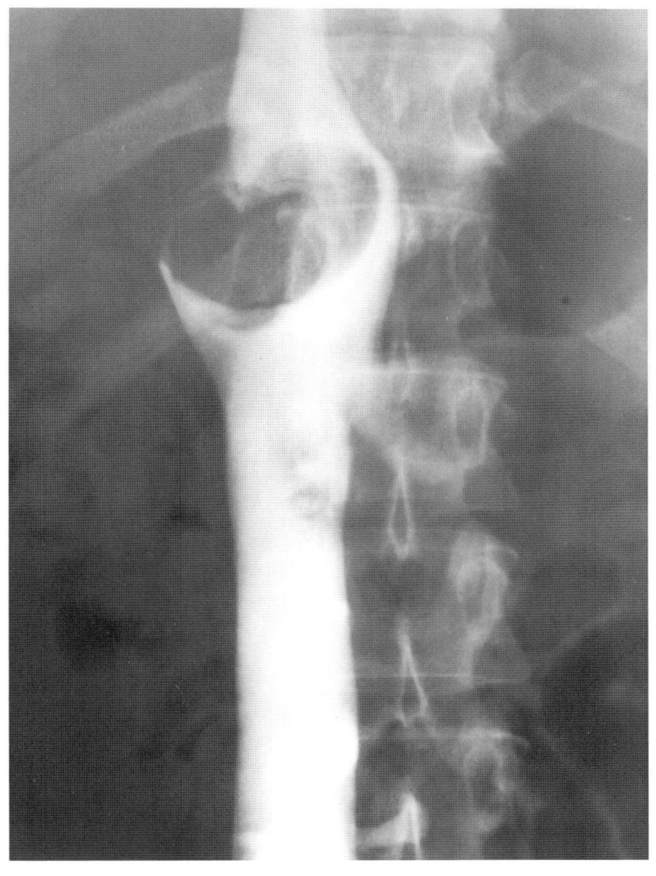

C

17.46C, clearly delineated tumor extension into the inferior vena cava, which not only expanded the IVC, but almost entirely occluded it. This explained his ankle swelling.

Deep venous thrombosis (DVT) of the lower extremities and pelvis may extend proximally up into the IVC, producing caval thrombosis. In Figure 17.47, note the large lobulated filling defect within the contrast-opacified inferior vena cava. The cava above the renal veins is clear. The immediate danger to the patient is dislodgment of the thrombus, which would lead to a massive pulmonary embolism.

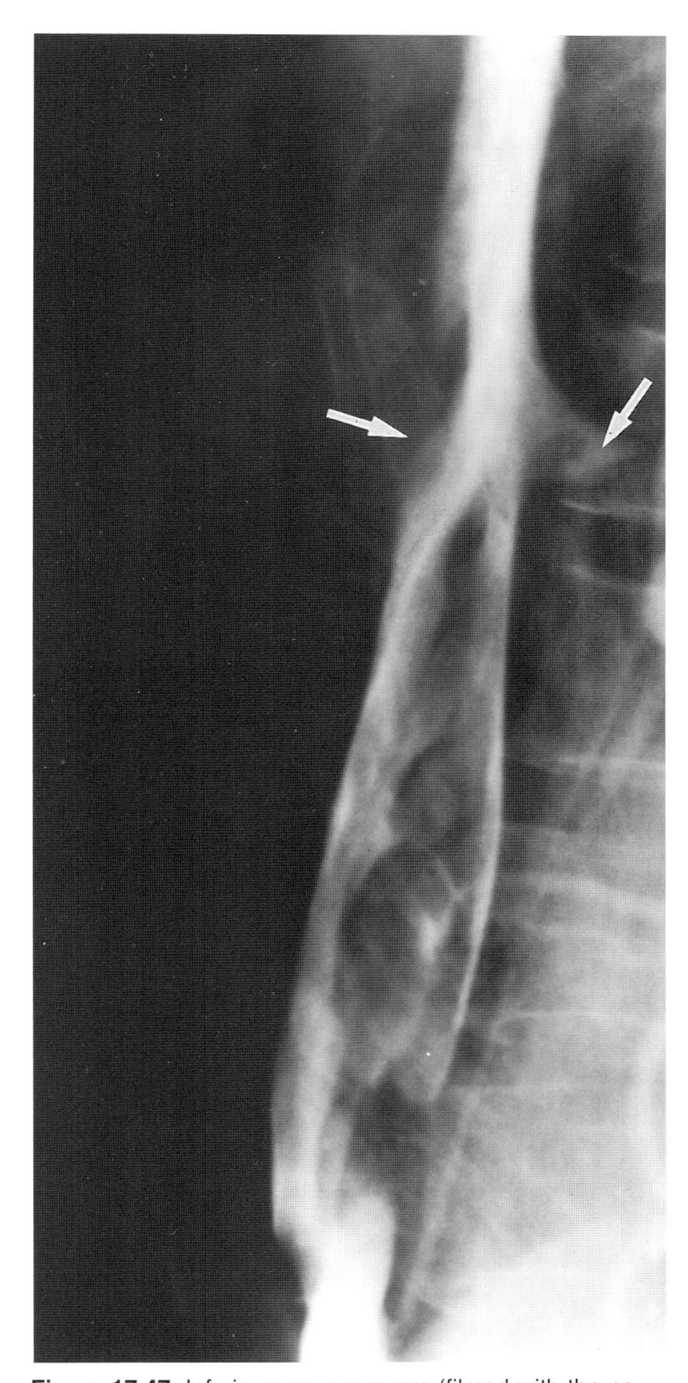

Figure 17.47. Inferior vena cavagram (filmed with the patient rotated into a steep oblique position) showing extension of lower extremity and pelvic venous thrombosis up into the cava. The *arrows* indicate the level of the renal veins. Unopacified blood is entering from the right renal vein; a small amount of contrast material is refluxing into the left renal vein.

Deep Venous Thrombosis of the Lower Extremities

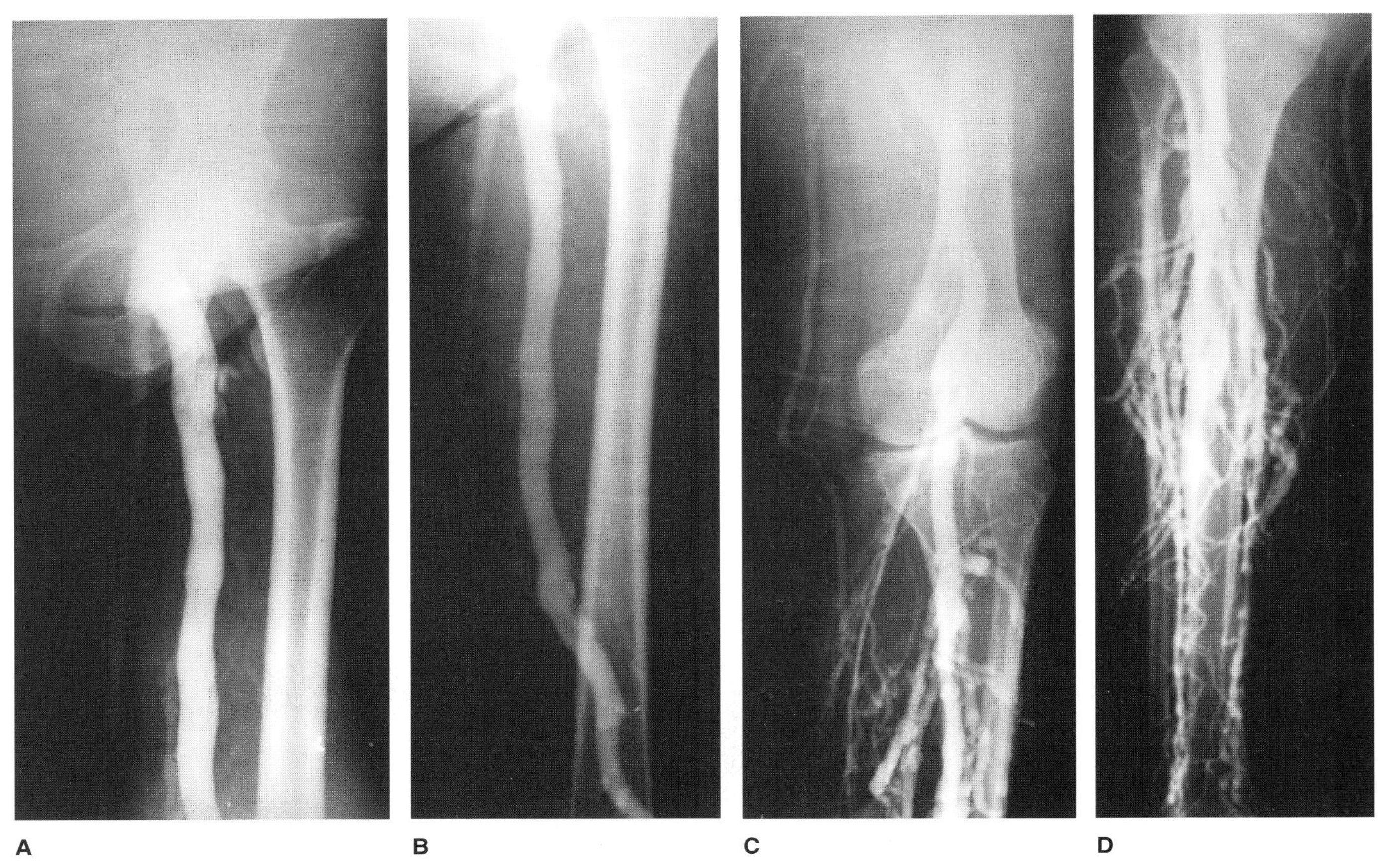

Figure 17.48. Normal left leg venogram. A: Pelvic segment. B: Upper leg segment. C: Knee segment. D: Lower leg segment.

One of the most common clinical conditions for which you may request vascular imaging is suspected deep venous thrombosis of the lower extremities. The occurrence of DVT is often related to venous stasis and a hypercoaguable state. The risk factors include prolonged immobilization, trauma, congestive heart failure, pregnancy, and neoplastic disease. In addition, certain surgical procedures are associated with an increased risk of DVT, especially hip and knee replacement surgery and vascular surgery of the aorta and its branches.

Patients with acute DVT may present with pain and swelling of the affected limb, and physical examination of such patients may reveal warmth, distention of superficial veins, and calf tenderness. But many patients with acute DVT have asymptomatic lower extremities, and seek medical attention for symptoms of pulmonary emboli that have arisen because of DVT.

Deep venous thrombosis has traditionally been diagnosed by *leg venography,* which requires a needle stick of a dorsal foot vein with a small-gauge needle and hand injection of contrast material. The contrast material is carried passively with the venous blood flow proximally to opacify the entire deep venous system of the leg. Compare the normal leg venogram in Figure 17.48 with the anatomic drawing of the deep venous system of the leg in Figure 17.43. On the leg venogram find the common femoral vein, (superficial) femoral vein, and popliteal vein. The anterior and posterior tibial veins and the peroneal veins of the lower leg are paired and are shown superimposed over their muscular tributaries, which drain the calf muscles. Compared with the leg arteries, which you saw earlier, the veins are larger in caliber and have characteristic bulbous dilatations where valves are located. The greater and lesser saphenous superficial veins are

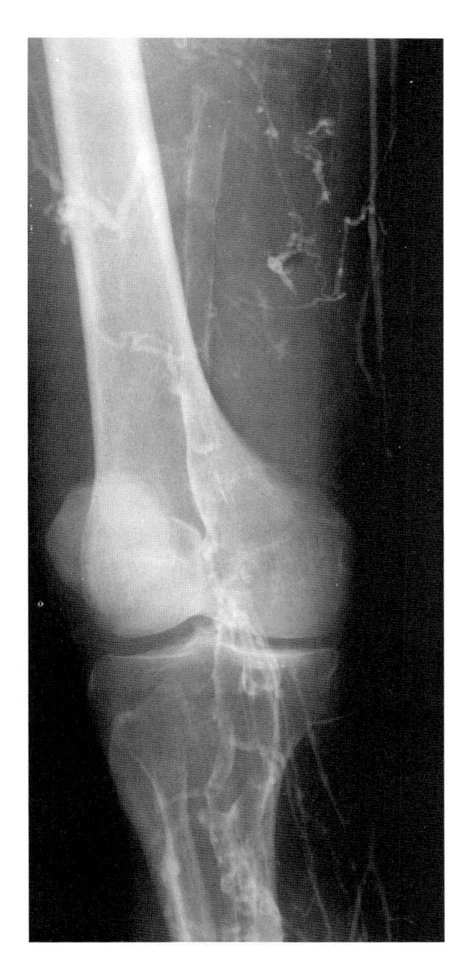

Figure 17.49 ***(left).*** Film from the leg venogram of a patient with deep venous thrombosis; it shows extensive thrombus (long filling defects) within the femoral vein, the popliteal vein, and the veins of the lower leg.

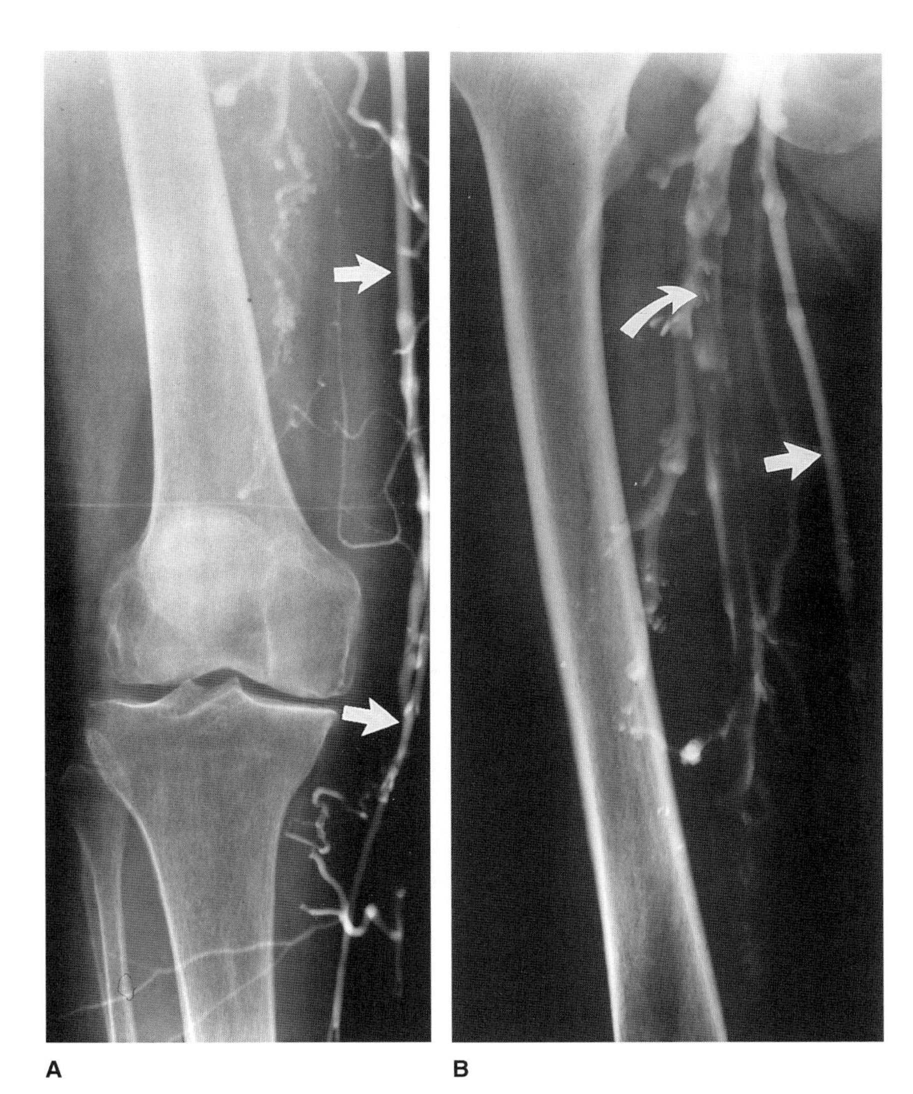

A B

Figure 17.50. Leg venogram of a patient whose deep venous thrombosis has almost completely occluded the *deep* venous system. A: Contrast material is shown flowing proximally in the greater saphenous vein *(straight arrows),* the major venous drainage of the *superficial* venous system. B: Slightly higher, thrombus can be seen within the deep venous system *(curved arrow).* The *straight arrow* again indicates the greater saphenous vein.

not opacified on normal venographic examinations because normal venous flow proceeds from the superficial into the deep venous system. But the superficial venous system can be opacified at venography when the deep venous system is occluded and consequently the major venous drainage of the leg is provided by the superficial venous system.

At venography thrombus will appear as filling defects within the deep veins, as you can see in Figure 17.49. The popliteal and femoral veins are filled with thrombus, outlined by a tracing of contrast material flowing in the thin spaces between the thrombus and the vein walls. DVT may also be shown as venous occlusion with contrast flow into the superficial venous system (Figure 17.50).

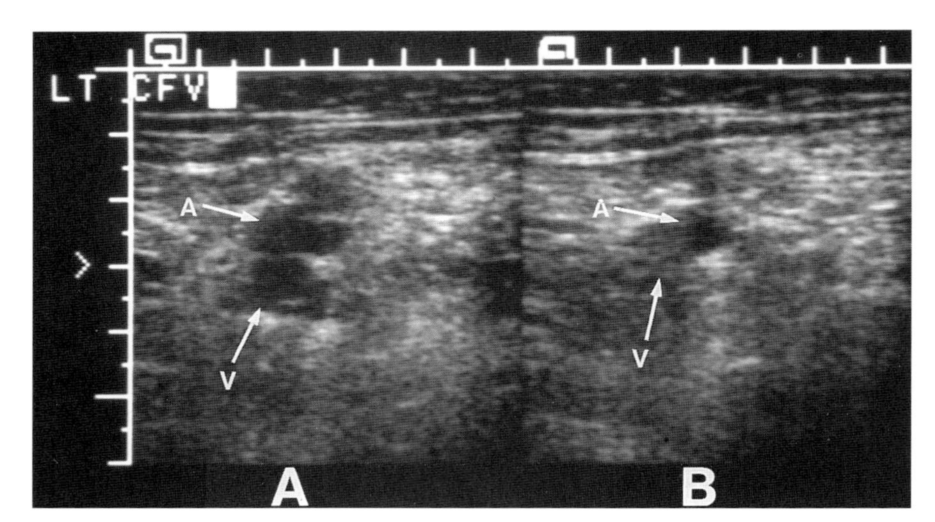

Figure 17.51. Normal venous ultrasound of the left common femoral vein. A: *Without compression.* The common femoral artery *(A)* and vein *(V)* are shown as two round anechoic/hypoechoic areas. B: *With compression.* The femoral artery *(A)* is essentially unchanged; the femoral vein *(V)* completely compresses (winks).

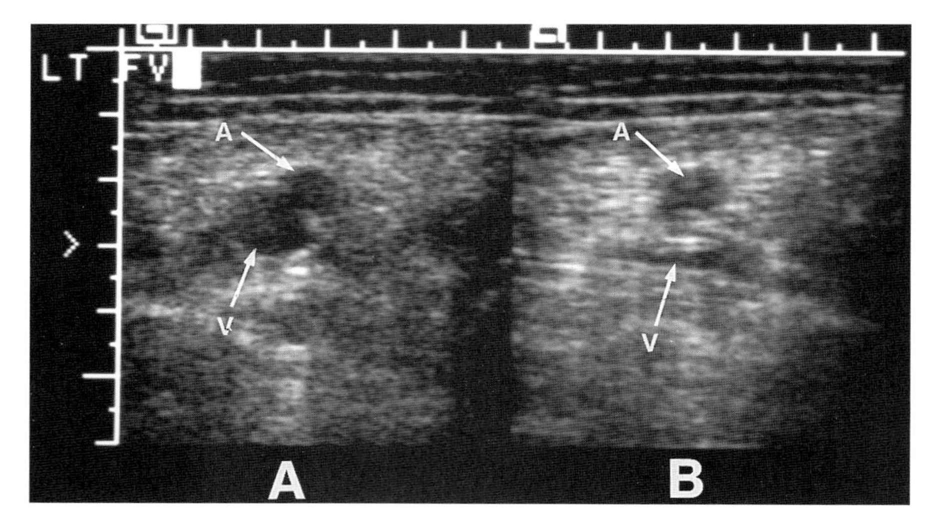

Figure 17.52. Normal lower-level venous ultrasound of the left femoral vein of the patient in Figure 17.51. A: *Without compression.* The femoral artery *(A)* and vein *(V)* are shown as two round anechoic/hypoechoic areas. B: *With compression.* The femoral artery *(A)* is essentially unchanged; the femoral vein *(V)* almost completely compresses.

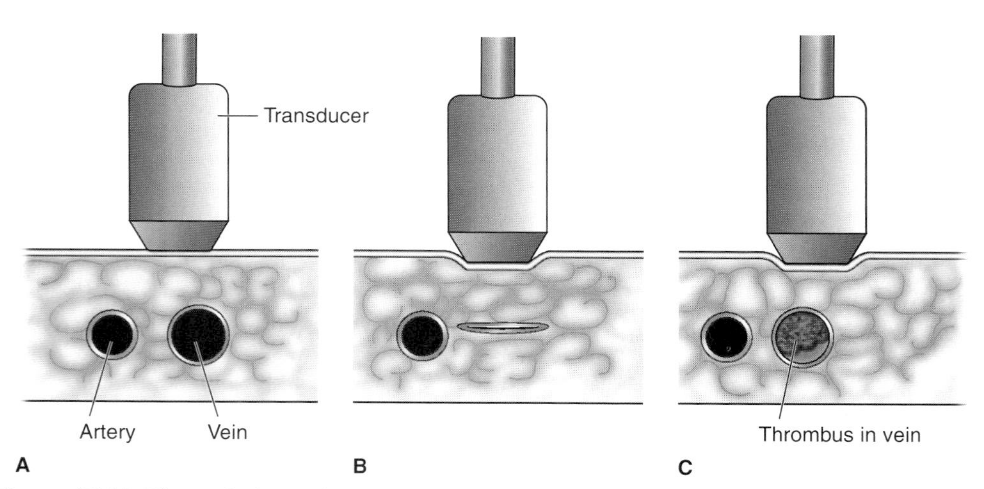

Figure 17.53. The technique of venous ultrasound. A: Normal vein without compression. B: Normal vein with compression. The vein completely compresses (winks) because no thrombus is present. C: Thrombosed vein with compression. Thrombus within the vein prevents its compression and its round shape does not change.

Today, deep venous thrombosis is diagnosed almost entirely by *venous ultrasound.* For this examination, an ultrasound transducer is held over the groin to show the femoral artery and vein in cross section; they appear as two round black anechoic/hypoechoic areas (because they contain liquid blood) (Figures 17.51A and 17.52A). The key feature of this examination is to determine whether the vein is compressible with slight pressure from the ultrasound transducer. In a normal examination, compression applied to the transducer compresses the *femoral vein* until the lumen disappears (the vein *"winks"*), but the *femoral artery,* because it has a thicker and tougher wall, and high internal pressure, does *not compress and wink* (Figures 17.51B and 17.52B).

By contrast, if the patient has DVT and thrombus within the vein, which will interfere with its compressibility, the vein will not compress or wink. Figures 17.54 and 17.55 show absence of compression in the femoral and popliteal veins in a patient with extensive venous thrombosis. Ultrasound may also show that a thrombosed venous lumen contains echogenic material rather than anechoic fluid. And, finally, a Doppler ultrasound examination can show absence of Doppler flow signals in the thrombosed vein (Figure 17.56). In black and white photographs, Doppler ultrasound shows blood flow as white, the lack of flow as black (no signal); compare the color Doppler images in Figure 17.12.

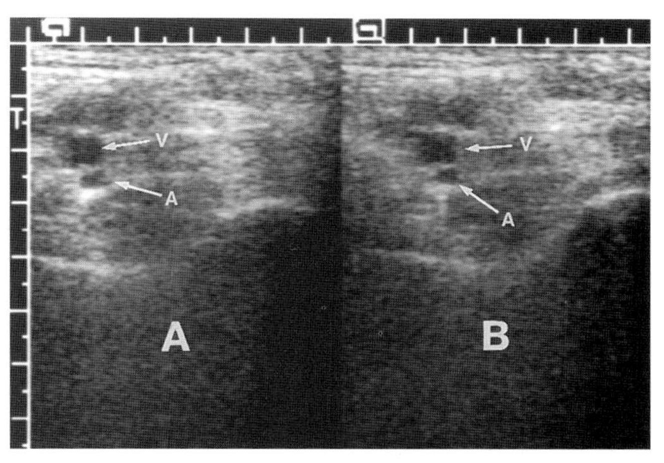

Figure 17.55. Abnormal venous ultrasound of the left popliteal vein, at the level of the knee, of the patient in Figure 17.54. A: *Without compression.* The popliteal artery *(A)* and vein *(V)* appear as two round anechoic/hypoechoic areas. B: *With compression.* The popliteal artery *(A)* and vein *(V)* remain unchanged. The popliteal vein does not compress because it is filled with thrombus.

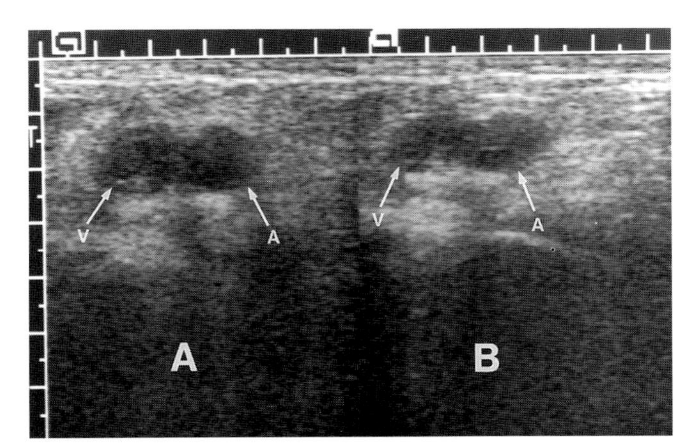

Figure 17.54. Abnormal left common femoral venous ultrasound showing the presence of deep venous thrombosis. A: *Without compression.* The femoral artery *(A)* and vein *(V)* appear as two round hypoechoic areas. B: *With compression.* The femoral artery *(A)* and the femoral vein *(V)* remain unchanged. The femoral vein does not compress because it is filled with thrombus.

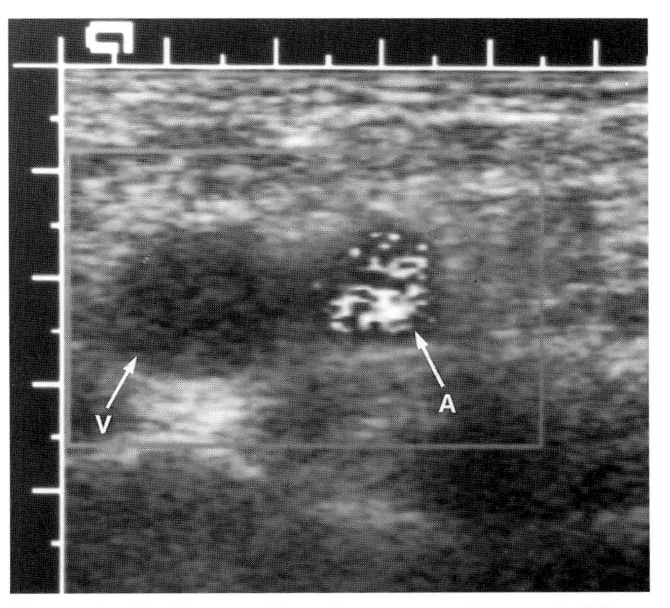

Figure 17.56. Abnormal Doppler venous ultrasound of the left femoral vein of the patient in Figures 17.54 and 17.55. Note the bright Doppler flow signals (appearing white on this black and white photograph) within the femoral artery *(A),* where flowing blood is present. These signals would appear in color on a color photograph, like those in Figure 17.12. The absence of Doppler flow signals in the femoral vein *(V)* is consistent with venous occlusion due to deep venous thrombosis.

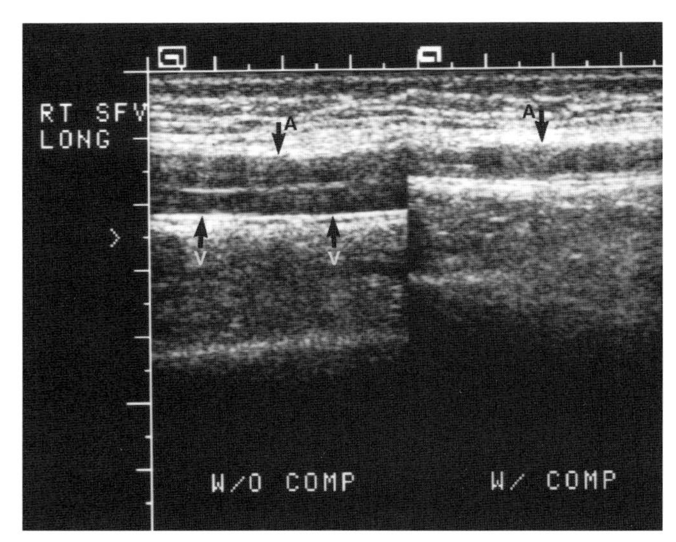

Figure 17.57. Normal longitudinal venous ultrasound of the femoral vein of the thigh. *Without compression (left):* The femoral artery *(A)* and vein (*V*'s) appear as two tubular anechoic/hypoechoic areas. *With compression (right):* The femoral artery *(A)* is essentially unchanged; the femoral vein completely compresses and is no longer visible.

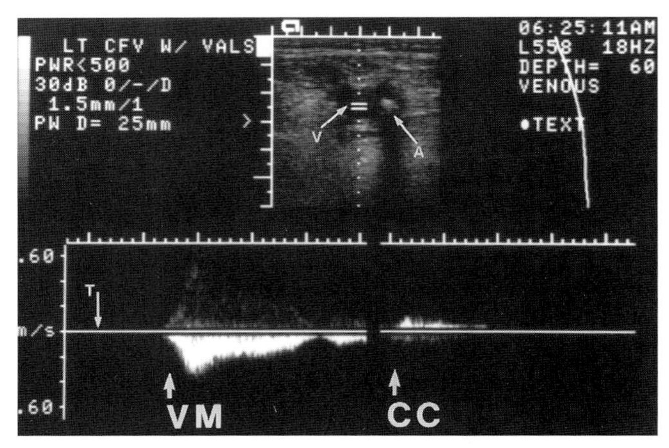

Figure 17.58. Normal common femoral vein Doppler flow tracing *(T)* with Valsalva maneuver *(VM)* and calf compression *(CC)*. At the top you can see the Doppler measurement site overlying the left femoral vein *(V)*, next to the femoral artery *(A)*. Note that as the Valsalva maneuver *(VM)* commences, the Doppler flow drops below the baseline tracing, because this maneuver decreases flow into the inferior vena cava and subsequently flow into the femoral vein in a normal patient; if the inferior vena cava or iliac veins proximal to the femoral vein were occluded with thrombus, the Valsalva maneuver would have no effect. Note also that the flow increases with calf compression; if the calf veins were occluded, compression would have no effect.

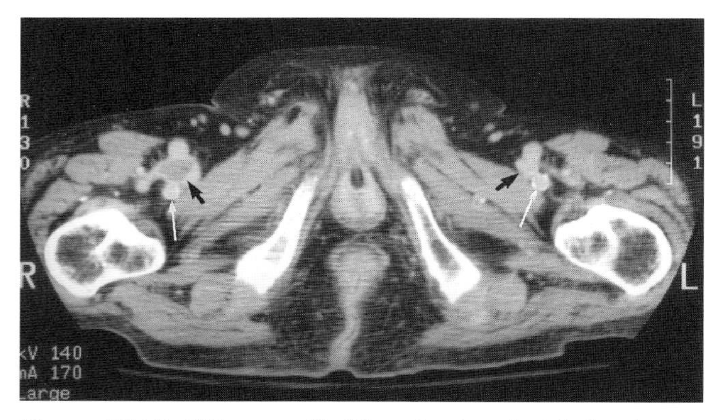

Figure 17.59. CT scan with IV contrast material showing right femoral vein thrombus (see text). The *black arrows* indicate femoral veins and the *white arrows* indicate femoral arteries.

In examining a lower extremity with venous ultrasound the ultrasonographer can search for thrombus within the deep venous system at multiple levels from the groin to the calf. By rotating the ultrasound transducer, the sonographer can image the veins longitudinally (Figure 17.57) as well as in cross section. A positive study will show lack of compressibility of the deep veins or lack of intraluminal venous thrombus at all the sites examined. Using Doppler techniques the examiner can observe the presence or absence of venous flow over the femoral vein (Figures 17.56 and 17.58). In a patient without thrombus, venous flow should stop when the patient performs a Valsalva maneuver and increase when the patient's calf is squeezed (which will cause flow augmentation). The flow effects caused by these techniques will be altered by the presence of thrombus, central to the transducer with the Valsalva maneuver and peripheral to the transducer with calf compression.

You may be wondering whether deep venous thrombosis can be seen on cross-sectional examinations such as CT and MR. Most definitely, yes, as you can see in Figure 17.59, which depicts a patient with DVT of the right leg. This slice from a contrast CT scan of the abdomen, obtained for another reason, shows thrombus in the right femoral vein *(black arrow on the right)*. Compare the normal left femoral vein *(black arrow on the left)*. The *white arrows* point to the femoral arteries, which are distinguished in this older patient by rims of atherosclerotic calcification. Remember, however, that you should not study a patient with suspected DVT with CT when the diagnosis can be made with the less expensive and simpler ultrasound examination. But whenever CT is requested to diagnose another condition, scans of the veins should be checked for signs of DVT.

Lymphangiography

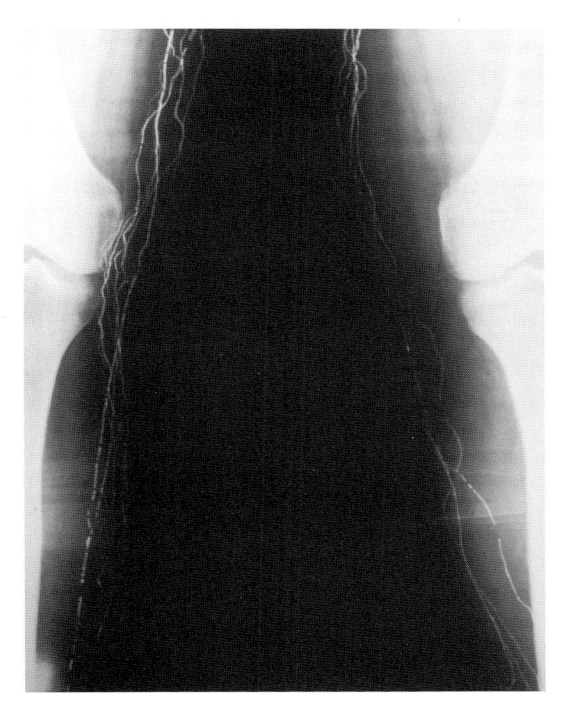

A

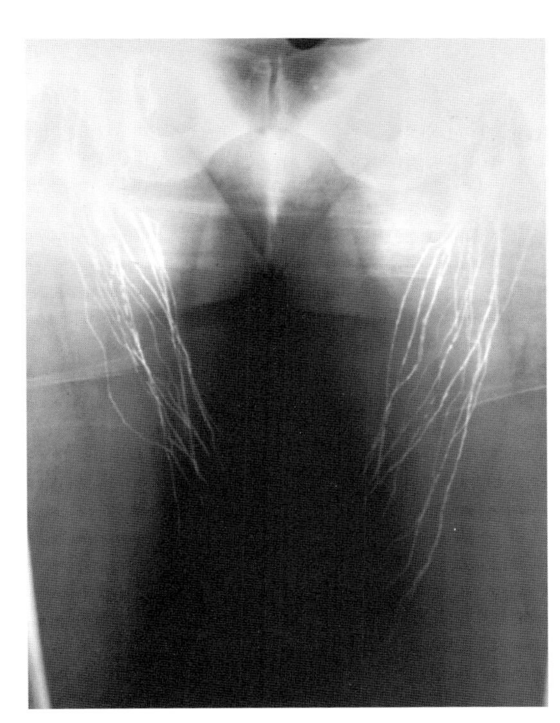

B

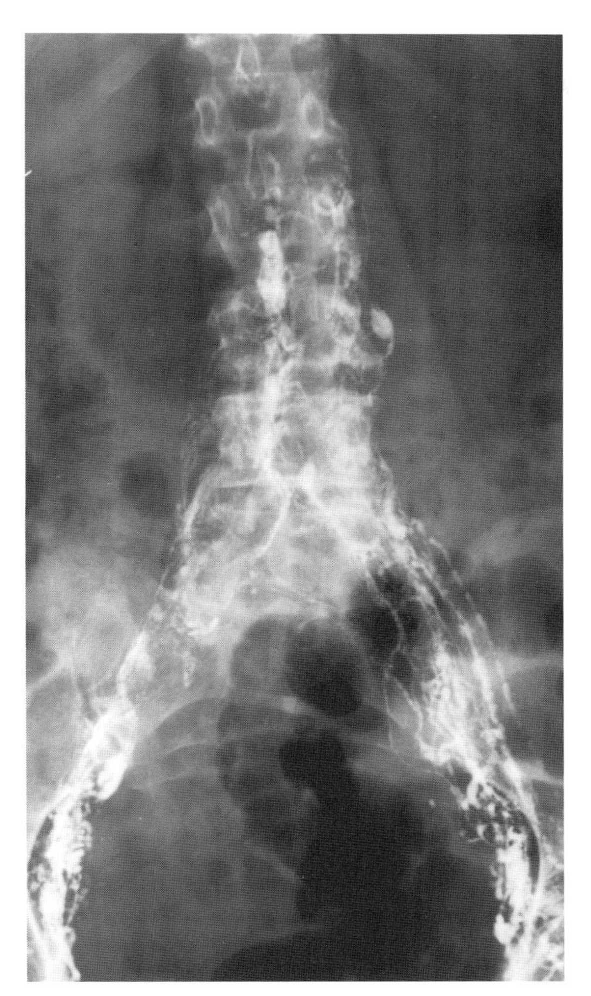

C

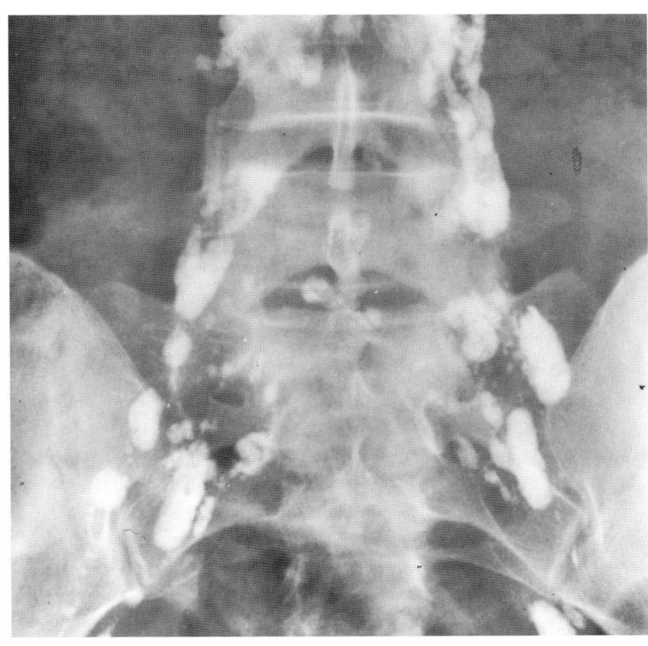

D

Figure 17.60. Normal lymphangiogram. A, B, and C: Initial lymphatic, channel phase films taken over the knees, upper legs, pelvis, and abdomen. Note the multiple thin lymphatic channels that follow the course of the greater saphenous vein. D: Nodal phase film, obtained 24 hours later. (See text on the next page.)

Lymphangiography is a delicate procedure requiring the efforts of both the patient and the radiologist. In order to radiographically opacify the iliac and para-aortic nodes, the radiologist must cannulate a lymphatic channel on the dorsum of each foot with a tiny 30-gauge needle via a small cutdown. A few milliliters of an iodinated, oily contrast medium are then slowly injected. This agent is transported passively, with the normal lymphatic flow, through the lymphatic channels of the legs, pelvis, and retroperitoneum, which drain into the thoracic duct. Within the first hour or two, only the lymph channels are opacified *(channel phase)*; later the nodes are opacified *(nodal phase)*.

But how does the radiologist find a transparent lymph channel to cannulate within each cutdown? After sterilely prepping and draping each foot, the radiologist injects a tiny volume of blue dye intradermally into the first and fourth toe web spaces. Within 10 to 20 minutes the blue dye is picked up by the lymphatic system of the foot, visually revealing the lymph channels for cannulation and injection of x-ray contrast medium. The entire procedure takes about 2 hours. Channel phase films are taken during the infusion of contrast medium or immediately afterward; nodal phase films are taken the next day.

Figure 17.60 illustrates the two stages in the lymphangiogram filming of a patient with normal lymph nodes. Figures 17.60A, B, and C are channel phase films, taken when the lymphatic channels are best opacified, but the lymph nodes are barely seen. Note the multiple small-diameter iliac lymph channels that follow the course of their associated blood vessels. At the level of the first lumbar vertebra the para-aortic channels drain into the cisterna chyli. Figure 17.60D is a nodal phase film taken 24 hours later. The lymphatic channels have cleared of contrast medium and the lymph nodes are better seen. Normal nodes are oval in shape with a homogeneous distribution of contrast medium. You will remember that the thoracic duct empties into the left subclavian vein, so that any excess oily contrast medium not retained by the nodes will flow into the venous system to the pulmonary capillary bed. Ultimately, the residual oily material will be coughed up by the patient. This is no problem for normal, healthy individuals, but you should be aware

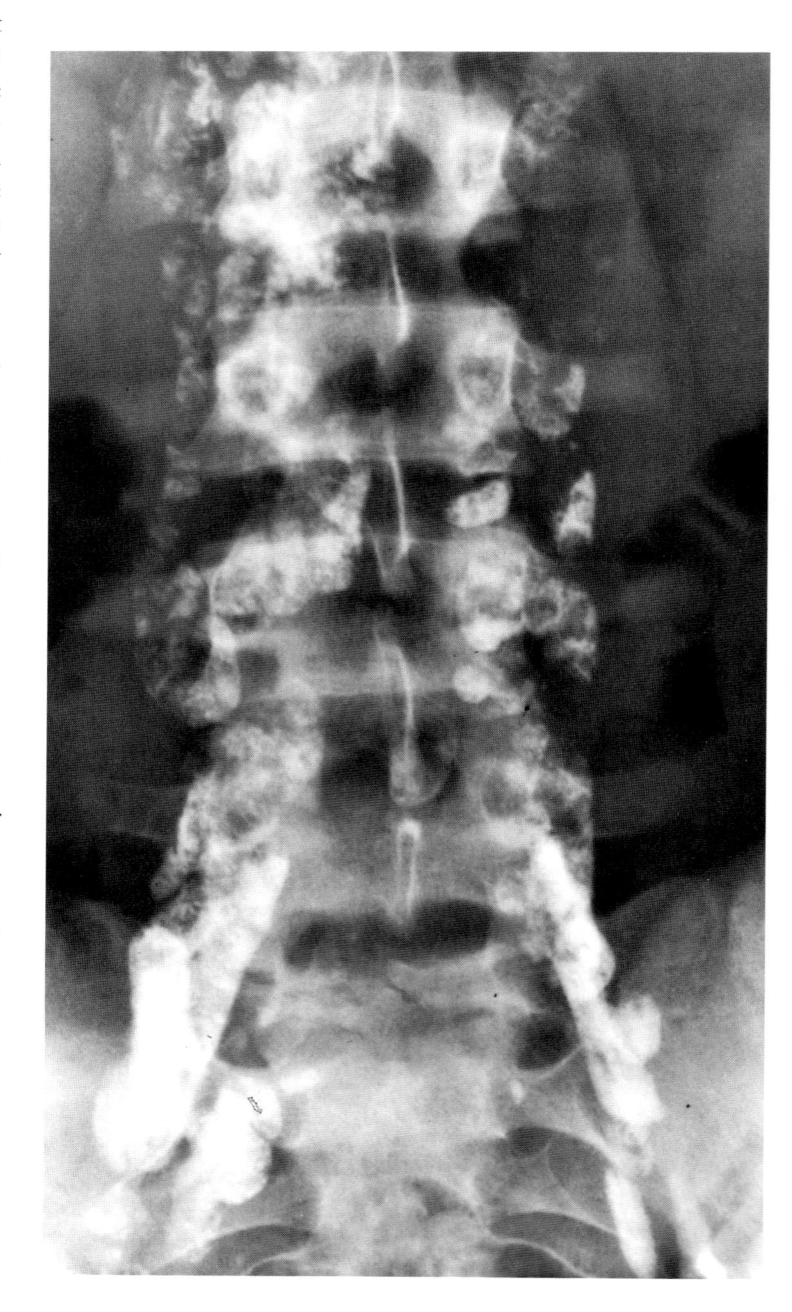

Figure 17.61. Abnormal nodal phase lymphangiogram showing enlarged para-aortic lymph nodes containing filling defects. This patient had widespread lymphoma.

that lymphangiography *is* hazardous for patients with severely compromised pulmonary function and those with a right-to-left cardiac shunt, which carries the risk of "paradoxical" systemic embolization. The radiologist performing the procedure always calculates the amount of contrast medium according to the needs of the patient.

Lymph nodes with lymphomatous involvement are uniformly enlarged, with the "foamy" appearance you see in Figure 17.61. With metastatic involvement, single or multiple "punched-out" focal defects are seen in normal-sized or enlarged nodes. In addition, metastases may occlude lymph channels so that blocked channels with collateral circulation may be seen on the channel phase films, and the nodes supplied from the blocked channels may not be opacified at all. In the staging of malignant disease, if an abnormal nodal defect is identified it can be biopsied percutaneously by the radiologist for cytological examination. Of course, nonneoplastic conditions such as infections may also cause lymph node enlargement.

Enlarged lymph nodes may also be imaged by CT. Because CT is faster, easier to perform, less expensive, and better tolerated by patients than lymphangiography, it has replaced lymphangiography as the initial procedure for staging malignancies that involve the retroperitoneal nodes. But although *CT can readily demonstrate enlarged nodes (Figure 17.62),* it *cannot detect lymphomatous or metastatic disease involving normal-sized nodes.* Consequently, lymphangiography is still indicated in staging lymphomas and genitopelvic malignancies, such as cervical and testicular carcinomas, when the CT examination is normal. Generally speaking, retroperitoneal lymph nodes measuring less than 1.0 centimeter in diameter at CT are considered to be normal in size.

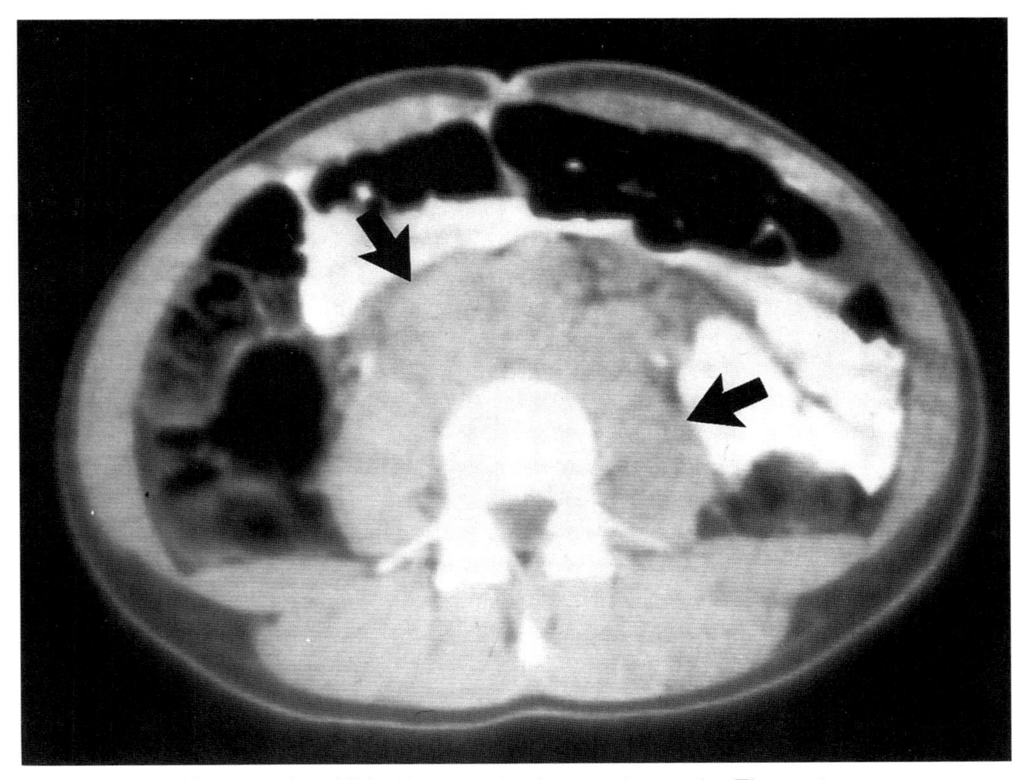

Figure 17.62. CT scan of a child with extensive lymphadenopathy. The aorta, inferior vena cava, and psoas muscles are surrounded by a thick layer of enlarged lymphomatous nodes *(arrows).*

Problems

Unknown 17.1 (Figure 17.63)

PA (A) and lateral (B) chest films of a 67-year-old man with an asymptomatic lung mass *(arrows)*. Would you schedule this patient for a percutaneous aspiration biopsy for diagnosis or would you recommend another imaging examination?

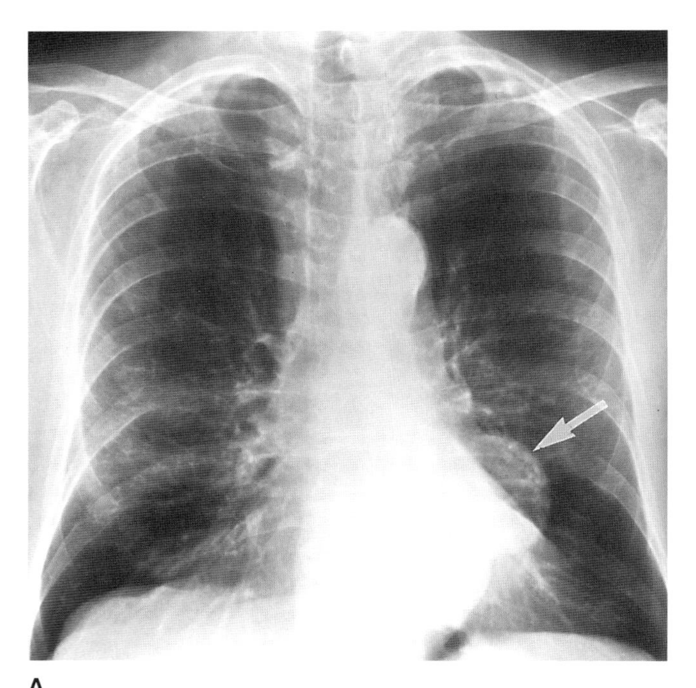

A

Figure 17.63 ***(Unknown 17.1)***

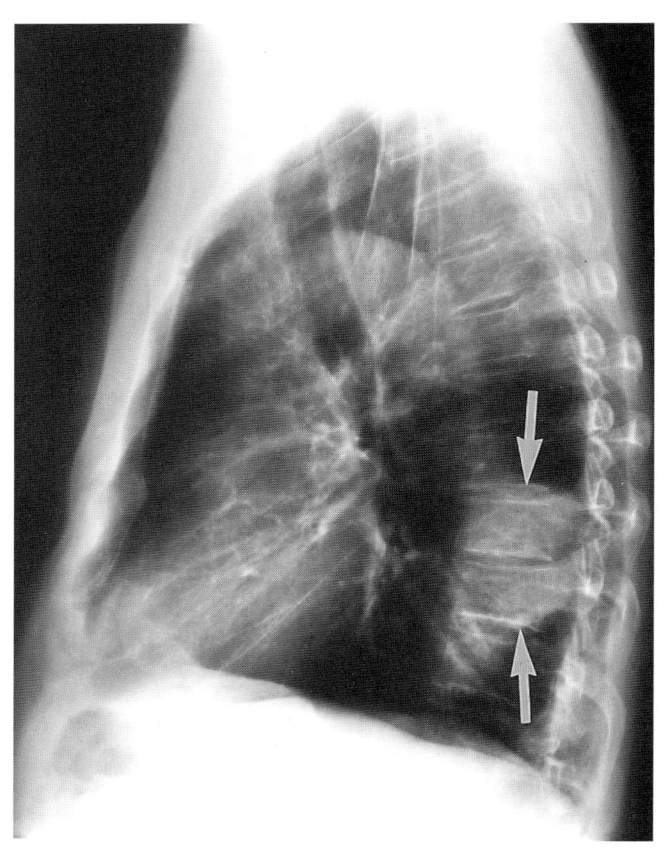

B

Unknown 17.2 (Figure 17.64)

This 63-year-old woman presented to the emergency ward with chest and back pain. What are your conclusions after reviewing her chest film (A) and chest CT scans (B-D)?

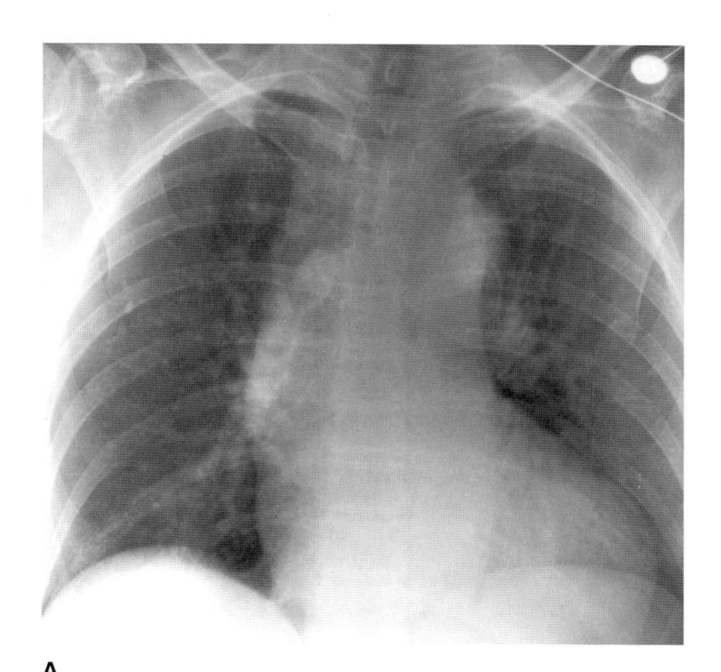

A

Figure 17.64 ***(Unknown 17.2)***

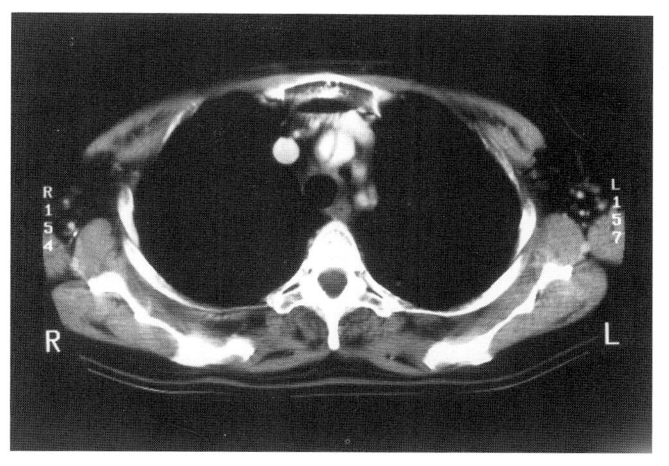

B

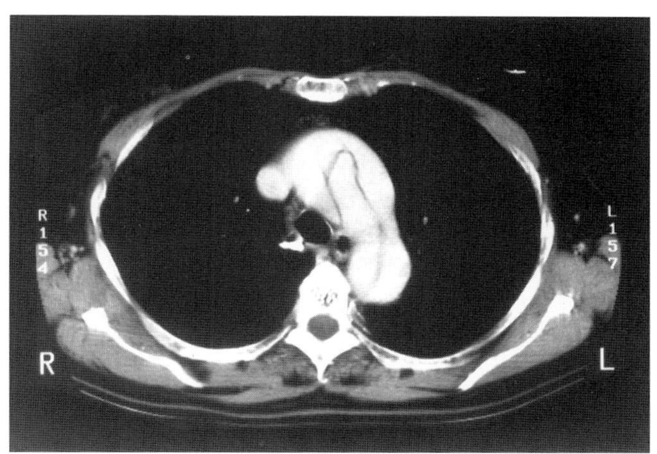

C

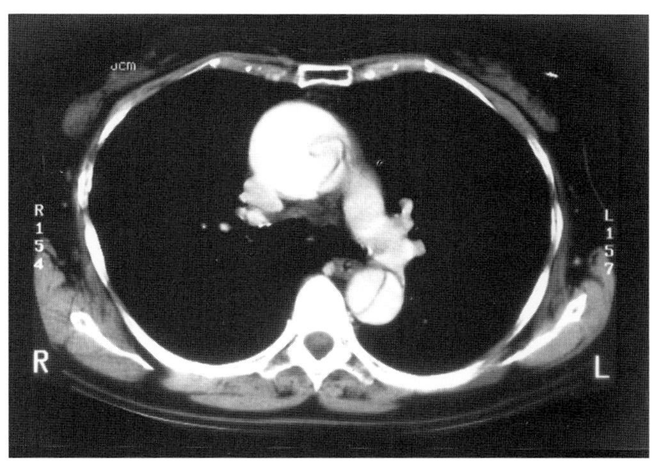

D

18 The Central Nervous System

The central nervous system (CNS) is very well organized, and shows little anatomic variation from patient to patient. The anatomy is complex, but once you learn the basics you will be comfortable with the imaging techniques in common use. In the process you will see how the anatomy of the brain and its coverings is altered by the conditions that imaging can diagnose.

In this chapter we will review the major anatomic structures shown on skull films, face films, CT scans, MR scans, cerebral arteriograms, and myelograms, and discuss the diagnostic imaging of the more common central nervous system conditions.

Figure 18.1. Mona Lisa positioned for a lateral skull film.

Imaging Techniques

The imaging workup of CNS conditions, in contrast to conditions affecting other parts of the body, usually begins with higher-tech studies such as CT and MR. Plain skull films are seldom requested today, because they show only the bony skull, and do not reveal abnormalities of the brain, except for intracranial calcifications and rare cases of pneumocephalus. CT and MR show the brain in exquisite detail and can image an enormous variety of brain pathologies. In fact, CT and MR scans show brain anatomy so well that these images are often used to teach neuroanatomy; you will be able to refresh your knowledge of neuroanatomy from the CT and MR images shown in this chapter. The imaging workup of your patients with acute and chronic CNS conditions will no doubt begin with a CT or MR scan.

Because skull films are still in use and studying them will increase your knowledge of imaging, we will review them for you. Only 75 percent of U.S. hospitals have a CT scanner; in the remaining institutions, mostly smaller rural facilities, skull films may be obtained even today for CNS emergencies such as head trauma. Any patient in one of these smaller hospitals who shows clinical signs of a serious head injury (a neurological deficit, for example) or whose plain skull film shows evidence of trauma (such as a skull fracture) should be transferred as quickly as possible to a facility with a CT scanner and a neurosurgical service.

The usual skull film series consists of five views: AP, PA, both laterals (each side in turn close to the film cassette), and a Towne's view (AP, with the x-ray tube angled caudad to show the occipital bone). Study the skull drawings (Figures 18.2 and 18.3) and the lateral and PA skull films on the following pages. On

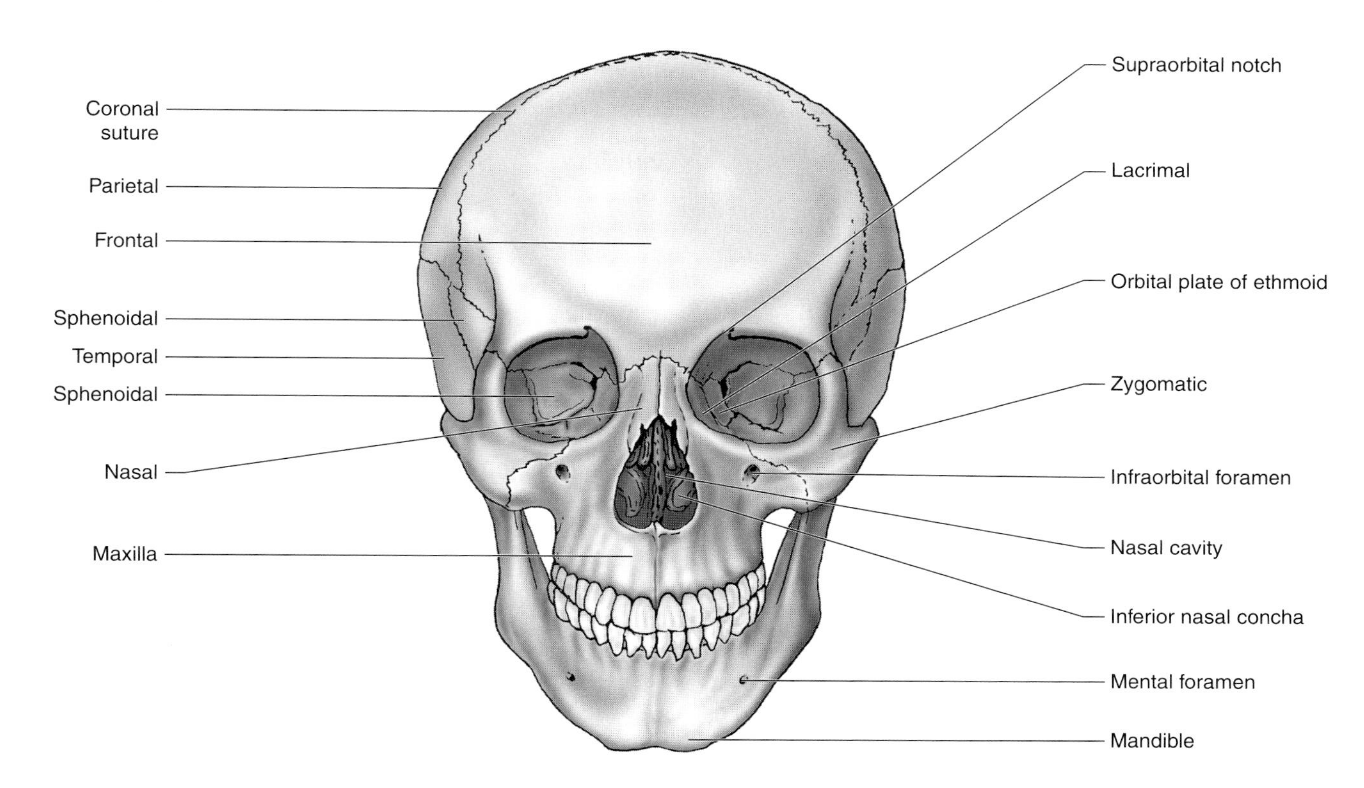

Figure 18.2. Frontal view drawing of the skull.

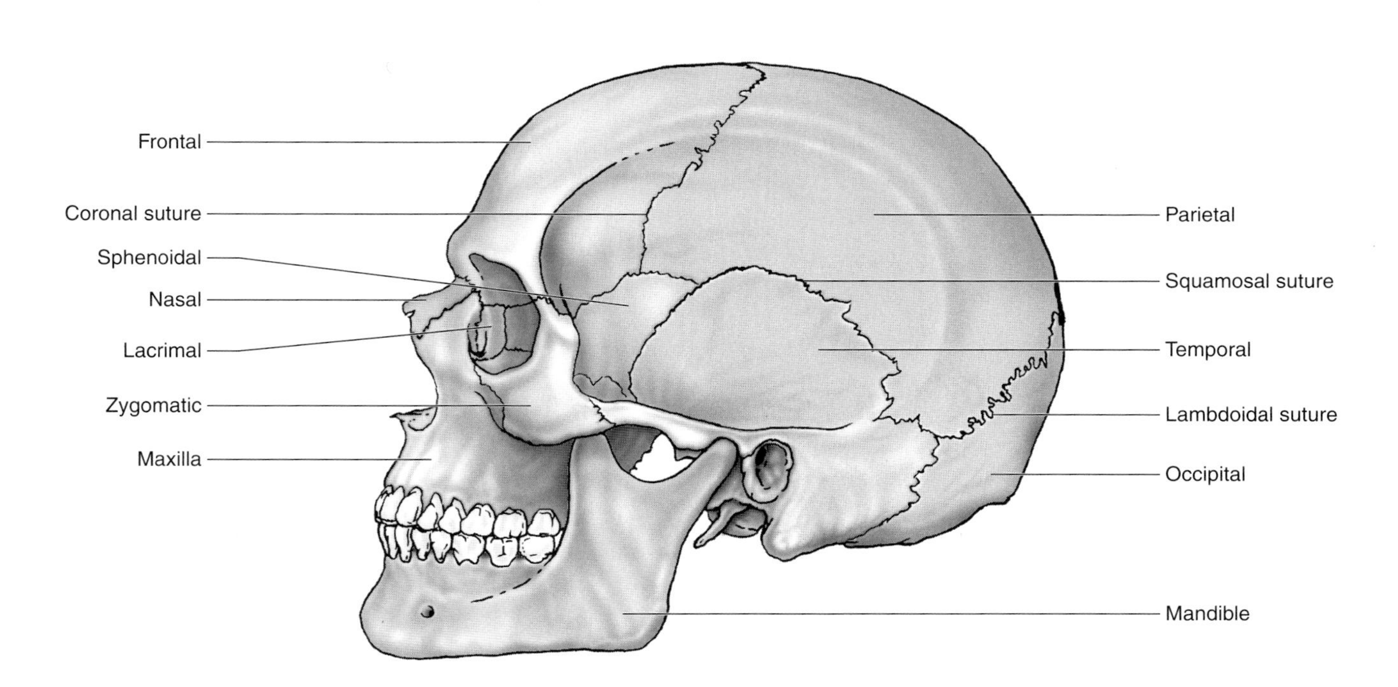

Figure 18.3. Side view drawing of the skull.

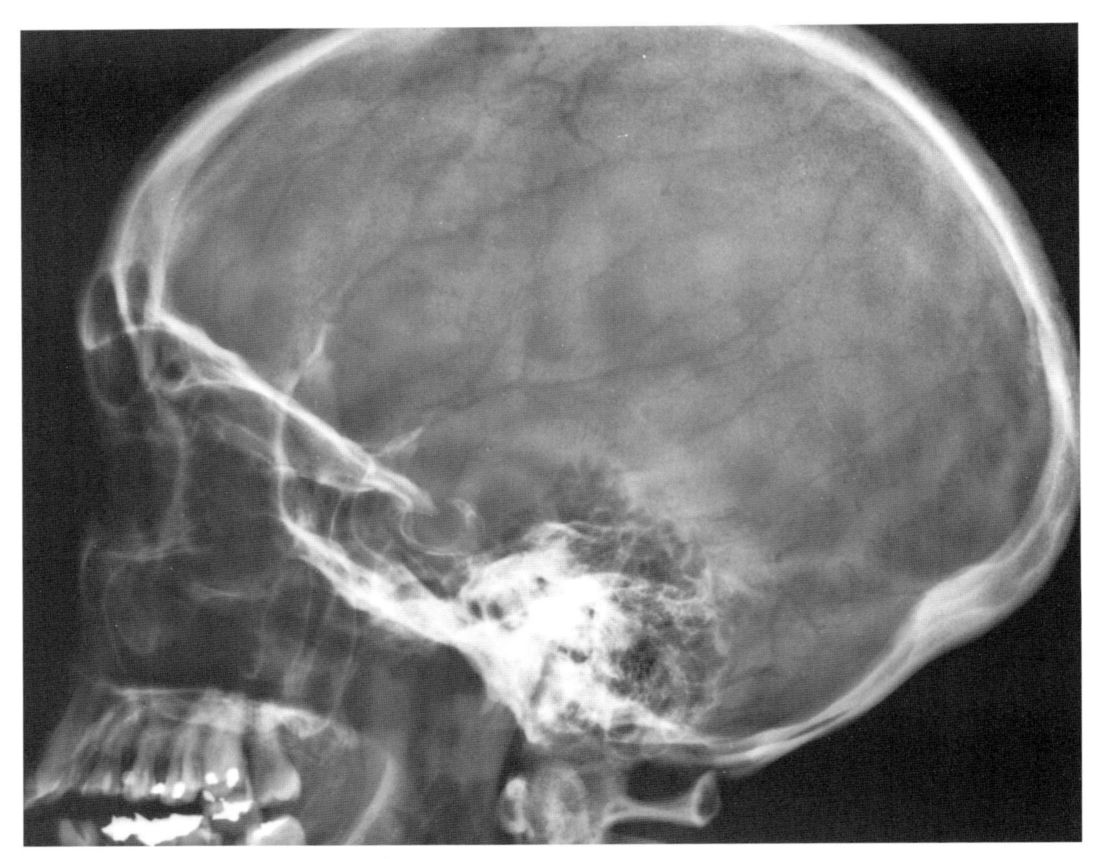

Figure 18.4. Normal lateral skull film.

the lateral view (Figure 18.4) the face is "burned out" by the exposure necessary to penetrate the calvaria (brain case). You can see that the calvaria is composed of two layers of compact bone, an inner table and an outer table, separated by the diploic space. The vascular markings you see include the meningeal artery grooves on the inner table and the venous channels within the diploic space. The calvaria is composed of the frontal bone (located anteriorly), two parietal bones (laterally), the occipital bone (posteriorly), the temporal bones (inferior laterally), and sphenoid bone (inferiorly). The coronal and lambdoidal sutures are superimposed on the lateral skull film and are barely visible as fuzzy lucencies along the expected course of the sutures; the lambdoidal sutures are easiest to see posteriorly in Figure 18.4.

Skull films made PA and AP look different, because in one the orbits are close to the film while in the other they are projected from far away and therefore are magnified, showing larger round circles of bone. Each depicts best the details of the bony structures closest to the film. Figure 18.6 is a PA film of a patient with a skull fracture having both depressed and linear components. Note that the fracture line is much blacker than either the vascular markings or the suture lines.

Sutures usually remain visible throughout life, distinguishable from fracture lines by their serpiginous character and white margins of compact bone. A frac-

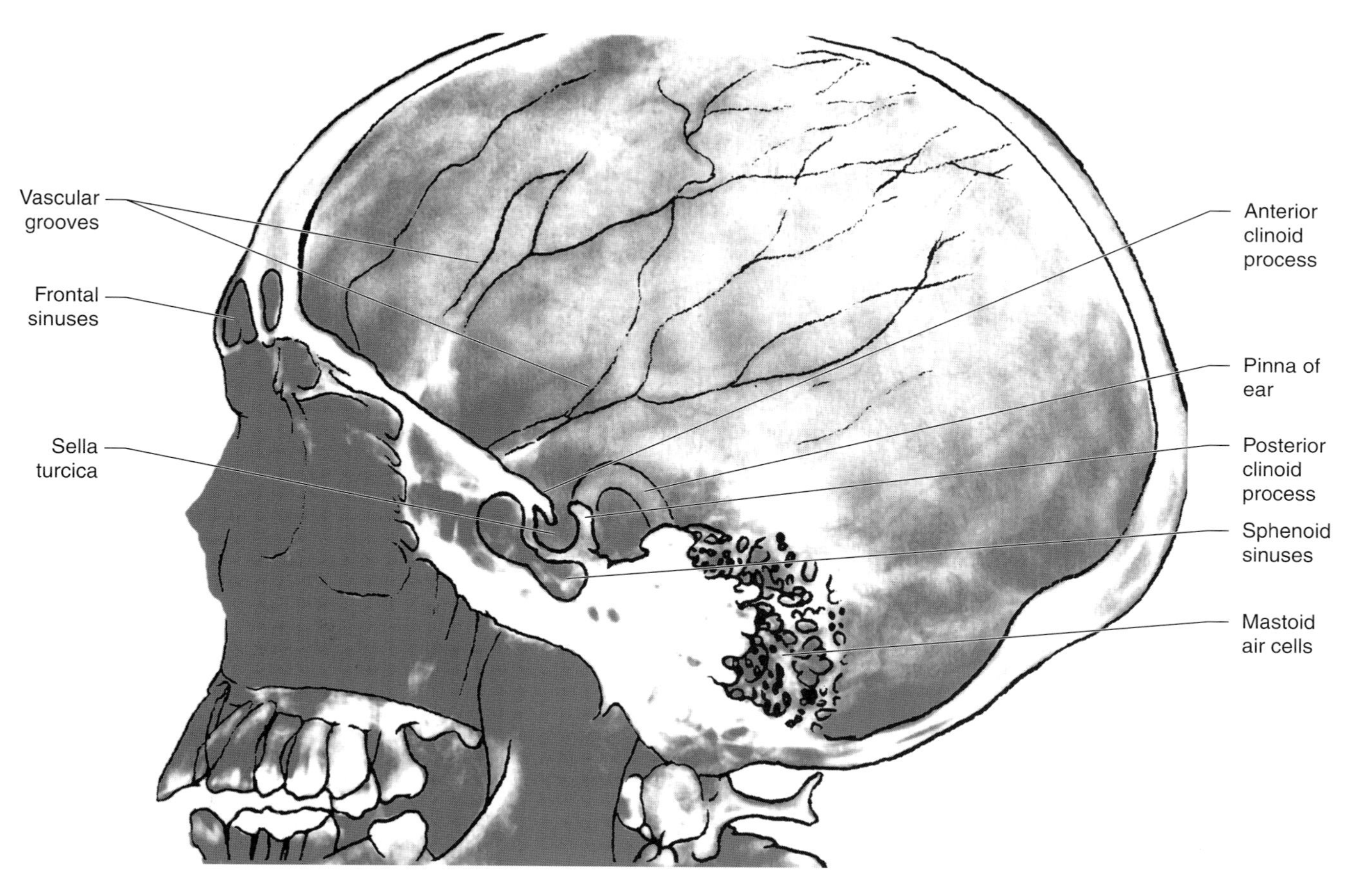

Figure 18.5. Labeled diagram to match Figure 18.4.

ture appears more linear and not marginated because there is no border of compact bone along a fracture line. Vascular grooves, such as impressions from the middle meningeal artery branches, appear in their expected anatomic locations and are not as black as fracture lines, which break both tables of bone and interrupt the diploic space.

In the neonate, as you have already seen, the brain can be visualized with ultrasound through the still-open fontanels and thin temporal bone; this is of course not possible in the older child or the adult. Ultrasound is the screening test of choice when perinatal brain injury or hemorrhage is suspected. *Cerebral arteriography,* with injection of an iodinated contrast medium into the carotid and vertebral arteries, is most commonly used to evaluate cerebrovascular disease and to better delineate cerebral aneurysms, arteriovenous malformations, and the vascular supply of brain tumors. Radioisotope brain scans, in which brain metabolism is studied by monitoring the photon or positron emission from intravenously injected radioisotopes, are performed much less frequently today because CT and MR yield more detailed information. Today radioisotope scans are useful in studying tumor metabolism and in determining brain death.

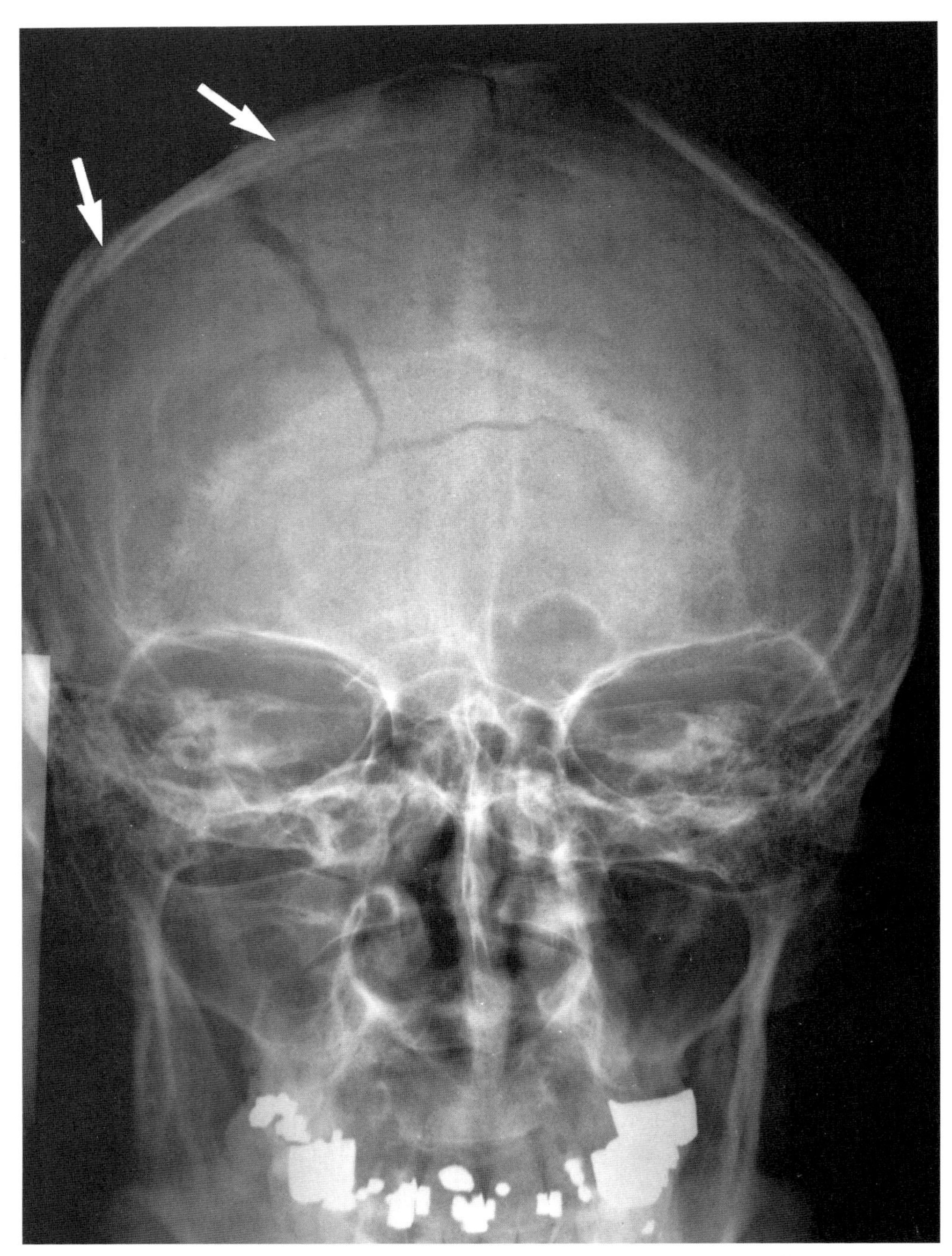

Figure 18.6. PA skull film with fractures that are both linear and depressed. The fracture fragment (between the *arrows*) is slightly depressed. This is not a simple fracture, therefore, but a comminuted one. Note the metallic dental fillings and crowns. Identify the sagittal and lambdoidal sutures, orbits, frontal sinuses, and mandible.

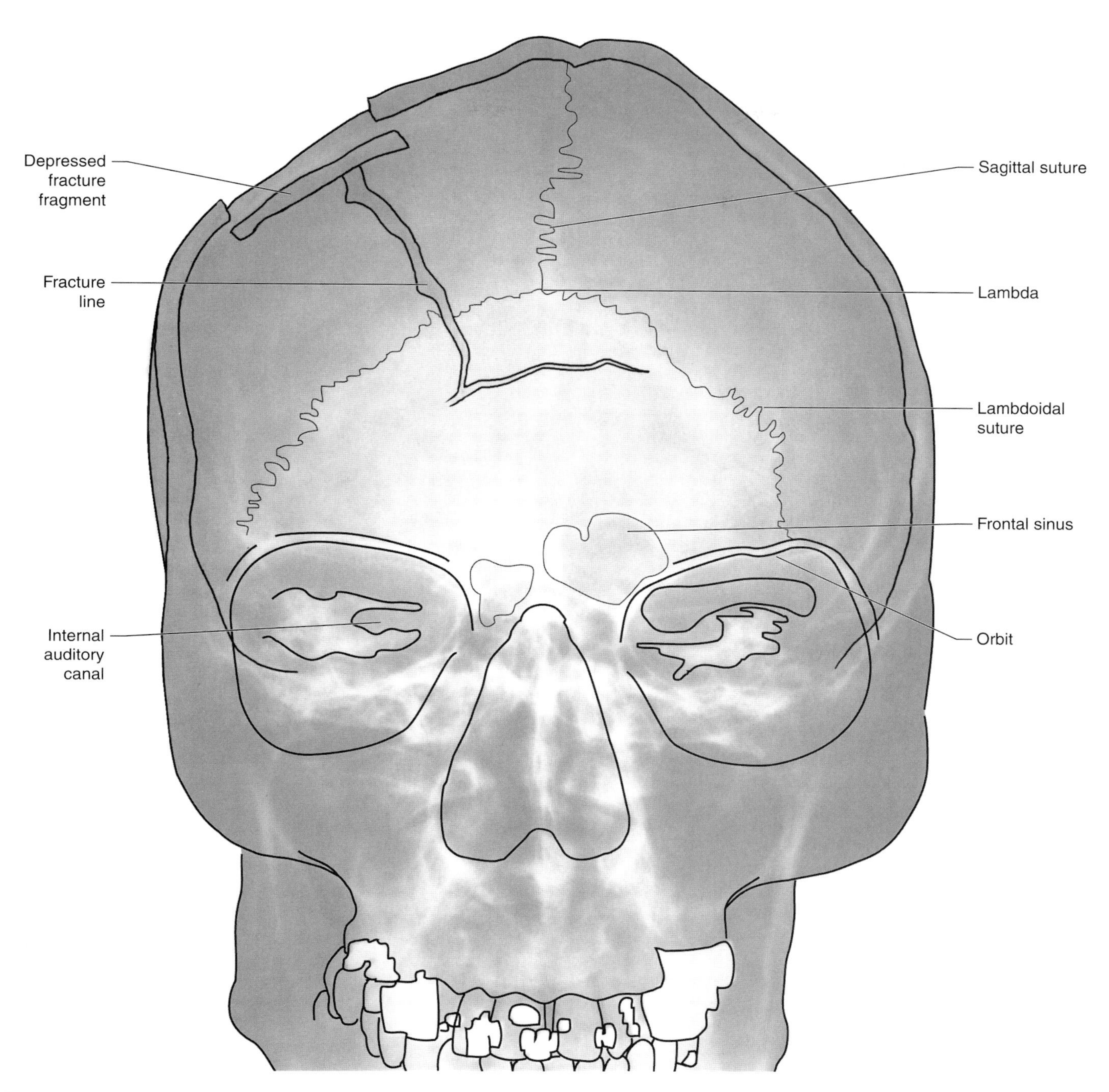

Figure 18.7. Labeled diagram to match Figure 18.6.

CT Anatomy of the Normal Brain

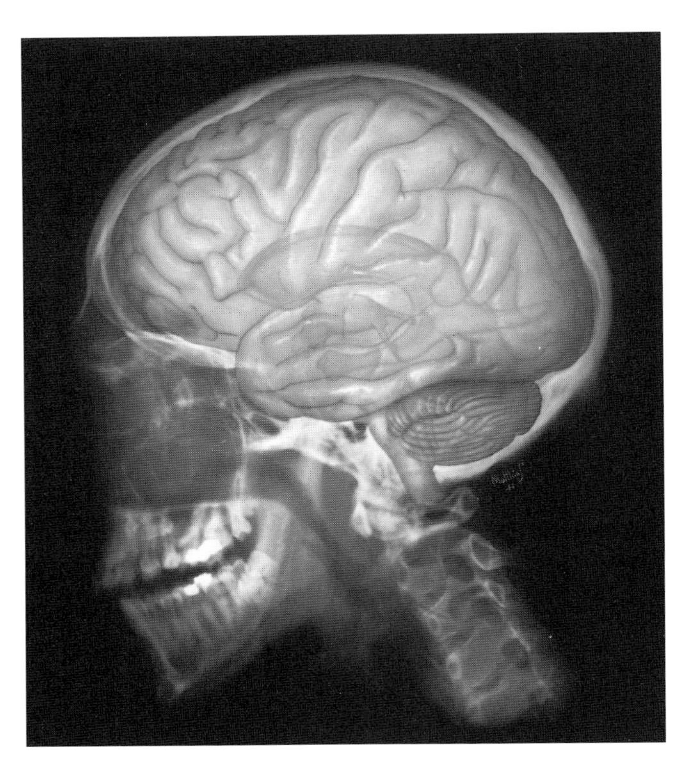

Figure 18.8. Superimposition of the apparently transparent brain and its ventricular system upon a lateral radiograph of the skull. The ventricular system is not as well seen here as in Figure 18.9.

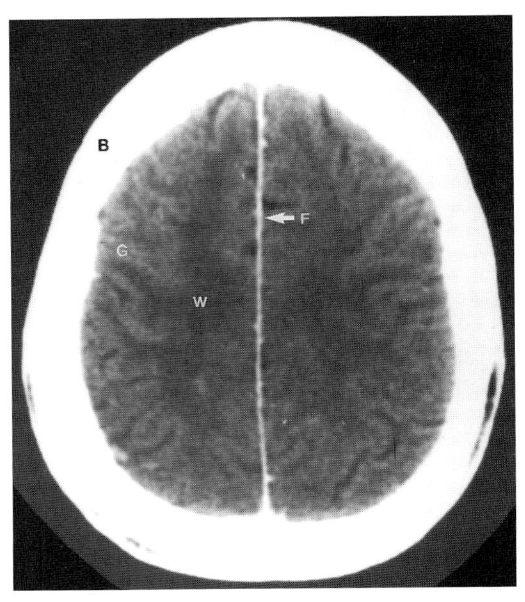

A. Upper scan obtained above the lateral ventricles. The falx cerebri *(F)* separates the two cerebral hemispheres. White matter *(W)* appears darker than gray matter *(G)*. Note the sulci and gyri. The bony calvaria *(B)* looks very white.

Figure 18.10A-E. Normal axial head CT scans obtained with intravenous contrast material and filmed with brain window settings.

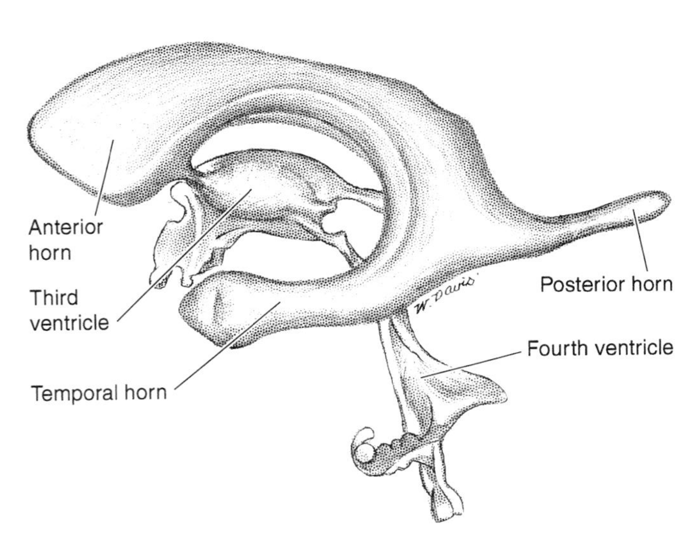

Figure 18.9. The ventricular system. Identify the two large lateral ventricles (superimposed in this drawing and not labeled), the third ventricle, the cerebral aqueduct (aqueduct of Sylvius) between the third and fourth ventricles, and the fourth ventricle. Note the anterior, posterior, and temporal horns of the lateral ventricle closest to you.

Study the CT scans of the normal head, performed after intravenous administration of contrast material, shown in Figure 18.10. They are a series of noncontiguous scans, photographed with a brain window, a CT setting of width and level that provides optimal visualization of brain tissue. These scans were obtained at intervals from the vertex of the skull down to its base. Compare the scans with the illustrations shown in Figures 18.8 and 18.9.

At CT cerebral white matter (shown centrally in the hemispheres) appears slightly less dense (blacker) than gray matter (shown peripherally) because the white-matter tracts contain more fatty tissue (myelin sheaths around neural processes). Cerebrospinal fluid (CSF) in the ventricles and subarachnoid spaces looks nearly black. The gray matter appears gray.

Recognize also the structures that appear bright (white) on head CT scans. In a noncontrast scan white areas almost always represent calcification or acute blood. On examinations performed with intravenous contrast material, like those in Figure 18.10, not only do calcifications and acute blood appear white, but so

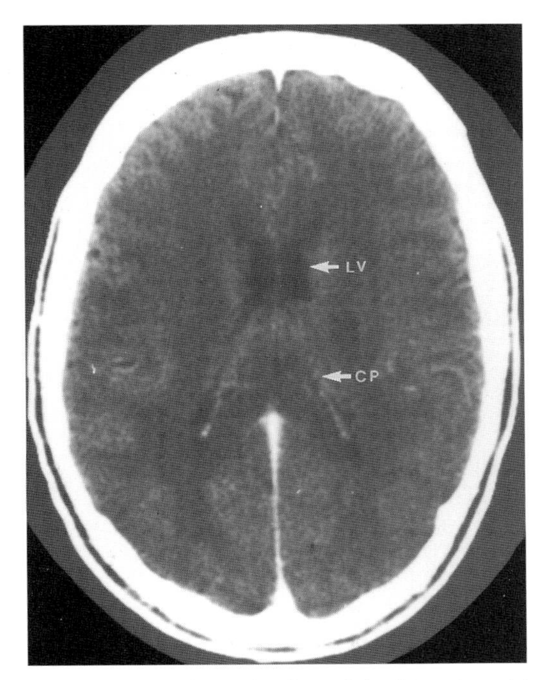

B. Scan through the bodies of the lateral ventricles. Cerebrospinal fluid within the lateral ventricles *(LV)* appears black, whereas the contrast-opacified choroid plexuses *(CP)* look white.

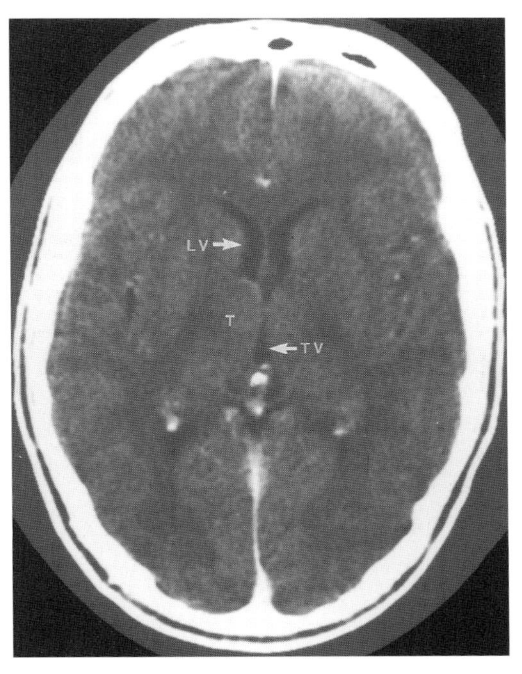

C. Scan through the level of the third ventricle *(TV)*. The thalami *(T)* are seen immediately lateral to the third ventricle. The anterior horns of the lateral ventricles *(LV)* can also be seen on this scan, separated by the thin septum pellucidum.

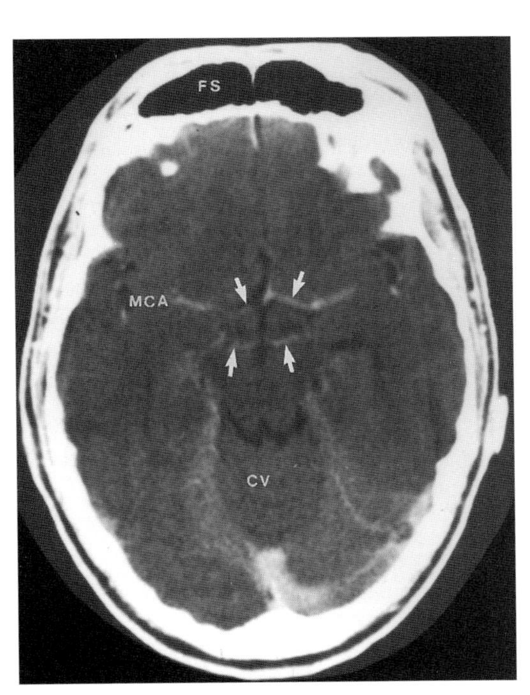

D. Scan through the intravenous contrast-opacified circle of Willis *(arrows)*. The middle cerebral arteries *(MCA)* can be seen coursing laterally within the sylvian fissures. This scan is low enough to show the cerebellar vermis *(CV)*. Air within the frontal sinuses *(FS)* is very black.

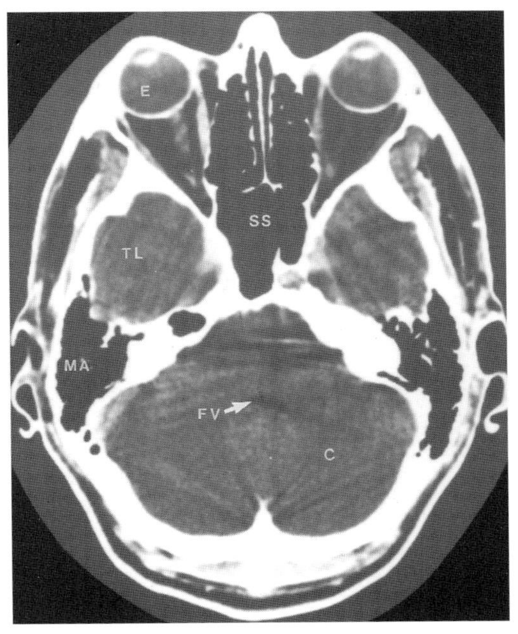

E. Scan through the skull base. The eyeballs *(E)* and optic nerves are visible, as are the sphenoid sinuses *(SS)*, temporal lobes *(TL)*, and mastoid air cells *(MA)*. The fourth ventricle *(FV)* can be seen anterior to the cerebellum *(C)*. You will soon see that MR provides better visualization of the posterior fossa structures (cerebellum, midbrain, and so on) than does a CT scan, in which the thick bone of the skull base produces artifacts (those streaklike densities coursing across the cerebellum) that appear to be originating from bone.

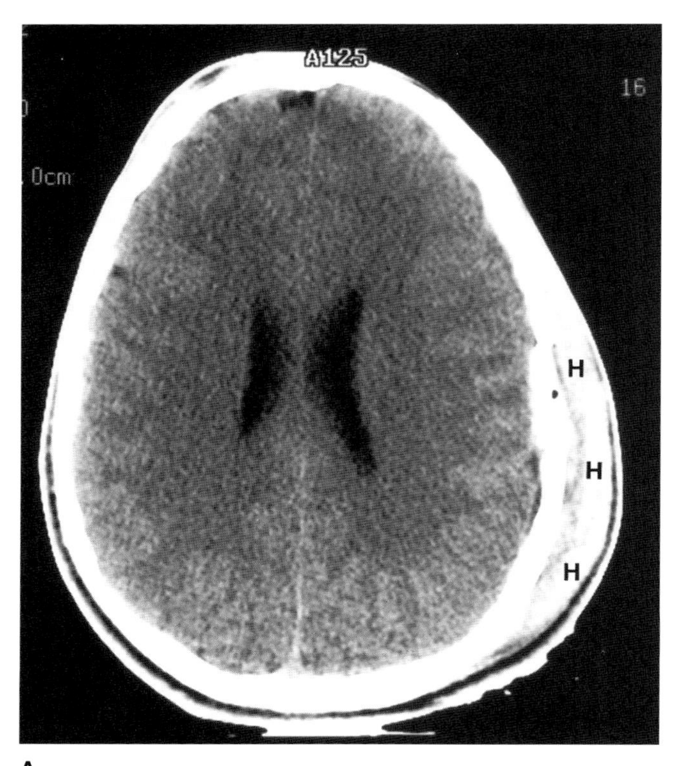

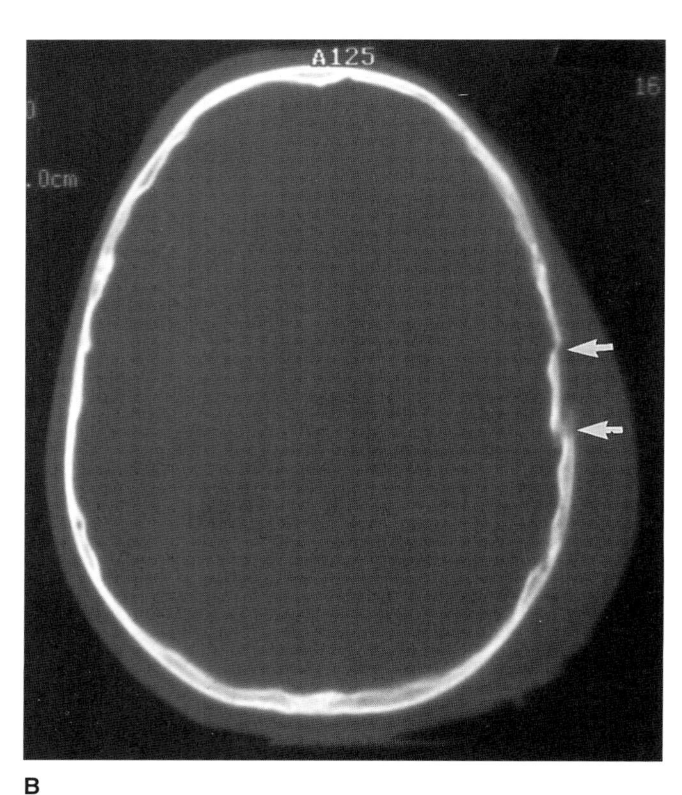

A B

Figure 18.11. Brain and bone window CT scans of a head trauma patient. Both scans were obtained at the same site. The only change is the CT window and level settings at which these images were photographed (window and level settings can be altered at the CT scanner control panel, even after scanning is completed). A: The brain window image shows a large subgaleal hematoma (*H*'s indicate hemorrhage between the skin of the scalp and the calvaria). No intracranial brain injury is seen. B: The bone window image, however, shows a depressed skull fracture at the site of the subgaleal hemorrhage *(arrows).*

also do vascular structures, such as the cerebral arteries and veins, the falx cerebri, and the choroid plexes of the lateral ventricles. The bones of the calvaria and face also appear white, of course, because of the calcium content of the bone. These structures are poorly defined on brain window scans; the bony detail, of course, would be better shown with bone window CT settings, as you can see in Figure 18.11.

MR Anatomy of the Normal Brain

MR scans of the brain allow even greater differentiation between gray and white matter than does CT. In addition, MR scanning allows direct imaging in the coronal and sagittal planes. Study the axial and midsagittal scans in Figures 18.12 and 18.13. At the MR settings used in these scans, the T1 settings, the CSF appears black, the gray matter gray, and the white matter white. Although cortical bone appears black (because of a lack of MR signal), fat (which has a strong MR signal) within the bone marrow and in the scalp appears very white. Compare the T1- and T2-weighted images of another patient in Figure 18.14. On the T2 settings, the CSF is white, and the fat-containing structures, such as the bone marrow and the white matter, are relatively darker.

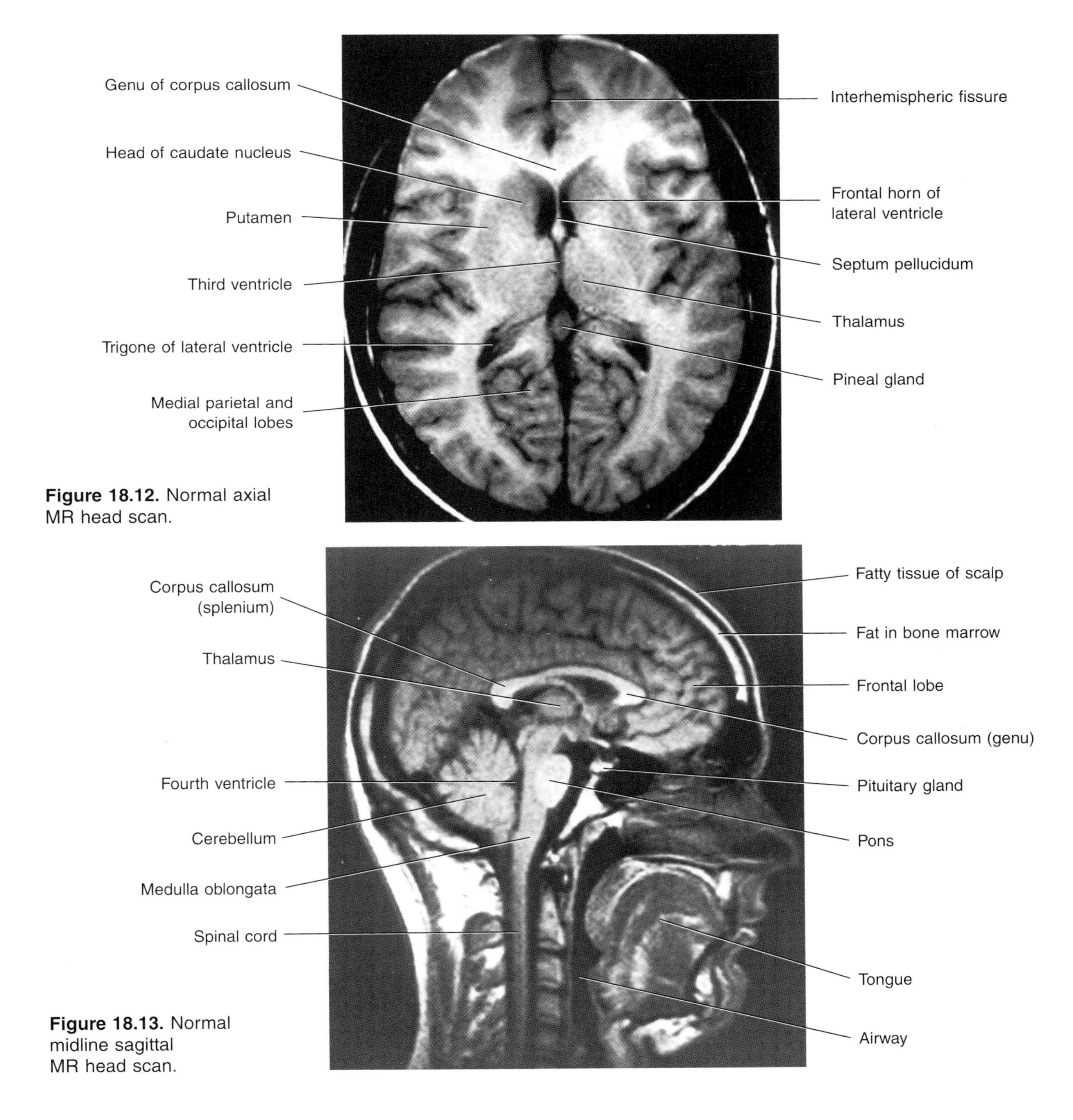

Figure 18.12. Normal axial MR head scan.

Figure 18.13. Normal midline sagittal MR head scan.

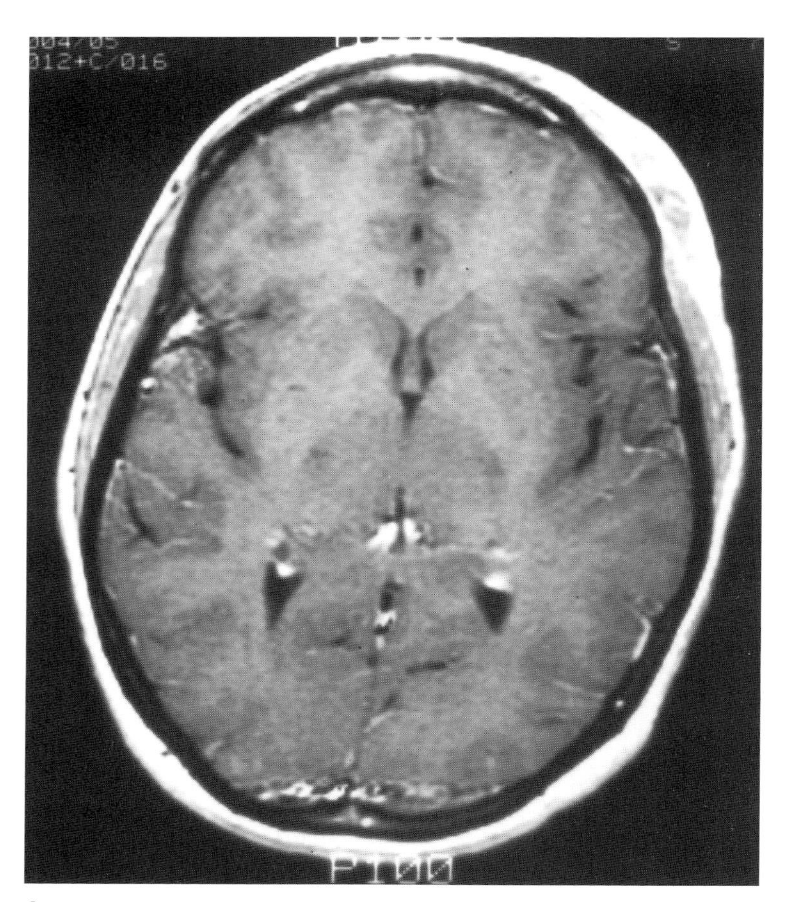

A

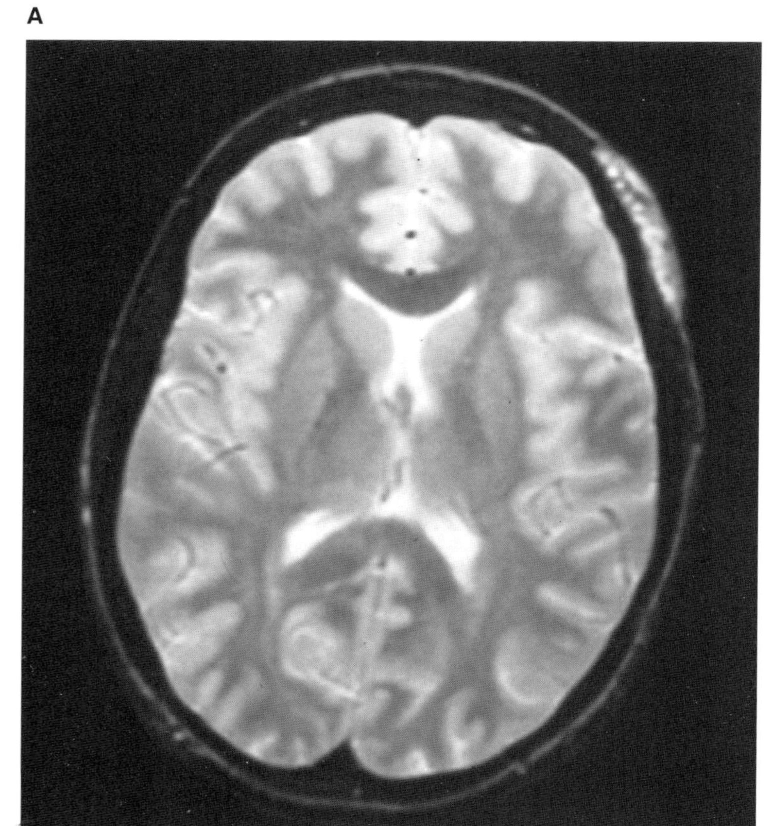

B

Figure 18.14. T1- and T2-weighted axial MR brain images taken of the same patient at the same level. A: T1-weighted image shows black CSF, white subcutaneous fat (beneath the scalp), and white matter brighter than gray matter. This image was obtained with intravenous contrast material. Note the enhancement of the veins and choroid plexus. B: T2-weighted image shows white CSF, black subcutaneous fat, and white matter darker than gray matter.

CT and MR Compared

Contrast the appearance of a posterior fossa tumor on the CT and MR scans in Figure 18.15. On the MR scan, not only are the resolution and the differentiation of white and gray matter improved, but fewer artifacts are seen than with CT. Linear streaks and other artifacts from thick bony structures interfere with the CT scan but not with the MR scan. Remember that cortical bone is black on the MR scan and white on the CT scan. On the T1-weighted MR scan, the subcutaneous fat is white, while on the CT scan it is black; the scalp is only barely visible on the CT scan, as a thin white line seen beyond the skull. The contrast resolution of MR is much greater than that of CT.

Intravenous contrast materials can improve visualization of many pathological conditions during CT and MR scanning. Iodinated contrast agents are used for CT scanning and gadolinium contrast agents for MR scanning. Iodinated agents increase the CT density of blood vessels and vascular structures so that they appear brighter and whiter on scans. Gadolinium, a rare-earth, metallic, paramagnetic substance, shortens the T1 and T2 relaxation times at MR scanning and thus enhances the appearance of pathological conditions.

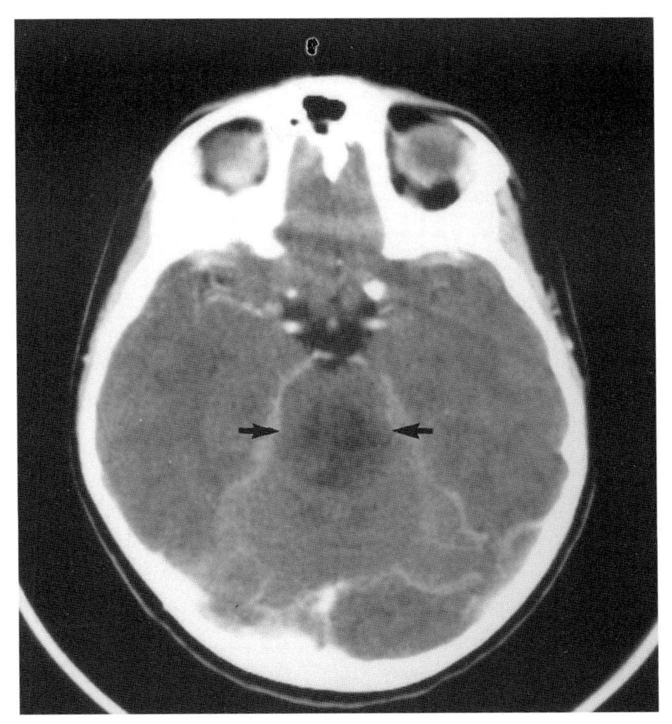

A

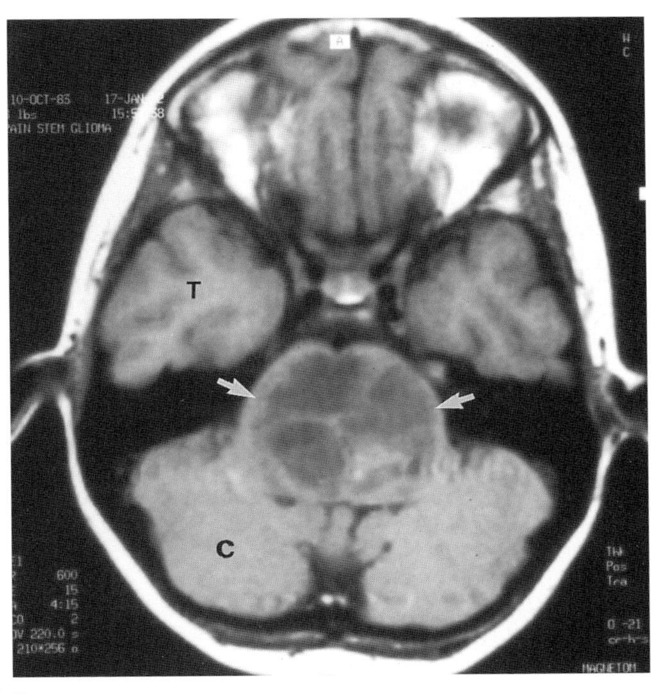

B

Figure 18.15. CT and MR scans of a patient with a midbrain tumor. A: CT scan with intravenous contrast material shows an indistinctly margined, low-attenuation region *(arrows)* in the midbrain. Linear streak artifacts originating from bone course across the brain parenchyma. B: MR scan shows a low-signal, well-defined tumor *(arrows)* in the midbrain. The cerebellum *(C)* and temporal horns *(T)* are well depicted.

Study Figure 18.16, consisting of six images from the CT and MR examinations of a patient with a right occipital glioma brain tumor. Compare the CT scan with the MR images. Note the differences between the T1- and T2-weighted MR images, and compare the T1-weighted MR images with and without gadolinium contrast material with the parasagittal (no gadolinium) and coronal (with gadolinium) images. As you can clearly see, MR is capable of providing a great variety of images, depending on the imaging parameters chosen, which are usually referred to as *MR protocols*. These are custom-selected by the radiologist for each patient, based on the patient's clinical presentation and the pathological conditions suspected.

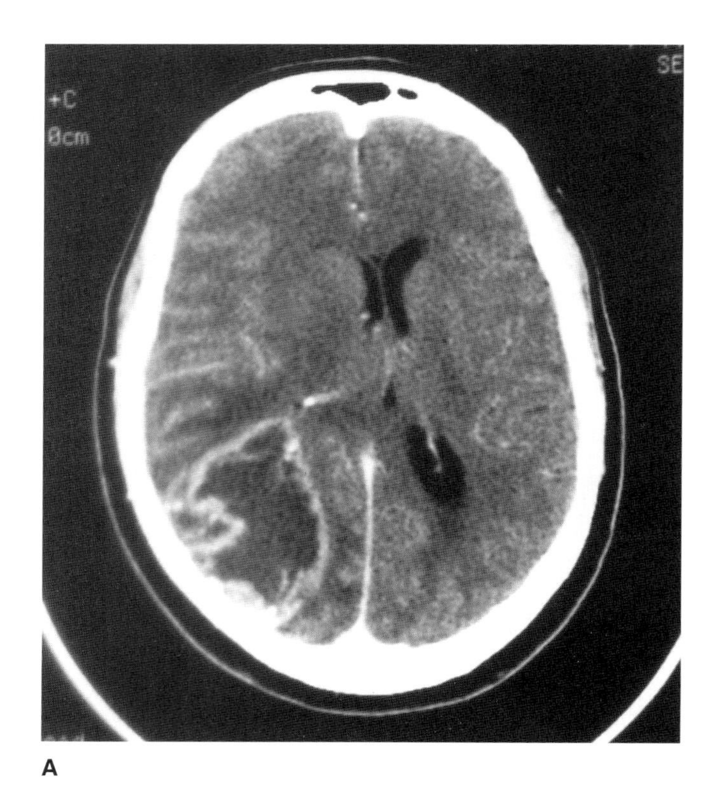

A

Figure 18.16. CT and MR scans of a patient with a right occipital glioma. Compare the appearance of the brain tumor on the various types of CT and MR scans. A: Axial CT scan with intravenous iodinated contrast material. The low-attenuation right occipital tumor has a hypervascular margin, and it compresses the right lateral ventricle and produces midline shift. B: T1-weighted axial MR scan without contrast material. An ill-defined right occipital mass is seen compressing the right lateral ventricle and producing midline shift. C: T2-weighted axial MR scan without contrast material. The tumor is better differentiated from the remainder of the brain than it is in B. D: T1-weighted axial MR scan with intravenous gadolinium contrast material. The margin of the tumor is brightly enhanced. E: T1-weighted parasagittal MR scan without contrast material shows an ill-defined mass *(arrows)* in the right occipital region. F: T1-weighted coronal MR scan with gadolinium shows the tumor with a brightly enhanced margin, within the right occipital lobe.

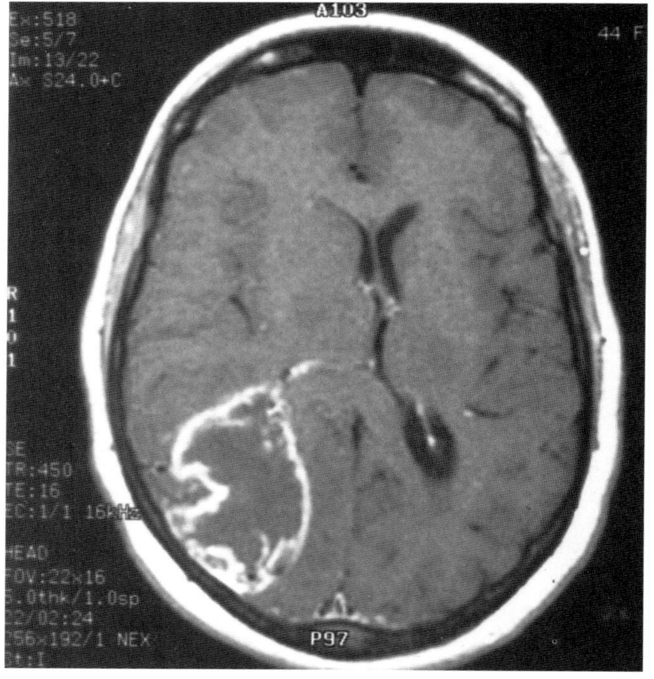

D

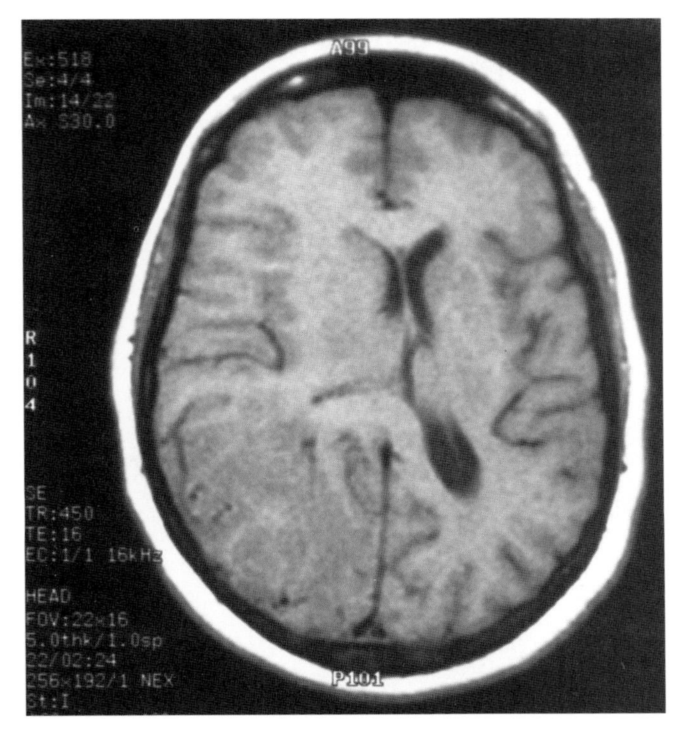
A99
Ex:518
Se:4/4
Im:14/22
Ax S30.0
R
1
0
4
SE
TR:450
TE:16
EC:1/1 16kHz
HEAD
FOV:22x16
5.0thk/1.0sp
22/02:24
256x192/1 NEX
St:I
P101

B

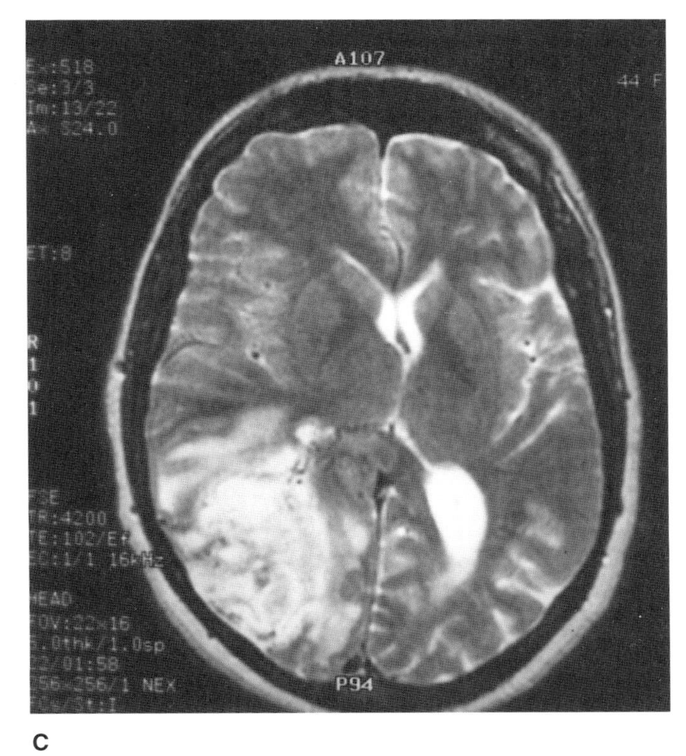
A107
Ex:518
Se:3/3
Im:13/22
Ax S24.0
44 F
ET:8
R
1
0
1
FSE
TR:4200
TE:102/Ef
EC:1/1 16kHz
HEAD
FOV:22x16
5.0thk/1.0sp
22/01:58
256x256/1 NEX
P94

C

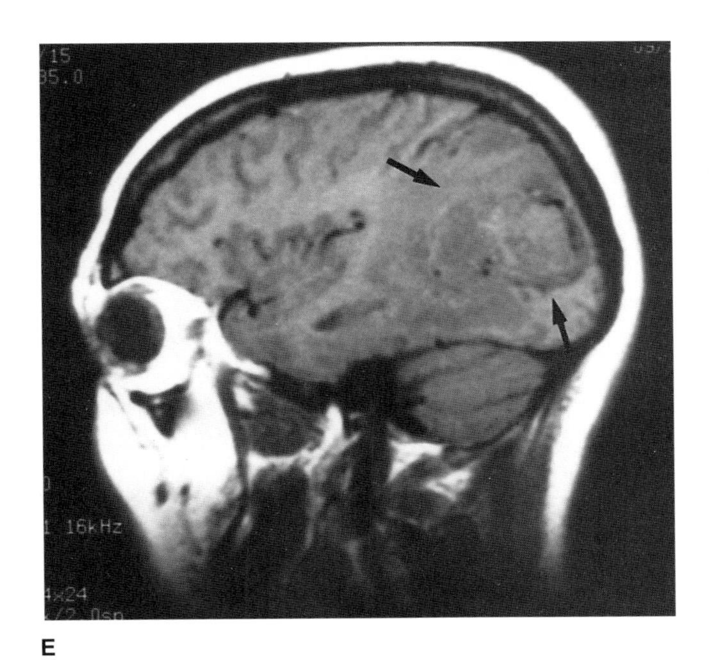
/15
35.0
1 16kHz
4x24

E

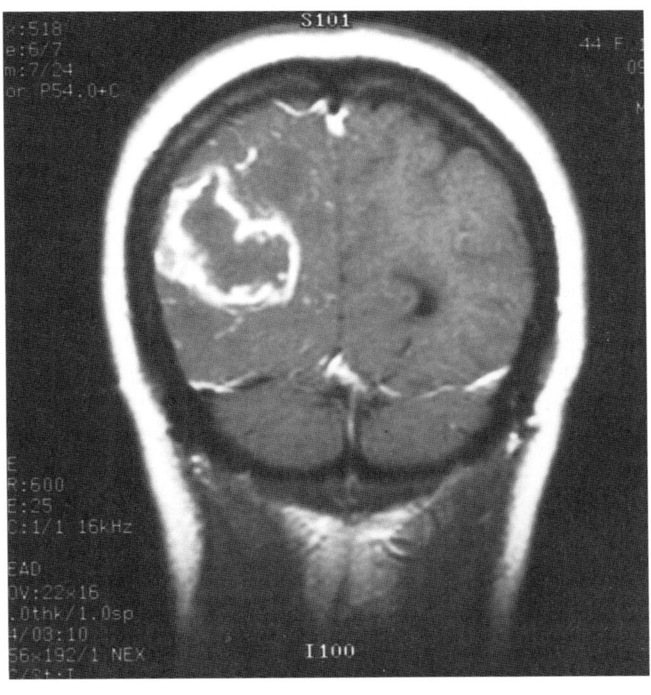
S101
x:518
e:6/7
m:7/24
or P54.0+C
44 F
R:600
E:25
C:1/1 16kHz
EAD
OV:22x16
.0thk/1.0sp
4/03:10
56x192/1 NEX
I100

F

Hydrocephalus, Brain Atrophy, and Intracranial Hemorrhage

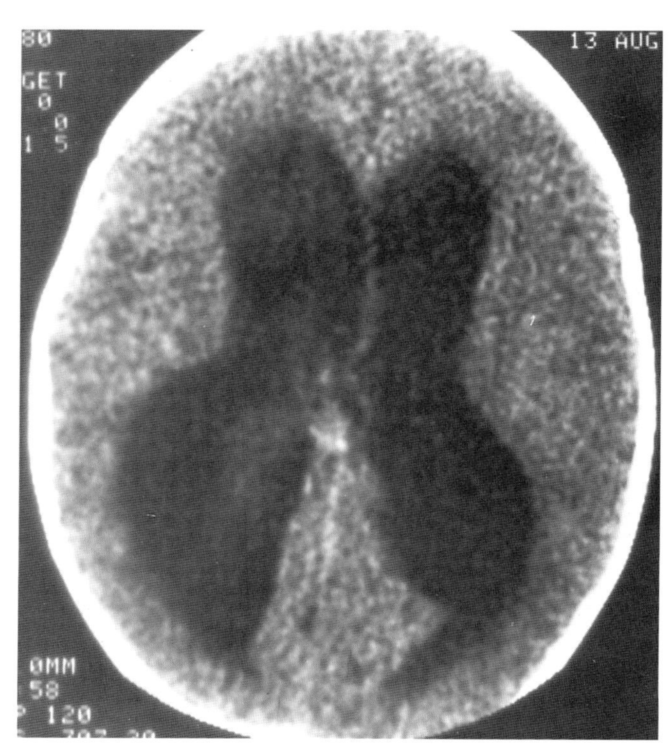

Figure 18.17. CT scan of an infant with obstructive hydrocephalus. The lateral ventricles are markedly dilated and the gyri and sulci are compressed.

Figure 18.18. MR scan of another child with severe obstructive hydrocephalus.

Cerebrospinal fluid is produced by cells of the choroid plexes of the lateral and third ventricles. From the production sites in these ventricles, CSF flows through the cerebral aqueduct (between the third and fourth ventricles at the level of the midbrain) into the fourth ventricle, which is the space between the pons and the cerebellum. The CSF then leaves the ventricular system via the two foramina of Lushka, at the sides of the inferior aspect of the cerebellum, and the foramen of Magendie, at the back of the inferior cerebellum. The CSF then flows over the cerebral hemispheres within the subarachnoid space, and is absorbed by arachnoid granulations at the superior sagittal sinus. At CT and MR we can see CSF within the ventricular system and over the cerebral hemispheres, extending within the sulci between the gyri.

Because CSF-containing spaces are clearly shown at CT as low-attenuation areas, it is easy to identify conditions that alter their size and shape, especially when these spaces are abnormally large or deformed. With *obstructive hydrocephalus,* the lateral and third ventricles are enlarged, while the fourth ventricle and the subarachnoid spaces remain small. This condition is usually caused by obstruction of the CSF flow at the cerebral aqueduct. Tumors of the midbrain are common causes of obstructive hydrocephalus. The child shown in Figure 18.17 had severe obstructive hydrocephalus caused by a tumor obstructing the cerebral aqueduct. In severe cases of obstructive hydrocephalus, the brain tissue becomes compressed symmetrically against the inside of the calvaria, effacing the gyri and sulci (Figure 18.18).

Communicating hydrocephalus, on the other hand, occurs when CSF is not reabsorbed from the subarachnoid space. In this condition the lateral and third ventricles are not dilated out of proportion with the fourth ventricle and subarachnoid spaces. Meningeal inflammation due to tumor cells or to hemorrhage is a common cause of communicating hydrocephalus.

In patients with *brain atrophy* the ventricles are enlarged, the sulci are prominent, and the brain is separated from the brain case, because the brain has become smaller. Brain atrophy, also called *hydrocephalus ex vacuo,* occurs with normal aging; you will regu-

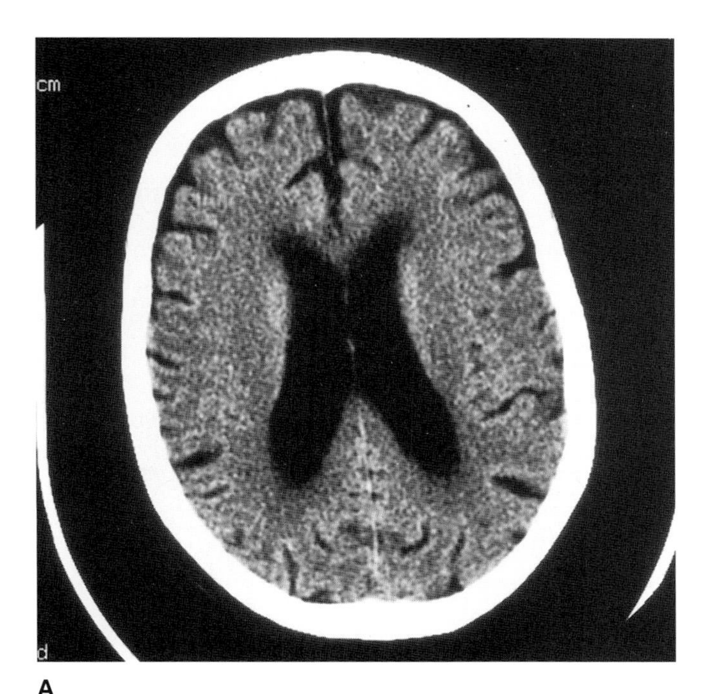

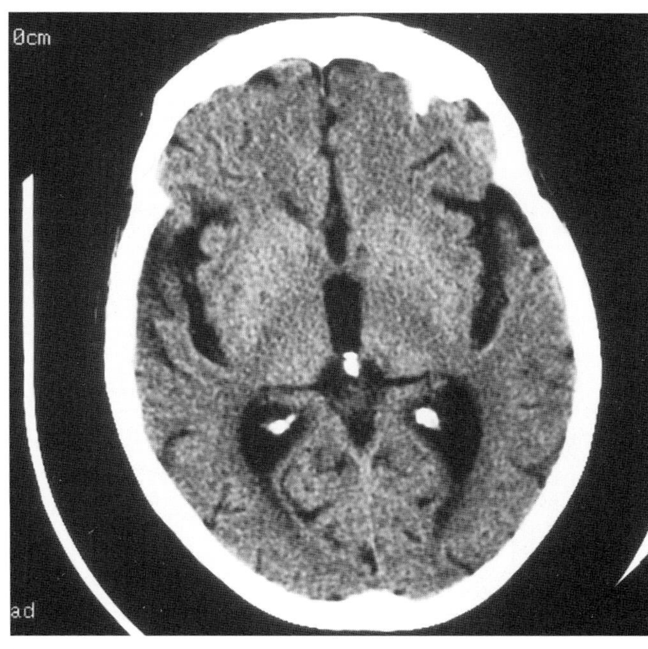

A B

Figure 18.19. CT scans of brain atrophy. Compare the normal CT scans in Figure 18.10. Note the increased size of the lateral ventricles and the prominence of the sulci, which is due to the decreased size of the gyri.

larly see it in the CT and MR scans of your elderly patients. Figure 18.19 shows a patient with age-related cerebral atrophy. The ventricles are enlarged and the sulci are prominent because of the decreased size of the gyri. Focal areas of hydrocephalus ex vacuo may often be seen in patients with localized areas of brain atrophy, which frequently occurs after a stroke.

Intracranial fresh blood is readily apparent at CT. Because of its high concentration of protein constituents, it appears denser (whiter) than the adjacent brain tissue. Figure 18.20 shows a CT scan of a hypertensive patient with a spontaneous intracerebral hemorrhage into the right thalamus. In Figure 18.21 (on the next page), a CT scan of another hypertensive patient, you can see a much bigger hemorrhage. There is a large collection of blood within the right cerebral hemisphere, and there is also blood within the ventricles.

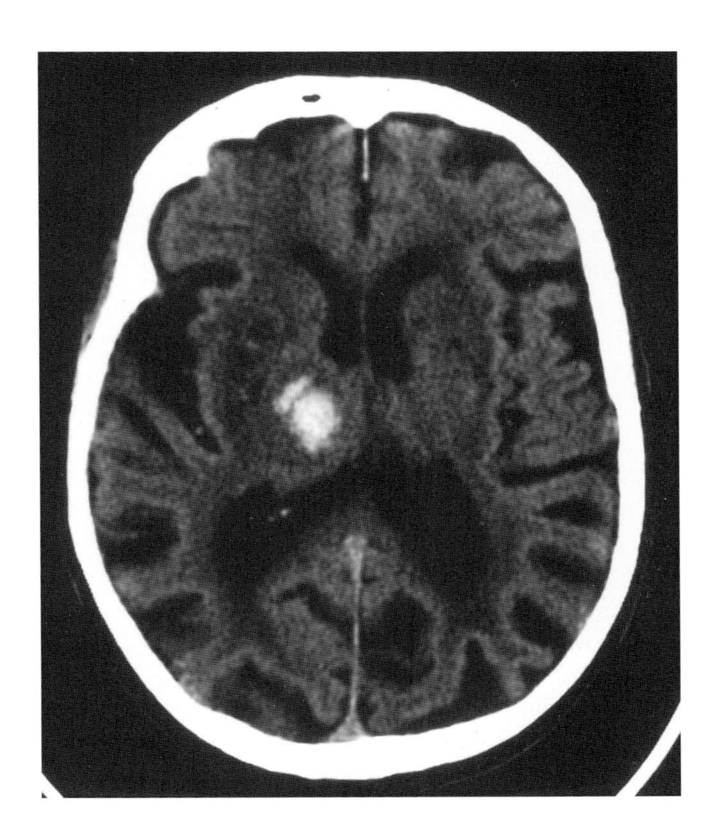

Figure 18.20. CT scan of a hypertensive patient with a thalamic bleed (hemorrhagic stroke). This patient also has brain atropy. Note the large ventricles and prominent sulci.

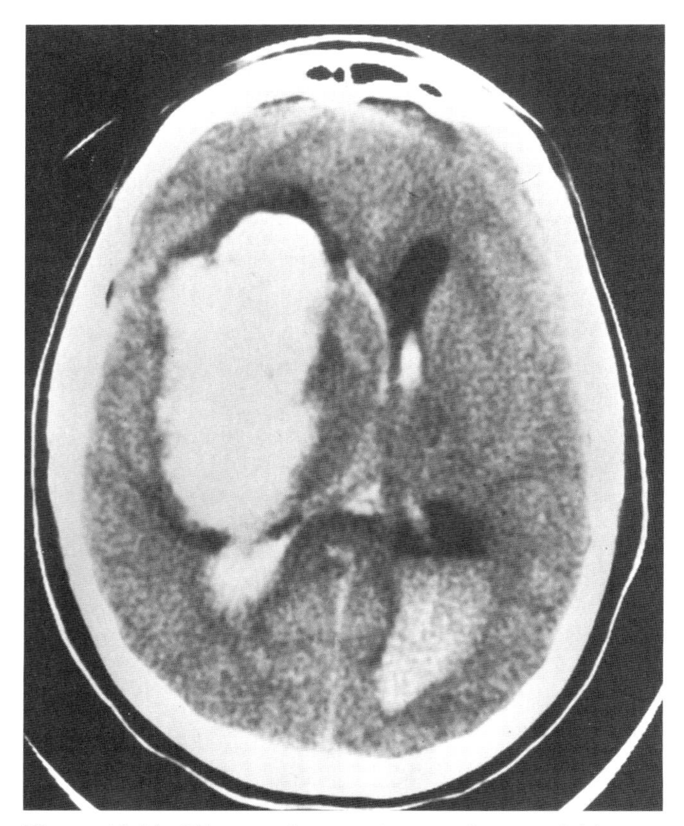

Figure 18.21. CT scan of a spontaneous intracranial hemorrhage in a hypertensive patient. Note the large collection of blood within the right cerebral hemisphere and smaller collections in the lateral ventricles, which are compressed and displaced to the left.

Both of these patients have suffered a *hemorrhagic stroke*. This type of stroke can mimic an *ischemic stroke*, which is caused by a perfusion deficit and is often treated with anticoagulation. Differentiating between these two conditions is extremely important because they require different treatment. An emergency CT scan can quickly and accurately determine which type of stroke has afflicted the patient. Patients suffering from hemorrhagic strokes should not be given anticoagulation treatment. Figure 18.22 shows CT scans of a patient with a spontaneous *subarachnoid hemorrhage* due to bleeding from a small anterior communicating artery aneurysm (not seen) of the Circle of Willis. Note the bright white blood in the basal cisterns *(arrows)* surrounding the midbrain in Figure 18.22A and higher up in the Sylvian fissures *(arrows)* in Figure 18.22B. Compare these images with the normal CT scan in Figure 18.10.

Figure 18.22 ***(below).*** Noncontrast CT scans of a subarachnoid hemorrhage from a circle of Willis aneurysm (not seen). A: The *arrows* indicate blood within the basal cisterns surrounding the midbrain. Blood can be seen in almost all the subarachnoid spaces. B: Higher-level scan shows more subarachnoid hemorrhage. The *arrows* indicate blood in the Sylvian fissures.

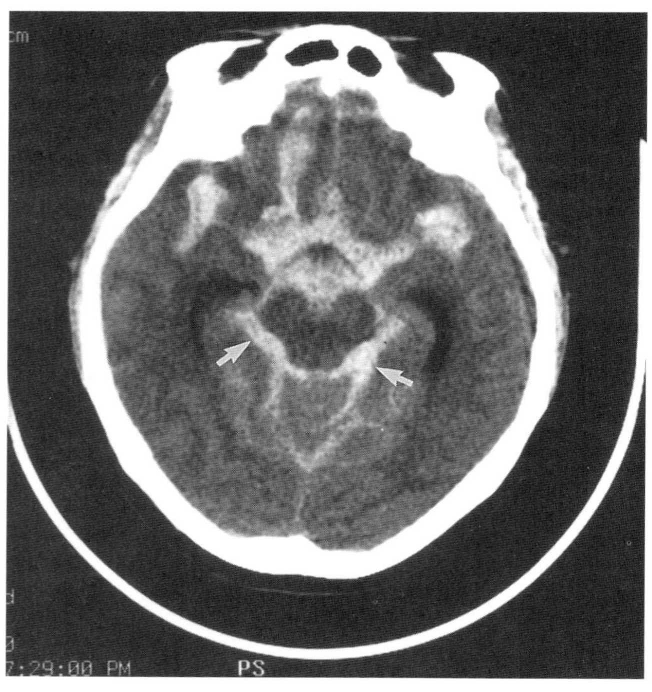

A

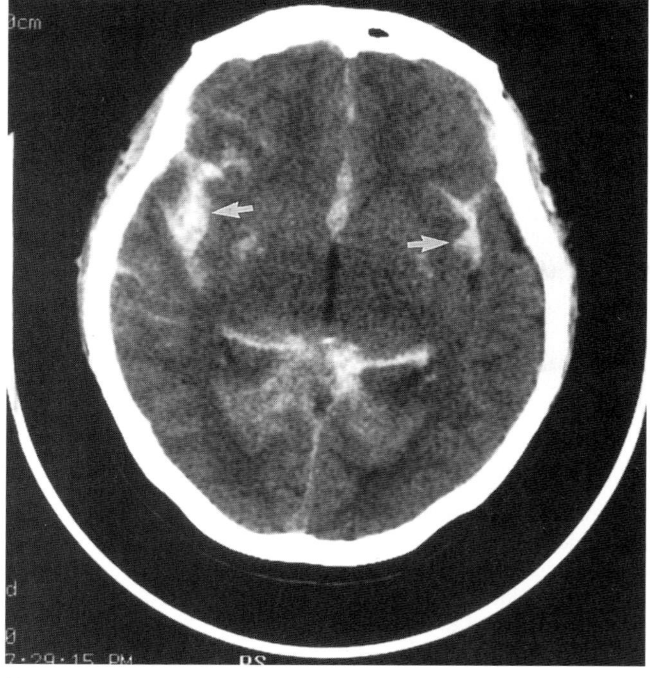

B

Normal Cerebral Arteriography

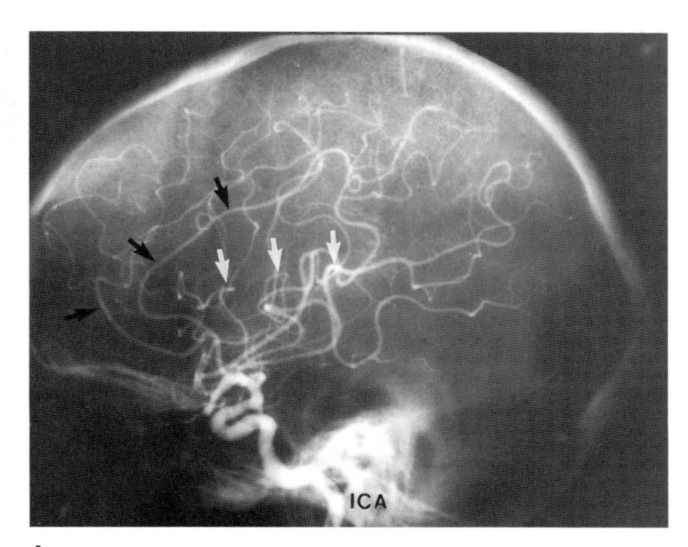

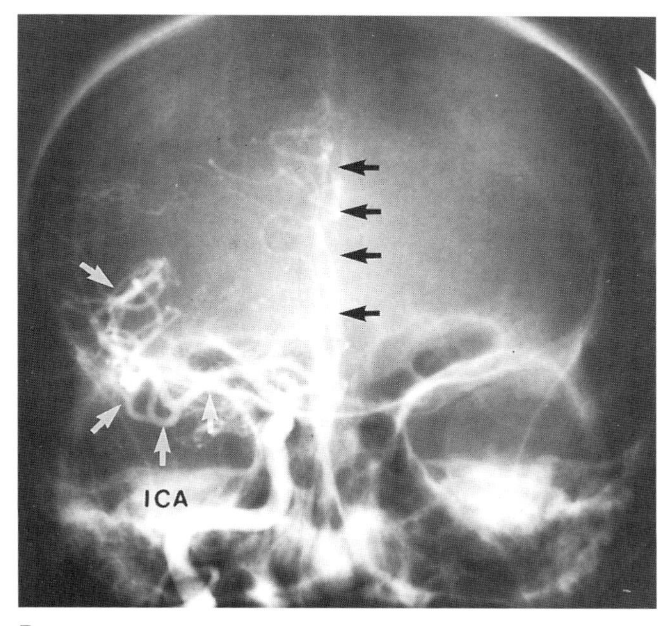

A B

Figure 18.23. Normal selective right internal carotid arteriogram. A: Lateral view. The *black arrows* indicate anterior cerebral artery branches; the *white arrows* indicate middle cerebral artery branches. *ICA* is the internal carotid artery. B: Frontal view. The *black arrows* indicate the more medial anterior cerebral artery branches; the *white arrows* indicate the laterally coursing middle cerebral artery branches.

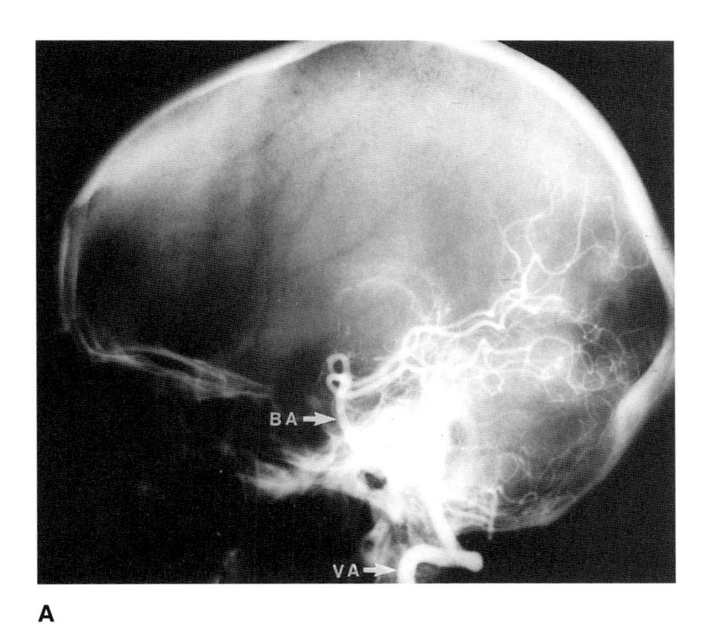

A

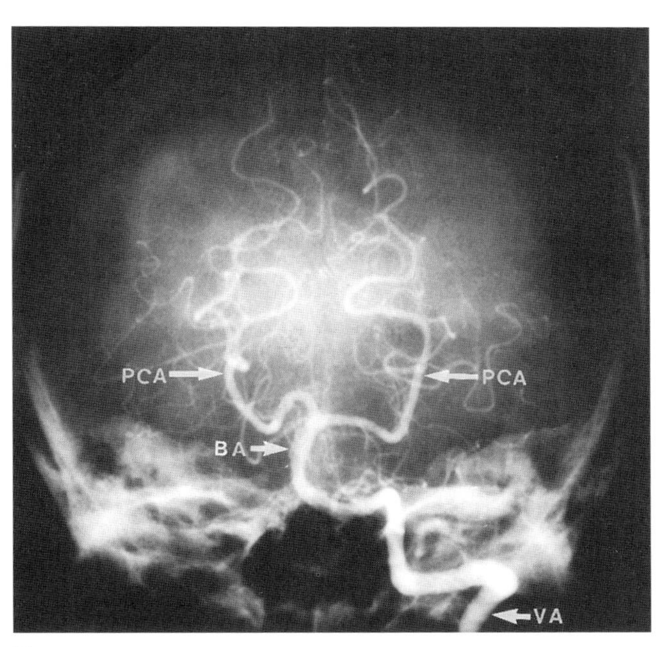

B

Figure 18.24. Selective left vertebral arteriogram. A: Lateral view. The opacified left vertebral artery *(VA)* ascends superiorly into the skull. The left and right vertebral arteries join to form the basilar artery *(BA),* which divides into the two posterior cerebral arteries (overlying each other on the lateral view). B: Frontal view. The left vertebral artery and the basilar artery are again seen. The two posterior cerebral arteries *(PCA)* are shown separated from each other.

Head Trauma

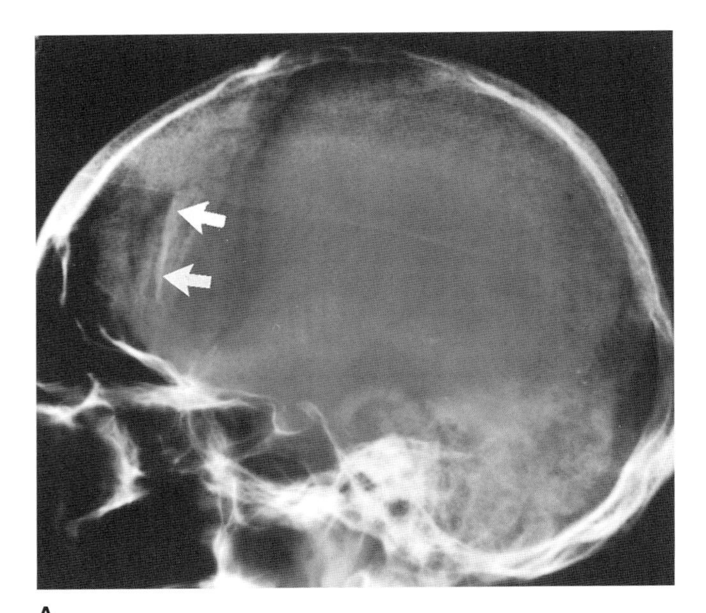

A

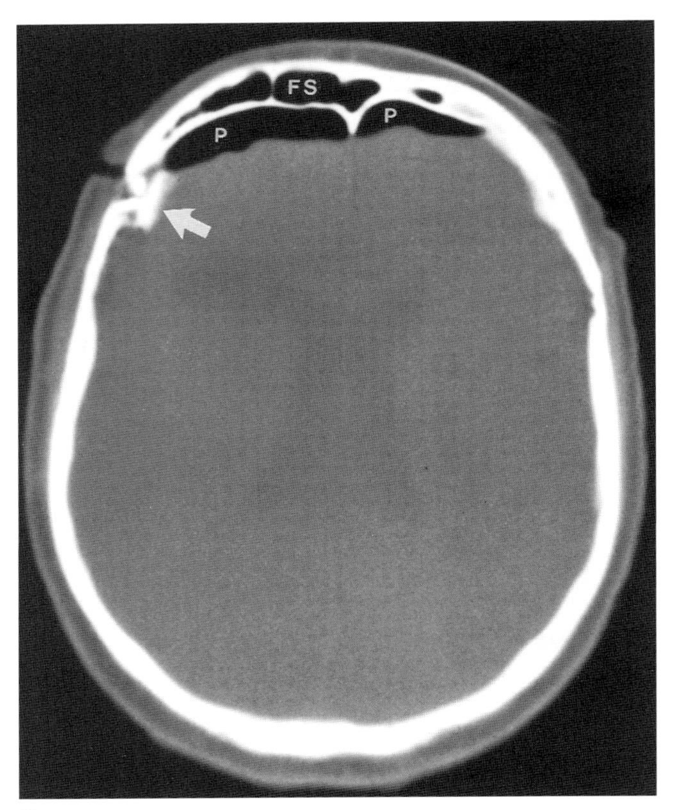

B

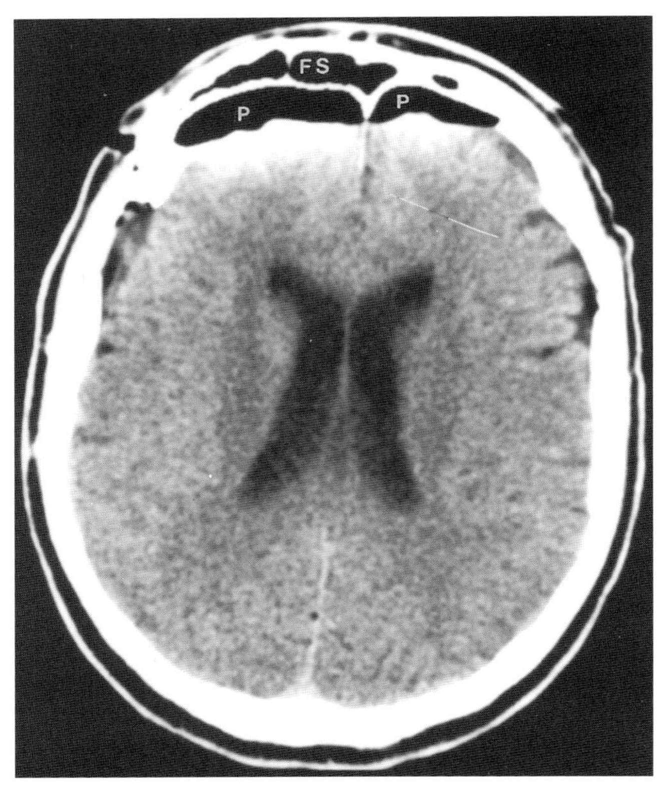

C

Figure 18.25. Depressed skull fracture with brain contusion. A: Lateral skull film shows a depressed fracture fragment *(arrows)*. B: Bone window CT scan again shows the depressed fracture fragment *(arrow)* and pneumocephalus *(P)*. Air is also seen within the frontal sinuses *(FS)*. C: Brain window CT scan at the same level shows brain contusion (increased whiteness) of the frontal lobe tips adjacent to the pneumocephalus.

CT has had a dramatic impact on the diagnostic workup of the head trauma patient. When intracranial injury is suspected, CT can diagnose intracranial hemorrhage, brain contusion, pneumocephalus, foreign bodies, and skull fractures in an examination that is quickly and easily performed. Furthermore, CT can identify secondary effects of trauma, such as edema, brain herniation, ischemic infarction, and hydrocephalus. In acute head trauma, intracerebral hemorrhage and extracerebral blood collections (subdural hematoma, epidural hematoma, and subarachnoid hemorrhage) can be diagnosed by CT with nearly 100 percent accuracy. CT allows serial scanning without risk, so that patients can safely be followed after trauma.

In any patient who has sustained head trauma and presents with signs of neurological dysfunction, CT is indicated. *Do not obtain plain films of the skull before the CT examination, because doing so would waste precious time and subject the patient to an unnecessary dose of radiation for little or no diagnostic gain.* The radiation exposure to the patient of a head CT is no greater than that of a plain skull film series.

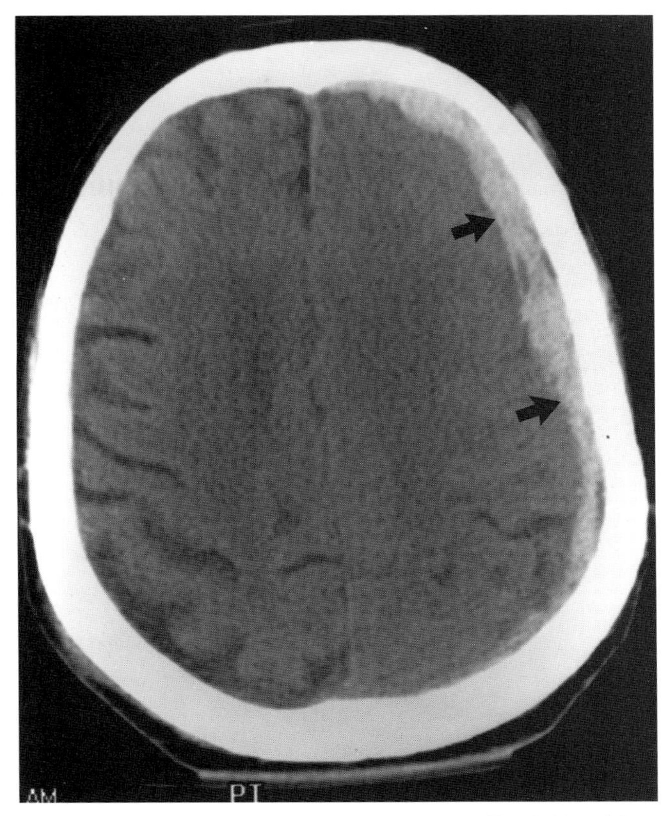

Figure 18.26A. Acute subdural hematoma. Fresh blood in the subdural space appears white *(arrows)* and takes on a crescent shape. Note the compression of the left sulci and gyri compared with those on the right side. Midline shift is also present.

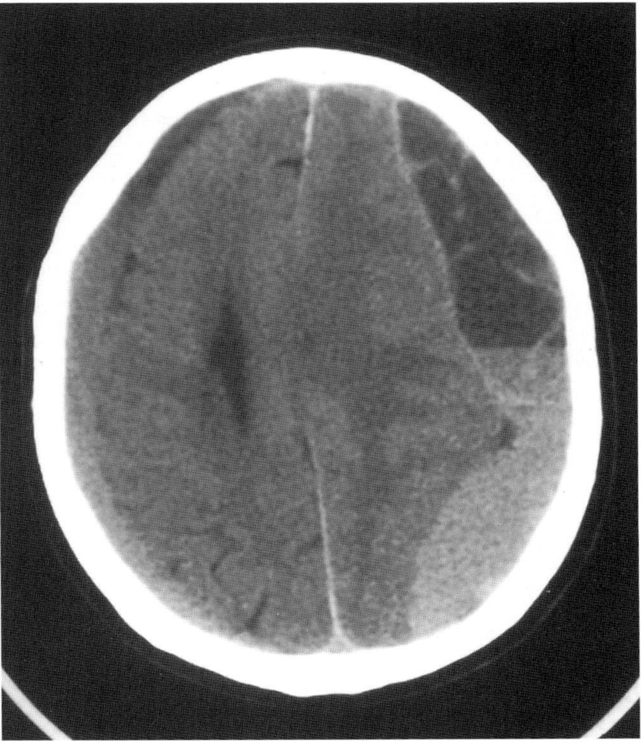

Figure 18.26B. Bilateral subacute subdural hematomas. Fluid collections are seen in both subdural spaces, compressing the brain medially. These are subacute hematomas, with settling of the blood cells, which produces a hematocrit effect, especially well seen on the left. The larger fluid collection on the left side completely compresses the left lateral ventricle and produces midline shift.

Remember that bright areas on a head CT scan are acute blood, calcium, or contrast material. When a CT scan is performed on a head trauma patient, intravenous contrast material *should not* be administered, because contrast enhancement of the cortical vessels can mask a diagnosis of subarachnoid hemorrhage. The images should be photographed at both brain and bone windows. Extravasated blood has a high protein concentration and appears white on the brain window images, compared with the surrounding brain tissue. The bone window images can show the presence and configuration of skull fractures. Care should always be taken during the transport of head trauma patients to the CT scanner to avoid further exacerbating any injuries of the cervical spine, which often occur in these patients.

CT can depict depressed and basilar skull fractures better than plain films. True, a horizontal, nondisplaced linear fracture in the axial plane of section may rarely be overlooked by CT; but this is not as important as the brain injuries that are *not* overlooked by CT. Two secondary signs of skull fracture at CT are pneumocephalus and opacification (due to blood) of the mastoid air cells and sphenoid sinuses (signs of a basilar skull fracture).

If a head trauma patient with no neurological dysfunction is initially studied with plain skull films at a medical center without a CT scanner, these should be scrutinized for findings that indicate the need for transfer to a hospital with a CT scanner. Such findings include displacement of a calcified pineal gland (suggesting a unilateral mass or hematoma), pneumocephalus, an air-fluid level in the sphenoid or frontal sinuses, a depressed skull fracture, or a linear fracture crossing a meningeal artery groove or major venous sinus. Today most patients whose plain films show an acute skull fracture are referred for a CT examination.

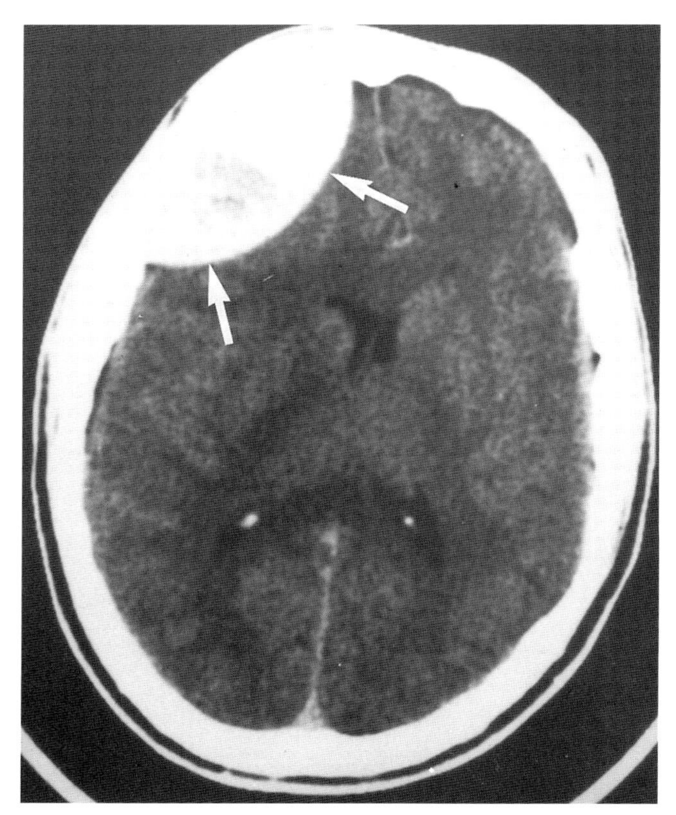

A

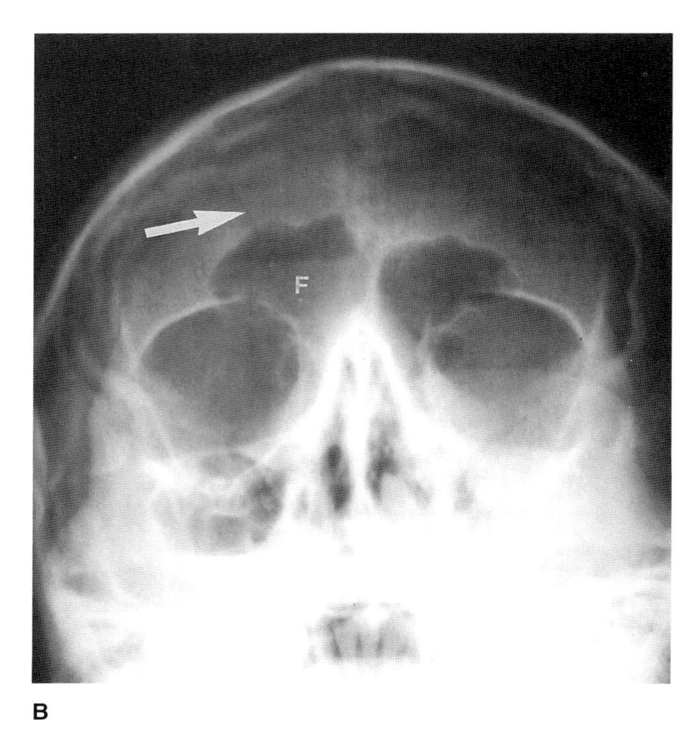

B

Figure 18.27. Acute epidural hematoma. A: The *arrows* indicate a lenticular-shaped, biconvex epidural hematoma. B: Upright skull film of the same patient shows a right frontal fracture *(arrow)* overlying the hematoma, as well as a collection of blood *(F)* in the right frontal sinus (note the fluid level).

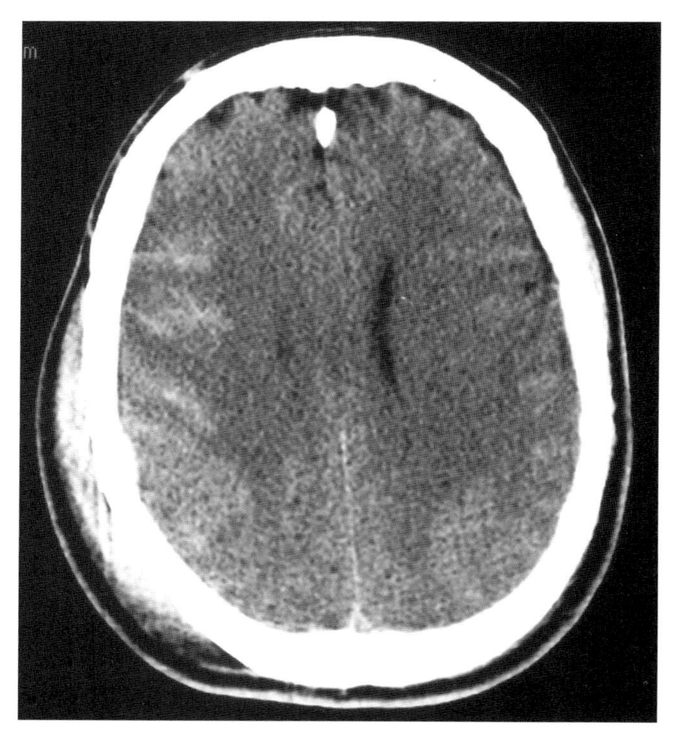

Figure 18.28. CT scan of intracerebral hemorrhage. Streaks of acute hemorrhage are seen within the right brain parenchyma. These represent shear-induced hemorrhage from the rupture of small intraparenchymal blood vessels. Note the compression of the right lateral ventricle. Compare the normal CT scan in Figure 18.10.

MR can also show acute brain injuries. But during an MR scan the patient's entire body is positioned within the bore of an enormous magnet, so that observation and careful monitoring of an acutely injured patient is much more difficult than with CT. CT is preferable for showing acute hemorrhage, skull fractures, and foreign bodies.

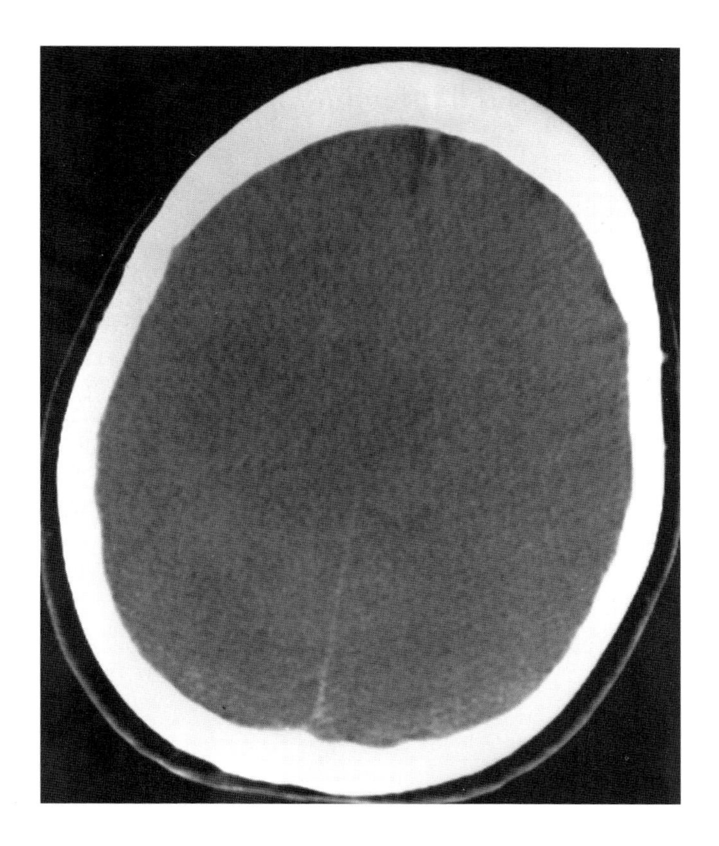

Figure 18.29. Traumatic cerebral edema in a child. This condition, usually caused by an increase in cerebral blood volume or an increase in tissue fluid content, can lead to brain herniation. The gyri and sulci are completely effaced. The compressed ventricular system is not visible.

Cerebrovascular Disease and Stroke

Cerebrovascular disease is a leading cause of death, and in the elderly it is a major cause of disability. One of the most common manifestations is a *stroke* characterized by the sudden onset of a neurological deficit associated with a perfusion deficit to a specific region of the brain. Strokes are usually caused by acute cerebral ischemia resulting from thrombosis of a cerebral artery or a cerebral artery embolism. Emboli may arise from the heart or from atherosclerotic plaques, frequently located in and narrowing the carotid circulation at the level of the common carotid artery bifurcation (Figure 18.30). Less frequently, strokes are caused by arterial spasm, which may occur following acute subarachnoid or intracerebral hemorrhage, or by carotid artery dissection following trauma. When a patient presents with symptoms of stroke, the first step is to obtain a CT scan

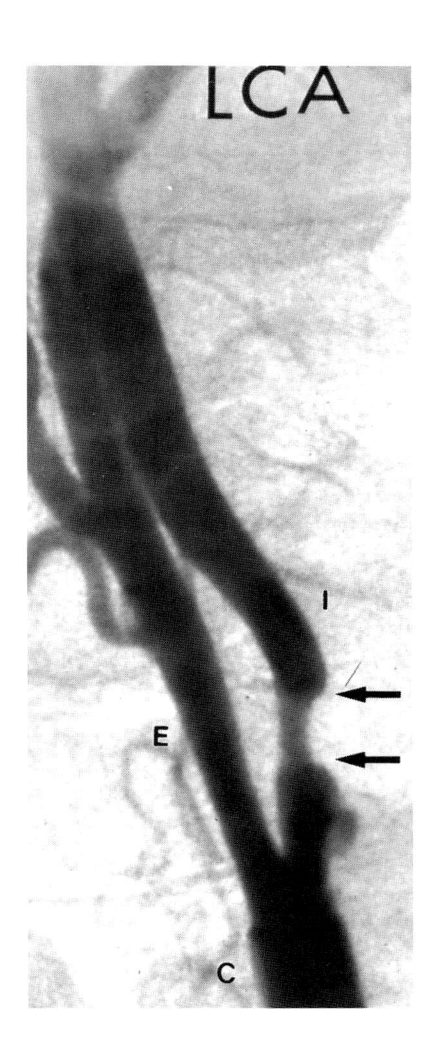

Figure 18.30. Selective left common carotid arteriogram (the arteries look black because of photographic subtraction) shows a stenosis *(arrows)* in the internal *(I)* carotid artery. The external *(E)* and common *(C)* carotid arteries are also visible.

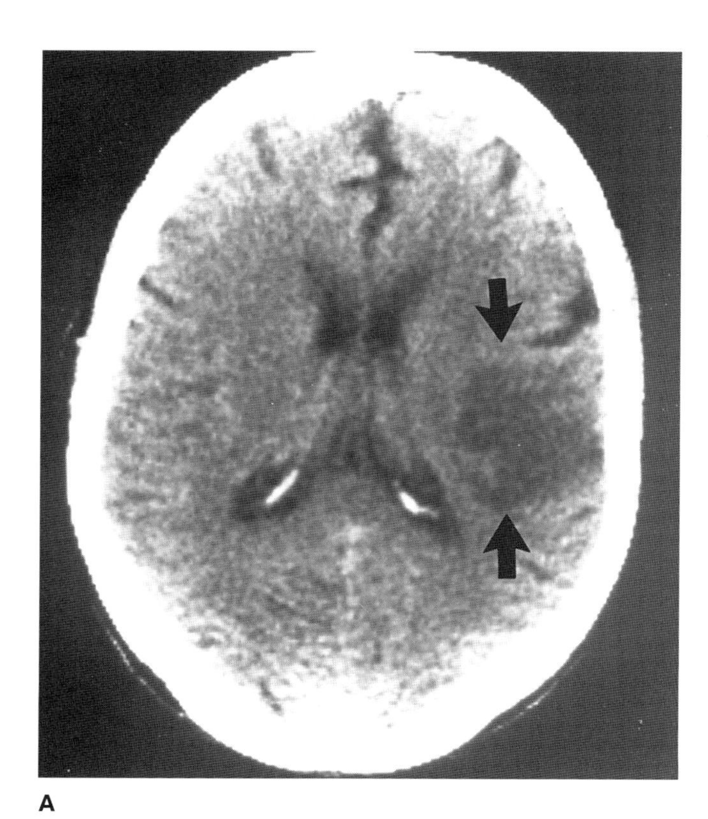
A

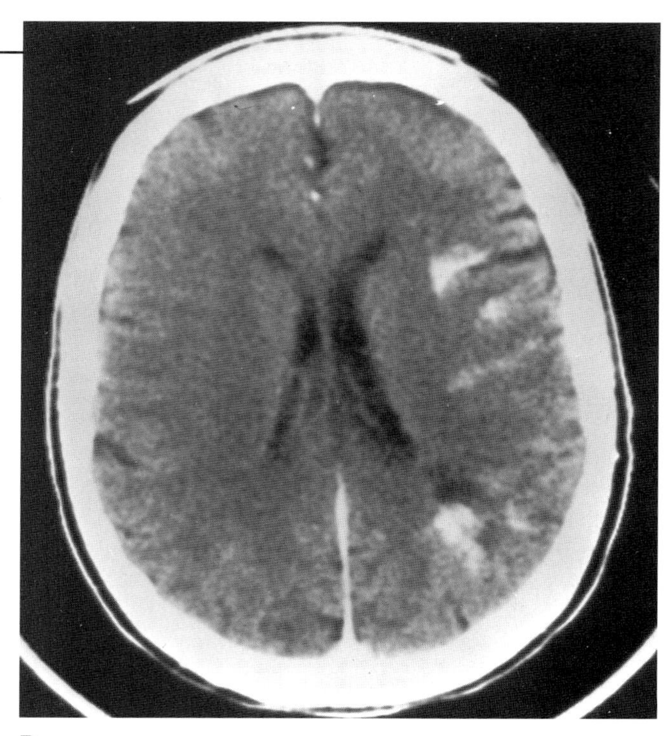
B

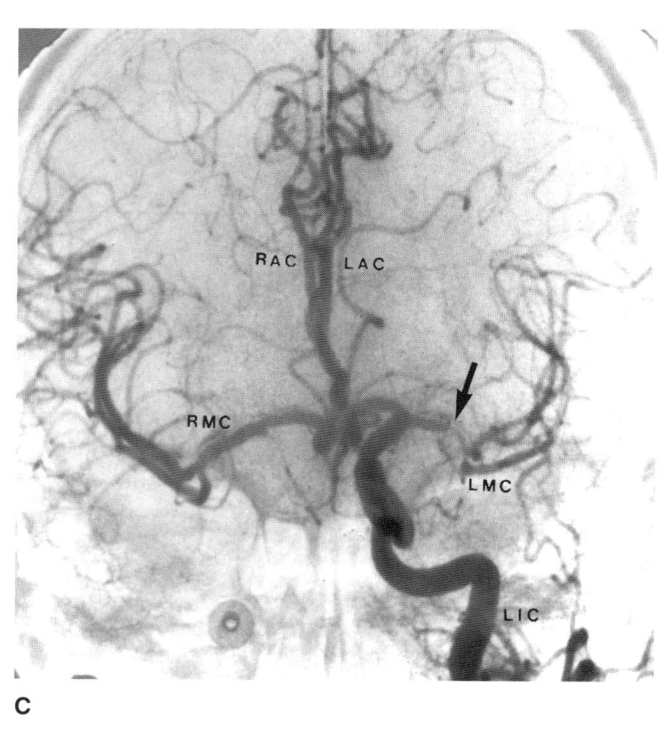

C

Figure 18.31. Stroke resulting from a left middle cerebral artery embolus. A: Early CT scan (12 hours after the stroke) shows a low-density area of infarction *(arrows)* in the left hemisphere. B: Repeat CT scan with intravenous contrast material one week later shows hyperdensity within the area of infarction. C: Frontal view of a selective left internal carotid arteriogram shows an embolus *(arrow)* almost totally occluding the left middle cerebral artery. The left internal carotid artery *(LIC),* right middle cerebral artery *(RMC),* left middle cerebral artery *(LMC),* right anterior cerebral artery *(RAC),* and left anterior cerebral artery *(LAC)* are clearly shown. The right middle and anterior cerebral arteries were opacified with contrast material as the result of an angiographic maneuver in which the right carotid artery was externally compressed during injection of contrast medium into the left internal carotid artery.

to exclude hemorrhage as the cause. With an acute ischemic stroke the CT scan is often normal. If the diagnosis is in doubt, an MR scan or angiography could confirm the stroke and determine its cause.

Ischemic strokes usually cause brain *infarcts,* permanent areas of brain damage that can be detected by CT and MR. It is important to note that the CT scan of a stroke patient may appear normal within the first 8 to 12 hours after the stroke. An MR scan usually indicates stroke damage before CT does. It is also important to know that ischemic strokes can turn into hemorrhagic strokes, usually 2 to 4 days after presentation, probably because of necrosis and reperfusion of the infarct. Therefore, any patient showing clinical deterioration in the poststroke period should be rescanned to rule out a new hemorrhage.

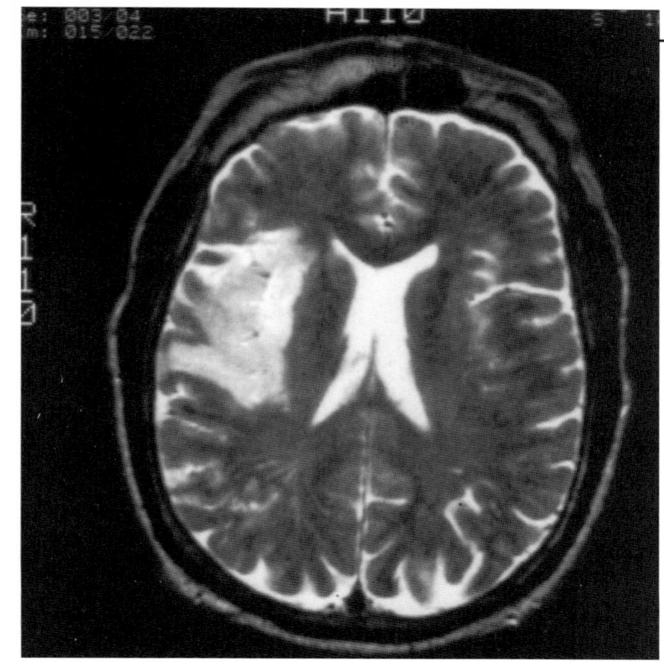

A

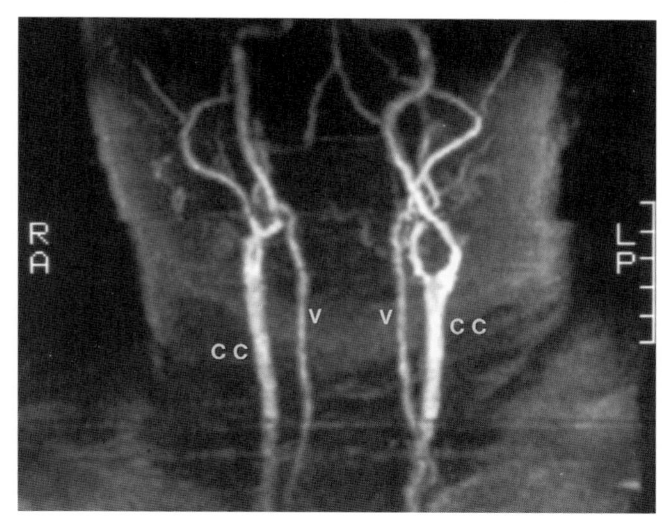

B

Figure 18.32. MR examination of an acute right middle cerebral stroke. A: Axial T2-weighted MR scan showing an increased signal in the right frontoparietal region. B: MR angiogram of the neck arteries. C: Detail view of the circle of Willis. B and C show patency of the common carotid *(CC),* internal carotid *(C),* vertebral *(V),* basilar *(B),* and middle *(M)* cerebral arteries. No major artery occluded.

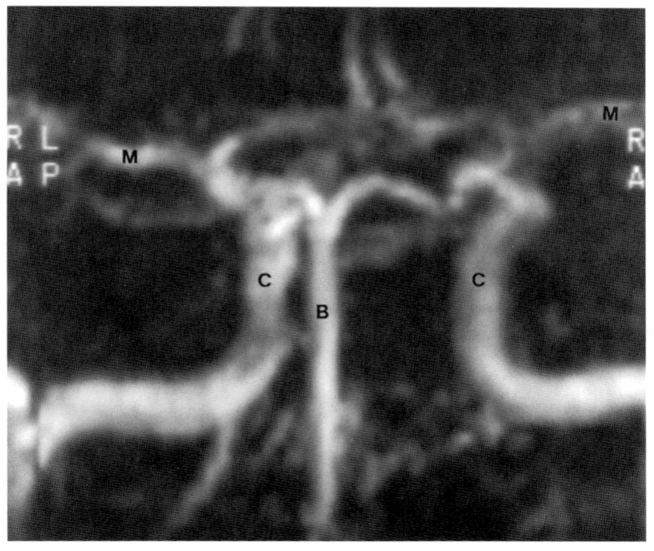

C

Strokes must be differentiated from *transient ischemic attacks (TIAs),* which are focal neurological deficits that develop suddenly, last for a brief period (minutes or hours), and then completely resolve. TIAs are thought to represent cerebral emboli that have dissolved before causing permanent damage. They are not associated with brain infarction and do not have detectable CT findings. TIAs are often a warning sign of an impending ischemic stroke, and they are regarded as cause for a diagnostic workup of correctable conditions leading to stroke, such as carotid artery stenosis or cardiac arrhythmia. The neurological deficits associated with strokes last 24 hours or longer; they may be partially or fully reversible, or they may be permanent.

CT can show the location and size of infarcts and can differentiate them from tumors, aneurysms, and other lesions that may mimic the clinical presentation of a stroke. The CT appearance of a brain infarct will vary with time. Although the initial CT scan may be normal, at 12 to 24 hours the infarct will appear as a mottled or homogeneous area of diminished CT attenuation (Figure 18.31A), representing both interstitial and intracellular edema; generally there is no enhancement with intravenous contrast material. At 1 to 4 weeks the majority of brain infarcts show patchy contrast enhancement (Figure 18.31B), representing breakdown in the blood-brain barrier (leak of contrast material from newly formed capillaries). One month later the density of the infarct will approach that of CSF, as the necrotic brain tissue is replaced by CSF.

Finally, note that an acute stroke can be shown earlier by MR than by CT (Figure 18.32A). Furthermore, to determine the cause of the stroke the arterial supply to the brain can be evaluated with MR angiography, which requires no vascular catheterization or injected contrast medium. The MR scanner computer can be directed to stack axial slices of the neck and brain to make three-dimensional models of the carotid and vertebral arteries and their branches (Figures 18.32B and C).

Brain Tumors

Your search for a brain tumor may be precipitated by a variety of signs and symptoms of central nervous system dysfunction, including new seizures, headaches, personality changes, alterations in consciousness, visual changes, motor and sensory deficits, and papilledema. These clinical presentations may occur with other neurological conditions, such as stroke and migraine, and consequently clinicians should request imaging examinations as part of the diagnostic workup of new central nervous system disorders, to rule in or rule out a tumor. Brain tumors are exceedingly sensitive to detection by CT and MR. Skull films are usually normal with brain tumors and are rarely obtained today; skull abnormalities are seen on plain films only if the tumor contains calcifications, causes bony erosion or bony proliferation, or produces signs of increased intracranial pressure (enlargement of the sella turcica, for example). Angiography may be useful after a CT or MR diagnosis of brain tumor to assist in surgical planning by determining the vascularity of the tumor and showing its relation to adjacent vascular structures.

CT and MR scanning not only can diagnose brain tumors with high accuracy, but can localize them, determine their size, show their effect on adjacent structures, and suggest the histologic type of the tumor. Brain tumors can arise either from the brain itself (an example being glioma arising from glial cells) or from the coverings (meninges) of the brain (meningioma, for instance). The brain is also a common site for metastatic disease, especially from lung and breast cancers. Not infrequently, a patient is symptomatic from a brain metastasis before the primary tumor is detected.

At CT most brain tumors show contrast enhancement in all or part of the tumor after the administration of intravenous contrast material (Figure 18.33). The appearance of primary tumors varies with the degree of malignancy. Grade 1 gliomas may appear as low-attenuation masses with minimal contrast enhancement, whereas grade 3 and 4 gliomas may appear cavitary with thick walls and an irregular bright contrast enhancement, called *ring enhancement*.

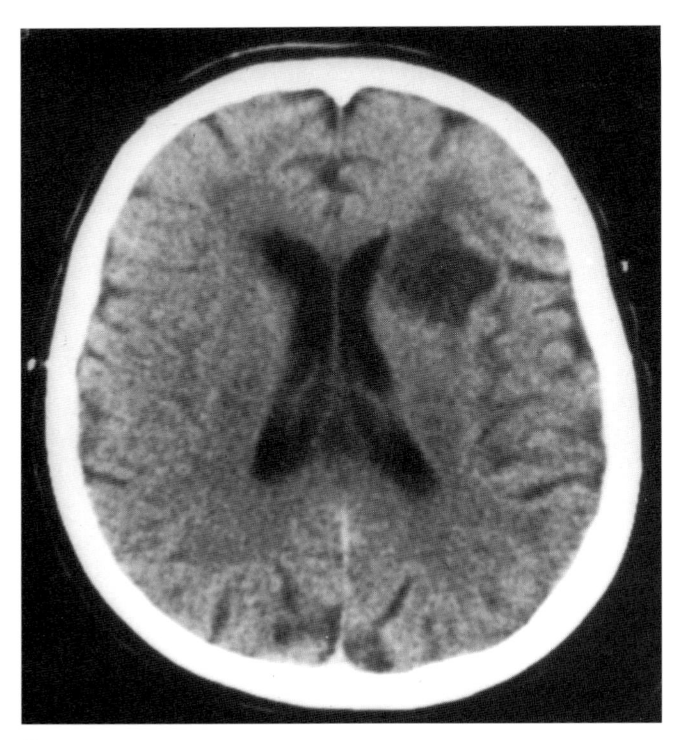

A

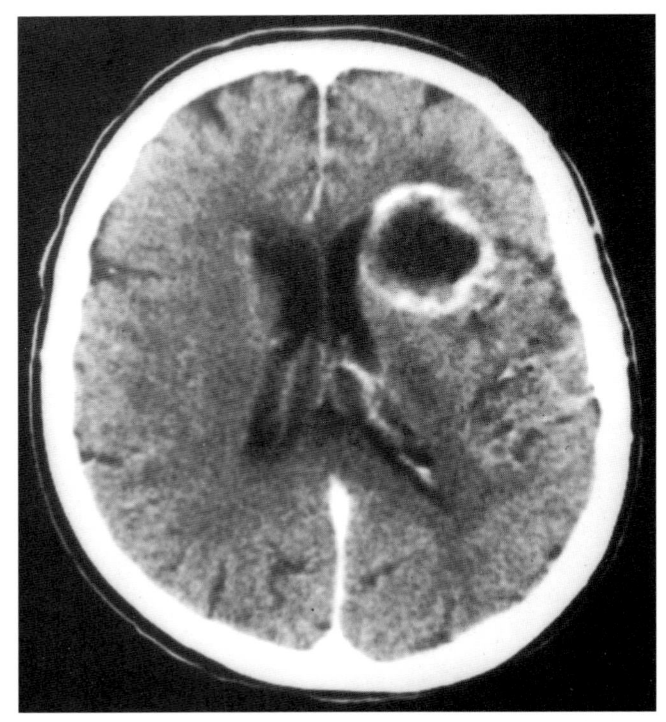

B

Figure 18.33. CT scans of a grade 3 astrocytoma. A: Noncontrast scan. B: Scan obtained at the same level after injection of intravenous contrast material, showing ring enhancement.

At CT brain tumors are frequently shown surrounded by a low-density area of edema, which may be massive in patients with metastatic tumors. Edema appears less dense than normal brain tissue (Figure 18.34), because the increase in tissue fluid reduces the CT attenuation nearly to that of CSF. Metastatic tumors are often found at the gray matter–white matter junction, and they are often multiple, whereas primary tumors are nearly always solitary. Brain abscesses, or foci of infections, such as *toxoplasmosis,* mimic brain metastases because they also appear as multiple, contrast-enhancing lesions; you will see examples in Chapter 20.

Figure 18.34. Contrast-enhanced CT scan showing a right frontal brain metastasis from a breast cancer. The *white arrow* points to the metastasis itself; the *black arrows* indicate a large area of low-attenuation edema surrounding the tumor.

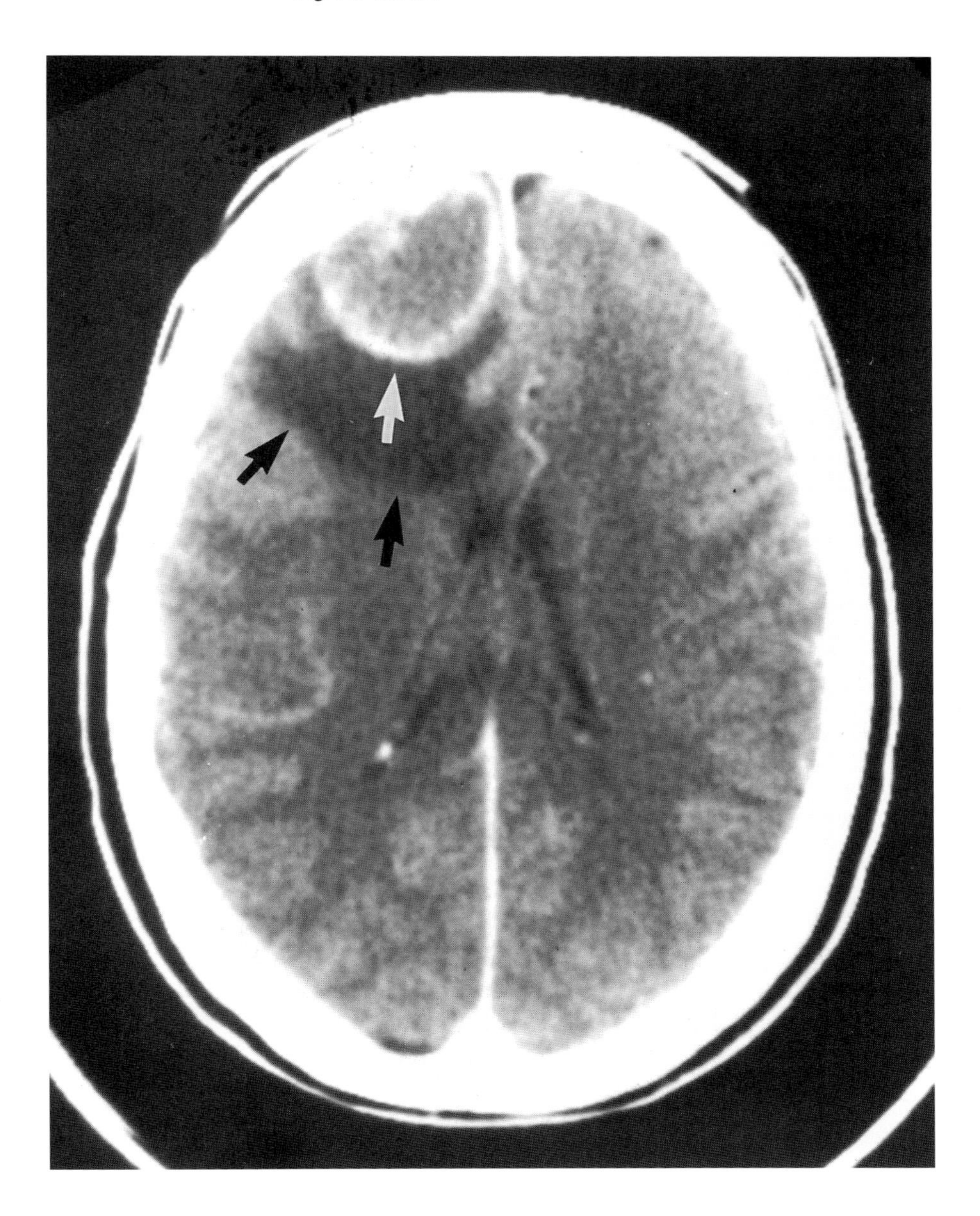

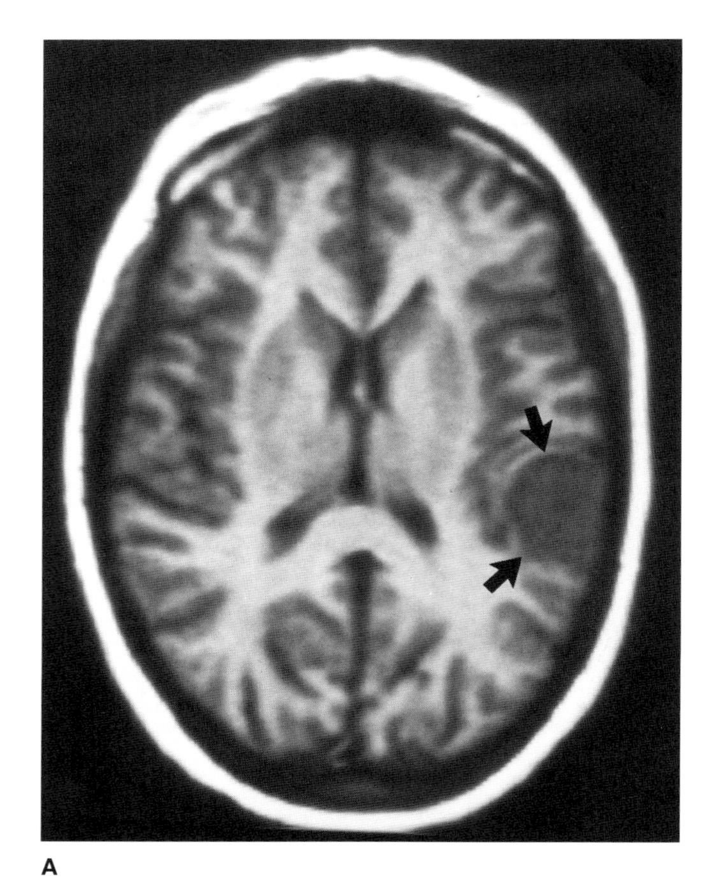

A

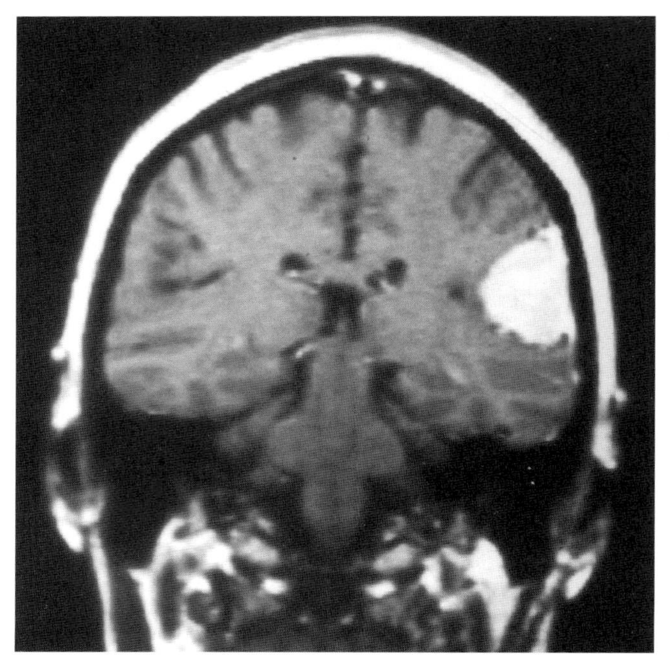

B

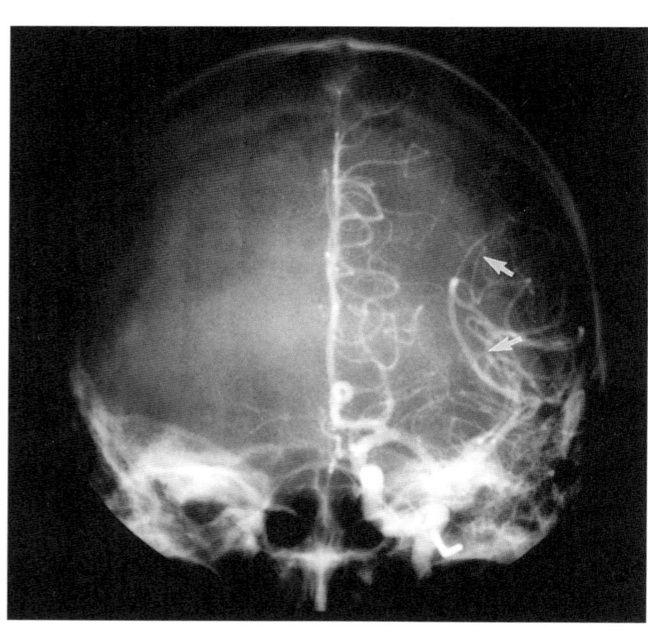

C

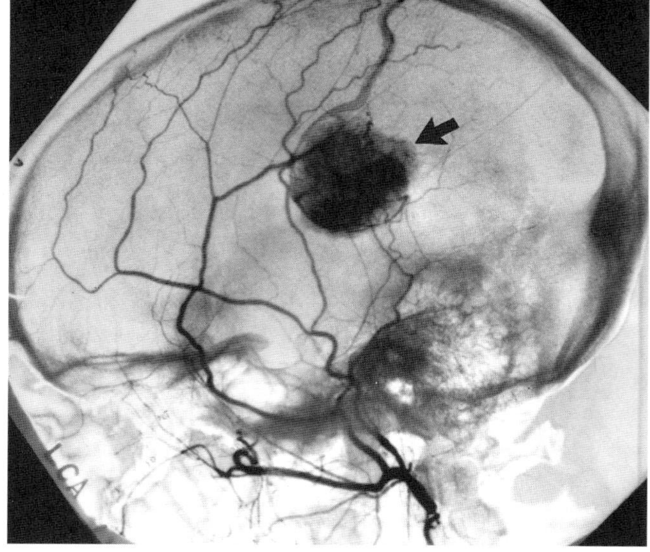

D

Figure 18.35. MR scans and arteriography of a left parietal meningioma. A: Axial MR scan shows a mass *(arrows)* in the left parietal region, abutting the skull. B: Coronal MR scan with gadolinium contrast material shows marked enhancement of the mass. C: Frontal view from a selective left *internal* carotid arteriogram shows that the blood supply to the mass is not provided by the intracerebral circulation; instead, the left middle cerebral arteries *(arrows)* are displaced around the mass. D: Lateral view from a selective left *external* carotid arteriogram shows that the mass *(arrow)* is supplied by meningeal branches and is therefore most probably a meningioma, which this proved to be.

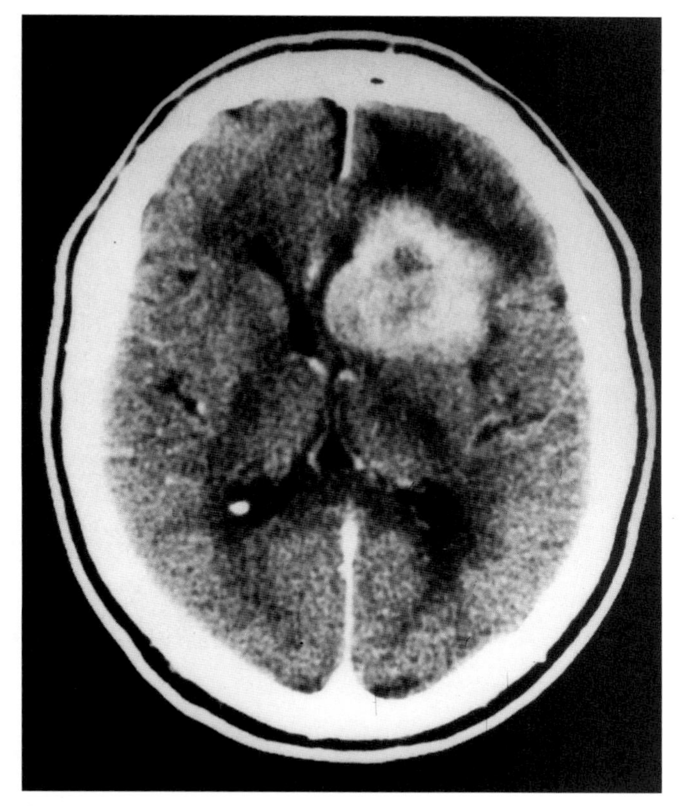

A

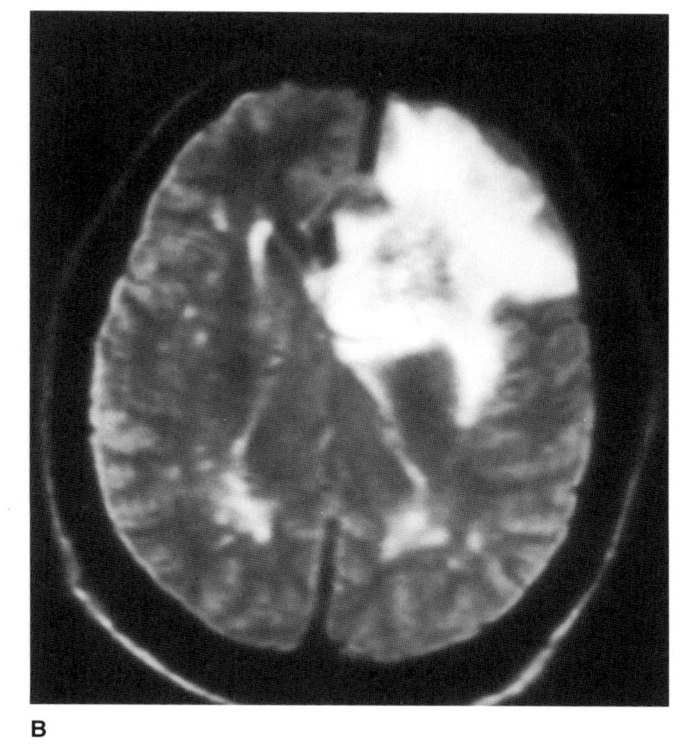

B

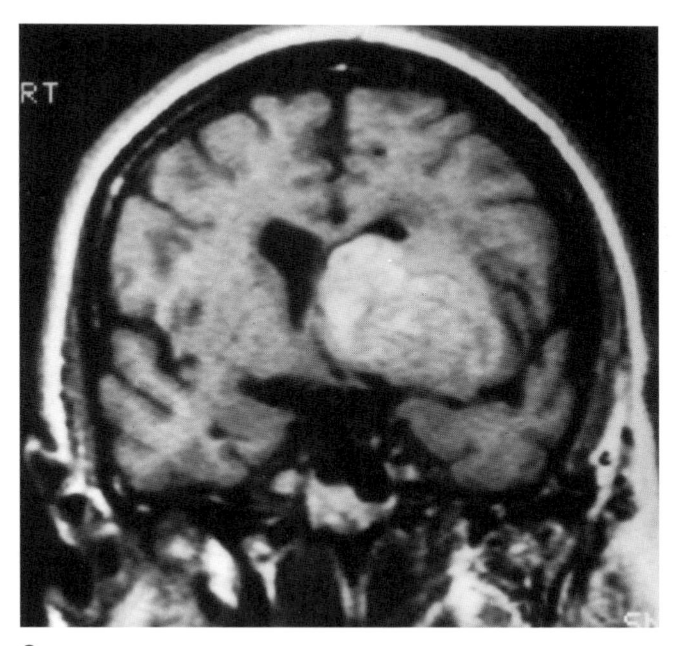

C

Figure 18.36. Malignant melanoma metastasis to the left basal ganglia. A: CT scan with intravenous contrast material shows a contrast-enhancing metastasis in the head of the left caudate nucleus with associated low-attenuation edema in the frontal region anterior to it. The mass compresses the left lateral ventricle and produces midline shift. B: Axial proton-density MR scan shows a markedly increased signal from both the metastasis and the edema; compare the CT scan. C: Coronal T1-weighted MR scan shows compression by the mass of the left lateral ventricle and midline shift.

At MR a brain tumor may appear whiter or blacker than the surrounding normal brain tissue, depending on the MR protocol used (Figures 18.35 and 18.36). The edema associated with metastases is also well shown with MR. Today, MR is nearly always obtained before therapy because of its ability to yield multiplanar images. Like other intracranial space-occupying lesions, brain tumors compress, distort, and displace surrounding structures, and the clinical findings are related to the location and size of the tumors. On an MR examination it is possible to see the effect of the tumor on surrounding structures in the sagittal, coronal, and axial planes (Figure 18.16, pages 518–519).

Cerebral Aneurysm and Arteriovenous Malformation

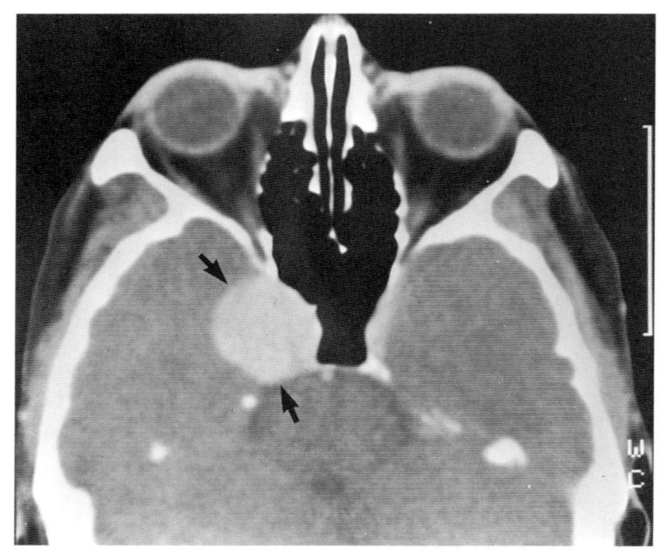

A

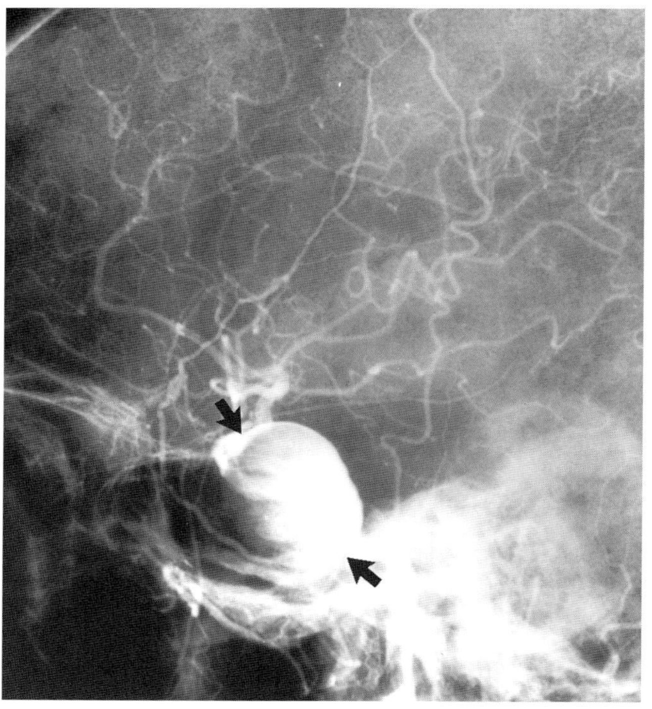

B

Figure 18.37. Cerebral aneurysm. A: CT scan with intravenous contrast material identifies a large aneurysm *(arrows)* just to the right of the sphenoid sinuses. B: Right internal carotid arteriogram shows the aneurysm *(arrows)* arising from this artery.

The physical signs and symptoms of brain tumors can be produced by two vascular lesions, cerebral aneurysm and cerebral arteriovenous malformation. Both can rupture and bleed intracranially, producing sudden severe headache or deterioration in a patient's central nervous system function. In addition, symptoms can be produced simply by the mass effect of the vascular lesion itself.

Both lesions can be diagnosed by CT and MR, although angiography is generally required for treatment planning. In recent years interventional angiography has been employed in the treatment of both lesions. A transcatheter approach makes it possible to occlude cerebral aneurysms and malformations with particulate material or detachable balloons. These transcatheter techniques have been especially useful in cases where surgical access to the lesion would be difficult; they will be discussed further in Chapter 19.

Cerebral aneurysms are congenital saccular outpouchings of arteries, usually occurring at the circle of Willis or, less commonly, at other major arterial branch points. Such an outpouching is less strong than the adjacent normal blood vessel wall and is very prone to rupture and leak at a later date. When an aneurysm leaks, it nearly always bleeds into the subarachnoid space, the compartment in which the vessels traverse. Blood is very irritating to the meninges, and even a small leak is sufficient to produce a severe headache, which patients inevitably describe as "the worst headache I've ever had." Although such a patient could be suffering from migraine, consider requesting a CT scan to rule out a subarachnoid hemorrhage when you are faced with this complaint. Now turn back to page 522 and look again at Figure 18.22, which shows hemorrhage in the subarachnoid space from a leaking aneurysm.

With the administration of intravenous contrast material an aneurysm, if large enough, will appear at CT as a smooth-margined, spherical mass of high CT attenuation (Figure 18.37). In fact, the attenuation will be the same as that measured in any major blood vessel, such as the patient's aorta. The aneurysm may contain thrombus, and subarachnoid hemorrhage may be seen around the aneurysm if leak or rupture has occurred. The aneurysm may have visible calcification within its wall. It is extremely important to differentiate an aneurysm from a brain tumor, because biopsy of an aneurysm would be disastrous. Most large aneu-

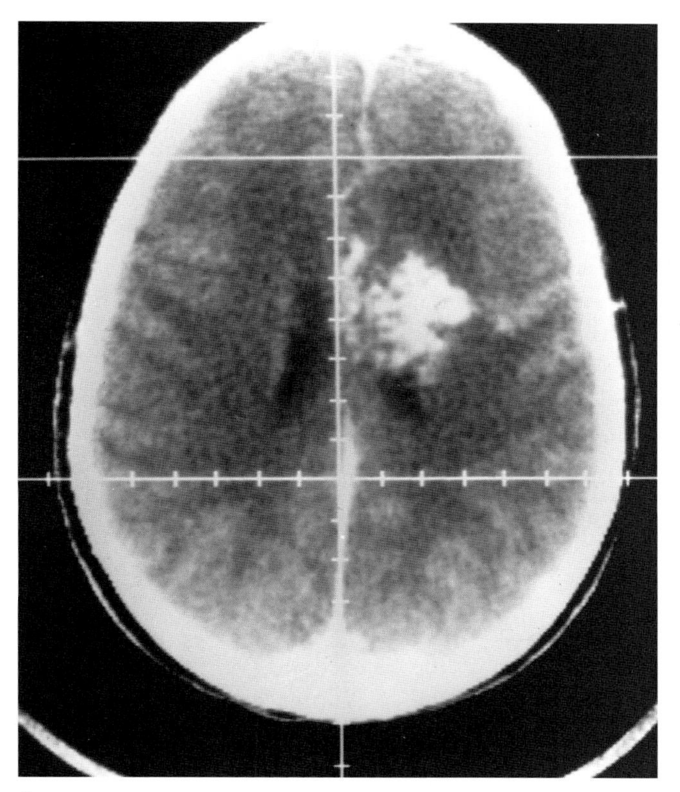

A

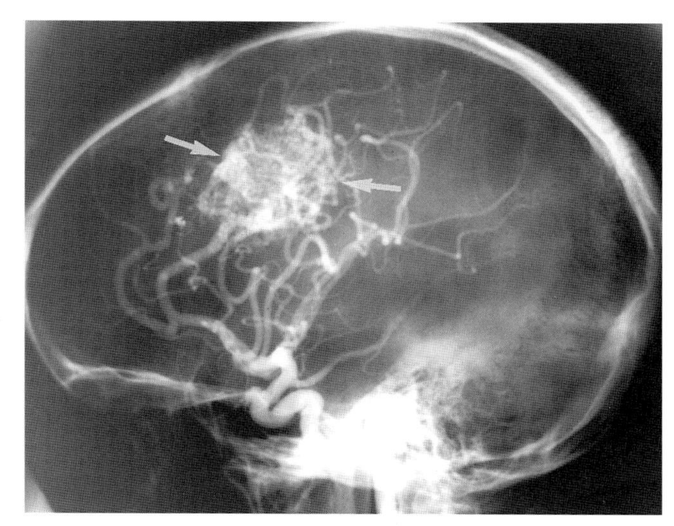

B

Figure 18.38. Cerebral arteriovenous malformation. A: CT scan, after administration of intravenous contrast medium, of a patient with diplopia shows a tangle of abnormal blood vessels in the left cerebral hemisphere. B: Selective left internal carotid arteriogram shows the arteriovenous malformation *(arrows)* opacified with contrast material. The large caliber of the cerebral arteries reflects the increased blood flow passing through them to supply the arteriovenous malformation.

rysms can be seen at CT. Small aneurysms, measuring only a few millimeters in diameter, may be overlooked by CT; arteriography is necessary to detect them. But in symptomatic patients with leak or rupture, the CT scan will show the subarachnoid hemorrhage, even if the aneurysm is too small for CT to detect. Once the diagnosis of subarachnoid hemorrhage is made, usually by CT, the patient should have angiography to determine the size, location, and geometry of the aneurysm. Traditionally, aneurysms have been surgically treated by craniotomy and aneurysm clipping; today radiological interventional techniques such as transcatheter embolization can supersede surgical intervention.

Arteriovenous malformations (AVMs) are tangles of abnormal blood vessels located within the brain. These lesions are congenital and have the capacity to bleed. In contrast to aneurysms, these lesions, because of their location, bleed within the brain tissue rather than within the subarachnoid space. Intraparenchymal blood usually does not cause a headache, but in most cases it does cause the patient to have a seizure. Arteriovenous malformations may produce local pressure effects on surrounding structures, may cause a "steal" of blood away from the adjacent brain.

At contrast-enhanced CT an arteriovenous malformation appears as a tangle of blood vessels—some seen on end as high-attenuation dots, some cut through the side and seen as high-attenuation serpiginous structures (Figure 18.38). On a noncontrast CT scan an AVM often appears only as a group of calcifications within the brain substance. Because of the arteriovenous shunting that takes place in AVMs, the blood flow itself is increased, and arteriography will show an increase in the caliber of the supplying arteries. Note the large-caliber left internal carotid artery branches in Figure 18.38B, compared with the normal internal carotid arteriogram in Figure 18.23A.

The Face

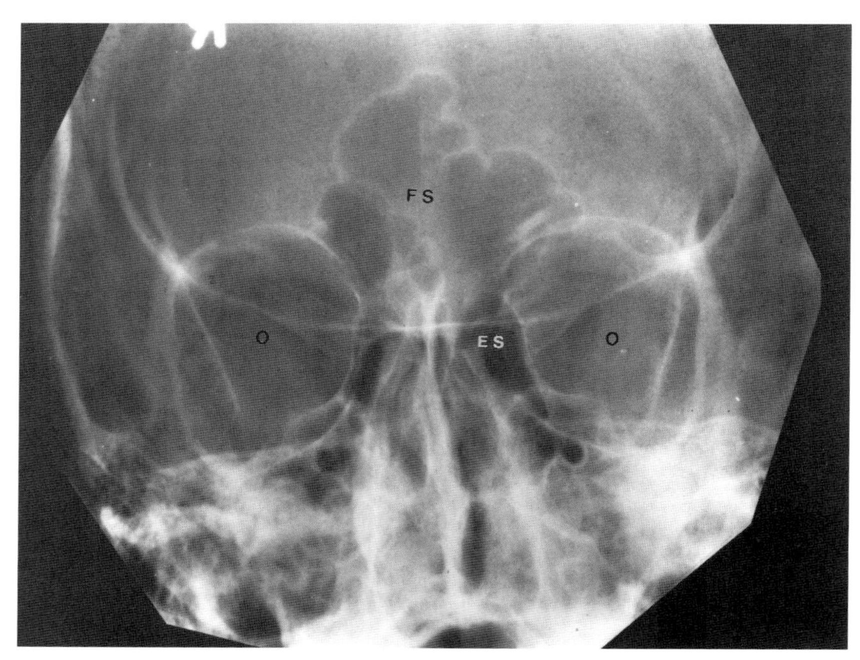

Figure 18.39. Normal Caldwell view of the face. The frontal sinuses *(FS)*, ethmoid sinuses *(ES)*, and orbits *(O)* are well shown.

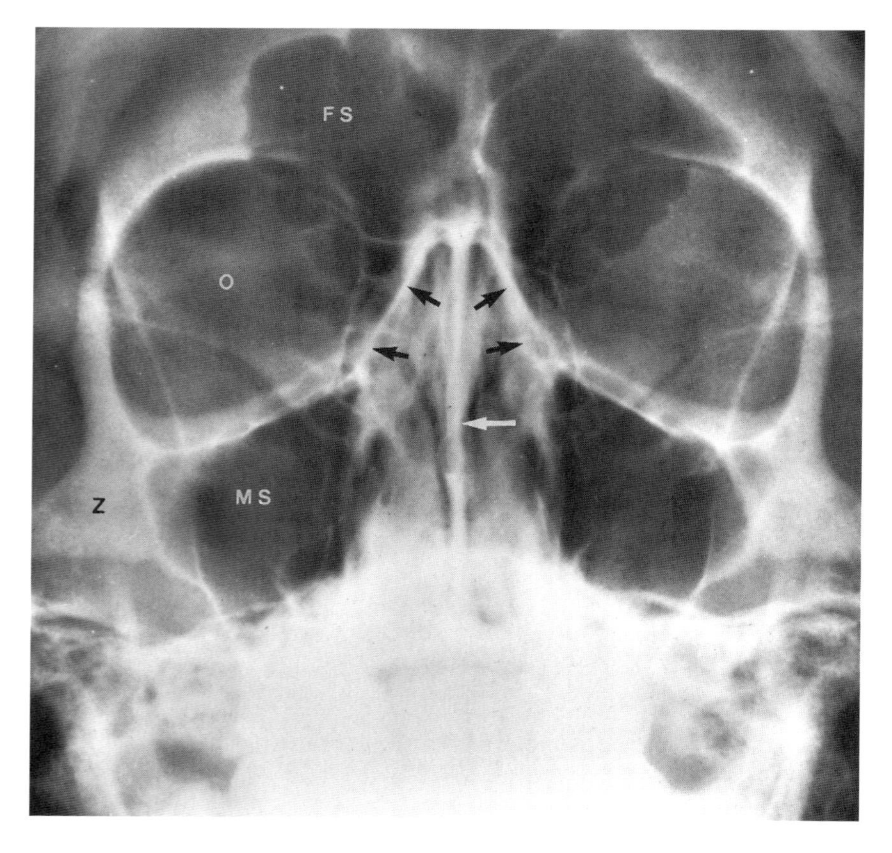

Figure 18.40. Normal Waters view of the face. The frontal sinuses *(FS)*, orbits *(O)*, maxillary sinuses *(MS)*, and zygoma bones *(Z)* can be identified. The *white arrow* indicates the nasal septum; the *black arrows* show the margins of the nasal fossa.

The facial bones are not well seen on a routine skull film series. In fact, they are burned out because of the higher x-ray exposure necessary to see the skull. When your patient has a clinical condition requiring plain film imaging of the face, you should request a *facial film series* that includes Caldwell, Waters, and lateral views of the face (Figures 18.39, 18.40, and 18.41). These are coned close to the margins of the face and taken with a PA technique (so that the facial bones are as close to the film as possible to obtain optimal facial bone detail).

Facial films are most frequently requested to search for facial fractures after trauma. They are also indicated for patients with suspected inflammatory conditions of the face, such as sinusitis and orbital cellulitis, or for patients with a suspected facial tumor. In addition to the routine facial film series, specialized series may be indicated for certain patients; a consultation with the radiologist will help determine the most appropriate examination for the patient's condition. These series include the nasal series (suspected nasal trauma), mandible series (suspected mandible trauma), and sinus series (specially coned down to provide the most optimal imaging of the paranasal sinuses).

The facial bones constitute one of the most complex arrangements of curving bony surfaces in the entire

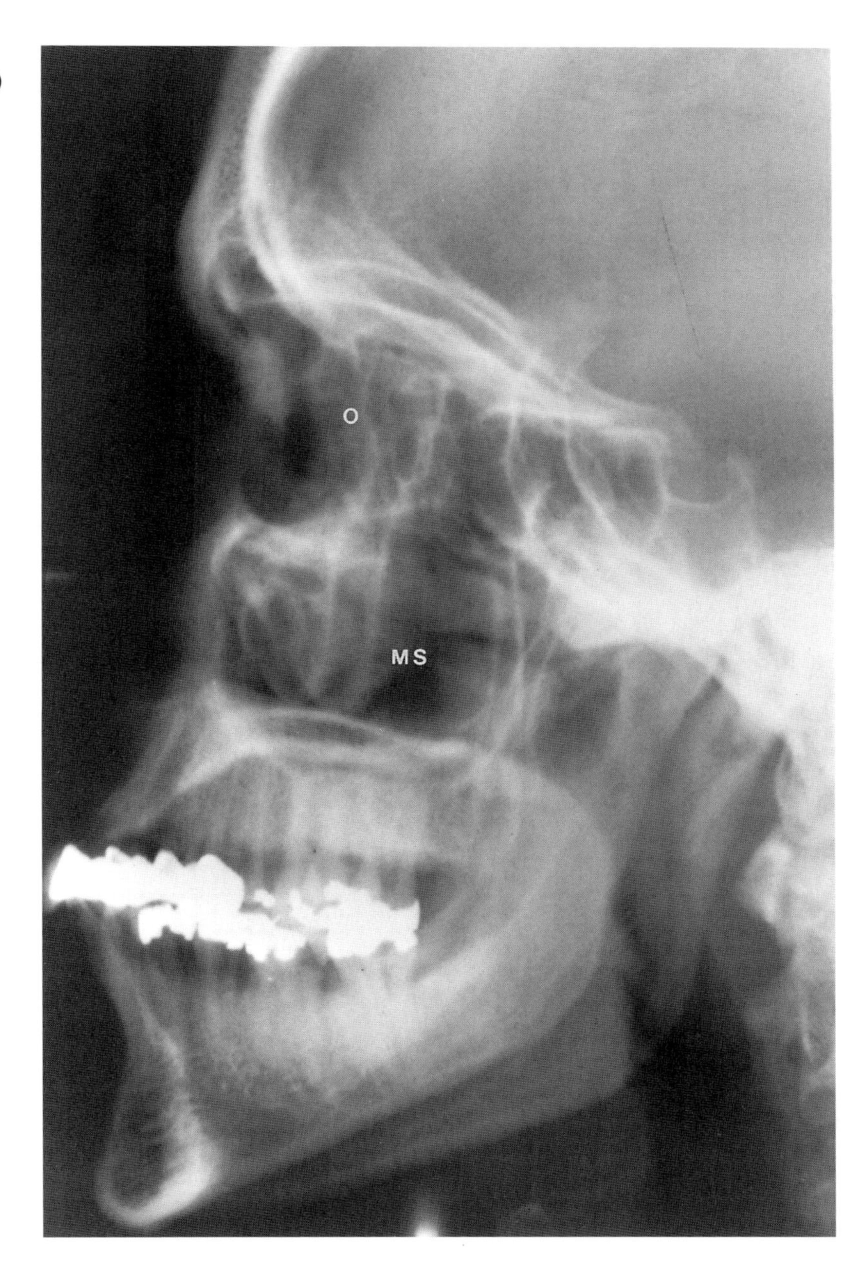

Figure 18.41. Normal lateral view of the face. The two orbits *(O)* and maxillary sinuses *(MS)* overlie each other. Note the metallic dental fillings and crowns.

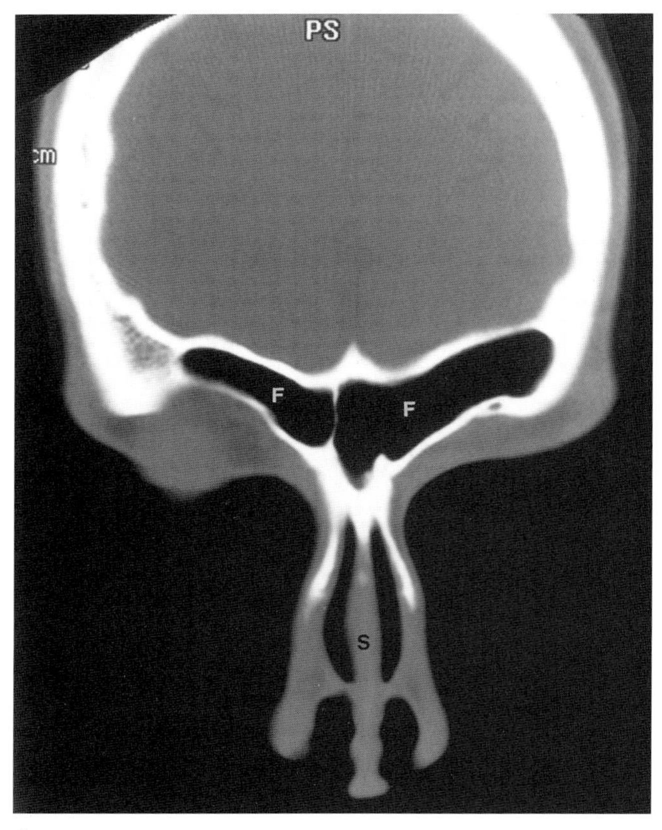

A

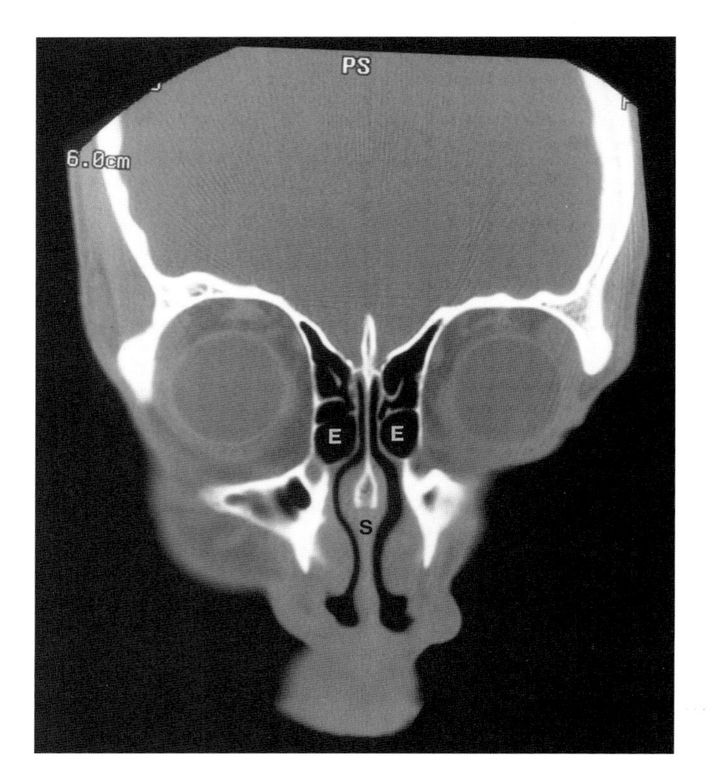

B

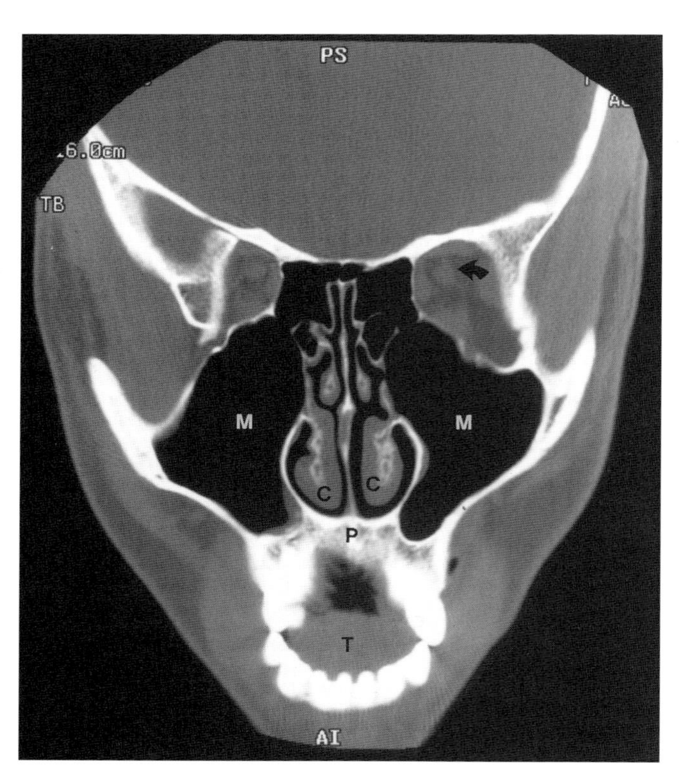

C

Figure 18.42. Normal face CT examination. A: Coronal scan of the anterior face through the frontal lobes of the brain, the frontal sinuses (*F*'s), and the nose; *S* indicates the nasal septum. B: Coronal scan through the midorbits showing the eyeballs, ethmoid sinuses (*E*'s), and nasal septum *(S)*. C: Coronal scan through the posterior orbits showing the optic nerve *(arrow)* surrounded by extraocular muscles (well shown because of surrounding orbital fat), the maxillary sinuses (*M*'s), nasal turbinates, or conchae (*C*'s) within the nasal cavity, hard palate *(P),* and tongue *(T)* within the oral cavity. D: Axial scan of the lower face through the maxillary sinuses (*S*'s), zygomatic arches (*Z*'s), and nasal septum *(S)*. E: Axial scan of the upper face through the midorbits showing the eyeballs, medial and lateral rectus muscles *(arrows),* ethmoid sinuses (*E*'s), and sphenoid sinuses (*S*'s). Note the opaque lenses within the anterior eyeballs.

body. These curving surfaces surround four spaces (oral cavity, nasal cavity, and two orbits) and the paranasal sinuses (frontal sinuses, maxillary sinuses, ethmoid sinuses, and sphenoid sinuses). The facial skeleton is often difficult to evaluate on plain film x-rays, which visually compress this anatomy into two-dimensional black and white images. CT however, can beautifully display this complex anatomy in a series of axial and coronal slices (Figure 18.42). CT can identify more fracture lines and fracture fragments than plain films, and can better show the position and orientation of fracture fragments (Figure 18.43). Three-dimensional

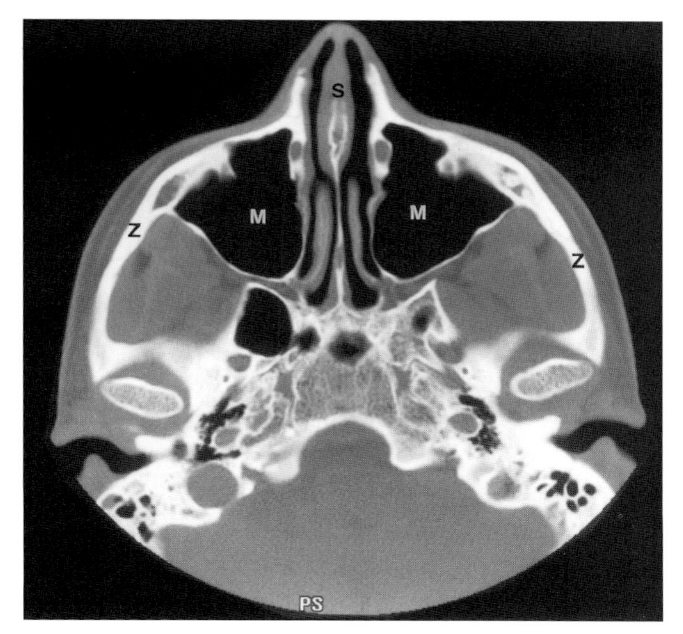

D

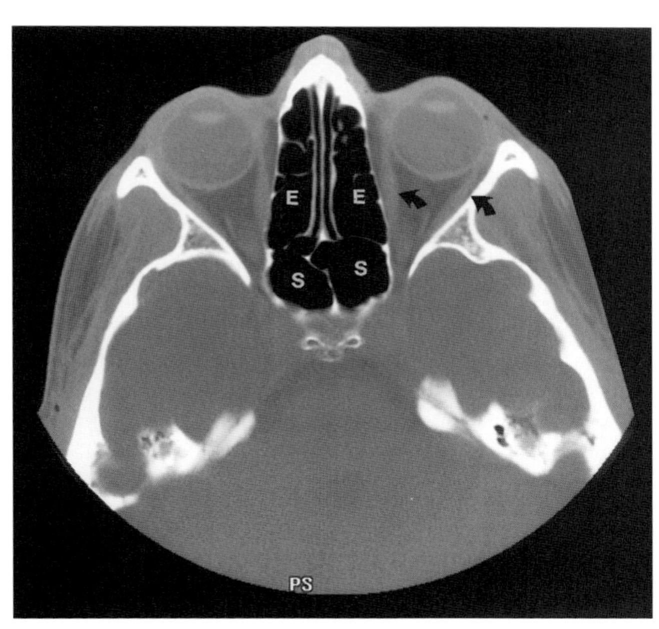

E

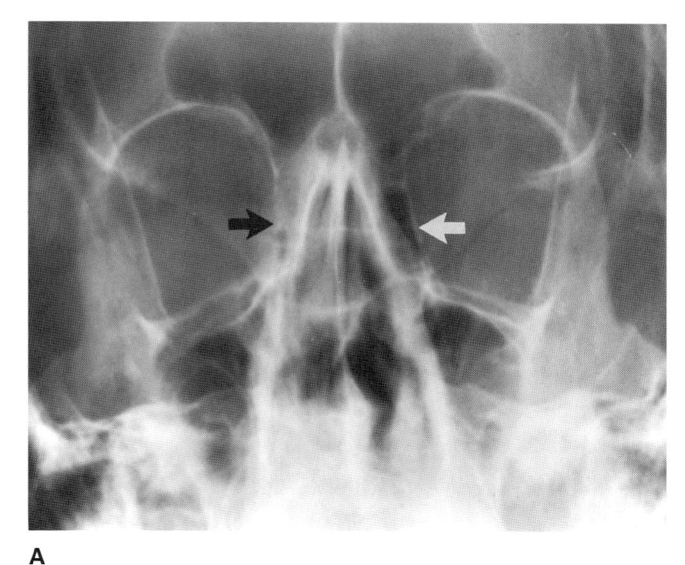

A

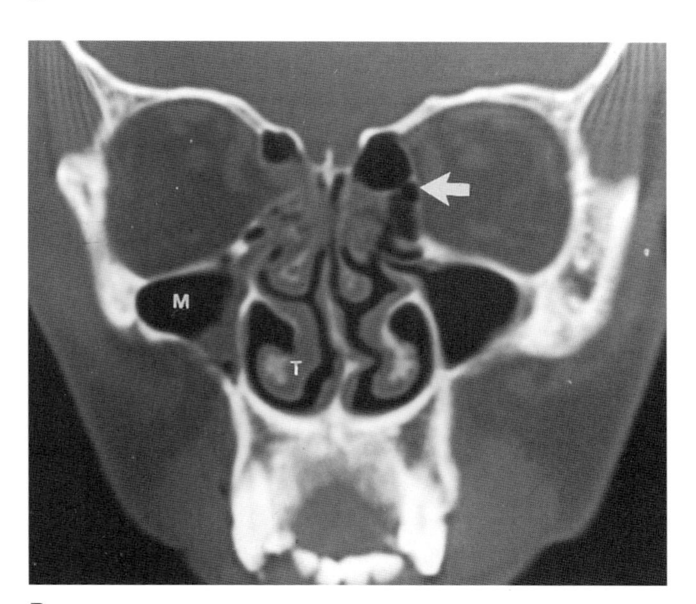

B

Figure 18.43 ***(above).*** Right orbit medial wall blowout fracture in a patient who had been struck in the right eye by an assailant. A: Plain film suggested a fracture *(black arrow)* of the medial wall of the right orbit. Because the medial wall is so poorly seen on this film, it must be displaced from its normal anatomic position. Note also the opacification (representing blood) of the underlying right ethmoid sinus. Compare the normal left medial orbital wall *(white arrow)* and normally aerated left ethmoid sinus. B: Coronal CT scan confirms medial displacement of the right medial orbital wall (therefore, a definite fracture). Note the position of the normal left medial orbital wall *(white arrow).* Opacification can be seen in the right ethmoid sinus. The nasal turbinates are well seen (*T* indicates the right inferior turbinate). You may have noticed on the plain film that the right orbital floor (also the roof of the right maxillary sinus, *M*) appears to be lower than the left; CT shows that this is not a fracture, but instead an anatomic variant.

reformations of facial CT scans (3DCT) are particularly good at displaying major displaced fracture fragments, which facilitates optimal surgical planning (Figure 18.44, on the next two pages).

CT can also visualize soft-tissue structures directly and therefore can show disorders of these structures that would not be possible to image with plain films or conventional tomography. For example, with orbital trauma CT can show soft-tissue injuries such as

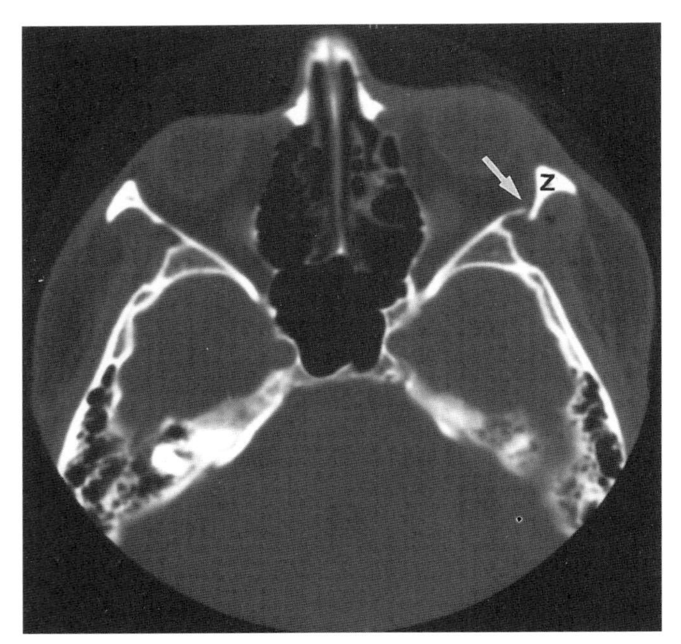

A. Axial CT scan through the midorbits shows a fracture of the zygomatic-sphenoid suture *(arrow),* along the posterior lateral margin of the orbit. The zygoma *(Z)* is posteriorly displaced.

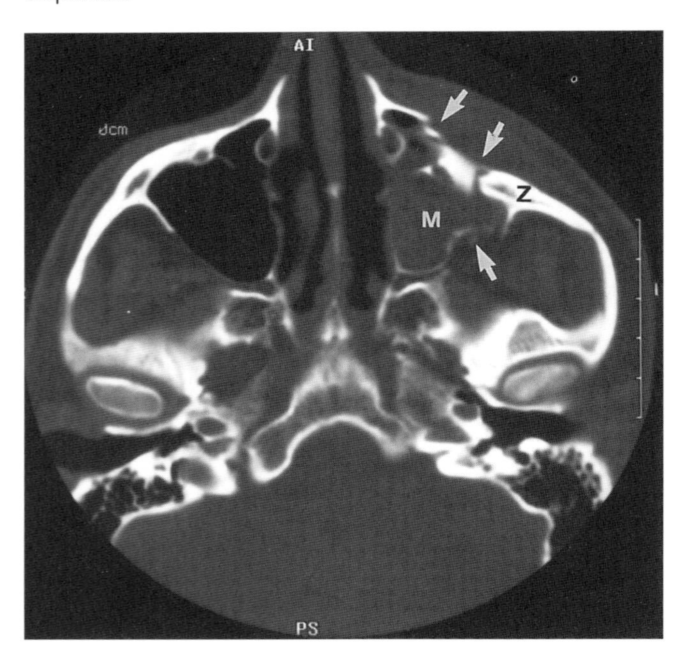

B. Axial CT scan at a lower level through the maxillary sinuses shows fracturing at the sutures between the zygoma and maxillary bones *(arrows)* along the anterior and posterior lateral walls of the right maxillary sinus. The left zygoma *(Z)* is displaced posteriorly and the left maxillary sinus *(M)* is opacified with blood.

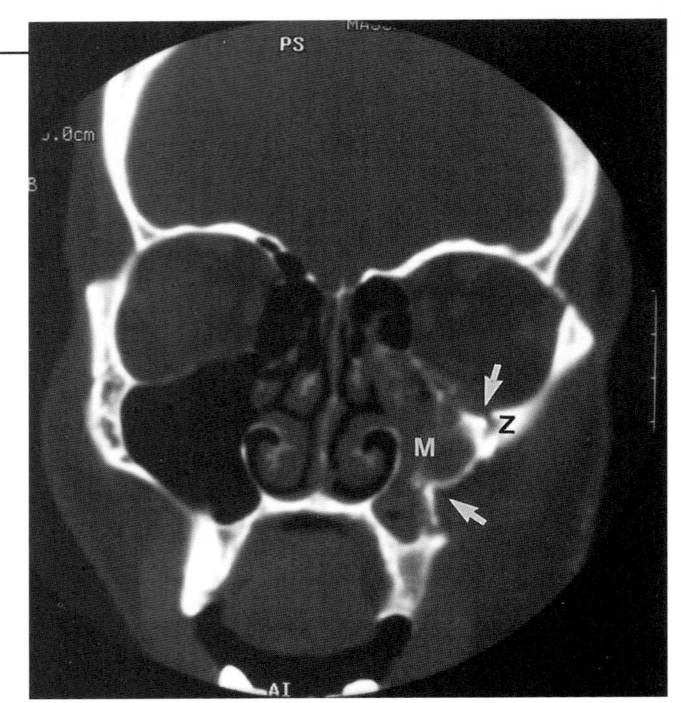

C

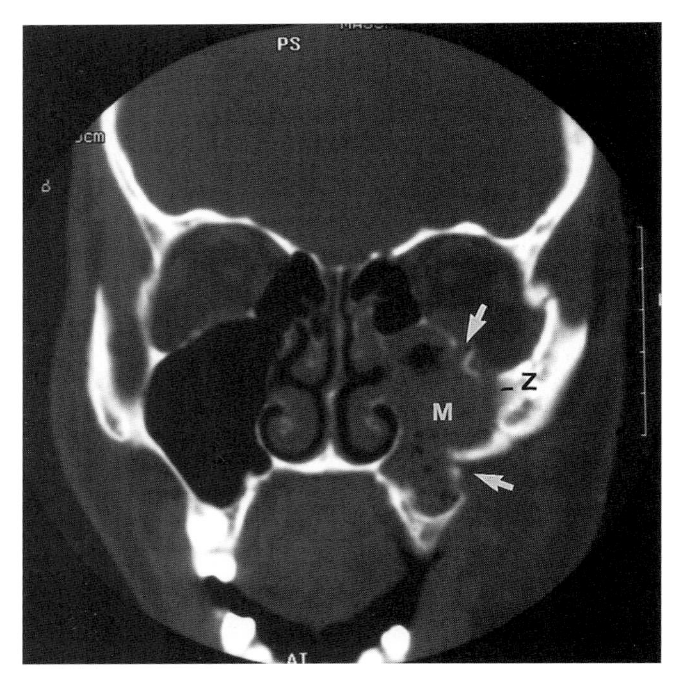

D

C-D. Anterior and posterior coronal CT slices show fractures *(arrows)* at the margins of the zygoma *(Z)* and maxillary bones involving the floor of the left orbit and the lateral wall of the left maxillary sinus. The maxillary sinus *(M)* is filled with blood.

Figure 18.44. Left zygoma fracture: conventional CT (axial and coronal slices) and three-dimensional CT (films taken from the front, base, and both sides). This young man was struck in the left cheek in an altercation; the blow caused fracturing at the suture lines between the left zygoma and its neighbors (maxillary bone, frontal bone, temporal bone, and sphenoid bone), with posterior and inferior displacement of the zygoma bone itself. You may wish to review the anatomy of the zygoma bone in Figures 3.3 and 3.5, pages 44, 45, before proceeding.

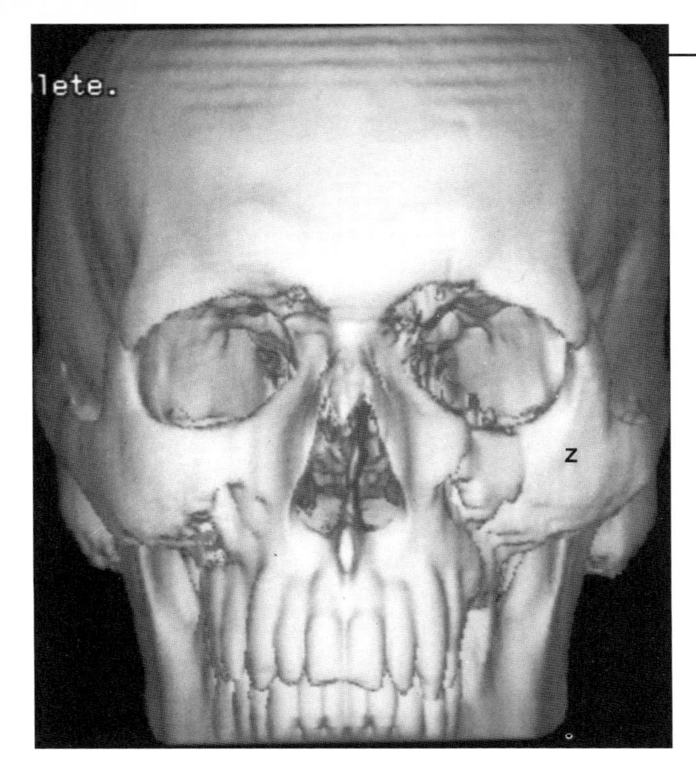

E

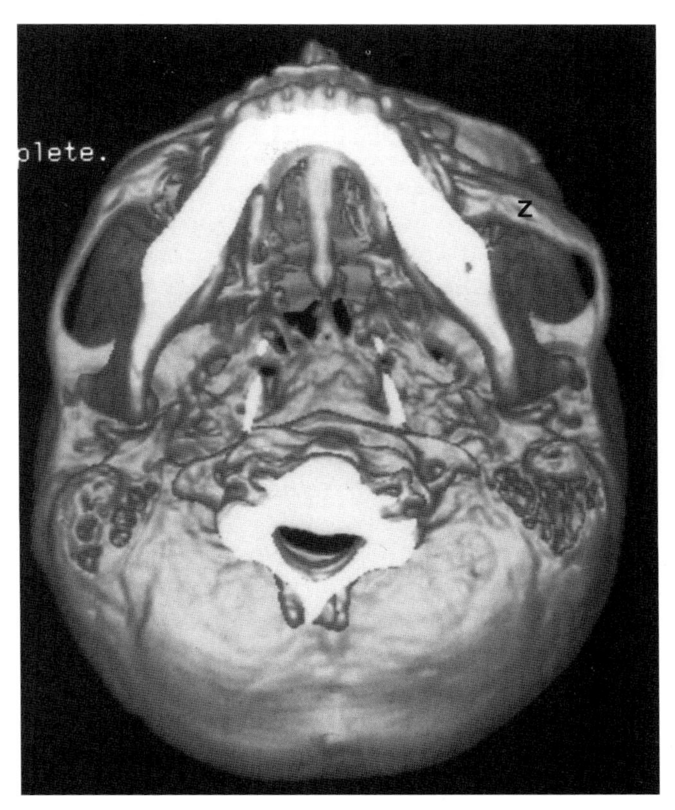

F

E-H. The axial slices were computer integrated to produce a three-dimensional model that was rotated and filmed from the front (E), base (F), and both right (G) and left (H) sides. Note how well this displaced fracture is displayed by 3DCT. The left zygoma *(Z)* is shown to be posteriorly and inferiorly displaced. Note the inferior zygoma displacement on the anterior view. The posterior displacement is very evident on the base view and in the comparison of the cheekbones (zygomas) on the two laterals. It is easy to understand from these images why the patient has clinical flattening of the left cheek.

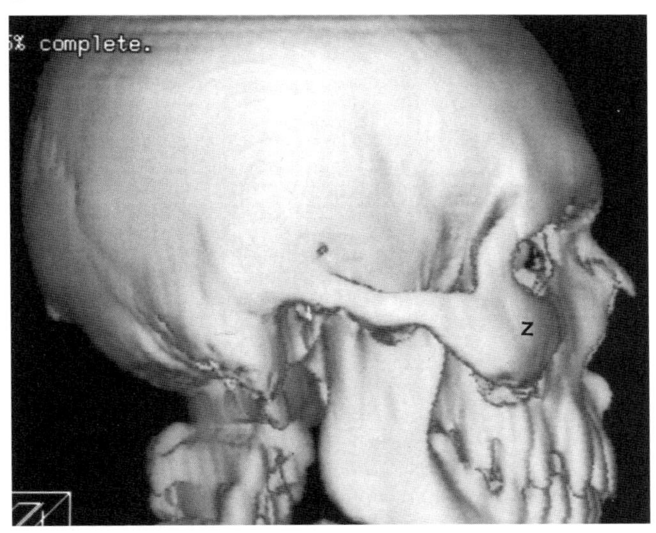

G

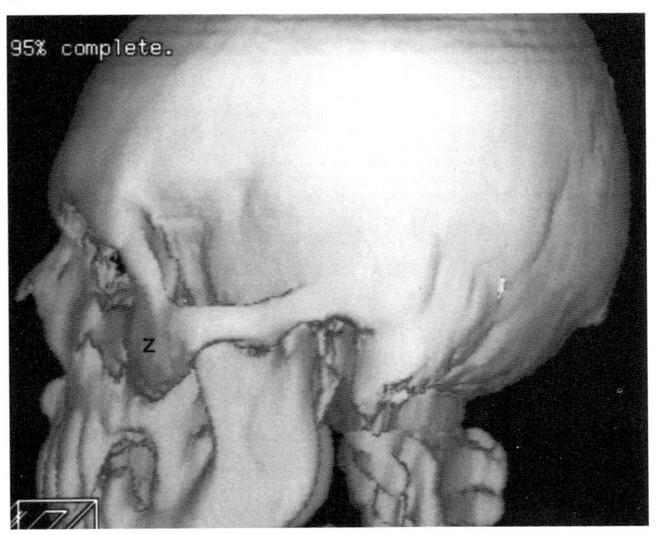

H

extraocular muscle entrapment and impingement, eyeball rupture, lens dislocation, the presence of foreign bodies, and injuries to the optic nerve. MR is also an excellent technique for imaging soft-tissue pathology of the face and neck, especially neoplastic conditions.

In patients with acute inflammatory or allergic sinusitis, plain films of the sinuses may show thickening of the sinus mucosal membranes, abnormal air-fluid levels, and—rarely, in advanced cases—evidence of bony destruction. But the preferred imaging technique for sinusitis is CT, which may visualize many disorders not detected by plain films, including changes indicative of early sinusitis. CT may show the cause of recurrent sinusitis, often a blocked sinus ostium (which may require surgical correction), as well as early complications of sinusitis.

Low Back Pain and Lumbar Disc Syndrome

Low back pain is a very common problem, and you will no doubt see a large number of patients with this complaint during your career. In your emergency department, they may present with acute symptoms, after trauma or lifting a heavy object; in your office or clinic, they may present with subacute or chronic back symptoms. There are many causes of low back pain, including muscle strain, arthritis, vertebral fracture, lumbar disc disease, and even bone tumors, and you should choose the imaging technique that best shows what you think may be the most likely cause. A plain film series of the lumbar spine (AP, lateral, and coned-down lateral views) is only helpful when the cause alters the radiographic appearance of the spine, as do fractures, primary and metastatic tumors, and arthritis. After trauma to the spine, for example, plain films are exceedingly helpful in ruling in or ruling out a vertebral fracture. But a plain film series does not help evaluate muscle strain; young patients with low back pain caused by muscle strain, and no neurological symptoms or signs, will have normal lumbar spine x-rays, and they should not be unnecessarily radiated.

When patients present with back pain and neurological symptoms or signs, such as sciatica, you should consider the diagnosis of lumbar disc syndrome (degenerative disc, herniated disc, bulging disc). Since plain films are helpful only in ruling out conditions that change the x-ray appearance of the bony spine, such as trauma, tumor, or arthritis, these patients may require more sophisticated imaging with MR or CT. Plain films may show a narrowed intervertebral disc space, but this finding is nonspecific because disc space may narrow as part of the normal aging process in a patient with some other cause of back pain.

Patients with sciatica often complain of pain like an electric shock that radiates down one or both legs. This *radicular pain* is caused by the irritation of one of the spinal nerve roots where it leaves the spinal canal. The usual causes of radicular pain are intervertebral disc herniation into the neural foramina and degenerative arthritis that narrows one of the vertebral foramina. Radicular pain most commonly occurs in the sides and the back of the legs, caused by the narrowing of the foramina of the L4–L5 and L5–S1 vertebral levels. The pain may be associated with muscle weakness, a sensory deficit, or abnormal reflexes.

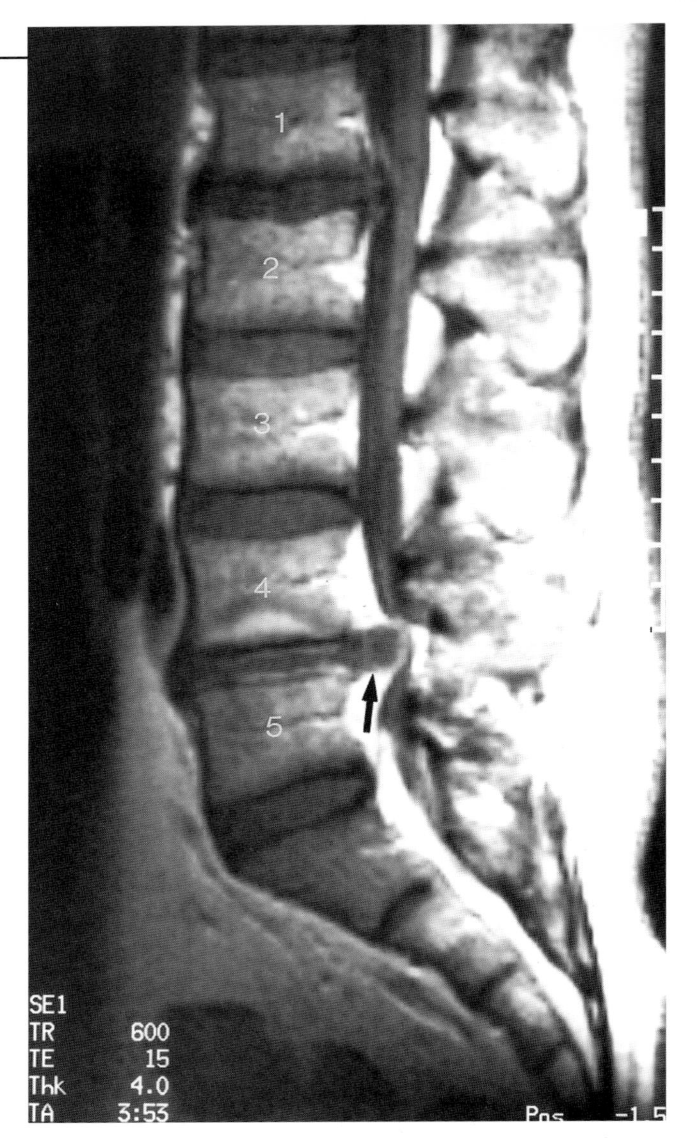

Figure 18.45. Sagittal MR scan of the lumbar spine of a 55-year-old man with a herniated L4–L5 intervertebral disc *(arrow). Numbers* overlie the five lumbar vertebral bodies. The dark vertical structure posterior to the vertebral bodies represents the thecal sac, containing the spinal cord and spinal nerves. Note how it is compressed by the herniated disc. You can easily imagine why this patient suffers pain and has motor weakness in both legs. Observe also that the L4–L5 disc space is narrower than the other lumbar intervertebral disc spaces.

The spinal cord and nerve roots are not seen on plain spine films but can be visualized by CT, MR, and myelography. Probably the best way to image the lumbar spine in patients with radicular pain is to obtain an MR scan. Figure 18.45 shows the sagittal MR scan of a patient with a herniated L4–L5 disc. MR can also show osteophytes associated with degenerative arthritis, which may also cause nerve root syndromes.

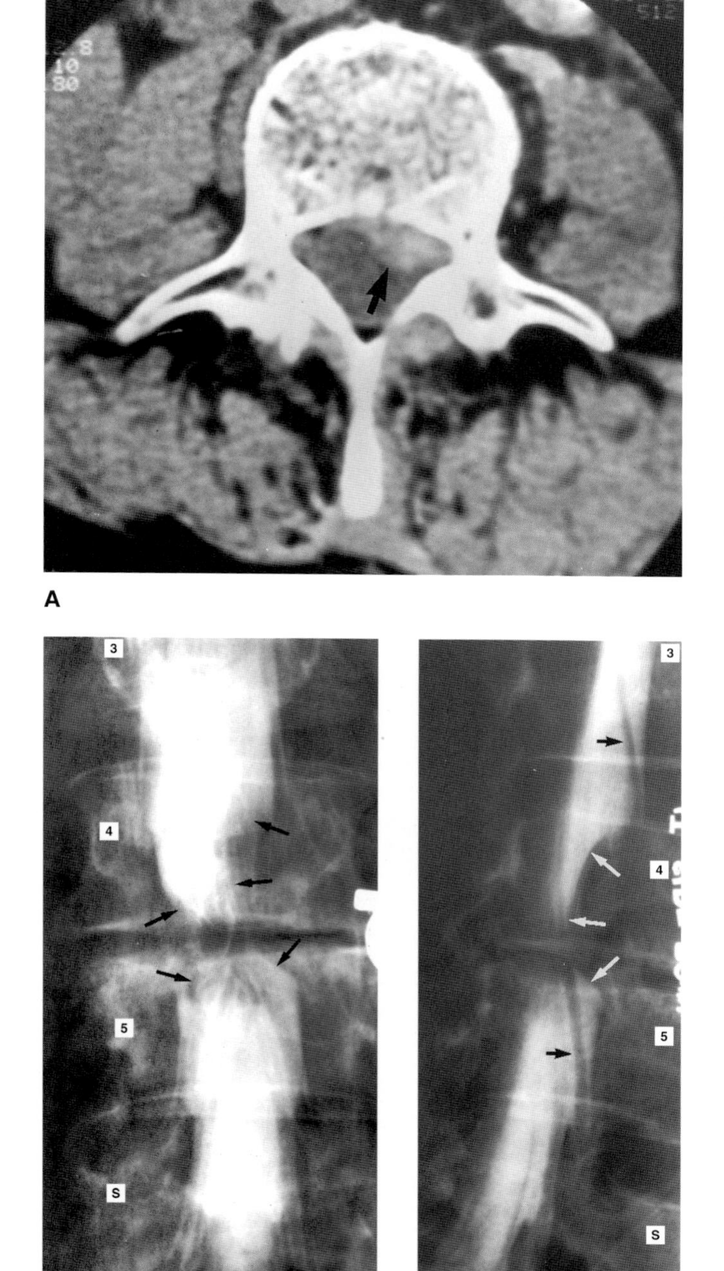

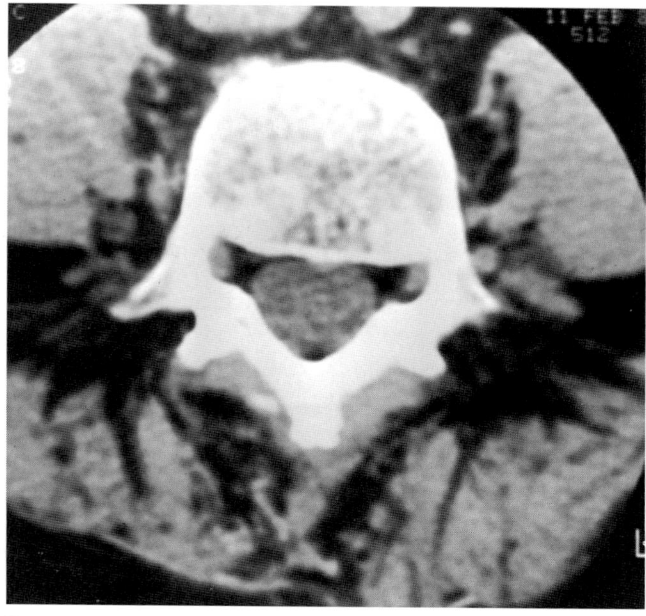

Figure 18.46. Herniated L4–L5 lumbar disc. A: CT scan through the pedicles of the L4 vertebra shows herniated disc material *(black arrow)* extending posteriorly on the left side into the neural canal. B: Lower-level CT scan through the pedicles of the L5 vertebra shows a normal portion of the neural canal, with black fatty tissue outlining the thecal sac centrally and nerve roots in both lateral recesses. C: Frontal view myelogram of the same patient shows bilateral compression of the opacified *(white)* thecal sac by herniated disc material *(black arrows)* at L4–L5. There is more compression from the left side than from the right. Lumbar vertebrae 3–5 are shown; the sacrum *(S)* can also be seen. D: Oblique myelogram of the same patient. The *white arrows* indicate the severe compression on the left side of the thecal sac. The *black arrows* indicate two radiolucent spinal nerves traversing the contrast-enhanced thecal space.

Lumbar disc herniation can also be demonstrated by CT and myelography (Figure 18.46). To perform a myelogram, the radiologist places a needle percutaneously into the lumbar spine, below the termination of the spinal cord (conus), with its tip in the thecal sac. Then, under fluoroscopic control, a myelographic contrast material is injected to allow visualization of the spinal cord, spinal nerve roots, and the cul de sac. Since the contrast material is radiopaque, the spinal cord and nerve roots will appear radiolucent. In addition, CT slices can be obtained after contrast injection (CT myelography) to provide improved detail.

The majority of patients with nerve root syndromes caused by lumbar disc herniation will recover with conservative treatment and may not require any diagnostic imaging. Surgical intervention is indicated only for the small number of patients who fail to recover on optimal conservative management or who suffer major disruption of motor functions, including bowel and bladder dysfunction. Such patients should be imaged with an MR scan; today myelography is seldom used, being reserved for patients whose clinical findings are unusual and whose scanning findings are indeterminate.

Spinal Tumors

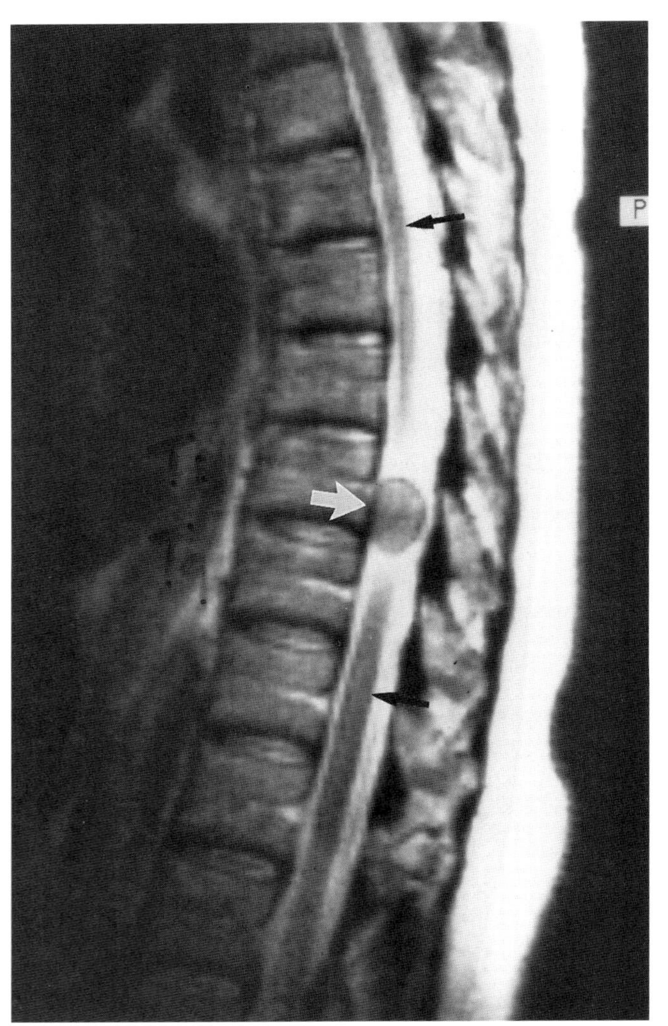

A

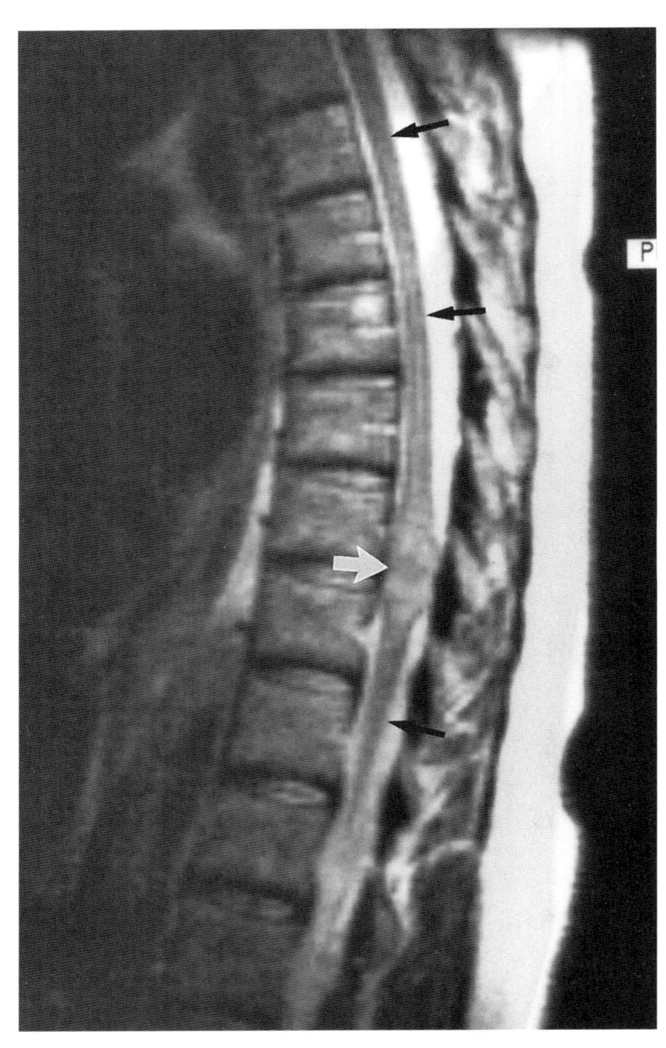

B

Radiology plays an important role in the diagnosis and evaluation of spinal tumors. Diagnostic imaging with myelography, CT myelography, and MR can determine whether a spinal tumor is *intramedullary* (located within the spinal cord itself), *intradural/extramedullary* (located within the dural sac, which contains spinal fluid, but outside the spinal cord), or *extradural* (located outside the dural sac). Establishing the presence of a neoplasm in one of these three "compartments" permits formulation of the differential diagnosis. Intramedullary tumors (spinal cord tumors) are usually ependymomas (the most common), astrocytomas, or hemangioblastomas. Meningiomas are the most common intradural/extramedullary tumors, but nerve sheath tumors may also occur in the dural sac. The most common extradural mass (impinging on the dural sac) is a disc herniation (as in Figure 18.46), but the most common extradural tumor is metastatic spread from cancer of the breast, lung, or prostate.

Study Figure 18.47, which shows the imaging examinations of a patient with an intradural/extramedullary spinal tumor that proved to be a meningioma.

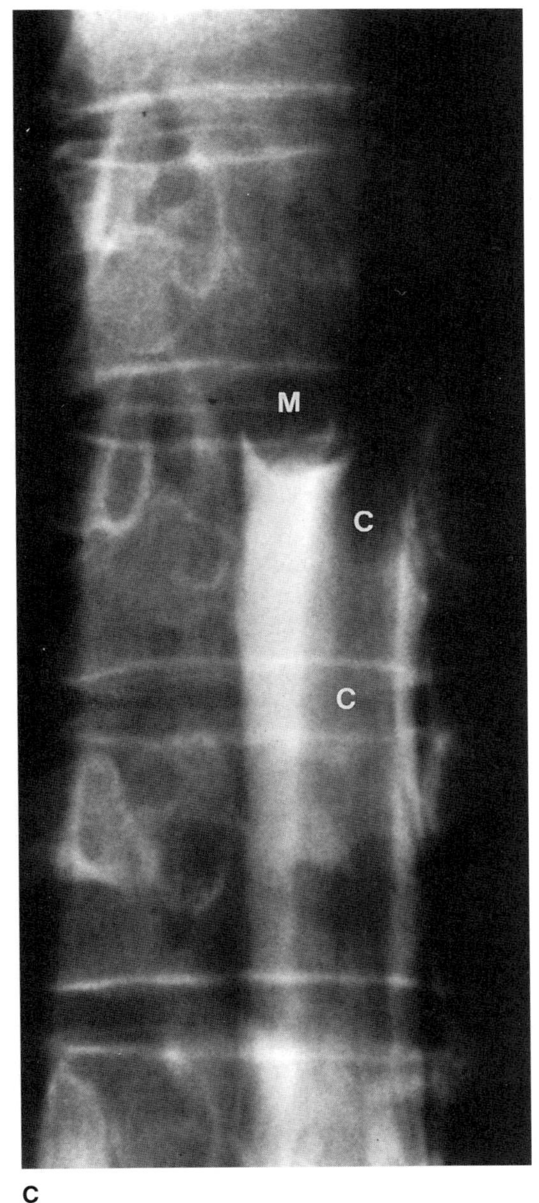

C

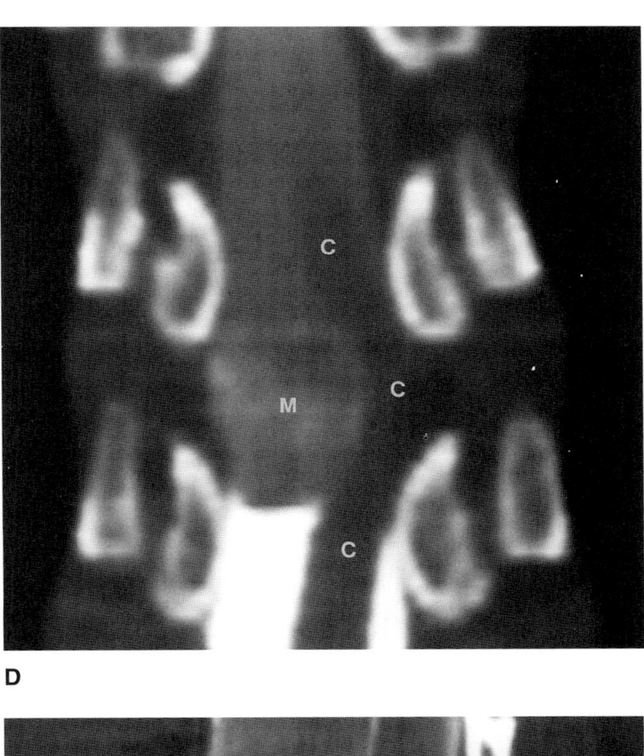

D

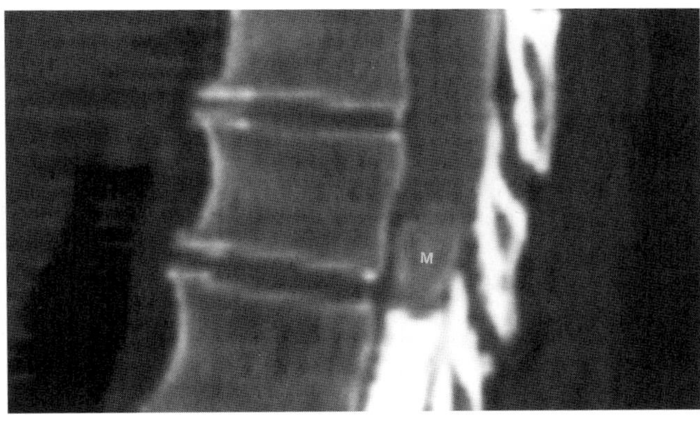

E

The MR scan (Figures 18.47A and B) identified a spinal tumor (of the thoracic spine) responsible for the patient's back pain and abnormal neurological examination. On the MR scan the neoplasm appears to be adjacent to the spinal cord. A myelogram obtained after contrast injection into the lumbar thecal sac was filmed with plain films (Figure 18.47C) and CT (Figure 18.47D, coronal reformation, and Figure 18.47E, sagittal reformation). The myelogram film and CT myelogram confirm an intradural/extramedullary tumor, inside the dural sac, but outside the spinal cord, which it compresses.

Figure 18.47. Intradural/extramedullary spinal tumor that proved to be a meningioma (see text). A-B: Midline adjacent slices from a sagittal MR scan show a small tumor *(white arrow)* adjacent to the spinal cord *(black arrows).* The fluid in the thecal sac appears white on this MR scanning protocol. C: Myelogram film shows obstruction of the flow of intrathecal contrast material (which was injected through a lumbar puncture and appears white) at the site of the mass *(M).* The spinal cord (*C*'s) is displaced and compressed by the mass. CT films obtained with the same myelographic contrast material (D, coronal reformation, and E, sagittal reformation) show obstruction of contrast material, and compression and displacement of the spinal cord (*C*'s) by the mass *(M).* The mass is clearly shown to be extramedullary (outside the cord).

Problems

Unknown 18.1 (Figure 18.48)

Study the MR scans of this previously healthy 52-year-old man presenting with headache, a left facial droop, and weakness in the left hand. What are your conclusions?

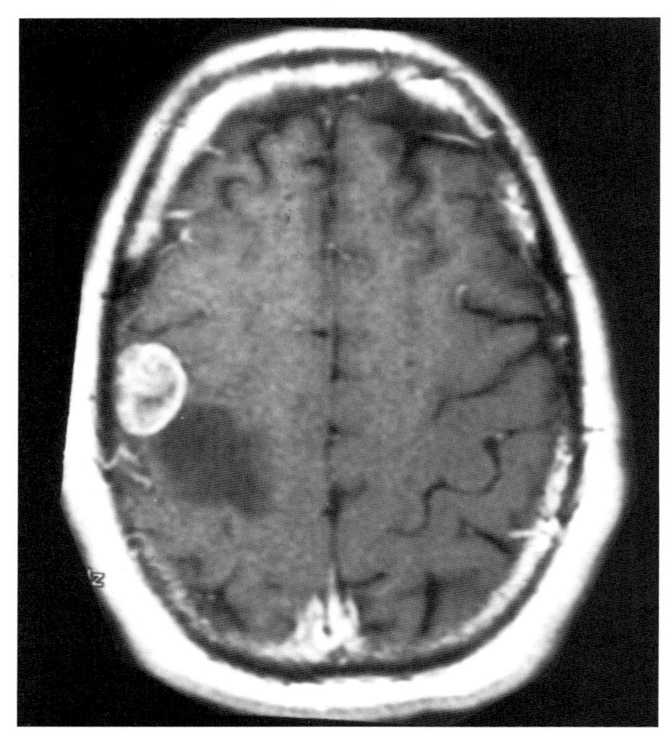

A. Upper axial scan.

Figure 18.48 ***(Unknown 18.1)***

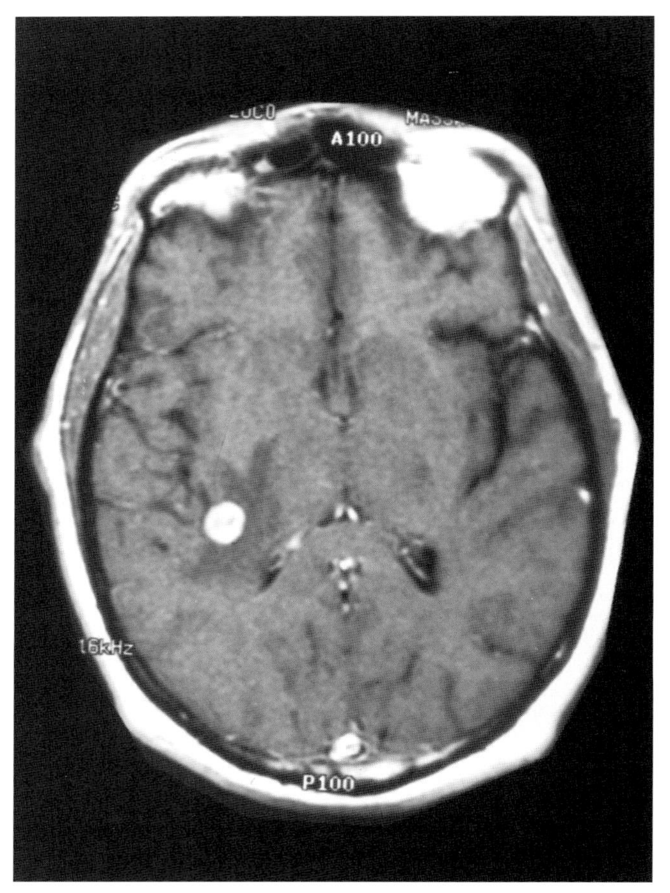

B. Mid-level axial scan.

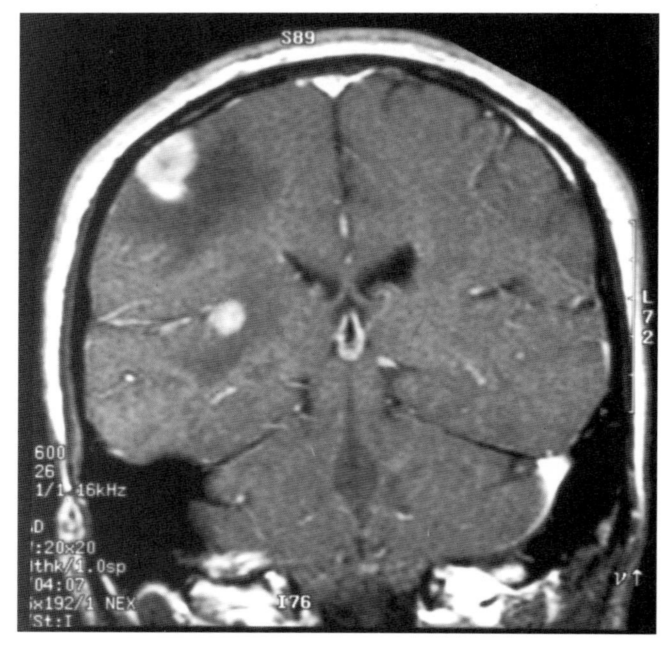

C. Coronal scan.

Unknown 18.2 (Figure 18.49)

This 9-year-old child with headache, nausea, and dizziness has a brain tumor. Where is it located on this sagittal MR scan? Compare the normal MR scan in Figure 18.13, page 515.

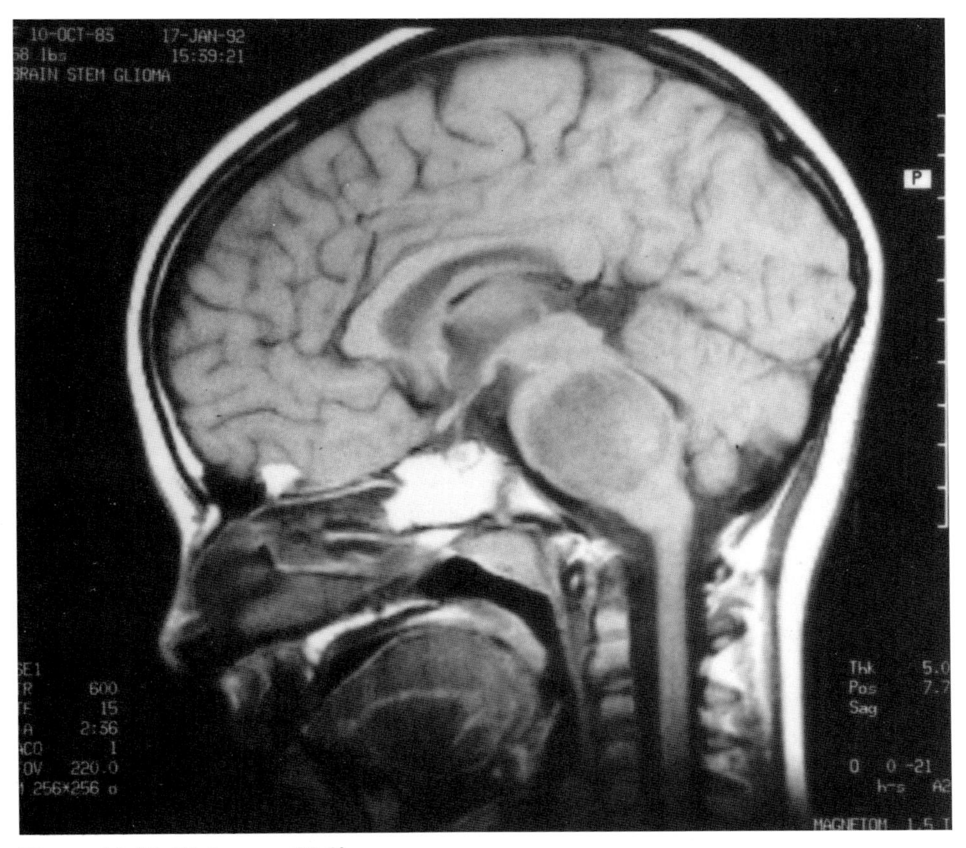

Figure 18.49 ***(Unknown 18.2)***

19 Interventional Radiology

Radiologists have, of course, appeared many times in these pages, usually as consultants on which imaging techniques to select or as interpreters of the findings generated by those techniques. In addition to fulfilling these roles—and generally advising physicians and reassuring patients—radiologists are often called upon to perform therapeutic and interventional procedures. There are quite a variety of these procedures, and here we will introduce you to those you are most likely to request for your patients.

Percutaneous Transluminal Angioplasty

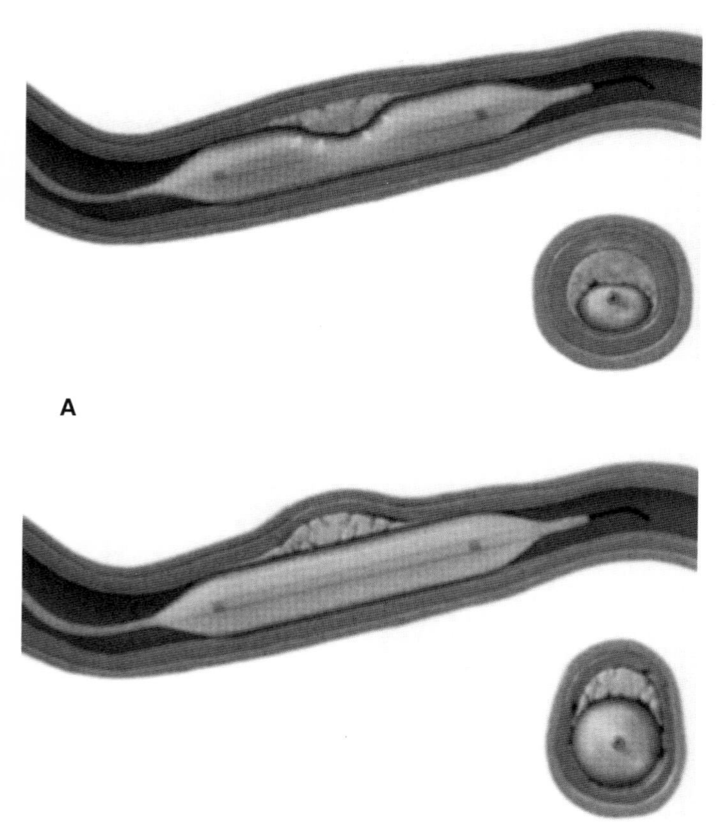

Percutaneous transluminal angioplasty (PTA) is the mechanical recanalization of an occluded or stenotic artery by displacement of the obstructing material within it, usually atherosclerotic plaque. Today the procedure is performed by radiologists with a percutaneously inserted balloon catheter (Figure 19.1) that is inflated within the stenotic or occluded artery. The procedure is quite safe, with low morbidity and mortality rates, and permits relief of a patient's claudication, angina, or other ischemia-related symptoms without surgery. PTA is most successful in the treatment of short single stenoses and occlusions; it is less successful in treating long-segment occlusions, diffuse multifocal stenoses, and severely calcified stenoses. In patients with claudication, rest pain, or nonhealing ulcers in a lower extremity due to short-segment atherosclerotic disease, PTA of the iliac, femoral, and popliteal and tibial arteries offers a safe alternative to arterial bypass surgery (Figure 19.2).

Patients are given aspirin before PTA and their diagnostic arteriograms are reviewed. During the procedure performed to relieve stenosis, arterial pressures are measured across the stenosis. A gradient of 20 millimeters or more is hemodynamically significant, a gradient that should benefit from angioplasty. The radiologist selects a balloon catheter that will overdilate the artery by 1 millimeter. After balloon inflation across the stenosis, pressures are again measured to document elimination of the pressure gradient and a repeat arteriogram is performed. A successful angioplasty is one that shows less than 20 percent residual stenosis. Arterial occlusions are treated in a similar fashion.

Figure 19.1. The mechanism of angioplasty. A: The percutaneously inserted balloon catheter is inflated within the arterial stenosis. A balloon "waist" is formed as the atherosclerotic plaque initially indents the inflating balloon. B: With further inflation the balloon waist disappears as the vessel caliber increases and the luminal narrowing is eliminated.

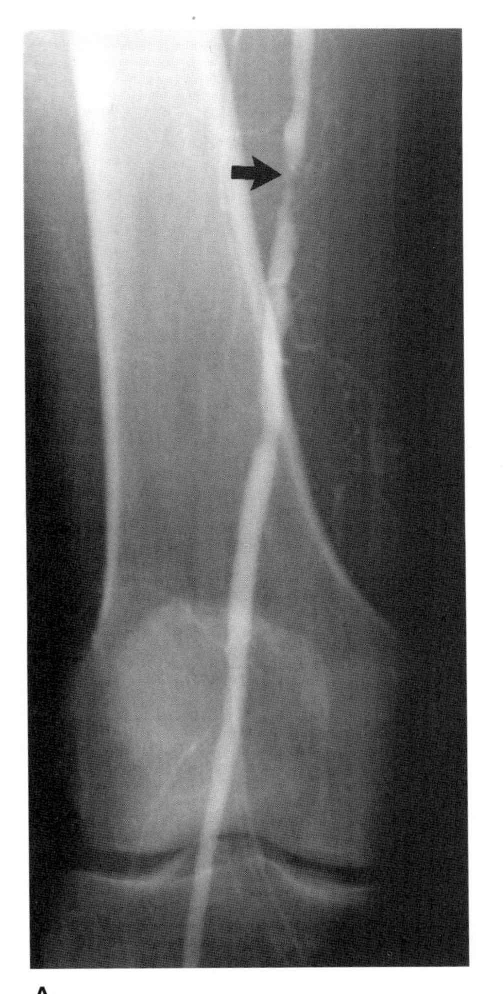

A

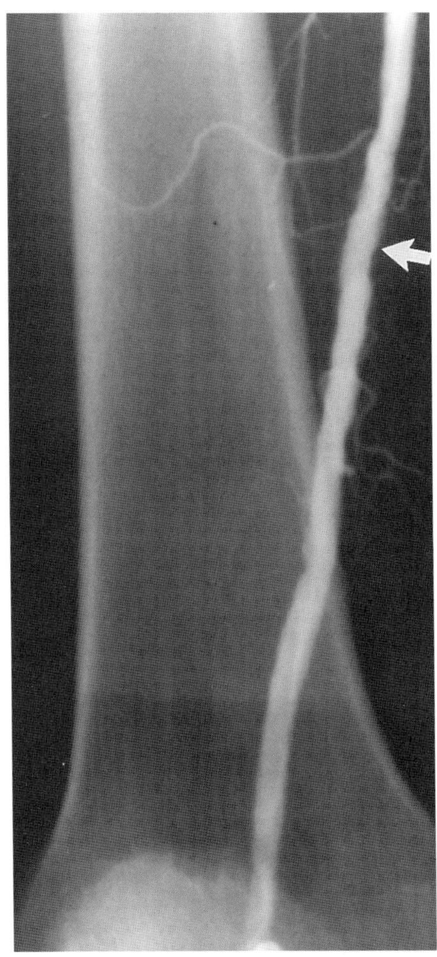

B

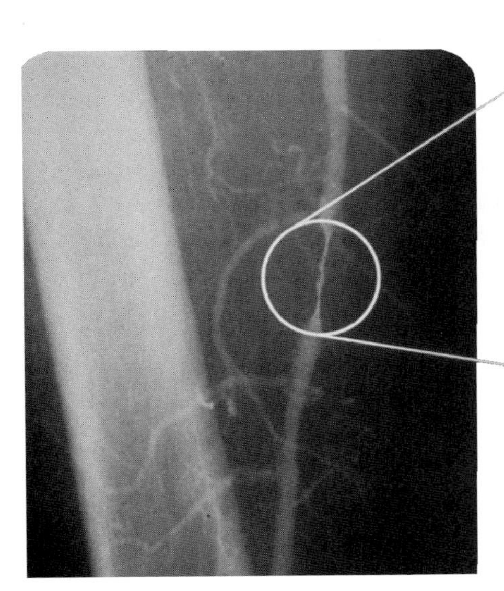

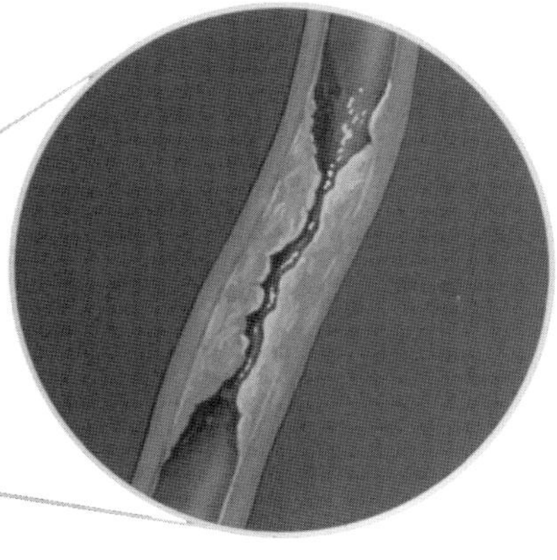

C

Figure 19.2. Angioplasty to relieve claudication of the right leg. A: Preangioplasty diagnostic arteriogram shows a tight stenosis *(arrow)* in the distal right superficial femoral artery, at the level of the adductor canal, just above the right knee. B: Post angioplasty arteriogram shows excellent dilatation of the stenosis (the *arrow* indicates the level of prior stenosis). The patient's claudication was markedly improved. C: Artist's depiction of a similar lesion in another patient.

A

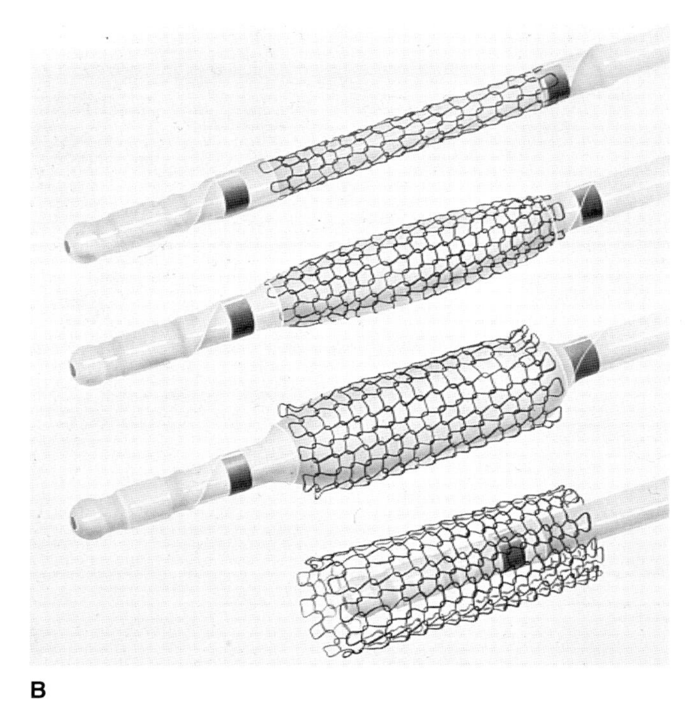

B

Figure 19.3. A: Arterial stent. B: Mechanism of stent placement *(top to bottom):* A special percutaneous balloon catheter preloaded with the stent is positioned across the site of the stenosis. Inflation begins within the stenosis. At full inflation the catheter sleeves release the stent. The insertion catheter is withdrawn.

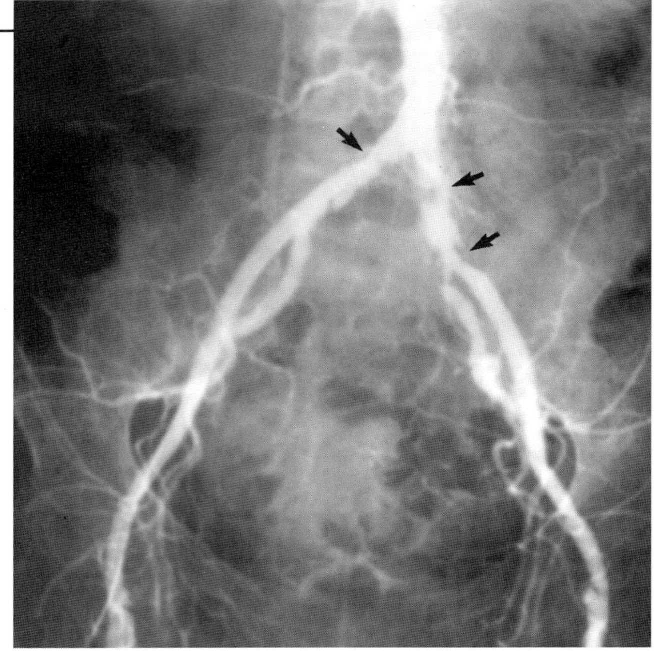

A

Figure 19.4. Angioplasty of iliac artery stenoses with multiple stent placement in an elderly patient with bilateral claudication. A: Preangioplasty arteriogram shows stenoses *(arrows)* of both common iliac arteries. B: The angioplasty balloon catheter, loaded with a stent, is positioned across the proximal left common iliac artery stenosis. C: The balloon is inflated, dilating the stenosis and expanding the stent. D: The balloon is deflated and the fully expanded stent is then in the proper position. E: Plain film taken after two other stenoses were dilated and stented. Can you find all three stents? F: Final arteriogram showing marked improvement in the caliber of the iliac arteries. G: Artist's depiction of similar lesions in another patient.

Recurrent stenoses and occlusions are common causes of angioplasty failure and they are often related to the degree of residual narrowing. Many of these lesions are treated with repeat angioplasty. An alternative treatment is the placement of an intravascular stent to maintain patency. One type is an expandable metal stent inserted, mounted on an angioplasty balloon catheter, as shown in Figure 19.3. The stents are several centimeters in length; several may be linked together for treatment of a long stenotic segment. They have been used in the iliac, coronary, renal, and femoral arteries (Figure 19.4). Vascular stents have also been used to relieve certain *venous* conditions, including malignant obstruction of the superior and inferior vena cavae and also benign venous occlusive disease.

Percutaneous transluminal renal angioplasty can improve or cure renovascular hypertension. As you no doubt remember, renovascular hypertension is one of the less common causes of hypertension, the most frequent being essential hypertension. When renal artery stenosis is detected in a patient with hypertension it may be helpful to determine whether the stenosis is only *coincidental* or whether the renal artery stenosis is the *cause* of the hypertension. The differentiation is made by measuring the level of renin in

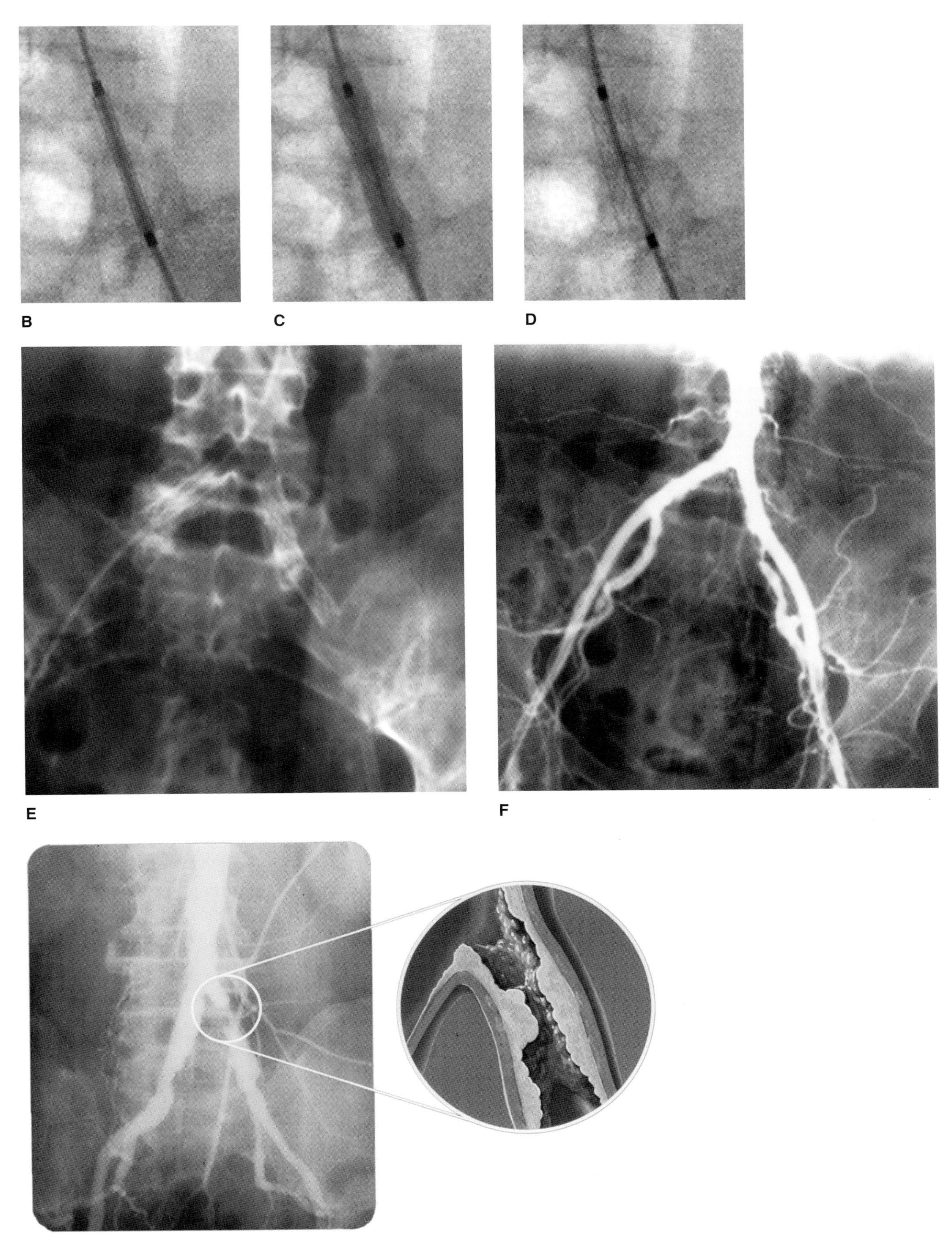
B
C
D
E
F
G

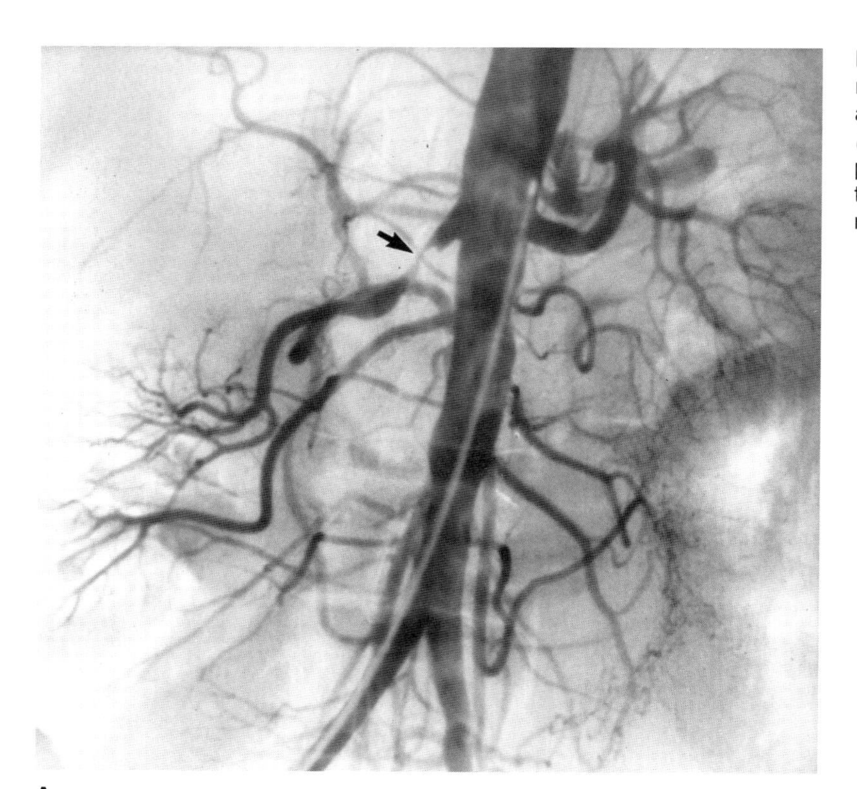

A

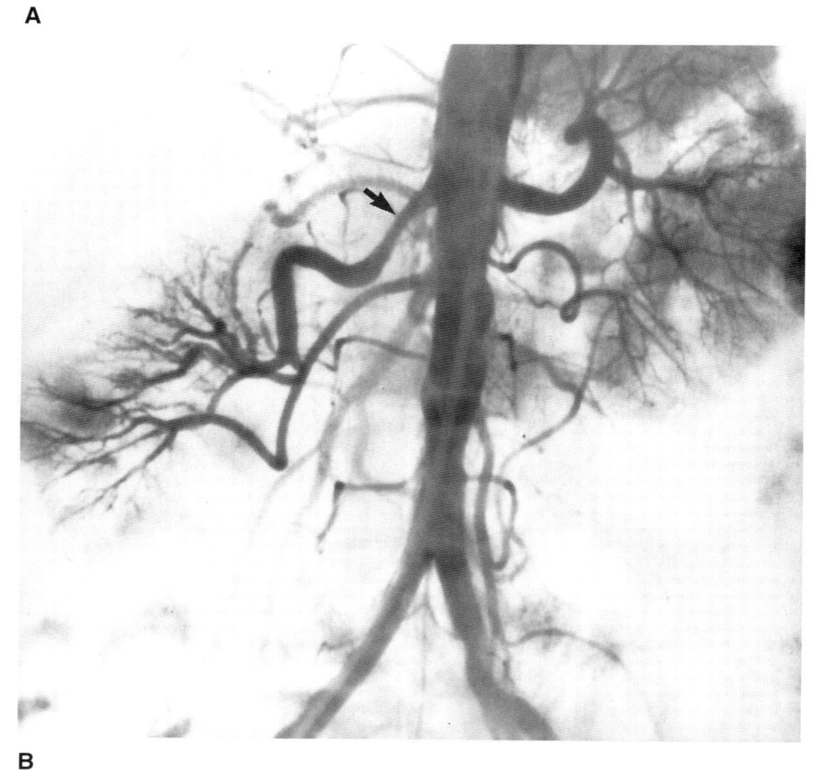

B

Figure 19.5. Renal artery angioplasty to relieve renovascular hypertension. A: Preangioplasty abdominal aortogram shows a tight stenosis *(arrow)* of the right renal artery. B: Postangioplasty aortogram shows excellent dilatation of the stenosis. The patient's hypertension was markedly improved by the procedure.

samples of venous blood obtained by selective renal venous catheterization. Renovascular hypertension is diagnosed when the renin level is 1.5 times higher in the venous blood from the kidney distal to the renal artery stenosis than it is in blood samples from the opposite renal vein or the inferior vena cava. Following a successful renal artery angioplasty there should be a drop in the level of renal vein renin from the affected side, and consequently an improvement in the patient's hypertension (Figure 19.5).

Transcatheter Embolization

Two angiographic techniques are used to decrease arterial blood flow, vasoconstrictor infusion and transcatheter embolization. An infusion of a vasoconstrictor through a selectively positioned arterial catheter can *reduce* blood flow to the structures supplied by that artery. You will soon read more about arterial infusion vasoconstrictor therapy in the treatment of gastrointestinal hemorrhage with intra-arterial vasopressin.

Transcatheter embolization techniques can also be used to cause a *stoppage* of arterial blood flow, which may be indicated in a variety of conditions. Embolization is used to treat uncontrolled bleeding when medical therapy fails or surgery is not appropriate, as in cases of pelvic bleeding after massive trauma and pelvic fracture (Figure 19.6).

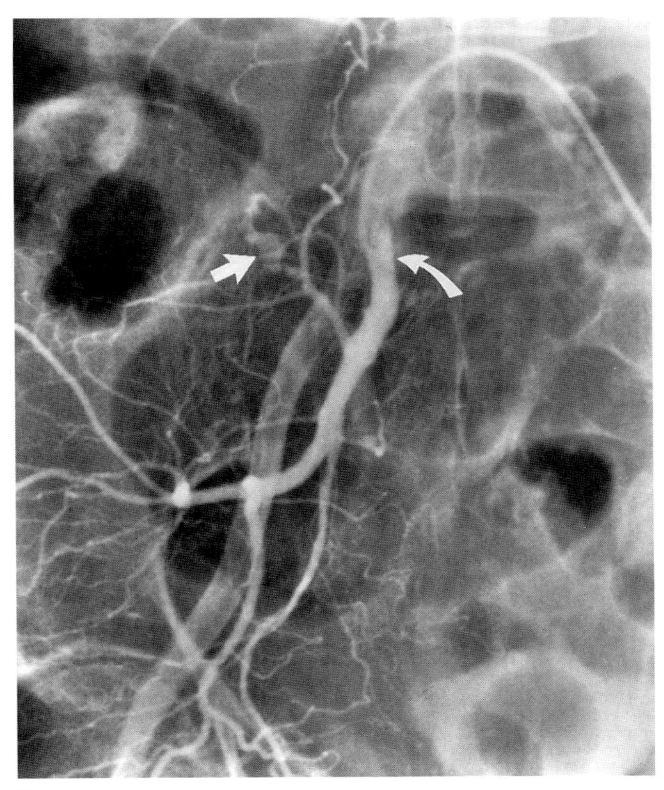

A

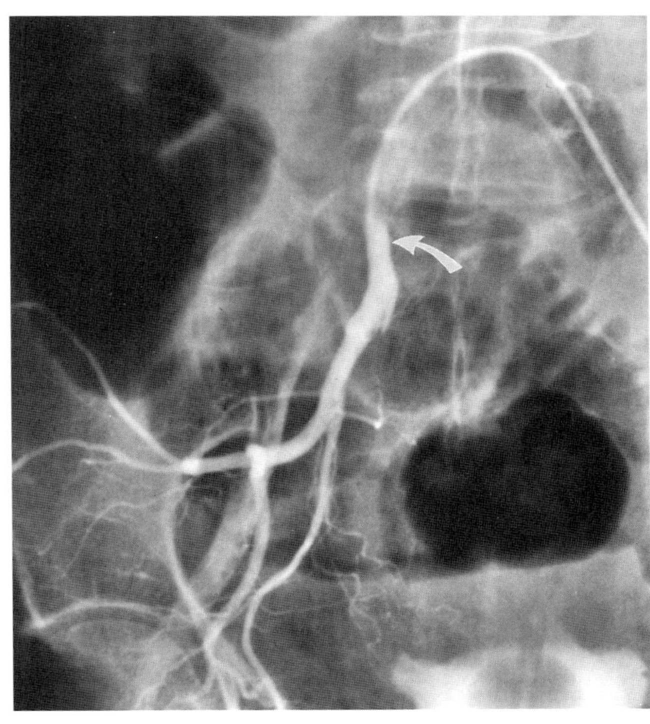

B

Figure 19.6. Embolization of a pelvic arterial bleeding site in a trauma victim with pelvic fractures. A: Preembolization right internal iliac arteriogram shows contrast extravasation *(straight arrow)* at the bleeding site, a laceration of the right iliolumbar artery. The *curved arrow* indicates the right internal iliac artery origin. B: Postembolization arteriogram (the *curved arrow* again indicates the right internal iliac artery) showing occlusion of the iliolumbar artery branch and no further bleeding. The patient did well clinically without requiring further blood transfusions.

As an adjunct to surgery, embolization can be performed just before the operation, to reduce blood flow to a malignancy so that there will be less blood lost in surgery. This technique is often used to occlude the renal artery preoperatively in patients with very vascular renal cell carcinomas (Figure 19.7). Through an arteriographic catheter positioned within the renal artery, the radiologist injects small pieces of particulate material, while watching on the fluoroscope, until the arterial bed is occluded. Figure 19.7C shows the right renal artery after embolic occlusion with Gelfoam (surgical gelatin); blood flow to the right renal parenchyma and tumor has been eliminated. Severe ischemic kidney pain follows the embolization procedure and therefore it is performed on the morning of surgery, just after preoperative sedation and just before general anesthesia and nephrectomy.

In cases of unresectable tumors, embolization of the arterial blood supply can decrease tumor bulk and palliate pain. Transcatheter embolization techniques are also used to treat arteriovenous malformations and fistulas in all parts of the body.

Not every artery may safely be embolized; if the organ the artery supplies is not to be removed, embolization is safer if there is a collateral blood supply. Embolization of arteries that supply end-organ arteries, such as the superior mesenteric artery branches supplying the vasa recta of the small and large bowel, is usually not performed because it may cause organ infarction.

Several embolization materials are in common use, including Gelfoam particles, steel coils, detachable balloons, and glues. They are selected based on the size of the artery to be occluded, technical factors involved in the selective catheterization of the artery, and the length of time that complete occlusion is necessary. Gelfoam particles, for example, produce only temporary occlusion, whereas glues and coils produce permanent occlusion. Detachable balloons are used to occlude cerebral aneurysms and arterial-venous fistulas.

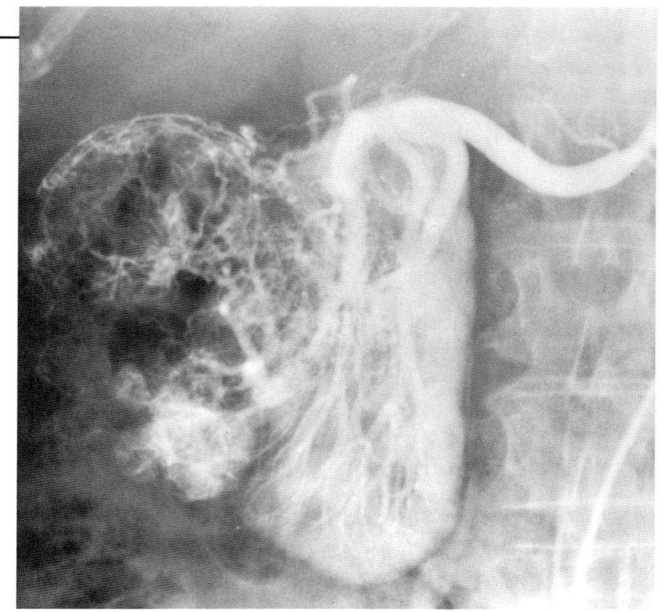

A

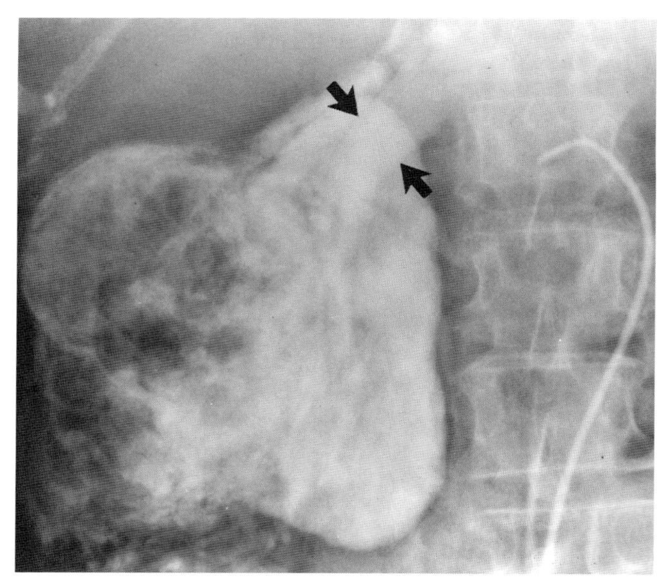

B

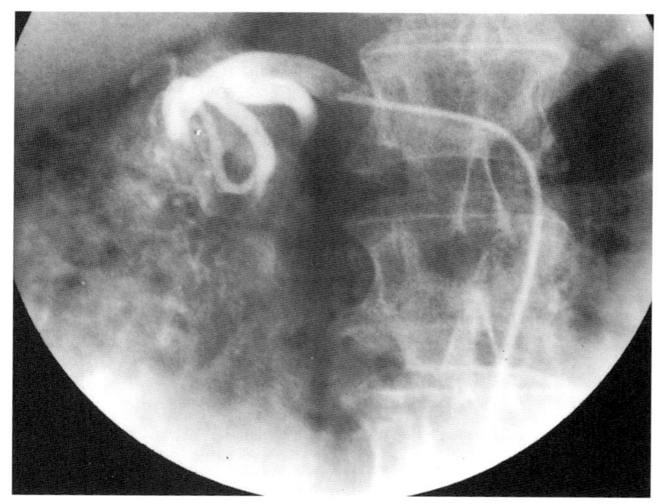

C

Figure 19.7. Preoperative embolization of a large renal cell carcinoma of the right kidney. A: Arterial phase of the initial arteriogram shows a large vascular tumor of the lateral portion of the right kidney. Note the bizarre tumor vasculature. B: Venous phase of the initial arteriogram shows a normal right renal vein *(arrows)* without evidence of tumor invasion. C: Postembolization arteriogram shows occlusion of all the right renal artery branches and cessation of blood flow to the renal parenchyma and tumor.

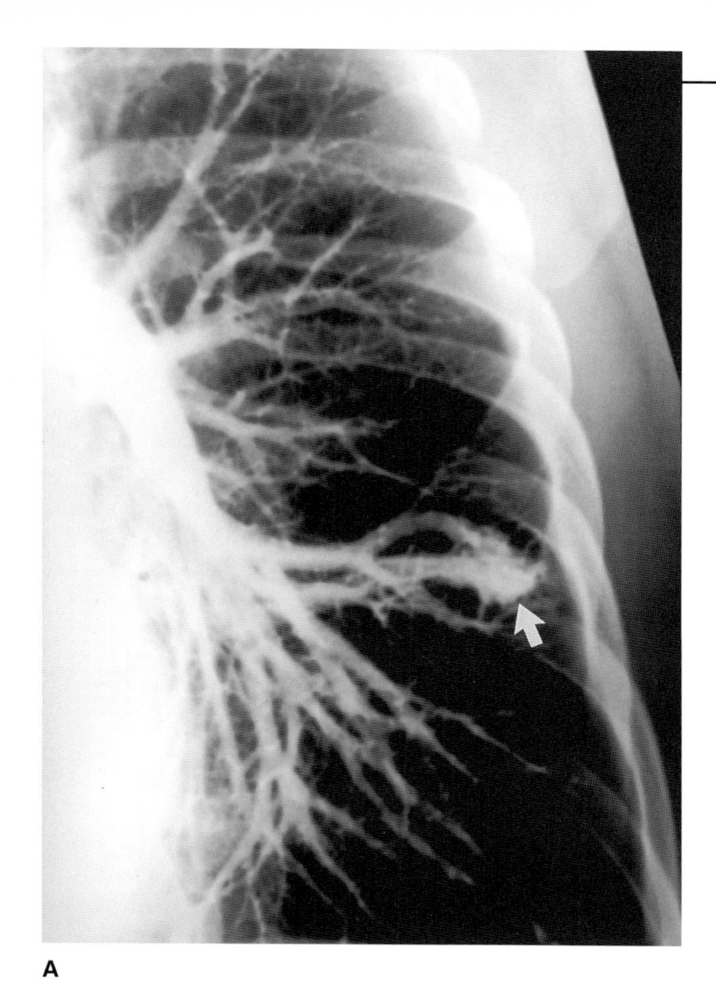

A

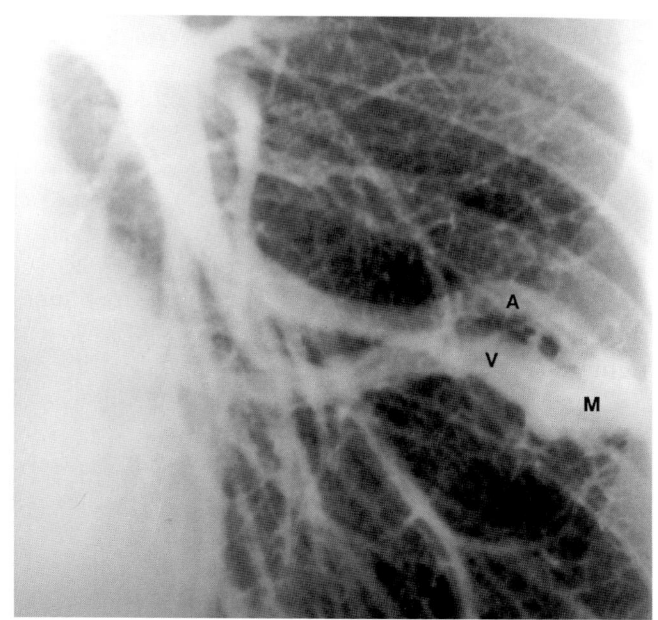

B

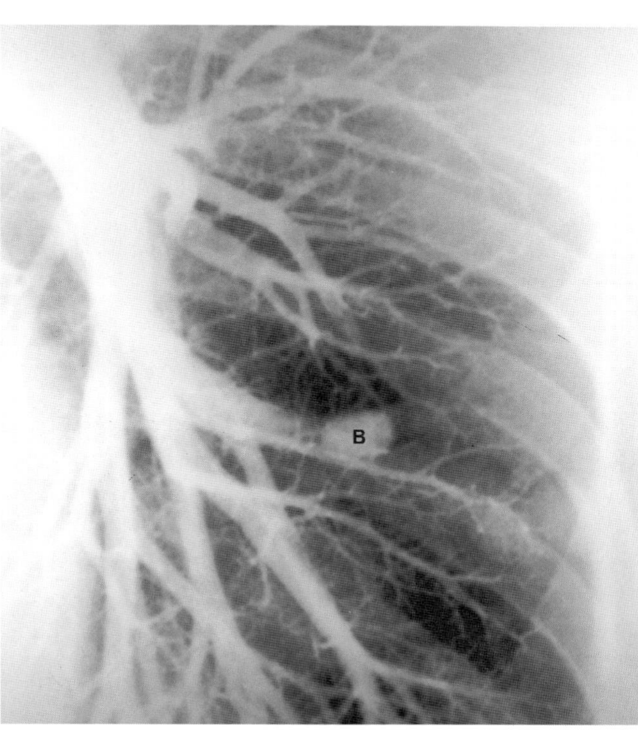

C

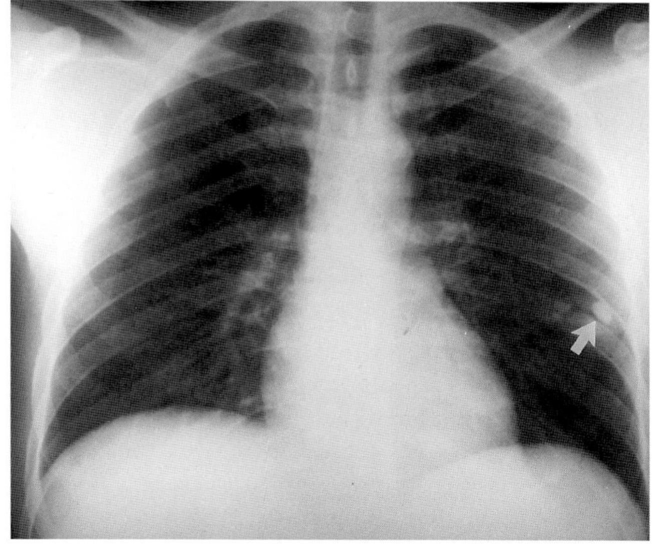

D

Figure 19.8. Therapeutic embolization of a congenital pulmonary arteriovenous malformation. This young man had suffered a small stroke due to a paradoxical embolus that passed through this malformation into the systemic circulation and subsequently the brain. A small peripheral lung nodule had been identified in the left lung on his chest film. A: Left pulmonary arteriogram determined that the left lung nodule was a pulmonary arteriovenous malformation *(arrow).* B: Magnification arteriogram showed the feeding artery *(A)* supplying the malformation *(M)* and its draining vein *(V).* C: The feeding artery was embolized with a detachable balloon *(B),* occluding blood flow through the malformation. D: Postembolization chest film showed that the detached balloon *(arrow)* had migrated slightly closer to the malformation. The patient did not suffer any recurrence of stroke.

Angiographic Diagnosis and Control of Acute Gastrointestinal Hemorrhage

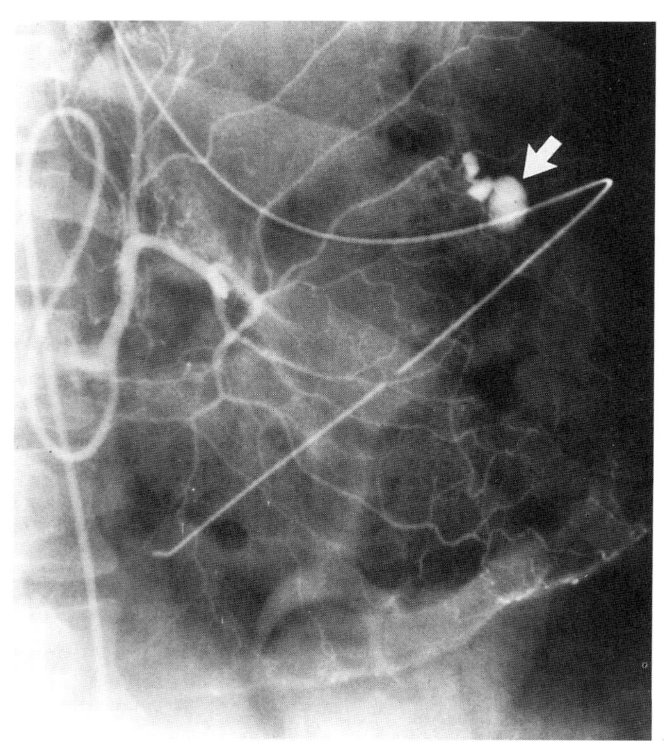

A

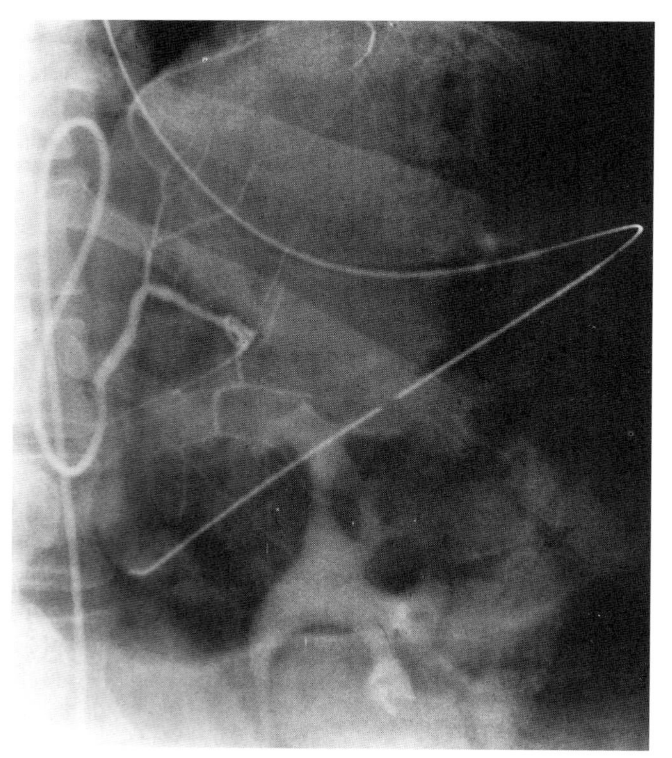

B

Figure 19.9. Left gastric arteriograms of a patient with an acutely bleeding gastric ulcer. A: The *arrow* indicates extravasation of contrast material at the bleeding site. B: Repeat arteriogram 20 minutes after treatment with a left gastric artery infusion of vasopressin shows decreased caliber of the left gastric artery and its branches, and no further bleeding. The infusion was continued for 24 hours, after which the catheter was withdrawn. The patient did well and experienced no rebleeding.

In patients with acute, unremitting, and life-threatening gastrointestinal hemorrhage, angiography offers a technique for both localization and treatment of the bleeding. After selective arterial catheterization of the artery supplying the responsible lesion, sites of GI hemorrhage can be determined by the injection of contrast material. Angiographic hemostasis is then accomplished through vasoconstrictor infusion or, in the case of some bleeding sites, embolic occlusion of the arterial supply. Figure 19.9 depicts a patient with an acutely bleeding gastric ulcer, and Figure 19.10 shows a patient with an acutely bleeding colonic diverticulum. Both patients were successfully treated with infusion therapy with vasopressin.

Angiography is indicated for patients who are *actively* bleeding, because the angiographic demonstration of hemorrhage requires radiographic depiction of contrast medium extravasation from the arterial lumen into the bowel. Documentation of *active upper GI bleeding* is generally provided by observation of the nasogastric tube aspirate; if vigorous iced saline lavage fails to clear a patient's bright red bloody aspirate, active bleeding is probably occurring proximal to the ligament of Treitz. Documentation of *active lower GI bleeding* is somewhat more difficult. A patient with slow bleeding from the small or large bowel may have intermittent movements of bloody stool; in the interim, whether or not bleeding has stopped may be unclear. The usual measures taken to determine whether bleeding has ceased include monitoring the patient's vital signs, blood requirements, and frequency of hematochezia.

In uncertain cases a radioisotope bleeding scan should be performed (Figure 19.11). For this procedure a sample of the patient's red blood cells is labeled with a radioisotope and reinjected into the patient. Scans of the abdomen are then obtained at frequent

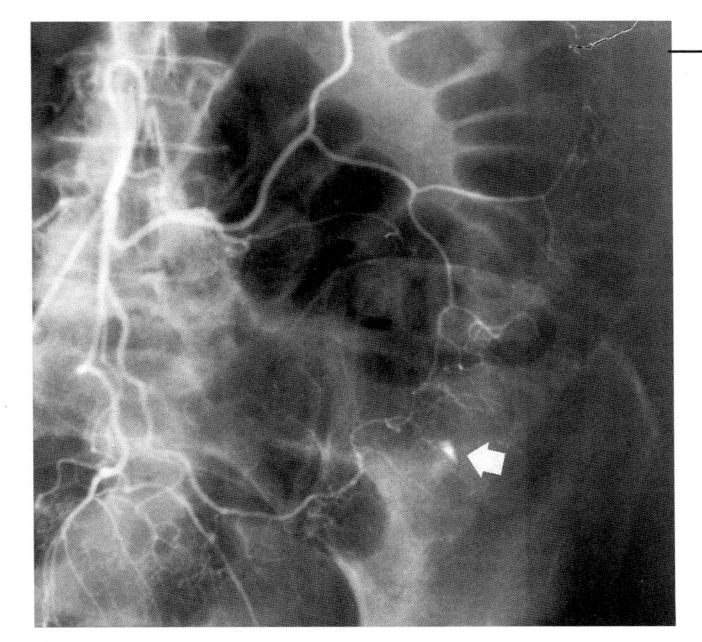

A

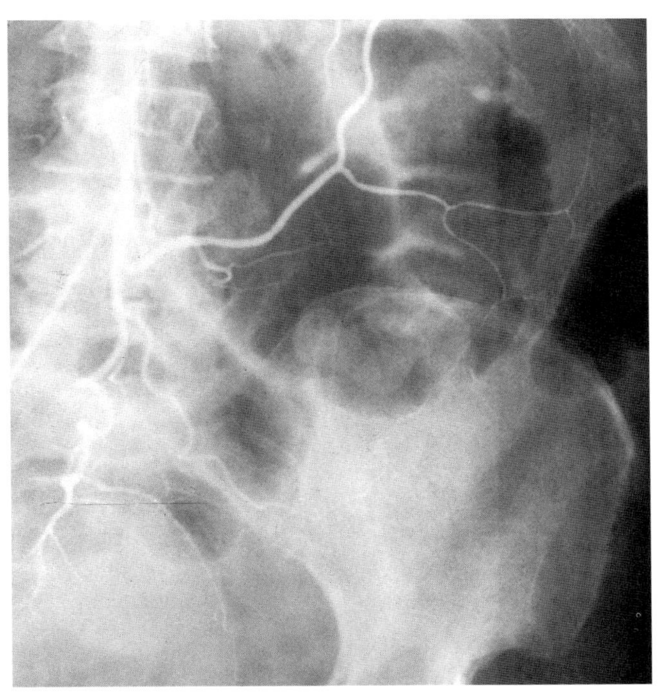

B

Figure 19.10. Inferior mesenteric arteriograms of a patient with a bleeding colonic diverticulum. A: The *arrow* points to the site of active bleeding (contrast medium extravasation) in the distal descending colon. B: Repeat arteriogram after 20 minutes of treatment with vasopressin shows no further bleeding. The infusion was administered for 24 hours and then was discontinued. No rebleeding occurred.

intervals. Normally, activity is shown only overlying the major blood vessels. Scans of a patient with continued bleeding are usually positive within 10 to 15 minutes and show pooling of radioactivity outside the vascular system and overlying the bowel. A positive bleeding scan calls for emergency angiography, whereas a negative scan indicates that angiography is not needed at that time.

Acutely bleeding *upper* GI lesions that can be localized with selective arteriography include esophagitis, Mallory-Weiss tears, esophageal tumors, gastritis, gastric tumors, ulcers, and duodenal diverticula. *Lower* GI lesions include colonic diverticula, neoplasms of the small and large bowel, and vascular malformations. Since angiography primarily localizes bleeding sites rather than characterizing the causes of bleeding, patients need endoscopic and barium examinations for definitive diagnosis once hemostasis is achieved. Although gastroesophageal varices can be identified at angiography, *extravasation* from varices cannot. Ideally, patients with acute upper GI bleeding should be endoscoped before angiography to rule out varices, and patients with acute lower GI bleeding should be sigmoidoscoped to rule out rectal causes, such as hemorrhoids. *Barium examinations should not be performed* initially in an acutely bleeding patient, because residual barium may interfere with angiographic visualization.

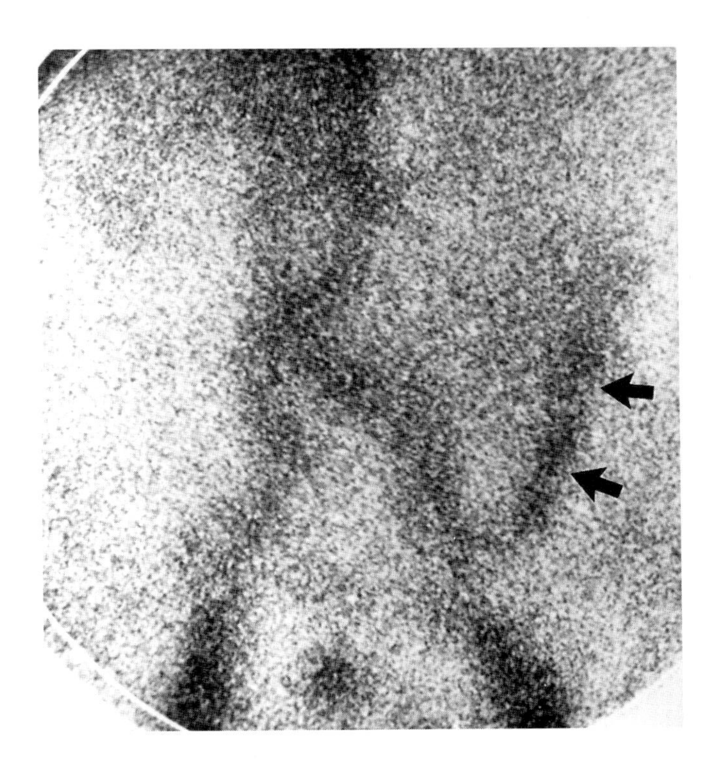

Figure 19.11 ***(right).*** Positive radioisotope bleeding scan of the patient in Figure 19.10, obtained before the initial arteriogram, showing accumulation of isotope overlying the course of the distal descending colon *(arrows).*

Inferior Vena Cava Filters

Inferior vena cava filters are placed in patients with documented pulmonary embolism or deep venous thrombosis who have contraindications to anticoagulation therapy or who have suffered recurrent pulmonary embolisms while being treated with optimal levels of anticoagulation. The filters are placed in the inferior vena cava just below the level of the renal veins, where they trap life-threatening thromboemboli originating in leg veins or pelvic veins, preventing them from reaching the heart and lungs. The filters are usually inserted through a femoral vein (Figure 19.12), but occasionally through a jugular vein, as in patients with iliac vein thrombosis. Filter placement via a femoral vein in a patient with thrombus in the iliac veins or inferior vena cava would result in dislodgment of the thrombi and additional pulmonary emboli. For this reason, all patients for whom this procedure is considered must have an inferior vena cavagram performed via contrast injection into a femoral vein to demonstrate a clear pathway before femoral vein filter placement.

The original filter applicators required a surgical venous cutdown, but improvements in technology now permit percutaneous filter placement. Several kinds of filters are in current use. One of the most common is the Greenfield filter (Figure 19.13), which appears radiographically as a cone-shaped arrangement of 6 metal tines. All filters in common use today are collapsible, and they are inserted via a venous catheter under fluoroscopic control.

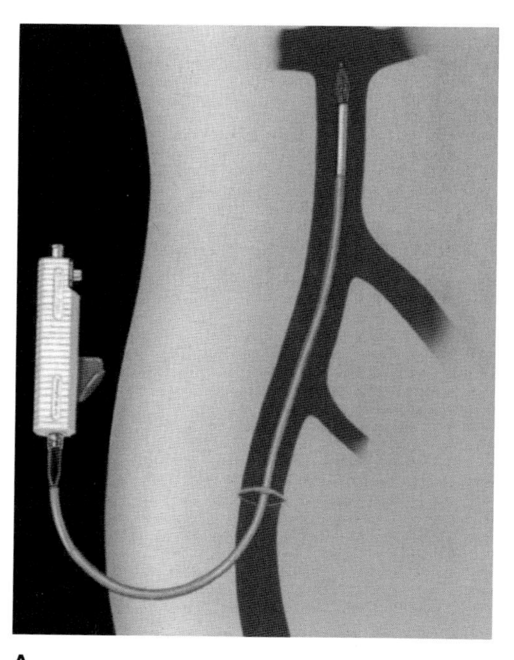

A

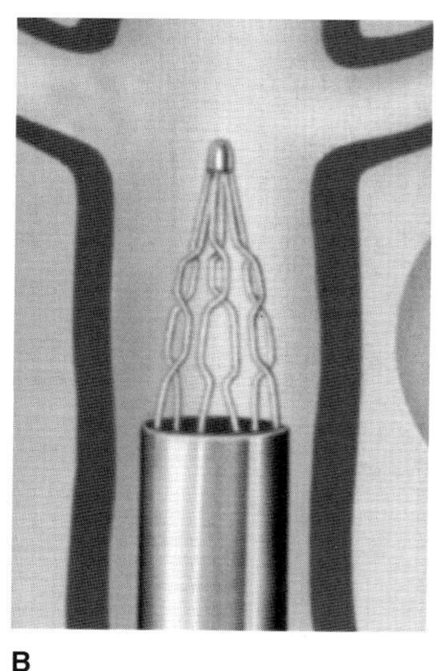

B

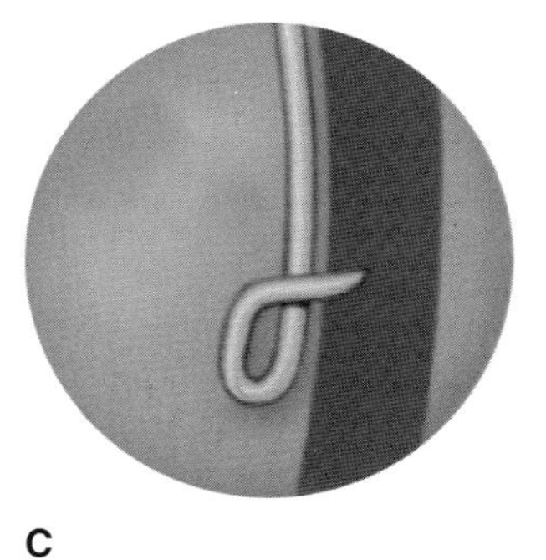

C

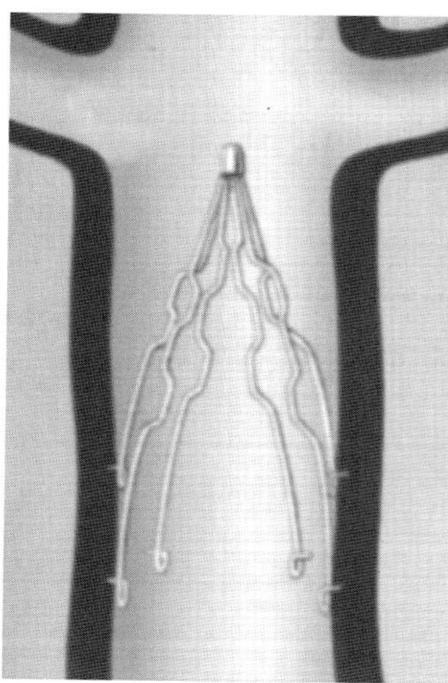

D

Figure 19.12. Insertion of a Greenfield inferior vena cava filter. A: Filter introduction catheter (containing the compressed filter) is inserted through a femoral vein; the tip of the catheter is advanced to the implantation site, just below the level of the renal veins. B: The filter is released at the implantation site, where it expands to its normal cone shape. C: Angled hooks partially penetrate the inferior vena cava wall to provide filter stability. D: Expanded filter in position.

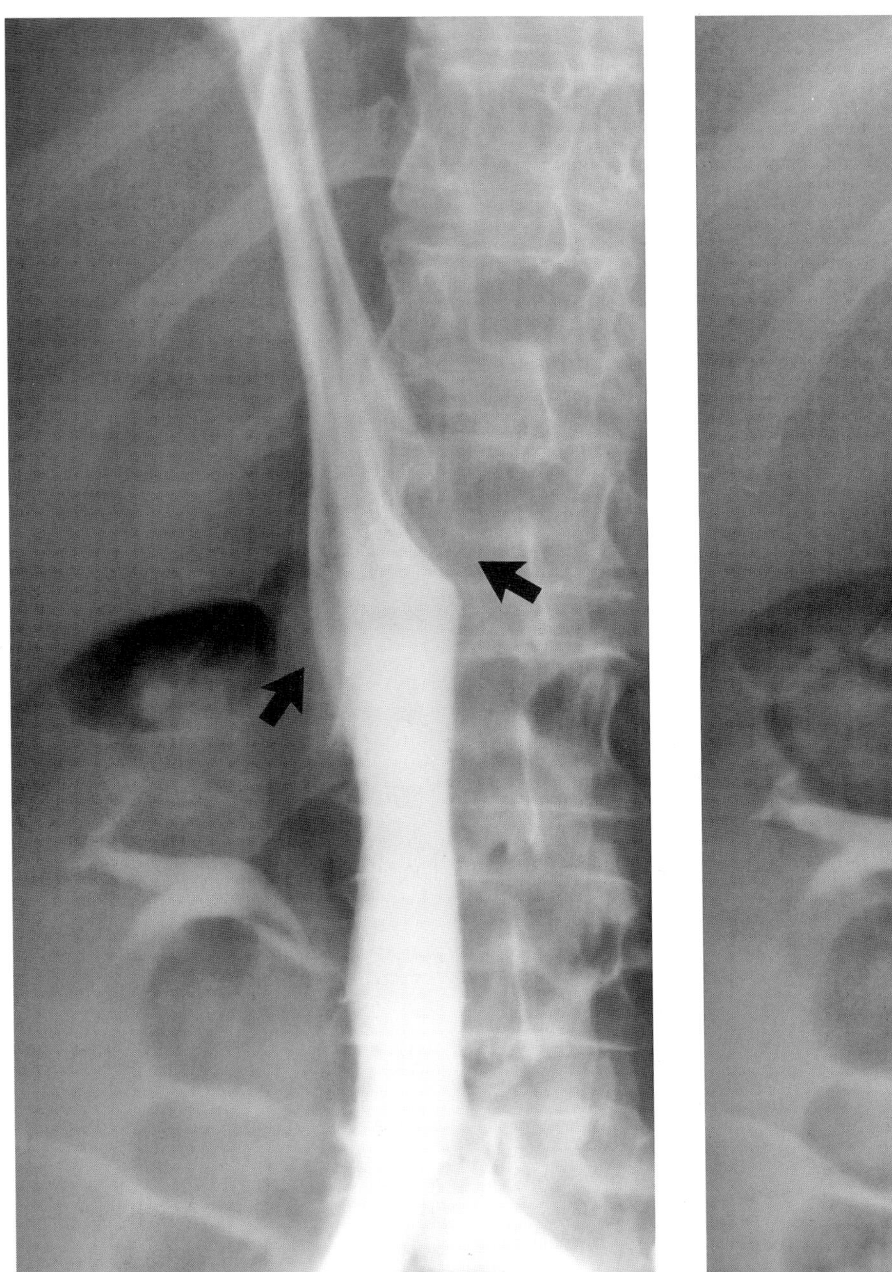

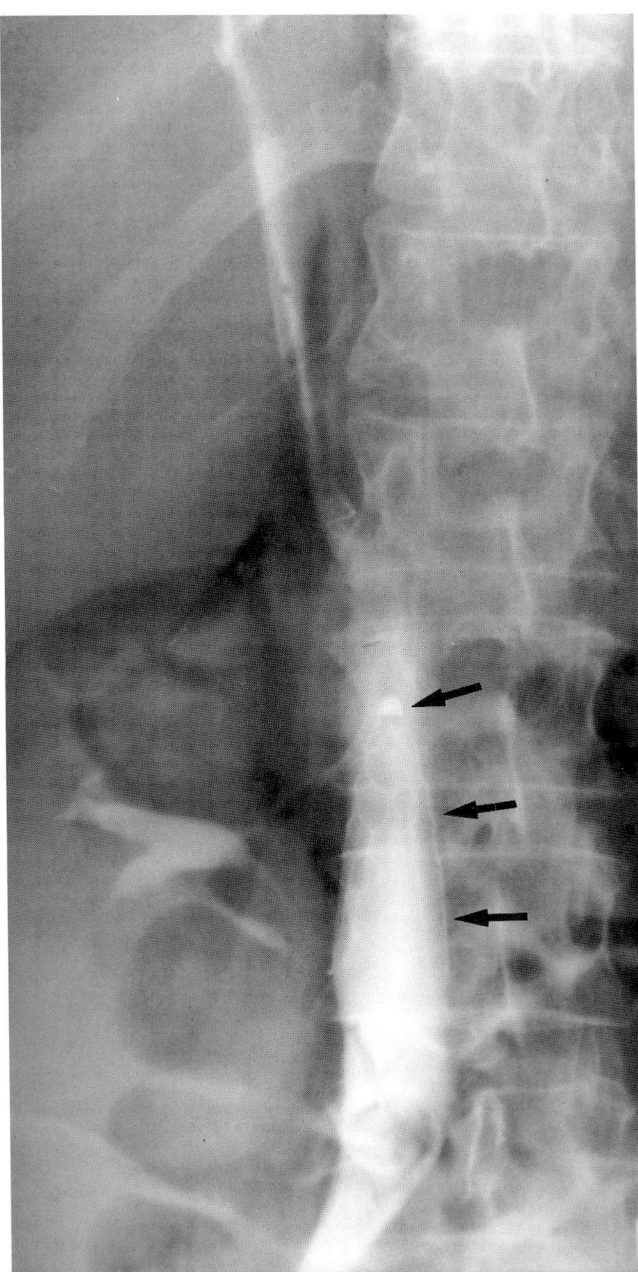

Figure 19.13. Inferior vena cavagrams of a Greenfield inferior vena cava filter, correctly located just below the level of the renal veins. A: Early cavagram film. The dense contrast material hides the filter; well shown are the sites of the right and left renal veins *(arrows),* where unopacified blood from the kidneys is entering the inferior vena cava. B: Slightly later cavagram film. The cone-shaped filter is now revealed *(arrows),* consisting of 6 metallic tines, joined superiorly, that have anchored themselves inferiorly into the caval wall.

Image-Guided Venous Access

Achieving venous access is an important and essential part of modern medical practice. The treatment of many medical conditions requires venous access for the administration of chemotherapeutic agents, antibiotics, nutritional supplements, and blood products. Traditionally, venous lines have been inserted without imaging guidance by "blind puncture." But the target of the venipuncture, for example the subclavian vein (Figure 19.14), can be imaged by ultrasound or fluoroscopy with injection of contrast material, in order to identify the exact site of the vein at the time of puncture. Imaging the vein can expedite the placement of Hickman catheters, portacaths, and other specialized venous lines. Image-guided venipuncture is especially recommended when blind puncture has been unsuccessful.

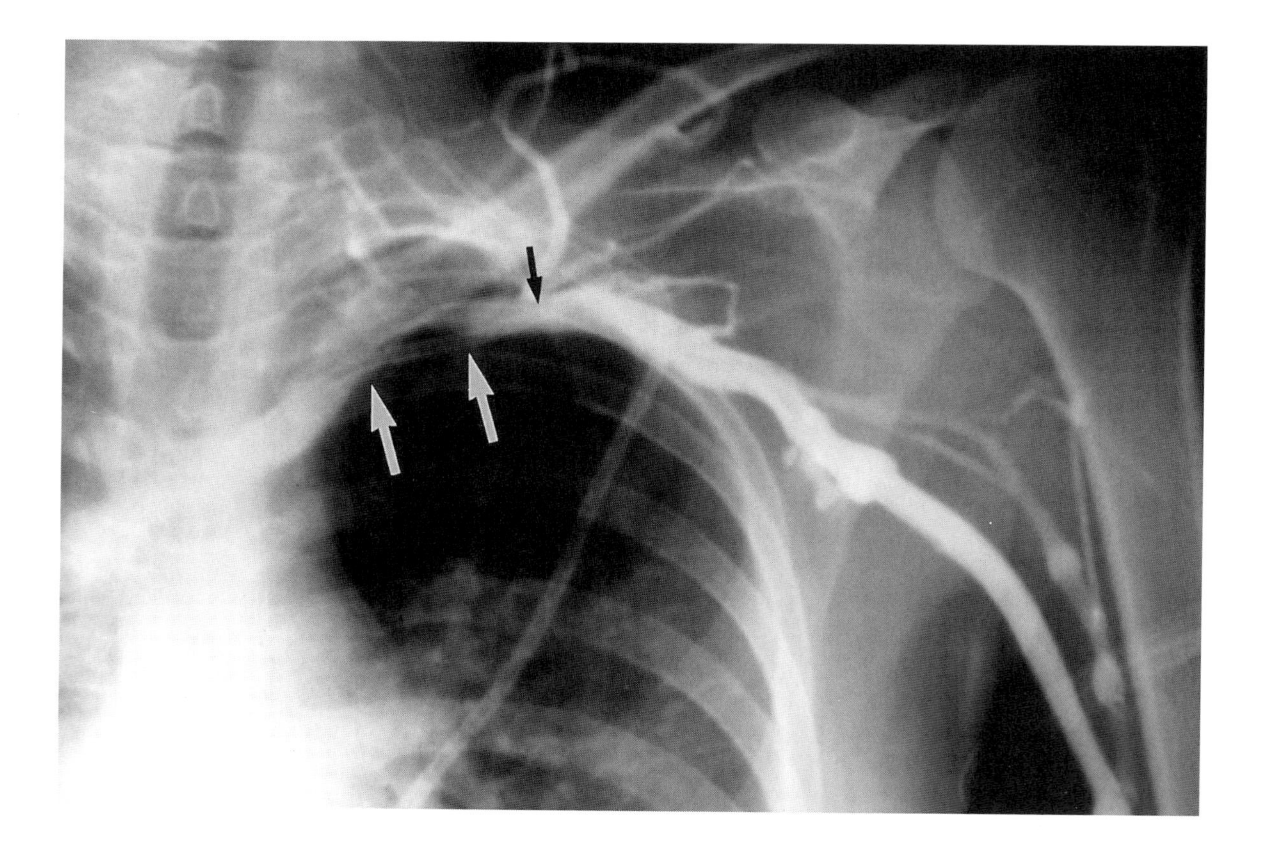

Figure 19.14. Arm venogram (contrast material injected in a peripheral arm vein) to opacify the left subclavian vein, in this case to identify a complication of subclavian line placement, thrombus formation. The space between the *white arrows* indicates a segment of vein where almost no contrast material is seen flowing around the catheter. The *black arrow* points to the site where the subclavian line enters the subclavian vein. This method of subclavian vein opacification can assist in line placement under fluoroscopic guidance.

Percutaneous Aspiration Needle Biopsy of the Thorax

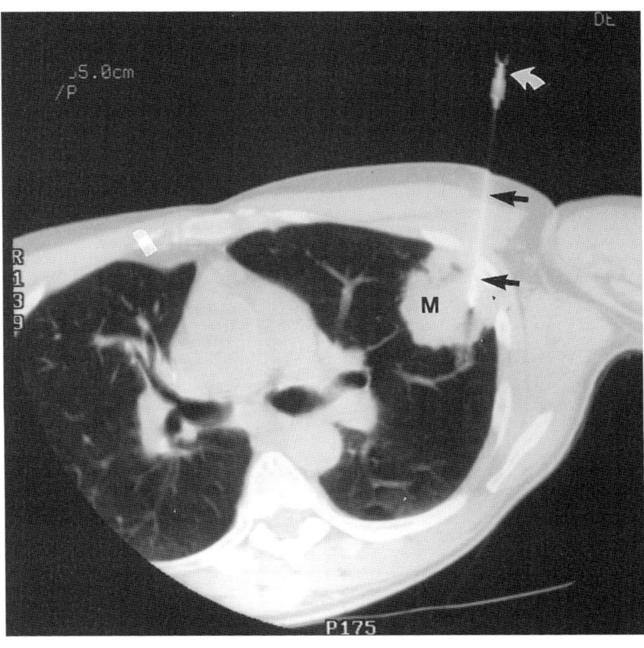

Figure 19.15. CT-guided lung biopsy. The biopsy needle *(arrows)* is percutaneously inserted through the anterior-lateral left chest wall and advanced into the lung mass *(M)* under CT guidance (repeat CT slices taken through the lesion as the needle is advanced into the patient). The *curved arrow* points to the needle's hub. When the tip is shown to be within the lesion, material is withdrawn for pathological examination.

When a new pulmonary nodule or mass is identified on your patient's chest film, your most important immediate task is to determine whether the lesion is cancer or not. A nodule that was present on prior chest films and has not changed over many years can be assumed to be a benign, probably old granulomatous lesion that does not require further diagnostic workup. A nodule with a dense central calcification may also be assumed to be benign. But if the nodule is a truly new finding, not present on prior films, or if no prior films exist for comparison, a diagnostic workup is mandatory.

Since CT can show smaller nodules than plain films, and also can show nodules in pulmonary sites that are difficult to see on plain films, a CT scan should be performed next. If additional nodules are seen on the CT scan, it can be assumed that the patient has pulmonary metastases from a cancer located elsewhere. If the nodule is single, a primary lung cancer should be suspected. When a single pulmonary nodule is cancerous, the CT scan may show enlarged nodes in the mediastinum, indicating that metastatic spread has occurred.

The most important component of the diagnostic workup is a pathological examination of material from the lesion. The possibilities include a percutaneous needle aspiration biopsy performed by the radiologist, bronchoscopy with bronchial washings, and surgical open biopsy. Bronchoscopy with cytological examination of bronchoalveolar lavage fluid is most useful in diagnosing central tumors involving the main or lobar bronchi. Today percutaneous aspiration biopsy performed under CT guidance is recommended for diagnosing peripheral nodules and masses; this technique can establish a diagnosis in over 90 percent of patients with lung cancer. Before the advent of CT, aspiration biopsy of lung nodules was performed under fluoroscopic guidance. But since CT can show smaller nodules than fluoroscopy, CT is definitely the preferred technique. In fact, with CT the radiologist can biopsy a nodule as small as 0.3 centimeters.

Percutaneous aspiration needle biopsies are performed with the patient either prone or supine, depending on the location of the lesion. Breath holding instruction is provided. Then the radiologist directs the tip of a special 22- to 23-gauge biopsy needle into the target nodule. First the needle's position within the nodule is confirmed by CT (Figure 19.15); then aspiration is performed. Immediately after aspiration, an on-site cytopathologist stains and reviews the biopsied material. Samples are also sent for culture and gram-staining when infection is suspected. The procedure may take from 30 minutes to 1 to 2 hours. As you might suspect, a potential complication is pneumothorax; therefore inspiration and expiration chest films or CT scans with lung window settings should be obtained immediately after biopsy. No treatment may be necessary for small pneumothoraces, but temporary treatment with a chest tube should be undertaken if the patient is symptomatic or if the pneumothorax is greater than 30 percent or is enlarging. A less common complication of needle aspiration biopsy is hemoptysis, which is usually self-limited, not requiring any treatment. Percutaneous biopsy can be performed on an outpatient basis.

Percutaneous Aspiration Needle Biopsy of the Abdomen

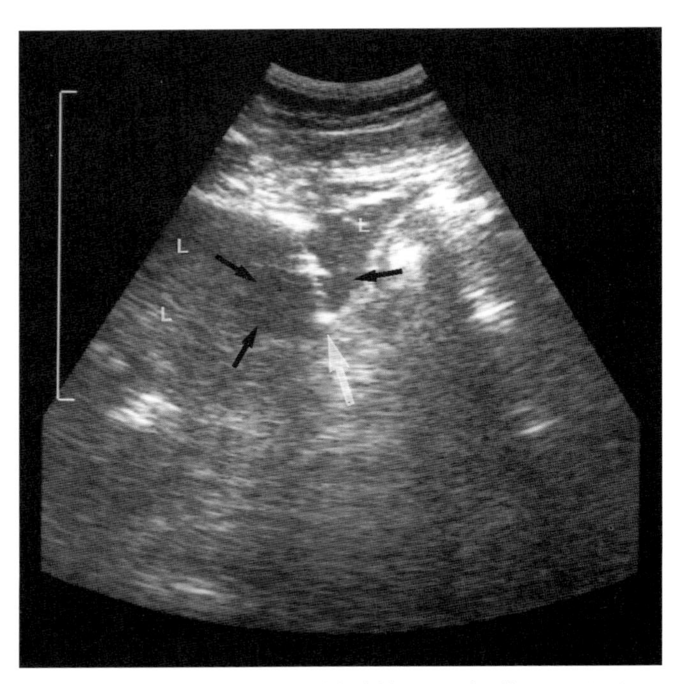

Figure 19.16. Ultrasound-guided biopsy of a liver metastasis. This transverse (axial plane) scan shows the left lobe of the liver (*L*'s) displayed as a triangular structure directed toward the left abdomen (compare the normal CT anatomy, Figures 3.62 to 3.73, pages 74–79), which contains a hypoechoic mass *(black arrows).* The *white arrow* indicates the linear echogenic tract of the percutaneously placed biopsy needle, which is now in position within the mass for biopsy.

A percutaneous abdominal aspiration needle biopsy performed under CT or ultrasound guidance permits pathological examination of abdominal lesions without subjecting the patient to a laparotomy and open abdominal biopsy. With this technique tumors of the liver, kidneys, pancreas, retroperitoneum, and other structures can be biopsied with minimal patient discomfort, and the procedure can be performed on outpatients. This biopsy technique is helpful in diagnosing a primary abdominal tumor, confirming suspected metastases, staging of neoplastic disease, and diagnosing benign lesions such as abdominal cysts and inflammations of the abdomen.

The patient is allowed no solid food for 12 hours before the procedure, but can have clear liquids up to 2 hours before the biopsy. The patient is given preprocedure anesthesia and analgesia. The skin entry site is anesthetized with lidocaine and the biopsy is performed with a 18- to 22-gauge biopsy needle. Larger 14- to 16-gauge needles may be used to aspirate larger pieces of tissue for subtyping certain conditions, such as lymphoma. When bowel has to be traversed, only fine (18- to 22-gauge) needles should be used.

The biopsy may be performed with either CT or ultrasound guidance. When ultrasound guidance is used the lesion is localized and the distance and angle of its biopsy path are determined by ultrasound.

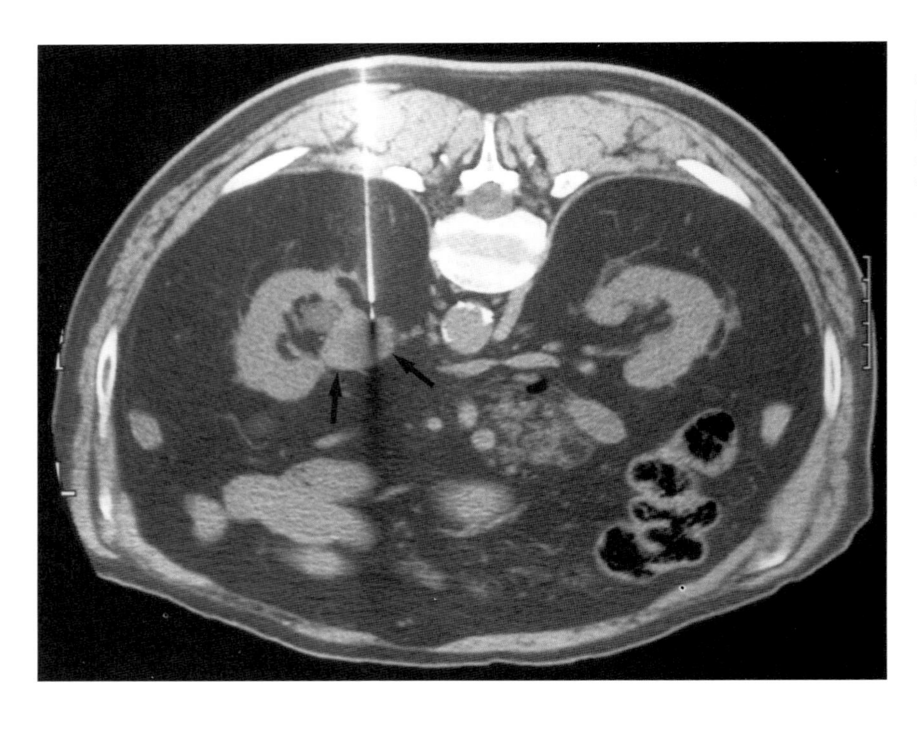

Figure 19.17. CT-guided biopsy of a small renal tumor. The patient is lying prone and the biopsy needle has been inserted from the back. The needle tip is shown within the mass *(arrows),* ready for biopsy. The tumor proved to be a renal cell carcinoma.

The radiologist then performs the biopsy, guiding the needle by hand or using a special ultrasound biopsy transducer guide, which has a built-in slot for the biopsy needle. The needle tip is visualized as a discrete echogenic complex within the target lesion (Figure 19.16). For CT-guided biopsy, needle tip placement within the lesion is confirmed by CT (Figure 19.17).

It is desirable but not mandatory to avoid certain structures during passage of an abdominal biopsy needle; these are the lung, pleura, gallbladder, small and large bowels, pancreas, and any dilated ducts (pancreatic, biliary). Following aspiration an immediate cytological examination is performed. Postprocedure patients are treated with bed rest, and their vital signs are monitored, for 1–2 hours; they are discharged after 2–4 hours. Positive tissue is recovered in approximately 85–90 percent of cases, depending upon the size and location of the target lesion. Complications are uncommon, occurring in less than 2 percent of cases; they include hemorrhage, infection, organ injury, pneumothorax, and pancreatitis (if the normal pancreas is biopsied or transgressed).

Percutaneous Abscess Drainage

Radiologically guided percutaneous drainage has become a common treatment for abscesses and noninfected fluid collections throughout the body. This technique obviates the need for surgical drainage and general anesthesia in the management of these collections. Initially, it was employed to drain postoperative abscesses in the peritoneal cavity, including the subphrenic and subhepatic spaces and paracolic gutters. Percutaneous abscess drainage is currently used to treat a wide variety of abscesses and fluid collections, including hematomas, lymphoceles, empyemas, lung and mediastinal abscesses, bowel fistulas, necrotic tumors, and benign cysts.

The procedure is usually performed under CT guidance; in some cases ultrasound guidance is used, especially for pleural fluid collections. A fine needle (20- or 22-gauge) is guided percutaneously into the fluid collection. Diagnostic aspiration can then be performed through this needle, including the withdrawal of material for bacteriological examination. A drainage catheter is then inserted either by trocar or the Seldinger technique (drainage catheter slid over an exchange wire), in tandem to the fine needle (Figure 19.18).

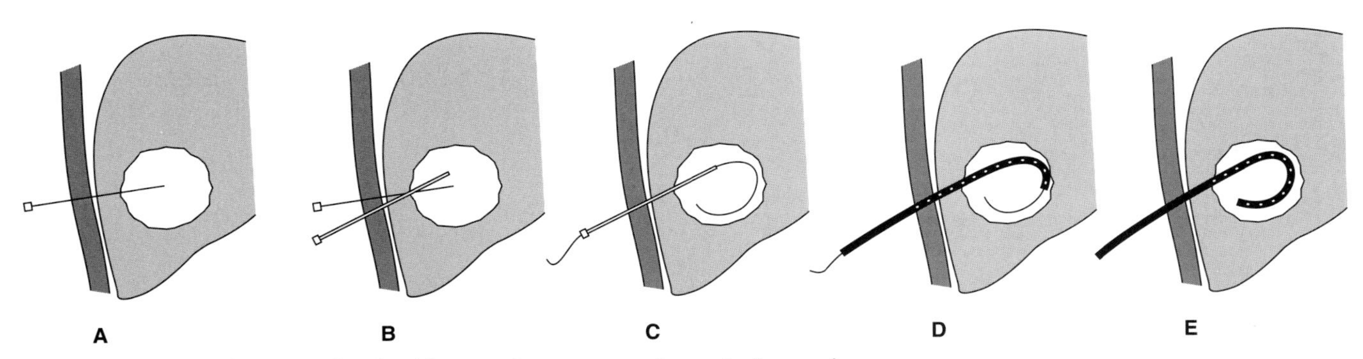

Figure 19.18. The five steps involved in percutaneous needle aspiration and catheter drainage of a hepatic abscess. A: The localization needle is positioned in the center of the abscess. B: A small-caliber catheter is percutaneously placed within the abscess. This catheter was advanced into the abscess over a trocar, which has been removed. C: A heavy-duty exchange wire is inserted through the catheter; the localization needle is removed. D: The small-caliber catheter is replaced with a large-caliber drainage catheter over the exchange wire. E: The wire is removed; the abscess is drained through the catheter.

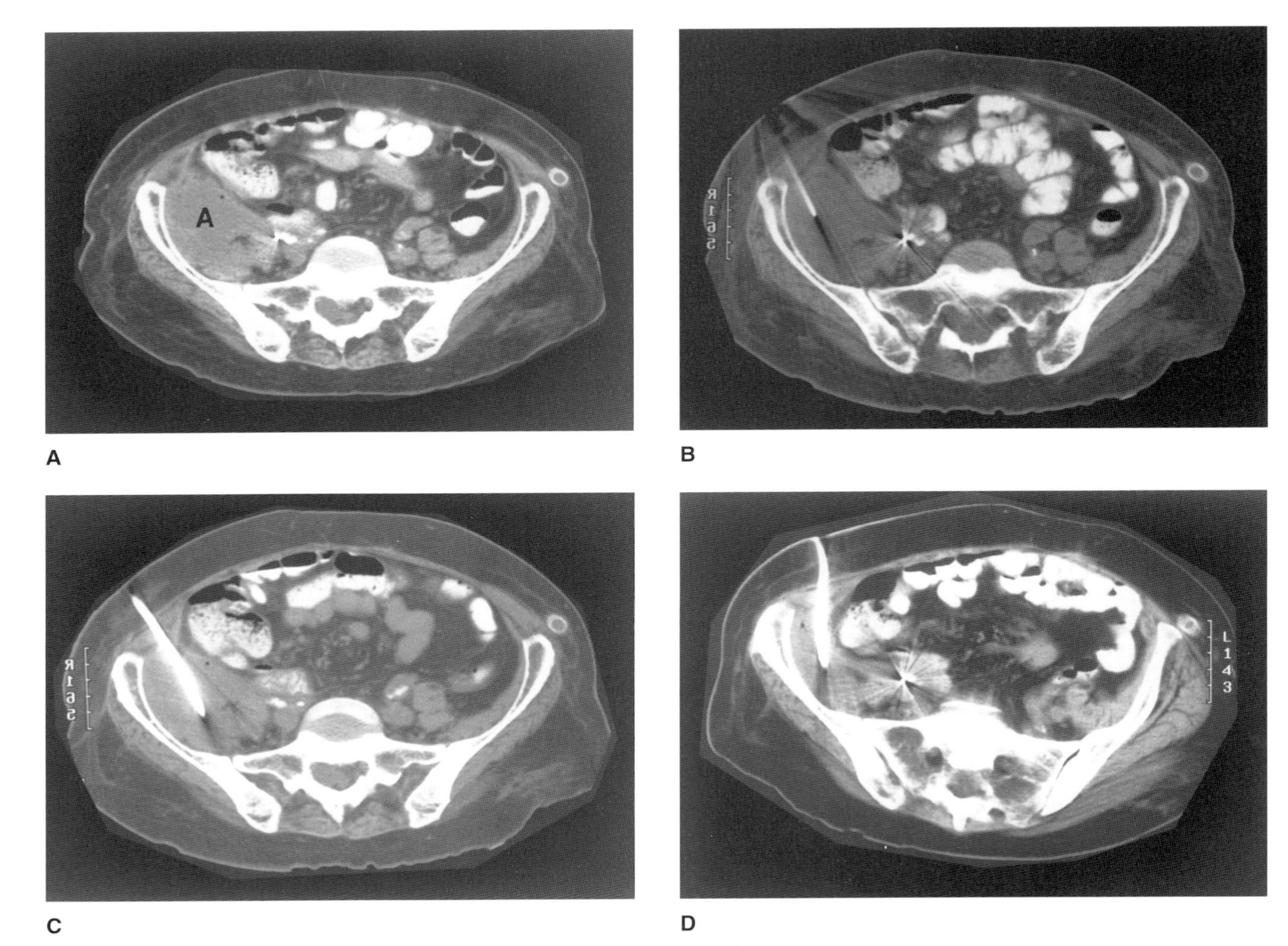

Figure 19.19. Percutaneous aspiration of a large abdominal abscess in an elderly woman with fever, chills, and abdominal pain. A: Initial CT scan shows a large abscess *(A)* in the right lower quadrant. B: Initial percutaneous needle puncture confirms the abscess by the aspiration of pus. C: Percutaneous placement of a large drainage catheter. D: Decrease in size of the abscess after initial drainage.

Large viscous purulent collections usually require a 12- to 16-French catheter to completely drain and lavage the abscess cavity (Figure 19.19). The patient is started on antibiotics and the clinical course is followed by repeat CT scans and contrast injections into the abscess cavity itself. The catheter is removed when drainage has ceased, the clinical course of the patient has improved, and resolution of the cavity is shown by CT. Percutaneous drainage is contraindicated in the absence of a safe percutaneous route.

Percutaneous Gastrostomy and Jejunostomy

Percutaneous gastrostomy is usually performed to feed patients with prolonged nasogastric nutritional needs after head injury, stroke, or major trauma. This procedure can be performed under local anesthesia and conscious sedation without subjecting the patient to a surgical procedure and general anesthesia. With the bowel opacified with previously administered barium, the patient fasting, and a nasogastric tube in place so that the stomach can be inflated with air, a special needle is percutaneously placed into the gastric lumen under fluoroscopic control. A guide wire is then inserted into the gastric lumen through a needle, over which a 12- to 16-French gastrostomy tube is placed, with its tip in the gastric lumen.

Percutaneous transgastric jejunostomy is a variation of percutaneous gastrostomy performed in patients who also require prolonged nutrition but who have a history of gastro-esophageal reflux or aspiration. To avoid reflux complications the feeding tube in these patients should terminate beyond the stomach, in the small bowel. The difference between this technique and gastrostomy, technically, is that the guide wire is advanced beyond the stomach into the proximal jejunum and then a jejunostomy tube is advanced over the guide wire.

Percutaneous Biliary Decompression

Percutaneous biliary drainage is a technique used to alleviate obstructive jaundice with its associated sepsis and pruritis in patients with biliary obstruction due to unresectable pancreatic cancer, cholangiocarcinoma (cancer of the bile duct), obstructing metastatic disease, or gallbladder cancer. It is an extension of percutaneous transhepatic cholangiography, which you read about in Chapter 14. Figure 19.20 illustrates the steps in this procedure. In A you can see a tumor encasing the common bile duct and causing mechanical biliary obstruction. A percutaneous biliary needle-catheter combination has been advanced into the biliary system in B. After removal of the needle in C, a small-caliber flexible guide wire is inserted through the biliary catheter and advanced down the biliary tree, through the obstructing stenosis, until the tip of this wire can be seen within the duodenum. In D the biliary catheter has been withdrawn from the guide wire and replaced with a larger-caliber, multi–side-hole drainage catheter which is advanced over the guide wire until its tip is within the duodenum. In E the guide wire has been removed. The drainage catheter is then clamped outside the skin surface. Now the patient has a route for internal biliary drainage. When unresectable tumors obstruct the biliary tree, this palliative treatment can eliminate symptoms due to jaundice.

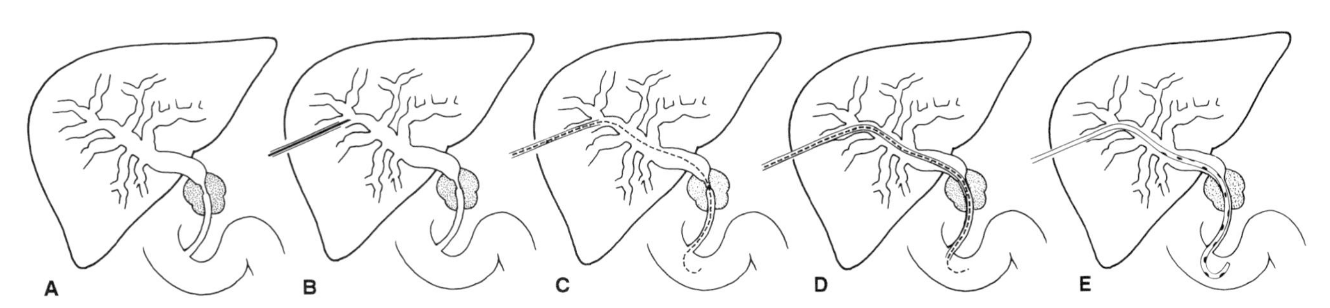

Figure 19.20. Technique for percutaneous transhepatic biliary drainage. (See text.)

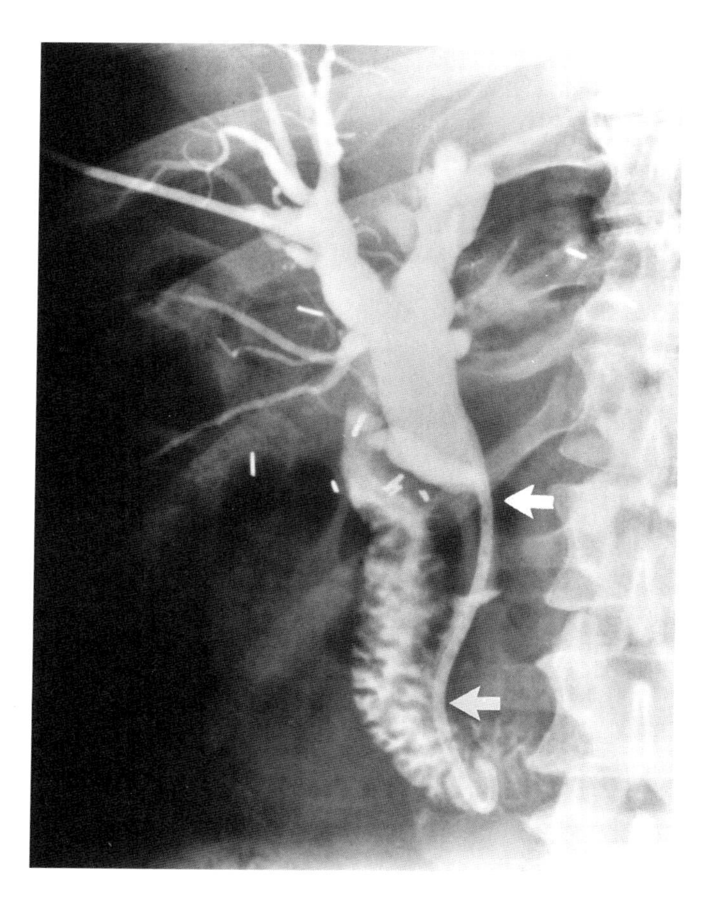

In Figure 19.21 you see the results of a percutaneous biliary drainage procedure. Water-soluble contrast material has been injected through the external portion of the drainage catheter, opacifying the intrahepatic biliary radicals and proximal common bile duct. Although the distal duct *(arrows)* is markedly narrowed around the tube by the unresectable pancreatic carcinoma, there is free flow of contrast material and bile into the duodenum through the drainage catheter. This patient's jaundice was entirely relieved.

Internal stents are now available that permit biliary drainage across sites of mechanical obstruction without any external catheter (Figure 19.22).

Figure 19.21 ***(left).*** Transhepatic cholangiogram of a patient with a percutaneous biliary drainage catheter transversing a pancreatic carcinoma that encases the common bile duct. (See text.)

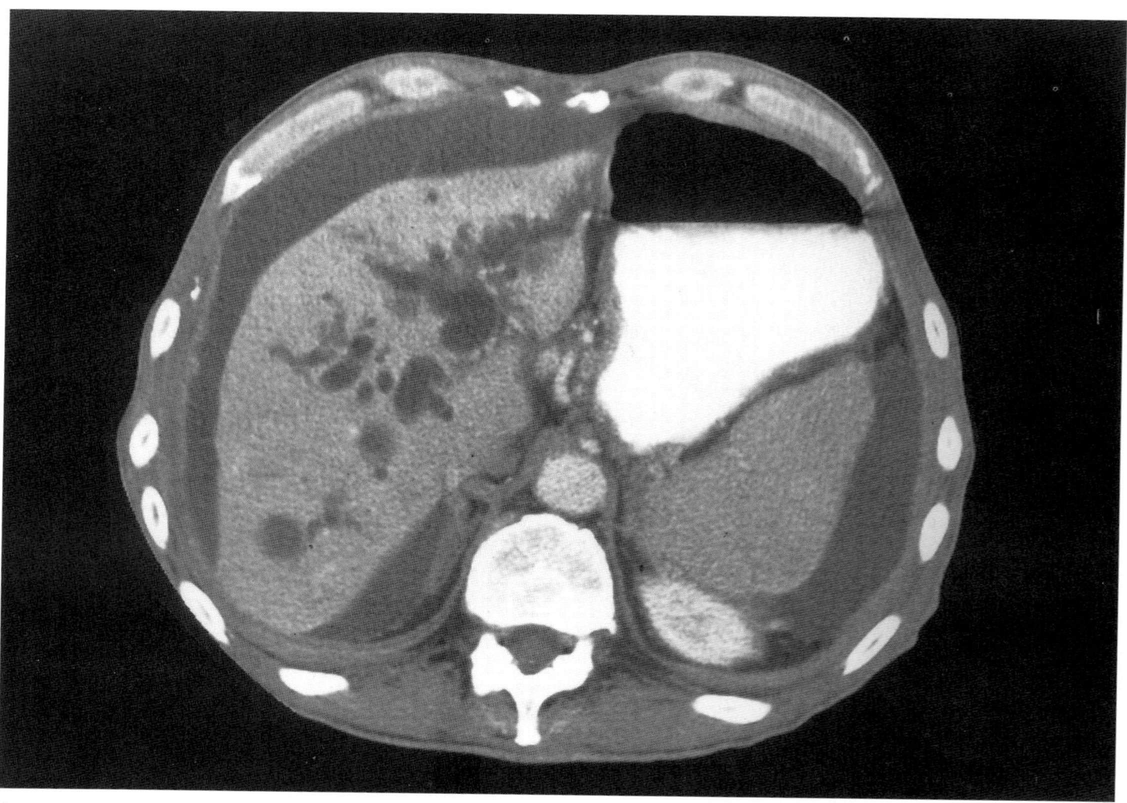

A

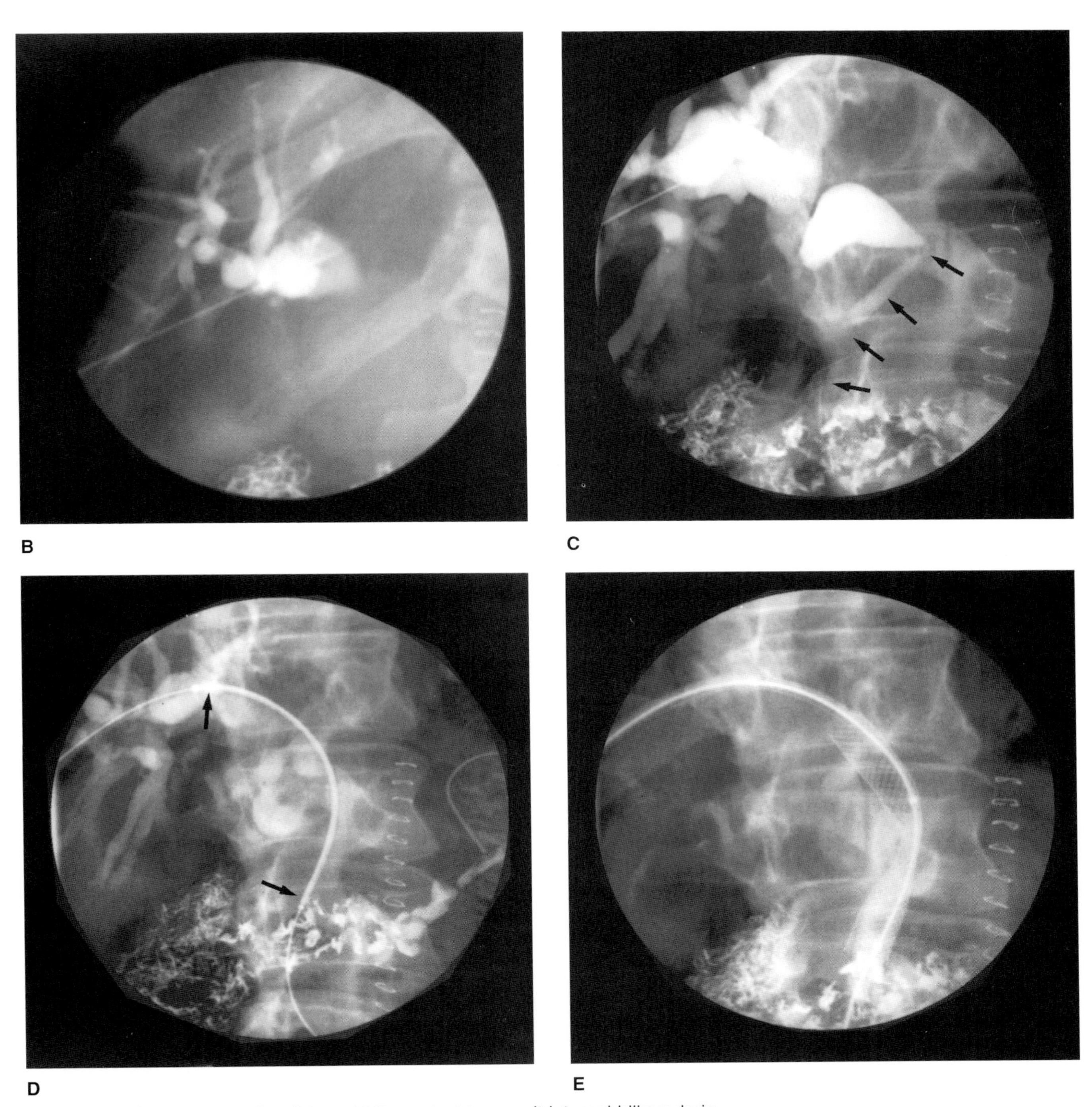

Figure 19.22. Placement of an internal biliary stent to permit internal biliary drainage in a patient with common bile duct obstruction caused by a metastatic tumor. A *(at left):* Initial abdominal CT scan shows marked dilatation of the intrahepatic bile ducts and (malignant) ascites around the liver and spleen. B and C: Percutaneous needle puncture of the biliary tree and contrast injection under fluoroscopic guidance shows marked dilatation of the intrahepatic biliary tree with narrowing and obstruction of the common bile duct (*arrows* in C) proximal to the duodenum. D: Percutaneous placement of a balloon dilatation catheter with a collapsed metallic stent *(between the arrows)* across the segment of obstructed common bile duct. E: Following dilatation of the balloon and expansion of the stent, the patient has internal drainage of his obstructed biliary tree. The dilatation catheter was removed at the end of the procedure and the patient's jaundice was relieved by the stent.

Percutaneous Cholecystotomy

Percutaneous cholecystotomy is a temporizing procedure that can relieve the symptoms of a patient critically ill with acalculus cholecystitis until a cholecystectomy can be performed. Percutaneous cholecystotomy is also useful in treating patients with sepsis from an unknown source, after a full fever workup. Under ultrasound guidance, the radiologist punctures the gallbladder, approaching transhepatically and entering the gallbladder along its attachment to the liver. This approach helps diminish the likelihood of bile spillage into the peritoneum. Using a trocar, the radiologist places an 8.3 French cholecystotomy catheter into the gallbladder with a single rapid controlled thrust of the catheter (Figure 19.23). The gallbladder is then aspirated and the catheter fixed into position. The catheter remains in place for 3 weeks to allow a tract to form.

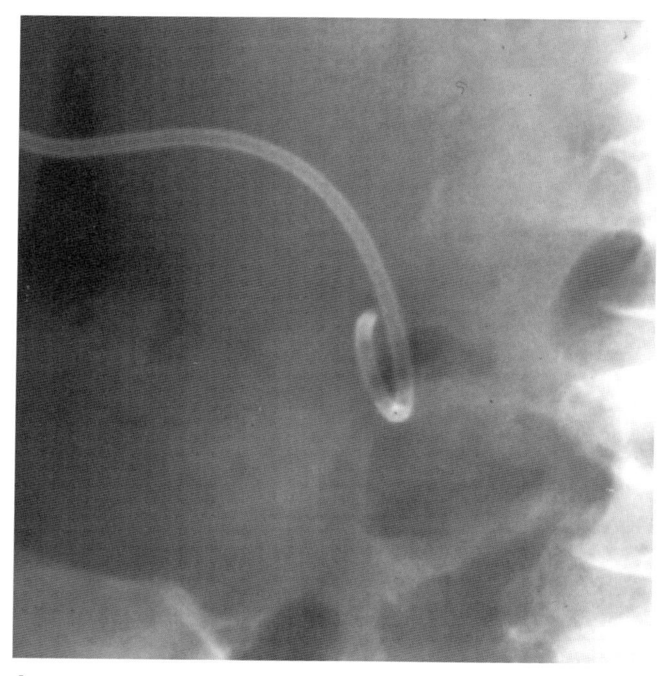

A

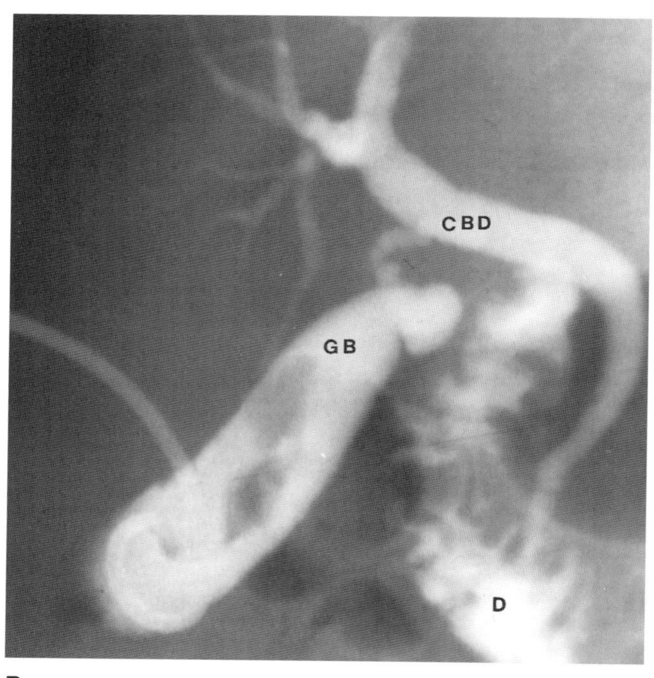

B

Figure 19.23. Percutaneous cholecystotomy. A: Detail of a plain film showing the percutaneously inserted pigtail catheter used for gallbladder drainage. B: Contrast material injected through the catheter opacifies the gallbladder *(GB)* and common bile duct *(CBD)*. There is free flow of contrast material into the duodenum *(D)*. C-E: Sequential CT scans following the course of the cholecystotomy tube *(arrows)* from its percutaneous puncture site to its termination in the gallbladder.

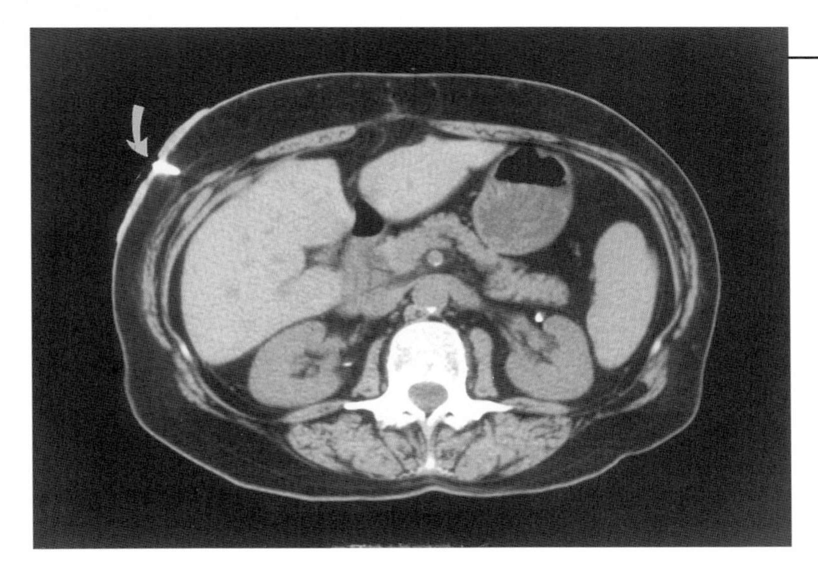

C

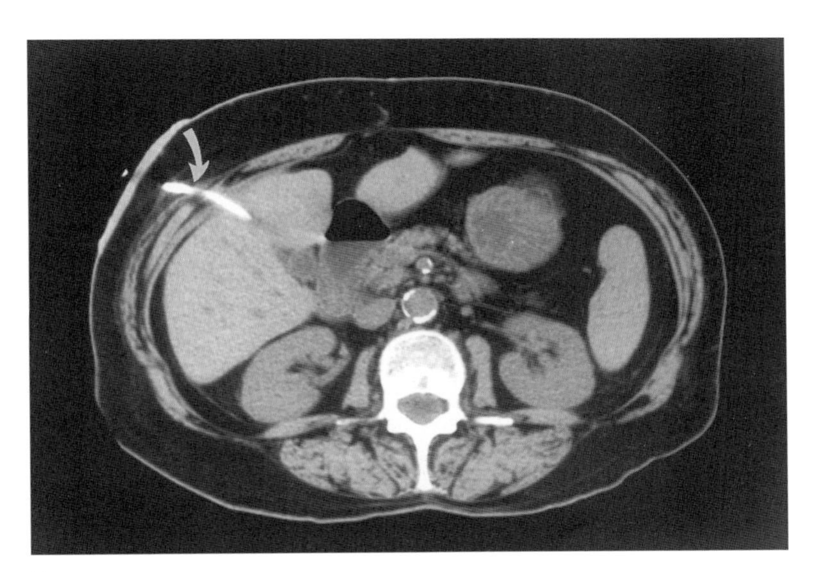

D

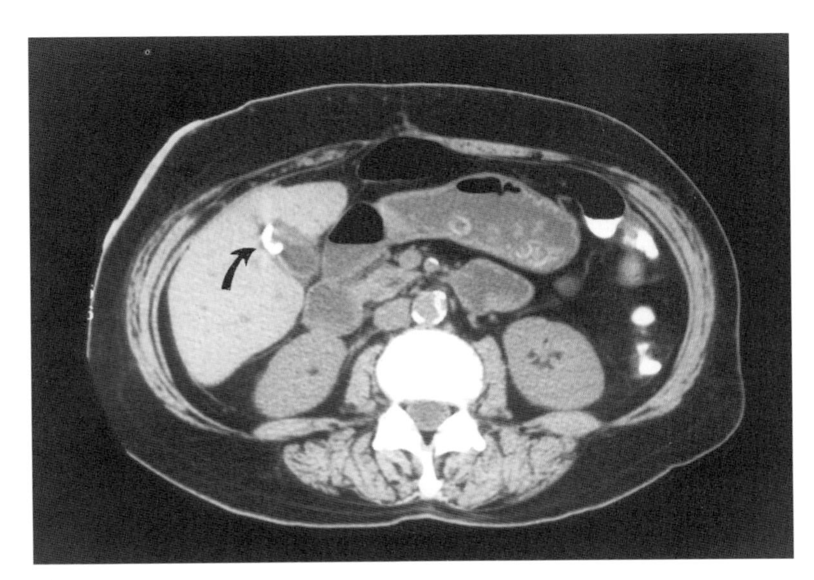

E

Radiological Management of Urinary Tract Obstruction

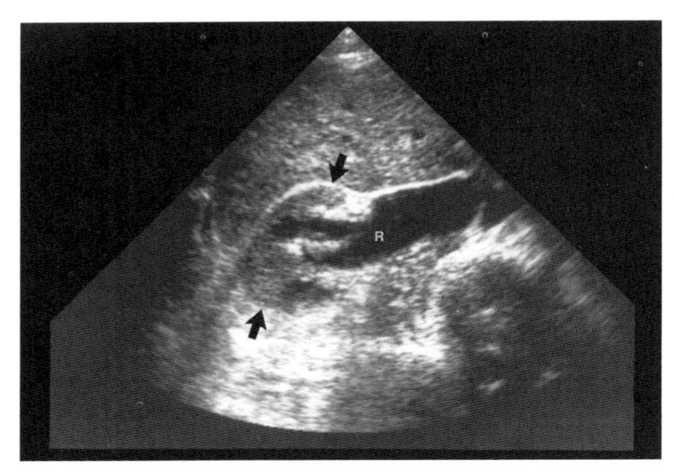

A

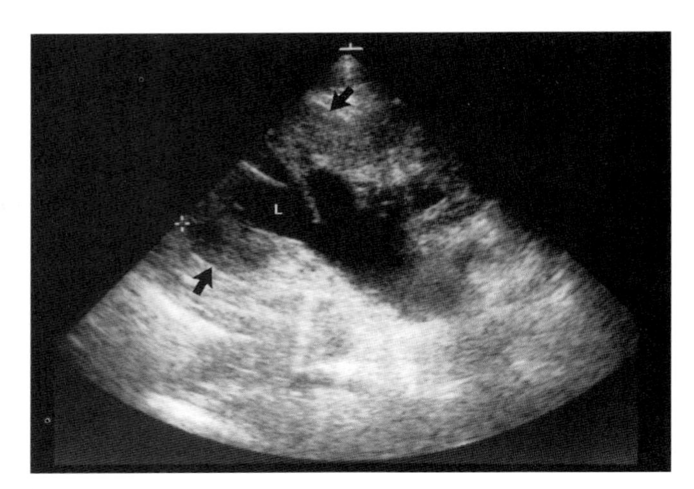

B

Figure 19.24. Bilateral percutaneous nephrostomy to relieve bilateral hydronephrosis in a man with metastatic prostate cancer. A and B: Right and left renal ultrasound scans show bilateral hydronephrosis. The *arrows* point to the kidneys; *R* and *L* indicate dilated right and left renal collecting systems (compare the normal renal ultrasound, Figure 14.53, page 332). C: Plain film shows bilateral percutaneous nephrostomy tubes in place; these now drain the urine produced by the kidneys. D: Contrast injection through the nephrostomy tubes opacifies portions of the urinary tract. The left ureter *(arrow)* is almost completely obstructed by metastatic tumor; the right ureter is narrowed at its junction with the bladder. A Foley catheter is seen within the bladder, which is indented by surrounding pelvic metastases. E: CT scan shows the courses of the percutaneous nephrostomy tubes, which terminate in the right *(R)* and left *(L)* renal collecting systems.

Urinary tract obstruction is a common problem requiring prompt diagnosis and treatment in order to prevent long-term injury of renal tissue and loss of renal function. As you learned in Chapter 14, obstruction of the urinary tract may be diagnosed by intravenous urography or ultrasound. Urography may show very early signs of obstruction, initially producing a delayed and increasingly dense nephrogram, with later films showing columnation of contrast media in a dilated ureter to the level of the obstruction. Hydronephrosis may take days or weeks to occur after the onset of obstruction. When hydronephrosis does develop, it will also be shown by urography. Ultrasound can detect hydronephrosis with a high degree of accuracy, but may overlook early obstruction.

Urinary tract obstruction may be temporary, and no intervention is necessary if the obstruction is likely to resolve spontaneously, as is the case with a partially obstructing stone in the distal ureter, which the patient will pass. Intervention is indicated if the obstruction is not likely to pass, as is the case with a large stone or malignant obstruction, or if the urinary tract becomes infected and the patient develops urosepsis. Common causes of chronic obstruction are malignant tumors of the urinary tract and surrounding structures, which may compress and obstruct the ureter. Impacted stones and benign ureteral strictures can also obstruct the urinary tract.

When urinary tract obstruction requires treatment, the obstruction can be relieved by percutaneous nephrostomy. With the patient in the prone position, the radiologist, using a posterolateral approach, inserts an 8- to 10-French nephrostomy drainage catheter into the dilated urinary tract, under either fluoroscopic or ultrasound guidance (Figure 19.24). The catheter is placed on external drainage and samples are sent for culture and cytology.

An extension of this procedure is the placement of an internal catheter or stent that transverses the site of obstruction and permits internal urinary drainage

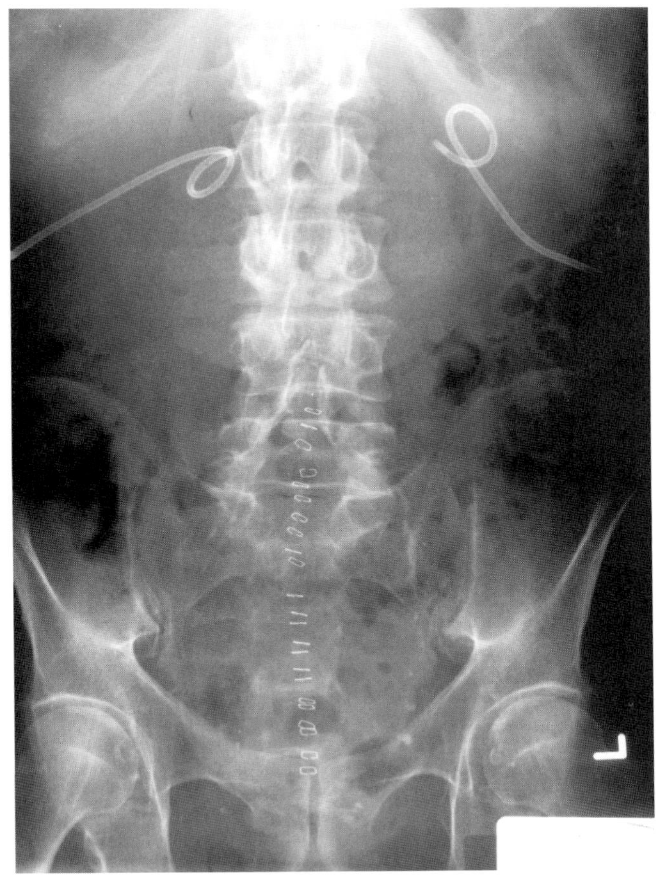

C

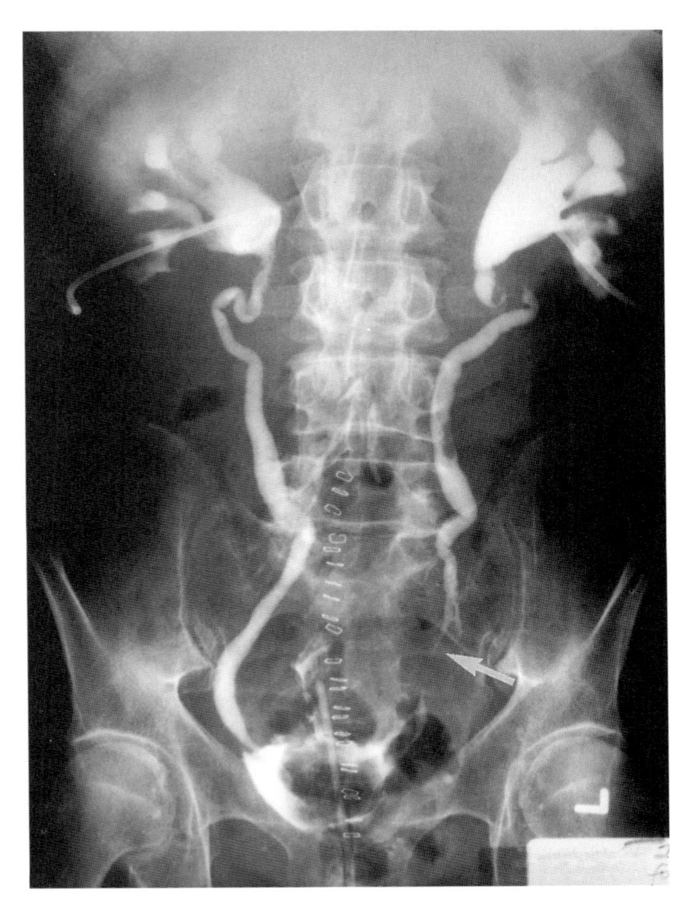

D

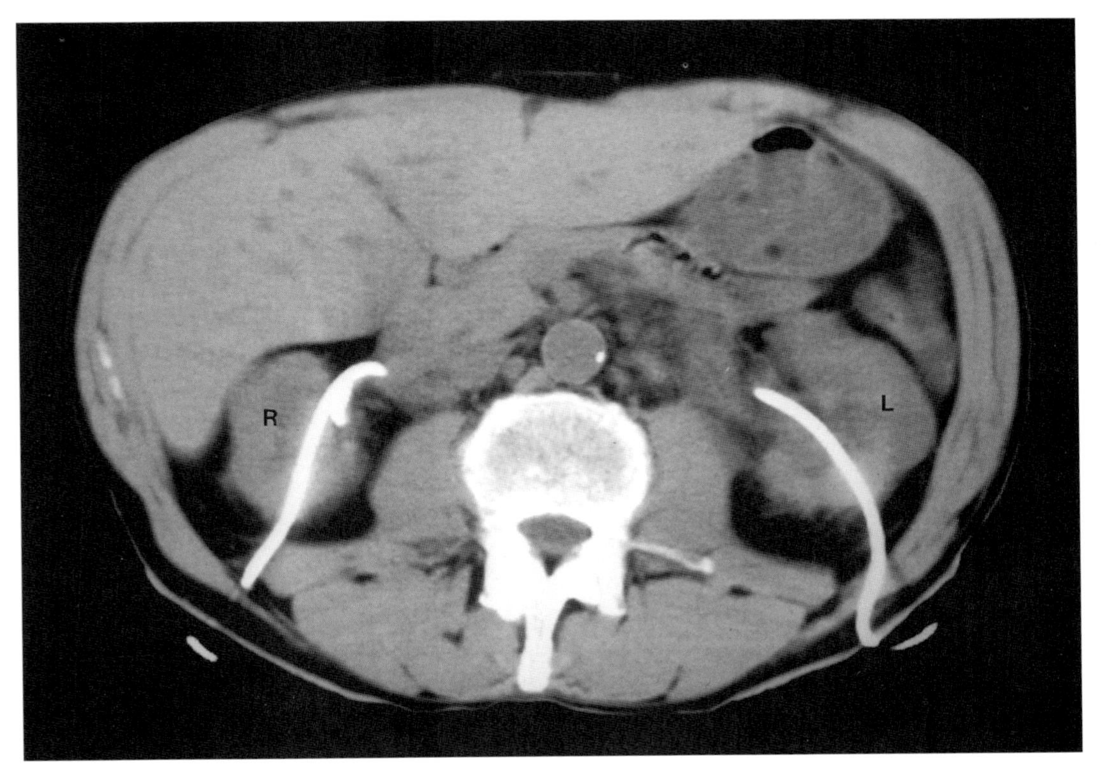

E

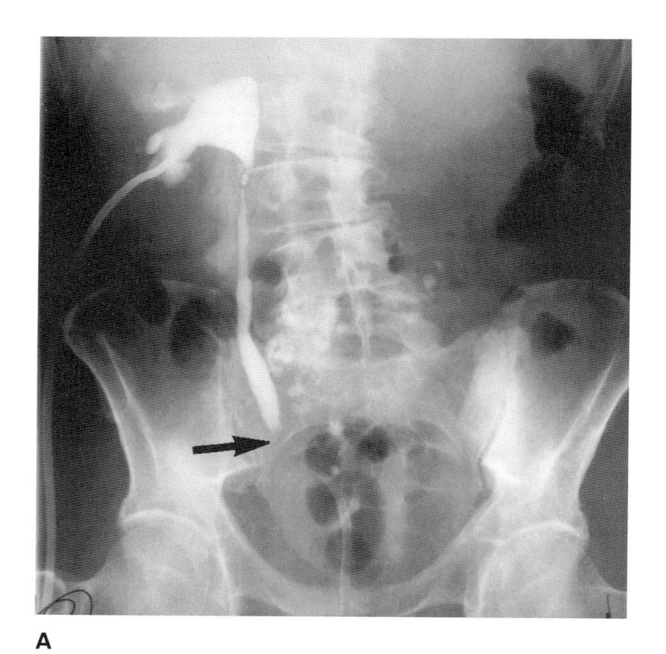

A

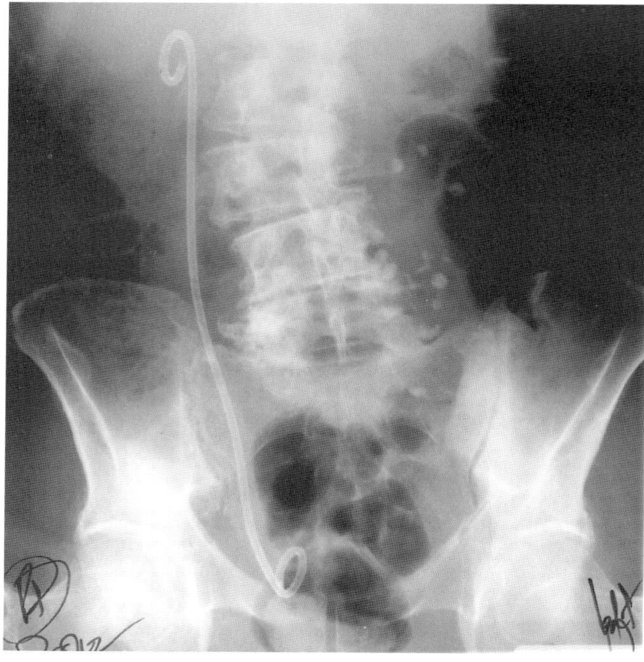

B

Figure 19.25. Percutaneous ureteral stent placement to provide internal drainage of an obstructed right kidney. A: Plain film of an initial percutaneous nephrostomy shows obstruction of the right ureter *(arrow)* by metastatic tumor within the pelvis. A second catheter can be seen within the bladder. B: Plain film shows the internal stent, which was placed to provide internal drainage across the ureteral obstruction. The proximal end of the stent lies within the right renal pelvis, the distal end within the bladder. Multiple side holes in the stent (not well shown) assist drainage of the urine.

from the kidney into the bladder. After percutaneous nephrostomy, the radiologist, using fluoroscopic guidance, negotiates a passage through the obstruction to gain access to the bladder (Figure 19.25). The stent, which bypasses the obstructing lesion, is left in place.

Ureterorenoscopy, or renal endoscopy, can be performed by percutaneously inserting a small-caliber endoscope. With this percutaneous technique uroepithelial tumors can be biopsied and urinary calculi can be removed. Although most small calculi are usually treated with extracorporeal shock wave lithotripsy, (ESWL), percutaneous renal endoscopy is indicated for the nonsurgical treatment of large calculi, such as staghorn calculi. The small calculi and fragments resulting from the endoscopic lithotripsy are removed directly via the endoscope.

Interventional Neuroradiology

Interventional radiological procedures have proven valuable in the treatment of two conditions affecting the central nervous system, cerebral aneurysms and cerebral arteriovenous malformations.

Cerebral aneurysms are not at all uncommon. It has been estimated that they occur in 2 percent of the general population, and nearly 25,000 people in North America suffer aneurysm rupture each year. About 50 percent of these die from the initial hemorrhage episode, and about 50 percent of those who survive die or suffer serious neurological consequences from rebleeding or vasospasm. It is therefore imperative that aneurysms be treated when detected.

The conventional surgical treatment for cerebral aneurysms is an open surgical craniotomy with clipping of the neck of the aneurysm. But certain cerebral aneurysms are located at sites that are extremely difficult for neurosurgeons to access safely. Today these aneurysms can be successfully treated with transcatheter embolization (Figure 19.26). Following catheterization of the cerebral arterial system by a femoral artery route, the tip of a special embolization catheter is navigated directly into the aneurysm to thrombose it with a detachable balloon or detachable metallic coils.

By similar angiographic approaches, cerebral arteriovenous malformations can also by thrombosed to prevent hemorrhage. The mechanics of occluding an arteriovenous malformation are made more complex by the anatomy of the malformation. The occluding agents chosen will depend upon the configuration of the malformation, and may include a variety of solid materials such as coils, spheres, and polyvinyl alcohol foam, or they may include liquid acrylic materials that are directed to harden within the malformation itself.

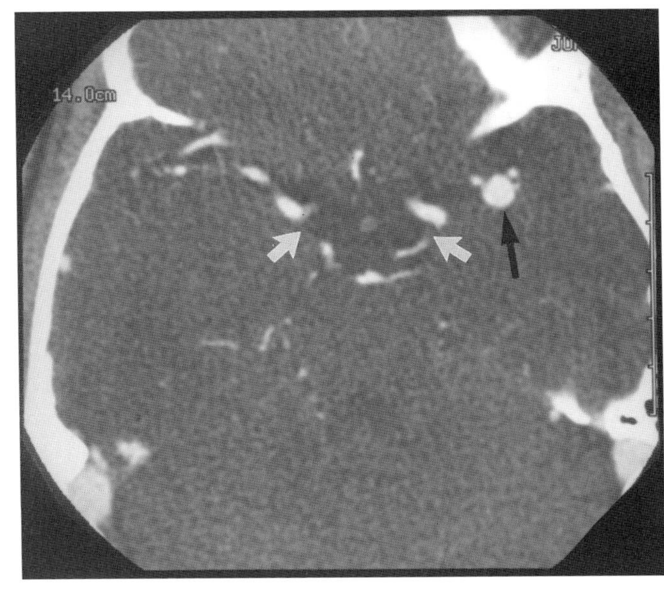

A

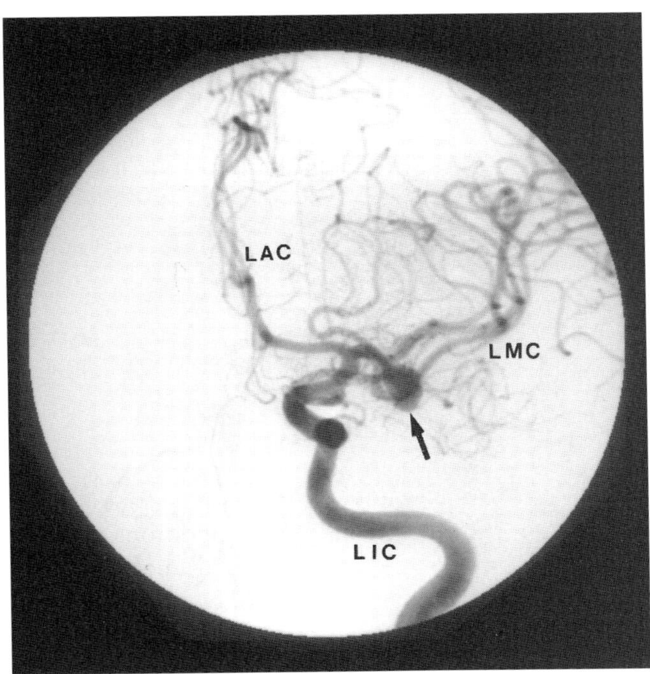

B

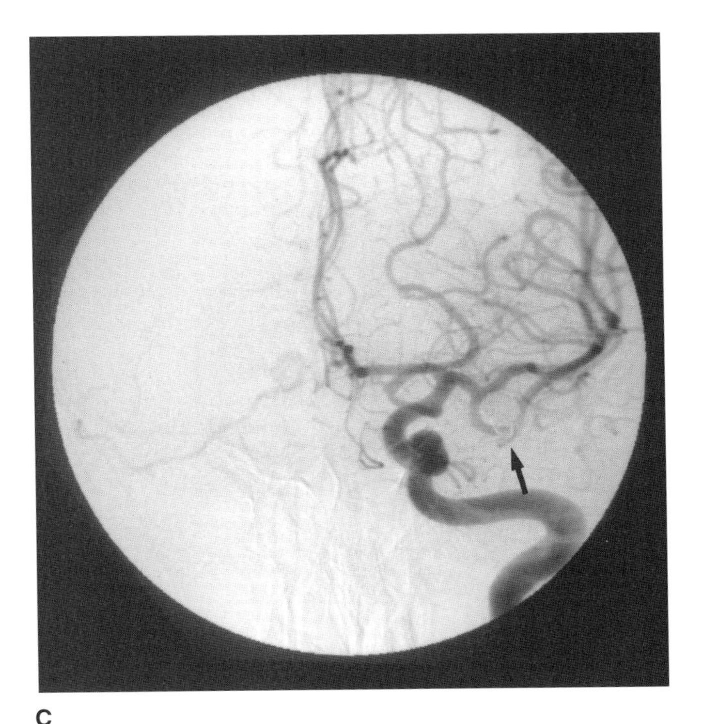

C

Figure 19.26. Transcatheter embolization of a left middle cerebral artery aneurysm in a 58-year-old woman who presented with dizziness. A: Her initial contrast CT scan shows a small contrast-filled aneurysm *(black arrow)* to the left of the circle of Willis *(white arrows)*. The aneurysm is closely associated with the course of the left middle cerebral artery. B: Anterior view of a left internal carotid *(LIC)* arteriogram shows a small aneurysm *(arrow)* arising from the left middle cerebral *(LMC)* artery. Also shown are left anterior cerebral *(LAC)* artery branches. The dark circle to the left of the aneurysm is not a second aneurysm; it is only overlap of tortuous portions of the internal carotid artery. C: Postembolization arteriogram shows the aneurysm now occluded with multiple small metallic coils.

20 Pathological Change over Time and Multisystem Disease: TB and AIDS

In the preceding chapters you have learned basic imaging principles and have been introduced to the wide range of clinically available imaging techniques. In addition, you have had the opportunity to observe the radiological appearance of many common and unusual conditions that you will encounter in your training and practice. In this chapter you will see how radiology can document and measure the gross pathological changes caused by a single disease over time or in many different parts of the body. We will use pulmonary TB to show how radiology can track the progress of a particular disease through the years, and AIDS to demonstrate how radiology can depict the presence or absence of disease in multiple systems.

Pulmonary TB: Change over Time

Pulmonary tuberculosis is caused by an aerobic acid-fast bacillus, *Mycobacterium tuberculosis.* It may take two forms, primary tuberculosis and reactivation, or post-primary, tuberculosis.

Primary TB commonly affects children. It is usually asymptomatic and there may not be any radiological sequelae of the infection. When the TB is symptomatic, it may manifest itself as pneumonia, which appears on plain films as segmental or lobar air-space opacification (Figure 20.1). This disease may mimic a

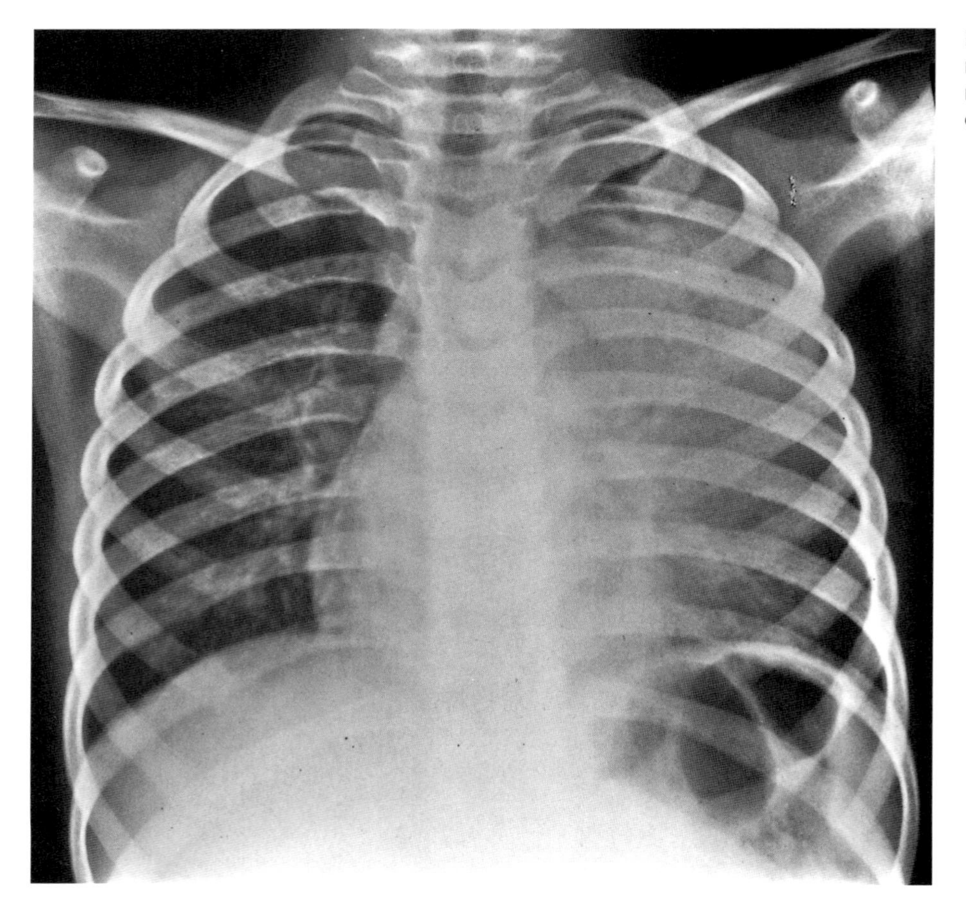

Figure 20.1. Primary TB in a child manifesting as left upper lobe pneumonia. Air-space consolidation has obliterated the left heart border.

bacterial pneumonia. A pleural effusion may also occur, associated with the parenchymal opacification. Unilateral or bilateral hilar adenopathy is common in children with primary TB. In some asymptomatic patients primary TB may produce a Ranke complex consisting of a parenchymal calcification (Ghon lesion) combined with nodal calcification (Figure 20.2). In most cases primary TB is eventually contained by a granulomatous response, but it may recur many years later as reactivation TB.

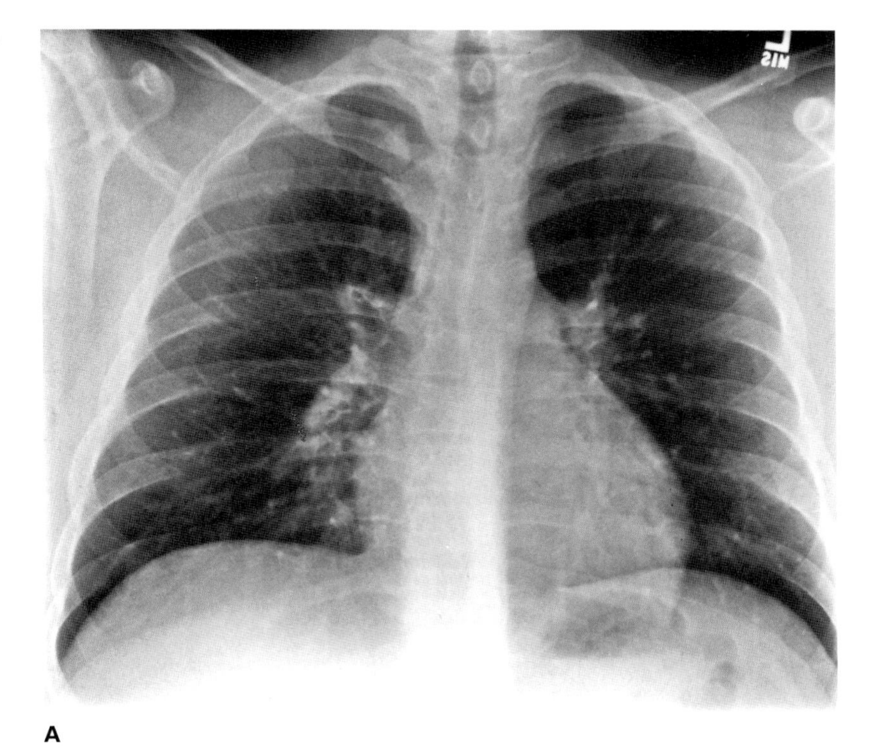

A

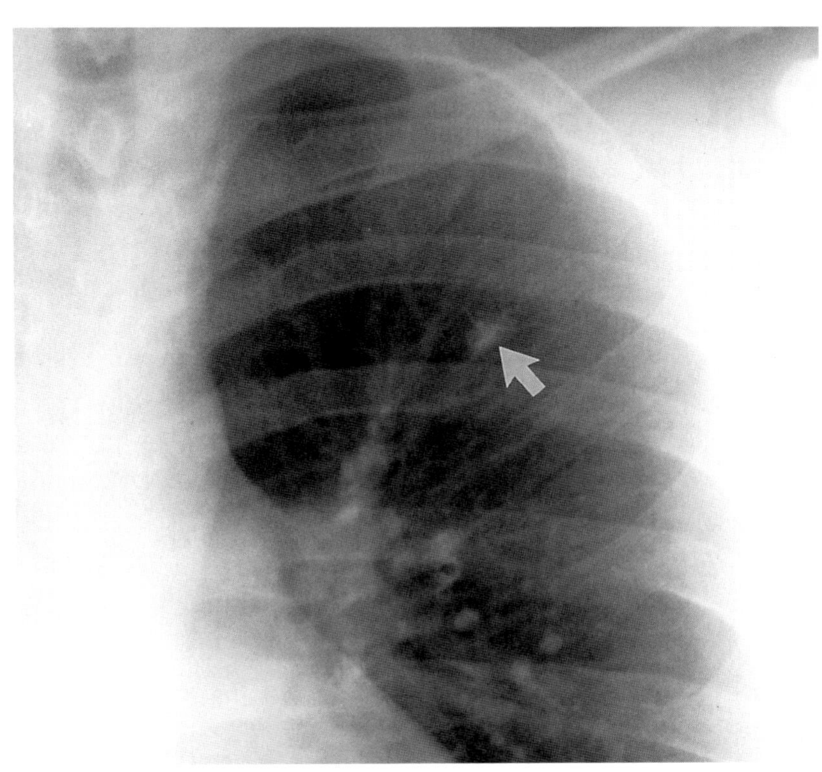

B

Figure 20.2. Ranke complex with a parenchymal calcification (Ghon lesion) in the left lung pulmonary parenchyma and calcified left hilar nodes. A: PA chest film. B: Magnification view of the left upper lung better shows the calcified nodule *(arrow)* and tiny white calcifications in the left pulmonary hilum.

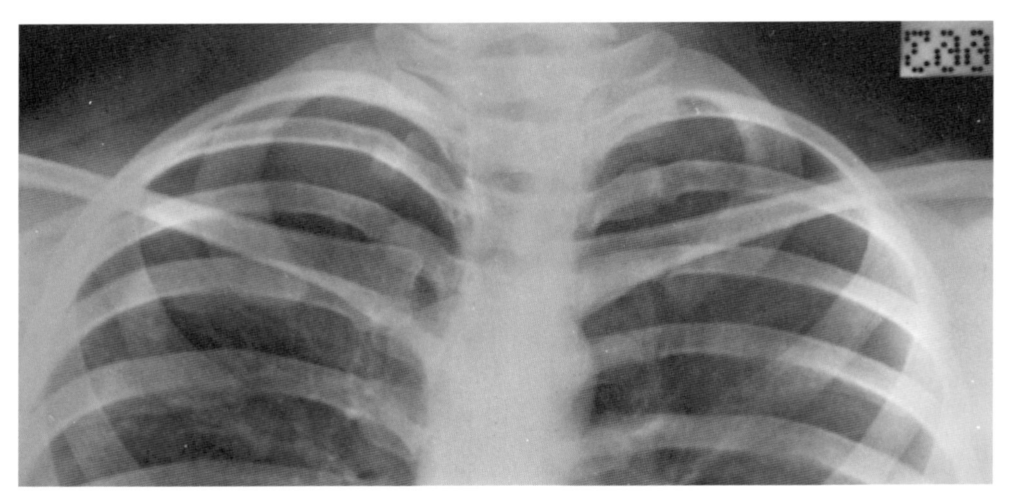

A

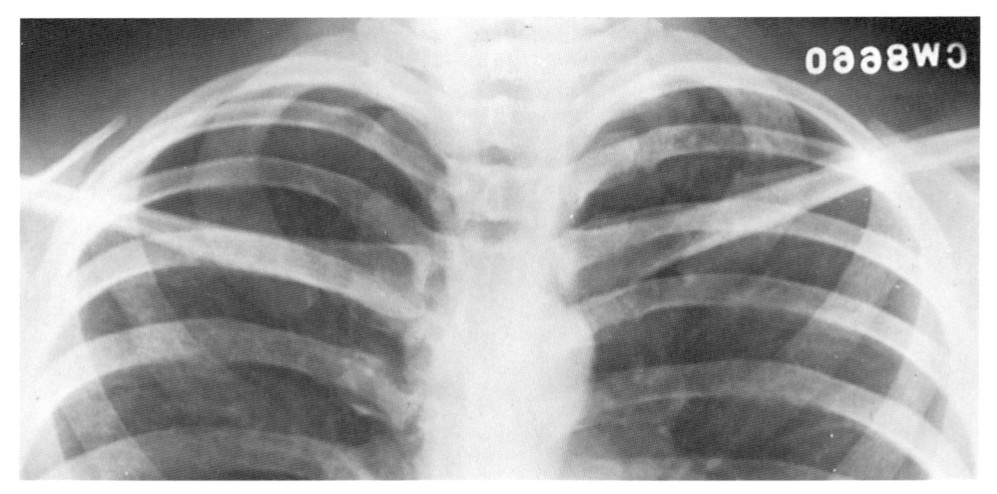

B

Figure 20.3. An example of minimal apical tuberculosis in details from two chest films made on the same patient. A: A routine film made in September is normal. B: A film made 5 months later shows numerous small, fluffy opacities in the lung tissue at the left apex, in the second and third interspaces, and superimposed on the first three ribs. Compare the interspaces.

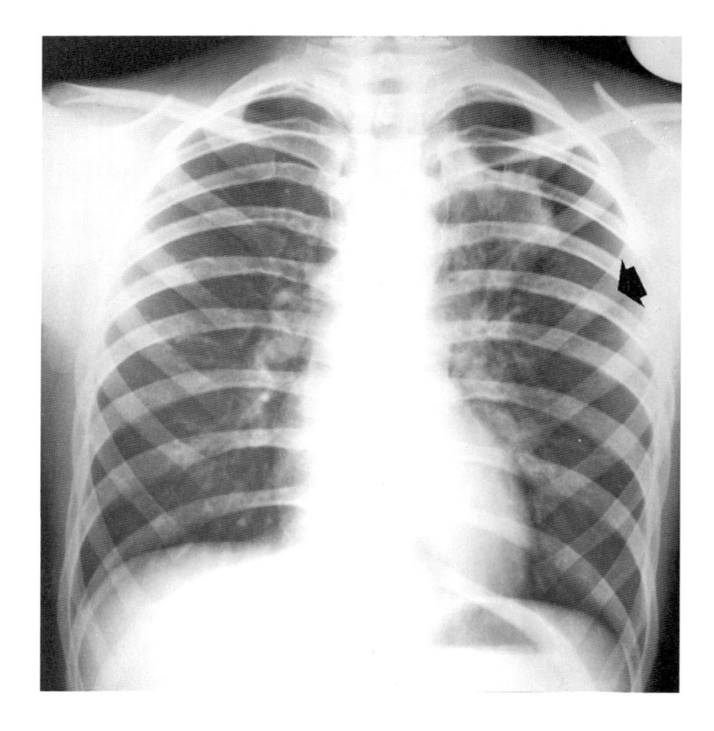

Figure 20.4. More extensive tuberculosis, in the left upper lobe, with a small pneumothorax (the *arrow* indicates the visible margin of the lung). The linear strands extending to the chest wall are pleural adhesions over the apex, restricting the degree of collapse.

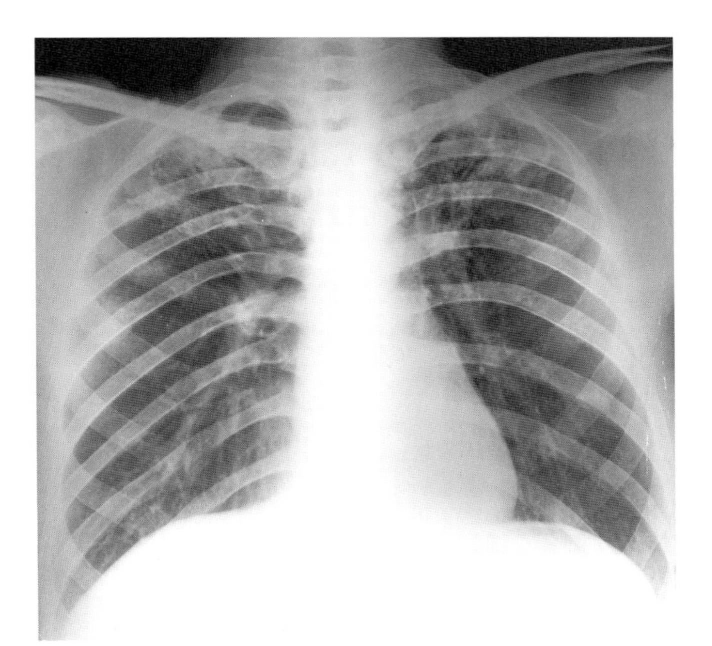

Figure 20.5. Bilateral chronic upper lobe tuberculosis. Note that, in addition to the obvious infiltrative parenchymal streaks in both upper lobes, ringlike cavities are present. The one on the right can be seen rising above the medial end of the right clavicle, and the one on the left overlies the fourth rib near the lateral chest wall. Note also the very characteristic pattern of vessels extending down into the lower lungs from the hila. These vessels are longer, straighter, and more vertical than the normal lower lung vessels and may be likened to the taut guy ropes of a tent. Their appearance is easy to remember when you realize that in this type of upper lobe tuberculosis there is so much scarring and retraction that the upper lobes are markedly reduced in size and both hila are drawn upward, stretching the vessels to the lower lobes.

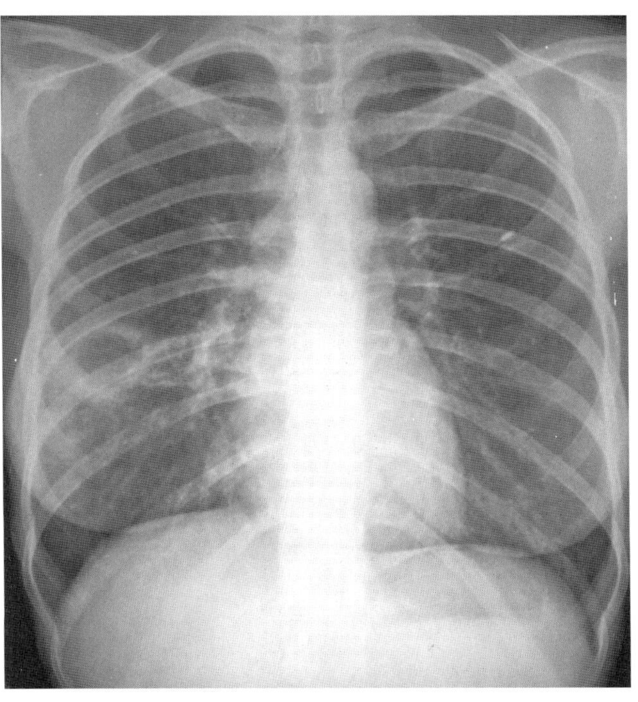

Figure 20.6. Cavitation in a relatively new tuberculous lesion. Considering the size of the cavity here, there is less extensive parenchymal density than might be expected in chronic involvement. When you compare carefully the segments of lung seen in the seventh, eighth, and ninth interspaces on the two sides, those on the left appear normal, while those on the right show scattered soft opacities. Think of the pathological process that is going on, rather than of the x-ray opacities. Consider that the balance between the resistance of the patient and the virulence of the disease must determine the rate of tissue breakdown. This being so, it must follow that the relation of the size of the cavity to the type of inflammatory opacity in the involved lung around it, as you see them in the x-ray, provides an index to the state of the host-disease balance. In a portion of lung showing many soft, fluffy opacities, the sudden appearance of a large, thin-walled cavity probably means rapid tissue breakdown in poorly resisting lung. A similar cavity in a segment of lung known to have been diseased for a long time, and showing instead the dense, discrete, and stringy opacities of healing fibrotic lesions, would not carry the same implications. Likewise, the progress of changes in a series of films made at intervals provides a useful index to the patient-disease relationship and all the factors that may influence it, as well as the success or failure of therapy.

Patients with *reactivation, or post-primary, TB* are symptomatic, presenting with cough, chills, night sweats, and weight loss. Both the apical and posterior segments of the upper lobes and the superior segments of the lower lobes are involved (Figures 20.3 and 20.4). Patchy air-space densities and nodular opacities are seen on plain films, as well as cavities, which usually indicate the presence of active and transmissible disease (Figures 20.5 and 20.6). Healing is associated

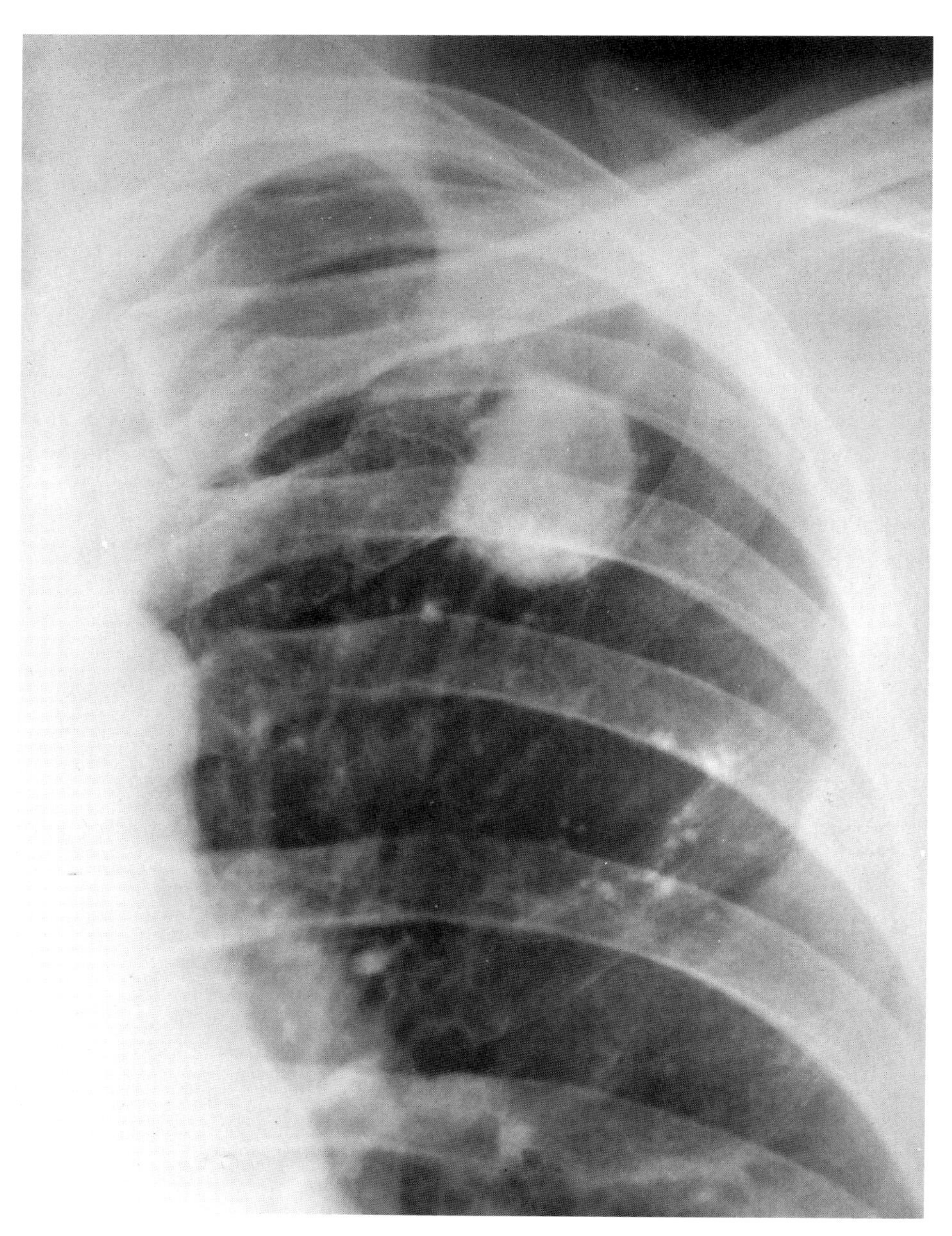

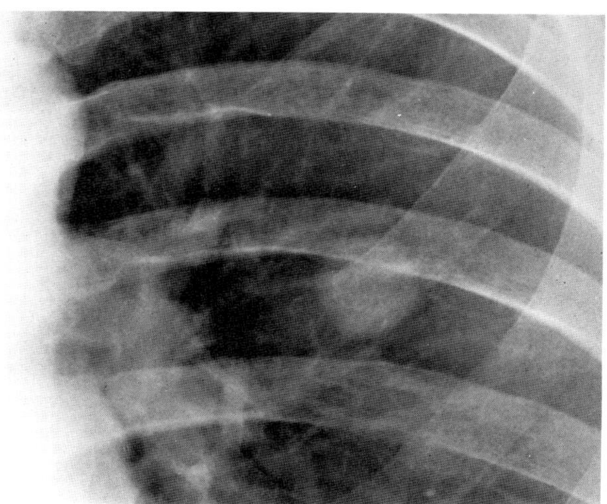

Figures 20.7 *(above)* and 20.8 *(left).* Tuberculomas in the lung. These lesions may closely simulate solitary tumors, either primary or metastatic. Some tuberculomas contain no calcium, while others calcify centrally (or expand around and engulf an earlier calcific focus). Still others calcify peripherally and appear to have a shell.

with scarring and volume loss in the upper lobes as well as bronchiectasis. Tuberculomas (Figures 20.7 and 20.8) are nodular opacities that may develop in either primary or reactivation TB. They may or may not calcify and may resemble solitary lung cancers. Miliary TB results from hematogenous spread of the organism and may complicate either primary or reactivation TB (Figure 20.9).

As an exercise, stop and review the *morphological changes* produced by the tubercle bacillus in the lung that you can recognize on radiographs and CT scans. To start with, of course, you know that identification of the etiological agent in tuberculosis is never settled by x-ray. Tuberculosis is a clinical diagnosis, not a radiological one, for all the morphological changes it produces can be closely mimicked by fungus, among other agents. Radiology can suggest a diagnosis of pulmonary tuberculosis but cannot establish it; only laboratory identification of the tuberculous bacillus can prove the disease is present.

Pleural effusion is one easily recognized manifestation of TB, the organism being identified by pleural biopsy even in the absence of visible changes in the lung. Consolidation with or without *cavitation* may be documented and followed on serial films, as may the progress of healing under therapy. Of course, you would be very suspicious of pulmonary consolidations and cavities *involving the pulmonary apices.* You have seen *enlarged hilar and paratracheal lymph nodes,* which although not at all specific for tuberculosis may be a part of the x-ray documentation, and you know that those nodes may calcify as they heal.

Tuberculosis may produce *pneumothorax,* and you know how to look for air in the pleural space. *Miliary tuberculosis* is recognizable on a chest film as soon as the minute interstitial granulomas reach 2 millimeters in diameter, although you know they cannot be differentiated radiographically from miliary carcinomatosis, disseminated in the same fashion to the interstitium via the capillary bed.

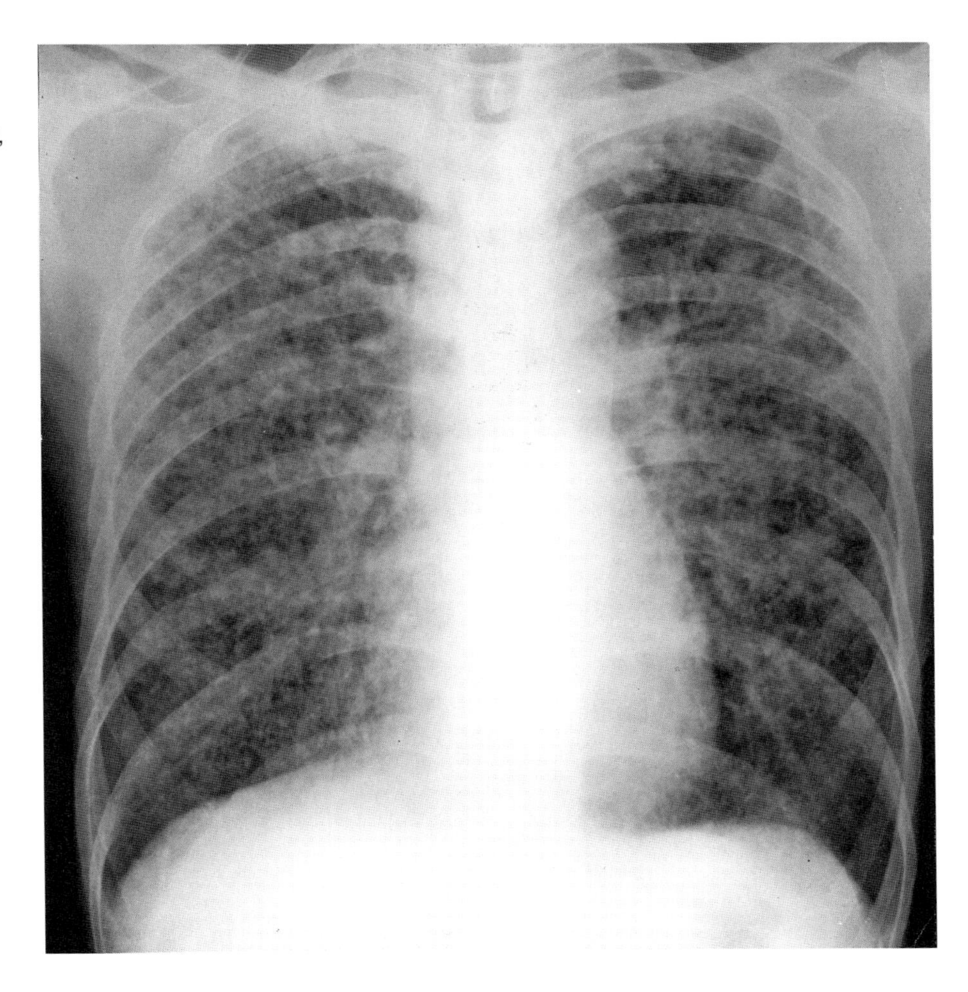

Figure 20.9. Miliary tuberculosis. The innumerable nodules scattered throughout the parenchyma here, resulting from hematogenous spread, have been superimposed upon the lungs, in which there was already some tuberculous infiltration in the upper lobes. In looking at any film, you have to consider that the opacities you see may represent acute changes superimposed on chronic ones.

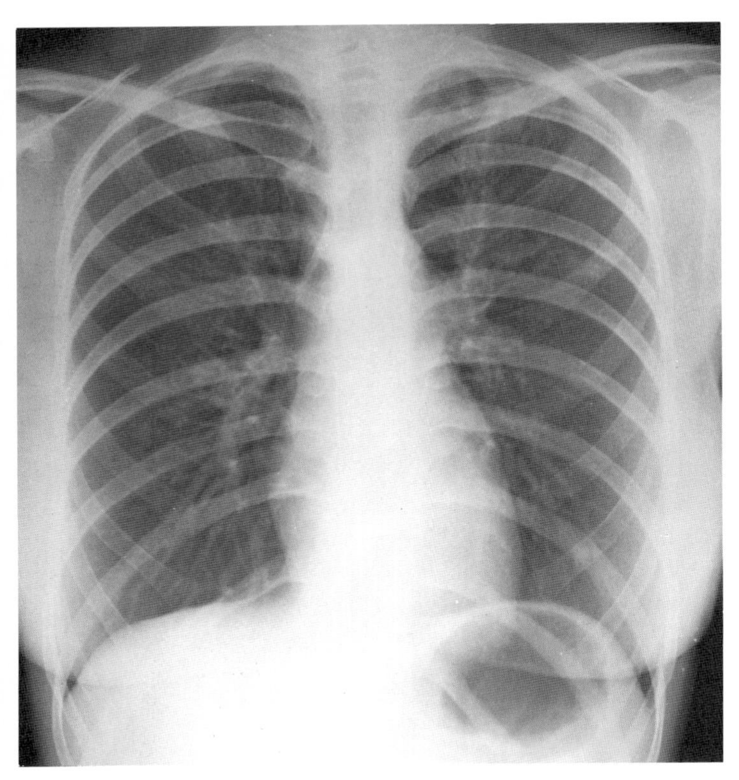

A. December 1943.

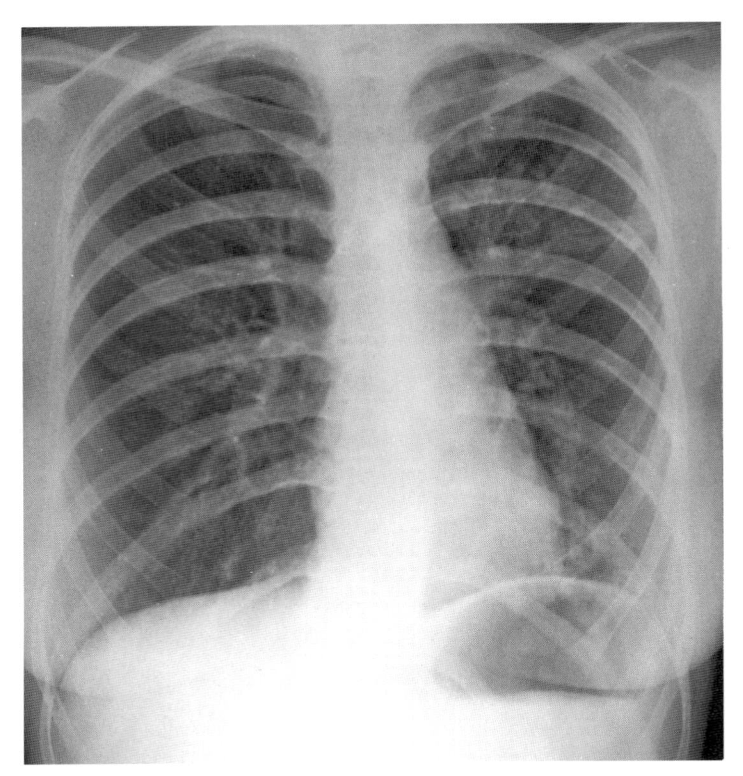

B. August 1944.

Figure 20.10. Serial films over a period of 3 years of a patient who had tuberculosis before modern therapy was available. In A the patient has a soft infiltrative opacity extending upward from the left hilum to the apex. In B, made 8 months later, there is progressive involvement of the left upper lobe and there are new areas of opacity extending down toward the left diaphragm, which may be either in the lingular segment of the upper lobe or in the lower lobe. In C, made 6 months later, there is involvement throughout the left lung, but with the development of scar tissue the opacities have taken on a harder, denser, and more discrete appearance. D, a year later, gives you radiographic indications that there is much more scar tissue retraction than on the earlier films. Note that the trachea and the heart have been drawn over to the left. Cavitation is obvious in the upper lobe. You see the profile of the diaphragm still. In E, there appears to be an immense cavitation replacing the upper lobe (absence of lung vessels). The left diaphragmatic profile and the profile of the left heart border have disappeared, indicating consolidation, and there is new spread of the disease to the right lung. Some pleural effusion on the left cannot be excluded. F is a radiograph of the postmortem specimen of the two air-inflated lungs, the vessels of which have been injected with an opaque substance.

Now study the series of chest films obtained in a patient with pulmonary tuberculosis over 3 years, before the advent of antituberculous drug therapy (Figure 20.10). Compare the disease progression in that patient with the course of TB in the patient in Figure 20.11 (on page 582), whose chest films document healing of pulmonary tuberculosis with drug therapy.

As well as showing you the varied morphological changes caused by a disease over time, this review of pulmonary TB should remind you of the many different radiological manifestations a single disease process can produce. TB, like many other diseases, can generate diverse symptoms at various stages of its development, and you should always remember this fact when you are confronted with a chest film showing pneumonia in a child, a cavitary lesion in an adult, or miliary nodules in patients of any age.

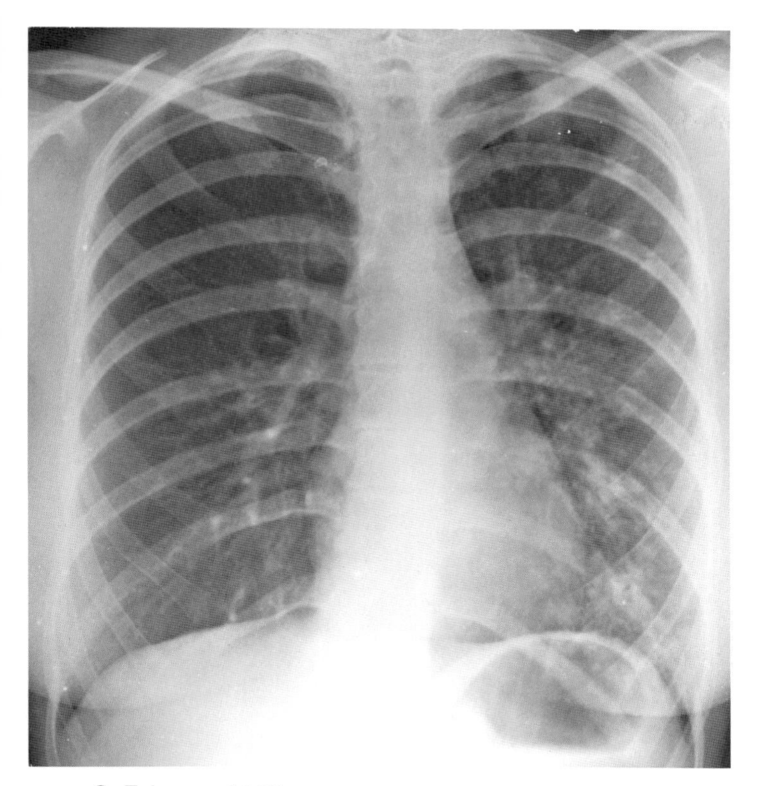

C. February 1945.

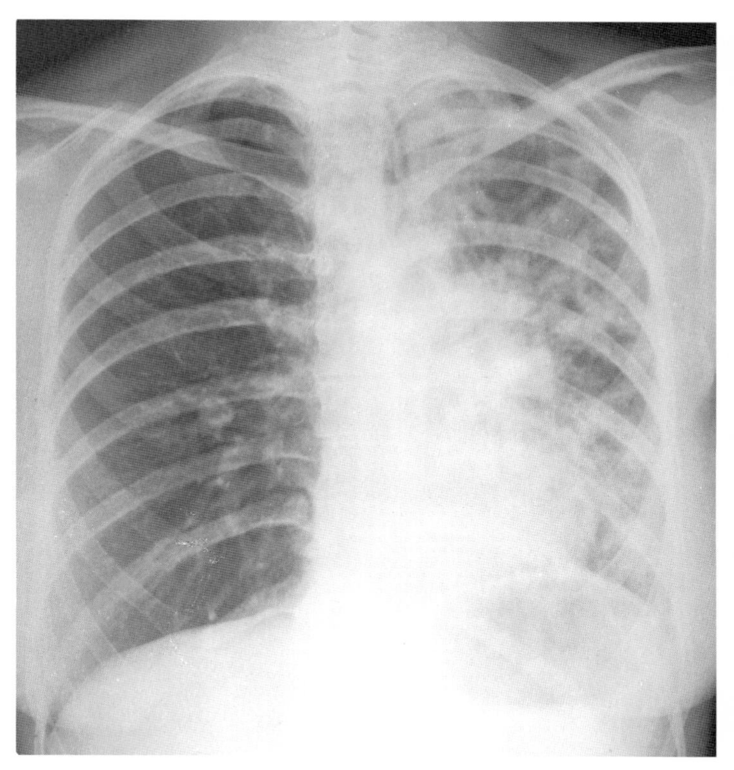

D. February 1946.

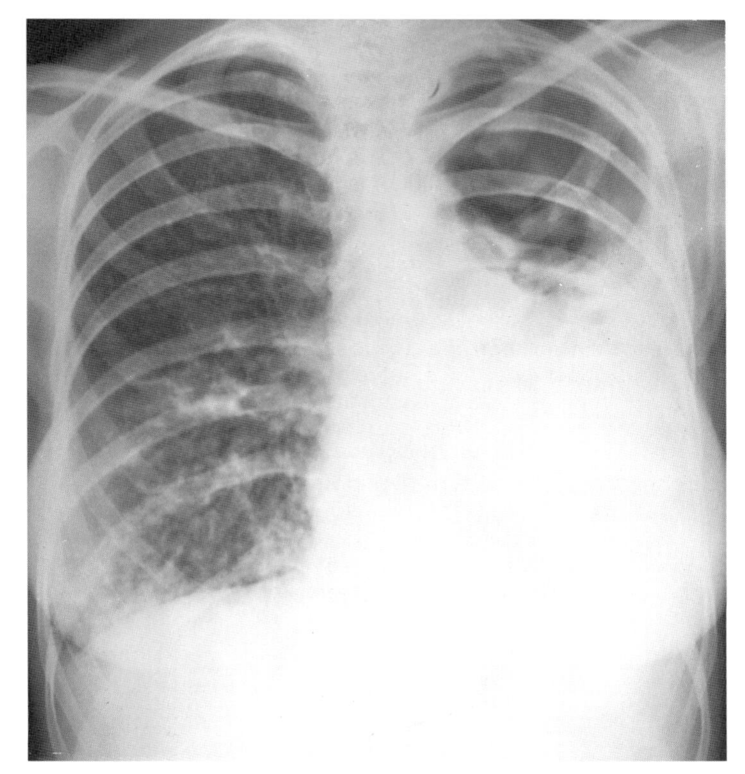

E. August 1946.

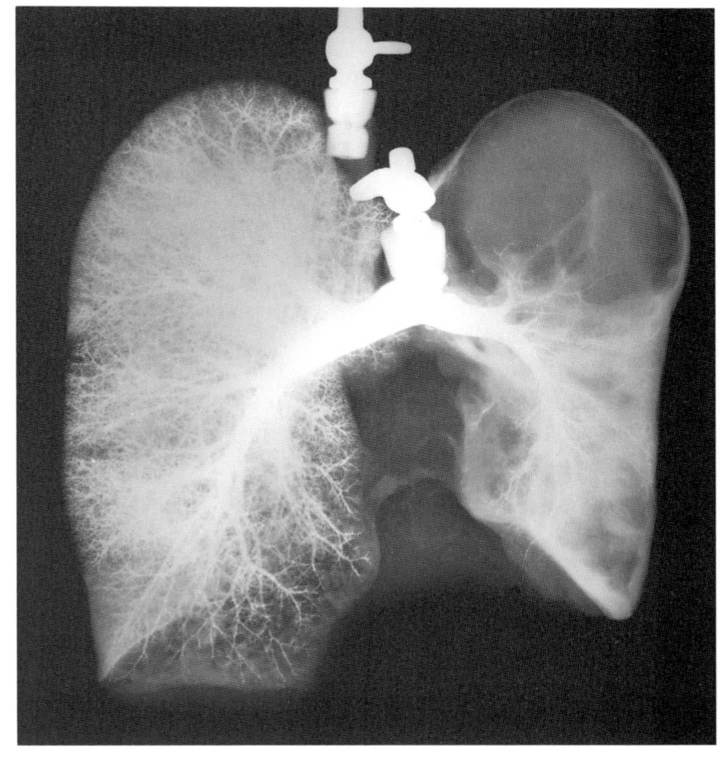

F. October 1946.

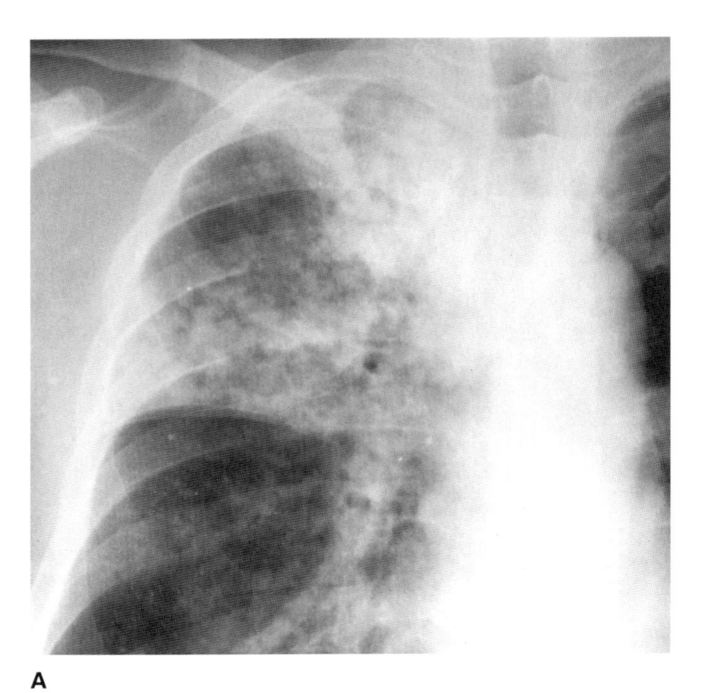

A

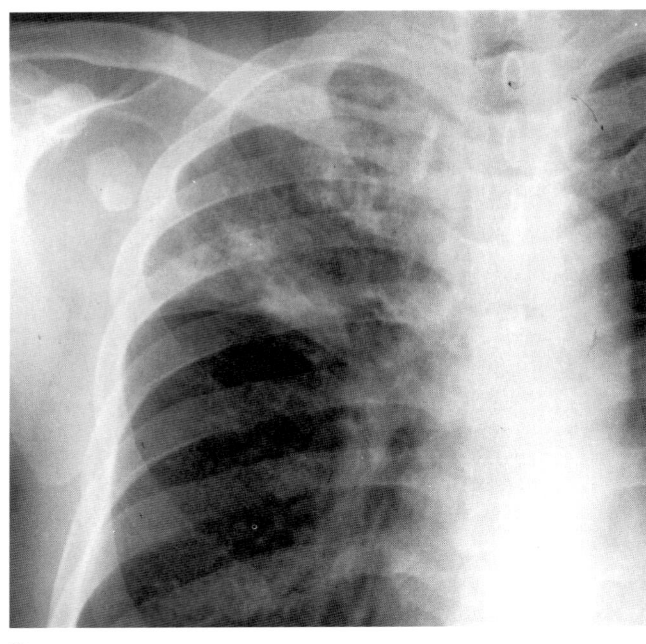

B

Figure 20.11. Healing in a patient with right upper lobe tuberculosis. A: The patient before treatment. B: The disease is resolving with drug therapy. Note the upward retraction of the horizontal fissure as the process heals. Ultimately, this patient became symptom free and sputum negative, and was considered cured.

AIDS: Multisystem Disease

Since the acquired immune deficiency syndrome (AIDS) was first described in 1981, it has spread dramatically. Caused by the human immunodeficiency virus (type 1) (HIV 1), a retrovirus, this infection has spread to every continent and is the leading cause of death of young men in the United States. More than 500,000 people have died and an estimated 1,000,000 are infected. AIDS usually involves multiple systems, including the respiratory, gastrointestinal, and central nervous systems. Radiology is often the first technique used to identify the extent of the disease.

Pulmonary Involvement in AIDS

The most common site for opportunistic infections in patients with HIV is the chest. Over 50 percent of AIDS patients develop a pulmonary infection or pulmonary neoplasm during the course of the disease. *Pneumocystis carinii pneumonia (PCP)* is the most common infection; AIDS patients frequently experience one or more episodes of this pneumonia. Patients present with a nonproductive cough, dyspnea, and fever. Chest films show a diffuse, bilateral interstitial abnormality with small, fine, reticular opacities. In the early stages, the chest film may demonstrate almost no abnormality, even when pneumocystis is present (Figure 20.12). The earliest radiological finding may be nothing more than a fine interstitial abnormality that could be confused with the chronic lungs seen in a longtime cigarette smoker. It is therefore always advantageous, when the radiologist is interpreting the chest films of patients with acute pulmonary symptoms and a fine interstitial abnormality, to have the old films available for comparison.

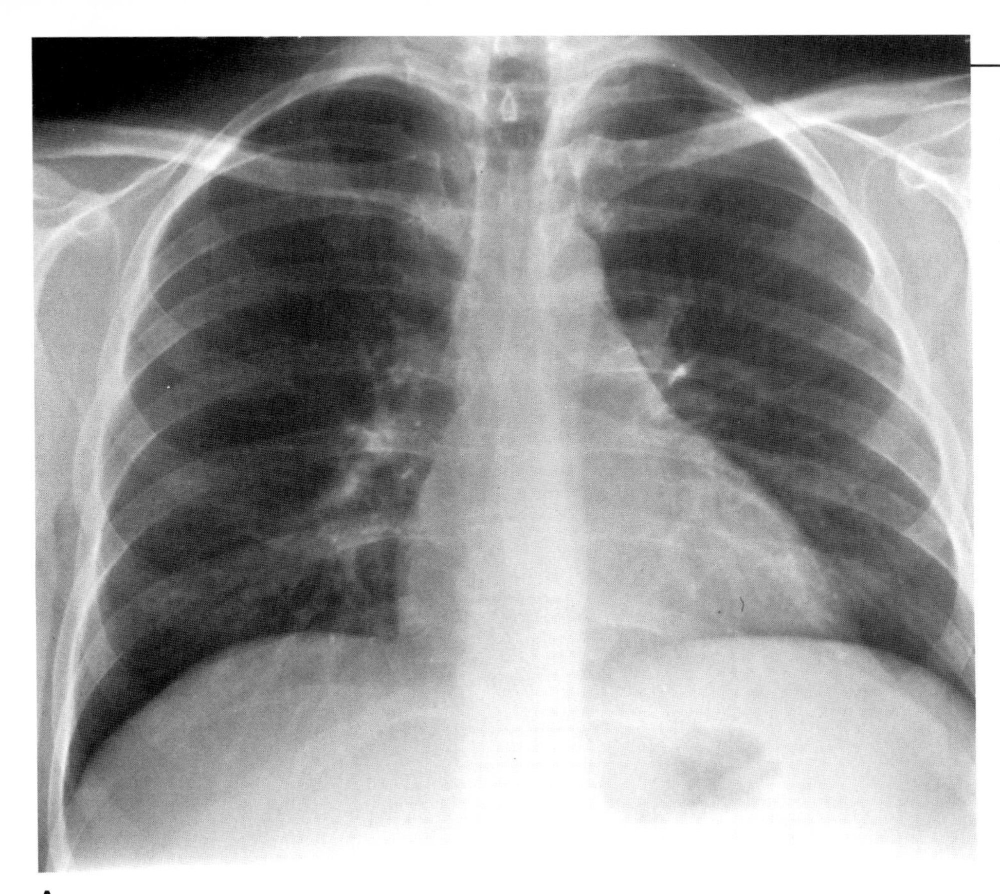

A

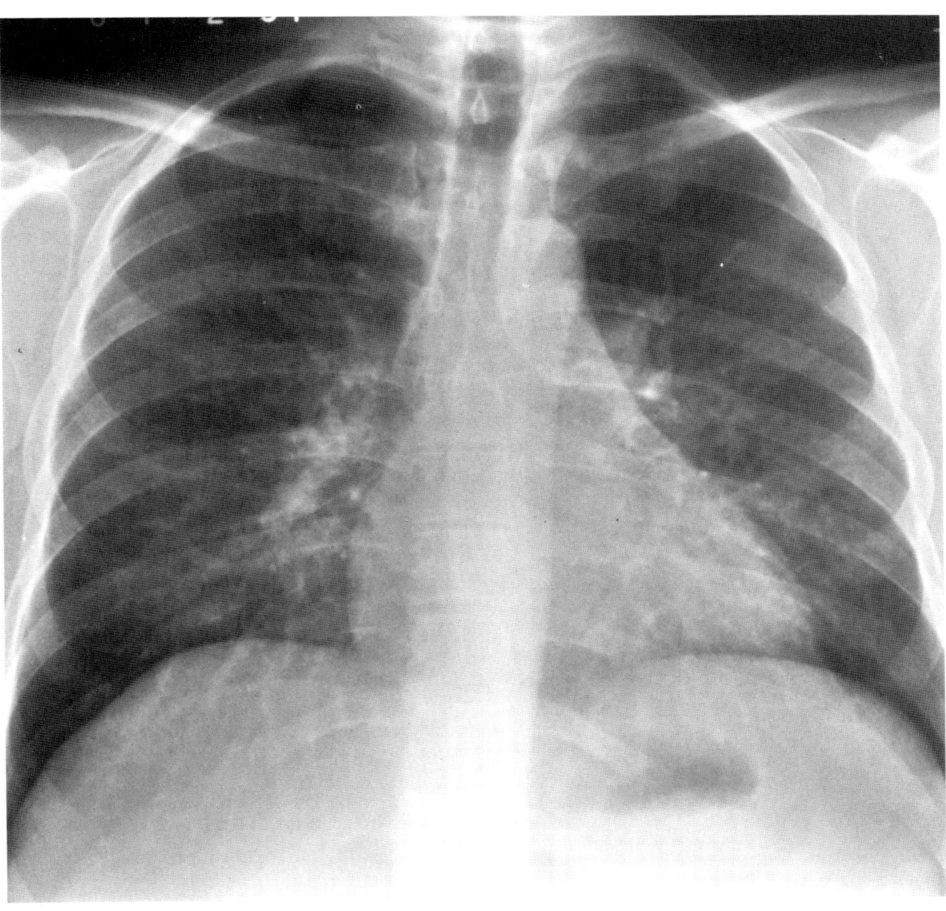

B

Figure 20.12. *Pneumocystis carinii* pneumonia (PCP) in a young man with HIV infection. A: Normal film taken one month before the patient presented. B: Film taken when the patient first presented with fever and cough. At this early stage of PCP only fine reticular opacities can be seen. Study the two films carefully to identify these reticular opacities in B overlying the pulmonary hila (making them appear shaggier than they did in A) and midlung zones.

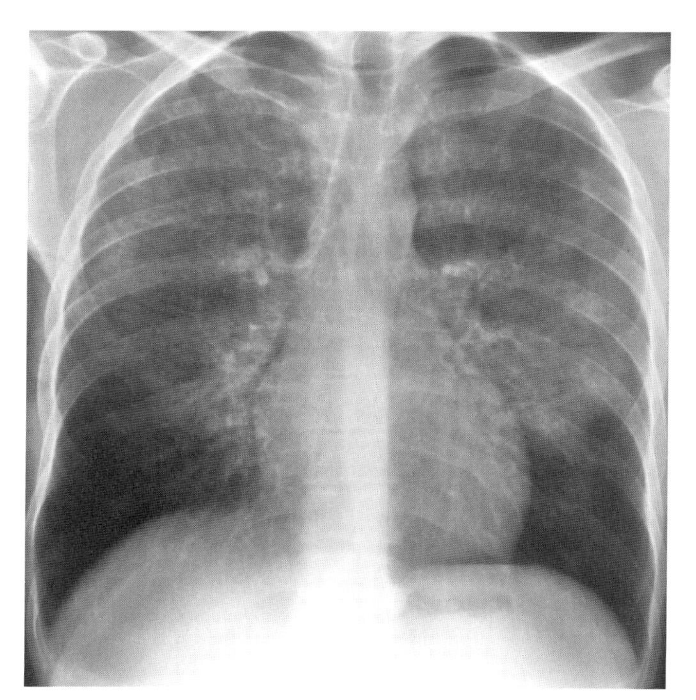

Figure 20.13. PCP involving the upper lobes with both reticular and ground-glass opacities. Upper lobe involvement is not infrequently seen in HIV patients who are taking prophylactic treatment with aerosolized pentamidine rather than systemic prophylaxis with Bactrim (often counterindicated because many patients are allergic to Bactrim); the inhaled drug is drawn by gravity to the lower lobes and thus the upper lobes are much more prone to infection. As you may have noticed, this distribution could be confused with the appearance of reactivation TB.

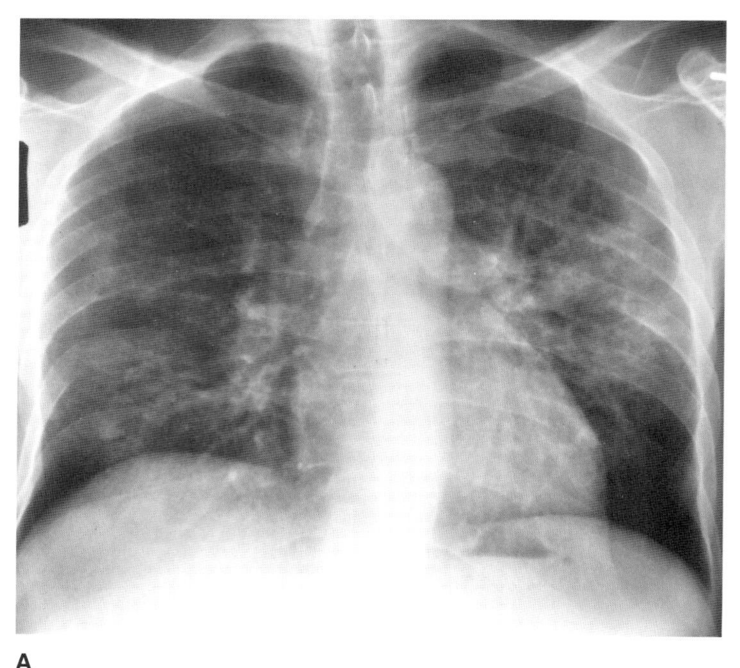

A

Figure 20.14. Rapid progression of severe PCP over 6 days. A: On the first day, when this AIDS patient presented with cough and fever, the chest film showed only fine reticular opacities with some ground-glass opacities in the left midlung. B *(top of facing page):* Four days later ground-glass opacities are obvious throughout both lungs. C *(top of facing page):* Six days later the lungs are completely consolidated with extensive air bronchograms. The patient died shortly after this film was taken.

When the infection is severe, alveolar opacities may be seen mixed with the interstitial pattern (Figures 20.13 and 20.14). Thin-walled cysts called pneumatoceles may develop in the pulmonary parenchyma; these may rupture, causing spontaneous pneumothorax. Approximately one-third of PCP patients also have a simultaneous infection caused by another agent, and consequently a mixed pulmonary pattern is not unusual, with findings such as hilar adenopathy and pleural effusion. A large pleural effusion usually indicates a concurrent bacterial infection, such as pneumococcal pneumonia.

Although chest films are the standard technique for imaging AIDS patients when pulmonary disease is suspected, CT is useful when the chest films are normal or equivocal in symptomatic patients. Early PCP pneumonia may show no abnormality on a chest film while a high-resolution CT scan is positive for early interstitial change and ground-glass opacities (Figure 20.15).

When pulmonary TB occurs in AIDS patients, it usually represents reactivation of a prior infection. In early HIV, when the patient's immune system is relatively intact and the CD4 count is above 400, the pulmonary pattern of *Mycobacterium* TB is similar to that of reactivation TB, described earlier, with cavitary lesions and opacities in the upper zones of the lung, and the patient will have a positive PPD test. The clinical presentation is also similar, including such symptoms as fever, chills, night sweats, cough, and weight loss.

In advanced stages of AIDS, with further impairment of the immune system, the PPD will be negative and the finding may be that of widely disseminated

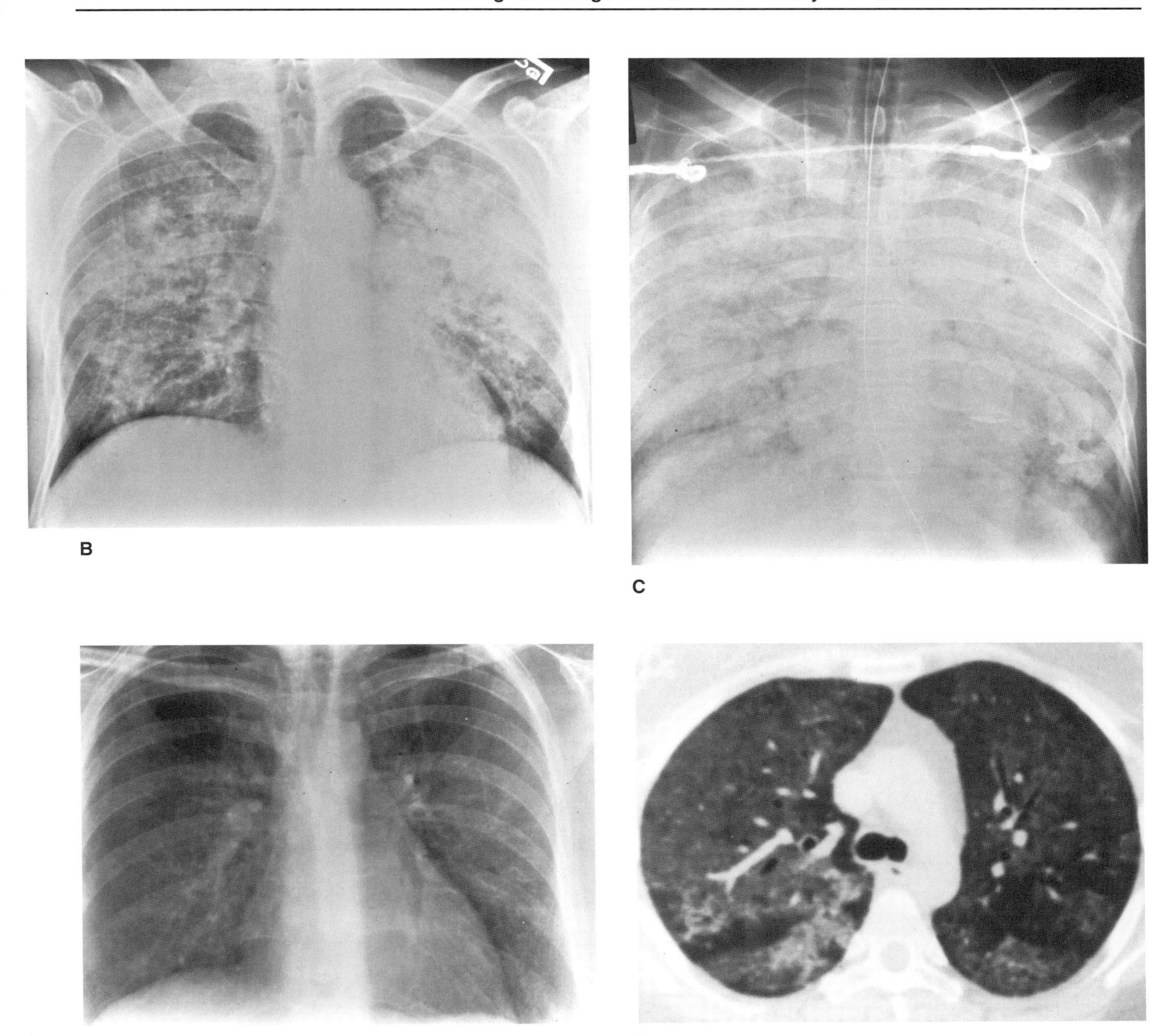

Figure 20.15. Plain film and CT scan of PCP. A: Plain film. B: The ground-glass opacities of PCP are better shown by CT than by the plain film obtained the same day. Compare the CT appearance of the pulmonary parenchyma with the normal chest CT scans in Chapter 3.

TB. The chest films may show a variety of patterns, including nodular and reticular interstitial patterns and adenopathy.

Atypical, nontuberculous mycobacteria may cause pulmonary disease in AIDS patients. *Mycobacterium avium complex (MAC)* produces infiltrates similar to those caused by reactivation tuberculosis, with opacities and cavities in the upper lobes. MAC infections in AIDS patients frequently involve extrapulmonary sites, including the abdominal organs, the lymphatic system, and the bone marrow.

Kaposi's sarcoma may occur in the lungs, lymph nodes, skin, and liver in patients with advanced AIDS. With pulmonary involvement the neoplasm invades the tracheobronchial mucosa as well as the peribronchial and perivascular interstitium of the lung. Chest films will show nonsegmental, nodular, and linear opacities extending out from the hila of the lungs into the midlung and lower lung regions (Figure 20.16A). CT will confirm the peribronchovascular distribution of opacities, with air bronchograms transversing areas of confluent disease (Figure 20.16B).

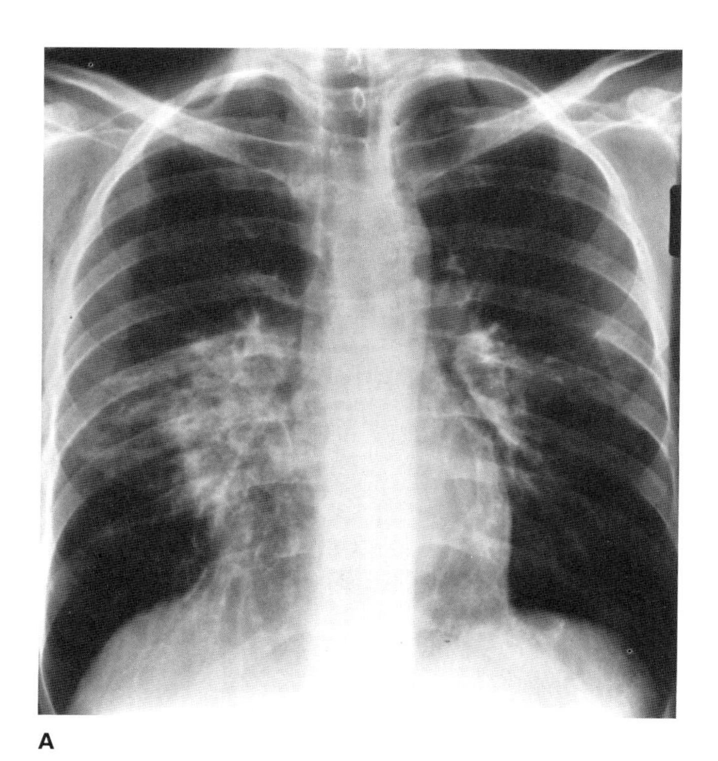

A

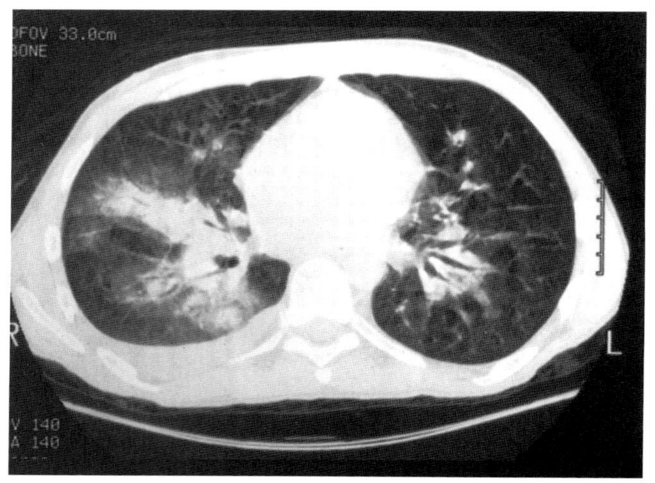

B

Figure 20.16. Kaposi's sarcoma of the lung. A: Chest film shows nodular and linear opacities extending out into the right lung from the right pulmonary hilum. Only minimal hilar opacities are seen on the left side. B: CT scan shows air bronchograms traversing bilateral, confluent, perihilar opacities.

Gastrointestinal Involvement in AIDS

The major feature of gastrointestinal involvement in AIDS is the occurrence of opportunistic infections. An opportunistic organism is, of course, an organism of low virulence that is often present in normal skin or mucosal flora which does not normally produce disease. Such organisms include *Candida albicans,* cytomegalovirus, herpes simplex virus, *Cryptosporidium* species, and *Mycobacterium* species. They may produce esophagitis, enteritis, and colitis in the immunocompromised host.

Esophagitis in patients with HIV is most commonly produced by an opportunistic infection with *Candida albicans,* a fungus. Oral involvement with this organism is known as *thrush,* which may be diagnosed on clinical examination. Patients with esophagitis may present with difficulty swallowing (dysphagia) or pain on swallowing (odynophagia). In the early stages of esophagitis, a barium esophagram will show mucosal irregularity. Later, an esophagram will reveal nodular thickening of the esophageal folds, as well as a shaggy or cobblestone pattern as the liquid barium fills linear spaces between plaques of monilial colonies and necrotic debris (Figures 20.17 and 20.18).

Esophagitis in HIV patients may also be caused by the herpes simplex virus, which produces multiple small focal ulcerations, often with a diamond shape. The esophageal mucosa between ulcerations may appear normal. Esophagitis caused by the cytomegalovirus also produces ulcerations, but these are usually larger and may appear in clusters.

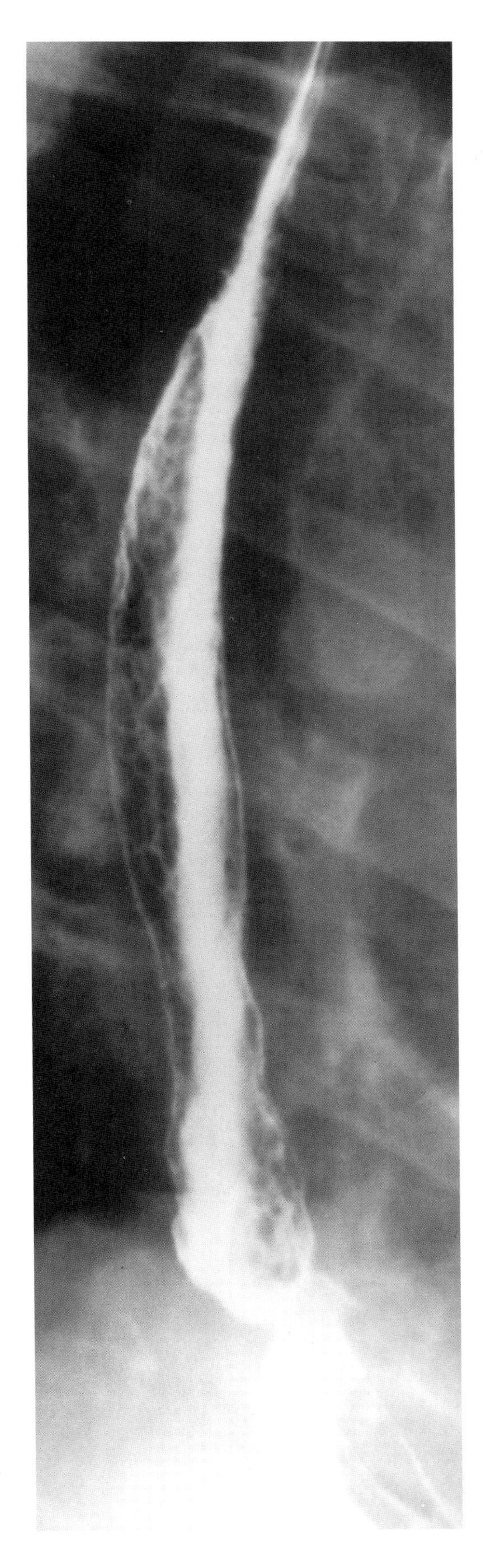

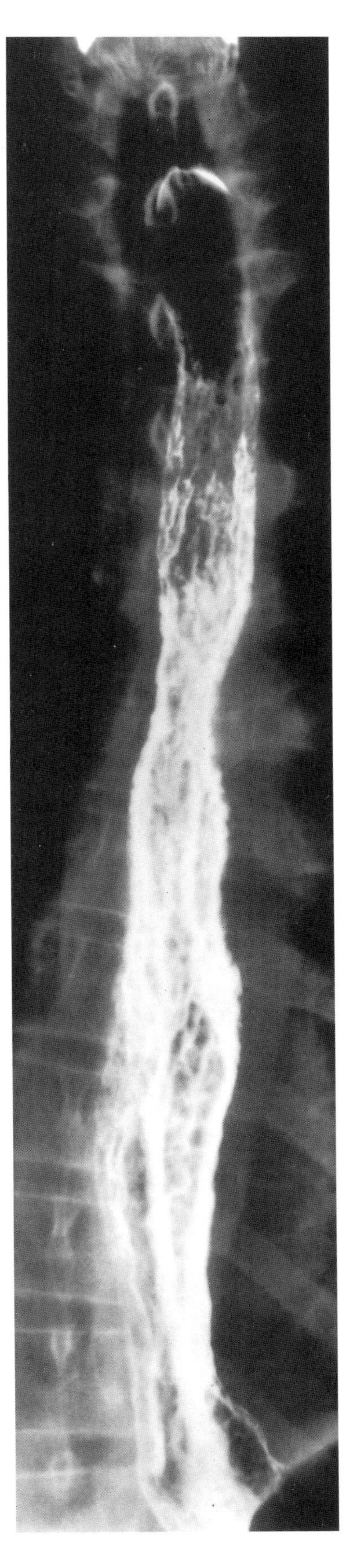

Figures 20.17 and 20.18. Barium swallow examinations of two AIDS patients with candida esophagitis. In both cases discrete, nodular, plaquelike lesions can be seen, with liquid barium filling in the linear spaces between the plaques of monilial colonies.

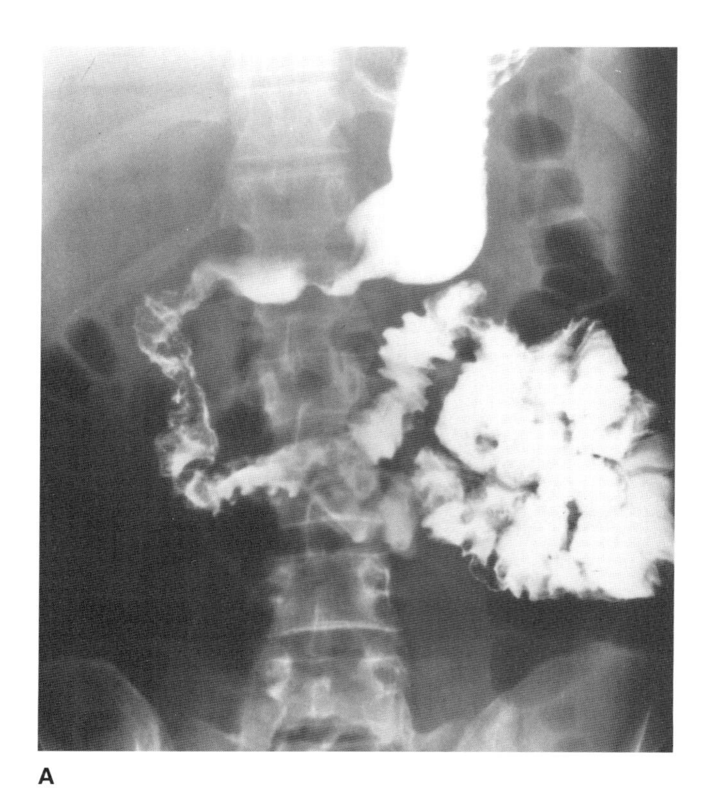

A

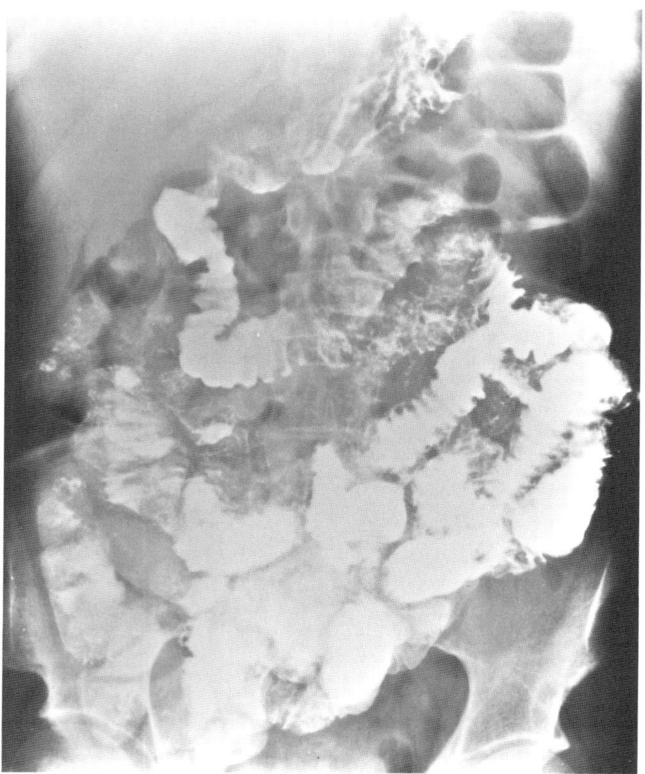

B

Figure 20.19. Upper GI series (A) and small-bowel examination (B) of an AIDS patient with *Cryptosporidium* enteritis. Marked thickening of the mucosal folds is evident in the duodenum and jejunum. The poor coating of the bowel mucosa results from the increased secretions associated with this infection.

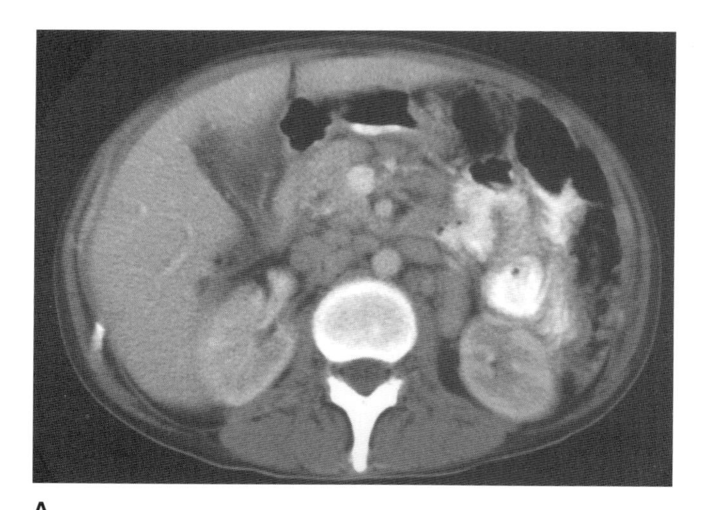

A

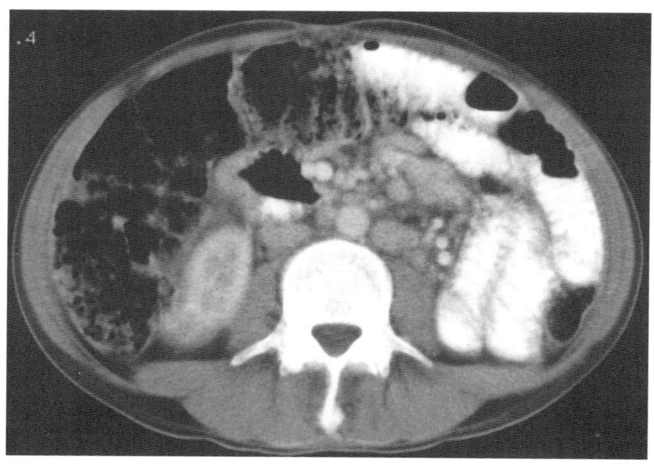

B

Figure 20.20. CT scans of an AIDS patient with an abdominal infection caused by MAC. The opacified loops of small bowel are dilated with thickened mucosal folds. Note the many dotlike densities around the aorta, inferior vena cava, and mesenteric blood vessels; they represent abnormal lymph nodes. Compare the normal abdominal CT scans in Chapter 3.

Enteritis in AIDS patients is commonly caused by any of several species of *Cryptosporidium,* small protozoan organisms that can produce severe diarrhea with dehydration. The proximal small bowel is usually involved (duodenum and jejunum). On barium examination you will see thickening of the mucosal folds and a marked increase in small-bowel secretions (Figure 20.19).

MAC may also produce enteritis, with thickening of the mucosal folds and bowel dilatation. MAC affects the mid- and distal small bowel more than the proximal. Mesenteric lymphadenopathy is common (Figure 20.20).

Kaposi's sarcoma is the most common abdominal neoplasm occurring in AIDS patients, often involving the stomach, bowel, and liver. On barium examination (Figure 20.21) multiple mural nodules are typically identified, each with a characteristic central umbilication. Some patients also have AIDS-related non-Hodgkin's lymphomas, which commonly involve extranodal sites in the bowel, abdominal viscera, and bones (Figure 20.22).

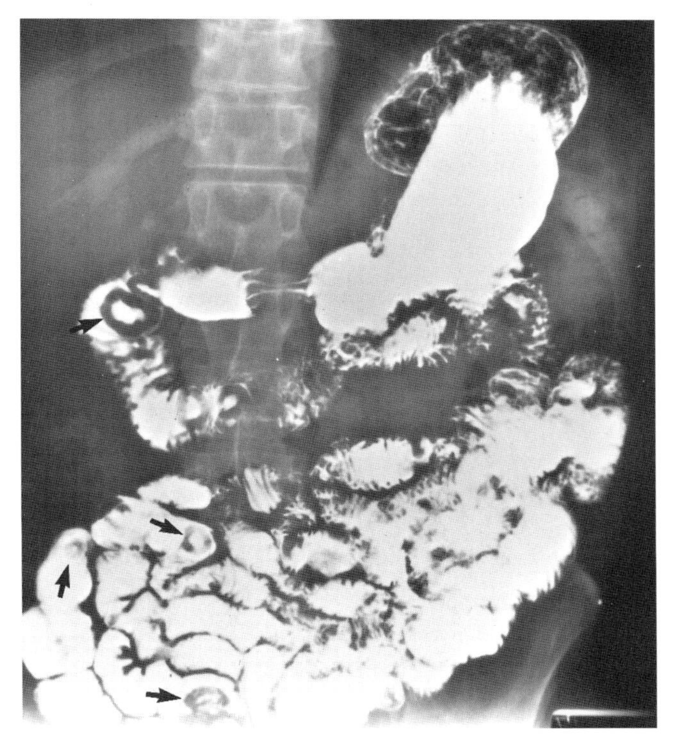

Figure 20.21. Barium examination of a patient with Kaposi's sarcoma involving the bowel. The *arrows* point to mucosal tumors; the liquid barium fills their central umbilications.

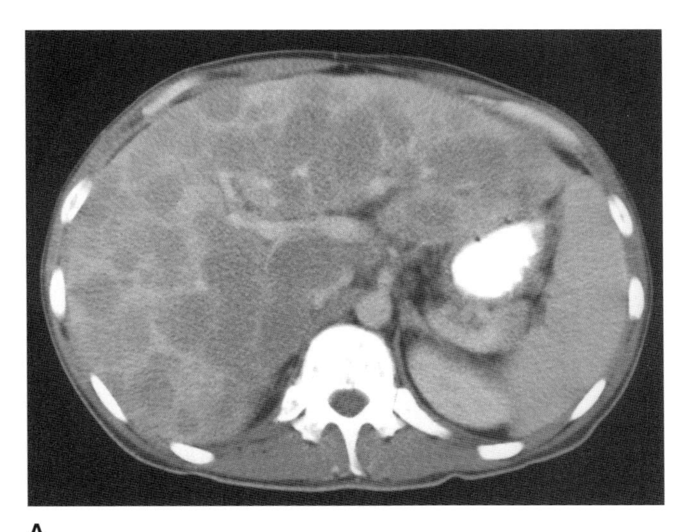

A

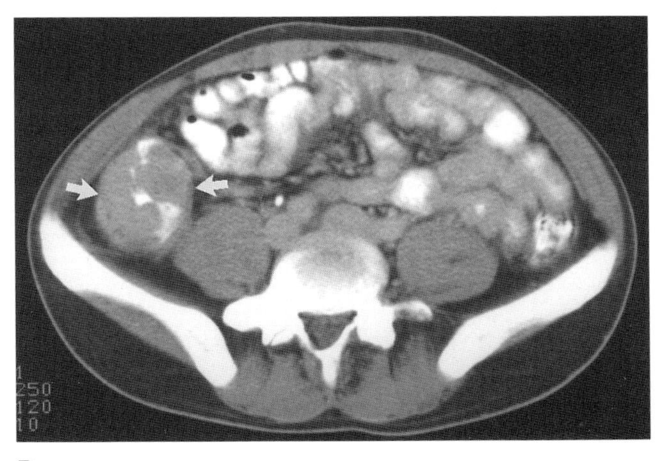

B

Figure 20.22. AIDS patient with B-cell, non-Hodgkin's lymphoma involving the liver and right colon. A: CT scan of the upper abdomen shows multiple tumor masses in the liver. B: CT scan of the lower abdomen shows a mass *(arrows)* involving the cecal portion of the right colon.

Central Nervous System Involvement in AIDS

A variety of opportunistic infections and neoplasms may affect the central nervous system of an AIDS patient. Furthermore, the HIV virus may damage the nervous system itself, causing a subacute encephalitis and meningitis. About 10 percent of AIDS patients present with CNS involvement; around 50 percent develop CNS complications during the course of their illness; and about 75 percent of these patients display evidence of CNS infection upon autopsy.

The most common finding in the brain of AIDS patients is progressive diffuse brain atrophy, which can be shown by CT, but is best demonstrated by MR scanning. Memory loss and cognitive impairment (AIDS dementia complex) may precede the appearance of abnormalities on CT or MR examinations.

Both viral and nonviral infections occur with AIDS. *Cryptococcus neoformans* and *Toxoplasma gondii* cause the most common *nonviral* infections. *Cryptococcus* produces a granular meningitis; patients present with headache, fever, meningeal signs, and seizures. This infection produces little in the way of radiological signs, and CT and MR examinations are often normal.

With *Toxoplasma,* however, the radiological findings are distinctive. This intracellular parasite, seen in 15 percent of HIV patients, produces mass lesions in the brain, causing focal neurological abnormalities and seizures. At CT with IV contrast material, these lesions display a characteristic ringlike or nodular enhancement associated with moderate edema. The lesions are usually multiple and bilateral, and vary in size; they commonly involve the basal ganglia. MR is a better technique for imaging these lesions than CT, because MR can detect very early lesions when the CT scan is still negative. Contrast-enhanced MR scanning of patients with toxoplasmosis also shows multiple ringlike enhancing lesions (Figure 20.23). You have already read in Chapter 18 that certain brain tumors can show ringlike enhancement, and as you may have correctly guessed, the presence of multiple lesions in an HIV-infected patient suggests an infectious rather than a neoplastic etiology. But any patient who does not respond to empirical treatment for toxoplasmosis after 2–4 weeks should have a biopsy performed to rule out a neoplasm.

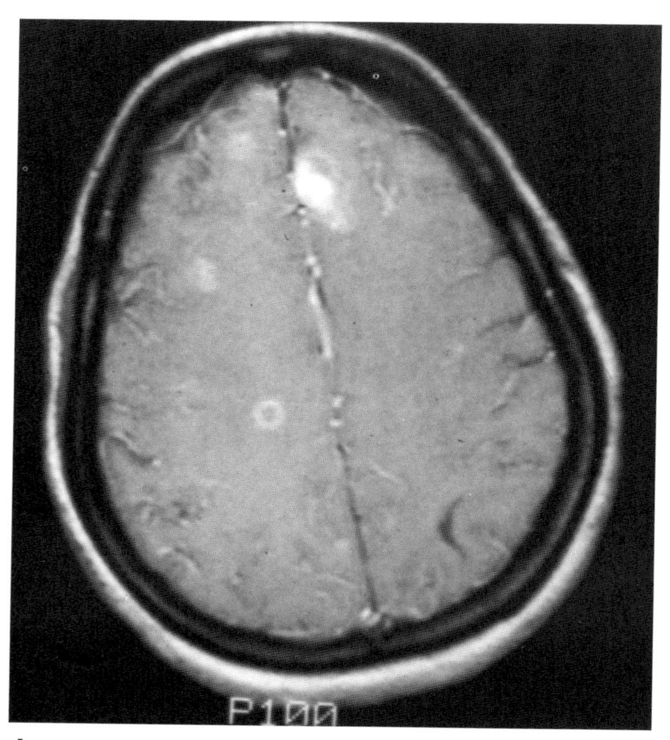

A

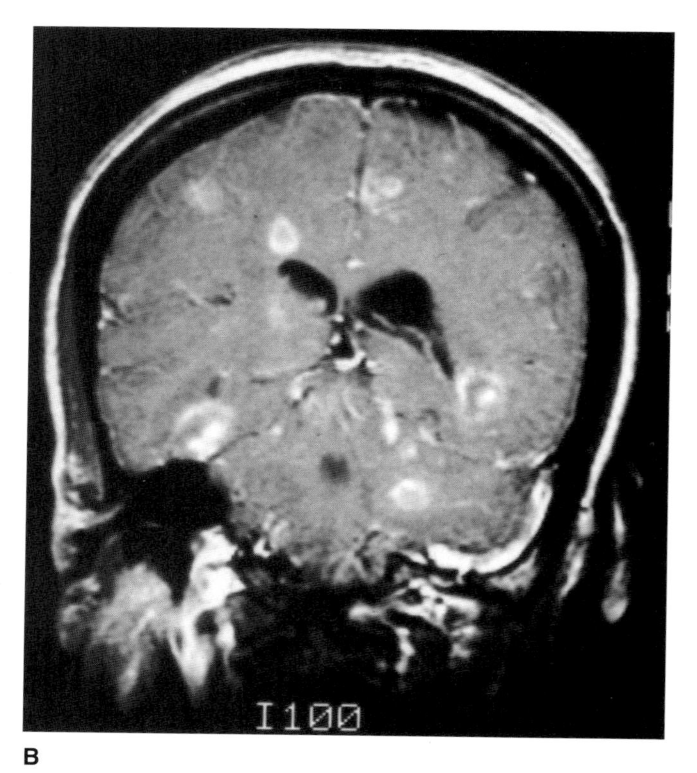

B

Figure 20.23. MR scans of an AIDS patient with toxoplasmosis. A: Axial view. B: Coronal view. Both views show the multiple, contrast-enhancing, ringlike lesions typical of this infection.

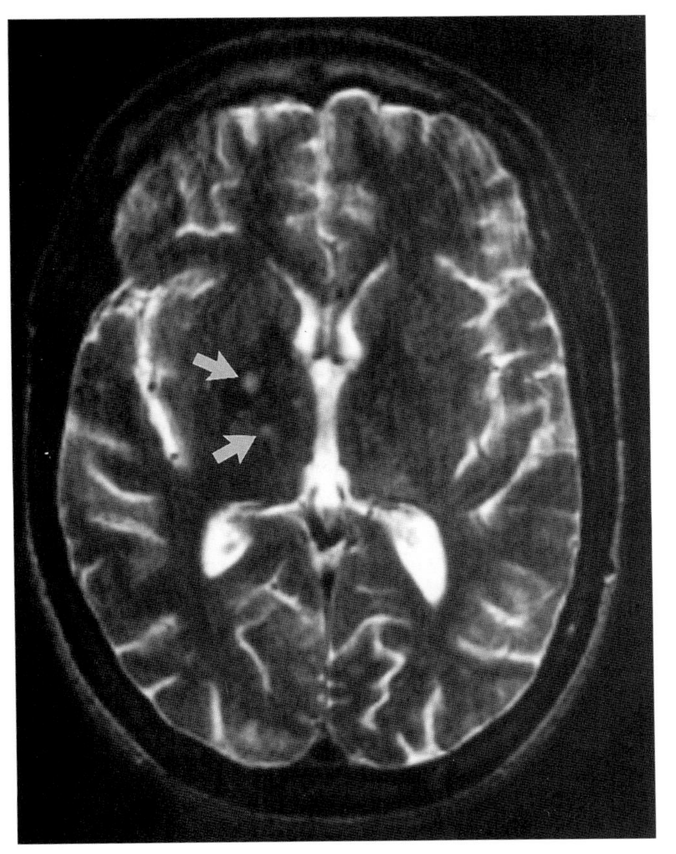

A

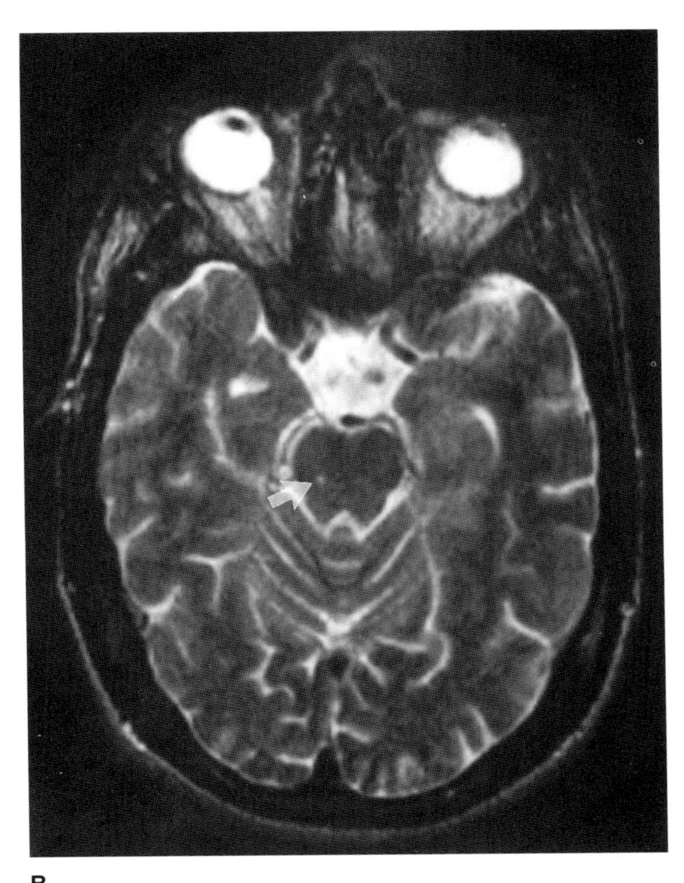

B

Figure 20.24. MR scan of an AIDS patient with progressive multifocal leukoencephalopathy (PML). T2-weighted axial MR images show multiple high-signal lesions *(arrows)* in the basal ganglia (A) and the midbrain (B).

Other agents that may cause nonviral CNS infections in AIDS patients include *Candida albicans,* mycobacterial species (both MAC and *Mycobacterium* tuberculosis), and *Coccidiodes.*

Among the common opportunistic *viral* infections seen in AIDS patients are *progressive multifocal leukoencephalopathy (PML)* caused by a papovavirus, *cytomegalovirus (CMV).* PML produces demyelination and edema, resulting in headache, dementia, and focal neurological abnormalities. CT and MR will show white matter abnormalities (Figure 20.24). CMV produces focal areas of necrotic encephalitis that appear as mass lesions on CT and MR scans. HIV directly affects the CNS, causing subacute encephalitis with mild confusion and memory loss. CT and MR scans may be normal or may show mild to severe brain atrophy.

Primary CNS lymphoma is a brain neoplasm that occurs in about 10 percent of HIV patients. Such lesions appear as brain masses, often greater than 4 centimeters in diameter; they enhance with IV contrast material at CT, and at MR they have a signal strength similar to that of brain parenchyma. Multiple lesions occur in about 50 percent of patients. The enhancement is variable but may show a characteristic double ringlike pattern. The appearance of these lesions may be confused with the appearance of lesions caused by toxoplasmosis. Toxoplasmosis is much more common than CNS lymphoma, but if a patient does not respond to toxoplasmosis therapy, CNS lymphoma must be considered. Another brain lymphoma that can occur in AIDS patients is systemic lymphoma, which involves the brain and spinal cord coverings, or meninges. Finally, Kaposi's sarcoma, common elsewhere in the body, does in rare cases occur in the brain.

Answers to Unknowns

Unknown 1.1 (Figure 1.14)

The pair of dice on the right have been loaded by boring holes into the substance of each die, filling with lead, recapping, and repainting the dots. Bits of lead wire have been used. Of the loaded pair, the die on the left has been x-rayed with the loaded face down, as it would tend to fall. The die on the right has been turned on its side and then x-rayed. You are now looking through it from side to side. The loaded face is down and very dense. The upper part of the die has been evacuated and left empty, which increases its tendency to fall with the 2 facing up—or the 5, depending on which is chosen by the tamperer.

Unknown 1.2 (Figure 1.15)

No, not an egg with a nail in it. The oval object could not be an egg because its radiodensity falls away at the edge and is much greater and fairly uniform in the center. This must therefore be a solid oval body of considerable density and homogeneous composition —except for the nail, which actually was in the center of it. The dark streaks are air in the interfaces after the object had been cracked open. It is a mineral bolus that was found in the stomach of a horse. The nail, typical of those used in shoeing horses, had undoubtedly been swallowed many years ago and remained in the stomach. The "stone," a concretion like a gallstone, had built up around it gradually.

Unknowns 2.1 and 2.2 (Figures 2.18 and 2.19)

Neither figure is a tomogram. You know this because the images are sharp and clear, unlike the superimposed blurry images you see in a conventional tomogram like the one in Figure 2.13B. These unknowns are both radiographs of cadaver slices. Figure 2.18 is a far-posterior coronal slice of the cadaver you have seen. This slice contains only the sacrum, posterior parts of the scapulae and ribs, the tips of the spinous processes of the vertebrae, and the muscles about them. Figure 2.19 is a radiograph of a midline sagittal slice of another cadaver, a woman, as you can determine from the structures you see within the pelvis. The uterus, cervix, and vagina are shown between the anteriorly positioned urinary bladder and the posterior rectum is shown just in front of the sacrum. This slice is approximately 1 centimeter thick, including the mid-portions of the vertebral bodies and their spinous processes. Can you find the liver, loops of bowel, and cardiac chambers?

Unknown 4.1 (Figure 4.13)

The ribs are numbered correctly. The structures indicated by the white arrows are cervical ribs arising from the seventh cervical vertebra, a not uncommon congenital variation.

Unknown 4.2 (Figure 4.14)

The eighth rib on the left is fractured posteriorly, close to its vertebral end. There is also a fracture of the lateral margin of the scapula.

Unknown 4.3 (Figure 4.15)

This patient has a fractured left clavicle. The dark streaks represent air that has infiltrated into the soft tissues through an overlying laceration sustained at the time of injury (subcutaneous emphysema).

Unknown 4.4 (Figure 4.19)

The left shoulder girdle is missing, removed surgically (forequarter amputation) because of a malignant bone tumor in the arm. Note that the medial half of the left clavicle is still present.

Unknown 4.5 (Figure 4.20)

Except for an old, healed fracture of the left clavicle (compare the normal right clavicle) this is a normal chest film of a young woman with normal breast shadows.

Unknown 4.6 (Figure 4.21)

The right breast shadow is missing. This woman underwent a right mastectomy two years before this film was taken. An incidental finding, unrelated to the breast cancer that occasioned the mastectomy, is the presence of multiple calcified left hilar lymph nodes, the result of prior exposure to histoplasmosis.

Unknown 4.7 (Figure 4.22A)

The breast shadows in this patient are exceedingly dense, and if you look carefully within the breast shadows, you will see large, round, well-circumscribed densities that represent silicone breast implants. This patient has undergone bilateral augmentation mammoplasties. Figure 4.22B shows you the lateral view.

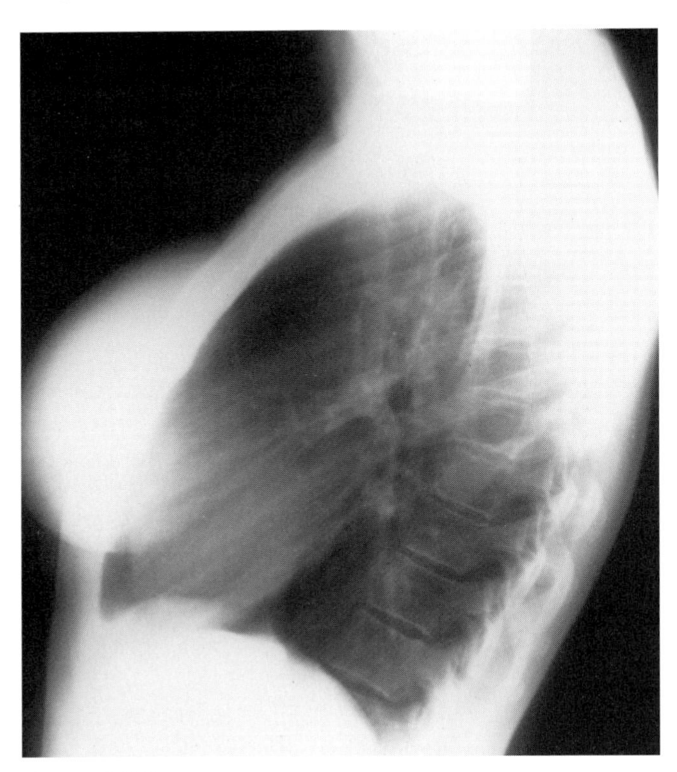

Figure 4.22B

Unknown 5.1 (Figure 5.32)

Multiple cavities with fluid-air interfaces in an intravenous drug abuser strongly suggest lung abscesses due to blood-borne bacteria (staphylococcus in this patient). This condition is often associated with infection of the heart valves (bacterial endocarditis).

Unknown 5.2 (Figure 5.33)

Compared with the normal lung HRCT, the pattern in this patient's HRCT shows subtle differences. Note the enlargement of and expansion of the distal air spaces seen in groups within the network of interstitial linear compartments. This patient has emphysema.

Unknown 6.1 (Figure 6.13)

The superior segment of the right lower lobe is densely consolidated. This is clearly air-space disease of one bronchopulmonary segment of a lobe.

Unknown 6.2 (Figure 6.14)

The right middle lobe is consolidated. Clinically, the patient appeared to have pneumonia.

Unknown 7.1 (Figure 7.7)

The right diaphragm is that horizontal linear density overlapping the right tenth rib with air below it, pneumoperitoneum having been an older method of "putting the lung to rest" in patients with tuberculosis. You are seeing the dome of the diaphragm in tangent, but the part curving anteriorly and posteriorly is not well seen. Blood vessels that seem to extend below the dome of the diaphragm are, of course, in the right lower lobe, posterior basal segment. The patient tried to take a deep breath but was unable to pull the diaphragm any farther down against the cushion of the subdiaphragmatic air.

Unknown 7.2 (Figure 7.18)

The bones are normal, as are the soft tissues, in this male patient. The heart and mediastinum appear to be slightly deviated to the left. There is a right pneumothorax with pronounced collapse of the right lower and middle lobes, which appear to be of soft-tissue density because they no longer contain any visible air. A lesser degree of volume loss is seen in the right upper lobe, which is still partially aerated and shows patchy pulmonary densities. Similar lesions are seen in the left lung. There is a short horizontal fluid level blunting the right costophrenic angle, clear indication that there must be both fluid and air in the right pleural space. Sputum examination revealed tubercle bacilli in this febrile patient.

Unknown 7.3 (Figure 7.19)

Hydropneumothorax several days after pneumonectomy. The mediastinum is slightly shifted toward the side of the absent lung. Note the missing sixth rib and the elevated stomach bubble, which is located under a high (unseen) diaphragm. Thus that side of the thoracic cage is much smaller than the other (missing lung, crowded ribs, high diaphragm).

Unknown 9.1 (Figure 9.33A)

Yes, the upper mediastinal mass to the right of the trachea certainly could be (and was) related to the small parenchymal lesion. In Figure 9.33B, a CT scan through the hilum, you can better see the mass *(M)*, which consisted of metastases to mediastinal and hilar nodes that were compressing *(arrow)* the superior vena cava *(S)*.

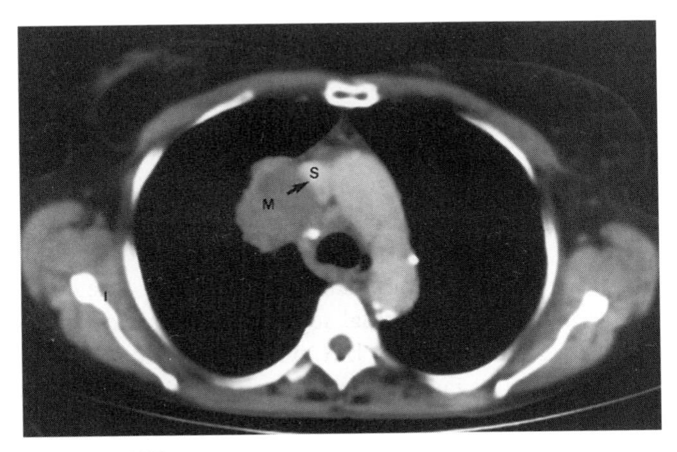

Figure 9.33B

Unknown 12.1 (Figure 12.11)

Plain film of a patient with small-bowel mechanical obstruction, showing multiple, air-filled, and dilated jejunal loops and a cleared colon.

Unknown 12.2 (Figure 12.12)

Plain film of another patient with mechanical small-bowel obstruction, although in contrast to the film in Figure 12.11 *(Unknown 12.1)* the air filling is minimal and the jejunal loops are primarily fluid filled.

Unknown 13.1 (Figure 13.4)

The small round filling defect near the pyloric canal on both spot films was consistently present at fluoroscopy and on films. At endoscopy it was found to be a benign gastric polyp.

Unknown 14.1 (Figure 14.81)

No doubt you noticed loops of opacified bowel extending outside the peritoneal cavity through a large gap in the linea alba between the two rectus abdominus muscles. In B you can see that the loops are located in the subcutaneous fascia just beneath the skin. This patient, of course, has a ventral hernia, at the site of a midline incision performed one year earlier. Did you also find the gallbladder stone in A?

Unknown 14.2 (Figure 14.82)

You should have noted a small, shrunken, and nodular liver as well as splenomegaly and low-attenuation peritoneal fluid representing ascites. This chronic alcoholic patient, of course, has cirrhosis. The cause of his abdominal pain is also apparent; the low-attenuation, irregularly margined mass in the liver is a tumor. If you considered the possibility of a hepatoma (which occurs more frequently in patients with cirrhosis), you should congratulate yourself for combining your image findings with your clinical knowledge.

Unknown 14.3 (Figure 14.83)

This CT scan was performed with both oral and intravenous contrast materials. You should have noted the large liver laceration and associated hemoperitoneum in this trauma patient. Did you identify the bright, white crescent shape that can be seen laterally within the hemoperitoneum? It is as bright as intravenous contrast material and that is exactly what it represents. The liver laceration has severed a right hepatic artery branch, and this white crescent represents extravasation of the intravenously injected contrast material into the hemoperitoneum. Intravenous contrast extravasation is a very important finding on trauma CT scans, and it tells you that the patient is actively bleeding on the CT scanner table and should be taken directly to the operating room for surgical management. This was done for this patient, who made an excellent recovery!

Unknown 15.1 (Figure 15.17)

Comminuted fracture of the humerus.

Unknown 15.2 (Figure 15.18)

Impacted comminuted fracture of the head and neck of the humerus.

Unknown 15.3 (Figure 15.19)

Radiograph of a fractured tibia several weeks after the injury, made the day the initial plaster cast was removed. The shadowy, flocculent white material (callus) is taking on mineral, which tells you that this is not a fresh fracture. The alignment and angulation of the tibial fracture fragment is not optimal.

Unknown 15.4 (Figure 15.20)

The fracture of the radial head is easier to see in B, an oblique view. Note the offset of white bone at the fracture point in A.

Unknown 15.5 (Figure 15.21)

Fracture of the distal end of the radial metaphysis at the wrist with displacement of its epiphysis (as seen in the lateral). Note the white overlap of bone in the AP view, often a clue to the presence of subtle fractures.

Unknown 15.6 (Figure 15.22)

No fracture is present. This is a normal immature wrist. A normal growth plate is always distinguished from a fracture by its smooth, dense margin.

Unknown 15.7 (Figure 15.23)

Impacted fracture of the radius 1.5 centimeters from the radiocarpal joint. The typical Colles' fracture adds a fracture of the styloid process of the ulna, not present here.

Unknown 15.8 (Figure 15.24)

Fracture of both bones of the forearm, with overriding due to muscle pull, which must be reduced, set end to end, and immobilized in good alignment in plaster.

Unknown 16.1 (Figure 16.74)

If you recognized the air-containing loops of bowel within the scrotum you probably had no difficulty in diagnosing a scrotal hernia.

Unknown 16.2 (Figure 16.75)

The highly echogenic linear structure that produces acoustic shadowing on the sagittal scan represents an intrauterine contraceptive device (IUD). Compare its appearance on the ultrasound examination with its appearance on the accompanying plain film (C). IUDs come in a variety of shapes and forms. One of the important uses of ultrasound imaging is to confirm intrauterine position. If the IUD is not seen within the uterus at ultrasound, a plain film of the abdomen should be obtained to determine whether the IUD is lying free within the peritoneal cavity (having perforated the uterine wall) or whether it has been expelled.

Unknown 16.3 (Figure 16.76)

The uterus is markedly enlarged for a woman of this age and the uterine cavity is filled with low-attenuation material. Did you consider a uterine neoplasm? This patient proved to have an endometrial carcinoma with hemorrhage. The uterine cavity was filled with blood clots.

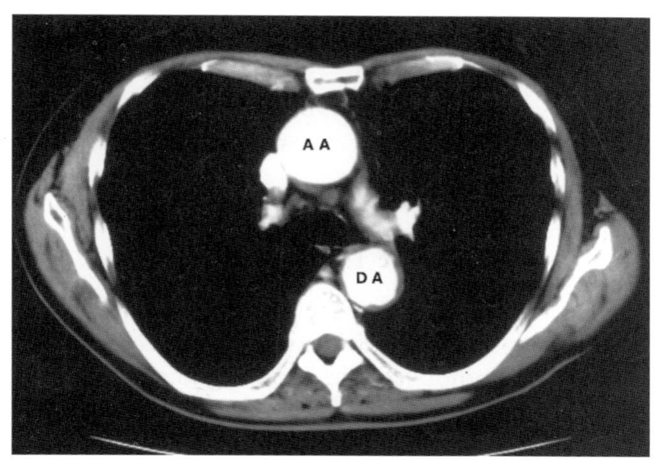

Figure 17.63C

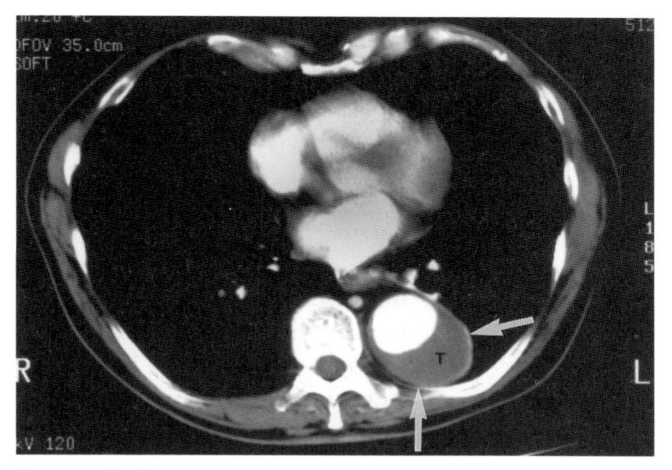

Figure 17.63D

Unknown 17.1 (Figure 17.63)

Any chest mass shown to be contiguous to, or adjacent to, the thoracic aorta could be an aortic aneurysm, which, of course, should not be biopsied under any circumstances. The patient had a chest CT examination with intravenous contrast material. You see the results in Figures 17.63C and 17.63D. C: An upper-level scan shows a normal-caliber ascending *(AA)* and normal-caliber descending aorta *(DA)*. D: A lower scan, at the level of the heart, shows that the mass represents an aneurysm *(arrows)* of the descending aorta that is laterally and posteriorly lined with thrombus *(T)*. Only the free-flowing lumen of the aneurysm is opacified with contrast material.

Unknown 17.2 (Figure 17.64)

The patient's chest film shows a dilated aorta and her CT scans a thin membrane (intimal flap) separating the aortic lumen into two compartments. This patient has an aortic dissection with patent true and false lumens. Note in B, an upper-level slice, just above the aortic arch, that the dissection extends up into the right brachiocephalic and left subclavian arteries, which are separated into two lumens. The intimal flap is well shown at the level of the aortic arch (C) and just below (D) in the dilated ascending and descending aorta.

Unknown 18.1 (Figure 18.48)

Multiple high-intensity (white) masses are shown, associated with areas of low-intensity (dark gray) edema. No doubt you thought of metastatic disease after seeing multiple lesions. This patient proved to have metastatic lung cancer.

Unknown 18.2 (Figure 18.49)

The tumor is located in the pons and it proved to be a glioma. Note that the pons is expanded.

Acknowledgments

My gratitude goes first and foremost to Lucy Frank Squire. She recognized early on the need for a textbook to introduce medical students to radiology and wrote the first edition in 1964. A model of clarity and a joy to read, it was the first user-friendly medical student textbook. From her years of student teaching, Dr. Squire had recognized the concepts students found difficult to grasp, and she presented them in a lucid, engaging, and easily comprehensible manner. A pioneer of small-group, interactive teaching, Dr. Squire captivated the students with unknowns to interpret and puzzles to solve. As a medical student I also was fascinated by *Fundamentals of Radiology.*

During my radiology residency at Massachusetts General Hospital, I had the good fortune to work with Dr. Squire in her Harvard Medical School Radiology Clerkship, which I now direct. This was the beginning of a 20-year collegial relationship during which we discussed and collaborated on radiology education for medical students. I was indeed honored to be chosen as co-author of the fourth edition of *Fundamentals of Radiology* and am further honored to be the author of the fifth edition.

My gratitude also goes to my medical students, who have helped in a multitude of ways. They have provided me with an astounding number of ideas and innovations for inclusion in the fifth edition, making possible a text more suited to their needs. Their enthusiasm, imagination, and friendship have been an inspiration to me.

I am grateful to all my good friends and colleagues who generously consulted with me on specific points or contributed illustrative cases to this edition of the book. My thanks go also to Francis Cunningham and Shelley Eshleman, who executed the drawings for the early editions, and to the many others, acknowledged individually below, who contributed to previous editions. For help in the preparation of the current edition, I would like to thank especially Drs. Oksana Baltarowich, Beryl Benacerraff, Robert T. Bramson, Philip Costello, Steven L. Dawson, Jill Dobbins, Dennis W. Foley, Carol Hulka, John A. Kaufman, J. Nash Lawrason, Mark Lerner, Joachim Lotz, Theresa C. McLoud, William E. Palmer, Franzisca A. Prosser, Carlos A. Rabito, Patrick M. Rao, James T. Rhea, Richard Sacknoff, Bruce Saffran, Jane C. Share, Arthur C. Waltman, and Isabel C. Yoder. I am also grateful to the ATL Company, DuPont Diagnostic Imaging, the Eastman Kodak Company, General Electric Medical Systems, and J/B Woolsey Associates.

I am indebted to my department chairman, Dr. James H. Thrall, for his support and encouragement. And I owe a special thanks to my secretary, Evelyn M. DeBerardinis, for her patience and her assistance in the preparation of the manuscript.

Chapter 1

Figure 1.1 The late Dr. Merrill Sosman brought this from Australia.

Figure 1.3 From *Medical Record* 149, February 15, 1896.

Figure 1.4 Courtesy Dr. W. Felts, Minneapolis, Minnesota.

Figures 1.5A, 1.11 From *Fundamentals of Radiography,* pp. 6, 25, published by Eastman Kodak Co., Rochester, New York.

Figure 1.5B Courtesy DuPont Diagnostic Imaging, Wilmington, Delaware.

Figure 1.8 Courtesy Dr. D. Eaglesham, Guelph, Ontario, Canada.

Figure 1.9 Courtesy Dr. E. Comstock, Wellsville, New York, and C. Bridgman, Rochester, New York.

Figure 1.10 Courtesy C. Bridgman, Rochester, New York, and S. Keck, New York, New York.

Figure 1.12 J/B Woolsey Associates.

Figure 1.14 Courtesy C. Bridgman, Rochester, New York.

Chapter 2

Figure 2.1 Courtesy Dr. A. Richards and the publisher, *Medical Radiography and Photography* (hereinafter *MR&P*) 32:28.

Figure 2.2B Courtesy Drs. W. Macklin, Jr., H. Bosland, and A. McCarthy and the publisher, *MR&P* 31:91.

Figure 2.3C Courtesy DuPont Diagnostic Imaging, Wilmington, Delaware.

Figures 2.3D, 2.20, 2.23, 2.24A, 2.30, 2.36, 2.41 Courtesy GE Medical Systems, Milwaukee, Wisconsin.

Figure 2.5 Courtesy C. Bridgman, E. Holly, and Dr. M. Zariquiey and the publisher, *MR&P* 32:49.

Figure 2.6 J/B Woolsey Associates.

Figures 2.7, 2.8 Courtesy Dr. H. Forsyth, Jr., and the publisher, *MR&P* 25:38.

Figures 2.10, 2.12 Courtesy Dr. C. Behrens, Bethesda, Maryland.

Figure 2.22 From *Fundamentals of Radiography,* p. 48, published by Eastman Kodak Co., Rochester, New York.

Figure 2.33 Courtesy Advanced Technology Laboratories, Bothell, Washington.

Chapter 3

Figures 3.1–3.3, 3.5, 3.8–3.11, 3.15, 3.16, 3.19, 3.20, 3.24, 3.25, 3.29, 3.34, 3.35, 3.39, 3.44–3.46 J/B Woolsey Associates.

Chapter 4

Figures 4.3, 4.18 J/B Woolsey Associates.

Figures 4.5, 4.6 Courtesy Dr. O. Alexander and the publisher, *MR&P* 30:35, 36.

Figure 4.7 Courtesy T. Funke and the publisher, *MR&P* 36:9, 29.

Figure 4.9 Courtesy S. Forczyk, Fall River, Massachusetts.

Figures 4.11, 4.12 Courtesy the late Dr. John Hope, Drs. E. O'Hare, T. Tristan, and J. Lyon, Jr., and the publisher, *MR&P* 33:30, 31.

Figure 4.13 Courtesy Dr. J. Atlee, Lancaster, Pennsylvania.

Figure 4.15 Courtesy A. Dini, Rochester, New York.

Figures 4.16, 4.17 From *Fundamentals of Radiography,* pp. 25, 30, published by Eastman Kodak Co., Rochester, New York.

Figure 4.19 Courtesy Dr. J. Jollman, Omaha, Nebraska.

Chapter 5

Figure 5.2 Courtesy R. Morrison and the publisher, *MR&P* 27:132.

Figures 5.9, 5.11 Courtesy Drs. B. Felson, F. Fleishner, J. McDonald, and C. Rabin and the publisher, *Radiology* 73:744.

Figure 5.10 Courtesy Dr. C. Dotter and the publisher, *MR&P* 34:48, and Dr. J. Reed, Detroit, Michigan.

Figures 5.12, 5.13 Courtesy Dr. B. Epstein and the publisher, *MR&P* 34:58, 62.

Figures 5.15, 5.22 Courtesy Dr. T. McLoud, Boston, Massachusetts.

Figures 5.16–5.18 Courtesy Drs. A. Bell, S. Shimomura, W. Guthrie, H. Hempel, H. Fitzpatrick, and C. Begg and the publisher, *Radiology* 73:566.

Figures 5.20B, 5.21E Courtesy Dr. R. Sherman, New York, New York.

Figure 5.21F Courtesy Dr. R. Wagner, Detroit, Michigan.

Chapter 6

Figures 6.1, 6.6, 6.9 J/B Woolsey Associates.

Figures 6.4, 6.5 Courtesy Dr. M. Zariquiey and the publisher, *MR&P* 33:68–76.

Figure 6.13 Courtesy the late Dr. G. Simon, London, England.

Figure 6.16 Courtesy Dr. J. Woodring and the publisher, *MR&P* 62:1.

Chapter 7

Figure 7.2 Courtesy Dr. M. Strahl, Brooklyn, New York.

Figure 7.3 Courtesy Dr. J. Hope et al. and the publisher, *MR&P* 33:26, 28.

Figure 7.5 Adapted from Sobotta-Uhlenhuth, *Atlas of Descriptive Human Anatomy,* 7th ed., Hafner Publishing Co., New York, New York.

Figure 7.6 Courtesy Dr. C. Dotter, Portland, Oregon.

Figure 7.7 Courtesy C. Brownell and the publisher, *MR&P* 27:114.

Figure 7.9 Courtesy Dr. W. Brosius, Detroit, Michigan.

Figure 7.11 J/B Woolsey Associates.

Figures 7.12, 7.17 Courtesy Dr. J. Petersen and the publisher, *Radiology* 74:36, 40.

Figure 7.16 Courtesy Dr. E. Carpenter, Superior, Wisconsin.

Figure 7.18 Courtesy Dr. G. Schwalbach, Rochester, New York.

Figure 7.19 Courtesy Dr. H. Forsyth, Jr., and the publisher, *MR&P* 31:129.

Figure 7.21 Courtesy Dr. M. Fisher and the publisher, *MR&P* 46:2.

Chapter 8

Figure 8.1 Courtesy Drs. I. Harris and M. Stuecheli and the publisher, *MR&P* 28:29.

Figure 8.2 Courtesy Dr. J. Hope et al. and the publisher, *MR&P* 33:26.

Figures 8.4, 8.5, 8.18, 8.21 Courtesy the late Dr. G. Simon, London, England.

Figures 8.7, 8.8, 8.15, 8.26 Courtesy Dr. J. Woodring and the publisher, *MR&P* 62:1 (back), etc.

Figures 8.9, 8.19, 8.21–8.24, 8.28–8.32 Courtesy Dr. D. Godwin, Seattle, Washington.

Figure 8.13 Courtesy Dr. W. Crandall, Sulphur, Oklahoma.

Figure 8.14 Courtesy Dr. H. Fulton and the publisher, *MR&P* 30:81, 82.

Figure 8.17 Courtesy Drs. E. Uhlmann and J. Ovadia and the publisher, *Radiology* 74:269.

Chapter 9

Figure 9.1 Courtesy Dr. J. Woodring and the publisher, *MR&P* 62:1.

Figures 9.3A, 9.4A Courtesy Dr. C. Dotter and the publisher, *MR&P* 30:70.

Figures 9.3B, 9.4B, 9.5B, 9.8B J/B Woolsey Associates.

Figures 9.5A, 9.8 Courtesy Drs. P. Markovits and J. Desprez-Curely and the publisher, *Radiology* 78:373, 377.

Figures 9.6, 9.11–9.14, 9.17–9.22, 9.24, 9.26–9.34 Courtesy Dr. J. Woodring and the publisher, *MR&P* 62.

Figure 9.7 Courtesy Dr. G. McDonnell and the publisher, *MR&P* 30:84.

Figure 9.9 Courtesy Dr. T. Newton and the publisher, *American Journal of Roentgenology* 89:277.

Figure 9.10 Courtesy Drs. R. Ormond, A. Templeton, and J. Jaconette and the publisher, *Radiology* 80:738.

Figure 9.23 Courtesy Dr. J. Hope et al. and the publisher, *MR&P* 33:35.

Figure 9.25 Courtesy Drs. A. Megibow and M. Bosniak, New York, New York.

Chapter 10

Figures 10.1, 10.5 Courtesy Dr. H. Forsyth, Jr., Rochester, New York.

Figure 10.10 Courtesy the late Dr. G. Simon, London, England.

Figure 10.19 Courtesy Drs. M. Klein and E. Walsh and the publisher, *Radiology* 70:674.

Figures 10.21, 10.22, 10.24 Courtesy Dr. R. Kubicka and the publisher, *MR&P* 61:14–39.

Figure 10.25 Courtesy Dr. B. Epstein and the publisher, *MR&P* 34:68, 69.

Figures 10.27, 10.30 J/B Woolsey Associates.

Figure 10.33 Courtesy Drs. A. Lieber and J. Jorgens and the publisher, *American Journal of Roentgenology* 86:1069–70.

Figure 10.43 Courtesy Dr. W. Tuddenham et al. and the publisher, *MR&P* 33:61.

Figure 10.46B Courtesy Dr. A. Strashun, Brooklyn, New York.

Figures 10.57, 10.64 Courtesy Dr. M. Lipton and Imatron, Inc., San Francisco, California.

Chapter 11

Figures 11.2, 11.3 Courtesy Dr. J. Hope et al. and the publisher, *MR&P* 33:37.

Figure 11.4 Courtesy Drs. B. Kalayjian and M. Sapula and the publisher, *MR&P* 30:57.

Figure 11.5 Courtesy Dr. J. McGillivray and the publisher, *MR&P* 30:27.

Figures 11.6, 11.27, 11.28, 11.40 J/B Woolsey Associates.

Figure 11.7B Courtesy Dr. C. Stevenson, Spokane, Washington.

Figure 11.11 Courtesy Dr. G. Thompson and W. Cornwell and the publisher, *MR&P* 25:43.

Figure 11.15 Courtesy V. Yamamoto and the publisher, *MR&P* 26:121.

Figure 11.18 Courtesy Drs. R. Salb and G. Burton and the publisher, *MR&P* 33:106.

Figure 11.19 Courtesy Dr. F. Nelans, Wagner, Oklahoma.

Figure 11.21 Courtesy Dr. L. Cole, Blossburg, Pennsylvania.

Figure 11.23 Courtesy Dr. J. Wainerdi, New York, New York.

Figure 11.29 Courtesy E. Holly and G. Weingartner and the publisher, *MR&P* 29:91.

Figure 11.31 Courtesy Dr. W. Tuddenham and the publisher, *Radiology* 78:697.

Figure 11.32 Courtesy C. Bridgman and the publisher, *MR&P* 26:12.

Figure 11.39 Courtesy Dr. W. Irwin, Detroit, Michigan.

Figure 11.42 Courtesy Drs. G. Stein and A. Finkelstein and the publisher, *MR&P* 31:5.

Figure 11.43 Courtesy Dr. G. Schwartz and the publisher, *MR&P* 30:55.

Figure 11.45 Courtesy Dr. J. Becker, Brooklyn, New York.

Figure 11.50 Courtesy Dr. N. Alcock and the publisher, *MR&P* 23:27.

Figure 11.56 Courtesy Dr. R. Filly and Acuson Corp., San Francisco, California.

Figure 11.57 Courtesy Dr. G. Grube and Acuson Corp., Loma Linda, California.

Figure 11.58 Courtesy Dr. E. Ferris and Acuson Corp., Little Rock, Arkansas.

Figure 11.60 Courtesy Dr. L. Berland and Acuson Corp., Birmingham, Alabama.

Chapter 12

Figure 12.2 Courtesy Dr. D. Haff, Northampton, Pennsylvania.

Figure 12.6 Courtesy Dr. C. Nice and the publisher, *Radiology* 80:44.

Figure 12.21 Courtesy Dr. J. McCort and the publisher, *Radiology* 78:51.

Figure 12.28 Courtesy Dr. E. Schultz and the publisher, *Radiology* 70:728.

Figure 12.29 Courtesy Dr. H. Welsh and the publisher, *MR&P* 34:78.

Chapter 13

Figure 13.3 Courtesy Dr. C. Nice and the publisher, *Radiology* 80:43.
Figure 13.4 Courtesy Dr. R. Sherman and Eastman Kodak Co., Rochester, New York.
Figures 13.8, 13.36 Courtesy T. Funke, Lorain, Ohio.
Figure 13.9 Courtesy Dr. S. Wyman, Boston, Massachusetts.
Figure 13.10 Courtesy Drs. W. Horrigan, H. Atkins, and N. Tapley and the publisher, *Radiology* 78:440.
Figure 13.11 Courtesy Dr. G. Joffrey, Olean, New York.
Figure 13.14 Courtesy Drs. J. Spencer and J. Schaeffer and the publisher, *MR&P* 29:22.
Figure 13.15 Courtesy Dr. M. Kulick, Brooklyn, New York.
Figure 13.19 From *Fundamentals of Radiography,* p. 6, published by Eastman Kodak Co., Rochester, New York.
Figure 13.20 Courtesy Dr. T. Orloff and the publisher, *MR&P* 35:53.
Figures 13.26, 13.27 Courtesy Dr. S. Prather, Jr., Augusta, Georgia.
Figure 13.30 Courtesy Dr. G. Jaffrey, Santa Rosa, California.
Figure 13.31 Courtesy Dr. E. Ahern and the publisher, *MR&P* 30:9.
Figure 13.33 Courtesy Dr. J. Higgason and the publisher, *MR&P* 30:60.
Figure 13.34 Courtesy Dr. J. Dulin, Iowa City, Iowa.
Figure 13.35 Courtesy Dr. J. Reed, Detroit, Michigan.
Figure 13.40 Courtesy Dr. C. Cimmino and the publisher, *MR&P* 30:45.

Chapter 14

Figure 14.11 Courtesy Dr. J. Ferrucci, Boston, Massachusetts.
Figure 14.31 Courtesy Dr. J. Pepe, Brooklyn, New York.
Figures 14.33, 14.34 Courtesy Dr. K. McKusick, Boston, Massachusetts.
Figure 14.37 Courtesy Drs. G. Stein and A. Finkelstein and the publisher, *MR&P* 31:12.
Figure 14.38 Courtesy Dr. E. Pirkey, Louisville, Kentucky.
Figure 14.40 Courtesy Dr. T. Van Zandt, Rochester, New York, and Eastman Kodak Co.
Figure 14.42 Courtesy Dr. S. Dallemand, Brooklyn, New York.
Figure 14.61 Courtesy Dr. J. Hope et al. and the publisher, *MR&P* 33:49.
Figure 14.62 Courtesy Dr. K. Fengler and the publisher, *Journal of Mount Sinai Hospital,* New York, New York.
Figure 14.66 Courtesy Drs. M. Bosniak and J. Becker, New York, New York.
Figure 14.68 Courtesy Dr. W. Foley and General Electric Company.
Figure 14.72 Courtesy Drs. A. Waltman and C. Athanasoulis, Boston, Massachusetts.
Figure 14.75 Courtesy Drs. J. Edeiken, G. Strong, and J. Khajavi and the publisher, *Radiology* 79:88.
Figure 14.76 Courtesy Drs. J. Hope and C. Koop and the publisher, *MR&P* 38:47.

Chapter 15

Figure 15.2 Courtesy Chicago Museum of Natural History, Chicago, Illinois.
Figure 15.3 J/B Woolsey Associates.
Figures 15.4, 15.9 Courtesy C. Bridgman and the publisher, *MR&P* 26:4, 27:72.
Figure 15.5 Courtesy Dr. G. Mitchell and the publisher, *MR&P* 34:6.
Figures 15.6, 15.31 Courtesy Drs. H. Isard, B. Ostrum, and J. Cullinan and the publisher, *MR&P* 38:97, 101.
Figure 15.8 Courtesy J. Cahoon and the publisher, *Radiography and Clinical Photography* 22:4, 6.
Figure 15.12 Courtesy Dr. L. Hilt, Eugene, Oregon.
Figures 15.17, 15.19–15.21 Courtesy T. Funke and the publisher, *MR&P* 36:9, 29.
Figure 15.25 Courtesy Dr. W. Irwin, Detroit, Michigan.
Figure 15.26 Courtesy Dr. D. Wilner, Atlantic City, New Jersey.
Figure 15.39 Adapted from A. W. Ham, *Histology,* 3rd ed., p. 295, published by J. B. Lippincott, Philadelphia, Pennsylvania.
Figure 15.40 Courtesy Dr. P. Lacroix, University of Louvain, Belgium.
Figure 15.41 Courtesy Dr. M. Urist, *Bone as a Tissue,* p. 37, and the publisher, McGraw-Hill Book Co., New York, New York.
Figure 15.42 Courtesy Dr. L. Luck, *Bone and Joint Diseases,* 1st ed., and the publisher, Charles C Thomas, Springfield, Illinois.
Figures 15.43, 15.44 Courtesy Dr. J. Feist, Pittsburgh, Pennsylvania.

Chapter 16

Figure 16.2 Harriet Greenfield.

Figure 16.5 Courtesy Dr. I. Andersson and the publisher, *MR&P* 62:cover.
Figures 16.7, 16.8, 16.33, 16.34, 16.44, 16.46, 16.59, 16.64, 16.66, 16.68, 16.72A Illustrations: J/B Woolsey Associates.
Figure 16.13 Courtesy J. Hill, Lancaster, England.
Figure 16.24 Courtesy Advanced Technology Laboratories, Bothell, Washington.

Chapter 17

Figure 17.9, 17.11 Courtesy Advanced Technology Laboratories, Bothell, Washington.
Figures 17.17–17.22, 17.30, 17.41–17.43, 17.53 J/B Woolsey Associates.

Chapter 18

Figures 18.2, 18.3, 18.5, 18.7 J/B Woolsey Associates.
Figure 18.8 Courtesy R. Matthias and the publisher, *MR&P* 28:cover.
Figure 18.9 W. Davis.
Figures 18.10, 18.23, 18.24, 18.27 From R. Novelline and L. F. Squire, *Living Anatomy,* published by Hanley and Belfus, Philadelphia, Pennsylvania.
Figures 18.39, 18.40 Courtesy Drs. G. Potter and R. Gold and the publisher, *MR&P* 51:2.

Chapter 19

Figure 19.1, 19.2C, 19.3, 19.4G, 19.12 Courtesy Medi-tech/Boston Scientific Corporation, Natick, Massachusetts.
Figure 19.7 Courtesy Drs. A. Waltman and C. Athanasoulis, Boston, Massachusetts.
Figure 19.18 J/B Woolsey Associates.
Figure 19.20 Courtesy Dr. C. Athanasoulis et al. and the publisher, *Interventional Radiology,* W. B. Saunders Company, Philadelphia, Pennsylvania.

Chapter 20

Figure 20.4 Courtesy Dr. G. Schwalbach, Rochester, New York.
Figure 20.6 Courtesy R. Bottin, Indianapolis, Indiana.
Figure 20.9 Courtesy Dr. G. Baron, Rochester, New York.
Figure 20.10 Courtesy Dr. W. Brosius, Detroit, Michigan.
Figure 20.11 Courtesy Dr. B. Suster, Brooklyn, New York.

Index

A

B

D

E

F

G

H

L

M

N

O

P

R

S

T

U

V

W

X